31st International Conference on Photonic, Electronic, and Atomic Collisions (ICPEAC2019)

Journal of Physics: Conference Series Volume 1412

Deauville, France
23 – 30 July 2019

Editors:

Lorenzo Ugo Ancarani
Christian Bordas
Franck Lepine
Dominique Vemhet

ISBN: 978-1-7138-1591-4
ISSN: 1742-6588

Printed from e-media with permission by:

Curran Associates, Inc.
57 Morehouse Lane
Red Hook, NY 12571

proceedings.com

Some format issues inherent in the e-media version may also appear in this print version.

Institute of Physics
Dirac House, Temple Back
Bristol BS1 6BE UK

Phone: 44 1 17 929 7481
Fax: 44 1 17 920 0979

techtracking@iop.org

Additional copies of this publication are available from:

Curran Associates, Inc.
57 Morehouse Lane
Red Hook, NY 12571 USA
Phone: 845-758-0400
Fax: 845-758-2633
Email: curran@proceedings.com
Web: www.proceedings.com

TABLE OF CONTENTS

PLENARY

IUPAP YOUNG SCIENTIST PRIZE

PHOTONS

LEPTONS

HEAVY PARTICLES

ATTOSECOND PHYSICS

HIGH-HARMONIC GENERATION

STRONG FIELD

WEAK FIELD / FIELD ASSISTED COLLISIONS

FREE ELECTRON LASER

LOW ENERGY TO ULTRACOLD PHYSICS

STRUCTURE / SPECTROSCOPY

SINGLE / MULTIPLE (DE)EXCITATION

SINGLE / MULTIPLE IONIZATION

CHARGE-EXCHANGE

E- ATTACHMENT

E- SCATTERING

PLASMAS & PARTICLE ACCELERATION

CLUSTERS / NANOSYSTEMS / SURFACES / SOLIDS

BIOSYSTEMS

ANTIMATTER & EXOTIC PARTICLES

STORAGE RING

INSTRUMENTATION

Journal of Physics: Conference Series **1412** (2020) 011001 doi:10.1088/1742-6596/1412/1/011001

PREFACE

The 31st International Conference on Photonic, Electronic and Atomic Collisions (XXXI ICPEAC) was held in the region of Normandy, France, starting with the tutorials at the University of Caen Normandie on July 23, 2019, and followed by the main conference on July 24-30, 2019 at the Deauville International Conference Centre. Held biannually, ICPEAC is one of the largest conferences on the atomic and molecular physics calendar, and covers the recent progress in photonic, electronic, and atomic collisions with matter. With a history dating back to 1958, the ICPEAC has its second coming to France, the first of which being the 1977 edition in Paris. The organization of ICPEAC 2019, the tutorials, and some of the satellite meetings involved 15 French institutes belonging to CNRS and various universities or institutions, and located all over the country. With the selection of France as the ICPEAC XXXI venue the local organizers began their preparations six years ahead in anticipation of the event.

The conference brought together 615 participants from 43 countries with 32% of young researchers (PhD students or doctors who defended their PhD less than 4 years ago) and 28% of female scientists.

ICPEAC XXXI aroused the submission of 808 abstracts among which 41% were submitted by young scientists. In April 2019, the ICPEAC General Committee has reviewed 761 Regular Abstracts submitted before 10 March 2019 and has selected "hot topics" as Special Reports. The Local Organizing Committee has examined 47 post-deadline abstracts submitted before 15 June, 2019. The Scientific Conference Program included 749 abstracts that were scheduled for presentation in one of the five poster sessions organized according to 18 different topics representing the main fields of ICPEAC 2019. The full poster list is published in the Conference Handbook available on line and was distributed in printed form to the participants. Among these abstracts, 469 are presented in issues 6-23 of this volume of the Journal of Physics Conference Series.

In addition, five plenary lectures by Joachim Burgdörfer (Institute of Theoretical Physics, TU-Wien, Austria), Klaus Bartschat (Drake University, USA), Christoph Bostedt (Paul Scherrer Institute & Ecole Polytechnique Fédérale de Lausanne, Switzerland), Astrid Bergeat (Institut des Sciences Moléculaires, University of Bordeaux, France), and Hidetoshi Katori (RIKEN & University of Tokyo, Japan) have been presented along with 62 progress reports and 28 special reports. Detailed write-ups of some of the invited lectures and reports are presented in issues 1-5 of this volume, constituting a tangible record of the meeting.

Preceding the main conference, three distinguished scientists, Gordon Drake (University of Windsor, Canada), Anne L'Huillier (Lund Attosecond Science Center, LTH, Sweden) and Thomas Stöhlker (Helmholtz Institute Jena, Germany) gave tutorial lectures specifically designed for students and early-career researchers.

During the conference, the 2019 IUPAP Young Scientist Prize was awarded to Andreas Mooser (Max-Planck Institute for Nuclear Physics & CERN) and Christian Smorra

(CERN). The Sheldon Datz Prize for an Outstanding Young Scientist attending ICPEAC was awarded to Vincent Lemelin (Department of Nuclear Medicine and Radiobiology, Sherbrooke University, Canada).

The Local Organizing Committee would like to express its gratitude to the "Société Française de Physique", the International Union for Pure and Applied Physics (IUPAP), the European Physical Society, the City of Deauville, the Normandy Region, the University of Caen Normandie and all the conference sponsors for their financial aid which made the event possible, offered the tutorials to all the young scientists, and supported 45% of them to attend the main conference.

Finally, the Local Organizing Committee would like to thank all the ICPEAC participants for their support and contributions to the conference.

The Editors,

For the proceedings of the ICPEAC XXXI papers:
Lorenzo Ugo Ancarani, LPCT, CNRS-Université de Lorraine
Christian Bordas, ILM, CNRS-Université de Lyon
Franck Lépine, ILM, CNRS-Université de Lyon
Dominique Vernhet, INSP, CNRS-Sorbonne Université

For the proceedings of the ICPEAC XXXI abstracts:
Henri Bachau, CELIA, CNRS- Université de Bordeaux
Richard Brédy, ILM, CNRS- Université de Lyon
Olivier Dulieu, LAC, CNRS- Université Paris-Saclay
Francis Penent, LCPMR, CNRS- Sorbonne Université
Dominique Vernhet, INSP, CNRS-Sorbonne Université

ICPEAC Committees

Executive Committee

Kiyoshi Ueda - *Scientific Chair* (Japan)
Emma Sokell - *Secretary* (Ireland)
Stefan Schippers - *Treasurer* (Germany)
Fritz Aumayr - *Vice Chair* (Austria)
Hossein Sadeghpour - *Past Chair* (USA)
Alexander Dorn (Germany)
Akiyoshi Hishikawa (Japan)
Ken-Ichi Ishikawa (Japan)
Eva Lindroth (Sweden)

ICPEAC XXX (Cairns, Australia, 2017)
Anatoli Kheifets - *Chair*
Igor Bray - *Co-chair*

ICPEAC XXXI (Deauville, France, 2019)
Dominique Vernhet - *Chair*
Lamri Adoui - *Co-chair*
Olivier Dulieu - *Co-chair*

ICPEAC XXXII (Ottawa, Canada, 2021)
Tom Kirchner - *Chair*
André Staudte - *Co-chair*

ICPEAC XXXII (Hamburg, Germany, 2023)
Robin Santra - *Chair*
Michael Meyer - *Co-chair*

General Committee

Argentina
D.Arbo

Australia
M. Bromley
R. Sang

Austria
C. Lemell

Belgium
F. Remacle

Brazil
R. Costa
M. Varella

Canada
Z.-C. Yan

China
X. Cai
D. Ding
B. Wei

Czech Rep.
J. Fedor

Finland
E. Kukk

France
P. Martin

Germany
A. Gumberidze
H. Kreckel
M.Meyer
O. Smirnova

Hungary
K. Varjú

India
L.Tribedi
M. Vinodkumar

Italy
P. Bolognesi

Japan
N.Nakamura
K. Okada
M. Tachikawa
M. Takahashi

Poland
M. Pajek

Russia
A. Grum-Grzhimailo
O. Vasyutinskii

Spain
A. Palacios
R. Prosmiti

Sweden
H. Zettergren

Switzerland
F. Merkt

Turkey
Z. Nur Ozer

United Kingdom
V. Averbukh
G. Gribakin

United States
L. Argenti
I. Fabrikant
T. Killian
H. Lewandowski

LOCAL ORGANIZING COMMITTEE

CHAIR

Dominique Vernhet (INSP - Paris)

CO-CHAIRS

Lamri Adoui (CIMAP - Caen)
Danielle Dowek (ISMO - Orsay)
Olivier Dulieu (LAC - Orsay)
Emily Lamour (INSP - Paris)

MEMBERS

INSP-Paris
C. Prigent
S. Steydli
M. Trassinelli

LCPMR-Paris
T. Marchenko
F. Penent

IPN-Orsay
M. Chabot

IPCMS-Strasbourg
P.-A. Hervieux

ILM-Lyon
C. Bordas
R. Brédy
F. Lépine

CIMAP-Caen
A. Domaracka
J. Douady
P. Rousseau

PIIM-Marseille
S. Ferri

LPCT-Metz
L. U. Ancarani

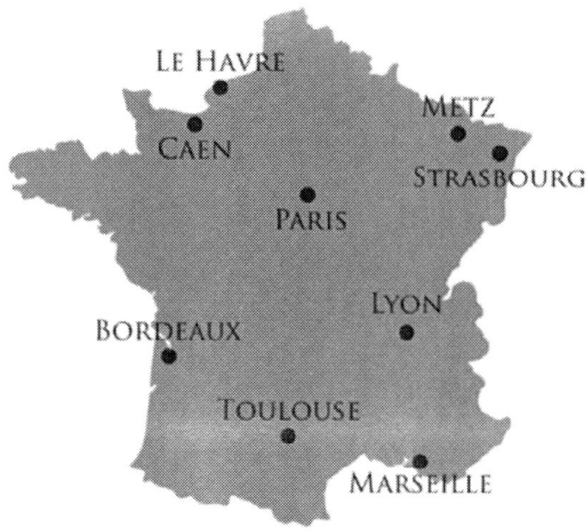

CELIA-Bordeaux
H. Bachau
V. Blanchet
F. Catoire
B. Fabre
B. Pons
A. Sopena

LOMC-Le Havre
I. Schneider

LPT-Toulouse
P. M. Dinh

ICPEAC XXXI SPONSORS

The Local Organizing Committee is very thankful to the institutions, the scientific societies, the laboratories and all the private companies and exhibitors that contributed to the success of the International Conference on Photonic, Electronic, and Atomic Conference, held in Normandy, France.

INSTITUTIONS, SCIENTIFIC SOCIETIES, LABORATORIES

Journal of Physics: Conference Series **1412** (2020) 011003 doi:10.1088/1742-6596/1412/1/011003

COMPANIES & EXHIBITORS

IOP Publishing

PANTECHNIK

Newport

Amplitude

edp sciences

APS physics

PFEIFFER VACUUM

Femto Easy

4D-3D

VACOM

FICHOU

EDWARDS

CAEN Tools for Discovery

RoentDek Handels GmbH

FASTLITE

Springer

Peer review statement

All papers published in this volume of *Journal of Physics: Conference Series* have been peer reviewed through processes administered by the proceedings Editors. Reviews were conducted by expert referees to the professional and scientific standards expected of a proceedings journal published by IOP Publishing.

60 years of attosecond physics at ICPEAC: from collisions to ultrashort pulses

Joachim Burgdörfer[1], Christoph Lemell[1] and Xiao-Min Tong[2]

[1] Institute for Theoretical Physics, Vienna University of Technology, Vienna, Austria, EU
[2] Center for Computational Sciences, University of Tsukuba, Tsukuba, Japan

E-mail: burg@concord.itp.tuwien.ac.at

Abstract. The field of attosecond physics has seen an almost explosive growth since the early 2000's and represents by now an increasing fraction of contributions to the bi-annual series of International Conferences of Photonic, Electronic, and Atomic Collisions (ICPEAC). The latter is anything but a coincidence as many of the underlying concepts of electronic and photonic dynamics are closely intertwined with atomic-scale collision processes. We illustrate this fruitful connection and its implications with the help of a few prototypical examples of current topical interest.

1. Introduction

Since the beginning of this millennium the field of "attosecond physics" or, more generally, of "attosecond science" has seen a near-exponential growth. Taking the number of recent research articles accounted by *Web of Science* with the term "attosecond" just in the title as an admittedly superficial indicator, this number rose from near zero before 1995 to 2275 as of July 2019 [1]. This disruptive development has its origin in the appearance of novel light sources with pulses as short as ~ 50 attoseconds (1 as $= 10^{-18}$s) [2, 3]. With the advent of chirped pulse amplification [4] and the frequency-comb technology [5], the generation of high-intensity laser beams as well as phase-stabilized few-cycle IR pulses, attosecond pulse trains [6] and isolated attosecond pulses [7] could be realized and are now routinely employed in many labs around the world.

The importance of well-characterized light fields on the time scale of attoseconds originates from the fact that this is the natural time scale of electronic dynamics in valence shells of atoms, molecules and solids. As already used by Niels Bohr in his atomic model [8], the orbital period of the atomic electron in hydrogen, $T_n = 2\pi n^3$ (in atomic units), is $T_1 \simeq 150$ as for the ground state (principal quantum number $n = 1$). Attosecond pulses thereby offer the opportunity to observe, to interrogate and, eventually, to actively control and manipulate electronic dynamics. A few examples for the realization of this promise involving state-of-the art light fields will be briefly discussed in the following.

Exploitation of electromagnetic pulses on the attosecond time scale to induce electronic transitions in atoms, molecules and solids has a long history in the field of atomic collisions for almost a century. Following up on Bohr's work on stopping of charged particles in matter [9], Weizsäcker [10] and Williams [11] described in the thirties of the 20th century the interaction of a fast charged particle passing by an atom in terms of a time dependent electromagnetic field pulse of "virtual photons" whose duration is on the attosecond scale. Consequently, the probability of Coulomb excitation peaks when the Fourier spectrum of the virtual photon field, $\omega \approx 1/\tau_c = v/2a$ (v is the collision velocity, a the typical linear dimension of the atomic target and τ_c the collision time), matches the atomic excitation energy $\Delta E = \varepsilon_f - \varepsilon_i$. This celebrated Massey criterion [12] often referred to as the velocity-matching criterion has been one of the corner stones of atomic collision physics, the overarching theme of ICPEAC now celebrating its 60th anniversary. (Sir Harrie Massey was involved in the third edition of ICPEAC in London in 1963). Fig. 1 illustrates the physics of the virtual photon field and of the resultant Massey criterion using the data for the total excitation cross section of the prototypical collision system of protons colliding with hydrogen ($H^+ + H \rightarrow H^+ + H(n=2)$). Converting the collision velocity scale to a time scale clearly shows that the excitation probability is highest for a collision time of about 50 as.

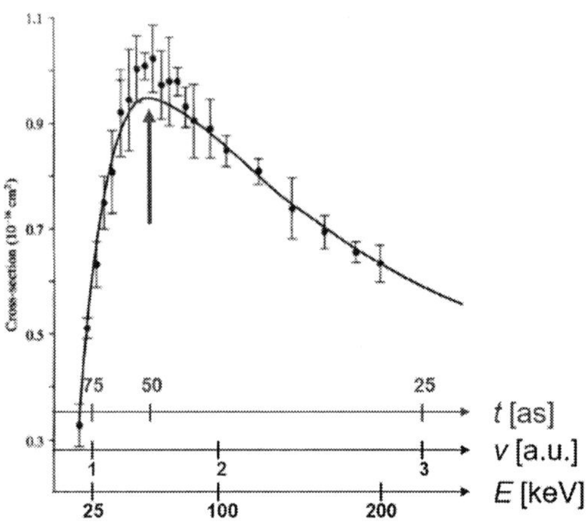

Figure 1. Excitation cross section for proton impact on hydrogen, $H^+ + H(1s) \rightarrow H^+ + H(n=2)$, as a function of projectile velocity v and collision time $\tau_c = v/2a$ (a: atomic radius). Experimental data [13], solid line: atomic close coupling calculation [14].

Since at that time the rapidly developing accelerator technologies, mostly for nuclear applications, were the driving force for the burgeoning field of atomic collisions, the primary focus was on collision energy or collision velocity rather than on collision time as key parameter for classifying and describing the underlying processes. This may have been one reason why attosecond physics, even though ubiquitously present, was not yet explicitly recognized as such. A deeper reason was likely the conceptual difficulty in mapping time onto a physical variable going back to the early days of quantum physics [15, 16]. As will be illustrated in the following, recent progress in the field of attosecond physics has helped to explore and to elucidate the role of time in quantum dynamics.

The aim of this presentation given as a plenary talk at ICPEAC XXXI is to demonstrate the close conceptual interconnection between the fields of atomic collisions and of light-field driven attoscience with the help of a few prototypical examples. These examples are chosen to highlight the interplay between photonic and collision processes on ultrashort timescales and are not meant to be representative or a review of the diverse and rapidly growing field of attosecond physics. Atomic units ($e = \hbar = m_e = 1$) are used throughout the text unless otherwise stated.

2. Synthesizing collision-like fields: the half-cycle pulses

The temporal shape of the collisional virtual photon pulse (Fig. 2) analyzed by Weizsäcker [10] and Williams [11],

$$F_\perp(t) = q\,\frac{b}{[b^2+(vt)^2]^{3/2}}, \tag{1}$$

$$F_\parallel(t) = q\,\frac{vt}{[b^2+(vt)^2]^{3/2}}, \tag{2}$$

(q: charge of projectile, b: impact parameter) form a half-cycle pulse (HCP) perpendicular to the velocity vector (Eq. 1, Fig. 2a) and a distorted single-cycle pulse along the velocity vector (Eq. 2, Fig. 2b). For fast collisions ($v \gg 1$ a.u.) the temporal width of the collisional HCP, τ_c, can be as short as a fraction of an attosecond [17]. Correspondingly, their Fourier spectra,

$$F_\perp(\omega) = \frac{q\omega}{v^2}\cdot K_1(\omega b/v), \tag{3}$$

$$F_\parallel(\omega) = \frac{q\omega}{v^2}\cdot K_0(\omega b/v), \tag{4}$$

where K_0 and K_1 are modified Bessel functions of the second kind, extend to the XUV and, for relativistic collisions, well into the X-ray regime. In fact, because of the time (or phase) shift between the two orthogonal components the vectorial field of the passing-by charged particle forms an elliptically polarized half-cycle pulse capable of transferring orbital angular momentum to the electrons of the target. This rotational coupling results in orientation and alignment of the atomic charge cloud and has been a frequently explored topic at many ICPEACs in the 70's and 80's [18, 19]. Collisions are thus the ideal source of ultrashort pulses in terms of both their absolute duration and the minimum number of cycles subtended. However, in collisions, it is difficult to independently control and vary v and b while suppressing competing processes occurring at close distances, in particular charge transfer.

Synthesizing collision-like pulses under well-controlled conditions, in particular HCPs, was therefore pursued to study the non-linear electronic response to HCPs. Freely propagating HCPs

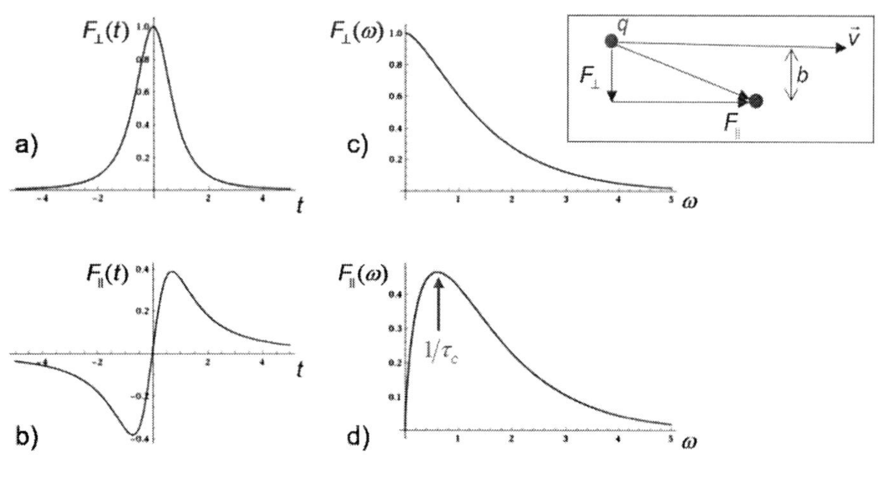

Figure 2. Temporal evolution of the projectile Couplomb field acting on the target perpendicular (a) and parallel (b) to the collision velocity $\vec{v}$ (insert: straight-line trajectory with impact parameter b). Fourier spectrum of the collisional field pulse perpendicular (c) and parallel (d) to $\vec{v}$.

with durations τ_{HCP} of a picosecond or less can be generated using a photoconducting switch triggered by a femtosecond laser [20, 21]. This is still slow compared to the classical orbital period T_n of ground-state atoms ($n = 1$) but ultrafast ($\tau_{\text{HCP}} \ll T_n$) for high Rydberg states. Impulsively driven ultrafast electronic dynamics was therefore first explored in the realm of Rydberg physics [22].

One difficulty in using freely propagating HCPs is that they are not inherently truly unidirectional, i.e., do not exactly match Eq. 1. They rather comprise a short intense unipolar electric field pulse followed by a weaker pulse of opposite polarity and much longer duration. The resulting pulse shape has been referred to as an "asymmetric monocycle" and resembles somewhat Eq. 2. The effects of the reverse polarity pulse must be considered in analyzing data. Nevertheless, significant momentum transfer to a Rydberg electron can be achieved because of the very different time scales associated with the unipolar pulse and the ensuing opposite-polarity tail.

Truely unipolar but non-propagating HCPs can be created by applying voltage pulses from a fast pulse generator(s) to an electrode. This approach has the advantage that by using a number of pulse generators in conjunction with pulse combiners and splitters it is possible to engineer complex pulse sequences that can be measured directly using a fast probe and sampling oscilloscope. The minimum pulse durations that can be generated, however, are typically limited to $\gtrsim 0.5 - 1$ ns meaning that the true impulsive limit can only be reached for atoms with very high $n \gtrsim 300 - 400$.

In the limit $\tau_{\text{HCP}} \ll T_n$, a single pulse $\vec{F}_{\text{HCP}}(t)$ simply delivers a collision-like impulsive momentum transfer or "kick"

$$\Delta\vec{p} = -\int_{-\infty}^{\infty} F_{\text{HCP}}(t)\, dt \tag{5}$$

to the atomic electron. The response of the electron becomes independent of the detailed shape of the pulse as only the integral matters. Classically, the application of such an HCP to an atomic electron with initial momentum $\vec{p}_i$ and energy E_i results in an energy transfer

$$\Delta E = \varepsilon_f - \varepsilon_i = \frac{(\vec{p}_i + \Delta\vec{p})^2}{2} - \frac{p_i^2}{2} = \frac{\Delta p^2}{2} + \vec{p}_i \cdot \Delta\vec{p}, \tag{6}$$

where ε_f is the final electron energy.

Quantum mechanically, the initial atomic state $|\phi_i\rangle$ is shifted (or "boosted") in momentum space by the application of the HCP, $|\psi(t = 0)\rangle = |\phi_i^B\rangle = e^{i\Delta\vec{p}\cdot\vec{r}}|\phi_i\rangle$. The corresponding expectation values of energy and momentum are

$$\langle\varepsilon\rangle_{t=0} = \langle\phi_i^B|H_{\text{at}}|\phi_i^B\rangle = \langle\phi_i|H_{\text{at}}|\phi_i\rangle + \frac{\Delta p^2}{2} + \langle\phi_i|\vec{p}\cdot\Delta\vec{p}|\phi_i\rangle\,, \tag{7}$$

$$\langle\vec{p}\rangle_{t=0} = \langle\phi_i^B|\vec{p}|\phi_i^B\rangle = \Delta\vec{p} + \langle\phi_i|\vec{p}|\phi_i\rangle\,, \tag{8}$$

where H_{at} is the atomic Hamiltonian. Since for bound states $\langle\phi_i|\vec{p}|\phi_i\rangle = 0$, $\langle\varepsilon\rangle_{t=0} = \varepsilon_i + \Delta p^2/2$, in agreement with the classical prediction. The "boosted" final electronic wavefunction can be expanded in terms of the field-free atomic basis as

$$|\psi(t)\rangle = \sum_j a_j(0)e^{-i\varepsilon_j t}|\phi_j(0)\rangle \tag{9}$$

and comprises a coherent superposition of many states having a broad distribution in $j = (n, \ell, m)$. Eq. 9 represents a nonstationary wavepacket whose evolution is governed by the atomic, i.e., field-free Hamiltonian H_{at}. The evolution of $|\psi(t)\rangle$ has been observed by applying a second probe HCP of opposite polarity after some time delay τ_D (inset Fig. 3). This pulse

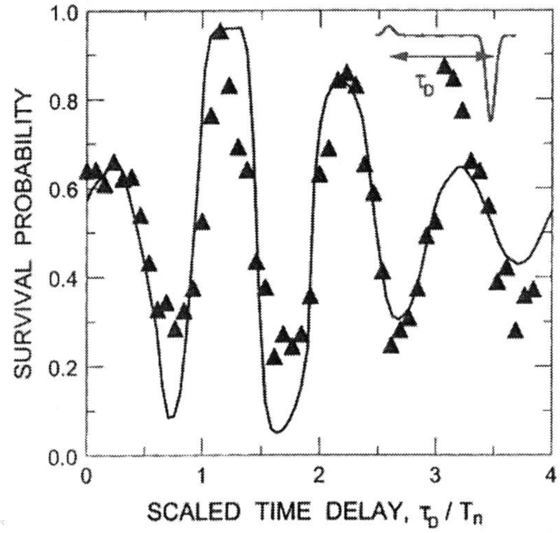

Figure 3. Measured survival probability for polarized K($n = 351$) atoms following application of two oppositely directed HCPs as a function of scaled time delay τ_D/T_n. The HCP sequence is illustrated in the inset. The first HCP delivers a scaled impulse $\Delta p_0 = \Delta p/p_n = n\Delta p = -0.1$, the second probe pulse a scaled impulse $\Delta p_0 = 0.9$. Solid line: results of CTMC simulations (adapted from [22]).

sequence resembles, in fact, the longitudinal component of the collisional pulse (Eq. 2). Unlike the collisional pulse, however, it features exquisite tunability of amplitude and delay between the two HCPs. Pronounced quantum beats are observed in the survival probability of K($n = 351$) Rydberg atoms [23] well reproduced by quasi-classical CTMC simulations (Fig. 3). This close classical-quantum correspondence for beats is an instructive illustration of Bohr's correspondence principle and points to periodic charge fluctuations of the revolving electron as their origin.

HCP-driven Rydberg atoms, discussed at several ICPEACs (see, e.g., [24]), can be viewed as a precursor to current attosecond pulse interactions with atoms. Several fundamental concepts related to the non-linear electronic response to ultrashort pulses could be explored, however, only on the nano-to-picosecond time scales and not yet for valence electrons. Only with the advent of attosecond pulses, watching and probing electronic motion near the electronic ground state became feasible.

3. Attosecond time delays

The availability of carrier-envelope phase (CEP) controlled IR pulses and of attosecond pulses has allowed to "clock" fundamental electronic processes. Observing quantum dynamics in the time domain, often referred to as attosecond chronoscopy [25], can be viewed as source of information complementary to that of spectroscopy.

One of the first major breakthroughs of attosecond chronoscopy was the observation of a finite time delay of photoemission relative to the arrival of the crest of the ionizing XUV pulse [26]. These time delays, typically of the order of about ten attoseconds, are directly related to the Eisenbud-Wigner-Smith (EWS) time delay [27, 28, 29] originally introduced for resonant elastic scattering,

$$\tau_{\mathrm{EWS}} = i\hbar S^{\dagger}(E)\frac{\partial}{\partial E}S(E), \tag{10}$$

in terms of the scattering matrix $S(E)$. Felix Smith, member of the organizing committee of the first ICPEAC in 1958, presented the time-delay operator in matrix form for multi-channel problems at ICPEAC II in 1961 [30]. The Hermitian operator τ_{EWS} features real eigenvalues, the eigentime delays, providing an explicit and constructive example of a bona-fide physical observable of quantum theory. The EWS time delay (Eq. 10) satisfies the correspondence principle of quantum to classical mechanics. τ_{EWS} can be shown to converge in the limit of small deBroglie wavelength $\lambda_{\mathrm{dB}} \to 0$ to the classical time delay τ_{cl} given, e.g., for a 1d scattering problem for a potential $V(x)$, by

$$\tau_{\mathrm{cl}} = \lim_{X\to\infty}\left[\int_{-X}^{X} dx \frac{1}{\sqrt{2m(E-V(x))}} - 2\frac{X}{\sqrt{2mE}}\right]. \tag{11}$$

Measurement of EWS time delays for photoionization has been realized by two alternative pump-probe settings involving an attosecond pulse (or pulse train) as a pump ionizing the atom and a few-cycle IR pulse as probe, both temporally near-perfectly correlated with each other through the process of high-harmonic generation (HHG). An intense IR pulse generates high-frequency radiation by strongly non-linear up-conversion [31, 32, 33, 34]. The underlying mechanism is nicely captured by the instructive semiclassical three-step model [35, 36] of tunneling ionization, acceleration by the strong IR field, and rescattering at the ionic core resulting in coherent radiative recombination. It reveals the close connection to the physics of electron-ion recombination, a major topic at many editions of ICPEAC's (see, e.g., [37, 38]).

The same IR pulse, suitably attenuated, provides in addition a temporally well correlated moderately strong IR probe field ($\sim 10^{11}$ W/cm^2 for streaking). The latter ponderomotively shifts the momentum of the outgoing photoelectron $\vec{p}_f$ relative to its field-free value, $\vec{p}_0$, by the vector potential of the IR field $\vec{A}_{\mathrm{IR}}(t)$,

$$\vec{p}_f(t) = \vec{p}_0 - \vec{A}_{\mathrm{IR}}(t)/c \tag{12}$$

at the instant of time t of its appearance in the continuum. By varying the temporal overlap (or relative phase) between the IR and the attosecond XUV pulses and, hence, by the sinusoidal modulation of the energy of the outgoing electron $E(t) = p_f^2(t)/2$, this attosecond streaking protocol [26, 39] allows to time the photoemission with attosecond precision. Attosecond streaking can be viewed as a "classical clock" mapping time onto energy. It can also be viewed as the (multi-photon) classical limit of a "quantum clock", the interferometric RABBITT technique (reconstruction of attosecond bursts by interference of two-photon transitions [6, 40, 41]). The latter utilizes one-photon transitions between continuum states induced by weaker IR pulses to determine phase differences in side bands between the photoelectron wavepackets ejected by adjacent odd harmonics of the XUV attosecond pulse train.

It is worth recalling that the current attosecond streaking technique closely mirrors a streaking method on the femtosecond scale employed in the field of atomic collisions then referred to as post-collision interaction (PCI) effect. Numerous PCI studies were presented at ICPEACs in the 80's. Streaking by PCI was, for example, used to time resolve the collective motion of autoionizing states coherently excited in Li^{+}+ He $\to$ He^{**}+ Li^{+} collisions [42, 43]. The energy of the electron emitted by the autoionizing states (lifetime $\sim$ 20 fs) was modulated relative to its field-free value by its Coulomb interaction $\sim 1/R(t)$ with the slowly receding Li^{+} projectile still in the vicinity of the target when autoionization occurs. Thereby, the emission time of the electron was mapped onto its energy shift relative to the kinetic energy of autoionization electron in free space.

The pioneering attosecond streaking experiment [26] for the relative EWS time delay between photoionization of the 2s and 2p electrons of neon by photons with energies of about 110 eV yielded a time delay of $\Delta\tau_{2\mathrm{s}-2\mathrm{p}} \approx 20$ as. Accompanying calculations [26] as well as a large number of subsequent theoretical studies [44, 45, 46, 47] could confirm that the formation of the outgoing 2p wavepacket is, indeed, delayed relative to the 2s wavepacket but found considerably smaller values (Fig. 4). As attosecond streaking features excellent time resolution but only limited energy resolution given by the width of the streaking modulation (see Eq. 12), $\delta E = |p_0 A_{\mathrm{IR}}|$, the streaking trace of the main line of the 2s electron overlaps with the emission from shake-up transitions accompanying the 2p emission which may distort the observed time delay [47].

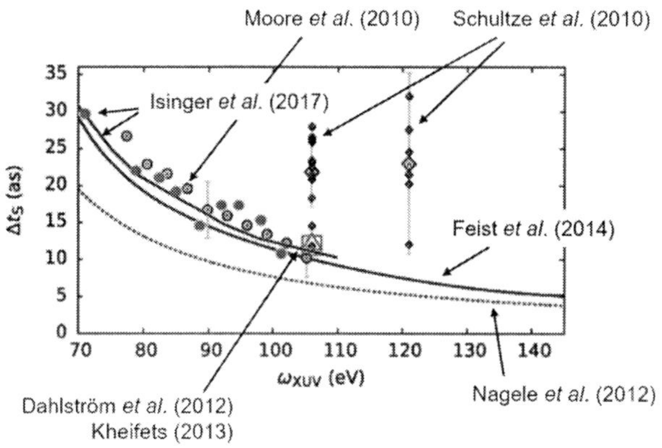

Figure 4. Relative time delay between the photoelectron emitted from Ne(2s) and from Ne(2p). Experiments [26, 49] and theory [26, 44, 45, 46, 47, 49]

Unlike for neon, photoemission from helium allows resolving shake-up transitions from main lines by streaking. Measurements and calculations demonstrated with one attosecond precision large differences between τ_{EWS} of the main line and shake-up transitions [48]. Exploiting the considerably higher energy resolution of RABBITT, Isinger *et al.* [49] could subsequently determine the relative time delay between the 2s and 2p main lines of neon without distortion by shake-up admixture and found good agreement with theory [44, 45, 46, 47] at somewhat lower photon energies.

Extracting τ_{EWS} for photoionization from attosecond streaking or RABBITT measurements requires the correction by an additional time shift, τ_{CLC}, due to the Coulomb-laser coupling (CLC) for streaking or τ_{CC} due to IR induced one-photon continuum-continuum (CC) transitions for RABBITT. Both τ_{CLC} and τ_{CC} are of the same origin, the interaction of the IR probe field with the outgoing electron in the atomic Coulomb field and their values have been found to be nearly identical, $\tau_{\mathrm{CLC}} \simeq \tau_{\mathrm{CC}}$ [25, 50, 51]. Only recently, these additional time shifts often referred to as measurement related effects have been recognized to be of the same physical origin as τ_{EWS} for photoionization, i.e., for a bound-continuum transition. τ_{CC} (or τ_{CLC}) can be viewed as the EWS time delay for continuum-continuum scattering by absorption or emission of IR photons [52].

In the current generation of experiments, only relative time delays either between different ionization processes of the same atomic species (e.g., Ne(2p) vs. Ne(2p) ionization) or different atoms in a gas mixture (e.g., pairs of rare gases [53]) are accessible. To establish an absolute time scale for delays, the reference to (virtually) exact solutions of the Schrödinger equation is required that has become available, e.g., for helium [54, 55].

Timing of the photoelectric effect, i.e., the determination of "time zero" for emission from solid surfaces poses an even bigger challenge as it represents a true many-body problem in an extended system. Observation of the photoelectron implies tracing out the "environmental" degrees of freedom of such a many-body system as concomitant electronic excitation or phonon scattering are not observed. The resulting time delay is therefore associated with partially decoherent dynamics rather than a fully coherent wavepacket. The multi-step process of primary excitation to the conduction band, subsequent transport and scattering and, eventually, transmission through the solid-vacuum boundary is accessible to classical transport simulations [56]. Time resolved photoemission depends not only on the electronic structure of the topmost layers but, at the same time, also probes the non-linear electronic response of the target surface to (moderately) strong IR fields and the electronic transport on the attosecond time scale.

The first measurement of the relative timing between emission from the conduction band (WCB) and from the 4f level (W4f) of tungsten [57] resulted in remarkably long time delays of

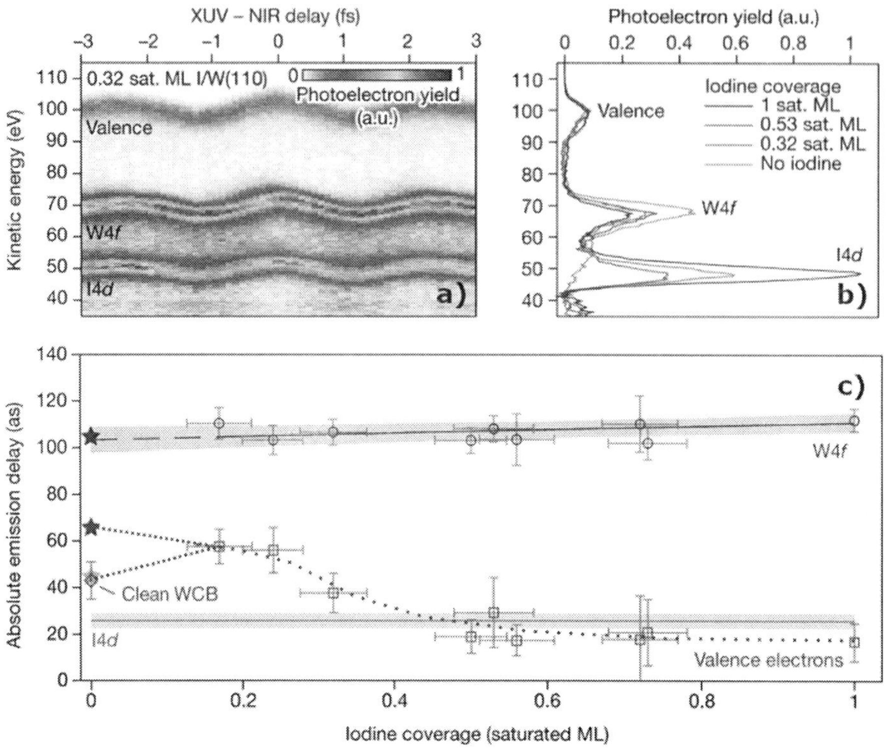

Figure 5. Measurement of absolute time delay of photoemission from tungsten using iodine as chronoscope. a) streaking trace b) photoelectron spectrum c) delay of the core level W4f and valence band as a function of the coverage of the W surface by chronoscope atoms. Delay of the W conduction band (WCB) for a clean surface is marked. Stars result from transport simulations: blue (W4f), green (WCB without surface states) and yellow (WCB including surface states). Dotted lines to guide the eye. (adapted from [61]).

$\sim$ 100 as and stimulated a large number of theoretical investigations (e.g. [56, 58, 59]). The origin of such time delays has remained an open question for a long time. Considerable progress has been made by controlled deposition of Mg adlayers on the W surface allowing for the accurate relative timing between the 2p-core level of Mg adlayers and the substrate 4f level of W [60]. A large fraction of the relative time delay could be unambiguously attributed to the transport from the substrate through the top adlayer. Very recently, the absolute timing of the photoelectric effect could be experimentally determined for the first time [61]. To this end, iodine was used as an atomic chronoscope. By a reference measurement of the relative time delay between the chronoscope atom and helium for which the absolute time delay τ_{EWS} can be determined ab-initio with (sub)attosecond precision, the time delay for 4d emission of the chronoscope iodine atom can be placed on an absolute scale. In turn, by depositing I on the tungsten surface, the absolute timing of both 4f core level and conduction band emission could be determined (Fig. 5). Furthermore, quenching the contribution from localized surface states by increasing the surface coverage by adsorbate atoms results in a non-monotonic time delay for conduction band electrons. This observation allowed to identify the important contribution of surface states which feature a very short time delay $\tau_{\mathrm{SS}} \sim 10$ as. Surface states are, in part, responsible for the short mean time delay of the conduction band of clean W of $\tau_{\mathrm{CB}} \sim 40$ as. These results highlight the sensitivity of the observable time delay to both the local electronic structure and to transport.

4. Quantum beats and decoherent dynamics

Another class of time resolved observables is associated with the appearance of quantum beats. A sudden perturbation of the initial state creates a wavepacket composed of a coherent superposition of excited eigenstates (see Eq. 9). An ultrashort pulse, in particular an attosecond pulse with duration τ_p of a few hundred attoseconds causes such impulsive perturbation thereby coherently exciting many electronically excited bound states and continuum states within the reach of its spectral width $\sim \hbar/\tau_p$ of several eV. The reduced density matrix of the system resulting from tracing out unobserved degrees of freedom (the "environment" E) [62],

$$\rho(t) = \mathrm{Tr}_E \left(|\psi(t)\rangle\langle\psi(t)| \right) , \tag{13}$$

describes the time evolution of the resulting partially decoherent wavepacket [63, 64]. Attosecond/IR pump-probe pulse sequences are well suited to generate and to observe the ensuing (de)coherent electronic dynamics that manifests itself in (damped) oscillations, the quantum beats, in, e.g., the one-particle density $\langle\vec{r}|\rho(t)|\vec{r}\rangle$ (see also section 2).

Quantum beat spectroscopy has a long history in the field of electronic and atomic collisions going back (at least) to beam-foil spectroscopy in the late 60's [65]. Fast atomic or ionic species were impulsively and coherently excited near the exit surface after traversing self-supporting thin carbon foils. Quantum beats were monitored in the time resolved down-stream photoemission from fast projectiles with the time-resolution severely limited by the spatial resolution of the detection. Accordingly, beats on the nanosecond (or GHz) scale such as fine-structure quantum beats in light ions [66] and Lamb shift s-p coherence beats in hydrogenic systems [67, 68] could be resolved. Collisions are an ideal tool for providing impulsive and sudden perturbations and, thus, for generating coherences over a remarkably wide range of energy and time scales extending from nanoseconds for fine-structure quantum beats in hydrogen to the (sub-) attosecond time scale for quasi-molecular X-ray emission in energetic highly-charged ion-atom collisions [69].

The availability of well-phase controlled XUV/IR pump-probe pulse sequences on the (sub-) femtosecond time scale has opened novel opportunities to generate and probe quantum beats in electronic systems on ultrashort time scales. One pioneering example was the observation of fine structure quantum beats in Kr^+ by attosecond transient absorption (ATA, [70]) where the $\mathrm{p}_{1/2}$-$\mathrm{p}_{3/2}$ beat period is 6.2 fs rather than $\sim$ 0.12 ns in hydrogen. Another is the charge migration in molecules [71, 72, 73], i.e., oscillations in the electron or hole density spreading over the molecule due to the impulsive excitation of a coherent superposition of many-electron states. These oscillations in the electronic system are, in general, fully reversible. A net irreversible charge transport occurs only through a "collapse of the wavepacket", a decoherent process induced, e.g., by coupling to vibronic degrees of freedom resulting in molecular rearrangement or dissociation.

The build-up of a Fano resonance in the time domain gives rise to quantum beats in the continuum [74, 75] by interference between the direct photoionization path with an amplitude $\sim \langle c|\vec{r}|i\rangle$ and the indirect ionization path through the quasi-bound resonance ($|R\rangle$) $\sim \langle R|\vec{r}|i\rangle$ which subsequently decays to the continuum state $|c\rangle$ via autoionization $\sim \langle c|V_{ee}|R\rangle$. The beat amplitude at energy E and time t after impulsive excitation is of the form

$$A(E,t) \sim e^{-\Gamma/2\,t} \cos[(E - E_R)t] \,. \tag{14}$$

The crests of the beat "wave" forms hyperbolas with $(E - E_R)t = n\pi$ (Fig. 6) which converge to the energetic position of the resonance E_R in the limit $t \to \infty$ while the amplitude is damped by the finite lifetime of the resonance ($\sim 1/\Gamma$). Quantum beats in autoionization provide a prototypical example of intrinsic partially decoherent dynamics. The recent first experimental observation of the build-up of Fano resonances in helium by ATA [76] and by RABBITT [77] has been a significant achievement of attosecond physics. It provided novel insights into the temporal

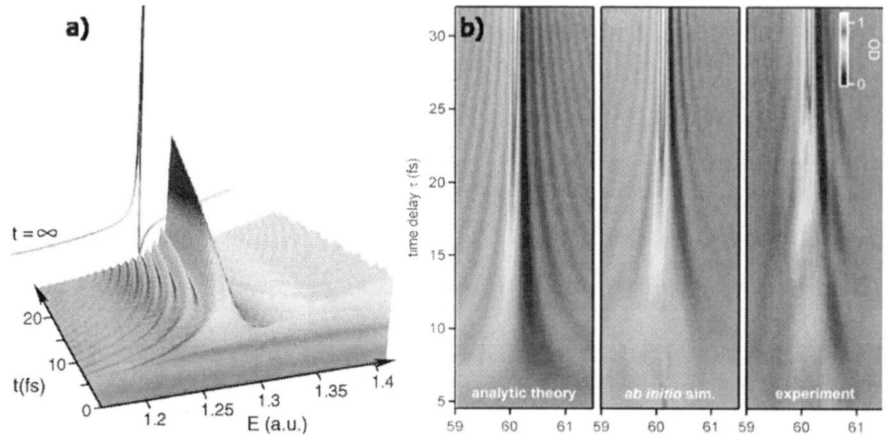

Figure 6. Build-up of Fano resonance in the time domain. a) Evolution of the photoelectron spectrum and asymptotic Fano profile, b) comparison between experiment, ab-initio simulation and simplified analytical model (adapted from [75, 76]).

evolution of wavepackets in the structured continuum and direct observation of quantum beats otherwise not visible in the well-established stationary ($t \to \infty$) Fano profile of the resonance.

The recent break-through in synthesizing intense optical attosecond (OAS) pulses [78] resembling half-cycle pulses with an effective temporal width below one femtosecond and a spectral distribution extended from $\sim$ 1 eV to about $\sim$ 4 eV has enabled for the first time to impulsively generate electronic bound-state wavepackets in atoms involving electronic inter-n coherences without being overshadowed by ionization. With a time-delayed weaker optical attosecond pulse as a probe, the electronic wavepacket motion can be mapped onto a quantum beat signal in photoemission from the excited states [79]. First experimental tests are underway [80]. The Ly_α and Ly_β emission spectra of hydrogen simulated for an OAS pump-probe pulse sequence as a function of delay time (Fig. 7) displays, indeed, an intricate beat pattern that results from strong-field induced interferences between a multitude of excitation paths.

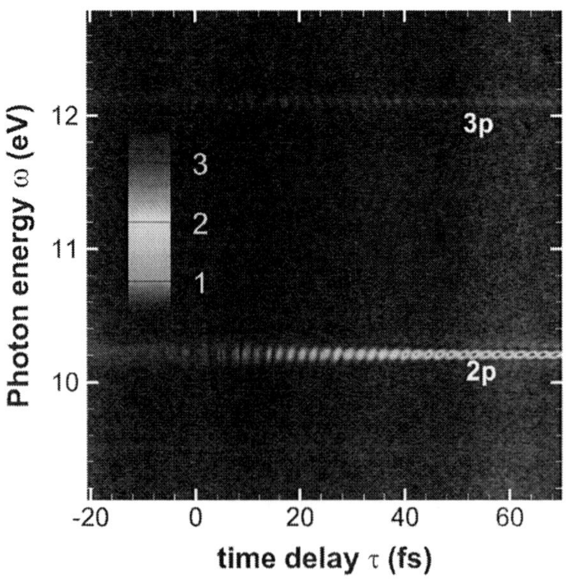

Figure 7. VUV emission spectrum as function of time delay between the OAS pump ($I = 5 \times 10^{13}$ W/cm^2) and OAS probe pulse ($I = 2 \times 10^{12}$ W/cm^2). The Ly_α and Ly_β emission lines are labelled by the initial state of the transition (from [79]).

5. Time resolved holography and diffraction

Tunneling ionization by a strong IR pulse creates a time-periodic sequence of ionization bursts confined to a small fraction of the optical cycle each emitted near a maximum of the driving field. This sequence of bursts gives rise to an ionization spectrum with approximately equispaced peaks in energy [81] with the peak spacing given by the IR photon energy often referred to as above threshold ionization (ATI). It was realized early on [82] that ATI can be described by a semiclassical two (or three) step model that could also account for HHG (see Sect. 3), however, with elastic electron-ion scattering rather than radiative recombination as the third step. While the ATI peaks are, semiclassically, the signature of temporal intercycle interferences, additional modulations not equispaced in energy result from intracycle interferences [83].

A well-controlled few-cycle IR pulse generates only few ionization bursts each of which extending typically only $\sim$ 100 as. Their relative strength can be controlled by the carrier-envelope phase (CEP). For a pulse with effectively only two subsequent ionization bursts, interferences in the photoelectron spectrum signify a temporal double slit [84] with spacing between the slits of $T/2$ and a width of each slit of $\sim$ 100 as. The resulting 2d momentum distribution of the outgoing electron along the laser polarization axis, $p_{\parallel}$, and perpendicular to it, $p_{\perp}$, displays a remarkably complex diffraction pattern (Fig. 8f) that carries both Ångstrom-scale spatial and attosecond-scale temporal information. The interference of two electron microbursts emitted from an atomic point source can be viewed as holographic image of the potential landscape the receding electron traverses [85]. This can be easily visualized by considering a single optical cycle consisting of two HCPs of opposite polarity (Fig. 8a,d) resembling the collisional field (c.f. Fig. 2). After the first half-cycle the tunneling ionized 2d momentum distribution is smooth and structureless centered about the mean momentum $\langle p_{\parallel}\rangle$ corresponding to the momentum transfer by a "kick" (see Eq. 5 and Fig. 8c). The second HCP reverses the direction of this wavepacket and, moreover, creates a second ionization burst traveling in the same direction (Fig. 8e). This second wavepacket serves as the reference wave for the hologram (Fig. 8f). The information of the potential landscape, in the present case of the atomic hydrogen

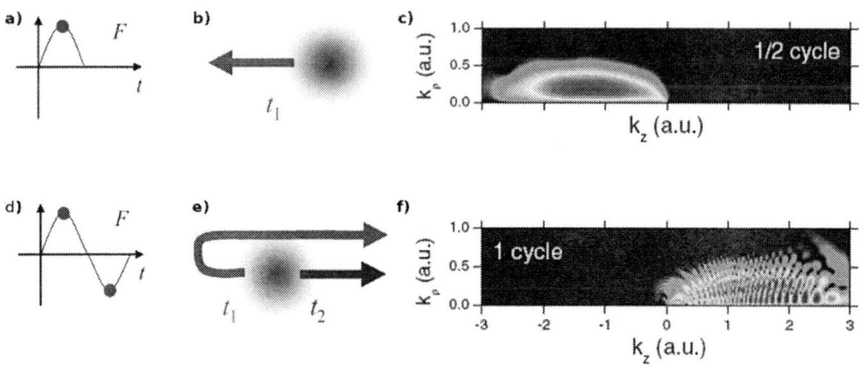

Figure 8. Temporal double slit for a single-cycle IR pulse ionizing hydrogen ($I = 2 \times 10^{14}$ W/cm^2). a-c: first half cycle. a) temporal distribution of the electric field $F(t)$ with maximum of tunneling ionization at $t = t_1$. b) orientation of unidirectional emission c) resulting vectorial momentum distribution, $P(k_z, k_\rho)$, along (k_z) and perpendicular (k_ρ) to the laser polarization axis. d-f: as a-c but after the full cycle featuring the rescattered first and a tunnel-ionized second ionization burst. The resulting diffraction pattern can be viewed as a hologram of the time-dependent potential landscape traversed by the wavepacket liberated during the first half cycle between times t_1 and t_2 (adapted from [85]).

Coulomb field and the IR field, traversed by the first wavepacket between the two subsequent field extrema is imprinted on the phase of the rescattered wavepacket. The development of the full potential of this holographic imaging technique with sub-fs temporal and Ångstrom-scale spatial resolution, e.g., for molecules, is still in its infancy. First proof-of-principle experiments for metastable xenon strong-field ionized by a mid-IR free electron laser (FEL) pulse [86] and for simple molecules [87] have demonstrated its feasibility.

The maximum energy of the redirected and returning electron is of the order of $3U_p$ where $U_p = F_0^2/(4\omega_{\rm IR}^2)$ (F_0 peak field strength) is the ponderomotive energy associated with the quiver motion of a free electron in the laser field. For large U_p, i.e., strong mid-IR pulses, the returning wavepacket can be approximated by a plane wave of a free electron. Accordingly, the interaction with the residual ion can be viewed as elastic or inelastic electron-ion scattering and, for HHG, as radiative recombination. This is the conceptual underpinning of the three-step model for HHG [33, 34, 35, 36] and of the quantitative rescattering theory [88, 89]. It also reveals another close link to the field of electronic and atomic collisions. Cross sections for electron diffraction, electron-impact ionization and radiative recombination can be directly employed in estimates for strong-field processes. As the electron "beam" in the rescattering process is temporarily confined to sub-femtoseconds, ultrafast diffraction and microscopy are coming into reach.

6. Outlook

With further progress in the development of novel light sources, the rapidly developing field of attosecond physics is expected to expand and widen its scope. It is tempting to speculate which directions this field will take in the future. Certainly, Yogi Berah's famous dictum "*It is always difficult to make predictions – in particular about the future*" applies here. While some of the destinations are already well within reach, others may be longshots.

Attosecond chronoscopy is increasingly complementing conventional spectroscopy, providing information not easily accessible in the frequency domain. One prominent example is relaxation and decoherence dynamics [90]. Analyzing the "arrow of time" of irreversibility is greatly facilitated by observing processes in the time domain and likely to become an important tool, in particular for more complex systems [91].

Another topic already well underway is ultrafast magnetic and spin dynamics as well as dichroism [92] accessible now with cirularly polarized attosecond XUV pulses. First observations of ultrafast demagnetization on the (sub-) femtosecond time scale have been reported [93, 94]. The formation of structured attosecond light beams carrying high orbital angular momenta [95] and electron vortices by broad-band circular attosecond pulses [96] appear to be in reach.

One obvious direction for future progress will be the further increase in intensity of attosecond XUV pulses so that highly non-linear attosecond processes such as multi-photon ionization and attosecond pump–attosecond probe scenarios can be realized. Further improvement of HHG sources as well as sub-fs pulse control of FEL will be key. One challenging long-term goal would be to reach the high-intensity high-frequency regime of atomic stabilization where the intense light field inhibits rather than stimulates photoelectron emission [97, 98].

Another tempting direction to be pursued is the quest for even shorter pulses on the sub-attosecond time scale reaching $\sim 10-100$ zeptoseconds (1 zs $= 10^{-21}$ s). While this has already been accomplished by high-energy collisions [17], HHG by mid-IR pulses reaching keV photon energies [99] provide the first steps towards its realization for light fields. It is, however, still a wide open question which novel physical observable can be explored with zeptosecond pulses.

Acknowledgments

JB and CL were supported by Austrian Science Fund FWF (Proj. Nos. SFB-F41, W1243), XMT was supported by a Grants-in-Aid for Scientific Research (JP16K05495) from the Japan Society for the Promotion of Science.

[1] Web of Science, http://apps.webofknowledge.com
[2] Krausz F and Ivanov M Yu 2009 *Rev. Mod. Phys.* **81** 163
[3] Guinness World Records, https://www.guinnessworldrecords.com/world-records/shortest-flash-of-light
[4] Strickland D and Mourou G 1985 *Opt. Comm.* **56** 219
[5] Udem Th, Reichert J, Holzwarth R and Hänsch T W 1999 *Opt. Lett.* **24** 881
[6] Paul P M, Toma E S, Breger P, Mullot G, Auge F, Balcou P, Muller H G and Agostini P 2001 *Science* **292** 1689
[7] Drescher M, Hentschel M, Kienberger R, Tempea G, Spielmannn C, Reider G A, Corkum P B and Krausz F 2001 *Science* **291** 1923
[8] Bohr N 1913 *Philos. Mag.* **26** 1
[9] Bohr N 1913 *Philos. Mag.* **25** 10
[10] von Weizsäcker C F 1934 *Z. Phys.* **88** 612
[11] Williams E J 1935 *Kgl. Danske Videnskab. Selskab Mat.-fys. Medd.* **13** No. 4
[12] Massey H S W and Smith R A 1933 *Proc. R. Soc. Lond. A* **142** 142
[13] Park J T, Aldag J E, George J M and Peacher J L 1976 *Phys. Rev. A* **14** 608
[14] Ford A L, Reading J F and Hall K A 1993 *J. Phys. B: At. Mol. Opt. Phys.* **26** 4537
[15] Pauli W 1926 *Handbuch der Physik* **23** 1
[16] Cohen-Tanoudji C, Diu B and Laloë 1977 *Quantum Mechanics* (New York: Wiley-VCH)
[17] Moshammer R, Schmitt W, Ullrich J, Kollmus H, Cassimi A, Dörner R, Jagutzki O, Mann R, Olson R E, Prinz H T, Schmidt-Böcking H and Spielberger L 1997 *Phys. Rev. Lett.* **79** 3621
[18] Fano U and Macek J H 1973 *Rev. Mod. Phys.* **45** 553
[19] Andersen N, Gallagher J W and Hertel I V 1986 *Proc. ICPEAC XIV (Palo Alto)* (Amsterdam: North-Holland) 57
[20] Jones R R, You D and Bucksbaum P H 1993 *Phys. Rev. Lett.* **70** 1236
[21] You D, Jones R R, Bucksbaum P H and Dykaar D R 1993 *Opt. Lett.* **18** 290
[22] Dunning F B, Lancaster J C, Reinhold C O, Yoshida S and Burgdörfer J 2005, *Adv. At. Mol. Opt. Phys.* **52** 49
[23] Arbo D G, Reinhold C O, Burgdörfer J, Pattanayak A K, Stokely C L, Zhao W, Lancaster J C and Dunning F B 2003 *Phys. Rev. A* **67** 063401
[24] Yoshida S, Reinhold C O, Persson E, Burgdörfer J, Tannian B E, Stokely C L and Dunning F B 2004 *Book of invited papers ICPEAC XXII, Stockholm 2003, Phys. Scr.* **T110** 424
[25] Pazourek R, Nagele S and Burgdörfer J 2015 *Rev. Mod. Phys.* **87** 765
[26] Schultze M, Fiess M, Karpowicz N, Gagnon J, Korbman M, Hofstetter M, Neppl S, Cavalieri A L, Komninos Y, Mercouris T, Nicolaides CA, Pazourek R, Nagele S, Feist J, Burgdörfer J, Azzeer A M, Ernstorfer R, Kienberger R, Kleineberg U, Goulielmakis E, Krausz F and Yakovlev V S 2010 *Science* **328** 1658
[27] Eisenbud L 1948 *PhD Thesis* (Princeton Univ.)
[28] Wigner E P 1955 *Phys. Rev.* **98** 145
[29] Smith F T 1960 *Phys. Rev.* **118** 349; ibid. **120** 1058
[30] Smith F T 1961 *Book of Abstracts, ICPEAC II, Boulder, CO.* Abstract B2
[31] McPherson A, Gibson G, Jara H, Johann U, Luk T S, McIntyre I A, Boyer K and Rhodes C K 1987 *J. Opt. Soc. Am. B* **4** 595
[32] Ferray M, L'Huillier A, Li X F, Lompre L A, Mainfray G and Manus C 1988 *J. Phys. B: At. Mol. Opt. Phys.* **21** L31
[33] L'Huillier A and Balcou Ph 1993 *Phys. Rev. Lett.* **70** 774
[34] Lewenstein M, Balcou Ph, Ivanov M Yu, L'Huillier A and Corkum P B 1994 *Phys. Rev. A* **49** 2117
[35] Kulander K C, Schafer K J and Krause J L 1993 *Super Intense Laser-Atom Physics* ed B. Piraux, A. LHuillier and K. Rzazewski (Berlin: Springer) p 95
[36] Corkum P 1993 *Phys. Rev. Lett.* **71** 1994
[37] Peart B, Walton D S and Dolder K T 1970 *J. Phys. B: At. Mol. Opt. Phys.* **3** 1346
[38] Müller A, Schennach S, Wagner M, Haselbauer J, Uwira O, Spies W, Jennewein E, Becker R, Kleinod M, Pröbstel U, Angert N, Klabunde J, Mokler P H, Spädtke P and Wolf B 1991 *Phys. Scr.* **T37** 62
[39] Itatani J, Quéré F, Yudin G L, Ivanov M Yu, Krausz F and Corkum P B 2002 *Phys. Rev. Lett.* **88** 173903
[40] Veniard V, Taïeb R and Maquet A 1996 *Phys. Rev. A* **54** 721
[41] Klünder K, Dahlström J M, Gisselbrecht M, Fordell T, Swoboda M, Guénot D, Johnsson P, Caillat J, Mauritsson J, Maquet A, Taïeb R and L'Huillier A 2011 *Phys. Rev. Lett.* **106** 143002
[42] Van der Straten P and Morgenstern R 1986 *Comm. At. Mol. Phys.* **19** 243
[43] Burgdörfer J and Morgenstern R 1988 *Phys. Rev. A* **38** 5520
[44] Moore L, Lysaght M A, Parker J S, van der Hart H W and Taylor K T 2011 *Phys. Rev. A* **84** 061404
[45] Dahlström J M, Carette T and Lindroth E 2012 *Phys. Rev. A* **86** 061402(R)

[46] Kheifets A S 2013 *Phys. Rev. A* **87** 063404
[47] Feist J, Zatsarinny O, Nagele S, Pazourek R, Burgdörfer J, Guan X, Bartschat K and Schneider B I 2014 *Phys. Rev. A* **89** 033417
[48] Ossiander M, Siegrist F, Shirvanyan V, Pazourek R, Sommer A, Latka T, Guggenmos A, Nagele S, Feist J, Burgdörfer J, Kienberger R and Schultze M 2017 *Nature Phys.* **13** 280
[49] Isinger M, Squibb R J, Busto D, Zhong S, Harth A, Kroon D, Nandi S, Arnold C L, Miranda M, Dahlström J M, Lindroth E, Feifel R, Gisselbrecht M and L'Huillier A 2017 *Science* **358** 893
[50] Nagele S, Pazourek R, Feist J, Doblhoff-Dier K, Lemell C, Tökesi K and Burgdörfer J 2011 *J. Phys. B: At. Mol. Opt. Phys.* **44** 081001
[51] Dahlström J M, Guénot D, Klünder K, Gisselbrecht M, Mauritson J L'Huillier A, Maquet A and Taïeb R 2013 *Chem. Phys.* **414** 53
[52] Fuchs J, Douguet N, Donsa S, Martin F, Burgdörfer J, Argenti L, Cattaneo L and Keller U 2019 *preprint* arXiv:1907.03607
[53] Palatchi C, Dahlström J M, Kheifets A S, Ivanov I A, Canaday D M, Agostini P and DiMauro L F 2014 *J. Phys. B: At. Mol. Opt. Phys.* **47** 245003
[54] Pazourek R, Feist J, Nagele S and Burgdörfer J 2012 *Phys. Rev. Lett.* **108** 163001
[55] Pazourek R, Nagele S and Burgdörfer J 2013 *Faraday Discuss.* **163** 353
[56] Lemell C, Solleder B, Tökesi K and Burgdörfer J 2009 *Phys. Rev. A* **79** 062901
[57] Cavalieri A L, Müller N, Uphues Th, Yakovlev V S, Baltuska A, Horvath B, Schmidt B, Blümel L, Holzwarth R, Hendel S, Drescher M, Kleineberg U, Echenique P M, Kienberger R, Krausz F and Heinzmann U 2007 *Nature* **449** 1029
[58] Kazansky A K and Echenique P M 2009 *Phys. Rev. Lett.* **102** 177401
[59] Zhang C-H and Thumm U 2009 *Phys. Rev. Lett.* **102** 123601
[60] Neppl S, Ernstorfer R, Cavalieri A L, Lemell C, Wachter G, Magerl E, Bothschafter E M, Jobst M, Hofstetter M, Kleineberg U, Barth J V, Menzel D, Burgdörfer J, Feulner P, Krausz F and Kienberger R 2015 *Nature* **517** 342
[61] Ossiander M, Riemensberger J, Neppl S, Mittermair M, Schaffer M, Duensing A, Wagner M S, Heider R, Wurzer M, Gerl M, Schnitzenbaumer M, Barth J V, Libisch F, Lemell C, Burgdörfer J, Feulner P and Kienberger R 2018 *Nature* **561** 374
[62] Blum K 1981 *Density Matrix Theory and Applications* (Berlin: Springer)
[63] Gabriel H 1969 *Phys. Rev.* **181** 506
[64] Bosse J and Gabriel H 1974 *Zeitschrift f. Physik* **266** 283
[65] Bashkin S 1968 *Beam-Foil Spectroscopy* (New York: Gordon and Breach)
[66] Andrä H J 1974 *Phys. Scr.* **9** 257
[67] Sellin I A, Liljeby L, Mannervik S and Hultberg S 1979 *Phys. Rev. Lett.* **42** 570
[68] Burgdörfer J 1979 *Phys. Rev. Lett.* **43** 505
[69] Schuch R, Meron M, Johnson B M, Jones K W, Hoffmann R, Schmidt-Böcking H and Tserruya I 1988 *Phys. Rev. A* **37** 3313
[70] Goulielmakis E, Loh Z-H, Wirth A, Santra R, Rohringer N, Yakovlev V S, Zherebtsov S, Pfeifer T, Azzeer A M, Kling M F, Leone S R and Krausz F 2010 *Nature* **466** 739
[71] Cederbaum L S and Zobeley J 1999 *Chem. Phys. Lett.* **307** 205
[72] Calegari F, Ayuso D, Trabattoni A, Belshaw L, De Camillis S, Anumala S, Frassetto F, Poletto L, Palacios A, Decleva P, Greenwood J B, Martin F and Nisoli M 2014 *Science* **346** 336
[73] Kraus P M, Mignolet B, Baykusheva D, Rupenyan A, Horny L, Penka E F Grassi G, Tolstikhin O I, Schneider J, Jensen F, Madsen L B, Bandrauk A D, Remacle F and Wörner H J 2015 *Science* **350** 790
[74] Wickenhauser M, Burgdörfer J, Krausz F and Drescher M 2005 *Phys. Rev. Lett.* **94** 023002
[75] Argenti L, Pazourek R, Feist J, Nagele S, Liertzer M, Persson E, Burgdörfer J and Lindroth E 2013 *Phys. Rev. A* **87** 053405
[76] Kaldun A, Blättermann A, Stooss V, Donsa S, Wei H, Pazourek R, Nagele S, Ott C, Lin C D, Burgdörfer J and Pfeifer T 2016 *Science* **354** 738
[77] Gruson V, Barreau L, Jimenez-Galan A, Risoud F, Caillat J, Maquet A, Carre B, Lepetit F, Hergott J F, Ruchon T, Argenti L, Taïeb R, Martin F and Salieres P 2016 *Science* **354** 734
[78] Hassan M Th, Luu T T, Moulet A, Raskazovskaya O, Zhokov P, Garg M, Karpowicz N, Zheltikov A M, Pervak V, Krausz F and Goulielmakis E 2016 *Nature* **530** 66
[79] Jiang W-C, Tong X-M, Pazourek R, Nagele S and Burgdörfer J 2019 (submitted)
[80] Goulielmakis E 2019 *private communication*
[81] Kruit P, Kimman J, Muller H G and van der Wiel M J 1983 *Phys. Rev. A* **28** 248
[82] Kuchiev M Yu 1987 *JETP Lett.* **45** 404
[83] Arbo D, Ishikawa K, Schiessl K, Persson E and Burgdörfer J 2010 *Phys. Rev. A* **82** 043426

[84] Lindner F, Schätzel M G, Walther H, Baltuska A, Goulielmakis E, Krausz F, Milosevic D B, Bauer D, Becker W and Paulus G G 2005 *Phys. Rev. Lett.* **95** 040401
[85] Arbo D G, Persson E and Burgdörfer J 2006 *Phys. Rev. A* **74** 063407
[86] Huismans Y, Rouzee A, Gijsbertsen A, Jungmann J H, Smolkowska A S, Logman P S W M, Lepine F, Cauchy C, Zamith S, Marchenko T, Bakker J M, Berden G, Redlich B, van der Meer A F G, Muller H G, Vermin W , Schafer K J, Spanner M, Ivanov M Yu, Smirnova O, Bauer D, Popruzhenko S V and Vrakking M J J 2011 *Science* **331** 61
[87] Wolter B, Pullen M G, Le A-T, Baudisch M, Doblhoff-Dier K, Senftleben A, Hemmer M, Schroter C D, Ullrich J, Pfeifer T, Moshammer R, Gräfe S, Vendrell O, Lin C D and Biegert J 2016 *Science* **354** 308
[88] Morishita T, Le A-T, Chen Z and Lin C D 2008 *Phys. Rev. Lett.* **100** 013903
[89] Le A-T, Lucchese R R, Tonzani S, Morishita T and Lin C D 2009 *Phys. Rev. A* **80** 013401
[90] Pabst S, Greenman L, Ho P J, Mazziotti D A and Santra R 2011 *Phys. Rev. Lett.* **106** 053003
[91] Jordan I, Huppert M, Brown M A, van Bokhoven J A and Wörner H J 2015 *Rev. Sci. Instr.* **86** 123905
[92] Beaulieu S, Comby A, Clergerie A, Caillat J, Descamps D, Dudovich N, Fabre B, Geneaux R, Legare F, Petit S, Pons B, Porat G, Ruchon T, Taïeb R, Blanchet V and Mairesse Y 2017 *Science* **358** 1288
[93] Siegrist F, Gessner J A, Ossiander M, Denker C, Chang Y P, Schröder M C, Guggenmos A, Cui Y, Walowski J, Martens U, Dewhurst J K, Kleineberg U, Münzenberg M, Sharma S and Schultze M 2019 *Nature* **571** 240
[94] Dewhurst J K, Elliott P, Shallcross S, Gross E K U and Sharma S 2018 *Nano Lett.* **18** 1842
[95] Popminchev D, Hernandez-Garcia C, Dollar F, Mancuso C, Perez-Hernandez J A, Chen M-C, Hankla A, Gao XH, Shim B, Gaeta A L, Tarazkar M, Romanov D A, Levis R J, Gaffney J A, Foord M, Libby S B, Jaron-Becker A, Becker A, Plaja L, Murnane M M, Kapteyn H C and Popmintchev T 2015 *Science* **350** 1225
[96] Ngoko Djiokap J M, Hu S X, Madsen L B, Manakov N L, Meremianin A V and Starace A F 2015 *Phys. Rev. Lett.* **115** 113004
[97] Vos R J and Gavrila M 1992 *Phys. Rev. Lett.* **68** 170
[98] Jiang W-C and Burgdörfer J 2018 *Opt. Expr.* **26** 19921
[99] Popminchev T, Chen M-C, Popmintchev D, Arpin P, Brown S, Alisauskas S, Andriukaitis G, Balciunas T, Mücke O D, Pugzlys A, Baltuska A, Shim B, Schrauth S E, Gaeta A, Hernandez-Garcia C, Plaja L, Becker A, Jaron-Becker A, Murnane M M and Kapteyn H C 2012 *Science* **336** 1287

Precision Measurements of the Fundamental Properties of the Proton and Antiproton

C. Smorra

RIKEN, Fundamental Symmetries Laboratory, 2-1 Hirosawa, Wako, Saitama, 351-0198, Japan
CERN, Esplanade des Particules 1, 1217 Meyrin, Switzerland

E-mail: christian.smorra@cern.ch

A. Mooser

RIKEN, Fundamental Symmetries Laboratory, 2-1 Hirosawa, Wako, Saitama, 351-0198, Japan
Max-Planck-Institut für Kernphysik, Saupfercheckweg 1, D-69117, Heidelberg, Germany

E-mail: andreas.mooser@mpi-hd.mpg.de

Abstract. Precision measurements comparing the fundamental properties of conjugate particles and antiparticles constitute stringent tests of CPT invariance. We review recent precision measurements of the BASE collaboration, which improved the uncertainty of the proton and antiproton magnetic moments and the comparison of the proton-to-antiproton charge-to-mass ratio. These measurements constitute the most stringent tests of CPT invariance with antiprotons. Further, we discuss the improved limit on the antiproton lifetime based on the storage of a cloud of antiprotons in the unique BASE reservoir trap. Based on these recent advances, we discuss ongoing technical developments which comprise a coupling trap for the sympathetic cooling of single (anti-)protons with laser-cooled beryllium ions, a transportable trap to relocate antiproton measurements into a high-precision laboratory, and a new experiment to measure the magnetic moment of helium-3 ions, which will improve absolute precision magnetometry.

1. Introduction

In the development of the Standard Model of particle physics, symmetries have played an important role [1]. In the early last century, the discrete symmetries of charge- (C), parity- (P), and time-reversal (T), and the combined CP-symmetry were believed to be individually conserved. However, experiments showed that, for example, the P-symmetry is violated in the nuclear beta decay [2], and that the combined CP-symmetry is violated in the decay of neutral kaons [3]. These experimental observations were included into the Standard Model as the V-A theory of the weak interaction and a CP-violating phase in the quark-mixing matrix, respectively. At present, the interactions in the Standard Model are described by local Lorentz-invariant quantum-field theories. These are expected to be invariant under the combined CPT transformation [4]. Therefore, CPT invariance is regarded as one of the fundamental symmetries in the Standard Model and should hold exactly. As consequences of this symmetry, conjugate particle-antiparticle pairs are created and annihilated in pair processes and have identical fundamental properties except for signs. This theoretical understanding is in conflict to our astronomical observations, which indicate that our Universe consists almost exclusively of matter [5]. This indicates that our understanding of the fundamental interactions is incomplete, since the Standard Model can neither explain the matter-antimatter asymmetry in our Universe, nor reproduce the observed matter excess based on the CP violation in the quark sector [6]. Searching for additional sources of symmetry violation or additional interactions may provide important hints to improve our understanding of the matter-antimatter asymmetry. So far, CPT invariance is holding up to all conducted experimental tests [7]. Therefore, it is essential to make more stringent CPT tests with increased experimental precision or to test CPT invariance in new systems.

Experimental tests of CPT invariance with antiprotons are conducted so far at the antiproton decelerator (AD) of CERN [8], which is presently the world's only low-energy antiproton facility. Here, high-precision spectroscopy is conducted on antiprotons [9, 10, 11, 12], antihydrogen [13, 14, 15, 16], and antiprotonic helium [17]. Opportunities on measurements in other exotic systems are presently discussed, such as on the positively-charged antihydrogen ion $\overline{H}^+$[18, 19] or the antihydrogen molecular ion $\overline{\mathrm{H}}_2^-$ [20, 21]. Of particular interest is also the search for a modified antimatter gravitation [19, 22, 23, 24]. The ongoing experimental efforts will benefit from the new deceleration stage ELENA [25], which reduces the antiproton energy for the experiments by a factor ~ 50 compared to the AD, down to 100 keV.

In the experiments of the BASE collaboration, we compare the fundamental properties of protons and antiprotons, such as their lifetimes $\tau_{p/\overline{p}}$, charge-to-mass ratios $(q/m)_{p/\overline{p}}$, and magnetic moments $\mu_{p/\overline{p}}$. The magnetic moment of the proton μ_p was known with a relative precision of 10 parts per billion (ppb) from a measurement using a hydrogen maser in 1972 [26], which provided the best measurement of this fundamental property for more than four decades. The antiproton magnetic moment $\mu_{\overline{p}}$ was until recently only known with a relative uncertainty on the 0.1 % level from the spectroscopy of antiprotonic helium atoms [27]. Experimental techniques for high-precision measurements of magnetic moments of single particles with $\sim$ 1 ppb precision were already developed in the 1970s and 1980s in the concourse of electron and positron measurements by the group of Dehmelt [28, 29, 30, 31]. These methods provided a promising basis to conduct also high-precision magnetic moment measurements of protons and antiprotons with comparable precision [32]. Extending the methods for protons and antiprotons required however to considerably advance several experimental techniques, since their magnetic moment is about a factor 660 smaller than those of the leptons. Essential steps to realize such measurements for protons and antiprotons were the development of the double-trap technique to suppress line-broadening effects from magnetic field gradients [33], the development of a Penning-trap system to detect spin-transitions of a single proton at first with a statistical method [34], and to reach the sensitivity of detecting single spin transitions with protons [35] and antiprotons [36]. These techniques were implemented in two experiments: the proton g-factor experiment at the University of Mainz (BASE-Mainz) [37], and the BASE experiment at the antiproton decelerator of CERN [38]. These developments resulted in the first direct measurements of the proton and antiproton magnetic moments using a single-trap method with an intermediate level of precision of the order 10^{-6} [10, 37, 39, 40]. The application of the double-trap technique with the sensitivity on detecting single spin-transitions resulted in the first parts-per-billion measurements of the proton and antiproton magnetic moments [12, 41], and recently even in a 0.3 ppb measurement of the proton magnetic moment [42]. The proton measurements constitute the most precise measurement of a nuclear magnetic moment and decrease the uncertainty of the proton magnetic moment by a factor of 33 compared to the previous best measurement [26], and the antiproton measurement improves the uncertainty by more than six orders of magnitude compared to exotic atom spectroscopy [27], and a factor of 350 compared to the best single-trap measurement [40], which was also performed by the BASE collaboration. By combining these measurements, a 10^6-times more stringent test of CPT invariance in the baryon sector was realized and improved limits on CPT-odd antiproton coefficients in the non-minimal Standard Model Extension [43] and on CPT-odd dimension-five operators [44], which would cause a splitting of the proton/antiproton magnetic moments were reported. The developed methods also facilitated improved comparisons of the proton and antiproton charge-to-mass ratio [9, 11] and improved limits on the antiproton lifetime from laboratory experiments [45].

In the future, further improvements on testing CPT invariance in the baryon sector require improved experimental techniques for single protons and antiprotons. To this end, we are developing a sympathetic-cooling method for (anti)protons using a common-endcap electrode trap system [46, 47], and the BASE collaboration also targets to implement a quantum-logic method to improve the spin state determination [48]. In addition, we target to lower the ambient magnetic field fluctuations by building a transportable antiproton container to relocate our precision measurements into the low-noise environment of a high-precision laboratory. Based on these recent developments, we are also building an experiment to measure the magnetic moment of $^3He^{2+}$ with 1 ppb precision, which requires a 10-fold more-sensitive apparatus compared to (anti)proton magnetic moment measurements. This measurement will allow to establish hyperpolarized helium-3 as independent high-precision magnetometer.

2. The basic experimental techniques

Penning traps are presently the tool of choice for high-precision measurements of magnetic moments [12, 41, 42, 49, 50] and charge-to-mass ratios [11, 51, 52] of single particles. For these measurements, a superposition of a strong magnetic field $\vec{B} = B_0\hat{e}_z$ ($B = 1.9\,\mathrm{T}$) for radial confinement and a weak electric quadrupole field $\Phi = V_0C_2(z^2 - \rho/2^2)$ ($V_0 \sim 7\,\mathrm{V}$) confines a single charged particle, where C_2 is a trap specific geometry factor, and V_0 the voltage applied to the ring electrode of the trap. The trapped particle has three eigenmotions, the axial mode is a harmonic oscillation along the magnetic field lines, and the modified cyclotron mode and magnetron mode comprise the motion in the radial plane [53]. In this configuration, we make high-precision measurements of the frequencies of single trapped particles. The magnetic moment of the proton and antiproton are measured by determining the frequency ratio of the Larmor (spin-precession) frequency $\omega_L = (g/2)\,(q/m)\,B$ and the cyclotron frequency $\omega_c = (q/m)\,B$:

$$\frac{\omega_L}{\omega_c} = \frac{g_{p/\overline{p}}}{2} = \pm\frac{\mu_{p/\overline{p}}}{\mu_N}, \tag{1}$$

where $g_{p/\overline{p}}$ denotes the dimensionless magnetic moment of the proton/antiproton (g-factor), and μ_N the nuclear magneton. The cyclotron frequency ν_c is determined by measuring the three eigenfrequencies of the trapped particle using the relation $\nu_c^2 = \nu_+^2 + \nu_z^2 + \nu_-^2$ [53]. The Larmor frequency is determined by driving spin-transitions with an oscillating magnetic field and measuring the spin-transition probability as a function of the drive frequency.

To determine ν_c, the motional frequencies of the trapped particle are measured non-destructively by using image-current detection methods [54, 55, 56]. To this end, the current resulting from the movement of the image charge induced in the trap electrodes is picked up by a superconducting coil which is attached to one of the trap electrodes. This forms a parallel tank circuit, which acts on its resonance frequency as a large resistor. Matching the axial frequency of the particle to the resonance frequency of the tank circuit converts the image-current signal into a detectable voltage drop, which is read-out using cryogenic low-noise amplifiers and a Fast-Fourier Transform (FFT) analyzer. The detection systems reach single particle sensitivity for (anti)protons in traps with $d < 10\,\mathrm{mm}$ diameter, and provide measurements of the axial frequency ν_z with $\sim$ 20 mHz resolution in $\sim$ 1 min averaging time [56]. To measure the cyclotron frequency, we use a sideband method [57], where we couple the radial and axial modes and determine the cyclotron frequency from the axial and sideband frequencies [55]. Using this method, such measurements reach a precision of a few ppb in about 2 minutes averaging time [11, 58].

An essential technique in magnetic measurements is to detect spin transitions for the spectroscopy of the Larmor resonance. For this purpose, we utilize the continuous Stern-Gerlach effect and make quantum non-demonlition measurements of the spin state of single trapped particles [28]. To this end, a strong magnetic bottle $B_z(z,\rho) = B_0 + B_2(z^2 - \rho^2/2)$ is superimposed to one of our traps, the *analysis trap*. The magnetic potential of the bottle couples the orbital and spin angular momentum to the axial mode, such that quantum transitions of the spin or in the radial modes change the axial frequency:

$$\Delta\nu_z = \frac{1}{4\pi m\nu_z}\frac{B_2}{B_0}h\nu_+\left(\frac{g}{2}m_s + (n_+ + \frac{1}{2}) + \frac{\nu_-}{\nu_+}(n_- + \frac{1}{2})\right), \tag{2}$$

where B_2 the curvature of the magnetic field, and n_+, n_-, and $m_s = \pm 1/2$, are the cyclotron, magnetron, and spin quantum numbers, respectively. Single quantum transitions, Δn_+=1, Δn_-=1, and Δm_s=1, cause frequency shifts of about 65 mHz, 20 μHz and 180 mHz, respectively. Therefore, the challenge for the magnetic moment measurement is to distinguish spin transitions and cyclotron transitions that are driven by residual electric field noise in the trap. In the cryogenic 3.6 mm diameter trap of the BASE apparatus at CERN, the electric field noise reaches levels corresponding to a few cyclotron transitions per hour for particles below 50 mK cyclotron energy [59]. This enables identifying single spin transitions with high fidelity [35, 36]. A second challenge is the line broadening imposed by the magnetic bottle. Whereas electron $(g-2)$ measurements can be performed in a single trap with a moderate B_2 of 150 T/m^2 [53], the (anti)proton requires to go with the magnetic bottle strength close to the technical limit by using a ferromagnetic ring electrode, which results in $B_2 \sim$ 300000 T/m^2. This causes a significant line broadening of the spin and cyclotron resonance in this trap and limits the uncertainties of magnetic moment measurements of a few times 10^{-7} [40]. To reach a higher level of precision, it is essential to use the double-trap technique, where the spin-state identification takes place in the strongly inhomogeneous analysis trap, and the spectroscopy of the Larmor frequency is spatially separated into a second homogeneous trap, the *precision trap*, where the residual magnetic gradients and their line-broadening effects are at least a factor 10^5 smaller. Using this technique, the latest measurement of the proton magnetic moment reached down to 0.3 ppb in relative precision [42].

3. Precision measurement of the proton magnetic moment

The Penning trap system of the BASE collaboration which was used to perform the 0.3 ppb proton magnetic moment measurement is shown in Fig. 1 [42]. A single proton, loaded in-trap with a cryogenic field emission point, is prepared for spin-transition spectroscopy by cooling of the cyclotron mode in the precision trap with a resonant cyclotron detector with an effective temperature of 3.2 K. We select a low-energy thermal state ($E_+/k_B < 0.6$ K) by detuning the image-current detector, and transporting the proton adiabatically to the analysis trap, where the cyclotron energy is determined by measuring the axial frequency shift, see Eq. (2). The cooling cycle is repeated until the cyclotron energy is sufficiently low to observe single spin transitions. We identify the initial spin state of the cold proton by driving resonantly spin transition and measuring the resulting axial frequency shift [35]. Subsequently, the proton is transported back into the precision trap, where we measure the cyclotron frequency ν_c with the sideband method described in [57] and drive a spin transition at the frequency ratio $\Gamma = \nu_{\rm rf}/\nu_c$. Subsequently, we cool the cyclotron mode of the proton, since the sideband measurement couples the cyclotron

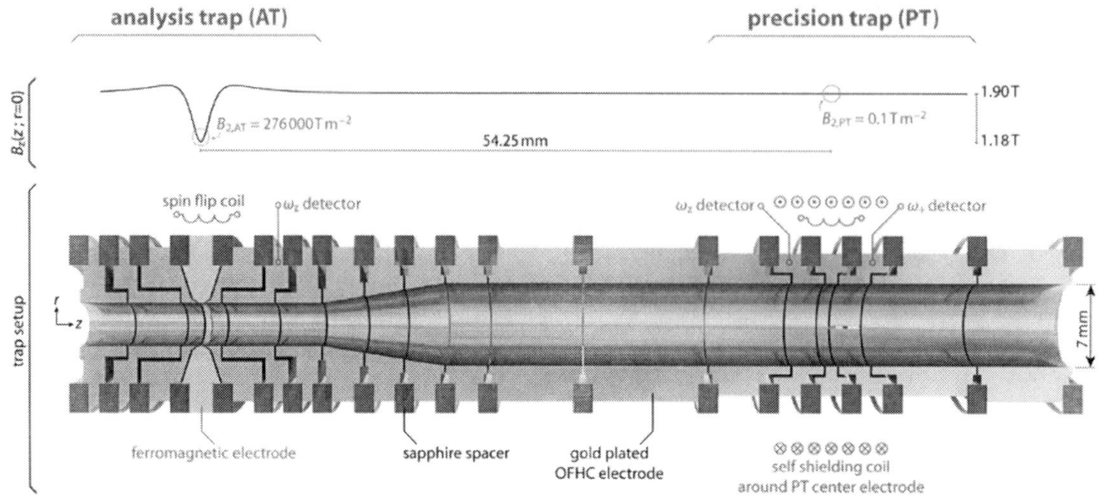

Figure 1. Double Penning-trap system for the proton magnetic moment measurement. The electrode stack in combination with the magnetic field pointing along the symmetry axis of the electrode form the Penning traps for the proton magnetic moment measurement. The gold-plated OFE copper electrodes are shown in gold and the sapphire rings providing the insulation in blue. The spin-state analysis trap has a ferromagnetic ring electrode shown in green to form the strong magnetic bottle required to apply the continuous Stern-Gerlach effect. The analysis trap and the homogeneous precision trap for high-precision frequency measurements are spatially seperated by several transport electrodes by a distance of 54.25 mm to decrease the residual inhomogeneity due to the magnetic bottle in the precision trap. A self-shielding coil stabilizes the magnetic field in the precision trap. Both traps have an axial image-currect (ω_z) detector, and spin-flip coil in their vicinity to irradiate oscillating magnetic fields to drive spin flips. The precision trap has in addition a cyclotron ω_+ detector for resistive cooling of the cyclotron mode.

mode to an axial image-current detector with an effective temperature of $T_+ \sim 350\,\mathrm{K}$. Once the proton is cooled below the energy threshold for single spin-flip detection, we shuttle the proton back to the analysis trap and determine the final spin state. This enables determining whether the spin in the homogeneous magnetic field of the precision trap was flipped. This spin flip determination also initializes the next measurement cycle. For the latest measurement, about 1300 of these measurement cycles were performed and the proton g-factor was extracted from a maximum-likelihood analysis by matching the line-shape function [60] to the measured spin-flip probability $P_{\mathrm{SF,PT}}(\Gamma)$, as shown in Fig. 2.

The most significant uncertainty of this measurement is due to measurement statistics (268 ppt) resulting from the slow cycle time due to the inefficient cooling procedure for the cyclotron mode. The largest systematic limits resulted from the image-charge shift, i.e. the back-interaction of the image-charge in the trap electrodes with the proton itself [61, 62], which causes a shift of the measured cyclotron frequency with about -98(3) ppt. An additional uncertainty of 80 ppt was added to account for potential shifts from the fitting procedure of the axial frequency. Including all systematic effects, the result of this proton magnetic moment measurement is

$$\mu_p/\mu_N = 2.792\,847\,344\,62(83), \tag{3}$$

which has a relative uncertainty (1 s.d.) of $\sim 300\,\mathrm{ppt}$. Further details on the experiment, the proton g-factor measurement and the data analysis are presented in reference [42].

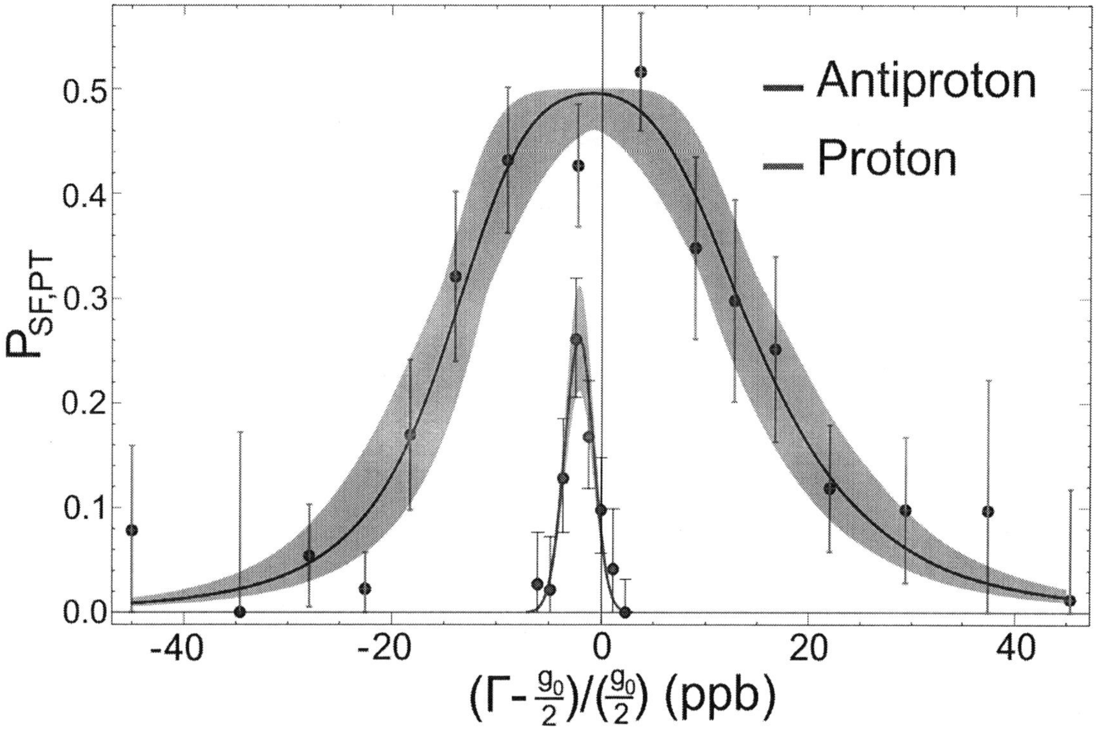

Figure 2. Spin-flip resonances of the antiproton and proton magnetic moment measurements [12, 42]. The lineshape function $P_{\mathrm{SF,PT}}(\Gamma)$ of the spin-flip resonance in the precision trap as function of the tested frequency ratio Γ is shown for protons and antiprotons in red and blue, respectively. The solid line corresponds to the parameters with the maximum likehood, and the shaded area corresponds to the largest change when the lineshape parameters are changed within 1 standard deviation. The antiproton resonance uses an improved lineshape function compared to reference [12]. The measurement is referenced to the BASE proton magnetic moment measurement in 2014: $g_0/2 = 2.792847350(9)$ [41].

4. Precision measurement of the antiproton magnetic moment

The antiproton g-factor measurement [12] was conducted in the BASE Penning-trap system at CERN [38], which is shown in Fig. 3. It consists of four Penning traps, two of them are the analysis trap and the precision trap for the double-trap measurement. In addition, there is an antiproton reservoir trap and a cooling trap (not shown in Fig. 3) to accelerate the cooling of the cyclotron mode [38]. The reservoir trap serves as a single antiproton source for the precision measurements [63]. A cloud of antiprotons initially prepared from a single antiproton decelerator pulse is stored in this trap and single antiprotons are extracted non-destructively by splitting the potential well with voltage ramps [64]. This technique was developed by the BASE collaboration, and was one of the key methods for the fast progress of the antiproton measurements. The BASE collaboration managed to run single particle measurements in the trap system continuously for 405 days without reloading - even during shutdown periods of the antiproton decelerator [45].

The antiproton magnetic moment measurement faced additional challenges compared to the one of the proton due to the higher radiofrequency noise level in the AD. The noise increases the cyclotron transition rate in the analysis trap, so that the cyclotron energy limit to reach a

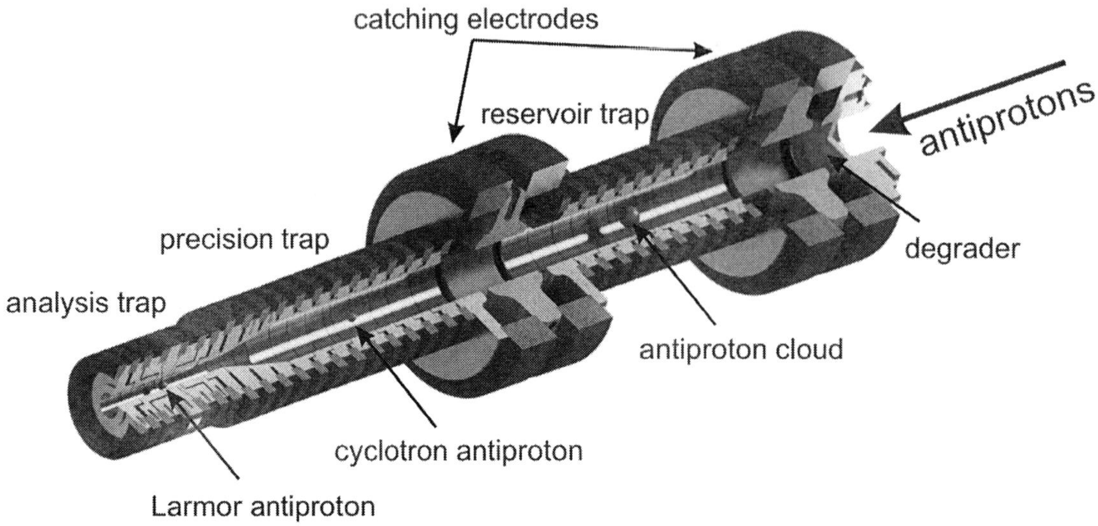

Figure 3. Trap system for the antiproton magnetic moment measurement. The trap electrodes and the sapphire rings forming the Penning traps are shown in gold and blue, respectively. A cloud of antiprotons in trapped in the reservoir trap, which supplies the other traps with single antiprotons for the precision measurements. It is surrounded by the catching electrodes, which confine antiprotons after they passed through the degrader by the application of high voltage pulses after the initial injection from the AD. The cold "Larmor antiproton" is shown in the anaysis trap, which has a ferromagnetic ring electrode to identify the spin state, and the second "cyclotron antiproton" is shown in the center of the precision trap, where it is used for cyclotron frequency measurements. More details about the trap system are found in reference [38].

sufficient axial frequency stability for the single spin-state detection is lower, $E_+/k_B < 0.2\,\mathrm{K}$ [59]. Further, the effective temperature of the cyclotron detector of 12.8 K for cyclotron cooling was also limited by radiofrequency noise. Both effects increased the cyclotron cooling time to about 12 hours per cooling cycle. To circumvent a significant decrease in the measurement statistics, the BASE collaboration invented a new two-particle measurement method, where the first particle is cooled to $E_+/k_B < 50\,\mathrm{mK}$ and is exclusively used for the spin-transition spectroscopy, whereas the second particle is only used for the cyclotron frequency measurement in the precision trap [12]. This novel scheme has been the essential step to realize the antiproton magnetic moment measurement, since it prevents the cyclotron mode of the first particle from being heated during the measurement cycle. In this measurement, cyclotron energy changes of the first particle occur only due to the residual interaction with the detuned cyclotron detector while the particle is in the precision trap. This results in a random-walk of the cyclotron energy with a standard deviation of $\Xi_{\Delta E_+}/k_B \sim 22\sqrt{N}\,\mathrm{mK}$, where N is the number of measurement cycles. In this way, 75 measurement cycles were performed on average before cyclotron mode cooling became necessary. In total, 933 data points were recorded, which resulted in the blue resonance shown in Fig. 2 and extracted the antiproton g-factor with a statistical uncertainty of $\sim 1.1\,\mathrm{ppb}$. The two-particle method boosts the measurement statistics by more than a factor of 5, but as a trade-off one has to consider systematic uncertainties due to the different orbits of the two particles in the precision trap. The most significant shifts scale with the residual linear and quadratic magnetic field gradients and impose in total a systematic shift of 0.44 ppb and a systematic uncertainty of $\sim 1.0\,\mathrm{ppb}$. In total, considering all systematic effects in reference [12] and from using an improved lineshape model which has recently been developed (publication in progress), the resulting antiproton magnetic moment is:

$$\mu_{\overline{p}}/\mu_N = -2.7928473443(46), \tag{4}$$

which differs by less than 5% of the uncertainty from the result in reference [12], and has a slightly increased relative uncertainty of 1.7 ppb for the 68% C.L., and 2.9 ppb for the 95% C.L.

5. Interpretation

The recent proton/antiproton magnetic moment measurements have improved the uncertainty of the proton magnetic moment by a factor of 33 compared to the hydrogen maser measurement [26], and the antiproton magnetic moment by more than a factor 10^6 compared to the spectroscopy of antiprotonic helium spectroscopy [27], and more than a factor 3000 compared to the best other competing Penning trap measurement [10]. By combining the two latest magnetic moment measurements, we obtain the following limit on the time-average difference of the proton and antiproton magnetic moments:

$$\frac{\delta\mu}{\mu_N} = \left(\frac{\mu_p}{\mu_N} + \frac{\mu_{\overline{p}}}{\mu_N}\right) = 0.3(8.3) \times 10^{-9}, \tag{5}$$

where the uncertainty in brackets is for 95 % C.L. Our result shows no significant deviation from the zero value predicted by CPT invariance and support CPT symmetry within the resolution of our measurements. Consequently, we can set limits on all CPT-odd interactions which produce a time-average difference of the Larmor frequency or the magnetic moments, such as the non-minimal Standard Model Extension (SME) [43] and a certain CPT-odd dimension-five operator [44]. We obtain the following constraints (95 % C.L.) on the combination of six non-minimal SME coefficients from the uncertainty of δg of our two latest measurements:

$$\left|\tilde{b}_p^Z\right| < 8.1 \times 10^{-25}\,\text{GeV}, \tag{6}$$

$$\left|\tilde{b}_{F,p}^{XX} + \tilde{b}_{F,p}^{YY}\right| < 4.6 \times 10^{-9}\,\text{GeV}^{-1}, \tag{7}$$

$$\left|\tilde{b}_{F,p}^{ZZ}\right| < 3.3 \times 10^{-9}\,\text{GeV}^{-1}, \tag{8}$$

$$\left|\tilde{b}_p^{*Z}\right| < 1.5 \times 10^{-24}\,\text{GeV}, \tag{9}$$

$$\left|\tilde{b}_{F,p}^{*XX} + \tilde{b}_{F,p}^{*YY}\right| < 3.1 \times 10^{-9}\,\text{GeV}^{-1}, \tag{10}$$

$$\left|\tilde{b}_{F,p}^{*ZZ}\right| < 1.1 \times 10^{-8}\,\text{GeV}^{-1}. \tag{11}$$

The last three combinations of coefficients represent the most stringent constraints for these antiproton coefficients [7]. In addition, we set the following limit (95 % C.L.) on a possible magnetic moment splitting for protons and antiprotons f_p^0 caused by a certain dimension-five operator described in reference [44]:

$$f_p^0 = \frac{\delta g}{2}\frac{\mu_N}{2} < 4.5 \times 10^{-12}\mu_B. \tag{12}$$

Despite the factor 3000 improvement in the sensitivity of testing CPT invariance in the baryon sector, there is still no hint how the matter-antimatter asymmetry was created, neither in the measurements of the BASE collaboration nor in the results of many other high-precision tests of the fundamental interactions [65]. Currently, the BASE collaboration is targeting to increase the measurement precision by a factor 10 to 100 in the next years. The ongoing technical developments to reach this improved sensitivity are described in the sections 8 and 9. These technical advances will allow revisiting tests of the CPT invariance in the baryon sector with one or two orders improved sensitivity.

6. High-precision comparisons of the proton-to-antiproton charge-to-mass ratio

A second particle property which can be measured with high precision in Penning traps is the charge-to-mass ratio. To this end, one measures the cyclotron-frequency ratio of two particles in the same magnetic field, which results in a direct comparison of their charge-to-mass ratios. In our case, we measure the antiproton-to-proton cyclotron frequency ratio:

$$\frac{\nu_{c,\bar{p}}}{\nu_{c,p}} = \frac{(q/m)_{\bar{p}}}{(q/m)_p}. \tag{13}$$

Consequently, this measurement provides a second stringent CPT invariance test in the baryon sector. Such measurements were already performed by the TRAP collaboration [66], with the latest result reported in 1999 having 90 ppt relative uncertainty [9]. One key technique is to use the negative hydrogen ion as proxy for the proton, since the conversion factor $R = \nu_{c,p}/\nu_{c,H^-}$ is known with 0.2 ppt uncertainty [11, 58] and since this greatly reduces systematic uncertainties related to inverting the trap voltage.

The BASE Penning trap system has also been utilized to conduct such a high-precision comparison of the charge-to-mass ratios of protons and antiprotons [11]. Compared to these earlier measurements, the cyclotron frequency was measured at lower cyclotron energy in thermal equilibrium with the axial detection system resulting in an effective temperature of about 350 K. Further, the BASE collaboration first realized in this measurement a scheme using the fast adiabatic transport to exchange the particles in the measurement trap. This method enabled measuring the cyclotron-frequency ratio with a cycle time of 4 minutes, about a factor 60 faster than in earlier measurements. The cycle time was limited by the necessity to average the magnetic field fluctuations over one deceleration cycle of the antiproton decelerator. This fast and entirely non-destructive measurement scheme is meanwhile also used for other precision charge-to-mass ratio measurements [51].

In 2014, the BASE collaboration recorded about 6500 cyclotron frequency ratios with this measurement scheme in 35 days, which resulted in a statistical uncertainty of 62 ppt. The major systematic uncertainty resulted from the interplay of the residual magnetic field gradient in the measurement trap with a tiny change of ~ 30 nm of the equilibrium position of the trap caused by a trap voltage adjustment for the negative hydrogen ion. This resulted in a correction of 119(20) ppt. The final result of the charge-to-mass ratio comparison was:

$$\frac{(q/m)_{\bar{p}}}{(q/m)_p} + 1 = 1(69) \times 10^{-12}, \tag{14}$$

which is also consistent with CPT invariance.

The charge-to-mass ratio comparison is sensitive to gravitational anomalies of antiprotons, since any difference in the gravitational binding energy would result in a different gravitational redshift of the proton and antiproton cyclotron frequencies [67]. The cyclotron-frequency ratio including the gravitational redshift by the gravitational potentials U and $\alpha_g U$ for protons and antiprotons, respectively, is given as:

$$\frac{\nu_{c,\bar{p}}}{\nu_{c,p}} = \frac{(q/m)_{\bar{p}}}{(q/m)_p}(1 + 3(\alpha_g - 1)Uc^{-2}). \tag{15}$$

Here, α_g is the parameter characterizing the modified gravitational acceleration of antiprotons, m_p the proton mass, and $m_{\bar{p}}$ the antiproton mass. Limits derived from such comparisons are however ambiguous, since they are simultaneously a test of CPT invariance and gravitational anomalies. Consequently, an intrinsic charge-to-mass ratio difference from breaking the CPT invariance cannot be disentangled from an anomalous gravitation. Assuming CPT invariance to be valid and using for U the gravitational potential of the local galactic supercluster, which is an approach consistent with earlier literature [67], but which has been controversially discussed [68], limits on the gravitational anomaly parameter for antiprotons were extracted from this measurement: $(\alpha_g - 1) < 8.7 \cdot 10^{-7}$ [11].

7. Improved antiproton lifetime limits

Another fundamental property of antiprotons is their lifetime $\tau_{\bar{p}}$, which is required by CPT invariance to be identical to the proton lifetime τ_p. Searching for proton decays is of high interest in the scope of finding a baryon-number violating process, therefore searches based on huge samples with more than 1000 tons of water were made looking for a disappearance signal. These experiment found $\tau_p > 10^{29}\,\mathrm{y}$ [69], and even higher constraints have been obtained for specific decay channels [70]. On the contrary, due to the difficulty in producing and storing antiprotons, $\tau_{\bar{p}} > 0.3\,\mathrm{y}$ was until recently the best limit for an antiproton disappearance signal [71].

The trap system of the BASE collaboration enables to search for a disappearance signal by storing clouds of antiprotons in the reservoir trap, and counting the amount of confined particles over time by making use of the lineshape of the image-current signal [45]. Based on experiments from July 2014 until December 2016, the BASE collaboration has not observed any antiproton loss due to annihilation with residual gas. This enabled our collaboration to set a limit of $\tau_{\bar{p}} > 10.2\,\mathrm{yrs}$ (68 % C.L.) on the total lifetime of antiprotons, which represents the best laboratory measurement so far [45]. Other laboratory constraints were set by the APEX collaboration, which was operating at the Fermilab antiproton storage ring. APEX searched for antiproton decays in the storage ring producing negative leptons as decay product, and obtained limits for 13 decay channels in the range of 200 yrs to $7 \cdot 10^5$ yrs [72]. In contrast, the measurement of our collaboration constrains all possible decay channels, including also decays into negative kaons [73, 74] and other "dark decay channels", which were invisible to the APEX detector. However, astrophysical bounds on the antiproton lifetime suggest that $\tau_{\bar{p}} \gtrsim 10^6\,\mathrm{yrs}$ [75], so that improved experimental techniques are required to exceed these bounds in laboratory experiments. The existing methods may further improve the present experimental limit by a factor 100 by increasing the number of antiprotons in the reservoir trap. Further increases require new approaches for efficient non-destructive counting of a large cloud of trapped particles. In case annihilation events are observed at a higher lifetime and vacuum sensitivity, we also require an independent vacuum measurement, which could be based on the recombination rate of highly-charged ions to constrain the annihilation rate with residual gas [45].

8. Coupling trap

One of the major limitations in proton/antiproton g-factor measurements is the impact of the cyclotron cooling time on the measurement statistics. Recooling protons and antiprotons below the energy threshold for single spin-flip detection requires on average 2 hours and 12 hours, respectively, in the latest measurements. The resistive cooling technique is limited by the detector temperature to a few Kelvin, whereas laser cooling of trapped ions has already reached the motional ground state of cyclotron mode in a Penning trap [76]. Atomic ions without suitable laser-cooling transition can be sympathetically cooled by co-trapped ions, however we require also the cooling of a single antiproton, which cannot be performed with co-trapped ions.

To reduce the cooling limits for single protons and antiprotons, the BASE collaboration follows an early idea by Heinzen and Wineland [77] and developed a "common-endcap" two-trap system, where trapped particles in separate potential wells interact through image currents induced onto a shared trap electrode. The principle is illustrated in Fig. 4. By matching the axial frequencies of the trapped particles, the image-current interaction leads to a periodic exchange of the energy for both particles in the shared axial mode. The coupling trap system consists of two orthogonal compensated Penning traps [78] where the electrodes in between the two ring electrodes transmit the image-current signal to couple protons to laser-cooled beryllium ions [47]. The trap system is optimized to have a low capacitance on the common endcap electrode (5 pF) and a short particle-electrode distance (4 mm) to obtain a short energy exchange time. We expect to reach an exchange time of 55 s by coupling a cloud of 100 beryllium ions to a single proton [47]. If this goal can be reached, (anti)protons will be deterministically cooled to a cyclotron energy of a few 10 mK with a significantly reduced cooling time. The lower temperature also ensures that the spin-flip detection fidelity is close to 100 %, and that frequency measurements can be performed at lower temperatures and lower amplitude-dependent frequency shifts. Consequently, the sympathetic cooling in the common-endcap trap system will be one of the essential improvements to improve the precision of proton/antiproton magnetic moment measurements.

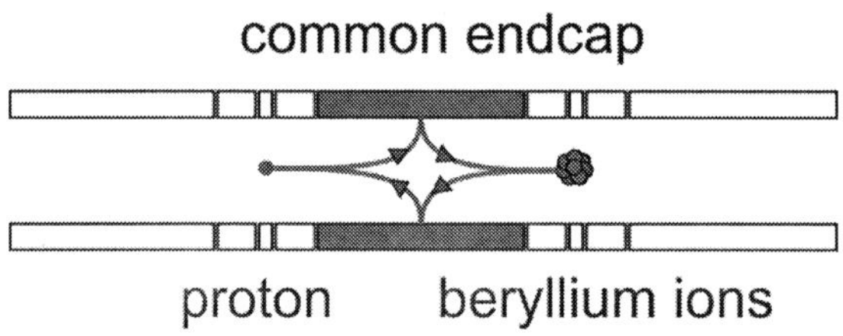

Figure 4. Image-current interaction in a common-endcap two-trap system. The proton (red) and a cloud of beryllium ions (green) interact by image-currents though a common-endcap electrode (blue). The blue lines visualize the Coulomb interaction with the image currents in the common-endcap electrode.

9. A Transportable Trap for Antiprotons

Another important limitation in the antiproton measurements is the magnetic field and radiofrequency noise imposed by the operations in the antiproton decelerator facility. The operation of the antiproton decelerator causes about 1 μT peak-to-peak fluctuations in the BASE experimental area, which would, if not suppressed, cause relative shifts of the cyclotron frequency of order 5×10^{-7}. Presently, the BASE collaboration makes use of the magnetic shielding provided by self-shielding superconducting solenoids [58], which were recently improved [79] to suppress cyclotron-frequency ratio fluctuations to below 2 ppb. However, on the long-term, we target to perform antiproton precision measurements in a low-noise environment, such that we can reach an even higher magnetic field stability. To this end, we target to develop a transportable trap for antiprotons, which can be loaded at the antiproton decelerator/ELENA facility, and can be transported to a high-precision laboratory to conduct single-antiproton measurements with a transportable antiproton reservoir.

Compared to earlier attempts [82, 83, 84] and presently active projects [85], we target to develop a transportable trap for single particle experiments, which requiring only a low amount of antiprotons. The transportable trap will form a single antiproton source for single-particle precision measurements based on the reservoir trap technique [64]. We have demonstrated the long-term storage of antiprotons up to more than a year [45], which is sufficient to conduct precision measurements in an offline laboratory.

The possibility to make antiproton charge-to-mass ratio and magnetic moment measurements in a high-precision laboratory will lead to further improvements in precision beyond limits imposed by the noisy environment at the production site. Further, the possibility to increase the number of antiproton precision experiments enables new kinds of experiments based on simultaneous antiproton cyclotron frequency measurements. This will allow testing Lorentz-invariance within the framework of the Standard Model Extension [43], and enable searches for topological dark-matter with a network of antimatter clocks [80], such as presently performed with magnetometers [81] and atomic clocks [65]. Another possibility is to search for changes in the gravitational redshift of the antiproton cyclotron frequency by an active modification of the gravitational potential by measuring at the cyclotron-frequency ratio of antiprotons and negative hydrogen ions at different altitudes. However, this requires a further advance in measurement precision by a factor of ~ 70 to reach the sensitivity to probe antiproton gravity at a height difference of 300 m.

10. Magnetic Moment Measurements of He-3 ions

Based upon the developments for the proton and the antiproton magnetic moment measurements we started the construction of a new experiment, which aims at the first direct high-precision measurement of the nuclear magnetic moment of $^3He^{2+}$ and a high-precision measurement of the ground-state hyper-fine splitting in a magnetic field of $^3He^+$. To this end sympathetic laser-cooling, a novel analysis-trap as well as a novel schema for the detection of the spin-state will be applied for the first time.

The measurement of the nuclear magnetic moment of $^3He^{2+}$ will establish hyperpolarized ^3He as an independent magnetometer, which compared to H_2O probes has smaller systematic uncertainties concerning sample shape, impurities, environmental dependencies and exhibits no line-shape distortion with a poorly defined center [86, 87, 88, 89]. Furthermore, the diamagnetic shielding correction of ^3He is the one which is most precisely known for all atoms [90]. Making use of optical pumping and long relaxation times in low-pressure ^3He gas cells, relative precisions of 10^{-12} within seconds were already achieved [91]. However, limited by theoretical diamagnetic shielding corrections to 12 ppb, the ^3He probes do not provide an absolute calibration independent of H_2O, because the magnetic moment of ^3He was only determined by measuring the ^3He NMR frequency relative to the NMR frequency of H_2O [92]. Thus, so far, any measurement of a given magnetic field using ^3He and H_2O cannot be considered as independent and uncorrelated but rather represents a revaluation of existing NMR frequency-ratio measurements used to determine the $^3He^{2+}$ magnetic moment. In addition, deviations in NMR frequency comparisons compared to accepted values are reported [93], which can be explained by inconsistencies in proton shielding corrections or a shift of the nuclear magnetic moment of ^3He at the 100 ppb level. The planed measurement will thus eventually overcome the limitation of a calibration correlated to H_2O NMR probes, resolve deviations in reported nuclear magnetic measurements data and allow for a ppb test of theoretical diamagnetic shielding corrections.

The measurement of the ground-state hyper-fine splitting in a magnetic field of $^3He^+$ will provide high-precision determinations of the bound electronic and nuclear magnetic moments in a light hydrogen like atom. In addition, it will give access to the zero-field ground-state hyper-fine splitting, which strongly depends on nuclear structure effects, e.g. nuclear polarizability and the Zeemach radius. Currently, the most precise measurement results from spectroscopy in a radio-frequency quadrupole ion trap [94]. The measurement was limited to relative precision of 1 ppb by the second-order Doppler effect in a room temperature apparatus. We aim to improve the precision by making use of single ion spectroscopy in a cryogenic Penning-trap system.

11. Conclusion

We have presented the recent improvements of the BASE collaboration in comparing the fundamental properties of protons and antiprotons. The measurements presented here support the CPT invariance as fundamental symmetry of the Standard Model up to the present measurement resolution. New methods are presently being developed to make further improvements in the sensitivity of testing CPT invariance, and to conduct a measurement of the He-3 magnetic moment.

Acknowledgments

We acknowledge all former and present members of the BASE collaboration for their contributions to the results presented here. In particular we would like to thank Klaus Blaum and Stefan Ulmer for their continuous support and their comments to this manuscript. We acknowledge financial support by RIKEN, the RIKEN Foreign Postdoctoral Researcher program, RIKEN Incentive Research Project Funding, the Max-Planck Society, the Max-Planck-RIKEN-PTB Center for Time, Constants and Fundamental Symmetries, and the CERN fellowship program. We acknowledge support from CERN, in particular, from the AD operation team and the CERN cryolab team.

References

[1] Weinberg S 2018 Essay: Half a Century of the Standard Model *Phys. Rev. Lett.* **121** 220001
[2] Wu C S, Ambler E, Hayward R W, Hoppes D D and Hudson R P 1957 Experimental Test of Parity Conservation in Beta Decay *Phys. Rev.* **105** 1413-5
[3] Christenson J H, Fitch V L, Cronin J W and Turlay R 1964 Evidence for 2π decay of K_2^0 meson *Phys. Rev. Lett.* **13** 138-40
[4] Lüders G 1957 Proof of the TCP theorem *Ann. Phys.* **2** 1-15
[5] Cohen A G, De Rujula A and Glashow S L 1998 A Matter-Antimatter Universe? *Astroph. J.* **495** 539-49
[6] Dine M and Kusenko A 2004 Origin of the matter-antimatter asymmetry *Rev. Mod. Phys.* **76**, 1-30
[7] Kostelecky V A and Russell N 2019 Data table for Lorentz and CPT violation *arXiv* 0801.0287v12 [hep-ph]
[8] Hemery J Y and Maury S 1999 The Antiproton Decelerator: Overview *Nucl. Phys.* A **655** 345-52
[9] Gabrielse G, Khabbaz A, Hall D, Heimann C, Kalinowsky H and Jhe W 1999 Precision mass spectroscopy of the antiproton and proton using simultaneously trapped particles *Phys. Rev. Lett.* **82** 3198-201
[10] DiSciacca J *et al.* 2013 One-Particle Measurement of the Antiproton Magnetic Moment *Phys. Rev. Lett.* **110** 130801
[11] Ulmer S *et al.* 2015 High-precision comparison of the antiproton-to-proton charge-to-mass ratio *Nature* **524** 196-9
[12] Smorra, C. *et al.* 2017 A parts-per-billion measurement of the antiproton magnetic moment *Nature* **550** 371-4
[13] Ahmadi M *et al.* 2018 Chacterization of the 1S-2S transition in antihydrogen *Nature* **557** 71-4
[14] Ahmadi M *et al.* 2017 Observation of the hyperfine spectrum of antihydrogen *Nature* **548** 66-9
[15] Kuroda N *et al.* 2014 A source of antihydrogen for in-flight hyperfine spectroscopy *Nat. Commun.* **5** 3089
[16] Malbrunot C *et al.* 2018 The ASACUSA antihydrogen and hydrogen program: results and prospects *Phil. Trans. Roy. Soc.* **376** 20170273
[17] Hori M, Aghai-Khozani H, Soter A, Barna D, Dax A, Hayano R, Kobayashi T, Murakami Y, Todoroki K, Yamada H, Horvath D and Venturelli L 2016 Buffer-gas cooling of antiprotonic helium to 1.5 to 1.7 K, and antiproton-to-electron mass ratio *Science* **354** 610-4
[18] Walz J, Pittner H, Herrmann M, Fendel P, Henrich B and Hänsch T W 2003 Cold antihydrogen atoms *Appl. Phys.* B **77** 713-7
[19] Perez P *et al.* 2015 The GBAR antimatter gravity experiment *Hyp. Int.* **233** 21-7

[20] Dehmelt H 1995 Economic synthesis and precision spectroscopy of anti-molecular hydrogen ions in Paul trap *Phys. Scr.* **T59** 423
[21] Myers E G 2018 CPT tests with the antihydrogen molecular ion *Phys. Rev.* A **98** 010101
[22] Aghion S *et al.* 2014 A moire deflectometer for antimatter *Nat. Commun.* **5** 4538
[23] Bertsche W A 2018 Prospects for comparison of matter and antimatter gravitation with ALPHA-g *Phil. Trans. Roy. Soc.* **376** 20170265
[24] Charman A E *et al.* 2013 Description and first application of a new technique to measure the gravitational mass of antihydrogen *Nat. Commun.* **4** 1785
[25] Bartmann W *et al.* 2018 The ELENA facility *Phil. Trans. Roy. Soc.* **376** 20170266
[26] Winkler P F, Kleppner D, Myint T and Walther F G 1972 Magnetic moment of the proton in Bohr magnetons *Phys. Rev.* A **5**, 83-114
[27] Pask T *et al.* 2009 Antiproton magnetic moment determined from the HFS of (p)over-barHe(+) *Phys. Lett.* B **678** 55-9
[28] Dehmelt H and Ekström P 1973 Proposed g-2 delta-omegaz experiment on single stored electron or positron *Bull. Am. Phys. Soc.* **18** 727
[29] Wineland D, Ekström P and Dehmelt H 1973 Monoelectron Oscillator *Phys. Rev. Lett.* **31** 1279-82
[30] Van Dyck R S, Schwinberg P B and Dehmelt H G 1987 New High-Precision Comparison of Electron and Positron g Factors *Phys. Rev. Lett.* **59** 26-9
[31] Dehmelt H, Mittleman R, Van Dyck R S and Schwinberg P 1999 Past electron-positron g-2 experiments yielded sharpest bound on CPT violation for point particles *Phys. Rev. Lett.* **83** 4694-6
[32] Quint W, Alonso J, Djekic S, Kluge H J, Stahl S, Valenzuela T, Verdu J, Vogel M and Werth G 2004 Continuous Stern-Gerlach effect and the magnetic moment of the antiproton *Nucl. Instr. Meth.* B **214** 207-10
[33] Häffner H, Beier T, Djekic S, Hermanspahn N, Kluge H J, Quint W, Stahl S, Verdu J, Valenzuela T and Werth G 2003 Double Penning trap technique for precise g factor determinations in highly charged ions *Eur. Phys. J.* D **22** 163-82
[34] Ulmer S, Rodegheri C C, Blaum K, Kracke H, Mooser A, Quint W, and Walz J 2011 Observation of Spin Flips with a Single Trapped Proton *Phys. Rev. Lett.* **106** 253001
[35] Mooser A, Kracke H, Blaum K, Bräuninger S A, Franke K, Leiteritz C, Quint W, Rodegheri C C, Ulmer S, and Walz J 2013 Resolution of Single Spin Flips of a Single Proton *Phys. Rev. Lett.* **110** 140405
[36] Smorra C *et al.* 2017 Observation of individual spin quantum transitions of a single antiproton *Phys. Lett.* B **769** 1-6
[37] Rodegheri C C, Blaum K, Kracke H, Kreim S, Mooser A, Quint W, Ulmer S and Walz J 2012 An experiment for the direct determination of the g-factor of a single proton in a Penning trap *New J. Phys.* **14** 063011
[38] Smorra C *et al.* 2015 BASE - The Baryon Antibaryon Symmetry Experiment, *Eur. Phys. J. Special Topics* **224** 3055-108
[39] DiSciacca J and Gabrielse G 2012 Direct Measurement of the Proton Magnetic Moment *Phys. Rev. Lett.* **108** 153001
[40] Nagahama H *et al.* 2017 Sixfold improved single particle measurement of the magnetic moment of the antiproton *Nat. Commun.* **8** 14084
[41] Mooser A, Ulmer S, Blaum K, Franke K, Kracke H, Leiteritz C, Quint W, Rodegheri C C, Smorra C and Walz J 2014 Direct high-precision measurement of the magnetic moment of the proton *Nature* **509** 596-9
[42] Schneider G *et al.* 2017 Double-trap measurement of the proton magnetic moment at 0.3 parts per billion precision *Science* **358** 1081-4
[43] Ding Y and Kostelecky V A 2016 Lorentz-violating spinor electrodynamics and Penning traps *Phys. Rev.* D **94** 056008
[44] Stadnik Y V, Roberts B M and Flambaum V V 2014 Tests of CPT and Lorentz symmetry from muon anomalous magnetic dipole moment *Phys. Rev.* D **90** 045035
[45] Sellner S *et al.* 2017 Improved limit on the directly measured antiproton lifetime *New J. Phys.* **19** 083023
[46] Heinzen D J and Wineland D J 1990 Quantum-limited Cooling and Detection of Radio-frequency Oscillations by Laser-cooled Ions *Phys. Rev.* A **42** 2977-94
[47] Bohman M *et al.* 2018 Sympathetic cooling of protons and antiprotons with a common endcap Penning trap *J. Mod. Opt.* **65** 568-76
[48] Meiners T *et al.* 2018 Towards sympathetic cooling of single (anti-)protons *Hyp. Int.* **239** 26
[49] Hanneke D, Fogwell S and Gabrielse G 2008 *Phys. Rev. Lett.* **100** 120801
[50] Sturm S, Vogel M, Köhler-Langes F, Quint W, Blaum K and Werth G 2017 High-Precision Measurements of the Bound Electrons Magnetic Moment *Atoms* **5** 4
[51] Heiße F, Köhler-Langes F, Rau S, Hou J, Junck S, Kracke A, Mooser A, Quint W, Ulmer S, Werth G, Blaum K and Sturm S 2017 High-Precision Measurement of the Protons Atomic Mass *Phys. Rev. Lett.* **119** 033001

[52] Myers E G 2019 High-Precision Atomic Mass Measurements for Fundamental Constants *Atoms* **7** 37
[53] Brown L S and Gabrielse G 1986 Geonium theory: Physics of a single electron or ion in a Penning trap *Rev. Mod. Phys.* **58** 233-311
[54] Wineland D J and Dehmelt H G 1975 Principles of the stored ion calorimeter *J. Appl. Phys.* **46** 919-30
[55] Ulmer S, Blaum K, Kracke H, Mooser A, Quint W, Rodegheri C C and Walz J 2011 Direct Measurement of the Free Cyclotron Frequency of a Single Particle in a Penning Trap *Phys. Rev. Lett.* **107** 103002
[56] Nagahama H *et al.* 2016 Highly sensitive superconducting circuits at 700 kHz with tunable quality factors for image-current detection of single trapped antiprotons *Rev. Sci. Instrum.* **87** 113305
[57] Cornell E A, Weisskoff R M, Boyce K R and Pritchard D E 1990 Mode coupling in a Penning trap: pulses and a classical avoided crossing *Phys. Rev.* A **41** 312-5
[58] Higuchi T *et al.* 2018 Progress towards an improved comparison of the proton-to-antiproton charge-to-mass ratios *Hyp. Int.* **239** 27
[59] Borchert M J *et al.* Measurement of Ultralow Heating Rates of a Single Antiproton in a Cryogenic Penning Trap *Phys. Rev. Lett.* **122** 043201
[60] Brown L S 1985 Geonium lineshape *Ann. Phys.* **159** 62-98
[61] Porto J V 2001 Series solution for the image charge fields in arbitrary cylindrically symmetric Penning traps *Phys. Rev.* A **64** 023403
[62] Schuh M, Heiße F, Eronen T, Ketter J, Kohler-Langes F, Rau S, Sega T, Quint W, Sturm S and Blaum K 2019 Image charge shift in high-precision Penning traps *Phys. Rev.* A **100** 023411
[63] Ulmer S, Yamazaki Y, Smorra C, Blaum K, Franke K, Matsuda Y, Nagahama H, Quint W, Walz J, Mooser A and Schneider G 2013 Technical Design Report BASE *CERN Document Server* **CERN-SPSC-2013-002** SPSC-TDR-002
[64] Smorra C *et al.* 2015 A reservoir trap for antiprotons *Int. J. Mass Spectr.* **389** 10-3
[65] Safronova M S, Budker D, DeMille D, Kimball D and Derevianko A 2018 Search for new physics with atoms and molecules *Rev. Mod. Phys.* **90** 025008
[66] Gabrielse G 2006 Antiproton mass measurements *Int. J. Mass Spectr.* **251** 273-80
[67] Hughes R J and Holzscheiter M H 1991 Constraints on the gravitational properties of antiprotons and positrons from cyclotron-frequency measurements *Phys. Rev. Lett.* **66** 854-7
[68] Nieto M and Goldman T 1991 The argument against Antigravity and the gravitational acceleration of antimatter *Phys. Rep.* **205** 221-81
[69] Ahmed S N *et al.* 2004 Constraints on Nucleon Decay via Invisible Modes from the Sudbury Neutrino Observatory *Phys. Rev. Lett.* **92** 102004
[70] Abe K *et al.* 2017 Search for proton decay via $p \to e^+\pi^0$ and $p \to \mu^+\pi^0$ in 0.31 megaton years exposure of the Super-Kamiokande water Cherenkov detector *Phys. Rev.* D **95** 012004
[71] Gabrielse G, Fei X, Orozco L A, Tjoelker R L, Haas J, Kalinowski H, Trainor T A and Kells W 1990 Thousandfold improvement in the measured antiproton mass *Phys. Rev. Lett.* **65** 1317-20
[72] Geer S H *et al.* 2000 New Limit on CPT Violation *Phys. Rev. Lett.* **84** 590-3
[73] Kobayashi K *et al.* 2005 Search for nucleon decay via modes favored by supersymmetric grand unification models in Super-Kamiokande-I *Phys. Rev.* D **72** 052007
[74] Babu K S, Pati J C and Wilczek F 1998 Suggested new modes in supersymmetric proton decay *Phys. Lett.* B **42** 3337-47
[75] Geer S H and Kennedy D C 2000 A New Limit on the Antiproton Lifetime *Astrophys. J.* **53** 2648-52
[76] Goodwin F, Stutter G, Thompson R C and Segal D M 2016 Resolved-Sideband Laser Cooling in a Penning Trap *Phys. Rev. Lett.* **116** 143002
[77] Wineland D J and Heinzen D J 1990 Quantum-limited cooling and detection of radio-frequency oscillations by laser-cooled ions *Phys. Rev.* A **42** 2977-94
[78] Gabrielse G, Haarsma L, Rolston S L 1989 Open-endcap Penning traps for high precision experiments *Int. J. Mass Spec.* **88** 319-32
[79] Devlin J A *et al.* 2019 A superconducting solenoid system with adjustable shielding factor *Phys. Rev. Appl.* accepted
[80] Derevianko A and Pospelov M 2014 Hunting for topological dark matter with atomic clocks *Nature Physics* **10** 933-6
[81] Wu T *et al.* 2019 Search for Axionlike Dark Matter with a Liquid-State Nuclear Spin Comagnetometer *Phys. Rev. Lett.* **122** 191302
[82] Tseng C H *et al.* 1993 Portable Trap Carries Particles 5000-Kilometers *Hyp. Int.* **76** 81-93
[83] Wada M and Yamazaki Y 2004 Technical developments toward antiprotonic atoms for nuclear structure studies of radioactive nuclei *Nucl. Instr. Meth.* B **214** 196-200
[84] Lewis R A 1997 Antiproton portable traps and medical applications *Hyp. Int.* **109** 155-64
[85] Aumann T *et al.* 2019 Experiment Proposal for PUMA: antiprotons and radioactive nuclei *CERN Document*

Server CERN-SPSC-2019-033 / SPSC-P-361
[86] Reid R 1975 Nuclear Magnetic Shielding in Hydrogen Molecule *Phys. Rev.* A **11** 403-8
[87] Grange J *et al.* 2015 Muon (g-2) Technical Design Report *arXiv*:1501.06858
[88] Fei X, Hughes V W and Prigl R 1997 Precision measurement of the magnetic field in terms of the free-proton NMR frequency *Nucl. Instr. Meth.* A **394** 349-56
[89] Chupp T and Swanson S 2001 Medical imaging with laser-polarized noble gases *Adv. At. Mol. Opt. Phy.* **45** 41-98
[90] Rudzinski A, Puchalski M and Pachucki K 2009 Relativistic, QED, and nuclear mass effects in the magnetic shielding of He-3 *J. Chem. Phys.* **130** 244102
[91] Nikiel A, Blümler P, Heil W, Hehn M, Karpuk S, Maul A, Otten E, Schreiber L M and Terekhov M 2014 Ultrasensitive ^{3}He magnetometer for measurements of high magnetic fields *Eur. Phys. J.* D **68** 330
[92] Flowers J L, Petley B W and Richards M G 1993 *Metrologia* A Measurement of the Nuclear Magnetic Moment of the He-3 Atom in terms of that of the Proton **30** 75-87
[93] Jackowski K, Jaszunski M and Wilczek M 2010 Alternative Approach to the Standardization of NMR Spectra. Direct Measurement of Nuclear Magnetic Shielding in Molecules *J. Phys. Chem.* A **114** 2471-5
[94] Schuessler H A, Fortson E N and Dehmelt H G 1969 Hyperfine Structure of Ground State of He-3+ by Ion-Storage Exchange-Collision Technique *Phys. Rev.* **187** 5-38

The attoclock and tunnelling time

A.W. Bray

Theoretical Physics, Research School of Physics, The Australian National University, Canberra, Australia

E-mail: alexander.bray@anu.edu.au

Abstract. Ultrafast laser pulses have enabled experiments probing the temporal dynamics of an electron tunnelling out of an atom. A quintessential example of such is known as the 'attoclock'. While the investigation is ongoing, the window for non-zero 'tunnelling times' in this context appears to have largely closed.

The role of time in quantum mechanics is under ever greater scrutiny due to the development and application of ultrafast pulsed lasers. Similar to a snapshot camera, by interacting with a system on a timescale comparable with its dynamics you gain access to its underlying temporal nature. With pulse durations now commonly in the few-femtosecond (10^{-15} s) or even attosecond (10^{-18} s) domain, the motion of electrons within atoms is within reach. Such systems are sufficiently fundamental that experiments provide unprecedented tests of theory at the heart of quantum mechanics.

A quintessential example of this is a class of experiments known as the 'attoclock'. In such, the rotating electric field vector of a (near)-circularly polarised laser pulse applied to a target atom or molecule maps the timing of the resultant tunnel ionisation onto the photoelectron momentum. From the distribution of this momentum a measurement of the 'tunnelling time' for the process is derived. However, despite a decade passing since the techniques inception [1], lack of consensus persists over its interpretation, in no small part due to the notion of tunnelling time itself.

First examined by MacColl [2] in 1932 the concept of a tunnelling time has received great attention and is the subject of a recent review [3]. A quantity using this moniker has been defined in several different contexts, but the definitions are often mutually incompatible as each are fundamentally different methods of quantifying the underlying physical process. The question is, does the attoclock contain any physics attributable to a tunnelling time, and if so, is it compatible with an existing definition or is yet of another form? While this question remains somewhat open, there is an increasing volume of evidence that the answer to the precondition is, in fact, "no".

For an attoclock experiment to conclude an observation of tunnelling time the rotation observed in the photoelectron momentum distribution must be unexplained by classical trajectories assuming instantaneous tunnelling. For the earliest reported attoclock measurements this was not found to be the case [4, 5]. In stark contrast, the results of Landsman *et al.* [6] observed rotations far beyond what could be explained by such trajectories. However, to this date, the data remains unexplained by *ab initio* theory despite considerable efforts [7, 8]. The remaining suggestion of tunnelling times in attoclock problems is from Camus *et al.* [9].

This work relies on the finding that the difference in rotations from two atomic targets is not well explained by trajectories with instantaneous tunnelling and zero initial velocity.

On the theoretical side, for a simplified attoclock problem[1] using near single-cycle pulses the nature of the rotation has been found to be of entirely Coulombic origin [10, 11]. To experimentally confirm this finding it has been suggested to examine negative ions [12]. Further evidence is provided via backpropogation [13] or through application of an additional linearly polarised field [14]. Most convincing however, is the recent work of Sainadh *et al.* [15] which presents agreement between theory and experiment for an atomic hydrogen target, and concludes that they see no rotations attributable to a delay in the tunnelling process.

For the attoclock, the window for any physics that would be attractively described by a tunnelling time appears to have all but closed. What keeps the door ajar is the aforementioned not yet fully understood experiments of Camus and Landsman.

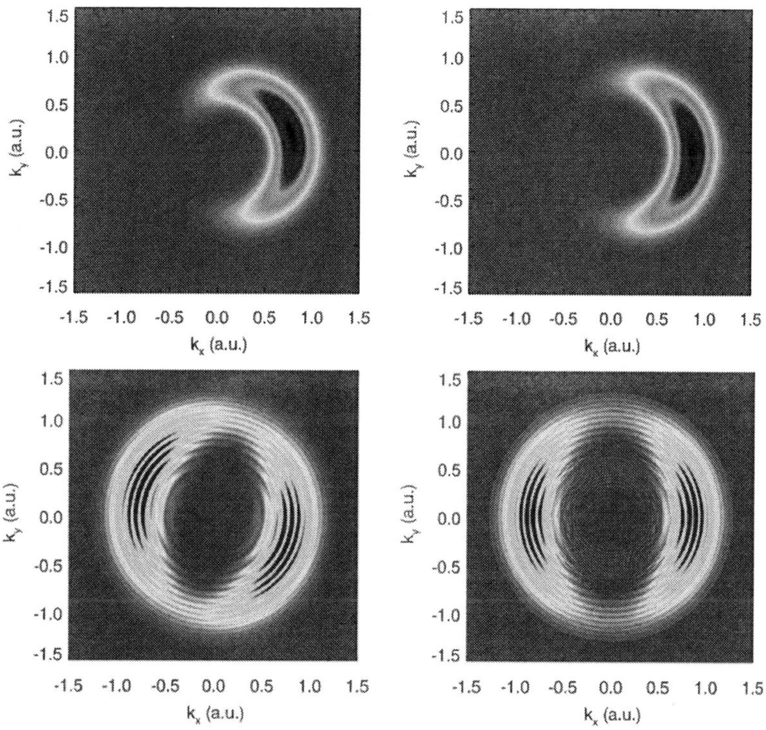

Figure 1. Left column: 3D-TDSE calculated photoelectron spectra from hydrogen for short (top) and long (bottom) pulses. The short pulse is approximately 1.6 fs FWHM in intensity of peak 0.86×10^{14} W/cm^2, 800 nm, ellipticity 1.0, and anti-clockwise helicity. Its corresponding distributions assume CEP stability. The long pulse is approximately 6 fs FWHM in intensity of peak 1.5×10^{14} W/cm^2, 770 nm, ellipticity 0.85, and clockwise helicity. Its corresponding distributions are averaged over CEP. Right column: As for left but from a Yukawa potential (screening parameter $a = 1$) of hydrogenic binding energy. The colouration of probability is linear and normalised from red (min) to black (max).

[1] It should be noted that the simplified problem has the advantage of perfectly symmetric photoelectron distributions as there are no effects of inter-cycle interference (see Figure 1). Accordingly there is no ambiguity in characterising said distribution by a single angle and corresponding trajectory.

References

[1] Eckle P, Smolarski M, Schlup P, Biegert J, Staudte A, Schöffler M, Muller H G, Dörner R and Keller U 2008 *Nature Physics* **4** 565
[2] MacColl L A 1932 *Phys. Rev.* **40**(4) 621–626
[3] Landsman A S and Keller U 2015 *Physics Reports* **547** 1–24
[4] Eckle P, Pfeiffer A N, Cirelli C, Staudte A, Dörner R, Muller H G, Büttiker M and Keller U 2008 *Science* **322** 1525–1529
[5] Pfeiffer A N, Cirelli C, Smolarski M, Dimitrovski D, Abu-Samha M, Madsen L B and Keller U 2012 *Nature Physics* **8** 76
[6] Landsman A S, Weger M, Maurer J, Boge R, Ludwig A, Heuser S, Cirelli C, Gallmann L and Keller U 2014 *Optica* **1** 343–349
[7] Ivanov I and Kheifets A 2014 *Physical Review A* **89** 021402
[8] Majety V P and Scrinzi A 2017 *Journal of Modern Optics* **64** 1026–1030
[9] Camus N, Yakaboylu E, Fechner L, Klaiber M, Laux M, Mi Y, Hatsagortsyan K Z, Pfeifer T, Keitel C H and Moshammer R 2017 *Physical review letters* **119** 023201
[10] Torlina L, Morales F, Kaushal J, Ivanov I, Kheifets A, Zielinski A, Scrinzi A, Muller H G, Sukiasyan S, Ivanov M and Smirnova O 2015 *Nature Physics* **11** 503
[11] Bray A W, Eckart S and Kheifets A S 2018 *Physical review letters* **121** 123201
[12] Douguet N and Bartschat K 2019 *Physical Review A* **99** 023417
[13] Ni H, Saalmann U and Rost J M 2018 *Physical Review A* **97** 013426
[14] Han M, Ge P, Fang Y, Yu X, Guo Z, Ma X, Deng Y, Gong Q and Liu Y 2019 *Physical Review Letters* **123** 073201
[15] Sainadh U S, Xu H, Wang X, Atia-Tul-Noor A, Wallace W C, Douguet N, Bray A, Ivanov I, Bartschat K, Kheifets A, Sang R T and Litvinyuk I V 2019 *Nature* **568** 75

XUV+IR photoionization of argon atoms: selection of sideband orders

R Della Picca[1], A A Gramajo[1], S López[2] and D G Arbó[2]
[1] CONICET and Centro Atómico Bariloche (CNEA), 8400 Bariloche, Argentina
[2] Institute for Astronomy and Space Physics IAFE (UBA-Conicet), Buenos Aires, Argentina
E-mail: renata@cab.cnea.gov.ar

Abstract. We present a theoretical study of atomic ionization due to an XUV pulse in the presence of an infrared laser. Within the strong field approximation and considering the periodicity and symmetry of the transition matrix we show that the photoelectron spectrum can be described from the contribution during only one (or half) infrared cycle. These symmetry and periodicity properties impress selection rules which destructively cancel certain sideband orders favoring others. In particular, we analyze the photoionization of Argon in four geometrical arrangements of the polarization vector and the photoelectron momentum direction.

1. Introduction

Laser assisted photo emission (LAPE) processes take place when typically extreme ultraviolet (XUV) radiation and infra red (IR) intense laser fields overlap in space and time. Since the first theoretical prediction of "sidebands" peaks [1] a lot of experiments and theoretical studies have been performed in this area, see for example [2, 3, 4, 5, 6, 7, 8] and references therein. The simultaneous absorption of one high-frequency photon together with the exchange of several additional photons from the laser field lead to equally spaced "sideband" peaks in the photoelectron (PE) spectra, located on each side of the photoionization line. From the theoretical point of view, the formation of these peaks can be explained as the constructive interference between electron waves emitted at different optical cycles of the IR laser field [9, 10]. In our previous works [10, 11, 12] we have employed the semiclassical model (SCM) based on the strong field approximation (SFA) to identify electron trajectories and to describe the photoelectron (PE) spectrum as product of inter- and intracycle interferences factors, where the former accounts for sidebands formation and the latter as a modulation.

Despite the interpretation in terms of these interferences was previously deduced employing the saddle point approximation in the temporal integration of the transition matrix, in this work we show that this approximation is not strictly necessary to describe the PE spectrum as the contribution of intra- and intercycle factors. In the special case that the electron emission is perpendicular to the laser polarization vector, an alternative factorization of the PE spectrum in terms of intra-half and inter-half-cycle interferences can be obtained due to the initial state symmetry [12]. This results in that the even orders of sideband peaks are canceled and the PE spectra present characteristic sideband peaks separated by twice the IR photon energy. Experimentally, this kind of peaks suppression was confirmed in one color above treshold ionization (ATI) of Xe at 800 nm [13]. However, to the best of our knowledge there is no

LAPE experiments to confirm the selection of sideband orders at particular emission directions. In the present work we explore the photoionization of argon from different initial states and at different geometrical arrangements to theoretically establish under what situation the selection of certain sideband production can be expected. High resolution experiments under the mentioned conditions would be desirable in order to corroborate the present study.

The paper is organized as follows: In Sec. 2, we briefly resume the SFA theory and analyze the properties of the temporal integral of the transition matrix. In Sec. 3 we present the results for Argon LAPE from shell 3 and for four geometrical arrangements. Concluding remarks are presented in Sec. 5. Atomic units are used throughout the paper, except when otherwise stated.

2. Analysis and properties of the temporal integral of the transition matrix

The theoretical approach based on the SFA is presented in detail in [10]. The model describes the ionization of an atomic system (initially in the state φ_i with ionization potential I_p) interacting with an XUV pulse assisted by a IR laser, linearly polarized along $\hat{\varepsilon}_X$ and $\hat{\varepsilon}_L$, respectively. The emitted electron with momentum $\vec{k}$ and energy $E = k^2/2$ is described by the Volkov wave function [14]. Then, the photoelectron momentum distributions can be calculated as the square of the transition matrix T_{if}

$$\frac{dP}{d\vec{k}} = |T_{if}|^2 \qquad \text{where} \qquad T_{if} = -\frac{i}{2} F_{X0} \int_{t_0}^{t_0+\tau_X} \hat{\varepsilon}_X \cdot \vec{d}[\vec{k} + \vec{A}(t)]\, e^{iS(t)}\, dt. \tag{1}$$

Here the generalized action is defined as

$$S(t) = -\int_t^{\infty} dt' \left[\frac{(\vec{k} + \vec{A}(t'))^2}{2} + I_p - \omega_X \right] \tag{2}$$

and the dipole element is $\vec{d}(\vec{v}) = \langle e^{i\vec{v}.\vec{r}} | \vec{r} | \varphi_i(\vec{r}) \rangle (2\pi)^{-3/2}$. F_{X0} and ω_X are the respective amplitude and frequency of the XUV pulse. F_{X0} is nonzero only during the temporal interval $(t_0, t_0 + \tau_X)$ where τ_X is the XUV pulse duration. During this lapse the IR electric field can be modeled as a cosine-like wave, hence, the vector potential can be written as

$$\vec{A}(t) = \frac{F_{L0}}{\omega_L} \sin(\omega_L t)\, \hat{\varepsilon}_L \tag{3}$$

with F_{L0} the amplitude and ω_L the frequency of the laser electric field.

In the following we analyze some properties of the transition matrix in Equation (1). First, we examine the action S defined in Equation (2), which can be written as

$$S(t) = S_0 + at + b\cos(\omega_L t) + c\sin(2\omega_L t), \tag{4}$$

where $a = k^2/2 + I_p + U_p - \omega_X$, $b = -\hat{\varepsilon}_L \cdot \vec{k} F_{L0}/\omega_L^2$, $c = -U_p/2\omega_L$, the ponderomotive energy $U_p = (F_{L0}/2\omega_L)^2$ and S_0 is a constant that can be omitted. We observe that $[S(t) - at]$ is a time-oscillating function with the same period of the laser $T = 2\pi/\omega_L$, then

$$S(t + NT) = S(t) + aNT \tag{5}$$

with N an integer number. Furthermore, since the potential vector Equation (3) is T-periodic, the dipole element is also periodic: $\vec{d}[\vec{k} + \vec{A}(t + NT)] = \vec{d}[\vec{k} + \vec{A}(t)]$.

Let us introduce the quantity $I(t)$ as the contribution from zero to time t as

$$I(t) = \int_0^t \ell(t')\, e^{iS(t')}\, dt' \qquad \text{with} \qquad \ell(t) = -\frac{i}{2} F_{X0} \hat{\varepsilon}_X \cdot \vec{d}[\vec{k} + \vec{A}(t)] \tag{6}$$

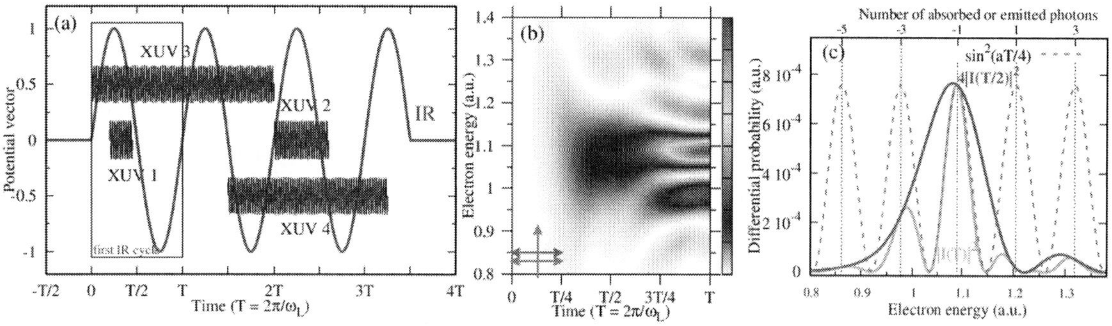

Figure 1. (a) Scheme of different XUV+IR photoionization cases. (b) Square modulus of the integral I Eq. (6) as a function of time and electron energy for the case of photoionization of Ar(3s) with electronic emission direction (yellow arrow) perpendicular to both polarization vectors (red -IR- and blue -XUV- horizontal arrows). (c) Cuts of plot (b) at $t = T/2$ and T in arbitrary units. The laser parameters are $F_{L0} = 0.041$, $\omega_L = 0.057$, $F_{X0} = 0.01$, $\omega_X = 41\omega_L$ and $\tau_X = 3T$.

for $0 \leq t \leq T$. When the integral (6) covers all the first IR cycle, it coincides with the photoionization transition matrix for a XUV pulse that holds exactly one IR cycle [see Equation (1)]. For this reason we call $|I(T)|^2$ as the "intra-cycle" contribution.

We are interested in considering an arbitrary situation (i.e. arbitrary delay t_0 and XUV duration τ_X) and to express the transition matrix in Eq. (1) in terms of the integral $I(t)$ in Equation (6). Therefore, we analyze four cases schematized in figure 1(a):

XUV 1 (streaking): the XUV pulse is shorter than T and acts during the first IR cycle. The integration of the transition matrix from t_0 to $t_0 + \tau_X$ can be written as the subtraction of two integrals in the intervals $[0, t_0]$ and $[0, t_0 + \tau_x]$:

$$T_{if} = \int_0^{t_0+\tau_X} \ell(t) e^{iS(t)}\, dt - \int_0^{t_0} \ell(t) e^{iS(t)}\, dt = I(t_0 + \tau_X) - I(t_0).$$

provided that $t_0 + \tau_X \leq T$.

XUV 2 (delay of several cycles): the XUV pulse starts at the beginning of the Nth IR cycle ($t_0 = NT$) and it is shorter than T, *i.e*, $\tau_X \leq T$. Performing the transformation $t = t' + NT$ and keeping in mind the T-periodicity of ℓ and S [see Equation (5)] it is easy to see that

$$T_{if} = \int_{NT}^{NT+\tau_X} \ell(t) e^{iS(t)}\, dt = \int_{t'=0}^{t'=\tau_X} \ell(t' + NT) e^{iS(t'+NT)} dt' = I(\tau_X) e^{iaNT}. \quad (7)$$

XUV 3 (duration of several cycles - intercycle contribution): when the XUV covers several IR cycles ($\tau_X = NT$), the integral over each cycle can be summed up using Eq. (7) to obtain

$$\begin{aligned} T_{if} &= \int_0^{NT} \ell(t) e^{iS(t)} dt = \sum_{n=0}^{N-1} \int_{nT}^{nT+T} \ell(t) e^{iS(t)} dt = \sum_{n=0}^{N-1} I(T) e^{ianT} \\ &= I(T) \frac{\sin(aTN/2)}{\sin(aT/2)} e^{[iaT(N-1)/2]}. \end{aligned} \quad (8)$$

Then, the PE momentum distribution is

$$|T_{if}|^2 = \underbrace{|I(T)|^2}_{\text{intracycle}} \underbrace{\left[\frac{\sin(aTN/2)}{\sin(aT/2)}\right]^2}_{\text{intercycle}}, \quad (9)$$

where we can identify the intra- and intercycle factors. This kind of factorization was previously obtained within the semi classical model (SCM) in LAPE [10] and the simple man's model in ATI [15]. In these works each contribution was recognized as the interference stemming from trajectories within the same optical cycle (intracycle interference) and from trajectories released at different cycles (intercycle interference). In the present work, we show in a more general way that these interferences are a consequence of the periodicity of the transition matrix.
The zeros of the denominator in the intercycle factor in Equation (9), *i.e.*, the energy values satisfying $aT/2 = n\pi$, are preventable singularities since the numerator also cancels, however maxima are reached at theses points. Such maxima are recognized as "sideband peaks" in the PE spectrum that occur when

$$E_n = n\omega_L + \omega_X - I_p - U_p \tag{10}$$

and they correspond to the absorption (positive n) or emission (negative n) of n IR photons following the absorption of one XUV photon. In fact, when $N \to \infty$ the intercycle factor becomes a series of delta functions, i.e., $\sum_n \delta(E - E_n)$.

XUV 4 (general case): we consider an arbitrary XUV pulse of duration $\tau_X = NT + \Delta$, where $\Delta \leq T$ and delay $t_0 = MT + \delta$, where $\delta \leq T$ (with N and M nonzero integer numbers). Then, using the previous results we can obtain

$$\begin{aligned} T_{if} &= \int_{MT+\delta}^{MT+\delta+NT+\Delta} \ell(t)e^{iS(t)}dt = e^{iaTM}\int_{\delta}^{\delta+NT+\Delta} \cdots = e^{iaTM}\left[\int_0^{NT} + \int_{NT}^{NT+\delta+\Delta} - \int_0^{\delta}\right] \quad (11) \\ &= e^{iaTM}\left[I(T)\frac{\sin(aTN/2)}{\sin(aT/2)}e^{iaT(N-1)/2} + e^{iaTN}I(\delta+\Delta) - I(\delta)\right] \quad \text{if} \quad \delta+\Delta \leq T \\ &= e^{iaTM}\left[I(T)\frac{\sin(aT(N+1)/2)}{\sin(aT/2)}e^{iaTN/2} + e^{iaT(N+1)}I(\delta+\Delta-T) - I(\delta)\right] \quad \text{if} \quad \delta+\Delta \geq T. \end{aligned}$$

When the XUV covers an integer number of IR cycles ($\Delta = 0$), Equation (11) results in:

$$T_{if} = e^{iaT\frac{2M+N-1}{2}}\frac{\sin(aTN/2)}{\sin(aT/2)}\left[I(T) + I(\delta)\,2i\sin(aT/2)\,e^{(iaT/2)}\right],$$

where the intercycle contribution can be factorized as in the previous case. Here we note that at the sideband position $aT = 2n\pi$ the second term inside the bracket is zero, then the emission probability is independent of the delay δ and the PE spectrum becomes equal to Equation (9). This independence was observed in previous works (see Figure 6 of [11] and 5 of [12]) as a blurred SCM discontinuity in the PE spectrum as a function of the delay between XUV and IR pulses. Anyway, in the general case ($\Delta \neq 0$) we note that the first term inside the brackets in Equation (11) determines the predominant contribution to the PE spectrum when $N >> 1$ due to the increase of the intercycle interference term for $aT = 2n\pi$ [see discussion below Equation (10)]. In such case, the PE spectrum approximately behaves like Equation (9).

Throughout the analysis of the previous four cases, we have covered two LAPE regimes: the streaking (cases 1 and 2) and sideband regimes (cases 3 and 4). In all these cases, we found that the temporal integration over the XUV duration with however delay could be written as function of the magnitude $I(t)$ defined during the first IR cycle. As an example, in Figure 1(b) we show $|I(t)|^2$ for photoionization of Ar($3s$). From its definition in Equation (6), it is clear that it increases from zero at time zero and it depends on the electron energy and the geometrical arrangement between $\hat{\varepsilon}_X$, $\hat{\varepsilon}_L$ and $\hat{k}$ (see the arrows in the bottom left corner of the figure). With this information and employing equations (7) to (11), the PE spectra can be constructed for several configurations of the XUV+IR fields. Furthermore, we note that the precedent analysis can also be done not only within the SFA but also within others approaches, such as the Coulomb-Volkov approximation, as long as its dipole elements $\vec{d}[\vec{k}+\vec{A}(t)]$ maintain the temporal T-periodicity.

Now, we consider the particular situations in which $\hat{\varepsilon}_L \perp \vec{k}$. Because of this, $b=0$ in Equation (4) and $[S(t)-at]$ has not only T- but also $T/2$-periodicity. If under this circumstance, the dipole element also satisfies $\hat{\varepsilon}_X \cdot \vec{d}[\vec{k}+\vec{A}(t+T/2)] = \pm\hat{\varepsilon}_X \cdot \vec{d}[\vec{k}+\vec{A}(t)]$, *i.e.*, it is symmetric or antisymmetric with respect to the middle of the IR cycle, the integration over all the IR cycle can be written as a sum over the two half cycles. Then, depending on the symmetry (+) or antisymmetry (−) of the dipole element with respect to $T/2$ we have

$$I(T) = \int_0^{T/2} \ell(t)e^{iS(t)}dt + \underbrace{\int_{T/2}^{T} \ell(t)e^{iS(t)}dt}_{\pm e^{iaT/2}I(T/2)} = I(T/2)(1 \pm e^{iaT/2}) \qquad (12)$$

$$|I(T)| = |2\, I(T/2)\, \cos(aT/4)| \text{ if } + \text{ (symmetric)} \qquad (13)$$

$$|I(T)| = |2\, I(T/2)\, \sin(aT/4)| \text{ if } - \text{ (antisymmetric)}\,. \qquad (14)$$

The factors $\cos(aT/4)$ or $\sin(aT/4)$ in equations (13) and (14) cancel out odd or even sideband peaks in the intercycle contribution, respectively. In consequence, the PE spectrum presents structures corresponding to absorption or emission of only even or odd number of IR photons and the energy difference between two consecutive sideband peaks is 2ω instead of ω as for the general conservation energy rule in Equation (10). With this in mind, for antisymmetric dipole elements the PE spectrum of Equation (9) becomes

$$|T_{if}|^2 = \underbrace{4\, \underbrace{|I(T/2)|^2}_{\text{intrahalfcycle}}\, \sin^2(aT/4)}_{\text{intracycle}} \underbrace{\left[\frac{\sin(aTN/2)}{2\sin(aT/4)\cos(aT/4)}\right]^2}_{\text{intercycle}} \qquad (15)$$

$$= \underbrace{|I(T/2)|^2}_{\text{intrahalfcycle}} \underbrace{\left[\frac{\sin(aTN/2)}{\cos(aT/4)}\right]^2}_{\text{interhalfcycle}}, \qquad (16)$$

which reaches maxima only for odd n in Equation (10) and it is suppressed at energy values E_n with even n. In particular, the absorption of only one XUV photon alone (in the absence of absorption or emission of IR photons) is forbidden. Alternatively, the odd sideband orders are canceled whereas the even orders stay on for symmetric dipole elements. In correspondence with our previous analysis within the SCM (see Eq. (18) in Ref. [12]), the equations (15) and (16) indicate that the PE spectrum can be factorized in two different ways: (i) as the product of intra- and intercycle interference factors and (ii) as the product of intrahalf- and interhalf-cycle interference contributions. Obviously, the two different factorizations give rise to the same results.

As an example, we examine again the Ar(3s) photoionization presented in figure 1 (b) and (c). The length-gauge dipole element for an hydrogen-like 3s state is written in Appendix A. When both lasers are polarized in the same direction $\hat{z}$ and the electronic emission is perpendicular to them ($\vec{k}=k\hat{x}$), the term $\ell(t) \sim \hat{z}\cdot[k\hat{x}+A(t)\hat{z}]f_1[k^2+A^2(t)] = A(t)f_1[k^2+A^2(t)]$ can be written as the product of the vector potential amplitude $A(t)$ and a function f_1 which depends

Table 1. Component of dipole element matrix $\hat{\varepsilon}_X \cdot \vec{d}(\vec{v})$ where $\vec{v} = \vec{k} + \vec{A}(t)$, and its symmetries respect $T_L/2$, depending on the initial states and the geometrical arrangements.

	F $v^2 = (k+A)^2$ $v_z = k + A$	P $v^2 = k^2 + A^2$ $v_z = A$	OF $v^2 = k^2 + A^2$ $v_z = k$	OP $v^2 = (k+A)^2$ $v_z = 0$
$\vec{d}_{3s} \cdot \hat{z} = v_z f_1(v^2)$ $3s$	$= (k+A) f_1((k+A)^2)$ no symmetry all SB	$= A f_1(k^2 + A^2)$ antisymmetric odd SB	$= k f_1(k^2 + A^2)$ symmetric even SB	$= 0$ not SB
$\vec{d}_{3p0} \cdot \hat{z} = f_2(v^2) + f_3(v^2) v_z^2$ $3p0$	$\sim (k+A)^2$ no symmetry all SB	$\sim A^2$ symmetric even SB	$\sim A^2$ symmetric even SB	$= f_2((k+A)^2)$ no symmetry all SB
$\vec{d}_{3p1} \cdot \hat{z} = v_z (v_x + i v_y) f_4(v^2)$ $3p1$	$= 0$ not SB	$= A k f_4(k^2 + A^2)$ antisymmetric odd SB	$= A k f_4(k^2 + A^2)$ antisymmetric odd SB	$= 0$ not SB

on time through $A^2(t)$. Hence, $\ell(t)$ results antisymmetric respect to the middle of the IR cycle and Equation (14) is verified.

In figure 1 (c) we plot the cuts of figure 1 (b) at times $t = T/2$ and T. We observe that, according to Equation (14), the intracycle contribution $|I(T)|^2$ (orange line) could be written as the product of the intrahalfcycle $|I(T/2)|^2$ (green line) and the interference factor $\sin^2(aT/4)$ (dashed line) which is null at even orders of the sideband peaks.

3. LAPE of argon atoms

In this section we want to explore in what situations some sideband orders are canceled in view of the preceding discussion. For that, we consider the argon photoionization from the shell 3, and analyze the symmetries of the states $3s$, $3p_0$ and $3p_1$ for different geometrical arrangements between the momentum direction $\hat{k}$ and the polarization vectors of both linearly polarized fields. We fix $\hat{\varepsilon}_X$ along the $\hat{z}$ axis, whereas the IR polarization vector $\hat{\varepsilon}_L$ can be parallel or orthogonal to it. For the sake of simplicity, we restrict our analysis to the case where the XUV pulse duration is a multiple of the IR optical cycle, i.e., $\tau_X = NT$. Then, the PE momentum distribution is given by Equation (9), and we examine the emission parallel and perpendicular to the XUV polarization direction. Different combinations of these geometrical arrangements lead to four study cases, namely:

Forward (F): Both polarization vectors and electronic emission direction are parallel to the $\hat{z}$ axis, *i.e.*, $\hat{\varepsilon}_L = \hat{\varepsilon}_X = \hat{k} = \hat{z}$.

Perpendicular (P): Both polarization vector are parallel and the electron emission direction is perpendicular to them, *i.e.*, $\hat{\varepsilon}_L = \hat{\varepsilon}_X = \hat{z}$ and $\hat{k} = \hat{x}$.

Orthogonal Forward (OF): The laser polarizations are orthogonal and the electron emission is parallel to the XUV one, *i.e.*, $\hat{\varepsilon}_X = \hat{k} = \hat{z}$ and $\hat{\varepsilon}_L = \hat{x}$.

Orthogonal Perpendicular (OP): The laser polarizations are orthogonal and the electron emission is parallel to the IR one, *i.e.*, $\hat{\varepsilon}_X = \hat{z}$ and $\hat{\varepsilon}_L = \hat{k} = \hat{x}$.

The hydrogen-like dipole matrix elements for the subshells $3s$, $3p0$ and $3p1$ are described in Appendix A. In table 1 we present the analysis of the expected sideband (SB) orders depending on the geometrical arrangement (columns) and the dipole element $\hat{z}$-component (rows) evaluated in $\vec{v} = \vec{k} + \vec{A}(t)$. For each geometrical configuration the magnitudes v^2 and v_z are shown in the

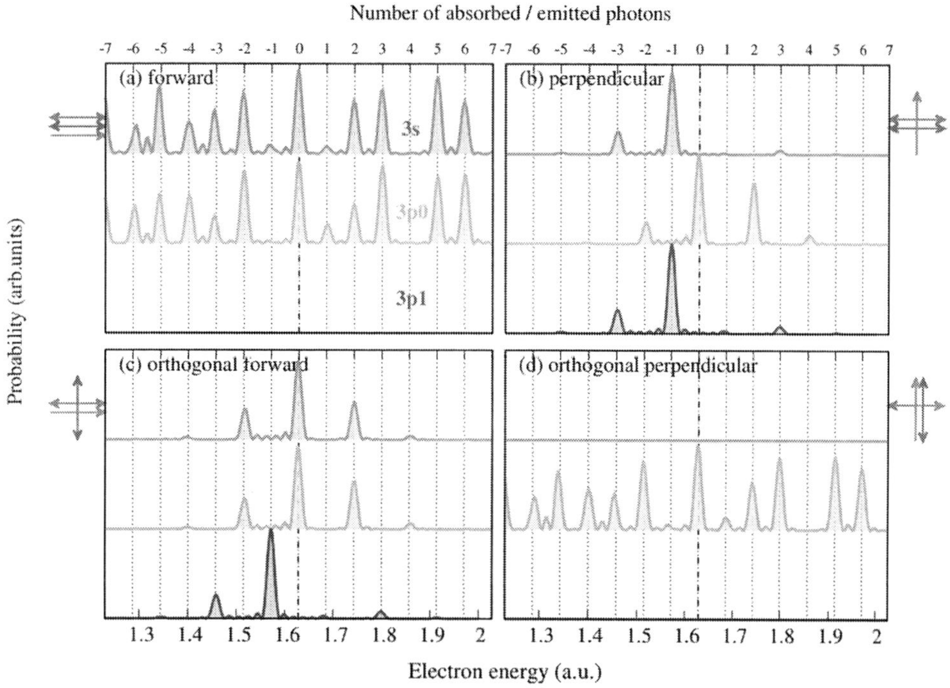

Figure 2. PE spectra in arbitrary units for LAPE of Argon from shell 3. The arrows represent the geometrical arrangement between $\hat{\varepsilon}_X$ (blue), $\hat{\varepsilon}_L$ (red) and $\hat{k}$ (orange). The laser parameters are the same as in Fig. 1.

first row. In F and OP configurations, the temporal dependence of $v^2 = [k + A(t)]^2$ does not have any particular symmetry for what we would expect all sideband orders according to Equation (9). In particular, (F $3p1$), (OP $3s$) and (OP $3p1$) have null dipole elements. Instead, the P and OF configurations (center columns) verify the conditions $b = 0$ (since $\hat{\varepsilon}_L \perp \vec{k}$) and there is a defined symmetry of the dipole element (symmetric or antisymmetric) with respect to $T/2$. Thus, even or odd SB orders are expected.

In order to corroborate the theoretical predictions of table 1, we also present in Fig. 2 the SFA calculation of Eq. (1). The four geometrical configurations P, F, OF and OP are considered in figures 2 (a) to (d), respectively, and for the initial states $3s$ (in green) $3p_0$ (yellow) and $3p_1$ (red). The numbers n of absorbed or emitted photons are indicated as vertical dashed lines in the figure. Since the difference of ionization potential I_p for the different argon initial states considered (see Appendix A), the $3s$-PE spectrum is shifted 0.48 a.u. to high energies in order to match the same sideband order for all states. In the figure we observe that the sideband peaks have different heights, because they depend on the modulation of the intracycle factor [11, 12]. Furthermore, in agreement with the analysis of table 1 we note that: (i) in P and OF configurations only odd or even sideband are present, whereas (ii) the F and OP configurations exhibit all SB orders, except when the emission is forbidden (F $3p1$, OP $3s$ and $3p1$) or when the intracycle modulation suppress some particular SB orders ($n = -1$ and 4 for example). Besides, (iii) the energy range of the emitted electrons is much smaller in P and OF arrangements than in the F and OP cases, strongly depending on the IR intensity [10]. In addition, we want to emphasize that TDSE calculations for Argon (P $3s$) corroborate the selection of only odd sideband peaks allowed in the PE spectrum, as it can be observed from Fig. 1 of Ref. [16].

4. Conclusions

We have studied the electron emission produced by an XUV pulse assisted by an IR laser field emphasizing on the analytic properties deduced from the SFA transition matrix element. We have shown that in several XUV+IR configurations, the PE spectrum can be described as a function of the time integral $I(t)$ during the first IR optical cycle not only in the sideband but also in the streaking regime. In particular, we have shown that intra-, inter-, intrahalf- and interhalfcycle interferences are a consequence of the periodicity and symmetry of the transition matrix element. For the case of photoionization of argon from the third shell, we have analyzed the symmetries of the states s, p_0 and p_1 in four different geometrical arrangements and have corroborated the corresponding selection rules that determine the presence of all, none, odd or even sideband orders in the PE spectra.

Appendix A. Dipole elements

The dipole element is defined by $\vec{d_i}(\vec{v}) = (2\pi)^{-3/2}\int d\vec{r}\,\exp[-i\vec{v}\cdot\vec{r}]\,\vec{r}\,\phi_i(\vec{r})$. Depending on the initial hydrogen-like state ϕ_i we have that the $\hat{z}$-component of the dipole elements are

$$\begin{aligned}
\vec{d}_{3s}\cdot\hat{z} &= v_z f_1(v^2) & f_1(v^2) &= -2i\,(2\alpha)^{5/2}\,(3v^4+11\alpha^4-18\alpha^2v^2)/\pi(v^2+\alpha^2)^5\\
\vec{d}_{3p0}\cdot\hat{z} &= f_2(v^2)+v_z^2 f_3(v^2) & f_2(v^2) &= \sqrt{3}\,2^4\alpha^{7/2}\,(v^4-\alpha^4)/\pi(v^2+\alpha^2)^5\\
 & & f_3(v^2) &= -\sqrt{3}\,2^5\alpha^{7/2}\,(3v^2-5\alpha^2)/\pi(v^2+\alpha^2)^5\\
\vec{d}_{3p1}\cdot\hat{z} &= v_z(v_x+iv_y)f_4(v^2) & f_4(v^2) &= \sqrt{3}\,2^{9/2}\alpha^{7/2}\,(3v^2-5\alpha^2)/\pi(v^2+\alpha^2)^5
\end{aligned}$$

where $\alpha = \sqrt{2I_p}$ and we have introduced the functions f_j (for $j = 1$ to 4) to indicate explicitly the dependence on v^2. In order to consider the photoionization of argon, we have set the ionization potential $I_p = 15.78$ eV for the $3p$ subshells and $I_p = 28.84$ eV for the $3s$ state.

References

[1] Véniard V, Taïeb R and Maquet A 1995 *Phys. Rev. Lett.* **74**(21) 4161–4164
[2] Itatani J, Quéré F, Yudin G L, Ivanov M Y, Krausz F and Corkum P B 2002 *Phys. Rev. Lett.* **88**(17) 173903
[3] Drescher M and Krausz F 2005 *Journal of Physics B Atomic Molecular Physics* **38** S727–S740
[4] Maquet A and Taïeb R 2007 *Journal of Modern Optics* **54** 1847–1857
[5] Radcliffe P, Arbeiter M, Li W B, Düsterer S, Redlin H, Hayden P, Hough P, Richardson V, Costello J T, Fennel T and Meyer M *New Journal of Physics* **14** 043008
[6] Meyer M, Radcliffe P, Tschentscher T, Costello J T, Cavalieri A L, Grguras I, Maier A R, Kienberger R, Bozek J, Bostedt C, Schorb S, Coffee R, Messerschmidt M, Roedig C, Sistrunk E, Di Mauro L F, Doumy G, Ueda K, Wada S, Düsterer S, Kazansky A K and Kabachnik N M 2012 *Phys. Rev. Lett.* **108**(6) 063007
[7] Mazza T, Ilchen M, Rafipoor A J, Callegari C, Finetti P, Plekan O, Prince K C, Richter R, Danailov M B, Demidovich A, De Ninno G, Grazioli C, Ivanov R, Mahne N, Raimondi L, Svetina C, Avaldi L, Bolognesi P, Coreno M, O'Keeffe P, Di Fraia M, Devetta M, Ovcharenko Y, Mller T, Lyamayev V, Stienkemeier F, Dsterer S, Ueda K, Costello J T, Kazansky A K, Kabachnik N M and Meyer M 2014 *Nature Communications* **5** 3648
[8] Düsterer S, Hartmann G, Bomme C, Boll R, Costello J T, Erk B, Fanis A D, Ilchen M, Johnsson P, Kelly T J, Manschwetus B, Mazza T, Meyer M, Passow C, Rompotis D, Varvarezos L, Kazansky A K and Kabachnik N M 2019 *New Journal of Physics* **21** 063034
[9] Kazansky A K, Sazhina I P and Kabachnik N M 2010 *Phys. Rev. A* **82**(3) 033420
[10] Gramajo A A, Della Picca R, López S D and Arbó D G 2018 *Journal of Physics B: Atomic, Molecular and Optical Physics* **51** 055603
[11] Gramajo A A, Della Picca R, Garibotti C R and Arbó D G 2016 *Phys. Rev. A* **94**(5) 053404
[12] Gramajo A A, Della Picca R and Arbó D G 2017 *Phys. Rev. A* **96**(2) 023414
[13] Korneev P A, Popruzhenko S V, Goreslavski S P, Yan T M, Bauer D, Becker W, Kübel M, Kling M F, Rödel C, Wünsche M and Paulus G G 2012 *Phys. Rev. Lett.* **108**(22) 223601
[14] Wolkow D 1935 *Zeitschrift für Physik* **94** 250–260 ISSN 0044-3328
[15] Arbó D G, Ishikawa K L, Schiessl K, Persson E and Burgdörfer J 2010 *Phys. Rev. A* **82**(4) 043426
[16] Della Picca R, Gramajo A A, López S D and Arbó D G *J. Phys. Conf. Series* Poster ICPEAC. In press

Machine learning for analysing ab initio molecular dynamics simulations

Florian Häse,[1,2,3,4] **Ignacio Fdez. Galván,**[5] **Alán Aspuru-Guzik,**[2,3,4,6] **Roland Lindh**[5] **and Morgane Vacher**[7]

[1] Department of Chemistry and Chemical Biology, Harvard University, Cambridge, MA 02138, USA
[2] Department of Chemistry, University of Toronto, Toronto, ON M5S 3H6, Canada
[3] Department of Computer Science, University of Toronto, Toronto, ON M5S 3H6, Canada
[4] Vector Institute for Artificial Intelligence, Toronto, ON M5S 1M1, Canada
[5] Department of Chemistry – BMC, Uppsala University, Box 576, 75123 Uppsala, Sweden
[6] Canadian Institute for Advanced Research (CIFAR) Senior Fellow, Toronto, ON M5S 1M1, Canada
[7] Department of Chemistry – Ångström, The Theoretical Chemistry Programme, Uppsala University, Box 538, 751 21 Uppsala, Sweden

E-mail: morgane.vacher@kemi.uu.se

Abstract. Post-calculation analyses are often required to extract physical insights from ab initio molecular dynamics simulations. In the present work, we use different machine learning classifiers to take a new perspective on the decomposition reaction of dioxetane. Upon thermally activated decomposition, dioxetane can form products in an electronically excited state and can thus chemiluminesce. Simulated dynamics trajectories exhibit both successful and frustrated dissociations. As an exhaustive and systematic study of the decomposition mechanism "by hand" is beyond feasibility, machine learning models have been employed to study the relevant nuclear distortions governing molecular dissociation. According to all classifiers used in the study, the two sets of geometries differ by the in-phase planarisation of the two formaldehyde moieties. New insights are obtained from this analysis: if both moieties are not planar enough when the dissociation is attempted, it is frustrated and the molecule remains trapped. The postponing of the decomposition reaction by the so-called entropic trap enhances the chemiexcitation efficiency.

1. Introduction

Ab initio molecular dynamics simulations provide a complete picture of the evolution of the electrons and the nuclei along a chemical reaction. However, post-calculation analyses are often required to understand *why* a molecule reacts the way it does [1, 2, 3]. Recently, we have studied the thermally-activated decomposition of dioxetane [4, 5, 1]. The reaction occurs in two steps leading to two formaldehyde fragments: first the O–O bond breaks and then the C–C bond breaks (Figure 1). This reaction is responsible for the non-adiabatic population of electronic excited states, a process called chemiexcitation [6, 7]. The subsequent radiative relaxation is called chemiluminescence, or bioluminescence when occurring in living organisms [8].

It was suggested that an "entropic trap" regulates the outcome of the dissociation (instead of a transition state) [9]. By delaying the exothermic ground-state dissociation, the entropic

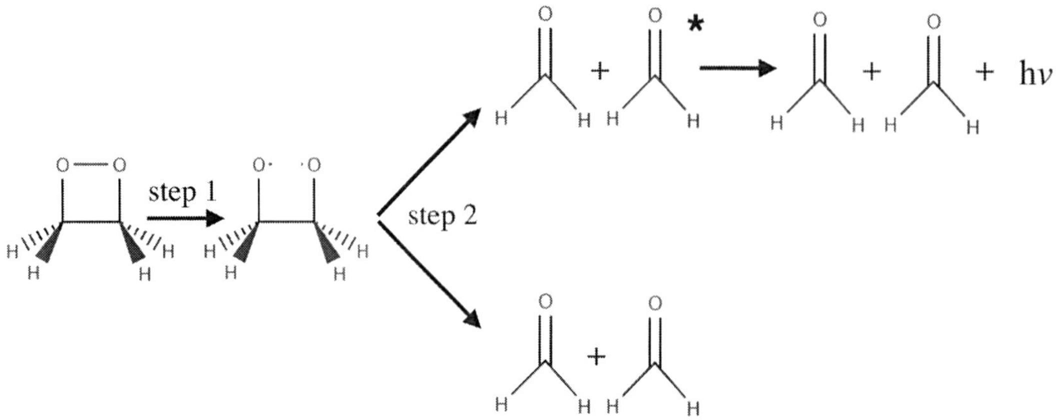

Figure 1. Chemiluminescent (top) and dark (bottom) decomposition reactions of 1,2-dioxetane into two formaldehydes.

trap would give the molecule time to explore other routes and access electronic excited states. More precisely, it was recently demonstrated that the yield of chemiexcitation is determined by the time scale of the decomposition reaction [5]: the slower the reaction, the higher the chemiexcitation yield. Ab initio molecular dynamics trajectories that dissociate late in time exhibit several attempts to dissociate before the final successful dissociation. We call these attempts *frustrated dissociations* [4]. Understanding the yield of chemiexcitation thus relates to understanding why some dissociations are frustrated and others are successful.

In the present work, we use statistical models frequently employed in machine learning to analyse the data produced from ab initio molecular dynamics simulations. Based on our analyses, physical insights into the dissociation mechanism, the entropic trap and thus the chemiexcitation yield are extracted.

2. Theoretical methods

We will below in some details describe the specifics of the ab initio molecular dynamics that produced the data, and the subsequent machine learning algorithms for the post-simulation analysis.

2.1. Data

The analysis is performed on an ensemble of 250 Born-Oppenheimer trajectories, initiated around the first transition state (corresponding to the O–O bond breaking) and heading towards the products [10, 4]. The vibrational ground state is reproduced by sampling initial nuclear positions and velocities (along all nuclear coordinates except the reaction coordinate) from a Wigner distribution [11, 12]. The electronic structure is described using the complete active space self-consistent field (CASSCF) method [13] with an active space of 12 electrons in 10 orbitals, and the ANO-RCC-VTZP basis set [14]. The dynamics simulations are performed using the OpenMolcas package [15].

Geometries along the trajectories which correspond to local maxima in the time evolution of the C–C bond length are assigned the label *frustrated*; the largest local maximum reaches 1.76 Å. *Successful* geometries are geometries at which the C–C bond length reaches 1.76 Å and keeps on increasing monotonically until successful dissociation occurs. Among the ensemble of trajectories, 420 frustrated and 250 successful geometries are extracted, respectively.

2.2. Machine learning algorithms

We aim to identify the nuclear distortions that distinguish best the frustrated and the successful geometries. To do that, we suggest the use of standard machine learning classifiers to elucidate the relevant nuclear distortions that enable the maximal distinction of frustrated and successful geometries. In this work, we employ three different statistical models: linear discriminant analysis (LDA), linear support vector classification (SVC) and support vector machines (SVM) with a linear kernel.

LDA assumes that the probabilities $p(x|y = 0)$ and $p(x|y = 1)$ follow normal distributions with identical covariances, and aims to identify a predictor for y by maximizing the likelihood ratios between the two distributions [16]. Under these assumptions, the decision criterion for x belonging to either $y = 0$ or $y = 1$ is based on a linear combination of the input features.

SVC and SVM [17] with linear kernels provide a more flexible approach to separating the space. These models create classifiers by constructing linear combinations of the input features and minimizing the quadratically smoothed hinge loss [18]. The two approaches used in this study differ in their regularization and penalty terms. SVC were implemented with an L2 regularization on the learned coefficients [19] while no further penalty terms were added to SVM classification.

The linearity of all three models simplifies the interpretation of the trained classifiers as learned feature transformations are based on linear combinations of the provided input features.

3. Results

In this section, we will present the results on the two sets of successful and frustrated geometries obtained from the ab initio molecular dynamics simulations. First, we will present how the two sets compare in normal mode coordinates. Then, we will show how the machine learning algorithms perform in separating the two sets of geometries, before finally interpreting the new suggested nuclear coordinate.

3.1. Normal modes as starting points for nuclear coordinates

The geometries are expressed in terms of normal modes (calculated at the initial transition state structure), which form a complete set of linearly independent nuclear coordinates. The normal modes are numbered from 0 to 17, where 0 refers to the reaction coordinate. The relevant normal modes are represented in a simplified manner in Figure 2.

Figure 3 presents the two sets of successful and frustrated geometries in normal mode coordinates. The upper subplot shows the average position with the standard deviation represented by the black line. First, we see that both sets of geometries do not show significant average distortions along modes 2, 3, 5, 9, 10, 12, 14 and 17. Then, along most other modes, the successful geometries show larger average distortions than the frustrated ones. The difference between both sets of geometries is quantified by the Kolmogorov–Smirnov (KS) statistics [20] plotted in the lower part of the figure. The modes that exhibit the largest differences between the two sets of geometries are modes 0, 8 and 13, with a KS statistic larger than 0.5, 0.6 and 0.5, respectively. The distribution of successful and frustrated geometries along normal mode 8 (which exhibits the highest KS statistics) is presented in Figure 4a. Normal mode 8 corresponds mainly to the planarisation of the formaldehyde moieties, with only the "inner" hydrogen atoms moving (and not the "outer" ones). Normal mode 0 corresponds mainly to the O-C-C-O dihedral angle while normal mode 13 corresponds mainly to the in-phase bending of the two H–C–H angles (Figure 2).

3.2. Performance of the machine learning analysis

The three different classifiers used suggest new combined coordinates to maximally separate the two sets of geometries. The composition of these new coordinates in terms of normal modes is

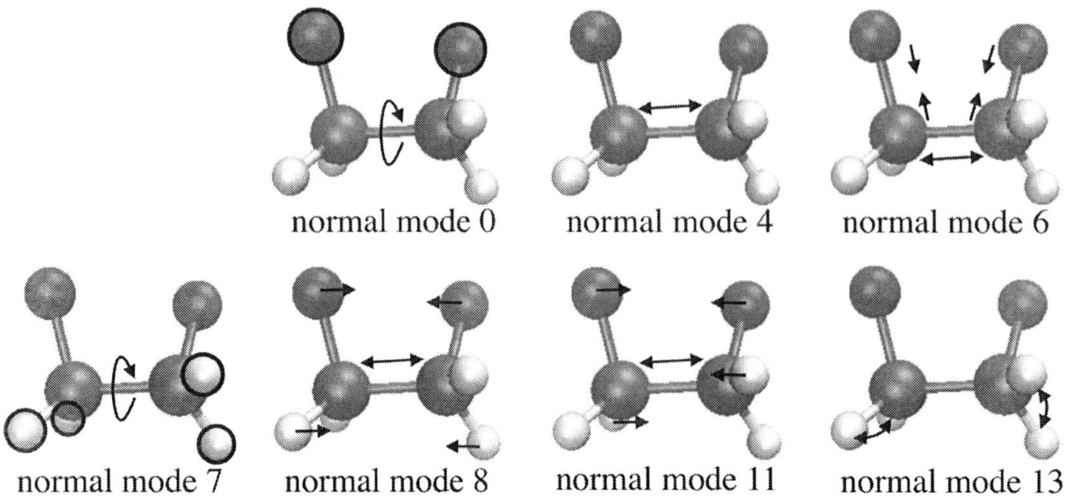

Figure 2. Simplified representation of the relevant normal mode coordinates (calculated at the O–O bond breaking transition state structure). Normal mode 0 corresponds mainly to the OCCO dihedral angle while normal mode 7 corresponds mainly to the HCCH dihedral angles.

represented in Figure 5. The distributions of successful and frustrated geometries along these combined coordinates are presented in Figures 4b-d, where the KS statistics is also indicated for each case. Regarding the ability to separate the two sets of geometries, all three algorithms perform very well since they reach a KS statistics higher than 0.99, indicating that the two sets of geometries barely overlap along these new nuclear coordinates.

3.3. Interpreting the machine learning analysis

What new nuclear coordinate has been identified for the separation? Looking at the coefficients in Figure 5, the new coordinates suggested by the three classifiers are quite similar. The angle of the LDA coordinate vector is only 11 and 10 degrees with the linear SVC and SVM (with a linear kernel) coordinate vectors, respectively; the angle between the two SVC and SVM coordinate vectors is only 4 degrees. Based on a visual inspection of the new coordinate, it mainly corresponds to the in-phase planarisation of the two formaldehyde moieties.

In all classifier cases, the normal mode with the largest amplitude is normal mode 11. Normal mode 11 corresponds mainly to the planarisation of the formaldehyde moieties, with only the "outer" hydrogen atoms moving and not the "inner" ones as in normal mode 8 (Figure 2). It is noted that along normal mode 11 alone, the two sets of geometries do not differ much (KS $<$ 0.2, see Figure 3). The second most important normal mode is mode 8, already identified as a good separative coordinate (Figure 3).

Taking the linear SVC classifier for instance, the five most important modes (11, 8, 6, 4 and 7) out of the 18 normal modes account for more than 90% of the suggested new nuclear coordinate. Using only their coefficients in creating the new coordinate (and setting the others to zero) gives a KS of 0.896. When re-optimising these five coefficients, a KS of 0.989 is obtained.

To test the robustness of the analysis procedure, we repeated the training of the linear SVC classifier with 1.8 Å and 1.9 Å for the threshold value used to detect the successful geometries. The angle between the two new coordinate vectors is only 7 degrees. It is on the same order of magnitude as the angle obtained for different classifiers and it is satisfactorily small. It is also noted that, although it might be interesting to study in detail, we do not expect the results to depend much on the choice of initial basis of nuclear coordinates, for example when using

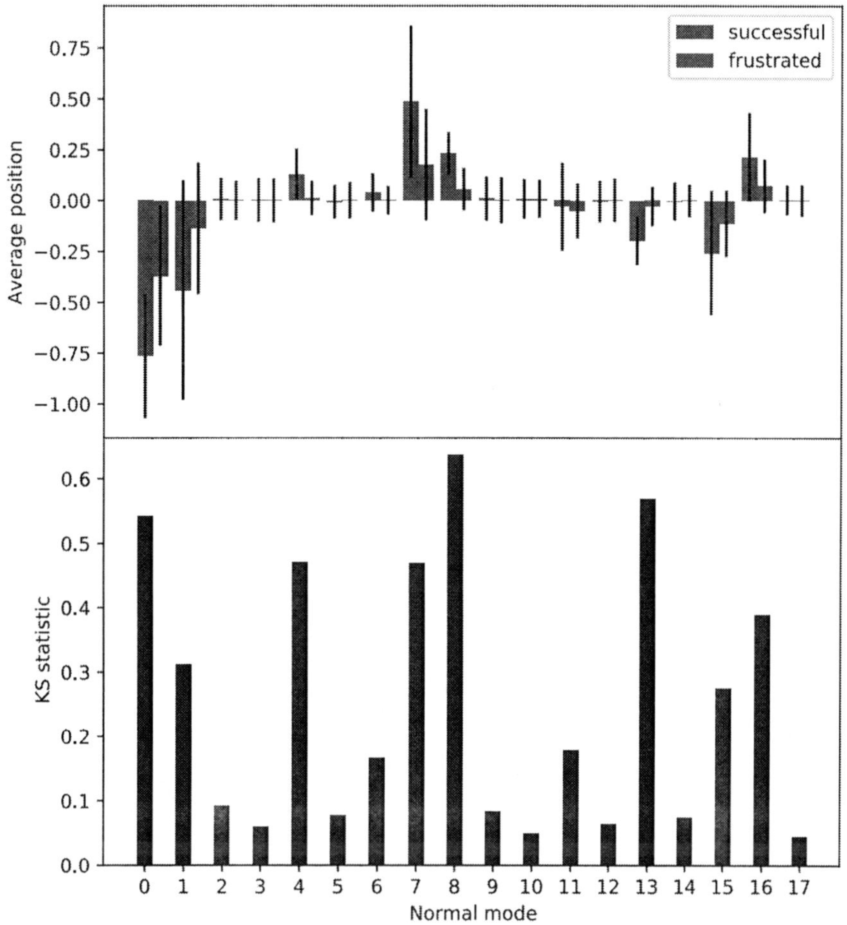

Figure 3. Frustrated (red) and successful (green) geometries in normal mode coordinates. The upper subfigure shows the average positions and standard deviations. The lower one shows the KS statistics which quantifies the difference between both sets of geometries.

localised modes [21] instead of normal modes.

4. Conclusions and discussions

In this work, we have used three different machine learning classifiers to suggest a nuclear coordinate that distinguishes between frustrated and successful dissociations (identified along ab initio molecular dynamics simulations). The three algorithms perform very well in separating the two sets of geometries and all three suggest a similar new nuclear coordinate. The latter mainly corresponds to the in-phase planarisation of the two formaldehyde moieties.

We note that it was previously suggested that the nuclear distortions responsible for the frustration of dissociations were a small O-C-C-O dihedral angle and large O-C-C angles [4]. The quantification of how much the two sets of geometries differ along those previously suggested coordinates gives a KS of slightly larger than 0.2 and slightly lower than 0.5 along the O-C-C angle and O-C-C-O dihedral angle, respectively. The machine learning classifiers are thus much better at finding separative nuclear coordinates than humans. Also, it is interesting to observe that none of these coordinates seem to be directly identified as important by the machine learning classifiers. (Normal mode 7 corresponds mainly to the H-C-C-H dihedral angles, and

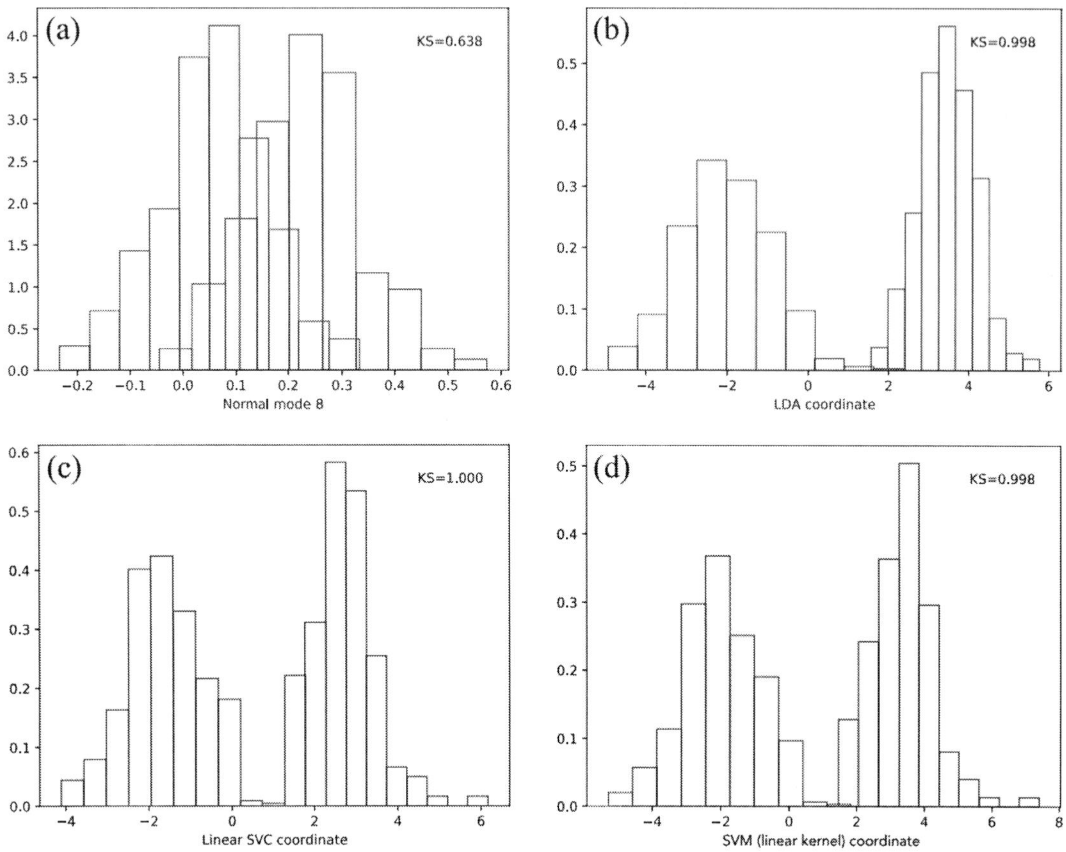

Figure 4. Frustrated (red) and successful (green) geometries along (a) normal mode 8, and the coordinates optimised with (b) LDA, (c) linear SVC and (d) SVM with a linear kernel. The KS statistics is indicated to quantify the difference between both sets of geometries.

not the O-C-C-O angle.) The planarisation distortions must indirectly imply the reduction of the O-C-C angles though.

The present analysis of the ab initio molecular dynamics simulations brings a fresh and more accurate look onto the dissociation mechanism and the entropic trap. Contrary to what was previously suggested [4], the geometrical condition that is necessary for a trajectory to escape the entropic trap and for dissociation to be successful is the in-phase planarisation of the two formaldehyde moities. If both moities are not planar enough when the dissociation is attempted, the molecule remains trapped. This way, the so-called entropic trap leads to frustrated dissociations, postponing the decomposition reaction. A chemical distortion that would hamper the in-phase planarisation of the two formaldehyde moities may thus enhance the efficiency of chemiexcitation. Performing the same type of analysis on excited state dynamics might reveal complementary relevant nuclear coordinates.

References

[1] Häse F, Fdez Galván I, Aspuru-Guzik A, Lindh R and Vacher M 2019 *Chem. Sci.* **10**(8) 2298–2307 URL `http://dx.doi.org/10.1039/C8SC04516J`

[2] Grazioli G, Roy S and Butts C T 2019 *Journal of Chemical Information and Modeling* **59**

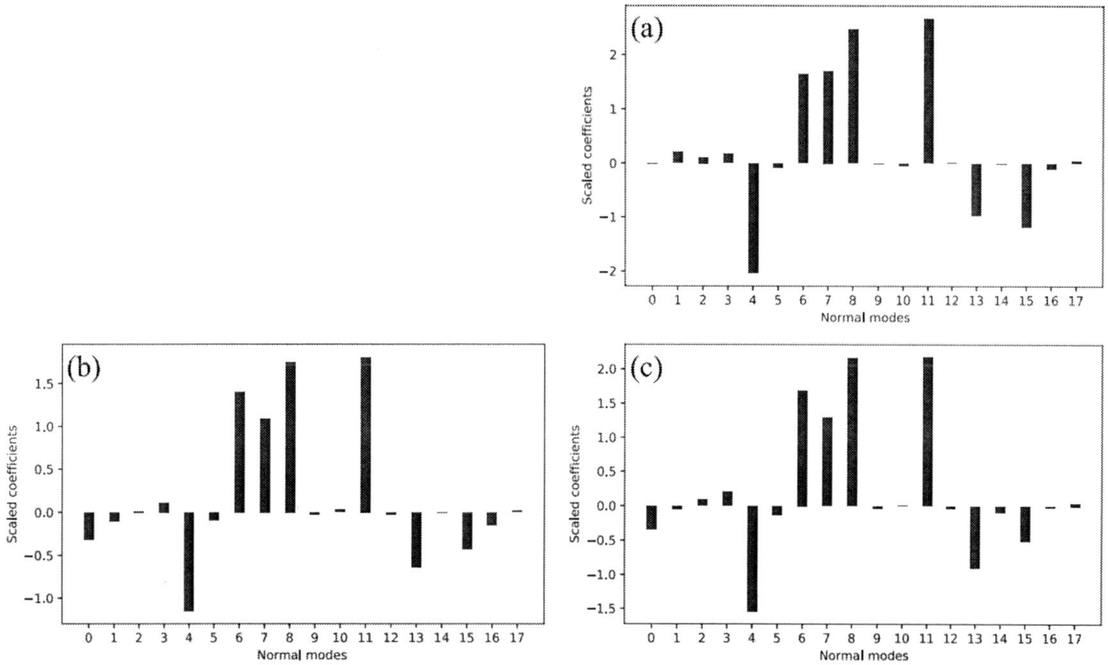

Figure 5. Combined coordinates (expressed in terms of normal modes) suggested by (a) LDA, (b) linear SVC and (c) SVM with a linear kernel.

2753–2764 pMID: 31063694 (*Preprint* `https://doi.org/10.1021/acs.jcim.9b00134`) URL `https://doi.org/10.1021/acs.jcim.9b00134`

[3] Berishvili V P, Perkin V O, Voronkov A E, Radchenko E V, Syed R, Venkata Ramana Reddy C, Pillay V, Kumar P, Choonara Y E, Kamal A and Palyulin V A 0 *Journal of Chemical Information and Modeling* **0** null pMID: 31276400 (*Preprint* `https://doi.org/10.1021/acs.jcim.9b00135`) URL `https://doi.org/10.1021/acs.jcim.9b00135`

[4] Vacher M, Brakestad A, Karlsson H O, Fdez Galván I and Lindh R 2017 *J. Chem. Theory Comput.* **13** 2448–2457 pMID: 28437611 (*Preprint* `https://doi.org/10.1021/acs.jctc.7b00198`) URL `https://doi.org/10.1021/acs.jctc.7b00198`

[5] Vacher M, Farahani P, Valentini A, Frutos L M, Karlsson H O, Fdez Galván I and Lindh R 2017 *J. Phys. Chem. Lett.* **8** 3790–3794 pMID: 28749694 (*Preprint* `https://doi.org/10.1021/acs.jpclett.7b01668`) URL `https://doi.org/10.1021/acs.jpclett.7b01668`

[6] FdezGalván I, Gustafsson H and Vacher M 2019 *ChemPhotoChem* **3** 957–967 (*Preprint* `https://onlinelibrary.wiley.com/doi/pdf/10.1002/cptc.201800232`) URL `https://onlinelibrary.wiley.com/doi/abs/10.1002/cptc.201800232`

[7] Vacher M, Fdez Galván I, Ding B W, Schramm S, Berraud-Pache R, Naumov P, Ferré N, Liu Y J, Navizet I, Roca-Sanjuán D, Baader W J and Lindh R 2018 *Chem. Rev.* **118** 6927–6974 pMID: 29493234 (*Preprint* `https://doi.org/10.1021/acs.chemrev.7b00649`) URL `https://doi.org/10.1021/acs.chemrev.7b00649`

[8] Navizet I, Liu Y J, Ferré N, Roca-Sanjuán D and Lindh R 2011 *ChemPhysChem* **12** 3064–3076 ISSN 1439-7641 URL `http://dx.doi.org/10.1002/cphc.201100504`

[9] Feyter S D, Diau E W G, Scala A A and Zewail A H 1999 *Chemical Physics Letters* **303** 249 – 260 ISSN 0009-2614 URL `http://www.sciencedirect.com/science/article/pii/S0009261499002638`

[10] Lourderaj U, Park K and Hase W L 2008 *International Reviews in Physical Chemistry* **27** 361–403 (*Preprint* `http://dx.doi.org/10.1080/01442350802045446`) URL `http://dx.doi.org/10.1080/01442350802045446`

[11] Wigner E 1932 *Phys. Rev.* **40**(5) 749–759 URL `http://link.aps.org/doi/10.1103/PhysRev.40.749`

[12] Barbatti M, Granucci G, Lischka H, Persico M and Ruckenbauer M 2006 Newton-x: a package for newtonian dynamics close to the crossing seam, version 0.11b www.univie.ac.at/newtonx

[13] Roos B O, Taylor P R and Siegbahn P E 1980 *Chemical Physics* **48** 157 – 173 ISSN 0301-0104 URL http://www.sciencedirect.com/science/article/pii/0301010480800450
[14] Roos B O, Lindh R, Malmqvist P Å, Veryazov V and Widmark P O 2004 *The Journal of Physical Chemistry A* **108** 2851–2858 (*Preprint* http://dx.doi.org/10.1021/jp031064+) URL http://dx.doi.org/10.1021/jp031064+
[15] Fdez Galván I, Vacher M, Alavi A, Angeli C, Aquilante F, Autschbach J, Bao J J, Bokarev S I, Bogdanov N A, Carlson R K, Chibotaru L F, Creutzberg J, Dattani N, Delcey M G, Dong S S, Dreuw A, Freitag L, Frutos L M, Gagliardi L, Gendron F, Giussani A, González L, Grell G, Guo M, Hoyer C E, Johansson M, Keller S, Knecht S, Kovačević G, Källman E, Li Manni G, Lundberg M, Ma Y, Mai S, Malhado J P, Malmqvist P Å, Marquetand P, Mewes S A, Norell J, Olivucci M, Oppel M, Phung Q M, Pierloot K, Plasser F, Reiher M, Sand A M, Schapiro I, Sharma P, Stein C J, Sørensen L K, Truhlar D G, Ugandi M, Ungur L, Valentini A, Vancoillie S, Veryazov V, Weser O, Wesołowski T A, Widmark P O, Wouters S, Zech A, Zobel J P and Lindh R *Journal of Chemical Theory and Computation* PMID: 31509407 (*Preprint* https://doi.org/10.1021/acs.jctc.9b00532) URL https://doi.org/10.1021/acs.jctc.9b00532
[16] Franklin J 2005 *The Mathematical Intelligencer* **27** 83–85 ISSN 0343-6993 URL https://doi.org/10.1007/BF02985802
[17] Aizerman M A 1964 *Automation and remote control* **25** 821–837
[18] Rosasco L, Vito E D, Caponnetto A, Piana M and Verri A 2004 *Neural Computation* **16** 1063–1076 (*Preprint* https://doi.org/10.1162/089976604773135104) URL https://doi.org/10.1162/089976604773135104
[19] Ng A Y 2004 *Proceedings of the Twenty-first International Conference on Machine Learning* ICML '04 (New York, NY, USA: ACM) pp 78– ISBN 1-58113-838-5 URL http://doi.acm.org/10.1145/1015330.1015435
[20] 2008 *Kolmogorov–Smirnov Test* (New York, NY: Springer New York) pp 283–287 ISBN 978-0-387-32833-1 URL https://doi.org/10.1007/978-0-387-32833-1_214
[21] Jacob C R and Reiher M 2009 *The Journal of Chemical Physics* **130** 084106 (*Preprint* https://doi.org/10.1063/1.3077690) URL https://doi.org/10.1063/1.3077690

Journal of Physics: Conference Series **1412** (2020) 042004 doi:10.1088/1742-6596/1412/4/042004

Time-dependent effective potential for the ultrafast correlated electron dynamics of molecules in intense laser fields: Application to anisotropic ionization of CO

H Kono[1], S Ohmura[2], T Kato[3], H Ohmura[4] and S Koseki[5]

[1] Department of Chemistry, Graduate School of Science, Tohoku University, Sendai 980-8578, Japan
[2] Department of Physical Science and Engineering, Nagoya Institute of Technology, Nagoya, Aichi 466-8555, Japan
[3] Department of Chemistry, School of Science, The University of Tokyo, 7-3-1 Hongo, Bunkyo-ku, Tokyo 113-0033, Japan
[4] National Institute of Advanced Industrial Science and Technology (AIST), 1-1-1 Higashi, Tsukuba, Ibaraki 305-8565, Japan
[5] Department of Chemistry, Graduate School of Science, Osaka Prefecture University, Osaka 599-8531, Japan

Author to whom any correspondence should be addressed:
hirohiko.kono.d6@tohoku.ac.jp

Abstract. By using the multiconfiguration time-dependent (TD) Hartree-Fock (MCTDHF) method, we simulated the multielectron dynamics of a CO molecule irradiated by near-IR two-cycle pulses with different carrier envelope phases. The ionization rate calculated is higher when the laser electric field $\varepsilon(t)$ points from the nucleus C to O than the opposite case, in agreement with the results of two-color ionization experiments. The mechanism of anisotropic tunnel ionization in CO was examined by converting the obtained multielectron dynamics to the representation in terms of TD natural orbitals $\{\phi_j(t)\}$. Within the framework of MCTDHF, we derived the unitary equations of motion for $\{\phi_j(t)\}$. From the derived equations, we defined the TD effective potentials $\{v_j^{\text{eff}}(t)\}$ that govern the dynamics of $\{\phi_j(t)\}$. In $v_{5\sigma}^{\text{eff}}(t)$ for the 5σ HOMO, a narrow hump that originates from two-body electron-electron repulsion is formed on the top of the field-induced distorted barrier near the nucleus C when $\varepsilon(t)$ points from C to O, which is responsible for the directional anisotropy of tunnel ionization. For 4σ HOMO-2, a high barrier to suppress ionization is formed in $v_{4\sigma}^{\text{eff}}(t)$ when $\varepsilon(t)$ points from C to O, in correlation with the electron-electron interaction with a 5σ electron on route to ionization. For the opposite phase, $v_{4\sigma}^{\text{eff}}(t)$ becomes barrierless, which enhances high-harmonic generation through $\phi_{4\sigma}(t)$.

1. Introduction

Electronic motion induced by external fields is one of the fundamental processes that trigger various physical or chemical phenomena in a wide range of timescale. Intense, ultrashort laser pulses are utilized to induce/manipulate large amplitude electronic motions such as tunnel ionization and subsequent high-order harmonic generation (HHG). Ti:Sapphire near-infrared (IR) laser systems can now provide ultrafast few cycle pulses of focused intensity greater than 10^{14} Wcm^{-2}. A number of experimental and theoretical researches have been devoted to profoundly understanding the intense-field induced electron dynamics in atoms [1], molecules [2], solids [3], and biological systems [4]. The electronic response of an atom or a molecule can be categorized into adiabatic and nonadiabatic regimes with respect to temporal change in the applied laser electric field [5]. Accordingly, ionization induced by intense fields is usually characterized by the Keldysh parameter $\gamma = (\omega/f_{max})(2I_P)^{1/2}$ [6] (in atomic units), where ω is the angular frequency of the applied laser field, f_{max} is the maximum of the electric field envelope, and I_P is the ionization potential. The ionization mechanism is classified as adiabatic "quasistatic" tunnel ionization or multiphoton ionization according as $\gamma < 1$ or $\gamma > 1$.

Diverse theoretical approaches have been developed to investigate the field-induced ionization of atoms and molecules and to quantify the ionization rate or probability. When the light intensity of the laser is beyond a certain limit, conventional perturbation theory is insufficient for the description of ionization. Epoch-making in the development of nonperturbative approaches is the one proposed by Keldysh [6], in which intense-field ionization is formulated as an electronic transition from the ground state of an atom to Volkov states (quantum states of an electron freed in an oscillating laser electric field). In this approach, the ground state is not perturbed by the applied field and the detailed atomic energy structure is not taken into account in determining the ionization probability. Another defect in Keldysh's approach was its neglect of the Coulomb interaction of the electron released in the laser field with the parent ion. To remedy the defect, Perelomov et al. developed a method, called the Perelemov-Popov-Terent'ev (PPT) approach [7], where the long-range Coulomb interaction is included as the first-order correction in the quasi-classical action of the electron. These types of approaches have been generalized by *S*-matrix theory for obtaining the transition probability of a quantum process. An element of the *S*-matrix is in general equivalent to the transition amplitude obtained by the standard prescription of projection of the total wave function of the system onto the final state. Since the early days of nonperturbative ionization of atoms or molecules in intense laser fields, several versions of the S-matrix approach have been introduced which are together known as the Keldysh-Faisal-Reiss (KFR) theory or Strong Field Approximation (SFA) [8].

As the field intensity becomes higher and its optical period $2\pi/\omega$ becomes longer (i.e. $\gamma \to 0$), an electron penetrates (or goes beyond) the distorted "quasistatic" barrier for ionization before the sign of the laser electric field changes. In this case, quasistatic tunneling theory, namely the Ammosov–Delone–Krainov (ADK) model [9], is most commonly used. The ADK model is the quasistatic limit ($\gamma \to 0$) of the PPT approach. Tong et al. extended the tunnelling ionization ADK model to molecules by considering the symmetry property and the asymptotic behaviour of the molecular electronic wave function (named MO-ADK theory) [10]. Adiabatic theory of intense field ionization for finite-range potentials has now been completed by Tolstikhin and Morishita [11].

The above approaches are developed within the framework of the single active electron (SAE) approximation where only the atomic or molecular orbital under consideration is allowed to interact with the applied field and the other orbitals are kept fixed at the shapes obtained for the ground-state calculation. There are ways of including the effects of field-induced dynamic polarization on other orbitals. It is possible to estimate dynamic polarization effects by constructing an "effective potential" semiempirically [12] or by systematically expanding the S-matrix of a multielectron system to different ionization processes. In fact, only the leading term associated with an ionization process is evaluated in a systematic series expansion of the S-matrix [13].

Time-dependent processes can be explicitly described beyond the SAE approximation by numerically solving the time-dependent (TD) Schrödinger equation of a multielectron system. Among such methods are the time-dependent Hartree-Fock (TDHF) method and the multiconfiguration time-dependent Hartree-Fock (MCTDHF) method [14-16]. In the TDHF, the multielectron wave function, as a function of time t, is approximated by a single Slater determinant and the time evolution of orbitals is formulated under the mean field approximation. This basic approach takes into account the electron-electron interaction but not correlated electron motion. On the other hand, the total wave function $\Phi(t)$ in the MCTDHF method is expanded in terms of many Slater determinants corresponding to various electron configurations, by which the correlated electron motion can be described beyond the mean field approximation. In this method, both the orbitals and the coefficients of Slater determinants are optimized in compliance with the Dirac-Frenkel variational principle. For the simulation of the electron dynamics of large systems, real-time real-space time-dependent density functional theory (TDDFT) is utilized [17, 18], where the quantification of electron correlation depends on the specific property of the exchange-correlation functional employed. The variations of the electron density and correlation energy in time could be analyzed, but the description of correlated electron motion by TDDFT is incomplete because only the total electron density is determined.

More reliable and detailed information can be obtained as the level of the theory employed is higher. Accordingly, the results obtained becomes complicated to analyze. For example, the TD orbitals in the MCTDHF method reflect the effects of electron-electron (two-body) interaction or electron correlation, but it is not straightforward to extract those effects from the time evolution of orbitals. The temporal change in orbitals cannot be linearly decomposed into those due to the one-body interaction (interaction of an electron with nuclei and external fields) and due to the two-body interaction. In previous papers, we have proposed a way of converting the multielectron wave function $\Phi(t)$ obtained by MCTDHF to a single orbital picture [19, 20]. For this purpose, using the TD natural spin-orbitals $\{\phi_j(t)\}$ [21], which diagonalize the first-order reduced density matrix of electrons constructed from $\Phi(t)$, we derived the corresponding effective potentials $\{v_j^{\rm eff}(\boldsymbol{r},t)\}$ for $\{\phi_j(t)\}$ as a function of the spatial coordinates of an electron, $\boldsymbol{r}$ [21]. The TD effective potential $v_j^{\rm eff}(t)$ for $\phi_j(t)$ can be decomposed into $v_j^{\rm eff}(t) = v_1(t) + v_{2,j}(t)$, where $v_1(t)$ is the one-body interaction and $v_{2,j}(t)$ arises from the two-body interaction. Modern laser technology enables us to tailor the form and phase of a laser field for manipulation of the dynamics of molecules on a femtosecond or even attosecond time scale. The two-color field can also be used to answer the fundamental question, *i.e.*, to investigate the anisotropy in electron dynamics, for example, unidirectional ionization of heteronuclear diatomics such as CO. The emission direction of C^+ or O^+ from CO molecules in two-color ionization experiments indicates that ionization is enhanced when the electric field points from the nucleus C to the nucleus O [22]. This type of experiment can now be combined with the field-free molecular orientation technique applied to CO [23]. By using MCTDHF, we calculated the effective potentials of natural orbitals for CO in an intense near-IR field to investigate the mechanism of anisotropic ionization of CO [20]. The results indicated that when the laser electric field points from C to O, tunnel ionization from the C atom side is enhanced and that a hump structure originating from two-body electron-electron repulsion is then formed *near* the C nucleus on the top of the field-induced distorted barrier of the effective potential for the 5σ highest occupied molecular orbital (HOMO). We attributed this hump formation to the directional anisotropy of tunnel ionization.

Effective potential is a versatile tool to analyze intrinsic multielectron nature in electronic dynamics. Correlation in dynamics between natural orbitals can be revealed by examining the temporal change in the corresponding effective potential, though the effective potential is a one-body quantity. In the present paper, we show that the dynamics of the HHG of the 4σ (HOMO-2) orbital in CO is correlated with the 5σ ionization dynamics: When the laser electric field points from O to C, HHG from 4σ is enhanced while the ionization from 5σ is reduced. In the following, we present a brief summary of the effective potential approach and the numerical results for the effective potentials of the 4σ and 5σ orbitals.

2. Effective potential for natural orbitals in the MCTDHF framework

The equation we solve is the TD Schrödinger equation for an N-electron system in intense laser fields

$$i\hbar\,\partial\Phi(t)/\partial t = \hat{H}(t)\Phi(t), \qquad (1)$$

where $\Phi(t)$ is the TD wave function and $\hat{H}(t)$ is the electronic Hamiltonian including the interaction with the applied field $\varepsilon(t)$. We solve (1) by using the MCTDHF method where $\Phi(t)$ is expanded in terms of electron configurations $\{\Phi_I(t)\}$ (Slater determinants or configuration state functions) [15,16]:

$$\Phi(t) = \sum_{I=1}^{M} C_I(t)\Phi_I(t), \qquad (2)$$

where M is the number of $\Phi_I(t)$ and $C_I(t)$ are the configuration interaction (CI) coefficients for $\Phi_I(t)$. Each electron configuration is constructed of N spin orbitals (SOs) $\{\psi_k(t)\}$. A SO $\psi_k(t)$ is the product of a one-electron spatial orbital and a one-electron spin function; $\psi_k(t) = \langle \boldsymbol{x} | k(t)\rangle$, where $\boldsymbol{x} = (\boldsymbol{r}, \mu)$ denotes the spatial coordinates $\boldsymbol{r}$ and the spin coordinate μ of an electron. The number of SOs used, $N_{\rm O}$, is larger than N ($N_O = N$ in TDHF). The working equations of motion (EOMs) of $\{\psi_j(t)\}$ and $\{C_I(t)\}$ have been derived by applying the Dirac-Frenkel TD variational principle to possible variations of $\Phi(t)$.

The resulting EOMs for $\psi_k(t)$, from which the effective potentials $v_j(\boldsymbol{r},t)$ are derived, are expressed as

$$i\hbar\frac{\partial}{\partial t}|\psi_k(t)\rangle = \hat{h}(t)|\psi_k(t)\rangle + \sum_{ij}^{N_O} A^{-1}(t)_{ki}\left[\hat{V}_{ij}(t) - \Lambda_{ij}(t)\right]|\psi_j(t)\rangle, \quad (3)$$

where $\hat{h}(t)$ is the one-body Hamiltonian including the electric dipole interaction $\boldsymbol{r}\cdot\varepsilon(t)$ with laser electric fields $\varepsilon(t)$, $\hat{V}_{ij}(t)$ is the orbital coupling between i and j mediated by other orbitals (depending on $\{C_I(t)\}$), $A_{km}(t) = \langle\Phi(t)|\hat{a}_k^\dagger(t)\hat{a}_m(t)|\Phi(t)\rangle$ is the (k,m) matrix element of the 1st order reduced density matrix (1RDM), where $\hat{a}_j(t)$ and $\hat{a}_j^\dagger(t)$ denote, respectively, the annihilation and creation operators for an electron in $\psi_j(t)$. $\Lambda_{ij}(t)$ is the following Lagrange multiplier

$$\Lambda_{ij}(t) = \sum_k^{N_O}\langle\psi_j(t)|\left[\hat{V}_{ik}(t) + A_{ik}(t)\hat{g}(t)\right]|\psi_k(t)\rangle, \quad (4)$$

where $\hat{g}(t)$ is to be determined so that the orthonormalization condition $\langle\psi_j(t)|\psi_k(t)\rangle = \delta_{jk}$ is maintained.

Equation (3) can be rewritten as

$$i\hbar\partial|\psi_k(t)\rangle/\partial t = \left[\hat{h}(t) - \hat{g}(t)\right]|\psi_k(t)\rangle + \hat{Q}(t)\left[\hat{g}(t)|\psi_k(t)\rangle + \sum_{ij}^{N_O} A^{-1}(t)_{ki}\hat{V}_{ij}(t)|\psi_j(t)\rangle\right] \quad (5)$$

where $\hat{Q}(t) = 1 - \sum_s^{N_O}|s(t)\rangle\langle s(t)|$ is the projector to the complement to the current orbital space. Since

$$i\hbar\langle\psi_j(t)|\frac{\partial}{\partial t}|\psi_k(t)\rangle = \langle\psi_j(t)|\left[\hat{h}(t) - \hat{g}(t)\right]|\psi_k(t)\rangle, \quad (6)$$

the time propagation is unitary if $\hat{g}(t)$ is a Hermitian operator like $\hat{g}(t) = \sum_{mn}^{N_O}|\psi_m(t)\rangle g_{mn}(t)\langle\psi_n(t)|$. To ensure the numerical stability in solving the EOMs for $\{\psi_j(t)\}$ and $\{C_I(t)\}$, we choose $\hat{g}(t) = \hat{h}(t)$; then, $\langle\psi_i(t)|\partial\psi_j(t)/\partial t\rangle = 0$ (for $i = j$ and $i \neq j$), i.e., the orthonormalization condition holds in the time-propagation of $|\psi_j(t)\rangle$. Equation (5) are coupled with the EOMs for $\{C_I(t)\}$.

The 1RDM at t is calculated from $\{\psi_j(t)\}$ and $\{C_I(t)\}$, which is given by

$$\rho(\boldsymbol{x},\boldsymbol{x}',t) = \sum_{ij}^{N_O} A_{ij}(t)\psi_i^*(\boldsymbol{x},t)\psi_j(\boldsymbol{x}',t). \quad (7)$$

For $\hat{g}(t) = \hat{h}(t)$, $A_{ij}(t) \neq 0$. The rhs in (7) can be diagonalized by a unitary transformation from $\{\psi_j(t)\}$ to $\{\phi_j(t)\}$

$$\rho(\boldsymbol{x},\boldsymbol{x}',t) = \sum_j^{N_O} A_{jj}(t)\phi_j^*(\boldsymbol{x},t)\phi_j(\boldsymbol{x}',t). \quad (8)$$

In this case, $A_{ij}(t) = 0$ for $i \neq j$ and the orbitals that satisfy (8) are called natural orbitals [21]. The diagonal elements for natural orbitals, $n_j(t) = A_{jj}(t)$, is the occupation number of $\phi_j(t)$. The expectation value $\boldsymbol{d}(t)$ of any one-particle operator $\hat{\boldsymbol{d}}$, such as the field-induced dipole moment, is expressed as $\boldsymbol{d}(t) = \sum_j^{N_O} n_j(t)\boldsymbol{d}_j(t)$, where $\boldsymbol{d}_j(t) = \langle\phi_j(t)|\hat{\boldsymbol{d}}(t)|\phi_j(t)\rangle$. We have shown for LiH that the dynamics of $\phi_j(t)$ is correlated with the "time-dependent chemical potential" of $\phi_j(t)$ [19]. The natural orbital representation provides a bridge connecting the multielectron dynamics to a single electron picture.

The matrix elements of the 1RDM obey in general the following equation [20]:

$$dA_{km}(t)/dt = -(i/\hbar)\left[W_{mk}(t) - W_{km}^*(t) + g_{mk}(t)\left(n_k(t) - n_m(t)\right)\right]. \quad (9)$$

Here, $W_{km}(t)$ is defined as $W_{km}(t) = \left[\psi_k(t)\psi_q(t)|\psi_r(t)\psi_p(t)\right]\langle\Phi(t)|\hat{a}_m^\dagger(t)\hat{a}_r^\dagger(t)\hat{a}_p(t)\hat{a}_q(t)|\Phi(t)\rangle$, where $[\cdots|\cdots]$ represents the conventional two-body integral (in atomic units)

$$\left[\psi_k\psi_q|\psi_r\psi_p\right] = \int dx_1 dx_2 \psi_k^*(x_1)\psi_q(x_1)|r_1 - r_2|^{-1}\psi_r^*(x_2)\psi_p(x_2). \quad (10)$$

Suppose that $\psi_j(t_0) = \phi_j(t_0)$ at $t = t_0$. To satisfy $A_{ij}(t) = 0$ for $i \neq j$, we set $g_{mk}(t)$ in (9) as

$$g_{mk}(t) = \left[W_{mk}(t) - W_{km}^*(t)\right]/\left[n_m(t) - n_k(t)\right]. \quad (11)$$

Substitution of (11) into (5) yields the EOMs for the natural orbitals

$$i\hbar\frac{\partial}{\partial t}|\phi_k(t)\rangle = \left[\hat{T} + v_1(r,t)\right]|\phi_k(t)\rangle + \sum_j^{N_O}\hat{U}_{kj}(t)|\phi_j(t)\rangle, \quad (12)$$

where $\hat{T}$ is the kinetic energy operator of an electron, the one body interaction term $v_1(\boldsymbol{r},t)$ consists of the electric dipole interaction $\boldsymbol{r}\cdot\varepsilon(t)$ and the Coulomb attractive interaction between an electron and the nuclei, and the coupling between $\phi_k(t)$ and $\phi_j(t)$, $\hat{U}_{kj}(t)$, is given by

$$\hat{U}_{kj}(t)=\left(1-\delta_{kj}\right)\frac{W_{jk}(t)-W_{kj}^{*}(t)}{n_k(t)-n_j(t)}+Q(t)\frac{\hat{V}_{kj}(t)}{n_k(t)}. \quad (13)$$

In the present treatment, the spatial part of the natural SO is independent of its spin state. Equation (12) can then be formally rewritten as

$$i\hbar\,\partial\phi_k(r;t)/\partial t=\left[\hat{T}+v_k^{\text{eff}}(\boldsymbol{r},t)\right]\phi_k(\boldsymbol{r},t), \quad (14)$$

where the one-body quantity $v_k^{\text{eff}}(\boldsymbol{r},t)$ is considered the effective potential that determines the dynamics of $\phi_k(t)$ under multielectron interaction

$$v_k^{\text{eff}}(\boldsymbol{r},t)=v_1(\boldsymbol{r},t)+\sum_j^{N_O}\left\langle \boldsymbol{r}\left|\hat{U}_{kj}(t)\right|\phi_j(t)\right\rangle\Big/\phi_k(\boldsymbol{r},t)=v_1(\boldsymbol{r},t)+v_{2,k}(\boldsymbol{r},t). \quad (15)$$

Multielectron dynamics can be analyzed by examining separately the temporal change in the one-body term $v_1(\boldsymbol{r},t)$ and that in the two-body term $v_{2,k}(\boldsymbol{r},t)$. We designate the expectation value of $\hat{T}+v_k^{\text{eff}}(\boldsymbol{r},t)$ over $\phi_k(t)$ as the real-valued effective energy $\varepsilon_k(t)$. In practice, we obtain the "instantaneous" natural orbitals $\{\phi_j(t)\}$ at time t from $\Phi(t)$ that are the numerical solutions of (5) for $\hat{g}(t)=\hat{h}(t)$. $v_k^{\text{eff}}(\boldsymbol{r},t)$ can be plotted as function of $\boldsymbol{r}$ by inserting $\{\phi_j(t)\}$ and the corresponding $\{C_I(t)\}$ (different from those for $\hat{g}(t)=\hat{h}(t)$) into (15).

3. Results and discussion

We briefly introduce the results [20] related to the ionization and HHG dynamics of a CO molecule interacting with an intense near-IR two-cycle pulse and present the TD effective potentials for the 4σ and 5σ natural orbitals in the dynamics. In the numerical simulation, we considered 10 spatial orbitals constructed mainly from 1s, 2s, 2p orbitals of C and O atoms, that is, 1σ, 2σ, 3σ, 4σ, 2×1π, 5σ, 2×2π (LUMO) and 6σ. We chose the high lying six orbitals of 2×1π, 5σ, 2×2π and 6σ as the active orbitals that construct the complete configuration space to accommodate 14–8=6 electrons. The lower lying SOs are treated to be occupied by one electron but the shapes of orbitals vary according to the EOMs for the SOs. This calculation scheme is referred to as the time-dependent complete active space self-consistent field (TD-CASSCF) method, developed in [16]. The cylindrical coordinates z and ρ are used for numerical grid point representation of the spatial orbitals; z is chosen parallel to the molecular axis. The nuclei C and O are placed on the z axis ($\rho=0$): $z=-1.066\ a_0$ for C and $z=1.066\ a_0$ for O, where a_0 is the Bohr radius (the difference is equal to the equilibrium internuclear distance $R_e=2.132\ a_0$).

The polarization direction of the applied electric field $\varepsilon(t)$ was assumed to be parallel to the C-O axis. The frequency of $\varepsilon(t)$ is $\hbar\omega=0.06E_h$ = 1.64 eV (wavelength λ= 760 nm, optical period $T_c=2\pi/\omega=2.53$ fs), where E_h is the Hartree. The electric field $\varepsilon(t)$ is assumed to have a form

$$\varepsilon(t)=f(t)\cos(\omega t-\varphi), \quad (16)$$

where φ is the carrier-envelope phase (CEP) and the envelope function $f(t)$ takes the form of $f(t)=f_{\max}\sin^2\left(\pi t/2T_c\right)$ so that $f(t)$ attains its maximum value of $f_{\max}=0.06\ E_h/(ea_0)$ =3.1×10^{10} V/m (the corresponding light intensity $I_{\max}=1.3\times10^{14}$ W/cm^2) at $t=T_c$ and then returns to zero at $t=2T_c$ (two-cycle pulse). The electric field $\varepsilon(t)$ for $\varphi=-0.5\pi$ is shown in figure 1. The sign of the field is reversed when $\varphi=0.5\pi$. The conditions of the numerical simulation such as grid intervals to describe MOs are given in [20].

3.1. Ionization of CO

The Keldysh parameter γ is ~1.0 in the present condition of the applied pulse. The simulated ionization processes are thus categorized as tunnel ionization. According to the results of the TD-CASSCF numerical simulation [20], ionization exclusively occurs from the 5σ HOMO natural orbital when $\varepsilon(t)>0$, i.e., the laser electric field points from the nucleus C to O (for instance, around $t=t_1$ in figure 1). The electric field exerts a force on electrons toward the side of $z<0$ (from O to C) when $\varepsilon(t)>0$. We have also confirmed that ionization from 5σ is significantly reduced when $\varepsilon(t)<0$. In the present condition, the calculated ionization probability around the second peak of $\varepsilon(t)$ (at $t = t_1$) is five times smaller for $\varphi = 0.5\pi$ than for $\varphi = -0.5\pi$. The ionization components from 4σ and 1π are orders of magnitude less than that of 5σ. This is in accord with previous experimental and theoretical works, where the ionization rate takes the maximum when $\varepsilon(t) > 0$ and takes the minimum when $\varepsilon(t) < 0$. This is attributed to the fact that the 5σ orbital has a large lobe protruding from the C atom. The total ionization probabilities for the two-cycle pulses of $\varphi = \pm 0.5\pi$ obtained by the TDHF method are about 20 times smaller than those obtained by the TD-CASSCF.

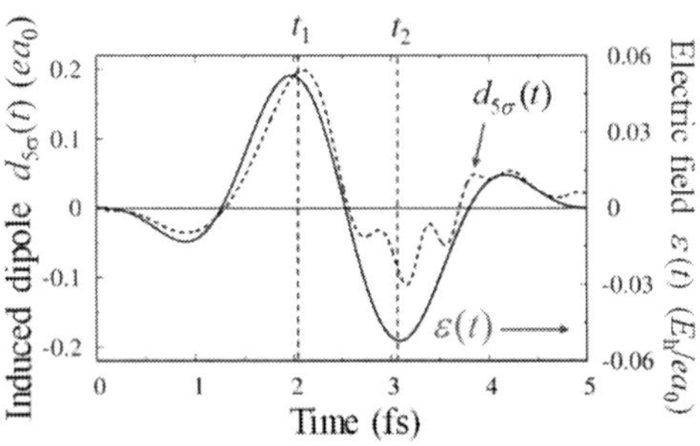

Figure 1. Electric field profile $\varepsilon(t)$ of the two-cycle pulse of (16) for CEP $\varphi = -0.5\pi$ and the corresponding induced dipole moment $d_{5\sigma}(t)$ of the 5σ natural orbital: $\omega = 0.06 E_h/\hbar$ (λ=760 nm), maximum field strength ε_{max} = 0.052 $E_h/(ea_0)$ = 2.7×10^{10} V/m at $t \approx t_1$.

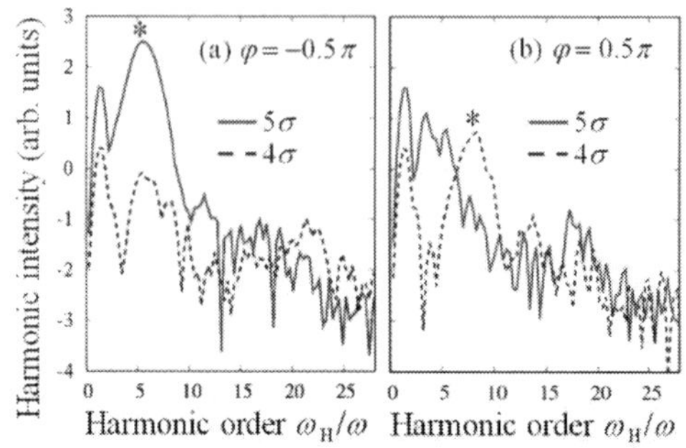

Figure 2. High-harmonic spectra $S(\omega_H)$ of the 4σ (broken lines) and 5σ (solid lines) paths of CO for the two-cycle pulses of (a) $\varphi = -0.5\pi$ and (b) $\varphi = 0.5\pi$. The polarization direction is parallel to the C-O axis. * denotes the highest peak in each path.

3.2. HHG

HHG spectra $S(\omega_H)$ are also affected by anisotropic ionization, where ω_H is the frequency of HHG. The HHG spectrum of a single molecule can be calculated from the Fourier transform of the dipole acceleration $\ddot{d}(t)$ (See $d_{5\sigma}(t)$ in figure 1). $S(\omega_H)$ of CO for two-cycle pulses of (16) are shown in figure 2. We decompose the HHG spectrum into different paths associated with respective natural orbital (i.e., $\ddot{d}(t) \approx \sum_j^{No} n_j(t)\ddot{d}_j(t)$). As shown in figure 2(a) for $\varphi = -0.5\pi$, the HHG generated by $\phi_{5\sigma}(t)$, denoted as the 5σ path, is much larger than those of the other paths, especially in the range of $\omega_H/\omega \le 9$ (The HHG cutoff energy for 5σ is $\approx 18\hbar\omega$). This is due to the recombination of the ionization component from 5σ born at $t = t_1 \approx 2$ fs (around the second peak of $\varepsilon(t)$), which demonstrates that ionization leading to HHG mainly proceeds from 5σ when $\varepsilon(t) > 0$. The ionization component from 5σ born at the third peak of $\varepsilon(t)$ (<0 at $t = t_2 \approx 3$ fs) is relatively small and the fourth peak is much weaker (confirmed by wavelet analysis). Only the ionization component born at $t \approx t_1$ contributes to the HHG.

The results for $\varphi = 0.5\pi$ are shown in figure 2(b). The HHG of the 5σ path for $\varphi = 0.5\pi$, which originates from the electron freed at $\varepsilon(t_2) > 0$, is weaker than for $\varphi = -0.5\pi$. This is because the fourth peak of $|\varepsilon(t)|$ is weaker than the third peak at $t \approx t_2$ (for $\varphi = -0.5\pi$, $\varepsilon(t_1) \approx |\varepsilon(t_2)|$). The HHG of the 5σ path is generated in either case by a 5σ electron freed when $\varepsilon(t) > 0$. The opposite trend is observed for the 4σ path in figure 2. For the 4σ path, the 5th order peak for $\varphi = -0.5\pi$ originates from ionization at $t \approx t_2$ and the 8th order peak for $\varphi = 0.5\pi$ originates from ionization at $t \approx t_1$. Both are due to the ionization component of 4σ generated when $\varepsilon(t) < 0$. For $\varphi = 0.5\pi$, the 8th order peak of the 4σ path is a prominent peak, which is much higher than the 8th order peak of the 5σ path.

3.3. Effective potential

We here investigate the mechanism of anisotropic near-IR induced ionization of CO by analysing multielectron (correlated) dynamics in terms of the TD effective potentials of natural orbitals defined by (15). The intrinsic multielectron nature in ionization of each natural orbital can be revealed by examining the temporal change in the corresponding effective potential, i.e., by resorting to the language of a single electron picture. Plotted in figure 3 are effective potentials $v_{5\sigma}^{\text{eff}}(t) = v_1(t) + v_{2,5\sigma}(t)$ of the

5σ at $t_1 \approx 2$ fs, where ρ is fixed at 0.755 a_0. The potentials in figure 3(a) and 3(b) are obtained by TDHF and TD-CASSCF, respectively. At $t=t_1$, $\varepsilon(t)$ points from C to O. The change in $v_1(t)$ is governed by the dipole interaction, i.e., $v_1(t) \approx v_1(0) + z\varepsilon(t)$; thus, $v_1(t) + v_{2,5\sigma}(0)$ linearly slants. The change in $v_{2,5\sigma}(t)$ differs between the two cases. In the z-range shown in figure 3, while $v_{2,5\sigma}(t) - v_{2,5\sigma}(0) > 0$ for TDHF, $v_{2,5\sigma}(t) - v_{2,5\sigma}(0)$ for TD-CASSCF has a hump structure formed on the top of the field-induced distorted barrier $v_1(t) + v_{2,5\sigma}(0)$. As a result, the total one of $v_{5\sigma}^{\text{eff}}(t) = v_1(t) + v_{2,5\sigma}(t)$ has a hump thin barrier for tunnelling, formed at $z \approx -3$ a_0 near the C nucleus. This is favourable for ionization toward negative z. When the phase of $\varepsilon(t)$ is reversed at $t \approx t_2$, the tunnel barrier formed is far away from the nuclei ($z \approx 15$ a_0), which is responsible for suppressed tunnel ionization when $\varepsilon(t) < 0$. These features of $v_{5\sigma}^{\text{eff}}(t)$ are consistent with the directional anisotropy of tunnel ionization of CO. As shown in figure 3(a), hump formation cannot be described by TDHF. The formation of the hump is hence attributed to correlated electron motion. As expected from the comparison between figure 3(a) and (b), the ionization probability for TDHF is less than 1/20 of the TD-CASSCF value [20].

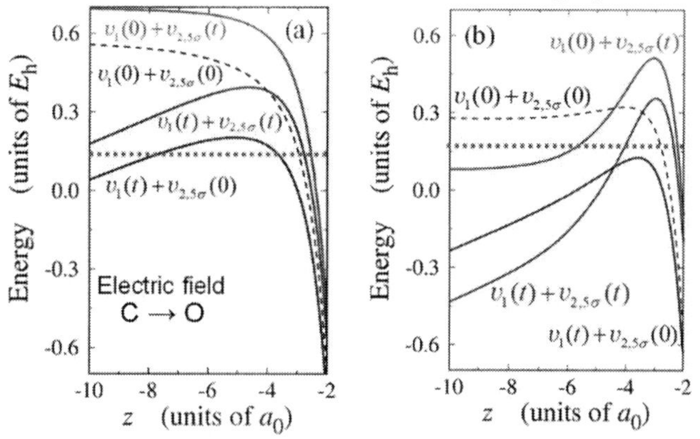

Figure 3. Effective potentials $v_{5\sigma}^{\text{eff}}(t) = v_1(t) + v_{2,5\sigma}(t)$ of the 5σ at $t=t_1 \approx 2$ fs of the two-cycle pulse (16): (a) TDHF and (b) TD-CASSCF (M=400). $v_1(t)$ is the one-body part in $v_{5\sigma}^{\text{eff}}(t)$ and $v_{2,5\sigma}(t)$ is the two-body part. The horizontal dotted line denotes the effective energy $\varepsilon_{5\sigma}(t)$.

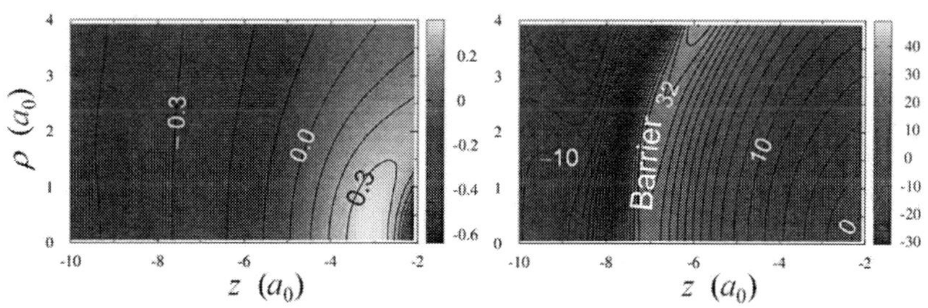

Figure 4. 2D plots of TD-CASSCF effective potentials at $t = t_1 \approx 2$ fs of the two-cycle pulse (16) of $\varphi = -0.5\pi$: (a) $v_{5\sigma}^{\text{eff}}(t)$ and (b) $v_{4\sigma}^{\text{eff}}(t)$. The heights of contour lines are given in units of E_h.

The TD-CASSCF effective potentials of $v_{5\sigma}^{\text{eff}}(t)$ and $v_{4\sigma}^{\text{eff}}(t)$ are depicted in figure 4 as functions of z and ρ. Figure 4(a) shows that the barrier around $z=-3$ a_0 of $v_{5\sigma}^{\text{eff}}(t)$ lowers as ρ increases, which reflects the fact that the electron mostly moves along the C-O axis in the case where the polarization direction is parallel to the C-O axis. At $t \approx t_1$, the electron density is peaked near the position of $z=-3$ a_0 and $\rho=0$, which we call an exit zone of ionization. $v_{5\sigma}^{\text{eff}}(t)$ has thus a hump in the same exit zone of ionization, which results from interactions between orbitals through electron-electron repulsion. As shown in figure 4(b), $v_{4\sigma}^{\text{eff}}(t)$ has a high barrier originating from $v_{2,4\sigma}(t)$, in correlation with the electron-electron interaction with a 5σ electron on route to ionization. The high barrier formed at a radius of ~7a_0 from the nuclei suppresses the birth of a 4σ electron that contributes to HHG (See figure 2(a)). On the other hand, for $\varphi = 0.5\pi$, $\upsilon_{4\sigma}^{\text{eff}}(t)$ at $t \approx t_1$ ($\varepsilon(t_1) < 0$) is nearly barrierless [20]. The electron released then has more space for rescattering, which enhances the HHG of the 4σ path as shown in figure 4(b).

4. Concluding remarks

We proposed an effective potential approach to analyze the multielectron dynamics of molecules in external fields. Within the framework of MCTDHF, we defined the TD effective potentials $\{\upsilon_j^{\text{eff}}(t)\}$ that govern the dynamics of natural orbitals $\{\phi_j(t)\}$: $\upsilon_j^{\text{eff}}(t) = \upsilon_1(t) + \upsilon_{2,j}(t)$, where $\upsilon_1(t)$ is the one-body interaction including the interaction with the applied laser field $\varepsilon(t)$ and $\upsilon_{2,j}(t)$ is due to the two-body electron-electron repulsion. The mechanism of intense near-IR field ionization in CO was investigated by examining the temporal change in $\upsilon_{5\sigma}^{\text{eff}}(t)$ for the 5σ HOMO. We found that for CO, the hump formed near the C atom (an exit zone of ionization), which is due to the change in electron-electron interaction (or electron correlation), is responsible for the anisotropic ionization observed. Two-body interaction can significantly modify the shape of the tunnel barrier in the effective potential. We also found that the suppression of 5σ ionization is correlated with the enhancement of 4σ HHG, which can be experimentally confirmed by combining two-cycle near-IR pulses of different carrier envelope phases with orientation experiments of CO [23].

Another approach in line with the concept of effective potential is developed by Kato and Yamanouchi [24], where an effective potential itself, from which SOs are obtained, is formulated as an optimization target in the framework of multiconfiguration expansion. Further development of effective potential theory is inevitable to grasp the core of multielectron correlation dynamics.

[1] Joachain C J, Kylstra N J and Potvliege R M 2012 *Atoms in Intense Laser Fields* (Cambridge: Cambridge University Press)
[2] Nisoli M, Decleva P, Calegari F, Palacios A and Martín F 2017 *Chem. Rev.* **117** 10760
[3] Schiffrin A et al. 2013 *Nature* **493** 70
[4] Dharmadhikari A K, Bharambe H, Dharmadhikari J A, D'Souza J S and D. Mathur D 2014 *Phys. Rev. Lett.* **112** 138105
[5] Kawata I, Kono H and Fujimura Y 1999 *J. Chem. Phys.* **110** 11152
[6] Keldysh L V 1965 *Sov. Phys. JETP* **20** 1307
[7] Perelemov A M, Popov V S and Terent'ev M V 1966 *Soviet Phys. JETP* **23** 924
[8] Faisal F H M 1973 J. Phys. B **6** L89; Reiss H R 1980 *Phys. Rev.* A **22** 1786
[9] Ammosov M V, Delone N B and Krainov V P 1986 *Soviet Phys. JETP* **64** 1191
[10] Tong X M, Zhao Z and Lin C D 2002 *Phys. Rev.* A **66**, 033402
[11] Tolstikhin O I and Morishita T 2012 *Phys. Rev.* A **86** 043417
[12] Zhang B, Yuan J and Zhao Z 2014 Phys. Rev. A **90** 035402
[13] Becker A and Faisal F H M 2005 *J. Phys. B: At. Mol. Opt. Phys.* **38** R1
[14] Zanghellini J, Kitzler M, Fabian C, Brabec T and Scrinzi A 2003 *Laser Phys.* **13** 1064
[15] Kato T and Kono H 2004 *Chem. Phys. Lett.* **392** 533; Kato T and Kono H 2008 *J. Chem. Phys.* **128** 184102
[16] Miyagi H and Madsen L B 2013 *Phys. Rev.* A **87** 062511; Sato T and Ishikawa K 2013 *Phys. Rev.* A **88** 023402.
[17] Eds., Marques M A L, Ullrich C, Nogueira F, Rubio A, Burke K, Gross E K U 2006 *Time-Dependent Density Functional Theory* (Berlin Heidelberg: Springer)
[18] Yabana K 2012 *Phys.Rev.* B **85** 045134
[19] Ohmura S, Kono H, Oyamada T, Kato T, Nakai K and Koseki S 2014 *J. Chem. Phys.* **141** 114105
[20] Ohmura S, Kato T, Oyamada T, Koseki S, Ohmura H and Kono H 2018 *J. Phys. B: At. Mol. Opt. Phys.* **51** 034001; Kato T, Yamanouchi K and Kono H 2018 *Attosecond Molecular Dynamics* (London: Royal Society of Chemistry)
[21] Löwdin P 1955 *Phys. Rev.* **97** 1474
[22] Ohmura H, Saito N and Morishita T, 2011 *Phys. Rev. A* **83** 063407
[23] Kraus P M, Baykusheva D and Wörner H J 2014 *J. Phys. B: At., Mol. Opt. Phys.* **47** 124030
[24] Kato T and Yamanouchi K 2018 *Phys. Rev.* A **98** 023405

Bright single-cycle terahertz source based on gas cells irradiated by two-color laser pulses

T.V. Liseykina
Institute of Physics, University of Rostock, D-18051 Rostock, Germany
Institute of Computational Mathematics and Mathematical Geophysics SB RAS, Lavrent'ev ave. 6, 630090 Novosibirsk, Russia

S.V. Popruzhenko
Prokhorov General Physics Institute RAS, Vavilova str. 38, 119991 Moscow, Russia
Voronezh State University, Universitetskaya pl. 1, 394036 Voronezh, Russia
E-mail: sergey.popruzhenko@gmail.com

Abstract. We study the excitation of electron currents in a transparent cell of sub-millimeter size filled by an atomic gas and illuminated by an intense two-color femtosecond laser pulse. The pulse consists of a strong fundamental component and its second harmonic of low intensity, both circularly polarized. We show that for a sufficiently small 20 μm-size interaction volume the plasma oscillation excited by asymmetric ionization is almost spatially homogeneous. This coherent dipole plasma oscillation results in a remarkably efficient conversion of the electron energy into that of radiation emitted in the terahertz frequency domain. Simultaneously, strong quasi-static electric fields of maximal strength $E_{\rm m} \simeq 10$ MV/cm are shown to exist inside the plasma during several hundred femtoseconds after the ionizing two-color laser pulse has gone.

1. Introduction

Electromagnetic radiation in the terahertz (THz) frequency domain is known to have a considerable potential for utilization both in fundamental experimental physics and in various applications. This includes THz spectroscopy of molecular gases and solids, control of magnetism in complex materials, noninvasive diagnostics, imaging and more, see [1, 3, 2, 4, 5] for review and references. Sources of high-intensity short THz pulses are of particular interest as they provide nonlinear regime of interaction and can be used for control of fast processes. Several efficient methods for the generation of strong THz waves are presently known and being applied in laboratories [4, 5]. Optical rectification in crystals [6] provides the currently highest convergence efficiency from optical or infrared (IR) to THz radiation at frequencies $\nu < 3$ THz. Such sources can deliver THz pulses of $\simeq 10$ μJ energy and $\simeq 1$ MV/cm electric field strength. Their main limitations are the relatively narrow spectral width determined by material absorption and the intensity damage threshold of crystals which becomes particularly restrictive at high repetition rates of pump lasers. Another family of perspective methods for the THz generation is based on laser frequency conversion in plasmas. This includes in particular emission of relativistic laser plasmas at solid-state densities [7, 8] and several schemes based on nonlinear ionization of gases [9, 10, 11, 12, 13]. These ionization schemes are technically simple and can employ ambient air at normal conditions as a target. The emitted spectrum is broad and typically extends up to $10-30$ THz. In combination with the absence of a damage threshold and the possibility of using high-repetition pump lasers, this makes ionization-based sources of THz radiation a promising alternative to nonlinear crystals.

Two-color laser radiation consisting of a strong pulse of frequency ω and its relatively weak second harmonic (SH) is conventionally used to excite asymmetric ionization currents emitting THz waves [10, 14]. By now, employing linearly polarized 800-nm pulses and their co-polarized SH, THz waves with an electric field strength of the order of 10 MV/cm [15] have been obtained providing the presently strongest THz source operating in the high-repetition-rate regime. Over the past years the $\omega - 2\omega$ scheme has been extensively studied employing different field polarization states and wavelengths [16, 17, 18, 19]. According to recent theoretical predictions [20, 21], application of mid-infrared two-color pulses instead of conventionally used 800-nm radiation may help increasing the THz emitted energy by one order of magnitude or even more. This foreseeing enhancement can make THz sources of interest for fundamental experiments in atomic and molecular physics and for other applications requiring nonlinear regimes of interaction.

The physical mechanism of $\omega - 2\omega$ THz generation is well understood on the single-atom level. A two-color laser field generally generates asymmetric photoelectron momentum distributions, so that after the averaging over fast oscillations with the frequencies ω and 2ω and over the distribution, the electron current does not vanish, in contrast to the case of a quasi-monochromatic field [22, 23, 24, 25]. The collective response of media ionized by bichromatic fields is less studied owing to a high complexity of the problem. Most of the theory developed so far used simple models for the photo-induced current or nonlinear susceptibility which enter the right-hand side of an exact or reduced inhomogeneous wave equation (see examples in [13, 22, 24]). Such approaches allow to compute the response of spatially extended media including filaments, but they suffer from model simplifications and do not take self-consistently into account the back reaction of THz radiation created in the plasma on the electron motion. Meanwhile, radiation can significantly influence the electron dynamics as the number of coherently emitting electrons may reach giant values of $\sim 10^{10}$ and higher.

In this paper, we address the problem of the macroscopic THz response of laser-driven plasma in a complimentary way by employing a particle-in-cell (PIC) simulation of the plasma dynamics during and after the interaction with a two-color ionizing laser pulse. Our simulation includes the ionization step and allows for a fully self-consistent calculation of the electron current including the back reaction of the coherently emitted radiation on the plasma dynamics. We employ the PIC code UMKA developed for studying laser-plasma interactions in the strong-field regime and adapted to include ionization [26, 27]. High numerical costs of PIC simulations restrict the interaction volume in all dimensions by size ~ 100 μm, so that instead of considering filaments created in open air we examine radiation emitted from a spatially restricted target. Such targets can be realized with thin gas jets or small gas-filled cells, the latter were used in experiment [19]. We analyze the plasma dynamics, calculate distributions of the quasi-static electric field inside the plasma and the radiation spectra emitted in the forward direction, and estimate the efficiency of the electron energy conversion into that of THz radiation. Our results show a relatively high energy conversion which appears more efficient for a smaller interaction volume and indicate the presence of strong long-living quasi-static electric fields. On this basis, we suggest that the application of small gas-filled cells or thin gas beams where a two-color laser field can induce an almost homogeneous oscillating dipole, may lead to a much higher IR-to-THz energy conversion efficiency than that presently achieved in the regime of filamentation.

2. Basic equations and numerical model

We consider a two-color circularly polarized pulse described by the vector potential

$$\mathbf{A}(\mathbf{r},t) = \mathbf{A}_{\omega}(\mathbf{r},t) + \mathbf{A}_{2\omega}(\mathbf{r},t) \tag{1}$$

whose form is determined by its time dependence in the plane $(x = 0, y, z)$:

$$\mathbf{A}_{\omega}(x=0,y,t) = \frac{E_0}{\omega} e^{-\frac{y^2}{2R^2}} \Big(0, f(\varphi)\cos(\varphi), f(\varphi-\pi)\sin(\varphi)\Big)\ , \quad f(\varphi) = \sin^2\left(\frac{\varphi}{2N}\right)\ . \tag{2}$$

Here E_0 and ω are the electric field amplitude and the carrier frequency of the fundamental pulse with duration of N periods, $R = 16$ μm corresponding to the focal spot diameter (FWHM) of 27 μm and the phase $\varphi = \omega(t - x/c)$ with c being the speed of light. The phase variable $\varphi \in [0, 2\pi N]$, and $\mathbf{A} = 0$ outside of this interval. The SH field has the same spatio-temporal envelope

$$\mathbf{A}_{2\omega}(x=0,y,t) = \epsilon\frac{E_0}{2\omega} e^{-\frac{y^2}{2R^2}} \Big(0, f(\varphi)\cos(2\varphi-\alpha), f(\varphi-\pi)\sin(2\varphi-\alpha)\Big) \tag{3}$$

and is shifted by phase α with respect to the fundamental. The fundamental field amplitude was taken $E_0 = 0.17$ at.u. $= 0.87$ GV/cm, corresponding to intensity $\mathcal{I} = 2 \times 10^{15}$ W/cm^2 of circularly polarized radiation. The number of cycles $N = 100$ for 0.8 μm wavelength and $N = 40$ for 2 μm corresponds to the fixed pulse duration $\tau = 267$ fs. The relative electric field amplitude in Eq. (3) is taken $\epsilon = 0.22$ corresponding to a 5% relative intensity of SH. The choice of circularly polarized radiation is based on its higher efficiency in the excitation of the net photoelectron current responsible for emission of THz waves [19, 21]. For an isotropic gas target, the phase shift α between the two fields in Eqs. (2) and (3), determines the orientation of the symmetry axis in the polarization plane, but otherwise does not affect the plasma dynamics [21]. In our calculations we use a rectangular two-dimensional cell which makes results α-dependent. To focus on the physically relevant situation when plasma oscillations are exited along the direction where the plasma size is spatially restricted, we adjust the value of α to direct the net photoelectron momentum after ionization along the y axis. In the case of linearly polarized pulses, the plasma dynamics is expected to be generally similar, but with a clear dependence on the phase shift, as was observed in [19].

The rectangular gas cell is restricted in the (x, y) plane, $0 \le x \le L, -L/2 \le y \le +L/2$, with $L = 20$ μm or 50 μm and homogeneous in z direction which makes the calculation two-dimensional (2D) in the position space, while electron momenta and electromagnetic fields are calculated in the full dimension. The incoming laser pulse propagating along the x axis is determined by Eqs. (2), (3) on the front side of the cell $x = 0$, while the electromagnetic field in the whole space is calculated numerically from the system of Vlasov-Maxwell equations. With the phase shift α fixed as described above, translation invariance along the z axis is not expected to introduce unphysical effects. Spatial and temporal resolution were $\Delta x = \Delta y = (2\pi c/\omega)/40$ and $\Delta t = (2\pi/\omega)/80$, correspondingly, and 64 macroparticles per species per numerical cell were used.

The gas is assumed initially neutral, and its ionization proceeds along the field tunneling mechanism. We describe ionization events probabilistically, with the distribution function determined by the tunneling ionization rate taken from [28, 29]. We only consider single-electron ionization, so that all the atomic species are either neutral or single-charged. When an ionization event happens, the charge state of a chosen ion changes from 0 to +1, and a free electron is created at rest at the position of the ion. The energy required for ionization is subtracted from the field via the work of the "ionization current" $\mathbf{j}_{\rm ion}$ parallel to the electric field at the ion location. The energy conservation is secured by the condition that the value $(\mathbf{j}_{\text{ion}} \cdot \mathbf{E})\Delta t$ is equal to the energy spent on ionization per time step Δt [30, 31]. If the field energy in a cell is insufficient for further ionization, this cell is not considered anymore during the current time step [32]. For target and laser parameters we consider below, a small fraction of the laser energy is sufficient to singly ionize all atoms in the interaction volume, so that effects of saturation play a minor role. In order to check the role of cell boundaries we have also made several runs for bigger cells with the sizes up to 150 μm in the longitudinal and lateral directions. Results of these calculations show that the effective lateral size of the plasma is determined by that of the focal waist $\simeq 30$ μm and does not depend on the cell width and on its shape in the lateral direction. Instead, the longitudinal cell size has a direct effect on the plasma response.

3. Numerical results

Figure 1 shows the vector potential, Eq. (1), the target geometry, distributions in electron velocity v_y taken at a time instant $t = 333$ fs approximately equal to the time when the laser pulse leaves the cell after the interaction, and spectra $P(\nu) \sim \mathbf{E}(\nu)\mathbf{E}^*(\nu)$ of radiation emitted in the forward direction. These distributions demonstrate a clear dependence on the cell size, on the laser wavelength, and on the gas density. Firstly, at intensity $\mathcal{I} = 2 \times 10^{15}$ W/cm^2, single-electron tunnel ionization of argon happens within a couple of optical cycles on the front edge of the pulse, so that during the interaction and after the laser pulse is gone the electron concentration of the plasma n_e is close to that of atoms n_0. As a consequence, the plasma frequency $\omega_p = \sqrt{4\pi e^2 n_e/m_e}$ approximately equals to 5×10^{13} s^{-1} for $n_0 = 10^{18}$ cm^{-3} and to 1.5×10^{14} s^{-1} for $n_0 = 10^{19}$ cm^{-3}. These numbers agree with the positions of maxima in the radiation spectra. Note that the plasma wavelength, $\lambda_p = 2\pi c/\omega_p$, which is equal to 40 μm and 13 μm for 10^{18} cm^{-3} and 10^{19} cm^{-3}, respectively, determines the number of oscillations per cell size L in the electron distribution. Secondly, the laser wavelength determines the initial photoelectron velocity (see e.g. [21] for details) $v_0 \approx keE_0\lambda/(2\pi m_e c)$ with $k \approx 0.1$. These parametric scalings are clearly seen on the distributions shown in Fig. 1(c-f).

Below we focus on the distribution of the electric field inside the plasma and on the total energy emitted in the THz domain. Analysis of the magnetic fields and of the temporal structure of THz pulses will be presented elsewhere. Figure 2 shows spatial distributions of the electric field E_y taken at times $t \approx 400$ fs after the interaction has started. The fields are averaged over the fundamental period aiming to smear out fast oscillations induced by the two-color pump pulse. Cuts of E_y alog the x-axis for $y = 0$ and $y = L/2$ are depicted by black and white lines, respectively. The plots allow estimating the peak values of the oscillating electric fields inside the plasma and in the near-field zone. On the cell axis, the magnitude of this quasi-static electric field reaches $E_{\text{m}} \approx 0.01 E_0 = 8.6$ MV/cm for the 50 μm cell at concentration $n_0 = 10^{19}$ cm^{-3} and 0.8 μm wavelength. For the 20 μm cell and 2 μm wavelength the peak electric field inside the plasma is even higher despite of the 10-times lower electron density.

Finally, we compute the total energy emitted in the THz domain. To this end, we notice that at $t > 400$ fs THz pulses are already well localized in space relatively far from the plasma, so that their linear energy density (i.e. energy per unit length in z direction) can be estimated as

$$W_{\text{THz}} \approx \int_S dx dy \frac{\bar{\mathbf{E}}^2}{4\pi} \qquad (4)$$

where the integral is taken over the 2D area S in the (x, y) plane where the THz wave is localized. Here $\bar{\mathbf{E}}$ denotes the electric field averaged over fast oscillations with frequencies ω and 2ω. The averaging procedure rules out the contribution of the pump laser field. Results are summarized in Table 1 for the same set of parameters as used for Fig. 2.

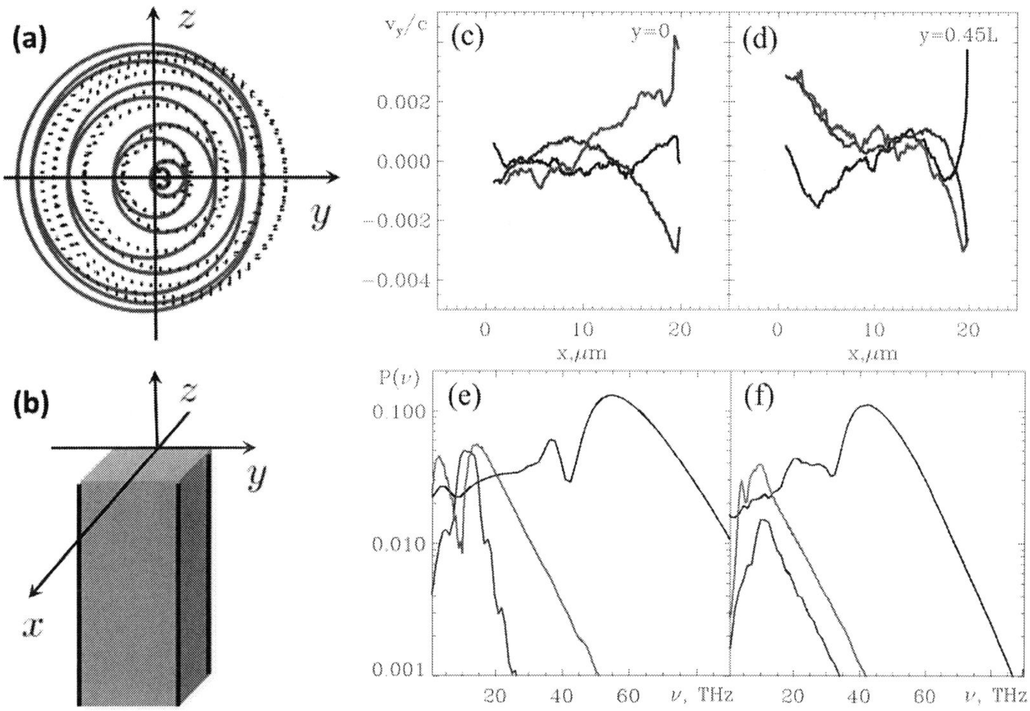

Figure 1. (color online) Vector potentials of the single-color $\mathbf{A}_\omega(0,0,t)$ (black dotted line) and the two-color $\mathbf{A}(0,0,t)$ (solid red line) laser pulses defined by Eqs. (1), (2) and (3) (a); geometry of the 2D target (b); distributions in the electron velocity $v_y(x)$ normalized to the speed of light c at $t = 333$ fs and $y = 0$ (c) and $y = 0.45L$ (d) for the 20 μm cell; spectral radiation power $P(\nu)$ in the forward direction measured in relative units for the 50 μm (e) and 20 μm (f) cells. Black lines correspond to parameters ($\lambda = 0.8$ μm), $n_0 = 10^{19}$ cm^{-3}, blue lines – to ($\lambda = 0.8$ μm), $n_0 = 10^{18}$ cm^{-3} and red lines – to ($\lambda = 2$ μm), $n_0 = 10^{18}$ cm^{-3}. On panel (a), the number of cycles in the pulse is reduced to $N = 10$ for better visibility.

Table 1. THz pulse energy per unit length calculated from Eq. (4) for parameters of Fig. 2, and normalized to its value for $n_0 = 10^{19}$ cm^{-3}, $\lambda = 0.8$ μm and $L = 50$ μm.

Cell size	$n_0 = 10^{19}$ cm^{-3} $\lambda = 0.8$ μm	$n_0 = 10^{18}$ cm^{-3} $\lambda = 0.8$ μm	$n_0 = 10^{18}$ cm^{-3} $\lambda = 2$ μm
50μm	1.0	0.21	2.0
20μm	0.44	0.16	0.88

4. Discussion and outlook

Results reported in the previous section show several clear tendencies. First, in all cases the THz pulse duration is shorter than that of the pump pulse. The main spike of the field lasts about 100 fs for $n_0 = 10^{19}$ cm^{-3} and 150 fs for $n_0 = 10^{18}$ cm^{-3}. These times are approximately equal to the plasma periods $T_p = 2\pi/\omega_p$ at the electron concentration equal to n_0. At the same time, the THz pulse duration appears almost independent on the cell size and the pump

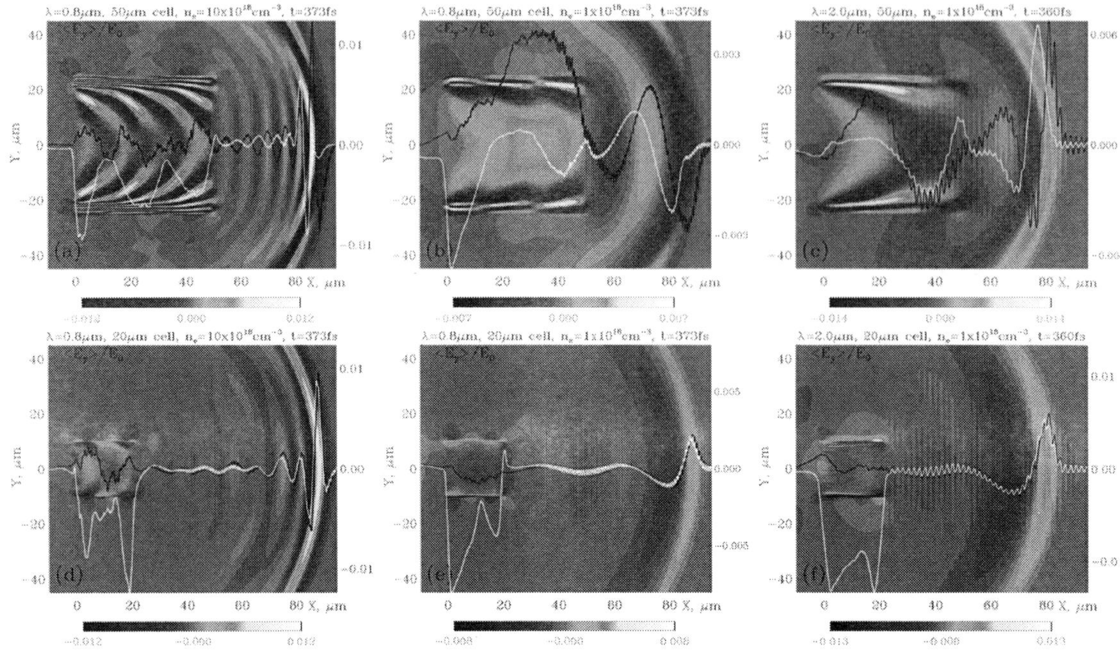

Figure 2. (color online) Distributions of the time-averaged electric field $E_y(x,y)$ and cuts of these distributions taken on the cell axis $y = 0$ (black lines) and the cell boundary, $y = L/2$ (white lines) at times $t \simeq 400$ fs. The cell size, atomic concentration, laser wavelength and time instants t are shown on the panels headlines.

wavelength. The short duration of the THz burst contrasts with the fact that plasma oscillations are clearly present in the cell long after the pump laser pulse is gone. A possible mechanism, which limits the pulse duration, can be connected to an extremely strong radiation damping of the plasma oscillations. In the long wavelength limit, the radiation reaction force acting on an electron is proportional to the number of coherently radiating electrons. For THz wavelengths and plasma concentrations considered here, this number can be as large as $N_e \simeq 10^{10}$ leading to an almost prompt damping of that component of the plasma oscillation, which emits radiation. The following plasma evolution remains time-dependent but does not lead to significant emission anymore. This behavior can be interpreted as resulting from a weak coupling of the survived plasma oscillation mode to the field of radiation [33] or, equivalently, as a formation of a specific non-radiating dynamic configuration. The latter has been extensively studied in nanoplasmonics and optics of metamaterials [34]. A quantitative description of this radiation damping effect on the plasma dynamics at long wavelengths will be given elsewhere.

Second, the quasi-static electric field excited inside and in a close vicinity of the cell reach the value $E_{\rm m} \approx 0.013E_0 = 11.6$ MV/cm for parameters of Fig. 2(d). The peak amplitude practically does not depend on the cell size and grows with the electron concentration and the laser wavelength. Assuming a homogeneous plasma oscillation with electron velocity distribution shown on Fig. 1(c-d), one may estimate the electric field amplitude as $E_{\rm m} \simeq E_0 v_0/v_p \sim \sqrt{n_e}\lambda$ where v_0 is the characteristic value of the initial velocity v_y and $v_p = eE_0/m_e\omega_p$. Estimating v_0 from the data of Fig. 1 one obtains values of the electric field amplitude close to those shown on Fig. 2. Thus, the quasi-static electric field induced inside the cell by the plasma oscillation grows with the atomic concentration and the laser wavelength, the latter agrees with the known data and calculations of two-color THz emission in mid-infrared fields [18, 20, 21]. The dependence of the electric field distribution on the cell size is less trivial. At 50 μm the field is mostly concentrated near the cell edges where it oscillates spatially in the y-direction with a relatively small period. These lateral oscillations can be explained as resulting from an non-homogeneous ionization on the shoulders of the pump laser pulse whose width in the transverse direction is determined by the parameter $R = 16$ μm and is smaller than the cell size. Instead, for the smaller cell the field ionization is quickly saturated in the whole volume making the system behaving similarly to a plain capacitor with an almost homogeneous time-dependent electric field inside.

Finally, we consider the energy emitted in the THz domain. Its relative value given in Table 1 shows that the smaller cell emits with a much higher efficiency per electron than that of 50 μm size. Indeed, taking into account that for our 2D calculation the linear concentration of electrons in the 20 μm cell is 6.25 times smaller, we obtain, from the numbers shown in Table 1, that this cell emits $2.7 - 4.7$ times more energy per electron, at other parameters fixed. This enhanced efficiency is the result of higher coherency of the electron motion inside the smaller cell. While the 50 μm cell supports $2 - 3$ plasma wavelengths, the smaller cell shows almost a pure dipole oscillation of the electron gas. As a result, the electrons emit in-phase maximizing both the emission power and the radiation damping. The higher radiation damping is also responsible for a much weaker tail of the THz pulse in the case of the smaller cell. For a 3D plasma this quasi-dipole oscillation would make THz emission almost isotropic in the (x, z) plane. In our 2D model, a considerable emission happens in the backward direction. Its contribution is not seen on Figs. 2(e,f) for $t \simeq 400$ fs because the back propagated THz wave at this time is located at $x \approx -80 \div -60$ μm. Backward THz emission from two-color ionization of air has been recently observed in experiment [35] where the laser pulse was tightly focused to create a small spike instead of a conventional elongated filament.

Effects considered in this paper essentially depend on the plasma size (particularly on that in the longitudinal direction). A cell restricted by sharp boundaries is a numerically convenient model, but it might appear a setup difficult for experimental realizations. The problem of cell material, which will unavoidably interact with a strong laser pulse, becomes particularly severe for the smallest considered cell size of 20 μm. Alternative realizations of the scheme with a spatially restricted interaction volume can be achieved by using either tight focusing or thin gas jets. In the tight focusing scheme, the lateral plasma size will also be limited to the value of several microns, highly reducing the total THz energy output. Instead, in a gas jet flowing from a thin linear nozzle, the longitudinal and lateral plasma dimensions can be independently controlled by the jet thickness and the focal spot size.

In conclusion, we studied emission of THz waves from a small gas cell where a non-equilibrium plasma is created by means of strong-field two-color ionization of atoms. Our main funding is the high relative efficiency of the electron energy conversion into THz radiation for smaller plasma emitters. Quasi-static electric fields are shown to achieve values > 10 MV/cm and can be further enhanced by applying longer mid-infrared wavelengths and using higher concentrations in the target. Consequently, a gas jet, gas-filled cell or fiber of size of $10 - 20$ μm can serve as a highly efficient source of quasi-static electric and magnetic fields both in the near-filed and far-field zones.

Acknowledgments

Authors acknowledge useful discussions with V. Bukin, A. Gopal, G.G. Paulus and V. Strelkov. SVP acknowledges financial support of the Russian Science Foundation through grant No.18-12-00476 in the part related to the development of the model and analysis of numerical results. The development of numerical algorithms was partially supported by the Russian Science Foundation through the grant No.19-71-20026. TVL acknowledges the Siberian Supercomputer Center

ICMMG SB RAS (Novosibirsk) for providing the computing time on NKS-1P cluster. Most of the numerical simulations were performed using the computing resources granted by the John von Neumann-Institut für Computing (Research Center Jülich) under the project HRO04.

References

[1] Pickwell E and Wallace V P 2006 *J. Phys. D* **39** R301
[2] Dexheimer S L 2008 *Terahertz Spectroscopy: Principles and Applications* (New York: Taylor & Francis Group)
[3] Lee Y-S 2009 *Principles of Terahertz Science and Technology* (New York: Springer)
[4] Hafez H A, Chai X, Ibrahim A, Mondal S, Férachou D, Ropagnol X and Ozaki T 2016 *J. Opt.* **18** 093004
[5] Dhillon S D *et al.* 2017 *J. Phys. D* **50** 043001
[6] Yeh K-L, Hoffmann M C, Hebling J and Nelson K A 2007 *Appl. Phys. Lett.* **90** 171121
[7] Gopal A *et al.* 2012 *New J. Phys.* **14** 083012
[8] Gopal A *et al.* 2013 *Phys. Rev. Lett.* **111** 074802
[9] Bartel T, Gaal P, Reimann K, Woerner M and Elsaesser T 2005 *Opt. Lett.* **30** 2805
[10] Kim K Y, Glownia J H, Taylor A J and Rodriguez G 2007 *Opt. Express* **15** 4577
[11] Silaev A A and Vvedenskii N V 2009 *Phys. Rev. Lett.* **102** 115005
[12] Bergé L, Skupin S, Köhler C, Babushkin I and Herrmann J 2013 *Phys. Rev. Lett.* **110** 073901
[13] Kostin V A and Vvedenskii N V 2018 *Phys. Rev. Lett.* **120** 065002
[14] You Y S, Oh T I and Kim K Y 2012 *Phys. Rev. Lett.* **109** 183902
[15] Oh T I, Yoo Y J, You Y S and Kim K Y 2014 *Appl. Phys. Lett.* **105** 041103
[16] Dai J, Karpowicz N, and Zhang X.-C. 2009 *Phys. Rev. Lett.* **103** 023001
[17] Wen H, and Lindenberg A M 2009 *Phys. Rev. Lett.* **103** 023902
[18] Clerici M *et al.* 2013 *Phys. Rev. Lett.* **110** 253901
[19] Meng Ch, Chen W, Wang X, Lü Zh, Huang Y, Liu J, Zhang D, Zhao Z, and Yuan J 2016 *Appl. Phys. Lett.* **109**, 131105
[20] Fedorov V Yu and Tzortzakis S 2018 *Phys. Rev. A* **97** 063842
[21] Tulsky V A, Baghery M, Saalmann U and Popruzhenko S V 2018 *Phys. Rev. A* **98** 053415
[22] Kim K Y 2009 *Phys. Plasmas* **16**, 056706
[23] Kotelnikov I A, Borodin A V, Shkurinov A P 2011 *J. Exp. Theor. Phys.* **139** 1081
[24] Johnson L A, Palastro J P, Antonsen T M, and Kim K Y 2013 *Phys. Rev. A* **88** 063804
[25] Popruzhenko S V, Tulsky V A 2015 *Phys. Rev. A* **92**, 033414
[26] Vshivkov V A, Naumova N M, Pegoraro F, and Bulanov S V 1998 *Phys. Plasmas* **5**, 2727
[27] Liseykina T, Bauer D 2013 *Phys. Rev. Lett.* **110**, 145003
[28] Popov V S 2004 *Usp. Fiz. Nauk* **147**, 921 [2004 *Phys. Usp.* **47**, 855]
[29] Popruzhenko S V 2014 *J. Phys. B: At. Mol. Opt. Phys.* **47** 204001
[30] Rae S C and Burnett K 1992 *Phys. Rev. A* **46**, 2077
[31] Mulser P, Cornolti F, and Bauer D 1998 *Phys. Plasmas* **5**, 4466
[32] Kemp A J, Sentoku Y, Cowan T, Fuchs J, and Ruhl H 2004 *Phys. Plasmas* **11**, L69
[33] Thiele *et al.* 2018 *Optica* **5** 1617
[34] Yang Y and Bozhevolnyi S 2019 *Nanotechnology* **30** 204001
[35] Ushakov A A, Panov N A, Chizhov P A, Shipilo D E, Bukin V V, Savel'ev A B, Garnov S V and Kosareva O G 2019 *Appl. Phys. Lett.* **114**, 081102

Positron collisions with targets of biological and technological relevance

Alessandra Souza Barbosa[1,2], **Sergio D'A Sanchez**[1] **and Márcio H F Bettega**[1]

[1]Departamento de Física, Universidade Federal do Paraná, Caixa Postal 19044, 81531-990 Curitiba, Paraná, Brazil
[2] Laboratório de Colisões Atómicas e Moleculares, CEFITEC, Departamento de Física, FCT–Universidade Nova de Lisboa, Caparica, Portugal

E-mail: alessandra@fisica.ufpr.br

Abstract. In this work we present calculated cross sections for elastic scattering of low-energy positrons by molecules, obtained with the Schwinger multichannel method. We extended a previous study on the role of polarization effects in the positron-allene scattering and also present some preliminary results for positron collisions with cyclopentane. We also compare the differential cross sections of benzene and its azaderivative pyrimidine at selected energies, where similar angular behaviour was found.

1. Introduction

Positron physics attracts great interest due to its fundamental, technological and biological applications. Since many of its applications are based on basic interactions of positrons with molecules, positron-scattering studies have become a cornerstone towards understanding the underlying physics of these interactions [1, 2]. In recent years, thanks to the advent of better theoretical methodologies and computers to perform the simulations, theoretical positron-molecules studies have gained new breath. At the same time, experimental advances have considerably improved the available data [3].

However, positron scattering calculations can be a difficult task due to some intrinsic problems. For example, besides the inelastic process already observed in electron-scattering, the incoming positron can capture one of the electrons of the target to form positronium (Ps). Moreover, Ps formation can be responsible for a large amount of the total cross section for positron impact energies over the positronium formation threshold [4]. Even more troublesome, at energies below the Ps formation and electronic excitation thresholds, the agreement between the theoretical calculations and experimental data is far from good. This is mainly because the description of the polarization potential, which is crucial in the positron scattering due to the absence of exchange effects, is an extremely hard task for theoreticians.

In recent years, our group has put a lot of effort in order to provide some reliable results for positron scattering by molecules and systems of technological and biological relevance. For example, employing the Schwinger multichannel method [5, 6], we have performed extensive studies in order to improve the description of the polarization potential for two small non-polar molecules, allene [7] and silane [8]. In both works we have shown that the inclusion of extra functions in chargeless centers seems to improve the description of the polarization effects. This was corroborated by the analysis of the *s*-wave cross section and eigenphase. It is worth mentioning that for the first time a bound state was predicted by an *ab initio* positron-molecule scattering calculation [7]. We have also worked on the description of positron interactions with biological systems, such as the pyrimidine molecule [9] and tetrahydrofuran [10].

In this contribution we employed the Schwinger multichannel method to obtain calculated cross sections for positron scattering by molecules of technological and/or biological relevance. We will review and present additional results on the description of the polarization effects in the positron-allene scattering, and present some preliminary results for positron scattering by cyclopentane. The present cyclopentane cross sections will be compared with previous results for tetrahydrofuran (THF) due to the similarity of both molecules. Indeed, THF can be seen as a cyclopentane where one of the CH group is replaced by an O atom. Besides, we will also present a comparison between the differential cross sections (DCS) for positron scattering by benzene and pyrimidine.

The remainder of this paper is as follows. In the next section we will provide some of the theoretical details of the Schwinger multichanel method, then in section 3 the results will be presented and discussed. The last section will present a brief summary of our findings.

2. Methodology and theoretical details

The elastic cross sections presented here were calculated with the Schwinger multichannel method (SMC) as implemented for positron-molecule scattering. Since the method has been described in detail in several publications [5, 6], here we will only discuss those aspects of the SMC method that are relevant to the present calculations.

The working expression for the scattering amplitude is:

$$f(\vec{k}_f, \vec{k}_i) = -\frac{1}{2\pi} \sum_{m,n} \langle S_{\vec{k}_f} | V | \chi_m \rangle (d^{-1})_{mn} \langle \chi_n | V | S_{\vec{k}_i} \rangle \qquad (1)$$

where $d_{mn} = \langle \chi_m | A^{(+)} | \chi_n \rangle$ and the $A^{(+)}$ operator is given by $A^{(+)} = Q\hat{H}Q + PVP - VG_P^{(+)}V$. $|S_{\vec{k}_{i,f}}\rangle$ is a solution of the unperturbed Hamiltonian H_0 (the target Hamiltonian plus the kinetic energy of the incoming positron) and is given by a product of a target state and a plane wave; V is the interaction potential between the incident positron and the electrons and nuclei of the target molecule; $|\chi_m\rangle$ is a set of $(N+1)$-particle configuration state functions used in the expansion of the trial scattering wave function, $\hat{H} = E - H$ is the collision energy minus the full Hamiltonian of the system ($H = H_0 + V$), P is a projection operator onto the open-channel space defined by the target eigenfunctions, and $G_P^{(+)}$ is the free-particle Green's function projected onto the P-space. Finally $Q = (\mathbb{1} - P)$ is the projector onto the closed electronic channels of the target.

In the static plus polarization approximation (SP), the direct space is composed by configuration state functions (CSFs) of the form:

$$|\chi_j\rangle = |\Phi_1\rangle \otimes |\varphi_j\rangle \oplus |\Phi_i\rangle \otimes |\varphi_j\rangle, \qquad (2)$$

where $|\Phi_1\rangle$ represents the ground state of the molecule obtained at the Hartree-Fock (HF) level, $|\varphi_j\rangle$ is a single-particle orbital used to expand the positron scattering orbital (see below) and $|\Phi_i\rangle$ is obtained from virtual single excitations of the target from the hole (occupied) orbitals to a set of particle (unoccupied) orbitals. Thus, it is worth mentioning that the description of the polarization effects depends on the basis functions employed in the target description as well as on the configurations employed in the SP calculations. With the aim of making the reading clear, the specific details of each calculation will be provided in the next section along with the results.

3. Results

3.1. Small apolar molecules

Recently we have shown that the description of the polarization effects in positron scattering by allene [7] and silane [8] can be improved by the inclusion of extra functions in chargeless centers. In particular, for the allene molecule the inclusion of extra functions placed in the vertices of two hexagons lying in planes perpendicular to the molecular axis, and the inclusion of four extra centers along the CH bond, as shown in the left side of figure 1, improves the description of the polarization effects [7]. We have found that as the extra centers were included in the calculations the polarization effects seem to be better described, once the minimum present in the s-wave cross sections moved to higher energies. Moreover, the abrupt increase in the integral cross section (ICS) as the impact energy goes towards zero was initially characterized as a virtual state and then as a bound state, as the number of extras centers and of the configurations space functions were increased [7]. Here we present an extension of this work, where similar calculations have been carried out but with the extra functions placed in chargeless centers located at the vertices of two squares lying in planes perpendicular to the molecular axis, as shown in the right side of figure 1.

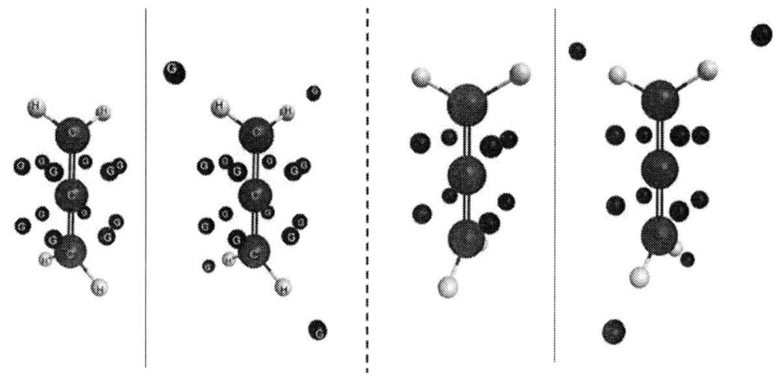

Figure 1. Geometrical structure of allene molecule and the position of the extra centers employed in the previous [7] (left) and present calculations.

The scattering calculations were carried employing the same geometry and basis set used in the previous study [7]. In all calculations, the 65 first modified virtual orbitals (MVO) [11] were employed as particle and scattering orbitals. The 11 valence orbitals were employed as hole orbitals and, with the 65 MVOs, as scattering orbitals. This is labelled as 65p76s along this text. The 0c-65p76s, hex-65p76s and hex+4-65p76s calculations correspond respectively to no extra centers, 12 extra centers placed in the vertices of two hexagons and 16 extra centers (the previous 12 and additional four extra centers placed along the CH bond). These calculations correspond to the SP-1, SP-3 and SP-5 presented previously [7]. Similarly, two other calculations labelled squa-65p76s and squa+4c-65p76s employed extra centers placed in the vertices of the two squares and with four additional centers along the CH bond, respectively. Since our aim is to discuss the low-energy physics of positron-allene scattering, we will present only the s-wave cross sections and eigenphases, which are presented in figure 2.

By inspection of figure 2 it is seen that the inclusion of the extra centers brings the minimum present in the cross section to higher energies, shifting it from 2.7 eV in the 0c-65p76s calculation to at around 3.3 eV. It is also noted that the cross sections obtained in the calculations employing extra centers, seem to overlap in the 0.5-5.0 eV energy region. Moreover, at the same energy the s-wave eigenphase changes sign from positive to negative. This is particularly related to the net potential felt by the incoming positron, which is a combination of the repulsive static potential and the attractive polarization potential. These two contributions cancel each other at the particular energy where the s-wave eigenphase changes sign (thus, the net potential changes from attractive to repulsive). Besides the minimum, the dramatic increase in the cross section as the impact energy goes to zero is also noticeable. This is a common feature in positron scattering and can be related either to a virtual or a bound state, depending on the behaviour of the s-wave eigenphase. Or, equivalently, the virtual or bound state is corroborated by the calculation of the scattering length (SL) defined as [12]

$$\mathrm{SL} = -\lim_{k\to 0}\frac{1}{k}\tan\delta_0(k) \tag{3}$$

where δ_0 and k are the s-wave eigenphase and the positron momentum, respectively. Thus, if the SL is positive (negative) a bound (virtual) state takes place.

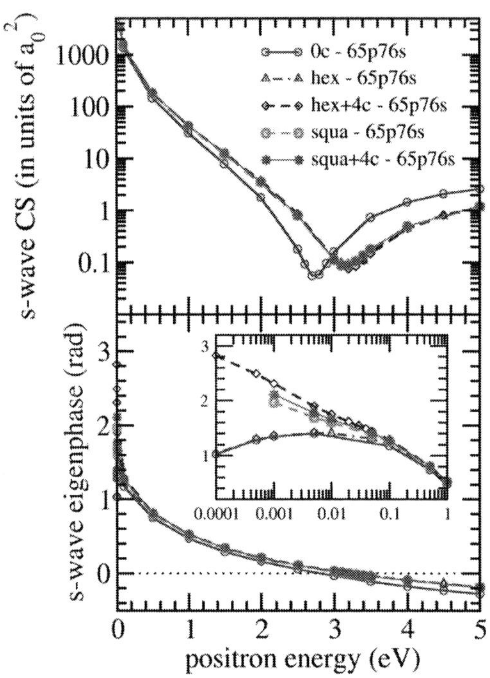

Figure 2. Positron scattering by allene: (top panel) s-wave cross section (CS) and (bottom panel) respective eigenphase calculated employing different schemes to account for the polarization effects. The inset presents the same data within a logarithmic energy scale, in order to highlight their very low-energy behavior. (hex = 2 hexagons = 12 extra centers; squa = 2 squares = 8 extra centers)

As discussed previously, the 0c- and hex- calculations support a virtual state, whereas the hex+4- calculation supports a bound state [7]. Moreover, we have also seen that only by increasing the configuration space by including more virtual single excitations of the target, the calculations employing 12 extra centers lead to a bound state [7]. Regarding the results obtained in the squa-65p76s and squa+4c-65p76s calculations, it is noted that at lower energies the s-wave eigenphase increases, presenting a similar behaviour to the hex+4c- case. The squa- and squa+4- calculations lead to SL equal to 256 a_0 and 183 a_0, respectively, supporting a bound state in both cases. That is, even employing a smaller basis set than previously, in the hex-65p76s calculation, the effective potential seems to be more attractive once a bound state is supported in squa-65p76s calculations. These results reveal the importance on the choice of the positions of the extra functions. This is noted by comparing the hex-65p76s and squa-65p76s results, which reveal that a reduced basis set (due to the lower number of extra functions employed in the latter calculation) can provide a better description of the polarization effects, in particular at impact energies lower than 0.1 eV. However, it should be noted that for higher impact energies, all calculations employing the extra functions seem to provide similar results regardless of the position of the extra functions.

3.2. Tetrahydrofuran and cyclopentane

Here we present preliminary calculated cross sections for low-energy positron scattering by cyclopentane (C_5H_{10}) and compare with the previous results for tetrahydrofuran (C_4H_8O). Both molecules are cyclic with five atoms in the ring and, as shown in figure 3, THF can be obtained by the replacement of a CH_2 group of cyclopentane by an oxygen atom. Besides this similarity it is important to note that whereas cyclopentane presents a dipole moment that is approximately zero, THF presents a dipole moment of 2.03 D [10]. Thus this difference shall play an important role in the positron scattering dynamics.

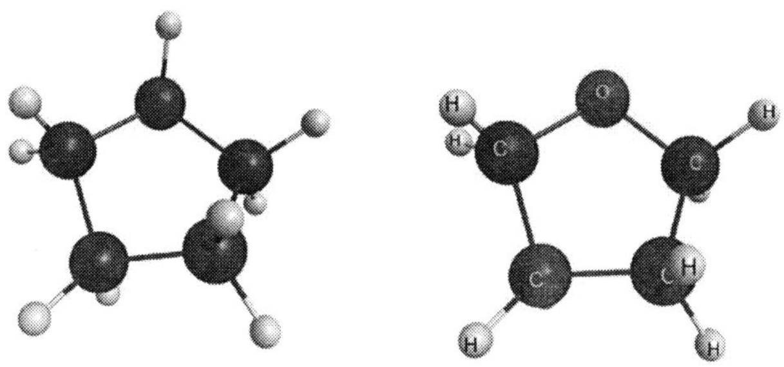

Figure 3. Geometrical structure cyclopentane (left) and tetrahydrofuran (right) molecules.

The scattering calculations for cyclopentane were carried out in the C_2 symmetry group following the same procedure as in our previous work on THF [10]. That is, the 15 outermost occupied orbitals were employed as hole orbitals, the 35 lowest MVOs [11] as particle orbitals and these MVOs along the 20 occupied orbitals were used as scattering orbitals. Besides this calculation, labelled SP-1, we also carried out two others where we have optimized the molecular geometry in the C_{2v} symmetry group. At this symmetry group all the carbon atoms lie in the same plane. Although not working in the ground state geometry, using a larger symmetry group permits the use of a larger number of CSFs in the calculation.

In figure 4 we present preliminary calculated integral cross sections for cyclopentane obtained in the C_2 and C_{2v} symmetry groups. In the C_{2v} symmetry group we carried out two calculations, the first one employing the same polarization scheme employed in the C_2 calculations (SP-1) and the second one (SP-2) employing the 15 valence orbitals as hole, the 55 lowest MVOs as particle orbitals and all the occupied orbitals with these 55 MVOs as scattering orbitals. The present results are compared with the available results for THF.

From figure 4 it is noted that both SP-1 calculations for the cyclopentane present an overall similar behaviour, where the main differences (e.g. the minimum at around 0.5 eV and the ICS magnitude at impact energies higher than 2.0 eV) are due to the different geometry employed in each calculation. This is better seen in figure 5 where the symmetry decomposition, according to the C_2 symmetry group, of the SP-1 calculated ICSs are presented. It should be noted that the only difference among these calculations is the employed molecular geometry. That is, the description of the polarization effects employed the same number of hole, particle and scattering

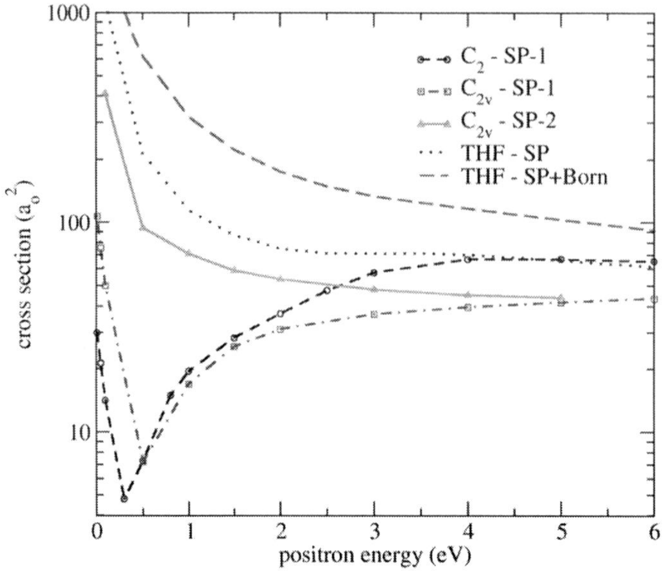

Figure 4. Calculated integral cross sections for elastic scattering of positrons by cyclopentane, in the static plus polarization approximation, obtained in the C_2 and C_{2v} symmetry groups. The present results are compared with the calculated results for THF in the SP and SP+Born approximations [10].

orbitals in each calculation. When comparing SP-1 and SP-2 calculations, carried out in the C_{2v} symmetry group, we observe that the ICS obtained in SP-2 scheme presents larger magnitudes. Moreover, the former results present a minimum at around 0.5 eV, whereas this minimum is not seen in the SP-2 results. From the comparison of the symmetry decomposition of the cross sections, in the C_{2v} symmetry group, presented in figure 6 it is seen that this is not only due to an improvement in the totally symmetric irreducible representation, A_1, but also in the B_1 and B_2 symmetries where a pronounced structure at around 1.0 eV is observed. As shown in the literature, this structure is due to an angular momentum barrier effect [13].

3.3. Pyrimidine and benzene

We have recently devoted some effort to the study of positron scattering by molecules of biological relevance including pyrimidine, one of the building blocks of the nitrogenated bases of DNA. Pyrimidine can be seen as a derivative of benzene with the replacement of two CH groups of benzene molecule by a N atom in the 1 and 3 positions. Thus, this section will be devoted to a comparison of the SMC results, in particular the differential cross sections, for pyrimidine and benzene. The scattering calculations were carried out in the optimized geometry of both molecules, as discussed previously [9, 14], in the C_{2v} symmetry group for pyrimidine and D_{2h} for benzene (although benzene belongs to the D_{6h} symmetry group, once the SMC method only deals with abelian groups). For the description of the polarization effects we have employed the 15 valence orbitals as hole orbitals, 50 MVOs as particle orbital and these 50 MVOs along the 21 occupied orbitals as scattering orbitals, resulting in a total of 53612 CSFs for pyrimidine and 53424 CSFs for benzene.

It have been shown in several references (see for example [15] and references therein) that the differential cross sections for electron scattering by benzene and the azaderivatives present similar oscillatory behaviour. Thus, it could be important to look for similarities and/or differences in

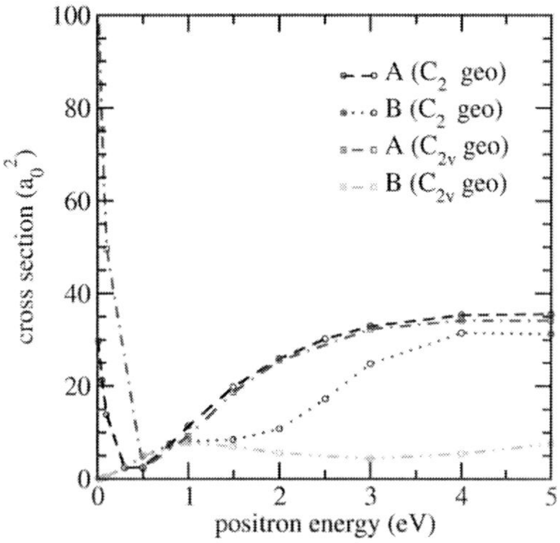

Figure 5. Symmetry decomposition, according to the C_2 symmetry group, of the SP-1 calculated integral cross sections for cyclopentane. It should be noted that the only difference among the calculations is the molecular geometry employed in the calculations.

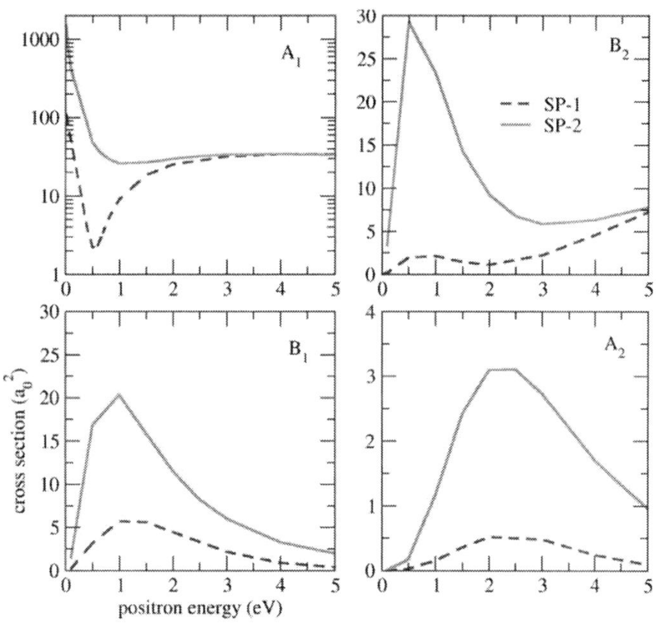

Figure 6. Symmetry decomposition, according to the C_{2v} symmetry group, of the calculated integral cross sections for elastic scattering of positrons by cyclopentane in the static plus polarization approximation, employing two different schemes (different numbers of CSFs).

the comparison of the DCS for positron scattering by benzene and its azaderivatives, such as pyrimidine. Figure 7 presents the differential cross sections for elastic scattering of positrons by benzene and pyrimidine at selected energies. The dashed black line corresponds to the SP calculations for benzene, whereas the solid cyan and the dot-dashed magenta lines correspond to the results for the pyrimidine molecule, obtained in the SP and SP+Born, where in the latter the long range effects of dipole moment potential are taken into account through the well known

Born-closure relation.

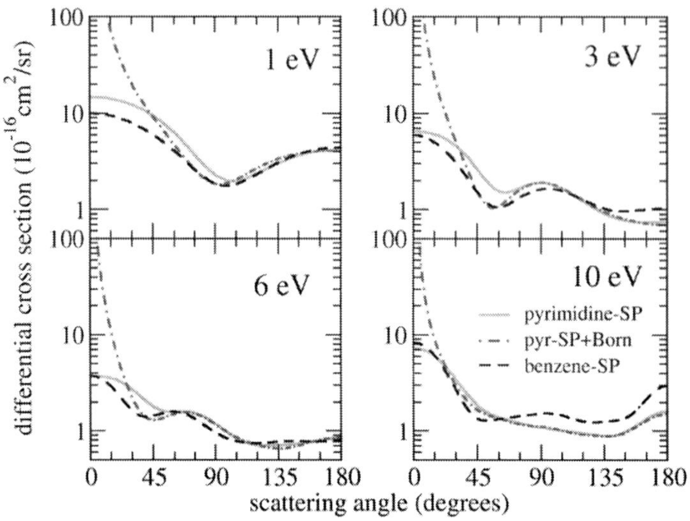

Figure 7. Calculated differential cross sections for elastic scattering of positrons by benzene and pyrimidine at selected energies.

From figure 7 it is seen that the main differences in the pyrimidine DCS in the SP and SP+Born approximations are in the low-angle region, which is a signature of the long range dipole moment potential. Moreover, when comparing benzene and pyrimidine cross sections, it is seen that the overall angular behaviour of the DCSs are very similar for both molecules at the SP approximations. On the other hand, they present differences at angles typically lower than 30 degrees when the Born correction for the long-range potential due to the dipole moment is included. It could be further interesting to study the DCS angular behaviour for positron scattering by other azaderivatives, such as pyridine where only one CH group in benzene ring is replaced by a N atom.

4. Summary

In this paper we have reviewed and extended some of our results in positron scattering by molecules obtained by employing the Schwinger multichannel method. We have discussed how the description of the polarization effects can be improved, either by including extra functions placed in chargeless centers or by enlarging the configuration space. In particular, we have seen that for the allene molecule, the position of the extra functions seems to play an important role in the description of the polarization effects, in particular at impact energies lower than 0.1 eV. This is based on the fact that the calculation employing functions in only eight extra centers (in the vertices of two squares) supports a bound state in positron-allene scattering, whereas the previous calculation employing functions in twelve extra centers (vertices of two hexagons) [7] supported a virtual state. For cyclopentane we have carried out calculations in the C_2 and C_{2v} symmetry groups (and, thus, employing molecular geometries optimized in these symmetry groups), where the latter calculations were carried out employing two different schemes for the SP approximation. It was noted that the C_2 and C_{2v} cross sections present an overall similar behavior when the same polarization scheme is employed, while the main differences, in particular the minimum at around 0.5 eV and the ICS magnitudes at impact energies higher than 2 eV, are due to the molecular geometry. Besides that, by enlarging of the CSFs space, in the SP-2 calculations, the ICS presents greater magnitude due to an angular momentum barrier effect. Finally, we have presented a comparison of the differential cross sections, at selected energies, for benzene and its azaderivative pyrimidine. We have seen that for both molecules the overall behaviour are quite similar, presenting more relevant differences in the low-angle region where the pyrimidine DCSs present an abrupt increase due to the long-range nature of the dipole moment potential. This study could be further extended by carrying out calculations for other azaderivatives, as pyridine.

Acknowledgments

ASB acknowledges support from the Brazilian agency CNPq, from UFPR (Brazil), from the European Physical Society and from the Portuguese National Funding Agency FCT-MCTES through the research grant UID/FIS/00068/2019 (CEFITEC). MHFB and SS acknowledge support from the Brazilian agency CNPq. MHFB acknowledges support from Finep (under project CT-Infra). The authors acknowledge the computational support from Professor Carlos M. de Carvalho at LFTC-DFis-UFPR and at LCPAD-UFPR and from CENAPAD-SP.

References

[1] Danielson J R, Weber T R and Surko C M 2008 *AIP Conf. Proc.* **1037** 84
[2] Surko C M, Gribakin G F and Buckman S J 2005 *Journ. of Phys. B* **38** R57
[3] Brunger M J , Buckman S J and Ratnavelu K 2017 *J. Phys. Chem. Ref. Data* **46** 023102-1
[4] Kauppila W E, Miller E G, Mohamed H F M, Picpinos K, Stein T S and Surdotovich E 2004 *Phys. Rev. Lett.* **93** 113401
[5] Germano J S E and Lima M A P 1993 *Phys. Rev.* A **47** 3976
[6] Silva E P, Germano J S E and Lima M A P 1994 *Phys. Rev.* A **49** R1527
[7] Barbosa A S, Sanchez S d'A and Bettega M H F 2017 *Phys. Rev.* A **96** 062706
[8] Barbosa A S and Bettega M H F 2017 *Phys. Rev.* A **96** 042715
[9] Barbosa A S, Pastega D F and Bettega M H F 2015 *J. Chem. Phys.* **143** 244316
[10] Barbosa A S and Bettega M H F 2019 *J. Chem. Phys.* **150** 18430
[11] Bauschlicher C W 1980 *J. Chem. Phys.* **72** 880
[12] Joachain C J 1975 *Quantum Collision Theory* (Amsterdam: North-Holland)
[13] Lopes A R, Sanchez S d'A and Bettega M H F 2011 *Phys. Rev.* A **83** 062713
[14] Barbosa A S, Blanco F, García G and Bettega M H F 2019 *Phys. Rev.* A **100** 042705
[15] Barbosa A S, Pastega D F and Bettega M H F 2013 *Phys. Rev.* A **88** 022705

Electron transfer to phenyl boronic acid upon potassium collisions

F Ferreira da Silva[1,*], **B Pamplona**[1], **M Mendes**[1], **G García**[2], **and P Limão-Vieira**[1]
[1] Atomic and Molecular Collisions Laboratory, CEFITEC, Department of Physics, Universidade NOVA de Lisboa, 2829–516 Caparica, Portugal
[2] Instituto de Física Fundamental, Consejo Superior de Investigaciones Científicas, Serrano 113-bis, 28006 Madrid, Spain

E-mail: f.ferreiradasilva@fct.unl.pt

Abstract. We present negative ion formation from collisions of neutral potassium atoms with neutral phenylboronic acid $C_6H_5B(OH)_2$ molecules in the lab frame energy range from 10 to 1000 eV. From the assignment of the negative ion time-of-flight (ToF) mass spectra, BO^- is the main fragment detected at energies below 100 eV, however above 100 eV BO_2^- is the predominant fragment accounting on average for 30% of the total anion yield for collision energies above 250 eV. The rich fragmentation pattern results in the formation of more than thirty fragment anions, with twenty boron containing ions.

1. Introduction

Low-energy electron (LEE) interactions studies with biomolecules have attracted the attention of the international scientific community for almost two decades. Sanche and co-workers have described the interaction of such electrons with plasmid DNA, in an energy range up to 20 eV [1,2], showing that LEE interactions can be very efficient in inducing single and double strand breaks (SSB and DSB) as well as loss of supercoiled DNA. The underlying molecular mechanisms of such reactions [2] have been attributed to dissociative electron attachment (DEA) interactions, showing that DNA damage in condensed films upon LEE irradiation occurs at the same energy as resonance capture from isolated constituents, such as water, cytosine and thymine and sugar unit [1,3,4]. Gas phase electron capture has been studied in a wide range of target molecules leading to dissociation through resonant processes. In the case of thymine and cytosine, the anions formed with the highest yield have been assigned to the dehydrogenated parent anions; however, due to ring opening above 4 eV other fragments, such as CNO^-, are formed with considerable intensities [5]. Additionally, we note that comprehensive characterisation of the fragmentation pathways induced by free electrons was reported for uracil RNA base [3], valine [6,7] in bare and hydrated environments, and the simplest amino acid glycine [8]. Moreover, H^- formation from thymine and uracil upon DEA has been shown to be not only bond selective i.e. (C-H) versus (N-H), but site selective (N1 versus N3 site) [9]. Site selective fragmentation at hydrogen sites has been also shown for different functional groups, in the particular case of carboxylic acids, alcohols and amines [10]. The effect of hydration on pyrimidine [11] by surrounding water molecules has been shown to be an efficient quenching mechanism regarding fragmentation of such a relevant biomolecule. Furthermore, different environments, such as helium nanodroplets matrix, have been extensively investigated in order to give insight on the role of the surrounding matrix dictating the dissociation channels in electron capture processes. Simple organic molecules such as formamide [12], acetic acid [13] and some amino acids [14,15] have been studied in such cold environments and differences compared with DEA in bare target molecules. Moreover, LEE studies on molecular targets that may act as potential drug cancer agents have been performed, namely cis-platinum [16] and bromo pyruvic acid [17]. Moreover we note electron transfer studies yielding negative ion formation in potassium collisions with several biological molecules, in particular for DNA/RNA bases showing that the underlying molecular decomposition mechanism leading to H^- formation is site and bond selective [18]. Furthermore, and within the scope of electron transfer studies, the anionic fragmentation pattern of aliphatic amino acids [19–21] and halogenated uracil derivatives used as potential radiosensitizers have been comprehensively investigated [22]. Upon electron transfer in neutral potassium-neutral molecule collisions, the presence of a potassium cation in the vicinity of a temporary negative ion leads to a relevant coulombic complex which induces new fragmentation pathways that are not allowed in free electron capture [23].

In recent years, boronic acid-containing molecules have been considered to be very promising compounds with potential biomedical applications. Boron is the fifth element of the periodic table and has an electronic structure similar to carbon, conferring to it a high potential for use in pharmaceutical drug design as well as in medicinal chemistry [24]. Amongst the different applications of boronic acids, we stress its relevance as fluorescent cellular dyes [25], saccharide sensors (particularly important for the controlled release of drugs in vivo) [26], dopamine sensors and several enzymatic inhibitors [27]. Furthermore, they have been proven helpful in the treatment of HIV, cancer, diabetes and obesity. The characteristics that allow such a wide range of applications are mostly chemical, namely their reactivity and stability [28]. These acids continue to prove their importance in medical practices, although some fundamental chemical properties and action mechanisms still remain unknown. In the present communication we describe recent electron transfer experiments in potassium collisions with phenylboronic acid (figure 1).

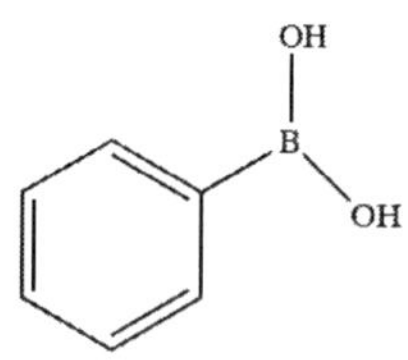

Figure 1. Phenyl boronic acid chemical structure

2. Experimental set-up

The phenyl boronic acid, $C_6H_5B(OH)_2$, sample was purchased from Sigma-Aldrich with a minimum stated purity of 99% and was used as delivered. The sample was heated up to 423 K and temperature controlled by a PID unit. In order to test for thermal decomposition products in the target beam, mass spectra were recorded at different temperatures. No differences were observed in the relative peak intensities as a function of temperature. The negative ion time-of-flight mass spectra were obtained from neutral potassium atoms collisions at different energies. The present crossed beam set-up configuration was described by Antunes *et al.* [29] and further modified as described in Mendes *et al.* [30]. In brief, K^+ ions produced in a commercial source form Heatwave lab, US, were accelerated from 10 to 1000 eV. The accelerated K^+ ions pass through an oven where they resonantly charge-exchanges with neutral potassium to produce a beam of fast (hyperthermal) atoms. Before and after collecting time-of-flight (ToF) mass spectra, the hyperthermal K beam is monitored by a Langmuir-Taylor detector. The phenyl boronic acid effusive beam was introduced into the collision chamber by a 1 mm diameter aperture, where it crossed with the neutral hyperthermal K beam in between two parallel electrostatic plates. The produced anions where extracted into a reflectron ToF (KORE R-500-6) by a pulsed electrostatic field (~290 Vcm^{-1}) and detected by a MCP device. The base pressure in the collision chamber was typically 6×10^{-5} Pa and the working pressure was 4×10^{-4} Pa. From the obtained mass spectra (resolution $m/\Delta m$

$\approx$ 700) background signal was subtracted (without the heated sample) and the mass calibration was performed based on well-known nitromethane negative ions formed in potassium collisions [29].

3. Results and discussion

Figure 2 presents a ToF mass spectra of phenyl boronic acid. In the upper panel (A), the mass spectrum was obtained from K collisions at 43 eV available energy (70 eV collision energy in the lab frame) whereas in the lower panel (B) the mass spectrum was obtained at 200 eV available energy (300 eV in the lab frame), with more than thirty fragment anions assigned. The present measurements were carried out at the lowest centre-of-mass collision energy of 2.5 eV (6.8 eV collision energy); however, only when the available energy rises above 12.8 eV negative ion formation from phenylboronic acid was observed. In table 1 the assigned negative anions formed after potassium collisions are listed.

Figure 2. ToF mass spectra obtained for phenyl boronic acid after collisions with potassium atoms at 43 eV (A) and 200 eV (B) available energy (70 and 300 eV collision energy, respectively). The most intense fragment anions are assigned; all fragments are listed in table 1.

3.1 BO^- and BO_2^-

The most intense fragment anions detected in the wide range of collision energies investigated in these experiments where assigned to BO^- and BO_2^-, with the sum of their yields accounting for more than 40% of the total anion intensities. In both cases, the isotopic ratio between the lighter (^{10}B) and the heavier (^{11}B) anions is ~ 1/4. Boron monoxide (BO^-) m/z 26 (and m/z 27) is observed at the lowest available energy, 12.8 eV, whereas for 6.2 eV no traces of such anion have been detected. Boron dioxide (BO_2^-) m/z 42 (and m/z 43) was only observed above 16.3 eV available energy. Boron monoxide and boron dioxide have high electron affinities, 2.51 and 4.46 eV [31], which is not surprising considering they are formed with high intensity in the ToF mass spectra. Figure 3 shows the branching ratio of these anions as function of the collision energy. For available energies below 100 eV, boron monoxide is the most predominant anion, whereas with increasing available energy (above 100 eV), its yield is less than 20% of the total anion yield. Of relevance, boron dioxide (BO_2^-) follows a opposite trending. For available energy above 250 eV, BO_2^- formation is predominant reaching a plateau. In conclusion, the formation of these negative ions is strongly dependent on the collision energy. It is clear that for low collision energies, excision of the B-O bond against the O-H bond is the predominant pathway, while at available energies above 100 eV, O-H bond rupture becomes dominant (see figure 4).

Figure 3. Branching Ratios for BO^- (m/z 26 and m/z 27) and BO_2^- (m/z 42 and m/z 43) ions as a function of the collision energy. Green line with triangles is for boron dioxide. Purple line with full circles is for boron monoxide with the contribution of the isobaric fragment $C_2H_2^-$. The dashed line represents the branching ratio of boron monoxide without the $C_2H_2^-$ contribution and was obtained out of the isotopic ratio of ^{11}B and ^{10}B.

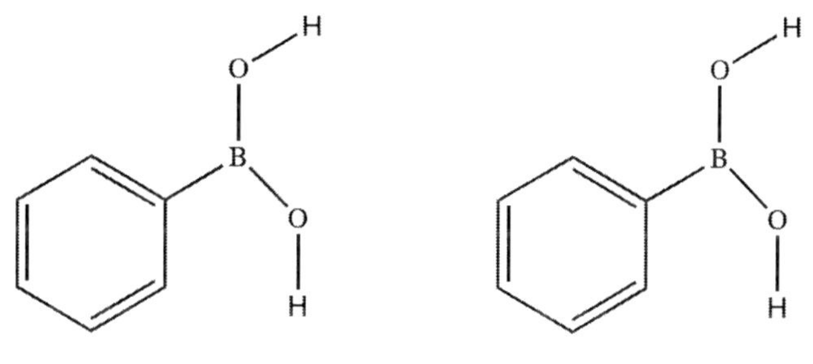

a b

Figure 4. a) Schematic bond rupture leading to BO^- formation; b) schematic bond rupture leading to BO_2^- formation.

Table 1. Assignment of the negative ions formed in potassium collisions with phenyl boronic acid

Anion	Mass (a. m. u.)
$C_6H_4BO^-$	103
$C_5H_4{}^{11}BO^-$	91
$C_5H_4{}^{10}BO^-$	90
$C_5H_2{}^{11}BO^-$ / $C_6H_6{}^{11}B^-$	89
$C_6H_5^-$	77
$C_6H_3^-$	75
$C_3H_3{}^{11}BO^-$	66
$C_5H_5^-$	65
$C_5H_4^-$ / $C_3H^{11}BO^-$ / $C_4H_5{}^{11}B^-$	64
$C_5H_3^-$ / $C_3H^{10}BO^-$ / $C_4H_5{}^{10}B^-$	63
$C_2{}^{11}BO^-$ / $C_3H_3{}^{11}B^-$	51
$C_2{}^{10}BO^-$ / $C_3H_3{}^{10}B^-$	50
$C_3H_2{}^{11}B^-$	49
$^{11}BO_2H_3^-$	46
$^{11}BO_2^-$	43
$^{10}BO_2^-$	42
$C^{11}BOH^-$ / $C_3H_4^-$	40
$C^{11}BO^-$ / $C_3H_3^-$	39
$C_2H_3{}^{11}B^-$ / $C_3H_2^-$	38
$C_2H_2{}^{11}B^-$ / C_3H^-	37
O_2^-	32
CH_2O^-	30
$^{11}BO^-$	27
$^{10}BO^-$ and $C_2H_2^-$	26
C_2H^-	25
$CH^{11}B^-$ / C_2^-	24
OH^-	17
O^-	16
CH_2^-	14
CH^-	13
C^-	12

3.2 Other boron containing fragments

Apart from boron monoxide and boron dioxide, twenty-one other boron-containing negative ions have been assigned. The heavier fragment formed is assigned to $C_6H_4BO^-$ m/z 103; this is the only fragment that does not result from the opening of the ring. In the unimolecular decomposition, the reaction leading to $C_6H_4BO^-$ formation may yield $H_2O + H^\bullet$, or $OH^\bullet + H_2$ (or even $OH^\bullet + H + H$). Given that such anion yield is particularly low (see figure 2), its isotopic anion at mass 102 m/z is not discernible above the background noise. The fragments m/z 91 and m/z 90 were assigned to $C_5H_4{}^{11}BO^-$ and $C_5H_4{}^{10}BO^-$.

Formation of such fragment anions requires the opening of the ring with loss of CH + H_2O. The fragments m/z 89 and m/z 66 were assigned to $C_5H_2{}^{11}BO^-$ and $C_3H_3{}^{11}BO^-$ and no traces of the isotopic contributions were detected. However, for the latter, such formation requires ring opening leading to the loss of three carbon atoms and four hydrogen atoms, as well as the loss of one oxygen atom. Fragment anion m/z 64 was assigned to $C_3H^{11}BO^-$ / $C_4H_5{}^{11}B^-$ while m/z 63 to the respective isotopic forms ($C_3H^{10}BO^-$ / $C_4H_5{}^{10}B^-$). Although the ToF mass intensity of the latter anion is discernible from the background noise it also shows the isotopic contribution of $C_3H^{11}BO^-$ / $C_4H_5{}^{11}B^-$, the current mass resolution it is not sufficient to unambiguously assign such isobaric contributions. The same rational applies to anions m/z 51 and m/z 50, $C_2{}^{11}BO^-$ and $C_3H_3{}^{11}B^-$ and the corresponding isotopic contributions $C_2{}^{10}BO^-$ and $C_3H_3{}^{10}B^-$. Fragment anion m/z 49 was assigned to $C_3H_2{}^{11}B^-$; the mechanism leading to the formation of such anion implies the rupture of both B-O bonds. The negative ion m/z 46 was assigned to ${}^{11}BO_2H_3{}^-$; the reaction producing this anion involves the break of the B-C bond and an H transfer from the phenyl ring to the BO_2H_2 end . The anionic species m/z 40 and m/z 39 were assigned to $C^{11}BOH^-$ and $C^{11}BO^-$; it is interesting to note that, although these fragments are formed with low intensities, the C-B bond is preserved. Finally, fragments m/z 38 and m/z 37 were assigned to $C_2H_3{}^{11}B^-$ and $C_2H_2{}^{11}B^-$, where the C-B bond is preserved against the B-O bond. The lightest boron-containing fragment anion has been assigned to $CH^{11}B^-$, m/z 24.

3.3 Other relevant fragments

Molecular oxygen anion, m/z 32, is formed for collision energies above 23 eV (available energy), despite its low electron affinity (0.4480±0.0060 eV) [32]. The branching ratio of this anion amounts to 10% of all fragments formed. Fragment C_2H^- (m/z 25) is observed for the available energy above 23 eV with a branching ratio consistently between 8% and 11% of the total fragmentation pattern. Fragment $C_2H_2^-$ (m/z 26) is isobaric with BO^-; the contribution of these anions to the m/z 26 yield is observed for available energy above 100 eV, when the ratio between isotopic fragments increases, indicating that m/z 26 results also from the collision induced dissociation mechanism that opens a new channel leading to $C_2H_2^-$ formation. Boron-free fragments $C_5H_5^-$ (m/z 65) and $C_5H_4^-$ (m/z 64) are formed through ring opening, the former via rupture of just two C-C bonds, the latter through an additional C-H bond excision. Finally, we note that $C_5H_3^-$ (m/z 63) may also be related to two isobaric contributions, viz. C_3HBO^- and/or $C_4H_5B^-$ but due to the limited mass resolution it is not possible to unambiguously assign such anions, as well as the fact that there is no information about the thresholds of formation of such isobaric contributions.

3.4 Other fragments

The $C_6H_5^-$ and $C_6H_3^-$ ions are the ring counterpart anions to $B(OH)_2$ and $B(OH)_2 + H_2$ or $B(OH)_2 + H +$ H. Although the intensities of these two fragments are not particularly relevant in comparison with other anions, they are formed through the excision of the B-C bond, and in the case of $C_6H_3^-$ via releasing of two extra hydrogen atoms. The electron affinities of C_6H_5, BO and BO_2 are (1.0960 ± 0.0060 eV) [33], (2.510 ± 0.015 eV) [31] and (4.460 ± 0.030 eV) [31], respectively. Several low intensity fragment anions have been assigned to: $C_3H_4^-$, $C_3H_3^-$, $C_3H_2^-$, C_3H^- (m/z 40, m/z 39, m/z 38, m/z 37, respectively). These anions are isobaric with boron containing $C^{11}BOH^-$ and $C^{11}BO^-$, in the case of m/z 40 and m/z 39, and isobaric with $C_2H_3{}^{11}B^-$ and $C_2H_2{}^{11}B^-$ in the case of m/z 38 and m/z 37; however, due to their low yields in the ToF mass spectra it is difficult to precisely assign the corresponding anion. Fragment m/z 30 was assigned to CH_2O^- and m/z 25 to CH_2^-. The anion m/z 24 was assigned to C_2^-, although it can also be isobaric with $CH^{11}B^-$. Other lighter fragments have been assigned to OH^-, O^-, CH_2^-, CH^- and C^- with m/z 17, m/z 16, m/z 14, m/z 13 and m/z 12, respectively.

4. Conclusions

The present study describes negative ion formation in collisions of potassium atoms with phenylboronic acid. The ToF mass spectra show a very rich fragmentation pattern leading to the formation of several different boron-containing anions, mainly boron oxides (BO^- and BO_2^-). In the present work, it is shown that the energetic threshold of formation of these anions lies at a few eV and the trending of the branching ratios with the collision energy was analysed. The main dissociation pathways for available energies below 100 eV lead to the formation of BO^-, but for energies above 100 eV these dissociation pathways are supressed, resulting in the predominant formation of BO_2^-.

Acknowledgments
MM and FFS acknowledge the Portuguese National Funding Agency FCT-MCTES through PD/BD/106038/2015 and researcher position IF-FCT IF/00380/2014, respectively, and together with PLV the Research Grants UID/FIS/00068/2019 (CEFITEC), PTDC/FIS-AQM/31215/2017 and PTDC/FIS-AQM/31281/2017. This work was also supported by Radiation Biology and Biophysics Doctoral Training Programme (RaBBiT, PD/00193/2012); UID/Multi/04378/2013 (UCIBIO). G.G. acknowledges partial support from the Spanish Ministerio de Ciencia, Innovación y Universidades, Project No. FIS2016-80440.

References
[1] Martin F, Burrow P D, Cai Z, Cloutier P, Hunting D and Sanche L 2004 DNA strand breaks induced by 0-4 eV electrons: The role of shape resonances *Phys. Rev. Lett.* **93** 068101
[2] Boudaïffa B, Cloutier P, D Hunting D, Huels M A and L S 2000 Resonant Formation of DNA Strand Breaks by Low-Energy (3 to 20 eV) Electrons *Science* **287** 1658
[3] Denifl S, Ptasińska S, Hanel G, Gstir B, Probst M, Scheier P and Märk T D 2004 Electron attachment to gas-phase uracil. *J. Chem. Phys.* **120** 6557
[4] Simons J 2006 Electrons Cause DNA-Strand Breaks ? *Acc. Chem. Res.* **39** 772
[5] Denifl S, Ptasińska S, Probst M and Hrus J 2004 Electron Attachment to the Gas-Phase DNA Bases Cytosine and Thymine *J. Phys. Chem. A* **108** 6562
[6] Denifl S, Flosadóttir H D, Edtbauer A., Ingólfsson O, Märk T D and Scheier P 2010 A detailed study on the decomposition pathways of the amino acid valine upon dissociative electron attachment *Eur. Phys. J. D* **60** 37
[7] Denifl S, Mähr I, Ferreira da Silva F, Zappa F, Märk T D and Scheier P 2008 Electron impact ionization studies with the amino acid valine in the gas phase and (hydrated) in helium droplets *Eur. Phys. J. D* **51** 73
[8] dos Santos J S, da Costa R F and Varella M T D N 2012 Low-energy electron collisions with glycine. *J. Chem. Phys.* **136** 084307
[9] Ptasińska S, Denifl S, Grill V, Märk T, Illenberger E and Scheier P 2005 Bond- and Site-Selective Loss of H- from Pyrimidine Bases *Phys. Rev. Lett.* **95** 093201
[10] Prabhudesai V S, Kelkar A H, Nandi D and Krishnakumar E 2005 Functional group dependent site specific fragmentation of molecules by low energy electrons *Phys. Rev. Lett.* **95** 143202
[11] Neustetter M, Aysina J, Ferreira da Silva F and Denifl S 2015 The Effect of Solvation on Electron Attachment to Pure and Hydrated Pyrimidine Clusters *Angew. Chemie Int. Ed.* **54** 9124
[12] Ferreira da Silva F, Denifl S, Maerk T D, Doltsinis N L, Ellis A M and Scheier P 2010 Electron Attachment to Formamide Clusters in Helium Nanodroplets *J. Phys. Chem. A* **114** 1633
[13] Ferreira da Silva F, Jaksch S, Martins G, Dang H M, Dampc M, Denifl S, Maerk T D, Limao-Vieira P, Liu J, Yang S, Ellis A M and Scheier P 2009 Electron attachment and electron ionization of acetic acid clusters embedded in helium nanodroplets *Phys. Chem. Chem. Phys.* **11** 11631
[14] Ferreira da Silva F, Denifl S, Maerk T D, Ellis A M and Scheier P 2010 Electron attachment to amino acid clusters in helium nanodroplets: Glycine, alanine, and serine *J. Chem. Phys.* **132** 214306
[15] Ferreira da Silva F, Bartl P, Denifl S, Maerk T D, Ellis A M and Scheier P 2010 Formation of the Magic L-Serine Octamer in Helium Nanodroplets *Chem. Phys. Chem.* **11** 90

[16] Kopyra J, Koenig-Lehmann C, Bald I and Illenberger E 2009 A single slow electron triggers the loss of both chlorine atoms from the anticancer drug cisplatin: Implications for chemoradiation therapy *Angew. Chemie - Int. Ed.* **48** 7904
[17] Ferreira da Silva F, Márcio M T, Jones N C, Vrønning Hoffmann S, Denifl S, Bald I and Kopyra J 2019 Electron-Induced Reactions in 3-Bromopyruvic Acid *Chem. - A Eur. J.* **25** 5498
[18] Almeida D, Ferreira da Silva F, García G and Limão-Vieira P 2013 Selective bond cleavage in potassium collisions with pyrimidine bases of DNA *Phys. Rev. Lett.* **110** 023201
[19] Ferreira da Silva F, Lança M, Almeida D, García G and Limão-Vieira P 2012 Anionic fragmentation of glycine upon potassium-molecule collisions *Eur. Phys. J. D* **66** 78
[20] Ferreira da Silva F, Rafael J, Cunha T, Almeida D and Limão-Vieira P 2014 Electron transfer to aliphatic amino acids in neutral potassium collisions *Int. J. Mass Spectrom.* **365–366** 238
[21] Ferreira da Silva F, Meneses G, Ingólfsson O and Limão-Vieira P 2016 Side chain effects in reactions of the potassium-tyrosine charge transfer complex *Chem. Phys. Lett.* **662** 19
[22] Ferreira da Silva F, Almeida D, Antunes R, Martins G, Nunes Y, Eden S, Garcia G, Limao-Vieira P, Ferreira da Silva F and Limão-Vieira P 2011 Electron transfer processes in potassium collisions with 5-fluorouracil and 5-chlorouracil. *Phys. Chem. Chem. Phys.* **13** 21621
[23] Ferreira da Silva F, Matias C, Almeida D, Garcia G, Ingolfsson O, Flosadottir H D, Omarsson B, Ptasinska S, Puschnigg B, Scheier P, Limao-Vieira P and Denifl S 2013 NCO-, a Key Fragment Upon Dissociative Electron Attachment and Electron Transfer to Pyrimidine Bases: Site Selectivity for a Slow Decay Process *J. Am. Soc. Mass Spectrom.* **24** 1787
[24] Ban H S and Nakamura H 2015 Boron-based drug design *Chem. Rev.* **15** 616
[25] Santos F M F, Rosa J N, Candeias N R, Carvalho C P, Matos A I, Ventura A E, Florindo H F, Silva L C, Pischel U and Gois P M P 2016 A Three-Component Assembly Promoted by Boronic Acids Delivers a Modular Fluorophore Platform (BASHY Dyes) *Chem. - A Eur. J.* **2** 1631
[26] Mader H S and Wolfbeis O S 2008 Review Boronic acid based probes for microdetermination of saccharides and glycosylated biomolecules *Microchim Acta* **162** 1
[27] Fu H, Fang H, Sun J, Wang H, Liu A, Sun J and Wu Z 2014 Boronic Acid-based Enzyme Inhibitors: A Review of Recent Progress *Curr. Med. Chamistry* **21** 3271
[28] Cambre J N and Sumerlin B S 2011 Biomedical applications of boronic acid polymers *Polymer (Guildf).* **52** 4631
[29] Antunes R, Almeida D, Martins G, Mason N J, Garcia G, Maneira M J P, Nunes Y and Limão-Vieira P 2010 Negative ion formation in potassium-nitromethane collisions. *Phys. Chem. Chem. Phys.* **12** 12513
[30] Mendes M, Pamplona B, Kumar S, Ferreira da Silva F, Aguilar A, García G, Bacchus-Montabonel M-C and Limao-Vieira P 2019 Ion-Pair Formation in Neutral Potassium-Neutral Pyrimidine Collisions: Electron Transfer Experiments *Front. Chem.* **7** 1
[31] Zhai H-J, Wang L-M, Li S-D and Wang L-S 2007 Vibrationally Resolved Photoelectron Spectroscopy of BO- and BO2-: A Joint Experimental and Theoretical Study *J. Phys. Chem. A* **111** 1030
[32] Ervin K M, Anusiewicz I, Skurski P, Simons J and Lineberger W C 2003 The Only Stable State of O2- Is the X 2Πg Ground State and It (Still!) Has an Adiabatic Electron Detachment Energy of 0.45 eV *J. Phys. Chem. A* **107** 8521
[33] Guniona F R, Gillesa K M, Polaka L M and Lineberger W C 1992 Ultraviolet photoelectron spectroscopy of the phenide, benzyl and phenoxide anions, with ab initio calculations *Int. J. Mass Spectrom. Ion Process.* **117** 601

Resonances in molecules and molecular clusters

Jimena D Gorfinkiel
School of Physical Sciences, The Open University, Milton Keynes, MK7 6AA, UK
E-mail: J.Gorfinkiel@open.ac.uk

Abstract. Resonances play an important role in a number of atomic and molecular processes. Identifying and characterising resonances in electron scattering is essential as they can both enhance a number of processes (e.g. electronic and vibrational excitation) and are crucial in others like dissociative electron attachment and dissociative recombination. We discuss recent theoretical studies of shape and core-excited resonances, both in isolated molecules of biological relevance and in small molecular clusters. The latter are investigated to understand the effect of the environment, in particular hydration, in electron collisions in biological matter.

1. Introduction

Temporary anion states (also known as resonances) can enhance most electron-induced processes. More interestingly, they can be chemistry initiators by leading to the break-up of the molecule into at least one reactive species. Resonances are also important in other molecular processes, for example, photodetachment and associative detachment. Theoretical research on their identification and characterization is therefore an active area particularly, but not uniquely, associated to the understanding of electron-induced processes in molecules. Besides their fundamental interest, resonance mediated or enhanced processes are relevant in a number of applied fields, from the modelling of radiation for medical uses to focused electron beam induced deposition and nanofabrication [1].

Scattering methods are an effective tool for the accurate description of resonances [1]: the modelling of shape resonances can be done using several of these approaches, both for small and larger molecules. Core-excited resonances are harder to model accurately, because of the importance of correlation in their description. However, both types of resonances (and vibrational Feshbach resonances) play an important role in, among others, biological radiation induced damage and electron transfer reactions.

Recent joint theoretical and experimental work on pyrimidine [2] and thiophene [3] using electron energy loss spectroscopy [4] and the R-matrix method [5] has confirmed the ability of this theoretical approach to accurately describe core-excited resonances in mid-size molecules in the gas phase.

Nonetheless, a lot of the scattering processes of applied relevance happen in a condensed environment, where the target molecule is surrounded by other molecules. In particular, in cellular environments, biomolecules (DNA, proteins, etc.) are surrounded by water. To bridge the gap between our understanding of resonance formation in molecules in the gas phase and that in a condensed environment, molecular clusters are being investigated, both theoretically and experimentally [6].

Here, we briefly describe results obtained with the R-matrix method both for isolated molecules (pyrimidine) and small molecular clusters.

2. R-matrix method

The R-matrix method [5] is based on the separation of space into an inner region, containing the electronic density of all the target (molecule or cluster) electronic states of interest and an outer region where the scattering electron is distinguishable from the bound electrons. A sphere of radius a, known as the R-matrix sphere, separates both regions.

In the inner region, an accurate description of exchange and correlation is necessary and the wave functions describing the N+1-electron system (N-electron target + scattering electron) can be written as:

$$\Psi_k^{N+1} = \mathcal{A}\sum_{i=1}^{n}\sum_{j=1}^{n_c}\Phi_i(\mathbf{x}_N;\hat{r}_{N+1};\sigma_{N+1})\frac{u_{ij}(r_{N+1})}{r_{N+1}}a_{ijk} + \sum_{i=1}^{m}\chi_i(\mathbf{x}_{N+1})b_{ik} \tag{1}$$

where $\mathbf{x}_N$ and $\mathbf{x}_{N+1}$ stand for the spin-space coordinates of all N/N+1 electrons, respectively. r_{N+1} and $\hat{\mathrm{r}}_{N+1}$ correspond to the radial and angular coordinates of the scattering electron respectively and σ_{N+1} stands for its spin. The wave function Φ_i describes the ith electronic state of the target, as well as the angular and spin behaviour of the scattering electron. The functions $\frac{u_{ij}(r_{N+1})}{r_{N+1}}$ describe the radial behaviour of the scattering electron while the L^2-integrable functions χ_i are crucial for a good description of the short-range polarisation-correlation effects. $\mathcal{A}$ is the antisymmetrization operator, and the coefficients a_{ijk} and b_{ik} are determined by the requirement that the functions Ψ_k^{N+1} diagonalise the N+1 non-relativistic Hermitian Hamiltonian in the inner region [5].

In the outer region, the interaction between the scattering electron and the target molecule can be described using a single-centre multipole potential expansion. The basis functions Ψ_k^{N+1} and their associated eigenvalues are used to construct the R-matrix at the boundary between the regions. This R-matrix is propagated to an asymptotic distance where, by matching with known asymptotic expressions, the K-matrix can be determined. Resonance energies and lifetimes can then be obtained in several ways, in particular: (i) by fitting the eigenphase sum, $\delta_{sum}(E)$, obtained from diagonalizing the K-matrix using the well known Breit-Wigner expression:

$$\delta_{sum}(E) = \delta_r + \delta_{bg} = -\arctan\frac{\Gamma/2}{E - E_r} + \delta_{bg} \tag{2}$$

where E_r and Γ are the energy and width of the resonance respectively (and therefore its lifetime is $\tau = 1/\Gamma$); (ii) by calculating the S-matrices and, from them, the time-delay matrices:

$$\mathbf{Q}(E) = i\hbar\mathbf{S}\frac{d\mathbf{S}}{dE} \tag{3}$$

Peaks in the largest eigenvalues of $\mathbf{Q}(E)$ can be fitted to Lorentzian functions that depend on the resonance energy and width thus enabling their determination.

3. Results

Use of the R-matrix method and the UKRmol/UKRmol+ [7, 8] suites has enabled the investigation of shape and core-excited resonances in small and medium size molecules as well as the study of shape resonances in molecular clusters.

Below, we show two illustrative examples: pyrimidine, in which most core-excited resonances identified have been confirmed experimentally and thymine-$(H_2O)_5$ clusters, in which we have investigated the properties of hydration effects on resonance formation.

3.1. Core-excited resonances in pyrimidine

The nucleobases are one of the DNA constituents more widely studied, both experimentally and theoretically [9, 6]. However, whereas several works measured the anion yields due to dissociative electron attachment (DEA) both in the lower energy range, where shape resonances are responsible for the DEA, and at higher-energies, where core-excited resonances play an important role, not a lot is known about these core-excited resonances. The R-matrix approach has been applied to their study in both pyrimidinic and purinic DNA bases [10, 11, 12, 13] and (some of) the core-excited resonances of uracil have been investigated with other methodologies [14]. However, due to the lack of experimental information on the resonances themselves and the difficulty of linking one in a usually fairly dense spectrum of resonances with specific DEA products, it is difficult to ascertain the accuracy of the computational results.

A more satisfactory situation is found for pyrimidine, a model for pyrimidinic nucleobases: electron energy loss (EEL) spectra of it are available and enable a comparison with sophisticated R-matrix calculations. The comparison [2] shows excellent agreement between the resonances described by the calculation and those visible in the EEL spectra, if one accounts for the shift to higher energies in the computational results. This shift is due to an incomplete representation of the polarization effects in the R-matrix close-coupling calculations: for highly polarizable molecules (the spherical polarizability of pyrimidine is over 60 a_0^3) the standard close-coupling expansion is insufficient to fully describe the polarizing effect of the scattering electron. The amount of polarization described can not be quantified in this approach, but a qualitative indication of how well it is described can be obtained from comparing calculated shape resonance positions to experimental values. Use of pseudostates is likely to resolve the problem of underrepresentation [15, 16] but, until very recent software developments [8], the size of the calculations made them unfeasible.

Of all the resonances identified and characterized in the R-matrix calculations, 9 manifest themselves in the cross sections for exciting those states of pyrimidine that contribute to the energy losses ΔE=4.26 eV and ΔE=5.2 eV for which we have performed a comparison. Of these 9, two are not seen in the excitation cross sections, but the other 7 are; these results are summarized in table 1. The resonances that are not visible correspond to smaller peaks that would be 'washed out' in the experiment due to a broad Franck-Condon envelope. We note that these resonances, unlike all the others, are described by configurations in which an electron from a non-bonding orbital has been promoted to one of the lower π^* orbitals [2].

This kind of agreement between calculated and measured core-excited resonances has also been found for thiophene [3]: for this target, the cross sections for excitation into the first and second excited states calculated for two scattering angles (90° and 135°) agree very well with the EEL data and four core-excited resonances are clearly observed in both. These results demonstrate the capability of the R-matrix approach to model core-excited shape resonances in mid-size molecules of biological interest.

3.2. Effect of hydration

In order to understand the effect of hydration on resonance formation, we performed two sets of detailed investigations: ones for clusters of pyridine and another for thymine both with up to 5 water molecules. The aim was to gain insight into how the low energy π^* shape resonances in these molecules are affected by the presence of a few water molecules in their environment.

There were two main purposes of these studies: (i) to confirm, for larger molecules, the findings of Freitas *et al.* [18] in their study of HCOOH clusters with one and two water molecules; (ii) to extend the investigation to larger clusters. We note that whereas in the case of thymine additional water molecules will hydrogen-bond to the thymine itself, in the case of pyrimidine, these molecules hydrogen-bond to one another at the nitrogen 'end' of the molecule. This difference enabled us to investigate different effects [19, 20].

Table 1. Positions and calculated widths when available [17] (both in eV) of the core-excited resonances of pyrimidine identified in the calculations and the EEL spectra for Ranges I and II (see details in Regeta *et al.* [2]); note that the first resonance listed is actually of mixed shape and core-excited character.

Symmetry	Calculated E	Calculated Γ	EEL
$^2B_1$	4.78	0.38	4.35
$^2A_1$	5.96	0.18	-
$^2A_1$	6.15	0.18	-
$^2A_2$	6.11	0.51	5.55
$^2B_1$	6.37	0.58	5.55
$^2B_1$	7.11	0.48	6.52
$^2A_2$	7.33	0.43	6.52
$^2B_1$	8.47	1.69	7.45
$^2B_1$	12.3	-	10.3

The calculations presented were performed at static-exchange level to avoid uncertainty in the relative description of polarization effects. Nonetheless, calculations were also performed at static-exchange plus polarization level and the results obtained were qualitatively the same. The cc-pVDZ basis set was used to generate Hartree-Fock orbitals; due to the size of some of the clusters, an R-matrix radius of 18 a_0 was necessary. The geometries of the thymine clusters investigated were obtained from that determined by Smyth *et al.* [21] for the thymine-$(H_2O)_5$ cluster using sophisticated DFT calculations.

We interpreted the effects of hydration as arising from two distinct contributions, that we labelled indirect and direct effects. The indirect effects are due to the *distortion of the molecular geometry* caused by the presence of the water molecules; the direct effect is that due to the *presence* of the water molecules itself. In other words: the indirect effect can be computed in calculations by investigating the molecule of interest isolated, first in its equilibrium geometry and then in the geometry it adopts in the cluster; the direct effect is computed by comparing the resonances of the isolated molecule *in the cluster geometry* and those in the clusters themselves. The total effect of hydration is the sum of these two effects. We note that the effects can both go in the same direction (e.g. both lowering the energy of the resonance or both raising it) or not. In addition, the direct and indirect effects, as well as the total effect depend (weakly, for these systems) on the site to which the water molecule attaches via hydrogen bonding.

Figure 1 shows the time-delay for isolated thymine together with that for 5 different thymine-H_2O clusters. The geometries of the clusters correspond to taking the lowest energy geometry of thymine-$(H_2O)_5$ in its electronic ground state (shown in figure 2) and removing four of the water molecules. Three well defined peaks are visible in the time-delay: the two lower ones (for all targets) correspond to the first and second π^* resonances of thymine, of pure shape character. The third one corresponds to the third π^* resonance, that is of mixed shape core-excited shape character (and, for this reason, given that our calculation does not describe inelastic channels, is excluded from the analysis).

Two of the dimers (A and D) involve water being the hydrogen acceptor in the bond: in agreement with Freitas *et al*'s conclusions, these lead to a 'destabilization' of the resonances, i.e. a shift to higher energies. Conversely, dimers B, C and E correspond to situations in which water is the hydrogen donor. In this case, the resonances are 'stabilized', i.e. shifted to lower energies.

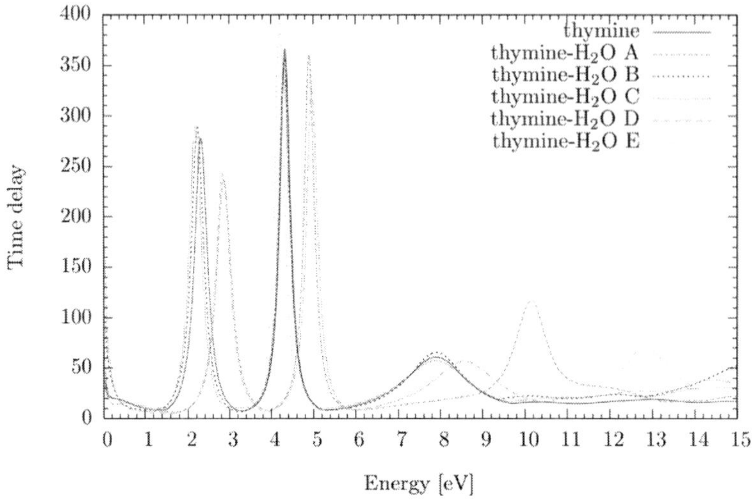

Figure 1. Largest eigenvalue of the time-delay matrix showing the first and second shape π^* resonances as well as the third mixed character π^* one in isolated thymine (red line) and 5 different thymine-H_2O clusters. The clusters are labelled according to which single H_2O molecule in the thymine-$(H_2O)_5$ cluster is taken into consideration. Note that the yellow line corresponding to cluster E is hardly visible under the red line and that the lines for clusters A and D overlap almost fully for the first two peaks.

The destabilizing shift is bigger than the stabilizing one. In addition, small differences can be observed between the resonance position for different geometries of the cluster corresponding to donor/acceptor situations: this shows that there is a (small) effect due to the specific site (atom) the water molecule binds to.

We have also observed that the effect of adding more water molecules, at least those that will form a hydrogen bond with the target of interest, is qualitatively additive; that is, it is possible (in general) to correctly predict whether resonances will be stabilized or destabilized but the shifts due to individual water molecules don't add up to the total shift. The prediction from figure 2 is therefore that the combined effect of the 5 water molecules would be small: the addition of two destabilizing water molecules and three stabilizing ones (whose effect is smaller), should lead to small shifts in the resonances towards lower energies. Indeed, the resonance shifts in thymine-$(H_2O)_5$ are 0.075 and 0.29 eV for the first and second π^* resonance respectively. Interestingly, the fact that one of the shifts is more than three times bigger than the other is not obvious from the shifts that occur when individual water molecules are present.

4. Conclusions

Theoretical and, more significantly, computational improvements over the last decades, have enable the accurate investigation of shape and core-excited resonances in molecules of biological relevance. Comparison with EEL spectra confirm that core-excited shape resonances in pyrimidine are well described when sophisticated models are used in R-matrix calculations. This paves the way for the study of DEA via these resonances, for which little is known beyond the measurement of anion production.

The effect of an aqueous environment on resonance formation and DEA can also be studied computationally, but in a less accurate manner. In this paper, we show that it is possible to calculate resonance positions (and widths) in an *ab initio* fashion and that the calculations enable

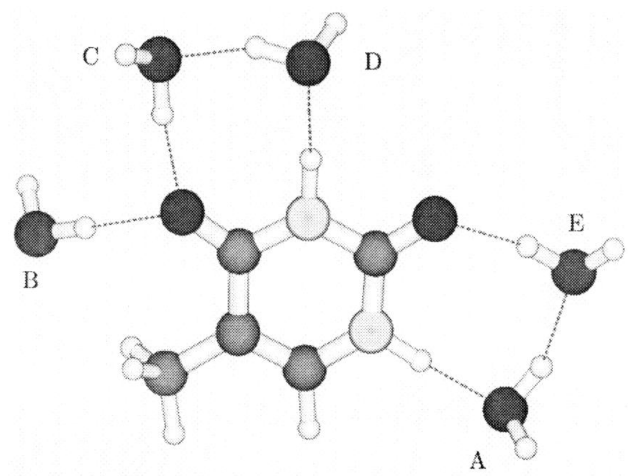

Figure 2. Lowest-energy geometry of the thymine-$(H_2O)_5$ cluster as determined by Smyth *et al.* [21]. The bottom right-hand molecule corresponds to 'A' and the labelling continues clockwise from it.

us to gain insight into the effect of hydrogen bonding on resonances. This insight could help us predict how resonances will be affected for systems for which calculations are still unfeasible.

Nonetheless, it is important to highlight that the microhydration effects on resonance formation do not provide the whole picture. Comparison with DEA experiments indicate that knowledge of the changes in resonance characteristics are not sufficient to provide even a correct qualitative picture of the effect on DEA. Recent experiments have shown this for the case of hydrogen loss from uracil and thymine in small clusters where experiments identified a complete quenching of this DEA channel at low energies upon hydration [22]. The calculations demonstrate that it is the presence of water that hinders H loss from hydrated nucleobases [23], and that this is not a resonance-related effect.

Acknowledgements

I gratefully acknowledge the contribution of many collaborators to the work presented in this paper, in particular Drs. Z. Mašín, A. Sieradzka, K. Regeta and Prof. M. Allan and Ms. Alexandra Loupas. This work used the ARCHER UK National Supercomputing Service (http://www.archer.ac.uk).

References

[1] Ingólfsson O 2019 *Low-Energy Electrons: Fundamentals and Applications* (Jenny Stanford Publishing)
[2] Regeta K, Allan M, Mašín Z and Gorfinkiel J D 2016 *J. Chem. Phys.* **144** 024302
[3] Loupas A, Regeta K, Allan M and Gorfinkiel J D 2018 *J. Phys. Chem.* A **122** 1146
[4] Regeta K and Allan M 2015 *Phys. Rev.* A **91** 012707
[5] Tennyson J 2010 *Phys. Rep.* **491** 29
[6] Gorfinkiel J D and Ptasińska S 2017 *J. Phys. B: At. Mol. Opt. Phys.* **50** 182001
[7] Carr J M, Galiatsatos P G, Gorfinkiel J D, Harvey A G, Lysaght M A, Madden D, Mašín Z, Plummer M, Tennyson J and Varambhia H N 2012 *Eur. Phys. J.* D **66** 20653
[8] Mašín Z, Benda J, Gorfinkiel J D, Harvey A G and Tennyson J *Comp. Phys. Commun.* submitted
[9] Baccarelli I, Bald I, Gianturco F A, Illenberger E and Kopyra J 2011 *Phys. Rep.* **508** 1
[10] Dora A, Tennyson J, Bryjko L and van Mourik T 2009 *J. Chem. Phys.* **130** 164307

[12] Dora A, Bryjko L, van Mourik T and Tennyson J 2012 *J. Phys. B: At. Mol. Opt. Phys.* **45** 175203
[13] Dora A, Bryjko L, van Mourik T and Tennyson J 2012 *J. Chem. Phys.* **136** 024324
[14] Fennimore M A and Matsika S 2016 *Phys. Chem. Chem. Phys.* **18**(44) 30536–30545
[15] Gorfinkiel J D and Tennyson J 2004 *J. Phys. B: At. Mol. Opt. Phys.* **37** L343
[16] Gorfinkiel J D and Tennyson J 2005 *J. Phys. B: At. Mol. Opt. Phys.* **38** 1607–1622
[17] Mašín Z and Gorfinkiel J D 2012 *J. Chem. Phys.* **137** 204312
[18] Freitas T C, Coutinho K, Varella M, Lima M A P and Canuto S 2013 *J. Chem. Phys.* **138**
[19] Sieradzka A and Gorfinkiel J D 2017 *J. Chem. Phys.* **147** 034302
[20] Sieradzka A and Gorfinkiel J D 2017 *J. Chem. Phys.* **147** 034303
[21] Smyth M, Kohanoff J and Fabrikant I I 2014 *J. Chem. Phys.* **140** 184313
[22] Kočišek J, Pysanenko A, Fárník M and Fedor J 2016 *J. Phys. Chem. Lett.* **7** 3401
[23] McAllister M, Kazemigazestane N, Henry L, Gu B, Fabrikant I, Tribello G and Kohanoff J 2019 *J. Phys. Chem.* B **123** 1537

Selective bond scission in formic acid by low-energy electrons

Brandon Griffin,[1] Ali Moradmand,[2] Joshua Williams,[1] Ali Belkacem,[3] Thomas N Rescigno,[3] Cynthia Trevisan,[2] C William McCurdy,[3,4] Thorsten Weber[3] and Daniel S Slaughter[3]

[1]Department of Physics, University of Nevada Reno, Reno, Nevada 89557, USA
[2]Department of Sciences and Mathematics, California Maritime Academy, Vallejo, California 94590, USA
[3]Chemical Sciences Division, Lawrence Berkeley National Laboratory, 1 Cyclotron Rd, Berkeley, CA 94556, USA
[4]Department of Chemistry, University of California, Davis, CA 95616

E-mail: DSSlaughter@lbl.gov

Abstract. We report recent results of mass-resolved anion fragment momentum imaging experiments to investigate dissociative electron attachment to formic acid, for incident energies between 5 eV and 9 eV. A remarkable site-selectivity is found for a resonance at 8.5 eV by comparing anion fragment yields for two deuterated isotopologues of formic acid. This results in an H^- fragment from the O-H bond of the transient anion, with negligible contribution from C-H break. In contrast, a lower-energy resonance at 7.1 eV dissociates by C–H or O–H break to produce H^- and the neutral radicals HOCO or HCOO.

1. Introduction

Low energy electrons are well-established to have importance in chemical processes initiated by ionizing radiation [1]. This is because electrons are produced in high abundance with energies in the range for resonances to drive molecular dissociation by inelastic scattering and dissociative electron attachment (DEA). In DEA, the energy of a free electron is efficiently converted into vibrational kinetic energy and dissociation into two or more atomic or molecular fragments [2]. The process necessarily involves a strong coupling between electronic and nuclear motion, resulting in rich nonadiabatic dynamics such as conical intersections [3] or electron transfer [4] between electronic states of the transient anion, that are very difficult, if not impossible to model without detailed experimental data. In molecules with few degrees of symmetry, site selectivity may be observed where dissociation favors one moiety within the transient anion over another, depending on the electron attachment energy.

Formic acid (HCOOH), being the simplest organic acid, is a useful model system in developing a fundamental understanding electron-driven chemistry. It also has several important applications in nature and industry. The first organic acid to be observed in the interstellar medium [5, 6, 7], formic acid continues to attract intense interest as a chemical precursor or an intermediate species in the abiotic synthesis [8, 9] of biologically-relevant molecules. In Earth's atmosphere, the chemistry of formic acid is relevant to solubilities and reaction rates of natural and synthetic pollutants [10]. In energy transport applications, formic acid has been identified as a high-efficiency fuel for direct liquid fuel cells [11].

A low-energy shape resonance in formic acid has been the subject of several investigations, primarily focusing on the mechanism of electron attachment around 1.8 eV. Several theoretical studies produced two possible mechanisms: electron attachment to a σ^* resonance [12, 13, 14, 15] that directly dissociated to $HCOO^-$ + H; or attachment to a π^* resonance [16, 17, 18], followed by C–H bending out-of-plane, to stabilize the anion electronically and couple to a σ^* state to produce $HCOO^-$ + H. The subsequent experimental investigation of Janeckova *et al.* [19] measured a relatively weak dependence of the DEA cross section on substitution of the C–H hydrogen with deuterium, supporting the direct mechanism with no out-of-plane C–H bending. In this work, we examine two different resonances between 5 eV and 9 eV that produce H^-, O^- and OH^-. The resonances have been studied previously by measurements of the mass-resolved ion yields and their incident electron energy-dependence [20, 21, 22, 23]. In this contribution we focus on the site selectivity of DEA leading to H^- loss from the formyl or hydroxyl moieties.

2. Experimental methods

A DEA reaction microscope was used to measure the mass-resolved anion fragment yields and initial momentum after dissociative electron attachment. The technique has already been described elsewhere [24, 3], so only the most important details are included here. The experiments employ an energy-tunable, pulsed electron beam that is directed orthogonally with the time-of-flight axis of a momentum imaging spectrometer. Deuterated formic acid (HCOOD or DCOOH, 98% isotopologue purity) was obtained from Sigma Aldrich and degassed in the vacuum chamber sample manifold by a series of freeze-pump-thaw cycles. Formic acid vapor was admitted into the ultra-high vacuum chamber by heated (80°C) stainless tubing and collimated by a long stainless steel capillary of 0.3 mm inner diameter to form the gas target. The tubing and vacuum apparatus were baked to >120° for 48 hours beforehand to reduce the presence of water below significant concentrations, as determined by the rate of H^- or D^- from H_2O or D_2O, respectively, at the $^2B_1$ resonance energy of 6 eV. The pure formic acid effusive beam was maintained at low pressure conditions, <2 Torr within the tubing upstream of the capillary, to ensure negligible contributions from formic acid dimers [25]. The collision region, defined by the intersection of the molecular and electron beams, was aligned between the first two parallel electrodes of the anion fragment momentum imaging spectrometer. The first electrode of the anion spectrometer was pulsed following a delay from the electron gun pulse, extracting negative ion fragments into the spectrometer at a system repetition rate of 50 kHz. A magnetic field coaxial to the electron gun allowed the separation of anions from the scattered electron background and assisted in the low energy electron beam transport and collimation. The electron beam was characterized, using the sharp increase in O^- from CO_2 around its 4.0 eV DEA threshold, to determine the energy spread of 0.5 eV (full width at half maximum). The electron beam mean energy was calibrated by the same O^- onset curve and determined within ±0.1 eV for each experiment.

The momentum-imaging anion spectrometer produced 12 V/cm anion extraction and acceleration fields, followed by a position-focusing lens to map the initial momentum of anion fragments onto the position and time-sensitive anion detector. The detector consisted of a grid electrode for anion post-acceleration and a pair of 75 mm diameter chevron microchannel plates that amplified each detected particle onto a two-layer delay line anode, allowing the event-by-event acquisition of the 3-dimensional momentum of each ion, encoded in the time and position of each ion hit. Extensive shielding of the detector and spectrometer prevented most of the background scattered electrons from entering the spectrometer or hitting the detector. In the list-mode data record, the 3-dimensional momentum of each detected ion fragment was stored, allowing for both on-the-fly and offline analysis.

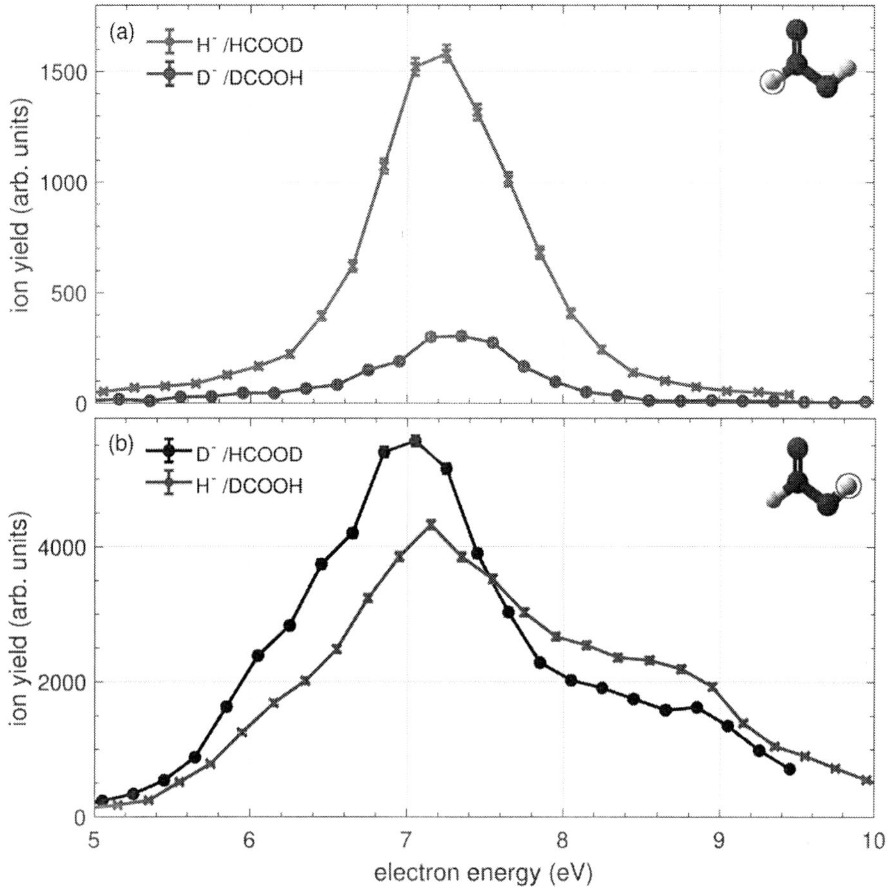

Figure 1. Incident electron energy dependence of relative yields of anion fragments for dissociative electron attachment to deutrated formic acid. (a) C–H break measured by H^- from HCOOD and D^- yield from DCOOH. (b) O–H break measured by H^- from DCOOH and D^- yield from HCOOD. Error bars represent plus/minus one standard deviation and, where they not visible, they are smaller than the data point symbols.

Momentum imaging experiments require several hours of data collection to achieve acceptable statistical uncertainties in the momentum-differential data. To produce mass-resolved anion fragment yields, the same experimental approach was employed for a much shorter time of 120 s per incident electron energy, to reduce the systematic uncertainties originating from instabilities in the molecular beam density. The anion fragment yields are not calibrated to an absolute scale and the relative yields for each of the two target molecules, HCOOD and DCOOH, were measured in different experimental runs under very similar conditions. Therefore the yields for each target are only approximately normalized to the same arbitrary scale, with the scale uncertainty for each target molecule estimated to be $\pm 50\%$.

3. Results

The relative yields of anions produced by C–H and O–H break, measured as a function of incident electron energy are presented in figure 1.

A single narrow peak in the H^- (D^-) yield is measured for C–H (C–D) bond scission, having

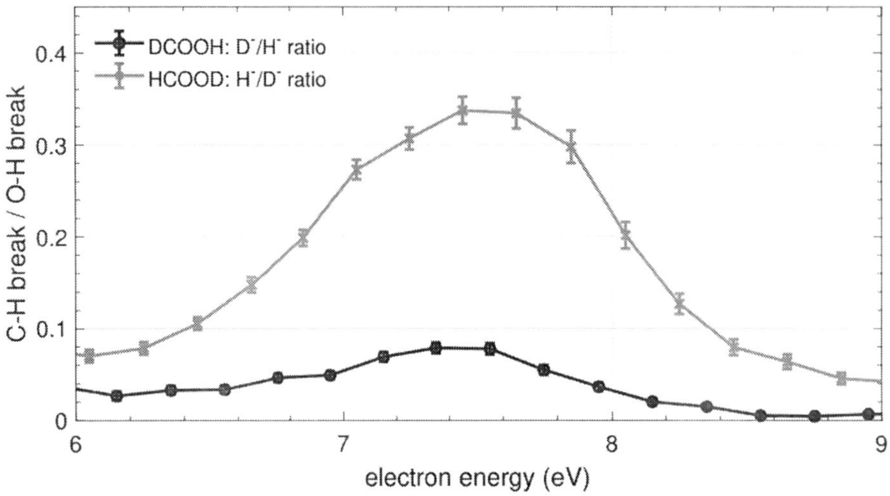

Figure 2. Yield fractions at the 7.1 eV resonance, D^-/H^- fraction from DCOOH and H^-/D^- fraction from HCOOD, showing a higher rate of O–H (or O–D) break compared to C–H (or C–D) break in both formic acid isotopologues.

a width of about 1 eV and symmetric shape about its maximum at 7.1 eV (figure 1 (a)). A peak at the same energy also occurs in the O–H break product yield (figure 1 (b)), however the shape is considerably different and about 0.5 eV broader than the corresponding C–H peak. A higher energy shoulder in the H^- (D^-) yield for O–H (O–D) scission is prominent at about 8.5 eV. Determining the absolute DEA cross sections from the present anion yields is not possible in the present experiments, however the relative yields of anion fragments reveal a significant isotope effect for C–H dissociation. On comparison of the DCOOH data in figures 1 (a) and (b), C–D scission is considerably less-favorable than O–H scission, and this is also the case, to a lesser extent, for HCOOD. This effect is quantified in the relative yields shown in figure 2, where we compare the ratio of C–H to O–H dissociation for each molecule. Specifically, the fractions (C–D break)/(O–H break) for DCOOH and (C–H break)/(O–D break) for HCOOD show that, over the entire energy range, O–H break is favored over C–D break, and O–D break is favored over C–H break. Near the 7.1 eV resonance, C–H break contributes to a fraction of about 1/3 to the total ion yield for formic acid deuterated at the hydroxyl moiety (HCOOD), C–D break contributes to 1/12 of the total ion yield for the formyl-deuterated isotopologue DCOOH.

To better identify the transient anion resonances associated with the incident electron energy-dependent structures in the DEA product yields, we examine the measured three-dimensional momentum images in figure 3 for the fragment anions produced at a few relevant incident electron energies. Each image displays the momentum of D^-, originating from the site indicated by the inset cartoon of each panel, in the laboratory frame defined by the electron beam direction. The longitudinal or transverse coordinates correspond to anion fragment dissociation in the same direction or orthogonal to the electron beam, respectively. To display ions emitted within a uniform solid angle, each image is extracted from the full momentum-space Newton sphere by extracting an inverted conical 30° slice [24].

4. Discussion

The anion yields of figure 1 show structures corresponding to DEA involving at least two distinct transient anion resonances. At least 11 excited electronic states [26] exist between 7.5 eV and

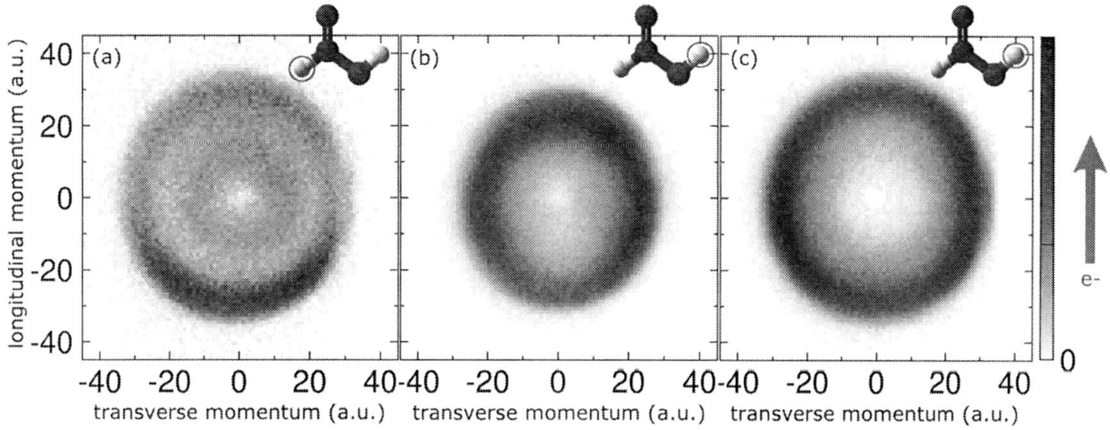

Figure 3. Momentum images of the D^- fragment for formic acid deuterated at (a) the formyl moiety (DCOOH), for 7.6 eV incident electron energy, and the hydroxyl moiety (HCOOD) for (b) 7.2 eV and (c) 8.5 eV incident electron energies. The incident electron direction is shown by an arrow on the right and the units of the linear intensity scale are arbitrary.

9 eV above the ground electronic state of formic acid. Any of them could be considered as possible parents of core-excited shape or Feshbach resonances. The minimum energy geometry of neutral formic acid in the ground electronic state is the *cis* (or *anti*) isomer, and the *trans* (or *syn*) isomer is 169 meV higher in energy [26]. For the moderate molecular beam temperatures of the present experiments, any contributions from the *trans* isomer are expected to be negligible, however other structural transformations may occur in the excited anion states. An investigation of these dynamics will require challenging resonance electronic structure calculations, which are beyond the scope of the present work. The peak yields for C–H and O–H scission are both around 7.1 eV, indicating a common resonance that produces different products, hydrocarboxyl radical (HOCO) and formyloxyl radical (HCOO), respectively, in two distinct reaction channels:

$$\mathrm{HCOOH} + e^- \rightarrow \mathrm{HCOOH}^- \rightarrow \mathrm{H}^- + \mathrm{HOCO} \quad (1)$$

$$\mathrm{HCOOH} + e^- \rightarrow \mathrm{HCOOH}^- \rightarrow \mathrm{H}^- + \mathrm{HCOO} \quad (2)$$

The adiabatic dissociation limits have already been determined to be 3.43±0.1 eV for reaction 1 and 3.79±0.23 eV for reaction 2 [20, 22]. The total kinetic energy release (KER) in the assumed two-body fragmentation can be extracted from the momentum images of figure 3, and it was found that the KER distributions span 2 to 4 eV for reaction 1 and 3 to 4 eV for reaction 2. The equilibrium geometry of *cis*-HOCO is somewhat similar to formic acid [27, 26], however it is known to have small barriers of about 1 eV to dissociation into CO + OH or CO_2 + H [28]. HCOO has at least three low-lying excited electronic states [29, 30] that could be accessible by 7 eV and 8.5 eV dissociative electron attachment, and they may decay dissociatively to H + CO_2.

The momentum image in figure 3(a), corresponding to the HOCO product channel (reaction 1), displays a broad peak in the backwards direction, opposite to the incident electron direction. In contrast, the HCOO product channel (reaction 2) has a minimum in the backwards direction at the lower-energy resonance in figure 3(b), and is more isotropic for the 8.5 eV resonance in figure 3(c). While we cannot rule-out three distinct electron attachment resonances contributing to these observations, the present fragment momentum measurements are consistent with electron attachment to a narrow core-excited resonance at 7.1 eV, which dissociates by C–H scission (reaction 1), and in a more complex process, to dissociation by O–H scission. The latter process may involve a more rapid stabilization of the transient anion, allowing DEA to compete more favorably with autodetachment and produce higher anion yields. The second resonance around 8.5 eV is expected to have a completely different molecular-frame electron attachment amplitude [3], resulting in the distinctly isotropic momentum ring in figure 3(c). An additional diffuse ring is also revealed for momenta <15 a.u. in figure 3(a). The shape of this feature was found to be insensitive to incident electron energies, suggesting that the origin may be a 3-body dissociation channel, possibly due to one or more higher-energy excited anion states.

5. Summary and outlook

Site-selectivity is observed in dissociative electron attachment in deuterated formic acid (HCOOD or DCOOH), where H^- or D^- is produced by breaking the C–H or C–D bond, respectively, in one DEA resonance centered at 7.1 eV. A second higher-energy resonance at about 8.5 eV produces no significant anions by C–H break, but both resonances produce H^- or D^- by breaking the O–H or O–D bond in DCOOH or HCOOD, respectively. Electron scattering calculations are needed to determine the molecular-frame electron attachment amplitudes in the molecular frame. Paired with the measurements presented here, a more complete analysis will then be possible to unravel the possible dissociation pathways involving hydrocarboxyl or formyloxyl neutral radicals.

Acknowledgments

This work is supported by the U.S. DOE Office of Basic Energy Sciences, Division of Chemical Sciences, Geosciences and Biosciences under contract DE-AC02-05CH11231. BG and JW acknowledge support from the National Science Foundation Grant No. NSF-PHY-1807017.

References

[1] Alizadeh E, Orlando T M and Sanche L 2015 *Annu. Rev. Phys. Chem.* **66** 379–398
[2] Fabrikant I I, Eden S, Mason N J and Fedor J 2017 *Adv. At. Mol. Opt. Phys.* vol 66 ed Arimondo E, Lin C C and Yelin S F (Academic Press) pp 545–657
[3] Slaughter D S, Belkacem A, McCurdy C W, Rescigno T N and Haxton D J 2016 *J. Phys.* B*: At. Mol. Opt. Phys.* **49** 222001
[4] Rescigno T N, Trevisan C S, Orel A E, Slaughter D S, Adaniya H, Belkacem A, Weyland M, Dorn A and McCurdy C W 2016 *Phys. Rev.* A **93** 052704
[5] Zuckerman B, Ball J A and Gottlieb C A 1971 *Astro. J.* **163** L41
[6] Bennett C J, Hama T, Kim Y S, Kawasaki M and Kaiser R I 2010 *Astrophys. J.* **727** 27
[7] Millar T J, Walsh C and Field T A 2017 *Chem. Rev.* **117** 1765–1795
[8] Becker S, Thoma I, Deutsch A, Gehrke T, Mayer P, Zipse H and Carell T 2016 *Science* **352** 833–836
[9] Orgel L E 2010 *Crit. Rev. Biochem. Mol. Biol.* **39** 99–123
[10] Millet D B, Baasandorj M, Farmer D K, Thornton J A, Baumann K, Brophy P, Chaliyakunnel S, de Gouw J A, Graus M, Hu L, Koss A, Lee B H, Lopez-Hilfiker F D, Neuman J A, Paulot F, Peischl J, Pollack I B, Ryerson T B, Warneke C, Williams B J and Xu J 2015 *Atmospheric Chem. Phys.* **15** 6283–6304
[11] Ong B C, Kamarudin S K and Basri S 2017 *Int. J. Hydrogen Energy* **42** 10142–10157
[12] Gallup G A, Burrow P D and Fabrikant I I 2009 *Phys. Rev.* A **79** 042701
[13] Rescigno T N, Trevisan C S and Orel A E 2009 *Phys. Rev.* A **80** 046701
[14] Gallup G A, Burrow P D and Fabrikant I I 2009 *Phys. Rev.* A **80** 046702
[15] Gallup G A 2013 *Phys. Rev.* A **88** 052705
[16] Gianturco F A and Lucchese R R 2004 *New J. Phys.* **6** 66
[17] Trevisan C S, Orel A E and Rescigno T N 2006 *Phys. Rev.* A **74**
[18] Rescigno T N, Trevisan C S and Orel A E 2006 *Phys. Rev. Lett.* **96**
[19] Janekov R, Kubala D, May O, Fedor J and Allan M 2013 *Phys. Rev. Lett.* **111** 213201
[20] Pelc A, Sailer W, Scheier P, Probst M, Mason N J, Illenberger E and Märk T D 2002 *Chem. Phys. Lett.* **361** 277–284
[21] Pelc A, Sailer W, Scheier P, Mason N and Mrk T 2002 *Eur. Phys. J.* D **20** 441–444
[22] Prabhudesai V S, Nandi D, Kelkar A H, Parajuli R and Krishnakumar E 2005 *Chem. Phys. Lett.* **405** 172–176

[23] Ptasińska S, Bass A D and Sanche L 2008 *J. Phys.: Conf. Ser.* **115** 012018
[24] Adaniya H, Slaughter D S, Osipov T, Weber T and Belkacem A 2012 *Rev. Sci. Instrum.* **83** 023106
[25] Allan M 2006 *J. Phys.* B*: At. Mol. Opt. Phys.* **39** 2939–2947
[26] Leach S, Schwell M, Dulieu F, Chotin J L, Jochims H W and Baumgärtel H 2002 *Phys. Chem. Chem. Phys.* **4** 5025–5039
[27] Clements T G, Continetti R E and Francisco J S 2002 *J. Chem. Phys.* **117** 6478–6488
[28] Johnson C J, Harding M E, Poad B L J, Stanton J F and Continetti R E 2011 *J. Am. Chem. Soc.* **133** 19606–19609
[29] Kim E H, Bradforth S E, Arnold D W, Metz R B and Neumark D M 1995 *J. Chem. Phys.* **103** 7801–7814
[30] Zou L, Li J, Wang H, Ma J and Guo H 2015 *J. Phys. Chem.* A **119** 7316–7324

Dissociative electron attachment to SO_2 near 7.5 eV resonance and axial-recoil approximation

Irina Jana[1], Sumit Naskar[2] and Dhananjay Nandi[1]

[1]Department of Physical Sciences, Indian Institute of Science Education and Research Kolkata, Mohanpur-741246, India

[2]Department of Chemical Sciences, Indian Institute of Science Education and Research Kolkata, Mohanpur-741246, India

E-mail: ij14rs047@iiserkol.ac.in

Abstract. We present experimental results for dissociative electron attachment to sulfur dioxide near the 7.5 eV resonance. The dissociation channels leading to O^-, S^- and SO^- ions formation have been investigated with a velocity map imaging technique. While the O^- and SO^- ions formed by dissociative electron attachment show angular distributions characteristic to a combination of A_1+B_2 symmetry, the S^- ions show angular distributions resulting from changes in the orientation of the dissociating bond due to bending mode vibrations implying the failure of axial-recoil approximation.

1. Introduction

Chemical reactions initiated by collisions with low-energy electrons play an important role in atmospheric chemistry, astro-chemistry, synthetic chemistry, and biology, amongst other areas. Understanding the mechanisms of these reactions in detail offers new insight into a range of vital physical and chemical processes, ranging from the breaking of a single chemical bond all the way through to complex multi-step processes. Low-energy electron attachment to molecules resulting in the formation of a temporary negative ion (TNI) and then dissociation of the TNI into negative ion and neutral fragments is known as dissociative electron attachment (DEA). Since the path-breaking works of Chandler and Houston [1] and Eppink and Parker [2], it is a well-known practice to measure the complete scattering distribution of the resulting negative ion fragment using the velocity-map ion imaging (VMI) technique [3, 4]. VMI combines time-of-flight mass spectrometry with imaging, separating ions of different mass and mapping their velocity distributions onto a position sensitive detector. Having controlled the initial velocity of the reactants and measured the final velocities of the products, one can 'join the dots' to learn about the forces and energetics that drive the chemical reaction under study. The kinetic energy and angular distribution of the negative ion fragments can thus be revealed from the velocity map images. Such experiments yield great insight into the physics underlying chemical reactivity [5, 6, 7, 8].

DEA to SO_2 produces three negative fragment ions O^-, S^- and SO^- at two resonances near 4.5 and 7.5 eV incident electron energies [9]. We report here DEA to SO_2 for the formation of O^-, S^- and SO^- ions close to the second resonance near 7.5 eV. According to the axial-recoil approximation, the symmetry of the TNI can be found from the negative

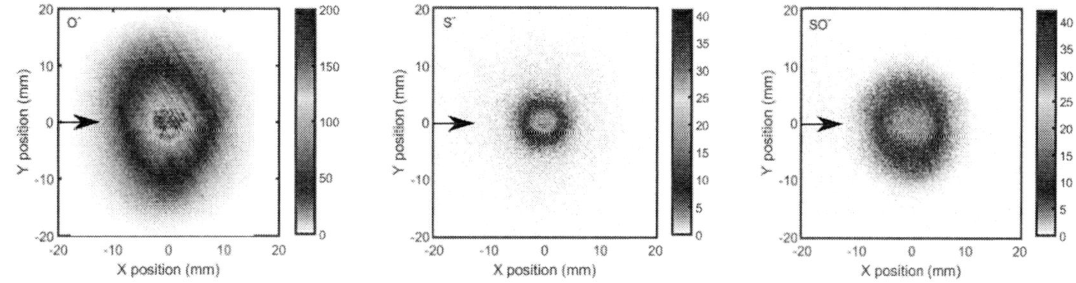

Figure 1. Time-sliced images for (a) O^-, (b) S^- and (c) SO^- ions at 7.5 eV incident electron energy. The incident electron beam is from left to right as shown by the arrow for all the figures.

ions' velocity-mapped images provided the molecule does not rotate or undergo any structural changes before the dissociation of the TNI into respective negative and neutral fragments takes place. Thus, the negative fragment ion retains the information on the orientational dependence of the electron capture process [10]. This methodology can be used to extract the angular distribution information, related to partial waves s, p, d,... in the attachment process, by fitting the experimental data with the theoretical formula given by O'Malley and Taylor and Azria *et al.* [11, 12]. However, this methodology completely fails in the absence of the axial-recoil approximation thus making it impossible to extract information about the symmetry of the TNI.

It can be noted from the existing literature that for some molecules, such as CO_2, the axial-recoil approximation can be seen to fail implying a strong nonaxial recoil effect due to the bend and stretch contributions to the dissociation dynamics [13, 14]. CO_2 is known to have a 4 eV shape resonance and 8 eV Feshbach resonance. The negative fragments ions' angular distributions around both the 4 and 8 eV resonances are observed to show remarkable difference from the axial-recoil prediction giving a clear evidence of an axial-recoil approximation breakdown.

In this article we report DEA studies to SO_2 around the second resonance at 7.5 eV for O^-, S^- and SO^- ions to present an interesting case where the axial-recoil approximation is seen to be valid and failing simultaneously for the same molecule depending on the negative ion fragment produced after the dissociation. The O^- and SO^- fragments are known to be produced from the asymmetric stretching of the molecule. The S^- ions, on the other hand, come from symmetric stretching simultaneously accompanied by symmetric bending with O_2 produced as the neutral counter-part [15].

2. Experimental apparatus

The experimental technique has been described in detail previously, so we will provide only a brief overview here [16, 17, 18]. The momentum of the final-state anionic fragments O^-, S^- and SO^- following DEA to SO_2 molecules were measured using a velocity slice imaging negative ion spectrometer. A gaseous target beam is effused from a narrow stainless steel capillary to intersect with a magnetically collimated electron beam pulsed at 10 kHz repetition rate forming a 'Newton sphere' of negative ions. The electron energy spread is typically 0.8 eV full width at half maximum. Ion focusing optics within the ion spectrometer allows the effective interaction and ion momentum images are recorded from a MCP-based position- and time-sensitive detector to an event list, so two-dimentional momentum distributions and time-of-flight information can be determined for a full 2π solid angle of detection. The central slice of the Newton sphere is selected for a time-window of 50 ns containing complete kinematic information of the negative ions. The kinetic energy distributions are obtained by integrating the ion counts over the entire ejection angle of the negative ion fragments and plotting them as a function of ion kinetic energy. On the other hand, the angular distributions are obtained by integrating the ion counts for a small ejection angle range θ to $\theta + d\theta$ of the ion fragments and plotting them as a function of ejection angle.

3. Results and discussion

The measured O^-, S^- and SO^- time-sliced velocity distributions for 50 ns width at 7.5 eV incident electron energy are shown in figure 1. All the images show an almost uniform central blob, with the diameters varying according to the respective kinetic energies of the negative ion fragments.

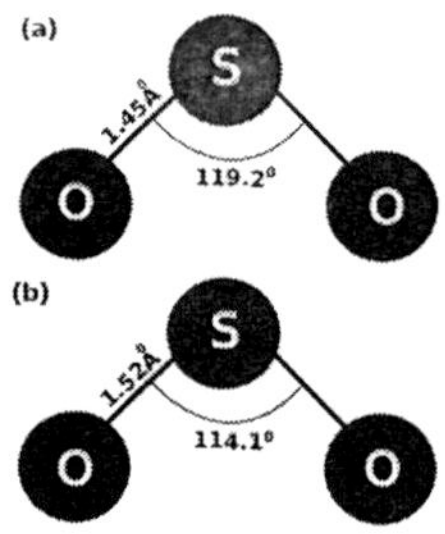

Figure 2. The calculated optimized ground state geometries of: (a) SO_2 and (b) SO_2 ion both showing the C_{2v} symmetry.

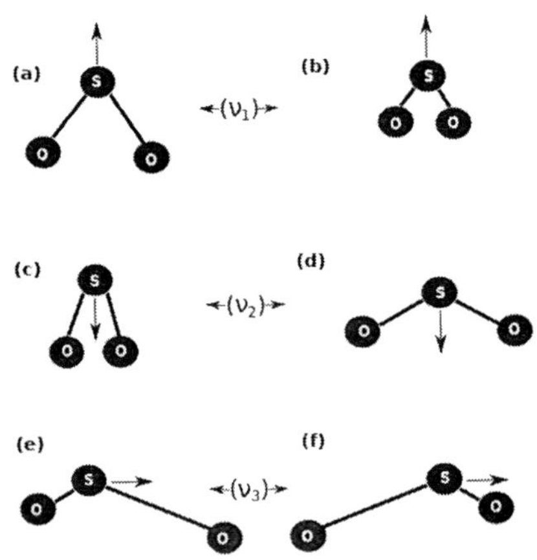

Figure 3. Depiction of computed vibration modes for SO_2 molecule. (a) and (b) represent symmetric stretching mode; (c) and (d) represent symmetric bending mode; (e) and (f) represent anti-symmetric stretching mode. The arrow denotes the dipole derivative unit vector in all the cases.

The ground state geometries of SO_2 and SO_2^- molecules optimized with Mϕller-Plesset perturbation theory up to second order (MP2) using aug-cc-pVQZ basis set to obtain a stable structure using the Gaussian 09 program package are shown in figure 2 [19]. The neutral SO_2 molecule has C_{2v} symmetry with a S-O bond length of 1.45 $\AA$ and 119.20^0 O-S-O bond angle in the optimized state (figure 2 (a)) [20]. After the attachment of an electron, the anion retains C_{2v} symmetry but the S-O bond length increases by 0.07 $\AA$ while the O-S-O bond angle decreases by 5.10^0 (figure 2 (b)) [16]. The three normal modes of vibration: symmetric stretching (ν_1), symmetric bending (ν_2) and asymmetric stretching (ν_3) for SO_2^- are shown in figure 3. It can be inferred from figure 3 that the symmetric stretching will give rise to S^- fragment formation. As the stretching may produce S^-, it has to be followed by the bending mode simultaneously such that the O-atoms can come close enough to form O_2. The same will be true for the formation of O_2^- fragment which has a very low cross-section as compared to S^- and hence is generally not reported [15]. The antisymmetric stretching mode alone may result in the formation of O^- and SO^- negative ions [15]. Thus, the dissociation (recoil) axis must rotate between the time of electron attachment to the neutral target molecule and subsequent dissociation of SO_2^{-*} for the production of S^- fragment. This structural change in the TNI undergoing bending mode vibration into $S^- + O_2$ fragments results in failure of the axial recoil approximation. On the other hand, the axial recoil approximation is valid for the production of O^- and SO^- fragments.

Using this approximation, the symmetry of the TNI can be found provided the symmetry of the neutral molecule and the angular distribution of the negative ion fragment is known by following the methodology of O'Malley and Taylor and Azria *et al.* [11, 12]. The general equation for describing angular distribution was given by Tronc *et al.* [21] which was then modified for polyatomic molecules by Azria *et al.* [12].

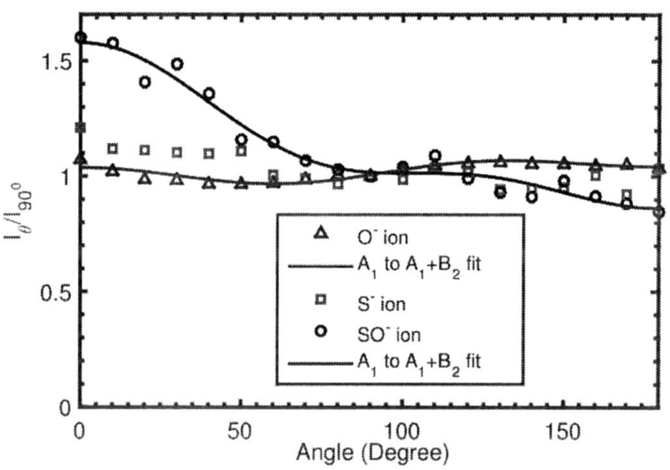

Figure 4. Angular distribution data along with necessary fits for A_1 to A_1+B_2 transition for O^-, S^- and SO^- ions for 7.5 eV incident electron energy at the second resonance.

The measured angular distribution for O^-, S^- and SO^- ions are shown in figure 4. The fit corresponding to A_1 to $A_1 + B_2$ transition is also shown in figure 4 with solid lines for O^- and SO^- ions only. Indeed, as mentioned before, the S^- cannot be fitted due to the failure in axial recoil approximation. Equations for A_1 to $A_1 + B_2$ transitions due to a combination of (s + p + d)- partial waves showing the scattering intensities coming to play for the C_{2v} point group are explained in details by Jana and Nandi [16].

Time-dependent density functional theory (TD-DFT) calculations are carried out to get the theoretical vertical energy values and symmetries of corresponding excited states for the SO_2^- molecule lying within the Franck-Condon region. Then the symmetry states lying near the second resonance observed in the range 7.0 - 7.6 eV by different groups [7, 8, 16, 17] and having symmetries with considerable value of oscillator strength (f) are noted. The table showing the symmetry states for the two resonances with considerable f-value are shown in table 1 in order of descending f-value.

Six symmetry states can be observed with high f-values and are reported in table 1. Amongst

Table 1. Symmetry states of excited TNI with corresponding oscillator strengths in descending order which are expected to contribute to DEA to SO_2 from TD-DFT calculations with SO_2^- ground state optimized geometry with the aug-cc-pVQZ basis and B3LYP exchange-correlational functional.

Energy range (eV)	Symmetry of TNI	Oscillator strength (f)	Ref[16]
7.0 - 7.6 eV	A_1	0.0516	
	B_2	0.0445	$A_1 + B_2$
	B_2	0.0383	
	B_2	0.0319	
	A_1	0.0240	
	B_2	0.0111	

these, four are B_2 and two are A_1 symmetry states. From this observation, it can be concluded that the second resonance occurs due to an A_1 to $A_1 + B_2$ transition. This result matches excellently with the recent experimental works on DEA to SO_2 [16, 17].

4. Conclusion

We have performed DEA to SO_2 at the second resonance at 7.5 eV incident electron energy. Although the neutral molecule and the TNI both have C_{2v} symmetry, from the normal vibration modes it is clear that O^- and SO^- ions are formed keeping the angle between the dissociation axis of the molecule and incident electron beam direction the same as that of the TNI, thus validating the axial recoil approximation. On the other hand, the S^- ion formation takes place with a rotation of the molecular dissociation axis resulting in a complete failure of the axial recoil approximation. Thus the axial recoil approximation can be seen to be validated and failed simultaneously for the same molecule with regards to the fragment ions formed. Also, the analysis of the angular distributions of O^- and SO^- ions and TD-DFT calculations show that the 7.5 eV peak is formed predominantly due to the A_1 to $A_1 + B_2$ transition.

Acknowledgment

The authors acknowledge support of M. Das (Indian Institute of Science Education and Research Kolkata) for discussions and included *ab initio* calculations using Gaussian 09 program package from the computational facility under the project SB/FTP/CS-164/2013 and financial supports from Science and Engineering Research Board (SERB) for supporting this research under the Project EMR/2014/000457.

References

[1] Chandler D W and Houston P L 1987 *J. Chem. Phys.* **87** 1445-1447
[2] Eppink A T and Parker D H 1997 *Rev. Sci. Instrum.* **68** 3477-3484
[3] Nandi D, Prabhudesai V S, Krishnakumar E and Chatterjee A 2005 *Rev. Sci. Instrum.* **76** 053107
[4] Nandi D, Prabhudesai V S and Krishnakumar E 2006 *Radiat. Phys. Chem.* **75** 2151-2158
[5] Nag P and Nandi D 2016 *Phys. Rev.* A **93** 012701
[6] Chakraborty D, Nag P and Nandi D 2016 *Phys. Chem. Chem. Phys.* **18** 32973-32980
[7] Krishnakumar E, Prabhudesai V S and Mason N J 2018 *Nat. Phys.* **14** 149
[8] Gope K, Tadsare V, Prabhudesai V S, Mason N J and Krishnakumar E 2016 *Eur. Phys. J.* D **70** 134
[9] Orient O J and Srivastava S K 1983 *J. Chem. Phys.* **78** 2949-2952

[10] Ram N B, Prabhudesai V S and Krishnakumar E 2009 *J. Phys.* B **42** 225203
[11] O'Malley T F and Taylor H S 1968 *Phys. Rev.* **176** 207
[12] Azria R, Abouaf R and Teillet-Billy D 1982 *J. Phys.* B **15** L569
[13] Moradmand A, Slaughter D S, Landers A L and Fogle M 2013 *Phys. Rev.* A **88** 022711
[14] Moradmand A, Slaughter D S, Haxton D J, Rescigno T N, McCurdy C W, Weber T, Matsika S, Landers A L, Belkacem A and Fogle M 2013 *Phys. Rev.* A **88** 032703
[15] Gope K, Prabhudesai V S, Mason N J and Krishnakumar E 2017 *J. Chem. Phys.* **147** 054304
[16] Jana I and Nandi D 2018 *Phys. Rev.* A **97** 042706
[17] Jana I and Nandi D 2019 *J. Phys.* B, **52** 185202
[18] Nag P and Nandi D 2015 *Meas. Sci. Tech.* **26** 095007
[19] Frisch M J, Head-Gordon M and Pople J A 1990 *Chem. Phys. Lett.* **166** 275-280
[20] Jana I, Naskar S, Das M and Nandi D Probing the molecular structure and properties of neutral and anionic ground states of SO_2 and CO_2 *Eur. Phys. J.* D (Accepted article)
[21] Tronc M, Schermann C, Hall R I and Fiquet-Fayard F 1977 *J. Phys.* B **10** 305

Electron attachment and quantum coherence in molecular hydrogen

Vaibhav S Prabhudesai[1], Nigel J Mason[2] and E Krishnakumar[3]

[1] Tata Institute of Fundamental Research, Homi Bhabha Road, 400005 Mumbai, India
[2] School of Physical Sciences, University of Kent, Canterbury, CT2 7NH, UK
[3] Raman Research Institute, Sadashiva Nagar, 560080 Bengaluru, India

Email: vaibhav@tifr.res.in

Abstract. Single electron attachment to a molecule may invoke quantum coherence in different angular momentum transfer channels. This has been observed in the 14 eV dissociative electron attachment resonance in molecular hydrogen where a coherent superposition of two negative ion resonant states of opposite parity is created, with the *s* and *p* partial waves of the electron contributing to the attachment process. Interference between the two partial wave contributions leads to a forward – backward asymmetry in the angular distribution of the product negative ions. Since these two resonant states dissociate to the same n = 2 state of H and H^-, this asymmetry is further modified due to interference between the two paths of the dissociating molecular negative ion along different potential energy curves. This interference manifests as a function of the electron energy as well as isotopic composition. This case is akin to the quantum interference observed in photodissociation by one-photon vs two-photon absorption.

1. Introduction

Dissociative electron attachment (DEA) is a dominant process in any medium in which low energy electrons are present. It is an efficient way of converting kinetic energy into chemical energy in a medium and has been exploited by modern laser technology and play a role in many planetary and astrophysical phenomena. A major characteristic of the DEA process is its ability to induce site selective bond breaks in molecules [1-4]. It has been shown that electron attachment to molecules depends on the functional groups present in them, leading to localization of both energy and charge at a given site in the molecule [1]. The subsequent dissociation of the molecular negative ion resonance (NIR) state takes place selectively at a given bond or site. Furthermore, since electron attachment occurs at distinct energies and leads to distinct molecular fragmentation pathways, DEA provides a way of controlling the chemistry by controlling the creation of reactive species in a medium. This is complementary to the coherent control of chemical reaction using lasers. However, electrons being ubiquitous, the possibility of chemical control using electrons may have wider applicability than coherent control of chemical reactions using lasers and is a more practical technology. Indeed, it is now being adopted in plasma technology and nanotechnologies [5].

Coherent control using lasers has been used to demonstrate control in photodissociation. In this, the quantum coherence manifests itself in terms of quantum interference of two or more channels that start from an initial state of a target molecule and lead to the same dissociation products. Selection of one of these channels has been demonstrated by controlling the coherent light fields. Breaking of inversion symmetry in the photodissociation using one photon vs two photon absorption is the most robust example of demonstration of such coherence and its control [6].

In many aspects, the DEA process is very similar to photodissociation. The kinetic energy of the attaching electron is transferred to the target molecule which causes its excitation and this excited state may then undergo dissociation. However, the most distinctive feature of DEA is the transfer of charge which results in the formation of a NIR. Depending on the electron affinity of the fragments, the DEA process invokes additional energy release which makes bond rupture in molecules possible for electron energies even below their dissociation limits. Also, unlike photoabsorption where in the visible frequency regime the dipole transition plays a dominant role, in electron attachment more than one value of angular momentum may be transferred with comparable strength. However, the most intriguing aspect of the DEA process is the rich electron-electron correlations that result due to interaction of the attached electron with those from the molecular target. This makes theoretical calculations of the electron attachment and DEA process more challenging. Experimentally, the DEA process has been studied through measurements of the absolute cross sections of various dissociation channels as well as measurements of the kinetic energy and angular distributions of the fragment anions [7-11]. Fragment kinetic energy distribution yields information about the energy partitioning in the dissociation process whereas angular distribution reflects the NIR state symmetries involved, provided the dissociation takes place on a fast enough time scale. Together such measurements allow modeling of the DEA dynamics.

We have recently observed in molecular hydrogen coherence in electron attachment leading to the formation of two negative ion resonances in coherent superposition and subsequent decay of the system through dissociation [12]. That this process is observed in the simplest molecular system makes it all the more interesting. It may be pointed out that no theoretical prediction or any observation has been made for such a process in any other system before. What we present here is a more detailed analysis of the data and discuss the implications of the results for understanding the NIRs in molecular hydrogen around 14 eV.

2. Negative ion resonances in molecular hydrogen

Before we discuss present results, it is appropriate to give a brief review of the status of negative ion resonances in molecular hydrogen. Being the simplest molecule, electron scattering from H_2 (and its isotopomers) has been the subject of considerable theoretical and experimental work. The NIRs in molecular hydrogen have been observed in DEA studies, electron transmission experiments, elastic and inelastic scattering experiments and various theoretical calculations. All these are summarised in several reviews [11, 13-17]. Recent measurements [18] on absolute cross sections for DEA on H_2 and D_2 have removed the disparity that existed in previous data [19, 20]. The DEA process in the molecule has been shown to have strong isotope effect [18, 19, 21, 22], due to the considerable mass difference in the nuclei. All the isotopomers are also found to have strong initial vibrational state dependence in the DEA cross sections [23, 24]. This has been explained by the combined effect of the Franck– Condon factors and the survival factor appropriate to the calculated potential energy curves [25, 26].

DEA to molecular hydrogen and its isotopomers is characterised by two relatively narrow resonances at 4 eV and 14 eV respectively and a broad one between 7 eV and 13 eV. Based on the measurements and the theoretical calculations [27-29] the symmetries of the three resonances seen in the DEA channel have been assigned as ${}^2\Sigma_u^+$ (4 eV), ${}^2\Sigma_g^+$ (10 eV) and ${}^2\Sigma_u^+$ (14 eV). The H^- angular distribution measurements in DEA at 4 eV and the 10 eV peaks were found to be consistent with these assignments [30]. Preliminary measurements [31] on the H^- angular distribution at 14 eV resonance were found to be isotropic and it was assumed that the symmetry of the resonance was the same as that determined by inelastic electron scattering measurements [32]. However, an interesting feature of measurements of ion yield curves is the observation of oscillation in the energy differential cross sections between 11 eV and 13 eV [31] and between 14 eV and 16 eV [30] for H_2 and the absence of such structures in D_2. No clear explanation has been made for both these structures, though it was proposed that the observed structure between 11 eV and 13 eV may involve an attractive ${}^2\Sigma_g^+$ NIR either through predissociation or a two-step process [33] in which the resonant elastic scattering from the attractive ${}^2\Sigma_g^+$ is followed by DEA through the repulsive $^2\Sigma_g^+$ state [31]. The structure in the 14 eV resonance was proposed as due to predissociation from a $^2\Delta_g$ state where the coupling takes place through rotations [30]. Since there is a requirement of unit angular momentum change between the two states in the predissociation process, Tronc *et al.* [30] proposed a breakdown of Born-Oppenheimer approximation with the Δ_g state having a dissociation limit of $H^-(1s^2)$ + H(3d) which is bound in the relevant electron energy range.

3. Experimental set-up

We have used the velocity slice imaging (VSI) technique [18, 34] to measure the angular distribution of the fragment anions produced in the DEA to H_2. In brief, an effusive molecular beam crosses a magnetically collimated pulsed electron beam at right angle. The molecular beam is coincident with the axis of the VSI spectrometer. The spectrometer consists of the interaction region followed by a single element electrostatic lens and 100mm long flight tube. The ions are detected using micro channel plates and phosphor screen-based position sensitive detector (PSD). The ions formed in the interaction region are extracted using a delayed pulsed electric field applied to the extraction electrodes. The electrostatic lens assembly provides the velocity focusing condition where the entire Newton sphere of the extracted fragment anions is projected onto the 2-dimensional PSD. The detector is subjected to an 80nsec wide voltage pulse, which is timed appropriately coinciding with the arrival of the central slice of the Newton sphere. The impacts of the ions on the detector are recorded using CCD camera and analyzed after adding several such slices in an off-line analysis. The energy resolution of the electron beam in the experiment is about 0.6 eV. The angular resolution depends on the electron energy and the collimating magnetic field strength as discussed in detail previously [34]. We estimate that it is close to 8° at 4 eV and 4° at 14 eV for a magnetic field of 50 Gauss used in the present measurements.

4. Results

Of the three resonances in H_2, the present discussion is focused upon the 14 eV resonance. We have obtained VSI images of H^- produced from H_2 and D^- from D_2 at a number of electron energies across this resonance. For comparison we have also measured the momentum images around the 4 eV resonance in H_2.

4.1. 4 eV resonance

The 4eV peak in DEA to H_2 is known to be due to ground NIR state of H_2^-. The momentum image obtained at 4.5 eV along with the angular distribution obtained from the image is shown in figure 1.

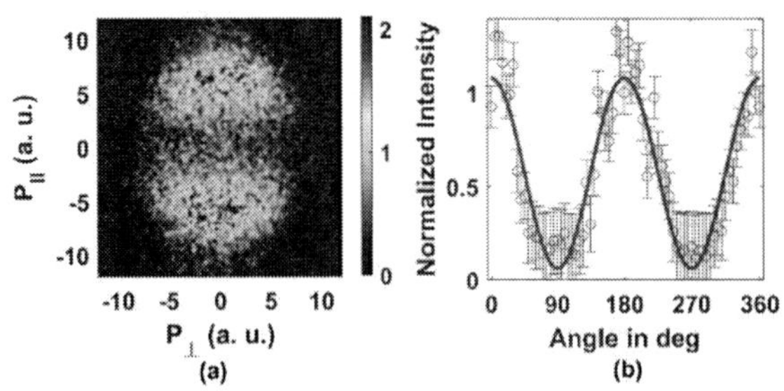

Figure 1. (a) Momentum image obtained for H^- from H_2 at 4.5 eV electron energy. (b) Angular distribution (normalized at $\theta = 180^0$) obtained from the image. Measured data (circles) and the fitted distribution (solid line). (Part (a) is reproduced from figure 3(a) of reference 12).

The ion signal peaks symmetrically in forward and backward directions with respect to the incoming electron beam. As the kinetic energy release is low for this resonance, the angular distribution was obtained from the outer edge of the momentum image. The angular distribution follows the $\cos^2\theta$ form with constant background. The exact fit function is of the form,

$$\mathrm{I}(\theta) = A^2 + B^2\cos^2(\theta) \tag{1}$$

with $A = 0.2457$ and $B = 1.01$. This is consistent with the transition to a NIR state of Σ_u symmetry from the target state of Σ_g symmetry. The dominant angular momentum transfer channel is $l = 1$ which is the lowest allowed partial wave for this transition in a homonuclear diatomic molecule. Ideally, the value of A should be zero. In the present case, it appears as due to a small background signal and imperfections in the momentum imaging. This background, which can be ignored in this case, becomes a limitation in the analysis of the 14 eV resonance, as will be seen later. We could not measure the D^- angular distribution at this energy due to its two orders of magnitude lower cross section [20].

4.2. 14 eV resonance

The VSI image obtained at 14.5 eV for both H_2 and D_2 along with the angular distributions are shown in figure 2. As can be seen from the figure, the H^- momentum image shows a clear forward-backward asymmetry with respect to the electron beam with more intensity in the backward direction. The VSI image of D^- from D_2 shows substantially different distribution with a different degree of asymmetry.

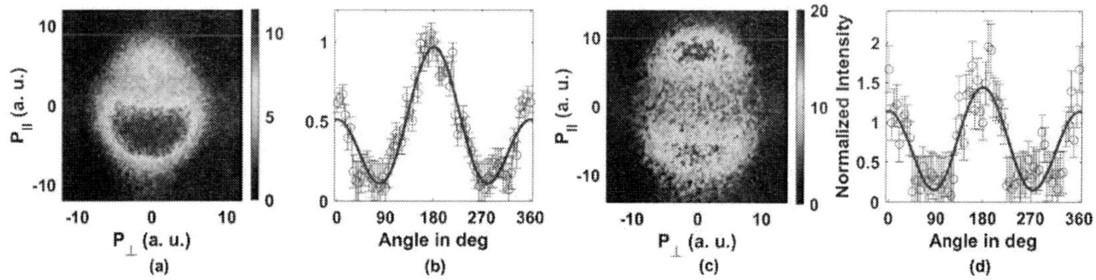

Figure 2. (a) Momentum image and (b) angular distribution (circles - measured data and solid line fit) for H^- from H_2 at 14.5eV electron energy and that ((c) and (d) respectively) for D^- from D_2. The angular distribution is normalized with respect to the count at $\theta = 180^0$. For the details of the fit, please see the text below. (Parts (a) and (c) are reproduced from figure 3(b) and (f) of reference 12).

5. Discussion

5.1. Manifestation of quantum coherence in DEA

For homonuclear diatomic molecules, due to the inversion symmetry present in the system, the dissociation process resulting from a given NIR state must also show inversion symmetry. As the initial neutral target state and the anion state formed after electron attachment has the inversion symmetry, the electron attachment must proceed with transfer of only odd or only even partial waves to the target [35]. The angular distribution thus obtained is given as

$$\mathrm{I}(k,\theta,\phi) = \left|\sum_{l=|\mu|}^{\infty} a_{l\mu}(k)\mathrm{Y}_{l\mu}(\theta,\phi)\right|^2 \tag{2}$$

where l has either only odd or only even values. As the target neutral state of H_2 is with even parity, the transfer of even partial wave will result in the formation of NIR with even parity and that of odd partial wave will generate an odd parity NIR state. Consequently, a single NIR state cannot explain the observed forward-backward asymmetry of the angular distribution where the inversion symmetry of the system is broken. However, if we have more than one resonance taking part in the electron attachment, the angular distribution of the product ion is given by

$$\mathrm{I}(k,\theta,\phi) = \sum_{|\mu|}\left|\sum_{l=|\mu|}^{\infty} a_{l\mu}(k)\mathrm{Y}_{l\mu}(\theta,\phi)\right|^2. \tag{3}$$

Please note that here the individual cross sections for each resonant state are simply added and will not break the inversion symmetry. A single electron on attachment creates one resonant state with a certain probability and in an ensemble it is possible to get different states by different electrons. The resultant observable angular distribution will be an incoherent addition of these signals. However, if a single electron upon attachment gives rise to a coherent superposition of different resonant states, the resultant angular distribution stems from the coherent addition of transition amplitudes for dissociation via each constituent state. In this case, the angular distribution will be given by

$$\mathrm{I}(k,\theta,\phi) = \left|\sum_{|\mu|} \sum_{l=|\mu|}^{\infty} a_{l\mu}(k) \mathrm{Y}_{l\mu}(\theta,\phi)\right|^2. \tag{4}$$

For example, in the case of H_2, considering only the lowest order allowed partial waves (angular momentum transfer) in the formation of coherent superposition of ${}^2\Sigma_g^+$ and ${}^2\Sigma_u^+$ states from the neutral ground state (${}^1\Sigma_g^+$), we can write this equation as

$$\mathrm{I}(k,\theta,\phi) = |a_{00}(k)\mathrm{Y}_{00}(\theta,\phi) + a_{10}(k)\mathrm{Y}_{10}(\theta,\phi)|^2. \tag{5}$$

Based on this, the functional form of the electron attachment probability with respect to angle will be

$$\mathrm{I}(k,\theta) = (a_g(k))^2 + (a_u(k))^2 \cos^2\theta + 2a_g(k)a_u(k)\cos\theta\cos\delta \tag{6}$$

where a_g and a_u are proportional to the magnitude of the two transition amplitudes to the ${}^2\Sigma_g^+$ and ${}^2\Sigma_u^+$ NIR states respectively with δ as the relative phase between the transferred s and p waves. From this, one can immediately see that due to the $\cos\theta$ term there will be a forward-backward asymmetry.

In the present experiment what we observe is the decay of this state through dissociation and under the axial recoil approximation the angular distribution in the attachment probability will manifest itself in the angular distribution of the H^- fragment. However, this by itself does not explain the observed results and we need to consider the evolution of the coherent states leading to dissociation. As the resonances progress towards the dissociation limit along two different potential energy curves, the phase between the two changes and, at the dissociation limit, there will be a phase difference between the two. This phase difference modifies the angular distribution as,

$$\mathrm{I}(k,\theta) = {A_g}^2 + {A_u}^2 \cos^2\theta + 2A_g A_u \cos\theta \cos\varphi \tag{7}$$

where A_g and A_u are the amplitudes of each resonant state at the dissociation limit after taking into account the respective autodetachment rate, which is a function of the inter-nuclear separation. We will discuss this point later. The phase φ is the sum of δ (equation (6)) and the phase difference arising from the propagation of the wavepacket along the two dissociation paths. As can be seen from equation (7), the extent of asymmetry observed in the differential cross section will explicitly depend on the relative phase between the two paths that contribute to the DEA channel as well as on the relative amplitudes in the two paths.

The observed difference in asymmetry in D_2 can then be understood as a manifestation of both the change in phase difference as well as change in relative contribution of the two paths. D_2 being a heavier molecule, the dissociation time would be $\sqrt{2}$ times longer than that in H_2. This will change the phase difference between two paths by a factor $\sqrt{2}$ resulting in a change in the interference intensity. On the other hand, each NIR state during dissociation competes with the autodetachment and the lifetime of this process varies as a function of inter-nuclear distance resulting in its survival probability as

$$\mathrm{p_i} = \int_{R_c}^{R_\varepsilon} \frac{\Gamma_a(R)}{\hbar v(R)} dR = \int_{t_{R_c}}^{t_{R_\varepsilon}} \frac{dt_R}{\tau(t_R)} \tag{8}$$

where $\Gamma_a(R)$ is the width of the anion potential energy curve, $v(R)$ is the speed of separation of the dissociating atoms, R_c is the inter-nuclear separation at which electron capture takes place and R_ε is the effective inter-nuclear separation beyond which the molecular anion is considered to be dissociated. $\tau(t_R)$ is the corresponding lifetime of the NIR. Assuming that the capture probability for each of the state is identical, one can estimate the forward backward asymmetry in the angular distribution using the potential energy curves of the two NIR states and their widths. As mentioned earlier, the relative amplitudes of each path will also depend on the lifetime of the individual NIR.

5.2. Angular distribution fits

Based on the model of coherent excitation of two resonances, we have analyzed the momentum images for the angular distributions, in an effort to quantify the relative contributions of the two resonances in the DEA signal and the relative phase, φ arising from the two dissociation paths. The angular distribution clearly shows the contributions due to both Σ_g^+ and Σ_u^+ states. The functional form for the angular distribution under the above model is given by equation (7). The result of the fits is given in table 1. We find that the best fit values for the A_g and A_u do not provide a consistent picture as far as the relative intensities of the two channels are concerned. We believe that this is due to unsatisfactory resolution in the momentum imaging, the main contribution to which is the electron energy spread (about 0.6eV FWHM) and background noise, as seen above in the case of 4 eV resonance. Although, the best momentum resolution obtainable from the spectrometer used is about 10% ($\Delta p/p$), the relatively large width of the slice used (80nsec) in comparison with the overall time spread of the Newton sphere of ions (about 200 nsec) makes it worse. We have tried overcoming this effect by obtaining the angular distribution for the outermost edge of the image as explained earlier. Moreover, due to low cross section, a longer data acquisition time results in the substantial contribution from the background noise spoiling the contrast of the image. This highlights the need for better measurements using improved electron energy resolution.

Table 1. Relative amplitudes and phase from the angular distributions H^- from H_2 and D^- from D_2.

For H_2				For D_2			
Electron energy (eV)	A_g	A_u	φ(rad)	Electron energy (eV)	A_g	A_u	φ(rad)
14.5	0.36±0.06	0.78±0.04	4.3±0.08	14.0	0.48±0.03	0.87±0.02	4.70±0.03
15.0	0.26±0.08	0.81±0.04	4.0±0.24	14.5	0.40±0.15	1.06±0.09	4.53±0.13
15.5	0.21±0.08	0.73±0.04	3.14±0.05	15.0	0.68±0.02	0.67±0.04	4.58±0.03

Since reliable quantitative estimates of the relative amplitudes could not be obtained from the fits of the data, we use the integrated intensities in the forward and backward hemispheres of the images to obtain the forward-backward asymmetry factor, η defined as,

$$\eta = \frac{(I_{forward} - I_{backward})}{(I_{forward} + I_{backward})} . \tag{9}$$

The values of η are given in table 2 and are compared with the results of simulations described below.

Table 2. Forward-backward asymmetry factor obtained for H^- and D^- signal. (Reproduced from table 1 from reference 12).

H_2		D_{21}	
Electron energy (eV)	Asymmetry factor η	Electron energy (eV)	Asymmetry factor η
14.5	-0.19 ± 0.02	14.0	0.02 ± 0.02
15.0	-0.17 ± 0.02	14.5	-0.03 ± 0.02
15.5	-0.12 ± 0.02	15.0	-0.08 ± 0.02

5.3. Simulation of η using potential energy curves

Around 14eV, the electron scattering experiments have identified a NIR state with ${}^2\Sigma_g^+$ and estimated its width to be 90meV [31]. Taking the potential energy curve for this state from the literature [26] and assuming a potential energy curve that runs close to some of those available in this energy range we estimate the asymmetry factor η as

$$\eta = \frac{\sqrt{3}\exp\left[-\left(\frac{t_g}{2\tau_g}+\frac{t_u}{2\tau_u}\right)\right]}{\exp\left[-\left(\frac{t_g}{\tau_g}\right)\right]+\exp\left[-\left(\frac{t_u}{\tau_u}\right)\right]}\cos\varphi \tag{10}$$

where τ_g and τ_u are the average lifetimes of the Σ_g and Σ_u NIR states involved in DEA, t_g and t_u are the dissociation times for the parent anion along the respective potential energy curves for electron attachment with a specified energy. The denominator of equation (10) is derived from using the cross section expression for DEA by each resonant state given by

$$\sigma_{DEA} = \sigma_c \times p_i \tag{11}$$

where the survival probability (p_i) for each resonant state is calculated using equation (8) and σ_c is the capture cross section. Although the width of the resonant states is strongly dependent on the inter-nuclear separation of the molecule, we have ignored this variation and replaced the lifetime with its average value over the dissociation path. We have also assumed that the two resonances show equal capture cross section at a given electron energy. This to some extent can be justified as the two potential energy curves considered for the simulations run very close to each other in the Frank-Condon region making their Franck-Condon factors almost equal for the transition from the vibrational and electronic ground state of the target. The numerator of equation (10) arises from the cross term of equation (7) integrated over the upper and lower half of space. The phase difference, φ, between the two paths is given by

$$\varphi = \frac{1}{\hbar}\int_{R_c}^{\infty}\left[\left(2\mathrm{m}\left(\mathrm{E}-\mathrm{V_u}(R)\right)\right)^{1/2} - \left(2\mathrm{m}\left(\mathrm{E}-\mathrm{V_g}(R)\right)\right)^{1/2}\right]\mathrm{d}R + \frac{\pi}{2} \tag{12}$$

where m is the reduced mass of the dissociating molecule, E is the electron energy and V(R) is the potential energy of the NIR states involved. The first term accounts for the phase difference arising from the two quantum paths of the dissociating wave packets. The additional phase of $\pi/2$ was obtained from the relative phase between the s and p partial waves of the plane wave description used for the incoming electron along the z-axis of the lab system. Ideally, in the presence of the target, the complex amplitudes of the partial waves will get distorted leading to an additional phase difference between the s and p waves, which will depend on the incident electron energy. However, as we use the zeroth order approximation, we have ignored this. We have taken the average lifetime for the $^2\Sigma_g^+$ state as 8fs based on its reported width and that for the $^2\Sigma_g^+$ as a parameter. The results obtained are shown in figure 3. The asymmetry factors are close to the observed values given in table 2.

As mentioned in the introduction, the symmetry of the NIR state active at 14eV was found to be a $^2\Sigma_g^+$ from electron scattering experiments [31]. On the other hand, the present angular distribution measurements of DEA indicate a strong contribution from the $^2\Sigma_u^+$ state. In fact, in the case of D_2 the slower dissociation results in greater loss of $^2\Sigma_g^+$ state amplitude to autodetachment, which makes the angular distribution more symmetric, dominated by the $^2\Sigma_u^+$ amplitude.

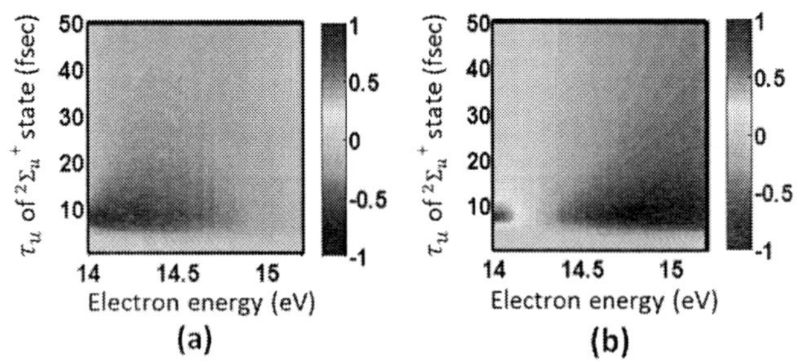

Figure 3. Estimated asymmetry factor as a function of electron energy and lifetime (τ_u) of the $^2\Sigma_u^+$ state in fsec for (a) H_2 and (b) D_2 where the lifetime of the $^2\Sigma_g^+$ state is taken as 8 fsec. (Reproduced from figure 4(b) and (d) of reference 12).

5.4. Signatures of quantum coherence in the scattered electron channel

We have seen above that the electron attachment process leads to a break in inversion symmetry at the 14 eV resonance due to coherent excitation of the two resonances of opposite parity. One would expect that the angular distribution of the autodetaching electrons from these resonances may also show corresponding asymmetry. It was the angular distribution of the ejected electrons at the incident energy of 14 eV that provided the identification of the symmetry of the resonance at this energy as $^2\Sigma_g^+$ [32]. However, a closer look at the data [32] shows a clear asymmetry in the angular distribution, which the authors acknowledged, but was left unexplained since a homonuclear diatomic molecule was expected to provide a symmetric distribution. The limited angular range they used in the backward direction also did not help them.

5.5. *Interferences in the ion yield curves*

As discussed in Section 1, equally spaced structures in the ion yield curve from H_2 in the 14 eV resonance were observed earlier, while no such structures were observed in D_2 [30]. The measurements were performed at 90^0 with respect to the incoming electron beam. The observed oscillation in the ion signal was found to be at the interval of 0.2eV. It was argued that the spacing might correspond to the vibrational spacing of a $^2\Delta_g$ state, which was predissociating due to mixing with the $^2\Sigma_g^+$ NIR contributing to the DEA [30]. Since the difference in angular momenta of the two states is more than 1 unit, the mixing could happen only under breakdown of Born-Oppenheimer approximation. One may wonder if there is a connection with their observation and the present results.

We may conjecture whether the interference observed in the present experiment in terms of angular distribution gives rise to that observed in the ion yield curve at 90°. At $\theta = 90°$, the interference term vanishes in the expression for $I(k, \theta)$ given above. So strictly, the interference that we observe could not have given rise to the oscillation in the ion yield curve. With the electron energy resolution of about 0.6 eV and reasonable imaging resolution ($\Delta p/p \sim 10\%$), we expect to observe the structures separated by 0.2 eV as reported by Tronc *et al.* [30] in the 90° direction. We are unable to see these structures, probably due to a large width of the time slice (80 nsec) in comparison with the overall time spread of the Newton sphere of ions (200nsec). Thus neither do we find a connection with those structures nor do we have an interpretation for it.

5.6. *Lifetimes of resonances in coherent superposition*

As can be seen from the logic of decay of a resonance, overlapping resonances with coherent excitation will not decay exponentially, although each of them shows exponential decay independently. Frishman and Shapiro have shown that such a superposition can be used to completely supress the decay of the resonance by suitably tweaking the relative phase between the two superposed states [36]. In light of this, it is interesting to see which way the coherent superposition of resonances affects the dynamics of the process. Our model, which gives reasonable match with the observed forward-backward asymmetry assumes that both the resonances retain their independent lifetime. That means they are understood to be decaying exponentially independent of one another. However, this is a crude approximation and it will be interesting to see a better model to describe the process.

6. Summary and conclusion

DEA to molecular hydrogen at the 14 eV resonance is found to be the result of coherent creation of two resonances of opposite parity leading to forward-backward asymmetry in the angular distribution of the fragment negative ion. As the wave packet propagates through two different paths based on the potential energy curves of the two resonances and leads to the same dissociation limit, we observe interference in the forward-backward asymmetry factor as a function of the electron energy. This interference is also dependent on the isotopic composition as seen in the case of D_2. Simulation using model potential energy curves shows fair agreement with the observed asymmetry. Fitting the angular distribution data as a function of electron energy using $^2\Sigma_g^+$ and $^2\Sigma_u^+$ symmetries qualitatively shows the contributions from both resonances. However, due to limited momentum resolution, quantitative estimates of the amplitudes of the two resonances could not be obtained with any reliability. We do not observe any signatures of the structures seen in an earlier report in the ion yield curves around 14 eV. It is clear that further experimental and theoretical work on NIR states are needed even for the simplest molecule.

Acknowledgement

E.K. acknowledges a Marie Curie Fellowship during the course of the measurements and Raja Ramanna Fellowship, Department of Atomic Energy India. V.S.P. acknowledges the Department of Atomic Energy, India for financial support. N.J.M. recognizes support from the European Union Framework 7 programme LASSIE Marie Curie ITN Grant Agreement 238258 and VAMDC INFRA-2008-1.2.2 Scientific Data Infrastructure. Grant Agreement no.: 239108.

References

[1] Prabhudesai V S, Kelkar A H, Nandi D and Krishnakumar E 2005 *Phys. Rev. Lett.* **95** 143202
[2] Prabhudesai V S, Nandi D, Kelkar A H and Krishnakumar E 2008 *J. Chem. Phys.* **128** 154309
[3] Ptasinska S *et al.* 2005 *Phys. Rev. Lett.* **95** 093201
[4] Ibănescu B C and Allan M 2009 *Phys. Chem. Chem. Phys.* **11** 7640
[5] Samukawa S 2005 *Japan. J. Appl. Phys.* **45** 2395
[6] Sansone G *et al.* 2010 *Nature* **465** 763
[7] Krishnakumar E and Nagesha K 1992 *J. Phys.* B*: At. Mol. Opt. Phys.* **25** 1645
[8] Rawat P *et al.* 2007 *J. Phys.* B*: At. Mol. Opt. Phys.* **40** 4625
[9] Rawat P *et al.* 2008 *Int. J. of Mass Spectrom.* **277**, 96
[10] Krishnakumar E and Prabhudesai V S 2019 *Springer Proc. In Phys.* **230** 20
[11] Fabrikant I I, Eden S, Mason N J and Fedor J 2017 *Adv. Atom. Mol. Opt. Phys.* **66** 545
[12] Krishnakumar E, Prabhudesai V S and Mason N J 2018 *Nat. Phys.* **14** 149
[13] Christophorou L G, McCorkle D L and Christodoulides A A 1984 *Electron-Molecule Interactions and Their Applications* vol 1 ed Christophrous L G (New York: Academic) ch 6
[14] Yoon J-S *et al.* 2008 *J. Phys. Chem. Ref. Data* **37** 913
[15] Yoon J-S *et al.* 2010 *Rep. Prog. Phys.* **73** 116401
[16] Schulz G J 1973 *Rev. Mod. Phys.* **45** 423
[17] Zecca A, Karwasz G P and Brusa R S 1996 *La Rivista del Nuovo Cimento* **19** 1
[18] Krishnakumar E, Denifl S, Čadež I, Markelj S and Mason N J 2011 *Phys. Rev. Lett.* **106**, 243201
[19] Schulz G J 1959 *Phys. Rev. Lett.* **113** 816
[20] Rapp D, Sharp T E and Briglia D D 1965 *Phys. Rev. Lett.* **14** 533
[21] Schulz G J and Asundi R K 1967 *Phys. Rev.* **158** 25
[22] Fedor J, Kočišek J and Janečková R 2015 *J. Phys.: Conf. Ser.* **635** 072093
[23] Allan M and Wong S F 1978 *Phys. Rev. Lett.* **41**, 1791
[24] Čadež I *et al.* 1988 *J. Phys.* B*: At. Mol. Opt. Phys.* **21** 3271
[25] Celiberto R *et al.* 2011 *Atom. Data Nucl. Data Tables* **77** 161
[26] Fabrikant I I, Wadehra J M and Xu Y 2002 *Phys. Scr.* **2002** 45
[27] Sharp T E 1971 *Atom. Data Nucl. Data Tables* **2** 119
[28] Eliezier I, Taylor H S and Williams Jr J K 1967 *J. Chem. Phys.* **47** 2165
[29] Stibbe D T and Tennyson J 1998 *J. Phys.* B*: At. Mol. Opt. Phys.* **31** 815
[30] Tronc M, Hall R I, Schermann C and Taylor H S 1979 *J. Phys.* B*: At. Molec. Phys.* **12** L279
[31] Tronc M, Fiquet-Fayard F, Schermann C and Hall R I 1977 *J. Phys.* B*: At. Molec. Phys.* **10** 305
[32] Weingartshofer A, Ehrhardt H, Hermann V and Linder F 1970 *Phys. Rev.* A **2** 204
[33] O'Malley T F 1966 *Phys. Rev.* **150** 14
[34] Nandi D, Prabhudesai V S, Krishnakumar E and Chatterjee A 2005 *Rev. Sci. Instrum.* **76** 053107
[35] O'Malley T F and Taylor H S 1968 *Phys. Rev.* **176** 207
[36] Frishman E and Shapiro M 2001 *Phys. Rev. Lett.* **87** 253001

Absolute cross sections for production of molecular dications by electron impact

L Sigaud[1] and E C Montenegro[2]

[1] Universidade Federal Fluminense, Instituto de Física, ZIP 24210-346, Niterói, RJ, Brazil
[2] Universidade Federal do Rio de Janeiro, Instituto de Física, P.O. 68528, ZIP 21941-972, Rio de Janeiro, RJ, Brazil

E-mail: lsigaud@if.uff.br

Abstract. Molecular dications are important metastable species present in a variety of environments including, but not restricted to, planetary atmospheres and the human body when submitted to radiation. On the other hand, symmetric molecular dications cannot be separated from the cation corresponding to the molecule broken in half (e.g. N_2^{++} and N^+) in stantard mass spectrometers. The DETOF technique allows one to identify the symmetric molecular dication, obtaining absolute cross sections related to their production. Results for dications of molecular nitrogen, molecular oxygen, ethylene and benzene are discussed here, giving emphasis on the mechanisms behind their production. It is seen that, while most metastable molecular dications result from a combination of direct double ionization and Auger-like processes, N_2^{++} and O_2^{++} represent so-called pure cases, where the dication is produced only by, respectively, double direct ionization or post-collisional Auger-like decay.

1. Introduction

Molecular dications have been the object of many different studies dedicated to their production, stability and subsequent fragmentation [1, 2, 3, 4, 5, 6, 7]. The most interesting aspect of dications is that, while singly ionized fragmentation process produce only singly-charged and neutral species, resulting in low kinetic energy releases (KER), they can dissociate into fragments with high kinetic energy due to Coulombic repulsion [8, 9, 10], leading to KER of the order of tens of eV. Their presence in planetary atmospheres have been predicted by many authors and the KER resulting from their fragmentation can lead to a local increase of the atmospheric temperature, which in turn is directly connected to planetary escape [1, 11, 12].

Although doubly charged molecular ions are possibly a crucial factor for the understanding of physical-chemical processes in energized media [13], such as planetary atmospheres, plasma environments and the human body when submitted to radiotherapy treatments, little is known regarding the absolute cross sections for the dication production in symmetric molecules, due to the intrinsic impossibility of separating moieties with the same mass to charge ratio in a time of flight (TOF) spectrum [1, 14]. As a consequence, they have never been detected in planetary atmospheres yet, although both molecular nitrogen and oxygen are the main components of Earth's atmosphere, where, on the other hand, O^{++} has already been detected [15, 16], showing that double ionization processes are present.

Besides, the mechanisms and pathways for dications formation and possible stability are still not known, raising the question if post-collisional processes play a significant role and, if that is the case, what would their relevance be for the production of these metastable species. In this work, the following symmetric molecular dications were investigated: nitrogen, oxygen, benzene and ethylene. In order to disentangle the dication from the fragment representing the breakup of the molecule in two identical parts (e.g. N_2^{++} and N^+), the well-established Delayed Extraction Time-Of-Flight (DETOF) technique was employed. The DETOF technique allows one to identify different kinetic energy distributions present in recoil ions. Since the fragmented molecules acquire kinetic energy in the fragmentation process, while dications retain their original Maxwell-Boltzmann velocity distribution, the identification of the latter becomes straightforward [17, 18, 19, 20].

A comparison between the production mechanisms of these four molecules and other non-symmetric molecular dications found in the literature show that a combination of direct double ionization and Auger-like post-collisional relaxation are usually present (as, for example, are the cases of ethylene and benzene). On the other hand, nitrogen and oxygen dications are produced by pure processes, where only one mechanism is predominantly present in their formation.

2. Experimental setup

The experimental setup is detailed elsewhere [21], thus only a summarized description is included. It consists mainly of a pulsed electron gun, operating in the 25 to 900 eV energy range, a gas cell and a time-of-flight mass spectrometer. The pressured inside the gas cell is controlled and kept below 5×10^{-4} torr in order to ensure a single-collisional regime, the electron beam is collected by a Faraday cup and the recoil ions, after passing through the TOF drift tube, are detected by a Micro-Channel Plate (MCP), allowing therefore absolute cross sections to be obtained. Both the electron beam and an 21 V/mm electrostatic extraction field used to guide the produced positively-charged ions into a TOF drift tube are synchronously pulsed with a frequency of 20 kHz and pulse durations of 50 ns and 100 ns, respectively. This electrostatic extraction field signal and the MCP signal are time-resolved by a time-to-digital converter analyzer in order to produce the TOF spectra, and the overall efficiency of the apparatus is determined by a calibration procedure described before [21].

The DETOF technique is based on this synchronization between the electron beam and extraction field pulses. If the extraction field is turned on exactly after the electrons pass through the interaction region inside the gas cell, all recoil ions produced in a collision would be guided through the collimator at the entrance of the TOF drift tube. On the other hand, if some time delay exists between these two pulses, the fragments after molecular breakup are allowed some free flight time before the extraction field is applied. Thus, species produced with high velocities after fragmentation may leave the collection region and will not be detected - in other words, this time delay works as a velocity selector. This means that consecutive measurements, by systematically and gradually increasing this time delay between the electron beam and extraction field pulses, provide a profile of the kinetic energy distributions present for a single peak in a TOF spectrum. For the particular case of dication production, since a molecular dication does not undergo fragmentation, it retains its original Maxwell-Boltzmann distribution, while all other higher kinetic energy distributions must correspond to the fragmented species with the same mass-to-charge ratio, which acquire momentum in the molecular breakup (e.g. N_2^{++} and N^+). This allows the disentanglement between the two different species present in the same peak in the TOF spectra, and since their relative proportions are known, their production absolute cross sections can be obtained. It must be noted that the time-frame of a measurement is in the order of microseconds [18, 20, 21] and, since the metastable lifetimes of all dications discussed here are orders of magnitudes higher than that, we can assume that only a negligible amount of them will dissociate inside the spectrometer before being detected.

3. Results

Absolute cross section values for the molecular dication production of nitrogen, oxygen and ethylene were already published before using the DETOF technique [18, 19, 22, 23]. However, in order to evaluate the predominant mechanisms for their production, the most important property to be analyzed is the dication-parent ion ratio, meaning the ratio between the doubly and singly-ionized non-fragmented molecule, as can be seen in figure 1. It is evident there that the behavior for all three molecules, specially when looking at the high impact energy range of the data, is very different. For nitrogen, the ratio steadily decreases with the increase of the impact energy, while for oxygen, after the increase above the dication threshold up to $\sim$ 100-200 eV, the ratio remains constant throughout the whole impact energy range. Ethylene, however, seems to exhibit both features: there is a marked decrease with the increase of the impact energy after the maximum value for the ratio has been reached, but above $\sim$ 500 eV impact energy the ratio reaches a constant value up to the highest impact energy observed here.

It is well known that, at high projectile velocities (with corresponding impact energy E), single ionization has a cross section dependency with a $1/E$ trend, while for direct double ionization, meaning the removal of two electrons in the same collisional process, it follows a $1/E^2$ behavior [24, 25]. Therefore, a direct double-to-single ionization ratio should follow a $\sim 1/E$ shape, since it is a comparison from a second order to a first order process. On the other hand, if the dication is produced by a single electron removal followed by a post-collisional Auger-like auto-ionization, the primary collisional process results in a single ionization. In this case, the doubly-to-singly ionized species ratio should be constant, since both cases come from a first order process.

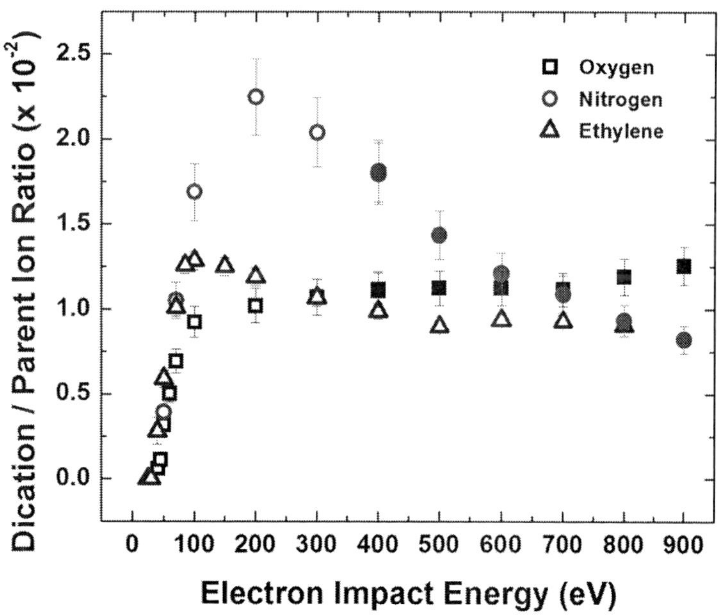

Figure 1. Ratio between dication and parent ion absolute cross sections for nitrogen (red circles - empty symbols [18] and full symbols [22]), oxygen (black squares - empty symbols [20] and full symbols [22]) and ethylene (blue triangles [23]).

In order to investigate this, a further analysis of the behavior of the ethylene dication-to-parent ion ratio is shown in figure 2. Once again the ratio between the dication and its parent ion is shown, but now the ratio between the K-shell ionization cross section from Talukder *et al.* [26] and our experimental parent ion cross section [23] are also included. A rough estimate of the contribution of both double direct ionization and K-shell ionization followed by Auger decay to the dication formation can be performed using the expression $R \sim A/E + B\sigma_K/\sigma_{C_2H_4^+}$, where R is the dication-to-parent ion ratio, σ_K is the K-shell ionization cross section from Talukder *et al.*, $\sigma_{C_2H_4^+}$ is the parent ion absolute cross section and A and B are the relative weights of each contribution. Using $A = 2.368$, in order to normalize it to the dication-to-parent ion ratio at 200 eV impact energy, and $B = 0.6$, a reasonably good agreement with the measured ratio is obtained for impact energies above 300 eV, shown in figure 2. This shows not only that both processes contribute significantly to $C_2H_4^{++}$ formation, but also that, since the value for B is below unity, single K-shell ionization followed by Auger decay must also contribute to the fragmentation of the molecule.

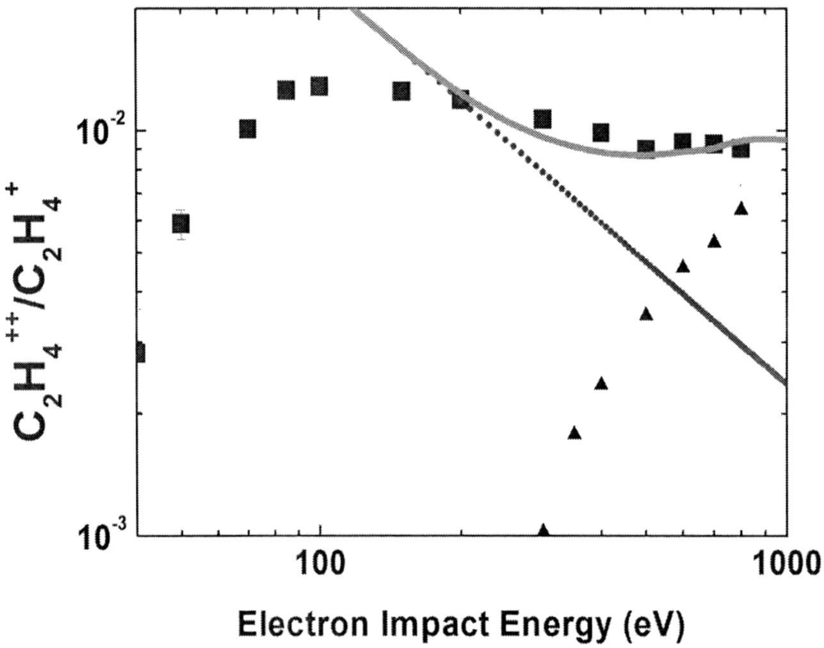

Figure 2. Ratio between ethylene dication and parent ion absolute cross sections (blue squares [23]). Theoretical points for a $\sim 1/E$ dependence, where E is the impact energy, normalized with the experimental ratio data at 200 eV impact energy (red circles). Ratio between the data for K-shell ionization of carbon by Talukder *et al.* [26] and our parent ion experimental data [23] (black triangles). A good agreement between the experimental ratio and the sum of the other two with adequate weights can be reached and is shown as a solid orange line here.

On the other hand, when one looks at the behavior for nitrogen and oxygen dications, shown in detail in figure 3, it can be observed that both molecules do not exhibit these two simultaneously competitive mechanisms. Figure 3 shows, along with the experimental dication-to-parent ion ratio, the K-shell ionization data for nitrogen and oxygen from Tawara *et al.* [27] and the double ionization fragmentation data (i.e. $N^+ + N^+$ and $O^+ + O^+$) from Tian and Vidal [28], all normalized to the parent ion cross section. It can be readily seen that K-shell ionization, particularly for the oxygen molecule case, is negligible in comparison to the ratios observed. This means that even if a relatively large contribution of post-collisional Auger decay comes after K-shell ionization, this value is still small enough so as to not play a significant role in the dication production. For nitrogen, in particular, it can be seen when comparing to the N^{+}+N^{+} data of Tian and Vidal [28], whose ratio also present a $\sim 1/E$ shape, that the main mechanism for producing metastable molecular nitrogen dication comes from direct double ionization (known as TS-2). Auger-like processes involving inner valence shell primary ionization are not energetically (from a molecular orbital's point of view) allowed, and therefore cannot be present. Thus, nitrogen dication represents a **pure** double direct ionization mechanism for dication production in the projectile energy range studied.

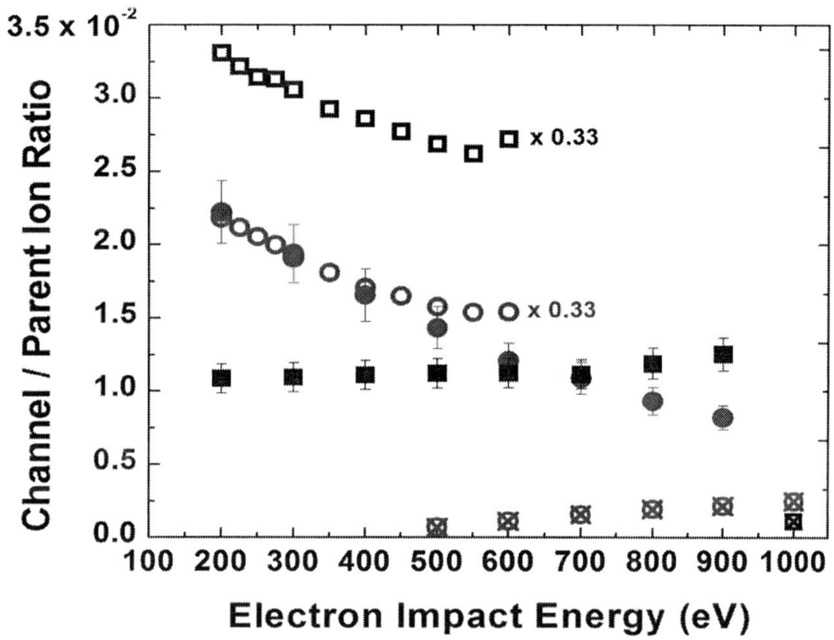

Figure 3. Ratio between different channels' - described below - and the parent ion cross sections as a function of the electron impact energy. Oxygen molecule, black squares: filled symbols - O_2^{++}, this work; open symbols - O^{+}+ O^{+}, Tian and Vidal [28]); crossed symbols - O K-shell ionization cross section, Tawara *et al.* [27]. Nitrogen molecule, red circles: filled symbols - N_2^{++}, this work; open symbols - N^{+}+ N^{+}, Tian and Vidal [28]); crossed symbols - N K-shell ionization cross section, Tawara *et al.* [27]. The data from Tian and Vidal have been multiplied by 0.33 in order to facilitate comparison in the same plot.

Meanwhile, the metastable molecular oxygen dication exhibits a completely different behavior from that of nitrogen. After reaching its maximum value at around 100-200 eV, the dication-to-parent ion ratio remains constant throughout the whole measured impact energy range. This leads to the conclusion that direct double ionization is negligible in comparison to post-collisional auto-ionization processes following single ionization, depicted by the ratio's constant dependency with the impact energy. It is important to notice as well that not only K-shell ionization of oxygen is negligible when compared to the ratio values, but also that its onset is at a much higher impact energy than what is observed in the present experiment. Therefore, the main mechanism behind metastable molecular oxygen dication production must be inner valence shell single ionization followed by an Auger-like ionization, leaving the molecule in a doubly-charged state. The interesting question that arises in this case is: why does O_2 not allow double direct ionization as a mechanism for its dication formation?

The final state of an oxygen molecular dication implicates the absence of its two outermost unpaired electrons. The probability of the two outermost electrons being removed in the same collisional process must be so small as to completely suppress this dication formation pathway. A possible explanation is that the molecular π orbitals of the unpaired electrons absent in the final dicationic state are spatially separated from each other in the oxygen molecule. Therefore, the trajectory of the projectile has to be a very particular one in order to access exclusively these two electrons in order to interact with them, which means that the geometry of the molecule itself suppresses this mechanism [22]. Therefore, it can also be assumed that direct double ionization of other than the two outermost electrons probably results in a fragmented state, as can be seen in figure 3 from the O^++O^+ data of Tian and Vidal. O_2^{++} is, hence, a **pure** Auger-like post-collisional auto-ionization following single ionization mechanism for dication formation.

A further evidence for this can be seen when comparing these results with collisions between electrons and O_2^+ [29]. Although the fragmentation channel still dominates, the stable O_2^{++} is now produced by direct ionization of the cation, without any geometrical constraint to inhibit the stable dication production. On the other hand, for stable N_2^+ ionization [30], the formation of the stable N_2^{++} is more probable than its fragmentation, which resembles what was observed [22].

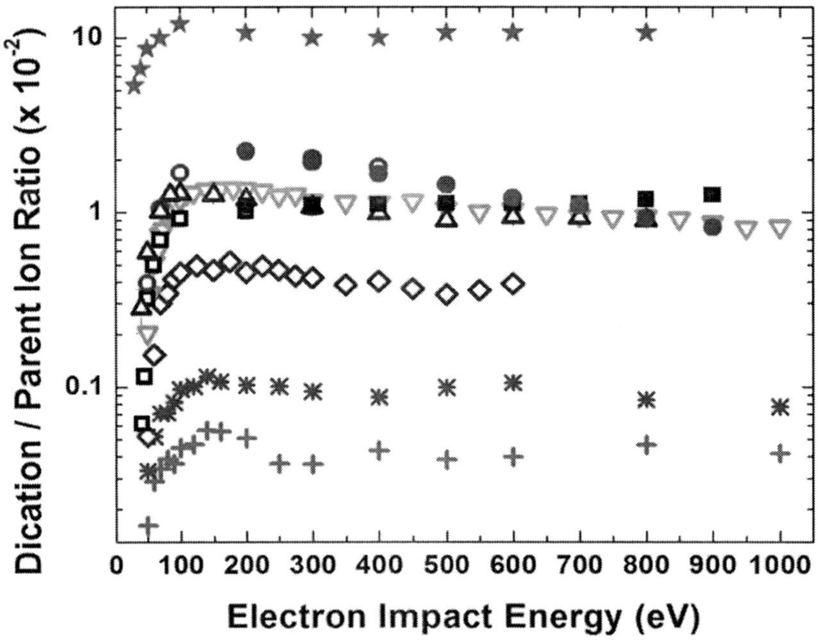

Figure 4. Ratio between dication and its parent ion cross sections as a function of the electron impact energy for different molecules: O_2 (black squares [20, 22]), N_2 (red circles [18, 22]), C_2H_4 (ethylene - blue up triangles [23]), C_6H_6 (benzene - pink stars, present work), NH_3 (gold crosses [31]), ND_3 (green asterisks [31]), CO (purple lozenges [32]) and CO_2 (orange down triangles [33]).

Comparison, in figure 4, of these results with several other molecules - namely benzene (present data), NH_3 (Rejoub *et al.* [31]), ND_3 (Rejoub *et al.* [31]), CO (Tian and Vidal [32]) and CO_2 (Straub *et al.* [33]) - shows that for almost every molecule single and double vacancy production processes (Auger-like post-collisional ionization and TS-2) are competitive mechanisms for the dication production. CO_2^{++} and ND_3^{++} data are still inconclusive, but seem to exhibit the same trend. The data for benzene shows that not only benzene dication has a much larger cross section than any other studied dication - meaning it is a much more stable molecule, since after the removal of two electrons the probability that it does not fragment is much larger than for other molecules - but also that the its main dication production mechanisms are exactly like ethylene's, presented above, with only different weights between them. It must be remembered that benzene has triple the amount of carbon atoms than ethylene, which increases the importance of carbon K-shell ionization. Thus, it is interesting to observe that the only two guaranteed pure mechanisms are, yet, N_2^{++} and O_2^{++}.

4. Conclusions

To sum it up, the DETOF technique allows the disentanglement of species with the same mass-to-charge ratio, isolating the dication of a symmetric molecule from its symmetric fragments. Therefore, it became possible now to determine the absolute cross sections for the formation of symmetric molecular dications such as N_2^{++}, O_2^{++} and $C_2H_4^{++}$ (prior to that, only dications of such molecules could only be measured by looking at hetero-isotopic nuclei, breaking the molecule's symmetry [14]).

The ratio of dication to parent ion absolute cross sections shows an important aspect of the dication formation - the energy dependence between first (single ionization) and second (direct double ionization) order processes. It is shown that ethylene and benzene, along many other molecules found in the literature, present a contribution of both processes to their dication formation: for low impact energies the direct double ionization dominates, while for high impact energies the dication production comes primarily from post-collisional Auger-like processes. This means that, for most molecules, dications, a doubly-ionized species in its final state, are not negligible in the high projectile velocities regime when compared to the singly-ionized moieties. Also, it must be noted that, since post-collisional processes depend only on the molecular orbital from which an electron was removed in the collision, theses processes must be present for dication formation regardless of the projectile (albeit with different cross section values).

Lastly, two cases are very important as benchmarks: (i) N_2^{++}, which shows a pure direct double ionization process because inner valence shell Auger-like decays are not energetically allowed and K-shell ionization is negligible in comparison; and (ii) O_2^{++}, which shows a pure post-collisional Auger-like decay following single inner valence shell ionization, in which direct double ionization is suppressed due to geometrical reasons. It must be mentioned that these behaviors could only be identified thanks to the DETOF technique, since N_2^{++} and O_2^{++} were not clearly observed prior to that in a TOF spectrum.

Acknowledgments

The authors would like to acknowledge the Brazilian funding agencies CNPq, CAPES and FAPERJ, as well as INCT-FNA (Proc. No. 464898/2014-5).

References

[1] Thissen R, Witasse O, Dutuit O, Simon Wedlund C, Gronoff G and Lilensten J 2011 *Phys. Chem. Chem. Phys.* **13** 18264
[2] Roithová J and Schröder D 2007 *Phys. Chem. Chem. Phys.* **9** 2341
[3] Mathur D 2004 *Phys. Rep.* **391** 1
[4] Price S D 2007 *Int. J. Mass Spec.* **260** 1
[5] Franceschi P, Thissen R, Zabka J, Roithová J, Herman Z and Dutuit O 2003 *Int. J. Mass Spectrom.* **228** 507
[6] Slattery A E, Field T A, Ahmad M, Hall R I, Lambourne J, Penent F, Lablanquie P and Eland J H D 2005 *J. Chem. Phys.* **112** 084317

[7] Alagia M, Candori P, Falcinelli S, Lavollée M, Pirani F, Richter R, Stranges S and Vecchiocattivi F 2009 *J. Phys. Chem.* A **113** 14755
[8] Leach S 1986 *J. Elec. Spec. Rel. Phenom.* **41** 427
[9] Nicolaides C A 1989 *Chem. Phys. Lett.* **161** 547
[10] Dolejsek Z, Farnik M and Herman Z 1995 *Chem. Phys. Lett.* **235** 99
[11] Simon C, Lilensten J, Dutuit O, Thissen R, Witasse O, Alcaraz C and Soldi-Lose H 2005 *Ann. Geophys.* **23** 781, and references therein
[12] Avakyan S V 2005 *J. Opt. Technol.* **72** 602
[13] Price S D 2003 *Phys. Chem. Chem. Phys.* **5** 1717
[14] Märk T D 1975 *J. Chem. Phys.* **63** 3731
[15] Šmilauer J, Truhlík V, Bošková J, Tríska P and Shultchishin J 1996 *Adv. Space Res.* **17 (10)** 135
[16] Witasse O, Slanger T and Thissen R 2011 *Ann. Geophys.* **29** 2235
[17] Natalia Ferreira, Sigaud L and Montenegro E C 2014 *Journal of Physics: Conference Series* **488** 012042
[18] Natalia Ferreira, Sigaud L, de Jesus V L B, Rocha A B, Coutinho L H and Montenegro E C 2012 *Phys. Rev.* A **86** 012702
[19] Sigaud L, Natalia Ferreira, Coutinho L H, de Jesus V L B and Montenegro E C 2012 *J. Phys.* B: *At. Mol. Opt. Phys.* **45** 215203
[20] Sigaud L, Natalia Ferreira and Montenegro E C 2013 *J. Chem. Phys.* **139** 024302
[21] Sigaud L, de Jesus V L B, Natalia Ferreira and Montenegro E C 2016 *Rev. Sci. Instrum.* **87** 083112
[22] Sigaud L and Montenegro E C 2018 *Phys. Rev.* A **98** 052701
[23] Sigaud L and Montenegro E C 2017 *Phys. Rev.* A **95** 012704
[24] Sant'Anna M M, Montenegro E C and McGuire J H 1998 *Phys. Rev.* A **58** 2148
[25] McGuire J H 1982 *Phys. Rev. Lett.* **49** 1153
[26] Talukder M R. Bose S and Takamura S 2008 *Int. J. Mass Spectrom.* **269** 118
[27] Tawara H, Harrison K G and de Heer F J 1973 *Physica* **63** 351
[28] Tian C and Vidal C R 1998 *J. Phys.* B: *At. Mol. Opt. Phys.* **31** 5369
[29] Cherkani-Hassani H, Belic D S, Jureta J J and Defrance P 2006 *J. Phys.* B: *At. Mol. Opt. Phys.* **39** 5105
[30] Bahati E M, Jureta J J, Belic D S, Cherkani-Hassani H, Abdellahi M O and Defrance P 2001 *J. Phys.* B: *At. Mol. Opt. Phys.* **34** 2963
[31] Rejoub R, Lindsay B G and Stebbings R F 2001 *J. Chem. Phys.* **115** 5053
[32] Tian C and Vidal C R 1998 *J. Phys.* B: *At. Mol. Opt. Phys.* **31** 895
[33] Straub H C, Lindsay B G, Smith K A and Stebbings R F 1996 *J. Chem. Phys.* **105** 4015

Role of methylation in electron scattering on $X(CH_3)_4$ (X= C, Si, Ge) molecules

Sylwia Stefanowska-Tur[1], Czesław Szmytkowski[1], Elżbieta Ptasińska-Denga[1], Paweł Możejko[1]

[1]Department of Atomic, Molecular and Optical Physics, Faculty of Applied Physics and Mathematics, Gdańsk University of Technology, Gabriela Narutowicza 11/12, Gdańsk, 80-233, Poland

E-mail: sylstefa1@student.pg.edu.pl

Abstract. To investigate how replacement of H atom with methyl group (CH_3) – in tetrahedral compounds of carbon, silicon and germanium – affects electron scattering process, total cross sections (TCS) for electron scattering from $C(CH_3)_4$, $Si(CH_3)_4$ and $Ge(CH_3)_4$ molecules have been compared with data for CH_4, SiH_4 and GeH_4 molecules. All examined data have been obtained with the same experimental setup. The shape of all discussed TCS energy dependences is very similar and is characterized by a dominant maximum peaked below 10 eV. For methylated compounds a gentle structure is also visible on high energy slope of main enhancement, between 10 – 20 eV. A simple formula for TCS evaluation for partially methylated carbon, silicon and germanium compounds is also proposed.

1. Introduction

In recent years, modern techniques of nanostructure production, like focus electron beam induced deposition (FEBID) method [1], have been of great interest. The precursors used in FEBID method often contain methyl groups, which must be removed from precursor prior to deposition of metallic compound onto substrate. Therefore it is important to understand how the CH_3 groups affect processes occurring during electron collisions with chemical compounds.

Among quantities which characterize electron collisions with molecules, total cross section is the one which contains overall information on scattering phenomena and can be measured in an absolute way and/or evaluated in theoretical calculations. In this work, we present our investigations on the role of methylation based on comparison of electron-scattering total cross section results for $C(CH_3)_4$, $Si(CH_3)_4$, $Ge(CH_3)_4$, CH_4, SiH_4 and GeH_4 molecules. The comparison for these molecular compounds can also help to find some regularities in the TCS energy dependence. These correlations may allow to estimate the TCS for methylated compounds of carbon, silicon and germanium for which there is no experimental or theoretical collisional data. Up to date, some efforts have been made to find correlations between total cross section value and physico-chemical parameters of numerous molecular targets (e.g. [2, 3, 4, 5]). The specific trends in electron-scattering total cross section energy dependence have been previously observed for hydrocarbon and perfluorocarbon target molecules: *isomer effect* [6] and *perfluorination effect* [7]. Moreover, *group additivity rule* has been formulated for nitrides and cyclic ethers [8]. Methylation effect has been also noticed in our previous experimental studies in which we have examined TCS

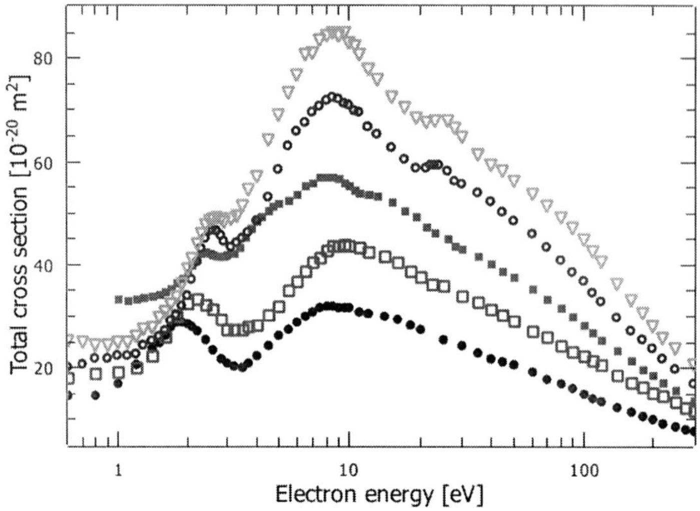

Figure 1. TCS for electron scattering from: 2,3-dimethyl-2-butene (▽) , 2-methyl-2-butene (○), 2-methylpropene (■), propene (□), ethylene (●) [9].

for selected alkenes (ethylene, propene, 2-methylpropene, 2-methyl-2-butene and 2,3-dimethyl-2-butene [9]) and dienes (allene, 1,2-butadiene and 2-methyl-1,3-butadiene molecules [10]). In figure 1 we have compared experimental TCS for selected alkenes [9], which have respectively 0, 1, 2, 3, and 4 methyl groups. The curves have very similar energy dependence over the whole investigated energy range. One of the most important observation is that, at energies above 4 eV, the TCS value increases by nearly the same value with the number of methyl groups attached around C=C bond. It is also clear that the magnitude and position of low-energy resonant peak is affected by presence of methyl groups, mainly due to redistribution of the electric charge in the methylated compounds via interaction of methyl group orbitals with the C=C orbitals [9]. The substitutional and additivity effects have been also observed and indicated for selected alkyne molecules: 1-butyne, propene and acetylene [11]. It was shown [11] that semi-empirical TCS for propyne, TCS(C_3H_4), and 1-butyne, TCS(C_4H_6), molecules can be easily obtained from the following simple rules:

$$\mathrm{TCS}_{(\mathrm{C_3H_4})}(E) = \sigma_{\mathrm{C_2H_2}}(E) + \frac{1}{2}\sigma_{\mathrm{C_2H_4}}(E) \tag{1}$$

and

$$\mathrm{TCS}_{(\mathrm{C_4H_6})}(E) = \sigma_{\mathrm{C_2H_2}}(E) + \sigma_{\mathrm{C_2H_4}}(E), \tag{2}$$

where $\sigma_{\mathrm{C_2H_2}}(E)$ and $\sigma_{\mathrm{C_2H_4}}(E)$ are the experimental TCS data [12, 13] and E denotes the incident electron energy. The obtained TCSs are in reasonable agreement with experimental data in the intermediate energy range (50-250 eV) [11]. It is worth also noting here that two theoretical methods: additivity rule [14] and independent atom method representation (IAM) [15] in combination with the screened additivity rule (SCAR) including interferences (+I)[16] are based on the general assumption that at intermediate- and high-impact electron energies the question of electron-molecule scattering can be reduced to the electron-atom collision problem.

2. Experimental details

All discussed experimental TCS data were obtained in our laboratory using electrostatic 127° electron spectrometer working in a linear transmission mode, described elsewhere [17]. Briefly, the electron beam with given energy E is directed into scattering chamber filled with studied molecules at a constant pressure p. The energy resolution of electron beam, $\Delta E = 100$ meV (FWHM). The energy scale was calibrated with respect to the well known resonant oscillatory structure visible around 2.3 eV when the target was mixed with N_2 molecules. The electrons that have not been scattered are detected by the Faraday cup collector. TCS can be determined from measured quantities using the Beer-Lambert-Bouguer (BLB) law:

$$TCS(E) = \frac{k\sqrt{T_m T_t}}{pL} \ln \frac{I_0(E)}{I_t(E)}, \tag{3}$$

where k denotes the Boltzmann constant, $I_t(E)$ and $I_0(E)$ are the intensities of the electron beam, passing the length L through the scattering chamber, taken in the presence and absence of the target molecules, respectively. T_m and T_t mean the temperature of manometer head and scattering cell. The target vapour pressure, p, was measured using a capacitance manometer and was kept within 80-180 mPa range, which ensured single-scattering conditions. The temperature of the scattering cell, T_t, is usually lower than $T_m = 322$ K by 10-20 K. Thus, recorded values of the target pressure, p, have been corrected taking into account the thermal transpiration effect [18]. The final absolute TCS value at each electron energy is a weighted mean of results from several measurement series.

3. Results

To investigate how substitution of CH_3 groups affects TCS energy dependence, we compared results for tetrahedral molecules XH_4 (X=C, Si, Ge) [19, 20, 21] with their fully methylated derivatives $X(CH_3)_4$ [22]. TCS curves for methylated compounds are shown in figure 2.

It is clearly visible that the shapes of the TCS curves for all targets are very similar to each other. Therefore we can conclude that substitution of H atoms with the CH_3 units does not influence the shape of the TCS energy dependence too much. TCSs for both molecular families are characterized by one dominant resonant-like maximum peaking at energies between 3 and 8 eV. For $Si(CH_3)_4$ and $Ge(CH_3)_4$ molecules the main maxima are shifted toward higher energies by 2.6 and 1.8 eV in comparison with SiH_4 and GeH_4 molecules, respectively. However, the TCS maximum for $C(CH_3)_4$ is shifted toward lower energy by 1.5 eV with respect to the CH_4 maximum. That results in higher TCS values around 2 eV of smaller SiH_4 and GeH_4 molecules compared to their methylated counterparts, while the TCS for CH_4 is lower than that for $C(CH_3)_4$ in the entire examined energy range. Table 1 summarizes the energies at which the TCSs have maxima.

Table 1. Electron energies [eV] at which TCSs have maxima.

Molecule	E_{max}
CH_4	8.0
SiH_4	3.0
GeH_4	3.8
$C(CH_3)_4$	6.5
$Si(CH_3)_4$	5.6
$Ge(CH_3)_4$	5.6

In general, the methylation causes a significant increase in the TCS value in the entire energy range. That is mainly related to increase in the geometrical size of the molecules [3, 23, 24].

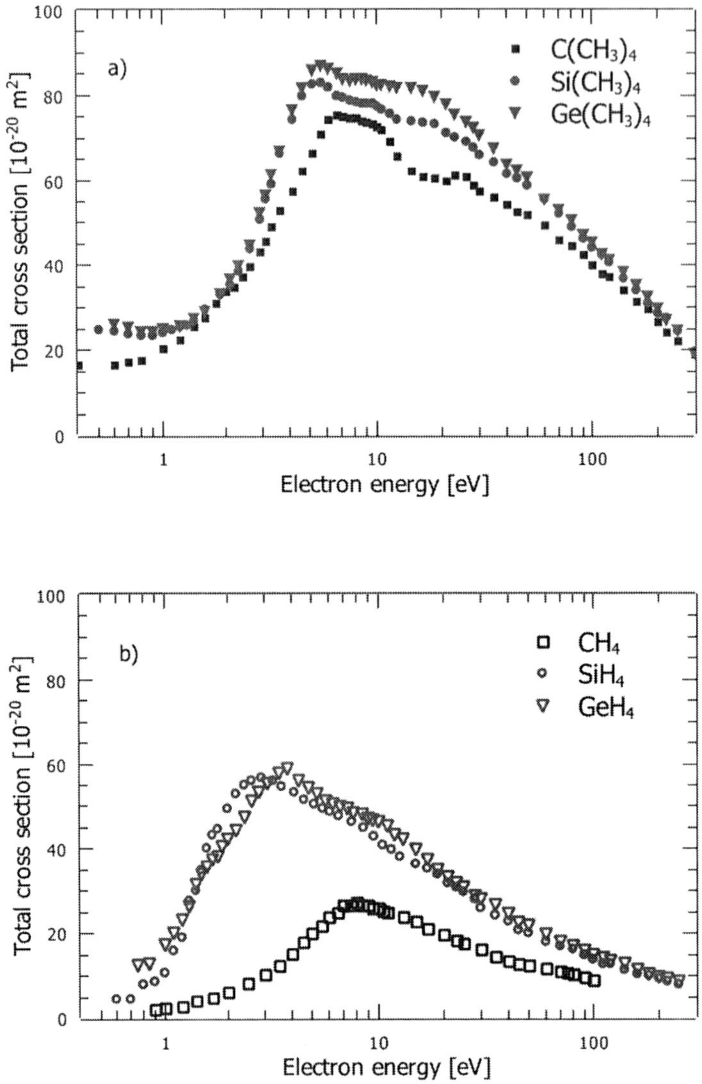

Figure 2. (a): TCS for electron scattering from: $C(CH_3)_4$ (■) [22], $Si(CH_3)_4$ (●) [22] and $Ge(CH_3)_4$ (▼) [22]. (b): TCS for electron scattering from: CH_4 (□) [19], SiH_4 (○) [21] and GeH_4 (▽) [20].

Influence of methylation is visible also above 10 eV, where for fully methylated compounds some gentle structure appears. These additional enhancements extend between 12 and 25 eV.

Figure 3 shows the experimental TCS curves for CH_4 and $C(CH_3)_4$ and the estimated TCS curves for selected methylated carbon compounds, which were obtained using a simple empirical additivity rule:

$$\mathrm{TCS}_{\mathrm{CH}_n(\mathrm{CH}_3)_m}(E) = \sigma_{\mathrm{CH}_4}(E) - m\sigma_{\mathrm{H}}(E) + m\sigma_{\mathrm{CH}_3}(E), \qquad (4)$$

where n denotes the number of hydrogen atoms bounded to the central atom and m indicates the number of methyl groups. TCS values for methyl group, σ_{CH} (E), have been determined as

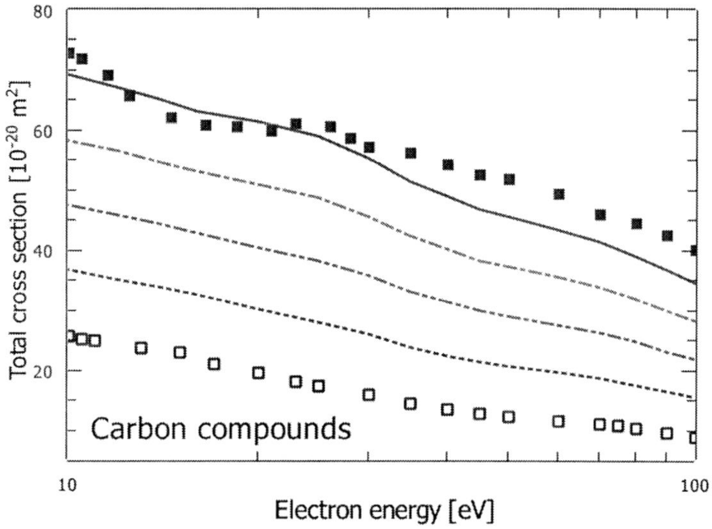

Figure 3. Experimental TCS for electron scattering from: CH_4 (□) [19] and $C(CH_3)_4$ (■) [22]. Estimated TCS (simple additivity rule) for electron scattering from: $C(CH_3)_4$ (——), $CH(CH_3)_3$ (— · —), $CH_2(CH_3)_2$ (— · · —), $CH_3(CH_3)$ (· · · · · ·).

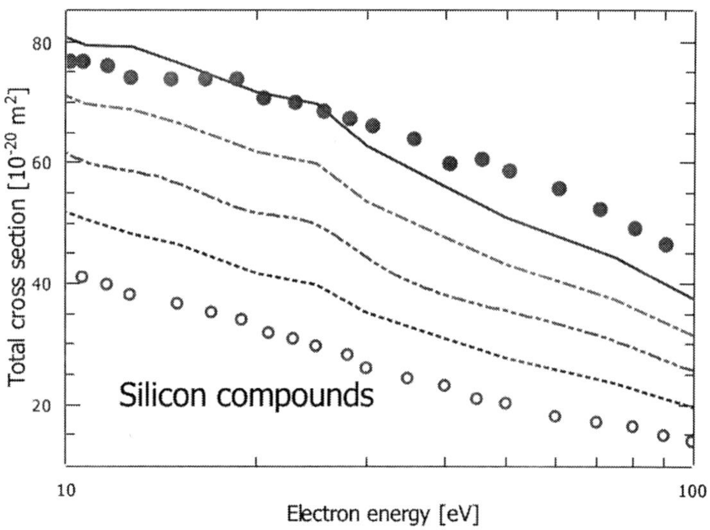

Figure 4. Experimental TCS for electron scattering from: SiH_4 (○) [21] and $Si(CH_3)_4$ (●) [22]. Estimated TCS (simple additivity rule) for electron scattering from: $Si(CH_3)_4$ (——), $SiH(CH_3)_3$ (— · —), $SiH_2(CH_3)_2$ (— · · —), $SiH_3(CH_3)$ (· · · · · ·).

a half of the experimental TCS of ethane (H_3C–CH_3) molecules [25], while TCS for hydrogen atom, σ_H, has been approximated as half of TCS measured for H_2 molecules [26]. The TCS for CH_4 molecules, $\sigma_{CH_4}(E)$, has been taken from [19]. In the above semi-empirical method of total cross section evaluation molecular target is treated as an assembly of independent subunits. Figure 4 shows estimated TCS energy dependence for silicon compounds. Similarly, the estimated

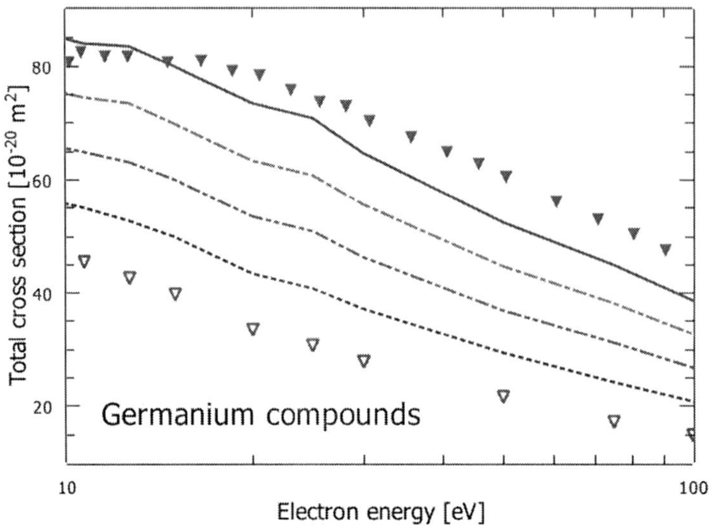

Figure 5. Experimental TCS for electron scattering from: GeH_4 (▽) [20] and $Ge(CH_3)_4$ (▼) [22]. Estimated TCS (simple additivity rule) for electron scattering from: (——) $Ge(CH_3)_4$, (— · —) $GeH(CH_3)_3$, (— · · —) $GeH_2(CH_3)_2$, (· · · · · ·) $GeH_3(CH_3)$.

TCS curves for selected methylated silicon compounds, were obtained using a simple empirical additivity rule:

$$\mathrm{TCS}_{\mathrm{SiH}_n(\mathrm{CH}_3)_m}(E) = \sigma_{\mathrm{SiH}_4}(E) - m\sigma_{\mathrm{H}}(E) + m\sigma_{\mathrm{CH}_3}(E), \qquad (5)$$

where the TCS for SiH_4 molecules, $\sigma_{\mathrm{SiH}_4}(E)$, has been taken from [21]. Figure 5 presents estimated TCS energy dependence for germanium compounds. TCSs for methylated Ge compounds were calculated as the sum of experimental TCS for molecular fragments:

$$\mathrm{TCS}_{\mathrm{GeH}_n(\mathrm{CH}_3)_m}(E) = \sigma_{\mathrm{GeH}_4}(E) - m\sigma_{\mathrm{H}}(E) + m\sigma_{\mathrm{CH}_3}(E), \qquad (6)$$

where the TCS for GeH_4 molecules, $\sigma_{\mathrm{GeH}_4}(E)$, has been taken from [20].

Evaluated this way TCSs for $C(CH_3)_4$, $Si(CH_3)_4$ and $Ge(CH_3)_4$ molecules are in reasonable agreement with experimental data [22] in investigated energy range. Discrepancies between measured and evaluated data are less than 15% in the entire energy range. Taking into account declared potential experimental uncertainties it is possible to state that the data obtained from equations (4), (5) and (6) should provide TCS evaluation with about 20% accuracy. As can be seen from figures 3, 4 and 5, the presence of extra methyl groups in studied methylated compounds ($CH(CH_3)_3$, $CH_2(CH_3)_2$, $CH_3(CH_3)$, $SiH(CH_3)_3$, $SiH_2(CH_3)_2$, $SiH_3(CH_3)$, $GeH(CH_3)_3$, $GeH_2(CH_3)_2$ and $GeH_3(CH_3)$) results mainly in general increase of the TCS magnitude. Also, the structure in TCS function seen above 10 eV for methylated compounds of silicon and germanium becomes more visible with increasing number of methyl groups. This additional structure can be associated with increasing efficiency in the ionization channel at this energy range. Although TCS results obtained this way are rather rough, they can serve as the first approximation of total cross section for electron scattering from methylated tetrahedral compounds of carbon, silicon and germanium studied in this work.

4. Conclusions

Analysis of experimental TCSs for electron scattering from tetrahedral compounds of carbon, silicon and germanium and their fully methylated derivatives has shown that methylation affects significantly the electron scattering processes. The methylation causes mainly increase in the TCS value in the entire energy range, which can be associated with growing geometrical size of the molecule. It was shown that applying the well-known additivity rule, the TCSs for $C(CH_3)_4$, $Si(CH_3)_4$, and $Ge(CH_3)_4$ molecules can be approximated quite well by a simple empirical formula obtained from the addition of cross sections of individual molecular sub-units. TCSs for partially methylated methane, silane and germane molecules are predicted for electron energies between 10 eV and 100 eV.

Acknowledgments

This work has been supported in part by the Polish Ministry of Science and Higher Education (MNiSzW Project 2017- 2018). This research was part of the COST Action CM1301 (CELINA). Sylwia Stefanowska-Tur kindly acknowledges the support of the Polish Ministry of Science and Higher Education within the Diamond Grant program (Project No. DI2015 018945).

References

[1] Huth M, Porrati F and Dobrovolskiy O V 2018 *Microelectron. Eng.* **185-186** 9
[2] Szmytkowski Cz *Z. Phys.* D **13** 69
[3] Nishimura H and Tawara H 1991 *J. Phys.* B **24** L363
[4] March N H, Zecca A and Karwasz G P 1994 *Z. Phys.* D **32** 93
[5] Garcia G and Manero F 1997 *Chem. Phys. Lett.* **280** 419
[6] Szmytkowski Cz and Kwitnewski S 2002 *J. Phys.* B **35** 2613
[7] Szmytkowski Cz and Ptasińska-Denga E 2001 *Vacuum* **63** 545
[8] Domaracka A, Możejko P, Ptasińska-Denga E and Szmytkowski Cz 2008 *Publ. Astron. Obs. Belgrade* **84** 35
[9] Szmytkowski Cz, Stefanowska S, Zawadzki M, Ptasińska-Denga E and Możejko P 2015 *J. Chem. Phys.* **143** 064306
[10] Szmytkowski Cz, Stefanowska S, Zawadzki M, Ptasińska-Denga E and Możejko P 2016 *Phys. Rev.* A **94** 042706
[11] Możejko P, Szmytkowski Cz, Zawadzki M, Ptasińska-Denga E 2014 *SPIG 2014 Contributed Papers of the 27th Summer School and Int. Symp. on the Physics of Ionized Gases (Belgrade)* (Belgrade: Institute of Physics) p 70
[12] Szmytkowski Cz, Możejko P, Zawadzki M, Maciąg K and Ptasińska-Denga E 2014 *Phys. Rev.* A **89** 052702
[13] Szmytkowski Cz, Kwitnewski S and Ptasińska-Denga E 2003 *Phys. Rev.* A **68** 032715
[14] Raj D 1991 *Phys. Lett.* A **160** 571
[15] Mott N F and Massey H S W 1965 *The Theory of Atomic Collisions* (Oxford: Oxford University Press) pp 194-196
[16] Blanco F and Garcia G 2015 *Chem. Phys. Lett.* **635** 321
[17] Szmytkowski Cz and Możejko P 2001 *Vacuum* **63** 549
[18] Knudsen M 1910 *Ann. Phys.* **31** 205
[19] Zecca A, Karwasz G, Brusa R S and Szmytkowski Cz 1991 *J. Phys.* B **24** 2747
[20] Możejko P, Kasperski G and Szmytkowski Cz 1996 *J. Phys.* B **29** L571
[21] Szmytkowski Cz, Możejko P and Kasperski G 1997 *J. Phys.* B **30** 4363
[22] Stefanowska-Tur S, Możejko P, Ptasińska-Denga E and Szmytkowski Cz 2019 *J. Chem. Phys.* **150** 094303
[23] Ariyasinghe WM, Powers D 2002 *Phys. Rev.* A **66** 052716
[24] Szmytkowski Cz, Domaracka A, Możejko P, Ptasińska-Denga E 2007 *Phys. Rev.* A **75** 052721
[25] Szmytkowski Cz and Krzysztofowicz A 1995 *J. Phys.* B **28** 4291
[26] Szmytkowski Cz, Maciąg K and Karwasz G 1996 *Phys. Scr.* **54** 271

Journal of Physics: Conference Series **1412** (2020) 052009 doi:10.1088/1742-6596/1412/5/052009

Many-body-theory calculation of positronium scattering by atomic hydrogen

A R Swann

School of Mathematics & Physics, Queen's University Belfast, University Road, Belfast BT7 1NN, United Kingdom

E-mail: a.swann@qub.ac.uk

Abstract. The recently developed many-body-theory approach for studying low-energy positronium-atom interactions is applied to calculate positronium scattering by atomic hydrogen. Calculations are carried out for the case where the two electrons are in a spin-triplet state. The elastic scattering phase shifts and cross sections are compared with existing coupled-channel, stochastic-variational, and Kohn-variational results, and excellent agreement is obtained.

1. Introduction

Positronium (Ps) is a light 'atom' consisting of an electron and its antiparticle, the positron. Calculations of Ps scattering by atoms are challenging, since both the target atom and Ps projectile are composite objects. The net electrostatic interaction between the two objects is zero. The presence of the exchange interaction between the target electrons and the electron in Ps results in repulsion for closed-shell targets, since the part of the Ps scattering wave function corresponding to the electron is required to be orthogonal to the occupied target orbitals. Mutual polarization of the target and projectile leads to the long-range attractive van der Waals interaction, which can significantly reduce the effect of the short-range Pauli repulsion [1]. Accurate calculations therefore need to account for the dynamical distortion of both objects during the collision.

A many-body-theory approach to calculating elastic scattering of Ps by closed-shell atoms has recently been developed [2]. The method accounts for virtual excitation of both the Ps and the target in an *ab initio* manner. In Ref. [2], elastic scattering phase shifts and cross sections were calculated for He and Ne targets and found to be in agreement with previous close-coupling [3] and model-potential [1] calculations. The dimensionless pickoff annihilation parameter ${}^1Z_{\mathrm{eff}}$ was also calculated for He and Ne and found to be in near-perfect agreement with experimental data [4].

In this work, the many-body-theory method is used to calculate scattering of Ps by atomic hydrogen for Ps momenta $K \leq 1$ a.u. We again consider only the case of elastic scattering, neglecting the possibility of excitation or ionization of the Ps (which is possible for $K \geq \sqrt{3}/2 \approx 0.866$ a.u.). Excitation or ionization of the H atom is only possible for $K \geq \sqrt{3/2} \approx 1.22$ a.u., which is outside the momentum range considered. We also neglect the inelastic process of Ps spin conversion, which is a relativistic effect that is suppressed at low momenta [5].

Unlike the noble-gas targets, the H atom is not closed-shell, and the elastic scattering phase shifts and cross sections depend on the total electron spin, which can have a value of 0 (singlet) or 1 (triplet). Here only the triplet case is considered, since the $1s$ orbital of the H atom is then effectively closed to the additional electron, and the many-body theory can be applied in the usual way.

The paper is structured as follows. In Sec. 2, a description of the salient features of the many-body-theory method for Ps-atom scattering, developed in Ref. [2], is given. The phase shifts and cross section for Ps-H scattering are presented in Sec. 3 and compared with the results of existing accurate *ab initio* calculations. A brief summary and outlook is given in Sec. 4.

Atomic units (a.u.) are used throughout.

2. Theory and numerical implementation

2.1. Many-body theory of electron- and positron-atom interactions

Although conventional treatments of scattering problems are usually based on the Schrödinger equation or Lippman-Schwinger equation, in many-body theory one instead uses the Dyson equation [6]. Before considering the Ps scattering problem, we first solve the Dyson equation for an electron interacting with the target atom and the Dyson equation for a positron interacting with the target atom. These equations are given by

$$(H_0^\pm + \Sigma_\varepsilon^\pm)\psi_\varepsilon^\pm(\mathbf{r}) = \varepsilon\psi_\varepsilon^\pm(\mathbf{r}). \tag{1}$$

Here, $H_0^\pm$ is the zeroth-order Hamiltonian, e.g., that of the electron $(-)$ or positron $(+)$ in the field of the atom, described in the Hartree-Fock approximation, and ε and $\psi_\varepsilon^\pm$ are the energy and quasiparticle wave function, respectively. The operator $\Sigma_\varepsilon^\pm$ is the nonlocal, energy-dependent, many-body correlation potential, which is equal to the electron or positron self-energy in the field of the atom; it acts on $\psi_\varepsilon^\pm$ as an integral operator, viz.,

$$\Sigma_\varepsilon^\pm\psi_\varepsilon^\pm(\mathbf{r}) \equiv \int \Sigma_\varepsilon^\pm(\mathbf{r},\mathbf{r}')\psi_\varepsilon^\pm(\mathbf{r}')\, d^3\mathbf{r}'. \tag{2}$$

As a result of the spherical symmetry of the problem, (1) can be solved separately for each electron or positron orbital angular momentum l, with the wave function in the form

$$\psi_\varepsilon^\pm(\mathbf{r}) = \frac{1}{r}\chi_{\varepsilon l}^\pm(r)Y_{lm}(\Omega), \tag{3}$$

where Y_{lm} is a spherical harmonic, m is the magnetic quantum number, Ω denotes solid angle, and $\chi_{\varepsilon l}^\pm$ is a radial function to be determined. Rather than working with the self-energy operator $\Sigma_\varepsilon^\pm$ in coordinate space, it is usually more convenient to work with its matrix elements in the basis of eigenstates $\varphi_\varepsilon^\pm$ of the Hartree-Fock Hamiltonian $H_0^\pm$. We have

$$\langle\varepsilon'|\Sigma_E^\pm|\varepsilon\rangle = \int_0^\infty\int_0^\infty P_{\varepsilon' l}^\pm(r')\Sigma_{El}^\pm(r,r')P_{\varepsilon l}^\pm(r)\, dr\, dr', \tag{4}$$

where $H_0^\pm\varphi_\varepsilon^\pm = \varepsilon\varphi_\varepsilon^\pm$, $\varphi_\varepsilon^\pm(\mathbf{r}) = r^{-1}P_{\varepsilon l}^\pm(r)Y_{lm}(\Omega)$, and $\Sigma_{El}^\pm$ is the self-energy for partial wave l.

The matrix elements $\langle\varepsilon'|\Sigma_E^\pm|\varepsilon\rangle$ are calculated using the diagrammatic technique [6]. Figure 1 shows the main contributions to the electron and positron self-energy. The first diagram for both the electron (top row) and positron (bottom row) accounts for the attractive long-range polarization potential $-\alpha/2r^4$, where α is the dipole polarizability of the atom. The remaining diagrams only contribute at short range. The second diagram for the positron represents the important process of virtual Ps formation, with the hatched Γ block denoting an infinite series of electron-positron Coulomb interactions [7, 8]. The electron and positron self-energy matrix

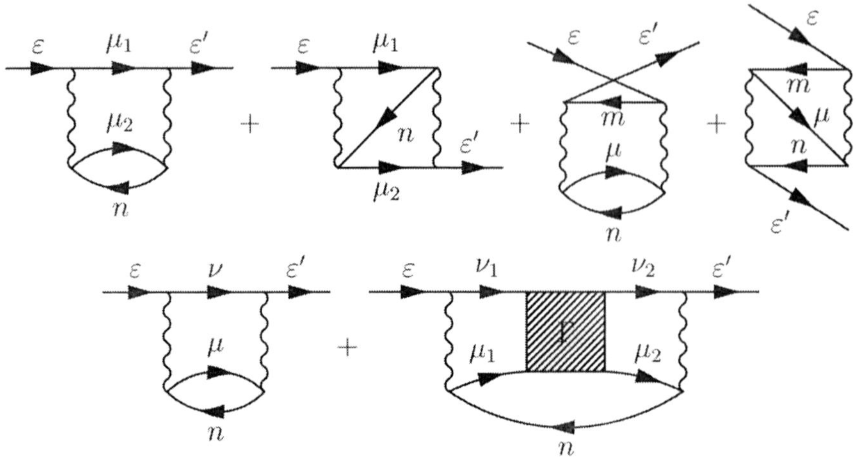

Figure 1. The main contributions to the self-energy of the electron (top row) and positron (bottom row) in the field of the atom. Lines labelled ε or ε' represent electron or positron Hartree-Fock wave functions. Lines labelled ν, ν_1, and ν_2 (μ, μ_1, and μ_2) represent positron (excited electron) states, which are summed over. Lines labelled m and n represent holes in the atomic ground state. Wavy lines represent Coulomb interactions. The shaded Γ block represents the sum of the electron-positron ladder-diagram series [7, 8], which accounts for virtual Ps formation.

elements are calculated as described in Ref. [8], using a B-spline basis with 40 splines of order 6, defined over an exponential knot sequence, in a spherical cavity of radius 30 a.u.

In Ref. [2], it was found that the electron and positron self-energy matrix elements $\langle\varepsilon'|\Sigma_E^{\pm}|\varepsilon\rangle$ depend rather weakly on the energy E at which they are calculated. Consequently, all calculations for the He and Ne targets were carried out for a fixed energy of $E = 0$ [2]. In the present calculations for the H target, the matrix elements $\langle\varepsilon'|\Sigma_E^{\pm}|\varepsilon\rangle$ are again calculated for a fixed energy of $E = 0$.

After the self-energy matrix elements have been calculated, (1) is solved to obtain the energy eigenvalues ε and wave functions $\psi_\varepsilon^{\pm}$ in the spherical cavity. For the electron, there are negative-energy states, which correspond to holes in the atomic ground state, along with positive-energy excited pseudostates that span the continuum. For the positron, there are only positive-energy 'continuum' pseudostates.

2.2. Many-body theory of Ps-atom interactions

The wave function Ψ of Ps satisfies the two-particle Dyson equation

$$\left(H_0^- + \Sigma_E^- + H_0^+ + \Sigma_E^+ + V + \delta V_E\right)\Psi = \mathcal{E}\Psi, \tag{5}$$

where V is the electron-positron Coulomb interaction, and δV_E is the energy-dependent screening correction that arises from polarization of the atom. As with the electron and positron self-energy, we calculate matrix elements of δV_E at a fixed energy of $E = 0$. Figure 2 shows the diagrammatic representation of the matrix elements of V and δV_E. Diagram (a) is the bare Coulomb interaction V. Diagram (b) is the main contribution to δV_E; it cancels the long-range electron and positron r^{-4} polarization potentials and results in the asymptotic R^{-6} van der Waals potential, where R is the distance from the Ps centre of mass to the atom. The exchange diagrams (c) and (d) are typically much smaller and partially cancel each other, so they can be neglected.

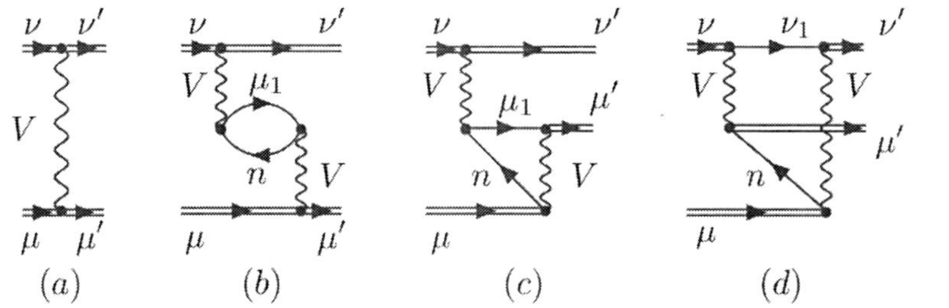

Figure 2. The main contributions to the electron-positron interaction in Ps: (a) the bare Coulomb interaction V; (b)–(d), screening δV_E with exchange contributions (with mirror images). Double lines labeled ν (μ) represent positron (electron) Dyson states in the field of the atom.

A Ps wave function with angular momentum J and parity Π can be constructed from the single-particle electron and positron Dyson states $\psi_\varepsilon^\pm$ as [1, 9]

$$\Psi_{J\Pi}(\mathbf{r}_-, \mathbf{r}_+) = \sum_{\mu,\nu} C_{\mu\nu}^{J\Pi} \psi_\mu^-(\mathbf{r}_-)\psi_\nu^+(\mathbf{r}_+), \tag{6}$$

where only products of electron and positron states whose orbital angular momenta satisfy $|l_\mu - l_\nu| \le J \le l_\mu + l_\nu$ and $(-1)^{l_\mu + l_\nu} = \Pi$ are included in the expansion. To ensure orthogonality of the electron part of the Ps wave function to the occupied target orbitals, the negative-energy electron states are excluded from the expansion. The energy eigenvalues $\mathcal{E}$ and expansion coefficients $C_{\mu\nu}^{J\Pi}$ are found by solving the matrix eigenvalue problem for the Hamiltonian matrix

$$\langle \nu'\mu'|H|\mu\nu\rangle = (\varepsilon_\mu + \varepsilon_\nu)\delta_{\mu\mu'}\delta_{\nu\nu'} + \langle \nu'\mu'|V + \delta V_E|\mu\nu\rangle, \tag{7}$$

where ε_μ and ε_ν are the energies of the Dyson states ψ_μ^- and ψ_ν^+, respectively. We carry out calculations for $J^\Pi = 0^+$, 1^-, and 2^+ to investigate S-, P-, and D-wave scattering of Ps(1s), respectively. To ensure the accurate description of Ps states by (6), we confine the constituent electron and positron of the Ps to a cavity of radius $R_c = 10$–16 a.u. [1, 9]. The reason for using a smaller cavity here than when calculating the self-energy matrix elements for the electron and positron is to assist convergence of the expansion (6) with respect to the maximum orbital angular momentum of the states included: if the electron-positron pair were allowed to move far away from the centre of the cavity, then the solid angle subtended by the electron-positron pair would be small, and electron and positron states with large orbital angular momenta would need to be included in (6) to accurately resolve the internal structure of Ps. To accurately represent the positive-energy 'continuum' in the cavity, we use a second B-spline basis of 60 splines of order 9, defined over a quadratic-linear knot sequence [1]. Calculations are performed with a range of different numbers of radial states and orbital angular momenta included in (6), up to $n_{\max} = 20$ and $l_{\max} = 20$ and extrapolated to $n_{\max} \to \infty$ and $l_{\max} \to \infty$ [9]. The inclusion of such high angular momenta in the expansion of the Ps wave function is necessary to ensure convergence of the single-centre expansion about the target nucleus.

The elastic scattering phase shifts η_L (L = 0, 1, 2) are obtained from the discrete Ps(1s) energy eigenvalues in the cavity as [1]

$$\eta_L = \tan^{-1} \frac{J_{L+1/2}(K[R_c - \rho])}{Y_{L+1/2}(K[R_c - \rho])} \quad (\mathrm{mod}\,\pi), \tag{8}$$

where K is the Ps centre-of-mass momentum (which is related to the Ps energy eigenvalue by $\mathcal{E} = -1/4 + K^2/4$), J_α and Y_α are the Bessel and Neumann functions, respectively, and ρ is

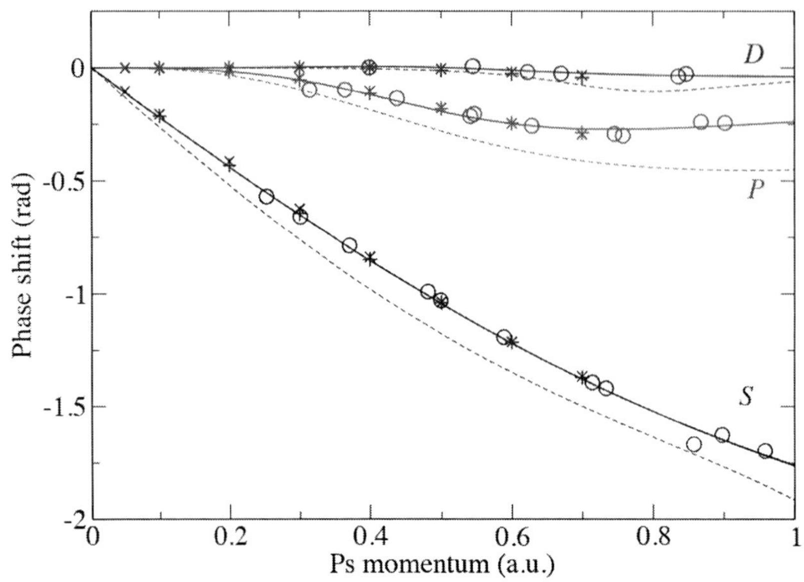

Figure 3. Phase shifts for elastic scattering of Ps by H, with the two electrons in the spin triplet state. Circles, phase shifts calculated at discrete value of the Ps centre-of-mass momentum; solid curves, effective-range-theory fits to the phase shifts; crosses, coupled-channel calculations [10]; plusses, Kohn-variational calculations [11]. The dashed curves are the corresponding frozen-target phase shifts, obtained by setting $\Sigma_E^{\pm} = 0$ and $\delta V_E = 0$. Black, red, and blue symbols and curves are for the S, P, and D waves, respectively.

the collisional radius of Ps at the wall of the cavity [1, 9]. Effective-range-theory fits are used to interpolate the phase shifts calculated at the discrete values of the Ps centre-of-mass momentum. The elastic cross section σ is then obtained as

$$\sigma = \frac{4\pi}{K^2} \sum_L (2L+1) \sin^2 \eta_L. \tag{9}$$

3. Results for elastic Ps-H scattering

The S-, P-, and D-wave phase shifts for elastic scattering of Ps on H, with the two electrons in the spin-triplet state, are shown in figure 3. The S-wave phase shift is a negative and decreasing function of K for low K, which indicates that the scattering length is positive and the effective Ps-H interaction is repulsive. This is similar to the picture of low-energy Ps scattering from He and Ne [2].

Also shown in figure 3 are coupled-channel calculations of Blackwood *et al.* [10], which included 14 Ps and 14 H states in the expansion of the wave function for the S wave, and 9 Ps and 9 H states for the P and D waves. Additionally, complex-Kohn-variational calculations of Woods *et al.* [11] are shown. We see that there is excellent agreement between the present calculations and those of Blackwood *et al.* [10] and Woods *et al.* [11] for all three partial waves.

A frozen-target calculation of the phase shifts, where distortion of the Ps is accounted for but the H atom is kept in its ground state, can be carried out by setting $\Sigma_E^{\pm} = 0$ and $\delta V_E = 0$ in (1) and (6). The results of such a calculation are also shown in figure 3. These phase shifts are lower (i.e., more negative) than the many-body-theory phase shifts. This is as expected: the removal of the attractive correlation effects, which at long range give rise to the attractive van der Waals interaction, leads to the Ps-H interaction becoming even more repulsive.

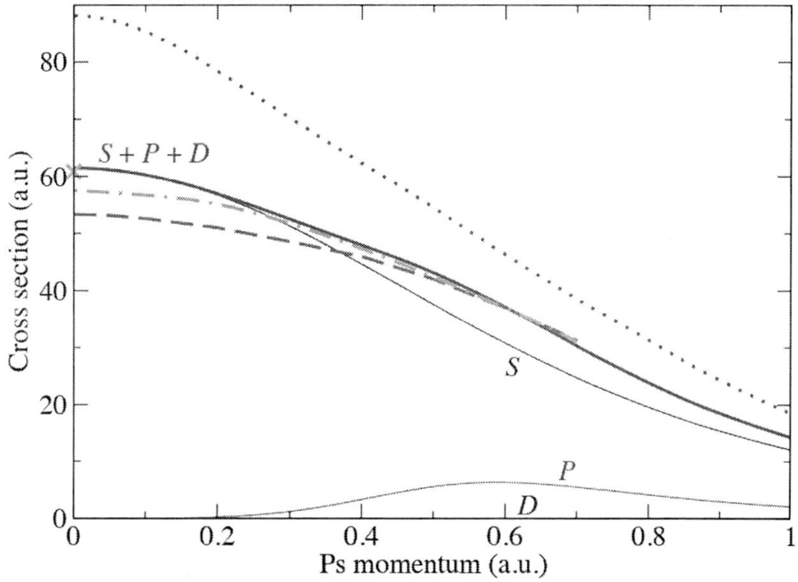

Figure 4. Cross section for elastic scattering of Ps by H, with the two electrons in the spin-triplet state. Solid black curve, S-wave contribution; solid red curve, P-wave contribution; solid blue curve (barely visible), D-wave contribution; thick solid green curve, total. Dashed magenta curve, coupled-channel calculation [10]; dot-dashed brown curve, Kohn-variational calculation [11]; orange cross, stochastic-variational calculation [12]. The dotted green curve is the frozen-target calculation, obtained by setting $\Sigma_E^{\pm} = 0$ and $\delta V_E = 0$.

The elastic scattering cross section is shown in figure 4. Also shown are the corresponding frozen-target result, the coupled-channel calculation of Blackwood *et al.* [10], the Kohn-variational calculation of Woods *et al.* [11], and the zero-energy result of a stochastic-variational calculation [12]. The scattering is almost purely due to the S wave for momenta up to $K \approx 0.2$ a.u., after which the P-wave partial cross section becomes visible. The D wave barely contributes in the momentum range considered, and consequently, the cross section is converged across the momentum range considered with respect to the number of partial waves L included in (9).

The frozen-target value of the cross section at zero momentum is 40% larger than the many-body-theory value, which emphasizes the need to account for distortion of the target to obtain accurate results. The many-body-theory value at zero momentum is in very close agreement with the stochastic-variational result [12]; indeed, the many-body-theory value of the scattering length is 2.21 a.u., while the stochastic-variational value is 2.2 a.u.

At low momenta, the many-body-theory cross section is slightly larger than the coupled-channel [10] and Kohn-variational [11] calculations, which gave values of 2.06 a.u. and 2.14 a.u., respectively, for the scattering length. However, the present cross section comes into very close agreement with the coupled-channel [10] and Kohn-variational [11] calculations at momenta $K \approx 0.5$ a.u. and $K \approx 0.3$ a.u., respectively.

4. Conclusions

Many-body theory has been used to investigate elastic scattering of Ps by atomic hydrogen, for the case where the two electrons have the total spin of 1. The scattering phase shifts, cross section, and scattering length have been calculated and compared with existing *ab initio* calculations, with excellent overall agreement being achieved.

Although the many-body-theory approach to Ps-atom scattering had previously been successfully applied to the many-electron He and Ne targets [2], carrying out calculations for H provides a useful verification of the reliability of the results obtained, as for H there are highly accurate calculations in the literature with which to compare. The main source of uncertainty in the many-body-theory calculations stems from the single-centre expansion of the Ps wave function in terms of single-particle electron and positron wave functions. It is vital to ensure that sufficiently many radial states and angular momenta are included in the expansion to enable robust extrapolation of the Ps energies to the limits $n_{\max} \to \infty$ and $l_{\max} \to \infty$. Some further uncertainty arises from the dependence of $\Sigma_E^{\pm}$ and δV_E on energy, the neglect of higher-order contributions to $\Sigma_E^{\pm}$ and δV_E, and using effective-range theory to extrapolate the S-wave phase shift to the limit $K \to 0$ (which determines the value of the scattering length and zero-energy cross section). However, the excellent agreement of the present scattering length with the stochastic-variational [12] and Kohn-variational [11] values and the very good agreement of the cross section at higher momenta with the Kohn-variational [11] and coupled-channel [10] calculations gives confidence in the many-body-theory approach and its application to other targets.

Future work will entail calculating elastic Ps scattering on heavier noble-gas atoms, viz., Ar, Kr, and Xe, for which the best existing *ab initio* calculations are of the frozen-target type [1, 13, 14]. Preliminary calculations indicate that, as has been found for H, He, and Ne, the net Ps-atom interaction for these targets is repulsive at low Ps energies; this is in contrast to the results of recent experiments which predicted the low-energy Ps-atom interaction to be attractive for Ar and Xe [15]. For these targets, the pickoff annihilation parameter ${}^1Z_{\rm eff}$ can also be calculated, which should help resolve the uncertainty in the experimental measurements of this quantity for Kr and Xe due to the contribution of spin-orbit quenching to the total annihilation rate [4, 16].

Acknowledgments

I am grateful to D V Fursa for fruitful conversations at ICPEAC 2019 in Deauville and to G F Gribakin for helpful comments on the manuscript. This work was supported by the Department for Employment and Learning, Northern Ireland, UK, and by the EPSRC UK, Grant No. EP/R006431/1.

References

[1] Swann A R and Gribakin G F 2018 *Phys. Rev.* A **97** 012706
[2] Green D G, Swann A R and Gribakin G F 2018 *Phys. Rev. Lett.* **120** 183402
[3] Walters H, Yu A, Sahoo S and Gilmore S 2004 *Nucl. Instrum. Methods Phys. Res.* B **221** 149
[4] Charlton M 1985 *Rep. Prog. Phys.* **48** 737
[5] Mitroy J and Novikov S A 2003 *Phys. Rev. Lett.* **90** 183202
[6] Fetter A L and Walecka J D 2003 *Quantum Theory of Many-Particle Systems* (New York: Dover)
[7] Gribakin G F and Ludlow J 2004 *Phys. Rev.* A **70** 032720
[8] Green D G, Ludlow J A and Gribakin G F 2014 *Phys. Rev.* A **90** 032712
[9] Brown R, Prigent Q, Swann A R and Gribakin G F 2017 *Phys. Rev.* A **95** 032705
[10] Blackwood J E, McAlinden M T and Walters H R J 2002 *Phys. Rev.* A **65** 032517
[11] Woods D, Ward S J and Van Reeth P 2015 *Phys. Rev.* A **92** 022713
[12] Ivanov I A, Mitroy J and Varga K 2001 *Phys. Rev. Lett.* **87** 063201
[13] Blackwood J E, McAlinden M T and Walters H R J 2002 *J. Phys.* B **35** 2661
[14] Blackwood J E, McAlinden M T and Walters H R J 2003 *J. Phys.* B **36** 797
[15] Brawley S J, Fayer S E, Shipman M and Laricchia G 2015 *Phys. Rev. Lett.* **115** 223201
[16] Saito H and Hyodo T 2006 *Phys. Rev. Lett.* **97** 253402

Journal of Physics: Conference Series **1412** (2020) 052010 doi:10.1088/1742-6596/1412/5/052010

Dissociative electron attachment to polyatomic molecules

Xu-Dong Wang, Xiao-Fei Gao, Hao Li, Bin Wu and Shan X Tian

Hefei National Laboratory for Physical Sciences at the Microscale, Department of Chemical Physics, University of Science and Technology of China, Hefei 230026, China

Email: sxtian@ustc.edu.cn

Abstract. Dissociative electron attachment (DEA) to molecule plays a key role in atmosphere, interstellar space and ionization damages of biological tissue. Experimental DEA studies of polyatomic molecules in gas phase provide the dynamics details that are the fundamentals to establish the physicochemical models of the electron-induced reactions in complicated environments. Since 2012, we successively set up two ion-velocity-map-imaging apparatuses, and accomplished a series of experimental studies of the DEA dynamics. Here is a brief review about our progresses on polyatomic molecules.

1. Introduction

Essential roles of DEA to molecule are well known during the past decades, such as the bond cleavage of the biological molecules in ionization irradiation damages [1] chemical reactions in interstellar space [2], oxygen origin in Earth's primitive atmosphere [3]. Rich information about the DEA dynamics has been revealed by introducing the ion momentum imaging technique in the experiments [4-9]. This technique as well as its developed mode, ion velocity map imaging (VMI) [6,10], enables us to efficiently obtain the momentum distribution of the anionic product Y^- of the DEA to molecular XY, $e^- + XY \rightarrow XY^- \rightarrow X + Y^-$.

Quantum scattering theory has been successfully applied for elucidations of the DEA dynamics of diatomic molecules [11-13], in particular, providing an analytic formula to fit the angular distributions of Y^- recorded using the conventional turn-table detection [14] or the ion imaging technique. Recently, coherent interferences of the intermediate CO^- resonant states [15-17] and interference of two DEA pathways of H_2 [18] have been reported, which enriches the basic knowledges about DEA dynamics. If the two-body dissociation of the resonant-state anion for a polyatomic molecule is fast enough, the axial-recoil approximation [11-13] is applicable to interpret the angular distribution of the anionic product [19-23]. However, the axial-recoil approximation is invalid if the anion fragment recoil axis is allowed to rotate within the dissociation time or the anion fragment is produced in a multi-stepped many-body dissociation process. Moreover, with the improvements of energy resolution, the vibrational-state resolved angular distributions challenge the current quantum scattering calculations in which the ro-vibrational states of the products are not considered explicitly [11-13].

2. Experimental developments

Full angular and kinetic energy distributions of the anionic product can be obtained simultaneously from an ion momentum image recorded with the VMI technique [6], and dynamic information about electron attachment and the resonant-state properties of the intermediate XY^- can be further derived. Therefore, it is essential to record a clear momentum image. In 2012, we established an ion VMI apparatus in which the electrons were emitted from a thermal filament and had an energy spread about 0.5 eV [6]. Recently, to investigate the molecular vibrational effects, we set up a high-resolution VMI apparatus by introducing trochoidal electron monochromator (TEM). In this machine, the energy spread of the incident electrons is reduced to 100 meV or less, thus the DEA dynamics study can be achieved at the molecular vibrational state level [24].

2.1. Application of ion velocity map imaging

According to the VMI working principle [10], the ion velocity (or momentum) image has the higher signal-to-noise ratio and the better momentum resolution than those of the image recorded with the ion image technique [4]. The double (spatial and velocity) focusing conditions are prerequisite in the VMI technique [10], while that are not critical for the regular ion imaging [4]. To realize the double focusing, we should confine the reactions to happen in a volume as small as possible; in practice, the volume with a size of $2\times2\times2$ mm^3 is acceptable [6]. Without the specifically designed VMI lenses, the ion momentum image is doomed to be blurred for the larger reaction volume.

As discussed previously [16], the double focusing length is fixed in a designed VMI system, thus the time-of-flight tube cannot be arbitrarily elongated, and otherwise the momentum resolution also becomes worse. Meanwhile, the ion Newton sphere should not be distorted (compressed in the parallel or perpendicular direction) during the flight. The VMI design should be practiced carefully, to have the Newton sphere as large as possible for the central time-slicing [10]. Furthermore, a clear ion momentum image also benefits both from the narrow widths of the incident electron pulse and the time-slicing.

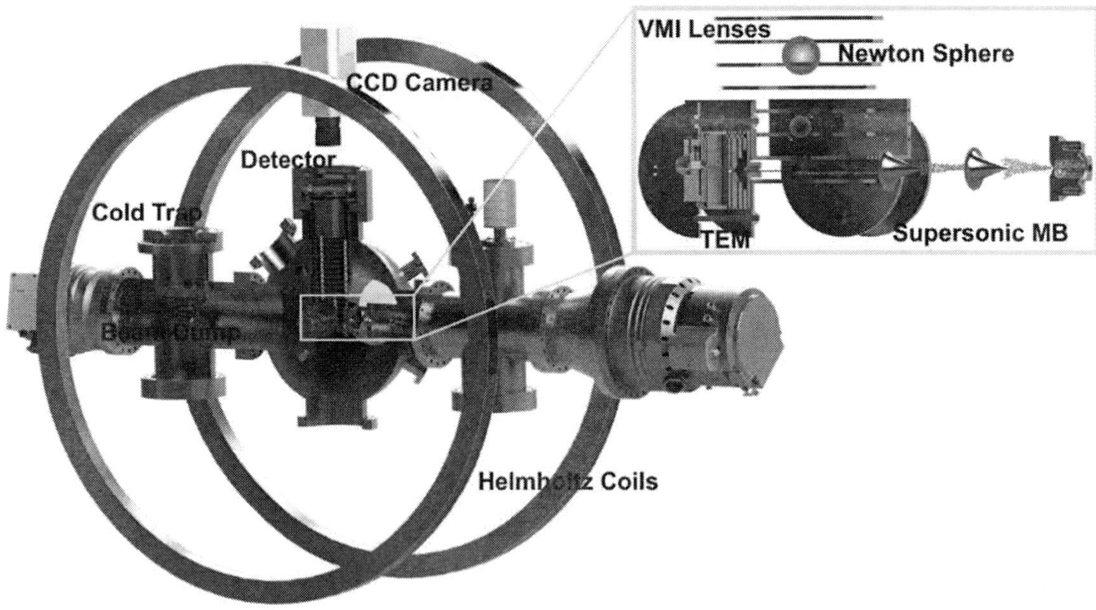

Figure 1. Schematic overview of our high-resolution experimental setup.

2.2. High-resolution apparatus

The TEM was used in the DEA studies several decades ago, primarily to record the production efficiency curves of the anionic yields. Overcoming the difficulties in combined usage of the TEM and VMI techniques, we successfully set up a high-resolution apparatus. An overview of the experimental arrangement is depicted in figure 1, where the key components are shown in an amplified panel. The electrons emit from a thermal filament, then they are pulsed and energetically monochromized with a homemade TEM (in which a homogenous magnetic field about 70 Gauss is produced with a pair of Helmholtz coils). This monochromator is set perpendicular to the continuous supersonic molecular beam

(MB) in the horizontal (*x-y*) plane. A retarding field analyzer and a Faraday cup installed at the downstream of the TEM are used to determine the electron energy spread (typically about 100 meV) and monitor the current of the monochromized electrons, respectively. Each bunch of the fragment anions produced in the DEA process goes through the VMI lens system which is perpendicular to the electron (along the *y* axis) and molecular (along the *x* axis) beams (sometime an effusive MB is set below the pusher lens and along the *z* axis), then these anions are detected with a pair of microchannel plates (MCPs) and a phosphor screen. A charge-coupled-device (CCD) camera is used to record the images.

The voltages on the VMI lenses are optimized for various anions with different velocities. The kinetic energy resolution ($\Delta E/E$) is about 2% for the O^- anions with the kinetic energies less than 3.5 eV, but $\Delta E/E$ reaches 5.5% at the very low kinetic energy (0.1 eV). This is better than our previous design ($\Delta E/E = 5\%$ when $E \leqslant 1.0$ eV) [6]. In our experiments, the width of a typical time-slicing pulse applied on the rear MCP is 30 - 45 ns which is thin enough to get a clear sliced image. When this apparatus is operated under the anion mass spectrometry mode, the DC high voltages are applied on the anion detector, while the pulsed voltage height of the pusher lens is larger than that used in the VMI mode, to improve the mass resolution.

3. Recent progresses

Using the low-resolution apparatus [6], we carried out a series studies for polyatomic molecules CO_2 [3,19], N_2O [20], OCS [25], NF_3 [22], CF_4 [26], $CFCl_3$ [21] and ethanol and acetaldehyde [27]. The DEA processes to the vibrational ground-state (cold) CO_2 (see figure 2) and vibrationally excited (hot) CO_2 [24] were investigated with the high-resolution apparatus. The following parts are classified with the different molecules.

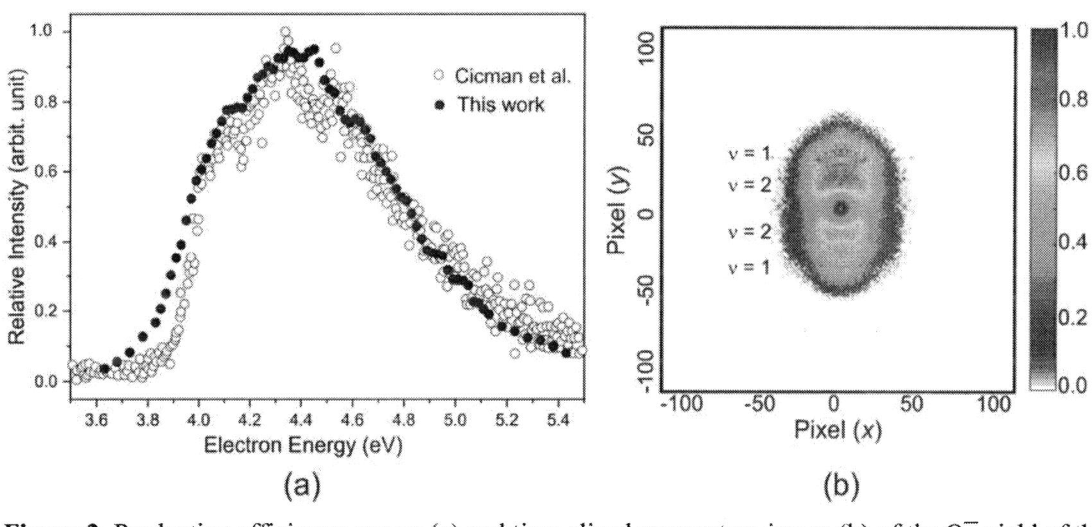

Figure 2. Production efficiency curves (a) and time-sliced momentum image (b) of the O^- yield of the DEA to CO_2. In (a),the present result (solid circles) is compared with that obtained by Cicman *et al.* [28] after renomalizations. In (b), the image is recorded at 4.55 eV, the electron incident direction (along y axis) is from top (backward) to bottom (forward) and through the image center, and the image size is scaled with the pixel numbers of the CCD camera.

3.1. *Triatomic molecules*

CO_2. Below the electron attachment energy of 10 eV, there are two resonant states of CO_2^- related to DEA processes of CO_2 [19]. Firstly, we examined the DEA of CO_2, $e^- + CO_2 \rightarrow CO_2^- \rightarrow CO + O^-$, around the second resonance $^2\Pi_g$ state at 8.2 eV of CO_2^-, and found a dynamic Renner-Teller effect on the angular distributions of the fast O^- product [19]. The slow O^- ions may be produced together with

CO at the high-lying vibrational states. Secondly, around the first resonance $^2\Pi_u$ state at 4.4 eV of CO_2^-, there were arguments about the assignments to the O^- production efficiency curve, namely, using the vibrational states of the intermediate CO_2^- or the product CO [28]. As shown in figure 2(a), we compare the production efficiency curves and find a good agreement between ours (recorded with the high-resolution apparatus) and that by Cicman *et al.* [28]. Furthermore, in the O^- image at the electron energy of 4.55 eV (see figure 2(b)), we find the fine structures in the forward-backward direction, namely, the outside signals correspond to CO at the vibrational state $n_v = 1$ while the inner ones are attributed to CO at $n_v = 2$ state. This image is much clearer than that reported previously [9], owing to the present higher resolution. The present assignment with CO vibrational states is in agreement with a previous conclusion [29], deserving further elucidations on its DEA dynamics.

At the electron energies around 18 eV, two DEA processes, $e^- + CO_2 \rightarrow 2O + C^-/O_2 + C^-$, could compete with each other, but the latter pathway was identified for the first time in our VMI measurements [3]. This work not only revealed the interesting DEA mechanisms but also demonstrated an important role of DEA in the origin of molecular oxygen on the planets and the early Earth [3]. The O_2 molecules produced in the above DEA process are populated at the high-lying vibrational states, which triggers our explorations of the molecular vibrational effects using the high-resolution apparatus.

Around the resonance $^2\Pi_u$ state of CO_2^-, the vibrational excitation of CO_2 is a competitive channel of DEA. Owing to the high cross sections and the mode selections of the CO_2 vibrational excitations nearby the resonant state, the high DEA cross sections are still observed below the threshold of $e^- + CO_2 \rightarrow CO + O^-$, and the angular distribution also indicates the strong dependences of the vibrational states of CO_2 [24]. Those are completely different from what we observed in figure 2 for the vibrational ground-state target.

N_2O & OCS. In the DEA process $e^- + N_2O \rightarrow N_2 + O^-$, we observed the evolution of momentum distributions of the O^- fragment in terms of the electron attachment energy, and attributed it to a competition between two distinctly different indirect pathways, namely, climbing over and bypassing the energy ridge after the molecular structure bending. These two pathways prefer leaving the N_2 fragment in the high vibrational (at the higher electron energies) and rotational (at the lower electron energies) states, respectively [20]. Trimodal speed distributions of S^- ion were found in $e^- + OCS \rightarrow CO + S^-$ in the electron attachment energy range of 6.2-7.0 eV, which arose from the different populations of the ro-vibrational states of CO fragment [25].

3.2. Other polyatomic molecules

NF_3. Dissociation dynamics of temporary NF_3^- was investigated in the electron attachment energy range of 0.5 - 4.5 eV. Six pathways can be accessed in energetics, but three of them were explored. The kinetic and angular distributions of the F^- yield indicated that this fragment was produced via two-body and three-body fragmentations, at the higher electron attachment energies. The anisotropic distributions of the fast F^- ions arose from the two-body dissociations and was relevant to the ^2E resonant state of NF_3^-, whereas the slow ions were produced via various three-body dissociations [22].

CF_4. The DEA processes $e^- + CF_4 \rightarrow CF_3 + F^-/ CF_3^- + F$ are two complementary pathways at a common energy (5.3- 6.0 eV). Although CF_4 is an apolar molecule, the electron attachment prefers to proceed along the local polar C-F bond, then the impulse dissociation of CF_4^- leads to the backward-distributed F^- (together with the forward-distributed CF_3) or the forward-distributed CF_3^- (together with the forward-distributed F). This spatial orientation characteristics was observed in the momentum images of F^- and CF_3^- anionic yields [26].

$CFCl_3$. The momentum images of the F^- ions produced in $e^- + CFCl_3 \rightarrow CCl_3 + F^-$ indicated image pattern evolutions with the electron attachment energy in a range of 2.0–7.0 eV. The angular and kinetic energy distributions of this anion were interpreted with two shape-resonant states of $CFCl_3^-$, while the F^- ions with the lower kinetic energies at higher attachment energies could be produced via three-body dissociations or due to the co-production of CCl_3 at an electronically excited state ($^2A''$) [21].

Ethanol (CH_3CH_2OH) & Acetaldehyde (CH_3CHO). In the DEA processes, the dominant anionic fragments were O^- and OH^- ions for ethanol while O^- and CH_3^- ions for acetaldehyde. Interestingly, we found that the O^- production in the DEA to ethanol was achieved via a cascade process, namely, experiencing the intermediate species $(CH_3CHO)^-$ formed by the double dehydrogenations [27]. This is quite similar to the alcohol dehydrogenases in human body.

3.3. Theoretical simulations
Since it is still a challenge in theory to describe the nuclear motions and internal energy disposals of the molecular anion prior to the dissociation, we developed a theoretical simulation strategy. Although most processes do not start in the lowest anionic state, the applicability of our method is based on the following assumptions: the DEA dynamic evolution finally undergoes on the potential energy surface of the lowest anionic state, if the ground-state fragments are produced in the asymptotic region; the coupling between the high and lowest anionic states is accomplished quickly after the vertical electron attachment. The DEA processes of N_2O [20], ethanol [27] and dialanine [30] have been qualitatively interpreted well, using our molecular dynamics simulations.

The computations with sophisticated quantum scattering methods are time-consuming, and the convergence is frequently problematic for the anionic resonant states of polyatomic molecules. We utilized a simplified quantum scattering method, the single-center expansion potential scattering, to study the shape resonant states formed in the electron attachments to cysteine and cystine. We found that the resonant states showed the novel selectivity in the bond cleavages, and the S–S bond cleavage of cystine anion was highly favored [31].

4. Remarks
During the past decades, different groups including ours made a lot of efforts to explore the DEA dynamics. There are some long-standing arguments about the spectral assignments of the anionic production efficiency curve, namely, using the vibrational states of the neutral target XY, the resonant-state XY^- or the polyatomic fragment X (or Y^-). Owing to the combination between VMI and TEM techniques, we can observe the molecular vibrational effects, not only of the products (as shown in figure 2(b)) but also the target molecules [24]. Further theoretical developments to reveal the nuclear motions induced by electron attachment are highly demanded.

Acknowledgement
Our project is supported by National Natural Sciences Foundation of China (Grant Nos. 21273213, 21573209, 21625301, 21727804).

References
[1] Boudaïffa B, Cloutier P, Hunting D, Huels M A and Sanche L 2000 *Science* **287** 1658
[2] Millar T J, Walsh C and Field T A 2017 *Chem. Rev.* **117** 1765
[3] Wang X-D, Gao X-F, Xuan C-J and Tian S X 2016 *Nat. Chem.* **8** 258
[4] Chandler D W and Houston P L 1987 *J. Chem. Phys.* **87** 1445
[5] Nandi D, Prabhudesai V S, Krishnakumar E and Chatterjee A 2005 *Rev. Sci. Instrum.* **76** 053107
[6] Wu B, Xia L, Li H-K, Zeng X-J and Tian S X 2012 *Rev. Sci. Instrum.* **83** 013108
[7] Adaniya H, Slaughter D S, Osipov T, Weber T and Belkacem A 2012 *Rev. Sci. Instrum.* **83** 023106
[8] Moradmand A, Williams J B, Landers A L and Fogle M 2013 *Rev. Sci. Instrum.* **84** 033104
[9] Moradmand A, Slaughter D S, Haxton D J, Rescigno T N, McCurdy C W, Weber T, Matsika S, Landers A L, Belkacem A and Fogle M 2013 *Phys. Rev.* A **88** 032703
[10] Eppink A T J B and Parker D H 1997 *Rev. Sci. Instrum.* **68** 3477
[11] O'Malley T F and Taylor H S 1968 *Phys. Rev.* **176** 207
[12] Cadez I, Tronc M and Hall R I 1975 *J. Phys.* B **8** L73
[13] Klar H and Morger H 1979 *J. Phys.* B **12** 2369
[14] Tronc M, Fiquet-Fayard F, Schermann C and Hall R I 1977 *J. Phys.* B **10** L459

[15] Tian S X, Wu B, Xia L, Wang Y-F, Li H-K, Zeng X-J, Luo Y and Yang J 2013 *Phys. Rev.* A **88** 012708
[16] Tian S X and Luo Y 2015 *Phys. Rev.* A **91** 056702
[17] Wang X-D, Xuan C-J, Luo Y and Tian S X 2015 *J. Chem. Phys.* **143** 066101
[18] Krishnakumar E, Prabhudesai V S and Mason N J 2018 *Nat. Phys.* **14** 149
[19] Wu B, Xia L, Wang Y-F, Li H-K, Zeng X-J and Tian S X 2012 *Phys. Rev.* A **85** 052709
[20] Xia L, Wu B, Li H-K, Zeng X-J and Tian S X 2012 *J. Chem. Phys.* **137** 151102
[21] Zeng X-J, Xia L, Li H-K, Lau K-C and Tian S X 2013 *Phys. Rev.* A **87** 012711
[22] Li H-K, Xia L, Zeng X-J and Tian S X 2013 *J. Phys. Chem.* A **117** 3176
[23] Azria R, Coat Y L, Lefever G and Simon D 1979 *J. Phys.* B **12** 679
[24] Li H, Gao X-F, Meng X and Tian S X 2019 *Phys. Rev.* A **99** 032703
[25] Li M-Y, Gao X-F, Wang X-D, Li H and Tian S X 2016 *Int. J. Mass Spectrom.* **404** 20
[26] Xia L, Zeng X-J, Li H-K, Wu B and Tian S X 2013 *Angew. Chem. Int. Ed.* **52** 1013
[27] Wang X-D, Xuan C-J, Feng W-L and Tian S X 2015 *J. Chem. Phys.* **142** 064316
[28] Cicman P, Senn G, Denifl G, Muigg D, Skalny J D, Lukac P, Stamatovic A and Märk T D, 1998 *Czech. J. Phys.* **48**, 1135
[29] Stamatovic A and Schulz G J 1973 *Phys. Rev.* A **7** 589
[30] Feng W-L and Tian S X 2015 *J. Phys. Chem.* A **119** 1838
[31] Wang Y-F, Tian S X and Yang J 2011 *Phys. Chem. Chem. Phys.* **13** 15597

Positronium collisions with atoms and molecules

R S Wilde[1] **and I I Fabrikant**[2]

[1]Department of Natural Sciences, Oregon Institute of Technology, Klamath Falls, OR 97601, USA
[2]Department of Physics and Astronomy, University of Nebraska-Lincoln, Lincoln, NE 68588, USA

E-mail: robyn.wilde@oit.edu

Abstract. We describe recent theoretical efforts to explain the observed similarity between electron and positronium (Ps) scattering with both atoms and molecules. In order to perform these calculations at a more ab initio level we have developed a Free Electron Gas (FEG) approximation to the exchange and correlation potentials similar to that used in electron-molecule scattering. The FEG results give good agreement with experiment above the Ps ionization threshold and exhibit resonance features below the threshold. To take account of the effect of orthogonality between the scattered electron and the target wavefunctions we have used an orthogonalizing pseudopotential (OPP). The FEG plus OPP results give good agreement with experiment for Ps-rare-gas scattering. Preliminary results for Ps-H_2 scattering also agree well with experiment and demonstrate the importance of including orthogonality for scattering below the Ps ionization threshold.

1. Introduction

The interaction between positronium (Ps) and atoms, molecules and solids is important in such practical applications as positron annihilation spectroscopy [1] and positron emission tomography (PET) [2]. In both applications the production of gamma rays due to annihilation of Ps is influenced by its local environment. A better theoretical understanding of the effect of these interactions may help with the interpretation of these applications.

The theoretical study of Ps scattering is also important for understanding the results of recent experiments in which a beam of Ps is scattered by various atomic and molecular targets [3-5]. These experiments have shown that when the total electron and Ps scattering cross sections are plotted as a function of projectile velocity they are very similar in shape and magnitude above the Ps ionization (fragmentation) threshold which occurs at a Ps velocity of 0.5 a.u. (6.8 eV). This similarity is somewhat surprising since it seems to suggest that the positron is not as important as the electron in Ps collisions with atoms and molecules in this velocity region.

In electron collisions with atoms and molecules the static interaction is very important, although the exchange interaction can also have an important effect especially near resonances and thresholds. Correlation effects are also important in electron collisions due to the polarization interaction. In Ps collisions, on the other hand, the static and polarization interactions disappear due to the equal and opposite charges of the positron and electron in Ps. Long-range correlation effects are due to the weaker van der Waals forces between Ps and the target atom or molecule. This suggests that the exchange interaction may be very important in Ps collisions. It also suggests that short-range correlation effects may be more important for Ps collisions with atoms and molecules than for electron collisions with the same targets.

The similarity between electron and Ps scattering above the Ps ionization threshold has been theoretically explained using the impulse approximation by Fabrikant and Gribakin [6] as mainly being due to the dominance of the exchange interaction in Ps-atom and Ps-molecule scattering. However, the impulse approximation is applicable only at high Ps velocities (energies). Therefore, other theoretical methods are needed for Ps velocities below the Ps ionization threshold. For these lower Ps velocities, the pseudopotential approach [7] has been developed. Both the impulse approximation and pseudopotential method use information on electron-atom and positron-atom scattering to develop a model for Ps-atom scattering. Specifically, in the pseudopotential method, a model potential (pseudopotential) is used in which the parameters of the potential are adjusted to reproduce the accurately calculated electron-atom and positron-atom phase shifts. The pseudopotential method also gives good agreement with experiment above the Ps ionization threshold [7,8] and again demonstrates the dominance of the exchange interaction in this region.

At velocities below the Ps ionization threshold the pseudopotential results for Ar, Kr and Xe exhibit a peak in the cross section [7,8]. This peak was explained [7] by a rapid decrease of the S-wave and the P-wave phase shifts in the low-velocity region. This behavior is very different from the results for the electron scattering cross sections for scattering from these atoms in this velocity region where a Ramsauer-Townsend minimum is observed.

On the other hand, the results of beam experiments that have recently been extended to velocities below the Ps ionization threshold [9] exhibit a very small cross section in this region which seems to suggest the possibility of a Ramsauer-Townsend minimum in Ps scattering with the heavy rare-gases. However, the pseudopotential method [7,8] as well as recent calculations by Green *et al.* [10] and Swann and Gribakin [11] have led to larger cross sections below the Ps ionization threshold for rare-gas atoms. It appears that there is some discrepancy between experiment and theory for Ps scattering by rare-gas atoms below the Ps ionization threshold. It should be noted that the beam experiments have thus far only been extended down to velocities of about 0.2 a.u. (1.1 eV) and only for the targets Ar and Xe. This means that there is a gap in the velocity (energy) range of about 0.06 a.u. (0.1 eV) - 0.2 a.u. (1.1 eV) for which practically no experimental data exists. Certainly, more experimental data is needed and there is a need for accurate theoretical calculations in this range.

Another interesting, but more complicated topic due to non-spherical interactions, is that of Ps-molecule collisions, for which the similarity between electron and Ps scattering has also been observed above the Ps ionization threshold in collisions with H_2, N_2, CO_2, O_2 and SF_6 [4, 5]. There have been relatively few theoretical studies of Ps-molecule collisions and these have been generally restricted to the simplest target H_2 [12-15].

We have previously applied the pseudopotential method to Ps collisions with H_2 by assuming that the Ps-molecule interaction is spherically symmetric [16]. While we obtained good results for H_2 in this way it turns out that severe difficulties are encountered when we attempt to use the pseudopotential approximation for more non-spherical systems like N_2. For this reason, we have recently developed a Free Electron Gas (FEG) approach to the calculation of exchange and correlation energies for Ps scattering [17]. From the target charge density, we can then use the Thomas-Fermi relation to obtain local exchange and correlation scattering potentials.

One deficiency of the local FEG potentials is that they cannot take into account the antisymmetric character of the total wave-function with respect to interchange of the projectile and target electrons. This deficiency can be rectified by enforcing orthogonality between the scattering wave-function and the bound orbitals of the target atom or molecule. Enforcing orthogonality has been shown to be important for electron-molecule scattering [18]. A method of enforcing orthogonality by using an orthogonalizing pseudopotential (OPP) for electron and Ps scattering has been proposed by Mitroy *et al.* [19,20]. An advantage of the OPP is that it does not require adjustable fitting parameters like the pseudopotential method of Refs. [7,8].

In the remainder of this report we describe our calculation of the local FEG potentials and nonlocal OPP. We then summarize our results of using these potentials to describe Ps scattering by rare-gas (Rg) atoms and several molecular targets.

2. Scattering potentials

A common method of including correlation in Ps scattering is to use the van der Waals potential with a short-range cutoff which is usually introduced in the form

$$V_{corr} = -\frac{C_W}{R^6}\left[1 - \exp[(R/R_c)^8]\right] \quad (1)$$

where C_W is the van der Waals constant, and R_c is an adjustable cut-off radius. In this way the long-range interaction is included properly, but the short-range correlation potential can depend significantly on the choice of cutoff radius [11].

In order to calculate the exchange and correlation effects in a more ab initio way we have computed the free-electron-gas exchange (FEG) and correlation energies, as a function of Fermi energy k_F using first and second order perturbation theory [17]. The energies obtained in this way can be converted to exchange and correlation potentials using the Thomas-Fermi model where the Fermi momentum depends on the charge density of the target atom or molecule through $k_F(r)=[3\pi^2\rho(r)]^{1/3}$.. The target charge density, $\rho(\boldsymbol{r})$, is calculated using accurate Hartree-Fock wave-functions. As described in Ref. [17] we evaluate the charge density at the Ps center of mass $\boldsymbol{R}$ so that $\boldsymbol{r} = \boldsymbol{R}$. The FEG potentials are similar to free electron gas approximations used in electron-molecule scattering [21,22]

One advantage of the current FEG correlation potential is that it can be matched smoothly to the long range van der Waals' form thus avoiding the necessity of an adjustable cutoff radius that affects the potential at small values of $\boldsymbol{R}$. We have found that the FEG correlation potential for Ps-rare-gas scattering is much more attractive at very small values of R than that of equation (1) [23].

For Ps-molecule scattering the charge density and local FEG potentials are not spherically symmetric. Therefore, in the case of molecules with Σ symmetry such as N_2 and CO_2, we expand the FEG potentials in Legendre polynomials $P_\lambda(\cos\theta)$ depending on the angle θ between the molecular axis and $\boldsymbol{R}$. Also due to the molecular symmetry only even values of λ contribute to this expansion. For H_2 we have included terms up to λ=12 and for N_2 and CO_2 we have included terms up to λ=24. All of this is very similar to electron scattering from these molecules and we can describe the partial cross sections and eigenphase sums in a way similar to electron-molecule scattering, by conserved projection of the Ps angular momentum on the molecular axis and parity e.g. Σ_g, Π_u etc.

As mentioned above, we use the OPP to enforce orthogonality between the electron in Ps and the occupied target atomic or molecular orbitals, φ_i. The OPP is defined by the projection operator [19,20]

$$\gamma\hat{P} = \gamma\delta(\boldsymbol{r}_p - \boldsymbol{r}'_p)\sum_i^N |\varphi_i\rangle\langle\varphi_i| \quad (2)$$

where $\boldsymbol{r}_p$ is the positron coordinate. This operator is added to the exchange and correlation potentials and leads to a nonlocal term. The δ function ensures that the OPP affects only the electron coordinate, emphasizing in this way the dominant role of the electron constituent of Ps in the scattering. The strength parameter γ in these equations is made large enough so that orthogonality is enforced and the scattering calculations are converged in the sense that further increase of γ leads to a negligible change in the phase shifts and cross sections. For velocities below 2 a.u. the phase shifts for Ps-rare-gas scattering are converged for values of γ=2 a.u. and above and we have used a value of γ=20 a.u. in these calculations. For the present Ps-H_2 calculations we have used a value of γ=3 a.u. The sum in equation (2) is over the N occupied orbitals of the target atom or molecule.

We have derived an expression for the nonlocal OPP kernel describing Ps-Rg atom scattering in [23]. In the case of Ps-molecule scattering the situation is more complicated due to the non-spherical nature of the molecular orbitals. However, if we expand the molecular orbitals in Legendre polynomials and retain only the first, spherically symmetric term (λ=0) the expression for the kernel reduces to that of [23]. In our present calculations for Ps-H_2 scattering described in section 4 we make this approximation.

3. Ps-rare-gas scattering

As mentioned in the introduction the beam results for Ps-Rg scattering with Ar and Xe indicate the possibility that a Ramsauer-Townsend type minimum may occur in Ps scattering by the heavy rare-gas atoms, due to the sharp decrease in the cross section for velocities just below the Ps ionization threshold. The recent pseudopotential results [7,8] as well as the calculations of Swann and Gribakin [11], however exhibit larger cross sections below threshold with no Ramsauer-Townsend minimum.

Electron and Ps scattering might not be expected to be very similar in the low velocity region due to the fact that at these velocities the long-range interaction between the projectile and the target is very important, and this interaction is different for the two projectiles: for electron scattering the polarization potential decays as $1/r^4$, while for Ps scattering it is the significantly weaker van der Waals interaction which decays as $1/R^6$, The results of a number of experiments and calculations for Ps scattering by rare-gases [24] have resulted in a positive scattering length which seems to preclude the existence of a Ramsauer-Townsend minimum in Ps scattering by rare-gas atoms. This conclusion is further supported by the results of swarm experiments [24] which have also led to the conclusion that the Ps-rare-gas scattering length is positive. The swarm measurements are generally restricted to very low energies up to about 0.1 eV. Recently Shibuya and Saito [25] have developed a method to extract scattering phase shifts for Ps-Xe scattering from the experimental swarm data and, using effective range theory, have extrapolated their results up to an energy of about 1 eV. They obtain a peak with a very large magnitude in the Ps-Xe scattering cross section that seems to be somewhat similar to the peak obtained in the pseudopotential results. However, it seems that the use of the effective range theory over such a large velocity (energy) range is inappropriate and the huge magnitude of this peak seems physically unreasonable [23].

In order to help clarify the situation for low velocity Ps-rare-gas scattering we have recently applied the local FEG approximation and OPP described in section 2 to the calculation elastic Ps scattering with rare-gas (Rg) atoms [23]. The use of just the attractive, local FEG potentials leads to shape resonances at low Ps velocities that are not seen experimentally. The addition of the OPP to the local potentials removes the resonances and generally leads to good agreement with beam experiments [3,9,26], especially for the heavier atoms Ar, Kr and Xe. For He and Ne the cross sections are too large compared with experiment in the low-velocity region which may be due to the FEG potential not being as appropriate for such smaller systems since it is based on the statistical Thomas-Fermi model.

At velocities below the Ps ionization threshold the cross section is very sensitive to the intermediate and long-range correlation potential. This is because the exchange and correlation potentials are masked by the OPP at small values of R. In fact, for Xe we can get a Ramsauer-Townsend type minimum in the cross section if we use a potential of the form of equation (1) and adjust the cutoff parameter R_c so that the correlation potential is attractive enough in the region of intermediate R. This sensitivity illustrates the necessity of accurate determination of the exchange and correlation potentials in this region. The use of the local FEG exchange and correlation potentials with the OPP, in general leads to smaller cross sections and better agreement with results of beam measurements than previous pseudopotential calculations. We illustrate these results for Ar and Xe in figure 1. In these calculations we have included partial waves up to l=10. Note, that above the Ps ionization threshold previous calculations have shown that the excitation of Ps is negligible compared to elastic scattering and Ps ionization [6,27]. Therefore, we have added Ps ionization cross sections calculated by using the binary-encounter approximation [8] to the elastic cross section to obtain total cross sections in this region.

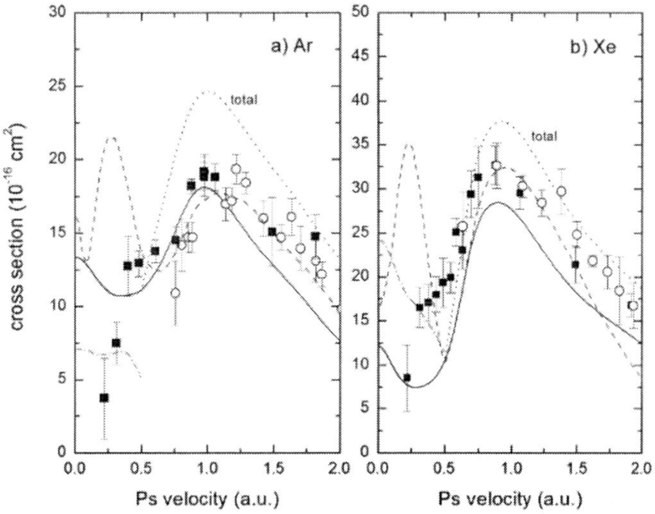

Figure 1. Cross sections for Ps scattering with a) argon and b) xenon. Solid line: elastic cross sections using the FEG potentials plus OPP of [23]. Dotted line: total (elastic plus Ps ionization) cross section. Blue dashed line: pseudo-potential results [8]. Red dash-dot line: results of [11] using the correlation potential of equation (1) with R_c=2.5 a.u. for Ar and R_c=3.0 a.u. for Xe. ■ [9]; ○ [3].

In the thermal energy region, inaccessible to beam experiments, our results compare well with swarm experiments measuring the Ps momentum transfer cross sections from observation of Ps moderation in gases. We obtain excellent agreement with measurements of Nagashima *et al.* [28] for He, Ne, and Ar, with Skalsey *et al.* [29] for Ne, and with Shibuya *et al.* [25,30] for Xe. However, the extrapolation of the Ps-Xe phase shifts, obtained by Shibuya and Saito [25] to higher energies leads to strongly overestimated (by an order of magnitude) total cross sections. As mentioned above this is due to the failure of the effective range expansion at higher energies.

It appears that our model provides a good description of Ps scattering by Rg atoms, especially the heavier atoms Ar, Kr and Xe. Our calculations confirm the positive sign of the scattering length and the absence of the Ramsauer-Townsend minimum in Ps scattering by heavier rare-gas atoms. However, the S- and P-wave scattering phase shifts decrease with energy significantly slower than in the previous pseudopotential calculations and significantly improves agreement with the beam experiments in the energy range below the Ps ionization threshold.

4. Ps-molecule scattering

The impulse approximation and pseudopotential method have also been applied with success to Ps collisions with molecular hydrogen [16]. In this work the Ps-H_2 interaction was taken to be spherically symmetric by retaining only the first term in the Legendre expansion of the e^--H_2 and e^+-H_2 scattering potentials. For molecules with less spherical symmetry we have found that the pseudopotential method is difficult to implement since it is challenging to reproduce the scattering *T*-matrix elements using a model potential. Therefore, we have used the FEG model to construct local potentials and employed them in calculating scattering cross sections for Ps-N_2 collisions [31]. The resulting calculation is in good agreement with experimental measurements [32] above the Ps ionization threshold. In the present calculations we use partial waves up to l=10 for H_2 and l=17 or N_2 and CO_2.

An interesting feature of the experimental results is that there is evidence [4,32] for a resonance Ps-CO_2 state similar to the Π_u resonance in e^--CO_2 collisions, and for a Ps-N_2 resonance state similar to the

Π_g resonance in e^--N_2 collisions near the Ps ionization threshold. Our recent results for Ps-N_2 scattering [31] also exhibit resonance feature in this region. The position of the resonance peak is slightly below the observed position which could be due to not including the effect of nuclear motion and/or orthogonality through the OPP. The effect of the nuclear motion on the Π_g resonance in electron scattering by N_2 is well known [33]. Therefore, a next step in the theoretical development should be incorporation of vibrational dynamics along the lines of the boomerang model [34]. This also opens an opportunity for the calculation of vibrational excitation cross sections in Ps-N_2 collisions. Below the Ps ionization threshold our cross sections for Ps-N_2 scattering [31] seem again to be larger than the experimental beam measurements [32]

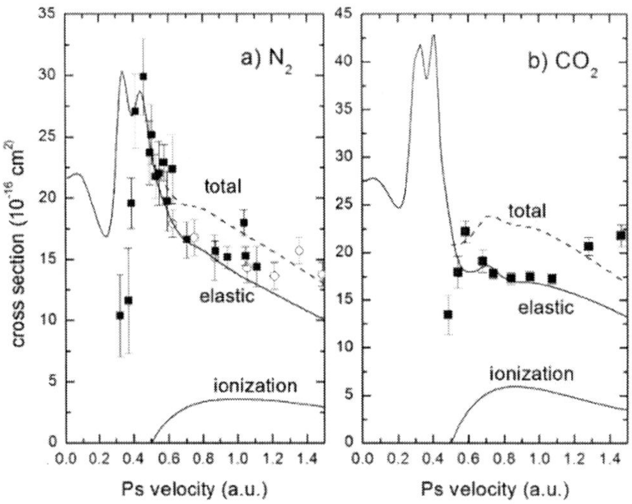

Figure 2. Cross sections for Ps scattering with a) N_2 of [31] and b) CO_2 using the FEG approximation. Solid lines: elastic cross sections. Dotted lines: total cross section (elastic plus Ps ionization). For N_2; ■ [3]; ○ [32]. For CO_2; ■ [4].

Calculations using the FEG exchange and correlation potentials for Ps-CO_2 scattering also show resonance features as demonstrated in figure 2. In this case the resonance position is further below the Ps ionization threshold and larger in magnitude. Above the Ps ionization threshold we add Ps ionization cross sections to the elastic FEG results using the binary-encounter approximation which was extended to Ps-molecule scattering in [31]. Our total cross sections in this region exhibit quite good agreement with the experimental measurements of [4]. From these results and those for N_2 it seems that the FEG model of the exchange and correlation potentials give a good description of Ps-molecule scattering above the Ps ionization threshold. Below the threshold we see resonance features, but the magnitude of the cross section is generally larger than the experimental beam results. We note that in our calculations of Ref. [31] we have used only the attractive local FEG potentials and have not included the repulsive OPP. Therefore, a next step for scattering below the Ps ionization threshold would be to develop the OPP for Ps-molecule collisions and see if it leads to better agreement with the experimental results in this region.

At this point we have made preliminary calculations in which the FEG plus OPP model is applied to Ps-H_2 scattering. In calculating the OPP in this case we have made the approximation of including only the first (spherically symmetric) term in the expansion of the occupied σ_g orbital. This approximation allows us to calculate the OPP in the same way that was done for Ps-He scattering in [23]. Although we have made the spherically symmetric approximation for the OPP, non-spherical terms have always been included in the calculation of the FEG potentials.

The results for Ps-H_2 in this case are shown in figure 3. Once again, we see a sharp resonance feature below the Ps ionization threshold when only the FEG potentials are used. When the OPP is added to the FEG potentials its repulsive nature suppresses the resonance and reduces the cross section below the threshold and gives good agreement with the measurements of Skalsey *et al.* [29,35] at low velocities and of Garner *et al.* [26] at velocities above the Ps ionization threshold. This is similar to Ps-Rg scattering and shows the importance of including the effect of orthogonality in Ps-molecule scattering. We are currently in the process of developing the OPP for Ps-N_2 and Ps-CO_2 collisions.

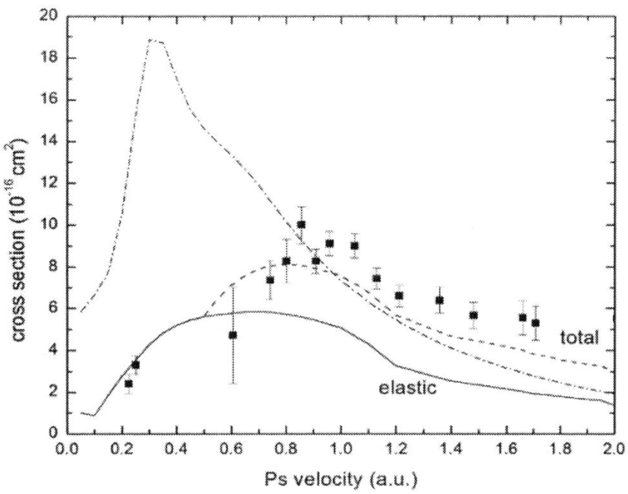

Figure 3. Cross sections for Ps scattering with H_2. Solid line: elastic cross section using the FEG potentials plus OPP. Dotted line: total cross section (elastic plus Ps ionization). Dash-dot line: Elastic cross section using the FEG approximation only. ■ Measurements of [29,35] (low velocities) and [26] (v=0.6 a.u. and above).

5. Conclusion and outlook

In summary, we have described our recent calculations of the total cross section for Ps scattering by rare-gas atoms and several molecules. In these calculations we have employed a FEG model for the calculation of exchange and correlation potentials. The FEG model gives good agreement with experimental results for the total cross section when we add Ps ionization cross sections calculated using the binary encounter approximation.

Below the Ps ionization threshold, the FEG results generally exhibit resonance behavior due to the attractive nature of the FEG potentials. For Ps-Rg scattering it is important to include the effect of orthogonality between the target and projectile electrons. We have developed the OPP to describe this effect. When the OPP is added to the FEG potentials we get relatively good agreement with experimental results for the heavy rare-gases, Ar, Kr and Xe, even at velocities below threshold. This illustrates the importance of including the effect of orthogonality at low velocities. Although we get better agreement with experiment, we do not see evidence for a Ramsauer-Townsend minimum.

For Ps-molecule scattering, when the FEG approximation is used, we again get good agreement with experiment above the Ps ionization threshold and see resonance structures near or below the threshold. For Ps-N_2 scattering the position and magnitude of resonance features agree quite well with experiment, but for Ps-CO_2 and Ps-H_2 collisions the resonance features have a large magnitude and appear at velocities significantly below the Ps ionization threshold.

We have begun the development of the OPP for Ps-molecule scattering by using a spherically symmetric approximation for the Ps-H_2 OPP kernel. When this is used along with the FEG potentials to describe Ps-H_2 scattering the resonance features are suppressed and the results agree well with experiment over the entire velocity range.

We are currently working on extending the OPP to more non-spherical molecules like N_2 and CO_2. Future work will also incorporate vibrational motion which will allow the calculation of vibrational excitation and other nuclear motion dependent processes.

Acknowledgments
IIF was supported by the US National Science Foundation, Grant No. PHY-1803744

References

[1] Tuomisto F and Makkonen I 2013 *Rev. Mod. Phys.* **85** 1583
[2] Champion C and Le Loirec C 2007 *Phys. Med. Biol.* **52** 6605
[3] Brawley S J, Armitage S, Beale J, Leslie D E, Williams A I and Laricchia G 2010 *Science* **330** 789
[4] Brawley S J, Williams A I, Shipman M and Laricchia G 2010 *Phys. Rev. Lett.* **105** 263401
[5] Brawley S J, Williams A I, Shipman M and Laricchia G 2012 *J. Phys.: Conf. Ser.* **338** 012018
[6] Fabrikant I I and Gribakin G F 2014 *Phys. Rev. Lett.* **112** 243201
[7] Fabrikant I I and Gribakin G F 2014 *Phys. Rev.* A **90** 052717
[8] Gribakin G F, Swann A R, Wilde R S and Fabrikant I I 2016 *J. Phys.* B **97** 064004
[9] Brawley S J, Fayer S E, Shipman M and Laricchia G 2015 *Phys. Rev. Lett.* **115**, 223201
[10] Green D G, Swann A R and Gribakin G F 2018 *Phys. Rev. Lett.***120** 183402
[11] Swann A R and Gribakin G F 2018 *Phys. Rev.* A **97** 012706
[12] Biswas P K and Adhikari S K 1998 *J. Phys.* B **31** L737
[13] Biswas P K and Adhikari S K 2000 *J. Phys.* B **33** 1575
[14] Comi M, Prosperi G M, and Zecca A 1983 *Phys. Lett.* **93A** 289
[15] Zhang J Y, Wu M S, Gao X, Yang Y J, Varga K, Yan Z C and Schwingenschlogl U 2019 *Phys. Rev.* A **100** 032701
[16] Wilde R S and Fabrikant I I 2015 *Phys. Rev.* A **92** 032708
[17] Fabrikant I I and Wilde R S 2018 *Phys. Rev.* A **97** 052707
[18] Morrison M A and Collins L A 1981 *Phys. Rev.* A **23** 127
[19] Mitroy J and Ryzhikh G F 1999 *Comput. Phys. Commun.* **123** 103
[20] Ivanov I A, Bromley M W J, and Mitroy J 2003 *Comput. Phys. Commun.* **152** 9
[21] Hara S 1967 *J. Phys. Soc. Jpn.* **22** 710
[22] O'Connell J K and Lane N F 1983 *Phys. Rev.* A **27** 1893
[23] Wilde R S and Fabrikant I I 2018 *Phys. Rev.* A **98** 042703
[24] See the summaries of these results in Refs. [11] and [22]
[25] Shibuya K and Saito H 2018 *Phys. Rev.* A **97** 052702
[26] Garner A, Laricchia G and Ozen A 1996 *J. Phys.* B **29** 5961
[27] Blackwood J E, McAlinden, and Walters H R J 2002 *J. Phys.* B **35** 2661; 2003 **36** 797
[28] Nagashima Y, Hyodo T, Fujiwara K and Ichimura A 1998 *J. Phys.* B **31** 329
[29] Skalsey M, Engbrecht J J, Nakamura C M Vallery R S and Gidley D W 2003 *Phys. Rev.* A **67** 022504
[30] Shibuya K, Kawamura and Saito H 2013 *Phys. Rev.* A **88** 042517
[31] Wilde R S and Fabrikant I I 2018 *Phys. Rev.* A **97** 052708
[32] Shipman M, Brawley S J, Sarkadi L and Laricchia G 2017 *Phys. Rev.* A **95** 032704
[33] Lane N F 1980 *Rev. Mod. Phys.* **52** 29
[34] Birtwistle D T and Herzenberg A 1971 *J. Phys.* B **4** 53
[35] Skalsey M, Engbrecht J J, Bithell R K Vallery R S and Gidley D W 1998 *Phys. Rev. Lett.* **80** 3727

Towards prediction of the rates of antihydrogen positive ion production in collision of antihydrogen with excited positronium

T Yamashita[1,2], Y Kino[2], E Hiyama[1,3], K Piszczatowski[4], S Jonsell[5], P Froelich[4]
[1]Nishina Center, RIKEN, Wako, 351-0198, Japan
[2]Department of Chemistry, Tohoku University, Sendai, 980-8578, Japan
[3]Department of Physics, Kyushu University, Fukuoka, 819-0395, Japan
[4]Department of Chemistry, Uppsala University, Uppsala, Box 518 751-20, Sweden
[5]Department of Physics, Stockholm University, Stockholm, SE-10691, Sweden

E-mail: takuma.yamashita@riken.jp

Abstract. We present a 4-body calculation of scattering between an antihydrogen atom ($\bar{\mathrm{H}}$) and a positronium (Ps) aiming at the prediction of cross sections for the production of antihydrogen positive ions ($\bar{\mathrm{H}}^+$). The antihydrogen positive ions are expected to be a useful source of ultra-cold anti-atoms for the test of matter-antimatter gravity. We convert the Schrödinger equation to a set of coupled integro-differential equations that involve intermediate states which facilitate the internal region description of the scattering wavefunction. They are solved using a compact finite difference method. Our framework is extended to scattering between an excited Ps and $\bar{\mathrm{H}}$. Cross sections of the reactions, Ps (1s/2s/3s) + $\bar{\mathrm{H}} \rightarrow e^- + \bar{\mathrm{H}}^+$, in s-wave collisions, are calculated. It is found that the reactions originating from Ps (1s/2s) + $\bar{\mathrm{H}}$ produce $\bar{\mathrm{H}}^+$ with a constant cross section within 0.05 eV above the threshold while the reaction cross section from Ps (3s) decreases as the collision energy increases in the same energy interval. Just above the threshold, the cross section of $\bar{\mathrm{H}}^+$ production from Ps (3s) + $\bar{\mathrm{H}}$ in s-wave collision is 7.8 times larger than that from Ps (1s) + $\bar{\mathrm{H}}$ in s-wave and 2.3 times larger than that from Ps (2s) + $\bar{\mathrm{H}}$ in s-wave. The near-threshold de-excitation reaction from Ps (3s) + $\bar{\mathrm{H}}$ occurs more rapidly than the $\bar{\mathrm{H}}^+$ production.

1. Introduction

The antihydrogen atom ($\bar{\mathrm{H}}$), consisting of an antiproton ($\bar{\mathrm{p}}$) and a positron (e^+), is a good probe for CPT symmetry and for physics beyond the standard model. Experimental development of sources of cold antihydrogen atoms [1–7] is aimed at a precise measurement of energy levels of $\bar{\mathrm{H}}$ and spectroscopy of $\bar{\mathrm{H}}$ has been reported recently [8–10].

Another application of the cold antihydrogen atoms is measurement of gravitational properties of antimatter [11] testing the weak equivalence principle. Some experiments have been planned to study the gravitational properties of $\bar{\mathrm{H}}$ [12–22]. The antihydrogen positive ion ($\bar{\mathrm{H}}^+ = \bar{\mathrm{p}}e^+e^+$) may be manipulated by electric fields and be used in studies of particle physics and atomic physics. $\bar{\mathrm{H}}^+$ ions are expected to be useful intermediates in the production of antihydrogen atoms [16, 17]. It would be prepared by sympathetic cooling of $\bar{\mathrm{H}}^+$ (with Be^+ ions) and the subsequent positron detachment. $\bar{\mathrm{H}}^+$ can also be utilized to develop an energy-tunable antihydrogen beam that will be used in atomic collision experiments.

Reaction between positronium (Ps = e^+e^-) and antihydrogen atom ($\bar{H}$ = $\bar{p}e^+$) has been featured recently as one of the promising production schemes of $\bar{H}^+$. A channel of $\bar{H}^+$ production,

$$\text{Ps (1s)} + \bar{\text{H}}\text{ (1s)} \rightarrow \text{e}^- + \bar{\text{H}}^+, \tag{1}$$

opens at the collision energy 6.05 eV in the center-of-mass system. A reaction between the second excited Ps (n = 3) and $\bar{H}$ (1s) is also promising because the positron transfer to $\bar{H}$ from Ps (n = 3),

$$\text{Ps } (n = 3) + \bar{\text{H}}\text{ (1s)} \rightarrow \text{e}^- + \bar{\text{H}}^+, \tag{2}$$

can occur from the collision energy 1.7 meV. The low-energy collision between Ps (n = 3) and $\bar{H}$ (1s) resulting in $\bar{H}^+$ production has been expected to have large cross sections [23–25]; however, these rearrangement reactions compete with several inelastic reactions of Ps (de-)excitation and require rigorous theoretical treatment.

The Ps-$\bar{H}$ system is the charge-conjugate system of Ps-H. Low-energy scattering of Ps (1s) by H has been studied in several works as one of the most fundamental scattering problems for the investigation of positronium-atom interaction [26–29]. A number of theoretical studies have produced H (1s) + Ps (1s) total/differential cross sections and scattering length [30–38]. On the other hand, the investigation of inelastic scattering has been limited to calculations using methods suitable for intermediate or high-energy collisions [23, 25, 39]. Calculations based on the close-coupling method [35, 40, 41] were performed to study slow inelastic collisions; however, the information on the reaction branches including a rearrangement reaction channel, namely $\bar{H}^+$ production channel, is still limited.

This work presents a four-body calculation of Ps-$\bar{H}$ scattering with an open channel of e^- + $\bar{H}^+$ as well as all possible competing inelastic channels. We revise our previous pilot calculation [42] and extend the method to scattering of excited Ps by $\bar{H}$. Numerical results are reported for s-wave collision between Ps (1s/2s/3s) and $\bar{H}$ (1s). Atomic units (a.u.; $m_e = \hbar = e = 1$) are used throughout this paper, except where mentioned otherwise.

2. Theory

In this paper we consider a reaction of Ps (n_0l_0) + $\bar{H}$ (1s) resulting in Ps ($n'l'$) + $\bar{H}$ (1s) or e^- + $\bar{H}^+$. For simplicity, we denote the two positrons with e_1^+ and e_2^+. The Schrödinger equation for the scattering state is written as

$$(H - E)\Psi = 0, \tag{3}$$

where the Hamiltonian H includes kinetic energy operators and all inter-particle Coulomb potential interactions expressed in the center-of-mass coordinate system.

The whole system can be characterized by its parity P, a total angular momentum J and its projection onto z-axis M, where the z-axis can be chosen to be in an arbitrary direction. The $\bar{H}^+$ production channel, e^- + $\bar{H}^+$, can open when the two positrons form a spin singlet under the Coulombic interactions. The ground state of $\bar{H}^+$ possesses positive parity and zero angular momentum in total; therefore, the parity of the $\bar{H}^+$ production channel is characterized by the angular momentum between e^- and $\bar{H}^+$. The parity satisfies $P = (-1)^J$, which means the antihydrogen positive ion production channel opens only in natural parity. The channels of Ps (nl) + $\bar{H}$ (1s) can be distinguished by a set of quantum numbers $\alpha \equiv \{n, l, \lambda\}$ for a given J and M, where λ is the angular momentum of the scattering partial wave and satisfies a condition of natural parity, $(-1)^{l+\lambda} = (-1)^J$. We denote the initial channel as $\alpha_0 \equiv \{n_0, l_0, \lambda_0\}$.

We construct the total wavefunction Ψ in equation (3) as

$$\Psi = \sum_\alpha \psi_\alpha + \sum_\upsilon b_\upsilon \Phi_\upsilon, \tag{4}$$

where Φ_υ describe the intermediate, square-integrable states of the scattering (obtained by the diagonalization of the full 4-body Hamiltonian, see below), b_υ are the expansion coefficients, and ψ_α

are the channel functions that describe the asymptotic behavior of the initial/final states. $\psi_\alpha(\mathbf{r},\mathbf{R},\boldsymbol{\rho})$ for Ps (nl) + $\bar{\mathrm{H}}$ (1s) is written as

$$\psi_\alpha(\mathbf{r},\mathbf{R},\boldsymbol{\rho}) = \frac{1}{\sqrt{2}}\left\{\mathcal{R}^{\mathrm{Ps}}_{nl}(r)\mathcal{R}^{\bar{\mathrm{H}}}_{1\mathrm{s}}(R)\mathcal{R}_\alpha(\rho)\left[Y_l(\hat{\mathbf{r}})Y_\lambda(\hat{\boldsymbol{\rho}})\right]_{JM} + (1\leftrightarrow 2)\right\}, \tag{5}$$

where $\left[Y_l(\hat{\mathbf{r}})Y_\lambda(\hat{\boldsymbol{\rho}})\right]_{JM} = \sum_{m_l}\sum_{m_\lambda} C(lm_l\lambda m_\lambda|JM)Y_{lm_l}(\hat{\mathbf{r}})Y_{\lambda m_\lambda}(\hat{\boldsymbol{\rho}})$ with $C(lm_l\lambda m_\lambda|JM)$ the Clebsh-Gordan coefficients. $(1\leftrightarrow 2)$ is a term where two positrons are exchanged. Since we consider the spin singlet, the spatial part should be symmetrized and thus $(1\leftrightarrow 2)$ is added in the positive sign. $\mathbf{r}$, $\mathbf{R}$ and $\boldsymbol{\rho}$ are relative vectors from e^- to e^+_1, from $\bar{\mathrm{p}}$ to e^+_2, and from the center-of-mass of Ps to that of $\bar{\mathrm{H}}$, respectively. $\mathcal{R}^{\mathrm{Ps}}_{nl}(r)$, $\mathcal{R}^{\bar{\mathrm{H}}}_{1\mathrm{s}}(R)$ and $\mathcal{R}_\alpha(\rho)$ are radial functions of Ps (nl), $\bar{\mathrm{H}}$ (1s), and the relative motion between them, respectively. It should be noted that the spherical harmonics of $\bar{\mathrm{H}}$ (1s), $Y_{00}(\hat{\mathbf{R}}) = 1/\sqrt{4\pi}$, is included in $\mathcal{R}^{\bar{\mathrm{H}}}_{1\mathrm{s}}(R)$ for brevity. The channel function for e^- + $\bar{\mathrm{H}}^+$ is written as

$$\psi_\alpha(\mathbf{r}',\mathbf{R}',\boldsymbol{\rho}') = \phi^{\bar{\mathrm{H}}^+}_{\Lambda=0}(\mathbf{r}',\mathbf{R}')\mathcal{R}_\alpha(\rho')Y_{JM}(\hat{\boldsymbol{\rho}}'), \tag{6}$$

where $\phi^{\bar{\mathrm{H}}^+}_{\Lambda=0}$ is a three-body wavefunction of $\bar{\mathrm{H}}^+$ that includes the exchange terms and $\mathcal{R}_\alpha$ is a radial function of the relative motion between e^- and $\bar{\mathrm{H}}^+$. $\phi^{\bar{\mathrm{H}}^+}_{\Lambda=0}$ reproduces the three-body energy of $\bar{\mathrm{H}}^+$ as $-0.527\,445$ a.u. which differs by $0.000\,001$ a.u. from the best variational result. $\mathbf{r}'$, $\mathbf{R}'$ and $\boldsymbol{\rho}'$ are relative vectors from $\bar{\mathrm{p}}$ to e^+_1, from the center-of-mass of $(\bar{\mathrm{p}},\mathrm{e}^+_1)$ to e^+_2, and from the center-of-mass of $\bar{\mathrm{H}}^+$ to e^-, respectively. $\Lambda = 0$ means the $\bar{\mathrm{H}}^+$ has zero total angular momentum.

This expression of the scattering state is basically the same as the one used in three-body muonic atom collisions [43,44] and recently used in four-body antihydrogen atom collisions [42,45]. Functions Φ_υ are expanded in finite range Gaussian functions and diagonalize the Hamiltonian according to $H\Phi_\upsilon = E_\upsilon\Phi_\upsilon$ with conditions $\langle\Phi_{\upsilon'}|\Phi_\upsilon\rangle = \delta_{\upsilon'\upsilon}$ and $\langle\Phi_{\upsilon'}|H|\Phi_\upsilon\rangle = E_\upsilon\delta_{\upsilon'\upsilon}$. The Schrödinger equation (3) is converted to a set of coupled equations with the following conditions, $\langle\mathcal{R}^{\mathrm{Ps}}_{nl}(r)\mathcal{R}^{\bar{\mathrm{H}}}_{1\mathrm{s}}(R)\left[Y_l(\hat{\mathbf{r}})Y_\lambda(\hat{\boldsymbol{\rho}})\right]_{JM}|H-E|\Psi\rangle_{\mathbf{r},\mathbf{R},\hat{\boldsymbol{\rho}}} = 0$, $\langle\phi^{\bar{\mathrm{H}}^+}_{\Lambda=0}(\mathbf{r}',\mathbf{R}')Y_{JM}(\hat{\boldsymbol{\rho}}')|H-E|\Psi\rangle_{\mathbf{r}',\mathbf{R}',\hat{\boldsymbol{\rho}}'} = 0$, and $\langle\Phi_\upsilon|H-E|\Psi\rangle = 0$. Here, $\langle\cdots\rangle_{\mathbf{r},\mathbf{R},\hat{\boldsymbol{\rho}}}$ means the integration over these coordinates leaving out integration over ρ. Expressing the radial functions of the relative motion in initial and final channels as $\mathcal{R}_\alpha(\rho) = \chi_\alpha(\rho)/\rho$, we obtain a set of coupled integro-differential equations for the channel functions, χ_α, that allow simultaneous determination of the expansion coefficients b_υ. We solve it by a compact finite difference method [46] under proper boundary conditions at $\rho\to\infty$

$$\chi_\alpha(\rho) \to \rho h^{(-)}_\lambda(k_\alpha\rho)\delta_{\alpha\alpha_0} - \sqrt{\frac{\upsilon_{\alpha_0}}{\upsilon_\alpha}}S_{\alpha\alpha_0}\rho h^{(+)}_\lambda(k_\alpha\rho) \tag{7}$$

for the initial channel α_0 and final channels dissociating into Ps (nl) and $\bar{\mathrm{H}}$ (1s). For the final channel dissociating into e^- + $\bar{\mathrm{H}}^+$, at $\rho'\to\infty$,

$$\chi_\alpha(\rho') \to -\sqrt{\frac{\upsilon_{\alpha_0}}{\upsilon_\alpha}}S_{\alpha\alpha_0}\rho' H^{(+)}_J(k_\alpha\rho'). \tag{8}$$

$h^{(\mp)}_\lambda$ is an incoming/outgoing spherical Hankel function and $H^{(\mp)}_\lambda$ an incoming/outgoing spherical Coulomb-Hankel function. υ_α is a speed of the relative motion between the fragments in channel α. The S-matrix elements $S_{\alpha\alpha_0}$ give partial cross sections $\sigma_{\alpha\alpha_0} = (2J+1)\pi|\delta_{\alpha\alpha_0} - S_{\alpha\alpha_0}|^2/k^2_{\alpha_0}$.

3. Results and discussion

Table 1. List of channels α = 1-7 in $J = 0$ opening just above the $e^- + \bar{H}^+$ dissociation threshold. In α = 1-6, the $\bar{H}$ is in 1s state.

	$\alpha = 1$	$\alpha = 2$	$\alpha = 3$	$\alpha = 4$	$\alpha = 5$	$\alpha = 6$	$\alpha = 7$
fragments	Ps (1s)+$\bar{H}$	Ps (2s)+$\bar{H}$	Ps (2p)+$\bar{H}$	Ps (3s)+$\bar{H}$	Ps (3p)+$\bar{H}$	Ps (3d)+$\bar{H}$	$e^-+\bar{H}^+$
partial wave	s-wave	s-wave	p-wave	s-wave	p-wave	d-wave	s-wave

Construction of the intermediate states $\{\Phi_\upsilon\}$ plays a primary role in accurate determination of the cross section. In this work we construct $\{\Phi_\upsilon\}$ using a Gaussian expansion method (GEM) [47–49], in terms of radial Gaussian functions and spherical harmonics, according to

$$\Phi_\upsilon = \sum_{c=1}^{3} \sum_{n_c N_c \nu_c} \sum_{l_c L_c \lambda_c} A^{(\upsilon)}_{c n_c N_c \nu_c l_c L_c \lambda_c} \left\{ r_c^{l_c} R_c^{L_c} \rho_c^{\lambda_c} \exp\left(-\frac{r_c^2}{r_{n_c}^2} - \frac{R_c^2}{R_{N_c}^2} - \frac{\rho_c^2}{\rho_{\nu_c}^2} \right) \right.$$
$$\times \left. \left[\left[Y_{l_c}(\hat{\mathbf{r}}_c) Y_{L_c}(\hat{\mathbf{R}}_c) \right]_{\Lambda_c} Y_{\lambda_c}(\hat{\rho}_c) \right]_{JM} + (1 \leftrightarrow 2) \right\}, \tag{9}$$

where $\{\mathbf{r}_c, \mathbf{R}_c, \rho_c\}$ is a set of Jacobian coordinates (see figure 1). The first coordinate set $c = 1$ is best suited to describe the four-body interactions in the Ps + $\bar{H}$ configuration and $c = 2$ and $c = 3$ are best suited to describe the four-body interactions in the $e^- + \bar{H}^+$ configuration. The linear coefficients $A_{cn_cN_c\nu_cl_cL_c\lambda_c}$ are determined by diagonalizing the Hamiltonian H in the Gaussian basis set specified in equation (9). The non-linear coefficients $\{r_{n_c}\}$, $\{R_{N_c}\}$ and $\{\rho_{\nu_c}\}$ are distributed according to a geometrical progression between chosen minimum and maximum values. The relative motion between Ps and $\bar{H}$ for $c = 1$ and that between e^- and $\bar{H}^+$ for $c = 2$ and $c = 3$ are described in terms of $\{\rho_{\nu_c}\}$.

We proceed with the discussion on the s-wave collision between Ps (ns) + $\bar{H}$ (1s) ($n \leq 3$) resulting in $e^- + \bar{H}^+$. The whole system is $J = 0$ and all open channels α = 1-7 just above the $e^- + \bar{H}^+$ dissociation threshold are listed in table 1. The maximum range of the $\{\rho_{\nu_c}\}$ is chosen to be 40 a.u. for $c = 1$ and 110 a.u. for $c = 2$ and $c = 3$. Each function Φ_υ is described in 44320 Gaussian basis functions. It should be emphasized that each of the functions in the set $\{\Phi_\upsilon\}$ contains contributions expressed in several sets of Jacobian coordinates, namely those natural for the initial channel, the final channel, all energetically opened channels and also the important virtual excitation channels [41].

The antihydrogen positive ion production cross sections ($\sigma_{\rm ion}$) of the s-wave collision of Ps (ns) + $\bar{H}$ (1s) are presented in figure 2. The energy of the scattering is expressed in two ways: as the relative collision energy being $E - \epsilon_{\rm Ps} - \epsilon_{\bar{H}}$ and as the antihydrogen impact energy on a Ps target at rest being $(E - \epsilon_{\rm Ps} - \epsilon_{\bar{H}})m_{\bar{H}}/m_{\rm Ps}$, where $\epsilon_{\rm Ps}$ and $\epsilon_{\bar{H}}$ are internal energy of Ps and $\bar{H}$ in the initial channel α_0. These cross sections are compared to each other and to other competing reactions, such as elastic scattering cross sections ($\sigma_{\rm ela}$) and Ps de-excitation cross sections ($\sigma_{\rm dex}$) for the same energy interval of the relative kinetic energy in final channel of $e^- + \bar{H}^+$. Note that the total cross sections of Ps (3s) + $\bar{H}$ (1s) are expected to be dominated by the s-wave scattering due to the small collision energy. The s-wave cross sections of Ps (1s/2s) + $\bar{H}$ (1s) are not enough to evaluate the total cross sections and we are working on inclusion of higher partial waves towards the total understanding of $\bar{H}^+$ production reactions. On the

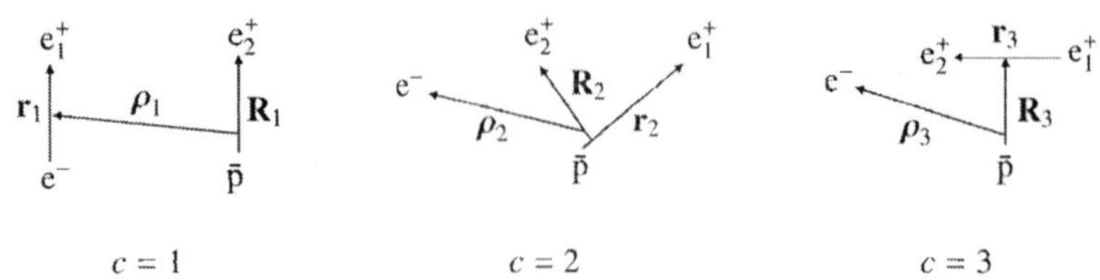

Figure 1. Three sets of Jacobian coordinate systems for intermediate states.

other hand, it is worthwhile to investigate the reaction Ps (ns) + $\bar{\mathrm{H}}$ (1s) → e^- + $\bar{\mathrm{H}}^+$ in s-wave collision for understanding the threshold behavior of the state-to-state cross sections.

Convergence of cross section against the maximum number of intermediate states is examined and we have found that cross sections almost converge with $\upsilon = 60$ ($E_{\upsilon=60} = 0.239\,68$ a.u.) from $\upsilon = 1$ (a ground state). For safety, the intermediate states are included from $\upsilon = 1$ to $\upsilon = 80$ ($E_{\upsilon=80} = 0.245\,75$ a.u.) so as to cover the collision energy sufficiently.

The $\bar{\mathrm{H}}^+$ production cross sections from Ps (1s/2s) + $\bar{\mathrm{H}}$ (1s) scattering, $\sigma_{\mathrm{ion}} = \sigma_{71}$ and σ_{72}, begin from a finite value at the threshold and show a steady value throughout the presented energy region (see figure 2 a and b). The steady behavior of the $\bar{\mathrm{H}}^+$ production cross section above the threshold in Ps (1s/2s) + $\bar{\mathrm{H}}$ (1s) (shown in figure 2 a and b) is consistent with Wigner's threshold law [50]. In the reaction Ps (1s/2s) + $\bar{\mathrm{H}}$ (1s) → e^- + $\bar{\mathrm{H}}^+$, the produced fragments attract each other through the Coulomb interaction and the speed of the relative motion is sufficiently smaller than that between the initial fragments, which results in a constant cross section above the threshold. The steady behavior of σ_{ion} is also consistent with the cross section behavior reported in a close-coupling (CC) calculation [35]. In that calculation, the $\bar{\mathrm{H}}^+$ production cross section from Ps (1s) + $\bar{\mathrm{H}}$ (1s) at the threshold was reported to show the steady behavior around 3.8 a_0^2. A similar behavior around $\pi\ a_0^2$ was reported in a coupled-pseudostate-approach (CPA) calculation [51,52]. For the reaction of Ps (2s) + $\bar{\mathrm{H}}$ (1s), the $\bar{\mathrm{H}}^+$ production cross section at the threshold was reported to be $\sim$ 60 a_0^2 in the CPA calculation [51, 52]. Note that our calculation only includes s-wave cross sections and shows smaller value than the CC and CPA calculations. This could be due to the lack of the higher partial wave contributions and the strictness of our method. A major difference in the calculation methods between this work and the CC approach is that the present one expresses the four-body correlation during the scattering by using the four-body basis functions which diagonalize the four-body Hamiltonian while the CC approach expresses the four-body correlation by linear combination of target wavefunctions. Besides, the CC calculation [35] included Ps excitation and $\bar{\mathrm{H}}^+$ formation but not $\bar{\mathrm{H}}$ virtual excitation effects.

The $\bar{\mathrm{H}}^+$ production cross section from the s-wave collision of Ps (3s) + $\bar{\mathrm{H}}$ (1s), $\sigma_{\mathrm{ion}} = \sigma_{74}$, decreases rapidly with increasing the collision energy (see figure 2 c), which is a different trend from those in Ps (1s/2s) + $\bar{\mathrm{H}}$ (1s) (see figure 2 a, b). In the reaction Ps (3s) + $\bar{\mathrm{H}}$ (1s) → e^- + $\bar{\mathrm{H}}^+$, both the initial and final fragments have similar relative speeds, which is out of the coverage of the Wigner's threshold law; therefore, the cross section $\sigma_{\mathrm{ion}} = \sigma_{74}$ does not show the steady behavior of σ_{ion}. The decrease of the cross section may be related to the fact that the inelastic cross sections are scaled by the factor of k_{i}^{-2} where k_{i} is the relative momentum in the initial.

In the low energy limit of the collision between Ps (3s) and $\bar{\mathrm{H}}$ (1s) in s-wave, the $\bar{\mathrm{H}}^+$ production cross section becomes $\sim$ 8.5 a_0^2. A previous calculation [23] based on a continuum distorted wave-final state (CDW-FS) theory predicted $\sim$ 18 a_0^2. Another previous work based on the first Born approximation [52] predicted that the $\bar{\mathrm{H}}^+$ production cross section was over 3000 a_0^2 just above the $\bar{\mathrm{H}}^+$ production threshold energy and decreased rapidly as the collision energy increased. One should note that these calculations included higher partial wave contributions. Our $\bar{\mathrm{H}}^+$ production cross section shows huge discrepancy from the first Born approximation and the behavior of the cross section against the collision energy obtained in this work is completely different from that obtained in CDW-FS. Namely, the present cross section rapidly decreases as the collision energy increases whereas the cross section in CDW-FS slowly decreases up to the collision energy $\sim$ 1 a.u.

The elastic scattering cross sections $\sigma_{\mathrm{ela}} = \sigma_{11}$ and σ_{22} for Ps (1s/2s) + $\bar{\mathrm{H}}$ (1s) are of the same magnitude as the $\bar{\mathrm{H}}^+$ production cross section; however, the elastic one in Ps (3s) + $\bar{\mathrm{H}}$ (1s) is around 100 times larger than the $\bar{\mathrm{H}}^+$ production cross section. Besides, the de-excitation cross section from Ps (3s) + $\bar{\mathrm{H}}$ (1s), $\sigma_{\mathrm{dex}} = \sigma_{14} + \sigma_{24} + \sigma_{34}$, is 10 times larger than the $\bar{\mathrm{H}}^+$ production cross section while the de-excitation cross section from Ps (2s) + $\bar{\mathrm{H}}$ (1s), $\sigma_{\mathrm{dex}} = \sigma_{12}$, shows smaller value than the $\bar{\mathrm{H}}^+$ production cross section.

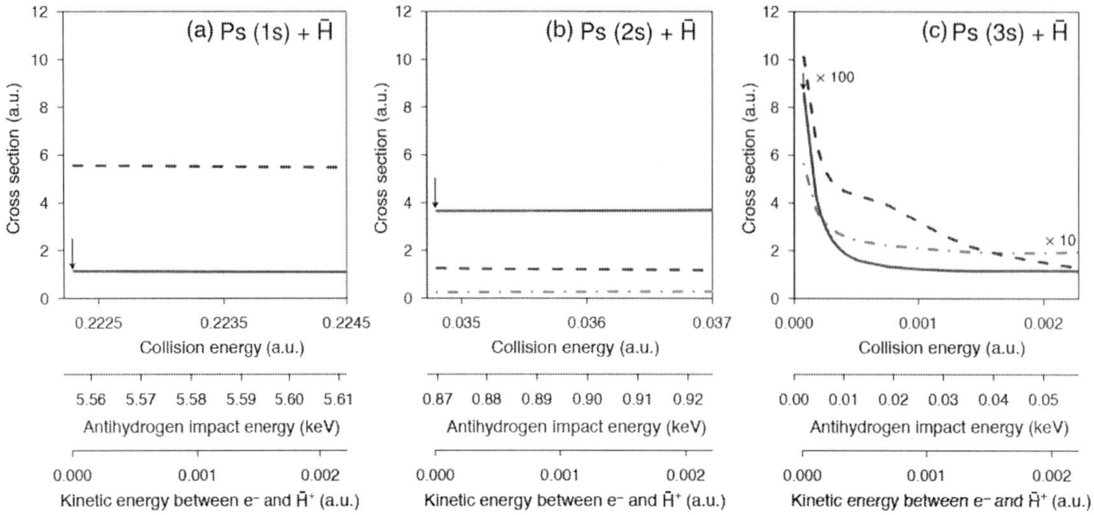

Figure 2. Cross sections of Ps (ns) + H̄ (1s) in s-wave collision are shown in (a) n = 1, (b) n = 2 and (c) n = 3. The arrow denotes the energy where e^- + $\bar{H}^+$ channel opens. Red lines are reaction cross sections $\sigma_{ion} = \sigma_{7\alpha_0}$ (α_0 = 1, 2 and 4) resulting in e^- + $\bar{H}^+$, black dashed lines are elastic cross sections $\sigma_{\alpha_0\alpha_0}$, and blue dotted-dashed lines are inelastic de-excitation cross sections σ_{dex} resulting in Ps ($n'l'$) + H̄ (1s) where $n' < n$. In (c) the elastic cross section should be multiplied by 100, and the inelastic de-excitation cross section by 10.

4. Summary

We have presented a four-body scattering calculation method, and applied it towards prediction of the production rate of $\bar{H}^+$ in reactions between Ps (nl) and H̄ (1s). The use of intermediate states obtained by the diagonalization of the full 4-body Hamiltonian projected onto the subspace spanned by the Gaussian basis functions facilitates the description of the four-body interactions in the internal scattering region, explicitly including the rearrangement reactions. S-wave scattering cross sections between Ps (ns) and H̄ (1s) are presented. We have found that the cross section for the reaction Ps (1s) + H̄ (1s) → e^- + $\bar{H}^+$ is consistent in magnitude with a previous close-coupling calculation [35]. The behavior of the cross section of the reaction Ps (2s) + H̄ (1s) → e^- + $\bar{H}^+$ shows a similar trend as that of Ps (1s) + H̄ (1s) → e^- + $\bar{H}^+$, both starting from a finite value and remain constant just above the threshold, which is in agreement with the Wigner's threshold law. On the other hand, the cross section of the reaction Ps (3s) + H̄ (1s) → e^- + $\bar{H}^+$ shows a drastic decrease with the collision energy just above the e^- + $\bar{H}^+$ threshold. We have found that the threshold value of the $\bar{H}^+$ production cross section from Ps (3s) + H̄ (1s) in s-wave becomes 7.8 times larger than that from Ps (1s) + H̄ (1s) and 2.3 times larger than that from Ps (2s) + H̄ (1s). It is also seen that the reaction Ps (3s) + H̄ (1s) → e^- + $\bar{H}^+$ competes with predominant elastic scattering and Ps de-excitation scattering.

Acknowledgments

T.Y. would like to thank Special Postdoctoral Researcher (SPDR) program in RIKEN, Y.K. would like to thank JSPS KAKENHI Grant Number 17K05592 and 18H05461, S.J. would like to thank the Swedish Research Council (VR) for financial supports. The computation was conducted on the supercomputers ITO at Kyushu University and HOKUSAI at RIKEN.

References

[1] Amoretti M *et al.* 2002 *Nature* **419** 456
[2] Enomoto Y *et al.* 2010 *Phys. Rev. Lett.* **105** 243401
[3] The ALPHA Collaboration 2011 *Nat. Phys.* **7** 558
[4] Kuroda N *et al.* 2014 *Nat. Commun.* **5** 4089
[5] Widmann E *et al.* 2018 *Hyperfine Interact.* **240** 5
[6] Kolbinger B *et al.* 2018 *EPJ Web Conf.* **181** 01003
[7] Malbrunot C *et al.* 2018 *Philos. Trans. Roy. Soc.* A **376** 20170273
[8] Ahmadi M *et al.* 2018 *Nature* **557** 71
[9] Ahmadi M *et al.* 2017 *Nature* **548** 66
[10] Ahmadi M *et al.* 2018 *Nature* **561** 211
[11] Scherk J 1979 *Phys. Lett.* B **88** 265
[12] Cesar C L *et al.* 2005 *AIP Conf. Proc.* **770** 33
[13] The ALPHA Collaboration 2013 *Nature Commun.* **4** 1785
[14] Indelicato P *et al.* 2014 *Hyperfine Interact.* **228** 141
[15] Pérez P *et al.* 2015 *Hyperfine Interact.* **233** 21
[16] Pérez P and Sacquin Y 2012 *Classical Quant. Grav.* **29** 184008
[17] Sacquin Y 2014 *Eur. Phys. J.* D **68** 31
[18] Kellerbauer A *et al.* 2008 *Nucl. Instrum. Meth. Phys. Res.* B **266** 351
[19] Scampoli P and Storey J 2014 *Mod. Phys. Lett.* A **29** 1430017
[20] Knecht A *et al.* 2014 *Hyperfine Interact.* **228** 121
[21] Kimura M *et al.* 2015 *J. Phys.: Conf. Ser.* **631** 012047
[22] Evans C *et al.* 2018 *EPJ Web Conf.* **182** 02040
[23] Comini P and Hervieux P A 2013 *New J. Phys.* **15** 095022
[24] Comini P and Hervieux P A 2013 *J. Phys.: Conf. Ser.* **443** 012007
[25] Comini P, Hervieux P A and Biraben F 2014 *Hyperfine Interact.* **228** 159
[26] McAlinden M T, MacDonald F G R S and Walters H R J 1996 *Can. J. Phys.* **74** 434
[27] Adhikari S K 2002 *Nucl. Instrum. Methods Phys. Res.* B **192** 74
[28] Fabrikant I I and Gribakin G F 2014 *Phys. Rev.* A **90** 052717
[29] Fabrikant I I and Gribakin G F 2014 *Phys. Rev. Lett.* **112** 243201
[30] Drachman R J and Houston S K 1976 *Phys. Rev.* A **14** 894
[31] Adhikari S K and Mandal P 2000 *J. Phys.* B: *At. Mol. Opt. Phys.* **33** L761
[32] Adhikari S K 2001 *Phys. Rev.* A **63** 054502
[33] Adhikari S K and Mandal P 2001 *J. Phys.* B: *At. Mol. Opt. Phys.* **34** L187
[34] Biswas P K 2001 *J. Phys.* B: *At. Mol. Opt. Phys.* **34** 4831
[35] Blackwood J E, McAlinden M T and Walters H R J 2002 *Phys. Rev.* A **65** 030502
[36] Reeth P V and Humberston J W 2003 *J. Phys.* B: *At. Mol. Opt. Phys.* **36** 1923
[37] Zhang J Y and Mitroy J 2008 *Phys. Rev.* A **78** 012703
[38] Woods D, Ward S J and Van Reeth P 2015 *Phys. Rev.* A **92** 022713
[39] Roy S and Sinha C 2008 *Eur. Phys. J.* D **47** 327
[40] Campbell C P, McAlinden M T, MacDonald F G R S and Walters H R J 1998 *Phys. Rev. Lett.* **80** 5097
[41] Blackwood J E, McAlinden M T and Walters H R J 2002 *Phys. Rev.* A **65** 032517
[42] Froelich P, Yamashita T, Kino Y, Jonsell S, Hiyama E and Piszczatowski K 2019 *Hyperfine Interact.* **240** 46
[43] Kamimura M 1988 *Muon Catal. Fusion* **3** 335
[44] Kino Y and Kamimura M 1993 *Hyperfine Interact.* **82** 45
[45] Piszczatowski K, Voronin A and Froelich P 2014 *Phys. Rev.* A **89** 062703
[46] Zhao J and Corless R M 2006 *Appl. Math. Comp.* **177** 271
[47] Hiyama E, Kino Y and Kamimura M 2003 *Prog. Part. Nucl. Phys.* **51** 223
[48] Hiyama E 2012 *Prog. Theo. Exp. Phys.* **2012**
[49] Hiyama E and Kamimura M 2018 *Front. Phys.* **13** 132106
[50] Wigner E P 1948 *Phys. Rev.* **73** 1002
[51] McAlinden M T, Blackwood J E and Walters H R J 2002 *Phys. Rev.* A **65** 032715
[52] Swann A R, Cassidy D B, Deller A and Gribakin G F 2016 *Phys. Rev.* A **93** 052712

Atomic processes with twisted electrons

V A Zaytsev[1], A Surzhykov[2,3], V G Serbo[4,5], V P Kosheleva[1,6,7], M E Groshev[1], V A Yerokhin[8], V M Shabaev[1] and Th Stöhlker[6,7,9]

[1]Department of Physics, St. Petersburg State University, Universitetskaya naberezhnaya 7/9, 199034 St. Petersburg, Russia
[2]Physikalisch-Technische Bundesanstalt, D-38116 Braunschweig, Germany
[3]Technische Universität Braunschweig, D-38106 Braunschweig, Germany
[4]Novosibirsk State University, RUS-630090 Novosibirsk, Russia
[5]Sobolev Institute of Mathematics, RUS-630090 Novosibirsk, Russia
[6]GSI Helmholtzzentrum für Schwerionenforschung GmbH, D-64291 Darmstadt, Germany
[7]Helmholtz-Institute Jena, D-07743 Jena, Germany
[8] Peter the Great St. Petersburg Polytechnic University, St. Petersburg 195251, Russia
[9]Institut für Optik und Quantenelektronik, Friedrich-Schiller-Universität, D-07743 Jena, Germany

E-mail: v.a.zaytsev@spbu.ru

Abstract. The present status of the fully-relativistic nonperturbative calculations of the fundamental atomic processes with twisted electrons is presented. In particular, the elastic (Mott) scattering, the radiative recombination, and for the very first time, the Bremsstrahlung processes are considered. The electron-ion interaction is accounted for in a nonperturbative manner, that allows obtaining reliable results for heavy systems. We investigate the influence of the "twistedness" of the incoming electron on the angular and polarization properties of the emitted electrons and photons for the elastic and inelastic scattering, respectively. It is found that these properties exhibit a strong dependence on the opening angle of the vortex electron beam in all processes considered.

1. Introduction

The electron is called twisted (or vortex) if it possesses a well-defined total angular momentum projection onto the propagation direction. Such particles being the solutions of the free Dirac equation with the imposed cylindrical boundary conditions were predicted theoretically [1] and realized experimentally [2, 3, 4] about a decade ago. From these articles, the extensive investigations dedicated to the creation, detection, and application of twisted electrons have started (see [5, 6, 7] for a review and relevant references). The interest in such particles is caused mainly by the magnitude of their total angular momentum (TAM) projection $\hbar m$. This projection can substantially exceed the one defined solely by the spin angular momentum of the conventional (plane-wave) electrons. Nowadays, the twisted electrons with $m \sim 1000$ can be routinely produced [8]. The magnetic dipole moment $\mu = m\mu_B$ (μ_B is the Bohr magneton) of the electrons with such a high TAM projection is by three orders of magnitude larger than those in the plane-wave case. This fact stimulates investigations of the utility of the twisted electrons for studying various subtle magnetic effects [9] and magnetic properties of the materials [10, 11, 12, 13].

Interaction of twisted electrons with ionic and atomic targets is of particular interest. Indeed, with the growth of TAM projection, the spin-orbit interaction increases thus providing a unique opportunity to get a better insight into the role of the complex interplay between spin and orbital angular momenta in various atomic processes. This fact has stimulated investigations of processes involving ionic (or atomic) targets and twisted electrons. In [14], the transfer of the orbital angular momentum from an incident vortex beam to the internal motion of the hydrogen electron was studied. The Rutherford potential scattering of electron vortices was investigated in [15]. Matula and co-authors [16] discussed the radiative recombination of twisted beams with bare nuclei into the ground states of corresponding H-like ions. The atom excitation in coarse of the inelastic electron-vortex-beam scattering was studied in [17]. In [18, 19], the investigation of the beam-size effects in the scattering of the twisted packets by the hydrogen atom was presented. Serbo *et al* [20] examined the Mott scattering of high-energy twisted electrons by neutral atoms whose electrostatic potential was approximated by a sum of three Yukawa potentials. The detailed description of the hydrogen atom ionization by the vortex beam was presented in the recent work by Harris and co-authors [21]. It is worth mentioning that only in [20] the relativistic formalism was utilized meanwhile in other investigations the nonrelativistic approach was used.

All these studies, however, were performed in the framework of the first Born approximation, i.e., the interaction of the twisted electrons with the target potential was taken into accounted in a perturbative manner. Such approximation is applicable only for the description of low-Z (Z is the nuclear charge number) atomic systems and relatively large projectile velocities. Meanwhile, the largest sensitivity to the "twistedness" is expected in high-Z systems where the spin-orbit interaction is strongly enhanced. To perform an accurate description of the processes with heavy ions and atoms, one needs to account for the interaction of the vortex electron with the target potential nonperturbatively. In [22] the fully-relativistic approach aimed at performing such descriptions was developed. In this formalism, the vortex electron is described by the wave function being the solution of the Dirac equation with the target potential and the twisted-wave asymptotic behavior. As a first application of the developed approach, in [22] the recombination of the vortex electron beams with heavy bare nuclei was described in details. Later, Kosheleva and co-authors [23] applied a similar method for the investigation of the higher-order effects beyond the first Born approximation in the process of the elastic (Mott) scattering of twisted electrons by neutral atoms. Additionally, in this work, the differential cross section and the degree of longitudinal polarization for scattering by such a heavy element as gold ($Z = 79$) were evaluated. We also mention another approach for the description of the twisted electrons in external fields which is based on the Foldy-Wouthuysen representation of the Dirac equation. This approach was successfully applied by Silenko and co-authors [24] to study the twisted electrons in electric and magnetic fields.

Here we apply the formalism developed in [22] for the investigation of the fundamental atomic processes with twisted electrons. In particular, the elastic (Mott) scattering, the radiative recombination, and the Bremsstrahlung processes are considered. It should be noted that the Bremsstrahlung from twisted electrons propagating in the field of ionic or atomic targets has not been examined so far. We pay special attention to the investigation of the influence of the "twistedness" of the incoming electron on the angular and polarization properties of the emitted electrons and photons for the elastic and inelastic scattering, respectively. It is found that these properties exhibit a strong dependence on the opening angle of the vortex electron beam in all processes considered.

2. Basic formalism

Let us start with the brief recall of the main properties of the free twisted electrons which we take here in the form of the Bessel waves. The vortex states are characterized by the following set of quantum numbers: the energy ε, the helicity μ, and the projections of the linear p_z and total angular m momenta onto the propagation direction which is chosen as the z-axis. Twisted electrons also possess a well-defined transversal momentum $\varkappa = \sqrt{\varepsilon^2 - 1 - p_z^2}$ and the so-called opening angle $\theta_p = \arctan(\varkappa/p_z)$. The explicit expression for the wave function of the state with the quantum numbers listed above is given by [20]

$$\psi_{\varkappa m p_z \mu}(\mathbf{r}) = \int d\mathbf{p} \frac{e^{im\varphi_p}}{2\pi p_\perp} \delta(p_\parallel - p_z)\delta(p_\perp - \varkappa) i^{\mu - m} \psi_{\mathbf{p}\mu}(\mathbf{r}). \quad (1)$$

Here $p_\parallel$ and $p_\perp$ are the longitudinal and perpendicular components of the momentum $\mathbf{p}$, respectively, and $\psi_{\mathbf{p}\mu}$ is the plane-wave solution of the free Dirac equation

$$\psi_{\mathbf{p}\mu}(\mathbf{r}) = \frac{e^{i\mathbf{p}\cdot\mathbf{r}}}{\sqrt{2\varepsilon(2\pi)^3}} u_{\mathbf{p}\mu}, \tag{2}$$

where $u_{\mathbf{p}\mu}$ is the Dirac bispinor [25, 26] which satisfies the normalization condition $u^\dagger_{\mathbf{p}\mu} u_{\mathbf{p}\mu'} = 2\varepsilon\delta_{\mu\mu'}$. Equation (1) implies that in the momentum space, the twisted electron can be represented as a coherent superposition of the plane-waves with the linear momenta $\mathbf{p}$ covering the surface of a cone with the opening angle θ_p. Additionally, from equation (1), it is seen that the vortex beam has an inhomogeneous probability distribution and an inhomogeneous flux density. In particular, the flux density equals

$$j_z^{(\mathrm{tw})}(\mathbf{r}) = \psi^\dagger_{\varkappa m p_z \mu}(\mathbf{r})\alpha_z \psi_{\varkappa m p_z \mu}(\mathbf{r}) = \frac{p}{\varepsilon(2\pi)^3} \sum_\sigma 4\mu\sigma \left[d^{1/2}_{\sigma\mu}(\theta_p)\right]^2 J^2_{m-\sigma}(\varkappa r_\perp), \tag{3}$$

where $p = |\mathbf{p}|$, α_z is the z component of the vector of Dirac matrices, $d^J_{MM'}(\theta)$ is the small Wigner matrix [27, 28], J_n is the Bessel function of the first kind [29, 30], and $r_\perp$ is the perpendicular component of $\mathbf{r}$.

Having discussed the basic properties of the free twisted electrons, we proceed to study their scattering (elastic and inelastic) by a single ion. First, we fix the z-axis along the propagation direction of the vortex beam. The position of the ion is given by the impact parameter $\mathbf{b} = (b_x, b_y, 0)$, which defines the relative position of the target and the incident electron. The necessity of this parameter is provided by the inhomogeneity of the probability flux density (3) of the vortex beam. To describe the scattering of the vortex beam by heavy ions accurately, one needs to account for the electron-atom interaction nonperturbatively. It can be performed via the construction of the wave function of the twisted electron as the solution of the Dirac equation in the external field of the ionic target with the following asymptotics [31]

$$\Psi^{(+)}_{\varkappa m p_z \mu}(\mathbf{r}+\mathbf{b}) \xrightarrow[r\to\infty]{} \psi_{\varkappa m p_z \mu}(\mathbf{r}+\mathbf{b}) + G^{(\mathrm{tw})}_{m\mu,\mathbf{b}}(\theta_p, \mathbf{n})\frac{e^{ipr}}{r}. \tag{4}$$

Here $G^{(\mathrm{tw})}$ is the bispinor amplitude and the transformed coordinate system $\mathbf{r} - \mathbf{b} \to \mathbf{r}$ (being convenient for practical calculations) is used. The wave function (4) is given by [22]

$$\Psi^{(+)}_{\varkappa m p_z \mu}(\mathbf{r}+\mathbf{b}) = \int d\mathbf{p}\frac{e^{im\varphi_p}}{2\pi p_\perp}\delta(p_\parallel - p_z)\delta(p_\perp - \varkappa) i^{\mu-m} e^{i\mathbf{p}\cdot\mathbf{b}} \Psi^{(+)}_{\mathbf{p}\mu}(\mathbf{r}), \tag{5}$$

where $\Psi^{(+)}_{\mathbf{p}\mu}$ describes the conventional (asymptotically plane-wave) electron propagating in the ionic target potential. The explicit form of this wave function can be found, e.g., in [26, 32, 33]. Utilizing (5) one can connect the amplitude of the twisted electron scattering with the plane-wave one as follows

$$\tau^{(\mathrm{tw})}_{\varkappa m p_z \mu}(\mathbf{b}) = \int \frac{e^{im\varphi_p}}{2\pi p_\perp}\delta(p_\parallel - p_z)\delta(p_\perp - \varkappa) i^{\mu-m} e^{i\mathbf{p}\cdot\mathbf{b}} \tau^{(\mathrm{pl})}_{\mathbf{p}\mu} d\mathbf{p}. \tag{6}$$

This amplitude accounts for the electron-ion interaction to all orders and describes the scattering process uniquely. Thus, the theoretical description is regarded as complete.

The process of the scattering of the vortex beam by a single ion is challenging for the experimental realization. Here, therefore, a more realistic scenario of the scattering by an infinitely extended (macroscopic) target is considered. We describe this target as an incoherent superposition of ions being randomly and homogeneously distributed. The differential cross section for this case is given by [20]:

$$d\sigma^{(\mathrm{tw})} = \int \frac{d\mathbf{b}}{\pi R^2}\left|\tau^{(\mathrm{tw})}_{\varkappa m p_z \mu}(\mathbf{b})\right|^2 = \frac{1}{\cos\theta}\int \frac{d\varphi_p}{2\pi} d\sigma^{(\mathrm{pl})}(\hat{\mathbf{p}}), \tag{7}$$

where (πR^2) stands for the cross section area (R is the radius of the cylindrical box) and the volumes of the phase spaces are omitted for brevity. The integrand in this expression designates the differential cross section for the scattering of the plane-wave electron incoming along $\hat{\mathbf{p}} = \mathbf{p}/|\mathbf{p}|$ direction. From equation (7) it is seen that the cross section for the scattering of the twisted electrons by the macroscopic target does not depend on the TAM projection m. It is worth mentioning that this dependence is restored if the incident electron is chosen as a superposition of two (or more) vortex states with different TAM projections [34, 35, 16, 22].

3. Results and discussions

We apply the approach which is described above for the investigation of the fundamental atomic processes with the twisted electrons. In particular, the elastic (Mott) scattering, the radiative recombination, and the Bremsstrahlung processes are studied. Here we restrict ourselves to the consideration of the scattering of 100 keV twisted electrons by the macroscopic target consisting of lead ($Z = 82$) nuclei. Additionally, we consider that the electrons and photons emitted in the course of the processes under investigation are asymptotically described by the plane waves. This assumption corresponds to the fact that in experiments the detectors of the conventional (plane-wave) particles are mostly utilized.

3.1. Elastic (Mott) scattering of twisted electrons

We start with the elastic (Mott) scattering of the twisted electrons. For this process, equation (7) takes the following form

$$\frac{d\sigma^{(\mathrm{tw})}_{\mu_f\mu_i}}{d\Omega_f} = \frac{1}{\cos\theta_p}\int \frac{d\varphi_p}{2\pi}\frac{d\sigma^{(\mathrm{pl})}_{\mu_f\mu_i}}{d\Omega_f}(\hat{\mathbf{p}}), \tag{8}$$

where the solid angle of the emitted electron Ω_f is defined by the azimuthal φ_f and polar θ_f angles, μ_f and μ_i are the helicities of the initial and final electron states, respectively. The explicit expression for the differential cross section of the Mott scattering of the plane-wave electrons can be found, e.g., in [36, 37, 38, 39]. On the left panel of figure 1, the differential cross section being averaged over μ_i and summed over μ_f is presented. The right panel of figure 1 displays the degree of the longitudinal polarization which is defined by

$$P = \frac{d\sigma_{1/2\,1/2} - d\sigma_{1/2\,-1/2}}{d\sigma_{1/2\,1/2} + d\sigma_{1/2\,-1/2}}, \tag{9}$$

where $d\sigma_{\mu_f\,\mu_i} \equiv \frac{d\sigma_{\mu_f\mu_i}}{d\Omega_f}$ and it is assumed that the incident electron is completely longitudinally polarized ($\mu_i = 1/2$). From this figure, it is seen that both the differential cross section (left panel) and the degree of the longitudinal polarization (right panel) exhibit a strong dependence on the opening angle of the twisted electron. Moreover, one can see that the angular distribution of the emitted electrons has a peak at the angle equal to the opening angle θ_p. This feature of the differential cross section can be exploited for analysis of kinematic properties of the vortex beam. For more details on the Mott scattering of the twisted electrons, we refer to [20, 23].

3.2. Radiative recombination of twisted electrons

We now turn to the consideration of the radiative recombination of the twisted electrons with lead ($Z = 82$) nuclei into the ground $1s$ state of the corresponding H-like ions. In accordance with (7), the differential cross section for this process expresses as

$$\frac{d\sigma^{(\mathrm{tw})}_{\chi m_f\mu_i}}{d\Omega_k} = \frac{1}{\cos\theta_p}\int \frac{d\varphi_p}{2\pi}\frac{d\sigma^{(\mathrm{pl})}_{\chi m_f\mu_i}}{d\Omega_k}(\hat{\mathbf{p}}). \tag{10}$$

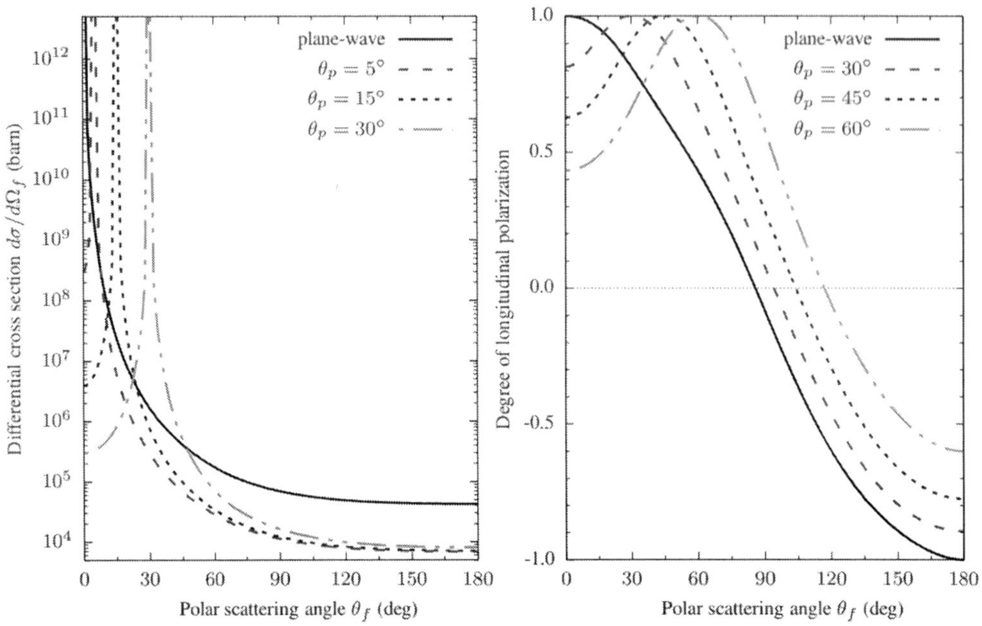

Figure 1. The angular and polarization properties of the electrons emitted in the course of the elastic (Mott) scattering of 100 keV twisted electrons by lead nuclei. On the left panel, the differential cross section (8) averaged over μ_i and summed over μ_f is presented. On the right panel, the degree of the longitudinal polarization (9) is depicted.

Here χ is the angle of the linear polarization of the emitted photon whose solid angle Ω_k is defined by the azimuthal φ_k and polar θ_k angles, and m_f is the TAM projection of the $1s$ electron state. The integral in equation (10) can be readily evaluated with the usage of the well-known expressions from the plane-wave case, which can be found, e.g., in the review [40]. The differential cross section averaged over μ_i and summed over m_f and χ is presented on the left panel of figure 2. From this panel, one can see that the "twistedness" of the incident electron results in qualitative changes of the angular distribution of the emitted photons. Namely, the photons are predominantly emitted in the forward direction, where the plane-wave differential cross-section tends to zero. On the right panel of figure 2 we present the degree of the linear polarization (first Stokes parameter)

$$P_l = P_1 = \frac{d\sigma_{90^\circ} - d\sigma_{0^\circ}}{d\sigma_{0^\circ} + d\sigma_{90^\circ}}, \tag{11}$$

where $d\sigma_\chi \equiv \frac{1}{2}\sum_{m_f\mu_i} \frac{d\sigma_{\chi m_f \mu_i}}{d\Omega_k}$. We note that the second and third Stokes parameters are identically equal to zero. From the right panel of figure 2, it is seen that at large opening angles, the degree of the linear polarization takes negative values. That corresponds to the photons polarized perpendicular to the reaction plane. A similar effect was predicted for the Vavilov-Cherenkov radiation from twisted electrons [41]. More detailed investigations of the radiative recombination of the twisted electrons performed in the first Born approximation and within the fully-relativistic nonperturbative treatment can be found in [16] and [22], respectively.

3.3. Bremsstrahlung from twisted electrons

Let us now briefly discuss the Bremsstrahlung from the twisted electrons scattered by the macroscopic target consisting of bare lead ($Z = 82$) nuclei. With the usage of equation (7), one can obtain the double

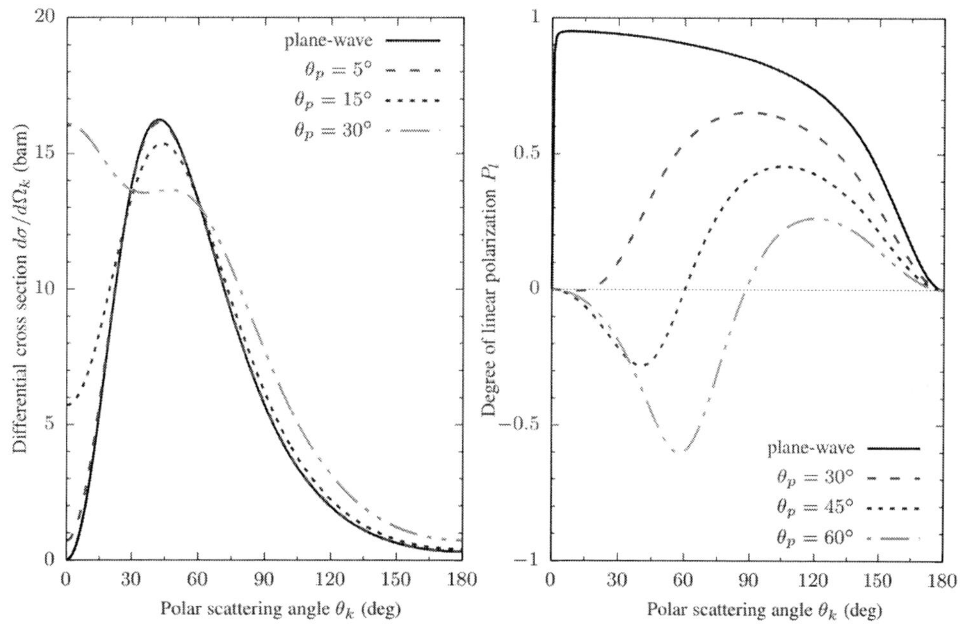

Figure 2. The angular and polarization properties of the photons emitted in the course of the radiative recombination of 100 keV twisted electrons with lead nuclei into the ground $1s$ state of the corresponding H-like ions. On the left panel, the differential cross section (10) averaged over μ_i and summed over m_f and χ is presented. On the right panel, the degree of the linear polarization (11) is depicted.

differential cross section (DDCS) for this process in the following form

$$\frac{d\sigma^{(\mathrm{tw})}_{\chi\mu_f\mu_i}}{d\omega d\Omega_k} = \frac{1}{\cos\theta_p}\int\frac{d\varphi_p}{2\pi}\frac{d\sigma^{(\mathrm{pl})}_{\chi\mu_f\mu_i}}{d\omega d\Omega_k}(\hat{\mathbf{p}}), \tag{12}$$

where ω stands for the energy of the emitted photon. In the present paper, for the evaluation of the plane-wave DDCS, we utilize the theoretical formalism and numerical algorithms developed in [42]. The results of our calculations for the case of 50 keV Bremsstrahlung are presented in figure 3. On the left panel of this figure, the DDCS averaged over μ_i and summed over μ_f and χ is displayed. The right panel of figure 3 demonstrates the degree of the linear polarization (first Stokes parameter), which is defined by a formula similar to (11) with $d\sigma_\chi \equiv \frac{1}{2}\sum_{\mu_f\mu_i}\frac{d\sigma_{\chi m_f\mu_i}}{d\omega d\Omega_k}$. As in the case of the radiative recombination process, the second and third Stokes parameters are identically equal to zero. From figure 3, it is seen that, the "twistedness" of the incoming electrons leads to qualitative changes of the angular and polarization characteristics. More detailed investigation of the process of the Bremsstrahlung from twisted electrons will be presented in a forthcoming publication.

4. Conclusion

In the present paper, the fundamental atomic processes with twisted electrons were considered. In particular, the elastic (Mott) scattering, the radiative recombination, and the Bremsstrahlung processes were studied within the formalism, which takes into accounts the electron-ion interaction nonperturbatively. This allows one to obtain reliable results even for such heavy systems as lead ($Z = 82$) nuclei. Special attention was paid to the investigation of the influence of the "twistedness" of the incoming electron on the angular and polarization properties of the emitted particles. It was found

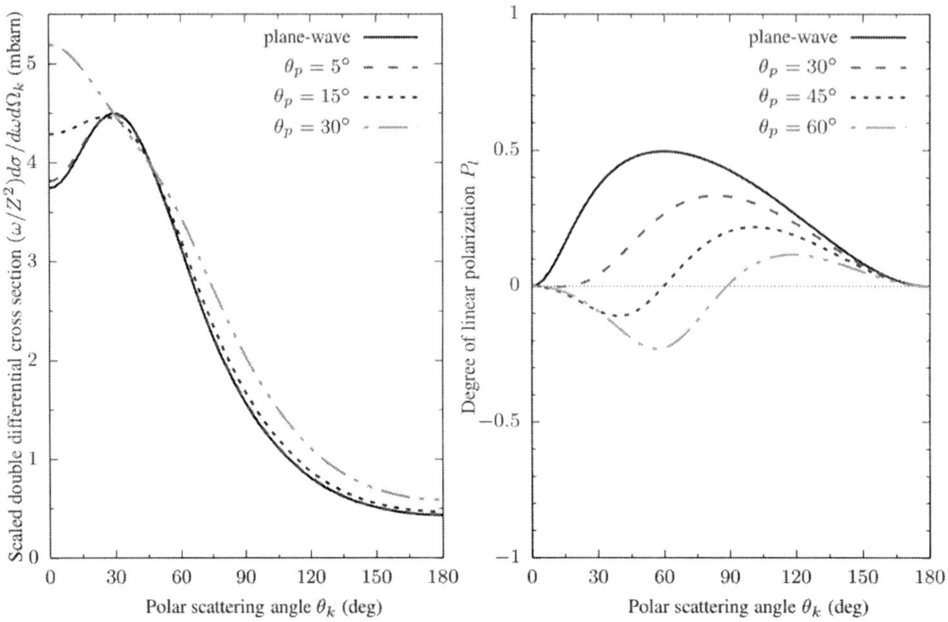

Figure 3. The angular and polarization properties of 50 keV Bremsstrahlung from 100 keV twisted electrons scattered by lead nuclei. On the left panel, the scaled double differential cross section (12) averaged over μ_i and summed over μ_f and χ is presented. On the right panel, the degree of the linear polarization is depicted.

that these properties exhibit a strong dependence on the opening angle of the vortex electron beam in all considered processes.

Acknowledgments

This work was supported by the grant of the President of the Russian Federation (Grant No. MK-4468.2018.2), by RFBR (Grant No. 18-32-00602), and by SPbSU-DFG (Grants No. 11.65.41.2017 and No. STO 346/5-1). V.A.Y. acknowledges support by the Ministry of Education and Science of the Russian Federation Grant No. 3.5397.2017/6.7.

References

[1] Bliokh K Y, Bliokh Y P, Savel'ev S and Nori F 2007 *Phys. Rev. Lett.* **99** 190404
[2] Verbeeck J, Tian H and Schattschneider P 2010 *Nature* **467** 301
[3] Uchida M and Tonomura A 2010 *Nature* **464** 737
[4] McMorran B J, Agrawal A, Anderson I M, Herzing A A, Lezec H J, McClelland J J and Unguris J 2011 *Science* **331** 192
[5] Bliokh K Y *et al* 2017 *Phys. Rep.* **690** 1
[6] Lloyd S M, Babiker M, Thirunavukkarasu G and Yuan J 2017 *Rev. Mod. Phys.* **89** 035004
[7] Larocque H, Kaminer I, Grillo V, Leuchs G, Padgett M J, Boyd R W, Segev M and Karimi E 2018 *Contemp. Phys.* **59** 126
[8] Mafakheri E *et al* 2017 *Appl. Phys. Lett.* **110** 093113
[9] Ivanov I P and Karlovets D V 2013 *Phys. Rev. Lett.* **110** 264801
Ivanov I P and Karlovets D V 2013 *Phys. Rev.* A **88** 043840
[10] Rusz J and Bhowmick S 2013 *Phys. Rev. Lett.* **111** 105504
[11] Béché A, Van Boxem R, Van Tendeloo G and Verbeeck J 2013 *Nat. Phys.* **10** 26
[12] Schattschneider P, Löffler S, Stöger-Pollach M and Verbeeck J 2014 *Ultramicroscopy* **136** 81
[13] Edström A, Lubk A and Rusz J 2016 *Phys. Rev. Lett.* **116** 127203
[14] Lloyd S, Babiker M and Yuan J 2012 *Phys. Rev. Lett.* **108** 074802
Lloyd S, Babiker M and Yuan J 2012 *Phys. Rev.* A **86** 023816

[15] Van Boxem R, Partoens B and Verbeeck J 2014 *Phys. Rev.* A **89** 032715
[16] Matula O, Hayrapetyan A G, Serbo V G, Surzhykov A and Fritzsche S 2014 *New J. Phys.* **16** 053024
[17] Van Boxem R, Partoens B and Verbeeck J 2015 *Phys. Rev.* A **91** 032703
[18] Karlovets D V, Kotkin G L and Serbo V G 2015 *Phys. Rev.* A **92** 052703
[19] Karlovets D V, Kotkin G L, Serbo V G and Surzhykov A 2017 *Phys. Rev.* A **95** 032703
[20] Serbo V G, Ivanov I P, Fritzsche S, Seipt D and Surzhykov A 2015 *Phys. Rev.* A **92** 012705
[21] Harris A L, Plumadore A and Smozhanyk Z 2019 *J. Phys.* B **52** 094001
[22] Zaytsev V A, Serbo V G and Shabaev V M 2017 *Phys. Rev.* A **95** 012702
[23] Kosheleva V P, Zaytsev V A, Surzhykov A, Shabaev V M and Stöhlker Th 2018 *Phys. Rev.* A **98** 022706
[24] Silenko A J, Zhang P and Zou L, 2018 *Phys. Rev. Lett.* **121** 043202
Silenko A J, Zhang P and Zou L, 2019 *Phys. Rev. Lett.* **122** 063201
Silenko A J, Zhang P and Zou L, 2019 *Phys. Rev.* A **100**, 030101(R)
[25] Berestetsky V B, Lifshitz E M and Pitaevskii L P 2006 *Quantum Electrodynamics* (Butterworth-Heinemann, Oxford)
[26] Eichler J and Meyerhof W 1995 *Relativistic Atomic Collisions* (Academic, San Diego)
[27] Rose M E 1957 *Elementary Theory of Angular Momentum* (New York: Wiley)
[28] Varshalovich D A, Moskalev A N and Khersonskii V K 1988 *Quantum Theory of Angular Momentum* (Singapore: World Scientific)
[29] Watson G N 1922 *A Treatise on the Theory of Bessel Functions* (London: Cambridge)
[30] Abramovitz M and Stegun I A (eds.) 1964 *Handbook of Mathematical Functions* (Washington D.C.: U.S. Govt. Printing Office)
[31] Schweber S S 1961 *An Introduction to Relativistic Quantum Field Theory* (New York: Row Peterson)
[32] Rose M E 1961 *Relativistic Electron Theory* (New York: Wiley)
[33] Pratt R H, Ron A and Tseng H K 1973 *Rev. Mod. Phys.* **45** 273
Pratt R H, Ron A and Tseng H K 1973 *Rev. Mod. Phys.* **45** 663(E)
[34] Ivanov I P 2011 *Phys. Rev.* D **83** 093001
Ivanov I P 2012 *Phys. Rev.* A **85** 033813
[35] Guzzinati G, Schattschneider P, Bliokh K Y, Nori F and Verbeeck J 2013 *Phys. Rev. Lett.* **110** 093601
[36] Mott N F 1932 *Proc. R. Soc. London* **135** 429
[37] Dogget J A and Spencer L V 1956 *Phys. Rev.* **103** 1597
[38] Sherman N 1956 *Phys. Rev.* **103** 1601
[39] Gluckstern R L and Lin S R 1964 *J. Math. Phys.* **5** 1594
[40] Eichler J and Stöhlker Th 2007 *Phys. Rep.* **439** 1
[41] Ivanov I P, Serbo V G and Zaytsev V A 2016 *Phys. Rev.* A **93** 053825
[42] Yerokhin V A and Surzhykov A 2010 *Phys. Rev.* A **82** 062702

Collisional cooling of trapped ions with cold atoms: results and insights

Sourav Dutta[1,2], **Rahul Sawant**[1,3] **and S A Rangwala**[1]
[1]Raman Research Institute, C. V. Raman Avenue, Sadashivanagar, Bangalore 560080, India.
[2]Tata Institute of Fundamental Research, Navy Nagar, Colaba, Mumbai 400005, India.
[3]Midlands Ultracold Atom Research Centre, School of Physics and Astronomy, University of Birmingham, Edgbaston, Birmingham B152TT, United Kingdom.

E-mail: sarangwala@rri.res.in

Abstract. We present a series of recent experimental and numerical results which allow us to propose and explain the mechanisms for collisional cooling of a trapped ion by sequential binary collisions with atoms. Our experiments in hybrid atom-ion traps study the cooling of ions when the reservoir of the coolant atoms is spatially localized at the centre of the ion trap. Under these conditions the widely and long held expectation that for the ion to collisionally cool, the atom must be of lighter mass (i.e. $m_I > m_A$) is overturned. Instead we show ion cooling for $m_I \leq m_A$ in addition to $m_I > m_A$, and explain why earlier work establishing mass ratios does not apply to the experiments in hybrid traps. Further, for cooling of an ion A^+ by the parent atom A, the mechanism of resonant charge exchange (RCE) allows for extremely efficient cooling of the ion A^+. This mechanism is demonstrated by comparing the cooling rate of an ion with a localized ensemble of its parent atoms and with a lighter atomic species, where in the latter case the RCE mechanism does not exist. The difference in measured cooling rates for a given number density of the atoms and the theoretically calculated difference between elastic and RCE cross sections are used to show that the cooling efficiency per collision with RCE is much greater than the cooling by elastic collision. We conclude with some perspectives and prospects of future experiments.

1. Introduction

Ever since the pioneering trapping experiments with ions [1], a primary objective has been to cool them for precision measurements [2]. As typical ion trap potentials are very deep, the trapped ion can be very energetic, which is not useful for precision work. It has therefore always been a simultaneous requirement to have a cooling mechanism in place, wherein the energetic ions can be cooled or thermalized to a fraction of the trap depth. In the era before laser cooling and other trap specific cooling developed, the available cooling mechanism was multiple sequential two-body collisions of a single trapped ion with neutral atoms or even molecules, which were at much colder temperature. Typically, the ion trap volume was flooded with a neutral cool gas, such that the ion trap was immersed in the cold gas. The resulting collisional cooling of ions by atoms, by flooding the ion trap with a neutral atom gas was called buffer gas cooling [3].

Subsequently other ion cooling methods have been developed, such as; resistive cooling with and without feedback, radiative cyclotron or synchrotron cooling and the very versatile and effective laser cooling [4]. While direct laser cooling is convenient only for a small number of ions with suitable electronic structure, its utility is greatly expanded by sympathetic cooling of ions that are hard to laser cool [4]. The effectiveness of this cooling depends on the ratio of the mass of the two ions.

With the expansion of laser cooling to neutral atoms [5], trapped dilute gas ensembles of localized atoms were readily prepared, as a Magneto-Optical Trap (MOT) or a degenerate quantum gas of Bose or Fermi particles. Science with cold trapped ions and atoms progressed independently and rapidly for approximately two decades. Since the mid 2000's a number of new experiments developed, which combined trapped ions with cold and ultracold atoms, in hybrid trap atom-ion experiments [6–14]. The hybrid traps combine radio frequency Paul traps with atom traps. Penning traps for ions have been avoided as the magnetic field required to trap ions causes nonlinear perturbation of atomic states, which hampers the laser cooling of atoms. The focus in these experiments shifted from precision manipulation of each individual system to understanding the interactions between atoms and ions at the coldest temperatures. The study of collisions and reactive processes in these systems is of contemporary interest.

Here we consider the case of an atom-ion collision, when both the atoms A and ion B^+ are trapped. In particular, the atoms A are laser cooled and localized at the ion trap centre. Both A and B are alkali atoms. For all collision energies elastic collisions between the atoms and ions occur. Further, when the ion species is the same as the atom, resonant charge exchange (RCE) collisions also occur. We consolidate several of our previous experimental and numerical results on the collisional cooling of ions with atoms. The experiments are performed at collsion energies above the temperature equivalent of 10 K. The ion cooling mechanisms are studied numerically and explain ion cooling consistently all the way upto mK equivalent collision energies. In the results discussed below, two cooling mechanisms manifest for this system. The first is that localized atoms at the ion trap centre will always cool trapped ions in collisions through elastic collisions. This is true even if the trapped ion has significantly lighter mass than the trapped atom, which is different from the commonly held views originating from the buffer gas cooling experiments and analysis. Further we demostrate that the process of resonant charge exchange (RCE) is an extremely effective cooling mechanism for homonuclear atom-ion pairs at collisional energies equivalent to room temperature.

2. Two-body atom-ion interaction

In hybrid atom-ion traps figure 1, it is the low energy atom-ion collision, which is of interest. This implies that the long range part of the molecular ion potential plays a critical role in the collision. As the long range force is attractive, for an atom and ion in the ground state, the potential energy in this limit is $V(R)|_{R\gg R_e} \propto -\alpha_d/(2R^4)$, where R_e is the distance of minimum potential energy and α_d is the dipole polarizability. For excited states, the smallest contributing multipole will determine the long range interaction. As the collision energy decreases, lower partial waves contribute maximally to the collision. The total collision cross section for the polarizability interaction as a function of collision energy E, varies as $E^{-1/3}$ over a large range of collision energies with deviations at the few partial wave limit (very low energies) and at the very high energy end [15].

3. Collisional cooling of ions with a buffer gas

For a system of co-trapped atoms and ions, the ion number density is many orders of magnitude lower that atom number density enabling multiple collisions of each trapped ion with atoms. The collisional cooling of the trapped ions depends on how energy is exchanged between the ions and the atoms. The process of collisional cooling of ions with atoms is complex, because the ion is dynamically trapped in a Paul trap, which combines both static and oscillatory electric fields to fashion an effective trap for the charged particles. In such a trap, the trapping and energy of the ion post collision (PC) depends on how the PC velocity relates to the instantaneous rf field configuration. If the trapped ion maintains its synchronization PC with the field, it will

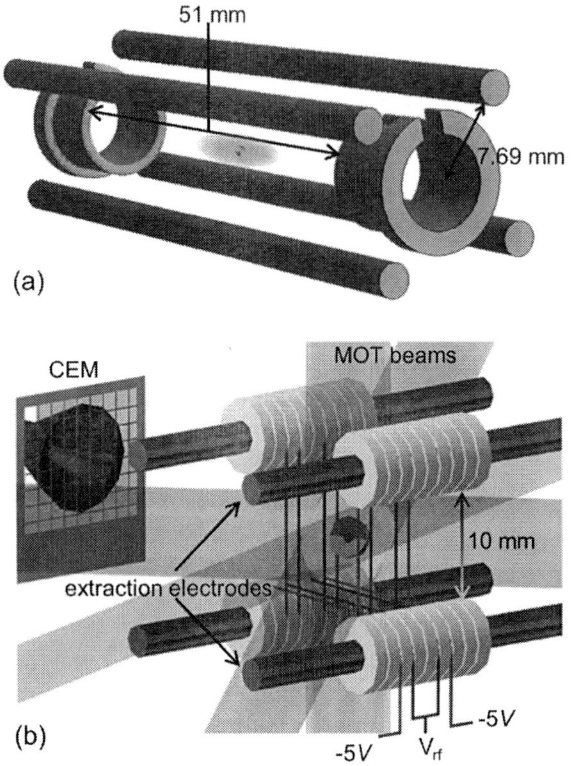

Figure 1. (a) Linear Paul trap with a MOT at the centre. The centred ellipsoid represents the volume within which the ion is trapped. Diagonal electrodes are equipotential and the two pairs of diagonal electrodes are 180° out of phase. The CEM is downstream from the end cap. along the trap axis. In (b) the spherical quadrupole trap with rf potentials on the two central wires and dc potentials on the outer pair of wires is shown. Grey represents the trapped ion volume extent and the dot in the centre is the MOT. Switching Voltage on the quadrupole rods extracts the trapped ions into the CEM, perpendicular to the trap.

stay trapped, but if the velocity and phase synchronization changes significantly, then the PC ion can heat and even exit the trap.

While buffer gas cooling makes relatively modest technical demands, there are subtleties in the cooling outcomes. When the direction of the ion velocity changes significantly per collision, after multiple collisions it will heat out of the ion trap, even if the atoms are at zero kinetic temperature. Kinematically, relatively large changes in the ion velocity will occur when the mass of the atom (m_A) is comparable to or greater than the mass of the ion (m_I). Indeed this is observed in many experiments [3], and therefore the practical operational condition for buffer gas ion cooling is when $m_A/m_I \ll 1$. Under these conditions the ion direction post collision suffers slight changes and so the ion remains trapped and cools rapidly. This gives rise to a critical mass ratio of the atom to ion mass, R_C above which it is theoretically impossible to cool a trapped ion with a buffer gas. Several theoretical analysis of trapped ion-atom collisions have arrived at different values of R_C (= 1 [3], 1.55 [16], 1.47 [17], 0.95 [18]) for buffer gas cooling (with our K^+ ion trap parameters).

4. Collisional cooling of ions in hybrid trap experiments: elastic collisions

Hybrid trap experiments differ from the buffer gas as the cooled and trapped atoms are tightly localized and co-centred with the ion trap, and so atom-ion collisions can occur only at the bottom of the ion trap. In the region around the trap centre, the trap field tends to zero and the motion of the ion approximates a force free trajectory near the trap centre. Under these circumstances, we experimentally show that ion cooling occurs irrespective of the mass of the atom. In particular, we show ion cooling when $m_A/m_I \geq R_C$ [12; 19].

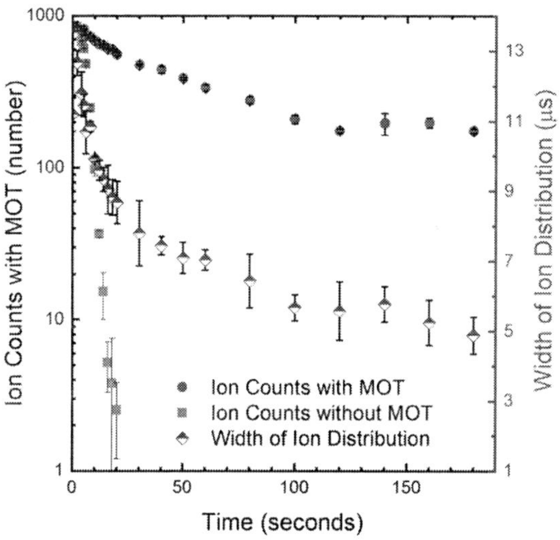

Figure 2. Rb^+ lifetime with MOT (blue, circle) and without MOT (grey, square) is shown as a function of time. The width of the extracted ion arrival time distribution is shown by the red diamonds referenced to the right vertical axis. Figure adapted from data in Ravi et. al. [12] with linear Paul trap of figure 1.

4.1. Cooling of ions by atoms of equal mass

In early experiments [12; 20], we showed that Rb^+ ions can be trapped for several minutes, in the presence of a Rb MOT. The laser cooled MOT atoms are precisely co-centred with the ion trap and the spatial extent of the MOT atoms is much less than of the trapped ions ($V_{MOT} \ll V_{IT}$). This is true of both the experimental geometries illustrated in figure 1, and results predominantly in atom-ion collisions proximate to the trap centre. In real traps, there is ion heating due to trap imperfections, which results in loss of the trapped ions. The lifetime of the trapped Rb^+ ion is extended dramatically in the presence of the Rb MOT (figure 2), whereas in the absence of the MOT the trapped ion population decays in the order of 10 seconds due to ion heating [12; 20]. Similar results were also seen in Na-Na^+ experiments [13]. The increase in lifetime is a signature of cooling of ions by atoms of equal mass. This observation is in departure from Major and Dehmelts work [3], where $R_C = 1$, and thus the ion is expected to neither heat nor cool on the average, when in collision with the buffer gas atoms of equal mass.

The experiment is performed by creating the ^{85}Rb MOT, while the ion trap is ON. The ionizing light is switched ON briefly and then either the MOT is kept ON or switched OFF. The trapped ions are held for variable duration of time in each case and are extracted out of the ion trap to a channel electron multiplier (CEM) with appropriate electrostatic lensing [21] to determine the number of trapped ions at different hold times. The rapid loss of trapped ions in the absence of the MOT allows the determination of the trapped ion heating rate. The persistence of trapped ions at long times, created under the same conditions, but held in presence of the MOT, then allows the determination of a cooling rate for the trapped ions in collision with the localized and trapped atoms. Further, when the trapped ion number stabilized between two and three minutes of hold time [12], the arrival time distribution of ions counted by CEM pulses continued to shrink, allowing the determination of ion temperature by application of the virial theorem. These results show unambiguously that trapped ions of Rb^+ are very efficiently cooled by localized and centred Rb MOT atoms, which have equal mass. While classically the maximum transfer of energy in an elastic collision is for equal masses, most collisions are glancing and therefore, a new cooling mechanism was invoked to explain the extremely efficient cooling of Rb^+. This mechanism is resonant charge exchange (RCE) for homonuclear atom-ion collisions. The discussion of cooling by RCE is deferred to section 5.

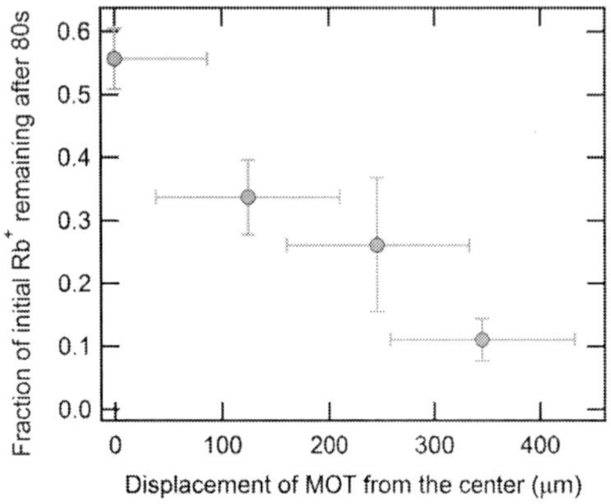

Figure 3. Rb^+ trapped fraction at a hold time of 80 s, as a function of displacement of the Rb MOT from the ion trap centre. The horizontal error bars are due to uncertainties in the determination of the precise displacement and MOT distortion when displaced.

In figure 3 we show how the precise centering of the MOT affects the trapped ion lifetime. This experiment has been performed in the spherical Paul trap shown in figure 1 (b). The MOT is displaced with respect to the ion trap centre, by shifting the magnetic centre for the MOT. The fraction of surviving Rb^+ ions is measured after a hold time of 80 seconds for different MOT displacement. The change in the effectiveness of the ion cooling is clearly seen in the dramatic fall in trapped ion lifetimes as a function of MOT displacement from the co-centred configuration. It is therfore standard experimental procedure to first centre the traps by maximizing the trapped ion lifetime and then take measurements.

4.2. Cooling of ions by atoms of higher mass

Given that $V_{MOT} \ll V_{IT}$ in hybrid traps where localized atoms are placed at the centre of the ion trap, the experiment to test the validity of the critical mass paradigm requires a heavier atom MOT with a lighter ion. In figure 4, we show cooling of: $^{85}Rb^+$ with ^{133}Cs MOT ($m_A/m_I = 1.565$), $^{133}Cs^+$ with ^{133}Cs MOT ($m_A/m_I = 1$) and $^{133}Cs^+$ by ^{85}Rb MOT ($m_A/m_I = 0.639$). These results are clearly consistent with cooling by localized and centred

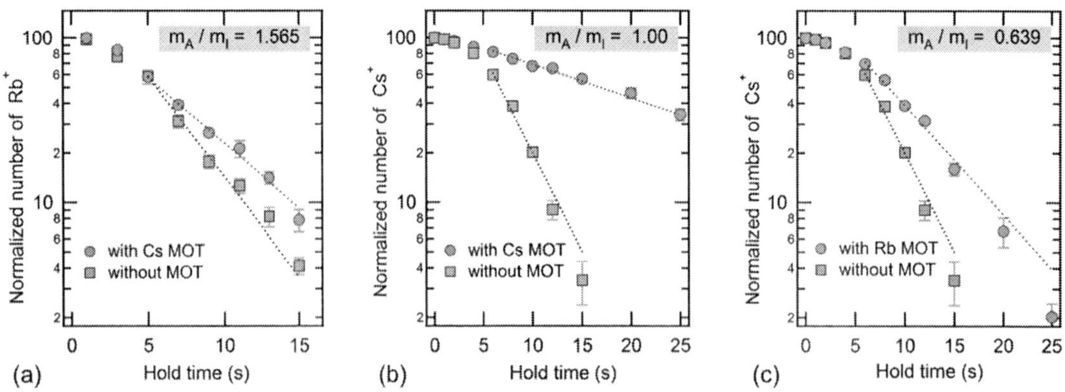

Figure 4. Panel (a) shows Rb^+ lifetime with (red, circle) and without (grey, square) Cs MOT. Panel (b) shows Cs^+ lifetime with (red, circle) and without (grey, square) Cs MOT. Panel (c) shows Cs^+ lifetime with (red, circle) and without (grey, square) Rb MOT. Figure adapted from data in Dutta et. al. [19] with linear Paul trap of figure 1. The equal mass case clearly differentiates from the unequal masses. In all cases, ion cooling is observed.

atoms $\forall\ m_A/m_I$, which stands it apart from buffer gas cooling [12; 22; 23].

In all two species experiments first a MOT is created, the ion trap is switched on and the MOT atoms are ionized from their excited states by briefly turning on a blue light, creating trapped ions. The MOT cooling light is turned off, which empties the first MOT within milliseconds (this instant is taken as t = 0). Then the MOT light for a different species is turned on (or not). The new MOT loads in 5 seconds to steady state. The trapped ions are held for a predetermined amount of time and then are extracted to a channel electron multiplier (CEM). The lifetimes of the trapped ions in the presence of and absence of MOT atoms is measured. The longer lifetime of ions in presence of MOT atoms indicate the cooling of ions by the MOT atoms.

To further check localized cooling, another experiment was performed with $^{39}K^+$ ions and a MOT of ^{85}Rb atoms, in the spherical Paul trap configuration of figure 1 (b). Once again, enhanced lifetimes were measured for the $^{39}K^+$ ion in the presence of the ^{85}Rb MOT as seen in figure 5(a), for an even higher ratio of $m_A/m_I = 2.179$. Here the cooling is present but less effective, than in the $^{85}Rb^+-^{133}Cs$ ion-atom case. The contribution of localization of cold atoms to ion cooling is tested by noting that displacement of the ion trap centre by even a few 100 microns from the ion trap centre completely eliminates the ion cooling. For the heteronuclear atom - ion experiments, no charge exchange collisions were seen in the control experiments.

4.3. Discussion

An intriguing observation is that while ion collisional cooling occurs, all the trapped ions eventually escape the trap, so there is significant heating of the trapped ion as well. This occurs because of the co-existence of two reservoirs of atoms with which the trapped ions collide and due to trap imperfection related losses. The localized cold atom reservoir cools the ion and the background vapour has the action of the buffer gas in the regular paradigm. In the case of Rb^+-Cs (K^+-Rb) experiment, the dominant background gases are Rb and Cs (K and Rb) i.e. one of the background gases is heavier than the trapped ion, and therefore leads to heating of the ion. The result is that there is both collisional heating (due to background atoms) and cooling (due to MOT atoms) present simultaneously, which results in the eventual emptying of the ion trap. This is further understood in a simulation of the experiment shown in figure 5.

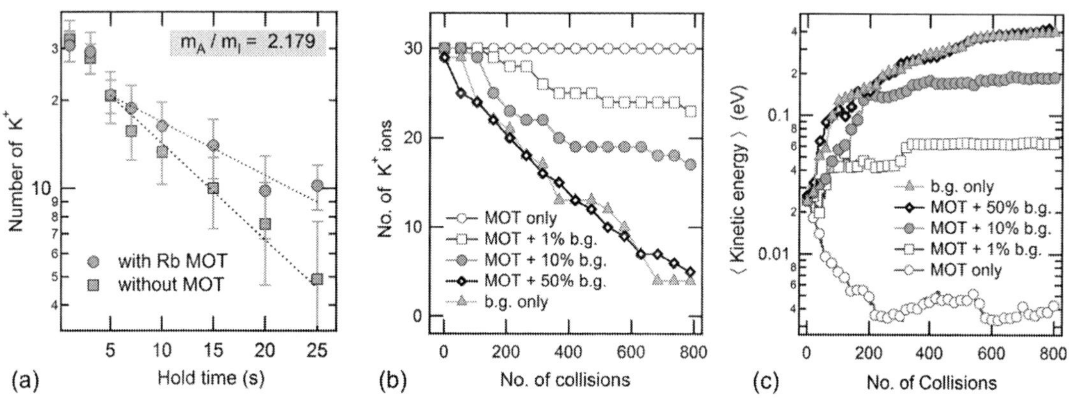

Figure 5. Experimentally measured K^+ ion lifetime with Rb MOT (red, circle) and without MOT (grey, square) is shown as a function of time in panel (a). The simulated lifetime of the same system as a function of number of ion-atom collisions is in (b) and the consequent heating or cooling is shown in (c). The background vapour plays a critical role in the eventual fate of the ion. Figure adapted from data in Dutta et. al. [19].

Here the trapped ion lives forever in the ideal trap when there is no heavy atom background. For any reasonable pressure of the higher mass atom, the ion trap empties out.

The important question which arises is, assuming a Gaussian density of cold atoms at the centre of a perfect ion trap, what is the final temperature and state of the ion, in the complete absence of the heavier mass background gas? Once again from simulations we can determine that, given a small but finite size of the cold atom distribution with full width at half maximum (FWHM) d, the final steady state temperature of a trapped ion will approximately be determined by the trap depth at a distance d even if the atoms are much colder than the kinetic temperature of the ion. This is so because at the centre of the ion trap, the ion effectively sees a buffer gas of heavier atoms, and only upon travelling radially outwards, the localization starts influencing the ion behaviour. The collisions at the trap centre do not heat the ion out of the trap as opposed to a uniform buffer gas which heats the ion in the outer reaches of the trap when the atom is heavier.

The ion can be put in thermal equilibrium with the heavy MOT atoms, if the ion trap depth at d is smaller than the atom temperature T_{MOT}. Under this situation the atom and ion can equilibrate, under the approximation of classical elastic collision, in the simulation.

5. Cooling of trapped ions by resonant charge exchange

We now focus on the homonuclear ion-atom collisions in figures 2 and 4 (b). In any collision, the probability of a head-on collision is negligible and a glancing collision is more likely. Thus irrespective of the condition of equal masses, the average (typical) elastic collision between ion and atom does not transfer much energy from the ion to the atom. The strength of the ion cooling, shown in figure 2 above, signals something else at work. The cooling by resonant charge exchange (RCE) was therefore proposed as a cooling mechanism. RCE is only active when the colliding atom-ion pair have a parent-daughter relation, i.e. the daughter ion is one electron removed from being the parent atom [3; 15]. In such a scenario, in a single glancing collision where little kinetic energy is exchanged, the electron from the cold atom can transfer to the incoming fast ion (which presents an identical ion core as that of the atomic nucleus), resulting in a fast atom leaving the trap and a very cold ion at the ion trap centre. Such cooling is called RCE or "swap" cooling and occurs when the atom and ion are identical species, since then the states available for the electron with the ion core do not extract an energy cost when the electron exchanges from atom to ion. While this mechanism was proposed and simulated to explain Rb^+ ion cooling with Rb MOT [12; 20], experimentally there was no discrimination between the elastic and RCE mechanisms.

To determine whether the RCE mechanism results in cooling of the trapped ion, we compare the collisional cooling of the Cs^+ ion with MOTs of Cs and Rb atoms. In the case of the Cs^+-Cs homonuclear ion–atom pair RCE is possible along with elastic collisions. On the other hand, for Cs^+-Rb the collisions are exclusively elastic. On considering only elastic collisions between the trapped Cs^+ ion and an atom of Rb and Cs, the cooling efficiency for the trapped Cs^+ ion with a localized Rb MOT is greater than for an identical Cs MOT.

To measure the cooling of trapped ions via RCE, we first load a Cs MOT, turn on the ion trap, ionize the MOT atoms as described earlier and then either (a) hold the Cs MOT, or (b) empty the Cs MOT and load the Rb MOT, followed by counting the ion number in the trap as a function of trap hold time in both cases. It can be seen easily in figure 6(a) that the same lifetime of trapped ions is achieved for a Cs MOT with 1/2 the peak atomic density and the same spatial volume as the Rb MOT. In addition, we also see (figure 6(b)) that the temporal width of the arrival time distribution of the extracted ions decreases in a nearly identical manner as the hold time increases in both these cases. This implies that the cooling rate for Cs^+ ions with Cs atoms k_{Cs} $(\equiv k^e_{Cs} + k^{\times}_{Cs})$, is equal to twice the Cs^+ cooling rate with Rb atoms $k_{Rb}(\equiv k^e_{Rb})$, i.e. $k_{Cs}(\equiv k^e_{Cs} + k^{\times}_{Cs}) \approx 2k^e_{Rb}$, where the superscript e and $\times$ refer to elastic and RCE respectively.

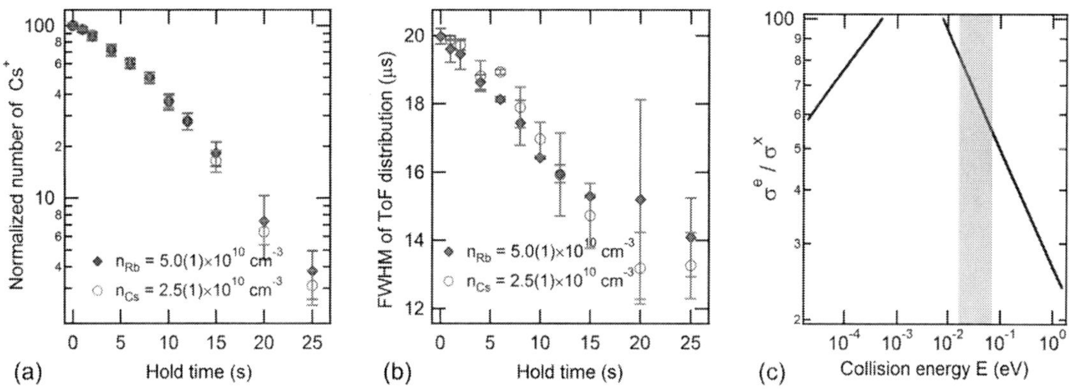

Figure 6. Panel (a) shows the Cs^+ ion lifetime with Rb and Cs MOT, where the MOT number density is controlled such that the lifetimes are well matched. In panel (b) the corresponding temporal width of the arrival time distributions that are also well matched is shown. The ratio of the elastic to RCE cross sections in the energy range of the experiment, marked in blue is shown in (c). Figure adapted from data in Dutta and Rangwala [24].

At energies much larger than s-wave collision energies, the cross sections for elastic and resonant charge exchange can be treated independently to a good approximation. The ratio of the elastic (σ^e) and RCE ($\sigma^\times$) cross sections is illustrated in figure 6(c). Clearly in the energy range of operation (shaded region in figure 6(c)) of the present experiment $\sigma^e \gg \sigma^\times$. Identifying $\delta E^e_{Cs}(\delta E^e_{Rb})$ as the effective energy lost by a Cs^+ ion per elastic collision with a Cs (Rb) atom, we use the simulations of Höltkemeier et al. [23; 25] to estimate that for a cold uniform buffer gas $\delta E^e_{Cs} = 0.5\delta E^e_{Rb}$ and that in the limit of a very tightly confined MOT at the trap centre, $\delta E^e_{Cs} = 1.04\delta E^e_{Rb}$. The total cooling rates relate the cross sections and the energy lost per collision by the ion as [24]

$$\frac{\delta E^\times_{Cs}}{\delta E^e_{Cs}} = \frac{\sigma^e}{\sigma^\times} \times \frac{k^\times_{Cs}}{k^e_{Cs}}.$$

Given that $\sigma^e \gg \sigma^\times$, which implies that the elastic collision is much more probable than RCE, and k^e_{Cs} $k^\times_{Cs}$, the experimental data in conjunction with above expression implies that $\delta E^\times_{Cs} \gg \delta E^e_{Cs}$ [24]. From figure 6 above, the ratio of the cooling efficiency between RCE and elastic collision was determined to be

$$39 \leq \frac{\delta E^\times_{Cs}}{\delta E^e_{Cs}} \leq 154.$$

The upper limit of the range represents the case of a uniform buffer gas and the lower limit denotes a spatially localized gas of infinitesimal extent at the trap centre. As the MOT in the experiment is of finite but small extent, the effectiveness of energy transfer in a single collision is contained within the range above. This underlines that ion cooling is very efficient when RCE is active.

6. Conclusions

Long held views on collisional cooling of trapped ions have been modified by the fact that in hybrid trap experiments, the location and distribution of the atomic cloud allows hitherto veiled mechanisms to manifest. Ion cooling by the localized and centred MOT uncovers that different parts of the reservoir of atoms have different effects on the energetics of the trapped and colliding ion. It opens up the possibility of an ion as a probe for finite and small systems, by measurement of the ion properties. Remarkably whether the ion will attain a steady state or equilibrate with a gas of atoms turns out to be position and size dependent. The ubiquitous atom-ion mass ratios for ion cooling are no longer the discriminant between collisional heating and cooling of the ion. A new cooling mechanism by resonant charge exchange, which is a purely quantum and symmetry enabled exchange mechanism, manifests in a dramatic manner so as to dominate the otherwise primary cooling channel mediated by elastic collisions. Further developments prime this class of experiments as an important and new method for probing quantum systems of both ions and neutrals.

Acknowledgments

The authors acknowledge K. Ravi, S. Lee, A. Sharma, G. Werth, T. Ray and S. Jyothi for key experimental and conceptual contributions. S. A. R. and R. S. acknowledge support from the Indo-French Centre for the Promotion of Advanced Research-CEFIPRA, Project No. 5404-1. S.D. acknowledges support through the DST INSPIRE Faculty Award IFA14-PH-114.

References

[1] Paul W, *Rev. Mod. Phys.* **62**, 531 (1990)
[2] Wineland D J, Drullinger R E and Walls F L, *Phys. Rev. Lett.* **40**, 1639 (1978)
[3] Major F G and Dehmelt H G, *Phys. Rev.* **170**, 91 (1968)
[4] Major F G, Gheorghe V N and Werth G 2005 *Charged Particle Traps* (Springer-Verlag, Berlin)
[5] Phillips W D, *Rev. Mod. Phys.* **70**, 721 (1998)
[6] Grier A T, Cetina M, Oručević F and Vuletić V, *Phys. Rev. Lett.* **102**, 223201 (2009)
[7] Zipkes C, Palzer S, Sias C, and Köhl M, *Nature*, **464**, 388 (2010)
[8] Schmid S, Härter A and Denschlag J H, *Phys. Rev. Lett.* **105**, 133202 (2010)
[9] Rellergert W G, Sullivan S T, Kotochigova S, Petrov A, Chen K, Schowalter S J and Hudson E R, *Phys. Rev. Lett.* **107**, 243201 (2011)
[10] Hall F H J, Aymar M, Bouloufa-Maafa N, Dulieu O and Willitsch S, *Phys. Rev. Lett.* **107**, 243202 (2011).
[11] Ravi K, Lee S, Sharma A, Werth G and Rangwala S A, *Appl. Phys. B* **107**, 971 (2012).
[12] Ravi K, Lee S, Sharma A, Werth G and Rangwala S A, *Nat. Commun.* **3**, 1126 (2012).
[13] Sivarajah I, Goodman D S, Wells J E, Narducci F A, and Smith W W, *Phys. Rev. A* **86**, 063419 (2012).
[14] Haze S, Hata S, Fujinaga M, and Mukaiyama T, *Phys. Rev. A* **87**, 052715 (2013).
[15] Côté R and Dalgarno A, *Phys. Rev. A* **62**, 012709 (2000)
[16] De Voe R G, *Phys. Rev. Lett.* **102**, 063001 (2009)
[17] Zipkes C, Ratschbacher L, Sias C and Köhl M, *New J. Phys.* **13**, 053020 (2011)
[18] Chen K, Sullivan S T and Hudson E R, *Phys. Rev. Lett.* **112**, 143009 (2014)
[19] Dutta S, Sawant R and Rangwala S A, *Phys. Rev. Lett.*, **118**, 113401 (2017)
[20] Ray T, Jyothi S, Ram N B and Rangwala S A, *Appl. Phys. B* **114**, 267 (2014)
[21] Jyothi S, Ray T and Rangwala S A, *Appl. Phys. B* **118**, 131 (2015)
[22] Rouse I and Willitsch S, *Molecular Physics*, DOI 10.1080/00268976.2019.1581952, (2019)
[23] Höltkemeier B, Weckesser P, López-Carrera H and Weidemüller M, *Phys. Rev. Lett.* **116** 233003 (2016)

[24] Dutta S and Rangwala S A, *Phys. Rev. A*, **97**, 041401(R) (2018)
[25] Höltkemeier B, Weckesser P, López-Carrera H and Weidemüller M *Phys. Rev. A*, **94**, 062703 (2016)

Photo-processing of astro-PAHs

C Joblin[1], G Wenzel[1], S Rodriguez Castillo[1,2], A Simon[2], H Sabbah[1,3], A Bonnamy[1], D Toublanc[1], G Mulas[1,4], M Ji[1], A Giuliani[5,6], L Nahon[5]

[1]Institut de Recherche en Astrophysique et Planétologie (IRAP), Université de Toulouse (UPS), CNRS, CNES, 9 Avenue du Colonel Roche, F-31028 Toulouse Cedex 4, France
[2]Laboratoire de Chimie et Physique Quantiques (LCPQ/IRSAMC), Université de Toulouse and CNRS,UT3-Paul Sabatier, 118 Route de Narbonne, 31062 Toulouse, France
[3]Laboratoire Collisions Agrégats Réactivité (LCAR/IRSAMC), Université de Toulouse (UPS), CNRS, 118 Route de Narbonne, F-31062 Toulouse, France
[4]Istituto Nazionale di Astrofisica (INAF), Osservatorio Astronomico di Cagliari, Via della Scienza 5, I-09047 Selargius (Cagliari), Italy
[5]Synchrotron SOLEIL, L'Orme des Merisiers, F-91192 Gif-sur-Yvette Cedex, France
[6]INRA, UAR1008, Caractérisation et Elaboration des Produits Issus de l'Agriculture, F-44316 Nantes, France

E-mail: `christine.joblin@irap.omp.eu`

Abstract. Polycyclic aromatic hydrocarbons (PAHs) are key species in astrophysical environments in which vacuum ultraviolet (VUV) photons are present, such as star-forming regions. The interaction with these VUV photons governs the physical and chemical evolution of PAHs. Models show that only large species can survive. However, the actual molecular properties of large PAHs are poorly characterized and the ones included in models are only an extrapolation of the properties of small and medium-sized species. We discuss here experiments performed on trapped ions including some at the SOLEIL VUV beam line DESIRS. We focus on the case of the large dicoronylene cation, $C_{48}H_{20}^+$, and compare its behavior under VUV processing with that of smaller species. We suggest that C_2H_2 is not a relevant channel in the fragmentation of large PAHs. Ionization is found to largely dominate fragmentation. In addition, we report evidence for a hydrogen dissociation channel through excited electronic states. Although this channel is minor, it is already effective below 13.6 eV and can significantly influence the stability of astro-PAHs. We emphasize that the competition between ionization and dissociation in large PAHs should be further evaluated for their use in astrophysical models.

1. Introduction

Polycyclic aromatic hydrocarbons (PAHs) are present in photodissociation regions (PDRs) associated with massive star-forming regions, such as the prototypical Orion Bar region [1]. Their interaction with vacuum ultraviolet (VUV) photons can trigger various molecular processes: -(i)- ionization resulting in gas heating by thermalisation of the emitted electrons [2], -(ii)- photodissociation limiting the survival of PAHs and producing molecules such as H_2 and C_2H_2 in PDRs [3–7], and -(iii)- radiative cooling leading to the well-known aromatic infrared (IR) emission bands between 3 and 15 μm, which constitute the only direct diagnosis we have so far for the presence of these large molecules in astrophysical environments as proposed in the initial PAH model [8, 9].

The chemical evolution of PAHs in PDRs has been modelled by several authors [3–6, 10]. These models determined a critical molecular size, typically of 50 – 60 carbon atoms, below which these PAHs are not expected to survive. In some models, the critical size is estimated from the ability of the molecule to lose C_2H_2 since in PDRs the absorption of UV photons is faster than the chemistry that could rebuild the carbon skeleton. The PAH experiencing C_2H_2 loss will therefore be ultimately destroyed if its fragments also experience C_2H_2 loss. Other models focus on the more common hydrogen loss. In this case the competition with rehydrogenation by reactivity with the abundant H and H_2 species needs to be considered (see for instance [5]).

All these chemical models rely on photophysical and chemical rates that are still only partly known for small and medium-sized species up to 24 carbon atoms. These species are easier to handle in gas-phase in the laboratory compared to large ones. However, their properties might differ, especially when one considers the interaction with VUV photons. In addition, one should keep in mind the extreme isolation conditions of astrophysical environments in which radiative cooling can play a key role in the photophysics of PAHs. The challenge for laboratory experiments is therefore to address the multiscale aspects of the photophysics of PAHs from very fast femtosecond processes such as ionization [11] to the very long timescale of IR cooling (see Fig.1). This challenge has to be addressed for large PAH species of $\sim 50-100$ carbon atoms, which are the best candidates to survive in PDRs. This article reports our methodology and first results towards the study of large PAH species and evidence for their specificity.

2. The photophysics of astro-PAHs

In PDRs, astro-PAHs are well isolated. Infrared emission, which is a slow process and can extend over seconds (cf. inset graph in Fig. 1), is therefore a major process in energy relaxation due to the lack of collisions that could enter into competition. Considering a typical gas (mainly hydrogen) density of $\sim 10^4\,\mathrm{cm}^{-3}$ in PDRs, the timescale for collisions is typically 28 hours taking an optimistic rate of collisions of $10^{-9}\,\mathrm{cm}^3\mathrm{s}^{-1}$. The timescale for the absorption of a VUV photon depends on the photon flux. It is several hours in the NGC 7023 PDR, which was studied by Montillaud *et al.* [5] (cf. Fig. 1), and tens of minutes in the brighter Orion Bar. This leaves time to lose most of the internal energy by infrared cooling. The energy of the available VUV photons is usually below 13.6 eV due to the ionization of hydrogen atoms but can be higher ($\sim$20 eV) in the ionized bubbles around massive stars [13].

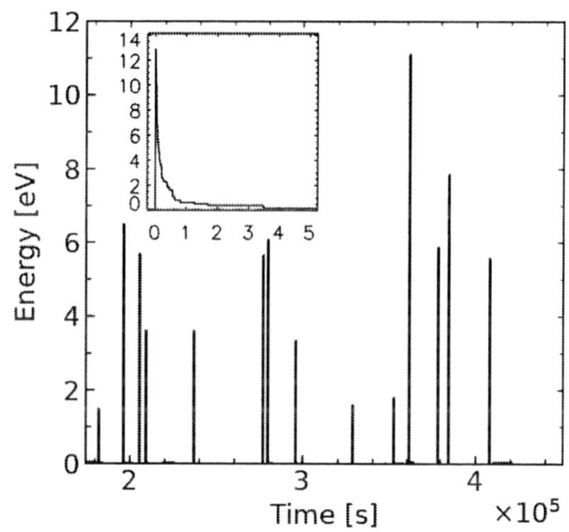

Figure 1. Timescale for the absorption of UV photons by a coronene cation $C_{24}H_{12}^+$ in the NGC 7023 NW PDR studied in the chemical model of Montillaud et al. [5]. An example of an IR cooling cascade is also provided in the inset, zoomed in over a 5 s time interval. The calculations were performed using a dedicated Monte Carlo code to describe the photophysics of PAHs [12].

Since the initial description of the photophysics of an astro-PAH [14] we can now draw a more complete scheme of all processes triggered by the interaction with a VUV photon and their molecular timescales (cf. Fig. 2). On the very short timescales is ionization. XUV fs experiments have shown that it occurs on a characteristic time of 40 fs for small PAHs [11] and this time was found to increase with size. Dissociation starts on longer timescales [11]. It is found, in the case of PAHs, to be well described by statistical theories. In particular the fitting of breakdown curves obtained in imaging photoelectron photoion coincidence (iPEPICO) spectroscopy at the Swiss Light synchrotron with a model based on the Rice-Ramsperger-Kassel-Marcus (RRKM) theory has been successful to quantify activation energies and dissociation rates by studies in the $1-100\,\mu s$ range [15]. In this time window, dissociation is not in competition with radiative cooling. It is only in the analysis of the experiments in ion traps and storage rings (e.g. in the PIRENEA setup described below) that this competition has to be taken into account considering the involved long timescales.

An interesting result in the last years of the photophysics of PAHs has been the confirmation in the laboratory of recurrent fluorescence as a main radiative cooling mechanism. Recurrent fluorescence, also called Poincaré fluorescence, has been predicted by Léger *et al.* [16]. Boissel *et al.* [17] reported evidence for its role in the cooling of trapped anthracene cations submitted to the radiation of a Xe lamp. The process was confirmed and quantified in the Mini-ring storage ring [18,19]. The use of storage rings has opened the possibility for dynamical studies over the ms window for the study of the fast radiative cooling of energized PAHs. Timescales longer than seconds are now becoming accessible with the new generation of cryogenic rings such as DESIREE at the University of Stockholm or the electrostatic cryogenic storage ring CSR at the Max-Planck Institute for Nuclear Physics and University of Heidelberg. These will be really valuable to address the infrared cooling of PAHs [20].

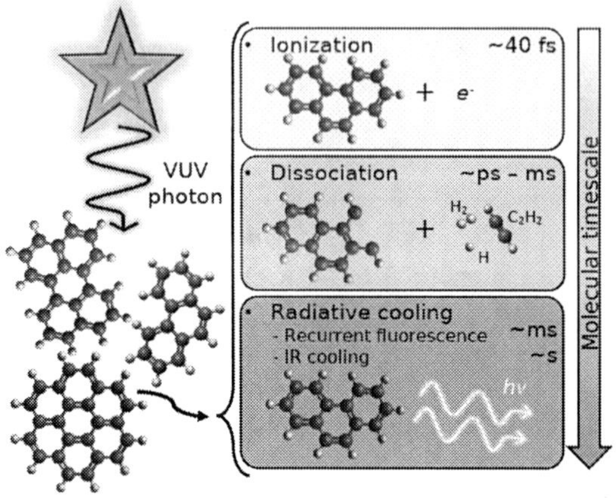

Figure 2. Schematic of the different relaxation processes of astro-PAHs upon VUV photon absorption and involved characteristic molecular timescales.

3. Methods

3.1. Experimental methods

We have used ion trap experiments to investigate the effect of size on the dissociation of PAHs. More specifically, we discuss here the importance of the C_2H_2 loss channel and the variation of the H dissociation rate with the PAH size. Two setups have been used: PIRENEA, a dedicated setup for astrochemistry, and the commercial linear ion trap available at the VUV DESIRS beamline. These setups are briefly discussed below.

• The PIRENEA setup for astrochemistry is a cryogenic Fourier transform ion cyclotron resonance mass spectrometer (FTICR-MS) that has been specifically designed to approach the conditions of the interstellar medium in terms of isolation of the trapped species. It is therefore best suited to study the photophysics of PAH ions on long timescales. It is unfortunately not interfaced with a tuneable VUV source, therefore a multiple photon absorption scheme with lower energy photons is used to achieve fragmentation. In this scheme, the photons are absorbed sequentially. Due to fast internal conversion (timescale typically less than ps [21]), the energy absorbed in an excited electronic state is rapidly converted into internal energy of the ground state before absorption of the next photon. Boissel *et al.* [17] have demonstrated that the use of a Xe lamp is convenient to control the heating of the ions and study their dissociation in competition with radiative cooling, which we call the dissociation at threshold. In these conditions, branching ratios between the different fragments can be obtained as illustrated in this article. In these experiments, one can easily follow the relative abundance of the parent and the different fragments as a function of the irradiation time. The fitting of these curves with a kinetic Monte Carlo model can be used to extract a dissociation rate (close to the threshold).

The model of Montillaud *et al.* [5] has been built on the analysis of data obtained for the coronene cation, $C_{24}H_{12}^+$.

• The VUV DESIRS beamline at the synchrotron SOLEIL is equipped with a Thermo Scientific LTQ XL^{TM} linear ion trap [22]. The parent ions are produced from a neutral precursor in a solution by an atmospheric pressure photo-ionization (APPI) source. Inside the trap, the species of interest (given m/z) can be isolated by ejecting other species. The isolated singly charged cations are then irradiated by the tuneable VUV synchrotron radiation and a few hundreds of mass spectra are recorded for each photon energy. The photon rate is in the range of $10^{12}-10^{13}$ photons s^{-1}. Over the studied 8−20 eV energy range, the irradiation time (between typically 0.2 and 0.8 s) and the opening of the exit slit are tuned in order to maximize the signal-to-noise ratio and to minimize multiple photon absorption events, which are therefore rare in our experimental conditions. The resulting photoproducts, *i.e.* fragments (H, H_2 and eventually C_2H_2 loss) and doubly charged cations, are mass-analyzed and action spectra can be built as a function of photon energy. Two campaigns were performed, one on small/medium species up to 24 carbon atoms [23, 24] and one on larger species up to 48 carbon atoms.

3.2. Numerical simulations

Molecular dynamics (MD) simulations are performed to study the dissociation of PAH radical cations in their ground electronic states. The electronic structure is described on-the-fly within the self-consistent-charge density functional-based tight binding (SCC-DFTB) scheme [25]. This methodology has been found to be promising to study isomerization effects during dissociation and trend in the H/C_2H_2 branching ratio as a function of internal energy [26]. As the C_2H_2 channel was found to be overestimated relative to H loss, we adopt here a similar procedure as in ref. [27] consisting in slightly modifying C-H and H-H interaction potentials by scaling the initial values of the parametrized atomic integrals $\langle \phi_\mu^{C,H} | \hat{h}[\rho_0] | \phi_\nu^{H} \rangle$ and $\langle \phi_\mu^{C,H} | \phi_\nu^{H} \rangle$ by 0.95 ($\phi_{\mu,\nu}$ are the atomic orbitals and $\hat{h}[\rho_0]$ refers to the monoelectronic Hamiltonian at the reference density). Using this modified potential, we performed hundreds of simulations at the lowest energy values to observe dissociation in a reasonable running time.

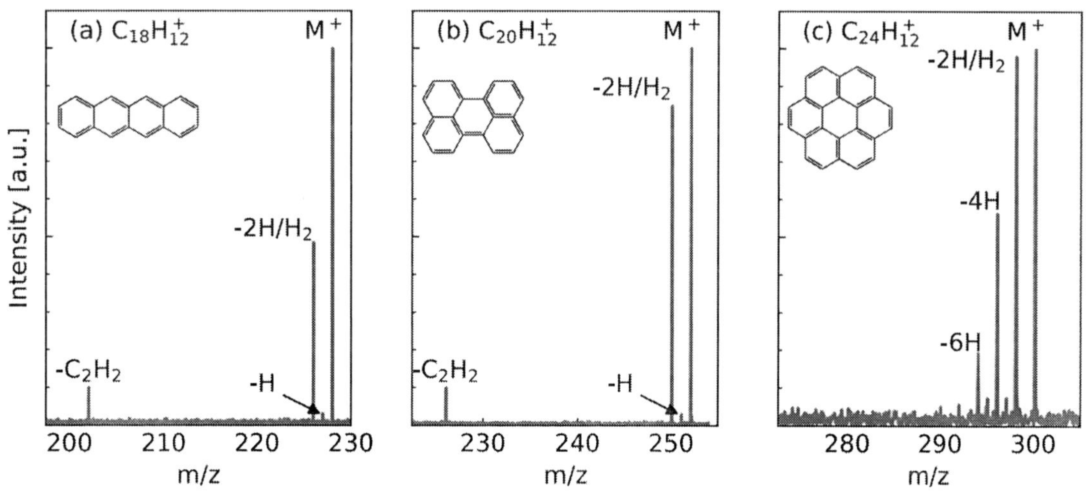

Figure 3. Fragmentation mass spectra of the cations of (a) tetracene, $C_{18}H_{12}^+$, (b) perylene, $C_{20}H_{12}^+$, and (c) coronene, $C_{24}H_{12}^+$. The ^{12}C isotopomer parent ions were isolated in the cryogenic ion cyclotron resonance cell of PIRENEA and exposed to the UV-visible light of a Xe arc lamp.

4. Results

4.1. Loss of C_2H_2 vs loss of hydrogen

The fragmentation of PAHs can involve H and H_2 loss but also carbonaceous fragments, in general C_2H_2 but in some cases also C_4H_2 or CH_3 (case of hydrogenated PAHs) [15]. The mass spectra measured with the PIRENEA setup following irradiation with the Xe lamp of the isolated ^{12}C isotopomers of three medium-sized molecules are shown in Fig. 3. The less compact structure tetracene, $C_{18}H_{12}^+$, shows the most intense peak for the C_2H_2 fragment, followed by perylene, $C_{20}H_{12}^+$. The compact ion coronene, $C_{24}H_{12}^+$, does not exhibit C_2H_2 loss, neither in PIRENEA nor in SOLEIL experiments at all VUV photon energies up to 20 eV [23]. A similar result was obtained for the larger molecules, dicoronylene, $C_{48}H_{20}^+$, but also the less compact ion dibenzo[fg,ij]phenanthro[9,10,1,2,3-pqrst]pentaphene, $C_{36}H_{18}^+$, whose structures are shown in Fig. 4.

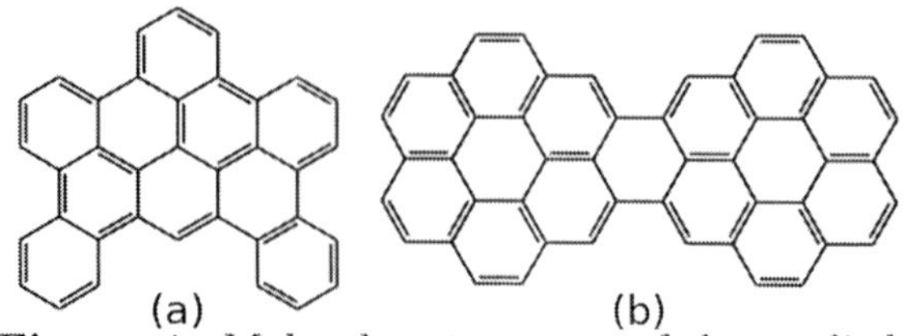

Figure 4. Molecular structures of the studied (a) dibenzophenanthropentaphene, $C_{36}H_{18}^+$, and (b) dicoronylene, $C_{48}H_{20}^+$, cations.

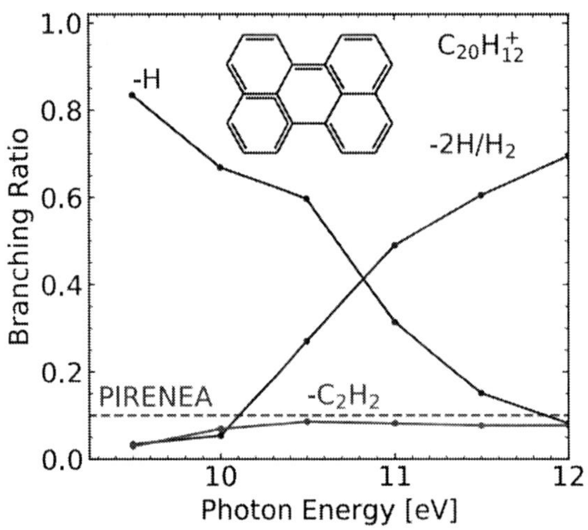

Figure 5. Fragmentation branching ratio (BR) for the perylene cation recorded at SOLEIL as a function of the VUV photon energy (first campaign, [23]). Also reported is the $C_2H_2/(H+H_2)$ BR derived from the PIRENEA experiments (cf. Fig. 3).

We report in Fig. 5 the evolution of the fragmentation branching ratio (BR) with the energy of the absorbed VUV photon for perylene, $C_{20}H_{12}^+$, up to 12 eV. In this range, the $C_2H_2/(H+H_2)$ BR does not vary with energy, whereas the opening of the $2H/H_2$ channel(s) proceeds. At higher energies, sequential fragmentation is observed and the results are more difficult to interpret. The value of BR recorded at SOLEIL is also consistent with the PIRENEA value, showing that it does not depend on the excitation scheme: multiple UV-visible photon absorption for PIRENEA compared to single VUV photon absorption at SOLEIL.

The above results appear in line with a statistical fragmentation of the hot ion. It is therefore of interest to compare them with the results of MD/SCC-DFTB calculations. A total energy of 24.8 eV is necessary to observe a reasonable number of fragmentation events for the perylene cation during the simulation time (1620 simulations of 500 ps). Averaging over all simulations, losses of H, H_2 and C_2H_2 were observed with BRs of 3.7, 0.6 and 0.7 %, respectively. In the case of the coronene cation at the same energy of 0.275 eV per mode (*i.e.* 28.1 eV of internal energy), the H, H_2 and C_2H_2 losses were observed with BRs of 6.2, 0.6 and 0.5 %, respectively. The values of 0.08 and 0.07 derived for the C_2H_2/H and $C_2H_2/(H+H_2)$ BRs, are therefore significantly lower than those for the perylene cation at 0.19 and 0.16, respectively. These simulations show the right trends compared to the experiments. Indeed, one should consider that, even with such a large number of simulations, statistics is probably not reached as the number of events at the energy dissociation threshold remains small. MD simulations can also be used to get structural information along the dynamics, which is not accessible in our experiments but can provide interesting insights into dissociation pathways and final products [28]. In particular it is interesting to investigate the role of isomerization upon dissociation [29]. This is an ongoing work which is not discussed in this article.

4.2. Dissociation rates

The results of the model by Montillaud *et al.* [5] rely on the dissociation rate for the H loss of the coronene cation that was obtained by fitting the experimental data of the PIRENEA experiment with a kinetic Monte Carlo model and assuming an activation energy of 4.8 eV based on DFT calculations. This rate is compared in Fig. 6 with the more recent rate derived from iPEPICO measurements [15]. J. R. Barker showed that it is possible to describe the RRKM rate with a simple analytical expression derived from the Laplace transform inversion of the Arrhenius thermal molecular rate [30]. The two adjustable parameters are the activation energy E_0 and a pre-exponential factor A in the equation:

$$k(U) = A\rho(U - E_0)/\rho(U) \quad (1)$$

where ρ is the density of vibrational states (DoS) of the parent ion and U its internal energy. We computed the DoS using the Beyer & Swinehart algorithm [31] and the list of harmonic vibrational modes listed in the theoretical spectral database of PAHs [32]. We then fitted the dissociation rate of the coronene cation from [15] using the mean value of 4.41 eV reported by the authors for E_0 and adjusting the value of A to 3.05 $10^{15}\,s^{-1}$.

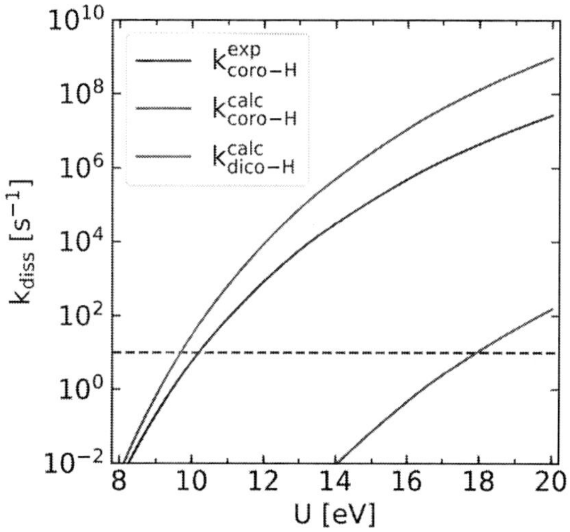

Figure 6. Dissociation rates of H loss for the coronene (coro) and dicoronylene (dico) cations. For coro, the experimental rate is taken from West *et al.* [15] and the calculated one from Montillaud *et al.* [5]. For dico, the calculated rate was obtained from Eq. 1 with values of $E_0 = 4.41$ eV and $A = 3.05 \cdot 10^{15}\,s^{-1}$ (cf. sect. 4.2). The horizontal dotted line represents a typical value for the rate of emission of IR photons.

It is now of interest to test whether such rates can be extrapolated to large PAH sizes. West *et al.* derived that the value of $E_0 = 4.41\,eV$ is typical for the loss of the first H in the studied PAHs up to 24 carbon atoms [15]. Assuming that A is independent of size, we can then calculate the dissociation rate of the dicoronylene cation, $C_{48}H_{20}^+$, using Eq. 1. The result is shown in Fig. 6 and shows that $C_{48}H_{20}^+$ is not expected to lose H below 17 − 18 eV. This prediction can be tested with our SOLEIL data.

Figure 7 (upper panel) compares the different photoproducts observed for coronene [23] and dicoronylene cations. For the latter, ionization leading to the dication is largely dominating, whereas there is more competition with hydrogen loss in the case of coronene. Zooming on the dicoronylene-H fragment channel (lower panel in Fig. 7), one can notice that the signal, although weak, starts to increase at around 12 eV, which strongly disagrees with the predicted dissociation rate (cf. Fig. 6). One could think of improper cooling of a small population of ions in the trap, possibly related to the sequential absorption of VUV photons over a short enough period so that these ions had not enough time to cool before the absorption of another VUV photon. However, a closer inspection of our data led us to notice that the fragmentation curve is following very closely the ionization curve until both curves split at ∼ 17 eV (lower panel in Fig. 7). This suggests that both channels are strictly in competition below 17 eV, which can be rationalized if dissociation is driven by electronic excited states. Thermal dissociation from the hot ground state of the cation would then start when both curves would deviate at ∼ 17 eV in reasonable agreement with the calculated dissociation rate. All this points to the occurrence of non-statistical dissociation through the same intermediate states leading to ionization. This non-statistical dissociation process is different from the one that has been reported for collisions with atoms at center-of-mass energies from a few tens to a few hundreds of eV [33]. In this case, specific non-statistical fragments are observed. On the contrary, in our experiments with VUV photons we observe a direct H loss at energies below those expected for statistical fragmentation following internal conversion. Direct dissociation in the excited states in a non-statistical process was reported earlier for large molecules, namely protonated peptides [34] and amino-acids [35] but never for PAHs.

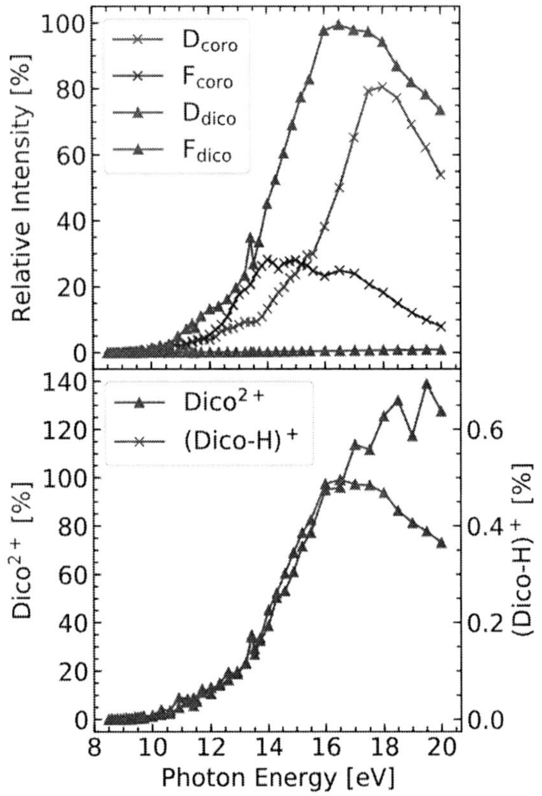

Figure 7. Upper panel: Dication, D, and fragment, F, photoproducts upon VUV photon absorption of coronene, $C_{24}H_{12}^+$ (coro) and dicoronylene, $C_{48}H_{20}^+$ (dico). Lower panel: Comparison of the dication and the H loss channel of dicoronylene.

This work combined with a previous study on the hexa-peri-hexabenzocoronene, $C_{42}H_{18}^+$, cation [36] shows that ionization is by far the dominant process in the processing of large PAHs by VUV photons. In addition, our SOLEIL experiments suggest that excited electronic states could play a role in dissociation. So far only statistical (thermal) dissociation has been considered in astronomical models, which requires energies over 13.6 eV for large PAHs and therefore rare multiple absorption events [5]. On the other hand, although of very low occurrence relative to ionization, dissociation from excited electronic states can proceed from the absorption of a single VUV photon of energy less than 13.6 eV. It can therefore be very competitive relative to a multiple absorption process.

5. Conclusion

The study of the photophysics and stability of PAHs in astrophysical environments has motivated a wealth of studies involving a large variety of experimental setups to address the multiscale dynamics of these systems. Major progresses have been achieved but quantitative studies should now be extended to larger species. We discuss here some results obtained in ion traps for species containing up to ~50 carbon atoms and came to the following conclusions. The loss of C_2H_2 is not expected to occur in the fragmentation of these large PAHs. Ionization largely dominates fragmentation and this competition has to be better evaluated to be used in astrophysical models. In addition, the possibility to dissociate at energies below 13.6 eV, as suggested by our SOLEIL experiments on the dicoronylene cation, has to be further investigated in particular through probing the relaxation of excited electronic states with ultra-fast diagnosis [11].

Acknowledgments
The research leading to this result is supported by the European Research Council under the European Union's Seventh Framework Programme ERC-2013-SyG, Grant Agreement no. 610256 NANOCOSMOS. G Wenzel is supported by the H2020-MSCA-ITN-2016 Program EUROPAH, Grant Agreement no. 722346. The computing mesocenter CALMIP (UMS CNRS 3667) is acknowledged for generous allocation of computer resources (p0059).

References

[1] Peeters E, Spoon H W W and Tielens A G G M 2004 *Astrophys. J.* **613** 986–1003
[2] Bakes E L O and Tielens A G G M 1994 *Astrophys. J.* **427** 822
[3] Allain T, Leach S and Sedlmayr E 1996 *Astron. Astrophys.* **305** 602–615
[4] Le Page V, Snow T P and Bierbaum V M 2001 *Astrophys. J., Suppl. Ser.* **132** 233–251
[5] Montillaud J, Joblin C and Toublanc D 2013 *Astron. Astrophys.* **552** A15
[6] Andrews H, Candian A and Tielens A G G M 2016 *Astron. Astrophys.* **595** A23
[7] Castellanos P, Candian A, Andrews H and Tielens A G G M 2018 *Astron. Astrophys.* **616** A167
[8] Léger A and Puget J L 1984 *Astr. Astrophys. Lett.* **137** L5–L8
[9] Allamandola L J, Tielens A G G M and Barker J R 1985 *Astrophys. J.* **290** L25–L28
[10] Visser R, Geers V C, Dullemond C P, Augereau J C, Pontoppidan K M and van Dishoeck E F 2007 *Astron. Astrophys.* **466** 229–241
[11] Marciniak A *et al.* 2015 *Nat. Commun.* **6** 7909
[12] Mulas G, Malloci G, Joblin C and Toublanc D 2006 *Astron. Astrophys.* **460** 93–104
[13] Compiègne M, Abergel A, Verstraete L, Reach W T, Habart E, Smith J D, Boulanger F and Joblin C 2007 *Astron. Astrophys.* **471** 205–212
[14] Léger A, d'Hendecourt L and Défourneau D 1989 *Astron. Astrophys.* **216** 148–164
[15] West B, Rodriguez Castillo S, Sit A, Mohamad S, Lowe B, Joblin C, Bodi A and Mayer P M 2018 *Phys. Chem. Chem. Phys.* **20**(10) 7195–7205
[16] Léger A, Boissel P and d'Hendecourt L 1988 *Phys. Rev. Lett.* **60** 921
[17] Boissel P and de Parseval P 1997 *J. Chem. Phys.* **106** 4973
[18] Martin S, Bernard J, Brédy R, Concina B, Joblin C, Ji M, Ortega C and Chen L 2013 *Phys. Rev. Lett.* **110** 063003
[19] Martin S *et al.* 2015 *Phys. Rev. A* **92** 053425
[20] Stockett M H, Björkhage M, Cederquist H, Schmidt H T and Zettergren H 2019 *Faraday Discuss.* **217** 126
[21] Ghanta S, Sivaranjana Reddy V and Mahapatra S 2011 *Phys. Chem. Chem. Phys.* **13** 14531–14541
[22] Milosavljević A R, Nicolas C, Gil J, Canon F, Réfrégiers M, Nahon L and Giuliani A 2012 *J. Sync. Rad.* **19** 174
[23] Zhen J, Castillo S R, Joblin C, Mulas G, Sabbah H, Giuliani A, Nahon L, Martin S, Champeaux J P and Mayer P M 2016 *Astrophys. J.* **822** 113
[24] Rodriguez Castillo S, Simon A and Joblin C 2018 *Int. J. Mass Spectrom.* **429** 189–197
[25] Elstner M, Porezag D, Jungnickel G, Elsner J, Haugk M, Frauenheim T, Suhai S and Seifert G 1998 *Phys. Rev. B* **58**(2092) 7260
[26] Simon A, Rapacioli M, Rouaut G, Trinquier G and Gadéa F X 2017 *Phil. Trans. R. Soc. A* **375** 20160195
[27] Simon A, Rapacioli M, Lanza M, Joalland B and Spiegelman F 2011 *Phys. Chem. Chem. Phys.* **13** 3359–3374
[28] Jusko P, Simon A, Wenzel G, Brünken S, Schlemmer S and Joblin C 2018 *Chem. Phys. Lett.* **698** 206–210
[29] Johansson H A B *et al.* 2011 *J. Chem. Phys.* **135** 084304
[30] Barker J R 1983 *Chem. Phys.* **77** 301–318
[31] Beyer T and Swinehart D F 1973 *Commun. ACM* **16** 379
[32] Malloci G, Joblin C and Mulas G 2007 *Astron. Astrophys.* **462** 627–635
[33] Stockett M H *et al.* 2015 *J. Phys. Conf.* **635** 012036
[34] Grégoire G, Kang H, Dedonder-Lardeux C, Jouvet C, Desfrançois C, Onidas D, Lepere V and Fayeton J A 2006 *Phys. Chem. Chem. Phys.* **8** 122–128
[35] Grégoire G, Lucas B, Barat M, Fayeton J A, Dedonder-Lardeux C and Jouvet C 2009 *Eur. Phys. J. D* **51** 109–116
[36] Zhen J, Castellanos P, Paardekooper D M, Ligterink N, Linnartz H, Nahon L, Joblin C and Tielens A G G M 2015 *Astrophys. J. Lett.* **804** L7

Cold collisions of C_2^- with Li and Rb atoms in hybrid traps

Milaim Kas[1], Jacques Liévin[1], Nathalie Vaeck[1] and Jérôme Loreau[1,2]
[1] Service de Chimie Quantique et Photophysique, Université libre de Bruxelles (ULB) CP 160/09, 1050 Brussels, Belgium
[2] Department of Chemistry, KU Leuven, Celestijnenlaan 200F, B-3001 Leuven, Belgium

E-mail: jerome.loreau@kuleuven.be

Abstract. We present a theoretical investigation of reactive and non-reactive collisions of Li and Rb atoms with C_2^- molecular anions at low temperatures in the context of sympathetic cooling in hybrid trap experiments. Based on accurate potential energy surfaces for the singlet and triplet states of the Li–C_2^- and Rb–C_2^- systems, we show that the associative electronic detachment reaction is inefficient but that it occurs with a large rate for collisions involving excited states of either Li/Rb or C_2^-. We also investigate rotationally inelastic collisions in order to explore the cooling of the translational and rotational degrees of freedom of C_2^- in hybrid ion-atom traps. The effect of micromotion is taken into account by considering Tsallis distributions of collision energies. We show that the translational cooling occurs much more rapidly than rotational cooling and that the presence of excited atoms leads to losses of anions on a timescale comparable to that of rotational cooling.

1. Introduction

Theoretical and experimental studies of low temperature reactive and non-reactive collisions involving molecular anions have drawn interest recently, both due to their importance in astrochemistry [1] as well as in the field of cold physics and chemistry [2]. In this context, one of the main experimental objectives is to produce samples of (ultra)cold molecular anions. While direct laser cooling anions is challenging [3], these species can be trapped efficiently using either radio-frequency (rf) or Penning traps. By co-trapping anions with ultracold atoms, it should therefore be feasible to sympathetically cool down anions by collisions. In order to investigate this possibility, a knowledge of the collisional properties of anions at low temperature is particularly important. In the case of anions the main reactive channel leading to losses from the trap is usually the associative electronic detachment reaction, $A + BC^- \rightarrow ABC + e^-$. This has been observed experimentally for collisions of ultracold Rb atoms with OH^- in a hybrid trap composed of a magneto-optical trap (MOT) combined with an octupole rf trap [4] and explained on the basis of accurate *ab initio* calculations [5]. In particular, we have recently shown that among the alkali and alkali earth series, the associative electronic detachment of OH^- at low temperature is expected to proceed efficiently only for collisions involving Rb and Cs atoms [6]. For the anions that are stable against collisional detachment, sympathetic cooling could be feasible.

Besides OH^-, other candidates for producing cold molecular anions have been suggested [7]. An interesting species that has been widely studied both theoretically and experimentally is

C_2^-, an open-shell species with $^2\Sigma_g^+$ symmetry and an electron affinity of 3.2 eV. Furthermore, it possesses several electronic excited states that could be used in laser-cooling schemes [8] as well as to explore excited-state dynamics. In a recent work, we have studied the interaction of C_2^- with Li and Rb atoms and concluded that the associative detachment reaction should proceed with a rate much smaller than the Langevin rate provided that the ultracold atoms are in their ground state [9]. In section 2 we discuss the potential energy surfaces (PESs) that were constructed for these systems. We then investigate reactive and non-reactive collisions of C_2^- with Li and Rb atoms in Section 3 and present a model for the translational and rotational cooling of C_2^- in hybrid traps in Section 4.

2. *Ab initio* calculations

Accurate *ab initio* calculations on the Li–C_2^- and Rb–C_2^- systems have been performed with the CASSCF/ic-MRCI+Q method and large basis sets [9]. The system is described using the Jacobi coordinates R (the distance of the atom to the center of mass of C_2^-) and θ (the relative angle), while the internuclear distance of C_2^- r is fixed to its equilibrium value, 1.268 Å. The interaction between an alkali atom and $C_2^-(X\,^2\Sigma_g^+)$ leads to $^{1,3}\Sigma^+$ states for linear geometries ($C_{\infty v}$ point group) or A' states at bent geometries (C_s point group). Two-dimensional PESs were obtained for both systems in singlet and triplet states, the former being the ground state of the collisional complex. The equilibrium geometry corresponds to the T-shaped configuration with a well depth of 2.30 eV for Li-C_2^- and 1.33 eV for Rb-C_2^-. An interesting feature of C_2^- is that it presents several bound excited electronic states, which opens the possibility of studying excited-state reactivity in anionic systems as well as the use of Doppler thermometry [10] to measure the translational temperature of C_2^-. We obtained PESs for the lowest excited molecular states of the complex, which correlate to the Li/Rb(^2S)+$C_2^-(^2\Pi_u)$ dissociation channel, located 0.50 eV above the ground state. Finally, we have also calculated the PESs for the corresponding neutral species (Li–C_2 and Rb–C_2) in order to investigate the associative detachment reaction. The neutral ground state ($X\,^2\Sigma_g^+$ or $X\,^2A'$) was incorporated along with the anionic states into the state-averaged procedure in order to obtain a common CASSCF wave function. The results of these *ab initio* calculations are shown in Figure 1 for Li–C_2^- and Rb–C_2^- for the linear geometry ($\theta = 0°$).

3. Collisional dynamics

3.1. Reactive Collisions: associative electron detachment

The associative electron detachment (AED) process A + BC$^-$ $\rightarrow$ ABC + e^-, in which a bond is formed while an electron is ejected, can proceed through two different mechanisms. The fastest one occurs when the system reaches a geometry at which the neutral and anionic PESs of the complex cross. The range of nuclear coordinates for which the anion PES lies higher in energy than the neutral PES defines the autodetachment region where the wave function associated to the entrance channel is embedded into the electronic continuum. At such geometries, the anionic state is unstable and the excess electron is spontaneously ejected. The rate at which electron detachment occurs depends on the probability that the electron tunnels through a centrifugal barrier and will therefore be larger when the excess electron occupies an orbital with a small orbital quantum number. This can be calculated using model potentials or more elaborated methods, such as R-matrix approaches [11], that allow to directly obtain the width, hence the lifetime, of these autodetaching states. Since it involves electronic motion, the detachment rate is usually very high ($10^{14} - 10^{15}$ s^{-1} [12]). The second mechanism involves an energy transfer between nuclear and electronic motions and occurs through non-adiabatic couplings. It was found [12] that the following characteristics lead to large detachment rates: *(i)* a very diffuse highest occupied molecular orbital (HOMO) that varies quickly with R; *(ii)* a small energy difference between anion and neutral PES; *(iii)* a small reduced mass. Although the calculation

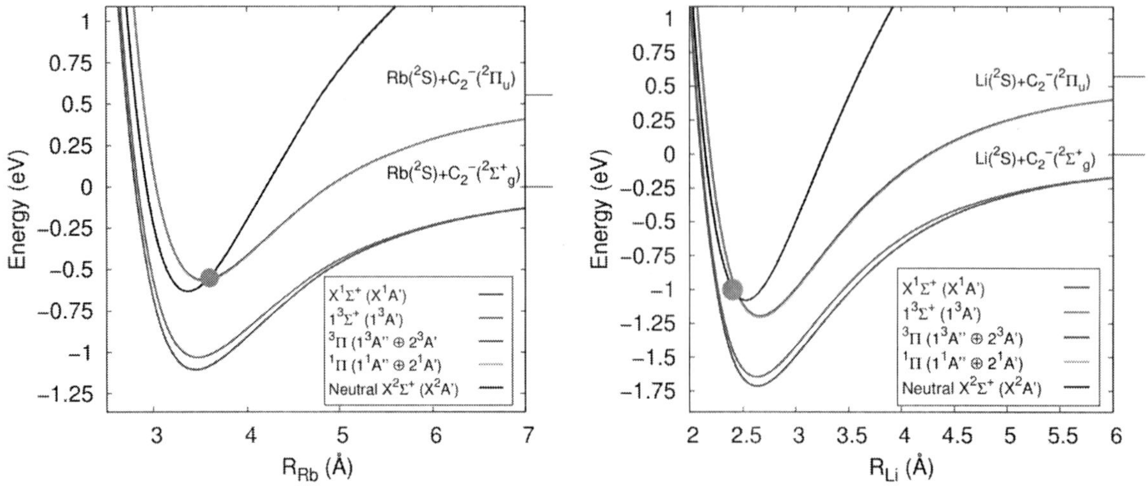

Figure 1. Potential energy curves for the lowest states of Rb–C_2^- (left) and Li–C_2^- (right) molecular species at $\theta = 0°$. The energy is given relative to the ground state dissociation energy. The two dissociation limits at which the molecular states correlate are shown. The full grey circle shows the crossing geometry between the anion and neutral potential energy curves.

of the detachment rate and is challenging, general trends and approximate rates can be obtained on the basis of the PESs as well as energetic considerations. Since the first mechanism ("curve jumping process" into the autodetachment region) leads to detachment rates that are usually several orders of magnitude larger than for the second mechanism (non-adiabatic driven electron detachment), one can neglect the latter if the neutral and anion PESs cross in an energetically accessible region of the nuclear coordinates.

From the first excited entrance channel, $M(^2S)$+$C_2^-(^2\Pi_u)$, a crossing with the neutral PES occurs close to the minimum of the well (shown by the gray circle in Figure 1). The first mechanism will therefore dominate the detachment process and, given the large rate of electron detachment when the crossing is reached, the rate of associative detachment is likely to be close to the capture rate. The entrance channel $M(^2P)$+$C_2^-(X\,^2\Sigma_g^+)$ lies even higher in energy and every collision will cause the collisional complex to enter the autodetachment region as well. Since $M(^2P)$ posses a quadrupole moment, the additional quadrupole-charge interaction will increase the capture rate compared to the Langevin one and we find that $k \approx 1 \times 10^{-8}$ cm^3s^{-1} below room temperature. On the other hand, for the lowest entrance channel, $M(^2S)$+$C_2^-(X\,^2\Sigma_g^+)$, no crossing occurs between the singlet or triplet PES of the anion and the neutral PES. As a result, associative detachment can only be triggered by the second mechanism, *i.e.* non-adiabatic couplings between nuclear and electronic motions. The detachment process from the ground state entrance channel should then be hindered. Differences between the singlet and triplet states should arise as the HOMO of the triplet state is more diffuse and varies more rapidly with R, as shown in Fig. 2.

3.2. Inelastic Collisions

Given that the AED reaction is predicted to be inefficient for collisions of C_2^- in the ground electronic state with Li and Rb atoms, we can investigate the rotational cooling dynamics of C_2^-. The close-coupling approach [13] was employed to calculate cross sections and rate coefficients for rotationally-inelastic collisions involving rotational states up to $j = 20$. The collision can occur on the singlet or the triplet PES, and the total cross section is given by the weighted sum with a ratio 1:3. For both Li and Rb, the cross sections are similar for the two spin symmetries, while the rate coefficients are larger for collisions involving Li than for Rb atoms [9].

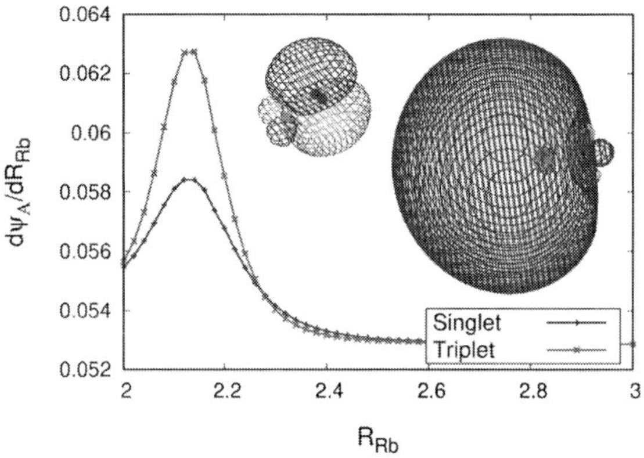

Figure 2. Derivative of the wave function along the R coordinate for Rb-C_2^- and shape of the HOMO for the singlet (left) and triplet (right) state. Note that the triplet state exhibits a strong dipole-bound character.

4. Cooling model

Based on the preceding results and discussion, we can develop a model to describe the simultaneous translational and rotational cooling of C_2^- through collisions with co-trapped, laser-cooled Li and Rb atoms.

The rotational cooling will be determined by the evolution of the population of the rotational states, which depend on the state-to-state inelastic rate coefficients:

$$\frac{dp_j(t)}{dt} = -n_a \sum_l k_{jl}(T)p_j + n_a \sum_l k_{lj}(T)p_l \tag{1}$$

where n_a is the density of laser-cooled atoms, p_j is the population of the rotational level j of C_2^-, and k_{jl} is the rate coefficient for inelastic collisions from level j to level l. The rate coefficients depend on the translational temperature T. The translational cooling of the anions is determined by elastic collisions and can be modeled by considering hard-sphere interactions as a function of the number N of collisions as $dT/dN = -(T - T_a)/\kappa$, where T_a is the temperature of the atoms and $\kappa = (m_i + m_a)^2/(2m_i m_a)$ [14]. This equation can be recast as

$$\frac{dT}{dt} = -\frac{R(T)}{\kappa}(T - T_a) \tag{2}$$

in terms of the collisional rate $R(T) = n_a \sigma_{el} v$, where the elastic cross section σ_{el} and the velocity v depend on the temperature, while the density of the atoms is assumed to be constant. This equation is solved assuming an initial temperature $T_0 = 300$ K.

A general characteristics of rf traps is that ions undergo micromotion, leading to non-thermal energy distributions [15] and limiting the minimal temperature of sympathetically cooled ions [16]. Several approaches can be employed in order to reduce the amount of micromotion and the residual heating, among which using systems with a small atom-ion mass ratio $\xi = m_a/m_i$ [17] or increasing the pole order of the trap. For instance, in the case of OH^- ions in an octupole trap cooled by ultracold Rb atoms, temperatures of less than 0.1 K should be reachable [18] despite the large value of ξ. Without taking into account the complete trap geometry, the effect of the non-thermal velocities can be investigated by computing the inelastic rate coefficients using a Tsallis distribution instead of a Maxwellian one. The characteristics of this distribution is a power-law tail that depends on a parameter $q > 1$ as well as on the pole order of the trap [18, 19]. The Maxwell-Boltzmann distribution is recovered for $q \to 1$ and large pole orders.

Another factor limiting the feasibility of cooling is the presence of excited states of the ultracold atom due to the MOT laser. This leads to two additional loss channels: inelastic collisions $M(^2P) + C_2^- \to M(^2S) + C_2^-$ in which the energy released is larger than the trap depth, and associative detachment $M(^2P) + C^- \to MC_2 + e^-$. As discussed in Section 3,

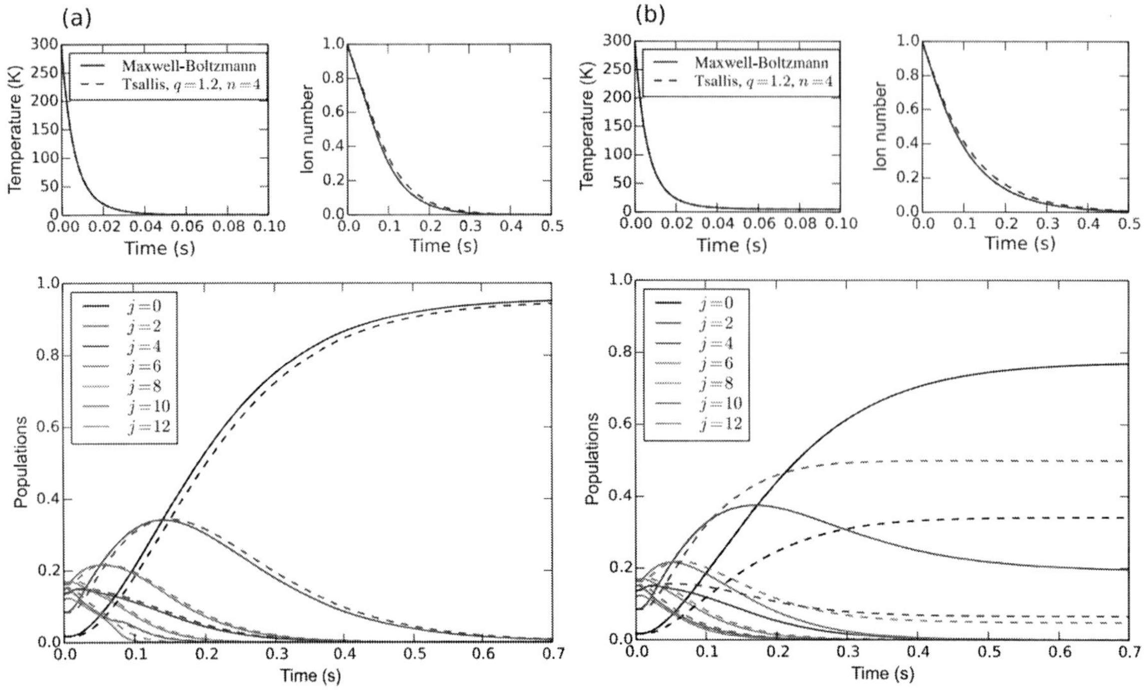

Figure 3. Evolution of the translational temperature, of the number of ions, and of the rotational populations as a function of time. (a) $T_a = 0.1$ K; (b) $T_a = 5$ K. Full lines: results obtained with a Maxwell-Boltzmann distribution of velocities; dashed lines: results obtained with a Tsallis distribution with parameter $q = 1.2$ for an octupole trap. The atom density is $n_a = 10^{10}\text{cm}^{-3}$ and the fraction of excited atoms is 5%.

the rate coefficient for associative detachment from the excited state M(^2P) is expected to be close to the value provided by capture theory. The impact of the excited states of the ultracold atoms can be somewhat mitigated by using a dark spontaneous force optical trap (dark SPOT), allowing the population of the excited state to remain below 10%.

The results of our simulations are shown for Rb-C_2^- in Fig. 3. For two atom temperatures ($T_a = 0.1$ K and $T_a = 5$ K) and two distributions of velocities (Maxwell-Boltzmann or a Tsallis distribution with parameter $q = 1.2$ in an octupole trap), we compare the evolution of the translational temperature, of the number of ions, and of the rotational populations. The atom density is fixed to $n_a = 10^{10}\text{cm}^{-3}$ and the fraction of excited atoms is 5%. We observe that the translational temperature reaches its limit much more rapidly than the rotational temperature. For $T_a = 0.1$ K the difference between the two distributions is small, as could be expected, while for $T_a = 5$ K the effect is more pronounced. With the Maxwell-Boltzmann distribution the rotational populations approach the value corresponding to a rotational temperature identical to that of the buffer gas since the rate coefficients satisfy detailed balance. For the Tsallis distribution, on the other hand, the power tail leads to a higher population of excited rotational states, and thus to a rotational heating. This heating is more pronounced for low values of the multipole order, as expected (not shown).

It is worth noting that even with a fraction of Rb(^2P) of 5%, the large rate coefficient for the AED reaction still suggests a rapid depletion of the molecular anions from the trap. This depletion occurs on a timescale longer than that of the translational cooling, but shorter than that of the rotational cooling. This will fundamentally limit the possibility of producing internally cold C_2^- anions in collisions with Rb atoms. Increasing the density of the Rb atoms will lead to a faster cooling, but also to more losses due to the associative detachment collisions with the excited atoms.

Finally, for Li atoms the mass ratio ξ is much more favorable so that the heating due to micromotion is reduced. On the other hand, the rate constant for associative detachment from the excited ^2P state is similar to that of Rb, so the loss of anions should be comparable.

5. Conclusions

The prospects for the sympathetic cooling of molecular anions by collisions with ultracold atoms in hybrid traps have been explored for C_2^-. In general, the efficiency of the cooling process is limited by factors such as heating through coupling with the micromotion in the rf trap and reactive collisions. The first concern can be addressed by using systems with low atom/anion mass ratios, high trap orders, or a localized buffer gas [18]. In order to limit the impact of associative detachment reactions, one general rule would be to choose a closed shell AB^- species with a large electron affinity, and systems other than OH^- and C_2^- have been proposed [7, 20]. However, the position of the associative detachment region depends on the atomic collision partner and must be investigated using accurate *ab initio* methods. In addition, the presence of excited states of the atoms in the MOT leads to a loss as these atoms will react with the anions on a timescale similar to that of the internal cooling. A complementary approach to cool anions is to use helium buffer gas cooling [14, 21], although in that case the temperatures are limited to a few kelvins. The dynamics of buffer-gas cooling can also be investigated based on the calculated inelastic cross sections [22, 23].

Acknowledgments

We acknowledge financial support from the Fonds de la Recherche Scientifique de Belgique-FNRS through the IISN 4.4504.10 grant. Computational resources have been provided by the Shared ICT Services Centre of the Université libre de Bruxelles.

References

[1] Millar T J, Walsh C and Field T A 2017 *Chem. Rev.* **117** 1765
[2] Tomza M, Jachymski K, Gerritsma R, Negretti A, Calarco T, Idziaszek Z and Julienne P S 2019 *Rev. Mod. Phys.* **91** 035001
[3] Cerchiari G, Yzombard P and Kellerbauer A 2019 *Phys. Rev. Lett.* **123** 103201
[4] Deiglmayr J, Göritz A, Best T, Weidemüller M and Wester R 2012 *Phys. Rev. A* **86** 043438
[5] Kas M, Loreau J, Liévin J and Vaeck N 2016 *J. Chem. Phys.* **144** 204306
[6] Kas M, Loreau J, Liévin J and Vaeck N 2017 *J. Chem. Phys.* **146** 194309
[7] Tomza M 2017 *Phys. Chem. Chem. Phys.* **19** 16512–16523
[8] Yzombard P, Hamamda M, Gerber S, Doser M and Comparat D 2015 *Phys. Rev. Lett.* **114** 213001
[9] Kas M, Loreau J, Liévin J and Vaeck N 2019 *Phys. Rev. A* **99** 042702
[10] Gianfrani L 2016 *Phil. Trans. R. Soc. A* **374** 20150047
[11] Tennyson J 2010 *Phys. Rep.* **491** 29–76
[12] Simons J 1998 *J. Phys. Chem. A* **102** 6035
[13] Arthurs A and Dalgarno A 1960 *Proc. R. Soc. A.* **256** 540
[14] deCarvalho R, Doyle J, Friedrich B, Guillet T, Kim J, Patterson D and Weinstein J 1999 *Eur. Phys. J. D* **7** 289
[15] Chen K, Sullivan S T and Hudson E R 2014 *Phys. Rev. Lett.* **112** 143009
[16] Cetina M, Grier A T and Vuletić V 2012 *Phys. Rev. Lett.* **109** 253201
[17] Haze S, Sasakawa M, Saito R, Nakai R and Mukaiyama T 2018 *Phys. Rev. Lett.* **120** 043401
[18] Höltkemeier B, Weckesser P, López-Carrera H and Weidemüller M 2016 *Phys. Rev. Lett.* **116** 233003
[19] Meir Z, Sikorsky T, Ben-shlomi R, Akerman N, Dallal Y and Ozeri R 2016 *Phys. Rev. Lett.* **117** 243401
[20] Kas M, Loreau J, Liévin J and Vaeck N 2019 *J. Phys. Chem. A* **123** 8893
[21] Gianturco F A, Lakhmanskaya O Y, Vera M H, Yurtsever E and Wester R 2018 *Faraday Discuss.* **212** 117
[22] Schullian O, Loreau J, Vaeck N, van der Avoird A, Heazlewood B R, Rennick C J and Softley T P 2015 *Mol. Phys.* **113** 3972
[23] Doppelbauer M J, Schullian O, Loreau J, Vaeck N, van der Avoird A, Rennick C J, Softley T P and Heazlewood B R 2017 *J. Chem. Phys.* **146** 044302

Theoretical studies of charged particle collisions

Åsa Larson[1], Patrik Hedvall[1], Michael Sahlin[1], Marjan Khamesian[2] and Ann E Orel[3]

[1] Department of Physics, Stockholm University, SE-106 91 Stockholm, Sweden
[2] Department of Chemistry, P. O. Box 518, S-751 20 Uppsala, Sweden
[3] Department of Chemical Engineering, University of California, Davis, CA 95616, USA

E-mail: aasal@fysik.su.se

Abstract. Results from quantum mechanical *ab initio* studies on mutual neutralization in collision of H^+ with H^- and H_2^+ with H^- are reported. The theoretical studies require computations of the potentials as well as the non-adiabatic interactions among highly excited electronic states. The electronic states are transformed to a diabatic representation and the coupled radial Schrödinger equation is solved using a logarithmic derivative method. For H_2^+ collisions with H^-, the process is studied using reduced dimensionality. Calculated cross sections and final state distributions are compared with measurements.

1. Introduction

In mutual neutralization oppositely charged ions collide, an electron is transferred and neutral atoms or molecules are formed. The reaction is by necessity a non-adiabatic process since there is a transition from an ion-pair state to a covalent state. Due to the long-range attraction of the ion-pair state potential there are in general avoided crossings at large internuclear distances ($\sim 10-40$ a_0)where the electron transfer will most probably take place. Here we will report from theoretical studies on the following mutual neutralization collisions:

$$H^+ + H^- \rightarrow H + H, \tag{1}$$

and

$$H_2^+ + H^- \rightarrow \begin{cases} H_2 + H \\ H + H + H \end{cases}, \tag{2}$$

where either the atomic or molecular fragments can be excited. The H^++H^- mutual neutralization reaction is important for understanding the formation of H_2 and the resulting cooling of the gas in the early universe [1]. The processes (1) and (2) are fundamental since the smallest of all atomic and/or molecular ions are involved. Still, these mutual neutralization reactions are non-trivial to study at low collisions energy both theoretically and experimentally. Experimental studies of mutual neutralization at low collision energies (or low temperatures) can be performed using *e.g.* the flowing afterglow Langmuir-probe technique [2], single-pass merged beam measurements [3] or using the double electrostatic storage ring DESIREE [4] at Stockholm University.

2. Structure calculations on H_2

We have previously performed series of theoretical studies [5, 6] on mutual neutralization in H^++H^- collisions by considering ${}^1\Sigma_g^+$ and ${}^1\Sigma_u^+$ electronic states of H_2 associated with the H($n \leq 3$)+H covalent limits as well as the ion-pair states. All non-adiabatic radial couplings among these states were included and the coupled Schrödinger equation for the radial wave function was solved in a strictly diabatic representation using a logarithmic derivative method [7]. In the theoretical model, the symmetry requirement due to the presence of identical nuclei was considered. The total and differential cross sections for mutual neutralization were calculated for energies between 0.001 and 100 eV. Reasonable agreement with previous calculated cross sections as well as with the measured cross sections above a few eV was obtained. Also the calculated final state distributions were compared with experimental results and satisfactory agreement was found. In the collision process between H^+ and H^- both electrons can be transferred. This double charge transfer process has also been studied [8], however, for this reaction the calculated cross section was found to be an order of magnitude larger than the measured cross sections for collision energies below 100 eV. This is not yet understood.

In order to test the approximations applied in the theoretical model and perform detailed comparisons with experiments, we have now returned to the system to improve the model and include effects previously neglected. New extensive structure calculations on the H_2 system have been performed in order to obtain consistent molecular data of all electronic states with asymptotic limits below the ion-pair H^++H^- threshold. These are the covalent states associated with the H($n \leq 4$)+H limits. The calculations were carried out using the DALTON program [9, 10]. The adiabatic potential energy curves for states of all symmetries as well as the radial non-adiabatic coupling elements, $\langle i\,|\partial_R|\,j\rangle$, and rotational coupling elements, $\langle i\,|iL_y|\,j\rangle$, were computed using the full configuration interaction method with a hydrogen basis set composed of (11s,8p,7d,6f) primitive functions contracted to (9s,8p,7d,6f). A basis set similar to the one used in our previous study [5] was used, however, with more diffuse basis functions added where the exponents of these functions were optimized in order to obtain correct asymptotic energies of covalent states and correct positions of the large distance avoided crossings due to the interaction between the ion-pair and the covalent states. An accurate description of these avoided crossings at large distances is a necessity in order to compute reliable cross sections for the mutual neutralization reaction. The calculated adiabatic potential energy curves are displayed with red lines in Fig. 1 and compared with previously calculated potential energy curves [11, 12, 13].

Mutual neutralization processes like (1) and (2) are in general driven by the non-adiabatic interactions between the ion-pair and covalent states occurring at large internuclear distances. Other types of interactions have in general been neglected. Rotational couplings originating from the matrix elements of the angular part of the nuclear kinetic energy operator have the form $\langle i\,|L_+|\,j\rangle\, J\,(J+1)\,/\,[2\mu R^2]$ and these will couple states with $\Delta\Lambda = \pm 1$. Bardsley pointed out [14] that even though these couplings are localized at small internuclear distances, the large angular momentum contributing in these types of processes, could make the rotational couplings important to consider for mutual neutralization reactions. We have here calculated these rotational couplings between all the ${}^1\Sigma_g^+$ and ${}^1\Pi_g$ as well as the ${}^1\Sigma_u^+$ and ${}^1\Pi_u$ set of states. Some of the rotational coupling elements of the lower lying electronic states have previously been calculated by Wolniewicz and Dressler [13] and they are found to be in reasonable agreement with our data.

3. H^+-H^- mutual neutralization

The cross section for mutual neutralization in $H^+ + H^-$ collisions was calculated by considering the 11 electronic states of ${}^1\Sigma_g^+$ symmetry and the 10 states of ${}^1\Sigma_u^+$ symmetry, all with threshold energies below the ion-pair limit. The radial non-adiabatic couplings among these states were

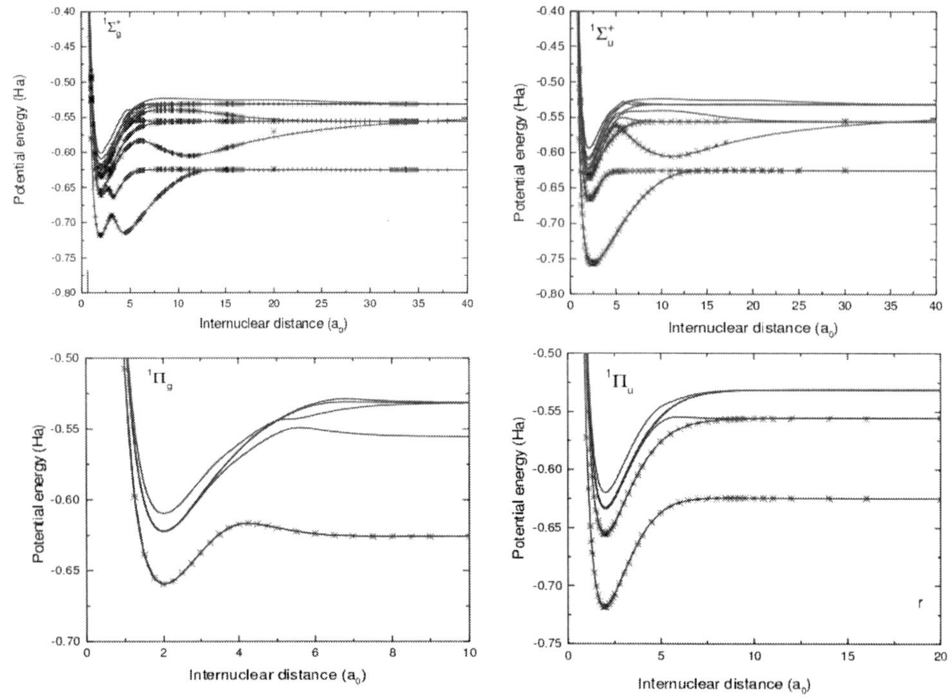

Figure 1. Calculated adiabatic potential energy curves (red lines) of H_2 in $^1\Sigma_g^+$, $^1\Sigma_u^+$, $^1\Pi_g$ and $^1\Pi_u$ symmetries are compared with results obtained by Detmer *et al.* (blue crosses) [11] and Dressler and Wolniewicz (black stars) [12, 13]. The potential associated with the ground state fragments is not displayed.

considered, but the rotational couplings to the $\Pi_{g/u}$ states have not yet been included in the model. The electronic states were diabatized and the coupled Schrödinger equation was solved using similar numerical parameters as in [5, 6]. The calculated total mutual neutralization cross section is displayed in Fig. 2 a) and compared with our previous calculation [5, 6] as well as with other calculated cross sections [15] and measurements [16, 17]. the electronic states associated with the H($n = 4$)+H limit do not significantly change the total neutralization cross section as can be seen by comparing the black and blue curves in the figure. There is a small increase of the cross section (by approximately 1% at 30 eV) when the $n = 4$ states are included. There is a reasonable comparison of the cross section computed here with the one computed by Fussen and Kubach [15] (red dashed curve) using an one-electron quantum approach. The calculated cross section agrees with measured ones (at higher collision energies, but not with the one measured by Moseley *et al.* [16] at low energies. Considering the small effect on the total neutralization cross section, it is interesting to note that by analyzing the final state contribution [see Fig. 2 b)], the H($n = 4$)+H fragments are formed with a ratio of about 10% at collision energies above 10 eV. At low collision energies, the H($n = 3$)+H fragments are mainly formed as the mutual neutralization reaction is driven by non-adiabatic interactions due to the avoided crossings between ionic and covalent states around 35 a_0. At larger energies, the $n = 2$ channels start to contribute when the avoided crossings around 10 a_0 start to become important.

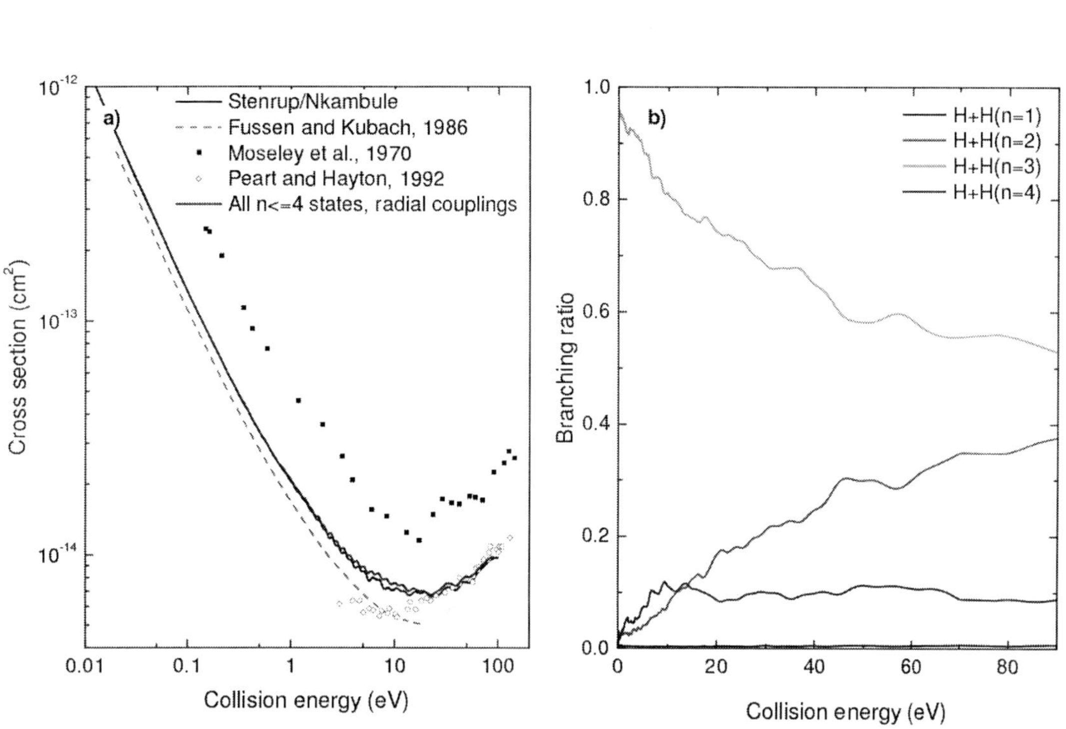

Figure 2. In a) the calculated cross section for H^+ + H^- mutual neutralization cross section (blue line) is compared with previously calculated cross sections [5, 6, 15] and measurements [16, 17]. In b), the calculated final state distributions are displayed.

4. Structure calculations on H_3

Performing an *ab initio* quantum mechanical study on the mutual neutralization reaction when H_2^+ collides with H^- as in reaction (2) is demanding. The potential energy surfaces and the non-adiabatic coupling elements have to be computed as functions of the three vibrational coordinates. The H_3 system with only three electrons allows for structure calculations using the full configuration interaction method and a hydrogen basis set of (11s,5p,1d) contracted to (6s,5p,1d) was used. Fig. 3 shows the calculated potential energy surfaces of H_3 in $^2A'$ symmetry where the H-H Jacobi coordinate is frozen at $r = 1.99$ a_0 and $\theta = 89.9°$. (This angle is chosen rather than 90° in order to run the calculations in C_s symmetry and avoid C_{2v}.) As can be seen there are avoided crossings at large internuclear distances (20-30 a_0) when the attractive ion-pair state crosses the covalent state potentials. There are additionally higher lying states with a long-range attractive Coulomb potential. These states have configurations mimicking continuum states of H_2^-+H^+. The electronic states involved in the long range avoided crossings that are believed to play an important role in the H_2^+ + H^- mutual neutralization reaction are identified [red curves in Fig. 3 a)] and the non-adiabatic coupling elements among these states are computed. The structure calculations of the potential energy surfaces and non-adiabatic coupling elements are repeated for different values of the r and θ Jacobi coordinates.

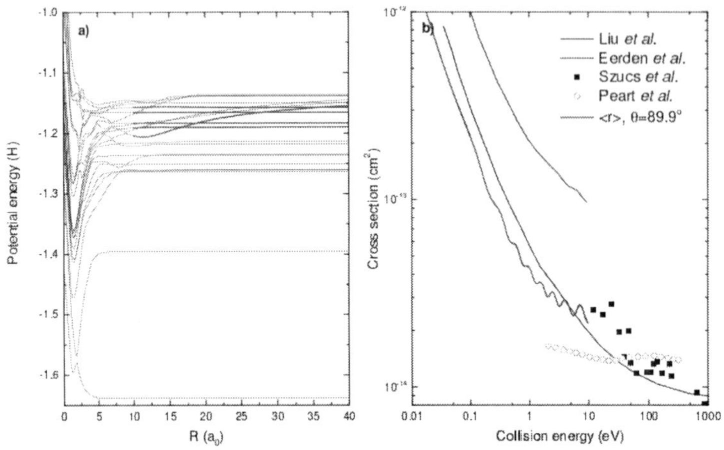

Figure 3. a) shows the potential energy surfaces of H_3 in $^2A'$ symmetry for fixed Jacobi coordinates r = 1.99 a_0 and θ = 89.9°. The electronic states that are believed to play an important role in the mutual neutralization reaction are shown with red curves and among these states the non-adiabatic coupling elements have been computed. In b), the calculated cross section for mutual neutralization in H_2^++H^- collisions is displayed (with the red curve) and compared with previous Landau-Zener calculations [20, 21] and measured cross sections [18, 19].

5. H_2^+-H^- mutual neutralization

An estimation of the mutual neutralization cross section was obtained by describing the H_2^+-H^- system as an one-dimensional system where the Jacobi (r, θ) coordinates are assumed to be frozen during the collision. The one-dimensional slices of the H_3 potential energy surfaces are diabatized and the coupled Schrödinger equation was solved using the same logarithmic derivative method as for the diatomic collision complexes. The cross section for mutual neutralization was then estimated by keeping θ = 89.9° frozen and from a weighted sum of the cross sections computed at the different values of the Jacobi coordinate r. The weights were given by the probability distribution obtained from the $H_2^+(v = 0)$ vibrational wavefunction. The resulting cross section is displayed with the red curve in Fig. 3 b). When the radial coordinate r changes, the neutralization cross section was changing which can be understood since the avoided crossings between the ionic and covalent states occur at other distances. The H_2^++H^- mutual neutralization cross section has previously been measured at collision energies above a few eV [18, 19]. The cross section has also been computed using multi-state semi-classical Landau-Zener calculations [20, 21]. In [20] the interactions to covalent states associated with H_2^*+H limit were considered, where the excited H_2 was assumed to be in an excited electronic state of gerade symmetry. In [21] only covalent states associated with H_2 in the electronic ground state and excited H-atoms were considered. Present *ab initio* calculation includes both types of states. By analyzing the partial cross sections it is found that mainly states associated with H_2^*+H are formed and the agreement between the present result with that of [20] is better than the agreement with [21].

6. Discussion and conclusion

The results presented here for processes (1) and (2) are preliminary. In the case of H^++H^-, the next step is to include the computed rotational couplings in the model. Some of the calculated non-adiabatic coupling elements between the covalent states are not completely zero asymptotically. This is a well-known problem that have been discussed and been analyzed in the case of inelastic atomic collisions [22]. We believe, the influence of the electron translation is not as dramatic in mutual neutralization reactions compared to other charge-transfer collisions due to the relatively small non-adiabatic couplings between ionic and covalent states in the asymptotic region. However, we would like to apply the correct asymptotic boundary condition in the logarithmic derivative code considering that asymptotically some of the electronic states are coupled. In the *ab initio* studies on H^++H^- performed so far, only bound electronic states have been considered. These bound states are known to interact with electronic resonant states imbedded in the ionization continuum of H_2^+. By connecting the bound states with the electronic resonant states, autoionization could be included in the model and processes such as associative ionization or penning ionization could be studied.

The next step in the H_2^++H^- calculations is to perform an averaging of the calculated cross section over the angle θ. We are also running new structure calculations on the H_3 system using the full configuration interaction method and with a larger basis set in order to obtain more of the covalent states that possess avoided crossings with the ion-pair state at large distances. The plan is to perform the scattering calculations in reduced dimensionality to determine what states of H_2^*+H that are populated and then in the subsequent step follow the fate of the H_2^* fragments to determine whether two fragments or three fragments are formed in the reaction. Studying the collision in full dimensionality is the ultimate goal. This, however, is a non-trivial task.

References

[1] Glover S 2003 *Astrophys J.* **584** 331
[2] Wiens J P, Shuman N S, Miller T M, and Viggiano A A 2016 *J. Chem. Phys.* **144** 204309
[3] de Ruette N, Dochain A, Launoy T, Nascimento R F, Kaminska M, Stockett M H, Vaeck N, Schmidt H T, Cederquist H, and Urbain X 2018 *Phys. Rev. Lett* **121** 0834012
[4] Schmidt H T *et al.* 2013 *Review Scientific Instruments* **84** 055115
[5] Stenrup M, Larson Å, and Elander N 2009 *Phys. Rev. A* **79** 012713
[6] Nkambule S M, Elander N, Larson Å, Lecointre J, and Urbain X 2016 *Phys. Rev. A* **93** 032701
[7] Johnson B R 1985 *Phys. Rev. A* **32** 1241
[8] Mezei J Zs *et al.* 2010 *Phys. Rev. A* **82** 014701
[9] Aidas K *et al.* "The Dalton quantum chemistry program system", 2014 *WIREs Comput. Mol. Sci.* **4** 269 (doi: 10.1002/wcms.1172)
[10] Dalton, a molecular electronic structure program, Release v2016.2 (2019), see http://daltonprogram.org.
[11] Detmer T, Schmelcher P, and Cederbaum L S 1998 *J. Chem. Phys.* **109** 9694
[12] Dressler K and Wolniewicz L 1984 *Can. J. Phys.* **62** 1706
[13] Wolniewicz L and Dressler K 1988 *J. Chem. Phys.* **88** 3861
[14] Bardsley J N 1982 *Recombination Processes in Atomic and Molecular Physics* in *Atomic and Molecular Collision Theory* ed. Gianturco F A (Plenum Press, New York)
[15] Fussen D and Kubach C 1986 *J. Phys. B* **19** L31
[16] Moseley J, Alberth W, and Peterson J R 1970 *Phys. Rev. Lett* **24** 435
[17] Peart B and Hayton D A 1992 *J. Phys. B* **25** 5109
[18] Peart B *et al.* 1997 *J. Phys. B* **30** 4955
[19] Szücs S *et al.* 1983 ICPEAC proceeding
[20] Liu C L, Wang J G, and R. K. Janev, J. Phys. B, 39, 1223 (2006).
[21] Eerden M J J, van der Sanders M C M, Otorbaev D K, and Schram D C 1995 *Phys. Rev. A* **51** 3362
[22] Grosser J, Menzel T, Belyaev A K 1999 *Phys. Rev. A* **59** 1309

Ultracold collisions in the Yb-Li mixture system

Florian Schäfer[1], Hideki Konishi[1], Adrien Bouscal[2], Tomoya Yagami[1], Matthew D Frye[3], Jeremy M Hutson[3] and Yoshiro Takahashi[1]

[1] Department of Physics, Graduate School of Science, Kyoto University, Kyoto 606-8502, Japan
[2] Département de Physique, École Normale Supérieure, PSL Research University, 24 rue Lhomond, 75005 Paris, France
[3] Joint Quantum Centre (JQC) Durham-Newcastle, Department of Chemistry, University of Durham, South Road, Durham, DH1 3LE, United Kingdom

E-mail: schaefer@scphys.kyoto-u.ac.jp

Abstract. We report our experimental results on the collisional physics between non-S-state atoms (ytterbium (Yb), effectively a two-electron system, in the metastable $^3P_2$ state) and S-state atoms (lithium (Li), an alkali metal, in the ground state). At low magnetic fields, by measuring inelastic interspecies collisional losses in the double quantum degenerate mixture we reveal the strong dependence of the inelastic losses on the internal spin states of both species and suppressed losses in stretched state configurations. Increasing the magnetic field up to 800 G we further investigate the magnetic field dependence of the collisional interactions. There, smoothly increasing inelastic losses are observed towards higher fields. The combined knowledge of both the magnetic field and the spin state dependence of the collisional losses of this prototypical mixture system of non-S-state and S-state atoms provides a significant step forward towards controllable impurity physics realized in the Yb-Li ultracold system.

1. Introduction

The understanding and control of the collisional properties of single-species ultracold atomic systems was and is instrumental to their application in the research of fundamental phenomena, quantum simulation and quantum computation. For example, elastic collisions enable us to reach ultracold temperatures via forced evaporative cooling [1], light-assisted collisions lead in photo-association processes to molecular states [2], manipulation of the scattering lengths by means of Feshbach resonances [3] facilitate the crossover from Bose-Einstein condensates to Bardeen–Cooper–Schrieffer superfluids [4, 5] and give insight to Efimov trimer physics [6], and controlled collisions between atoms form one possible basis for quantum gates in quantum computation schemes [7].

Going from single-species to two-species systems adds additional degrees of freedom to the experimental toolkit. For example, the physics of impurity systems leads to questions of both transport and localization of quantum matter [8, 9, 10]. Related questions in mixed dimensional systems range from basic phase diagram considerations [11] to the possibility of chiral p-wave superfluids [12, 13] and their application to topologically protected quantum computation [14, 15]. Also, polar molecules associated from two-species ultracold mixtures allow implementation of novel spin-lattice models as new platforms for quantum computation protocols [16].

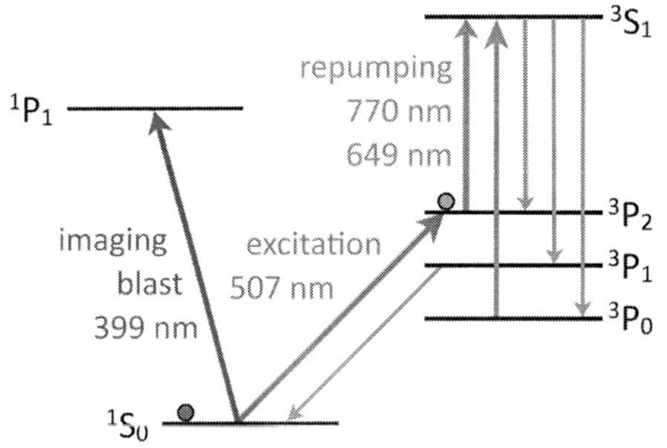

Figure 1. Yb level scheme of relevance to the experiments. Light at 507 nm connects the ground to the metastable $^3P_2$ state (green arrow). Fluorescence imaging at the strong $^1S_0 - {}^1P_1$ transition (purple arrow) reveals the remaining $^3P_2$ atoms after a blast of any remaining ground state atoms and transfer of the excited atoms to the ground state by two repumping lasers (red arrows) followed by spontaneous decay (gray arrows).

We here present and summarize our experiments on quantum degenerate mixtures of fermionic ^{6}Li and bosonic ^{174}Yb, where Yb is in the metastable $^3P_2$ state (see Fig. 1). First, at low magnetic bias fields we investigate inelastic interspecies collisional losses in the double quantum degenerate mixture that reveal the dependence of the inelastic losses on the internal spin states of both species. As a result, we discover two very distinct regimes: in collisions with ^{6}Li in the $F = 1/2$ hyperfine state overall constant inelastic collision rates are found. These results are in agreement with our previous report [17]. The collisional system with ^{6}Li in the $F = 3/2$ state, however, exhibits quite a different behavior. Now the magnetic sublevel of ^{174}Yb matters and we observe suppressed losses for stretched state configurations and up to 40-times increased relaxation rates in the other cases. Second, we measure the magnetic field dependence of the inelastic collision rates between ^{174}Yb, again in the metastable $^3P_2$ state, and ^{6}Li for several hyperfine levels of Li. Those experiments reveal a smooth but steady increase of the inelastic collision rates up to the highest investigated magnetic fields of 800 G and underline the importance of choosing collisional channels with suppressed inelastic dynamics.

This Progress Report is organized as follows: After this introduction we will review the necessary experimental details in Sec. 2. The individual results are then presented and explained in Sec. 3 and Sec. 4. Section 5 concludes the Report with a summary and a general discussion of our experimental findings.

2. Experiment

We here review the necessary experimental methods that are also discussed in more depth and detail in the related works [18, 19, 20]. Starting from a hot beam of thermal Yb and Li atoms from a dual-species oven, both species are sequentially Zeeman slowed and trapped in a magneto-optical trap (MOT). The atomic beam contains abundances of all stable isotopes of both species. By proper selection of the laser frequencies we choose to to slow down and cool only selected isotopes, namely bosonic ^{174}Yb and fermionic ^{6}Li. From the MOT the pre-cooled samples are transferred to a crossed far-off-resonant trap (FORT) created by two focused laser beams at wavelengths of about 1060 nm. Next, the intensities of the FORT lasers are decreased as to drive forced evaporative cooling of the ^{174}Yb atoms until quantum degeneracy, a Bose-Einstein condensate (BEC), is reached. Our setup is not designed to perform evaporative cooling of Li independently. Instead, Li is in thermal contact with Yb and is cooled to quantum degeneracy via sympathetic cooling [21].

During the initial phase of the evaporative cooling ramp we apply a 0.5 ms pulse of circularly polarized light resonant to either the Li $F = 1/2 \rightarrow F' = 1/2$ or the $F = 3/2 \rightarrow F' = 3/2$ D1-line transition. Together with a repumping light derived from the D2-line MOT light lasers we can effectively prepare a spin-polarized ^{6}Li sample in either of the four $F = 1/2$, $m_F = \pm 1/2$

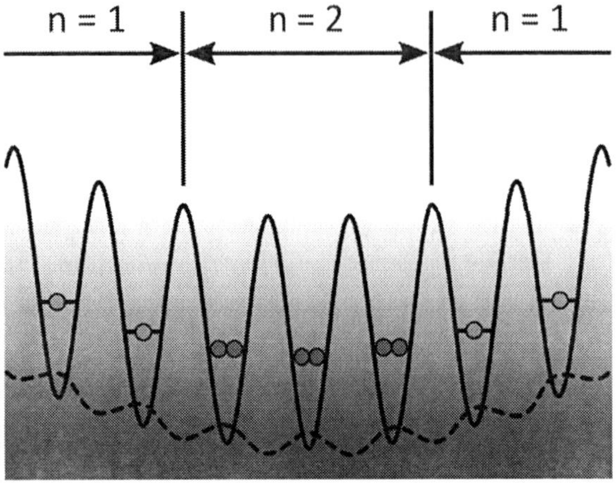

Figure 2. Quantum degenerate mixture of ^{174}Yb and ^{6}Li in a 3D optical lattice. Yb experiences a deep lattice potential (solid line) and forms a Mott-insulator shell structure (blue and green dots, only occupations numbers $n \leq 2$ shown). The effective potential for Li is shallow (dashed line) and the atoms remain unlocalized (red shaded area). For our investigation of inter-species scattering we focus on the interaction between ^{6}Li and ^{174}Yb(^{3}P$_2$) atoms (green dots) in $n = 1$ sites.

or $F = 3/2$, $m_F = \pm 3/2$ ground states. This polarization state remains with high fidelity until quantum degeneracy is reached and time-of-flight absorption imaging after application of a strong magnetic field gradient reveals above 90% of final spin-polarization. By choosing appropriate loading durations for the ^{174}Yb and ^{6}Li MOTs at this point we typically obtain a BEC of 10×10^4 ^{174}Yb atoms and a Fermi degenerate ^{6}Li gas of 3×10^4 atoms. The temperature of Li is about 300 nK or about 0.2 $T_{\rm F}$, where $T_{\rm F}$ is the Fermi temperature. Typical densities at this point are 5×10^{14} cm^{-3} for ^{174}Yb and 5×10^{12} cm^{-3} for ^{6}Li.

As we want to study the collisional dynamics between non-S-state atoms (^{174}Yb in the metastable ^{3}P$_2$ state) and S-state atoms (spin-polarized ^{6}Li in the ground state) it is necessary (i) to prepare the Yb atoms in the ^{3}P$_2$ state and (ii) to prevent fast losses stemming from inelastic Yb(^{3}P$_2$)-Yb(^{3}P$_2$) collisions [22]. To both ends we adiabatically load the quantum degenerate mixture into a three-dimensional (3D) optical lattice, formed by three pairs of counter-propagating laser beams of wavelength $\lambda_{\rm L} = 532$ nm. At $\lambda_{\rm L}$ the sign of the polarizabilities of Li and Yb are opposite. The chosen lattice depths are about 15 $E_{\rm R}^{\rm Yb}$ for ^{174}Yb and 0.7 $E_{\rm R}^{\rm Li}$ for ^{6}Li, where $E_{\rm R}^{\rm Yb}$ ($E_{\rm R}^{\rm Li}$) is the recoil energy of Yb (Li). Hence, while ^{6}Li only experiences a minor modulation of its wave function ^{174}Yb forms a Mott-insulator state (see Fig. 2 and Refs. [23, 17]). The difference in gravitational sag of the two species is addressed by an additional light field gradient, increasing the otherwise small spatial overlap between ^{174}Yb and ^{6}Li [17].

Using pairs of magnetic coils in Helmholtz configuration we then apply the desired bias fields, typically between 100 mG and 800 G. There, the metastable ^{3}P$_2$ state of Yb experiences significant Zeeman splitting of 2.1 MHz/G $\times\, m_J$. By application of a 0.5 ms long pulse of 507-nm light resonant to the ^{1}S$_0$ - ^{3}P$_2$ transition we then site and state selectively excite ^{174}Yb atoms only in singly occupied lattice sites ($n = 1$) and for a desired target Zeeman state $-2 \leq m_J \leq 2$. This effectively suppresses Yb(^{3}P$_2$)-Yb(^{3}P$_2$) collisional events. The excitation strength is chosen such that the number of excited ^{174}Yb atoms is about 10% of the total number of ^{6}Li atoms. In the following we can therefore assume that there is only a small number of non-S-state ^{174}Yb atoms immersed within a large Fermi sea of S-state ^{6}Li atoms. After the excitation a short light pulse at 399 nm removes the remaining ground state ^{174}Yb atoms from the system, ensuring no contributions from ground state Yb atoms to our results. During a variable holding time we then allow the excited ^{174}Yb atoms to interact with the background ^{6}Li gas. We note that losses of excited-state atoms due to collisions with hot background gas atoms and scattering from the trap lasers limit the lifetime of the ^{3}P$_2$ atoms to (850 ± 300) ms which is much longer than the typical timescales involved in interactions with ^{6}Li.

In the detection stage of the experiment we measure the remaining ^{174}Yb(^{3}P$_2$) atoms via the very sensitive fluorescence imaging method on the strong 399-nm transition (see Fig. 1 and

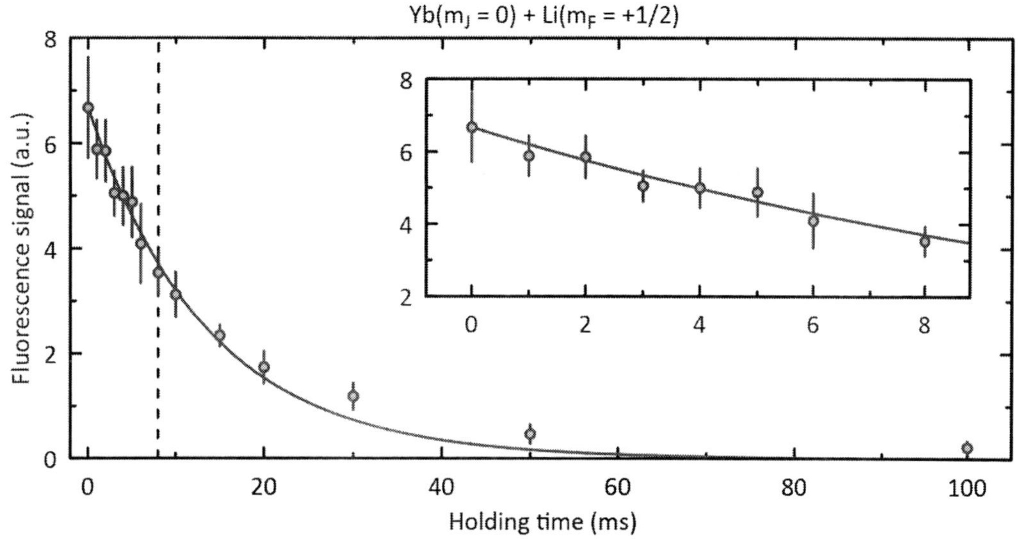

Figure 3. Decay of $^{174}Yb(^3P_2, m_J = 0)$ excited state atoms through inelastic collisions with $^6Li(^2S_{1/2}, F = 1/2, m_F = +1/2)$ ground state atoms, where we show the intensity of the obtained fluorescence signal as a function of the holding time. The data points (red) are the mean of independent measurements at each holding time. The error bars represent their standard deviations. An exponential decay (blue, solid line) is fitted to the initial time window (up to the dashed line) and shown enlarged in the inset. The lifetime obtained here is (13.5 ± 1.2) ms corresponding to an inelastic collisional rate coefficient of $(4.6 \pm 2.2) \times 10^{-11}$ cm^3/s.

Ref. [17]). For this we first apply a 399-nm light pulse to remove any accumulated $^{174}Yb(^1S_0)$ atoms from the system, followed by a short pulse of two repumping lasers (cf. Fig. 1) that transfers the remaining $^{174}Yb(^3P_2)$ atoms back to the ground state. There they are recaptured by a MOT operating on the 399-nm transition whose fluorescence light is integrated for 1 s on a sensitive charge-coupled device (CCD) camera to obtain a high-contrast signal of the MOT proportional to the number of $^{174}Yb(^3P_2)$ atoms. Imaging of the ^6Li atoms proceeds via standard time-of-flight absorption imaging.

3. Spin dependence

A first series of experiments is to establish the dependence of the inelastic collisions on the spin state of both the non-S-state $^{174}Yb(^3P_2)$ and the S-state $^6Li(^2S_{1/2})$ collisional partners. This is to gain more insight into anisotropy induced relaxation processes [24, 25, 26] in this collisional process.

The experiment proceeds as outlined in Sec. 2. During the first 100 ms of the adiabatic increase of the optical lattice depth we set a homogeneous bias magnetic field of 200 mG. This is to define a good quantization axis and to lift the degeneracy of the five $m_J = -2, \ldots, +2$ Zeeman states of $^{174}Yb(^3P_2)$. The experiment is executed for all combinations of Yb Zeeman states and accessible Li ground states, 20 configurations in total. For each combination we repeat the experiment several times, varying each time the holding time of the mixture in the optical lattice up to 100 ms. We point out that at the present bias magnetic field we could confirm that our method is sensitive to fine-structure changing collisions (e.g. $Yb(^3P_2) \to Yb(^3P_1)$) and hyperfine-structure changing collisions ($F = 3/2 \to F = 1/2$). In contrast, spin-changing collisions (m_J or m_F change) do not lead to sufficient gain in kinetic energy for the atoms to

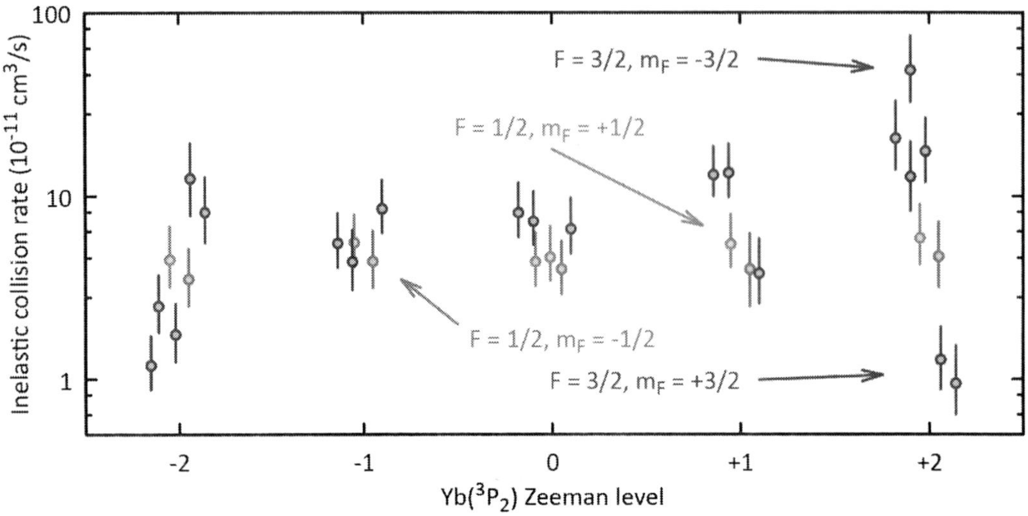

Figure 4. Inelastic collision rates between metastable ^{174}Yb(^{3}P$_2$) and ground state ^{6}Li in various hyperfine states at 200 mG bias magnetic field. Inelastic collisions with ^{6}Li(^{2}S$_{1/2}, F = 1/2, m_F = \pm 1/2$) (yellow and green symbols) show only a weak dependence on the m_J magnetic state of ^{174}Yb. In contrast, a clear m_J-state dependence in collisions with ^{6}Li(^{2}S$_{1/2}, F = 3/2, m_F = \pm 3/2$) is observed (blue and red symbols). The data points represent the median values, the error bars the $1-\sigma$ confidence interval of the determined collision rates. The data are slightly shifted horizontally for better visibility.

leave the optical lattice [19].

A representative decay of the remaining ^{3}P$_2$ atoms is shown in Fig. 3. The data qualitatively can be divided into two decay regimes. An initial, faster decay is followed by a slower decay dynamics. Having confirmed the linearity of our fluorescence light detection system, we interpret this as the crossover from a decay mostly governed by interspecies collisions to a mixed dynamics where also collisions of the Yb(^{3}P$_2$) atoms with background gas atoms combined with slightly reduced Li densities gain importance. We therefore typically limit the analysis of the data to the first few milliseconds (indicated by the dashed line and the inset in Fig. 3). The initial decay is then analyzed in terms of rate equations taking Yb(^{3}P$_2$)-Li inelastic losses and Yb(^{3}P$_2$) one-body losses into account. The details of the numerical analysis are described in our related work [19]. The analysis then gives access to a lifetime τ that, with knowledge of the atomic densities and their overlap, is translated into the inelastic loss parameter.

Figure 4 summarizes the results of this survey. The data fall into two categories. Looking at intermediate collisional rates we observe that the data for collisions with ^{6}Li(^{2}S$_{1/2}, F = 1/2, m_F = \pm 1/2$) all have very similar inelastic loss rates of about 4×10^{-11} cm^3/s. Collisions with ^{6}Li in the $F = 3/2$ hyperfine state, however, show a very pronounced and systematic dependence on the magnetic sublevel of both ^{174}Yb(^{3}P$_2$) and ^{6}Li. Especially for collisions between ^{174}Yb(^{3}P$_2, m_J = +2$) and ^{6}Li(^{2}S$_{1/2}, F = 3/2, m_F = \pm 3/2$) a 40-fold change of the loss rate is observed.

The findings are discussed in terms of the interspecies interaction operator $\hat{U}$ described in detail for the Yb(^{3}P$_2$)-Li(^{2}S$_{1/2}$) system in [27]. The anisotropy introduced into $\hat{U}$ by the non-S-state species leads to significant differences in the potentials of the four possible Yb(^{3}P$_2$)-Li(^{2}S$_{1/2}$) molecular states [28] and drives strong, spin dependent relaxation processes [26].

Indeed our inelastic loss rates are generally several orders of magnitude larger than those reported for systems where the anisotropy is suppressed by screening effects of outer s-shell orbitals [29]. The efficient suppression of inelastic losses observed in the configurations ^{174}Yb(^{3}P$_2, m_J = +2$)-^{6}Li(^{2}S$_{1/2}, F = 3/2, m_F = +3/2$) and ^{174}Yb(^{3}P$_2, m_J = -2$)-^{6}Li(^{2}S$_{1/2}, F = 3/2, m_F = -3/2$) stems from the lack of allowed inelastic decay channels for these stretched state configurations, assuming conservation of the total electronic angular momentum and its projection. According to [30] one can identify two contributions to $\hat{U}$, internal and external anisotropy. The internal anisotropy conserves the total electronic angular momentum and its projection. It therefore cannot cause inelastic changes in stretched state situations. In contrast, the external anisotropy couples to the rotational momentum of the nuclei and does not abide by the above conservation rule. In this light, our stretched state results allow direct access to the contribution by the external anisotropy that with inelastic collision rates of about 10^{-11} cm^3/s is in good agreement with earlier predictions [27]. The impact of the internal anisotropy is expressed by our findings for the non-stretched collisional systems where losses increase by one order of magnitude.

In order to understand the patterns of inelastic collision rates, we perform coupled-channel scattering calculations on Li+Yb(^{3}P$_2$). We use the same methods as described in Ref. [27], with all three spin-orbit states of Yb(^{3}P) included. The interaction potential for this system is known only from electronic structure theory [31] and is not accurate enough for quantitative predictions of scattering lengths or resonance predictions, but it should adequately represent the qualitative behaviour of the system. In order to sample the effects of different scattering lengths for the different electronic states, we construct 10 different interaction potentials. Each of these potentials is composed of 4 potential curves ($^2\Sigma^+$, $^2\Pi$, $^4\Pi$, and $^4\Sigma^+$), each of which is multiplied by a random scaling factor that varies over a range sufficient to randomise the scattering length of that potential completely.

We calculate loss rates at 200 mG for all 20 combinations of states for the experimental results shown in Fig. 4, for all 10 of our random potentials. There is a large variation of about 2 orders of magnitude in the overall magnitude of the loss rates between different potentials. This range encompasses the experimental results. For most interaction potentials, there is an approximate symmetry between calculated loss rates for (m_F, m_J) and $(-m_F, -m_J)$, although there are often differences of up 20%. Among the states of Li with $F = 1/2$, the differences between the loss rates for different Zeeman components are generally small, in agreement with the experiment.

For several of the potentials there is strong suppression of the loss rates for the spin-stretched states. Losses from these states require a change in L, the quantum number for end-over-end rotation, and so depend on the external anisotropy. In one view, this anisotropy is proportional to the difference between Σ and Π potentials. However, there may sometimes be a coincidental cancellation of the couplings, leading to suppression of loss from these states for some of our potentials. This cancellation may be conceptually similar to alkali-alkali systems with similar singlet and triplet scattering lengths [32, 33], where spin exchange is suppressed, but the details in this case are likely to be much more complicated. From the experimental results, it appears that the real system also exhibits this suppression.

4. Magnetic field dependence

We now turn our attention to the magnetic field dependence of the inelastic loss rates. Field strengths up to 800 G are considered. For field strengths > 300 G we directly ramp up the field, similar to the previous experiment, to the final value while preparing the optical lattice. For magnetic fields below 300 G, due to technical constraints of our setup, direct excitation is not efficiently possible. Instead we prepare and excite the sample at 300 G and then ramp down the field within 0.3 ms to the desired target field. Due to the strong Zeeman splitting of the Yb(^{3}P$_2$) state, the inevitable technical noise of the magnetic field and the narrow linewidth of

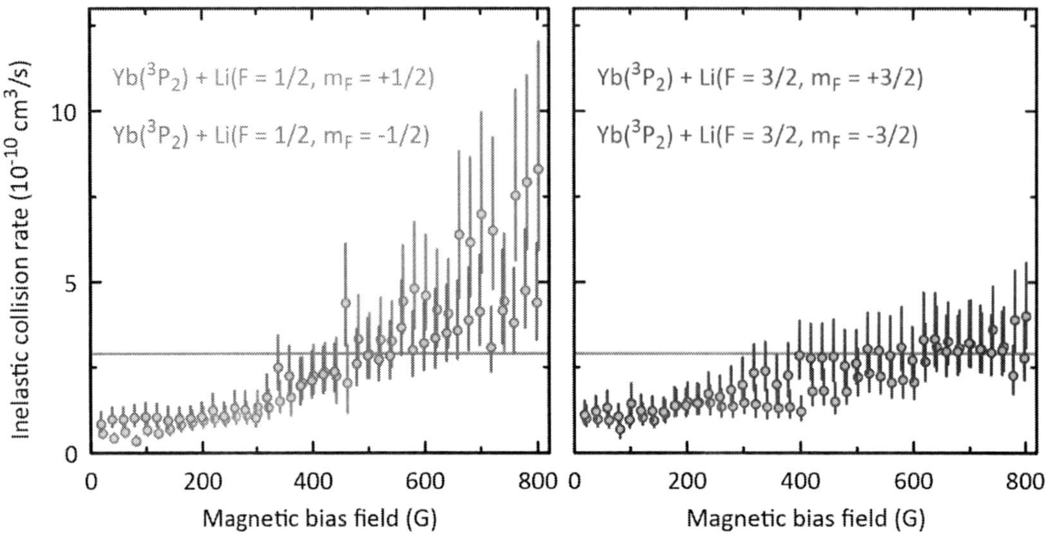

Figure 5. Summary of inelastic collision rates between ^{174}Yb(^{3}P$_2, m_J = 0$) and ^{6}Li(^{2}S$_{1/2}$) state atoms for bias magnetic fields between 20 and 800 G. The ^{6}Li atoms are either prepared in one of the two available $F = 1/2$ states (left panel) or the accessible $F = 3/2$ states (right panel). For easier comparison both panels are drawn to the same scale. Also indicated is the expected universal rate (gray lines). In all cases a smooth increase of the inelastic collision rate is found. No strong, resonance-like features indicative of possible Feshbach resonances are observed. The data is slightly offset horizontally for better readability.

the transition, we can reliably excite only the ^{174}Yb(^{3}P$_2, m_J = 0$) in this situation. Therefore, we limit this experiment to the $m_J = 0$ Zeeman state of ^{174}Yb(^{3}P$_2$). The data analysis is as in the low-field measurements of the preceding section.

The results are summarized in Fig. 5. Note, that even in this high-field situation we identify the states of Li by their low-field quantum numbers (F, m_F) at which they were prepared and that are adiabatically connected to the projections of the electronic and nuclear spin angular momenta (m_S, m_I) at high magnetic field. By using non-stretched state configurations the data include effects from both internal and external anisotropy and therefore should give a good overview of the general high-field behavior of the inelastic collisional loss rates. For all investigated spin states of ^{6}Li we find a smooth and general increase of the losses, starting with low fields at about 10^{-10} cm^3/s and staying below 10^{-9} cm^3/s when reaching 800 G. Especially for fields beyond about 600 G collisions involving ^{6}Li in the $F = 3/2$ hyperfine state have slightly lower losses then those with ^{6}Li in the $F = 1/2$ state. Overall, we find larger inelastic collisional rates than what was predicted in [27] for collisions involving Yb(^{3}P$_2, m_J = -2$). Also, for higher magnetic fields the losses approach and go beyond the predicted universal rate [27] of 2.9×10^{-10} cm^3/s (gray lines in Fig. 5); this corresponds to a model in which incoming flux can be reflected from the long-range potential, but all the flux that reaches short range is irreversibly lost with unit probability. [34]. This again exemplifies the importance of the anisotropy in $\hat{U}$ for ultracold collisions of ^{174}Yb(^{3}P$_2$) and ^{6}Li(^{2}S$_{1/2}$).

We have carried out further coupled-channel calculations to understand these results. We use the same interaction potentials as in the previous section, but now perform calculations up to 1000 G for the states shown in Fig. 5. The general behavior of the loss rates for these states is to increase from a small value at low field towards the universal loss rate [34], sometime rising above it at higher fields. There is resonant structure in most cases, which creates large peaks in loss rate. Some of which can have widths of hundreds of Gauss and are responsible for rates higher than universal. Many of the resonant features appear consecutively in the three lowest spin channels for Li ($F = 1/2, m_F = 1/2$; $F = 1/2, m_F = -1/2$; $F = 3/2, m_F = -3/2$; all of which correspond to electron spin $m_s = -1/2$) at progressively higher field, most likely due to a single state cutting upwards in energy through each of the thresholds in turn. Turning to the experimental results, the loss rates in the lowest Li channel ($F = 1/2\, m_F = 1/2$) rise strongly towards the upper end of the field range. The loss rates in the next states up rise less strongly, but one of them is still clearly above the universal rate. This behavior may be evidence of a broad resonance in the loss rate, centered above 800 G for the $F = 1/2, m_F = 1/2$ channel, and at even higher fields for the higher $F = 1/2, m_F = -1/2$ and $F = 3/2, m_F = -3/2$ channels. Further measurements at higher fields would be able to confirm if this is indeed the case.

5. Conclusions

We investigated the inelastic rates in the ^{174}Yb(^{3}P$_2$)-^{6}Li(^{2}S$_{1/2}$) system of non-S-state and S-state collisional partners. In contrast to S-state-only collisions effects of anisotropy become of significant importance here. This is evidenced by the strong dependence on the spin states we observed at low magnetic fields. At higher magnetic fields we find inelastic collision rates beyond the universal rate of complete loss at short distances. Furthermore only a smooth variation of the loss rates is found. While the general findings are supported by our coupled-channel calculations, the details are as yet mostly beyond our understanding. This renders experimental investigations all the more important and should encourage further theoretical studies. It would for example be interesting to extend our high-field studies to the numerically more investigated case of ^{174}Yb(^{3}P$_2, m_J = -2$) that might prove to be more collisionally stable.

In future experiments towards controllable impurity physics in the ultracold Yb-Li mixture system consideration of the anisotropy and its associated additional losses is necessary and stretched-state configurations should be preferred where possible to suppress unnecessary loss mechanisms. Our work thus provides a significant step towards a better understanding and control of ultracold Yb(^{3}P$_2$)-Li mixtures.

Acknowledgments

We thank Ruth Le Sueur for useful early discussions. This work was supported by the Grant-in-Aid for Scientific Research of JSPS Grants No. JP25220711, No. JP26247064, No. JP16H00990, No. JP16H01053, No. JP17H06138, No. 18H05405, No. 18H05228, JST CREST Grant No. JPMJCR1673, the Impulsing Paradigm Change through Disruptive Technologies (ImPACT) program by the Cabinet Office, Government of Japan, and MEXT-QLEAP. Further support was provided by U.K. Engineering and Physical Sciences Research Council (EPSRC) Grant No. EP/P01058X/1. HK achnowledges support from JSPS.

References

[1] Ketterle W and Druten N V 1996 *Advances In Atomic, Molecular, and Optical Physics* vol Volume 37 ed Benjamin Bederson and Herbert Walther (Academic Press) pp 181–236 ISBN 1049-250X
[2] Weiner J, Bagnato V S, Zilio S and Julienne P S 1999 *Rev. Mod. Phys.* **71** 1–85
[3] Chin C, Grimm R, Julienne P and Tiesinga E 2010 *Rev. Mod. Phys.* **82** 1225
[4] Bourdel T, Khaykovich L, Cubizolles J, Zhang J, Chevy F, Teichmann M, Tarruell L, Kokkelmans S J J M F and Salomon C 2004 *Phys. Rev. Lett.* **93** 050401
[5] Zwierlein M W, Stan C A, Schunck C H, Raupach S M F, Kerman A J and Ketterle W 2004 *Phys. Rev. Lett.* **92** 120403
[6] Zaccanti M, Deissler B, D'Errico C, Fattori M, Jona-Lasinio M, Müller S, Roati G, Inguscio M and Modugno G 2009 *Nature Physics* **5** 586–591 ISSN 1745-2473
[7] Daley A J 2011 *Quantum Information Processing* **10** 865–884 ISSN 1570-0755, 1573-1332
[8] Massignan P, Zaccanti M and Bruun G M 2014 *Rep. Prog. Phys.* **77** 034401 ISSN 0034-4885
[9] Chien C C, Peotta S and Ventra M D 2015 *Nat. Phys.* **11** 998–1004 ISSN 1745-2481

[10] Nishida Y 2016 *Phys. Rev. A* **93** 011606
[11] Nishida Y and Tan S 2008 *Phys. Rev. Lett.* **101** 170401
[12] Nishida Y 2009 *Annals of Physics* **324** 897–919 ISSN 0003-4916
[13] Wu Y J, He J, Zang C L and Kou S P 2012 *Phys. Rev. B* **86** 085128
[14] Kitaev A Y 2003 *Annals of Physics* **303** 2–30 ISSN 0003-4916
[15] Tewari S, Das Sarma S, Nayak C, Zhang C and Zoller P 2007 *Phys. Rev. Lett.* **98** 010506
[16] Micheli A, Brennen G K and Zoller P 2006 *Nat. Phys.* **2** 341–347 ISSN 1745-2473
[17] Konishi H, Schäfer F, Ueda S and Takahashi Y 2016 *New J. Phys.* **18** 103009 ISSN 1367-2630
[18] Hara H, Takasu Y, Yamaoka Y, Doyle J M and Takahashi Y 2011 *Phys. Rev. Lett.* **106** 205304
[19] Schäfer F, Konishi H, Bouscal A, Yagami T and Takahashi Y 2017 *New J. Phys.* **19** 103039 ISSN 1367-2630
[20] Schäfer F, Konishi H, Bouscal A, Yagami T and Takahashi Y 2017 *Phys. Rev. A* **96** 032711
[21] Myatt C J, Burt E A, Ghrist R W, Cornell E A and Wieman C E 1997 *Phys. Rev. Lett.* **78** 586–589
[22] Uetake S, Murakami R, Doyle J M and Takahashi Y 2012 *Phys. Rev. A* **86** 032712
[23] Bloch I, Dalibard J and Zwerger W 2008 *Rev. Mod. Phys.* **80** 885–964 ISSN 0034-6861
[24] Reid R H G and Dalgarno A 1969 *Phys. Rev. Lett.* **22** 1029–1030
[25] Mies F H 1973 *Phys. Rev. A* **7** 942–957
[26] Krems R V, Kłos J, Rode M F, Szcześniak M M, Chałasiński G and Dalgarno A 2005 *Phys. Rev. Lett.* **94** 013202 ISSN 0031-9007, 1079-7114
[27] González-Martínez M L and Hutson J M 2013 *Phys. Rev. A* **88** 020701
[28] Petrov A, Makrides C and Kotochigova S 2015 *New J. Phys.* **17** 045010 ISSN 1367-2630
[29] Hancox C I, Doret S C, Hummon M T, Krems R V and Doyle J M 2005 *Phys. Rev. Lett.* **94** 013201 ISSN 0031-9007, 1079-7114
[30] Krems R V, Groenenboom G C and Dalgarno A 2004 *J. Phys. Chem. A* **108** 8941–8948 ISSN 1089-5639
[31] Gopakumar G, Abe M, Das B P, Hada M and Hirao K 2010 *J. Chem. Phys.* **133** 124317 ISSN 0021-9606
[32] Julienne P S, Mies F H, Tiesinga E and Williams C J 1997 *Phys. Rev. Lett.* **78** 1880–1883
[33] Burke J P, Bohn J L, Esry B D and Greene C H 1997 *Phys. Rev. A* **55** R2511–R2514
[34] Idziaszek Z and Julienne P S 2010 *Phys. Rev. Lett.* **104** 113202

Negative ion relaxation and reactions in a cryogenic storage ring

H T Schmidt[1], S Rosén[1], R D Thomas[1], M H Stockett[1], W D Geppert[1], Å Larson[1], P Löfgren[1], A Simonsson[1], A Källberg[1], P Reinhed[1], M Björkhage[1], M Blom[1], J D Alexander[1], P K Najeeb[1], M Ji[1], N Kono[1], E K Anderson[1], G Eklund[1], M K Kristiansson[1], O M Hole[1], D Hanstorp[2], H Hartman[3], P S Barklem[4], J Grumer[4], K Hansen[5,2], M Gatchell[6,1], H Cederquist[1] and H Zettergren[1]

[1]Stockholm University, Department of Physics, 10691 Stockholm, Sweden
[2]University of Gothenburg, Department of Physics, 41296 Gothenburg, Sweden
[3]Malmö University, Department of Physics, 20506 Malmö, Sweden
[4]Uppsala University, Department of Physics and Astronomy, 75236 Uppsala, Sweden
[5]Tianjin University, Department of Physics, 300072 Tianjin, P R China
[6]Universität Innsbruck, Institut für Ionenphysik und Angewandte Physik, 6020 Innsbruck, Austria

E-mail: schmidt@fysik.su.se

Abstract. An overview of recent experimental results of studies of negative atomic and molecular ions in the Double ElectroStatic Ion-Ring ExpEriment, DESIREE is given. Metastable level lifetimes in atomic negative ions have been measured by time-dependent laser photodetachment. Rotational relaxation of diatomic anions is studied by near-threshold photodetachment. Spontaneous decays of small metal cluster anions by electron emission and fragmentation is studied with decay-channel specificity. Finally, mutual neutralisation of pairs of negative and positive ions has been investigated with initial and final state selectivity.

1. Introduction

The storage of fast charged particles have the general advantage that extended observation times and efficient detection of products with different mass-to-charge ratios than those of the stored ion beams are combined. When storage for few seconds is sufficient, this can be reached with room-temperature electrostatic rings and traps for few keV ion beams [1, 2, 3]. For light positive ions a very efficient, albeit costly, method exists to achieve much longer storage times. For these systems the dominating beam-particle loss process is often that of electron transfer to the stored ion from residual-gas collisions and by going to magnetic-confinement rings and much higher energies such losses are effectively eliminated due to the strong projectile velocity dependence of the electron-transfer process ($\sigma \propto v_P^{-11}$) [4] and very long storage lifetimes of many hours may result. For negative ions, the dominating loss mechanism is electron detachment in residual-gas collisions. That process does not have a very strong velocity dependence and thus the only way to eliminate residual-gas collisional losses and to reach very long storage lifetimes is to strongly reduce the residual-gas density. This is realised in the most recent generation of electrostatic ion-storage rings by means of cryogenic operation with several orders of magnitude reduction in residual-gas density. The cryogenic ion-beam storage rings currently in operation are DESIREE [5, 6], the Cryogenic Storage Ring, CSR [7], at Max-Planck Institut für Kernphysik in Heidelberg and the RIKEN Cryogenic E-ring, RICE [8], at RIKEN, Japan.

2. Brief description of DESIREE

Detailed descriptions of DESIREE are found in refs. [5] and [6], so here only its basic operating features are discussed. DESIREE consists of two electrostatic ion-beam storage rings of 8.7 m circumference, sharing a common straight section of 95 cm length where the two stored ion beams can be merged (see Fig. 1). After each of the three straight sections, position-sensitive micro-channel plate detectors are mounted to detect neutral products of reactions or spontaneous decays and after the common straight section a movable detector is mounted to detect charged fragments. The neutral detector after the merging section has a phosphor screen anode that is being observed in parallel by a CMOS camera and a segmented photomultiplier tube to allow coincident detection of two (or in principle more) neutral products with high time and position resolution. The merging and separation of two stored ion beams require that the two ion species have opposite charge polarity and a maximum ratio of the beam energies of 20. The latter is determined by the maximum voltage on the high-energy ion source platform of 100 kV and the practical lower limit for stable ion-beam storage of about 5 keV. For studies of reactions at close to zero center-of-mass energy this maximum energy ratio immediately translates to a maximum mass ratio of 20. Besides ion-ion reactions, studies are obviously possible of ions decaying spontaneously or after laser irradiation, as will be discussed below. To reach long storage lifetimes and to limit the background signal from ions colliding with residual gas molecules and losing their charge, a very good vacuum is required. In DESIREE this is achieved by applying a nested double vacuum system pumped by serially coupled standard turbomolecular ultrahigh vacuum pumps and cooling the entire inner vacuum volume to cryogenic temperatures (12-14 K) by means of four two-stage closed circuit helium cryogenerators. The vacuum surrounding the cooled inner chamber provides an excellent thermal insulation and in order to limit the effect of thermal radiation from the room-temperature environment, a copper screen covered in insulating material surrounds the inner vacuum chamber and is cooled to about 50 K by the first stages of the cryogenerators. Due to the cryogenic temperature of the inner chamber, residual-gas densities of the order of a few molecules per cubic mm is achieved resulting in storage lifetimes of keV anion beams of about 10^3 s [5, 9].

3. Lifetimes of metastable levels in atomic negative ions

Due to the absence of a long-range interaction between a free electron and a neutral atom, most atomic negative ions have only one or a few bound states. Some elements, as for example helium, do not have any stable bound states of the negative ion. However, metastable levels can exist if they are prevented from rapid disintegration by their symmetry. Such metastable negative ions are very efficiently studied with stored beams by just recording the production of neutrals as function of time after ion-injection [10, 11, 12, 13, 14]. For *bound* excited states of atomic negative ions a slightly more advanced experimental approach is necessary. One such approach is state-selective laser photodetachment which was developed at CRYRING [15]. Here the idea is simply that a laser wavelength is chosen so that the photon energy is less than the electron affinity, but higher than the binding energy of the electron in the excited state. In this way an increase in the rate of neutrals when the laser is applied can be ascribed exclusively to a population of ions in the excited state. Measuring such a signal as function of time after injection gives the decay of ions in the excited state due to the sought spontaneous decay and due to loss of the beam. By a series of careful control measurements the latter can usually be isolated and the real spontaneous lifetime deduced. At DESIREE the excellent vacuum condition has allowed us to use this method to measure longer lifetimes than what has previously been possible, reaching

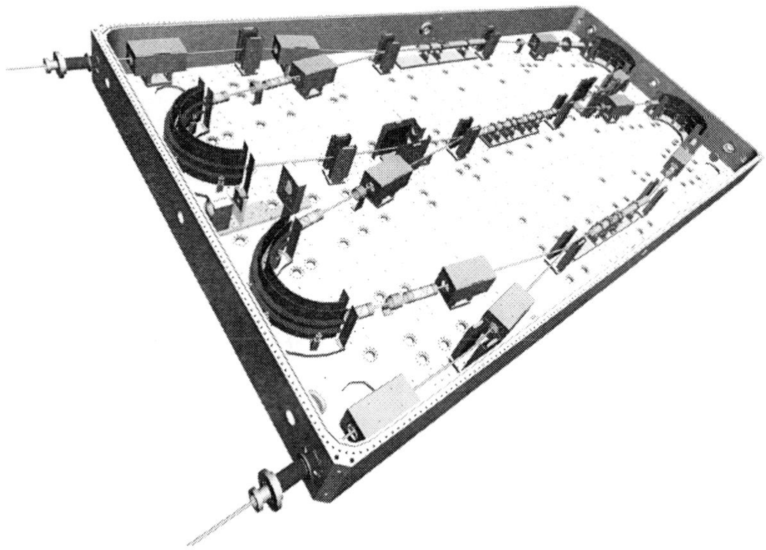

Figure 1. Drawing of the inner vacuum chamber with all the electrostatic ion-optical elements of DESIREE. Typical ion-beam trajectories are indicated by the light-blue lines.

the 503±54 s^{-1} lifetime of the upper fine structure level of S^- as the hitherto most longlived state studied [9]. In table 1 we give the results of the atomic negative ion excited state lifetimes measured so far at DESIREE [16, 17]. All the lifetimes listed in table 1 with the exception of

Table 1. Lifetimes of metastable levels of atomic negative ions measured in DESIREE.

Ion	*Level*	*Lifetime*	*Error*	*Reference*
S^-	$3p^5\,^2P_{1/2}$	503 s	54 s	[9]
Se^-	$4p^5\,^2P_{1/2}$	4.78 s	0.18 s	[9]
Te^-	$5p^5\,^2P_{1/2}$	0.463 s	0.008 s	[9]
Ni^-	$3d^94s^2\,^2D_{3/2}$	15.1 s	0.4 s	[16]
Pt^-	$5d^{10}6s\,^2S_{1/2}$	2.54 s	0.10 s	[17]
Pt^-	$5d^96s^2\,^2D_{3/2}$	0.05-0.2 s	—	[17]

the Pt^--level $5d^{10}6s\,^2S_{1/2}$ are for upper fine structure levels of the ground state. It was pointed out by Brage and Grumer [18] that for such states a scaling with the measured and calculated fine structure splittings should be fulfilled as the same radial matrix element is the most critical part of both these calculations. By their scaling method, they obtained a better agreement with theory for the case of sulfur. Surprisingly, however, applying the same method for nickel has not been successful and a 25% discrepancy remains for this system [19].

4. Rotationally cold molecular ions

A central motivation for applying cryogenic cooling in DESIREE and the other cryogenic electrostatic storage rings [7, 8] is that one can form and maintain a molecular ion beam with

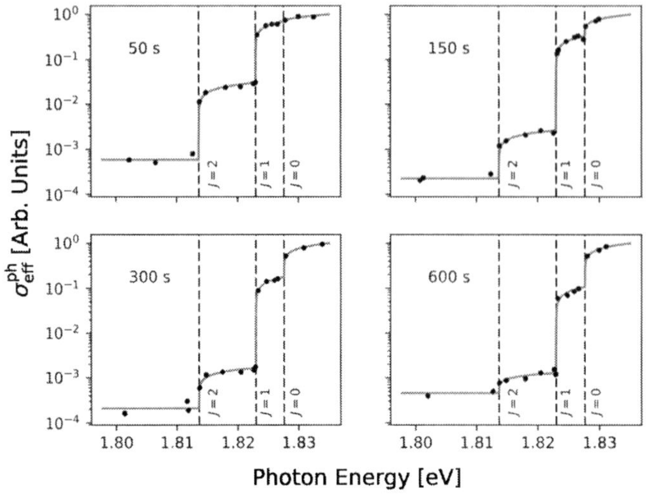

Figure 2. Measured effective relative photodetachment cross sections close to the $J = 0$, $J = 1$, and $J = 2$ detachment thresholds for different delays between ion injection and probing with a cw laser [21]. Reprinted figure with permission from [H. T. Schmidt *et al.*, Phys. Rev. Lett. **119**, 0733001 (2017)]. Copyright (2017) by the American Physical Society.

a dominance of the (few) lowest rotational level(s) only in a cryogenic environment. The OH^- molecular ion is a well-suited system for investigating the rotational cooling and such studies have been performed in CSR [20] and in DESIREE [21]. The experimental method was based on the rotational thermometry by near-threshold photodetachment [22]. Conceptually the idea is similar to the method used in the atomic negative ion lifetime measurements described above. In a narrow range of wavelengths corresponding to photon energies close to the electron affinity, photodetachment will be energetically allowed for all levels, all but $J = 0$, all but $J \leq 1$, all but $J \leq 2$ etcetera, depending on the exact value of the wavelength. In Fig. 2 the effective measured photodetachment cross section in the threshold range is shown for different storage times prior to the laser photodetachment. The staircase-like shape reflects the opening of the photodetachment channel for the three lowest rotational levels and the step heights are directly related to the population in the individual J-levels. The relative contributions from $J = 1$ and $J = 2$ decrease with time as the rotational cooling proceeds and thermal equilibirum is approached. Asymptotically, the rotational temperature of the stored ions reach 20.6 ± 0.5 K. By actively depleting the rotationally excited ions by photodetachment a rotational temperature of 12.3 ± 0.2 K was demonstrated for OH^- [21]. In a recent unpublished follow-up experiment using OD^- a rotational temperature of 7.4 ± 0.4 K was reached by the depletion method.

5. Decay of hot small metal cluster anions

In the previous section, the aim was to reach and characterize rotational level distributions of stored very cold molecular ions and cryogenic temperatures of the surroundings were needed. In the experiments described in this section, the focus is on decays of very hot metal cluster anions on the ms and s time scales, and it is not as obvious why these studies require cryogenic temperatures. It turns out, in fact, that the very low residual-gas density provides a tremendous improvement over earlier room-temperature studies. In pioneering work at the ELISA storage ring [1] it was established that ensembles of small, internally hot, cluster anions and many other complex systems decayed in the storage rings not by the well-known exponential law,

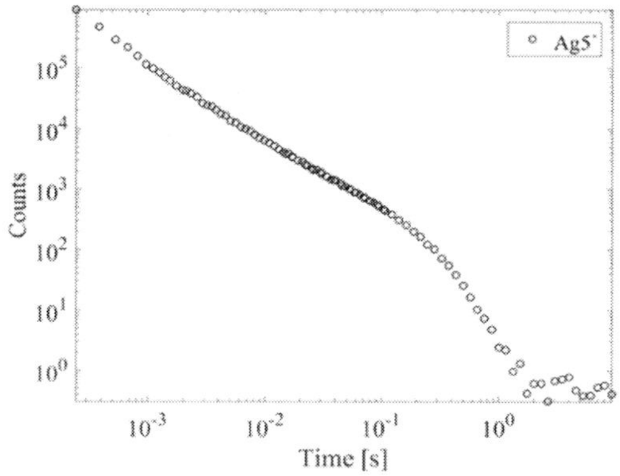

Figure 3. Rate of neutral products from the spontaneous decay of hot silver pentamer anions stored in DESIREE as function of time after ion production in a caesium sputter ion source.

but followed a power-law decay. This was ascribed to the large spread of internal energies in the ion ensemble. While all ions populating a particular quantum state would have the same constant decay rate leading to an exponential decay of those particular ions, very many different quantum states were populated in the stored ensemble of ions leading in practice to a continuum of internal energies and decay rates. The summation of all these different contributions leads to the observed power-law decay [23].

A limitation in studies of hot metal cluster decays is that ions with insufficient excitation energy to decay spontaneously may still contribute to the background neutral-particle count rate due to collisions with molecules of the residual gas. For this reason, room-temperature studies of such decays are typically only possible up to about 100 ms when the signal can no longer be separated from the background. So even if the storage lifetimes are often of the order of 1-10 s, the background collisions prohibit measurements beyond 100 ms. However, with the dramatically improved vacuum of DESIREE and other cryogenic systems [24], the background is dramatically decreased and decays in the second range may be studied. In Fig. 3 we show the measured decay when Ag_5^- is stored in DESIREE [25]. That same system was studied in the original ELISA work [23] but only until 50 ms, showing only the power-law decay and not the strong deviation from it, which only becomes visible after several hundred ms of storage and is ascribed to radiative cooling by infrared emission.

These much improved experimental conditions were used in measurements of decays for small clusters of copper [26] and silver [25]. In a recent study, metal dimer anions of silver and copper have been investigated with a slight addition to the detection scheme. Now not only the neutral products but also fragment ions can be detected. The new results confirm the conclusion of previous work from ELISA [27] that the dominating decay process for the tens of ms time range accessible in the ELISA experiment was fragmentation of highly rotationally excited dimer anions by tunneling through the centrifugal barrier. However, the new measurements also show that for times after injection in excess of 100 ms, the dominating decay is vibrational autodetachment of vibrationally highly excited dimers in lower angular-momentum states [28]. This is a surprising observation as no crossings are available between the initial and final states as function of the internuclear distance, which is the only parameter describing the dynamics on the Born-Oppenheimer surfaces of the dimer systems. Hence, the observation of vibrational autodetachment for this system constitutes a dramatic demonstration of the break-down of the Born-Oppenheimer approximation.

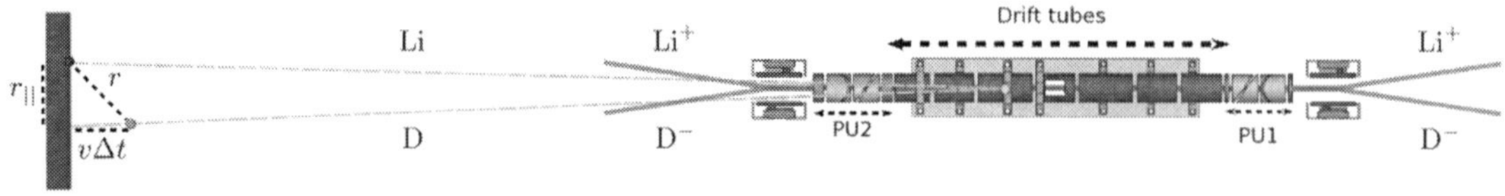

Figure 4. The straight section of DESIREE where the two stored beams can be merged is surrounded by sets of drift tubes. The split cylinders (yellow) marked PU1 and PU2 are position-sensitive electrostatic pickup electrodes applied to optimize the beam overlap. The seven green cylindrical electrodes can be put on bias voltages to adjust the center-of-mass energy. The ion beams are indicated in magenta and the two dashed lines are illustrating the trajectories of a pair of neutrals formed in an MN process inside one of the drift tubes.

6. Mutual neutralisation in DESIREE

As we have seen in examples above, the ability to store a beam of molecular ions for a long time in one of the storage rings inside the cryogenic environment of DESIREE yields excellent conditions for a range of different experiments. The construction of DESIREE as a *double* ring was, however, chosen in order to be able to investigate the interactions taking place between pairs of ions, one from each ring. In the mutual neutralisation (MN) process a pair of a positive and a negative ion react to form two neutral particles: $X^+ + Y^- \rightarrow X + Y$. The construction of DESIREE is optimized for this particular process. In Fig. 4, the merged-beams section and the path for a pair of neutrals towards the position and time-sensitive detector is shown. The distance in all three dimensions between the two neutrals is deduced from the measured detector impact positions and the time difference of the two neutrals. As this distance is given by the available kinetic energy in the relative motion of the two neutral products, its determination reveals the quantum states formed in the process. In a recent, as yet unpublished, study it was demonstrated that final states in neutral lithium of the configuration $1s^23s$ could be fully separated from states of the configurations $1s^23p$ and $1s^23d$ formed in the MN process: $\mathrm{Li}^+ + \mathrm{D}^- \rightarrow \mathrm{Li} + \mathrm{D}$ [29]. Evidently the distance between the two neutrals as they arrive at the detector will depend not only of their relative velocity but also on the distance from the detector at which the reaction took place. The two beams are merged over a distance of 95 cm and the distance to the detector from the center of the interaction region is 160 cm. In order to reach sufficient resolution, it is therefore necessary to run the experiment in such a way that the MN reactions taking place in only a small portion of the merged-beams section can be separated from other MN events. In the case of the $\mathrm{Li}^+/\mathrm{D}^-$ experiment mentioned above, this was achieved by storing the two beams at different energy/mass ratios and then tune in the ions to reach close to zero center-of-mass energy by applying a bias voltage on the three central drift tubes with a combined length of 25 cm. The Li^+ beam was stored at 17.0 keV, the D^- beam at 6.50 keV, and the three central electrodes were put on a bias voltage of -1278 V. In this way the center-of-mass energy was close to 0 eV for the collisions taking place in the biased section and 93 eV for collisions taking place in the non-biased drift tubes and in the pickup section.

Fig. 5 clearly demonstrates the separation of MN events of interest with $E_{CM} \approx 0$ eV taking place in the biased section and yielding events with $\Delta t < 50$ ns from those collisions taking place in the grounded sections before (right-side peak 350-450 ns) and after (peak 230-320 ns) the biased section. For the collisions at $E_{CM} \approx 93$ eV, the change in kinetic energies of the neutral products due to the collision's in-elasticity is relatively insignificant and the products continue with essentially un-changed velocities towards the detector. Therefore the value of Δt is mainly determined by the position along the beams' paths at which the reaction took place. This gives rise to the two-peak structure reflecting the two separated sections at which these higher energy

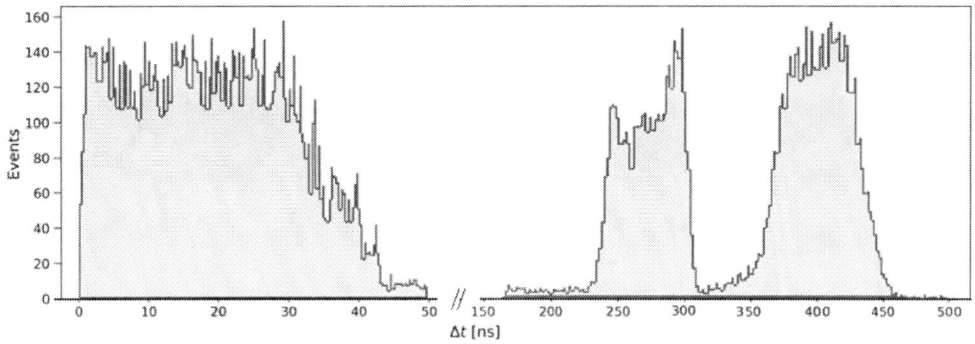

Figure 5. Histogram of time differences for pairs of neutral D and Li atoms formed in mutual neutralisation with the conditions described in the text. The reactions at $E_{CM} \approx 0\,\mathrm{eV}$ taking place in the biased three center electrodes result in $\Delta t < 50\,\mathrm{ns}$, while reactions in the grounded sections with $E_{CM} \approx 93\,\mathrm{eV}$ are easily distinguished through their much larger Δt.

collisions take place. For the biased section, on the other hand, the initial center-of-mass energy is of the order of tens of meV so that the final-state center-of-mass energy is dominated by the energy released in the reaction (1.26 eV for capture to $3s$ and 0.80 eV for capture to $3p$). For these events the time difference is primarily determined by the amount of kinetic-energy release and the random isotropic orientation of the interatomic vector after the collision. The motivation for studying this particular collision system comes from astrophysics. The mutual neutralisation process for metal ions and H^- in stellar atmospheres affects the line intensities in observations and a proper account of these processes is therefore essential for quantitative conclusions concerning the abundance of metals in stars. This triggered renewed theoretical efforts [30] and the soon to be published results from DESIREE will provide an experimental test of these calculations.

7. Summary and conclusions

During its first few years of operation, DESIREE has been used to study anions and their relaxation for a multitude of different systems. Atomic lifetimes have been determined, metal cluster decays have been investigated and small molecular ions have been shown to approach thermal equilibrium with the cryogenic surroundings. All these results have been made possible by the long storage lifetime and low rate of residual-gas collisions in the cryogenic extreme vacuum system. Recently, the first results of the studies of mutual neutralisation with ion beams stored in the two rings of DESIREE have been demonstrated. Also for these measurements the extreme vacuum is important to limit the background from residual-gas collisions even though such problems have been elegantly circumvented in a single-pass room-temperature experiment [31]. Experiments are now in the planning phase where DESIREE's full potential will be reached when ro-vibrationally relaxed molecular ions are being brought to interact, thereby combining the merged-beams capability of DESIREE with the long-time storage and the cryogenic conditions. To the best of our knowledge, DESIREE is the only available tool for studies of reactions between ro-vibrationally relaxed molecular ions at low and well-controlled center-of-mass energies.

Acknowledgments

This work was performed at the Swedish National Infrastructure, DESIREE (Swedish Research Council contract no. 2017-00621). HTS, HC, DH and HZ acknowledge their individual research grants (2018-04092, 2015-04990, 2016-03650 and 2016-04181) from the Swedish Research Council. MG, PSB, ÅL, RDT, HZ, HTS and HC acknowledge the support from the Knut and Alice Wallenberg Foundation (contract no. 2018.0028).

References

[1] Møller S P 1997 *Nucl. Instrum. Meth. Phys. Res.* A **394**, 281
[2] Zajfman D *et al.* 1997 *Phys. Rev.* A **55**, R1577
[3] Schmidt H T *et al.* 2001 *Nucl. Instrum. Meth. Phys. Res.* B **173**, 523
[4] Fischer D *et al.* 2010 *Phys. Rev.* A **81**, 012714
[5] Schmidt H T *et al.* 2013 *Rev. Sci. Instrum.* **84**, 055115
[6] Thomas R D *et al.* 2011 *Rev. Sci. Instrum.* **82**, 065112
[7] von Hahn R *et al.* 2016 *Rev. Sci. Instrum.* **87**, 063115
[8] Nakano Y *et al.* 2017 *Rev. Sci. Instrum.* **88**, 033110
[9] Bäckström E *et al.* 2015 *Phys. Rev. Lett.* **114**, 143003
[10] Andersen T *et al.* 1993 *Phys. Rev.* A **47**, 890
[11] Wolf A *et al.* 1999 *Phys. Rev. A* **59**, 267
[12] Pedersen U V *et al.* 2001 *Phys. Rev.* A **64**, 012503
[13] Reinhed P *et al.* 2009 *Phys. Rev. Lett.* **103**, 213002
[14] Schmidt H T *et al.* 2012 *J. Phys.: Conf. Ser.* **388**, 012006
[15] Andersson P *et al.* 2006 *Phys. Rev.* A **73**, 032705
[16] Kamińska M *et al.* 2016 *Phys. Rev.* A **93**, 012512
[17] Chartkunchand K C *et al.* 2016 *Phys. Rev.* A **94**, 032501
[18] Brage T and Grumer J 2017 *J. Phys.* B: *At. Mol. Opt. Phys.* **50**, 025001
[19] Si R *et al.* 2017 *Phys. Rev.* A **95**, 042504
[20] Meyer C *et al.* 2017 *Phys. Rev. Lett.* **119**, 023202
[21] Schmidt H T *et al.* 2017 *Phys. Rev. Lett.* **119**, 073001
[22] Otto R *et al.* 2013 *Phys. Chem. Chem. Phys.* **15**, 612
[23] Hansen K *et al.* 2001 *Phys. Rev. Lett.* **87**, 123401
[24] Menk S *et al.* 2014 *Phys. Rev.* A **89**, 022502
[25] Anderson E K *et al.* 2018 *Phys. Rev.* A **98**, 022705
[26] Hansen K *et al.* 2017 *Phys. Rev.* A **95**, 022511
[27] Fedor J *et al.* 2005 *Phys. Rev. Lett.* **94**, 113201
[28] Anderson E K *et al.* 2019 *Manuscript in preparation*
[29] Eklund G, Grumer J, Barklem P S *et al.* 2019 *Manuscript in preparation*
[30] Barklem P S *et al.* 2003 *Astron. Astrophys.* **409**, L1
[31] de Ruette N *et al.* 2018 *Phys. Rev. Lett.* **121**, 083401

Projectile Coherence Effects in Simple Atomic Systems

M Schulz[1], A Hasan[2], B Lamichhane[1], T Arthanayaka[1], M Dhital[1], S Bastola[1], L Nagy[3], S Borbély[3], and F Járai-Szabó[3]

[1] Physics Department and LAMOR, Missouri University of Science & Technology, Rolla, MO 65409, USA
[2] Dept. of Physics, UAE University, P.O. Box 15551, Al Ain, Abu Dhabi, UAE
[3] Faculty of Physics, Babes-Bolyai University, Kogalniceanu St. 1, 400084 Cluj, Romania

schulz@mst.edu

Abstract. Recent studies of projectile coherence effects in ion-atom collisions are presented. For intermediate-energy proton collisions an extensive literature provides strong support for the importance of such effects. In this regime coherence effects are now used as a tool to study the few-body dynamics very sensitively. In contrast, for high-energy ion impact the literature is much sparser and here an important role of coherence effects cannot be regarded as being established. In this context, a recent claim that in COLTRIMS experiments the coherence properties are determined only by the target beam is rebutted.

1. Introduction

Kinematically complete experiments on atomic collisions between charged particles and simple atoms and molecules have been performed for several decades [1,2] in order to study the fundamentally important few-body problem (FBP). The essence of the FBP is that the Schrödinger equation is not analytically solvable for more than two particles even when the forces are precisely known. Therefore, theory has to resort to numerical modelling efforts and the assumptions entering in these models have to be tested by detailed experimental data. One process that is particularly suitable to study the few-body dynamics is ionization because there the final state involves three unbound particles.

For electron impact a rich literature on kinematically complete experiments on ionization was developed over the last five decades [1,3-10] since the pioneering work of Ehrhardt et al. [11]. A qualitative understanding of the fully differential cross sections (FDCS) extracted from such experiments emerged relatively soon. Usually, the FDCS are dominated by two structures known as the binary and recoil peaks. The binary peak is due to a direct collision between the projectile and the electron and occurs at or near the direction of the momentum transfer **q**. The recoil peak also involves a projectile-electron interaction; however, here the electron subsequently backscatters off its parent nucleus so that it appears at or near the direction of –**q**. At large projectile energies and for light target atoms experimental data could to a large extent be well described by theory even quantitatively [12], however, for small projectile energies it took several decades before a (nearly) complete understanding emerged when several non-perturbative approaches were developed [12-14].

Studies on ion-impact ionization are significantly more challenging, both from an experimental and theoretical point of view. The difficulties are introduced by the much larger projectile mass compared to electron impact. This leads to extremely small scattering angles and energy losses (relative to the initial energy), which can be determined from a direct projectile measurement only for light ions at relatively small projectile energies [e.g. 15,16]. With the advent of cold target recoil-ion momentum spectroscopy (COLTRIMS) [17,18] kinematically complete experiments on ionization became feasible [e.g. 2,15,16,19-24]. Theoretically, the large projectile mass means that an enormous number of angular momenta has to be accounted for to describe the final projectile state. Only relatively recently were non-perturbative models developed [25-27]. However, for large perturbation parameters (projectile charge to speed ratio η) either large discrepancies to experiment were found [25] or the model has not been tested yet by comparison to experiment on a fully differential level [27].

It was very surprising that significant and qualitative discrepancies were found between experiment and theory even at very small η [19], where good agreement with perturbative and non-perturbative calculations was expected and routinely found in the case of electron impact [e.g. 28,29]. Even more surprising, very good agreement was achieved with a rather simple model, based on the first Born approximation (FBA), in which the nucleus-nucleus (NN) interaction (not included in the FBA) was accounted for retroactively by convoluting the FBA with classical Rutherford scattering [30].

This success led to the hypothesis that the problem with fully quantum-mechanical models could be that they all assume a fully coherent projectile beam [31], i.e. that the intrinsic momentum spread is zero. This is not a realistic assumption for fast heavy ions and as a result any interference predicted by theory may not be observable experimentally because fast heavy projectiles tend to be incoherent. In contrast, even very fast electrons usually have a much narrower intrinsic momentum spread due to their much larger deBroglie wavelength. A first indication that indeed measured cross section can be significantly affected by the projectile coherence properties were found in double differential cross sections (DDCS) for ionization of H_2 by 75 keV proton impact [31]. Later, numerous further experimental [e.g. 32-36] and theoretical [e.g. 37-40] studies supported this hypothesis.

In this article some of the most recent studies on the effect of the projectile coherence properties on FDCS are presented. While for intermediate energy proton collisions a large body of data and theoretical analysis lend strong support to the importance of coherence effects, the literature is much sparser for fast and heavy ion impact. In this regime, the hypothesis that the projectile coherence properties could be responsible for large discrepancies to theory is still awaiting ultimate evidence and did not go completely unchallenged. These critical views are also analyzed.

2. Experimental set-up

Kinematically complete experiments were performed on ionization of He and on dissociative capture (DC) from H_2 in collisions with 75 keV protons. Protons were generated with a hot cathode ion source and accelerated by a high voltage platform. The ions are usually generated with a large intrinsic momentum spread, i.e. with a small coherence length. However, in analogy to classical optics, the coherence length increases during the propagation of the projectile wave. Therefore, the coherence length Δr can be manipulated by placing a collimating slit with width a at a distance L from the target and is given by $\Delta r = L\lambda/2a$, where λ is the deBroglie wavelength. The experiments were performed for $a = 150$ μm and for two distances of 50 cm, corresponding to $\Delta r = 3.5$ a.u., and 6.5 cm. For the smaller distance the local collimation angle at the target is not restricted by the collimating slit and here $\Delta r = 1$ a.u., which is larger than $L\lambda/2a$.

The collimated projectile beam was crossed with a very cold target beam from a supersonic jet at a temperature of approximately 1 – 2 K. The recoiling target ions were extracted by a weak electric field of about 5 V/cm (40 V/cm) in the experiment on ionization (DC). The recoil ions (He^+ for ionization and the charged molecular fragment for DC) were detected by a two-dimensional position-sensitive multi-channel-plate (MCP) detector. After the collision, the projectiles were charge-state analyzed by a switching magnet. The beam component not charge-changed in the collision (ionization experiment) was decelerated by 70 keV and then energy-analyzed by an electrostatic parallel plate analyzer [41] and detected by a second MCP detector. For the DC experiment a third MCP detector was mounted at the 0° port of the switching magnet to detect the neutralized projectiles. The recoil-ion detector was set in coincidence with the respective projectile detector.

From the position information from each detector two momentum components of the corresponding particle were obtained. The third momentum component was obtained from the time-of-flight information contained in the coincidence time in the case of the recoil ion and from the energy loss in the case of the scattered protons. The momentum of the electrons ejected in ionization and the neutral molecular fragment in DC was determined from momentum conservation. In the case of projectiles neutralized in DC the third momentum component could not be measured directly. However, due to four conservation laws (energy in addition to momentum) only five of the nine momentum components are independent so that the measurement of 3 recoil-ion and 2 projectile components manifests a kinematically complete experiment. However, the resolution was not sufficient to determine the final state of the captured electron.

3. Results and Discussion

3.1. Ionization

For the analysis of the FDCS for ionization we use a coordinate system defined by the projectile momenta: the positive x-direction is given by the direction of the transverse component of **q** and the z-direction by the initial projectile momentum. This choice implies that q_x is always positive and q_y is always zero (within the experimental resolution). The FDCS were generated for an energy loss of ε = 30 eV, various fixed recoil-ion momentum components in the x-direction ranging from p_{recx} = 0.2 to 1.25 a.u., and fixed polar electron emission angles ranging from θ_{el} = 15° to 85° (measured relative to the projectile beam axis). The data were then analyzed as a function of the azimuthal electron emission angle φ_{el}. Here, φ_{el} = 90° is defined by the direction of q_x and φ_{el} = 270° by the direction of $-q_x$. At these angles, the binary peak and recoil peak, respectively, are expected in a first-order picture.

In analogy to classical optics the FDCS measured for a coherent beam $d\sigma_{coh}$ can be expressed as the one measured for an incoherent beam $d\sigma_{inc}$ multiplied by the interference term *I*. Therefore, if the coherence properties of the projectiles are indeed determined by the distance of the collimating slit to the target, the interference term can be extracted as the ratio R between the FDCS measured for large L relative to small L. These ratios are shown as a function of ϕ_{el} in figure 1 for a subset of the data, where the columns represent data for θ_{el} fixed at (from left to right) 25°, 45°, and 65°, and the rows data for p_{recx} fixed at (from top to bottom) 0.2 a.u., 0.7 a.u., and 1.25 a.u. Note that the regions in which no data are shown are kinematically prohibited because there q_x is negative, which is not possible in our coordinate system. The solid curves show a time-dependent calculation [42,43] in which the projectile is described by a wave packet with a width reflecting the coherence length. R was evaluated between FDCS calculated for a width of 3.5 a.u. and 1.0 a.u., respectively.

Theory predicts that for small p_{recx} and θ_{el} coherence effects are rather weak. Indeed, this prediction is confirmed by experiment, where R is nearly constant with a value close to 1. In contrast, for increasing p_{recx} and θ_{el} increasingly pronounced structures are found in R. Again, this finding is in very good qualitative accord with the experimental data, although there are some quantitative discrepancies. For the other kinematic settings similarly good agreement was found. These structures show that interference is present in the coherent FDCS. Here, different (non-observable) impact parameters leading to the same scattering angle interfere with each other, which is analogous to single-slit interference in classical optics and to which we refer as single-center interference.

Further theoretical analysis showed that the FDCS are very sensitive to the coherence length and that even at Δr = 3.5 a.u., which is substantially larger than typical impact parameters, the collision cannot be viewed as fully coherent [44]. This sensitivity makes the good qualitative agreement between theory and experiment the more significant. Overall, the strong departure of R from unity (except for small p_{recx} and θ_{el}) along with the reproduction of the experimental data by theory strongly support the hypothesis that cross sections for ion-atom collisions can be strongly influenced by the projectile coherence properties.

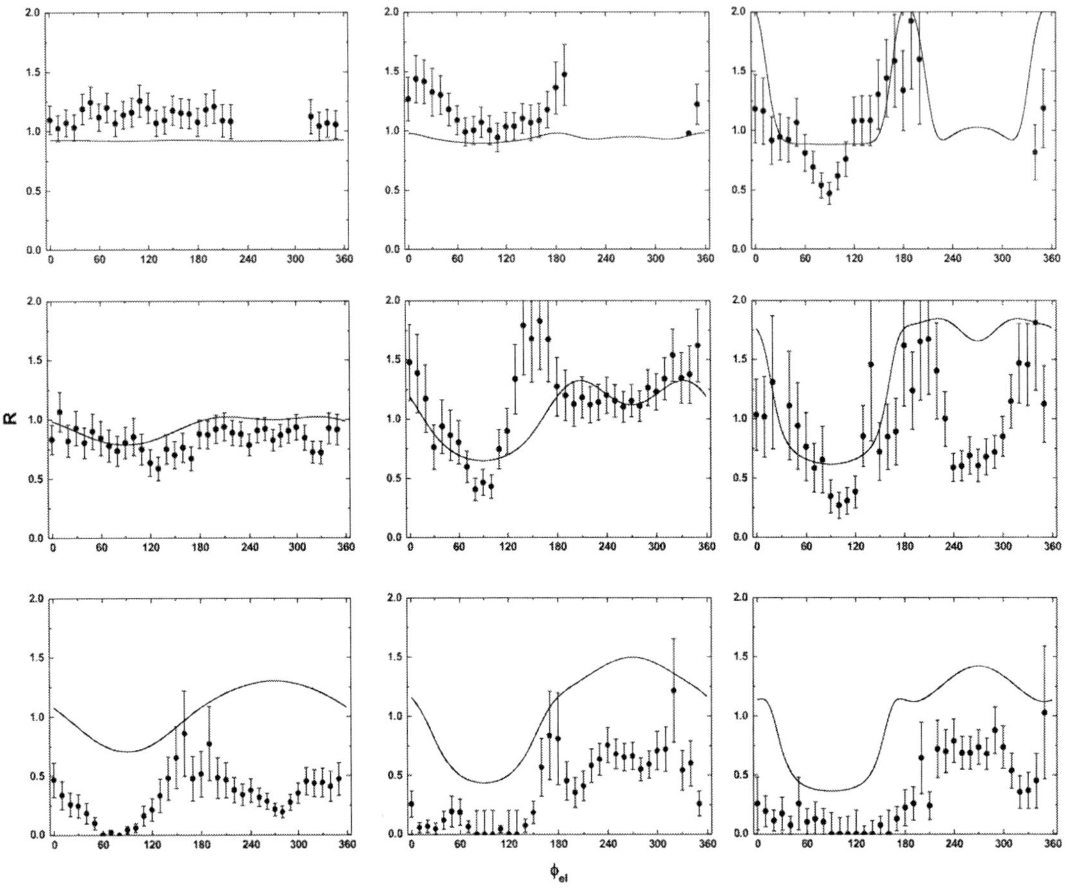

Figure 1. Ratio between fully differential cross sections for ionization of He by 75 keV proton impact measured at large and small distances between the collimating slit and the target as a function of the azimuthal electron ejection angle. The projectile energy loss was fixed at 30 eV. For data within a column θ_{el} was fixed at (from left to right) 25°, 45°, and 65° and for data within a row p_{recx} was fixed at (from top to bottom) 0.2, 0.7, and 1.25 a.u. The solid lines show time-dependent calculations in which the transition amplitude was convoluted with a wave packet describing the projectile with the width representing the transverse coherence length (3.3 and 1.0 a.u., respectively).

3.2. Dissociative Capture

In dissociative capture an electron from the target molecule is captured to the projectile and at the same time the residual molecular ion is excited leading to its fragmentation. There are two fragmentation channels which can be distinguished: in the first, the second electron residing with the molecule gets excited to a repulsive electronic state (i.e. the overall electronic process is transfer plus excitation). This channel leads to kinetic energy releases (KER, the sum kinetic energy of both fragments) ranging from approximately 5 to 25 eV. In the second fragmentation channel, the second electron remains in the ground state of the molecular ion, however, the nuclear motion is excited to a vibrational continuum state. This channel is known as ground state dissociation (GSD). Here the KER is significantly smaller (less than 2 eV). Therefore, the two fragmentation channels can be distinguished based on the KER.

From the momentum components of the molecular fragments extracted from the experiment the KER and the molecular orientation at the instant of the collision can be determined. In the following we will focus on GSD, which was selected by setting a condition on KER < 2 eV in the data analysis. Two molecular orientations were analyzed, both of which were perpendicular to the projectile beam axis. The first was also perpendicular to q_x while the second was parallel to q_x. As in the ionization experiment, the FDCS were measured at large and small L for each orientation and the ratios R between the corresponding $d\sigma_{coh}$ and $d\sigma_{inc}$ were analyzed [45].

The interference term for molecular two-center interference is given by

$$I_2 = 1 + \cos(\mathbf{p}_{rec} \cdot \mathbf{D}) = 1 + \cos(\mathbf{q} \cdot \mathbf{D}), \quad (1)$$

where **D** is the internuclear separation vector of the molecule. In this expression the dot product **q . D** is constant at 0 for the perpendicular orientation. Therefore, the ratio R for this orientation reflects single-center interference. For the parallel orientation the total interference term results from a combination of single- and two-center interference. Under the assumption that both are independent of each other we extracted the two-center interference term I_2 by dividing the ratio R by the single-center interference term obtained from the data for the perpendicular orientation.

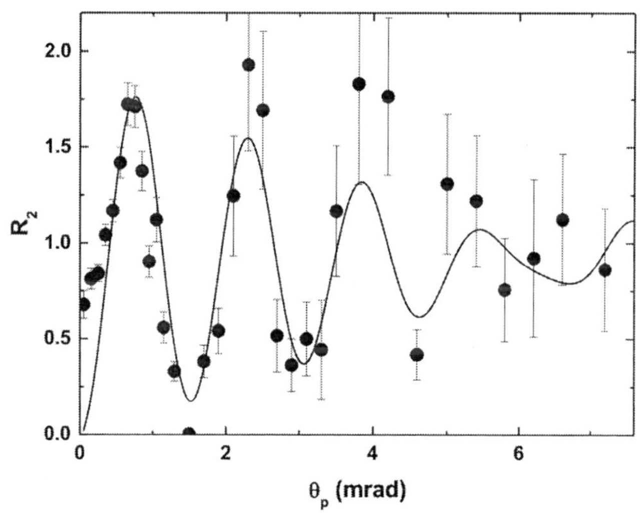

Figure 2. Ratio between fully differential cross sections for dissociative capture in 75 keV p + H_2 collisions measured at large and small distances between the collimating slit and the target as a function of the projectile scattering angle. The KER was fixed at 0 to 2 eV and the molecular orientation was perpendicular to the projectile beam axis and parallel to the transverse momentum transfer. The solid line represents equation (1) with a phase shift of π added to the argument of the cosine function.

The experimentally determined two-center interference term R_2 is plotted in figure 2 as a function of projectile scattering angle θ_p. A pronounced oscillating pattern can clearly be seen. For two reasons these data are particularly significant relative to the question whether the departure from R = 1 is a manifestation of coherence effects or due to experimental artifacts. First, in contrast to the ionization experiment the measurements for the large and small slit distance were performed simultaneously in the same experiment. This was accomplished by placing the x-slit at a small L and the y-slit at a large L and then selecting scattering in the x- and y-direction in the data analysis to obtain the incoherent and coherent FDCS, respectively. Thereby, the data for small and large L were taken under otherwise identical experimental conditions, i.e. differences between them cannot be explained by experimental artifacts. Second, the oscillation extrema in R_2 occur at exactly the same θ_p as in the absolute FDCS. This makes it very unlikely that the structures in R_2 are just artificial, which could result for example from inaccuracies in the calibration of θ_p affecting scattering in the x- and y-directions differently.

The two-center interference term of equation (1) predicts maxima where minima are found in the experimental data and vice versa. However, if a phase shift of π is added to the argument of the cosine function, I_2 (solid line in figure 2) is in nearly perfect agreement with the data. Such a phase shift was observed previously in electron capture to the dissociative $2p\sigma_u$ state of the projectile in H_2^+ + He collisions [46] and in dissociative ionization of H_2 by electron impact [47]. In [46] this shift was convincingly explained as a consequence of parity conservation: the switch in symmetry in the electronic wave function (from gerade in the ground state of He to ungerade in the final molecular state) has to be compensated by a corresponding switch in symmetry in the projectile wave function. However, the existing literature suggests that there are other factors which also need to be considered to fully understand the phase shift: Schmidt et al studied dissociative transfer excitation in p + H_2 collisions [48]. Although they also considered a case where the symmetry of the electronic wave function switches, they did not find any phase shift in the interference term. On the other hand, no switch in symmetry of the electronic wave function is involved neither in the present study nor in the one reported in [47], but yet in both cases a phase shift of π was found. At present, no explanation can be offered as to which factors apart from parity conservation may decide whether or not a phase shift occurs. However, we note that so far a phase shift was only observed in dissociative processes.

3.3. Coherence in Fast Collisions

For intermediate collision energies the presence of projectile coherence effects is supported by a large body of experimental data [e.g. 31,32,35,36,42,45]. However, for fast ion impact the literature is much sparser [33,34]. The aforementioned discrepancies between theory and experiment in FDCS for ionization in 1.2 GeV C^{6+} + He collisions were blamed on the very small transverse coherence length of the projectiles [31] and indeed data obtained for p + He collisions at the same perturbation and a coherence length larger by orders of magnitude were much better reproduced by theory [33]. On the other hand for somewhat slower proton collisions no significant differences between the FDCS for two different transverse coherence lengths were found [24]. However, there even the smaller coherence length was orders of magnitude larger than for 1.2 GeV C^{6+} impact and larger than the size of the target atom[1]. Under these conditions only weak coherence effects are expected [38] and the results of [24] are thus not inconsistent with the interpretation that the data for 1.2 GeV collisions are significantly affected by projectile coherence effects. Furthermore, the comparison between FDCS for ionization and transfer-ionization in 16 MeV O^{7+} + He also provided some strong indications for pronounced coherence effects [34].

Nevertheless, the interpretation that the failure of theory to reproduce measured FDCS for ionization in 1.2 GeV C^{6+} + He collisions is due to coherence effects was challenged by Kouzakov [49]. He argued that in an experiment where only the recoil ions and the ejected electrons are momentum-analyzed the FDCS can only be affected by the coherence properties of the target beam, but not the projectile beam. In the following, we argue that this conclusion is not consistent with what is generally known for coherence and interference effects.

[1] In the original publication the coherence length was calculated incorrectly. The correct value is at least 1.3 a.u. instead of 0.8 a.u.

First, it should be noted that the coherence properties of the collisions system are determined by the spread in the relative momenta of the two collision partners. The collision is coherent if this spread is narrow and incoherent if it is broad. But if the momentum distribution is broad for one of the collision partners, it must be broad for the relative momentum distribution because if an essentially constant value is added to a broad distribution the resulting distribution remains broad. Therefore, for interference structures to be observable (i.e. for the collision system to be coherent) *both* beams must be coherent. This is true regardless of which particle is actually momentum-analyzed in the experiment. To illustrate this important point consider the following scenario: suppose an experiment is performed in which both the projectile and the recoil-ion (and the ejected electrons) are momentum-analyzed and the coherence length of the projectiles is very small. In this scenario the analysis of Kouzakov yields that due to the incoherence of the projectiles (and therefore the collision system) no interference is observable. Then the experiment is repeated, but this time the projectiles are not detected. However, due to the constraints imposed by the conservation laws the experiment is nevertheless kinematically complete, i.e. exactly the same final state is observed in the experiment. In this case the analysis of Kouzakov yields that interference becomes observable. However, not detecting the projectiles obviously has no effect at all on the projectile coherence length, i.e. the collision system is still incoherent and no interference can be observed, contrary to the conclusion of Kouzakov. To put it in casual language: it is not possible to create an interference structure out of nothing simply by not looking at any incoherent collision partner in the experiment.

Finally, we note that this analysis demonstrates one important difference between incoherence and beam divergence, which constitutes part of the experimental angular resolution. Both are theoretically often treated by the same mathematical approach, namely by convoluting the transition amplitude for the coherent case with the momentum spread (either due to incoherence or experimental resolution) of the collision partners [e.g. 38-40,50]. However, the beam divergence will not have any effect on measured cross sections if the corresponding particle is not even detected, in which case it should not enter in the convolution. In contrast, the coherence properties are completely independent on whether or not the corresponding particles are detected and thus always need to enter in the convolution.

4. Conclusions

Over the last decade numerous studies have been performed demonstrating that the projectile coherence properties can have a major effect on measured cross sections, not because of imperfections in the experiment, but due to the inherent properties of the projectile wavefunction. This important point was overlooked for decades of scattering theory. The reason for this oversight is probably that FDCS on ionization were initially measured for electron impact, where the coherence properties do not play an important role. When such measurements became possible for ion impact, it seemed reasonable to apply the same theoretical techniques that worked well for electron impact to ion impact.

With the importance of coherence effects uncovered the literature on ion-atom collision studies should be revisited for previously unexplained discrepancies between experiment and theory. While it is unlikely that the coherence properties can offer a complete explanation for all of these cases, they may provide at least partial explanations in some cases. Furthermore, coherence effects can be used as a tool to study the few-body dynamics in collisions in greater detail. For example, the dependence of cross sections on the projectile scattering angle tends to be very steep. As a result, an interference structure superimposed on this steep dependence may be very difficult to identify. However, if the cross sections are measured for a coherent and an incoherent beam the interference term can be extracted as the ratio between these cross sections, in which the steep scattering angle dependence is removed. Thereby, a much more accurate analysis of the interference term is possible.

For fast ion impact the role of coherence effects is not completely settled yet. Although the claim that only the coherence properties of the target beam matter if the projectiles are not detected is unsubstantiated, this does not mean that the interpretation that discrepancies between experiment and theory are due to the coherence properties can be regarded as established. Further studies are needed, both theoretical and experimental.

5. Acknowledgements
This work was supported by the National Science Foundation under grant no. PHY-1703109.

References

[1] H. Ehrhardt, K. Jung, G. Knoth, and P. Schlemmer, Z. Phys. D 1, 3 (1986)
[2] M. Schulz and D.H. Madison, Intern. J. Mod. Phys. A 21, 3649 (2006)
[3] G. Stefani, L. Avaldi, and R. Camilloni, J. Phys. B 23, L227 (1990)
[4] A. Lahmann-Bennani, J. Phys. B 24, 2401 (1991)
[5] A.J. Murray and F.H. Read, Phys. Rev. Lett. 69, 2912 (1992)
[6] A. Dorn, R. Moshammer, C.D. Schröter, T.J.M. Zouros, W. Schmitt, H. Kollmus, R. Mann, and J. Ullrich, Phys. Rev. Lett. 82, 2496 (1999)
[7] J.F. Williams, P.L. Bartlett, and A.T. Stelbovics, Phys. Rev. Lett. 96, 123201 (2006)
[8] J. Röder, H. Ehrhardt, C. Pan, A.F. Starace, I. Bray, and D.V. Fursa, Phys. Rev. Lett. 79, 1666 (1997)
[9] D.S. Milne-Brownlie, M. Foster, J. Gao, B. Lohmann, and D.H. Madison, Phys. Rev. Lett. 96, 233201 (2006)
[10] X. Ren, S. Amami, O. Zatsarinny, T. Pflüger, M. Weyland, W.Y. Baek, H. Rabus, K. Bartschat, D.H. Madison, and A. Dorn, Phys. Rev. A 91, 032707 (2015)
[11] H. Ehrhardt, M. Schulz, T. Tekaat, and K. Willmann, Phys. Rev. Lett. 22, 89 (1969)
[12] T. Rescigno, M. Baertschy, W.A. Isaacs, and C.E. McCurdy, Science 286, 2474 (1999) [13] X. Ren, I. Bray, D.V. Fursa, J. Colgan, M.S. Pindzola, T. Pflüger, A. Senftleben, S. Xu, A. Dorn, and J. Ullrich, Phys. Rev. A 83, 052711 (2011)
[14] J. Colgan, M. Foster, M.S. Pindzola, I. Bray, A.T. Stelbovis, and D.V. Fursa, J. Phys. B 42, 145002 (2009)
[15] N.V. Maydanyuk, A. Hasan, M. Foster, B. Tooke, E. Nanni, D.H. Madison, and M. Schulz, Phys. Rev. Lett. 94, 243201 (2005)
[16] A. Hasan, S. Sharma, T.P. Arthanayaka, B.R. Lamichhane, J. Remolina, S. Akula, D.H. Madison, and M. Schulz, J. Phys. B 47, 215201 (2014)
[17] R. Dörner, V. Mergel, O. Jagutzski, L. Spielberger, J. Ullrich, R. Moshammer, and H. Schmidt-Böcking, Phys. Rep. 330, 95 (2000)
[18] J. Ullrich, R. Moshammer, A. Dorn, R. Dörner, L. Schmidt, and H. Schmidt-Böcking, Rep. Prog. Phys. 66, 1463 (2003)
[19] M. Schulz, R. Moshammer, D. Fischer, H. Kollmus, D.H. Madison, S. Jones, and J. Ullrich, Nature 422, 48 (2003)
[20] M. Schulz, R. Moshammer, A.N. Perumal, and J. Ullrich, J. Phys. B 35, L161 (2002)
[21] M. Schulz, R. Moshammer, A. Voitkiv, B. Najjari, and J. Ullrich, Nucl. Instrum. Meth. B 235, 296 (2005)
[22] M. Schulz, B. Najjari, A.B. Voitkiv, K. Schneider, X. Wang, A.C. Laforge, R. Hubele, J. Goullon, N. Ferreira, A. Kelkar, M. Grieser, R. Moshammer, J. Ullrich, and D. Fischer, Phys. Rev. A 88, 022704 (2013)
[23] R. Hubele, A.C. Laforge, M. Schulz, J. Goullon, X. Wang, B. Najjari, N. Ferreira, M. Grieser, V.L.B. de Jesus, R. Moshammer, K. Schneider, A.B. Voitkiv, and D. Fischer, Phys. Rev. Lett. 110, 133201 (2013)
[24] H. Gassert, O. Chuluunbaatar, M. Waitz, F. Trinter, H.-K. Kim, T. Bauer, A. Laucke, Ch. Müller, J. Voigtsberger, M. Weller, J. Rist, M. Pitzer, S. Zeller, T. Jahnke, L. Ph. H. Schmidt, J. B. Williams, S. A. Zaytsev, A. A. Bulychev, K. A. Kouzakov, H. Schmidt-Böcking, R. Dörner, Yu. V. Popov, and M. S. Schöffler, Phys. Rev. Lett. 116, 073201 (2016)
[25] M. McGovern, C.T. Whelan, and H.R.J. Walters, Phys. Rev. A 82, 032702 (2010)
[26] M.F. Ciappina, M.S. Pindzola, and J. Colgan, Phys. Rev. A 87, 042706 (2013)
[27] I.B. Abdurakhmanov, J.J. Bailey, A.S. Kadyrov, and I. Bray,Phys. Rev. A 97, 032707 (2018)
[28] P. Marchalant, C.T. Whelan, and H.R.J. Walters, J. Phys. B 31, 1141 (1998)

[29] S. Jones and D.H. Madison, Phys. Rev. A 62, 042701 (2000)
[30] M. Schulz, M. Dürr, B. Najjari, R. Moshammer, J. Ullrich, Phys. Rev. A 76, 032712 (2007)
[31] K.N. Egodapitiya, S. Sharma, A. Hasan, A.C. Laforge, D.H. Madison, R. Moshammer, and M. Schulz, Phys. Rev. Lett. 106, 153202 (2011)
[32] S. Sharma, A. Hasan, K.N. Egodapitiya, T.P. Arthanayaka, and M. Schulz, Phys. Rev. A 86, 022706 (2012)
[33] X. Wang, K. Schneider, A. LaForge, A. Kelkar, M. Grieser, R. Moshammer, J. Ullrich, M. Schulz, and D. Fischer, J. Phys. B 45, 211001 (2012)
[34] K. Schneider, M. Schulz, X. Wang, A. Kelkar, M. Grieser, C. Krantz, J. Ullrich, R. Moshammer, and D. Fischer, Phys. Rev. Lett. 110, 113201 (2013)
[35] S. Sharma, T.P. Arthanayaka, A. Hasan, B.R. Lamichhane, J. Remolina, A. Smith, and M. Schulz, Phys. Rev. A 90, 052710 (2014)
[36] T.P. Arthanayaka, S. Sharma, B.R. Lamichhane, A. Hasan, J. Remolina, S. Gurung, and M. Schulz, J. Phys. B 48, 071001 (2015)
[37] F. Jarai-Szabo and L. Nagy, Eur. Phys. J. D 69, 4 (2015)
[38] I. Fabre, F. Navarrete, L. Sarkadi, and R.O. Barrachina, Eur. J. Phys. 39, 015401 (2018)
[39] D.V. Karlovets, G.L. Kotkin, and V.G. Serbo, Phys. Rev. A 92, 052703 (2015)
[40] L. Sarkadi and R. Barrachina, Phys. Rev. A 93, 032702 (2016)
[41] A.D. Gaus, W. Htwe, J.A. Brand, T.J. Gay, and M. Schulz, Rev. Sci. Instrum. 65, 3739 (1994)
[42] T.P. Arthanayaka, B.R. Lamichhane, A. Hasan, S. Gurung, J. Remolina, S. Borbély, F. Járai-Szabó, L. Nagy, and M. Schulz, J. Phys. B 49, 13LT02 (2016)
[43] S. Borbély, J. Feist, K. Tőkési, S. Nagele, L. Nagy and J. Burgdörfer, Phys. Rev. A 90 052706 (2014)
[44] L. Nagy, F. Jarai-Szabo, S. Borbely, T. Arthanayaka, B.R. Lamichhane, A. Hasan, and M. Schulz in "Interdisciplinary Research on Particle Collisions and Quantitative Spectroscopy", World Scientific Publishing, Singapore, Dz. Belkic (ed.), in press (2019)
[45] B.R. Lamichhane, T. Arthanayaka, J. Remolina, A. Hasan, M.F. Ciappina, F. Navarrete, R.O. Barrachina, R.A. Lomsadze, and M. Schulz, Phys. Rev. Lett. 119, 083402 (2017)
[46] L. Ph. H. Schmidt, S. Schössler, F. Afaneh, M. Schöffler, K. E. Stiebing, H. Schmidt-Böcking, and R. Dörner, Phys. Rev. Lett. 101, 173202 (2008)
[47] A. Senftleben, T. Pflüger, X. Ren, O. Al-Hagan, B. Najjari, D. Madison, A. Dorn, and J. Ullrich, J. Phys. B 43, 081002 (2010)
[48] H.T. Schmidt, D. Fischer, Z. Berenyi, C.L. Cocke, M. Gudmundsson, N. Haag, H.A.B. Johansson, A. Källberg, S.B. Levin, P. Reinhed, U. Sassenberg, R. Schuch, A. Simonsson, K. Støchkel, and H. Cederquist, Phys. Rev. Lett. 101, 083201 (2008)
[49] K.A. Kouzakov, Eur. Phys. J. D 71, 63 (2017)
[50] J. Fiol, S. Otranto, and R. E. Olson, J. Phys. B 39, L285 (2006)

An introduction to Bayesian statistics for atomic physicists

M. Trassinelli

Institut des NanoSciences de Paris, INSP, CNRS, Sorbonne Université, F-75005 Paris, France

E-mail: martino.trassinelli@insp.jussieu.fr

Abstract. We present an introduction to some concepts of Bayesian data analysis in the context of atomic physics. Starting from basic rules of probability, we present the Bayes' theorem and its applications. In particular we discuss about the limits of classic statistics methods and in which cases a Bayesian analysis is mandatory. Moreover, we show how probability values can be assigned to different possible models (different numbers of peaks, type peak profile, etc.) from the analysis of experimental data.

1. Introduction

The standard analysis of atomic spectra consist on comparing n pairs $\{x_i, y_i\}$ of measured values (or $\{x_i, y_i, \sigma_i\}$ triplets if error bars are available) to the values of a function $F(x_i, \boldsymbol{a})$ used to model the data, where $\boldsymbol{a}$ are the function parameters. The common practice to determine the set of parameters $\boldsymbol{a}^{best}$ that describes at best the data is to find the maximum of the likelihood function

$$L(\boldsymbol{a}) = P(\{x_i, y_i, \sigma_i\}|\boldsymbol{a}) = \prod_{i=1}^{n} p(x_i, y_i, \sigma_i|\boldsymbol{a}), \tag{1}$$

where $p(x_i, y_i, \sigma_i|\boldsymbol{a})$ are the single conditional probability distribution for each channel to have x_i, y_i, σ_i for a given parameter values $\boldsymbol{a}$. If the data are normally distributed for each channel, we have that $p(x_i, y_i, \sigma_i|\boldsymbol{a}) = \frac{1}{\sqrt{2\pi}\sigma_i} \exp\left[-\frac{(y_i - F(x_i,\boldsymbol{a}))^2}{2\sigma_i^2}\right]$ and the equation above becomes

$$L(\boldsymbol{a}) = \prod_{i=1}^{n} \frac{1}{\sqrt{2\pi}\sigma_i} \exp\left(-\frac{[y_i - F(x_i, \boldsymbol{a})]^2}{2\sigma_i^2}\right) \propto \exp\left(-\sum_{i=1}^{n} \frac{[y_i - F(x_i, \boldsymbol{a})]^2}{2\sigma_i^2}\right). \tag{2}$$

Its maximization corresponds then to minimization of the function $\chi^2 = \sum_{i=1}^{n} [y_i - F(x_i, \boldsymbol{a})]^2/\sigma_i^2$ i.e. the chi-square function.

The maximum likelihood and chi-square minimization methods we just recalled are in mostly of the cases well adapted and sufficient for a correct data analysis. However they have some limitations that one should be aware:

i) Cause-effect inversion: The maximum likelihood method is based on the probability $L(\boldsymbol{a}) = P(\{x_i, y_i, \sigma_i\}|\boldsymbol{a})$ to obtain certain data values for a given set of parameters $\boldsymbol{a}$. But the final goal is just the opposite: for a given experimental data, one would like to evaluate the probability for having certain parameter values.

ii) Constraints on the parameter values: Hard constraints on parameter values (eg. a mass that cannot be negative) or simply prior parameter distribution from previous experimental results cannot be taken into account in a coherent way.

iii) Assignment of probability to models: If several models (hypotheses) can describe the experimental results, it is not possible assign probabilities to them from the analysis of the data set.

A remark on the last point, several classical statistics criteria are available to choose the most plausible model, like the χ^2-test, the likelihood-ratio test, etc. [1–6]. However, in the unfortunate case where there is no a clear propensity to a specific model and we are interested on a parameter common to all models (as the position of the a peak with undefined shape), the value of the most probable of the parameter of interest (and its associated standard deviation) cannot be extracted with classical statistics criteria.

To overcome these problems, a different approach has to be implemented with a new and more general definition of probability based on propositional logic. The probability is here a positive number that indicates the propensity that a certain proposition X is true, where X can indicates a numerical value, eg. "the peak position is between x_1 and x_2", or a more abstract concept, eg. "the model $\mathcal{M}$ correctly describe the data". Because of this general formulation, a probability function defined in this way can be applied to many different cases. In particular, as we will show in the next section, it can simply solve the three main problems of commonly used classic statistics methods (also called *frequentist* because of the definition of probability based on the repetitively of tries) mentioned above.

At present, Bayesian methods are routinely used in many fields: cosmology [7–9], particle physics [10], machine learning [11], ... They start to be applied only in recent years to nuclear physics [12] and there are very few applications in atomic physics (see eg. Refs. [13–15]) with almost no use in atomic spectroscopy. In the present article we would like to fill this gap and provide a short introduction of Bayesian data analysis methods in the context of atomic physics. A general and more extended introduction and application to physics cases can be found elsewhere [16–19].

2. Probability definition and Bayes' theorem

2.1. A general definition of probability

When basic logic and consistency are required, the probability function defined by propositions $X, Y, \ldots$ has to respect some basic properties [16, 20–22] (which are compatible to Kolmogorov probability axioms [23]):

$$0 \leq P(X|I) \leq 1, \tag{3a}$$
$$P(X|X, I) = 1, \tag{3b}$$
$$P(X + Y|I) = P(X|I) + P(Y|I) - P(X, Y|I), \tag{3c}$$
$$P(X, Y|I) = P(X|Y, I) \times P(Y|I) = P(Y|X, I) \times P(X|I). \tag{3d}$$

Here the probability $P(X|Y, I)$ means that X is true knowing that ("|") that Y is true and were I represents the current state of knowledge (knowledge of well tested physics laws as eg.). The joint probability $P(X, Y|I)$ means that both "X AND Y" are true (equivalent to the logical conjunction '$\wedge$'). The symbol '+' means the logical disjunction ($X + Y \equiv X \vee Y \equiv$ "X OR Y is true"). When a set of mutual exclusive assertions are considered $\{Y_i\}$, with $P(Y_i|Y_{j\neq i}) = 0$, a consequence of Eqs. (3) is the so-called *marginalization rule*

$$P(X|I) = \sum_i P(X, Y_i|I) = P(X|I) = \int_{-\infty}^{\infty} P(X, Y|I) dY. \tag{4}$$

where the last term is valid in the limit of continuous case $Y_{i+1} - Y_i \to dY$.

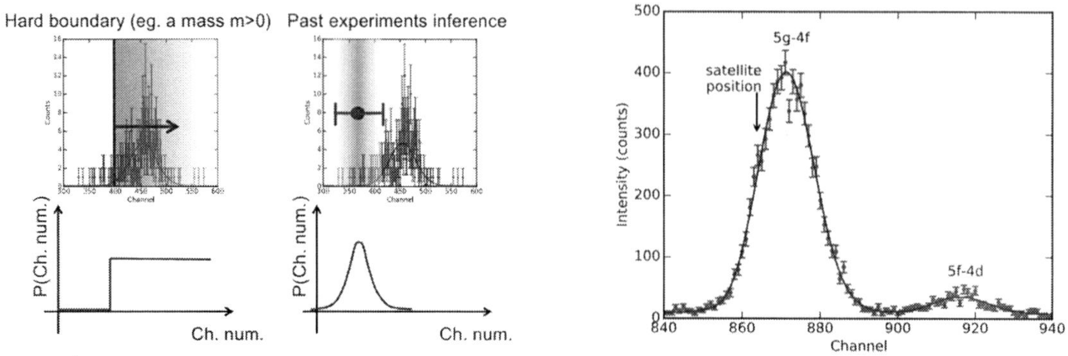

Figure 1. Left: Graphical representation of hard boundary from physical constraints (eg. a mass value that has to be positive and inference from past experiments in a given spectrum. In both cases, these constraints can be considered via the prior probability $P(a|I)$ of the different model parameter values (plots at the bottom). Right: Pionic nitrogen $5-4$ transitions. Possible additional transitions from the presence of one remaining electron in the K shell are indicated.

2.2. Bayes' theorem and posterior probability

An important corollary of Eq. (3d) the so-called called the Bayes' theorem

$$P(X|Y,I) = \frac{P(Y|X,I) \times P(X|I)}{P(Y|I)}, \tag{5}$$

named after Rev. Thomas Bayes, who first formulated theorems of conditional probability before 1761 [24], and rediscovered in 1774 and further developed by Pierre-Simon Marquis de la Laplace [25].

For a better insight in the implication of this theorem, we consider the particular case where X represent the hypothesis that the parameter values set $\boldsymbol{a}$ truly describes the data (via the function $F(x,\boldsymbol{a})$) and where Y correspond to the recorded data $\{x_i, y_i\}$. In this case, we have that the posterior probability of a certain set of parameter set is

$$P(\boldsymbol{a}|\{x_i,y_i\},I) = \frac{P(\{x_i,y_i\}|\boldsymbol{a},I) \times P(\boldsymbol{a}|I)}{P(\{x_i,y_i\}|I)} = \frac{L(\boldsymbol{a}) \times P(\boldsymbol{a}|I)}{P(\{x_i,y_i\}|I)}, \tag{6}$$

where I includes our available background information and where $P(\{x_i,y_i\}|\boldsymbol{a},I)$ is by definition the likelihood function $L(\boldsymbol{a})$ for the given set of data. In addition to the likelihood function, we have here the term $P(\boldsymbol{a}|I)$ that includes the prior probability on the parameters $\boldsymbol{a}$, which can includes possible boundaries. With the above formula, we solve in fact at once the two problems i) and ii) discussed in the introduction. As suggested in Fig. 1 (left), if we have hard boundaries a_{min}, a_{max}, $P(a|I) = 1/(a_{max} - a_{min})$ is considered. In the case of inference of previous measurement with result $a' \pm \sigma_{a'}$, the prior parameter probability is $P(a|I) = \frac{1}{\sqrt{2\pi}\sigma_{a'}} \exp\left[-\frac{(a-a')^2}{2\sigma_{a'}^2}\right]$.

2.3. Model testing and Bayesian evidence

An important consequence of the Bayes' theorem is the possibility to assign probabilities to different hypotheses (models) with a simple and well-defined procedure. In this case, X in Eq. (5) represents the hypothesis that the model $\mathcal{M}$ describes well the measured data and Y

represents the data themselves, as in the previous section. In particular case of atomic spectra, different hypotheses could correspond to different numbers of spectral component, presence of satellite lines or not, or different profiles modeling (Gaussian, Lorenzian, Voigt, ...). From Bayes' theorem, we have that the posterior probability of the model $\mathcal{M}$ is [8, 16, 21]

$$P(\mathcal{M}|\{x_i, y_i\}, I) \propto P(\{x_i, y_i\}|\mathcal{M}, I) \times P(\mathcal{M}|I), \tag{7}$$

where the first term of the right part is the so-called *Bayesian evidence* E of the model (also called *marginal likelihood* or *model likelihood*) and the second term is the prior probability assigned to the model from our background knowledge. Using the marginalization rule (Eq. (4)) and the probability properties (Eqs. (3a–3d)), we have

$$E(\mathcal{M}) \equiv P(\{x_i, y_i\}|\mathcal{M}, I) = \int P(\{x_i, y_i\}|\boldsymbol{a}, \mathcal{M}, I)P(\boldsymbol{a}|\mathcal{M}, I)d^N\boldsymbol{a} = \int L^{\mathcal{M}}(\boldsymbol{a})P(\boldsymbol{a}|\mathcal{M}, I)d^N\boldsymbol{a}, \tag{8}$$

where N is the number of the parameters of the considered model, and where we explicitly indicate the dependency of the likelihood function $L^{\mathcal{M}}(\boldsymbol{a})$ on the related model $\mathcal{M}$.

Considering equal priors, the probability of a model is higher if the evidence $E(\mathcal{M})$ is higher, which means that the likelihood function integral over the model parameter space is higher. This does not implies that the maximum of the likelihood function is larger. Considering maximal values of $L^{\mathcal{M}}(\boldsymbol{a})$ only can in fact introduce errors in the analysis, as recently discussed in the particular case of nuclear physics in Ref. [12]. Model selection criteria based on it, like the likelihood-ratio test and the Akaike information criterion, should be use then with precautions.

Probabilities evaluated from Eq. (7) are generally lower for models with higher number of parameters because of the higher dimensionality of the integral that corresponds to a larger parameter volume (and then to a lower average value of the likelihood function). This means that when the values of the likelihood function are similar, simple models (with a small number of parameters) are favored, accordingly to Ockham's razor principle.

The possibility to assigning probabilities to models has another important advantage. In the case we are interested to determine the probability distribution of a common parameter a_j without the need to identify the most probable model among the available choices $\mathcal{M}_\ell$, we can obtain the probability distribution $P(a_j|\{x_i, y_i\}, I)$ from the weighted sum

$$P(a_j|\{x_i, y_i\}, I) = \sum_\ell P(a_j|\{x_i, y_i\}, \mathcal{M}_\ell, I) \times P(\mathcal{M}_\ell|, I), \tag{9}$$

where $P(a_j|\{x_i, y_i\}, \mathcal{M}_\ell, I)$ are the probability distributions of a_j for each model and $P(\mathcal{M}_\ell|, I)$ are the probabilities of the different models. This is particularly useful when model probabilities are similar and classic statistics criteria are powerless.

3. A simple example: a possible satellite line presence

We consider a common case in spectroscopy where we would like to test the presence or not of a weak and unresolved spectral line close to an intense line, an example already discussed in details in Ref. [19]. In this specific example, we consider the $5g - 4f$ transition in pionic nitrogen, an hydrogen-like atom formed by a nitrogen nucleus and a negatively charged pion. During the formation of the pionic atoms, all electrons are expected to be ejected. The presence of a remaining electron in the K shell cannot however completely be excluded. Its presence can cause a shift of the main transition energy due to the Coulomb screening and then an appearance of a new component in the spectrum. To determine the probability of such a scenario, we have to calculate the evidence for the two possible models: Model 1 corresponding to the situation without remaining electrons (a pure hydrogen-like pionic atom) and Model 2 with the possible presence of one remaining electron. More details on the physics case can be found in Refs. [26,27].

The examined data consists in seven distinct spectra similar to the one represented in Fig. 1 (right) with a total of about 60000 recorded counts. Compared to model 1, model 2 has only one additional free parameter of the satellite line intensity, whose relative position with respect to the main line is fixed by theory.

The probabilities of the two hypotheses are calculated from the corresponding evidence E (Eq. (8)) of the models (calculated by Nested_fit program [19, 28]) and they are resumed in Tab. 1 together with some parameter model values.

Table 1. Evidence, probability and parameters values of the models 1 and 2 without and with a satellite line, respectively. The analysis is obtained with Nested_fit program [19,28] and partially reported in previous publications [26,27].

	Model 1	Model 2	Average value
log(Evidence)	-5031.00 ± 0.23	-5039.41 ± 1.82	
Probability	$99.98^{+0.02}_{-0.12}\%$	$0.02^{+0.12}_{-0.02}\%$	
Relative satellite amplitude	0.	$(9.8 \pm 2.2) \times 10^{-3}$	2.2×10^{-6}
Main line shift (ch.)	0.	0.079 ± 0.048	1.8×10^{-5}

As it can be observed, in the considered case the probability for having a satellite line is very low (only 0.02%). In addition using Eq. (9), the average of common parameter values can be calculated from the single most probable values relative to each model and the probabilities of the models themselves $P(\mathcal{M}_{1,2}|, I)$. This feature of Bayesian methods is, to the opinion of the author, one of the most useful tools for the analysis of complex or statistically-poor atomic spectra. Even if the probability of the two models was similar, a correct evaluation of the main peak position would have been possible.

4. Discussions and conclusions

In the previous sections we present the limitation of classic (frequentist) statistics analysis with respect to Bayesian statistics methods. Classic analysis can be safely used in many cases but one should be aware on its limitation. Classic methods cannot correctly take into account prior probabilities of parameters of the considered fit function (model/profile). Moreover, the study of maximal (minimal) value of likelihood (chi-square) function, instead of its behavior over the range of parameter values, can induce errors in the analysis as non-existing correlations and miss presence of multiple maxima (minima). These problems are avoided when the prior probabilities is included via Bayes' theorem (Eq. 6) and when the entire likelihood function $P(\boldsymbol{a}|\{x_i, y_i\}, I) = L(\boldsymbol{a})$ is considered.

For model selection, classic statistics methods provide only criteria to choose one model with respect to others, without the possibility to assign a probability to the different models. This is particularly annoying when these criteria fail to clearly indicate a preference for one particular model. With the Bayesian approach, this difficulty becomes trivial. The probability distribution (and mean, standard deviations, confidence intervals, etc.) of the different parameters that are common to the different models can be obtained from the probability distribution relative each model and the model probability.

Even if tremendously advantageous, Bayesian statistics methods have a drawback, partially responsible to the limited use among the scientific community. The evaluation of the Bayesian evidence and, more generally, the evaluation of the likelihood function $L(\boldsymbol{a})$ in the N-dimensional parameter space (where N is the number of the model parameter) can be very computationally costly. $L(\boldsymbol{a})$ is generally non analytic and its values and integral can be studied with Monte Carlo methods. In present days, when performing multicore computers are more and more available, the use of Bayesian data analysis libraries (eg. BAT [29], PyMC3 [30]) and programs (eg. CosmoMC [7], Multinest [9], Polychord [31], Nested_fit [19,28]) is more and more easy to implement and it should be used as much as possible. This message is addressed in particular to the atomic physics community, where Bayesian methods are still rarely used.

References

[1] Neyman J and Pearson E 1933 *Phil. Trans. R. Soc.* **231** 289–337
[2] Yates F 1934 *Sup. J. R. Stat. Soc.* **1** 217–235 ISSN 14666162
[3] Bevington P and Robinson D 2003 *Data reduction and error analysis for the physical sciences* (McGraw-Hill) ISBN 9780072472271 URL https://books.google.fr/books?id=0poQAQAAIAAJ
[4] Massey F J 1951 *J. Am. Stat. Assoc.* **46** 68–78 ISSN 0162-1459
[5] Akaike H 1974 *IEEE Trans. Automat. Contr.* **19** 716–723 ISSN 0018-9286
[6] Spiegelhalter D J, Best N G, Carlin B P and Van Der Linde A 2002 *J. R. Stat. Soc. B* **64** 583–639 ISSN 1467-9868 URL http://dx.doi.org/10.1111/1467-9868.00353
[7] Lewis A and Bridle S 2002 *Phys. Rev. D* **66** 103511
[8] Trotta R 2008 *Contemp. Phys.* **49** 71–104
[9] Feroz F, Hobson M P and Bridges M 2009 *Mon. Not. R. Astron. Soc.* **398** 1601–1614 ISSN 1365-2966
[10] Particle Data Group 2018 *Phys. Rev. D* **98** 030001
[11] Barber D 2012 *Bayesian Reasoning and Machine Learning* (Cambridge University Press) ISBN 9780521518147
[12] King G, Lovell A, Neufcourt L and Nunes F 2019 *Phys. Rev. Lett.* **122** 232502
[13] Stockton J K, Wu X and Kasevich M A 2007 *Phys. Rev. A* **76** 033613
[14] Calonico D, Levi F, Lorini L and Mana G 2009 *Metrologia* **46** 267 ISSN 0026-1394
[15] Mooser A, Kracke H, Blaum K, Bräuninger S A, Franke K, Leiteritz C, Quint W, Rodegheri C C, Ulmer S and Walz J 2013 *Phys. Rev. Lett.* **110** 140405
[16] Sivia D S and Skilling J 2006 *Data analysis: a Bayesian tutorial* 2nd ed (Oxford University Press)
[17] von der Linden W, Dose V and von Toussaint U 2014 *Bayesian Probability Theory: Applications in the Physical Sciences* (Cambridge University Press) ISBN 9781139952460
[18] von Toussaint U 2011 *Rev. Mod. Phys.* **83** 943–999
[19] Trassinelli M 2017 *Nucl. Instrum. Methods B* **408** 301–312 ISSN 0168-583X
[20] Cox R T 1946 *Am. J. Phys.* **14** 1–13
[21] Jaynes E and Bretthorst G 2003 *Probability Theory: The Logic of Science* (Cambridge University Press) ISBN 9781139435161
[22] Ballentine L E 1986 *Am. J. Phys.* **54** 883–889 URL http://scitation.aip.org/content/aapt/journal/ajp/54/10/10.1119/1.14783
[23] Kolmogorov A 1956 *Foundations of the theory of probability* (Chelsea Pub. Co.)
[24] Bayes M and Price M 1763 *Philosophical Transactions* **53** 370–418 URL http://rstl.royalsocietypublishing.org/content/53/370.short
[25] Laplace P 1825 *Essai philosophique sur les probabilités* (Bachelier) URL https://books.google.fr/books?id=pY9aAAAAcAAJ
[26] Trassinelli M, Anagnostopoulos D F, Borchert G, Dax A, Egger J P, Gotta D, Hennebach M, Indelicato P, Liu Y W, Manil B, Nelms N, Simons L M and Wells A 2016 *Phys. Lett. B* **759** 583–588 ISSN 0370-2693 URL http://www.sciencedirect.com/science/article/pii/S0370269316302787
[27] Trassinelli M, Anagnostopoulos D, Borchert G, Dax A, Egger J P, Gotta D, Hennebach M, Indelicato P, Liu Y W, Manil B, Nelms N, Simons L and Wells A 2016 *EPJ web conf.* **130** 01022
[28] Trassinelli M 2019 *submitted to MDPI Proceedings*
[29] Caldwell A, Kollár D and Kröninger K 2009 *Comp. Phys. Comm.* **180** 2197–2209 ISSN 0010-4655
[30] Salvatier J, Wiecki T V and Fonnesbeck C 2016 *PeerJ Comput. Sci.* **2** e55 ISSN 2376-5992
[31] Handley W J, Hobson M P and Lasenby A N 2015 *Mon. Not. R. Astron. Soc.* **453** 4384–4398 ISSN 0035-8711

Merged beam studies of mutual neutralization at subthermal collision energies

X. Urbain[1], N. de Ruette[2,3], A. Dochain[1], T. Launoy[4], R.F. Nascimento[2], M. Kaminska[2], M.H. Stockett[2], J. Loreau[4], J. Liévin[4], N. Vaeck[4], R.D. Thomas[2], H.T. Schmidt[2] and H. Cederquist[2]

[1] Institute of Condensed Matter and Nanosciences, Université catholique de Louvain, B-1348 Louvain-la-Neuve, Belgium
[2] Department of Physics, Stockholm University, Stockholm, SE-106 91, Sweden
[3] European Spallation Source ESS AB, P.O. Box 176, SE-211 00 Lund, Sweden
[4] Laboratoire de Chimie Quantique et Photophysique, Université Libre de Bruxelles, B-1050 Brussels, Belgium

E-mail: `xavier.urbain@uclouvain.be`

Abstract. We have measured the kinetic energy distributions for the mutual neutralization of a large ensemble of atomic anions and cations. Ions are accelerated to equal velocities in a merged beam setup, enabling measurements at collision energies as low as 5 meV. Three-dimensional momentum imaging is performed with two position sensitive detectors located several meters downstream from the few centimeter long region where the beams overlap. An unprecedented resolution in the kinetic energy release (KER) spectra allows us to identify the states of both reactants and products down to their fine structure. Knowing the angular distribution of the products in the laboratory and center-of-mass frame, allows for total, partial, and differential cross sections to be retrieved.

1. Introduction

Anions are present in a variety of astrophysical environments [1]. Their presence in molecular clouds has been confirmed quite recently with the detection of C_6H^- [2], soon to be followed by C_4H^- [3], C_8H^- [4], C_3N^- [5], C_5N^- [6], and lately CN^- [7]. They are thought to be the main reservoir of negative charges in cold and dense environments. While radiative and dissociative attachment are thought to be their dominant formation route, mutual neutralization would constitute an important destruction mechanism besides photodetachment [8].

Vuitton *et al* [9] characterized the abundance and role of CN^- and C_3N^- anions in the chemistry of Titan's upper atmosphere. More importantly, they identified associative detachment to radical species as the major destruction pathway. Closer to us, the role of oxygen anions in Earth's upper atmosphere is clearly established [10]. Electrons attach to oxygen on the night side, and ionospheric reactions take place that produce characteristic nightglows. Satellite images of the UV tropical nightglow of O I at 135.6 nm have been interpreted as resulting for a large part from the mutual neutralization reaction $O^- + O^+ \rightarrow O(^5P) + O$, followed by the radiative cascade $O(^5P) \rightarrow O(^5S) \rightarrow O(^3P)$ (note that the first emission at 777 nm is also observed) [11, 12]. Finally, high-resolution spectroscopy of trace metallic elements present in the outer layers of stellar objects is now widely available [13]. A quantitative interpretation of those spectra in terms of elemental abundances can only be reached through non-local thermodynamic equilibrium (non-LTE) stellar atmosphere modeling [14], which in turn requires full understanding of the ion-pair to covalent transitions responsible for collisional excitation between neutrals and mutual neutralization between ions [15, 16].

Of more practical application is the role played by negative ions in technical plasmas, the most obvious being the large scale production of high energy, high current negative deuterium beams as precursors of powerful neutral beams [17]. Their tangential injection into the toroidal plasma of tokamaks like ITER should increase the plasma temperature beyond the point where ohmic heating becomes inefficient. Volume production of D^- is normally impeded by the concurrent neutralization with D^+, D_2^+ and D_3^+ cations [18]. This limitation has been overcome by means of cesiated surfaces increasing the relative abundance of low energy electrons and the partitioning of the plasma volume between hot ionization and colder recombination regions, with the help of magnetic grids [19].

2. Mutual neutralization

Subthermal collisions, *i.e.*, collisions at velocities typical of a cold gas, do not necessarily proceed along weakly attractive potentials nor do they exhibit long-range potential barriers limiting the rate coefficient to some finite value. Indeed, anion-cation collisions are dominated by the long-range Coulomb attraction that captures increasingly large impact parameters as the collision velocity is reduced. Such processes have diverging rate coefficients as allowed by Wigner's threshold law. The reaction channels that are accessible at low energy are:

$$\begin{aligned} \mathrm{A^+ + B^-} &\rightarrow \mathrm{A^* + B^*} && \mathrm{MN} && (1)\\ &\rightarrow \mathrm{A + B^+} + e^- && \mathrm{TI} && (2)\\ &\rightarrow \mathrm{A^- + B^+} && \mathrm{CX} && (3)\\ &\rightarrow \mathrm{AB^+} + e^- && \mathrm{AI} && (4)\end{aligned}$$

where MN stands for mutual neutralization, TI for transfer ionization, CX for (double) charge exchange and AI for associative ionization (see [20] for a review). When dealing with polyatomic systems, e.g. $H_2^+ + O^-$ [21] and $CH^+ + O^-$ [22], one may further distinguish:

$$\begin{aligned} \mathrm{AB^+ + C^-} &\rightarrow \mathrm{AC^* + B} && \mathrm{RN} && (5)\\ &\rightarrow \mathrm{AC^+ + B} + e^- && \mathrm{RI} && (6)\\ &\rightarrow \mathrm{ABC} + h\nu && \mathrm{RA} && (7)\end{aligned}$$

where RN stands for reactive neutralization, RI for reactive ionization, and RA for radiative association. For the latter process to occur, a long-lived resonance must be formed, for the radiative stabilization to compete with the predissociation of the collisional complex.

Due to energy conservation, the energy imparted to mutual neutralization products reflects their degree of internal excitation. This exoergicity will be referred to as the (total) kinetic energy release (KER):

$$\begin{aligned} \mathrm{A^+ + B^-} &\rightarrow \mathrm{A^* + B^* + KER} && (8)\\ \mathrm{KER} &= [IP(\mathrm{A}) - E_{exc}(\mathrm{A^*})] - [EA(\mathrm{B}) + E_{exc}(\mathrm{B^*})] && (9)\end{aligned}$$

where IP is the ionization potential of atom A, E_{exc} its excitation energy, and EA the electron affinity of atom B. A schematic potential energy diagram is given in Figure 1 for the $O^+ + O^-$ reaction. As discussed below, the electron transfer will preferably occur at the crossing points

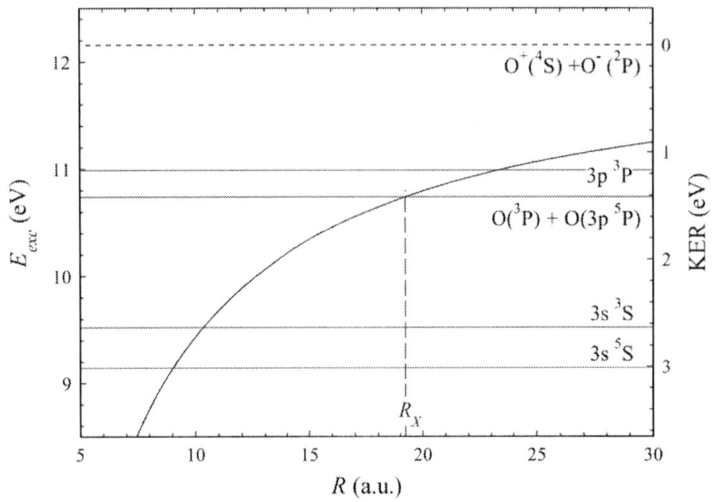

Figure 1. Diabatic potential energy diagram (schematic) for the O^+ + O^- mutual neutralization. The excitation energy E_{exc} of $O(2s^2 2p^3 nl\,^{2S+1}L)$ levels is converted to the kinetic energy release KER by means of Eq. 9. R_X is the internuclear distance at which electron transfer takes place.

constant potentials corresponding to the various covalent channels situated *below* the incoming asymptote. The experimental challenge will therefore be to energetically resolve these outgoing channels while collecting all mutual neutralization products, thereby obtaining total and partial MN cross sections. These will eventually benchmark theoretical calculations giving complete access to the full scattering matrix, *i.e.*, excitation, deexcitation, mutual neutralization and ion pair formation reactions. We describe hereafter some recent advances towards the completion of this task.

3. Single-pass merged beam experiment

In order to manipulate ion beams of different species and bring them to low relative collision velocities, we opt for the well-established merged beam arrangement (see [23] for a review). In this configuration, collimated beams of fast ions are brought into confluence at equal velocities, reducing the center-of-mass collision energy E_r to virtually zero. More precisely:

$$E_r = \mu\left[\frac{E_1}{M_1} + \frac{E_2}{M_2} - 2\sqrt{\frac{E_1 E_2}{M_1 M_2}}\cos\phi\right] \simeq \frac{\mu}{M_1} E_1 \phi^2 \qquad (10)$$

where E_1, M_1, E_2 and M_2 stand for the laboratory energies and masses of colliding ions, with $E_1/M_1 \simeq E_2/M_2$, μ is the reduced mass and ϕ is the average angle formed by the interacting beams. As a consequence, beam energies have to be scaled according to the mass ratio between the species under study, which limits the range of applicability of the method. In the set-up depicted on Figure 2 [24, 25], cations and anions are first accelerated and brought along parallel trajectories, to be electrostatically merged by two pairs of parallel plates mounted on a movable assembly. This motion is required to fulfil the merging criterion, *i.e.*, they must enter the last steering element at an angle that is inversely proportional to their kinetic energy for them to come out parallel to one another. Detuning the collision energy from near-zero to a finite and controlled value is performed by applying some electrical bias to the interaction cell located in the UHV section of the apparatus ($\sim 10^{-10}$ mbar). Reaction products may be charged or neutral. For mutual neutralization studies, both ion beams are dumped at the exit of the UHV region, while the neutrals fly unaffected towards the detection system located 3 to 5 m downstream. For the simple case of particles of equal velocity v and laboratory energy E, the kinetic energy release (KER) is related to the spatial (Δx) and temporal (Δt) separation of the

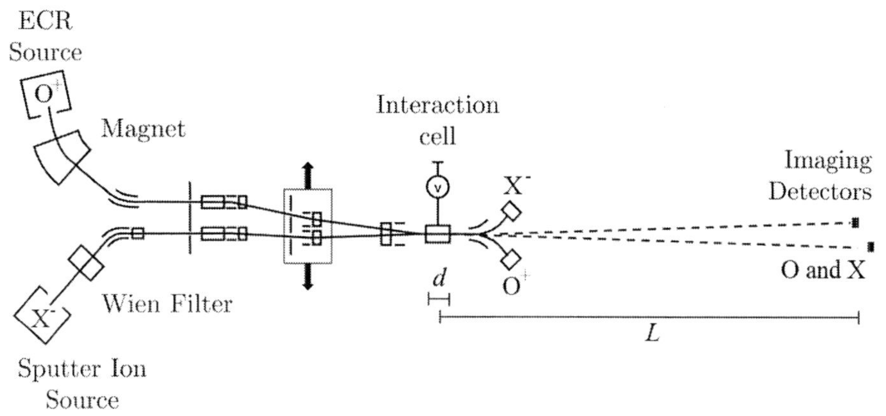

Figure 2. Schematics of the merged-beam apparatus for the study of $O^+ + X^-$ mutual neutralization. Ion sources are mounted on ± 20 kV platforms. A duoplasmatron source was used for H^- and D^- anions.

products detected at a distance L from the interaction region:

$$\mathrm{KER} = \frac{E}{2L^2}\left[\Delta x^2 + v^2 \Delta t^2\right] \tag{11}$$

The resolution of the KER measurement will strongly depend on the precision to which L is known, *i.e.*, on the ratio of the interaction length d to the average distance to the detectors. In the present case, $2 < d < 7.5$ cm and $L < 5$ m, producing a relative KER resolution of ~ 1 %.

Last, when dealing with merged electron and ion beams, it is common practice to describe their interaction in the co-moving frame by two temperatures, $T_{||}$ and $T_{\perp}$. The former is a direct measure of the longitudinal velocity spread, while the latter reflects the collimation of the beams. The distribution of relative velocities reads:

$$f(v_d, \vec{v}) = \frac{m}{2\pi k T_\perp}\sqrt{\frac{m}{2\pi k T_{||}}}\exp\left[-\frac{m v_\perp^2}{2\pi k T_\perp} - \frac{m(v_{||} - v_d)^2}{2\pi k T_{||}}\right] \tag{12}$$

where $v_{||} \simeq |v|(1 \pm \Delta E/2E)$ and $v_\perp \simeq |v| \sin\phi$, and v_d is the so-called detuning velocity, *i.e.*, the average velocity difference imparted to the beams by their acceleration and/or the interaction cell bias. For 12 keV C^- and C^+ beams with 5 eV energy dispersion, and 1 mrad angular spread, $T_{||} \simeq 8$ K and $T_\perp \simeq 70$ K, subthermal values indeed. Early attempts to measure MN cross sections were limited by the inherent difficulty of detecting pairs of particles hitting a single detector close in time and space. We circumvented this limitation by installing a pair of position sensitive detectors operating in coincidence. Their circular shape and the dead area between them requires some modeling to retrieve the actual momentum distribution.

4. Results and interpretation

A joint effort by UCLouvain and Stockholm University has allowed the measurement of MN reactions for a large number of systems: C^+, N^+ and O^+ in combination with D^-, C^-, O^-, Si^-, P^- and S^-. Figure 3 summarizes the findings in terms of kinetic energy release in the O^+ case, which directly correlate to the final electronic state of the oxygen atom. The anions are ordered according to the electron affinity, with tighter bound electrons in the anion occupying

MN with O^+ favors the $2s^22p^3\,3p\,^5P$ state with a branching fraction of 71 ± 9 % measured at 5 meV [25], in fair agreement with earlier theoretical predictions [10], *i.e.*, 42 % at 300 K. The ^3P state is however much less populated than originally estimated, *i.e.*, 15 ± 4 % against 46 %. Since the UV tropical nightglow of oxygen has two distinct components at 130.4 and 135.6 nm, corresponding to ^3S and ^5S respectively, a reevaluation of the actual role of MN may be in order.

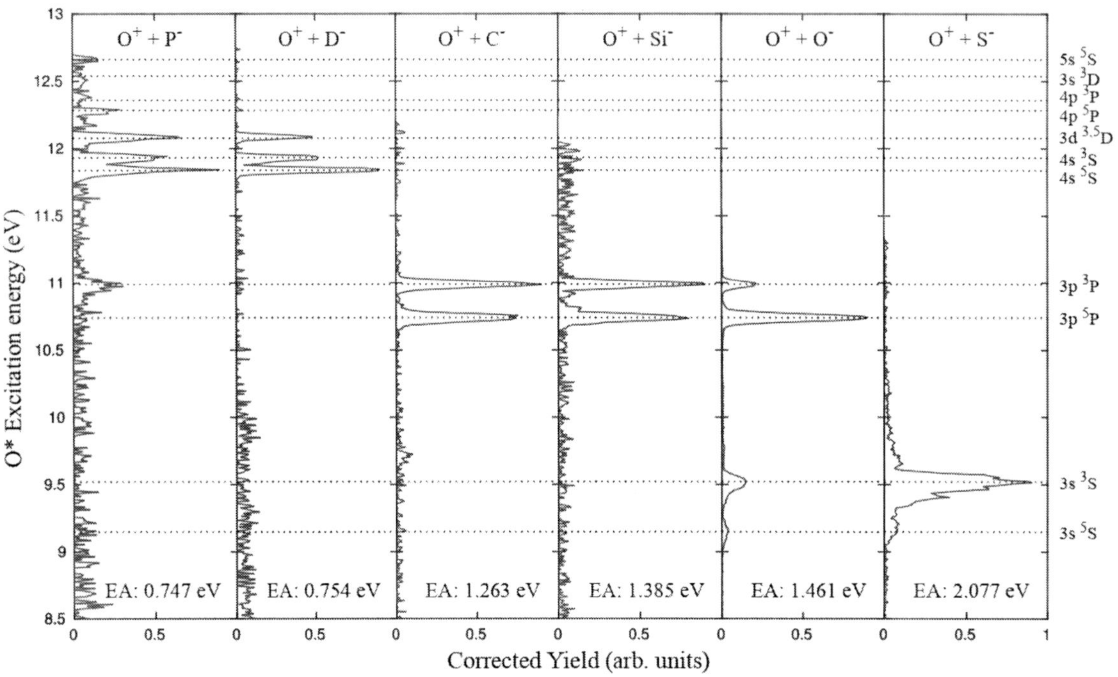

Figure 3. Kinetic energy release spectra for the $O^+ + X^-$ mutual neutralization. The energy scale is converted to the excitation energy of the $O(2s^22p^3nl\,^{2S+1}L)$ level by means of Eq. 9.

In order to rationalize the branching ratios measured for a wide variety of collisional systems, one is tempted to resort to a simplified treatment of the non-adiabatic interactions prevailing at the successive avoided crossings. A powerful approach is the Firsov-Landau-Herring asymptotic method [26] as applied by Zhou and Dickinson [27], that allows the evaluation of the one-electron exchange interaction $\Delta(R_X)$, where R_X is the crossing distance and $H_{if} \simeq \Delta(R_X)/2$ is approximately the coupling matrix element between states i and f, without explicitly treating the many-electron configuration interaction. These matrix elements are incorporated in a multi-state Landau-Zener calculation to generate total and partial cross sections. As shown by de Ruette *et al* [25], this single-electron treatment fails to reproduce the branching fractions observed in the case of $N^+ + O^-$, particularly for configurations like $N(2s2p^4\,^4P)$ that involve two-electron processes. This discrepancy is partially lifted by introducing some configuration interaction among terms sharing the same symmetry, as obtained by multiconfiguration Hartree-Fock (MCHF) calculations. What remains is a systematic underestimation of the high-KER, *i.e.*, short R_X channels. As shown by Mitrushchenkov *et al* [28] for the $Ca^+ + H^-$ MN reaction, the two-by-two treatment applied in the multi-state LZ approach underestimates the contribution of weak transitions to the cross section that may be active at smaller internuclear distances. The more reliable branching probability current method [29] still lies beyond a current computational to a full quantal treatment, as recently demonstrated by Launoy *et al* for the $Li^+ + H^-$ case [30]. Care must however be taken of the proper asymptotic description of the various channels, through the inclusion of diffuse orbitals.

5. Conclusions and perspectives

A single-pass merged beam apparatus has allowed us to collect a large ensemble of branching fractions among the various excited states produced by electron transfer in mutual neutralization reactions. We were able to assess the validity of earlier measurements and calculations, and to improve the theoretical treatment of such processes by including configuration interaction in the outgoing channels. Further experimental developments are needed to gain full access to the angular distribution of the products, which translates into state-specific differential cross sections. This total collection is needed to produce reliable cross sections. The forward detection is still impeded by the excessive background caused by rest-gas collisions. Such a limitation is absent in DESIREE, the double cryogenic storage ring facility located at Stockholm University [31, 32], where beams of anions and cations circulating in juxtaposed storage rings interact in a common straight section. Beams of anions and cations are stored for long times, allowing them to relax to their lowest quantum states [33]. This is particularly relevant for molecular species which can not be cooled in a continuous, single pass experiment. Such data are urgently needed for the astrochemical modeling of dense clouds and planetary atmospheres.

Acknowledgments

This work was supported by the Fonds de la Recherche Scientifique–FNRS (IISN Contract No. 4.4504.10) and by the Swedish Research Council (Contracts No. 2017-00621, No. 621-2014-4501, and No. 621-2015-04990). Computational resources have been provided by the Shared ICT Services Centre of the Université libre de Bruxelles and by the Consortium des Équipements de Calcul Intensif (CÉCI), funded by the Fonds de la Recherche Scientifique – FNRS under Grant No. 2.5020.11. The authors thank the Belgian State for the grant allocated by Royal Decree for research in the domain of controlled thermonuclear fusion. XU is Senior Research Associate of the Fonds de la Recherche Scientifique – FNRS. TL was funded by a fellowship of the Fonds pour la Formation à la Recherche dans l'Industrie et dans l'Agriculture–FRIA.

References

[1] Millar T J, Walsh C, Field T A 2017 *Chem. Rev.* **117** 1765
[2] McCarthy M C, Gottlieb C A, Gupta H and Thaddeus P 2006 *Astrophys. J.* **652** L141
[3] Cernicharo J, Guélin M, Agúndez M, Kawaguchi K, McCarthy M and Thaddeus P 2007 *Astron. Astrophys.* **467** L37
[4] Brünken S, Gupta H, Gottlieb C A, McCarthy M and Thaddeus P 2007 *Astrophys. J.* **664** L43
[5] Thaddeus P, Gottlieb C A, Gupta H, Brünken S, McCarthy M C, Agúndez M, Guélin M and Cernicharo J 2008 *Astrophys. J.* **677** 1132
[6] Cernicharo J, Guélin M, Agúndez M, McCarthy M, and Thaddeus P 2008 *Astrophys. J. Lett.* **688** L83
[7] Agúndez M, Cernicharo J, Guélin M, Kahane C, Roueff E, Klos J, Aoiz F J, Lique F, Marcelino N, Goicoechea J R *et al* 2010 *Astron. Astrophys.* **517** L2
[8] Wakelam V and Herbst E 2008 *Astrophys. J.* **680** 371
[9] Vuitton V, Lavvas P, Yelle R V, Galand M, Wellbrock A, Lewis G R, Coates A J, Wahlund J-E 2009 *Planet. Space Sci.* **57** 1558
[10] Olson R E, Peterson J R, Moseley J 1970 *Geophys. Res.* **76** 2516
[11] Sagawa E, Immel T J, Frey H U and Mende S B 2005 *J. Geophys. Res.* **110** A11302
[12] Qin J, Makela J J, Kamalabadi F Meier R R 2015 *J. Geophys. Res. Space Physics* **120** 10116
[13] Wys, R F G 2016 in *ASP Conf. Ser.* **507** *Multi-Object Spectroscopy in the Next Decade: Big Questions, Large Surveys, and Wide Fields* ed. Skillen I, Balcells M and Trager S (San Francisco, CA: ASP) 13
[14] Asplund M 2005 *Annu. Rev. Astron. Astrophys.* **43** 481
[15] Barklem P S 2016 *Astron. Astrophys. Rev.* **24** 9

[17] Toigo V *et al* 2017 *New J. Phys.* **19** 085004
[18] Fantz U, Franzen P and Wünderlich D 2012 *Chem. Phys.* **398** 7
[19] Bacal M and Wada M 2015 *Appl. Phys. Rev.* **2** *021305*
[20] Le Padellec A, Launoy T, Dochain A and Urbain X 2017 *J. Phys. B: At. Mol. Opt. Phys.* **50** 095202
[21] Staicu-Casagrande E M, Nzeyimana T, Naji E A, de Ruette N, Fabre B, Le Padellec A and Urbain X 2004 *Eur. Phys. J.* D **31** 469
[22] Le Padellec A, Staicu-Casagrande E M, Nzeyimana T, Naji E A and Urbain X 2006 *J. Chem. Phys.* **124** 154304
[23] Phaneuf R A, Havener C C, Dunn G H and Müller A 1999 *Rep. Prog. Phys.* **62** 1143
[24] Nzeyimana T, Naji E A, Urbain X and Le Padellec A 2002 *Eur. Phys. J.* D **19** 315
[25] de Ruette N, Dochain A, Launoy T, Nascimento R F, Kaminska M,Stockett M H, Vaeck N, Schmidt H T, Cederquist H and Urbain X 2018 *Phys. Rev. Lett.* **121** 083401
[26] Chibisov M I and Janev R K 1988 *Phys. Rep.* **166** 1
[27] Zhou X and Dickinson A 1997 *Nucl. Instrum. Methods* B **124** 5
[28] Mitrushchenkov A, Guitou M, Belyaev A K, Yakovleva S A, Spielfiedel A, and Feautrier N 2017 *J. Chem. Phys.* **146** 014304
[29] Belyaev A K 2013 *Phys. Rev. A* **88** 052704
[30] Launoy T, Loreau J, Dochain A, Liévin J, Vaeck N and Urbain X, accepted for publication in *Astrophys. J.* (*Preprint* arXiv:1906.06715)
[31] Thomas R D, Schmidt H T, Andler G, Björkhage M, Blom M, Brännholm L, Bäckström E, Danared H, Das S, Haag N *et al* 2011 *Rev. Sci. Instrum.* **82** 065112
[32] Schmidt H T, Thomas R D, Gatchell M, Rosén S, Reinhed P, Löfgren P, Brännholm L, Blom M, Björkhage M, Bäckström E *et al* 2013 *Rev. Sci. Instrum.* **84** 055115
[33] Schmidt H T, Eklund G, Chartkunchand K C, Anderson E K, Kaminska M, de Ruette N, Thomas R D, Kristiansson M K, Gatchell M, Reinhed P, Rosén S, Simonsson A, Källberg A, Löfgren P, Mannervik S, Zettergren H and Cederquist H 2017 *Phys. Rev. Lett.* **119** 073001

On the highly charged ion transmission spectroscopy applied to 2D materials

Richard A. Wilhelm
TU Wien, Institute of Applied Physics, Wiedner Hauptstr. 8-10/E134, 1040 Wien, Austria, EU

E-mail: wilhelm@iap.tuwien.ac.at

Abstract. The interaction of ions with solid materials may lead to charge exchange. Previous results show that charge exchange is confined to the uppermost surface layers of a solid or to the thickness of a freestanding 2D material, even when initially highly charged ions are used. Several processes are active during charge exchange and a clear distance or impact parameter dependence is observed. Here I show, that highly charged ion transmission can be used to determine the material structure of 2D materials especially when atomically resolved electron microscopy cannot be applied. The method is based on measurement of energy-, charge-, and angle-resolved ion transmission spectra in coincidence and a comparison with a local model for a time-dependent change in the projectile's electron occupation and consequently the scattering potential. Also necessary experimental procedures are discussed.

1. Introduction

The elastic and inelastic scattering of photons and charged particles is used in many different techniques for surface and bulk material characterizations [1, 2]. In angle-resolved photoelectron spectroscopy the absorption of photons leads to the emission of electrons and their angle-vs-energy spectrum reveals the material band structure [3]. Electron illumination of surfaces leads to electron emission used in secondary electron microscopy for image generation and the emission of characteristic electron energies due to an inner-shell Auger transition is used for elemental mapping in scanning Auger spectroscopy [4, 5]. Many more variations of electron and photon scattering techniques are used for a plethora of analytical purposes.

Elastic and inelastic ion scattering are also used in quantitative material analysis, i.e. low and medium energy ion scattering, Rutherford backscattering spectrometry, elastic recoil detection, particle induced x-ray and gamma ray emission and many more. However, in all established methods only the particle's energy, mass and/or angle of incidence (or angle of emission) is used for analysis. Heavy ions posses another degree of freedom: their charge state. Here I show, that the ion charge state and charge exchange is not just an obstacle when stopping forces are to be evaluated, but can in fact be used for sub-nm porosity determination in ultrathin materials. The recently emerged collection of organic and inorganic 2D materials contains different possible crystallographic phases, grain sizes and orientation, lateral and vertical heterostructures, and defect structures. Especially intrinsic defects in transition metal dichalcogenides (TMDs) demand atomic resolution in electron microscopy to be seen and density inhomogeneities in short-range ordered molecular nanosheets (SROMN) [6] stay hidden with current methods at hand. When materials become as thin as a single atom (graphene) or a few atoms (TMDs,

SROMN), single atomic defects, defect clusters and atomic scale variations in their homogeneity have a big impact on their properties. Hence, to control these defects one must detect them. As it turns out, charge exchange spectroscopy of ions is well suited for the task, especially in cases where atomic resolution in microscopy cannot be obtained. The reason lies in a strong interatomic distance dependence of charge exchange, which confines the important parts of the interaction process to the Å-length scale. From a spectroscopic analysis of the transmitted ions in comparison with a model simulation, detailed knowledge about the atomic scale structure of a material can be gained.

2. Charge exchange of highly charged ions

Studies on the interaction of highly charged ions revealed many interesting features. The formation of hollow atoms in front of the surface [7, 8], the emission of characteristic x-rays [9, 10], the emission of a large number of low energy electrons and a small number of high energy Auger electrons [11, 12, 13], the formation of nanostructures on susceptible materials and many more [14, 15]. One peculiarity emerged from the fact that ions scattered under grazing incidence from HOPG or transmitted through few nm carbon foils were found mainly in their neutral charge state [16, 17]. This pointed towards an extremely efficient charge exchange process, where several candidates were discussed in literature [18]. Some dispute arose around the sample preparation and vacuum conditions, where contaminations could severely affect the charge exchange. Experiments with C60 clusters in the gas phase, however, were very much in line with the previous findings on surfaces showing the same fast neutralization [19, 20]. Here, 'contaminations' can be ruled out.

A comprehensive explanation of the fast neutralization of HCI could recently be given by taking a multi-electron interatomic de-excitation channel into account: Interatomic Coulombic Decay (ICD) [21]. ICD is a process by which a donor (hollow atom) transfers its excitation energy to a neighbouring atom (or several of them). Electrons from the neighbouring sites are then ionized or excited. It turns out that ICD has an interatomic distance dependent rate, which agrees well with the observed de-excitation time scales. Currently, intensive work is dedicated to ab-initio calculation of the distance dependent ICD rate for extended systems [22, 23], well in agreement with results obtained from HCI transmission through 1 nm thick amorphous carbon nanomembranes (CNMs), single-, bi- and tri-layer graphene, single layer hBN, and single layer MoS_2 [24, 25]. The link between the ICD rate and HCI transmission spectroscopy results is discussed below.

3. Experimental procedures

Ion scattering from surfaces or molecules in the gas phase are typically performed in a standard ion beam setup using electrostatic [26], time-of-flight (TOF) [27] or magnetic spectrometers [28, 29] to determine the ion energy after interaction with the target. Especially, when it comes to charge exchange determination, additional demands must be met. The detector must be charge state sensitive, which is fulfilled by all the three concepts above, when ions are post accelerated by a DC voltage in the TOF measurement. However, also neutrals must be detected efficiently, which typically is challenging with electrostatic and magnetic spectrometers. When separate neutral counting in forward direction is performed, then special care must be taken to maintain the same solid angle for the neutral and the ion detector. Important for all concepts is the fact that detector efficiencies may depend on the ion charge state, when heavy ions at 10-100 keV energies are used, because here the electronic stopping contribution in the detector material is small and secondary electrons may be emitted with different probabilities depending on the charge state upon impact on the detector. Fig. 1 shows the pulse height distributions from a channeltron for different impinging Xe charge states after post amplification using a standard shaping amplifier. Even though the kinetic energy is not fixed here, one can still see a

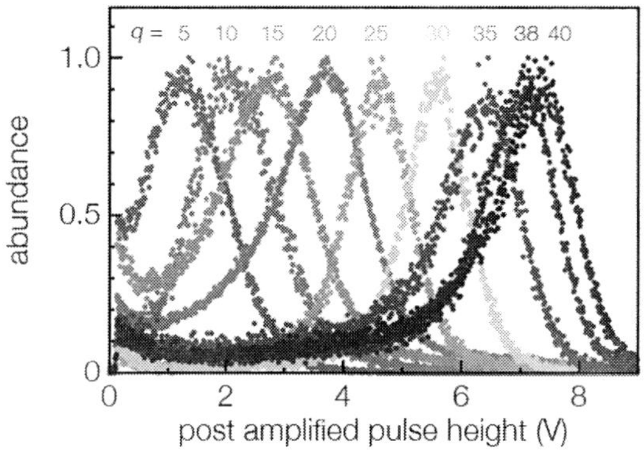

Figure 1. (color online) Post amplified pulse height distributions from a channeltron operated at 1900 V bias voltage for different incident Xe charge states and kinetic energies of $4.4\,\mathrm{kV} \times q$, i.e. $E_{kin} = 22 - 176\,\mathrm{keV}$.

charge state dependence, and especially low charge states and low kinetic energies may not be detected with $\sim 100\%$ efficiency. As a consequence, the detector efficiency must be calibrated for low ion charge states and for the correct kinetic energy range.

Recently it was shown that charge exchange depends strongly on the ion impact parameter, and consequently charge exchange observation depends on the observation (scattering) angle [30]. Therefore, it is highly advantageous to perform angle-resolved measurements with an angular resolution of $\sim 0.5°$ or better. One option is to use small apertures in electrostatic and magnetic spectrometers and perform subsequent measurements for different observation angles, i.e. move the detector stepwise. One crucial issue then is ion induced beam damage, which can be important especially for 2D materials and when highly charged ions are used [31, 32]. As sequential measurements of energy and angle increase the necessary exposure of the sample significantly, special care must be taken to collect enough statistics before the sample deteriorates too strongly.

One way to circumvent most of the issues here is to use a TOF based setup, which provides a high duty cycle, detects all charge states simultaneously, and uses a position sensitive ion detector providing angle resolution as well [33]. Using secondary electrons emitted from the interaction process with the target material as a start signal for the TOF measurement allows a duty cycle of 100% if the ion current is in the range of 10^5 ions/s. Even for somewhat higher intensities, measurements can be performed, when multi-start and multi-stop analysis is performed [34]. Electrostatic beam choppers, commonly used in TOF setups, provide a duty cycle of about $\sim 10^{-3}$ (chopper opening time ($\sim$ns) over TOF ($\sim \mu$s)). This may not be an issue if the ion source provides enough beam current. When sources for highly charged ions, i.e. typically electron beam ion traps (EBITs) [35, 36], are used, ion intensities of $10^3 - 10^5$ ions/s can be expected for the highest charge states ($q \gtrsim 30$) without chopping. A position sensitive multi-channel plate (MCP) with high timing resolution, e.g. a MCP with delay line anode [37], is favourable as a stop detector. Using a set of slits and a parallel plate deflector the impact positions on the MCP can be directly related to the charge state and the scattering angle. The timing information provides the stop trigger for the TOF measurement.

Since charge exchange spectroscopy is particularly powerful for 2D materials or the outermost surface layers of a bulk material, surface science methods for sample preparation and/or cleaning are necessary. Therefore, UHV conditions must be met as well as sample heating to $\gtrsim 400°$C should be possible.

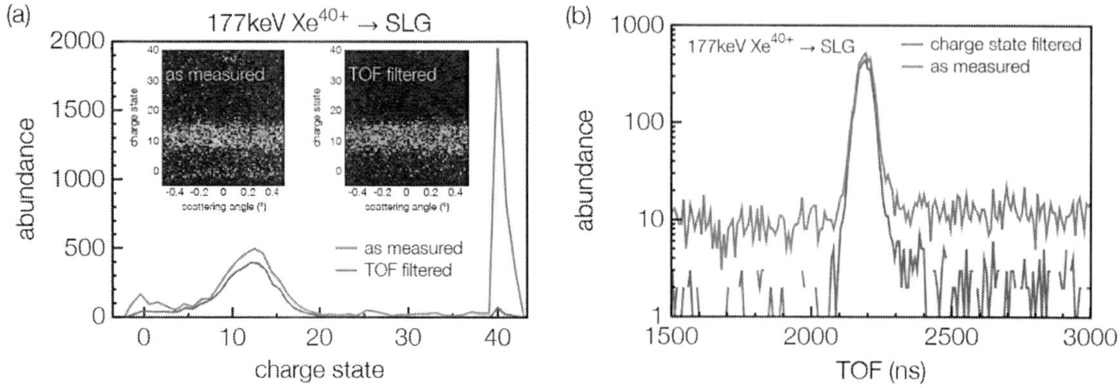

Figure 2. (color online) (a): Charge distribution extracted from the 2D position plot as measured and TOF filtered (c.f. text). The charge distributions are integrated over the scattering angle range covered by the detector. Insets show the 2D plots with and without TOF filtering. (b): TOF spectrum as measured and filtered by the charge-and-angle region selected in the 2D plot data. Filtering by charge-and-angle and/or TOF allows an almost background free measurement important at low count rates and reduces the influence of random coincidences. The discontinuities of the graph in (b) is due to 0 counts in the log-scale.

The following results are produced using an EBIT. HCI charge states are selected with a Wien filter. Ions are extracted from the source by a constant electrostatic field and decelerated in front of the target by a retarding field to adjust their final kinetic energy. Samples are kept at UHV conditions, i.e. below 5×10^{-9} mbar. Samples are prepared typically by chemical vapour deposition on suitable substrates and transferred onto standard TEM grids equipped with a Quantifoil support film without the use of a sacrificial polymer layer where-ever possible. Carbon Nanomembranes (CNM) are produced by cross-linking a self-assembled monolayer of nitro-biphenylthiol (NBPT) on a Au(111) surface. CNMs are subsequently transferred on a TEM grid. Some samples were treated with a specialized cleaning procedure to obtain large contamination free areas [38]. Electrons emitted from the ion interaction with the target are collected by a biased grid (+200-500 V) and accelerated to 30 keV to be measured in a surface barrier detector [39]. Ions are detected by a Roentdek delay line detector ~ 1.1 m behind the sample.

4. Ion transmission through 2D materials

Ion transmission through 2D materials reveals interesting details about the interaction of ions with solids. Before concrete conclusions can be drawn, a few things need to be considered. First, 2D materials can exist as freestanding membranes, but they can span over areas of a few μm in diameter at most. Therefore, support grid structures are typically used, e.g. holey or lacey carbon, or Quantifoil. Ion beams are typically focused to 0.1-1 mm in diameter if no focussed ion beams are used. Consequently, the ion beam probes large areas of the sample, i.e. the result may be influenced by inhomogeneities in the sample over the mm scale. Especially influences of a support film need to be considered as well as multilayer and contaminated areas. One way to deal with these effects is the use of the coincidence technique. Here, ions are measured with respect to their energy (loss) and charge state in a way, that either charge state spectra can be filtered by the TOF or vice versa. Fig. 2 shows the charge state spectrum of 177 keV Xe^{40+} ions transmitted through a single layer of graphene together with the corresponding TOF spectrum. The sample already shows large coverage and homogeneity, which is not always the case for 2D

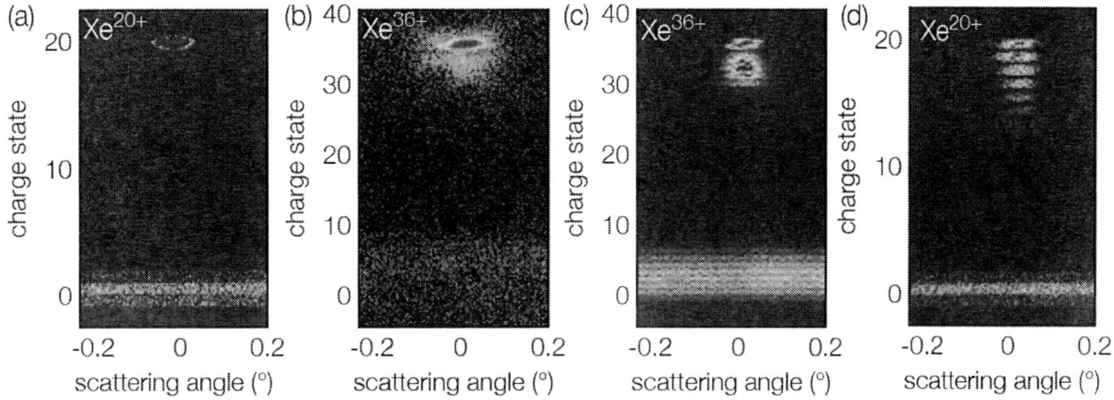

Figure 3. (color online) (a) and (d): Charge state patterns before and after exposure of a NBPT-CNM to 324 keV Xe^{36+}, respectively, obtained using 180 keV Xe^{20+} ions. (b) and (c): Charge state patterns of the same sample measured while exposure to 324 keV Xe^{36+}.

materials. However, the as-measured charge state spectrum shows a peak at $q = 0$, 12, and at 40. At the same time the TOF spectrum shows a background noise of about 10 counts per bin (bin size 0.3 ns), a peak at 2200 ns and slightly higher background at larger TOFs. Filtering the charge state spectrum for TOFs between 2100-2300 ns in Fig. 2(a) yields the red spectrum. The peak at $q = 40$ almost vanished and can be associated with ions passing through large uncovered areas in the sample. Those ions do not produce a start signal, because no electron emission takes place. Therefore, no TOF is measured for these ions and they are effectively filtered. Note, that the residual intensity at $q = 40$ is the result of random coincidences, i.e. electrons emitted from ion impinging on the TEM grid material itself (typically Cu or Au) and ions passing through uncovered areas. The number of ions impinging on the TEM grid and the number of ions transmitted through cracks are both large and therefore the probability of random coincidences is non-vanishing. In fact, the overall background signal in the TOF spectrum is the direct result of this.

Furthermore, the intensity of the peak at $q = 0$ in Fig. 2(a) is reduced as well, because these ions are transmitted through the Quantifoil support film. Since the film is 10-20 nm thick, ions are mainly in their neutral state. These ions have TOFs mostly larger than 2300 ns and contribute to the enhanced background at larger TOFs in Fig. 2(b). Finally, the distribution at $q = 12$ is the result of ions transmitted through the SLG itself and have a TOF in the region of 2100-2300 ns. The inset in Fig. 2(a) also shows how the noise level can be significantly reduced by TOF filtering.

The procedure can also be reversed and the TOF spectrum is filtered by the 2D angle-charge map in (a). Now the background in (b) is reduced almost to 0, i.e. the signal-to-noise ratio is large in order to determine energy losses precisely.

Data filtering by the means described above is helpful and even necessary when samples show low coverage or materials deteriorate quickly under the ion beam. In both cases only small statistics can be acquired in reasonable measurement time (also considering dark noise of the detector) or before the sample is destroyed.

TOF and charge state determination in a coincidence method enables also unambiguous identification of particular ion trajectories (e.g. through the support material), which would not be possible if both quantities are measured separately.

Ion transmission spectroscopy also allows to follow material damage in time. Fig. 3 shows different charge state patterns obtained from HCI transmission through a 1 nm thick NBPT-

CNM [40]. In (a) a spectrum is shown using 180 keV Xe^{20+} ions, where it was previously reported that this charge state is below a threshold for material perforation [41]. After the initial spectrum was measured, the sample was exposed to 324 keV Xe^{36+} ions and the charge state pattern is monitored continuously (c.f. Fig. 3(b)-(c)). While the charge state distribution at lower charge states below $q = 10$ remains constant, an upper charge state distribution develops between $q = 34 - 20$. As previously shown and also evident from the small angular range of these transmitted ions, the upper charge state distribution is a result of large impact parameter scattering. In fact, the impact parameter is above ~ 2 Å in order to maintain the high charge states for the ions (see discussion in the section below) [42]. It was also shown, that at the high charge state of $q = 36$ pores can be introduced efficiently into the material with diameters of up to 15 nm [41]. Ion transmission through previously damaged, i.e. perforated areas, will then lead to small charge exchange only. After the irradiation with $q = 36$, the sample was investigated with $q = 20$ again (c.f. Fig. 3(d)) and a significant change in the charge pattern can be observed. A pronounced distribution at higher exit charge states is visible, again in agreement with pore formation due to the high charge state irradiation.

5. Charge exchange mechanism in solids

In order to quantify charge exchange patterns from ion transmission and extract material porosity (c.f. Fig 3) a model calculation needs to be applied. Charge exchange of highly charged ions can be distinguished by two subsequent phases. The first phase is characterized by over-barrier charge transport from the surface to the ion. The classical-over-barrier (COB) model [43] describes quantitatively the distance of first electron transport and serves therefore as an input for the following full charge exchange model. One should notice that the COB model can only describe subsequent single electron transfers. At normal incidence conditions and with ion velocities in the range of 10^5 m/s or above, multi-electron transport must be considered. However, most quantities are determined by the critical distance for first electron transport, which can be reliably calculated.

The COB phase is followed by a quenching of the neutralized, but highly excited atom (hollow atom). The second phase consists of a competition of several de-excitation processes. Radiative de-excitation is charge conserving, but appears with lifetimes in the ns-μs range especially for the Rydberg-like states occupied in a hollow atom. Non-radiative de-excitation appears intra- and inter-atomic, whereas only the latter is charge conserving. The former leads to an efficient recharging of the hollow atom due to ionization of the Auger electrons in contrast to the latter, where the Auger electron is emitted from the target atom. It is important to consider, that inter-atomic Auger-type de-excitation has a distance dependent lifetime, which approaches about 1 fs for interatomic distances $R < 1$ Å and drops rapidly for larger R. In short, non-radiative de-excitation typically dominates and within this channel charge conserving inter-atomic Auger-type decay dominates for close collisions and recharging intra-atomic Auger-type decay dominates at distant collisions.

A quantitative formulation of the discussion above yields three rate equations for different electron occupations in the projectile

$$\frac{\mathrm{d}N_{core}}{\mathrm{d}t} = 0, \tag{1}$$

$$\frac{\mathrm{d}N_{cap}(t)}{\mathrm{d}t} = \lambda(t)\,(Z - N_{core} - N_{cap}(t) - N_{stab}(t)) - \gamma(t)N_{cap}(t), \tag{2}$$

$$\frac{\mathrm{d}N_{stab}(t)}{\mathrm{d}t} = \gamma(t)N_{cap}(t). \tag{3}$$

Here N_{core} is the number of frozen core electrons, $N_{cap}(t)$ the number of electrons captured from the surface and bound in highly excited Rydberg-like states, and $N_{stab}(t)$ the number of electrons

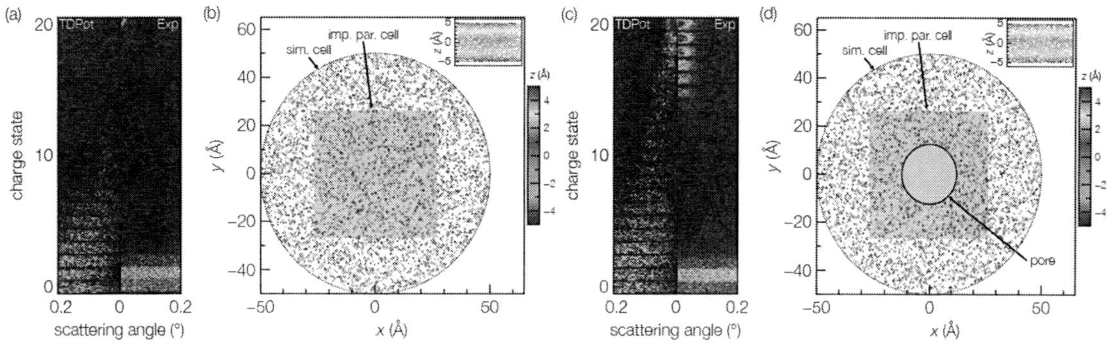

Figure 4. (color online) (a): Charge state patterns from 180 keV Xe^{20+} ions from a TDPot calculation and from experiment. (b): Respective simulated material structure. The simulation cell is indicated as well as the region in which impact parameters are chosen. The latter is selected to be smaller than the simulation cell to avoid edge effects (i.e. finite size effects). (c): Same as (a), but for the material after exposure to a Xe^{36+} beam, i.e. with ion induced pores. (d): Respective simulated material structure with a 25 Å pore.

quenched into lowly excited states such that they cannot undergo autoionization anymore. The rates $\lambda(t)$ and $\gamma(t)$ are the classical-over-barrier transport rate and the hollow atom quenching rate, respectively. An elaborated discussion about the determination of these rates can be found in [42]. In short, λ is assumed to be 0 for $R > R_c$, where R_c is the critical capture distance from the COB model, and 1 a.u. for $R \leq R_c$. The value of $\gamma = \gamma(R[t])$ is taken from the R-dependence of the Interatomic Coulombic Decay (ICD) process [21]. The occupation number of core, lowly excited and highly excited levels in the projectile modifies the screening of the interaction potential between ion and target atoms. In fact, the potential

$$\begin{aligned} V(R,t) &= \frac{(N_{core}+N_{stab}(t))Z_2}{R}\phi\left(\frac{R}{a_1((N_{core}+N_{stab}(t)),Z_2)}\right) \\ &+ \frac{N_{cap}(t)Z_2}{R}\phi^{cap}_{hollow}(R,r_0) \\ &+ \frac{(Z_1-N_{core}-N_{stab}(t)-N_{cap}(t))Z_2}{R}\phi\left(\frac{R}{a_2(Z_2)}\right), \end{aligned} \tag{4}$$

becomes explicitly time-dependent. The screening functions ϕ and ϕ^{cap}_{hollow} as well as the corresponding screening lengths a_1 and a_2 are described within the framework of the TDPot model [42]. With the help of Eqs. (1)-(4) the final charge state of the projectile as well as the nuclear and electronic energy losses can be calculated for every individual impact parameter. The electronic energy loss follows from the fact that for scattering in a time-dependent potential the center-of-mass (cms) kinetic energy is not conserved, whereas the momentum conservation still holds. The difference in cms kinetic energy before and after scattering is spent in the rearrangement of the electron configuration (alteration of the potential) and is here termed 'electronic energy loss'. Note, that the scattering is always inelastic when charge exchange occurs, i.e. nuclear energy loss can only be defined by the energy transfer to target nuclei, but not as the 'elastic' contribution. It is obvious that in case of charge exchange, nuclear and electronic energy loss are closely coupled and can not be treated as independent contributions to the stopping force anymore.

With a quantitative description of the charge exchange dynamics in a way that allows the determination of the final ion charge, energy, and angle, a material structure can be simulated

and the result compared to experiment. In Fig. 4(a) the charge pattern of 180 keV Xe^{20+} transmitted through a 1 nm thick amorphous CNM is shown (same as Fig. 3(a)), both as a result of a TDPot calculation (left) and the experiment (right). Fig. 4(b) shows the material structure as used for TDPot. A carbon layer was simulated, where the atom positions are chosen randomly within a cell of 100 Å in diameter and thickness of 10 Å. The atom areal density in TDPot is 5.54×10^{15} at$\times$cm^{-2}, which matches an estimated value from the production process of CNMs, i.e. the areal density of carbon atoms in the self-assembled monolayer of NBPT molecules on a Au(111) surface, which serves as a precursor for CNMs. The agreement between TDPot and experiment is good, i.e. the charge state distribution at $q < 5$ is well reproduced. TDPot shows additionally higher transmitted charge states ($5 < q < 12$) at small angles. A separated, somewhat higher charge state distribution was previously observed for CNMs also in experiment [30], but not for this particular sample here. A reason could be the presence of additional contaminations at the sample, which block the forward transmission of high charge states. Detailed analysis of the TDPot results show, that these higher transmitted charge states pass regions in the CNMs where the atom-atom distances are larger, i.e. low density regions in the membrane. This is in agreement with the small deflection angles both in TDPot (c.f. Fig. 4(a) $< 0.1°$) and previous experiments. If severe contaminations are present then they increase the areal density of the material, thus the probability to pass a low density area (channel) in the material decreases and higher transmitted charge states in the pristine case are absent.
Fig. 4(c) and (d) show the TDPot and experimental charge pattern as well as simulated structure for the same sample after it was exposed to a 324 keV Xe^{36+} beam. In both. TDPot and experiment a pronounced high charge state distribution at small deflection angles is observed. TDPot was simulated with a 25 Å pore in the center of the simulation cell. Pore formation for ions with $q > 20$ was observed previously utilizing transmission electron microscopy [31]. As expected, ions transmitted through the pore remain in high charge states and only ions passing the rim of the pore within the COB critical capture distance of ~ 10 Å undergo charge exchange but almost no ICD. Afterwards intra-atomic Auger decay takes place, i.e. the projectile recharges. Additionally, TDPot uses a finite simulation cell (100 Å diameter) and impact parameters are selected within a square of $37.5\,\text{Å}/\sqrt{2}$ edge length to avoid finite size effects in the simulation. Thus, the ratio between lower and higher charge state distribution is a direct measure for the average pore density in the membrane, i.e. $1\,\text{pore}/700\,\text{nm}^2 = 1.4 \times 10^{11}\,\text{pores}/\text{cm}^2$. This is in fair agreement with the fluencies typically applied for ion transmission measurements until severe material degradation (perforation) occurs. However, for a more detailed quantitative analysis further measurements and microscopy analysis are needed.

6. Outlook

Highly charged ion transmission spectroscopy is powerful not only for basic research in charge exchange of ions with solids, but also for material structure determination. As shown above, the methodology can be applied to materials, where the structure is not known and cannot be determined by transmission electron microscopy as it is the case for amorphous materials. Porosity can be determined on the sub-nm length scale, especially interesting for molecular sieving and desalination applications of these organic membranes. To further boost the quantitative comparison of experiments with simulations, material quality must be improved, i.e. cleaning procedures to reduce contaminations must be applied. Further, the ion transmission spectroscopy must be gauged against known structures, i.e. using crystalline 2D materials with different structures where the atomic positions are well known. This allows the determination of input parameters necessary for a simulation, such that the simulation can then be applied to unknown samples. A critical parameter is the distance dependent ICD rate, which is not known from ab-initio calculations at the moment for small atom separations. Currently, the hollow atom quenching rate (ICD rate) is estimated from an extrapolation of literature values at larger separations. Either adjusting the rate in the simulation to fit experimental results for various known sample structures or an approach to determine the rate at small separations from first principles must be done. In the not so far future, HCI based ion transmission spectroscopy may become a powerful addition for material structure determination for amorphous or short-ranged ordered molecular nanosheets.

Acknowledgments

Financial support from the Deutsche Forschungsgemeinschaft (WI 4691/1-1, project no. 322051344), the Austrian FWF (project no. Y 1174-N36), and the TU Wien (Innovative Projects) is gratefully acknowledged. I want to express my gratitude to F. Aumayr, S. Creutzburg, J. Schwestka, A. Niggas, and P.L. Grande for support of this work. I thank my collaborators M. Schleberger, L. Madauß, J. Kotakoski, H. Inani, A. Turchanin, and A. George for sample preparation and characterization. Parts of this research were carried out at IBC at the Helmholtz-Zentrum Dresden - Rossendorf e. V., a member of the Helmholtz Association. I would like to thank S. Facsko and R. Heller for assistance.

[1] Chang C C 1971 *Surf. Sci.* **25** 53–79
[2] Ferraioli L, Wang M, Pucker G, Navarro-Urrios D, Daldosso N, Kompocholis C and Pavesi L 2007 *J. Nanomater.* **2007** 1–5
[3] Damascelli A 2004 *Phys. Scr.* **T109** 61
[4] Seiler H 1983 *J. Appl. Phys.* **54** R1–R18
[5] Hembree G and Venables J 1992 *Ultramicroscopy* **47** 109–120
[6] Müller V, Hinaut A, Moradi M, Baljozovic M, Jung T A, Shahgaldian P, Möhwald H, Hofer G, Kröger M, King B T, Meyer E, Glatzel T and Schlüter A D 2018 *Angew. Chemie Int. Ed.* **57** 10584–10588
[7] Limburg J, Schippers S, Hoekstra R, Morgenstern R, Kurz H, Aumayr F and Winter H 1995 *Phys. Rev. Lett.* **75** 217–219
[8] Briand J P, de Billy L, Charles P, Essabaa S, Briand P, Geller R, Desclaux J P, Bliman S and Ristori C 1990 *Phys. Rev. Lett.* **65** 159–162
[9] Briand J P, de Billy L, Charles P, Essabaa S, Briand P, Geller R, Desclaux J P, Bliman S and Ristori C 1991 *Phys. Rev. A* **43** 565–567
[10] Schwestka J, Wilhelm R, Gruber E, Heller R, Kozubek R, Schleberger M, Facsko S and Aumayr F 2018 *Nucl. Instruments Methods Phys. Res. Sect. B Beam Interact. with Mater. Atoms* **422** 63–67
[11] Aumayr F and Winter H 2007 *Slow Heavy-Particle Induc. Electron Emiss. from Solid Surfaces* (*Springer Tracts in Modern Physics* vol 225) ed Winter H P and Burgdörfer J (Berlin, Heidelberg: Springer Berlin Heidelberg) chap 3, pp 79–112 ISBN 978-3-540-70788-2
[12] Köhrbrück R, Sommer K, Biersack J P, Bleck-Neuhaus J, Schippers S, Roncin P, Lecler D, Fremont F and Stolterfoht N 1992 *Phys. Rev. A* **45** 4653–4660
[13] Kost D, Facsko S, Möller W, Hellhammer R and Stolterfoht N 2007 *Phys. Rev. Lett.* **98** 225503
[14] Facsko S, Meissl W, Heller R, Wilhelm R, El-Said A S, Kowarik G, Ritter R and Aumayr F 2009 *J. Phys. Conf. Ser.* **194** 012060
[15] Heller R, Facsko S, Wilhelm R A and Möller W 2008 *Phys. Rev. Lett.* **101** 096102
[16] Winecki S, Cocke C L, Fry D and Stöckli M P 1996 *Phys. Rev. A* **53** 4228–4237
[17] Hattass M, Schenkel T, Hamza A V, Barnes A V, Newman M W, McDonald J W, Niedermayr T R, Machicoane G A and Schneider D H 1999 *Phys. Rev. Lett.* **82** 4795–4798
[18] Arnau A, Aumayr F, Echenique P, Grether M, Heiland W, Limburg J, Morgenstern R, Roncin P, Schippers S, Schuch R, Stolterfoht N, Varga P, Zouros T and Winter H 1997 *Surf. Sci. Rep.* **27** 113–239
[19] Martin S, Brédy R, Bernard J, Désesquelles J and Chen L 2002 *Phys. Rev. Lett.* **89** 183401
[20] Wilhelm R A, Gruber E, Schwestka J, Kozubek R, Madeira T I, Marques J P, Kobus J, Krasheninnikov A V, Schleberger M and Aumayr F 2017 *Phys. Rev. Lett.* **119** 103401
[21] Cederbaum L S, Zobeley J and Tarantelli F 1997 *Phys. Rev. Lett.* **79** 4778–4781
[22] Hemmerich J L, Bennett R and Buhmann S Y 2018 *Nat. Commun.* **9** 2934
[23] Bennett R, Votavová P, Kolorenč P, Miteva T, Sisourat N and Buhmann S Y 2019 *Phys. Rev. Lett.* **122** 153401
[24] Gruber E, Wilhelm R A, Pétuya R, Smejkal V, Kozubek R, Hierzenberger A, Bayer B C, Aldazabal I, Kazansky A K, Libisch F, Krasheninnikov A V, Schleberger M, Facsko S, Borisov A G, Arnau A and Aumayr F 2016 *Nat. Commun.* **7** 13948
[25] Wilhelm R, Gruber E, Schwestka J, Heller R, Fascko S and Aumayr F 2018 *Appl. Sci.* **8** 1050

[26] Markin S, Primetzhofer D and Bauer P 2009 *Nucl. Instruments Methods Phys. Res. Sect. B Beam Interact. with Mater. Atoms* **267** 634–637
[27] Klingner N, Heller R, Hlawacek G, von Borany J, Notte J, Huang J and Facsko S 2016 *Ultramicroscopy* **162** 91–97
[28] Vieluf M, Munnik F, Neelmeijer C, Kosmata M and Teichert S 2012 *Thin Solid Films* **520** 5900–5905
[29] Kosmata M, Munnik F, Hanf D, Grötzschel R, Crocoll S and Möller W 2014 *Nucl. Instruments Methods Phys. Res. Sect. B Beam Interact. with Mater. Atoms* **337** 27–33
[30] Wilhelm R A, Gruber E, Ritter R, Heller R, Facsko S and Aumayr F 2014 *Phys. Rev. Lett.* **112** 153201
[31] Ritter R, Wilhelm R A, Stöger-Pollach M, Heller R, Mücklich A, Werner U, Vieker H, Beyer A, Facsko S, Gölzhäuser A and Aumayr F 2013 *Appl. Phys. Lett.* **102** 063112
[32] Kozubek R, Tripathi M, Ghorbani-Asl M, Kretschmer S, Madauß L, Pollmann E, O'Brien M, McEvoy N, Ludacka U, Susi T, Duesberg G S, Wilhelm R A, Krasheninnikov A V, Kotakoski J and Schleberger M 2019 *J. Phys. Chem. Lett.* **10** 904–910
[33] Schwestka J, Melinc D, Heller R, Niggas A, Leonhartsberger L, Winter H, Facsko S, Aumayr F and Wilhelm R A 2018 *Rev. Sci. Instrum.* **89** 085101
[34] Jagutzki O, Cerezo A, Czasch A, Dorner R, Hattas M, Min Huang, Mergel V, Spillmann U, Ullmann-Pfleger K, Weber T, Schmidt-Bocking H and Smith G 2002 *IEEE Trans. Nucl. Sci.* **49** 2477–2483
[35] Schmidt M, Peng H, Zschornack G and Sykora S 2009 *Rev. Sci. Instrum.* **80** 063301
[36] Micke P, Kühn S, Buchauer L, Harries J R, Bücking T M, Blaum K, Cieluch A, Egl A, Hollain D, Kraemer S, Pfeifer T, Schmidt P O, Schüssler R X, Schweiger C, Stöhlker T, Sturm S, Wolf R N, Bernitt S and Crespo López-Urrutia J R 2018 *Rev. Sci. Instrum.* **89** 063109
[37] Czasch A, Milnes J, Hay N, Wicking W and Jagutzki O 2007 *Nucl. Instruments Methods Phys. Res. Sect. A Accel. Spectrometers, Detect. Assoc. Equip.* **580** 1066–1070
[38] Algara-Siller G, Lehtinen O, Turchanin A and Kaiser U 2014 *Appl. Phys. Lett.* **104** 153115
[39] Lakits G, Aumayr F and Winter H 1989 *Rev. Sci. Instrum.* **60** 3151–3159
[40] Turchanin A and Gölzhäuser A 2016 *Adv. Mater.* **28** 6075–6103
[41] Wilhelm R A, Gruber E, Ritter R, Heller R, Beyer A, Turchanin A, Klingner N, Hübner R, Stöger-Pollach M, Vieker H, Hlawacek G, Gölzhäuser A, Facsko S and Aumayr F 2015 *2D Mater.* **2** 035009
[42] Wilhelm R and Grande P 2019 *Commun. Phys.* **2** 89
[43] Burgdörfer J, Lerner P and Meyer F W 1991 *Phys. Rev. A* **44** 5674–5685

 1412 (2020) 072001 doi:10.1088/1742-6596/1412/7/072001

Temporal-spatial double slit interference of photoelectron

H Liang*, X R Xiao and L Y Peng†

State Key Laboratory for Mesoscopic Physics and Department of Physics, Peking University, Beijing, 100871, People's Republic of China

Synopsis The electron displacement induced by an ultrashort pulse is theoretically investigated. We propose and numerically demonstrate a scheme to accurately measure the electron displacement using a ruler formed by the interfering spirals in the photoelectron momentum distribution generated by two oppositely circularly polarized pulses. It can be understood as a temporal double-slit with an additional spatial shift on one slit.

Recently, the helical vortex structures in photoelectron momentum distributions (PMD) was numerically observed [1] from atom induced by two oppositely circularly polarized, time-delayed pulses and then verified experimentally [2, 3]. It was interpreted as the interference between $M = \pm L$ continue states induced by two count-rotate pulse.

Here [4], we show that this vortex structure can also be regarded as double-slit interference in time domain. Apply the pulse with configuration shown in Fig. 1, the electron ionized at a certain time t has the final momentum $\boldsymbol{p}$ opposite to the vector potential $\boldsymbol{A}(t)$ approximately. Consider the time t_1 and t_2 for two delayed pulses with same vector potential direction, those electron ionized at t_1 and t_2 would interference with each other with phase difference $\varphi_0 = (I_p+\boldsymbol{p}^2/2)(t_2-t_1)$, where I_p is the ionization potential of the atom. As t_1 gets smaller and t_2 gets larger, the corresponding vector potentials rotate the same angle $\delta\theta = \omega\delta t$ in the same direction. We have $\varphi_0(\theta+\delta\theta) - \varphi_0(\theta) = 2(I_p + \boldsymbol{p}^2/2)\delta\theta/\omega$, results evenly spaced fringes in photoelectron angular distribution with angular stripe spacing equals to $\pi\omega/(I_p + \boldsymbol{p}^2/2)$. By considering interference between different optical cycles, an additional condition $I_p + \boldsymbol{p}^2/2 \approx n\omega, n \in \mathbb{Z}$ is required, which corresponding to the n-photon process.

With such physics picture in mind, we now impose an additional attosecond laser pulse between these two opposite circularly pulses shown in Fig. 1. The electron ionized by the first pulse would gain a spatial displacement $\boldsymbol{r} = \int \boldsymbol{A}\,\mathrm{d}t$ from the intermediate pulse. Thus, a phase shift $\varphi_1 = \boldsymbol{p}\cdot\boldsymbol{r}$ is applied to the electron. Now, as the vector potential (momentum) rotate, we have $\delta\varphi_{\text{total}} = 2n\delta\theta + \frac{\delta\boldsymbol{p}}{\delta\theta}\cdot\boldsymbol{r}\delta\theta$, which approximately results a shift $\Delta\boldsymbol{p} = \omega\boldsymbol{n}\times\boldsymbol{r}/2$ in PMD. From the momentum shift in PMD, one can extract the electron displacement induced by the attosecond laser pulse.

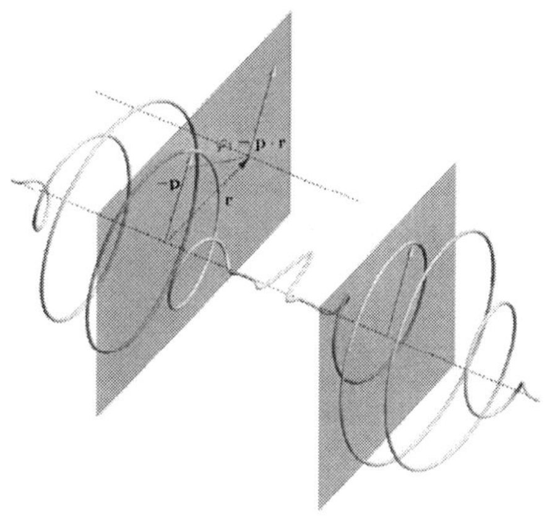

Figure 1. Demonstration for the present temporal-spatial double-slit interference.

References

[1] Djiokap J M N *et al* 2015 *Phys. Rev. Lett.* **115** 113004
[2] Pengel D *et al* 2017 *Phys. Rev. Lett.* **118** 053003
[3] Kerbstadt S *et al* 2019 *Nat. Commun.* **10** 658
[4] Xiao X R *et al* 2019 *Phys. Rev. Lett.* **122** 053201

*E-mail: haoliang@pku.edu.cn
†E-mail: liangyou.peng@pku.edu.cn

Angle-resolved RABBITT : from atoms to molecules

J Joseph[1 *], F Holzmeier[1 †], D Bresteau[2], C Spezzani[3], T Ruchon[2], J F Hergott[2], O Tcherbakoff[2], P D'Oliveira[2], R R Lucchese[4], J C Houver[1], B Carré[2] and D Dowek[1]

[1]ISMO, CNRS, Univ. Paris-Sud, Université Paris-Saclay, Orsay, 91405, France
[2]LIDYL, CEA, CNRS, Université Paris-Saclay, Gif Sur Yvette, 91191, France
[3]Elettra-Sincrotrone, Trieste, Basovizza, 34149, Italy
[4]Chemical Sciences Division, Lawrence Berkeley National Laboratory, Berkeley, California, 94720, USA

Synopsis We report recent results of photoelectron angular distributions (PADs) for photoionization (PI) of atomic and molecular targets, resolved in the time domain at the attosecond scale, combining electron-ion coincidence momentum spectroscopy and the RABBITT scheme. We propose a unified formalism synthetizing the $PAD(\theta,\tau)$ results in terms of a reduced set of coefficients.

Combining the XUV-IR RABBITT [1] interferometric technique at the 10kHz Attolab-FAB10 beamline, with electron-ion coincidence momentum spectroscopy, we investigate PADs for atomic (Ar) and molecular (NO, O_2) targets.
In this scheme, an attosecond XUV pulse train ionizes the target and a superimposed IR field with angular frequency ω and delay τ induces continuum-continuum (cc) transitions, generating sidebands (SB) in the photoelectron (PE) spectra. Analysing the 2ω modulation of the SB signal as a function of τ paved the way to studies of PI dynamics in real time, leading to the determination of Wigner time delays τ_W of the PE wavepacket generated by one-XUV-photon absorption [2]. This requires deciphering, within the atomic time delay τ_A, the τ_{cc} contribution due to the cc transition [3]. Advanced developments address angularly resolved RABBITT studies, with the goal to retrieve Wigner delays as a function of the ejection direction [4-9].
This experiment gives access to the PAD $I_{SB}(\theta,\tau)$, where θ is the emission angle relative to the light polarization **P**, and to the MFPAD $I_{SB}(\theta_e,\phi_e,\chi,\tau)$ for a molecule oriented at any χ angle relative to **P**, where (θ_e,ϕ_e) defines the emission direction in the molecular frame (MF).
In the RABBITT spectrogram:

$$I_{SB}(\theta,\tau)=A(\theta)+B(\theta)\cos(2\omega\tau+C(\theta))$$

the atomic PE delays $\tau_A(\theta)$ relative to $\tau_A(0°)$ are extracted from the C(θ) phase. Here we display an example of measured A(θ) (static) and B(θ) (dynamic) amplitudes for PI of Ar. A unified formalism synthetizing $I_{SB}(\theta,\tau)$ in terms of a reduced set of coefficients, which allows for generating e.g. the A(θ), B(θ) and C(θ) functions, will be presented. First MFPADs, obtained taking advantage of dissociative PI, will also be reported for PI of NO and compared with MCSCI calculations.

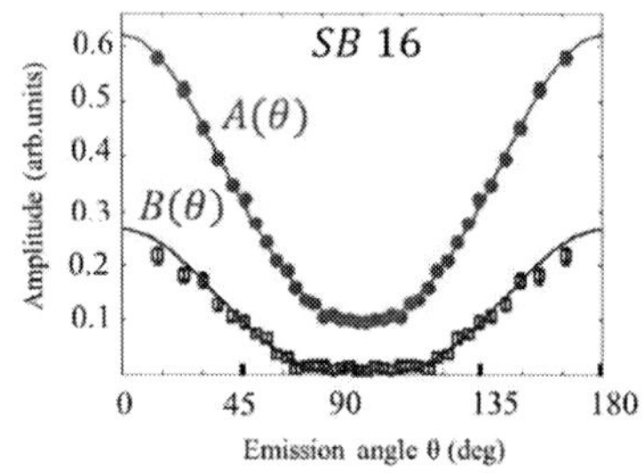

Figure 1 A(θ) and B(θ) for SB16 in XUV+IR PI of Ar.

Acknowledgments
We gratefully greet the memory of B Carré, coordinator of Attolab, deceased March 6, 2018. This research was supported by the ANR-10-LABX-0039-PALM, ANR-11-EQPX0005-ATTOLAB, ASPIRE Marie Sklodowska-Curie ITN (H2020 ID: 674960), and Lidex OPT2X-Paris-Saclay.

References
[1] Paul P M *et al* 2001 *Science* **292** 1689
[2] Klünder K *et al* 2011 *Phys. Rev. Lett.* **106** 143002
[3] Dahlström J M *et al.* 2013 *Chem. Phy.* **414** 53
[4] Heuser S *et al* 2016 *Phys. Rev. A.* **94** 063409
[5] Cirelli C *et al* 2018 *Nat. Commun.* **9** 955
[6] Bray A W *et al* 2018 *Phys. Rev. A.* **97** 063404
[7] Vos J *et al* 2018 *Science* **360** 1326
[8] Hockett P *et al* 2016 *J. Phys. B. At. Mol. Opt. Phys.* **49** 095602
[9] Baykusheva D and Wörner H J 2017 *J. Chem. Phys.* **146** 124306

* E-mail: jennifer.joseph@u-psud.fr
† E-mail: fabian.holzmeier@polimi.it

Broadband supercontinuum harmonics and isolated utrashort attosecond pulse generation by optimizing three-color laser fields

H-Y Yu, Z-H Jiao, X-X Zhou and P-C Li*

College of Physics and Electronic Engineering, Northwest Normal University, Lanzhou, 730070, China

Synopsis We present an efficient scheme for the generation of broadband supercontinuum harmonics and isolated ultrashort attosecond pulse by the optimized three-color laser fileds. The optimized laser pulse is synthesized by three one-color laser pulses with proper relative phases. The broadband supercontinuum harmonics are obtained by solving accurately the time-dependent Schrödinger equation using the time-dependent generalized pseudospectral method. Our scheme proposed produces a broadband supercontinuum spectrum with the photon energies near 0.3 keV. As a result, an isolated 17 as pulse can be obtained directly by superposing the supercontinuum harmonics.

Attosecond pulse can be used to probe the chemistry reaction and obtain the electronic dynamical informations of the atoms and the more details of the microcosmos. High-order harmonic generation (HHG) seems the most promising method to produce the attosecond pulse. According to the semiclassical three-step model, the electron can acquire additional kinetic energy after tunneling through the Coulomb potential, and then the electron recombines with the parent nucleus and emits radiation. The harmonic spectrum is characterized by a rapid drop at the beginning and a sharp cutoff around $I_p+3.17U_p$, where the I_p is the atomic ionization potential and U_p is the pondermotive energy.

In Fig 1, we present the HHG spectra of helium atom (He) driven by an optimized three-color laser field [1]. The harmonic spectrum of He can be calculated by solving the three-dimensional time-dependent Schrödinger equation by means of the time-dependent generalized pseudospectral method (TDGPS) [2]. The optimized three-color laser field, including two one-color mid-infrared driving pulse, is a good choice for the extension of the harmonic plateau and the generation of ultrashort attosecond pulse, because the ponderomotive energy is proportional to $I\lambda^2$, which implies that the harmonic cutoff can be extended using longer wavelength of the driving laser field. In our calculation, we use the laser intensity $I_1=3.0\times10^{14}$ W/cm^2 (800nm), $I_2=3.0\times10^{14}$ W/cm^2(2000nm), and $I_3=3.4\times10^{14}$ W/cm^2(2400nm), respectively. Three one-color laser pulses chosen are near 12-fs duration for the experimental realizations, and the relative phase is π ,0, and -0.5π, respectively.

By superposing the broadband supercontinuum harmonics from the 100th to 290th orders with the photon energies near 0.3 keV, an isolate 17as pulse can be generated directly. Although our simulation is a single-atom level, the previous studies indicate that the single-atom supercontinuum harmonics is similar even considering macroscopic phase matching.

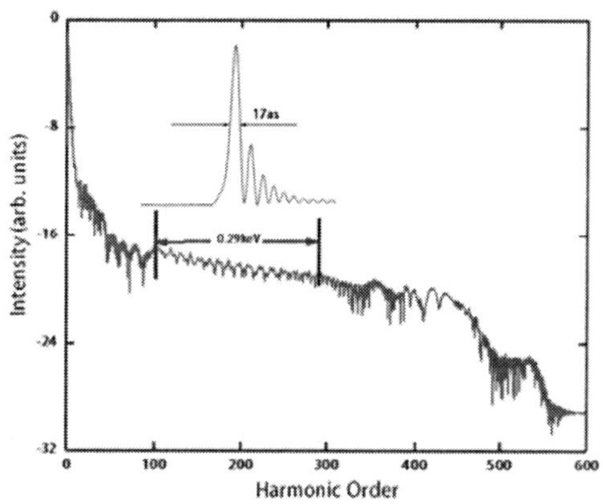

Figure 1. Supercontinuum harmonic spectra and ultrashort isolated attosecond generation of He atom driven by the optimized three-color laser field.

Acknowledgments

This work was supported by the National Natural Science Foundation of China (Grant No.11674268, No.11764038 and No.11765018).

References

[1] P. C. Li *et al.* 2011 *Opt. Express* **24** 23857
[2] X. M.Tong. *et al.* 1997 *Chem. Phys.* **217** 119

* E-mail: lipc@nwnu.edu.cn

Semiclassical interference in photoelectron spectra: Holographic structures

S D López[1] and D G Arbó[1 †]

[1]Institute for Astronomy and Space Physics IAFE (UBA-Conicet), Buenos Aires, 1428, Argentina

Synopsis We theoretically investigate on the holographic structure in doubly differential momentum distribution for atomic photoionization. We use the ab initio solution of the time dependent Schrödinger equantion and the two-step semiclassical model. We see how the holographic structures are formed due to the interference of the direct trajectories of electrons and the rescattered trajectories of electrons. We get rid of intra- and intercycle interferences by removing ionization in the second half of a one-cycle pulse.

When an intense short laser pulse interacts with an atom ionizing it, the photoelectron spectra present several structures that can be understood as double slit interferences in the time domain, namely intra- and inter-cycle interferences [1]. This type of structures is formed by electrons that emerge from the atom and follow directly to detector. There is another type of structures that requires interference of direct electron trajectories with others that interact with the parent core. In a classical picture, the rescattering trajectories return to the parent ion driven by the laser field and rescatters off to the detector. It is well known that a short range potential is sufficient to understand the rescattering rings at high energy, formed by very hard interaction among electron and ion [2]. Other structures can be explained only when long range coulomb interactions take place. This type of structures can be interpreted as holograms, i.e. the interference pattern between the direct (reference) and the rescattered (signal beam) electrons. In this way, the information of the interaction is encoded in the interference pattern between the reference and the signal [3].

In this work we present a theoretical analysis of interferences using a numerical solution of the time dependent Schrödinger equation (TDSE) and the semiclassical two step model (SCTS) [4]. We analyze the ionization of atomic hydrogen to characterize the role that the long-range Coulomb interaction plays in the holographic structures. With semiclassical approaches we can examine the trajectories that lead to different final conditions and get a deeper understanding of the electron kinematics associated to holographic structures.

We show in Figure 1 photoelectron distributions as a function of parallel k_z and perpendicular $k_\rightarrow$ momentum to polarization axis calculated with the SCTS model (upper pannel) and the TDSE (lower pannel). We observe interference patterns resembling a dispersion of an electron wave packet from a potential centered in the origin in both calculations.

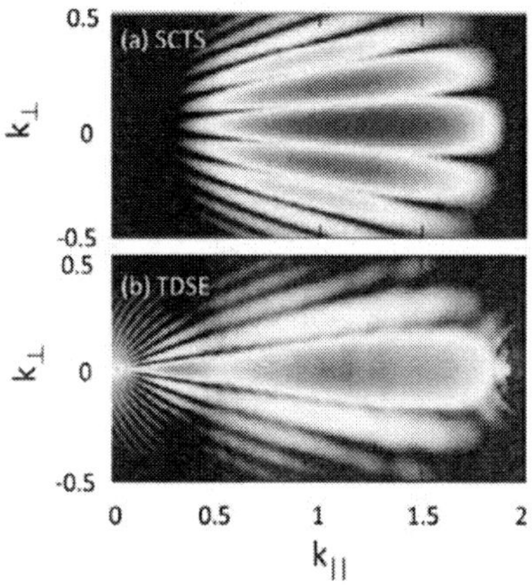

Figure 1. Momentum distribution after interaction of a near-infrared (λ = 800 nm) one-cycle laser pulse with a hydrogen atom. (a) SCTS, (b) TDSE. I = 10^{14} W/cm^2

This work has been supported by CONICET (Argentina PIP0386) and ANPCyT (Argentina PICT-2016-0296 and PICT-2014-2363).

References

[1] Arbó D G *et al* 2010 *Phys. Rev. A* **82** 043426; ibid **81** 021403
[2] Lewenstein M *et al* 1995 *Phys. Rev. A* **51** 1495
[3] Huismans Y *et al* 2011 *Science* **331** 61
[4] Shvetsov-Shilovski N *et al* 2016 *Phys. Rev. A* **94** 013415

† E-mail: diego@iafe.uba.ar

Journal of Physics: Conference Series **1412** (2020) 072005 doi:10.1088/1742-6596/1412/7/072005

Abnormal photoelectron angular distribution of Ar atoms in pulsed XUV and IR laser fields

XM Tong[1 *], and N Toshima[2]

[1]Center for Computational Sciences, University of Tsukuba, Tsukuba, 305-8577, Japan
[2]Institute of Materials Science, University of Tsukuba, Tsukuba, 305-8573, Japan

Synopsis We calculated the photoelectron angular distribution of Ar atoms in pulsed XUV and IR laser fields by solving the time-dependent Schrödinger equation and found that the angular distributions of sidebands are *not always* straight lines in the energy-angle plot as predicted by the strong field approximation. Further study shows that the abnormality is attributed to the photoelectron-nucleus Coulomb interaction. The abnormality depends on the IR intensity, IR pulse duration, XUV photon energy and the time-delay between the two pulses.

Combination of an extreme-ultraviolet (XUV) source generated from the high-order-harmonic generation (HHG) and the infrared (IR) laser opens a new door to investigate and control atomic photoabsorption processes.

Different from the XUV source generated from HHG, which is of the form of an attosecond pulse train or a single attosecond pulse, the photon energy of the XUV source radiated from a free-electron laser can be tuned almost continuously and the pulse width can be narrowed in a femto-second time-scale, so it opens another dimension to study the IR assisted photoabsorption as recently reported in both theory [1] and experiment [2,3]. The two experiments focused on the photoelectron angular distribution.

When we investigated the photoelectron angular distribution of Ar atoms ionized by an XUV assisted with IR fields as reported recently in the two experiments, we found that the sideband angular distributions, as shown in Fig. 1, are not iso-energy separated by one IR photon energy, or straight lines in energy-angle plot as predicted by the strong field approximation (SFA). We called this non-straight-line distribution in the energy-angle plot as an abnormal angular distribution.

To investigate the origin of the abnormality, we simulated the photoelectron angular distribution of Ar atoms in two-color XUV and IR laser pulses and confirmed that the angular distribution is *not always* iso-energy. Simply switching off the photoelectron-nucleus Coulomb interaction in the simulation, the SFA does predict iso-energy distributions, which are in consistent with our present knowledge of the angular distribution for a multi-photon absorption. This clearly shows that the abnormality is attributed to the Coulomb effect.

* E-mail: tong.xiaomin.ga@u.tsukuba.ac.jp

In this work, we also systematically investigated how the abnormal angular distribution depends on the IR intensity, pulse duration and the time-delay between the two pulses. The dependences may be used to extract the laser information by comparing measured angular distributions with simulated ones.

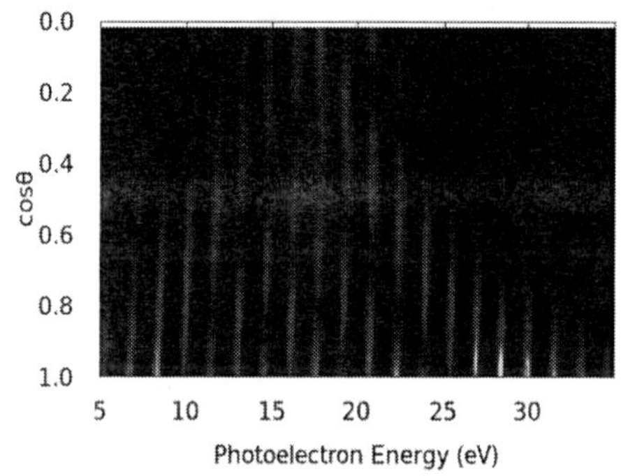

Figure 1. Photoelectron energy-angle distribution of Ar atoms ionized by a 33 eV, 8 fs pulsed XUV light in an 800 nm IR laser field. The laser intensity is 2×10^{13} W/cm^2 and the pulse duration is 30 fs.

Research was supported by a Grants-in-Aid for Scientific Research (JP16K05495) from the Japan Society for the Promotion of Science and by Multidisciplinary Cooperative Research Program in CCS, University of Tsukuba.

References

[1] Kazansky AK *et al* 2010 *Phys. Rev. A* **82** 033430
[2] Minemoto S *et al* 2018 *J. Phys. B* : **51** 075601
[3] Hummert J *et al* 2018 arXiv :1808 :04639

Opto-optical modulator (OOM) for extreme ultraviolet pulses

A Olofsson[1], E R Simpson[1], M Labeye[2], S Camp[2], N Ibrakovic[1], S Bengtsson[1], K J Schafer[2], M B Gaarde[2] and J Mauritsson[1*]

[1]Department of Physics, Lund University, P.O. Box 118, SE-221 00 Lund, Sweden
[2]Department of Physics and Astronomy, Louisiana State University, Baton Rouge, Louisiana, 70808, USA

Synopsis There is a profound, almost symbiotic relation between electrons and photons. When light interacts with matter the electrons in the material will start to move and oscillate. When a charged particle, such as an electron, oscillates it will act as a dipole and emit light. We utilize this symbiotic relation and extend the control of light to the extreme ultraviolet (XUV) region using the newly developed XUV opto-optical modulator.

To control light with short wavelengths is challenging since the tools available for visible light cannot easily be extended to shorter wavelength regions. We present a technique to control the phase of light in the extreme ultraviolet XUV wavelength region. We do this using the recently demonstrated opto-optical modulator (OOM), a technique for controlling the direction, duration and timing of XUV pulses using infrared (IR) control pulses [1,2].

In the OOM, coherent XUV light is used to promote an ensemble of atoms to a superposition of the ground state and a series of excited states. As the excited states decay [4,5], emission of the same XUV frequency light is reradiated, destructively interfering with the incident radiation to produce a normal absorption spectrum on-axis in the far-field. In this time, a second laser pulse can be used to control the target atoms, directly affecting the wavefronts, and consequently the direction, of any subsequent XUV emission. This is achieved through the application of a state-specific, spatial phase gradient across the radiating target focus, arising from the intensity-dependent, AC Stark-induced energy shifts that occur upon interaction of the target atoms with the control laser pulse. By offsetting the foci of the excitation and probe pulses, simple control over the spatial gradient of intensities applied to the target atoms can be achieved, allowing manipulation over the redirection angle of subsequent XUV emission.

Since the spatial profile of the IR probe in the interaction regioncan be easily controlled the method can be further extended to steer or split XUV pulses. We demonstrate this control through a which-way interference experiment where different parts of the XUV beam are made to overlap in the far-field producing interference. Control over the IR probe pulse may be further improved, for example, by using spatial light modulators to shape the IR beam and enable phase gratings and beam splitters for XUV pulses.

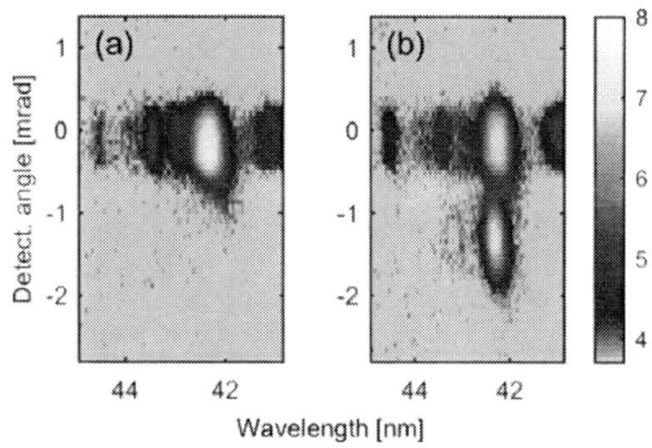

Fig. 1. a) The XUV spectrum transmitted through argon and b) the same spectrum partially re-directed by an IR control pulse. The IR pulse modulates the spatial phase of the excited dipoles, thus modifying the directionality of the emitted light.

References

[1] Bengtsson S *et al.* 2017 *Nature Photonics* **11** 252
[2] Bengtsson S and Mauritsson J 2019 *J. Phys. B* **52** 063002
[3] Ott C *et al* 2014 *Nature* **516** 374
[4] Brewer R G and Shoemaker R L 1972 *Phys. Rev. A* **6** 2001
[5] Hopf F A *et al* 1973 *Phys. Rev. A* **7** 2105

* E-mail: johan.mauritsson@fysik.lth.se

Analytical theory of attosecond transient absorption spectroscopy of perturbatively dressed systems.

D Kolbasova[1,2]* **and R Santra**[1,2]

[1] Center for Free-Electron Laser Science, DESY, Hamburg, 22607, Germany
[2] Department of Physics, Universität Hamburg, Hamburg, 20355, Germany

Synopsis The current work provides an elementary explanation of the signals one can obtain in attosecond transient absorption experiments. We perform a perturbative analysis of the transient absorption signal as a function of the time delay between the excitation pulse and the dressing pulse. We discuss an unusual kind of experiment, where the information is gained from the absorption spectrum of an attosecond XUV pump pulse, and the probe pulse gives a reference time, allowing to measure the time evolution of a system. We illustrate our concepts through numerical calculations for a model atom.

The advent of attosecond light pulses has opened up the possibility of studying fundamental questions related to the quantum dynamics of electrons on their natural time scale. Particularly, attosecond transient absorption (ATA) spectroscopy gives access to ultrafast dynamics of bound and autoionizing electronic states in rare gas atoms, molecules and solids. A theoretical analysis of the ATA signal is crucial for understanding and maximizing its information content.

In the current work we present an analytical theory of ATA spectroscopy for perturbatively dressed systems, which is applicable to the analysis of processes observed in pump-probe experiments if both pump and probe fields are sufficiently weak. We focus on an unusual kind of pump-probe experiment, where the information is gained from the absorption spectrum of an attosecond XUV pulse, which serves as a pump pulse at the same time. The optical probe pulse in this kind of experiment gives a reference time that provides a possibility to measure the time evolution of a system of interest. We build on the theoretical description of ATA spectroscopy developed in Ref. [1, 2] and extend it by performing a perturbative analysis of the impact of the dressing field. In contrast to Ref. [1, 2], where the optical field was assumed to be so strong that it causes tunnel ionization even in the electronic ground state, we here consider the situation where the optical dressing field only affects the electronic states that are reached after absorbing an XUV photon from the attosecond pulse. To this end, we employ a two-stage approach based on perturbation theory, which allows us to give an analytical expression for the transient absorption signal.

We focus on the situation in which the attosecond XUV pulse is used to create a coherent superposition of electronic states. As we explain, the resulting dynamics can be detected in the spectrum of the transmitted XUV pulse by manipulating the electronic wave packet using a carrier-envelope-phase-stabilized optical dressing pulse. In addition to coherent electron dynamics triggered by the attosecond pulse, the transmitted XUV spectrum encodes information on electronic states made accessible by the optical dressing pulse. The interference between the direct and the dressing-field-mediated excitation pathways gives rise to periodic modulations of the absorption peaks as a function of the pump-probe delay. We illustrate how the ATA signal reveals the real-time attosecond dynamics of the system through numerical calculations for a few-level model atom. The presented technique is suitable not only for atoms, but for more complex systems as well.

References

[1] Santra R *et al* 2011 *Phys. Rev. A* **83** 033405
[2] Pabst S *et al* 2012 *Phys. Rev. A* **86** 063411

*E-mail: daria.kolbasova@cfel.de

The single ionization of helium with energy-selected extreme ultraviolet photons

M Zhang[1,2], B Hai[1,2], D M Zhao[1], D P Dong[1,2], J T Lei[3], T Morishita[4], S F Zhao[5], S F Zhang[1]* and X Ma[1]†

[1]Institute of Modern Physics, Chinese Academy of Sciences, Lanzhou, 730000, China
[2]University of Chinese Academy of Sciences, Beijing, 100049, China
[3]School of Nuclear Science and Technology, Lanzhou University, Lanzhou, 730000, China
[4]Institute for Advanced Science, The University of Electro-communications, 1-5-1 Chofu-ga-oka, Chofu-shi, Tokyo 182-8585, Japan
[5]Key Laboratory of Atomic and Molecular Physics and Functional Materials of Gansu Province, College of Physics and Electronic Engineering, Northwest Normal University, Lanzhou 730070, China

Synopsis The platform dedicated to studies of extreme ultraviolet (EUV) laser on atom/molecule has been established recently at the Institute of Modern Physics, CAS, Lanzhou. The single ionization of helium with EUV photons of about 37.5 eV was conducted. Photoelectron angular distributions are obtained and compared with theoretical calculations from TDSE method.

Recently, the platform on studies of ultrafast dynamics of atomic processes has been established. The system consists of a reaction microscope [1] and an EUV laser system supported by KMLabs. The infrared laser ($\sim$ 800 nm, 25 fs, 3 kHz, $\sim$ 5 mJ) is focused into a 5-cm-long hollow fiber filled with noble gases to generate EUV light via High Harmonic Generation(HHG) method. EUV photon energy can be selected by the downstream grating and further focused at the crossing point of the target in the reaction microscope. The pulse duration of the EUV laser is estimated to be about 11 fs with the flux of 10^9 photons/s.

The first experiment on single ionization of helium has been conducted with a photon energy of about 37.5 eV. Ejected electrons and recoil ions are recorded in coincidence triggered by the laser pulse. Fig.1a. shows the momentum distribution of the photoelectrons. Dipole transition patterns can be clearly identified. Meanwhile Fig.1b. represents the photoelectron angular distribution with respect to the polarization direction of EUV photon where only one arm of the photoelectron in Fig.1a is selected. The measured angular distribution is in good agreement with theoretical results based on numerical calculations of Time Dependent Schrödinger Equation(TDSE) method [2]. Using the equation in [3] to fit the angular distribution, the fitted asymmetry parameter β =2.00 $\pm$ 0.07 is obtained. Further results will be represented at the conference.

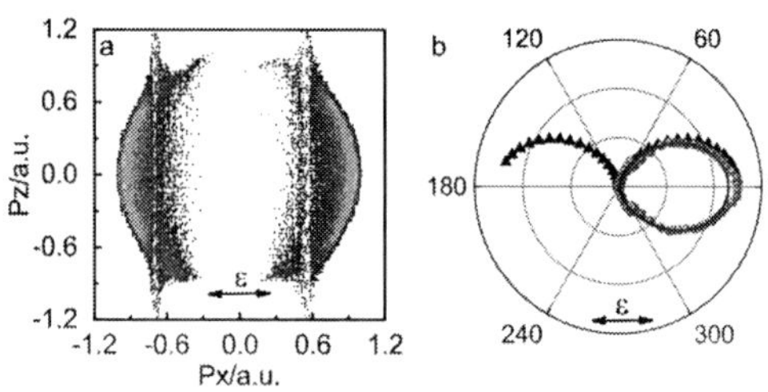

Figure 1. (a) Momentum distributions of the photoelectron. The x-axis is the polarization direction of the EUV photon and the y-axis is the direction of light propagation. The red vertical dashed lines represent the nodal line of electrons cyclotron motion in the magnetic field [1]. (b) Angular distribution of the photoelectron where black solid triangles represent results of TDSE, red solid spheres are the experimental results and blue lines are the fitting curves.

This work was supported by the National Key Research and Development Program of China(No.2017YFA0402300), and National Natural Science Foundation of China(No.11664035).

References

[1] Ullrich J *et al* 2003 *Rep Prog Phys* **66** 1463
[2] Morishita T *et al* 2007 *Phys Rev A* **75** 023407
[3] Manson S T *et al* 1982 *Rev Mod Phys* **54** 389

*E-mail: zhangshf@impcas.ac.cn
†E-mail: x.ma@impcas.ac.cn

Fano's propensity rule in angle-resolved attosecond interferometry

D Busto[1] *, J Vinblad[2], S Zhong[1], M Isinger[1], S Nandi[1], M Gisselbrecht[1], A L'Huillier[1], E Lindroth[2] and J M Dahlström[1]

[1]Department of Physics, Lund University, P.O. Box 118, SE-22 100 Lund, Sweden
[2]Department of Physics, Stockholm University, SE-106 91 Stockholm, Sweden

Synopsis Above-threshold ionization is a corner stone of attsecond science. In this work we extend Fano's propensity rule to two-photon above-threshold ionization and show that the asymmetry between absorption and emission of the second photon predicted by this propensity rule has strong implications for angle-resolved pump-probe experiments and in particular for attosecond photoelectron interferometry.

The majority of the techniques used in attosecond science rely on the photoionization of a target system upon absorption of an eXtreme Ultra-Violet (XUV) attosecond light field in the presence of a phase-locked Infra-Red (IR) pulse. In particular, the interferometric technique RABBIT (Reconstruction of attosecond beating by interference of two-photon transitions) is based on the interference of photoelectron wavepackets induced by absorption of an XUV attosecond pulse train followed by further absorption or emission of an IR photon. As the delay between the XUV and IR pulses is varied, the resulting sideband oscillates with a phase that carries information on the attosecond XUV field and on the photoionization process [1].

While most of the studies of photoionization dynamics performed until now have dealt with angle-integrated measurements, an increasing number of experiments have now access to angle-resolved information. This recent development has led to the experimental observation of unexpected effects such as the angular dependence of photoionization delays [2] or the delay dependence of the Photoelectron Angular Distribution (PAD) [3]. While some of these observations have been reproduced theoretically, their underlying origin is still not well understood.

In this work, we use channel-resolved many-body perturbation theory applied to rare gas atoms (He, Ne, Ar and Kr). We extend Fano's propensity rule [4], originally formulated for single photon ionization and stating that increasing (decreasing) angular momentum is favoured in the case of absorption (emission) of a photon, to two-photon above-threshold ionization. Subsequently, we focus on He and Ar and apply this result to the RABBIT technique which is based on quantum interference between two paths involving absorption or emission of an IR photon. The asymmetry between absorption and emission [Fig. 1(a)], predicted by Fano's propensity rule, results in incomplete interference which are responsible for the delay-dependence of the PAD [Fig. 1(b)] and the angular-dependence of the photoionization time delays. Our calculations are compared to experiments in Ar and show perfect agreement. Our results are very general and offer a new way to analyse and interpret time and angle-resolved results, beyond the traditional approximations.

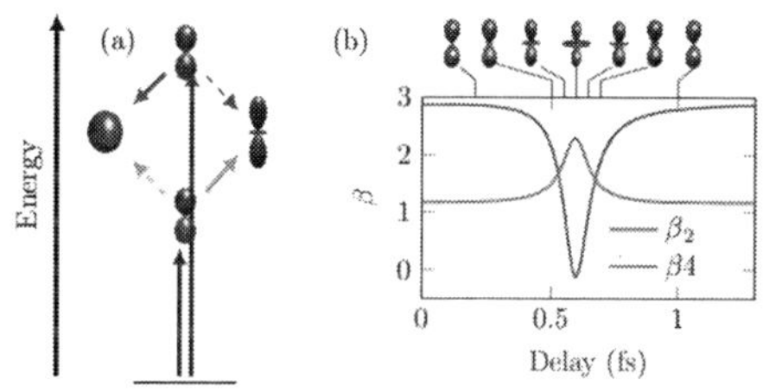

Figure 1. (a) Fano's propensity rule in the RABBIT scheme for He. Dashed arrows indicate weaker transitions. (b) Modifications of the PAD in He as a function of the delay (top) and corresponding asymmetry parameters (bottom).

*E-mail: david.busto@fysik.lth.se

References

[1] Klünder K *et al* 2011 *Phys Rev Lett* **106** 143002
[2] Heuser S *et al* 2016 *Phys Rev A* **94** 063409
[3] Cirelli C *et al* 2018 *Nat Commun* **9** 955
[4] Fano U 1985 *Phys Rev A* **32** 617

Disentangling spectral phases of interfering autoionizing states from attosecond interferometric measurements

C L M Petersson[1] *, L Barreau[2], M Klinker[1], A Camper[3], C Marante[1], T Gorman[3], D Kiesewetter[3], L Argenti[1], P Agostini[3], J González-Vázques[1], L F DiMauro[3], P Salières[2] and F Martín[1,4,5]

[1]Departamento de Química, Módulo 13, Universidad Autónoma de Madrid, 28049 Madrid, Spain
[2]LIDYL, CEA, CNRS, Université Paris-Saclay, 91191 Gif-sur-Yvette, France
[3]The Ohio State University, Department of Physics, Columbus OH 43210, USA
[4]Instituto Madrileño de Estudios Avanzados en Nanociencia (IMDEA-Nanociencia), Cantoblanco, 28049 Madrid, Spain
[5]Condensed Matter Physics Center (IFIMAC), Universidad Autónoma de Madrid, 28049 Madrid, Spain

Synopsis The spectral phase of the autoionising neon continuum has been probed via interferometric pump-probe measurements as well as ab initio, full-electron, time-dependent, theoretical calculations in energy intervals containing several auto-ionising. Theory and experiment are in excellent agreement. Despite the complex energy-dependence of the phase, a very simple model has been used to disentangle the contribution by different resonances. This work extends the applicability of reconstruction methods to multi-resonance regions of autoionising continua - a common scenario in multi-electron systems.

With the advent of attosecond science, the development of interferometric techniques has proved itself a highly useful avenue for the study of electron dynamics on their natural time scale. One such method is the pump-probe method known as reconstruction of attosecond beatings by interference of two-photon transitions (RABITT), which has been successfully used to study phenomena such as spectral phases [1]. Although this method has proven itself able to probe not just atomic [1] system, but also more complex system systems, such as molecules [2], solids [3]; it has so far been restricted to regions of autoionising continua containing one or no resonances. This is due to the difficulty in accurately describing electron correlation of multi-resonance regions, making accurate, full-electron, time-dependent ab initio theory, necessary to guide experiment inaccessible. This dearth of accurate calculations leaves simple models able to describe such regions a highly desired prospect.

In this communication, RABITT is used to probe neon below the second ionisation threshold, scanning the spectral phase both in a single- and a multi-resonance regions. Time-dependent, full-electron, ab-initio calculations were used to provide a theoretical prediction; enabled using an implementation of the XCHEM methodology [4], itself designed to accurately describe correlated processes in many-electron systems. Experimental data is also presented, with which the theoretical calculations show excellent result, providing further support for the XCHEM approach. In the multi-resonance region, the spectral phase has a complex dependence on the energy. Despite this, a very simple model based on a recent extension of the Fano model for two-photon ionisation induced by ultrashort pulses [5] is presented, capable of disentangling the contributions of different autoionising states and accurately predicting the energy-dependence of the spectral phase in the studied multi-resonance region.

This works extends the applicability of interferometric methods to multi-resonance regions of atomic and molecular systems, thus paving the way for reconstruction of electron wave packets coherently generated in systems previously unavailable due to their complexity.

References

[1] Cirelli C *et al* 2018 *Nat. Comm.* **9** 955
[2] Haessler S *et al* 2009 *Phys. Rev. A* **80** 011404(R)
[3] Cavalieri A L *et al* 2007 *Nature* **449** 1029
[4] Marante C *et al* 2017 *Phys. Rev. A* **13** 499
[5] Jiménez-Galán Á *et al* 2016 *Phys. Rev. A* **93** 023429

*E-mail: leon.petersson@uam.es

Spiral-like Structure: Unwinding Holographic Interference

A S Maxwell[1] *, X Y Lai[2], R P Sun[2], C Figueira de Morisson Faria[1] and X J Liu[2]

[1]Physics & Astronomy, University College London, Gower Street, London, WC1E 6BT, United Kingdom
[2]State Key Laboratory of Magnetic Resonance and Atomic and Molecular Physics and Center for Cold Atom Physics, Wuhan Institute of Physics and Mathematics, Chinese Academy of Sciences, Wuhan 430071, China

Synopsis A new spiral shaped holographic interference structure in above-threshold ionisation photoelectron momentum distributions is presented in experimental and theoretical results. Given a large Coulomb interaction and bound-state symmetries encoded in interfering trajectories we demonstrate its use for imaging.

Photoelectron holography has proven itself as a powerful tool for imaging atomic and molecular dynamics on an attosecond time scale [1, 2]. Electrons undergoing above-threshold ionization (ATI) take many paths to the detector, some of which will interact strongly with the core and others that will not, the interference between these trajectories can be used to image the target. This has led to the study of various holographic interference structures that can be used for imaging, including, the spider-like structure [1,2], between different forward-scattered trajectories and the fan-like structure [3, 4], resulting from direct- and forward-deflected trajectories.

Modelling these experimentally observed holographic patterns is difficult, requiring the Coulomb potential to be fully accounted for, ruling out models like the strong field approximation. This has led to the development of many other quantum trajectory models. However, recently a myriad of previously overlooked holographic patterns have been found using the Coulomb quantum-orbit strong field approximation (CQSFA) [4–6], including a spiral-shaped structure due to forward- and back-scattered paths [6], [Fig. 1(b)]. This was enabled as the CQSFA requires only a few trajectories per final momenta, while other Coulomb trajectory models require a total of around 10^9.

The spiral-like structure was not previously identified in any other theoretical or experimental work due similarity with ATI-rings and because it is most visible for a high energy low signal region. However, we show new experimental and theoretical work where, the spiral-like structure can be unambiguously identified [Fig. 1(c-d)] and perform a detailed study for different gases.

We find the spiral fringes can be related to the carpet-like interference structure [7], and argue the previous treatment, which neglects the Coulomb potential is over simplified. In addition, the symmetry the bound-state is shown to be encoded in the spiral fringes. The two trajectories that make up the spiral-like pattern undergo the largest interactions with the Coulomb potential, which affects the spiral patterns fringe positions. Hence, we fit an effective potential to the spiral-like structures in the experimental data to determine its efficacy as an attosecond imaging tool.

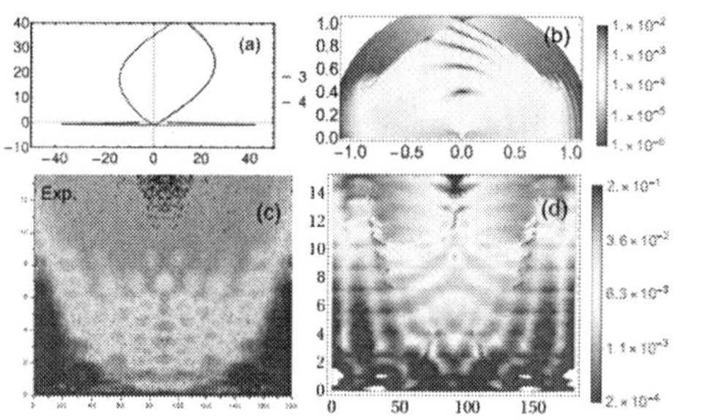

Figure 1. Panel (a), the two paths that make the spiral-like structure. Panel (b), spiral-like interference structure. Panel (c) Experiment and (d) CQSFA computation of ATI signal over energy and emission angle for xenon. For a laser intensity of $I = 7 \times 10^{13}$ W/cm^2 and wavelength $\lambda = 800$ nm.

References

[1] Huismans Y *et al* 2011 *Science* **331** 61
[2] Haertelt M *et al* 2016 *Phys. Rev. Lett.* **116** 133001
[3] Lai X Y *et al* 2017 *Phys. Rev. A* **96** 013414
[4] Maxwell A S *et al* 2017 *Phys. Rev. A* **96** 023420
[5] Maxwell A S *et al* 2018 *J. Phys. B* **51** 044004
[6] Maxwell A S *et al* 2018 *J. Phys. B* **51** 124001
[7] Korneev P A *et al* 2012 *Phys. Rev. Lett.* **108** 223601

*E-mail: andrew.maxwell.14@ucl.ac.uk

High-harmonic generation enhanced by dynamical electron correlation

Y Li[1] *, T Sato[1,2,3] and K L Ishikawa[1,2,3]

[1] Department of Nuclear Engineering and Management, Graduate School of Engineering, The University of Tokyo, Tokyo, 113-8656, Japan
[2] Photon Science Center, Graduate School of Engineering, The University of Tokyo, Tokyo, 113-8656, Japan
[3] Research Institute for Photon Science and Laser Technology, The University of Tokyo, Tokyo, 113-8656, Japan

Synopsis We theoretically investigate correlated electron dynamics in high-harmonic generation (HHG), using all-electron *ab initio* simulations for three-dimensional real alkali-metal atoms. The resulting harmonic spectra exhibit a plateau extended beyond the usual cutoff and a prominent resonance peak, which are clear manifestations of dynamical electron correlation.

Strong-field light-matter interaction has various applications. One prominent example is high harmonic generation (HHG), which serves not only as the standard tool for the generation of ultrashort laser pulses, but also as a promising way for the direct measurement and control of electron dynamics in atoms and molecules. Although the HHG process can be adequately treated as a one-electron problem in many cases, it fails when electron correlation plays a role [1], which is far from fully explored yet and is now attracting increasing interests.

Here, we report a fascinating mechanism leading to a drastic cutoff extension of the HHG spectra, induced by dynamical electron correlation between recolliding and inner-shell core electrons. This mechanism was first predicted for a one-dimensional multielectron model atom [2] and later extended to a one-dimensional two-electron model molecule [3]. Here we numerically simulate HHG from real alkali atoms using the muticonfiguration time-dependent Hartree-Fock (MCTDHF) method [4-6], which enables accurate description of many-electron atoms and molecules for a broad range of laser parameters. Taking sodium atom for example, we have successfully computed HHG spectra in an intense 1200-nm laser pulse, shown in Fig. 1. In comparison with single-active-electron (SAE) case, the multi-electron spectrum exhibits a second plateau with cutoff position at 133 eV, far beyond the first cutoff at 90 eV. Moreover, a promiment resonance peak shows up at 33 eV. These two remarkable features, which also exist in the HHG spectrum from other alkali atoms, is confirmed to be clear manifestations of multielectron effects. This demonstrates that high-harmonic spectroscopy provides new possibilities to explore dynamical electron correlation in strong laser pulses.

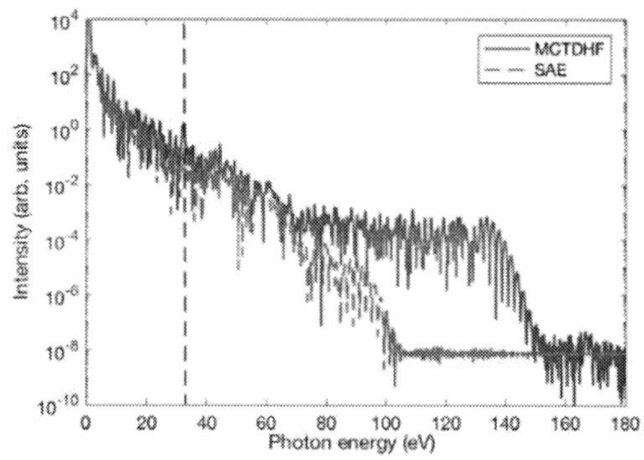

Figure 1. HHG spectra of Na computed by MCTDHF method (blue solid curve) and SAE approximation (red dashed curve). The laser pulse has a a wavelength of $\boldsymbol{\lambda} = \mathbf{1200}\ \boldsymbol{nm}$, a peak intensity of $\boldsymbol{I} = \mathbf{2 \times 10^{14}}$ W/cm^2. The vertical dashed line guides the position of resonance peak.

References

[1] Ishikawa K L and Sato T 2015 *IEEE J. Sel. Top. Quantum Electron.* **21** 8700916
[2] Tikhomirov I, Sato T, and Ishikawa K L 2017 *Phys. Rev. Lett.* **118** 203202
[3] Abanador P M *et al* 2018 *Phys. Rev. A* **97** 043414
[4] Zanghellini J *et al* 2003 *Laser Phys.* **13** 1064
[5] Kato T and Kono H 2004 *Chem. Phys. Lett.* **392** 533
[6] Sato T *et al* 2016 *Phys. Rev. A* **94** 023405

* E-mail: yangli@atto.t.u-tokyo.ac.jp

Attosecond time delay in multiphoton ionization

W Yang[1] *, X Liu[1], X Song[1], P Li[1] and J Chen[2]

[1]Department of Physics, Shantou University, Shantou, Guangdong 515063, China
[2] Institute of Applied Physics and Computational Mathematics, P. O. Box 8009, Beijing 100088, China

Synopsis Here with a semiclassical picture we show that, after emerging in the continuum state, the photoelectron can be temporarily re-trapped by the atomic potential in a quasi-bound state, stays there for hundreds of attoseconds before it gets free again, which is the origin of the remarkable time delay in above threshold ionization experiment. Our results build a bridge between the seemingly unrelated quantum EWS and classical Coulomb-induced time delays, and hold promise for intuitive interpreting fundamental electronic processes in photophysical and photochemical experiments.

Ultrafast pump-probe techniques using XUV pulses combined with a delayed infrared probe field have allowed measurement of timing and time delays during the process of one-photon ionization from atoms, molecules and solids, which opens a new perspective on real-time insights into fundamental photophysical and photochemical precesses. However, looking beyond the one-photon ionization and extracting the chronoscopic information from experimental observations in nonperturbative multiphoton ionization regime remains a challenge due to the complicated multiple above threshold ionization (ATI) channels.

Here we uncover the time information about ionization and scattering encoded in the recent phase measurements of argon atom in multiphoton ionization regime [1-3]. We show that the measured attosecond time delay is a result of statistic averages for different ionization pathways whose contributions can be clearly resolved. In the multiphoton ionization regime, the direct ionization electrons contributing to both the ATI and sideband peaks are emitted mainly at the nodes of the strong pump field with nonzero initial velocities. The sideband generation from the adjacent main ATI peaks by absorbing or emitting one probe photon for which the highest probability is at the peak of the probe field induces the observed π phase shift between the main ATI and sideband peaks. The rapid backward scattering contributing to the high energy of the spectrum corresponds to a negative time delay. Most importantly, we demonstrate direct real-time access to the re-trapped resonant scattering, a classical counterpart of quantum resonant scattering, and identify that it is the origin of the observed remarkably large time delay at lower momenta.

The good agreement of the semiclassical generalized quantum trajectory Monte Carlo simulations and statistic analysis with both the quantum TDSE and experimental results further consolidates the recent hints that the multiphoton and tunneling regimes share similar physical picture but display different characters [4]. Accurately measured time delay due to changes in the effective potential in atomic ionization of gas phase serve as a benchmark for the dynamic polarization response in more complicated systems in molecules or condensed phase. Our findings not only greatly advance the understanding of the the real-time dynamical information on the complex strong-field photoemission precesses encoded in the experimental phase measurements, but also provides unprecedented opportunities to explore the quantum-classical correspondence for timing and time delays, which is a fundamental problem in strong-field physics and quantum mechanism.

* E-mail: wfyang@stu.edu.cn

References

[1] Gong X *et al* 2017 *Phys. Rev. Lett.* **118** 143203
[2] Zipp L J, Natan A and Bucksbaum P H 2014 *Optica* **1** 361
[3] Song X *et al* 2018 *Phys. Rev. Lett.* **121** 103201
[4] Zimmermann H, Patchkovskii S, Ivanov M and Eichmann U 2017 *Phys. Rev. Lett.* **118** 013003

Surface-near-field reconstruction from streaked photoelectron spectra of Au(111) surfaces

M J Ambrosio* and U Thumm†

Department of Physics, Kansas State University, Manhattan, Kansas 66506, USA

Synopsis We calculate attosecond streaking traces from Au(111) surfaces and suggest their use for characterizing the plasmonic/dielectric response of solids at the nanoscale.

Recent theoretical work on attosecond streaking from nanoparticles demonstrated the reconstruction of plasmonic near-fields at metal-nanoparticle surfaces with atomic resolution in time and space [1, 2]. We here numerically model attosecond streaked photoelectron spectra to image the dielectric response of surfaces to intense IR laser pulses, indicating a method for characterizing the response at the solid-vacuum interface, which differs from the bulk response [3].

Time-resolved photoemission spectra and time delays from surfaces are sensitive to the target's dielectric response to the IR-streaking pulse and photoelectron transport in the few top crystalline layers of the substrate. We modeled these effects in quantum-mechanical simulations that describe the excitation of photoelectrons from localized initial states in the substrate by the attosecond XUV pulse into final excited states that satisfy outgoing-wave boundary conditions and include the interaction of the active electron with both, the surface-reflected IR-streaking field and lattice-periodic substrate potential [4] (Fig. 1, top graphs).

By accounting for the photoelectron final-state interaction with the Fresnel-reflected IR field of the streaking pulse at the substrate's optical surface and propagation of excited photoelectrons in the substrate (subject to electronic collisions), our numerical results show that the finite electron-transport time delay shifts the entire streaking trace without noticeable spectral distortion. In contrast, Fresnel reflection of the IR pulse induces spectral photoemission time delays that alter the shape of the streaking trace. Spectrally resolved Fresnel phase shifts, and thus surface-specific surface-dielectric-response properties, and can be retrieved by the fitting streaked spectra [5] (Fig. 1, bottom graphs).

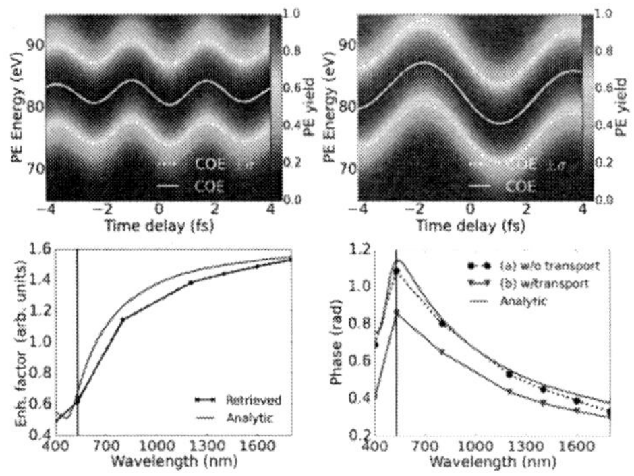

Figure 1. Top: Streaked traces from Au(111) surfaces for 800 (left) and 1600 nm (right) $10^{11}\,W/cm^2$ peak intensity IR pulses incident at 75° relative to the surface normal. Photoelectrons are emitted at −30°. **Bottom:** Total (= incident + reflected) electric-field amplitude (left) and phase retrieval (right) as a function of the IR wavelength. The blue vertical line shows the surface plasmon resonance at 530 nm.

Acknowledgements

We acknowledge the support from the US NSF and the Division of Chemical Sciences, Office of the Basic Energy Sciences, Office of Energy Research, US DoE.

References

[1] Li J *et al* 2018 *Phys. Rev. Lett.* **120** 223903
[2] Saydanzad E *et al* 2018 *Phys. Rev. A.* **98** 063422
[3] Shi N and RamprasadR 2007 *J. Computer-Aided Mater. Des.* **14** 133
[4] Ambrosio M J and Thumm U 2018 *Phys. Rev. A* **97** 043431
[5] Ambrosio M J und Thumm U *in preparation*

*E-mail: mj_ambrosio@phys.ksu.edu
†E-mail: thumm@phys.ksu.edu

Phase reconstruction of high-order harmonics by quantum path interference

A Olofsson*, N Ibrakovic, S Ek, S Bengtsson, E R Simpson and J Mauritsson

Department of Physics, Lund University, P.O. Box 118, SE-221 00 Lund, Sweden

Synopsis We present a method to retrieve the phase of XUV light generated through high-order harmonic generation, by observing the intensity dependence of the interference pattern in the far-field. This interference pattern arises from electron trajectories in the generation process. These trajectories spend different time in the continuum and the phase of the generated light will therefore depend differently on the instantaneous intensity of the driving field. By changing the intensity of the driving field the phase of the contributing trajectories can be extracted.

The process of high-order harmonic generation can be understood using a semi-classical three step model [1-4] in which an electron is first ionized via tunneling ionization, after which it is accelerated by the strong laser field, and finally it returns to the ion releasing its excess energy. This simple model predicts that there is more than one path [4] to the ion yielding the same photon energy (i.e. different electron trajectories) and that the phase of the light emitted from these different trajectories depends strongly on the instantaneous laser intensity.

We present an all optical method to retrieve the phase of electron trajectories and a simple model based on gaussian optics which explains the main features of the interferograms [5]. High-order harmonics are generated using a commercially available Yb:KGW (central wavelength 1030 nm) PHAROS laser (Light Conversion). The transform-limited pulse duration is 170 fs, and these long, multi-cycle driving pulses ensures a good spectral resolution in the experiment. The high-order harmonics are generated in Argon by focusing the laser into a gas-jet, and are detected using a flat-field spectrometer. The intensity of the pulses is varied either by stretching the duration or by attenuating the energy of the pulses.

In figure 1 we present the interferogram from a spectral lineout through harmonic 15 as a function of pulse duration. Two main structures can be observed in this figure: 1) for small divergences (<5 mrad) a rapid modulation with an invard curvature can be seen, and 2) for larger divergences a slower modulation with the opposite curvature can be observed. The on-axis interference pattern comes from the contributions from the short and the long trajectories. While the periodicity of the modulation is due to the relative phase between the trajectories, the curvature allows for an absolute measure of the phases. The second pattern (at larger divergences) comes from interference between contributions generated by the long trajectory, but at different spatial locations and thus at different intensities.

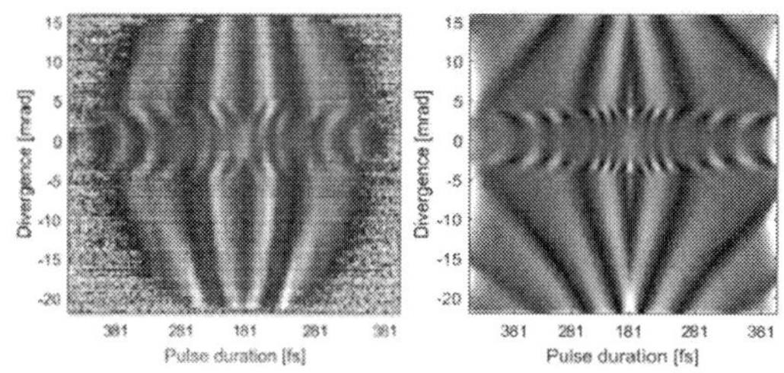

Figure 1. a) The interference structure of one harmonic as a function of the pulse duration of the driving pules. The left-right symmetry in the figure indicates that the sign of the chirp is not important for the process, while the change in intensity is. In b) a simple model based on gaussian beams is used to reproduce the main structures of the interference pattern, and retrieving the phase of the contributing paths.

The interference pattern is well reproduced using a simple gaussian model (see figure 1 b). This analytical model can be used to fit the interference pattern and extract the phases of the ingoing contributions. The model should allow an automated fitting procedure and extraction of the phase contribution for all the generated harmonics.

References

[1] Schafer K J *et al* 1993 *Phys. Rev. Lett.* **70** 1599
[2] Corkum P B 1993 *Phys. Rev. Lett.* **71** 1994
[3] Ferray M *et al* 1988 *J. Phys. B* **21** L31
[4] Lewenstein M *et al* 1994 *Phys. Rev. A.* **49** 2117
[5] Carlström S *et al* 2016 *New J. Phys.* **18** 123032

* E-mail: anna.olofsson@fysik.lth.se

Journal of Physics: Conference Series **1412** (2020) 072016 doi:10.1088/1742-6596/1412/7/072016

Sideband selection in XUV + IR photoionization

R Della Picca[1] *, A A Gramajo[1], S D López[2] and D G Arbó[2]

[1]CONICET and Centro Atómico Bariloche, CNEA, Bariloche, 8400, Argentina

[2]Institute for Astronomy and Space Physics IAFE (UBA-Conicet), Buenos Aires, Argentina

Synopsis We consider the ionization of argon atoms induced by a femtosecond laser pulse composed by overlapping extreme ultraviolet and infrared fields. We study several geometrical arrangements within the polarization vectors and emission direction that determinate only odd or even sideband peaks are present in the photoelectron spectrum.

The laser-assisted XUV photoelectric effect (LAPE) occurs when extreme ultraviolet (XUV) and infrared (IR) lasers overlap in space and time. The simultaneous absorption of one high-frequency photon and the exchange of several additional photons from the laser field lead to equally spaced sideband (SB) peaks in the photoelectron (PE) spectra [1] situated at energy values given by the conservation energy rule:

$$E_n = n\omega_L + \omega_X - I_p - U_P \qquad (1)$$

where $\omega_{X(L)}$ is the frecuency of XUV(IR) pulse, I_p is the ionization energy, U_p the ponderomotive energy and n is the number of absorbed or emitted IR photons. It is know that the temporal symmetry (established by the field repeatability cycle to cycle) is the responsible of the SB formation [2, 3]. Furthermore we have found in certain configurations and within the semiclassical model (SCM), there is an additional symmetry (inherent to each half cycle) that gives rise to a destructive interference resulting in only odd number of exchanged IR photons [3].

In the present work we investigate several geometrical arrangements between the polarization vectors and the emission direction, that lead to selective sideband orders in the ionization spectrum. Within the continuum-distorted wave Strong Field Approximation (SFA) supported by the ab initio solution of the time-dependent Schrdinger equation (TDSE) we study the angle-resolved momentum distribution of photoelectron for the case that both fields are linearly polarized. In the figure, we present the TDSE, SFA and SCM momentum distribution of the emitted electrons from hydrogen-like Argon 3s. In figures (b) and (c) we show the spectra for perpendicular and forward emission cases. In (b) it can be observed that the even sideband orders are selectively canceled whereas in the forward emission all sidebands are present.

During the conference we will theoretically analyze the conditions that give rise to a destructive interference resulting in only odd or even number of exchanged IR photons.

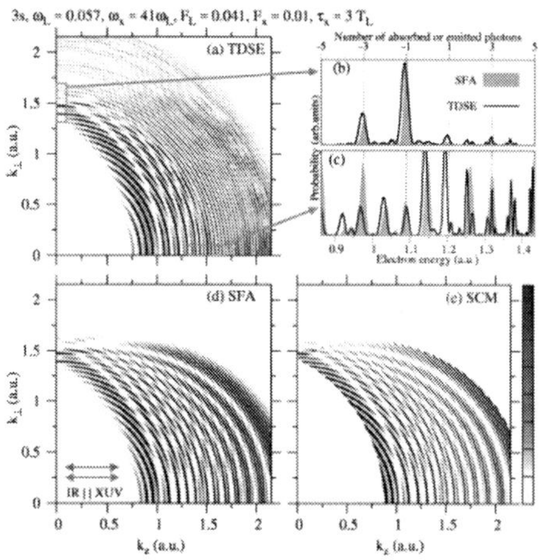

Figure 1. Photoelectron momentum distribution for LAPE process of Argon(3s) atoms in arbitrary units, (a) TDSE , (d) SFA and (e) SCM models. Differential emission probability in (b) perpendicular and (c) forward emission. The IR and XUV laser parameters are $\omega_L = 0.057$, the electric amplitude of the IR (XUV) is 0.041 (0.01) and $\omega_X = 2.3$ (all in atomic units). The XUV pulse duration is 3 IR cycles.

References

[1] Véniard V *et al* 1995 *Phys. Rev. Lett.* **74** 4161
[2] Maquet A *et al* 2007 *J. Mod. Opt.* **54** 1847
[3] Gramajo A A *et al* 2018 *J. Phys. B* **51** 055603

*E-mail: renata@cab.cnea.gov.ar

Atomic and molecular suite of R-matrix codes for ultrafast dynamics in strong laser fields and electron/positron scattering

J Wragg[3]*, J Benda[1], Z Mašín[2],†, G S J Armstrong[3], D D A Clarke [3], A C Brown[3] C Ballance[3], A G Harvey[4], K Houfek[2], A Sunderland[5], M Plummer[5], J D Gorfinkiel[1] and H van der Hart[3]

[1]School of Physical Sciences, The Open University, Walton Hall, MK7 6AA Milton Keynes, UK
[2]Institute of Theoretical Physics, Faculty of Mathematics and Physics, Charles University, V Holešovičkách 2, 180 00 Prague 8, Czech Republic
[3]Centre for Theoretical Atomic, Molecular and Optical Physics, School of Mathematics and Physics, Queen's University Belfast, Belfast BT7 1NN, United Kingdom
[4]Max-Born-Institut, Max-Born-Str. 2A, 12489 Berlin
[5]Scientific Computing Department, STFC Daresbury Laboratory, Sci-tech Daresbury, Cheshire WA4 4AD, United Kingdom

Synopsis We describe and illustrate a number of recent developments of the atomic and molecular ab initio R-matrix suites for both time-dependent calculations of ultrafast laser-induced dynamics and time-independent calculations of photoionization and electron scattering.

The field of attosecond science has presented computational physics with a set of new challenges to describe the dynamics of matter in strong femtosecond lasers. The recent move to study complex multielectron systems makes the need for a matching ab initio theoretical description even more pressing. In the more mature field of electron-molecule collisions, the need to study more complex targets has also required software improvements. Our R-matrix suites of codes [?, ?] have recently undergone a major upgrade and been made available to the community.

A set of atomic and the molecular time-independent codes form their core. On the atomic side these are the well-established RMATRXI and RMATRXII codes: both use a B-spline representation of the continuum; RMATRXI allows the inclusion of relativistic (Breit-Pauli) corrections. On the molecular side, a major development has led to the release of UKRmol+, based on code originally developed for electron/positron - molecule scattering. This suite has been extended to enable photoionization calculations and to include B-splines in the representation of the continuum via the new GBTOlib integral library. Significant improvements in performance of the molecular codes have been achieved developing a new parallel Hamiltonian build algorithm and a parallelized version of the R-matrix propagator.

The second layer of the R-matrix suite comprises R-matrix with time (RMT) [?] codes which can now be used to propagate in time both the atomic and molecular (in the fixed-nuclei approximation) wavefunctions. Recent developments allow the use of arbitrary polarized light and interfacing with the atomic RMATRXI suite thus enabling studies of spin-orbit dynamics in ultrafast laser fields.

The recent developments will be described, including illustrations of the results on generation of electron vortices from argon atoms, relativistic effects in the vicinity of the giant resonance in Xenon and validation tests of molecular RMT on photoionization of water. The benefits for the electron-molecule scattering community will be illustrated with the results for inelastic electron scattering from thiophene.

References

[1] https://gitlab.com/Uk-amor/RMT/rmt
[2] https://gitlab.com/Uk-amor/UKRMol
[3] Moore L R *et al* 2011 *J. Mod. Optics* **58** 1132

*E-mail: jack.wragg@qub.ac.uk
†E-mail: zdenek.masin@utf.mff.cuni.cz

Spin-orbit dynamics in atomic krypton observed through decay of autoionising states

J Wragg[1]*, G Armstrong[1] D Clarke[1] C Ballance[1] A Brown[1] and H van der Hart[1]

[1]Queen's University Belfast, Belfast, Northern Ireland

Synopsis We study the spin-orbit interaction within excited states of krypton by extending R-Matrix with time-dependence theory to include spin-orbit effects. We study an attosecond pump-probe scheme where we produce an autoionising state with the probe pulse. We observe dynamics through the subsequent decay of the autoionising state, from which we find a time-dependent signal when spin-orbit interactions are included. We examine the effect of the relative polarisation of the pulses, enabling an interpretation of the dynamics in terms of the quantum number m_ℓ of the core of the excited krypton atom.

In recent years, it has become possible to study spin-orbit dynamics experimentally from a time-dependent perspective (e.g. [1, 2]). To meet the need for corresponding theoretical methods, we have extended R-Matrix with Time-dependence (RMT) theory to include spin-orbit interaction. Over the past decade RMT has demonstrated its use in the ab initio study of ultrafast electron dynamics in atoms, and has recently been extended to describe dynamics in arbitrarily polarised light fields [3].

We extend RMT by enabling an interface with the RMatrixI package [4] such that spin-orbit effects are present in the resulting electron dynamics. To demonstrate these effects, we model a pump-probe setup on atomic krypton. The pump pulse (6 cycles, 10 eV) excites the $4s^2 4p^5 5s\ {}^1P$ state, after which the spin-orbit interaction will change the atom from the ${}^1P$ state to the ${}^3P$ state and back again. After a time delay, the probe pulse (6 cycles, 15 eV) will further excite the atom to the $4s4p^6 5s$ autoionising state. This state will subsequently decay to $4s^2 4p^5 k\ell$, which we observe through detection of the ejected electron.

Figure 1 shows the autoionisation signal from this model as a function of the time-delay between the two pulses. When the spin-orbit interaction is included in the model, we observe oscillations in the autoionisation signal with respect to the time delay between the two pulses. When the pulses are cross-polarised, we see the same oscillation, but out of phase.

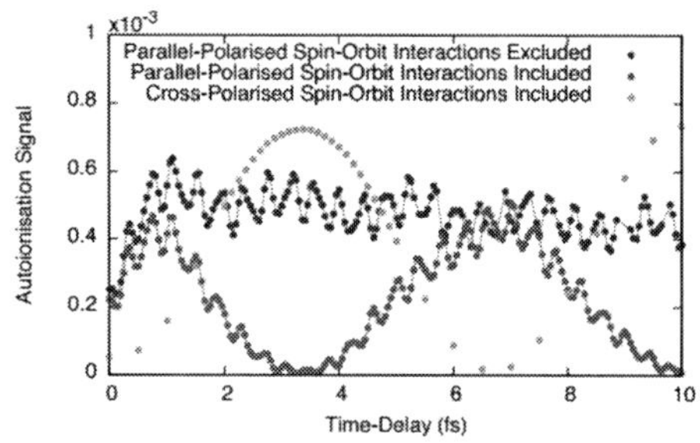

Figure 1. Signal from autoionisation of atomic krypton caused by a 10 eV pump pulse and a 15 eV probe pulse separated by a time-delay.

We will present an interpretation of this phase difference in terms of the spin-orbit induced dynamics of an electron between m_ℓ states, and the differing selection rules of the cross-polarised and parallel polarised pulses. We will also present a discussion of the interference effects present in Fig. 1 for both parallel-polarised cases, but absent in the cross-polarised case. Again, this difference can be attributed to the dipole-selection rules for the cross- and parallel-polarised pulses. This result may have actionable implications for improving the signal-to-noise ratio in experiments of this type.

References

[1] Jordan I *et al* 2017 *Phys. Rev. A* **95** 013404
[2] Wirth A *et al* 2011 *Science* **334** 6053 195
[3] Clarke D *et al* 2018 *Phys. Rev. A* **98** 053442
[4] Berrington K A *et al* 1995 *Comp. Phys. Com.* **92** 290

*E-mail: jack.wragg@qub.ac.uk

Amplification of a train of attosecond pulses in a plasma-based X-ray laser driven by an IR field

V A Antonov[1,2,*], K C Han[3], I R Khairulin[1] and O Kocharovskaya[3,†]

[1]Institute of Applied Physics of the Russian Academy of Sciences, Nizhny Novgorod, 603950, Russia
[2]Prokhorov General Physics Institute of the Russian Academy of Sciences, Moscow, 119991, Russia
[3]Department of Physics and Astronomy, Texas A&M University, College Station, TX 77843-4242, USA

Synopsis We suggest a technique to amplify a train of attosecond pulses, produced via high-harmonic generation of an infrared laser field, in active medium of a plasma-based X-ray laser driven by a replica of the same IR field as used to produce high harmonics forming a train of attosecond pulses.

In this contribution, we suggest a technique to amplify a train of attosecond pulses produced via high-harmonic generation (HHG) of an infrared (IR) laser field in the active medium of a plasma-based X-ray laser driven by a replica of the same IR field. We consider the X-ray lasers generating at the transition from the first excited to the ground-state energy levels of hydrogen-like ions [1]. Under the action of the IR field the frequencies of inverted transitions of the active medium vary in space and time along with variation of the IR field strength due to the linear Stark effect [2, 3]. As a result, the gain for the X-ray field appears not only at the frequency of the resonance, but also at the frequencies of the sidebands, separated from the resonance by an even multiple of the frequency of the driving IR field. If the last one coincides with the frequency of the field used for HHG, and the intensity of the driving field is high enough, then all the high-order harmonics incident to the medium can be amplified simultaneously. If the scattering of the harmonics into each other is suppresed due to the strong plasma dispersion of the active medium at the frequency of the IR field, while the intensity of the IR field is chosen in such a way, that the harmonics of different orders experience approximately the same gain, then there is a possibility for nearly uniform independent amplification of the harmonics, preserving their initial phase relations, and a temporal shape of the attosecond pulse train. We propose an experimental implementation of the suggested technique in active medium of C^{5+} ions dressed by an IR laser field at either 2.1 μm or 0.8 μm wavelength, and show the possibility to amplify by two orders of magnitude a train of attosecond pulses at a carrier wavelength 3.4 nm in the "water window" range with duration down to 100 as, see Fig. 1. We show also a possibility to isolate a single attosecond pulse from the incident attosecond pulse train during its amplification in modulated optically deep medium.

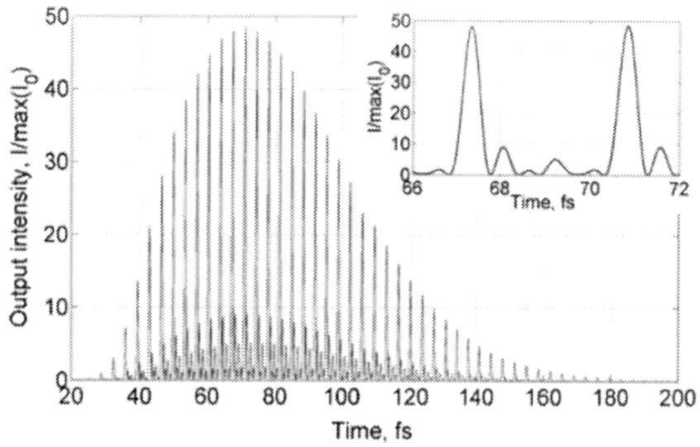

Figure 1. A train of amplified attosecond pulses. Output intensity is normalized to peak intensity of the incident attosecond pulse train. An insert shows two pulses from the train.

This work was partially supported by the Ministry of Education and Science of the Russian Federation under contract No.14.W03.31.0032 executed at the Institute of Applied Physics of the Russian Academy of Sciences.

References

[1] Suckewer S, Jaeglé P 2009 *Laser Phys. Lett.* **6** 411
[2] Antonov V A *et al* 2013 *Phys. Rev. A* **88** 053849
[3] Akhmedzhanov T R *et al* 2017 *Phys. Rev. A* **96** 033825

* E-mail: antonov@appl.sci-nnov.ru
† E-mail: kochar@physics.tamu.edu

Journal of Physics: Conference Series **1412** (2020) 072020 doi:10.1088/1742-6596/1412/7/072020

Many-body calculations of two-photon, two color matrix elements for attosecond delays

M Dahlström[1], A Ljungdahl[2], J Sörngård[2], J Vinbladh[2] and E Lindroth*[2]

[1]Department of Physics, Lund University, Box 118, SE-221 00 Lund, Sweden
[2]Stockholm University, Department of Physics, SE-106 91 Stockholm, Sweden

Synopsis Interferometric measurements based on the combination of extreme ultraviolet (XUV) attosecond pulse trains with weak infrared laser fields, have been successfully used to study time-resolved and recently angle-resolved photoemission in various atomic systems. These measurements rely on quantum interference between electron wavepackets created by absorption of XUV radiation, with further absorption or emission of laser photons. Here we present many-body calculations of the full two-photon process.

The dynamics, during for example a photoionization process, is encoded in the amplitude and phase of the released electron wave packet. A key quantity is the spectral derivative of this phase, the *group delay*. It reflects the delay or advance of the electron when it travels through the atomic potential to eventually emerge in the continuum. It has been established that it is meaningful to separate the interferometrically measured *atomic delay*, τ_A, into a *Wigner-like delay*, τ_W, associated with the one-photon extreme ultraviolet (XUV) ionization process and a contribution from the interaction with the infrared (IR) laser-field in the presence of the atomic potential, called the *continuum–continuum delay*, τ_{cc}. While the Wigner delay is strongly dependent on the atomic origin of the electron, the contribution from the IR photon has been found to be "universal" in the sense that it to a very good approximation depends only on the kinetic energy of the photoelectron, the photon energy of the laser field and the charge of the remaining ion. The universality of the τ_{cc} correction has been studied through many-body calculations of the two-photon process , benchmarked by state-of-the-art experiments [1], both for angular integrated measurements and for detection along the polarization axis [2, 3]. Although there are exceptions, in particular close to ionization thresholds and at resonances [3, 4], the universality of the contribution from the laser photon has hitherto proved to be a good approximation, with the important practical consequence that Wigner delays, τ_W, can be reliably extracted from measured atomic delays, τ_A.

We will present new studies investigating the calculations of the atomic delay in velocity gauge, the delays of heavy elements using relativistic many-body theory and additional possibilities with spherically polarized light.

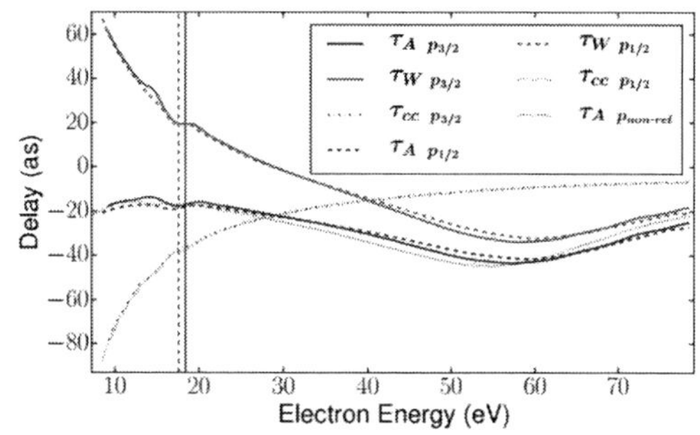

Figure 1. The atomic (blue solid and dashed line) and the Wigner (grey solid and dashed line) delays for the outermost shells, $4p_{1/2}$ and $4p_{3/2}$, in krypton. The difference for the two fine-structure components is clearly visible. The difference, $\tau_A - \tau_W$ is shown in red with dotted and dashed-dotted lines. Here the two fine-structure components cannot be distinguished, showing that the universality of τ_{cc} still holds.

References

[1] Isinger M *et al* 2017 *Science* **358** 893
[2] Dahlström J M *et al* 2012 *Phys. Rev. A* **86** 061402
[3] Dahlström J M and Lindroth E 2014 *J. Phys. B* **47** 124012
[4] Kotur M *et al* 2016 *Nat Commun.* **7** 10566

*E-mail: Eva.Lindroth@fysik.su.se

Polarisation Effects in Above-Threshold Ionisation of Excited Helium

A C Bray[1] *, A S Maxwell[1], C Figueira de Morisson Faria[1], Y Kissin[2], M Ruberti[2], M Khokhlova[2] and V Averbukh[2]

[1]Physics & Astronomy, University College London, Gower Street, London, WC1E 6BT, United Kingdom
[2]Department of Physics, Imperial College, Prince Consort Rd, London SW7 2BW, United Kingdom

Synopsis We present work on the disentanglement of core dynamics and binding potential influences for Above-Threshold Ionisation in Helium prepared in an excited p state, interacting with a mid-IR field. This was formed by comparing the Coulomb Quantum-Orbit Strong-Field Approximation (CQSFA) with B-spline Algebraic Diagrammatic Construction (ADC) simulations. This may lead to the identification of initial momentum distributions, for tunneling and re-scattering instances, from the CQSFA perspective. The latter method is an inverse approach; thus, the ADC simulations will allow comparison of findings.

Re-scattering effects in Above-Threshold Ionisation (ATI) are known for over twenty-five years and in general can be reasonably reproduced by using standard quantum orbit trajectory methods, such as the Strong-Field Approximation (SFA). For recent reviews see "The Plateau in Above-Threshold Ionisation: the Keystone of Re-scattering Physics" [1] and "Symphony on Strong Field Approximation" [2].

Thereby, a key issue in re-scattering is the shape of the electronic bound state from which the electron leaves or to which it returns. However, for certain orbital geometries, standard quantum trajectory methods such as the SFA erroneously lead to vanishing signals from re-scattered trajectories in the photoelectron spectra. This is because the SFA only considers re-scattering orbits that exactly pass through the origin and neglects any deflection caused by the residual Coulomb potential [3].

A striking example of this, is Helium prepared in an excited p state, whose orientation is orthogonal to that of a linearly polarised laser field [4]. Ab-initio methods, reveal that, in contrast to the SFA predictions, the ATI signal from this type of configuration is non-vanishing.

In this work, we have disentangled the different physical mechanisms that contribute to this effect; namely, the core dynamics and the interplay between the field and the residual binding potential for Helium. This is undertaken by comparing the Coulomb Quantum-Orbit Strong-Field Approximation (CQSFA) with ab-initio methods, namely B-spline Algebraic Diagrammatic Construction (ADC), that allow for core excitation and polarisation effects [5].

We also identify the main types of orbits responsible for a non-vanishing ATI signal within the CQSFA and possible initial momentum distributions of the instances of tunnelling and re-scattering by using the ADC model as a benchmark as the CQSFA is an inverse approach.

References

[1] Becker W *et al* 2018 *J. Phys. B* **51** 162002
[2] Amini K *et al* 2018 arXiv:1812.11447v1
[3] Maxwell A S *et al* 2017 *Phys. Rev. A* **96** 023420
[4] Averbukh V 2004 *Phys. Rev. A* **69** 043406
[5] Averbukh V *et al* 2018 *Royal Society of Chemistry* "First-principles Many-electron Dynamics Using the B-spline Algebraic Diagrammatic Construction Approach"

*E-mail: a.bray@ucl.ac.uk

Temporal metrology for advanced ultrashort XUV sources

T Sinyakova[1], C Bomme[2], C Bourassin-Bouchet[3], L Chopineau[2], D Cubaynes[1], L Dakroub[1], G Garcia[4], A Pandey[5], E Papagiannouli[5], O Guilbaud[5], F Quéré[2] and A Klisnick[1*]

[1]ISMO, CNRS, Université Paris-Saclay, Orsay Cedex, 9140, France
[2]LIDYL, CEA, CNRS, Université Paris-Saclay, Gif-sur-Yvette, 91191, France
[3]Laboratoire Charles Fabry, Institut d'Optique, CNRS, Université Paris-Saclay, Palaiseau Cedex, 91127, France
[4]Synchrotron SOLEIL, Gif sur Yvette Cedex, 91192, France
[5] LPGP, CNRS, Université Paris-Saclay, Orsay, 91400, France

Synopsis A new VMI spectrometer has been developed in order to enable the single-shot characterization of different ultrashort XUV sources through laser-dressed photoionization. The instrument has been commissioned with a plasma-based XUV laser and, then, implemented on a high harmonics source based on a plasma mirror.

The wide application of the ultra-short XUV sources for fundamental studies in many different areas requires the accurate measurement of their pulse temporal profile. Since the diversity of physical processes underlying their generation is directly reflected in their pulse temporal shapes, it is very important to develop an instrument that can cover a broad range of source characteristics, in terms of duration and pulse structure. A new temporal diagnostic based on angularly-resolved laser-dressed photoelectron spectroscopy is currently under development. It consists in a home-made velocity map imaging (VMI) spectrometer, involving a thick-lens configuration [1]. It was designed to allow the reconstruction of the XUV pulse temporal structure from the picosecond to attosecond time-scale, with a single shot capability.

The optimal characteristics of the VMI spectrometer, designed and constructed at ISMO, have been examined using a trajectory simulation of charged particles package SIMION. Commissioning experiments were performed at LASERIX facility (LUMAT, Université Paris-Sud). A plasma-based XUV laser emitting at 38 eV was used to photoionize Xe and Ne, in order to test the capabilities of the new VMI spectrometer and to optimize its performance.

Then, the VMI spectrometer was installed at UHI100 facility (LIDYL, CEA-Saclay) in order to characterize the temporal profile of high-order harmonics generated from a plasma mirror in presence of a IR-dressing field. Importantly, such sources have never been experimentally characterized in the attosecond time-scale so far. The sophisticated experimental set-up provides the temporal and spatial superposition between the XUV and dressing pulses. It, thus, allows one to complete single-shot-angularly-resolved measurements while maintaining the possibility of scanning the laser-XUV delay with attosecond precision.

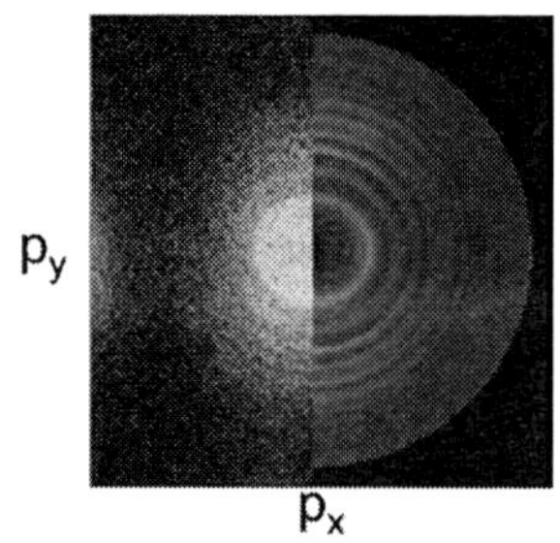

Figure 1. Raw 2D-momentum distribution of photoelectrons ejected from Ar by high harmonics generated from plasma mirror (left half) and slice of the 3D-momentum distribution after Abel inversion (right half).

Fig. 1 shows an example of obtained angularly-resolved VMI images, which will then be processed with an iterative algorithm in order to retrieve the XUV pulse temporal profile.

Acknowledgments :
This project has received funding from LabEx PALM (ANR-10-LABX-0039-PALM), LIDEX OPT2X, Sesame PULSE-X, Laserlab-Europe (EU-H2020 654148)

References :
[1] Kling N G *et al* 2014 *J. Instrum.* **9** P05005

* E-mail: annie.klisnick@u-psud.fr

Attosecond photoionization dynamics around Ar Cooper minimum

M Lejman[1 *], **C Alexandridi**[1], **M Turconi**[1], **L Barreau**[2], **D Platzer**[1], **A Borot**[1], **J-F Hergott**[1], **O Tcherbakoff**[1] **and P Salières**[1 †]

[1]Laboratoire Interactions, Dynamiques et Lasers, Commissariat à l'énergie atomique, 91191 Gif-sur-Yvette, France
[2]College of Chemistry, University of California, Berkeley CA, USA

Synopsis Photoionization time delays between the 3s and 3p subshells of argon have been measured over a large energy range (35-70eV) covering the Cooper minima in both subshells, providing stringent tests for interchannel correlation theories.

A Cooper minimum corresponds to a local decrease of the ionization cross section as a function of photon energy upon extinction of one ionization channel [1]. Although this phenomenon is well known in conventional spectroscopy, it is only recently that the ultrafast ionization dynamics around a Cooper Minimum (CM) have been investigated, especially on the attosecond time scale. The case of Argon is particularly interesting because the CM in the 3p channel around 50 eV [2] is also transferred to the 3s channel (around 43 eV) due to intershell correlation. Measuring attosecond ionization delays between the two subshells would thus provide a stringent test of multielectronic theoretical models trying to describe interchannel couplings accurately.

The RABBIT technique in principle allows the measurement of the spectral phase of the electron wave packets emitted in the two channels, from which the difference of ionization delays between the subshells can be deduced. However, two main difficulties arise for such measurements: i) the much smaller cross section for the 3s channel, in particular in the CM region, and ii) the overlap between the 3s sidebands and the 3p harmonic peaks in the RABBIT trace, have prevented access to the 3s channel spectral phase except at low energy, just below the CM [4, 5].

Thanks to the Rainbow RABBIT technique, we were able to overcome the above difficulties and to measure the difference between 3s and 3p ionization delays over a large energy range (35-70 eV) covering the Cooper minima in both subshells.

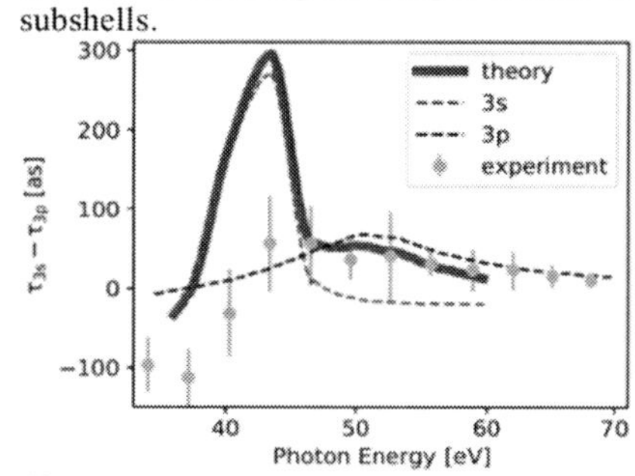

Figure 1. Difference between 3s and 3p ionization time delays measured as a function of the photon energy (dots) and compared with theoretical calculations [3].

The results (Figure 1) show strong variations of the delay difference, with a change of sign in the 3s Cooper minimum region. These results have been compared with previous measurements [4, 5, 6] and with theoretical calculations [3] for which the agreement is very good in the 3p Cooper minimum region (45-60 eV), but poor in the 3s Cooper minimum region (40-45 eV).

References

[1] Cooper J W 1962 *Phys. Rev.* **128** 681
[2] Houlgate R G *et al* 1974 *J. Phys. B* **7** L470
[3] J. M. Dahlström *et al* 2014 *J. Phys. B* **47** 124012
[4] Klünder K *et al* 2011 *Phys. Rev. Lett.* **106** 143002
[5] Guénot D *et al* 2012 *Phys. Rev. A* **85** 053424
[6] Schoun S B *et al* 2014 *Phys. Rev. Lett.* **112** 153001

* E-mail: mariusz.lejman@cea.fr
† E-mail: pascal.salieres@cea.fr

Subcycle transfer of photon momentum in nondipole tunneling ionization

H Ni[1]*, S Donsa[1] and J Burgdörfer[1]

[1]Institut für Theoretische Physik, Technische Universität Wien, 1040 Vienna, Austria, EU

Synopsis We study the subcycle transfer of photon momentum in tunneling ionization beyond the dipole approximation using the attoclock setup. We separate the momentum transfer during tunneling and propagation in continuum using the backpropagation method.

Strong-field ionization is usually studied using the dipole approximation thereby neglecting the photon momentum. The longitudinal momentum the departing electron gained directly from the photon is typically 1000 times smaller than the transverse momentum from the ionization process, and is thus difficult to resolve experimentally. With the advances of detecting technologies, this small momentum shift in the laser propagation direction becomes observable. Up to date, the studies of photon momentum transfer focus on the integral momentum transfer for the entire pulse. We, instead, simulate the subcycle transfer of photon momentum for a typical attoclock setup. The rotating vector potential of the elliptically polarized laser pulse acts as the clock of momentum transfer in the laser propagation direction. We find that the average transferred photon momentum shows a pronounced minimum near the pulse peak where the emission probability is maximal, as illustrated in Figure 1.

The (overall) nonzero shift of the longitudinal momentum is predicted to be $p_z = \frac{E}{c} + \alpha\frac{I_p}{c}$ with $\alpha \approx \frac{1}{3}$ [1] (for the entire pulse), where the first term comes from the propagation in the field after ionization while the second term comes from the tunneling process. This shift vanishes in the dipole limit. We use the backpropagation method [2, 3, 4] to explicitly separate these two contributions in order to test the expected shift [1], as well as to resolve these contributions on the subcycle scale.

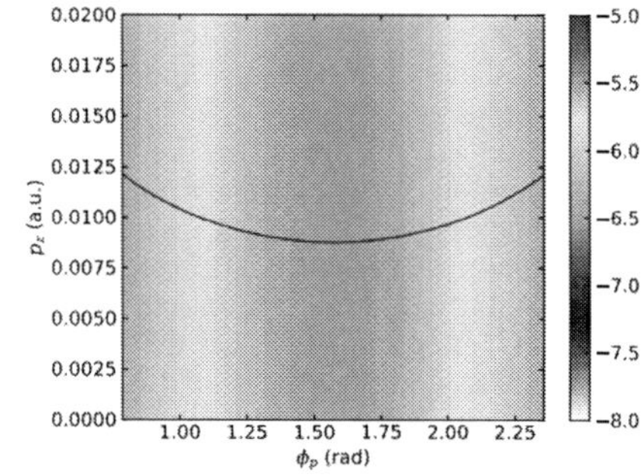

Figure 1. Electron momentum distribution around the maximal emission direction (logarithmic scale). The ϕ_p axis is the electron emission angle within the polarization plane and p_z is the longitudinal momentum in the laser propagation direction. The green solid line is the average longitudinal momentum as a function of emission direction $\langle p_z\rangle(\phi_p)$.

References

[1] Klaiber M *et al* 2013 *Phys. Rev. Lett.* **110** 153004
[2] Ni H *et al* 2016 *Phys. Rev. Lett.* **117** 023002
[3] Ni H *et al* 2018 *Phys. Rev. A* **97** 013426
[4] Ni H *et al* 2018 *Phys. Rev. A* **98** 013411

*E-mail: hongcheng.ni@tuwien.ac.at

Time-resolved interferometric photoemission from Cu(111) and Ni(111) surfaces

M J Ambrosio * **and U Thumm** †

Department of Physics, Kansas State University, Manhattan, Kansas 66506, USA.

Synopsis We calculated spectra and phases for photoemission from Cu(111) and Ni(111) surfaces. Our numerical model reproduces recently measured final-state shape resonances of the photoelectron in the substrate potential.

Interferometric photoemission by an XUV attosecond pulse train into the delayed field of an assisting IR laser pulse results in sideband-yield oscillations in the photoelectron spectrum that provide temporal (phase) information [1, 2, 3]. Recently measured interferometric (RABBITT) spectra reveal final-state resonances [2, 4] that we trace to the resonant enhancement of the final-photoelectron-state probability density in the substrate when the outgoing-electron de Broglie wavelength matches the substrate lattice spacing [Fig. 1 (a)].

We model the substrate's d-band in terms of linearly combinations of lattice-plane-localized orbitals and include the photoelectron-final-state interaction with the (i) Fresnel-reflected IR-laser pulse, (ii) periodic substrate potential along the surface-normal direction, and (iii) electronic scattering during the photoelectron transport from its excitation in the substrate to the surface [5].

Figure 1 (b) compares our simulated RABBITT phases with the experimental data for photoemission from Cu(111) of Ref. [4]. Our model reproduces the phase increase at high-harmonic order 20 which we interpret as due to the resonantly enhanced transport time in the solid. Figure 1 (c) shows our calculated RABBITT phases for emission from the $\Lambda_3^{\alpha,\beta}$ bands of Ni(111) together with the experimental results of Ref. [2].

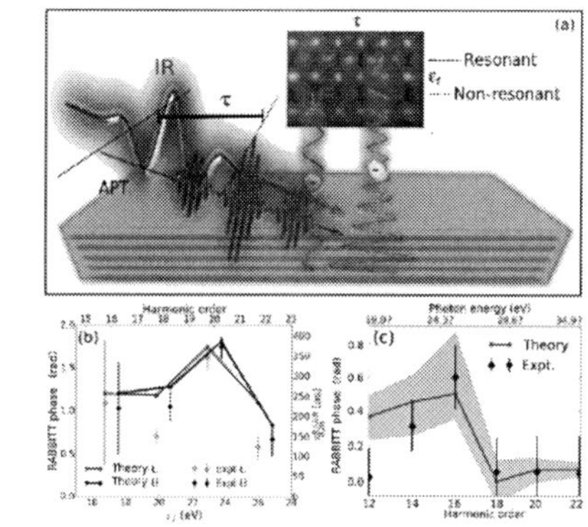

Figure 1. (a) Interferometric photoemission into resonant and non-resonant final states. (b) Calculated [5] and corresponding experimental [4] RABBITT phases for emission from the two 3d subbands of Cu(111). (c) RABBITT phases for emission from the $\Lambda_3^{\alpha,\beta}$ bands of Ni(111), compared with the experimental phases of Ref. [2].

Acknowledgements We acknowledge support from the US NSF and the Division of Chemical Sciences, Office of the Basic Energy Sciences, Office of Energy Research, US DoE.

References

[1] Locher R *et al* 2015 *Optica* **2** 405
[2] Tao Z *et al* 2018 *Science* **353**, 62
[3] Ambrosio M J *et al* 2018 *Phys. Rev. A* **97** 043431
[4] Kasmi L *et al* 2017 *Optica* **4** 1492
[5] Ambrosio M J *et al* 2019 submitted to *Phys. Rev. A*

*E-mail: mj_ambrosio@phys.ksu.edu
†E-mail: thumm@phys.ksu.edu

Atomic RABBITT-like experiments assisted by a circularly polarized laser

D I R Boll[1] and O A Fojón[1] *

[1]Instituto de Física Rosario, CONICET-UNR, Blvd. 27 de Febrero 210 bis, 2000 Rosario, Argentina

Synopsis We study theoretically the single ionization of atoms by the combined action of an attosecond pulse train with linear polarization and an assistant laser field with circular polarization. By means of a non-perturbative model, we obtain closed-form expressions for the angular distributions of photoelectrons. We show that such a configuration of light fields is similar to the usual polarization control technique where both the attosecond pulse train and the assistant laser field are linearly polarized with a controlled relative orientation.

In a previous study, we showed that angular distributions for an atomic RABBITT-like experiment, with collinear polarization for both the attosecond pulse train and the infrared laser field (IR), are asymptotically equivalent to those for the monochromatic ionization of an atomic dimer [1]. Moreover, we showed that the separation between these virtual emitters is governed by the photoelectron momentum and the properties of the infrared laser (wavelength, intensity and delay). The agreement of our results with the scarce experimental data available, was found to be comparable with that obtained through more complex methods. The strength of our model resides in the fact that it accounts for the interaction of the photoelectron with the IR to all orders, providing analytical expressions that reproduce the expected behaviour in the low IR intensity limit.

In this contribution, we go a step further. We demonstrate that the ionization of an atomic target with an attosecond pulse train, in combination with a circularly polarized infrared laser field, is asymptotically equivalent to a polarization control scheme [3, 4]. At variance with our previous results [1], where the delay dependence is mapped into different separations of the virtual emitters, in this case a delay variation translates into a different orientation of the virtual emitters while the distance between them is preserved. This equivalence can be understood by simple arguments. As the attosecond pulse trains have two opposite bursts per IR cycle, one of them will produce a photoelectron signal at times where the instantaneous polarization vector of the IR points in a direction having a polar angle Θ, as depicted in Fig. 1. The next attosecond burst, half an IR period later, will produce a photoelectron signal at times where the instantaneous polarization vector of the IR points in the opposite direction.

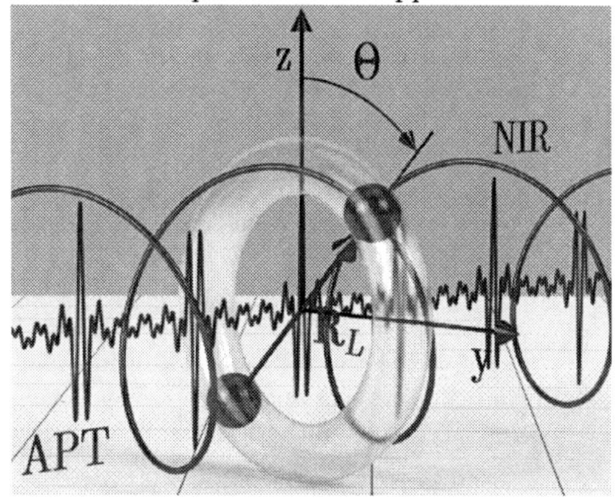

Figure 1. Pictorial representation of the reaction.

As these proccesses repeat over multiple IR periods, the photoelectron can only "feel" a linear variation of the IR.

Finally, we show that our analytical results allow us to define a "magic time": a delay such that angle-integrated signals for main- and sidebands are proportional to the total photoionization cross sections of the atomic target multiplied by the Cohen-Fano molecular interference factor [5].

*E-mail: ofojon@fceia.unr.edu.ar

References

[1] Boll DIR and Fojón OA 2016 *J. Phys. B: At. Mol. Opt. Phys.* **49** 185601
[2] Boll DIR and Fojón OA 2017 *J. Phys. B: At. Mol. Opt. Phys.* **50** 235604
[3] O'Keeffe P *et al* 2004 *Phys. Rev. A* **69** 051401
[4] Meyer M *et al* 2008 *Phys. Rev. Lett.* **101** 193002
[5] Cohen HD and Fano U 1966 *Phys. Rev.* **150** 30

Controlling sub-cycle instantaneous optical chirality in the photoionization of chiral molecules

S Rozen[1], A Comby[2], S Beauvarlet[2*], E Bloch[2], B Fabre[2], V Blanchet[2], D Descamps[2], S Petit[2], B Pons[2], N Dudovich[1] and Y Mairesse[2]

[1]Department of Physics of Complex Systems, Weizmann Institute of Science, Rehovot, 76100, Israel
[2]CELIA, Universite de Bordeaux-CNRS-CEA, Talence, 33405, France

Synopsis Photoionization of chiral molecules by circularly polarized light produces characteristic photoelectron angular distributions, showing a strong forward/backward asymmetry with respect to the light propagation axis. Up to now, this process, referred to as Photoelectron Circular Dichroism, has been studied using electric fields with constant ellipticity over the optical period. In this work, we show numerically and experimentally that the chiroptical photoelectron asymmetry can be controlled by engineering the subcycle instantaneous ellipticity and chirality of the ionizing electric field.

The photoionization of chiral molecules by circularly polarized light results in one of the most efficient ways to probe chirality in the gas phase. The photoelectron angular distribution indeed shows a forward/backward asymmetry with respect to the direction of light propagation, which is sensitive to both light helicity and enantiomeric form of the molecular target. This effect is referred to as PhotoElectron Circular Dichroism (PECD) [1, 2, 3].

Up to now, PECD has only been studied with laser or XUV fields whose ellipticity and chirality do not depend on time. The recent developments in attosecond spectroscopy have shown that two-color laser field enables the temporal properties of the ionization process to be controlled and measured with high accuracy [4].

Here we engineer an electric field whose subcycle temporal profile and thus instantaneous ellipticity and chirality can be tuned, by combining two phase-locked orthogonally polarized fundamental and second harmonic fields. We photoionize chiral molecules with this controlled chiral field and measure two-dimensional projections of the photoelectron angular distributions using a velocity map imaging spectrometer. We observe a clear forward/backward asymmetry which is opposite in the up and down parts of the detector.

This two-fold asymmetry, which we call Enantiomeric Subcycle Chiral Antisymmetric Response Gated by Orthogonal Two-color laser field (ESCARGOT), strongly depends on the relative phase between the orthogonal components of the fields and is indeed observed experimentally and reproduced numerically. Since changing this phase monitors the temporal evolution of instantaneous field ellipticity and chirality within one fundamental optical period, the ESCARGOT signal therefore provides a unique insight into the influence of instantaneous chirality in the dynamical photoionization process.

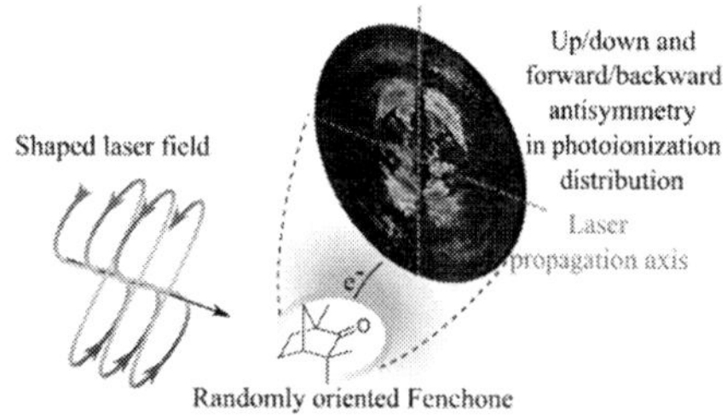

Figure 1. Principle of the experiment: the shaped electric field is focused into a jet of randomly oriented enantiopure fenchone molecules, in the interaction zone of a VMI spectrometer. The VMI measures the photoelectron momentum distribution, whose forward/backward antisymmetric component is shown in the figure. This component shows a clear up/down antisymmetry.

References

[1] Ritchie B 1976 *Phys. Rev. A* **13** 1411
[2] Bowering N *et al* 2001 *Phys. Rev. Lett.* **86** 1187
[3] Nahon L *et al* 2015 *JESRP* **204** 322–334
[4] Shafir D *et al* 2012 *Nature* **485** 343–346

*E-mail: sandra.beauvarlet@u-bordeaux.fr

NewStock, a time-dependent close-coupling program for atomic ionization

L Argenti[1*] **and E Lindroth**[2]

[1]Dep. Physics and CREOL, University of Central Florida, Orlando, 32816 Florida, USA
[2]Dep. Physics, Stockholm University, Stockholm, SE-10691 Sweden, EU

Synopsis We present NewStock, a time-dependent program to compute the photoionization of polyelectronic atoms by arbitrary light pulses. The program, which expands the capabilities of the Stock B-spline close-coupling structure code to the time-dependent domain, is applied to the pump-probe ionization of the neon atom.

Autoionizing states are a pervasive aspect of atomic ionization and a signature of electronic correlation. Thanks to attosecond (as) sources it became possible to resolve the evolution of these metastable states in real time, and to control them with light [1]. Only with the support of new dedicated *ab initio* tools, however, can the promises of as control come to fruition.

Ab initio time-dependent close coupling has affirmed itself as the leading tool to simulate the ionization of atoms and molecules under the influence of ultrashort pulses beyond the single-active-electron approximation [2]. Here we will present NewStock, a recently developed time-dependent programs that is able to compute the photoionization of small poly-electronic atoms by arbitrary light-pulse sequences. NewStock merges the capabilities of the Stock B-spline close-coupling structure code [3], based on the ATSP2K package [4], with those of state-of-the-art time-dependent two-active-electron codes [5]: it numerically solves the time-dependent Schrödinger equation in a space in which multi-reference parent ions are augmented with a spherical B-spline basis. Channel-specific mask functions periodically extract and analyze the outer part of the state vector, which allows us to use long moderately weak pulses while employing comparatively small quantization boxes. The channel- and energy-resolved photoelectron spectrum is computed by projecting the state function on multichannel scattering states.

The program is applied to compute the angularly-resolved photoelectron XUV-pump IR-probe spectrum of neon across the $2p^{-2}(^3\mathrm{P})3s(^2\mathrm{P}^e)$ threshold, where the $2p^{-2}(^1\mathrm{D})3s(^2\mathrm{D}^e)3p(^1\mathrm{P})$ configuration manifests itself as an intruder state [6]. The time- and angularly-resolved interference between one-photon and two-photon transitions reveals two key aspects of the autoionization dynamics of the intruder state: the initial fast decay to a metastable Rydberg wave packet of the $2p^{-2}(^3\mathrm{P})3s(^2\mathrm{P}^e)np(^2\mathrm{P}^o)$ series, as well as the slow decay of the latter to the continuum.

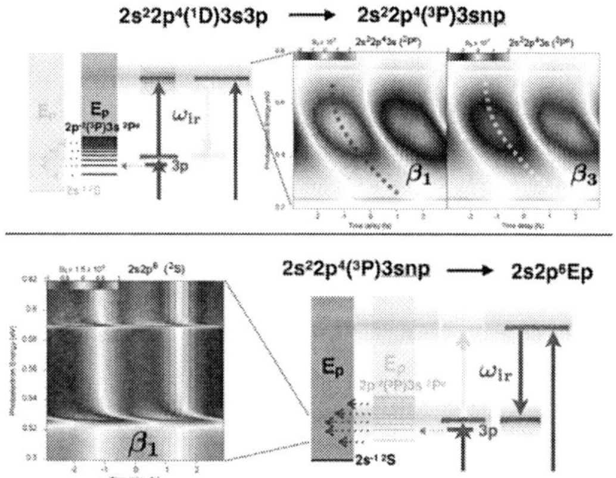

Figure 1. Ionization of neon across the $2p^{-2}(^3\mathrm{P})3s(^2\mathrm{P}^e)$ threshold with an as pulse train comprising isolated pairs of consecutive linearly-polarized harmonics of a fundamental IR frequency, in the presence of a weak IR pulse. The interference between one-photon and two-photon transitions amplitudes, in which either one or the other is resonant with the $2p^{-2}(^1\mathrm{D})3s(^2\mathrm{D}^e)3p(^1\mathrm{P})$ intruder state, gives rise to an anisotropic emission that incorporates the characteristic decay time of the intruder to the Rydberg series, and of each term of the series to the continuum.

References

[1] Gruson V *et al* 2016 *Science* **354** 734
[2] Argenti L *et al* 2015 *Phys. Rev. A* **91** 061403
[3] Carette T *et al* 2013 *Phys. Rev. A* **87** 023420
[4] Fröse Fischer C *et al* 2007 *Comp. Phys. Com.* **176** 559
[5] Argenti L *et al* 2010 *Phys. Rev. Lett.* **105** 053002
[6] Schulz K *et al* 1996 *Phys. Rev. A* **54** 3095

* E-mail: luca.argenti@ucf.edu

Polarization of attosecond pulses generated by two-color laser fields

M F Zhu[1], G L Wang[1 *], S F Zhao[1], X Y Li[2] and X X Zhou[1]

[1]College of Physics and Electronic Engineering, Northwest Normal University, Lanzhou, 730070, China
[2]Department of Experimental Teaching, Northwest Minzu University, Lanzhou, 730070, China

Synopsis We propose a circular-elliptical laser field, which consists of a circularly polarized field with its coplanar counter-rotating elliptical third harmonic pulse, to generate polarized high-order harmonics and attosecond pulses (APs) by He atom. We show that the polarization of APs depends evidently on the ellipticity ξ of harmonic laser field. Most importantly, nearly circular APs can be obtained with a proper ξ. Our results contribute to the generation of high elliptically polarized APs with s-state atom.

Although the so-called bicircular laser fields, which consist of two coplanar counter-rotating circularly polarized fields with angular frequencies of ω and 2ω or ω and 3ω, have been proposed in recent years in both experiments and theoretical simulations to generate circular harmonics, it is hard to obtain attosecond pulses (APs) with big polarization ellipticity, especially for the s-state atoms [1]. Here we demonstrate a feasibility of generating highly elliptically polarized APs by a named circular-elliptical laser field with He atom.

The electric field of driving laser pulse has the following form

$$\begin{cases} \boldsymbol{E}_{\omega}(t) = f(t)E_1[\cos(\omega t)\hat{x} + \sin(\omega t)\hat{y}]/\sqrt{2} \\ \boldsymbol{E}_{3\omega}(t) = f(t)\frac{E_2}{\sqrt{1+\xi^2}}[\cos(3\omega t)\hat{x} - \xi\sin(3\omega t)\hat{y}] \end{cases},$$

where $f(t)$ is a 8-optical-cycle trapezoidal envelope with linear ramps of one optical cycle of fundamental pulse, ξ is the ellipticity for harmonic pulse. Present investigation is based on the numerical solution of the time-dependent Schrödinger equation. Our simulations show that the polarization of third harmonic field almost has no influence on the harmonic spectra, and the helicity for $4q \pm 1$ harmonics are still the same for bicircular (ξ=1) and circular-elliptical ($\xi \neq$1) laser fields. However, compared with the case of bicircular field, the polarization of single harmonics generated by circular-elliptical laser changes from circular to elliptical state. By superposing several harmonics we obtain an attosecond pulse trains, which includes two attosecond bursts that approximately perpendicular with each other. With a proper time window we then pick out one burst.

Figure 1 shows the polarization dependence of one attosecond burst on the ellipticity of 3ω field. Here the total laser intensity is 2.5×10^{14} W/cm^2 and $E_1/E_2 = 1$, the wavelength of fundamental pulse is taken as 800 nm. It is obvious that the polarization of APs depends evidently on the ellipticity of third harmonic laser field. By choosing a reasonable ellipticity of the 3ω laser field we could obtain a bigger ellipticity for APs. With ξ=0.8, the ellipticity for this burst reaches about $\varepsilon \approx 0.9$.

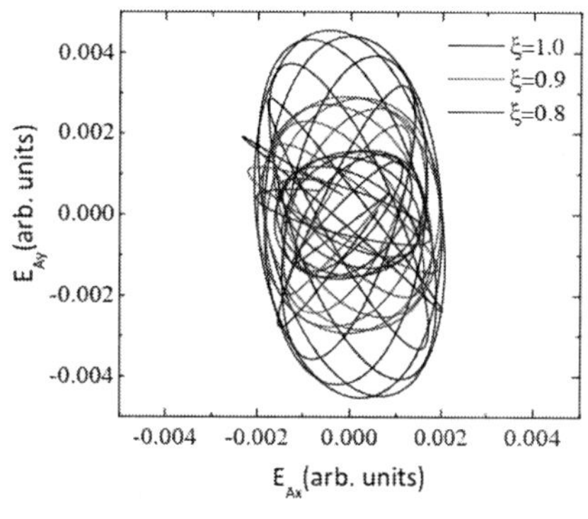

Figure 1. Electric field of APTs superposed from 30th to 36th harmonics generated by $\omega + 3\omega$ field with a time window of 3.7 to 3.95 optical cycle for different ξ.

Project supported by the National Natural Science Foundation of China (Nos. 11764038, 11864037, 11664035, 11564033), and the Foundation of Northwest Normal University [NWNU-LKQN-17-1].

References

[1] Milošević D B 2015 *Phys. Rev. A* **92** 043827

*E-mail: wanggl@nwnu.edu.cn

RABBITT attoclock for the direct measurement of ionization phases

S Donsa[1] *, N Douguet[2], J Burgdörfer[1], I Březinová[1] and L Argenti[2,3]

[1]Institute for Theoretical Physics, Vienna University of Technology, Vienna 1040, Austria, EU
[2]Department of Physics, University of Central Florida, Orlando, Florida 32186, USA
[3]CREOL, University of Central Florida, Orlando, Florida 32186, USA

Synopsis We use a circularly polarized attosecond pulse train (APT) together with a weak replica of the fundamental circularly polarized infra-red (IR) field to reconstruct the ionization phase of the dipole allowed $(2s2p)^1P^o$ resonance in atomic helium by analyzing the momentum distribution of the ionized photoelectrons within the polarization plane.

We propose the new attosecond spectroscopy technique "RABBITT attoclock" which generalizes RABBITT (Reconctruction of Attosecond Beating By Interference of Two-photon Transitions) to elliptically polarized light fields. Exploiting the properties of circularly polarized APTs it is possible to measure ionization phases within the photoelectron momentum distribution from a single shot measurement thereby avoiding the need of multiple time-delay measurements. The protocol uses an APT consisting of only $3n + 1$ and $3n + 2$ harmonics, where the two neighboring harmonics have left- or right-circular polarization, respectively, (such an APT was recently realized experimentally, e.g. in [1]) together with a weak replica of the fundamental IR field which can be either left- or right-circularly polarized. Within lowest-order perturbation theory this pulse setup leads to the superposition of the direct one-photon transition by absorbing one XUV photon, with the IR mediated two-photon transition from the neighboring harmonic by emission or absorption of an additional IR photon at each harmonic. Due to this superposition the momentum distribution of the photoelectrons shows a characteristic beating with ϕ or $3\,\phi$, respectively, depending on the polarization of the IR field (Fig. 1), with ϕ being the azimuthal angle.

We directly reconstruct the Fano phase variation caused by the optically allowed $(2s2p)^1P^o$ resonance in atomic helium. For the harmonic centered at the resonance we clearly observe the expected rapid phase jump in the momentum distribution. In the neighboring harmonic we observe the modulated Fano-phase variation caused by the two-photon transition with the resonance as intermediate state [2, 3]. In addition, the RABBITT attoclock technique allows to measure the phase variation caused by one-photon dipole forbidden resonances which can be reached via a two-photon transition such as the lowest energetic $^1D^o$ resonance in helium around 59.9eV. Considering the experimental difficulties in obtaining the full angular distribution at high energy we demonstrate that the methods works already if only 4 directions of the ionized electrons are measured (e.g. by a time-of spectrometer) and investigate the influence of finite energy resolution of the electron detector on the retrieved phase. Further we investigate how finite-pulse effects would influence the obtained phase-variation [3].

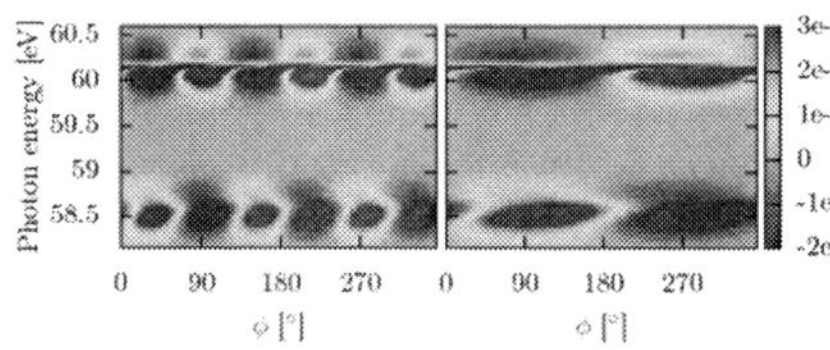

Figure 1. Photoelectron momentum distribution $P(\phi, \theta = \pi/2)$ of the APT together with a weak replica of the fundamental (left panel) left- and (right panel) right-circularly polarized IR, respectively. Note that the mean of the distribution was subtracted to make the oscillation visible.

References

[1] Kfir O *et al* 2015 *Nat. Photonics* **9** 99
[2] Gruson V *et al* 2016 *Science* **354** 734
[3] Jimenez-Galan A *et al* 2016 *Phys. Rev. A* **93** 023429

*E-mail: stefan.donsa@tuwien.ac.at

Generation and characterization of isolated attosecond pulses for coincidence spectroscopy at 100 kHz repetition rate

T Witting[1], F Furch[1], M Osolodkov[1], F Schell[1], C Menoni[2], C P Schulz[1]* and M J J Vrakking[1]

[1]Max-Born-Institut, Max-Born-Strasse 2A, 12489 Berlin, Germany
[2]Department of Electrical and Computer Engineering, Colorado State University, Fort Collins, CO 80523, USA

Synopsis An attosecond pump-probe beamline with 100 kHz repetition rate for coincidence experiments has been developed. It is based on non-collinear optical parametric chirped pulse amplification and delivers 100 μJ sub-4 fs to an high-harmonic generation source. Details on the generation and characterization of isolated attosecond pulses will be presented.

The generation of attosecond laser pulses in the extreme ultraviolet (XUV) via high-order harmonic generation (HHG) has enabled time-resolved studies of electronic processes in atoms, solids and molecules on the attosecond scale. We plan to combine an attosecond beamline with a reaction microscope. Detection of electrons and ions in coincidence will allow the retrieval of correlated momentum distributions of charged particles, giving access to channel-resolved photoelectron angular distributions and kinetic energy spectra. This wealth of information can help to better interpret experimental results when studying molecular systems, often involving coupling between electronic and nuclear degrees of freedom or multi-electron correlations. However, for detection in coincidence, only one atom or molecule should be ionized per laser shot. As a result, light sources with high repetition rate are required for pump-probe experiments to be completed in finite acquisition time.

Here we present an attosecond pump-probe beamline driven by a high power non-collinear optical parametric chirped pulse amplification (NOPCPA) system capable of running at a repetition rate of 100 kHz with a maximum output power of 24 W. The system delivers carrier-envelope phase (CEP) stable 7 fs pulses at 800 nm with pulse energies of 190 μJ after compression [1, 2]. We show further pulse compression to near single-cycle pulses with 100 μJ pulse energies (10 W) using the multiple plate continuum generation technique [3] and hollow core fibre compression. The CEP has been characterised on a single-and-every shot basis at full repetition rate using stereo ATI [4]. Using the compressed pulses we generate isolated attosecond pulses via HHG in krypton gas. We reach an XUV photon flux of 1.8×10^{13} photons/sec at the source. The emitted XUV attosecond pulses have been characterized by attosecond streaking. Using a time-domain ptychography algorithm we reconstruct the infrared streaking field and the isolated XUV attosecond pulse with a FWHM of 160 as. The XUV satellite (spaced at 1/2 a period of the IR driving field) pulse contrast is $1:10^{-2.4}$.

*E-mail: cps@mbi-berlin.de

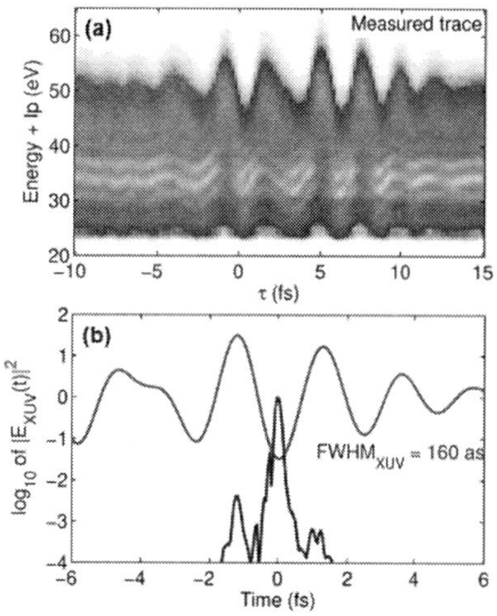

Figure 1. Recorded streaking trace (a), the retrieved infrared laser field (red line) and retrieved attosecond pulse intensity (b).

References

[1] Furch FJ *et al* 2017 *Opt. Lett.* **42** 2495
[2] Witting T *et al* 2018 *J. Opt.* **20** 044003
[3] Lu CH *et al* 2018 *Opt. Express* **26** 8941
[4] Hoff D *et al* 2018 *Opt. Lett.* **43** 3850

Comparison of strong field approximation and quantitative rescattering model in generating of attosecond pulse

X Y Li[1] *, J Fan[1] † and X L Wang[2]

[1]Department of Experimental Teaching, Northwest Minzu University, Lanzhou, 730030, China
[2] College of Electrical Engneering, Northwest Minzu University, Lanzhou, 730030, China

Synopsis In this paper, we want to investigate the differences between strong field approximation (SFA) and quantitative rescattering (QRS) model in macroscopically generating attosecond pulse. Our research shows that SFA can give comparable results to that from time-dependent Schrödinger equation (TDSE) in the cut-off range, but it encounters difficulties in reproducing the attosecond pulse of TDSE's in the low frequency range. However, the results from QRS agree well with that from TDSE method.

In the last two decades, attosecond science and technology have attracted significant attentions. Using attosecond pulse, scientists can measure photoemission time of tungsten [1], control of electron dynamics in carbon monoxide, measure tunneling delay time, etc. High-order harmonic generation (HHG) is one of the best method to synthesize attosecond pulse. HHG is generated by the nonlinear process of intense laser field interacting with atoms, molecules or other matters. In calculation, time-dependent Schrödinger equation is usually used to describe this interaction process, but it is time-consuming. Instead, strong field approximation (SFA) and quantitative rescattering (QRS) model are the two alternative methods. It is well know that SFA has failure to reproduce the harmonic spectra of TDSE's. QRS has been proved to calculate harmonic spectra as exact as TDSE, but it is as fast as SFA [2]. In this paper, we want to investigate whether QRS can generate attosecond pulse more exactly than that of SFA comparing the results of TDSE.

We used QRS, SFA and TDSE methods to calculate the single atom response of 1×10^{15} W/cm^2, 19.4 fs laser pulse interacting with 1 mm Ar gas jet, respectively. Through solving two three-dimensional (3D) Maxwell's wave equations of both laser field and harmonic field, we eventually can obtain the macroscopic attosecond pulse of these three methods.

Based on QRS, SFA and TDSE, Figure 1 and 2 show the attosecond pulse synthesized by different harmonic spectra ranges. They depict that QRS can obtain comparable results with TDSE all over the harmonic spectra, but SFA is fail on reproducting attosecond pulse of TDSE in the lower range of the harmonic spectra. Our reseach indicates QRS can solve the problem beyond SFA.

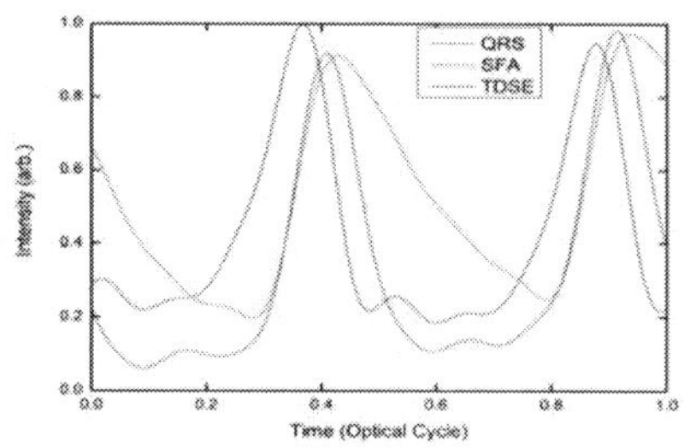

Figure 1. Attosecond pulse trains obtained by using QRS (black), SFA (red) and TDSE (blue). Synthesized by H17-H23.

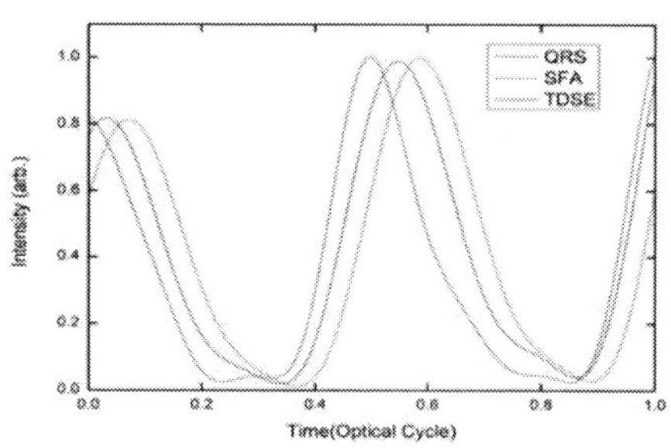

Figure 2. Attosecond pulse trains obtained by using QRS (black), SFA (red) and TDSE (blue). Synthesized by H23-H29.

Acknowledgment This work is supported by the National Natural Science Foundation of China under Grant No. 11564033, 11764036.

References

[1] Ossiander M *et al* 2018 *Nature* **561** 374
[2] Jin C *et al* 2009 *Phys. Rev. A* **79** 053413

* E-mail: lxycock@163.com
† E-mail: fanjuan@126.com

Angle-resolved time delays in photoionization with a high repetition rate laser source

H Srinivas[1] *, F Shobeiry[1] †, D Bharti[1], A Harth[1], T Pfeifer[1] and R Moshammer[1]

[1] Max-Planck Institut für Kernphysik, Saupfercheckweg 1, Heidelberg 69117, Germany

Synopsis Complete measurements of photoionization include both time-resolved and angle-resolved mapping of the atomic and molecular dynamics. The setup described in this work is suited to be used for attosecond pump-probe experiments as well as providing insights into the angular dependencies of photoionization.

Photoionization of atoms is a fundamental process in light-matter interaction, occuring at attosecond timescales. This process is not instantaneous and there is a finite time delay associated with the release of an electron. One popular method to measure such delays is an XUV-NIR pump-probe technique called RABBITT(Reconstruction of Attosecond beating by Interference of Two-Photon Transitions) [1] which was initially developed as a tool to characterize XUV attosecond pulse trains(APT) .

We present an experimental setup combining a source for XUV attosecond pulse trains and a *Reaction Microscope* [2]. This allows to not only temporally resolve atomic processes, but also to determine the angle of electron emission with respect to the polarization axis of the ionizing laser field. This is essential to analyze transitions between levels in the electronic continuum. Recent experiments [3] have shown that photoionization time delays vary as a function of the angle of emission due to an interference between various pathways in the electronic continuum.

The XUV attosecond pulse trains are generated using a laser with a central wavelength of 1030 nm, through High-Harmonic Generation (HHG). The laser pulses for the process of HHG are delivered at selectable repetition rates of 143 KHz, 71 KHz and 47 KHz. The pulse durations can also be chosen between 250 fs, 40 fs and 8 fs. An interferometric arrangement (Fig.1) is being implemented to perform pump-probe experiments. With high repetition rates, ionization events with a low cross-section can be measured with adequate statistics, in a comparitively shorter time frame.

To counter the effects of thermal drifts and external vibrations on the interferometer, a stabilzation scheme involving Fourier-transform interferometry [4] is being developed. This will be used to equalize the interferometric beam paths during delay scans in order to achieve temporal stability at the attosecond timescale.

Measurements on angle-dependent photoionization time delays in noble gases such as Argon are planned to be performed. The existence of features such as Cooper minima (a region of low ionization probabilities) in noble gases, results in a strongly anisotropic photoemission and thereby photoionization time delays of upto a 100 attoseconds [5].

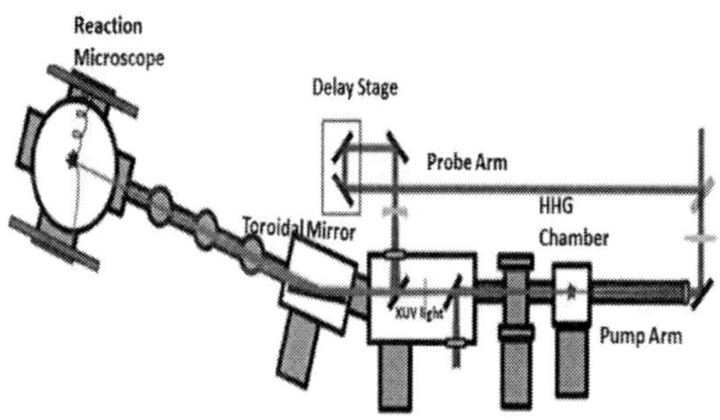

Figure 1. Schematic of the experimental setup

References

[1] Paul PM *et al* 2001 *Science* **292** 5522
[2] Dörner R *et al* 2000 *Phys. Rep.* **335** 192
[3] Cirelli C *et al* 2018 *Nat. Comm.* **9** 955
[4] Chini M *et al* 2009 *Opt. Exp.* **17** 24
[5] Pi L W and Landsman A S 2018 *Appl. Sci.* **8** 322

*E-mail: hemkumar@mpi-hd.mpg.de
†E-mail: shobeiry@mpi-hd.mpg.de

Can we break the symmetry along the polarization axis in photoionization?

Y-C Cheng[1†], S Mikaelsson[1], S Nandi[1], L Rämisch[1], C Guo[1], A Harth[1], J Vogelsang[1], M Miranda[1], C L Arnold[1], A L'Huillier[1] and M Gisselbrecht[1*]

[1]Department of Physics, Lund University, P. O. Box 118, SE-22100 Lund, Sweden

Synopsis Photoionization is a fundamental process in which an electron is emitted from an atom. The emission is traditionally considered to be symmetric with respect to the polarization axis, unless it is temporally confined to a period shorter than an optical cycle time. We demonstrate that this symmetry can still be broken by combining a train of a few attosecond pulses and a dressing laser field. The light fields act as temporal slits and phase modulator that releases electron wavepackets. The resulting photoelectron spectra differ for electrons emitted in opposite direction along the polarization.

The development of phase-controlled few-cycle infrared (IR) laser pulses opens the door to new control of the production of extreme attosecond ultraviolet (XUV) pulses, through the high-harmonic generation process [1]. In particular, it enables to investigate a regime of light matter interaction so far unexplored, and to bridge the gap between light-matter interaction with a single attosecond pulse[2] and a train of attosecond pulses[3]. Unexpectedly, in this intermediate regime, the photoionization dynamics exhibits a unique behaviour in a presence of a dressed laser field; the symmetry along the polarization axis can be broken.

The experiment was carried out using high order harmonic generated from a 200 kHz optical parametric chirped pulse amplification laser, emitting few-cycle pulses (≤6-fs), CEP stabilized, around 800 nm. The resulting attosecond pulse trains consist of 3-4 pulses with controllable phase difference [3]. When the ultrashort XUV light pulses impinge on atoms together with the generating IR field, multiple coherent ultrabroadband electron wave packets are created and characterized with a 3D momentum electron-ion spectrometer [4]. By scanning the CEP of the driving laser, the delay between the XUV and the IR fields remains constant while the relative phases between attosecond XUV pulse are changed. As a result, the individual phase of electronic wave packets emitted during an IR optical cycle time can be controlled, allowing thus to manipulate interferences between successive electron wave packets in the presence of the dressing field.

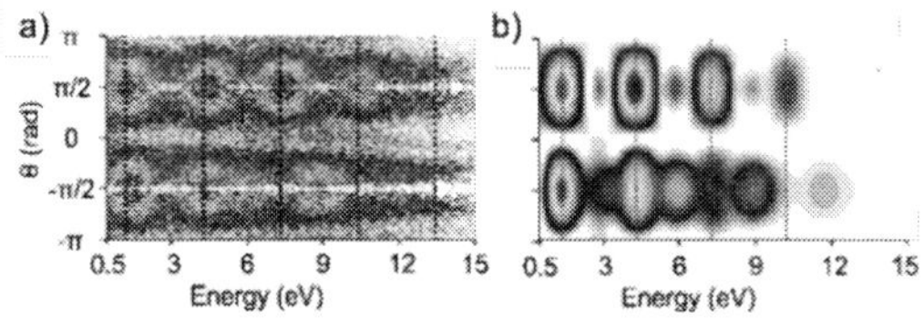

Figure 1. Experimental a) and simulated b) photoelectron kinetic energy distribution in helium as a function of the angle of emission with respect to the polarization axis.

Our measurements reveal a strong CEP- dependent asymmetry of the photoelectron spectra emitted along the polarization axis. A typical result of the experiment in helium is presented in Fig.1 together with time-dependent photoionization simulations based on the strong field approximation. We demonstrate that the amplitude and phase modulation of both IR and XUV fields are at the origin of the symmetry breaking, due to interferences between electronic wavepackets. We conclude that by combining weak intensity and very short pulse durations, it is possible to i) investigate and manipulate inter-/intra cycle interferences[5], ii) determine and control orientation in space. The latter is crucial for studying attosecond photoionization dynamics in molecules.

References

[1] Guo C *et al* 2018 *J. Phys. B.* **51** 034006
[2] Pazourek R *et al* 2015 *Rev. Mod. Phys.* **87** 765
[3] Dahlström J M *et al* 2013 *Chem. Phys.* **414** 53
[4] Gisselbrecht M *et al* 2005 *Rev. Sci. Instr.* **76** 013105
[5] Gramajo A A *et al* 2016 *Phys. Rev. A* **94** 053404

* E-mail: mathieu.gisselbrecht@sljus.lu.se
† E-mail: yu-chen.cheng@fysik.lth.se

Signatures of Coriolis effects in molecular bichromatic circularly polarized high-order harmonic generation

K J Yuan[1,2]* and A D Bandrauk[2]†

[1]Institute of Atomic and Molecular Physics, Jilin University, Changchun, 130012, China
[2]Laboratoire de Chimie Théorique, Faculté des Sciences, Université de Sherbrooke, Sherbrooke, J1K 2R1, Canada

Synopsis Circularly polarized high-order harmonic generation in molecules by bicircular pulses has been studied. We describe the laser induced electron dynamics in a rotating frame. The Coriolis effects give rise to orthogonal polarization harmonics. It is found that harmonic property depends on pulse frequencies and helicities, molecular orbital symmetries, and electron localizations, illustrating the importance of the field-molecule symmetry. The feature is described by dynamical symmetry theory and time profile analysis of harmonics.

High-order harmonic generation (HHG) opens up new frontiers in many areas of science [1]. Molecular systems due to additional degrees of nuclei have been shown to be good candidates for circularly polarized HHG [2]. We describe the circular HHG processes by circular polarization electron recollision model in a rotating frame [3, 4]. In a bichromatic (ω_1, ω_2) circularly polarized laser pulse with its polarization in the (x, y) plane, the field-molecule Hamiltonian can be rewritten as $H = H_0 + \bar{\omega} l_z + 2xE_0 \cos(\omega t)$ in a rotating frame at the rotating frequency $\bar{\omega} = (\omega_2 - \omega_1)/2$ with the mean frequency $\omega = (\omega_1 + \omega_2)/2$, as illustrated in Figure 1. The coplanar bichromatic circular pulses in the rotating frame $\bar{\omega}$ correspond to a linear driving field term $2xE_0 \cos(\omega t)$, in competition with a rotation angular momentum term $\bar{\omega} l_z$ (around the z axis) responsible for Coriolis forces. The resulting cutoffs are given by $N_m \hbar\omega_0 = I_p + 3.17U_p$, where the ponderomotive energy defines $U_p = (2eE_0)^2/4m_e\omega^2$.

We discuss how the symmetry of the initial electronic state of molecules determines the harmonic properties in the presence of bichromatic counter-rotating circularly polarized laser pulses. Comparison of HHG for different molecular electronic states shows that harmonic orders and polarizations are dependent on the molecular orbital symmetry. We also present the effect of the initial electron wave packet localization on HHG processes. Results show that the property of harmonic polarization depends on the molecular bond distance, i.e., the electron wave packet distributions. By tri-circular pulses, circularly polarized HHG can be produced with extended cut-offs, which depend on the intensity ratio of fields and the field-molecule symmetry. The resulting modulation of HHG spectra arises from a combination of two bicircular fields with counter-rotating and co-rotating components. The dependence of HHG properties thus allows to reveal molecular dynamical symmetries. The results can be extended to more complex molecular systems, and in principle provide a way to control HHG and study molecular electron dynamical symmetry by bicircular fields

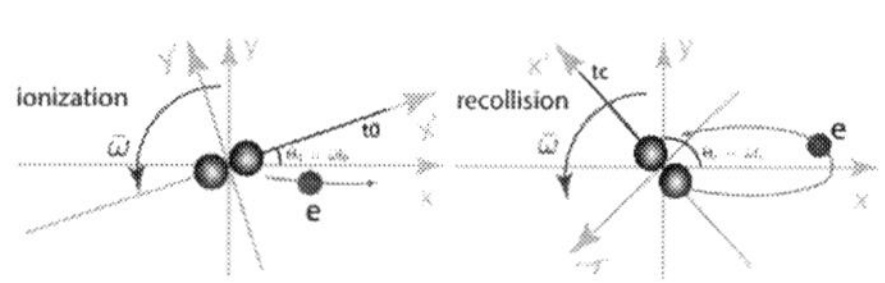

Figure 1. Circular polarization electron recollision of molecules (H_2^+ as a model) in a rotating (x', y') frame with frequency $\bar{\omega}$ by a bichromatic (ω_1, ω_2) circularly polarized laser pulse.

References

[1] Krausz F and Ivanov M 2009 *Rev. Mod. Phys.* **81** 163
[2] Yuan K J and Bandrauk A D 2013 *Phys. Rev. Lett.* **110** 023003
[3] Zuo T and Bandrauk A D 1995 *J. Nonlin. Opt. Phys. Mater.* **04** 533
[4] Yuan K J and Bandrauk A D 2018 *Phys. Rev. A* **97** 023408

*E-mail: kaijun.yuan@usherbrooke.ca
†E-mail: andre.bandrauk@usherbrooke.ca

Attosecond XUV-pump/XUV-probe spectroscopy of glycine

M Lara-Astiaso[1], A Palacios[1,2]*, J Delgado[5], J González-Vázquez[1], P Decleva[3] and F Martín[1,4,5]

[1] Departamento de Química, Universidad Autónoma de Madrid (UAM), 28049 Madrid, Spain
[2] Institute of Advanced Research in Chemical Sciences (IAdChem), UAM, 28049 Madrid, Spain
[3] Dipartimento di Scienze Chimiche e Farmaceutiche, Università di Trieste, 34127 Trieste, Italy
[4] Condensed Matter Physics Center (IFIMAC), UAM, 28049 Madrid, Spain
[5] Instituto Madrileño de Estudios Avanzados (IMDEA) en Nanociencia, 28049 Madrid, Spain

Synopsis We theoretically describe the outcome of an attosecond XUV pump-probe experiment in the smallest amino acid, glycine. We accurately evaluate the ionization amplitudes for the pump and the probe pulses. The molecular wave packet is then time-propagated using a trajectory-based Surface Hopping approach, accouting for the momentum distributions of the nuclei. Experimentally measurable photoelectron spectra are retrieved.

We investigate the ultrafast electron dynamics in ionized glycine using an attosecond XUV-pump/XUV-probe scheme. The existence of ultrafast charge migration was originally discussed by Weinkauf and collaborators after their pioneering studies on peptide chains [1]. Subsequent theoretical investigations by Cederbaum and collaborators demonstrated that sub-femtosecond charge fluctuations where observed in several biomolecules after sudden ionization, as a result of a superposition of cationic states [2]. The first experimental evidence was obtained in 2014 in the aminoacid phenylalanine, using an attosecond UV-pump/IR-probe scheme [3]. In this experiment, the large bandwidth of the attosecond pulse generates a superposition of one-hole states leading to a highly delocalized hole density that also fluctuates in the attosecond time scale.

A theoretical description of ultrafast charge migration in molecules first requires an accurate evaluation of the ionization amplitudes, which has only be achieved by a few theoretical approaches [3, 4]. Despite the theoretical efforts devoted to understand this phenomenon in large molecules, most prevoius studies assume the nuclei to be frozen, which is a valid approach for the first few femtoseconds [5], but unable to predict the subsequent molecular fragmentation. More importantly, existing methods to date have focused on describing the electron dynamics following the ionization induced by the pump pulse, but ignored the role of the probe pulse, except in a recent theoretical study in CO_2 [6]. The present work presents our progress to describe a complete XUV-pump/XUV-probe experiment in glycine. Ionization amplitudes are accurately evaluated using the static-exchange DFT approach within the time-dependent perturbation theory formalism [5], while the coupled electron-nuclear motion is then described using a Surface Hopping approach [7]. The final photoelectrum spectrum is retrieved for different probe pulses. Ongoing progress to retrieve different fragmentation paths will be presented.

*E-mail: alicia.palacios@uam.es

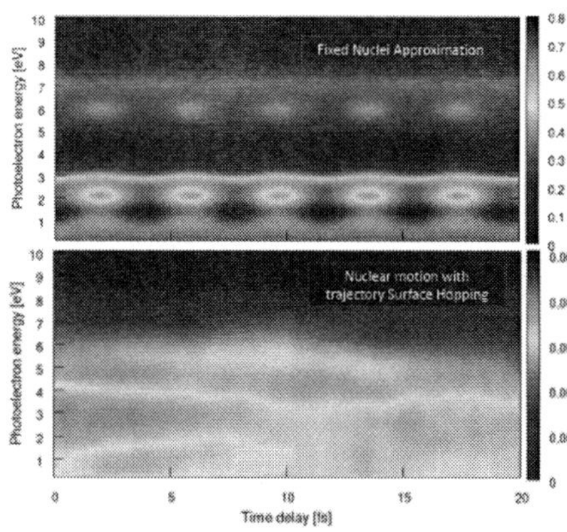

Figure 1. XUV-pump/XUV-probe time-resolved photoelectron spectra for glycine.

References

[1] Weinkauf *et al* 1995 *J. Phys. Chem.* **99** 11255
[2] Cederbaum LS *et al* 1999 *CPL* **307** 205
[3] Calegari F *et al* 2014 *Science* **346**336
[4] Mignolet B *et al* 2014 *J. Phys. B* **47** 124011
[5] Nisoli M *et al* 2017 *Chem. Rev.* **117** 10760
[6] Ruberti M *et al* 2018 *J. Chem. T. Comp* **14** 4991
[7] Richter M *et al* 2011 *J. Chem. Theo Comp* **7** 1253

UV-pump/UV-probe spectroscopy of N_2

V J Borràs[1], J González-Vázquez[1], M Klinker[1] and F Martín[1,2,3]

[1]Departamento de Química, Módulo 13, Universidad Autónoma de Madrid, 28049 Madrid, Spain
[2]Instituto Madrileño de Estudios Avanzados en Nanociencia (IMDEA-Nano), Campus de Cantoblanco, 28049 Madrid, Spain
[3]Condensed Matter Physics Center (IFIMAC), Universidad Autónoma de Madrid, 28049 Madrid, Spain

Synopsis UV-pump / UV-probe photoelectron spectra of N_2 are theoretically evaluated by solving the time-dependent Schrödinger equation in a basis of all-electron bound and continuum eigenstates provided by the XChem code.

The theoretical description of molecular photoelectron spectra in UV-pump/UV-probe experiments requires a good representation of the system's ionization continuum. For polyelectronic molecules, however, this is still a challenge due to the complicated short-range structure of correlated electronic wave functions. Whereas Quantum Chemistry Packages (QCP), implementing sophisticated methods to compute bound electronic molecular states, are nowadays standard in most laboratories, comparable tools for the continuum are not yet widely available. To tackle this problem a new approach developed in our group that uses a hybrid Guassian-B-spline (GABS) basis [1] (see Fig. 1), interfaced with existing QCPs via close-coupling scattering methods [2], has been used for the study of the multichannel ionization of molecular nitrogen.

In our group, single-photon ionization of N_2 has already been described within the fixed-nuclei approximation [3], and two-photon ionization of N_2 by assuming that the photons are absorbed sequentially [4], which are valid approximations when the pump and probe pulses are very short and do not overlap in time, respectively. The present study goes a step farther and includes a proper nuclear dynamics propagation as well as the possibility to describe absorption of more than one photon simultaneously. More specifically, we have solved the Time Dependent Schrödinger Equation (TDSE) in a basis of bound and continuum molecular states obtained from the XChem approach. To better handle the multichannel character of the ionization process while the nuclei move, we have developed a special methodology based on the projection of the XCHEM electronic states into the different parent ions and excited states of the neutral molecule. The calculated UV-pump / UV-probe photoelectron spectra show that the signature of nuclear dynamics is imprinted in the phases of the interferences observed in the temporal region where the pump and the probe pulses overlap.

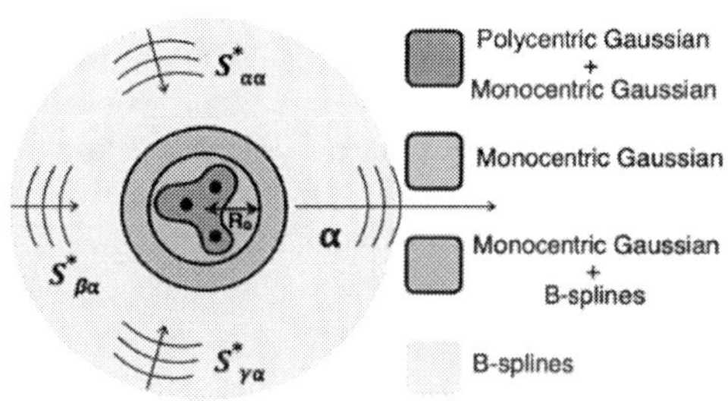

Figure 1. Schematic representation of the GABs basis.

References

[1] Marante C *et al* 2014 *Phys. Rev. A* **90** 012506
[2] Marante C *et al* 2017 *J. Chem. Theory Comput.* **13** 499
[3] Klinker M *et al* 2018 *J. Phys. Chem. Lett.* **9** 756
[4] Trabattoni A *et al* 2015 *Phys. Rev. X* **5** 041053

* E-mail: fernando.martin@uam.es

A 17-W, mid-IR parametric amplifier driven by a Yb:YAG thin-disk laser and its application to high harmonic generation in solids

T Kanai[1,2]*, Y Lee[1,2], M Seo[1,2] and D E Kim[1,2]

[1]Department of Physics, Center for Attosecond Science and Technology, Pohang University of Science and Technology, Pohang, 37673, South Korea

[2]Max Planck Center for Attosecond Science, Max Planck POSTECH/Korea Res. Init., Pohang, 37673, South Korea

Synopsis We developed a carrier-envelope phase-stable, KTA parametric amplifier delivering 17-W mid-IR pulses (signal: 1.4 micron, idler: 3.8 micron) at a 100-kHz rep. rate. The key aspect is stable generation of supercontinuum seed pulses with 1.4-ps pulses, which are the longest for supercontinuum generation on its longer wavelength side. This technology offers a robust architecture with characteristics of stable phases, simple dispersion management, wavelength tunability for 1.3-4.5 micron, and this high photon flux opens novel statistics-driven spectroscopy for rare events in strong field physics.

Driven by the λ^2 energy scaling law of strong field phenomena [1], high-power mid-IR optical parametric amplifiers (OPAs) [2] attract much attention as the next generation sources for attosecond science. Their longer wavelengths than that of traditional Ti:Sa lasers allow ones (i) to boost the energy scales of high field phenomena such as high harmonic generation [1] and plasma X-ray generation [3], and (ii) to realize novel non-perturbative phenomena in fragile and functional materials such as semiconductors [4], biomolecules [5], etc.. Here, we developed a novel mid-IR OPA driven by a 1.4 ps Yb:YAG thin-disk amplifier. The novelty of the development is in the design with supercontinuum seeding and in its high photon flux and by making full use of the latter aspect, novel signals at Van Hove singularities of ZnSe are clearly observed.

Figure 1 shows (a) the optical layout of the developed OPA, (b,c) typical harmonic spectra from ZnSe driven by the 3.8-μm and 3.5-μm pulses. One can clearly see harmonics up to the 8th order of the 3.8-μm wave below the bandgap (2.67 eV) as well as the novel two peaks around 250 nm and 400 nm above the bandgap [Fig. 1(c)]. Considering neither their separation nor the positions are integer multiples of the photon energy of 3.8-μm pulses, they cannot be explained as simple harmonic peaks. To understand the origins of the above peaks, we calculated the imaginary part of the dielectric function $\epsilon_2(\omega) := \mathrm{Im}[\epsilon(\omega)]$ based on the nonlocal pseudopotential model [6]. The positions of the two peaks were identified with those of $\epsilon_2(\omega)$ and their physical origins are in the so-called Van Hove singularities of the 2D exciton peak and one of the two 3D exciton ones. Depending on the time scale of these emission processes, which reflect their band structures, the present findings can serve as a novel route to observe transient band structures of solids.

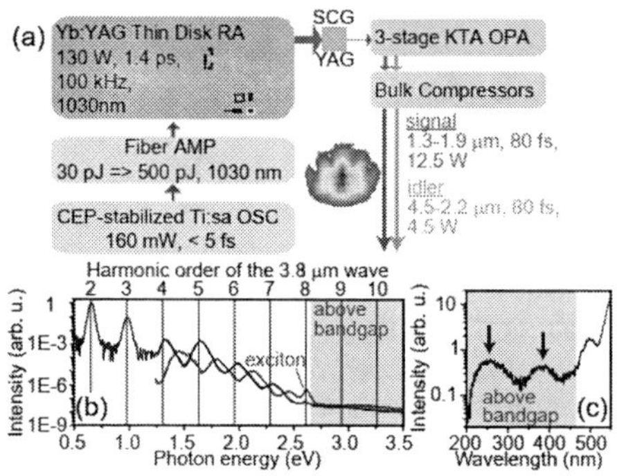

Figure 1. (a) The developed 17-W OPA. (b) Harmonic spectra driven by the 3.8-μm (blue) and the 3.5-μm (red) pulses. (c) That optimized for the novel two peaks at 250 nm and 400 nm.

References

[1] Popmintchev T *et al* 2012 *Science* **336** 1287
[2] Kanai T *et al* 2017 *Opt. Lett.* **42** 683
[3] Weisshaupt J *et al* 2014 *Opt. Lett.* **8** 927
[4] Ghimire S *et al* 2011 *Nat. Phys.* **7** 138
[5] Pupeza I *et al* 2015 *Nat. Photon.* **9** 721
[6] Chelikowsky J R and Cohen M L 1976 *Phys. Rev. B* **14** 556

*E-mail: goldwell74@gmail.com

Photoionization of neutral magnesium and sodium anion in C_{60} versus C_{240}

M Magrakvelidze[1] *, D Shields[2], R De[2], S T Manson[3] and H S Chakraborty[2] †

[1] Department of Physics, University of Mary Washington, Fredericksburg, Virginia 22401, USA
[2] Department of Natural Sciences, D.L. Hubbard Center for Innovation, Northwest Missouri State University, Maryville, Missouri 64468, USA
[3] Department of Physics and Astronomy, Georgia State University, Atlanta, Gergia 30303, USA

Synopsis We investigate the effects of confinement and electron correlation on the photoemission properties of Mg and Na^- endohedrally sequestered in C_{60} versus C_{240}. Cross sections and Wigner-Smith time delays for emissions from various electronic levels are computed and analyzed in time-dependent local density approximation.

The study of the photoemission in real time using short pulses is key in exploring electron correlations in matter. Technological advents in producing attosecond (*as*) pulses or pulse-trains made it possible to resolve electron dynamics in laser-matter interaction on *as* timescale [1]. For atomic and condensed-phase systems, relative time delays are measured between the photoemission from valence states [2]. Attempts were made to predict the time delays in photoionization process for Ar confined in C_{60} [3]. Here, we perform a comparative study of the effects of electron correlation on the photoionization of Mg and Na^- confined in C_{60} versus C_{240}.

The time-dependent local density approximation (TDLDA) method with the Leeuwen and Baerends (LB94) exchange-correlation functional is employed where the fullerene core is modelled by a spherical jellium shell [4]. We calculate photo cross sections and angle-integrated Wigner-Smith time-delays [3] for atomic-, fullerene- and atom-fullerene hybrid-type levels. We examine the size effects of the molecular cage on the plasmonically enhanced strength of the atomic ionization [5] and the on the routine cavity oscillations. Furthermore, the behavior of emission time delays in *as*, induced by this enhancement as well as by the cavity minima and confinement-modified Cooper minima, as a function of fullerene size is scrutinized in detail.

Figure shows TDLDA cross sections and delays for Mg valence 3*s* emission from C_{60} and C_{240} as a function of the photon energy. Cavity oscillations in the cross sections suggest a reduction in the separation between successive cavity minima from C_{60} to C_{240}. At these minima the correlation effects are dominant which makes the photomission Wigner-Smith delay important. The calculated delays exhibit structures at the minima which, however, are very different from the C_{60} to C_{240} confinement.

The study may open research in spectral and temporal measurements of A@C_{60} and A@C_{240}.

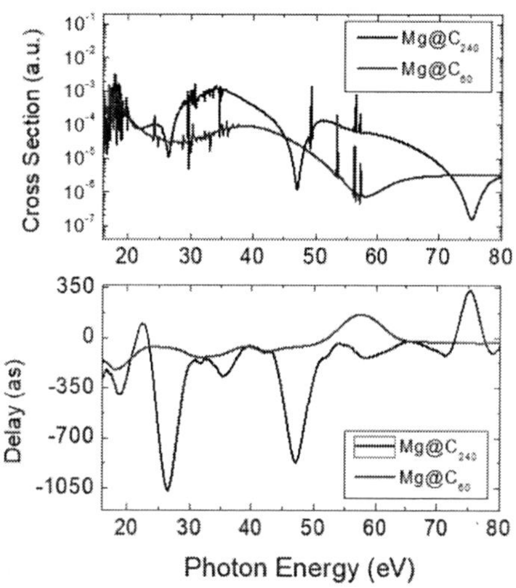

Figure 1. Cross Section (upper) and Wigner-Smith Delay (lower) for the Mg outer 3*s* electron @C_{240} vs @C_{60}.

Supported by the U.S. National Science Foundation grant PHY-1806206 and Chemical Sciences Division, US Department of Energy.

References

[1] Sansone G *et al* 2012 *J. Sel. Top. Quant.* **18** 507
[2] Neppl S *et al* 2012 *Phys. Rev. Lett.* **109** 87401
[3] Dixit G *et al* 2013 *Phys. Rev. Lett.* **111** 203003
[4] Choi J *et al* 2017 *Phys. Rev. A* **95** 023404
[5] Madjet M E *et al* 2017 *Phys. Rev. Lett.* **99** 243003

* E-mail: mmagrakv@umw.edu
† E-mail: himadri@nwmissouri.edu

Correlation drives a strong attractive force on plasmonic photoelectrons

M Magrakvelidze[1], **M E Madjet**[2 *] **and H S Chakraborty**[3 †]

[1]Department of Physics, University of Mary Washington, Fredericksburg, Virginia 22401, USA
[2]Qatar Environment and Energy Research Institute, Hamad Bin Khalifa University, Qatar Foundation, P.O. Box 5825, Doha, Qatar
[3]Department of Natural Sciences, D L Hubbard Center for Innovation, Northwest Missouri State University, Maryville, Missouri 64468, USA

Synopsis We show that the dynamical electron correlation suppresses the dipole field of incoming photon while inducing a strong attractive force at the spectral region of the giant plasmon resonance of C_{60}. A signature of this force is seen in the form of a binding well in the imaginary part of the induced potential. This attraction can significantly influence the emerging photoelectron's temporal delay behavior.

Fullerene molecules are a highly stable form of nanoscopic carbon allotrope that can exist at room temperature. Thus, they are routinely attractive for spectroscopic studies, including current interests in probing their photoresponse in real time [1]. Of particular interest is the spectral zone around 20 eV where a collective excitation occurs that decays *via* the electron continuum producing a giant plasmon resonance in the photoelectron intensity [2]. Theoretical studies of the electrons' collective motions at this resonance can thus be invaluable in developing deeper insights, particularly in understanding attosecond photoemission measurements in either RABITT or streaking spectrometry.

In the present study, a jellium potential representing 60 C^{4+} is constructed by smearing the total charge over a spherical shell of proper radius and thickness. The Kohn-Sham equations for a system of 240 electrons, made up of four valence electrons from each C atom, are solved to obtain the ground state structure in local density approximation [3]. Considering the dipole response of the system to the incoming photon, the energy dependent induced electron-density is then computed from the many body susceptibility in a linear response frame. This subsequently yields the complex radial induced potential $V_{\rm ind}(R,E)$ which, at plasmon energies, probes the collective response of the molecule.

At energies below/above the resonance peak $\mathrm{Re}[V_{\rm ind}]$ screens/anti-screens the external potential over the shell region of maximum wavefunction overlaps. The switch between these modes occures roughly at the plasmon peak where $\mathrm{Re}[V_{\rm ind}]$ is closely zero. In contrast, $\mathrm{Im}[V_{\rm ind}]$ in Figure 1 exhibits a deep minimum at the resonance peak, while falling off moving either side from the peak. This shape of $\mathrm{Im}[V_{\rm ind}]$ indicates a transient attractive force on electrons which may cause their time-delayed emergence.

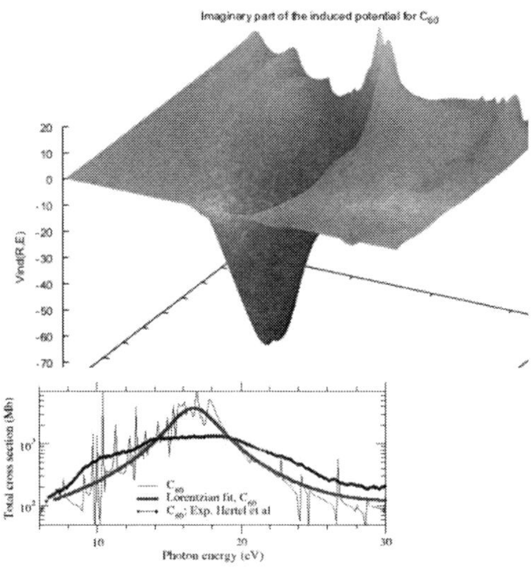

Figure 1. (Top) $\mathrm{Im}[V_{\rm ind}]$ as a function of C_{60} radius and photon energy. (Bottom) Calculated [3], Lorentzian fit, and measured [2] (red-shifted) ionization cross sections.

Supported by the National Science Foundation grant PHY-1806206.

References

[1] Barillot T J *et al* 2015 *Phys. Rev. A* **91** 033413
[2] Hertel I V *et al* 1992 *Phys. Rev. Lett.* **68** 784
[3] Choi J *et al* 2017 *Phys. Rev. A* **95** 023404

* E-mail: mmadjet@hbku.edu.qa
† E-mail: himadri@nwmissouri.edu

Numerical attoclock on atomic and molecular hydrogen

V Serov[1], **A W Bray**[2] **and A S Kheifets**[2]*

[1] Department of Theoretical Physics, Saratov State University, Saratov 410012, Russia
[2]Research School of Physical Sciences, The Australian National University, Canberra ACT 0200, Australia

Synopsis We simulate attosecond angular streaking on atomic and molecular hydrogen by solving the time-dependent Schrödinger equation (TDSE) driven by a single oscillation circularly polarized laser pulse. We compare results for this simulated "numerical attoclock" with predictions of a semiclassical model based on the saddle point method (SPM). This way we identify the essential physics behind the attoclock measurements. In molecular hydrogen, we observe a strong dependence of the width of the attoclock angular peak on the molecular orientation and attribute it to the two-center electron interference. See Ref. [1] for full detail.

We solve numerically a one-electron TDSE

$$i\partial\Psi(\boldsymbol{r})/\partial t = [\hat{H}_{\rm target} + \boldsymbol{A}(t)\cdot\hat{\boldsymbol{p}}]\Psi(\boldsymbol{r}) \qquad (1)$$

with $\boldsymbol{A}(t) = -A_0 f(t)[\cos(\omega t)\hat{\boldsymbol{e}}_x + \sin(\omega t)\hat{\boldsymbol{e}}_y]$. The pulse $f(t) = \cos^4(\omega t/4)$ for $-2\pi/\omega < t < 2\pi/\omega$ and zero elsewhere; ω corresponds to $\lambda =$ 800 nm. We also solve the saddle point equation

$$\partial S_{\boldsymbol{k}}(t_s)/\partial t = [\boldsymbol{k} + \boldsymbol{A}(t_s)]^2/2 + I_p = 0 \qquad (2)$$

with a semi-classical action $S_{\boldsymbol{k}}(t) = \int^t dt'\{[\boldsymbol{k} + \boldsymbol{A}(t')]^2/2 + I_p\}$. Both methods are used to obtain the photoelectron momentum distribution (PMD) in the polarization plane shown on the top panel of Fig. 1. The offset angle θ_A (mid panel) and the width of the attoclock peak $\mathcal{W}$ (bottom panel) are studied as functions of the driving field intensity for three molecular orientations: perpendicular to the polarization plane and parallel to the field $\boldsymbol{E}$ and potential $\boldsymbol{A}$ vectors at their peak values.

We apply the Keldysh-Rutherford (KR) model [2] to the offset angle due to the scattering in the Coulomb potential $\theta_A = (\omega^2/E_0)(Z/I_p)$. We scale the atomic hydrogen offset angle to that of H_2 $\theta_A^{\rm H} \times I_p^{\rm H}/I_p^{\rm H_2}$ to illustrate a nearly identical intensity dependance.

We find the attoclock offset angles θ_A of the H_2 molecule only weakly dependent on the molecular axis orientation relative to the polarization plane. This is so because the Keldysh tunnel width $b = I_p/E_0 \gg R$ is much larger that the inter-atomic distance at the field intensities under consideration. At the same time, the angular width $\mathcal{W}$ varies very significantly. Qualitatively, this can be understood as interference of the two trajectories obtained from the saddle point equation (2) in which the two-center molecular term $\pm\boldsymbol{E}(t_s)\cdot\boldsymbol{R}/2 = 0$ is added [3].

*E-mail: A.Kheifets@anu.edu.au

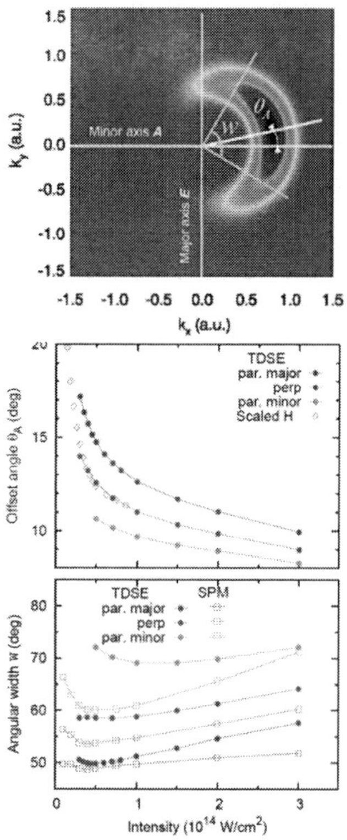

Figure 1. Top: Atomic hydrogen PMD in the polarization plane. Middle and bottom: attoclock offset angle θ_A and the angular width $\mathcal{W}$ of the H_2 molecule at three different orientiations.

References

[1] Serov V V, Bray A W and Kheifets A S 2008 arXiv:1812.09386
[2] Bray A W, Eckart S and Kheifets A S 2018 *Phys. Rev. Lett.* **121** 123201
[3] Chirilâ C C and Lein M 2006 *Phys. Rev. A* **73** 023410

H_2 photoionization induced by XUV pulses and X-ray free electron lasers

A Sopena[1,2]*, A Palacios[2], F Martín[2,3], F Catoire[1] and H Bachau[1]

[1]Centre des Lasers Intenses et Applications, Talence, 33405, France
[2]Universidad Autónoma de Madrid, Madrid, 28049, Spain
[3]Instituto Madrileño de Estudios Avanzados (IMDEA) en Nanociencia, Madrid, 28049, Spain

Synopsis We present a theoretical study on H_2 photionization taking into account coupled electron-nuclear motion and non-dipole effects. In the first part we study the effects of harmonic filtering in a pump-probe scheme using attosecond train pulses (ATP) in dipole approximation. In the second part we investigate nondipole effects in XUV and soft X-ray regimes.

The development of intense XUV sources through free electron lasers (FEL) and high-order harmonic generation (HHG) in the femtosecond (fs) and sub fs domains provides a unique tool to investigate non-linear laser matter interaction at ultra short time durations. Attosecond pulses provide the possibility for coherently and simultaneously control both electronic and nuclear dynamics, specifically populating different vibronic (electronic+vibrational) states [1], this makes mandatory the use of theoretical approaches beyond the Born-Oppenheimer approximation to be understood. On the other hand it is well-known that, in the XUV and soft X-rays domains, the dipole approximation must be carefully used. In particular, nondipole effects in atom and molecules result in asymmetries of the photoelectron distributions (see [2] for the cases of He or H_2, and references therein).

In this work, we present ab initio calculations performed on H_2 [3] to investigate the possibility of control excitation and ionization yields in two different cases. In the first case we investigated the effect of filtering different harmonics in a pump-probe scheme using VUV ATP. The higher state selectivity of ATP compared to single broad-bandwidth attosecond pulses together with the capabilities of IR sources to trace and drive the coupled electron and nuclear motion initiated by a VUV ATP makes this technique perfect to control excitation and ionization yields on attosecond timescales by means of electron wave packets interference [1]. For the second case we follow our recent research on stimulated Compton scattering in hydrogen [4] where we have shown that the diamagnetic term $\mathbf{A}^2$ (where $\mathbf{A}$ the vector potential of the field), usually neglected in dipole approximation, plays a crucial role in this context. These studies are now extended to the molecule H_2, results will be presented and discussed at the conference.

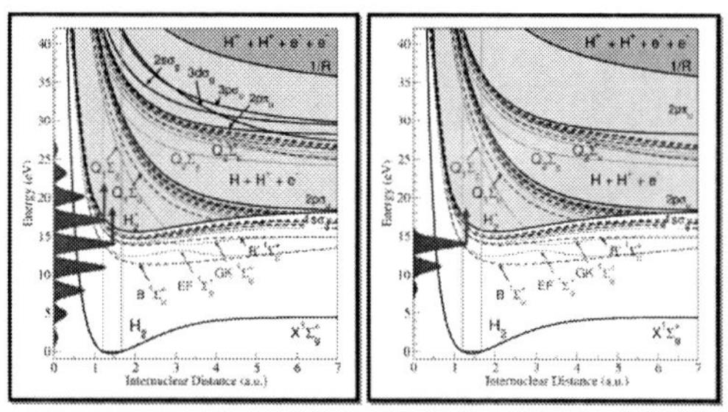

Figure 1. Schematics of the pump-probe simulation. Upper panels represent the temporal profile of the pump and probe laser fields. Lower panels represent the spectrum of said pulses superimposed to the energy diagram of H_2. Right figures correspond to the filtered case and left figures correspond to the unfiltered case.

References

[1] Ranitovic P *et al* 2014 *PNAS* **111** 912
[2] Zimmermann B *et al* 2015 *Phys. Rev. A* **91** 053410
[3] Palacios A *et al* 2015 *J. Phys. B* **48** 242001
[4] Bachau H *et al* 2014 *Phys. Rev. Lett.* **112** 073001

*E-mail: arturo.sopena@u-bordeaux.fr

Attosecond angular streaking on H_2 with all-ionic fragments detection

W Quan[1,2], M Zhao [1], V V Serov [3], M Z Wei [1], Y Zhou [1,2], X Y Lai[1,2], A S Kheifets [4] and X J Liu[1,2] *

[1]State Key Laboratory, Wuhan Institute of Physics and Mathematics, CAS, Wuhan 430071, China
[2] University of Chinese Academy of Sciences, Beijing 100049, China
[3] Department of Theoretical Physics, Saratov State University, Saratov 410012, Russia
[4]Research School of Physical Sciences, The Australian National University, Canberra ACT 0200, Australia

Synopsis We explore a novel attosecond angular streaking scheme on H_2 based on the all-ionic dissociative fragments detection. We aim to reveal the same information from the ionic attoclock as extracted from conventional attoclock measurements [1]. We examine the validity of the axial recoil approximation questioned in [2]. The advantage of the proposed scheme is that it does not rely on the adiabatic electron tunneling hypothesis which has raised the ongoing controversy of a finite tunneling time in the attosecond community [3]

The attoclock technique is based on measuring an offset angle of the peak photoelectron momentum distribution in the polarization plane of a close-to-circularly polarized few-cycle laser pulse. Tunnel ionization of H_2 gives rise to a photoelectron and an H_2^+ ion. Because the composite system $H_2^+ + e^-$ is neutral, its center of mass is not affected by the driving laser pulse. Hence, the offset angles of the photoelectron and the H_2^+ ion in the polarization plane should be similar. This similarity is illustrated in Fig. 1 where the calculated values of the photoelectron momentum distribution (left) and the experimental H_2^+ ion distribution (right) are displayed at various field intensities.

The conventional attoclock on H_2 [1] relies on the axial recoil approxiomation in which the molecule does not rotate between the tunnel ionization and dissociation instances. Hence the ionic fragment detection provides for the orientation of the molecular axis at the instant of tunneling. This approximation was questioned recently [2]. We also observe a noticeable molecular rotation and will need to figure it out in our determination of the photoelectron offset angle in the instantaneous molecular frame.

The experiment was conducted using a COLTRIMS set-up similar to previous studies [4]. The femtosecond laser pulses were generated from a commercial FEMTOPOWER Ti:sapphire laser system with a repetition rate of 5 kHz, a pulse duration of $\sim$ 30 fs, and a center wavelength of 800 nm.

*E-mail: xjliu@wipm.ac.cn

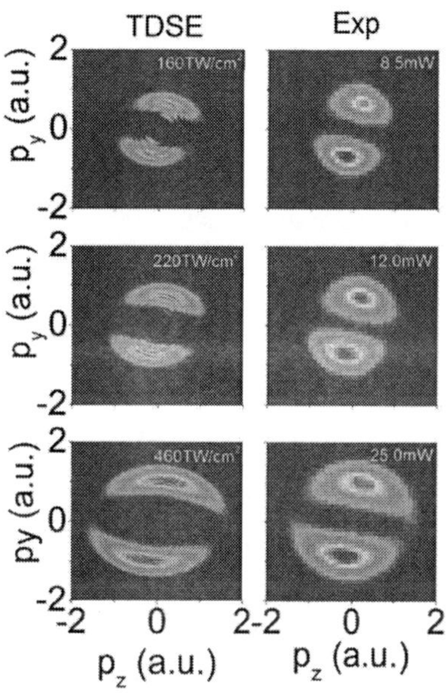

Figure 1. Photoelectron momentum distribution from solution of the time-dependent Schrödinger equation (TDSE) (left) and the H_2^+ momentum distribution from experiment (right) in the polarization plane at various field intensities.

References

[1] Sainadh U S, Ph.D. thesis, Griffith University, Brisbane, Australia, 2018.
[2] Shao Y *et al* 2017 *Phys. Rev. A* **95** 031404(R)
[3] Sainadh U S *et al* 2019 *Nature* (in press) arXiv:1707.05445
[4] Quan W *et al* 2016 *Opt. Express* **24** 023248

Journal of Physics: Conference Series **1412** (2020) 072044 doi:10.1088/1742-6596/1412/7/072044

Delayed relaxation of highly excited naphthalene cations

G Reitsma[1], J Hummert[1], J Dura[1], V Loriot[2], M J J Vrakking[1], F Lépine [2] and O Kornilov[1] *

[1]Max-Born-Institut für Nichtlineare Optik und Kurzzeitspektroskopie, Germany
[2]Institut Lumière Matière, Université Lyon 1, CNRS, France

Synopsis The efficiency of energy transfer in ultrafast electronic relaxation of molecules depends strongly on the complex interplay between electronic and nuclear motion. In this study we use wavelength-selected XUV pulses to induce relaxation dynamics of highly excited cationic states of naphthalene. Surprisingly, the observed relaxation lifetimes increase with the cationic excitation energy. We propose that this is a manifestation of a quantum mechanical population trapping that leads to delayed relaxation of molecules in the regions with a high density of excited states.

Relaxation of highly excited electronic states of large molecules follows a complex web of potential energy surfaces and involves transitions through many conical intersections. These dynamics pose a formidable challenge for modern quantum chemistry tools and therefore call for extensive experimental investigations. The complexity of the spectroscopic picture suggests that time-domain methods may bring insights, which are hard or impossible to reach using energy-domain tools only.

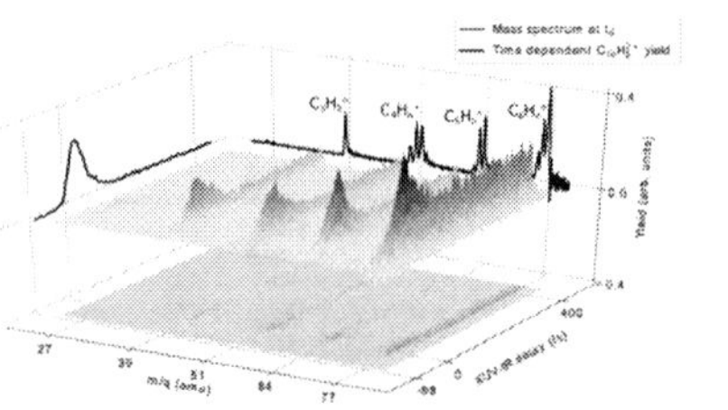

Figure 1. Fragment ion and di-cation yields for naphthalene molecules ionized by the XUV and IR pulses as a function of the time delay between the pulses.

The ultrafast XUV sources developed in the recent decades offer a perfect tool to study highly excited molecules, in particular, highly excited cations created upon photoionization with sufficiently large photon energy. Recently, XUV pulses produced by the method of high order harmonic generation were used to prepare highly excited cations of polyaromatic hydrocarbon molecules (PAH) [1, 2]. Surprisingly, it was observed that relaxation times of the states close to the double-ionization threshold increase both with the size of the molecule and with the energy of the excited state.

The above-mentioned experiments were carried out with broadband XUV sources exciting a broad range of cationic states without control over the excitation. Here we report time-resolved experiments on naphthalene molecules using an XUV time-delay-compensating monochromator beamline which delivers wavelength-selected XUV pulses and was previously used to study relaxation dynamics of highly excited nitrogen cations [3]. We record fragment ion and di-cation yields in a pump-probe XUV-IR configuration (see Figure 1) and observe that electronic states with higher excitation energy relax slower despite of having more total energy deposited into the molecule.

The striking similarity between our results and those of Refs [1, 2] leads us to formulating a conjecture, that relaxation of highly excited PAH molecules, which are known to very efficiently proceed towards the electronic ground state, exhibits the quantum mechanical effect of trapping and timescale separation, due to the large density of excited states and the low number of relaxation channels.

References

[1] Marciniak A *et al* 2015 *Nat. Comm.* **6** 7909
[2] Marciniak A *et al* 2019 *Nat. Comm.* **10** 337
[3] Eckstein M *et al* 2015 *J. Phys. Chem. Lett.* **6** 419

*E-mail: kornilov@mbi-berlin.de

Attosecond delays in photoionization of diatomic and polyatomic molecules

V Loriot[1], A Marciniak[1], S Nandi[1], M Hervé[1], G. Karras[1], E Constant[1] and F Lépine[1*]

[1]Institut Lumière Matière, UMR 5306 – UCBL – CNRS, Villeurbanne Cedex, 69622, France

Synopsis We have investigated the photoionization dynamics of several molecular systems using attosecond interferometric techniques based on ultrashort extreme ultraviolet (XUV) pulses. For diatomic molecules such as N_2, we can get a detailed picture of the photoionization process by measuring the attosecond delays. For polyatomic molecules such as naphthalene, such measurements can provide the average timescale over which photoionization processes at different electron energies can occur in complex quantum systems.

After photoionization of an atom, the resulting photoelectron evolves in the combined field of the ionic core and the remaining electrons. For molecular photoionization, in addition to these effects the interplay between the electronic and nuclear degrees of freedom can also influence the photoelectron. Attosecond interferometry techniques, e.g., RABBITT (Reconstruction of Attosecond Beating by Interference of Two-photon Transitions) [1] can provide snapshots of such electron dynamics in their natural timescale inside parent molecules [2].

In our experiment, one part of a near infrared (NIR) pulse (λ~800 nm) was used to generate high-order harmonics in the XUV regime. The remaining part of the NIR pulse was used as a dressing field either directly or after frequency-doubling (λ~400 nm). Following photoionization of the target molecule by a dressed XUV field, the photoelectrons were detected using a Velocity Map Imaging (VMI) spectrometer. The angle-resolved photoelectron spectra show redistributions of photoelectron signal which oscillate (Fig. 1) with the delay between the XUV and the dressing fields at twice the frequency of the fundamental. The phase of the oscillation contains the imprint of the dispersion dynamics of the photoelectron wavepacket scattered in the molecular potential.

For N_2, the 400 nm pulse was used as the dressing field (hence, Blue-RABBITT) to significantly reduce the spectral congestion in the photoelectron spectra [3], compared to standard RABBITT-measurements [2]. The corresponding phase-differences (Φ_A-Φ_X) between the A- and the X-states in the N_2^+ ions are shown in Fig. 1. The overall decreasing trend in (Φ_A-Φ_X) values along the measured photoelectron energies is qualitatively well reproduced by the theoretical calculations [4] for single photon ionization of N_2 molecules.

For polyatomic molecules such as naphthalene, many different ionization pathways lead to the same electron kinetic energy due to photoionization of several close-lying electronic states by the broadband XUV pulse. Using 800 nm pulse as the dressing field, the phase of the molecular photoionization can still be extracted. The phase provides a measure of the average timescale over which an electron with a given kinetic energy can escape the molecular potential. This shows the applicability of attosecond interferometry techniques for complex systems, and can serve as a starting point towards studying quantum transport phenomena in two-dimensional materials.

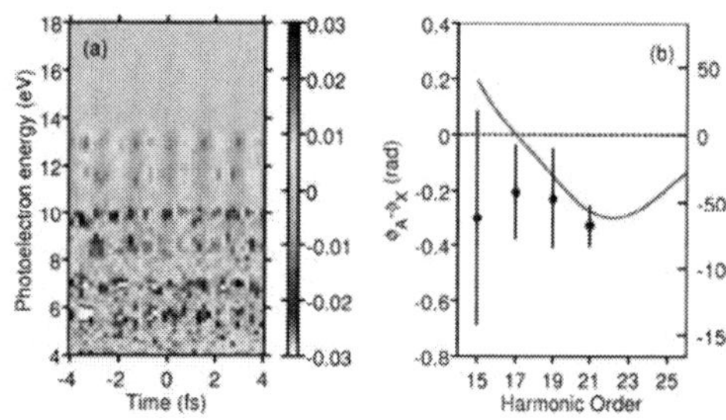

Figure 1. (a) Redistribution of photoelectron signal for photoionization of N_2. (b) Attosecond delay-differences between A and X-states of the N_2^+-ion, compared with that for single photon ionization of N_2, taken from [4].

References

[1] Paul P M *et al* 2001 *Science* **292** 1689
[2] Haessler S *et al* 2009 *Phys. Rev. A* **80** 011404 ; Huppert M *et al* 2016 *Phys. Rev. Lett.* **117** 093001
[3] Loriot V *et al* 2017 *J. Opt.* **19** 114003
[4] Hockett P *et al* 2016 *J. Phys. B* **49** 095602

* E-mail: franck.lepine@univ-lyon1.fr

Efficient table-top beamline for attosecond molecular and solid-state dynamics at the carbon K-edge

L Barreau[1] *, A D Ross[1], P M Kraus[1], S Garg[1], D M Neumark[1,2] and S R Leone[1,2,3]

[1]Department of Chemistry, University of California, Berkeley, California 94720, USA
[2]Chemical Sciences Division, Lawrence Berkeley National Laboratory, USA
[3]Department of Physics, University of California, Berkeley, California 94720, USA

Synopsis A harmonic continuum extending up to 375 eV is generated from 750 μJ, 9 fs mid-IR pulses centered at 1300 nm. The development of this source and its use for attosecond transient absorption spectroscopy experiments at the carbon K-edge are discussed.

Attosecond extreme ultra-violet transient absorption spectroscopy has established itself as a powerful tool for the observation of ultrafast electronic dynamics in atoms and materials [1,2]. Extending this method to the study of organic molecules and materials and biological systems requires the production of attosecond pulses covering the carbon K-edge spectral region (284 eV). Recent experimental developments have focused on producing short pulses around 1800 nm to drive the high-harmonic generation (HHG) process [3]. This approach makes use of the λ^2 scaling of the harmonic cutoff to increase the available photon energy, at the expense of the λ^{-5} - λ^{-6} scaling of the generation efficiency. Furthermore, very few systems can be optically excited with 1800 nm pulses.

Here we present the development of a new table-top HHG beamline dedicated to attosecond transient absorption experiments in carbon-containing gas-phase molecules and solid-state materials. The HHG is driven by a 750 μJ, sub-three-cycle mid-IR pulse centered at 1300 nm and the continuous harmonic spectrum extends to 375 eV (Fig. 1). Below 200 eV, the photon flux is an order of magnitude greater than the harmonics generated with the uncompressed 35 fs, narrowband 1300 nm pulses. We discuss the envisioned versatile pumping schemes: few-cycle pulses centered at 1300 nm and 800 nm, ultrashort visible pulses [4], and UV pulses produced by broadband sum-frequency generation [5].

This beamline will allow the study of ultrafast electronic dynamics such as passage through conical intersections in valence-excited organic molecules, carrier dynamics in graphene and organic semiconductors, or core-excited state dynamics.

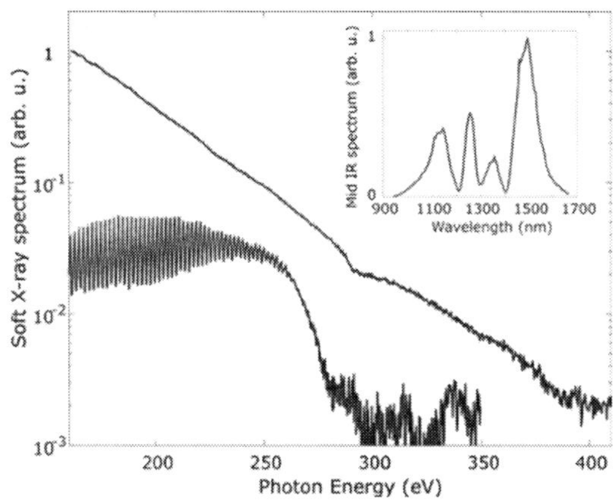

Figure 1. Harmonic continuum covering 150 to 375 eV produced with few-cycle mid-IR pulses (purple), and discrete harmonic spectrum generated with 35 fs, narrowband 1300 nm pulses (black). The inset shows the mid-IR spectrum after broadening in a hollow-core fiber, supporting 7 fs pulses.

References

[1] Ott C *et al* 2014 *Nature* **516** 374-378
[2] Schultze M *et al* 2014 *Science* **346** 1348
[3] Ren X *et al* 2018 *J. Opt.* **20** 023001
[4] Chang H -T *et al* 2016 *Opt. Lett.* **41** 5365
[5] Varillas R B *et al* 2014 *Opt. Lett.* **39** 3849

* E-mail: lou.barreau@berkeley.edu

Imaging an isolated water molecule with an attosecond electron wave packet

X Liu[1], K Amini[1], T Steinle[1], A Sanchez[1], M Shaikh[1], B Belsa[1], J Steinmetzer[2], A T Le[3,4], R Moshammer[5], T Pfeifer[5], J Ullrich[5,6], C D Lin[3], S Gräfe[2,7] and J Biegert[1,8,†]

[1]ICFO - Institut de Ciencies Fotoniques, The Barcelona Institute of Science and Technology, 08860, Castelldefels (Barcelona), Spain
[2]Institute of Physical Chemistry, Friedrich-Schiller University, Jena, 07743, Germany
[3]Department of Physics, J. R. Macdonald Laboratory, Kansas State University, Manhattan, 66506, USA
[4]Missouri University of Science and Technology, Rolla, 65409, USA
[5]Max-Planck-Institute für Kernphysik, Heidelberg, 69117, Germany
[6]Physikalisch-Technische Bundesanstalt (PTB), Braunschweig, 38116, Germany
[7]Abbe Center of Photonics, Friedrich-Schiller University, Jena, 07743, Germany
[8]ICREA, Pg. Lluís Companys 23, 08010 Barcelona, Spain

Synopsis We use laser-induced electron diffraction (LIED) to self-image the molecular structure of an isolated water molecular ion using its own retuning attosecond electron wave packet (EWP). Using LIED's sub-femtosecond and picometre spatio-temporal resolution imaging capabilities, we observe the symmetric stretching of the O-H and H-H internuclear distances with increasing laser field strength.

Many biologically important processes, such as protein folding dynamics in a water solvation shell [1], are strongly influenced by the geometric structure of water. In fact, the geometry of water molecules can be altered using electric fields on the order of 0.2 V/Å [2]. Here, we investigate the structural changes that a single water molecule undergoes in an intense (10^{13} – 10^{14} Wcm^{-2}), 97 fs (FWHM) 3.2 μm linearly laser pulse, corresponding to an electric field with field strengths of around 2 – 3 V/Å which mimics liquid water in its natural conditions. We use laser-induced electron diffraction [3] by Fourier Transform (FT-LIED) [4] to directly retrieve the geometric structure of isolated water molecules with attosecond and picometre spatio-temporal resolution. The LIED imaging process can be explained by the laser-induced electron recollision model [4]: an attosecond EWP is emitted by tunnel ionization, accelerated and returned by the driving laser field before elastically scattering against the parent ion. We employ the Reaction Microscope's electron-ion coincidence detection capabilities to isolate the high-energy LIED electrons that are detected in coincidence with H_2O^+ ions. Structural information, which is embedded within the high-energy electron momentum distribution, is directly retrieved without any prior knowledge nor the use of retrieval algorithms and *ab initio* calculations. We directly retrieve a symmetri with the ground electronic state structure of H_2O^+. We also observe that increasing the field strength of the laser pulse leads to a further stretched H_2O^+ structure. The observed increase in bond length could be related to the coupling between the molecular dipole and the electric field of the laser, thus permitting the simultaneous modification and measurement of the O-H bond length with LIED in field-dressed H_2O.

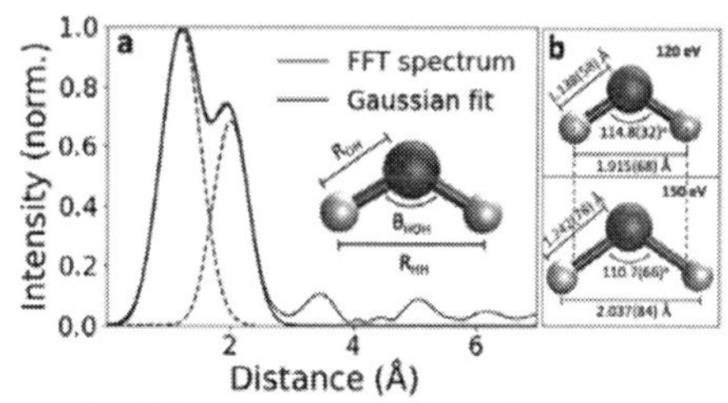

Figure 1. (a) Fourier spectrum (blue) from LIED data with Gaussian fits (black). The O-H and H-H internuclear distances are extracted using a sum of two Gaussians (black solid) along with the individual Gaussian fits (black dashed). (b) Experimentally retrieved molecular structures for laser intensities equivalent to a ponderomotive energy, U_p, of 120 and 150 eV.

References

[1] Zhang L *et al* 2007 *PNAS* **104** 18461
[2] Jung D H *et al* 1999 *Chem. Phys.* **244** 331
[3] Amini K *et al* 2018 arXiv:1805.06793. *PNAS* (*In press*)
[4] Pullen M G *et al* 2016 *Nat. Commun* **7** 11922

† E-mail: jens.biegert@icfo.eu

Ultrafast time-resolved photoelectron and ion fragment momentum imaging of vacuum ultraviolet-excited molecules

D S Slaughter[1] *, K A Larsen[1], F P Sturm[1], A Belkacem[1], P Ranitovic[1] and Th Weber[1]

[1]Chemical Sciences Division, Lawrence Berkeley National Laboratory, Berkeley, California 94720, USA

Synopsis New experimental results are presented on the dissociation dynamics of small neutral molecules in time-resolved photoexcitation-photoionization experiments by ultrashort vacuum- and extreme-ultraviolet laser pulses and charged particle momentum imaging.

Recent developments in tabletop high-order harmonic generation using ultrashort laser pulses are opening new possiblities in the measurement and control of excited electronic states of neutral molecules and molecular ions. For example, with ion fragment 3-dimensional momentum imaging to interrogate dissociation dynamics, we recently showed how an extreme ultraviolet attosecond pulse train synchronized to a near infrared (NIR) field allows precise control of the electronic wavepacket coupling two or more excited states of the oxygen molecular cation O_2^{+*}.[1]

We present recent progress on tracking the coupled electronic and nuclear dynamics in small molecules using spectrally-selected vacuum- and extreme-ultraviolet pulses to excite, control and probe transient molecules during photodissociation. Photoelectrons and fragment ions are analyzed by high-resolution cold target recoil ion momentum imaging (COLTRIMS)[2], to identify and investigate excited electronic states by photoelectron and fragment kinetic energy and angular distributions.

New results will be presented on the dynamics of dissociative photoionziation in H_2 and D_2 using extreme ultraviolet and NIR pulses, where analysis of the fragment momentum distribution (Figure 1) yields information on the few photon excitation path prior to dissociation. Progress towards photoelectron-photoion coincidence experiments, and improved selection, isolation and characterization of intense vacuum ultraviolet pulses for use with the COLTRIMS electron and ion imaging spectrometer (Figure 2) will also be discussed.

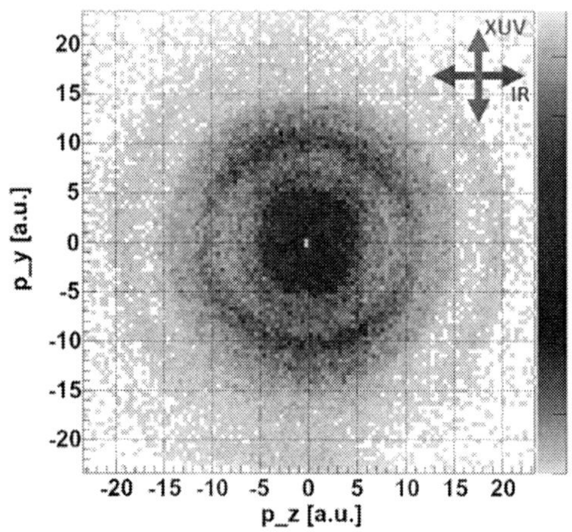

Figure 1. Slice of the ion fragment momentum distribution for dissociative photoionization of D_2 by a few-femtosecond laser pulse containing 89 nm and 800 nm photons with orthogonal polarization (see inset).

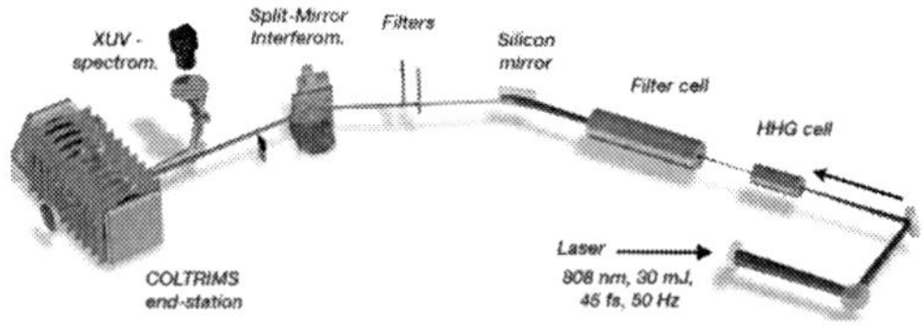

Figure 2. Schematic of the tabletop extreme ultraviolet laser system and momentum imaging experiment.

References

[1] Ranitovic P *et al* 2018 *Phys. Rev. A* **98** 013410
[2] Sturm F P *et al* 2016 *Rev. Sci. Inst.* **87** 063110

*E-mail: DSSlaughter@lbl.gov

Spatial distribution on high-order-harmonic generation of an H_2^+ molecule in intense laser fields

J Zhang, J Guo* and X S Liu†

Institute of Atomic and Molecular Physics, Jilin University, Changchun, 130012, People's Republic of China

Synopsis High-order-harmonic generation (HHG) for the H_2^+ molecule in a 3-fs, 800-nm few-cycle Gaussian laser pulse is investigated by solving the time-dependent Schrödinger equation. The spatial distribution on HHG is demonstrated and the results present the recombination process of the electron with the two nuclei. And there is little possibility of the recombination of the electron with the nuclei around the origin z = 0 a.u. and equilibrium internuclear positions z = ±1.3 a.u.. The electron-nuclear wave packet and ionization probability are investigated to explain the physical mechanism.

High-order-harmonic generation (HHG) via laser-matter interaction has been intensively studied in recent years because of its applications in generating isolated attosecond pulses [1]. The attosecond pulses are significant for probing, for example, nuclear dynamics and photoelectron wave packets [2].

In this work we investigate the spatial distribution on HHG of the H_2^+ molecule in an intense laser field, the dipole acceleration distribution in each electronic coordinate can be written as [3, 4].

$$d_A(z,t) = -\int_0^R \psi^*(z,R,t)\left(\frac{dV_C(z,R)}{dz} + (1+\frac{m_e}{2m_p+m_e})E(t)\right)\psi(z,R,t)dR$$

By using the dipole acceleration distribution in each electronic coordinate, the contribution of each electronic coordinate on HHG can be demonstrated as shown in Fig. 1, in which the horizontal axis represents the harmonic order, the vertical axis represents the electronic coordinate, and the depth of the color represents the intensity of the harmonics.

The results show that if the electric field reaches its negative peak, the electron is ionized along the positive-z direction. When the laser reverses, the electron slows down until it is pulled back and the contribution to the HHG mainly comes from the recombination of the electron with the nucleus along the positive-z direction. Meanwhile, if the electric field reaches its positive peak, the electron is ionized along the negative-z direction. When the laser reverses, the electron slows down until it is pulled back and the contribution to the HHG mainly comes from the recombination of the electron with the nucleus along the negative-z direction. We also illustrate that there is little possibility of recombination of the electron with the nuclei around the origin z = 0 a.u. and equilibrium internuclear positions z = ±1.3 a.u. from the spatial distribution of the HHG spectra. The underlying physical mechanism can be well understood by the three-step model and the time-dependent electron nuclear wave packet.

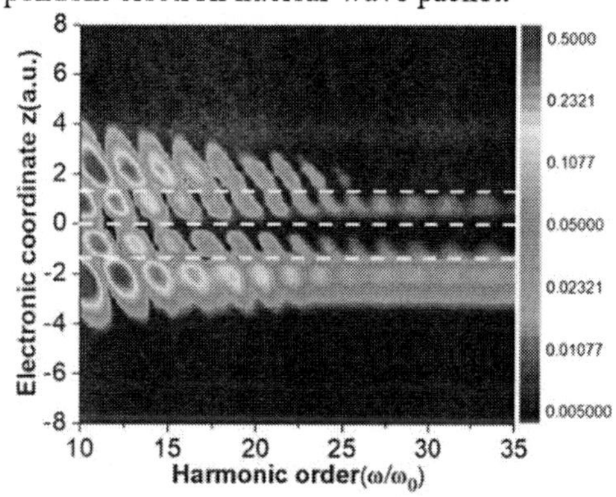

Figure 1. Spatial distribution of the HHG spectra as a function of the electronic coordinate and harmonic order for the 1D H_2^+ molecule in the Gaussian laser field. The white dashed lines indicate the positions of the electronic coordinate at the origin z = 0 a.u. and the equilibrium internuclear distance z = ±1.3 a.u., respectively.

References

[1] Paul P M 2001 *Science* **292** 1689
[2] Kienberger R *et al* 2002 *Science* **297** 1144
[3] Lein M *et al* 2002 *Phys. Rev. Lett.* **88** 183903
[4] Zhang J *et al* 2015 *Phys. Rev. A* **92** 013418

* E-mail: gjing@jlu.edu.cn
† E-mail: liuxs@jlu.edu.cn

Low-order harmonic generation of helium atom in an intense laser fields

J-W Zhou, Z-H Jiao, P-C Li [*] and X-X Zhou [†]

College of Physics and Electronic Engineering, Northwest Normal University, Lanzhou, 730070, China

Synopsis We calculate the low-order harmonic generation (LOHG) by solving three-dimensional time-dependent Schrödinger equation with an accurate model potential of helium (He) atom in an intense laser fields. The satellite-peak structures of the LOHG are obtained in the harmonic spectra of He atom. We analyze the emission properties of the LOHG by employing a synchrosqueezing transform technique. Our results show that the satellite-peak structures contain the information of the bound states and the Stark shift of He atom in an intense laser fields.

Recently, the low-order harmonic generation (LOHG) of atom in an intense laser fields has attracted considerable interest due to its application in high-intensity light source [1]. Compared to the plateau and the cutoff of high-order hamonic generation, the emission process of the LOGH is more complex [2]. Here, we calculate the LOHG spectra below the ionization threshould of helium (He) atom in an intense laser field by solving three-dimensional time-dependent Schrödinger equation (TDSE) with an accurate angular moment-dependent model potential of He atom by means of the generalized pseudospectral method. Our results show that the satellite-peak structures appear in the LOHG. To interpret the origin of the satellite peaks, we perform the time-frequency (TF) analysis on the time-dependent dipole moment of He in an laser fields by a novel synchrosqueezing transform technique [3].

In Fig 1(a), we present the LOHG spectra of He driven by a 470-nm laser pulse with 34 optical cycles, and the peak intensity is $I = 1.4\times10^{14}\,\mathrm{W/cm^2}$. One can see clearly that the satellite-peak structures of the 9th and 11th order can be found in the LOHG. Figure 1(b) shows the TF analysis of the LOHG. We find that the contribution for the satellite peak of the 9th order comes from the dress state of 2p state in the laser field and the 2p-1s transition in field-free case, respectively. While the contribution for the satellite peak of the 11th order only come from the transition between the dress states of 2p state and 1s state. The red shift of the 7th harmonic is attributed to the dress states of 2p state. We find that the energy interval between the satellite peak near the 8th order and the 9th order approximately equals to U_p(U_p is the electron ponderomotive energy), while energy interval between the adjacent satellite peaks at the same emission time is about 2ω. Here, U_x is the electron instant ponde-romonive energy.To confirm the contribution of 2p state to the LOGH, we projected the time-dependent wave function on the low excited state of He atom. We find that the population of 2p state is much lager than the other excited states in the falling edge of laser pulse.

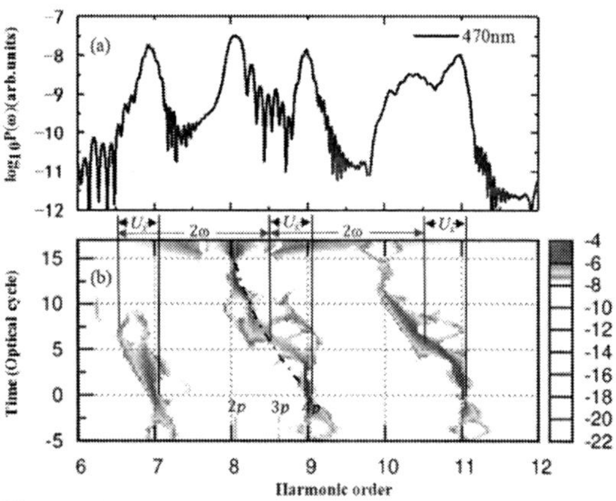

Figure 1. (a) LOHG spectra of He atom driven by a 470-nm laser pulse. (b) TF analysis for the LOGH.

Acknowledgments

This work was supported by the National Natural Science Foundation of China(Grant No.11674268, No.11764038 and No.11765018).

References

[1] Yost D C *et. al* 2009 *Nature Phys.* 5 815

[2] Xiong W H *et. al* 2017 *J Phys. B* **50** 032001

[3] Li P C *et. al* 2015 *Nature Commun.* **6** 7178

* E-mail: lipc@nwnu.edu.cn

† E-mail: zhouxx@nwnu.edu.cn

Gauge-invariant time-dependent configuration interaction singles method: applications to high-harmonic generation from atoms

T Teramura[1 *], T Sato[1, 2, 3] and K L Ishikawa[1, 2, 3]

[1]Department of Nuclear Engineering and Management, Graduate School of Engineering, The University of Tokyo, 7-3-1 Hongo, Bunkyo-ku, Tokyo 113-8656, Japan
[2]Photon Science Center, Graduate School of Engineering, The University of Tokyo, 7-3-1 Hongo, Bunkyo-ku, Tokyo 113-8656, Japan
[3]Research Institute for Photon Science and Laser Technology, The University of Tokyo, 7-3-1 Hongo, Bunkyo-ku, Tokyo 113-0033, Japan

Synopsis We present gauge-invariant reformulation of time-dependent configuration interaction singles (TDCIS) method and its applications to high-harmonic generation (HHG) from three-dimensional atoms. The conventional TDCIS method suffers from a violation of gauge-invariance, which prevents efficient simulations of high-field phenomena within the velocity gauge. In our reformulation, rotated velocity gauge transformed from the length gauge is employed instead of the velocity gauge. We simulate HHG from atoms and show gauge-invariance and performance of our rotated velocity gauge.

Modern laser technology generating intense laser pulses opened up a field of high-field physics. Perturbative treatment is no longer applicable to high-field phenomena therefore accurate theoretical methods are required.

Time-dependent configuration interaction singles (TDCIS) method is one of the promising *ab-initio* approaches which has been successfully applied to high-field phenomena, e.g. giant resonance in high-harmonic generation (HHG) [1,2]. The TDCIS method approximates the total wavefunction by a superposition of time-independent Slater determinants

$$\Psi(t) = \Phi_0 C_0(t) + \sum_i^{\mathrm{occ}} \sum_a^{\mathrm{vir}} \Phi_i^a C_i^a(t).$$

In this TDCIS wavefunction, Φ_0 is the Hartree-Fock ground state and Φ_i^a is a singly exited configuration replacing an occupied orbital ϕ_i with a virtual orbital ϕ_a. CI coefficients (C_0 and C_i^a) describe system dynamics. Utility of the TDCIS method has been enhanced by using time-dependent quantity called channel orbital defined by $\chi_i(t) \equiv \sum_a^{\mathrm{vir}} \phi_a C_i^a(t)$. This treatment removes a computational bottleneck, namely, calculation and storage of virtual orbitals [2]. Channel orbital χ_i can be interpreted as a electron wave packet from the corresponding occupied orbital ϕ_i.

However, the conventional TDCIS method suffers from a violation of gauge-invariance, as a conseqence of the truncated CI space. This prevents efficient simulations of high-field phenomena within the velocity gauge (VG). Instead of the conventional VG, we introduce *rotated* velocity gauge (rVG), which is transformed from the length gauge (LG) by an unitary operator $\hat{U}(t) \equiv \exp(-\mathrm{i}A(t) \cdot r)$ where $A(t)$ and r are the vector potential of laser field and electron position, respectively [3]. Figure 1 shows HHG spectra of He exposed to laser pulse whose wavelength is 400 nm and intensity is 4.0×10^{14} W/cm^2. This figure shows gauge-invariance between LG and rVG, as well as the breakdown of VG. Furthermore, we have succeeded in gauge-invariant simulations of HHG from Ne atom.

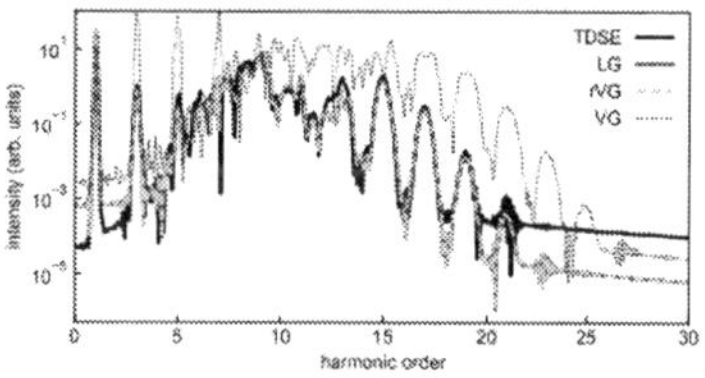

Figure 1. HHG spectra of He atom subject to a laser pulse with a wavelength of 400 nm and an intensity of 4.0×10^{14} W/cm^2. Black line shows the result of the time-dependent Schödinger equation (TDSE).

References

[1] Karamatskou A 2017 *J. Phys. B At. Mol. Opt. Phys.* **50** 013002
[2] Rohringer N *et al* 2006 *Phys. Rev. A* **74** 043420
[3] Sato T *et al* 2018 *Appl. Sci.* **8** 433

* E-mail: teramura@atto.t.u-tokyo.ac.jp

Dynamical origin of below-threshold harmonic generation of diatomic molecules

P-C Li[1,*], X-X Zhou[2], and S-I Chu[3]

[1]Department of Physics, College of Science, Shantou University, Shantou, Guangdong 515063, China
[2]College of Physics and Electronic Engineering, Northwest Normal University, Lanzhou, Gansu 730070, China
[3]Department of Chemistry, University of Kansas, Lawrence, Kansas 66045, USA

Synopsis We present an *ab initio* quantum study of below-threshold harmonic generation (BTHG) from H_2^+ molecules in the presence of an intense laser field by solving the time-dependent Schrödinger equation accurately in space and time. We find that multiple channels contribute to BTHG of H_2^+ molecules, which are related to the electron driven initially at specific time of the laser pulse from one nuclear core to another nuclear core or back to the parent core itself. The distinct contributions of these channels are distinguished by using a novel synchro squeezing time-frequency analysis.

High-order harmonic generation from atoms and molecules is a fundamental strong-field process attracting much attention due to its applications in ultrafast science and technology, such as the generation of attosescond pulse. The attosecond pulse further allows the direct detection and control of the electronic dynamics in atoms and molecules system. Recently considerable attention has been paid to the harmonic generation below the ionization threshold, where it is a potential way to produce vacuum-ultraviolet source. Most previous studies focused on the atomic system in below-threshold harmonic generation (BTHG) [1]. For the molecular systems, due to the presence of the extra internuclear degree of freedom [2], the BTHG becomes more complex.

In Figure 1(a), we present the below-threshold harmonics of H_2^+ molecular ions in an intense laser field. The black vertical dashed lines indicate the ionization threshold of the initial state marked by I_p of H_2^+ molecules. Figure 1(b) shows the Wavelet transform of time-frequency spectra of the BTHG. It is clearly seen that the Wavelet time-frequency spectra of H_2^+ molecules below ionization threshold are subject to some obscure features due to the broadband width in time-frequency distribution. However, as shown in Figure 1(c), the advantage of the synchrosqueezed transform [3] as comparing to other widely used the time-frequency transform is its capability to generate sharper and clearer time-frequency distributions which then allows to identify the individual role of multiple channels below the ionization threshold in time-frequency domain.

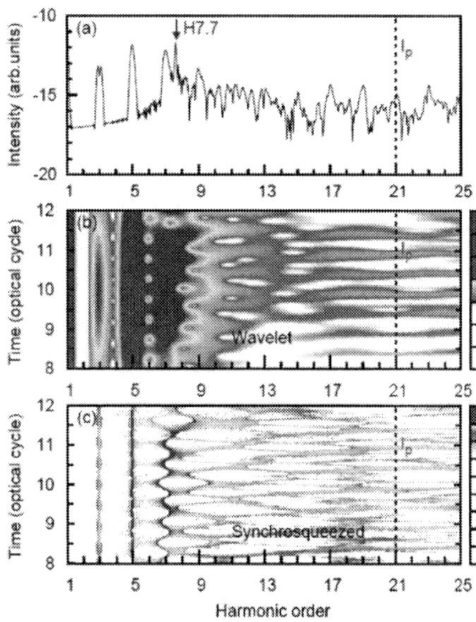

Figure 1. BTHG of H_2^+ molecules and time-frequency analysis by the Wavelet and Synchrosqueezed transform, respectively. The red arrow indicates that the resonant peak is located at the harmonic order 7.7 (H7.7).

Acknowledgments

This work was supported by the National Natural Science Foundation of China (Grant No.11674268, No.11764038 and No.11765018). It is also partially supported by the MoST of Taiwan and by US-DOE.

References

[1] Li P C *et al* 2015 *Nat. Commun.* **6** 7178
[2] Heslar J *et al* 2016 *Scientific Reports* **6** 37774
[3] Li P C *et al* 2014 *Phys.Rev.A* **90** 041401R

*E-mail: pchli@stu.edu.cn

Carrier-wave population transfer in semiconductors

R Zuo[1], X Song[1], T Meier[2] and W Yang[1,*]

[1]Department of Physics, Shantou University, Shantou, Guangdong 515063, China
[2]Department of Physics and Center for Optoelectronics and Photonics Paderborn (CeOPP), University of Paderborn, Paderborn, D-33098, Germany

Synopsis We show that a novel carrier-wave population transfer, i.e., interband excitations are confined to an extremely short time window due to their coupling to the intraband motion, is responsible for the enhancement of the interband HHG and the anomalous carrier-envelope phase dependence observed in recent experiments on semiconductors. The results give the answers of how inter- and intraband dynamics are coupled together and why the interband polarization can dominate the HHG emission in many semiconductors which is a highly debated topic in the community.

High order harmonic generation (HHG) in semiconductors has recently opened new regimes for strong field physics and attosecond science which allow to investigate and control fundamental physical processes in the time domain and enable various applications for light-based ultrafast electronics and attosecond pulse generation. However, until now, the underlying physical mechanisms are not yet fully understood and lively debated. It has been known that interband polarization and intraband current are the two basic sources for HHG in semiconductors. To date, numerous theoretical and experimental investigations have demonstrated that the interband contribution dominates the harmonics above band gap for ZnO, MgO, and GaAs exposed to mid-IR fields [1-6]. Whereas, little is known about why it dominates and how the two mechanisms couple to each other within sub-optical-cycle timescales.

Here we report a novel carrier-wave population transfer (CWPT) mechanism, i.e., the interband excitation is confined to an extremely short temporal window due the intraband motion [4]. This finding not only allows for a physically intuitive interpretation of the anomalous carrier-envelope phase dependence observed in recent HHG experiment from MgO [5], but also explains why the interband polarization dominates the coupled inter- and intraband dynamics. The validity of our results is confirmed by the consistency of our simulation with the experiment in [5] and results in [6]. Moreover, our results clarify the differences in essence of the HHG generation and control mechanisms between semiconductors and atoms: (i) in atomic HHG, the ionization process cannot be influenced by electron motion in the continuum state, whereas in semiconductions, intraband motion can dramatically influence the interband excitation. (ii) The attosecond pulse bursts are emitted about 3/4 of an optical cycle after the ionization in atoms, whereas in semiconductor, it is emitted just 1/4 of an optical cycle directly after the electronic excitation. Furthermore, motivated by the discovered mechanism, we demonstrate that the interband excitation can be controlled by an appropriately tailored two-color field. An ultrabroad supercontinuum spectrum covering the entire plateau region can be generated which directly creates an isolated-attosecond pulse even without phase compensation.

* E-mail: wfyang@stu.edu.cn

References

[1] Ghimire S *et al* 2011 *Nat. Phys.* **7** 138
[2] Vampa G *et al* 2014 *Phys. Rev. Lett.* **113** 073901
[3] Vampa G *et al* 2015 *Nature* **522** 462
[4] Song X *et al* 2019 *Opt. Express* **27** 2225
[5] You Y *et al* 2017 *Opt. Lett.* **42** 1816
[6] Schlaepfer F *et al* 2018 *Nat. Phys.* **14** 560

Journal of Physics: Conference Series **1412** (2020) 082006 doi:10.1088/1742-6596/1412/8/082006

Multi-channel contributions to infrared-driven high-harmonic generation in solids

F Navarrete[1], M F Ciappina[2] and U. Thumm[1 *]

[1]Department of Physics, Kansas State University, Manhattan, Kansas 66506, USA.
[2]Institute of Physics of the ASCR, ELI-Beamlines, Na Slovance 2, 182 21 Prague, Czech Republic.

Synopsis We numerical and analytically investigate the electron dynamics of high-order harmonic generation (HHG) in solids driven by strong mid-infrared few-cycle laser pulses. Our numerical model distinguishes contributions to the HH yield from (i) intra- and interband transitions and (ii) the central Γ−point of the first Brillouin zone (BZ) versus the entire first BZ of the initial valence band.

While HHG from atomic gases is comparatively well understood [1] and HHG from solids has been predicted theoretically decades ago [3], recent experiments [2] triggered many questions about the nature of HHG in solids and contributions from (i) the bulk and surface of the sample, (ii) different bands in the solid, (iii) intra- and interband transitions, and (iv) different initial states in the first BZ [2, 4, 5, 6].

Current numerical simulations focuss on HHG by IR-pulse-driven intra- and interband currents in the bulk and disagree on whether to include only initials states at the Γ−point of the initial valence band [4] or from the entire first BZ [5]. We developed a single-active-electron model to further scrutinize contribution to solid HHG. We solve the time-dependent Schrödinger equation in an (in the external driving field) adiabatic *Houston* basis in one spatial dimension [4] and compare exact numerical with approximate (but insightful) analytically derived HH yields.

Figure 1(a) shows our calculated interband yield for initial states with crystal momentum k in the first BZ. The V-shaped black line shows the maximum bandgap for each initial state that determines the HHG cutoff. Blue horizontal lines correspond to the range Δk of maximal contributions predicted by our analytical solution. Comparison with the total HH yield in Fig. 1(b) demonstrates that above the minimal bandgap (green line) the total yield is due to interband transitions, while intraband currents provide dominant contributions at HH energies below the minimal bandgap [Fig. 1(c)].

The HH spectrum in Fig. 1(d) is integrated over initial states from the entire first BZ. It mainly contains odd HHs and displays a cutoff (red line) at an intermediate initial k value between the maximum and minimum bandgap in the first BZ.

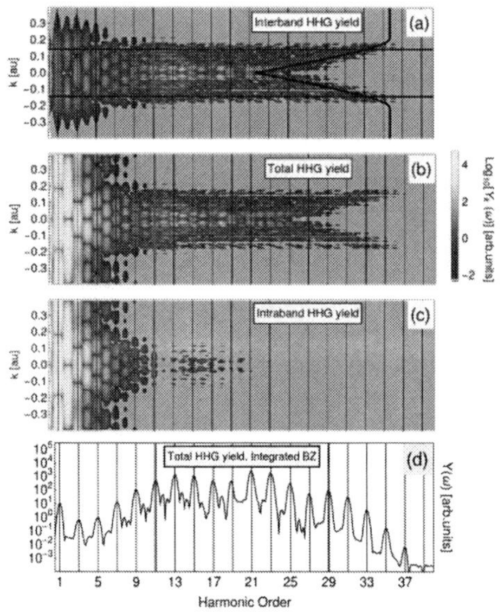

Figure 1. HHG spectra of MgO driven by a 10-cycle 3250 nm laser pulse with a peak field strength of 0.13 V/Å (2.25×10^{11} W/cm^2). Vertical thin black lines mark odd harmonics. The vertical green line indicates the minimum band gap at the Γ−point.

This work is supported by the US Air Force Office of Scientific Research and the US NSF.

References

[1] Le A T *et al* 2009 *Phys. Rev. A* **80** 013401
[2] Ghimire S *et al* 2011 *Nat. Phys.* **7** 138
[3] Plaja L *et al* 1992 *Phys. Rev. B* **45** 8334
[4] Wu M *et al* 2015 *Phys. Rev. A* **91** 043839
[5] Vampa G *et al* 2014 *Phys. Rev. Lett.* **113** 073901
[6] Navarrete F *et al* 2019 to be published.

*E-mail: thumm@phys.ksu.edu

Generation of high harmonics in solids

I Floss[1] *, C Lemell[1], K Yabana[2,3] and J Burgdörfer[1]

[1]Inst. for Theoretical Physics, Vienna University of Technology, Vienna, Austria, EU
[2]Center for Computational Sciences, University of Tsukuba, Tsukuba, Japan
[3]Graduate School of Pure and Applied Sciences, University of Tsukuba, Tsukuba, Japan

Synopsis We have developed an *ab-initio* multi-scale simulation to model light synthesis and propagation through transparent media accounting for the non-linear response of the solid. Using this approach we are able to investigate the build-up of high harmonics induced by strong and short laser pulses along the propagation direction inside a large band-gap dielectric, diamond. Detailed analysis of the results allows to identify minimum requirements necessary to realistically model laser-solid interactions.

Irradiation of crystalline insulators with strong and short laser pulses gives rise to non-linear response of the material and the generation of high harmonics (HHG). In recent experiments pronounced peaks at odd (and even) multiples of the laser frequency have been observed in the HHG spectra from crystals with (broken) inversion symmetry [1]. In contrast, simulated high harmonic spectra obtained from microscopic *ab-initio* models were found to be affected adversely by signals of coherent particle-hole pair excitations resulting in noisy spectra without clear harmonic peaks above the band-gap energy.

Several remedies for this problem have been proposed most of which only marginally footed on physical principles (e.g. extremely short decoherence times). As it turns out, laser-solid interaction is a true multi-scale problem which cannot be simplified below a certain level. We find that in order to reach at least qualitative agreement with experiment it is imperative to use both a sophisticated description of the electronic dynamics induced in a crystal cell on the microscopic level and the inclusion of mesoscopic effects of the extended system accounting for the pulse propagation in the dense medium and the inhomogeneous illumination of the crystal.

We combine refined single-cell simulations (time-dependent density functional theory and solution of semiconductor Bloch equations based on *ab-initio* band structures) with Maxwell's equations and include the transverse intensity distribution of the focal spot resulting in a clear and pronounced high harmonic spectrum [2]. Furthermore, we have accounted for decoherence in TDDFT on a phenomenological level [3].

*E-mail: isabella.floss@tuwien.ac.at

Analyzing the buildup of the final HHG spectrum within the crystal we are able to identify the ingredients necessary for realistic simulations of HHG from solids that yield agreement with experiments: 1) a fully 3-dimensional microscopic simulation of the crystal unit cell with an accurate band structure and a consistent set of coupling matrix elements, 2) the coupling of the microscopic response to the macroscopic pulse propagation which should also account for the transverse intensity distribution of the laser spot, and 3) accounting for decoherence with typical decoherence times $\tau_{\rm dec} > 10$ fs. Such simulations are able to model the transition from a noisy spectrum in the entrance layer to the clear HHG spectrum exiting the diamond layer after a propagation distance of 1 μm (Fig. 1).

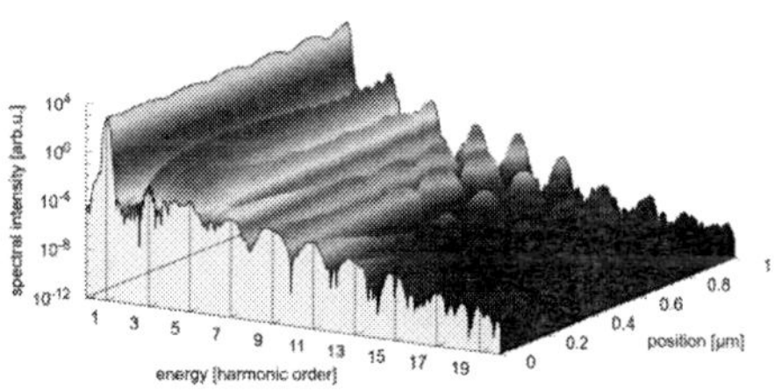

Figure 1. High harmonic spectrum (log-scale) induced in diamond by a linearly polarized few-cycle laser pulse with a carrier wavelength of 800 nm and peak intensity of 2×10^{13} W/cm^2 as a function of the position along the propagation direction inside a 1 μm thick diamond crystal.

References

[1] Ghimire S and Reis D A 2019 *Nat. Phys.* **15** 10
[2] Floss I *et al* 2018 *Phys. Rev. A* **97** 011401
[3] Floss I *et al* 2019 submitted

Coupling of autoionizing states by a chirped laser pulse

M Žitnik[1]*, A Mihelič[1], K Bučar[1], M Hrast[1,2], Ž Barba[1], Š Krušič[1,2], P Rebernik Ribič[3,4], J Urbančič[4], B Ressel[4], M Stupar[4], L Poletto[5], M Coreno[6], D Gauthier[7] and G De Ninno[4,3]

[1]J. Stefan Institute, Jamova Cesta 39, 1000 Ljubljana, Slovenia
[2]Faculty of Mathematics and Physics, University of Ljubljana, Jadranska 19, 1000 Ljubljana, Slovenia
[3]Elettra-Sincrotrone Trieste, Strada Statale 14-km 163,5, 34149 Basovizza, Trieste, Italy
[4]Laboratory of Quantum Optics, University of Nova Gorica, 5001 Nova Gorica, Slovenia
[5]Institute for Photonics and Nanotechnologies, IFN-CNR, 35131 Padova, Italy
[6]ISM, Consiglio Nazionale delle Ricerche, 34149 Basovizza, Italy
[7]LIDYL, CEA, CNRS, Université Paris-Saclay, CEA-Saclay, 91191 Gif-sur-Yvette, France

Synopsis We have observed the autoionization of the laser-coupled $2s2p\ ^1P^o$ and $2p^2\ ^1S^e$ resonances in helium. The ions were collected while varying the frequency and delay of the extreme-ultraviolet (EUV) excitation pulse with respect to the linearly chirped visible (VIS) laser pulse. From the measured frequency-delay map the Autler-Townes splitting, the EUV-VIS cross-correlation and the linear chirp parameter were extracted.

We have combined a high-harmonic-generation source (HHG) and an optical parametric amplifier laser source (OPA) to study the nature of short-lived quantum states. Upon simultaneous irradiation with the monochromatized HHG pulse and the OPA laser pulse, the initial state is excited and resonantly coupled to the neighboring state(s) having different properties in terms of excitation and decay probabilities. The ac-Stark map of the decay products provides information about properties of the hybrid photon-electron states created in the presence of the coupling field [1]. The approach is demonstrated for a paradigmatic case of a pair of autoionizing resonances in helium and can be applied to study the core-excited states in molecules and clusters.

We have solved a critical problem of how to measure a two-color ac-Stark map without relying on a laser frequency scan that may easily induce uncontrolled changes in the laser pulse properties such as focusing, timing, intensity and chirp. Instead of scanning the laser frequency, the delay between the HHG and the OPA pulse is varied, so that the OPA laser settings remain unchanged. We show by numerically solving the time-dependent Schrödinger equation that such a frequency-delay scan is a good substitute for the two-color map when the laser pulse is linearly chirped.

The method offers a possibility to systematically investigate selectively coupled pairs of short-lived atomic states with excitation energies from tens to hundreds of eV.

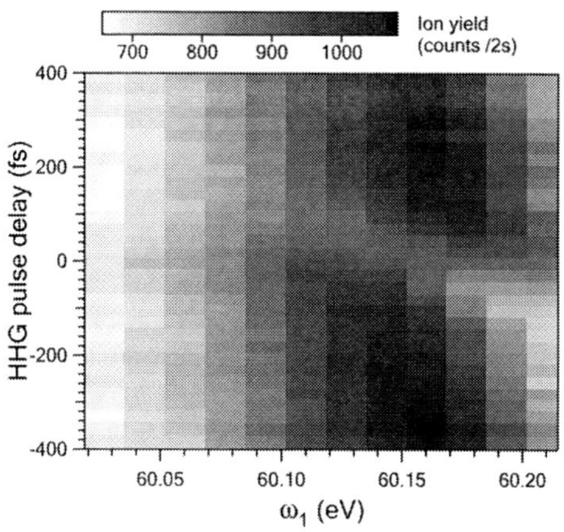

Figure 1. Measured ion yield map of the $2s2p\ ^1P^o$ resonance as a function of the EUV pulse photon energy and delay with respect to the chirped VIS laser pulse, tuned to the $2s2p\ ^1P^o - 2p^2\ ^1S^e$ transition at 1.938 eV.

References

[1] Autler S H and Townes C H 1955 *Phys. Rev.* **100** 703

*E-mail: matjaz.zitnik@ijs.si

High-energy ultrafast mid-IR optical parametric amplifier for strong-field science

D Faccialà[1]*, M Musheghyan[2], P Prasannan Geetha[1], A Pusala[3], G Crippa[3], A G Ciriolo[1], M Devetta[1], C Vozzi[1] and S Stagira[3]

[1]IFN-CNR, Piazza Leonardo da Vinci 32, Milano, Italy
[2]High Q Laser GmbH, Vienna, Austria and University of Kassel, Institute of Physics, Kassel, Germany
[3]Politecnico di Milano, Dipartimento di Fisica, Piazza Leonardo da Vinci 32, Milano, Italy

Synopsis We present a high-energy mid-infrared optical parametric amplifier source tunable between 2.8 and 3.8 μm for strong-field applications. We exploited this source in combination with a 800 nm source for generating high-order harmonics in atoms and molecules in a two-color pulse scheme.

Ultrafast laser sources in the mid-infrared region are of great interest for strong-field physics and spectroscopy. In particular, they enable the extension of the cut-off energy of the high-order harmonic generation (HHG) spectrum, thanks to the ponderomotive energy scaling of the rescattering electrons (Up $\propto \lambda^2$). Moreover, they allow to perform time-resolved vibrational spectroscopy of infrared active modes of many organic molecules [1, 2].

We present a 1-kHz broadband, tunable, mid-IR three-stage optical parametric amplifier (OPA) based on Potassium Titanyl Arsenate (KTA) crystals driven by a commercial 25-fs Ti:Sa laser. The mid-IR seed pulse is generated in the first stage via intra-pulse difference frequency generation and subsequently amplified in two stages. The output pulses have sub-90 fs duration, a maximum energy of 300 μJ at 3.25 μm, and they are tunable in the range 2.8-3.8 μm.

The mid-IR source was used in combination with the fundamental 800-nm source for generating high-order harmonics in atomic and molecular gases. The mid-IR pulses were tuned to 3.25 μm with a 200-nm bandwidth FWHM and had an on-target energy of 70 μJ, whereas the 800-nm pulses had an energy of about 170 μJ and a duration of about 50 fs. Fig. 1a shows HHG spectra in Krypton as a function of the delay between the mid-IR and the 800 nm pulses. In the region of temporal overlap between the pulses, a spectral modulation is observed. In particular, sidebands at harmonics of the mid-IR component appear at different delays, along with a strong suppression of the main 800-nm harmonics. In order to understand the origin of this spectral modulation, we performed simulations based on the strong-field approximation (SFA) [3]. The simulation results are reported in figure 1b.

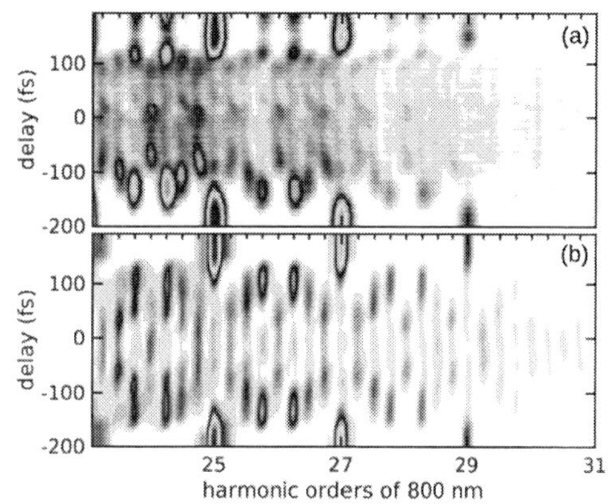

Figure 1. HHG spectra in Krypton as a function of the delay between the 800-nm pulse and the mid-IR pulse. (a) Experimental data; (b) theoretical simulation based on SFA.

The good agreement between the theoretical results and the experiment demonstrates that the observed sidebands in the region of the overlap can be ascribed to the perturbation of the electron trajectories induced by the 3.25 μm field.

Further investigations in both atomic and molecular systems are underway and will be presented.

*E-mail: davide.facciala@polimi.it

References

[1] Shan B and Chang Z *et al* 2011 *Phys. Rev. A* **65** 011804(R)
[2] Imahoko T *et al* 2007 *Appl. Phys. B: Lasers Opt.* **87** 629634
[3] Lewenstein M *et al* 1994 *Phys. Rev. A* **49** 2117

Ultrafast imaging of the Renner-Teller effect in a field-dressed molecule

K Amini[1,2], T Steinle[1], M Sclafani[1], M Shaikh[1], A Sanchez[1], X Liu[1], A T Le[3], J Steinmetzer[4], T Pfeiffer[5], R Moshammer[5], J Ullrich[5,6], M Lewenstein[1,7], R Moszynski[2], J G de Abajo[1,7], C D Lin[3], S Gräfe[4,8] and J Biegert[1,7*]

[1] ICFO - Institut de Ciencies Fotoniques, The Barcelona Institute of Science and Technology, 08860 Castelldefels (Barcelona), Spain
[2]Department of Chemistry, University of Warsaw, Warsaw, Poland
[3]Department of Physics, J. R. Macdonald Laboratory, Kansas State University, Manhattan, KS, USA
[4]Institute of Physical Chemistry, Friedrich-Schiller University, Jena, Germany
[5]Max-Planck-Institute für Kernphysik, Heidelberg, Germany
[6]Physikalisch-Technische Bundesanstalt (PTB), Braunschweig, Germany
[7] ICREA, Pg. Lluís Companys 23, 08010 Barcelona, Spain
[8]Abbe Center of Photonics, Friedrich-Schiller University, Jena, Germany

Synopsis We present experimental results of linear-to-bent transition of field-dressed molecules, mediated by Renner-Teller effect. Using the state-of-the-art laser-induced electron diffraction (LIED) technique, we image a bent and symmetrically stretched carbon disulfide (CS_2) molecule populating an excited electronic state under the influence of strong laser field. Our findings are well-supported by *ab initio* quantum mechanical calculations.

High-resolution ultrafast dynamics of structural changes are essential in the real-time observation of chemical reactions. Here, we observe ultrafast structural changes in CS_2 with a single-pulse excitation and imaging under the influence of strong laser field [1].

The experiment was performed using the LIED technique [2, 3] with the help of a passively CEP-stable mid-IR OPCPA laser system (3.2μm, 110fs, 160 kHz) [4] coupled with a reaction microscope [5], the latter of which is capable of measuring the full three-dimensional momenta of all charged particle species.

LIED probes an object's structure using its own electrons that are elastically re-scattered during strong-field induced re-collisions, as explained by the semi-classical re-collision model [6]. The excursion time of the rescattering electron and its high momentum transfer (~ 9.5 A^{-1}) provide a sub-femtosecond temporal resolution and picometre spatial resolution, respectively.

Our experimental findings show that the vibronic excitation of neutral CS_2 to a bent and symmetrically stretched structure occurs during the rising edge of our LIED laser field, as illustrated in Fig. 1. Our *ab initio* quantum dynamical calculations show that the measured bent structure in CS_2 in the presence of strong field is due to a linear-to-bent structural change mediated by the Renner-Telller effect, and is a consequence of the coupling of rovibronic states to the electronic states.

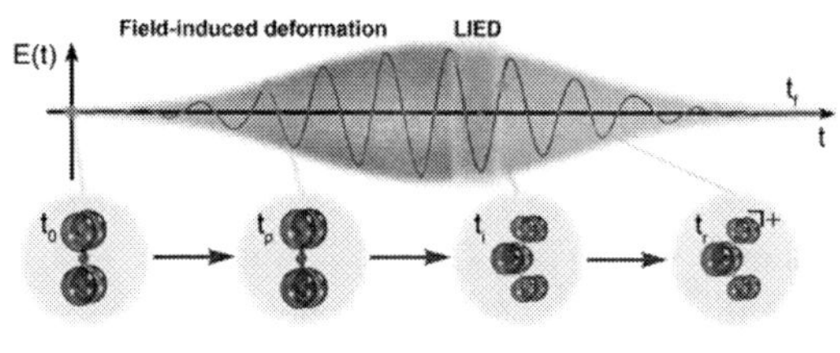

Figure 1. The CS_2 molecule is initially stretched and bent by 10° (at t_p) before leading to a bent CS_2 structure that is ionized (at t_i) and imaged by high-energy re-scattering electrons (at t_r).

In future work, LIED can be extended to two-pulse pump-probe studies to investigate the pathways of structural photo-induced chemical reactions suh as those in the photoisomerization of photoswitching molecules.

References

[1] Amini K *et al* 2019 arXiv:**1805**.06793 PNAS in print
[2] Meckel M *et al* 2008 *Science* **320** 1478
[3] Wolter B *et al* 2016 *Science* **354** 308
[4] Elu U *et al* 2017 *Optica* **4** 1024
[5] Ullrich J *et al* 2003 *Rep. Prog. Phys.* **66** 1463
[6] Corkum P B 1993 *Phys. Rev. Lett.* **71** 1994

* E-mail: jens.biegert@icfo.eu

Two-photon laser-assisted electron scattering on Hydrogen atom

A S Kornev* **and B A Zon**

Voronezh State University, Voronezh, 394018, Russia

Synopsis We considered the influence of light-dressed-atom effect on two-photon laser-assisted electron-atom scattering in the Born approximation. We performed calculations for the ground state of a Hydrogen atom. For small scattering angles, the main contribution to the cross section is made by the combined process, when one of the photons is absorbed by the atomic electron and the second one by the scattered electron. However, in the UV region, when two-photon resonances with highly excited atomic states arise, the main contribution is determined by the inverse polarization Bremsstrahlung absorption.

We considered the influence of light-dressed-atom effect on two-photon laser-assisted electron-atom scattering (LAES) in the Born approximation. We restricted ourselves to the LAES with the absorbtion of two photons and performed calculations for the ground state of a Hydrogen atom. The amplitude of the process is given by 10 diagrams (see figure 1). Here E_i is the incident electron energy, E_{1s} is the bound electron energy, ω is the laser frequency, $\boldsymbol{k}_i(\boldsymbol{k}_f)$ are the incident (scattered) electron momenta.

The M_0 diagrams correspond to the absorption of both photons by a free electron. This is the lowest order of the Bunkin–Fedorov–Kroll–Watson (BFKW) amplitude [1, 2]. In the M_1 diagrams, the Coulomb interaction of the unbound electron with the atom occurs with the transfer of the energy of ω from the atomic electron to the unbound one. The M_2 diagrams correspond to the absorption of both photons by the atomic electron. This is the inverse polarization Bremsstrahlung (PBs) absorption mechanism [3].

We performed calculations for the linearly polarized radiation and the 1 keV electron beam.

For the laser wavelength of 800 nm and small scattering angles, whitin the critical geometry ($\boldsymbol{qu} \approx 0$, $\boldsymbol{q} = \boldsymbol{k}_i - \boldsymbol{k}_f$, $\boldsymbol{u}$ is the polarization vector), the main contribution to the cross section is made by the $M_0 + M_1$ diagrams. Our results are in the qualitative agreement with those obtained in [4] with the Xe atom, where the deviation from the BFKW predictions was demonstrated.

However, in the UV region, when two-photon resonances with highly excited atomic states arise, the main contribution is determined by the inverse PBs absorption. The M_2 diagrams are dominant. As a result, in contrast to the BFKW predictions for the critical geometry, the LAES cross-section is non-zero.

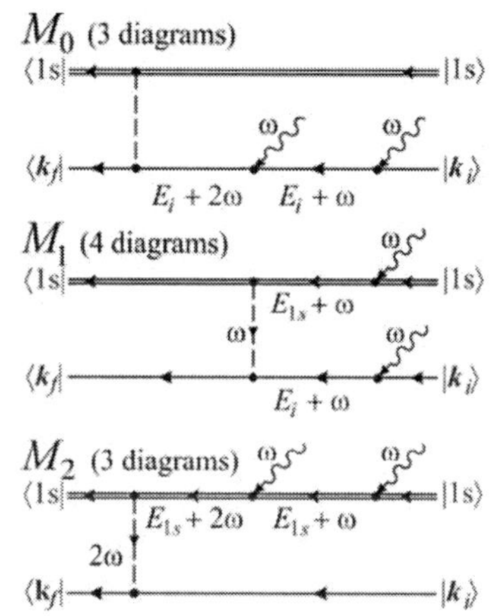

Figure 1. Diagrams for the LAES with the absorbtion of two photons. Single solid lines correspond to the scattered electron, double lines to the bound electron, dashed lines to the Coulomb interaction between electrons, wavy lines to the absorbed photons. The values of the energy of virtual states are indicated near the internal lines corresponding to the Green functions. Similar diagrams with transposed vertices are omitted here.

References

[1] Bunkin V F and Fedorov M V 1966 *Sov. Phys. – JETP* **22** 844

[2] Kroll N M and Watson K M 1973 *Phys. Rev. A* **8** 804

[3] Tsytovich V N and Oiringel I M 1992 *Polarization Bremsstrahlung* (New York: Plenum Press)

[4] Morimoto Y, Kanya R and Yamanouchi K 2015 *Phys. Rev. Lett.* **115** 123201

* E-mail: a-kornev@yandex.ru

Disentangling relativistic non-dipole effects in strong field atomic ionisation in non-relativistic regime

N Haram[1] *, I Ivanov[2], H Xu[1], K T Kim[2], A Atia-tul-Noor[1], U S Sainadh[1], I Litvinyuk[1] and R T Sang[1] †.

[1]Centre for Quantum Dynamics, Griffith University, Brisbane, QLD 4111, Australia
[2]Centre for Relativistic Laser Science, Institute for Basic Science, Gwangju-500-712, Republic of Korea

Synopsis Quantum mechanical description of the relativistic non-dipole effects is of fundamental interest in strong-field physics community, as they may have direct implications on the processes relying on re-collision dynamics. Here, we present a detailed experimental and theoretical study on the relativistic non-dipole effects in strong-field ionisation at moderate intensities. We compare our precise measurements with a truly ab-initio fully relativistic 3D model based on the time-dependent Dirac equation and demonstrate an excellent agreement between experimental results and theoretical predictions.

Relativistic non-dipole effects are known to have subtle effects on the re-collision electron dynamics in strong field ionisation and very important to study for a deeper insight into phenomena such as harmonic generation, atomic stabilisation, photo-electron holography etc [1]. We explore correlated electron dynamics obtained as a result of strong field ionisation of argon to study relativistic non-dipole effects in the non-relativistic intensity regime (10^{14} -10^{15} W/cm^2) using near-infrared linearly-polarised few-cycle laser pulses. These effects leave their signatures in the high-resolution, three-dimensional photo-electron momentum distributions recorded by a reaction microscope. In particular, they reveal themselves differently in the transverse electron momentum distribution (TEMD) as a function of laser intensity. We observe increasing counter-intuitive shifts of the peak of the TEMD with increasing laser intensity.

Furthermore, we support our experimental results by a truly ab-initio fully relativistic 3D model based on numerical solution of 3D-TDDE to investigate the relativistic non-dipole effects, which was always considered a formidable task so far. Owing to the importance of addressing relativistic non-dipole effects in strong field ionisation processes [2], such a relativistic development of theory is indispensable for the description of the ionization phenomena occurring for the currently available laser intensities of the order of 10^{18} W/cm^2 and beyond. Our approach based on the Dirac equation incorporates relativistic effects, such as the non-dipole effects, effects of the relativistic kinematics and spin-orbit interactions in the most natural way compared to numerous other theoretical models developed until now.

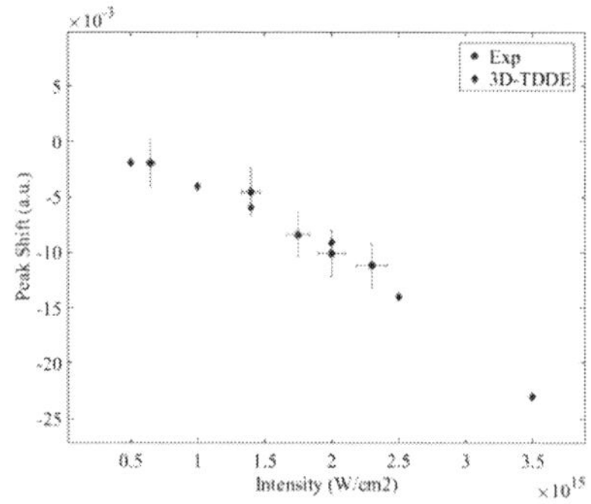

Figure 1. The peak shift of TEMD (the observable) extracted from theory and experiment agree quite well in the entire intensity range

Furthermore, for a better understanding of the underlying physics behind these non-dipole effects, we compared our existing results with the simulations done for a *Model Ar atom with Yukawa potential* to disentangle the combined effects of Coulomb potential and radiation pressure on the peak shift.

References

[1] Joachain C A *et al* 2002 *AIP Conference Proceedings* **611** 9
[2] Di Piazza A *et al* 2012 *Rev. Mod Phys.* **84** 1177

*E-mail: nida.haram@griffithuni.edu.au
†E-mail: r.sang@griffith.edu.au

Experimental comparison of frustrated tunnel ionisation in multi-cycle and few-cycle pulses

D Chetty[1]*, R D Glover[1], H Xu[1], A J Palmer[1], B A deHarak[2], P S Light[3], A N Luiten[3], I V Litvinyuk[1] and R T Sang[1]†

[1]Centre for Quantum Dynamics, Griffith University, Brisbane, 4111, Australia
[2]Physics Department, Illinois Wesleyan University, Bloomington, IL, 61702-2900, USA
[3]Institute for Photonics and Advanced Sensing and School of Physical Sciences, The University of Adelaide, Adelaide, 5005, Australia

Synopsis We present the first experimental evidence which qualitatively shows that frustrated tunnel ionization results in a larger number of excited neutrals per a tunnelling event for 6 fs pulses compared to 30 fs pulses. In addition, we also observe a less severe dependence on elliptical polarization for the 6 fs pulses. Furthermore, by varying the carrier envelope phase of the shorter pulse, we observe a modulation in the yield of excited neutrals which indicates possible control of electron trajectories.

The fact that a strong laser field can perturb the atomic potential of an atom allowing an electron to tunnel ionize lays the foundation for a rich set of phenomena. One such phenomenon is that of frustrated tunnel ionization (FTI) where the tunnelled electron is not ionized completely but is accelerated back towards the vicinity of the parent ion and recaptured into an excited state by the Coulomb field [1].

The detection of these excited neutrals requires a specific experimental set up and therefore only a few experimental studies investigating the creation of these states exist. Furthermore all experimental studies have been exclusive to multi-cycle laser pulses [1, 2, 3]. While FTI in few-cycle pulses has been investigated through the detection of radiation from the subsequent decay to the ground state [4], the only investigations dealing with the direct detection of the excited states have been through simulations [5]. Semi-classical simulations show that the FTI process depends significantly on the number of cycles that the liberated electron experiences. It predicts that as the number of cycles in the pulse increases, the ratio of excited neutrals to total number of tunnelling events decreases due to the higher probability of a high energy collision resulting in ionization.

In this work, we present the first experimental evidence that a larger number of tunnelled electrons are trapped in excited states as the pulse duration is decreased. We exploit the fact that a significant proportion of excited atoms decay into the metastable state which we then detect directly with a micro-channel plate [1]. Ions are collected and temporally displaced from metastable atoms using electric fields in a typical time of flight set up collinear with an atomic argon beam. Comparing the Ar^* and Ar^+ yields from 6 and 30 fs pulses centred at 800 nm with intensities ranging between 0.7 and 3×10^{14} W/cm^2, we find that the shorter pulse results in a larger ratio of Ar^*/Ar^+. In addition, we compare the dependence of ellipticity on the production of Ar^* and find a less severe dependence in the shorter pulse compared to the longer pulse. Furthermore, we explore the effect of varying the carrier-envelope phase of the 6 fs pulse to control the dynamics of the electron after tunnelling. We observe a modulation in Ar^* yields in the order of 8% with varying CEP, indicating subsequent control of electron trajectories in the FTI process.

References

[1] Nubbermeyer T *et al* 2008 *Phys. Rev. Lett.* **101** 233001
[2] Zimmerman H *et al* 2015 *Phys. Rev. Lett.* **114** 123003
[3] Zimmerman H *et al* 2017 *Phys. Rev. Lett.* **118** 013003
[4] Yun H *et al* 2018 *Nature Phot.* **12** 620
[5] Ortmann L *et al* 2018 *eprint arXiv* 1810.07164

*E-mail: dash.chetty@griffithuni.edu.au
†E-mail: r.sang@griffith.edu.au

Floquet formalism for multiphoton dynamics in superconducting quantum circuits

H Z Jooya* **and H R Sadeghpour**

ITAMP, Harvard-Smithsonian Center for Astrophysics, Cambridge, Massachusetts 02138, USA

Synopsis High-order nonlinear multiphoton dynamics of closed and open periodically driven quantum systems are studied within the semiclassical Floquet formalism. Multiphoton up/down frequency conversion in a superconducting circuit is realized and faithfully reproduced with calculations. The absorption spectra are suppressed at sharp avoided crossings of the Floquet states. We include the effect of coupling to cavity by time evolving the master equation. The characteristics of the time-averaged absorption spectra are investigated by probing details near and around the exceptional points where suppression occurs.

There has been rapidly growing interest in Floquet engineering, and successful implementations of this concept in various experiments, including the discrete time crystals, the Floquet-Bloch states on the surface of topological insulators, and the Floquet-Raman transition in driven solid-state spin systems, have necessitated a need for accurate theories to describe such concepts [1].

Within the many-mode Floquet formalism for closed systems, we first propose a versatile scheme to simulate the experimentally realized high order non-linear transitions, in the three-photon regime, when a transmon superconducting (SC) qubit is periodically driven by a pair of microwave fields in a cavity [2, 3]. As shown here, this method is used to implement a readily attainable scheme to up/down convert frequencies by multiphoton pumping between the microwaves, near and far way from the qubit resonance frequency, $\omega_q = 8.8356$ GHz. Effective Hamiltonians are derived which reveal the underlying physics of the Hermitian Floquet system. Population trapping in excited qubit states is manifest as the spectral suppression at specific frequencies [4].

Next, we extend our method to include relaxation by coupling to environment within the non-Hermitian theory, and study the spectral suppression with exceptional points (EP) [5]. We use the Floquet-Liouville supermatrix approach to investigate the appearance of single and multiphoton Floquet EPs. EPs are found by controlling the power and frequency of the driving microwave fields, and their effect on multiphoton-induced resonance fluorescence is demonstrated numerically. The connection between the observed spectral suppression near EPs and that obtained with the Hermitian Hamiltonian are discussed.

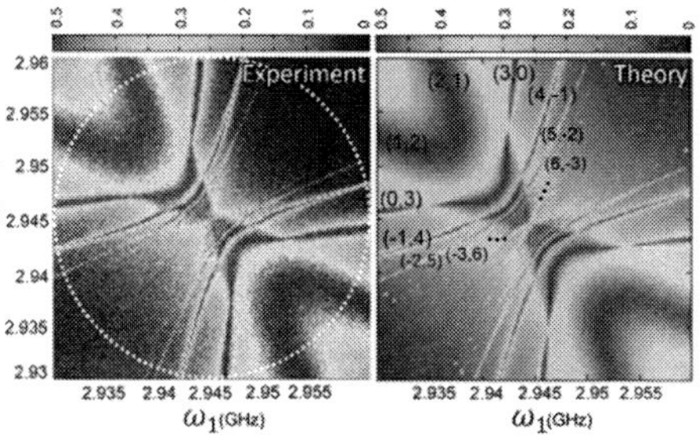

Figure 1. Three-photon transitions in a transmon SC qubit. The color map (z-axis) shows the transition probability from g to e. The number of photons involved in each transition is labeled. Various pumping possibilities can be identified, with $m_1\omega_1 + n_2\omega_2 = \omega_q$, when n+m =3. The frequency conversion occurs when m (or n)< 0, which implies that |m (or n)| number of photons is pumped from the first microwave to the second (or vice-versa).

References

[1] Kennes D M *et al* 2018 *Phys. Rev. Lett.* **120** 127601
[2] Jooya H Z *et al* 2016 *Sci. Rep.* **6** 174518
[3] Pan J *et al* 2017 *Phys. Rev. B* **96** 174518
[4] Jooya H Z *et al* 2018 arXiv.org/abs/1808.01303
[5] Pick A *et al* 2018 arXiv:1812.02005

*E-mail: hzjooya@cfa.harvard.edu

Dynamic photoemission interference of hydrogen atoms in intense ultra-fast elliptically polarized extreme ultraviolett pulses

G Li[1], N Wang[1], J Wang[1], A Liu[1]* and U Thumm[2]†

[1]Institute of Atomic and Molecular Physics, Jilin University, Changchun 130012, China
[2]James R. Macdonald Laboratory, Kansas State University, Manhattan, Kansas 66506-2604, USA

Synopsis We investigated dynamic interference (DI) in the photoionization of ground-state hydrogen atoms by intense extreme-ultraviolet (XUV) pulses with elliptical polarization. Numerically solving the time-dependent Schrödinger equation, we demonstrate that DI occurs not only in linearly but also in elliptically polarized pulses.

When subjected to the super-intense linear polarized laser field with high frequency, DI in the photoionization of hydrogen atoms from ground state can occur due to the interference of photoemission from equally ac Stark-shifted levels on the leading and trailing pulse side, separated in time by atomic stabilization against photoionization at the highest intensity of the temporal pulse profile [1-4].

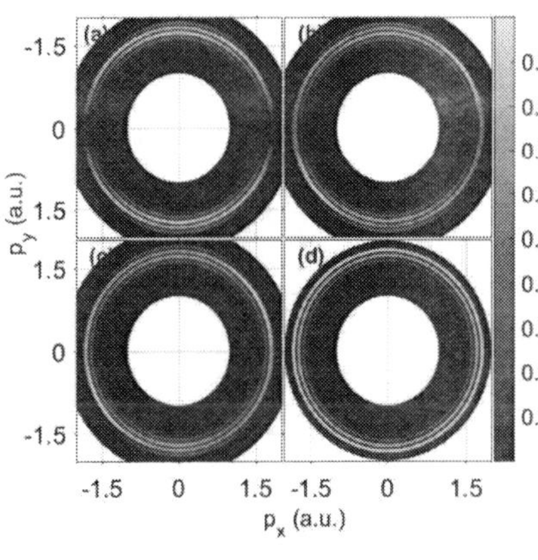

Figure 1. Photoelectron-momentum distributions for the ionization of ground-state hydrogen atoms by XUV pulses with a central photon energy of 53.6 eV, a pulse length of 1.5 fs, and a peak intensities of 3×10^{18}W/cm^2. (a) Linear polarization, (b,c) Elliptically polarized pulses with (b) $\eta = 0.25$ and (c) 0.5, and (d) Circular polarization.

By solving the time-dependent Schrödinger equation, we investigated DI for emission from ground-state hydrogen atoms in the presence of laser pulses with circular and elliptical polarization [5]. We found that DI from ground-state atomic hydrogen can (i) occur for photoionization by elliptically polarized ultra-fast laser pulses and (ii) be approximated within first-order perturbation theory for peak intensities up to 10^{18} W/cm^2, while higher intensities require a non-perturbative approach.

Figure 1 shows our full-dimensionality numerical results for the photoelectron-momentum distributions. It displays obvious DI patterns for different ellipticities, η=0, 0.25, 0.5 and 1. At 10^{18} W/cm^2 peak-pulse intensity, the shown angular distributions differ significantly from angular distributions we obtain in first-order perturbation theory [$\sim \cos^2(\theta)$] for pulses with much lower intensity ($I_0 \ll 10^{18}$ W/cm^2) and the same photon energy (not shown).

This work is supported by the NSFC under Grant Nos. 11774131, 91850114, 11604119 and, in part, by the US DOE under Award No. DE-FG02-86ER13491 and US NSF Grant PHY 1802085.

References

[1] Demekhin P V and Cederbaum L S 2012 *Phys. Rev. Lett.* **108** 253001
[2] Baghery M *et al* 2017 *Phys. Rev. Lett.* **118** 143202
[3] Guo J, Guo F, Chen J and Yang Y 2018 *Acta Phys. Sin.* **67** 073202
[4] Jiang W and Burgdörfer J 2018 *Opt. Express* **26** 19921
[5] Eberly J H and Kulander K C 1993 *Science* **262** 5137

*E-mail: aihualiu@jlu.edu.cn
†E-mail: thumm@phys.ksu.edu

Frustrated tunnel ionization of argon by intense few-cycle infrared laser radiation

T Pauly[1], N Smith[1], N Douguet[2] and K Bartschat[1*]

[1]Department of Physics and Astronomy, Drake University, Des Moines, Iowa 50311, USA
[2]Department of Physics, University of Central Florida, Orlando, Florida 32789, USA

Synopsis We report results for strong-field ionization and excitation of argon obtained by solving the time-dependent Schrödinger equation in a single-active-electron model. We analyze the process of "frustrated tunnel ionization" and its dependence on the potential and a variety of laser parameters, such as the intensity, carrier-envelope phase, and ellipticity. Our calculations support current experimental work at Griffith University.

We have carried out calculations for strong-field ionization and excitation of argon by solving the time-dependent Schrödinger equation in a single-active-electron model for the initially bound 3p electron. In particular, we are interested in the process of "frustrated tunnel ionization" [1, 2], where the field-ionized electron is driven back towards the nucleus by the changing laser field and ultimately recaptured into an excited bound state without the possibility of escaping again, due to the fact that the few-cycle pulse has already weakened too much. This process is currently being investigated experimentally at Griffith University [3]. We discuss the effect of different potentials, the carrier-envelope phase (CEP), the laser intensity, and the ellipticity of the radiation on the theoretical predictions.

Figure 1 shows the probabilities for survival of the Ar($3p^6$) ground state ($P_{\rm surv}$) as well as for excitation of any Rydberg state ($P_{\rm exc}$) and ionization ($P_{\rm ion}$). In order to increase the visibility regarding the dependence on the CEP of the linearly polarized laser field, the bottom part of the figure shows the probabilities normalized to their respective maximum. The current results were obtained with the static potential suggested by Green *et al.* [4], which was supplemented by a semi-empirical local potential to account for the electric dipole polarizability of the residual core.

Our calculations suggest a particularly strong CEP dependence of the probability for frustrated ionization, while the survival and ionization probabilities remain relatively constant when the CEP is varied. For the parameters chosen here, about 10% of the initially freed electrons are recaptured into bound states.

*E-mail: klaus.bartschat@drake.edu

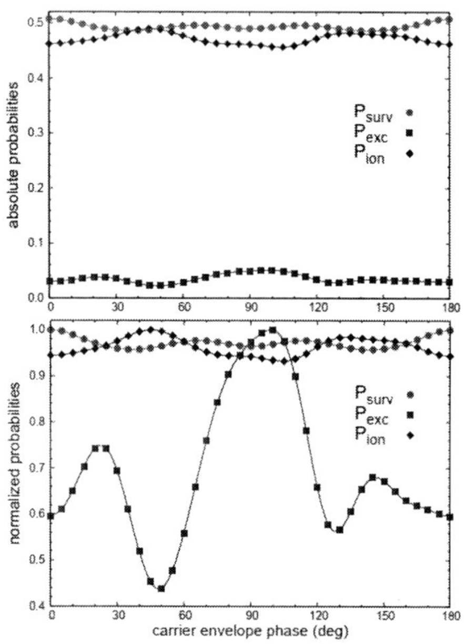

Figure 1. Absolute (top) and normalized (bottom) probabilities for survival of the ground state, excitation, and ionization of Ar(3p) as a function of the CEP for a 6 fs (FWHM) 800 nm laser pulse of peak intensity 4.0×10^{14} W/cm^2. See text for details.

This work is supported by the United States National Science Foundation under PHY-1803844 and XSEDE-090031.

References

[1] Nubbemeyer T *et al.* 2008 *Phys. Rev. Lett.* **101** 233001
[2] Popruzhenko S V 2018 *J. Phys. B* **51** 014002
[3] Chetty D *et al.* 2019 *private communication*
[4] Green A E S, Sellin D L and Zachor A S 1969 *Phys. Rev.* **184** 1

Photoabsorption of atoms in a strong elliptical laser field

X Gao * **and X-M Tong** †

Center for Computational Sciences, University of Tsukuba, 1-1-1 Tennodai, Tsukuba, Ibaraki 305-8577, Japan

Synopsis The photoabsorption of hydrogen atoms in a strong elliptical laser field was studied by solving the three-dimensional time-dependent Schrödinger equation. We found that the photoexcitation and ionization probabilities smoothly cross the ionization threshold and we can treate them in a unified way even in the strong elliptical laser field. Such findings may help us to understand how atomic structure affects the strong field ionization near the ionization threshold.

The interaction of light with atoms is of great scientific interest to understand the response of matter to an external electromagnetic field. In the weak field limit, the processes are treated by perturbation methods and atomic structures play an important role. In contrast, for strong laser fields (intensity ~10^{14} W/cm^2), most of the dynamic processes can be explained by the strong field approximation [1], in which, atomic structures are completely ignored. Recent studies [2] showed that the Coulomb interaction between the photoelectron and parent-core plays a crucial role for some processes.

There are two key processes for atoms in a strong laser field, photoabsorption including above threshold ionization (ATI) and high-order harmonic generation (HHG). By solving the three-dimensional time-dependent Schrödinger equation (TDSE3D) [3] for hydrogen atoms in a strong elliptical field, we investigated how atomic structure affects the photoabsorption and HHG.

Figure 1 shows the excitation (the negative photoelectron energy) and ionization probabilities of hydrogen atoms in elliptical laser fields as a function of the ellipticity (ε). By turning off the long range Coulomb interaction between the photoelectron and proton, we see that the Coulomb interaction or atomic structure does affect the photoabsorption process. If we convolute the absorption probabilities with a finite energy resolution, as shown in the inset, the photoabsorption probablilities cross the ionization threshold smoothly. Similar to case of the weak field [4], an intimate relation between the bound state excitation and the ionization probabilities above the threshold for the strong laser field ranging from the linearly polarization to the circularly polarization is found. While in the study, we found that the HHG yield and cutoff are almost the same if we take into account the Coulomb effect for the continumm states or not. The detailed results and discussion will be presented in the conference.

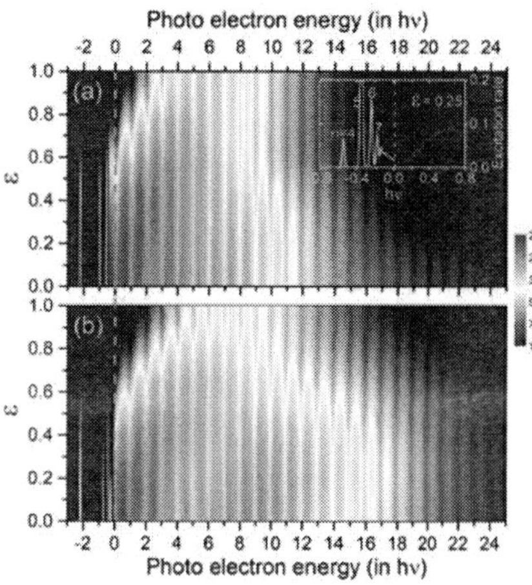

Figure 1. Photoabsorption of hydrogen atoms in elliptically laser fields as a function of ellipticities. The dashed lines indicate the ionization threshold. The laser parameters are: center wavelength 800 nm, pulse duration of full width at half magnitude 20 fs and laser intensity 1x10^{14} W/cm^2. (a): The TDSE3D calculations. The insertion exhibits the smooth connection of the excitation and ionization probabilities near the ionization threshold for laser ellipticity ε=0.25 as. (b): The TDSE3D calculations by turning off the long-range Coulomb potential.

References

[1] Lewenstein M *et al* 1994 *Phy. Rev. A* **49** 2117
[2] Blaga C I *et al* 2009 *Nature Phys.* **5** 335
[3] Tong X M 2017 *J. Phys. B:* **50** 144004
[4] Gao X *et al* 2009 *Chin. Phys. Lett.* **26** 013102

* E-mail: xgao@csrc.ac.cn,
† E-mail: tong.xiaomin.ga@u.tsukuba.ac.jp

Journal of Physics: Conference Series **1412** (2020) 092009 doi:10.1088/1742-6596/1412/9/092009

HILITE – A well-defined ion target for laser experiments

N Stallkamp[1,2†], S Ringleb[2*], B Arndt[3], M Kiffer[2], S Kumar[4], G G Paulus[2,5], W Quint[1,6], Th Stöhlker[1,2,5] and M Vogel[1]

[1]GSI Helmholtzzentrum für Schwerionenforschung, 64291 Darmstadt, Germany
[2]Institut für Optik und Quantenelektronik, Friedrich-Schiller Universität, 07743 Jena, Germany
[3]Goethe Universität Frankfurt, 60323 Frankfurt, Germany
[4]Inter-University Accelerator Centre, 110067 New Delhi, India
[5]Helmholtz-Institut Jena, 07743 Jena, Germany
[6]Physikalisches Institut, Ruprecht-Karls-Universität Heidelberg, 69120 Heidelberg, Germany

Synopsis We present a Penning-trap-based setup for the study of light-matter interactions in the high-power and/or high-intensity laser regime, such as multi-photon ionization and field ionization. The setup applies ion-cloud formation techniques to highly charged ions to the end of specific target preparation, as well as non-destructive detection techniques to identify and quantify the interaction educts and products.

The **H**igh-**I**ntensity **L**aser **I**on **T**rap **E**xperiment (HILITE) [1] aims to investigate the interaction of highly charged ions with extreme laser fields. To this end, we have built the setup based on an open-endcap Penning trap that facilitates the controlled formation of an ion target comprised of selected species, to superimpose it with a high-intensity and/or high-energy laser beam, and to non-destructively detect the confined ion species. To cover a broad range of laser parameters and hence different laser-matter interaction regimes, the setup is built in a compact fashion and allows transport to various large-scale laser facilities.

The Penning trap (see figure 1a) is located in the centre of a 6T superconducting magnet and cooled to cryogenic temperatures. The ions are provided by a dedicated source (EBIT), operated in a pulsed extraction mode to create a bunched beam of highly charged ions. These can be pre-selected by a velocity filter and decelerated by a pulsed drift tube. After deceleration, the ions are dynamically captured inside the trap by appropriate switching of electrode voltages. A single-pass 'charge counter' is placed on either side of the trap to determine the number, energy and structure of the incoming and outgoing ion bunches in a non-destructive way [2].

We will present the current status of the setup that has successfully demonstrated ion extraction and selection by the source and velocity filter, as well as transport, deceleration and storage inside the trap. So far, storage times of the order of several tens of minutes have been achieved. Furthermore, we will show the planned measurements at the free-electron laser FLASH@DESY (see figure 1b).

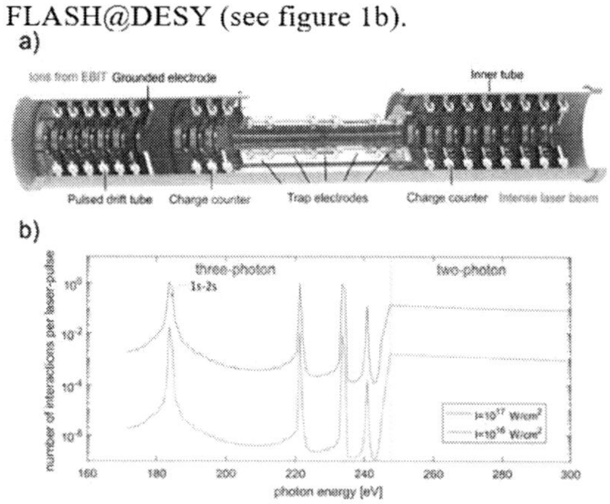

Figure 1. (a) Schematic of the Penning trap setup including deceleration and single-pass non-destructive ion detection devices. **(b)** Expected ion yield for hydrogen-like carbon for the proposed measurements in the X-ray photon-energy regime based on theoretical predictions [3].

References

[1] M Vogel *et al* 2012 *Nucl. Inst. Meth. B* **285** 65
[2] S Schmidt *et al* 2015 *Rev. Sci. Inst.* **86** 113302
[3] J Hofbrucker and A Volotka (HI Jena); *private communication*

† E-mail: n.stallkamp@gsi.de
* E-mail: stefan.ringleb@uni-jena.de

Branching ratio coherent control in ultrafast bond making

Z Amitay[1] *, L Liat[1], D M Reich[2], R Kosloff[3] and C P Koch[2]

[1] Schulich Faculty of Chemistry, Technion-Israel Institute of Technology, Haifa, 32000, Israel
[2] Theoretische Physik, Universität Kassel, Heinrich-Plett-Straße 40, Kassel, 34132, Germany
[3] Fritz Haber Research Centre and Institute of Chemistry, Hebrew University, Jerusalem, 91904, Israel

Synopsis Coherent control of bond making (photoassociation) is crucial for the realization of a new photochemistry, where a laser field is actively photo-driving the irradiated system in a coherent way from the initial state of reactants to the final state of desired products. In a significant step forward, we experimentally demonstrate here coherent control over the photoassociation branching ratio to different target states of the generated molecule. The model process is strong-field multiphoton femtosecond photoassociation of thermally hot magnesium atoms into bound magnesium dimer molecules.

One of the main inspirations for the birth of the coherent control field has been the innovative idea of generating a new type of photochemistry, where a laser field is actively photo-driving and accompanying the irradiated system in a coherent way from the initial state of reactants to the final state of desired products. This is conceptually different from the common photochemistry approach, where the reactants are individually photo-excited and then react in their "free time". The photonic tools of the corresponding bimolecular ultrafast coherent control scheme are shaped femtosecond pulses. First, an ultrafast coherent bond making (i.e., free-to-bound femtosecond photoassociation) is photo-induced between the free colliding reactants, and an excited bound molecular complex is generated in a desired coherent rovibrational wavepacket state. Then, subsequent photo-control over the molecular dynamics directs the system into target intermediate states. Last, these intermediate states serve as a platform for further coherent photo-control in order to selectively break bonds (i.e., femtosecond photodissociation) for generating the desired products.

This full scheme of binary reaction coherent control is still an unrealised dream. Its building block of photodissociation coherent control of bound molecules has been experimentally demonstrated extensively over the years. However, in strong contrast, its building block of photoassociation coherent control has been experimentally demonstrated only in a few works [1, 2, 3, 4]. In these works, we have studied strong-field multiphoton femtosecond photoassociation of thermally hot magnesium (Mg) atoms at a temperature of 1000 K, and have demonstrated femtosecond coherent control over the dynamics and yield of the photoassociation into bound magnesium dimer molecules (Mg_2).

Still, this demonstrated bond making control has been limited to driving the photoassociated molecule only into a single group of final molecular states. Here, in a significant step forward, we extend the demonstrated coherent control of bond making to experimentally controlling the photoassociation branching ratio (BR) to multiple groups of target states of the generated Mg_2 molecule. The BR control utilizes the finding that the photoassociation yields into the different target molecular states are all enhanced by positively chirped pulses, while the specific optimal (positive) chirp value for maximizing the yield into a given state depends on the state energy. The underlying control mechanism involves purification via Franck-Condon filtering, coherent de-excitation via chirp-dependent Raman transitions, and projection onto the final states.

Our work establishes the feasibility of high degree coherent control over the quantum dynamics of a molecule generated in thermal femtosecond photoassociation. This is another step toward the coherent control of photoinduced binary chemical reactions.

*E-mail: amitayz@technion.ac.il

References

[1] Rybak L *et al* 2011 *Phys. Rev. Lett.* **107** 273001
[2] Amaran S *et al* 2013 *J. Chem. Phys.* **139** 164124
[3] Levin L *et al* 2015 *Phys. Rev. Lett.* **114** 233003
[4] Levin L *et al* 2015 *J. Phys. B: At. Mol. Opt. Phys.* **48** 184004

Real-time reconstruction of the strong-field driven dipole response

V Stooß[1*], S M Cavaletto[1], S Donsa[2], A Blättermann[1], P Birk[1], C H Keitel[1], I Březinová, J Burgdörfer[2], C Ott[1] and T Pfeifer[1]

[1]Max Planck Institute for Nuclear Physics, Saupfercheckweg 1, 69117 Heidelberg, Germany, EU
[2]Institute for Theoretical Physics, Vienna University of Technology, Wiedner Hauptstraße 8, 1040 Vienna, Austria, EU

Synopsis Using ultrashort and phase-locked laser pulses, combined with an *in situ* measurement of a reference pulse, we present a method capable of retrieving the nonlinearly perturbed time-dependent response function of a quantum system driven by strong electric fields directly from spectroscopic absorption data with high precision.

Measurements of time-dependent processes in the quantum regime often employ pump–probe spectroscopy. A non-equilibrium state is prepared by a pump pulse and after a certain delay probed by a second probe pulse. Varying the time delay between pump and probe yields time-dependent information about the system under study.

By contrast, for linear light-matter interaction, the causality of observed spectroscopic signals defines the relationship between absorption and dispersion, thus the full temporal response function of a medium can be retrieved, with the prominent example of deducing lifetimes from the linewidths of isolated resonances.

We present a method which combines both principles in order to reconstruct the full temporal dipole response of a strongly driven time-dependent system from a single absorption spectrum [1] , see Fig 1a. The method relies on a sufficiently short pulse initializing the coherent excitation of the target system which subsequently interacts with a second field. The non-perturbative influence of the second field on the dynamics initialized by the coherent excitation can be directly reconstructed in real time from the absorption spectra. The reconstruction is applied to the observation of nonlinear dynamics between doubly excited states of helium. We observe Rabi-cycling and strong-field ionization during the interaction with the strong and short electric field in real time without varying the time-delay. Advanced ab-initio calculations together with a model using a limited basis set show the validity of the reconstructed response and show the breakdown of few-level quantum dynamics in the strong field regime.

Since the full time-dependent information of the response is encoded in the spectral line shape, it is crucial to precisely measure the absorption spectrum and minimize any systematic errors. This has been achieved with the *in situ* measurement of a reference spectrum of the XUV attosecond pulse which induces the dynamics. The new setup uses a transmission grating as beamsplitter in the XUV spectral range (Fig.1b) thereby improving the sensitivity to absorption changes by an order of magnitude.

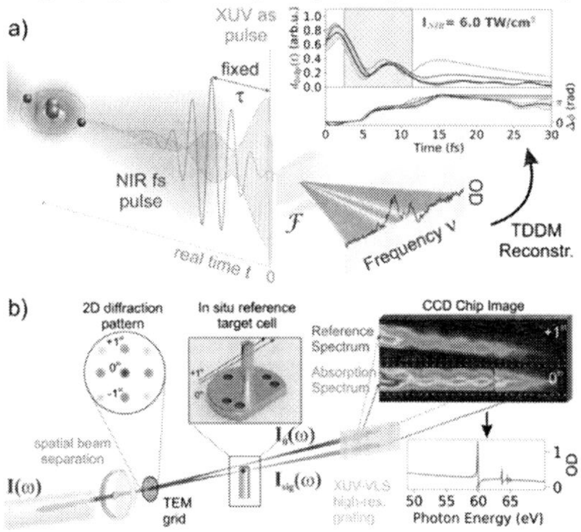

Figure 1. a) Principle of the response reconstruction

The presented methods allow the observation of complex real-time dynamics of excited states in the presence of a strong field.

References

[1] Stooß V *et al* 2018 *Phys. Rev. Lett.* **121** 173005

* E-mail: veit.stooss@mpi-hd.mpg.de

Generation of attosecond pulses in "water window" range by a plasma-based X-ray laser

I R Khairulin[1] *, V A Antonov[1,2] † and O Kocharovskaya[3]

[1]Institute of Applied Physics of the Russian Academy of Sciences, Nizhny Novgorod, 603950, Russia
[2]Prokhorov General Physics Institute of the Russian Academy of Sciences, Moscow, 119991, Russia
[3]Department of Physics and Astronomy, Texas A&M University, College Station, TX 77843-4242, USA

Synopsis We suggest a technique to generate a train of attosecond pulses in "water window" range by hydrogen-like C^{5+} plasma-based X-ray laser with two sequentical active plasma channels irradiated by two different optical laser fields with orthogonal polarizations. We show also the possibility to transform the radiation of a plasma-based X-ray laser dressed by an optical laser field into a train of attosecond pulses in a resonant absorber irradiated by the different optical field.

Plasma-based X-ray lasers are unique laboratory sources of short-wavelength coherent radiation with a relatively high pulse energy [1], which is typically much greater than that attainable via high-harmonic generation of an optical radiation. However, picosecond duration of the pulses produced by X-ray lasers prevents their application for study of the ultrafast processes unfolding on femto- and attosecond time scales.

Recently, we have shown that modulation of active medium of a hydrogen-like plasma-based X-ray laser by a strong linearly polarized optical laser field allows to transform a quasimonochromatic seeding X-ray field of the same linear polarization into a train of attosecond pulses [2]. The transformation occurs due to spectral broadening of the amplified X-ray field during its propagation through the medium, caused by sub-optical-laser-cycle modulation of frequency of the inverted transition by the optical field via the linear Stark effect.

In this contribution, we show that generation of the attosecond pulses can be implemented without using the seeding field. Instead, the active medium of the X-ray laser should be split into two parts, irradiated by the two different optical fields with mutially orthogonal polarizations. The first part of the active medium serves as a source of the linearly polarized seeding field, while the second part transforms it into an attosecond pulse train. Another option for the attosecond pulse formation is to replace the second part of the active medium by a resonant absorber dressed by an optical laser field. Our analytical and numerical results show the possibility to produce trains of attosecond pulses within "water window" range ($\lambda = 3.4$ nm) by C^{5+} plasma-based X-ray laser, see Fig. 1.

This work was partially supported by the Ministry of Education and Science of the Russian Federation under contract No.14.W03.31.0032 executed at the Institute of Applied Physics of the Russian Academy of Sciences.

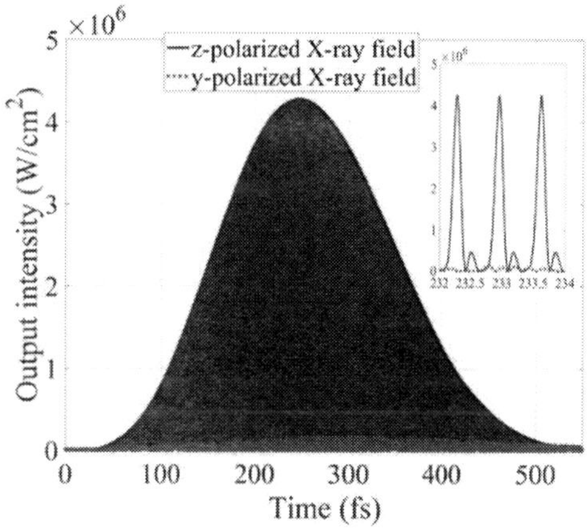

Figure 1. An X-ray field generated by a plasma-based X-ray laser after its transformation into a train of attosecond pulses in a resonant absorber dressed by an optical laser field.

References

[1] Suckewer S, Jaeglé P 2009 *Laser Phys. Lett.* **6** 411
[2] Akhmedzhanov T R *et al* 2017 *Phys. Rev. A* **96** 033825

* E-mail: lias_1_5@mail.ru
† E-mail: antonov@appl.sci-nnov.ru

doi:10.1088/1742-6596/1412/9/092013

Effects of relativistic interactions in photodetachment time delay of Br⁻

S Saha[1] *, J Jose[2], P C Deshmukh[3], A S Kheifets[4], V K Dolmatov[5] and S T Manson[6]

[1]Department of Physics, Indian Institute of Technology Madras, Chennai, 600036, India
[2]Department of Physics, Indian Institute of Technology Patna, Bihta, 801103, India
[3]Department of Physics, Indian Institute of Technology Tirupati, Tirupati, 517506, India
[4]Research School of Physics and Engineering, The Australian National University, Canberra ACT 0200, Australia
[5]Department of Earth Science, University of Northern Alabama, Florence, AL 35632, USA
[6]Department of Physics and Astronomy, Georgia State University, Atlanta, 30303, USA

Synopsis Wigner time delay in photodetachment of Br^- has been studied. The absence of Coulomb phase in photodetachment process makes the Wigner time delay more sensitive to the centrifugal barrier shape resonance and threshold effects. Important relativistic effects have been found near the threshold region.

In our previous study of photodetachment in Cl^- (Phys. Rev. A **99**, 043407, 2019), it was found that the shape resonance [1] and threshold effects [2] alter the time delay [3, 4] spectrum significantly, unlike in the case of photoionization of Ar which is isoelectronic with Cl^-. The differences occurred owing to the absence of Coulomb phase in photodetachment process. The time delay was found to be angle sensitive [5] only in case of photodetachment process in the near-threshold region. However, relativistic interactions had no effects on the time delay for those low Z systems. That prompted us to pursue the present study with high Z systems. In the present work, we report on the time delay in photodetachment of Br^- using the Wigner-Eisenbud time delay formalism [3, 4]. Photoionization of Kr has also been studied for a comparison. The calculations have been performed using relativistic random phase approximation (RRPA) [6].

The plots are shown in Fig. 1 for Br^- time delay for the $3d_{3/2}$ (top panel) and $3d_{5/2}$ (bottom panel) subshells for different ejection angles. The crossover between $d \rightarrow p$ and $d \rightarrow f$ [2] transitions near the threshold region makes the time delay spectrum very complex. However, the most important point is that the average time delay spectra for $3d_{3/2}$ and $3d_{5/2}$ of Br^- are qualitatively different only close to the threshold region. This clearly shows the importance of relativistic effects.

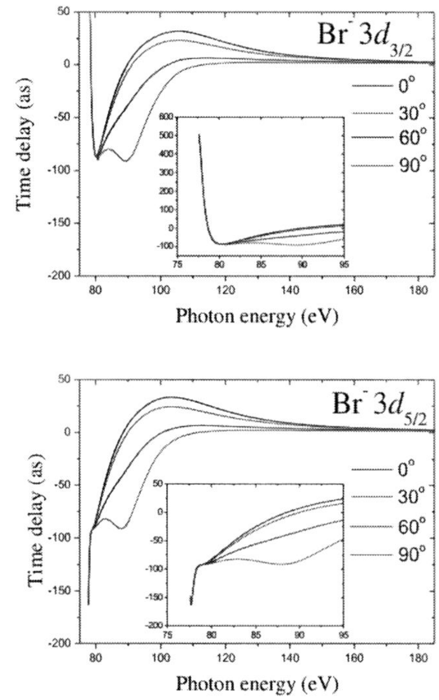

Figure 1. Photodetachment time delays of Br^- plotted for different electron ejection angles with respect to light polarization direction: $3d_{3/2}$ (top panel), $3d_{5/2}$ (bottom panel). Near-threshold regions magnified on the insets.

References

[1] Rau A R P and Fano U 1968 *Phys. Rev.* **167** 7
[2] Wigner E P 1948 *Phys. Rev.* **73** 1002
[3] Eisenbud L E 1948 *Ph. D. thesis*, Princeton Univ.
[4] Wigner E P 1955 *Phys. Rev.* **98** 145
[5] Kheifets A S *et al* 2016 *Phys. Rev. A* **94** 013423
[6] Johnson W R and Lin C D 1979 *Phys. Rev. A* **20** 964

* E-mail: soumyajit@physics.iitm.ac.in

Rescattering photoelectron spectroscopy of CO_2 molecule with an analytical returning electron wavepacket

M Okunishi[1] *, Y Ito[1], V Sharma[2], S Aktar[1], T Morishita[3], A I Dnestryan[4], O I Tolstikhin[4], R R Lucchese[5] and K Ueda[1]

[1]Institute of Multidisciplinary Research for Advanced Materials, Tohoku University, Sendai, 980-8577, Japan
[2]Department of Physics, Indian Institute of Technology, Hyderabad, India
[3]Institute for Advanced Science, The University of Electro-Communications, Chofu,182-8585, Japan
[4]Moscow Institute of Physics and Technology, Dolgoprudny 141700, Russia
[5]Lawrence Berkeley National Laboratory, Berkeley California, 94720-8196, United States

Synopsis Angle-resolved photoelectron momentum distributions of rescattering electron generated by intense infrared laser pulses were measured for CO_2 molecules. It is shown that the measured distributions along the outermost backward rescattering caustic are well reproduced by calculations based on a newly developed factorization formula with the analytical returning photoelectron wave packet.

In the past decade laser-induced electron diffraction (LIED) for simple gas phase molecules, which utilizes the elastically rescattered electrons generated by intense ultrafast laser pulses, was reported for extracting molecular geometry information [1]. The LIED is based on a proposed factorization formula for outermost photoelectron momentum distributions (PEMDs) induced by strong field ionization of the molecules. It states that PEMDs near the high-energy cutoff can be factorized into the differential cross section (DCS) for elastic scattering of a photoelectron on the parent ion and a returning photoelectron wave packet (RWP) [2]. Recently a new factorization formula for PEMDs along the outermost backward rescattering caustic was derived analytically [3], and it was demonstrated for strong-field ionization of NO and CO molecules, in which experimentally extracted PEMDs were compared with the theoretical calculations [4]. In the present study, we extend it to CO_2 to demonstrate the utility of the new formula.

The output of a Ti:Sapphire laser (800 nm, 100 fs, 1.5 mJ, 1 kHz) was injected into an optical parametric amplifier to convert the wavelength to 1300, 1450 and 1650 nm. The laser was introduced in a high vacuum chamber through a half waveplate and focused on CO_2 molecules. A field-free time-of-flight spectrometer was used to get the PEMD with waveplate orientation.

Figure 1 shows the PEMD measured with 1650 nm laser pulses along the outermost rescattered electron caustic, where the electron rescattering momentum is 3.5 atomic units. We compare the experimentally extracted PEMD with theoretical calculations. Three types of calculations are presented here, (i) independent atomic model (IAM) calculation of electron-molecule scattering, (ii) single active electron (SAE) model calculation, and (iii) *ab initio* calculations of electron-ion scattering.

*E-mail: m.okunishi@tohoku.ac.jp

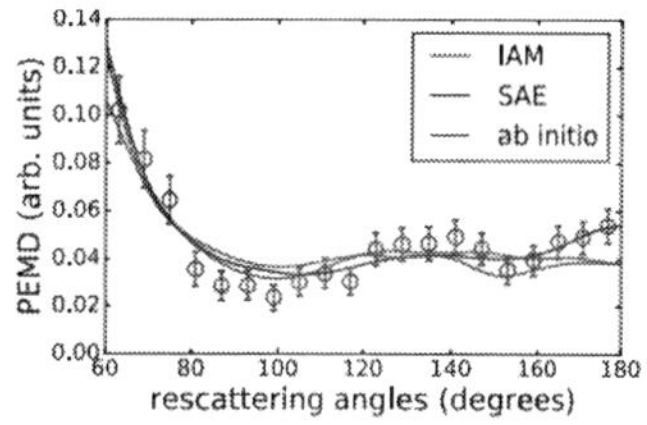

Figure 1. Experimentally extracted and theoretically calculated PEMD along the outermost electron caustic measured with 1650 nm laser pulses.

The SAE and *ab initio* calculations show good agreement with the experimental results for wide range of the rescattering angles, but IAM calculation fail to reproduce the modulation of the PEMD around 140-160 degrees.

References

[1] Ito Y *et al* 2017 *Phys. Rev. A* **96** 053414 and references therein
[2] Morishita T *et al* 2008 *Phys. Rev. Lett.* **100** 013903
[3] Morishita T and Tolstikhin O I 2017 *Phys. Rev. A* **96** 053416
[4] Ito Y *et al* 2018 *Phys. Rev. A* **97** 053411

Revealing the multiorbital and excitation effects in dissociative double ionization of CO molecules with angular streaking measurement

X Li[1], J Yu[1], X Yu[1], Y Yang[1], C Wang[1], S Luo[1 *] and D Ding[1 †]

[1]Institute of Atomic and Molecular Physics, Jilin University, Changchun, 130012, China

Synopsis We investigate the sequential double ionization dynamics for the dissociative channel $C^+ + O^+$ of CO molecules in strong circularly polarized laser fields. By using four-particle coincident measurement, the obtained ionic angular emissions and molecular frame photoelectron angular distributions of each ionization steps for various pathways reveals the significant effects of multiorbital ionization and laser-driven excitation.

Multiorbital effect exists widely in the strong-field tunneling ionization of molecules, where the direct tunnel from inner molecular orbitals populates the molecular cation to the electronic excited states. Distinguishing the multiorbital contributions helps us to understand the results of alignment-dependent harmonic phase interferometry, molecular angular-dependent ionization rate, multiple-pathway molecular dissociation dynamics, and so on [1,2]. Another effect that can also contributes to the molecular fragmentation is sequential field-driven excitation following strong-field ionization to the cationic ground state [3].

We perform a four-body coincident measurement for the two ionic fragments and two photoelectrons resulted from the charge-symmetric dossociative channel of CO in circularly polarized laser fields. Molecular frame photoelectron angular distributions (MFPADs) for various pathways that have distinct ionic kinetic energy releases (KERs), shown in Fig. 1(a), are reconstructed. The sequential ionization of two electrons is resulted in at the rising edge of the pulse envelope in the intense circular polarized laser with the intensity of 10^{15} W/cm^2. Considering that the drift momentum of electron follows the electrical field strength at ionization instant [4], we distinguish the orders of the ionization for the two electrons.

Figure 1(b) shows the KER-dependent ionic angular distribution in the plane perpendicular to the laser polarization plane. The substantial emissions out of the polarization plane (±90°) indicate the contributions from HOMO-1 (π-type orbital). Pathway-resolved dynamics are further extracted from the separated MFPADs of two electrons for three major pathways, P1, P2, and P3, shown in Fig. 1(c, d, e). The different MFPADs for P1 are attibuted to the participation of HOMO-1 ionization, whereas the MFPADs for P2 and P3, which comes from higher excited state of dication, reveal two ionization steps from HOMO, and therefore the sequential laser-driven excitation is essential for their emergence. The present work illustrates the feasibility of disentangling the ionization and excitation dynamics in molecular dissociative doubel ionization with multi-coincident angular streaking measurement.

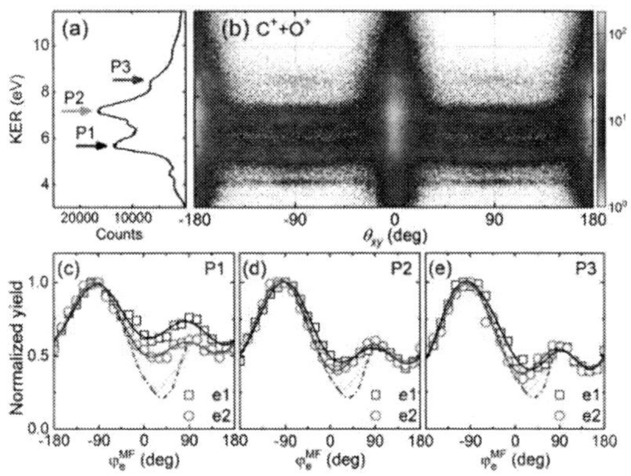

Figure 1. (a) Ionic KER and (b) angular distribution in the plane perpendicular to the polarization plane. (c)-(e) The separated MFPADs for the two electrons coincided with KER peaks P1, P2, and P3, respectively.

References

[1] Akagi H *et al* 2009 *Science* **325** 1364
[2] Luo S *et al* 2017 *Phys. Rev. A* **96** 063415
[3] Schell F *et al* 2018 *Phys. Chem. Chem. Phys.* **20** 14708
[4] Pfeiffer A N *et al* 2011 *Nat. Phys.* **7** 428

* E-mail: luosz@jlu.edu.cn
† E-mail: dajund@jlu.edu.cn

Dependence of the returning electron wave packet on the model potential for high-energy plateau photoelectron spectra

J H Chen[1] *, J Fan[2] and X Y Li[2] †
[1]School of Electronic and Information Engineering, Lanzhou City University, Lanzhou, 730070, China
[2] Department of Experimental Teaching, Northwest Minzu University, Lanzhou, 730030, China

Synopsis In the present work, we study the two-dimensional momentum distributions of high-energy plateau photoelectrons of a model-atom in a linearly polarized few-cycle laser pulse. Based on the improved strong-field approximation (ISFA), it is found that the returning electron wave packet is dependent of the model potential.

The above-threshold ionization (ATI) process in atomic, molecular, and other complex systems exposed to an intense laser field is probably the most fundamental and intriguing phenomenon in the field of laser-matter interactions. For the energy spectra of ATI electrons, the yield of photoelectrons with the energy above $2U_p$ has not changed with the electron energy rapidly until a new cutoff near about $10U_p$. Those are known as high-energy plateau photoelectrons. Here U_p is the electron ponderomotive energy. Recently, it is found that the two-dimensional momentum distributions (PMDs) of high-energy plateau photoelectrons is useful for imaging the atomic and molecular structures in the time scale of a few femtoseconds.

The main workhorse for the theoretical treatment of the high-energy plateau photoelectrons is the improved strong-field approximation (ISFA), in which the effect of the returning electron with the target ion is included. However, ISFA is not able to describe well the two-dimensional PMDs of high-energy plateau photoelectrons. Based on ISFA, two different theories have been developed recently, i.e., the quantitative rescattering model (QRS) [1] and the low-frequency approximation (LFA) [2,3]. Both theories have shown their successes by comparing with the *ab initio* results.

According to the QRS and LFA, the momentum distribution of the electrons on the back-rescattering ridge (BRR) can be factorized into the product of the returning electron wave packets and the electron-target ion elastic differential cross sections (DCSs), i.e.,

$$D(p,\theta)=W(p_r)\times\sigma(\theta_r,p_r)$$

where the wave packet $W(p_r)$ can be extracted from ISFA. And the DCSs $\sigma(\theta_r,p_r)$ can be obtained from the standard quantum-mechanical potential scattering theory.

In this work, we discuss the influence of the atomic model potential on the QRS model for the high-energy plateau photoelectron spectra in detail. Firstly, we give the high-energy plateau photoelectron spectra for a Yukawa potential, a zero-range potential, and a spherical shell potential in context of ISFA. The results obtained from numerically solving time-dependent Schrodinger equation (TDSE) are regarded as the benchmark results. Our results show that the energy distributions in high-energy region obtained from ISFA calculations using the spherical shell potential may either match or be better than those from ISFA using Yukawa potential and zero-range potential. And then, we extracted the electron wave packet from Yukawa potential, zero-range potential, and spherical shell potential respectively. It is found that the returning electron wave packet is dependent of the model potential. Our work may provide with a meaningful reference for the further study on the high-energy plateau electrons spectra of complex atoms.

Acknowledgment This work is supported by the National Natural Science Foundation of China under Grant No. 11564033, 11764036.

References

[1] Chen Z J *et al* 2009 *Phys. Rev. A* **79** 033409
[2] Cerkic A *et al* 2009 *Phys. Rev. A* **79** 033413
[3] Milošević D B 2014 *Phys. Rev. A* **90** 063423

* E-mail: chenyuwen1982@163.com
† E-mail: lxycock@163.com

Signatures of Light-Induced Potential Energy Surfaces in H_2^+

M Kübel[1,2,3]*, M Spanner[4], Z Dube[1], A Yu Naumov[1], M J J Vrakking[5], P B Corkum[1], D M Villeneuve[1] and A Staudte[1]†

[1]Joint Attosecond Laboratory, National Research Council and University of Ottawa, Ottawa, Ontario, Canada
[2]Ludwig-Maximilians-Universitt Munich, Am Coulombwall 1, D-85748 Garching, Germany
[3]Friedrich-Schiller Universitt Jena, Max-Wien-Platz 1, D-07743 Jena, Germany
[4]National Research Council, 100 Sussex Drive, Ottawa, Ontario K1A 0R6, Canada
[5]Max-Born-Institut, Max-Born-Strae 2A, D-12489 Berlin, Germany

Synopsis Using theory and Cold Target Recoil Ion Momentum Spectroscopy we find signatures of light-induced molecular potential energy surfaces in the 3-dimensional proton momentum distributions of dissociating H_2^+.

H_2 continues to provide fundamental insights into the mechanisms of intense light-matter interactions [1]. Recently, there has been significant interest in so-called light-induced conical intersections (LICI) that arise from the angle dependence of the single-photon coupling of electronic states. Analogously to regular conical intersections, electronic and nuclear motions are strongly coupled in the vicinity of LICIs. As a signature of such rovibronic dynamics, weak modulations in the angular distribution of protons emitted from H_2^+ have been reported [2]. More generally speaking, infrared (IR) laser pulses couple the σ_g and σ_u electronic states of H_2^+ typically through several pathways involving an odd number of photons [1], which can produce complex light-induced potential energy landscapes, and consequently even richer dynamics.

Here, we report the observation of strongly structured proton angular distributions obtained from the dissociative ionization of H_2. Our experiment relies on the STIER (Sub-cycle Tracing of Ionization Enabled by infraRed) technique [3], where an intense few-cycle visible pulse ionizes H_2 to H_2^+. The cation subsequently dissociates under the influence of a moderately intense mid-IR (2300 nm) pulse, which is polarized perpendicularly to the visible pulse and does not cause ionization of H_2 on its own. The 3-dimensional momentum of the ejected protons were recorded using COLTRIMS (cold target recoil ion momentum spectroscopy). Figure 1 a) shows the proton distribution produced in a few-cycle, 800 nm laser pulse. In Figure 1b) a cross-polarized, multi-cycle, 2300 nm pulse is superimposed. Strikingly, the weak mid-IR field does not only add a contribution along its polarization axis, but also produces angular features closer to the polarization of the visible laser pulse. Using numerical solutions of the time-dependent Schrdinger equation, we explain the features of the measured momentum distribution as an angle-dependent competition of different dissociation pathways in a complex light-induced potential energy landscape [4].

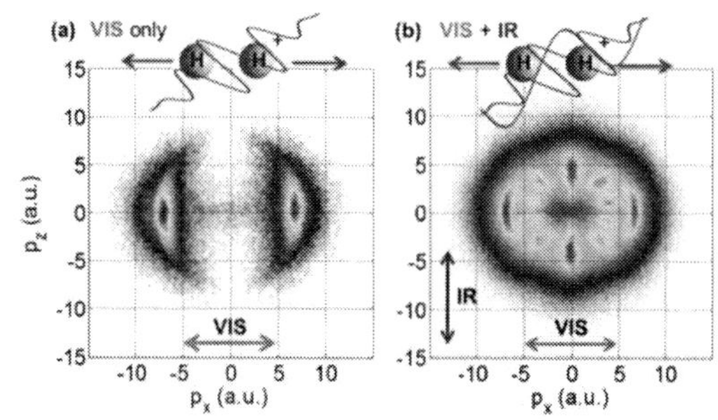

Figure 1. Measured proton momentum distributions produced by (a) a visible (730 nm) few-cycle pulse only and (b) and an additional mid-IR (2300 nm) dressing field. The arrows indicate the laser polarization.

References

[1] Ibrahim H *et al* 2018 *J. Phys. B* **51** 42002
[2] Natan A *et al* 2016 *Phys. Rev. Lett.* **116** 1
[3] Kbel M, *et al* 2017 *Phys. Rev. Lett.* **119** 183201
[4] Kbel M *et al* in preparation

*E-mail: matthias.kuebel@uni-jena.de
†E-mail: andre.staudte@nrc.ca

Dynamic interference in photoemission by chirped ultrashort extreme ultraviolett pulses

N Wang[1], G Li[1], A Liu[1] * and U Thumm[2] †

[1]Institute of Atomic and Molecular Physics, Jilin University, Changchun 130012, China
[2]James R. Macdonald Laboratory, Kansas State University, Manhattan, Kansas 66506-2604, USA

Synopsis We investigated dynamical interference (DI) in the photoionization of hydrogen atoms in intense ultrashort chirped extreme ultraviolett (XUV) pulses by numerically solving the time-dependent Schrödinger equation(TDSE) and examined its dependence on pulse intensity, duration, and chirp. We find the pulse chirp to red shift the photoelectron spectrum and weaken the DI effect.

DI in photoionization by intense extreme ultraviolett (XUV) pulses arises when electron emission from equally ac Stark-shifted levels at the raising and falling temporal pulse profile interfere and are temporally separated by a time interval of intense-field stabilization against photoemission [1, 2, 3]. In particular, DI from the *ground state* of hydrogen atoms has been the subject of a controversy between theoretical predictions [4, 5].

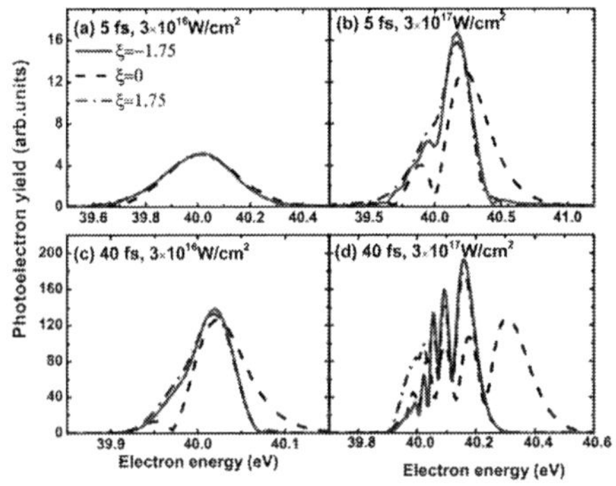

Figure 1. Photoelectron spectra from ground-state hydrogen atoms exposed to XUV pulses with 53.6 eV central photon energy, peak intensities of (a,c) 3×10^{16} and (b,d) 3×10^{17} W/cm^2, and pulse lengths of (a,b) 5 and (c,d) 40 fs for the chirp parameters ($\xi = 0, \pm1.75$).

Solving the TDSE in full dimensionality in dipole-approximation, we investigated DI in the photoemission from hydrogen atoms for a central XUV-pulse-photon energy of $\hbar\omega_{ctr} = 53.605$ eV and chirp parameters $\xi = 0, \pm1.75$. The chirped XUV pulse with pulse duration τ_0 is defined by the vector potential $A(t) = A_0(\xi)F(t)\sin[\omega(t)t]$ with $A_0(\xi) = \frac{1}{\omega}\sqrt{\frac{I_0/I_{au}}{\sqrt{1+\xi^2}}}$, $\omega(t) = \omega_0 + 2\ln 2\frac{\xi}{1+\xi^2}\frac{t}{\tau_0^2}$, and $F(t) = \exp[-2\ln 2\frac{1}{1+\xi^2}\frac{t^2}{\tau_0^2}]$. $I_{au} = 3.51\times10^{16}$ W/cm^2 is the atomic unit of intensity. Including ac Stark shifts and atomic stabilization, our numerical results confirm that dynamic interference is possible for single-XUV-photon ionization from the hydrogen ground state [6] (Fig. 1).

Compared with chirp-free pulses ($\xi = 0$), we find that the interference structure in photoemission spectra is increasingly red shifted for positively ($\xi = 1.75$) and negatively ($\xi = -1.75$) chirped pulses as the pulse peak intensity and duration increase, while the interference structure becomes either less pronounced [Fig. 1 (b,c)] or significantly narrower [Fig. 1 (d)].

This work is supported by the NSFC under Grant Nos. 11774131, 91850114, 11604119 and, in part, by the US DOE under Award No. DE-FG02-86ER13491 and US NSF Grant PHY 1802085.

References

[1] Eberly J H and Kulander K C 1993 *Science* **262** 5137
[2] Guo J *et al* 2018 *Acta Phys. Sin.* **67** 073202
[3] Jiang W and Burgdörfer J 2018 *Opt. Express* **26** 19921
[4] Demekhin P V and Cederbaum L S 2012 *Phys. Rev. Lett.* **108** 253001
[5] Baghery M *et al* 2017 Phys. Rev. Lett. **118** 143202
[6] Wang N *et al* to be published

*E-mail: aihualiu@jlu.edu.cn
†E-mail: thumm@phys.ksu.edu

Journal of Physics: Conference Series **1412** (2020) 092019 doi:10.1088/1742-6596/1412/9/092019

Super-interference in atomic photoionization: Moiré patterns in momentum maps

M Dran[1] and D G Arbó[1 †]

[1]Institute for Astronomy and Space Physics IAFE (UBA-Conicet), Buenos Aires, 1428, Argentina

Synopsis We theoretically investigate on the origin of lateral ring structures in the doubly differential momentum distribution for atomic ionization by laser pulses in the midinfrared spectral region. We demonstrate that such structures stems from the interplay between intra- and intercycle interference patterns which work as two separate grids in the two-dimensional momentum domain. When the periods of the two grids (intra- and intercycle) are similar, moiré patterns arise as concentric rings at high electron kinetic energy.

Moiré patterns appear in a myriad of situations not only in physics and mathematics but also in diverse disciplines such as art, fashion, digital photography and television [1]. For moiré interference pattern to appear, two slightly different patterns or grids must be present.

Photoelectron spectra have been described in the literature as an interference problem in the time domain. Trajectories stemming from different optical laser cycles give rise to *intercycle* interference energy peaks known as multiphoton (or ATI) peaks. Besides, another structure comes from the *intracycle* interference of the electron trajectories born during the same optical cycle [2]. These two interference patterns form two grids with different momentum-dependent spacing.

We theoretically investigate on the origin of lateral ring structures in the doubly differential momentum distribution for atomic ionization by laser pulses in the midinfrared spectral region [3]. We demonstrate that such structures stems from the interplay between intra- and intercycle interference patterns which work as two separate grids in the two-dimensional momentum domain. When the periods of the two grids (intra- and intercycle) are similar, moiré patterns arise as concentric rings at high electron kinetic energy in the forward and backward directions [4].

In Fig. 1 (a), the doubly differential momentum distribution calculated within the saddle point approximation (SPA) is displayed. One can see the intercycle interference pattern as iso-energetic concentric rings in agreement with the conservation of energy for the absorption of multiple photons of the laser. Furthermore, the intracycle pattern is displayed as symmetrical convex stripes with respect to the parallel momentum axis. Fig. 1 (b) shows the spacing of the two grids along the parallel momentum axis as a function of the parallel momentum. We observe that the spacing of the two intra- and intercycle grids coincide at parallel momentum ~ 0.85 a.u., just at the center of the lateral ring.

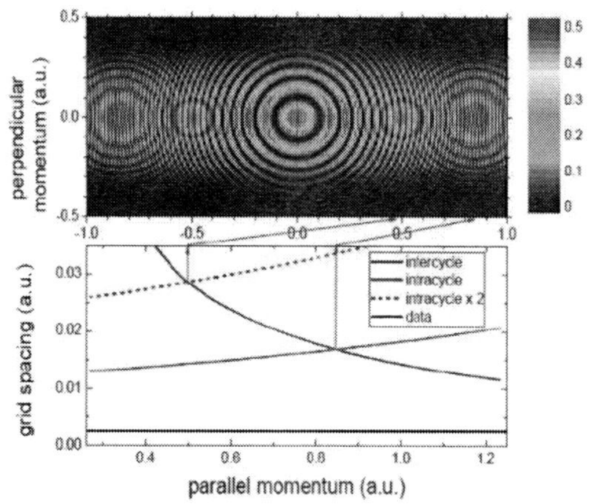

Figure 1. (a) Momentum distribution after interaction of a strong (I = 10^{14} W/cm^2) midinfrared (λ = 3200 nm) laser pulse with a hydrogen atom (semiclassical model). (b) Grid spacing as a function of the parallel momentum.

This work has been supported by CONICET (Argentina PIP0386) and ANPCyT (Argentina PICT-2016-0296 and PICT-2014-2363).

References

[1] Miao H *et al* 2016 *Nature Phys.* **12** 830
[2] Arbó D G *et al* 2010 *Phys. Rev. A* **82** 043426; ibid **81** 021403
[3] Lemell C *et al* 2013 *Phys. Rev. A* **87** 013421
[4] Dran M and Arbó D G 2018 *Phys. Rev. A* **97** 053406

† E-mail: diego@iafe.uba.ar

Classical theory of laser-assisted spontaneous bremsstrahlung

H B Ambalampitiya[1]* and I I Fabrikant[1]†

[1]Department of Physics and Astronomy, University of Nebraska-Lincoln, Lincoln, NE 68588, USA

Synopsis At low energies the classical trajectories for electron motion in a combined Coulomb and ac fields are affected by the Coulomb focusing. This leads to a strong enhancement of the laser-assisted bremsstrahlung cross section as compared to the field-free case.

Laser-assisted spontaneous bremsstrahlung is a process of creation of a photon with frequency Ω due to electron-atom scattering in the presence of an ac field of a lower frequency ω, $F = F_0 \cos(\omega t + \phi_0)$, i.e. the process

$$n\hbar\omega + e + A \to n'\hbar\omega + \hbar\Omega + e + A$$

which is equivalent to the harmonic generation in the continuum. The process is most efficient when the atomic system A is a bare nucleus or a positive ion.

This process has been studied since 1970s. Most of these and more recent papers, particularly those dealing with relativistic electrons, were treating the electron-ion interaction in the first order of perturbation theory. Others were concentrating on the resonance processes when the frequency of the emitted photon equals an integer times the laser frequency.

In the present paper we investigate another feature of the process due to the Coulomb focusing [1]. The action of the Coulomb potential, in combination with multiple electron returns due to the laser field, focuses parts of the electron wave function, increasing the efficiency of such processes as multiphoton ionization. Similar effects can occur in continuum-continuum transitions.

In contrast to the ionization problem, the bremsstrahlung problem can be treated purely classically with a very good accuracy [2]. In our treatment we launch electron trajectories parallel to the polarization of the ac field (z axis) at a distance from the Coulomb center, where the Coulomb interaction can be neglected, and calculate the radiation probability per unit frequency as a function of the impact parameter b. Integration over b allows determination of the bremsstrahlung cross section per unit frequency of the emitted photon $S(\Omega)$ for a given initial phase ϕ_0 of the ac field. The radiation spectrum is averaged then over ϕ_0. This averaging is equivalent to averaging over the initial starting point in the z direction [2]. The results for the probability as a function of b exhibit interesting features related to chaotic behavior of electron trajectories in a combined Coulomb and ac fields [3]. After integration over b and averaging over ϕ_0 these features are significantly smeared out.

As an example, we calculate radiation spectrum due to the Coulomb field of charge $Z = 1$ for the ac field intensity 100 TW/cm^2 (F_0 = 0.05338 a.u.) and the wavelength 3.1 μm (ω = 0.0147 a.u.), and compare it with the field-free $S(\Omega)$ for electron energy 6.35 a.u. corresponding to a typical kinetic energy of electron acquired by the ac field (about twice the ponderomotive energy $E_p = 3.30$ a.u.) when it approaches the Coulomb center.

Although the average over ϕ_0 decreases the cross section by an order of magnitude, we still observe a two order of magnitude gain as compared to the field-free cross section due to the Coulomb focusing. In particular, the laser-assisted cross section $S(\Omega)$ varies from 10^{-3} to 0.5×10^{-4} a.u. in the frequency range between 0.1 and 3 a.u. whereas the field-free cross section decreases from 1.1×10^{-5} to 0.18×10^{-6} a.u. in the same range.

References

[1] Brabec T *et al* 1996 *Phys. Rev. A* **54** R2551
[2] Fedorov M V and Ivanov M YU 1993 *Laser Physics* **3** 365
[3] Wiesenfeld L 1990 *Phys. Lett. A* **144** 467

*E-mail: harindranath@huskers.unl.edu
†E-mail: ifabrikant@unl.edu

Laser-assisted electron scattering and ionization processes in a quasi-Sturmian-Floquet approach

S A Zaytsev[1], A S Zaytsev[1], L U Ancarani[2]* and K A Kouzakov[3]

[1]Pacific National University, Khabarovsk 680035, Russia
[2]Université de Lorraine, CNRS, LPCT, F-57000 Metz, France
[3]Faculty of Physics, Lomonosov Moscow State University, Moscow 119991, Russia

Synopsis We demonstrate the robustness and efficiency of our recently formulated quasi-Sturmian-Floquet method to treat light-dressed electron states. For this purpose we study the archetypical processes such as laser-modified electron scattering on a Coulomb potential and laser-assisted electron-impact ionization of atomic hydrogen. For both processes we explore various regimes of laser radiation and identify the differences with respect to the corresponding laser-free situations.

We apply our recently formulated nonperturbative method [1, 2] to the theoretical treatment of laser-modified electron states in scattering and ionization processes. Our method is based on the Hermitian Floquet theory in the Kramers-Henneberger representation. We use quasi-Sturmian functions [3] to solve the system of coupled Lippmann-Schwinger-Floquet equations for the laser-modified Coulomb wave functions of an electron.

We investigate theoretically (i) a laser-assisted $(e, 2e)$ collision on atomic hydrogen,

$$\ell\omega + e^- + \mathrm{H}(1s) \to \mathrm{H}^+ + 2e^-, \tag{1}$$

and (i) a laser-modified nonrelativistic electron scattering on a proton target,

$$\ell\omega + e^- + p \to p + e^-, \tag{2}$$

where a net number ℓ of photons of frequency ω can be exchanged between the projectile-target system and the external field.

For process (1), we considered asymmetric $(e, 2e)$ kinematics with a laser field either perpendicular to the scattering plane [1, 2] or parallel to the incident electron momentum (present work). We investigated ejected-electron angular distributions for various ℓ values in the perturbative and nonperturbative regimes of laser radiation. Our calculations using the Born approximation in the projectile-target interaction converge fast with increasing number $\mathfrak{N}$ of Floquet components for the ejected-electron state. The modification due to the presence of laser radiation should be observable experimentally in the nonperturbative regime.

*E-mail: ugo.ancarani@univ-lorraine.fr

For process (2), we calculated the ratio of the laser-assisted angular differential cross section $d\sigma_\ell/d\Omega$ to the laser-free, Rutherford cross section $d\sigma_R/d\Omega \propto 1/\sin^4\frac{\theta}{2}$. We explored both the high-frequency ($E_e \ll \omega$) and low-frequency ($E_e \gg \omega$) regimes. For the latter case, Fig. 1 shows the convergence of our numerical results with increasing number $\mathfrak{N}$ of Floquet components of the electron state. An important feature is that the calculated ratio substantially differs, especially at larger angles, from unity, *i.e.*, from the value predicted by the customary Volkov and Coulomb-Volkov models.

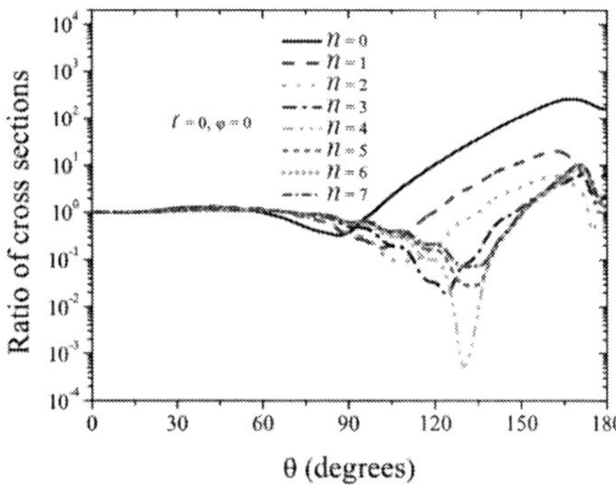

Figure 1. The ratio $\frac{d\sigma_{\ell=0}}{d\Omega} / \frac{d\sigma_R}{d\Omega}$ as a function of the electron scattering angle when $E_e = 0.43$ a.u., $\omega = 0.05$ a.u., and the laser electric field amplitude is $F_0 = 0.0125$ a.u. The laser field is linearly polarized and is perpendicular to the scattering plane.

References

[1] Zaytsev A S *et al* 2018 *Phys. Rev. A* **97** 043417
[2] Zaytsev S A *et al* 2019 *Eur. Phys. J. D* **73** 42
[3] Del Punta J A *et al* 2014 *J. Math. Phys.* **55** 052101

Photoelecton emission from silver nanoparticles after laser irradiation

L Budai[1], Z Márton[2,3], P Dombi[3,4] and K Tőkési[1,3*]

[1]Institute for Nuclear Research, Hungarian Academy of Sciences (Atomki), 4026 Debrecen Bem tér 18/c, Hungary
[2]Institute of Physics, University of Pécs, Pécs, Hungary
[3]ELI-ALPS, ELI-HU Non-profit Ltd., Dugonics tér 13, H-6720 Szeged, Hungary
[4]Wigner Research Centre for Physics, Budapest, Hungary

Synopsis We present theoretical study of photoelectron emission from silver nanoparticles after laser irradiation based on the classical simulation of the electron trajectories. The individual electron trajectories were analized with and without taken into account the image force between the emitted electron and its mirror charge. We show also the calculated energy spectra in comparision with the available experimental data. We observed a significant effect of the image acceleration to the calculated electron spectra.

Electron emission from nanoparticles induced by laser excitation has of great interest during the last years [1,2]. It is not only because it holds of very interesting and challenging fundamental and basic science both in experimental and theoretical points of views, but these researches also promise many technical applications. Among them it can be used in plasmonic biosensors, optoelectronics, photovoltaics, construction of ultrafast nanoemitters,

Along this line, we developed a computer simulation code based on the classical dynamics to calculate the electron trajectories emitted from a semi-ellipsoid protrusion of an Ag surface. At first the time dependent electric field was calculate (see fig.1) using the Kretschmann configuration [3]. The incident laser pulse was an 800 nm, 5 fs plane wave. The angle of incidence was 44.6°.

We performed a large number of classical trajectory simulations in the pre-calculated electric field obtained by the finite-difference time-domain simulations. During the simulations we are able to calculate the electron trajectories with and without taken into account the image force between the emitted electron and its mirror charge. We found that the image acceleration toward the surface plays a dominant role in shaping the electron trajectories [4]. We found also that the calculated energy distributions of photoelectrons are in agreement with the recent experiments. The experimentally obtained double peak structure was identified in our calculated spectra.

The work was support by the National Research, Development and Innovation Office (NKFIH) Grant KH126886

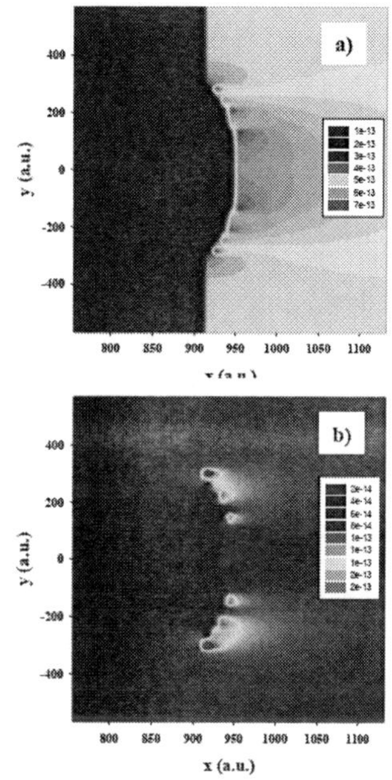

Figure 1. The E_x (a) and E_y (b) components of the electric field E in vacuum, in the vicinity of a semi-ellipsoid protrusion of an Ag surface layer, illuminated by an oblique incident, plane fs pulse through a transparent prism, in Kretschmann configuration. a) E_x at t = 110 fs, b) E_y at t = 110 fs.

References

[1] Rácz P *et al* 2017 *Nano Lett.* **17** 1181
[2] Budai J *et al* 2018 *Nanoscale* **10** 16261
[3] Kretschmann E and Raether H 1968 *Z. Naturforsch.* **23A** 2135
[4] Budai L *et al Eur. Phys J. D* to be published

* E-mail: tokesi@atomki.mta.hu

Journal of Physics: Conference Series **1412** (2020) 092023 doi:10.1088/1742-6596/1412/9/092023

Research on bremsstrahlung and Compton scattering in hot warm matters

Y Y Qi[1] * and J G Wang[2]

[1] School of Mathematics & Physics and Information Engineering, Jiaxing University, Jiaxing 314001, China
[2] Institute of Applied Physics and Computational Mathematics, P.O. Box 8009, Beijing 100088, China

Synopsis Bremsstrahlung and Compton scattering are the important processes including the continuum electrons in the hot warm matters. The distribution of the continuum electrons is affected by the plasma screening, which will influence the cross sections of Bremsstrahlung and Compton scattering. In the present work we will investigate the plasma influence on these processes.

The continuous radiation absorption process generally termed "free-free absorption" and its inverse process of bremsstrahlung are of considerable importance for a wide range of physical and astrophysical problems, from cosmic rays and nuclear physics applications to the study of gas discharge and radiative transfer phenomena [1-2].

The cross section for the free-free transition is [1-2]:

$$\sigma^{FF}(k_i,\omega)=\frac{8\pi^4}{3}\left(\frac{e^2}{\hbar c}\right)\frac{k_f}{m\hbar\omega}\sum_{l_i=0}^{\infty}\left\{(l_i+1)\left[M^2_{l_i+1,l_i}(r)+M^2_{l_i,l_i+1}(r)\right]\right\}$$

Adopting the dipole-acceleration of radial matrix element:

$$M^2_{l_f,l_i}=\frac{1}{\omega^2}\left|\int_0^{\infty}R_{\varepsilon_f l_f}(k_f,r)\frac{\partial V}{\partial r}R_{\varepsilon_i l_i}(k_i,r)r^2dr\right|^2$$

Here the continuous wave function can be solved by the the radial Schrödinger equation:

$$-\frac{1}{2r^2}\frac{d}{dr}\left(r^2\frac{dR_{El}(r)}{dr}\right)+\frac{l(l+1)}{2r^2}R_{El}(r)+V(r)R_{El}(r)=ER_{El}(r)$$

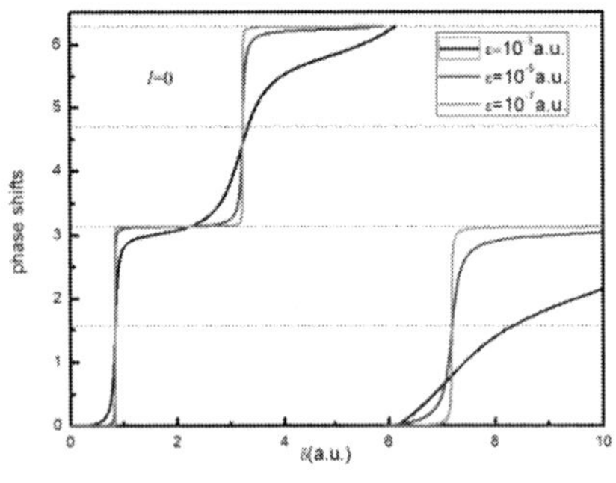

Figure 1. The s-wave phase shift with the plasma screening lengths for the debye potential.

Due to the plasma screening, the continuous wave function will be distorted, features of which can be reflected by their phase shifts. For example, the s-wave phase shift remarkablely changes even if the debye potential is adopted [3], as shown in Figure 1.

A loose bound will be produced when the phase shift jumps by $\frac{\pi}{2}$ or $\frac{3\pi}{2}$, the overlap matrix element will be enlarged, the cross section relevant to the continuous process will be involved.

The cross section for bremsstrahlung samples the influence of the debye potential as shown in figure 2.

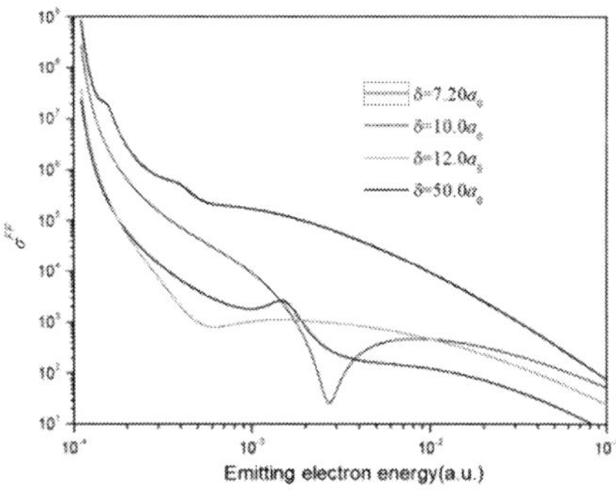

Figure 2. The free-free transition cross section with the emissing electron energy for the debye potential

References

[1] Wu J Y *et al* 2019 *Phys.Rev. A* **99** 012705
[2] Johnston R R 1967 *J.Quant.Spectrosc.Radiat. Transfer.* 7 815-835
[3] Qi Y Y *et al* 2015 *Chin. Phys. B* **24** 033403

* E-mail: yying_qi@zjxu.edu.cn

Journal of Physics: Conference Series **1412** (2020) 092024 doi:10.1088/1742-6596/1412/9/092024

Free-free experiments: dressed-atom effects during inelastic electron scattering

N L S Martin[1] *, B N Kim[1] †, C M Weaver[1] and B A deHarak[2]

[1]Department of Physics and Astronomy, University of Kentucky, Lexington, Kentucky 40506-0055, USA
[2]Physics Department, Illinois Wesleyan University, P.O. Box 2900, Bloomington, IL 61702-2900, USA

Synopsis Experiments and theory will be presented for the emission/absorption of 1.17 eV photons by electrons during the electron-impact excitation of a helium target in a laser field .

Free-free experiments investigate the absorption or emission of radiation during the collision of charged particles with atoms and molecules. The first experimental observation of dressed-atom effects – due to the electric field of the laser – during the elastic scattering of electrons by Xe were reported by Morimoto *et al* [1]. Their results were compared with an analytical expression by Zon that contains the electric dipole polarizability α of the target [2]. We are investigating dressing effects for *inelastic* electron scattering in the presence of a Nd:YAG laser; specifically we are investigating electron-impact excitation of the lowest excited states of He and Ar. Zon's expression is not valid for inelastic scattering so we have developed an equivalent inelastic expression. We will give the results of a simple calculation for dressing effects for the excited states of He (see Figure 1), and will give a progress report on our experiments.

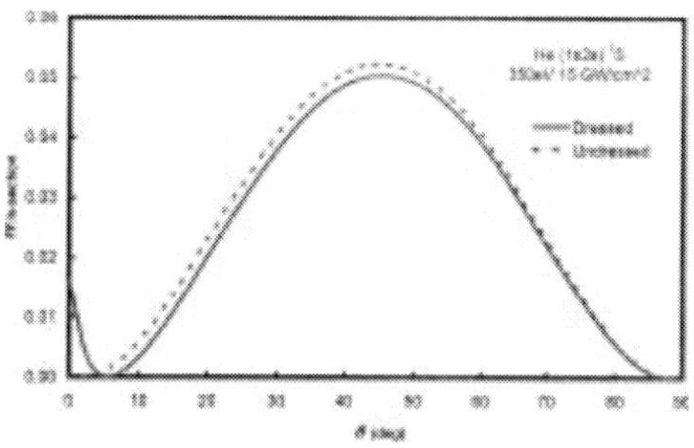

Figure 1. Calculated free-free signal for the excitation of He $(1s2s)\ ^1S$ by 350 eV electrons. The solid (green) curve includes the dressing of the atom by the 1.17 eV laser field, the dotted (red) curve does not.

References

[1] Morimoto Y, Kanya R and Yamanouchi K 2015 *Phys. Rev. Lett.* **115** 123201
[2] Zon B A 1977 *Sov. Phys. JETP* **46** 65

*E-mail: nmartin@uky.edu
†E-mail: brian.kim@uky.edu

Absolute cross sections of ultra-soft x-ray radiation in electron scattering upon atoms

A S Kornev[1], B A Zon[1]* and M Ya Amusia[2,3]

[1]Wave Processes Laboratory, Voronezh State University, Voronezh, 394018, Russia
[2]The Hebrew University of Jerusalem, 9190401, Israel
[3]A.F. Ioffe Physical-Technical Institute, St. Petersburg, 194021, Russia

Synopsis The soft-photon approximation (SPA) is shown to give quantitative interpretation of absolute measurement of bremsstrahlung (BrS) cross section, σ, on atomic targets in the region of large wavelengths λ, for which $\sigma \sim 1/\lambda^2$.

Refs. [1] presented results of absolute measurements of bremsstrahlung (BrS) cross sections values for ultra-soft x-ray photons ($\lambda = 7..18$ nm) generated in scattering of electrons with energies of 0.4..2 keV on Ar, Kr, and Xe atoms. The experimental data differ by a factor 3..4 from the calculations of Pratt *et al.* [2]. The important results reported in Refs. [1], have no theoretical explanation yet. Some of these results, namely, the values of BrS cross sections, including the limiting extrapolated values at infinitely large wavelengths of BrS photons, where the dependence $\sigma \sim 1/\lambda^2$ is valid without any doubt, we interpret on the basis of the soft-photon approximation (SPA), $\omega \ll E/\hbar$, where ω is the BrS photon frequency and E is the electron energy in the initial state. Previously the SPA was used to interpret ultra-soft x-ray spectra (*e.g.* [3] and refs. therein).

In the frame of SPA and in non-relativistic approximation, the double differential cross section of BrS for electron scattering on a certain target is presented as the product of the differential cross section of electron scattering on the same target without radiation, $d\sigma_e$, and the probability of photon radiation dw_γ [3],

$$d\sigma = d\sigma_e dw_\gamma, dw_\gamma = \frac{\alpha}{4\pi^2}\left[\frac{\mathbf{k}}{\omega}(\mathbf{v}-\mathbf{v}')\right]^2 \frac{d\omega}{\omega} d\Omega_\gamma. \quad (1)$$

where $\alpha \simeq 1/137$, $\mathbf{v}$ and $\mathbf{v}'$ are electron velocities in the initial and final states, respectively, $\mathbf{k}$ is the photon momentum, $d\Omega_\gamma$ is the solid angle of photon departure, and the system of units $\hbar = c = 1$ is used.

To calculate $\sigma_e(\theta)$, we used the data from Refs. [4, 5]. The Table below compares the theoretical calculations for spectral intensity $\lambda^2 \Delta I$ (in nm^{2}erg/s) with the experimental data [1]. The theoretical values lie within the range of experimental values for Kr and Xe and slightly out of them for Ar. Thus, the presented results demonstrate for the first time the quantitative agreement of theoretical and experimental BrS cross sections in electron-atomic collisions at low energies of bremsstrahlung photons.

Atom	Experiment	Theory
Ar	650 ± 260	921
Kr	1460 ± 584	1158
Xe	2300 ± 920	1725

References

[1] Gnatchenko E V, Tkachenko A A and Nechay A N 2007 *JETP Lett.* **86** 292; 2009 *Phys. Rev. A* **80** 022707
[2] Pratt R H *et al* 1977 *At. Data Nucl. Data Tables* **20** 175; 1981 Ibid **26** 477
[3] Gnatchenko E V, Tkachenko A A, Verkhovtseva E T and Zon B A 2005 *Phys. Rev. Lett.* **95** 023002
[4] Jablonski A, Salvat F and Powell C J 2004 *J. Phys. Chem. Ref. Data* **33** 409
[5] Powell C J, Jablonski A, Salvat F and Lee A Y 2016 NIST Electron elastic-scattering cross-section database, version 4.0, Natl Std. Ref. Data Series (NIST NSRDS-64)

*E-mail: zon@niif.vsu.ru

Photodetachment of C^- in the ground state

R Marion[1]*, K M Dunseath[2], M Terao-Dunseath[2] and X Urbain[1]

[1]Institute of Condensed Matter and Nanosciences, Université catholique de Louvain, B-1348 Louvain-la-Neuve, Belgium
[2]Univ Rennes, CNRS, IPR (Institut de Physique de Rennes)-UMR 6251, F-35000 Rennes, France

Synopsis Absolute total cross sections and asymmetry parameters for photodetachment of $C^-(^4S^o)$ have been measured using animated-crossed-beam and velocity map imaging techniques, for photon energies from threshold up to 6 eV. The cross sections are 15% to 25% larger than earlier experimental results above 2.2 eV. They are in very good agreement with those from a new *R*-matrix calculation using polarized pseudostates.

Negative ions, in particular those of carbon and oxygen, are of fundamental interest for the understanding of electron correlation and also play an important role in domains such as astrophysics and atmospheric physics. In the case of O^-, there was a long-standing discrepancy between theory and experiment, which was recently resolved by new measurements using an animated-crossed-beam technique to determine absolute total cross sections [1]. The new experimental cross sections are about 20% larger than the earlier values, and are in very good agreement with new theoretical results presented in the same publication. For C^-, the only experimental values near the photodetachment threshold date back more than 40 years [2, 3]. The measurements however were not absolute, and were normalized using the older O^- cross sections. Since the latter have now been superseded, it appears necessary to revise the experimental values for C^- by measuring them directly.

Here we report results for the photodetachment of C^-, determined using the same experimental techniques as in [1]. These are in good agreement with results of a new *R*-matrix calculation, which employs a basis set with polarized pseudostates to reproduce the polarizability of the carbon ground state. These give a polarizability of 11.58 a_0^3, which compares well with the value of 11.67 a_0^3 from an extensive coupled cluster calculation [4]. The electron affinity of the ground state is 1.258 eV, very close to the experimental value of 1.262 eV [5].

As shown in Figure 1, the agreement between the new theoretical and absolute experimental cross sections is very good. There is no sign of the small bump just above threshold (dashed line) present in the results of a previous *R*-matrix calculation [6]. Below 2.2 eV, the results are similar to those of earlier experiments but differences of the order of 15% to 25% appear at higher energies.

*E-mail: raphael.marion@uclouvain.be

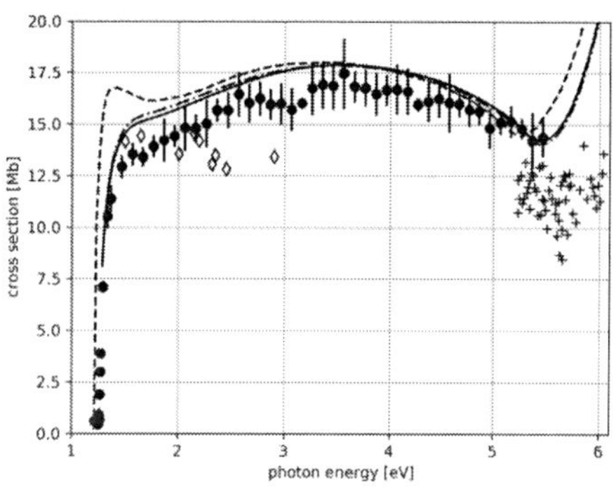

Figure 1. Experimental (circle: present work, diamonds: [2], crosses: [7]) and theoretical (solid: length form, dash-dot: velocity form) total cross sections for photodetachment of $C^-(^4S^o)$.

Results for the asymmetry parameter will also be presented at the conference. This work was supported by the Fonds de la Recherche Scientifique-FNRS through IISN grant No. 4.4504.10.

References

[1] Génévriez M *et al* 2018 *Phys. Rev. A* **98** 033410
[2] Seman M L and Branscomb L M 1962 *Phys. Rev.* **125** 1602
[3] Feldmann D 1977 *Chem. Phys. Lett.* **47** 338
[4] Das A K and Thakkar A J 1998 *J. Phys. B* **31** 2215
[5] Bresteau D *et al* 2016 *Phys. Rev. A* **93** 013414
[6] Zhou H L *et al* 2004 *Phys. Rev. A* **70** 022713
[7] Haeffler G *et al* 1996 *J. Phys. B* **29** 3017

Towards understanding the electronic structure and ion fragmentation patterns of indole and related compounds

H Sa'adeh[1, 2 *], O Plekan[2 †], A Ponzi[3], D Toffoli[4], P Bolognesi[5], J Chiarinelli[5], M C Castrovilli[5], C Vozzi[6], D Faccialà[6], M Devetta[6], A Ciavardini[7], L Avaldi[5], M Coreno[8], C Callegari[2], R Richter[2] and K C Prince[2]

[1]Department of Physics, The University of Jordan, Amman 11942, Jordan
[2]Elettra Sincrotrone Trieste, in AREA Science Park, 34149 Basovizza, Trieste, Italy
[3]Institut Ruđer Bošković, Bijenička cesta 54, 10000, Zagreb, Croatia
[4]Dipartimento di Scienze Chimiche, Università degli Studi di Trieste, Trieste, Italy
[5]CNR-Istituto di Struttura della Materia, Area della Ricerca di Roma 1, 00015 Monterotondo Scalo, Italy
[6]Istituto di Fotonica e Nanotecnologie, CNR, Piazza Leonardo da Vinci 32, 20133 Milano, Italy
[7]CERIC-ERIC, in Area Science Park, 34149 Basovizza, Trieste, Italy
[8]ISM-CNR, in Area Science Park, 34149 Basovizza, Trieste, Italy

Synopsis Electronic structure and VUV-induced fragmentation of selected molecules of the indole family have been studied using valence photoemission spectroscopy (PES), core X-ray photoelectron spectroscopy (XPS), near-edge X-ray absorption fine structure (NEXAFS) and Photo-Electron Photo-Ion COincidence (PEPICO).

Nitrogen-containing heterocyclic compounds are often termed prototypical, on the basis that a thorough understanding of their response to photoexcitation may offer a useful guide to the photochemistry and photophysics of more complex and more biologically relevant molecules. Here we are interested in indole (C_8H_7N, m=117 u) and other related molecules such as 2,3-Dihydro-7-azaindole ($C_7H_8N_2$, m=120 u) and 3-Formylindole (C_9H_7NO, m=145 u) due to their prevalence in naturally occurring and synthetic molecules displaying biological activity. For example, indole-based chromophores form the basic building blocks of the eumelanin pigments that occur frequently in nature and serve to protect organisms from the potentially damaging effects of UV light [1]. Despite the numerous studies dedicated to understanding the fundamental photophysical processes in nitrogen-containing heterocycles [2], up to now there is no experimental evidence concerning the full valence and core electronic structure of indoles under isolated conditions.

In order to obtain information about the VUV induced fragmentation patterns of indoles, PEPICO spectra were recorded (e.g. Figure 1). As well, molecular electronic structures were investigated by XPS and NEXAFS spectroscopies. Measurements were performed using the VG-TOF endstation at the GasPhase Photoemission beamline of Elettra-Sincrotrone Trieste, Italy. Results and their interpretation in the light of quantum-mechanical calculations will be presented.

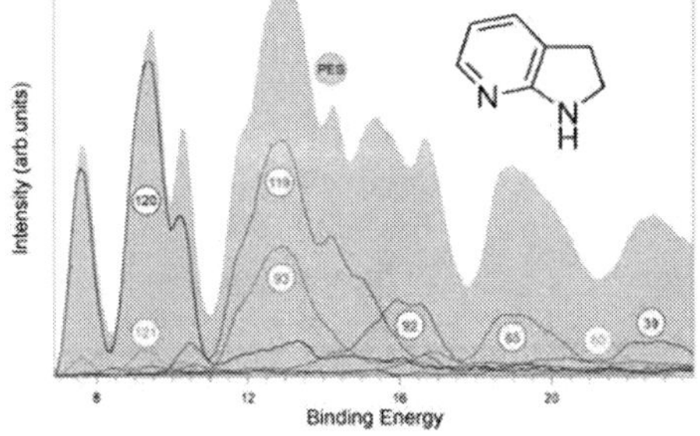

Figure 1. PES (grey full area) and PEPICO spectra (colored lines; numbers indicate masses) of main fragments of 2,3-Dihydro-7-azaindole measured at 60 eV photon energy [3].

Acknowledgements
HS acknowledges the TRIL fellowship awarded by the ICTP, Trieste, Italy, and the one-year scientific leave (2018-2019) granted by the University of Jordan, Amman, Jordan.

References
[1] Meredith P and Sarna T 2006 *Pigments Cell Res.* 19 572
[2] Marchetti B *et al* 2016 *Phys. Chem. Chem. Phys.* 18 20007
[3] Sa'adeh H *et al* in preparation

* E-mail: Hanan.Saadeh@ju.edu.jo
† E-mail: Oksana.Plekan@elettra.eu

Double photoionization of atomic carbon and neon

F L Yip[1] *, T N Rescigno[2] and C W McCurdy[2,3]

[1]Dept. of Sciences and Mathematics, California State University-Maritime Academy, Vallejo, CA 94590, USA
[2]Chemical Sciences Division, Lawrence Berkeley National Laboratory, Berkeley, CA 94720, USA
[3]Department of Chemistry, University of California-Davis, Davis, CA 95616, USA

Synopsis We consider the double photoionization of the $2p^2$ valence electrons of atomic carbon, which provides for distinct final-state symmetries depending on the three possible angular momentum couplings (^3P, ^1D, and ^1S) of the initially-bound p^2 electrons that are ejected into the continuum by a single photon. Comparison of this process with neon provides an analogous case for the resulting final states within the treatment of the double photoionization proceeding with the ejected electrons influenced by the remaining bound electrons.

Double photoionization (DPI) events provide a sensitive probe of electron correlation and focus on few-electron targets continues to reveal the consequences of electron correlation for atoms and molecules that possess several electrons. Results previously calculated for neon and argon [1] agree well with experimental measurements [2, 3] for various final LS states. We apply the same methodology to compute the DPI cross section from atomic carbon. Removal of the valence p^2 electrons yields similar final state continua connected by a single photon. We focus our comparisons with neon in the ^3P symmetry.

To represent the frozen-core $1s$ and $2s$ electrons in carbon, a basis of orbitals is constructed from an FEM-DVR grid [4] which allows the evaluation of the Coulomb and exchange interactions with the core electrons. Using this orbital-DVR grid, we solve the first-order driven Schrodinger equation in an expansion appropriate for the atomic symmetry (coupled spherical harmonics).

Good agreement has been found for the total and energy sharing cross section with previous results for the total double photoionization cross section and the largest two contributions to the energy sharing (single) differential cross section of ^3P carbon [5]. Comparing the triple differential cross section (TDCS) relative sizes and features of the TDCS lobes for carbon and neon (Figure 1) has also been done for excess energies that are the same percentage of the DPI threshold. Future work will examine the cross sections of these targets in other symmetries.

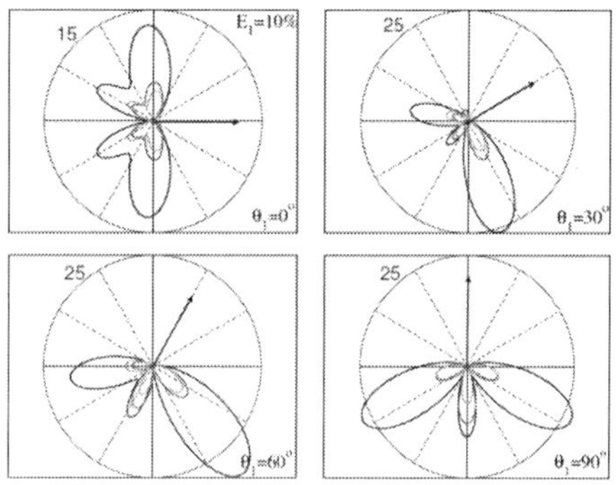

Figure 1. Triple differential cross sections (TDCS) for ^3P carbon (black curve) and neon (red and green curves) calculated in the velocity gauge. The carbon results and red neon curves are calculated at 10 eV excess photon energy (= 47.2 eV and = 76.0 eV, respectively), while the green neon result has the same percentage of excess energy in the fixed electron (= 83.7 eV) as the carbon black curve.

References

[1] Yip F L *et al* 2013 *Phys. Rev. Lett.* **110** 173001
[2] Krassig B *et al* 1996 *J. Phys. B* **29** 4255
[3] Bolognesi P *et al* 2004 *J. Phys. B* **37** 2285
[4] Yip F L *et al* 2010 *Phys. Rev. A* **81** 053407
[5] Carter S L and Kelly H P *et al* 1976 *J. Phys. B* **9** 1887

*E-mail: fyip@csum.edu

Benchmark calculations for photoionization of neutral iron

O Zatsarinny[1,*], K Bartschat[1], L Fernandez-Menchero[1] and S S Tayal[2]

[1]Department of Physics and Astronomy, Drake University, Des Moines, Iowa 50311, USA
[2]Department of Physics, Clark Atlanta University, Atlanta, GA 30314, USA

Synopsis The B-spline R-matrix method is used to investigate the photoionization of neutral iron from its ground and the lowest 43 excited states in the energy region from the ionization thresholds to 2 Ry. The sensitivity of the predictions is checked by comparing results obtained in different approximations. Including all terms from the $3d^64p$ and $3d^54s4p$ configurations considerably changes both the low-energy resonance structure and the background cross sections.

The enormous importance of iron-peak elements for astronomical observations is well known. Reliable photoionization cross sections for neutral iron, for example, are a requirement for accurate chemical abundances in late-type stars. The cross sections for ionization processes from both the ground and low-lying excited states are usually the minimum requirement for detailed nonlocal thermodynamic equilibrium modeling. Over several decades, calculations of photoionization cross sections for Fe I were carried out in a variety of approximations, with increasing size and sophistication. Presently, however, both electron-impact excitation rates and photoionization cross sections for Fe still represent a significant source of uncertainty in the models.

The purpose of the present work is to perform more elaborate and extensive calculations for photoionization of Fe I than what is currently available and, thereby, check the convergence of the predicted cross sections. Our improvements include both principal aspects of the collision calculations, namely the accuracy of the target states and the size of the close-coupling expansion.

The calculations were carried out with the B-spline R-matrix (BSR) code [1], where a B-spline basis is employed to represent the one-electron orbitals. Within this method, nonorthogonal sets of term-dependent orbitals are extensively used in both the target description and the representation of the scattering functions. That, in connection with multiconfiguration target expansions, provides a systematic way to account for a variety of correlation and relaxation effects. This feature was illustrated in detail in our recent calculations for electron collisions with Fe I [2] and Fe II [3], where the flexibility of the code allowed us to generate a target description of unprecedented accuracy for collision calculations.

The present close-coupling expansion contains all terms of the $3d^64s$, $3d^54s^2$, $3d^7$, $3d^64p$, and $3d^54s4p$ configurations of Fe II. This set of final target states covers all major channels in photo-ionization of neutral iron and thus provides a basis for approaching convergence in the predicted cross sections. The present photoionization cross sections exhibit significant differences compared to earlier results, in particular with the most recent R-matrix calculations of Bautista *et al.* [4]. We argue what none of the previous calculations can be considered converged due to the omissions of important states in the residual-ion expansions. For example, photoionization of the $3d^64s4p$ states of Fe I can occur in the one-electron approximation through ionization of the 3d, 4s, or 4p electron, respectively, thus leading to the $3d^54s4p$, $3d^64p$, and $3d^64s$ final ionic states of Fe II. To obtain converged results, all these channels should be considered on equal footing. In particular, we found that 3d-ionization becomes the dominant channel for higher energies. This leads to approximately the same value (~10 Mb) of the photoionization cross sections for all Fe I states.

We obtained total and partial photoionization cross sections for the first 44 bound states of Fe I. The photoionization of neutral iron exhibits numerous scattering channels. We performed a detailed analysis of these channels, showing that the relative population of the different ionic states changes considerably when varying the photon energy.

This work was supported by the United States National Science Foundation under grants PHY-1803844 and PHY-1520970, and the XSEDE allocation PHY-090031.

References

[1] Zatsarinny O 2006 *Comp. Phys. Commun.* **174** 273
[2] Wang K *et al* 2018 *ApJ* **867** 63
[3] Tayal S S and Zatsarinny O 2018 *Phys. Rev. A* **98** 012706
[4] Bautista M A *et al* 2017 *A&A* **606** A127

*E-mail: oleg.zatsarinny@drake.edu

Photoresponse studies of Ar and K^+ confined in neutral and charged Fullerenes

A Thuppilakkadan* **and H R Varma**†

School of Basic Sciences, Indian Institute of Technology Mandi, Kamand, H.P. 175005, India

Synopsis The present work we study the photoresponse of isoelectronic systems (IES) such as Ar and K^+ when they are confined inside neutral and charged fullerenes. The nature of the external cage (charged or neutral) play a significant role in determining the photoionization parameters along these isoelectronic members.

Atom encapsulated inside a fullerene ($A@C_{60}$) is a topic of significant interest across scientific disciplines because of its interesting physical and chemical properties. Response of an atom trapped inside a fullerene to ionizing radiation will differ from that of free atom. One of the interesting feature of $A@C_{60}$ is the confinement resonances [1] which appear in the photoionization spectrum. This arises because of the interference between the outgoing wave and the reflected photoelectron waves from the cage [2].

Here we report the theoretical photoionization studies of Ar and K^+ trapped inside neutral and charged fullerenes ($Ar@C_{60}$, $Ar@C_{60}^{-1}$, $K^+@C_{60}$ and $K^+@C_{60}^{-1}$). Our aim is to study the role of electronic correlation and relativistic effects on confinement resonances and its subsequent effects on the photoionization spectrum along the IES. The neutral and charged fullerenes are modeled by spherically attractive potential as reported in the literature [3, 4]. Well known relativistic random phase approximation [5] is used to obtain photoionization dynamics.

Figure 1 shows the 1s cross section of $Ar@C_{60}^q$ and $K^+@C_{60}^q$ (q=0 and -1). In all cases, confinement resonances are seen. Confinement enhances cross section at the threshold in the case of $Ar@C_{60}$ as compared to free Ar (figure not included), whereas the cross section is decreased at the threshold in $K^+@C_{60}$ as compared to free K^+ (figure not included). The situation changes when the neutral fullerene is replaced by a charged fullerene. Here the presence of negatively charged fullerene decreases the cross section in both cases. However the impact of the negative charge is different. In $Ar@C_{60}^{-1}$ cross section is almost zero and it then raises to a sharp resonance structure whereas in $Ar@C_{60}$ has its maximum value at the threshold. The sharp resonance appearing in $Ar@C_{60}^{-1}$ is called Coulomb confinement resonance (CCRs) [3] and its origin is attributed to the Coulomb potential barrier of the of charged shell.

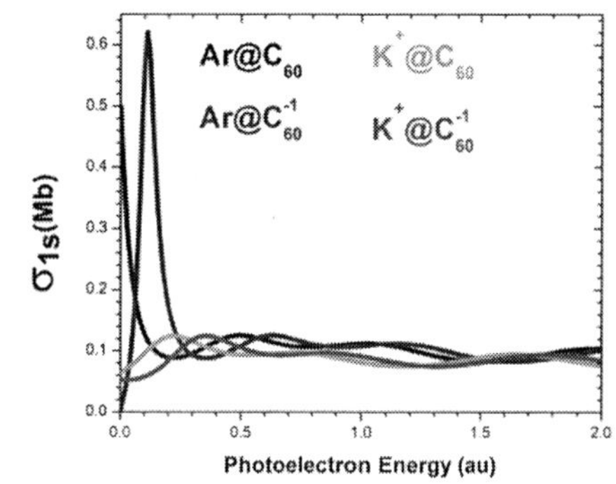

Figure 1. 1s photoionization cross section of $Ar@C_{60}$ (black), $Ar@C_{60}^{-1}$(blue), $K^+@C_{60}$ (green) and $K^+@C_{60}^{-1}$(red).

The photoionization spectrum of $K^+@C_{60}^{-1}$, however, is free from such resonances although the confinement oscillations are phase shifted. Detailed analysis of this will be carried out by studying the various photoionization parameters and also considering different type of systems and will be presented at the conference. This work is partially supported by the DST-SERB (India) grant EMR/2016/002695.

References

[1] Varma H R *et al* 2007 *Phys. Rev. A* **76** 012711
[2] Deshmukh P C *et al* 2014 *Phys. Rev. A* **89** 053424
[3] Kumar A *et al* 2017 *Phys. Rev. A* **94** 043401
[4] Dolmatov V K *et al* 2006 *Phys. Rev. A* **73** 013201
[5] Johnson W R and Lin C D 1979 *Phys. Rev. A* **20** 964

*E-mail: afsal_t@students.iitmandi.ac.in
†E-mail: hari@iitmandi.ac.in

Non-adiabatic transitions in the near-UV photodissociation of H_3^+

X Urbain[1]*, A Dochain[1], R Marion[1], T Launoy[1,2] and J Loreau[1,2]

[1]Institute of Condensed Matter and Nanosciences, Université catholique de Louvain, chemin du cyclotron 2, B-1348 Louvain-la-Neuve, Belgium
[2]Laboratoire de Chimie Quantique et Photophysique, Université libre de Bruxelles, Av. F. Roosevelt 50, B-1050 Brussels, Belgium

Synopsis The near-UV photodissociation of H_3^+ has been studied in a fast beam experiment. The kinetic energy release and branching ratio among the fragmentation channels were obtained by three-dimensional imaging of the dissociation products. It is found that the photodissociation of vibrationally excited H_3^+ mostly produces vibrationally cold H_2^+ and hot H_2. The non-adiabatic repopulation of the ground potential energy surface after excitation is accounted for by modelling the wavepacket dynamics on the intersecting potential energy surfaces.

The molecular ion H_3^+ is observed in many astrophysical environments [1], playing a central role in the insterstellar medium for the formation of hydrides and as a tracer of particle precipitation in the upper atmosphere of giant planets. H_3^+ is also a key species to understand deuterium fractionation.

The ground and first excited $^1A'$ potential energy surfaces of H_3^+ exhibit an avoided crossing seam at large $H_2 - H^+$ separation [2]. While the charge transfer reactions $H_2 + H^+ \rightleftharpoons H_2^+ + H$ probe this crossing seam in a full collision, the photodissociation of H_3^+ is actually probing it from within, as fragments depart from the classical turning point accessed via a vertical transition from the ground state potential well.

In this work, we consider the electronic excitation of the H_3^+ ground state to the first excited $^1A'$ state by UV photons of 3 to 5 eV, that triggers rapid dissociation into $H_2 + H^+$ and $H_2^+ + H$ products at wavelengths from 266 nm to 410 nm. Such wavelengths mainly probe the photodissociation from highly excited vibrational states of H_3^+ that are populated following its formation in the ion source plasma. This observation is corroborated by the storage ring experiments of Petrignani et al [3] that displayed a rapid decay of the photodissociation signal vs. storage time.

An analysis of the kinetic energy release of both channels was performed, showing a dominant contribution from the $H_2^+ + H$ channel with H_2^+ in the ground vibrational state. On the other hand, a strong vibrational excitation was observed for the H_2 fragments.

These findings were rationalised by performing time-dependent wavepacket simulations of the photodissociation process, demonstrating the impact of the non-adiabatic seam on the dynamics [4].

This work was supported by the Fonds de la Recherche Scientifique-FNRS through IISN grant No. 4.4504.10.

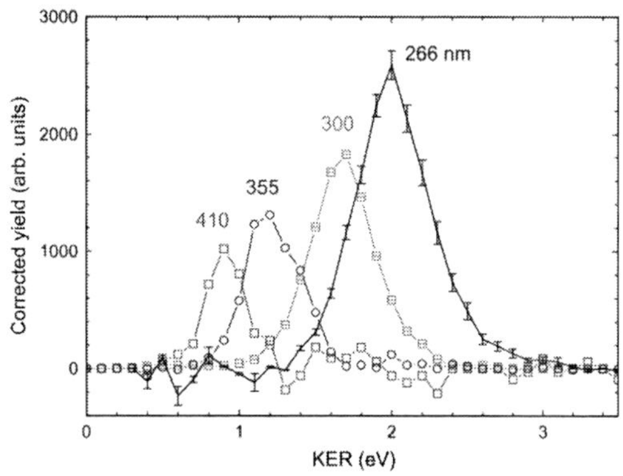

Figure 1. Measured kinetic energy release distributions for the $H_2^+ + H$ channel at various wavelengths. Integrated counts are corrected for the geometrical acceptance of the detection system and the contribution of rest gas collisions.

References

[1] Oka T 2013 *Chem. Rev.* **113** 8738-8761
[2] Preston R K and Tully J C 1971 *J. Chem. Phys.* **54** 4297
[3] Petrignani A *et al* 2010 *J. Phys. Chem. A* **114** 4864
[4] Urbain X *et al* 2019 submitted for publication in *Phil. Trans. R. Soc. A*

*E-mail: xavier.urbain@uclouvain.be

An experimental and theoretical investigation of XPS and NEXAFS of nicotine, nicotinamide, and nicotinc acid

H Sa'adeh[1, 2 *], F Backler[3], F Wang[3], A Ciavardini[4], S Piccirillo[5], M Coreno[6], R Richter[2] and K C Prince[2 †]

[1]Department of Physics, The University of Jordan, Amman 11942, Jordan
[2]Elettra Sincrotrone Trieste, in Area Science Park, 34149 Basovizza, Trieste, Italy
[3]Digital Chemistry Laboratory, Swinburne University of Technology, Melbourne, Australia
[4]CERIC-ERIC, in Area Science Park, 34149 Basovizza, Trieste, Italy
[5]Dipartimento di Scienze e Tecnologie Chimiche, Università di Roma, Rome, Italy
[6]ISM-CNR, in Area Science Park, 34149 Basovizza, Trieste, Italy

Synopsis The electronic structures of nicotine, nicotinic acid and nicotinamide have been studied by valence photoemission spectroscopy (PES), core X-ray photoelectron spectroscopy (XPS), and near-edge X-ray absorption fine structure (NEXAFS) and interpreted with the aid of quantum chemical calculations. Nicotinamide and nicotinic acid are closely related and show correspondingly similar spectral features, while nicotine is both structurally and spectroscopically diverse.

We report results on three biologically important derivatives of pyridine: nicotine ($C_{10}H_{14}N_2$, m=162 u), nicotinic acid ($C_6H_5NO_2$, m=123 u), and nicotinamide ($C_6H_6N_2O$, m=122 u). They are pharmacologically important chemicals for their psycho-active properties (nicotine), and as vitamins (nicotinic acid, vitamin B3, and nicotinamide, provitamin B3)

The ionization potentials and outermost valence ionic states of nicotinic acid and nicotinamide [1, 2] as well as calculations of the valence bands [2, 3] have been reported. No gas phase core level spectra are available so far and we report them here.

We have measured valence band (VB) photoelectron spectra (Figure 1), as well as core level carbon, nitrogen, and oxygen 1s X-ray photoemission and near edge X-ray absorption fine structure spectra. In addition, we have carried out quantum chemical calculations using the OVGF model for the valence bands and Core Electron Binding Energy model for the core levels [4]. XPS probes the chemical state of the molecule because the binding energy shifts depend on the chemical environment of the atom that is ionized, while VB spectra and NEXAFS provide complementary information about the occupied and unoccupied valence states. Measurements were performed at the GasPhase Photoemission beamline of Elettra-Sincrotrone Trieste, Italy. Results and their interpretation in the light of our quantum chemical calculations will be presented.

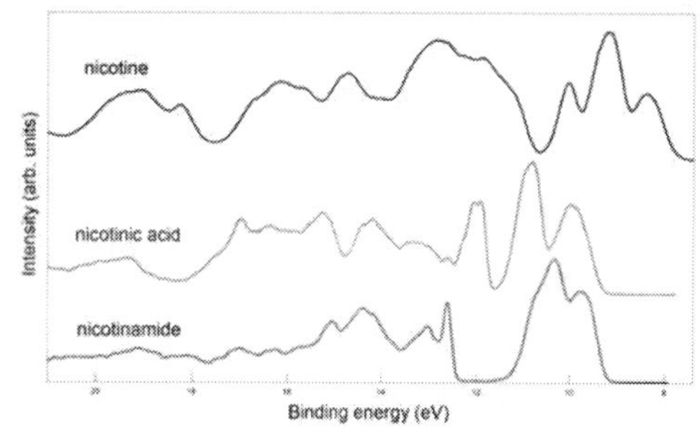

Figure 1. Valence band photoelectron spectra of nicotine, nicotinic acid, and nicotinamide measured at 100 eV photon energy [5].

Acknowledgements
HS acknowledges the TRIL fellowship awarded by the ICTP, Trieste, Italy, and the one-year scientific leave (2018-2019) granted by the University of Jordan, Amman, Jordan. FB acknowledges Research Training Scheme Program (RTSP) of Australian Governement.

References
[1] Dougherty D *et al* 1978 *J. Electron Spectrosc. Relat. Phenom.* **13** 379
[2] Opitz J 2007 *Int. J. Mass. Spectrom.* **265** 1
[3] Farrokhpour H and Ghandehari M 2013 *J. Phys. Chem. B* **117** 6027
[4] Islam S *et al* 2018 *J. Chem. Phys.* **149** 134312
[5] Sa'adeh H *et al* in preparation.

* E-mail: Hanan.Saadeh@ju.edu.jo
† E-mail: Prince@elettra.eu

Journal of Physics: Conference Series **1412** (2020) 102009 doi:10.1088/1742-6596/1412/10/102009

Effects of exchange-correlation functionals on the structure and the photoionization dynamics of Na_{40} versus Na_{92} cluster

R Shaik[1], H R Varma[1]* and H S Chakraborty[2]†

[1]School of Basic Sciences, IIT Mandi, Himachal Pradesh, 175005, India
[2]Department of Natural Sciences, D. L. Hubbard Center for Innovation, Northwest Missouri State University, Maryville, Missouri 64468, USA

Synopsis We study the photoionization properties of the Na_n (n=40 and 92) clusters in a spherical jellium frame using the Kohn-Sham density functional method. Two well known form of exchange-correlation (xc) functional in the framework of local density approximation (LDA) with Gunnarsson-Lundqvist parametrization[1] are employed: i) the electron self-interaction correction (LDA-SIC) [1] and ii) the van Leeuwen and Baerends model potential (LDA-LB94) [1].

The structure and photoionization dynamics of Na_n cluster are investigated using density functional theory. The primary objective of the current work is to study the impact of the form of xc in determining its properties by employing LDA-LB94 and LDA-SIC. To the best of our knowledge, not many work have been reported comparing these two schemes.

The real ionic core potential with tightly bound core electrons is replaced by a constant positive background of Na^{10+} ions smeared into a spherical jellium. The Kohn-Sham equation for the delocalized valence electrons ($3s^1$) from each Na atom is solved to obtain the ground state structure of the Na_n in the LDA. The unphysical electron self-interactions in LDA are corrected by using the above mentioned schemes. LDA-LB94 scheme being a prototype of the gradient-corrected class of approximations naturally produces the correct asymptotic behaviour. It is expected to be more realistic than LDA-SIC which removes self-interactions by orbit-by-orbit elimination resulting state-dependent potentials [1].

A time-dependent LDA (TDLDA) with the above two xc forms are used to study the photo response of Na_n clusters. In this method, the photoioinzation cross section is calculated by

$$\sigma_{PI}(\omega) = \sum_{nl} 2(2l+1)\left|\left\langle kl' \middle| \delta V(\vec{r}, \omega) \middle| nl \right\rangle\right|^2$$

where δV is self-consistent field potential including both the dipole interaction term (z), and the important electron correlation term. Here, nl denotes the occupied state and kl' are dipole-selected continuum.

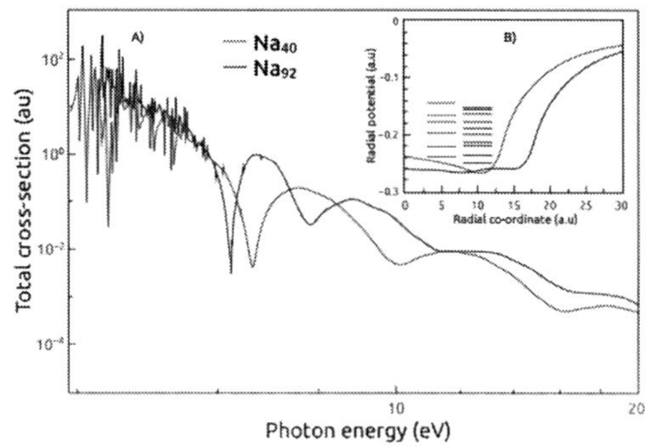

Figure 1. A) Total photoionization cross section B) Radial potentials and energy levels of Na_{92} and Na_{40} using LB94 scheme.

In Fig. 1 is shown the σ_{PI} for Na_{40} and Na_{92} along with the radial potentials and the energy levels (inset figure). The narrow structures at lower energies are the single electron excitations and the diffractive type broad oscillations are found at higher energies in both cases. It is to be noted that the oscillations shortened with increasing size of the cluster. These profiles will be compared with corresponding LDA-SIC results and will be presented in the conference. This work is partially supported by the NSF-USA grant PHY-1806206.

References

[1] Choi J *et al* 2017 *Phys. Rev. A* **95** 023404 and references therein.

*E-mail: hari@iitmandi.ac.in
†E-mail: himadri@nwmissouri.edu

An investigation of photo-double ionization in benzene

J Howard[1], E Sokell[1]*, P Bolognesi[2] and L Avaldi[2]

[1] School of Physics, University College Dublin Science Centre, Belfield, Dublin 4, Ireland
[2] CNR-Istituto di Struttura della Materia, Area della Ricerca di Roma 1, 00015 Monterotondo Scalo, Italy

Synopsis A photo-double ionization (PDI) coincidence technique has been used to study electron correlation as a function of photon impact energy on Benzene (C_6H_6). The simultaneous emission of two photoelectrons from the benzene π orbital has been measured using two sets of hemispherical electron energy analysers. Angle resolved single and coincidence photoelectron spectra are reported, providing new data on the PDI mechanism.

Benzene is a prime example of an aromatic hydrocarbon, a molecule with a planar arrangement where the π molecular orbitals overlap and the electrons are completely delocalised. Aromatic hydrocarbons have shown some interesting properties. like superconductivity at high temperatures [1] and excellent thermal properties. Hydrocarbons play a central role in biological process, astronomical process and combustion.

This cross-beam experiment was completed on the GasPhase Beamline at Elettra. The liquid benzene was placed in a test tube outside the vacuum chamber, freeze-pump-thawed to remove gases, and then admitted to the interaction region via a gas line and a gentle heating . The fully linearly polarised photon beam is delivered in a spot of about 200 μm diameter to the interaction region.

The 10 hemispherical electron energy analysers hosted in the multi-coincidence experimental chamber are mounted on two independent frames, holding respectively 3 and 7 analysers separated by 30 degrees of each other. In the present experiments, the two frames were placed in the perpendicular plane and used to detect electron coincidences from the PDI process. A detailed description of the apparatus and the related methods of the acquisition of the triple differential cross section (TDCS) and its analysis are described in [2].

At a resonance photon energy, it has been proposed [3] that, benzene allows the formation of a two-electron pseudo-particle, which has been likened to a 'Cooper pair'. This is only possible when the C-C distance matches the de Broglie wavelength. In this model the electron pair leaving the molecule breaks up and the electrons are emitted back-to-back. Previous results [4] have shown small enhancements in the single electron photoelectron spectra at angles of 70 and 138 degrees. However, other experiments [5] do not support this interpretation and the mechanism responsible for the enhancement is an open question. The present results, which are differential in angle and energy for each of the two electrons, provide new information on the PDI process.

An example of the measured TDCS is shown in figure 1. This TDCS, obtained at the resonance photon energy at which the enhancements were observed previously (~65 eV) [4], exhibits a difference from the equivalent He TDCS [6], which has been scaled to match the data.

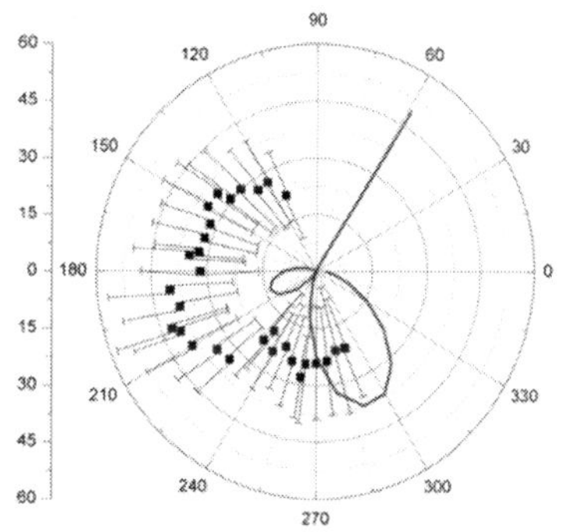

Figure 1. TDCS recorded at 65 eV (ke_1=ke_2=20 eV) (points) and equivalent He TDCS (solid red line). The arrow indicates the position of the fixed analyser.

References

[1] Rosseinsky M J and Prassides K 2010 *Nature* 464 39
[2] Bolognesi P *et al* 2004 *J. El. Spec. Rel. Phe.* **141** 105
[3] Wehlitz R 2016 *J. Phys. B.* 49 222004
[4] Wehlitz R *et al* 2012 *Phys. Rev. Lett.* **109** 193001
[5] Jänkälä K *et al* 2014 *Phys. Rev. Lett.* **112** 143005
[6] Kheifets A S *et al* 2000 *Phys. Rev. A* **62** 065402

* E-mail: emma.sokell@ucd.ie

Phase sensitive coherent control excitation schemes applied to photoionization microscopy of the hydrogen atom

P Kalaitzis[1 *], **S Danakas**[1], **K Ferentinou**[1], **F Lépine**[2], **C Bordas**[2] **and S Cohen**[1 †]

[1]Atomic and Molecular Physics Laboratory, Physics Department, University of Ioannina, 45110 Ioannina, Greece
[2]Institut Lumière Matière, Université Lyon 1, CNRS, UMR 5306, 10 rue Ada Byron 69622 Villeurbanne Cedex, France

Synopsis Photoionization microscopy (PM) is employed for imaging the flux of slow (meV) electrons ejected via atomic photoionization in the presence of a static electric field. Just above the ionization limit continuum and quasi-disrete states are degenerate. In hydrogen atom resonant excitation dominates over the continuum one, while the reverse holds for more complex atoms. We demonstrate theoretically that by using a bichromatic $\omega/2\omega$ laser field and by applying phase sensitive coherent control techniques it is possible to unbury and isolate otherwise obscured resonant features in PM images.

Photoionization microscopy (PM) is devoted to the measurement of the electron current probability density j in the case of photoionization of an atom in the presence of a static electric field of strength F. The spatial quantum interference structures exhibited by j are visible due to the low (meV) energy of the outgoing electrons. The PM studies are performed above the saddle point energy $E_{sp}=-2\sqrt{F}$ au. For $E_{sp}\leq E\leq 0$ quasi-bound Stark states (resonances) and continuum ones are degenerate. One of the goals of PM is to provide access to interferences stemming from the resonances which reflect intrinsic properties of the atomic system under study. In hydrogen atom the excitation of resonances is typically much stronger than the continuum one. Consequently, on-resonance PM images show specific resonant features, such as an abrupt increase of the number of fringes [1].

The Stark states of non-hydrogenic atoms are expressed as a mixture of quasi-bound and continuum hydrogenic states [2]. As a result, resonances and continua show comparable excitation strengths and the resonant image features are obscured. Here we show that, under certain conditions, it is possible to uncover the resonant signatures by applying one- and two-photon phase sensitive coherent control [3] over the excitation of the final Stark states. Then, the total excitation amplitude for each final state out of a common initial one is given by:

$$d_{tot} = E_{2\omega}\exp(\mathbf{i}\varphi_{2\omega})D^{(1)} + E_{\omega}^{2}\exp(\mathbf{i}2\varphi_{\omega})D^{(2)},$$

with $E_{2\omega}/\varphi_{2\omega}$ and $E_{\omega}/\varphi_{\omega}$ being the laser field amplitudes/phases and $D^{(1)}$ and $D^{(2)}$ one and two-photon excitation dipole matrix elements, respectively. Figure 1 shows the results of a theoretical hydrogenic calculation.

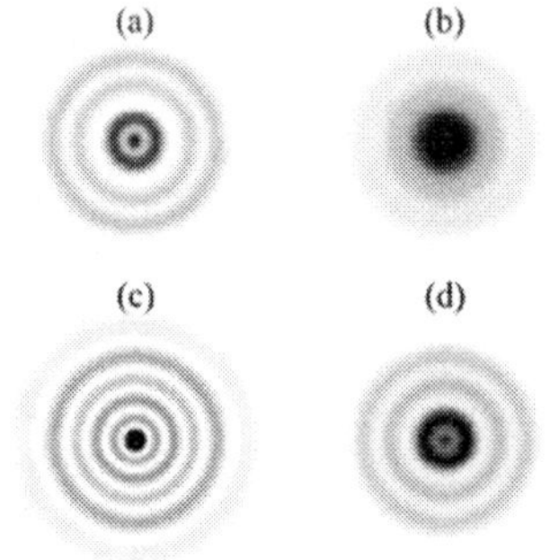

Figure 1. Hydrogenic PM images for ground state excitation and for F=680 V/cm and $E/|E_{sp}|$=-0.8943 (i.e. on the red size of the (n_1=5, m=0) shape resonance). (a) Single-photon (2ω) excitation only, (b) Two-photon (ω) excitation only, (c) Bichromatic $\omega/2\omega$ exctitation, $\Delta\varphi\equiv\varphi_{2\omega}-2\varphi_{\omega}=0$ and laser intensities optimized for bringing out the details of the resonance, (d) Same as in (c) but with $\Delta\varphi=\pi$.

With either the ω or the 2ω laser field alone the images fail to exhibit the five dark fringes of a weak resonance. This is achieved solely by applying both fields, setting the appropriate difference $\Delta\varphi\equiv\varphi_{2\omega}-2\varphi_{\omega}$ (here $\Delta\varphi=0$) and chosing appropriate values for $E_{2\omega}$ and E_{ω}. It turns out that the method appears to be quite efficient and the continuum excitation can be efficiently suppressed when the number of continua is small, i.e. just above E_{sp}.

References

[1] Stodolna A S *et al* 2013 *Phys. Rev. Lett.* **110** 213001
[2] Nicole C *et al* 2002 *Phys. Rev. Lett.* **88** 133001
[3] Shapiro M and Brumer P 2000 *Adv. At. Mol. Opt. Phys.* **42** 287

* E-mail: pkalaitz@cc.uoi.gr
† E-mail: scohen@uoi.gr

X-ray spectroscopy on ultrafast-decaying core-excited atomic ions

M Ilchen[1,2], T Mazza[1], T M Baumann[1], R Boll[1], A De Fanis[1], P Grychtol[1], V Music[1,2], Y Ovcharenko[1], D Rivas[1], N Berrah[3], B Erk[4], E Gryzlova[5], P Johnsson[6], C Küstner-Wetekam[2], L Marder[2], M Martins[7], Ch Ott[8], S Pathak[9], Th Pfeifer[8], D Rolles[9], P Schmidt[1,2], A Grum-Grzhimailo[5] and M Meyer[1*]

[1] European X-Ray Free-Electron Laser Facility GmbH, Holzkoppel 4, 22869 Schenefeld, Germany
[2]University of Kassel, Institute of Physics, Heinrich-Plett-Str. 40, 34132 Kassel, Germany
[3]University of Connecticut, Physics Department, Unit 3046 Storrs, CT 06269-3046, United States
[4]Deutsches Elektronen-Synchrotron DESY, Notkestrasse 85, 22671 Hamburg, Germany
[5]Lomonosov Moscow State University, Skobeltsyn Institute of Nuclear Physics, 119991 Moscow, Russia
[6]Lund University, Department of Physics, Box 117, 221 00 Lund, Sweden
[7]University of Hamburg, Institute of Experimental Physics, Luruper Chaussee 149, 22761 Hamburg, Germany
[8]MPIK, Quantum Dynamics & Control, Saupfercheckweg 1, 69117 Heidelberg, Germany
[9]Kansas State University, Dept. of Physics, 116 Cardwell Hall Manhattan, KS 66506-2601, United States

Synopsis Results from the first soft X-ray user experiment at the European XFEL on nonlinear photon-matter interaction will be presented. Angle-resolved electron time-of-flight spectroscopy employed at the AQS (Atomic-like Quantum Systems) endstation of the SQS (Small Quantum Systems) instrument reveals insight into the character of resonances in highly transient, core ionized neon ions, i.e. Ne^+ $1s^12s^22p^6 \rightarrow Ne^{+*}$ $1s^02s^22p^6np$, together with their respective relaxation dynamics. Enabled by the unique properties of the European XFEL, novel perspectives on efficient nonlinear spectroscopy will be discussed.

Highly intense X-ray pulses from X-ray free-electron lasers (XFELs) enable the study of nonlinear, photo-induced processes in atoms and molecules at short wavelengths. This includes sequential and nonsequential multiphoton ionization and excitation processes. Sequential ionization of core electrons is in strong competition with ultrafast relaxation via Auger emission [1]. At high X-ray irradiation, more than one photon can be absorbed before the Auger decay happens, i.e. typically within few femtoseconds. Consequently, the yield of this sequential photo-absorption by core electrons and accordingly the population of transient, highly excited ions increase significantly. Therefore, an efficient characterization of these short-lived ions becomes spectroscopically feasible. By using the intense pulses from the SASE3 soft X-ray undulators at European XFEL, we could, for the first time, study the core-hole excitation of transient, highly excited atoms by means of resonant Auger electron spectroscopy employing the Small Quantum Systems (SQS) instrument. Here, neon was ionized at the *1s* shell and, depending on the photon energy, either sequentially ionized to Ne^{2+} $1s^02s^22p^6$ or excited to different Ne^{+*} $1s^02s^22p^6np$ states within the same fs X-ray pulse. Thanks to the routinely available tunability of the SASE3 undulator, it was possible to scan the photon energy across the $1s^12s^22p^6 \rightarrow 1s^02s^22p^6np$ resonances with photon energy steps of 1 eV (much smaller than the bandwidth). The resulting Auger electron spectra were recorded by electron time-of-flight spectrometers. The results of these measurements are compared to previous studies of single-photon processes populating the same final states [2] as well as with results from calculations based on the Multiconfiguration Hartree-Fock (MCHF) method [3] including relativistic Breit-Pauli corrections (allowing non-orthogonal wave-functions). The results are benchmarking the performance of the SQS endstation at European XFEL and demonstrate the novel possibilities which have been opened up in particular for investigations of nonlinear phenomena and ultrafast decaying transient states in the X-ray domain.

* E-mail: michael.meyer@xfel.eu

References

[1] Young L *et al* 2010 *Nature.* **466** 56
[2] Goldsztejn G *et al* 2016 *Phys. Rev. Lett.* **117** 133001
[3] Froese Fischer C 1991 *Comput. Phys. Commun.* **64** 431

Two-color VUV+IR multiphoton ionization of the He^+ ion by circularly polarized light in the region of the 3p excited state

A N Grum-Grzhimailo[1]*, N Douguet[2] and K Bartschat[3]

[1] Skobeltsyn Institute of Nuclear Physics, Lomonosov Moscow State University, Moscow 119991, Russia
[2] Department of Physics, University of Central Florida, Orlando, Florida 32816, USA
[3] Department of Physics and Astronomy, Drake University, Des Moines, Iowa 50311, USA

Synopsis Circular dichroism in the two-color XUV + IR ionization of He^+ is considered theoretically in connection with the recent experiment by Ilchen *et al.* [1]. The sharp decrease of the circular dichroism with increasing intensity of the IR field when the XUV frequency is near the 1s – 3p resonant excitation is studied. It is confirmed that the main source of the effect is the axial component of the polarizability, which has opposite signs for magnetic substates with opposite projections.

In a recent experiment, intense circularly polarized extreme-ultraviolet (XUV) and near-infrared (NIR) laser pulses were combined to double ionize atomic helium via the oriented intermediate $He^+(3p)$ resonance state [1]. Specifically, the first XUV photon with energy $\omega_{XUV} = 48.37\,eV$ from the free-electron laser (FEL) FERMI in Trieste (Italy) ionized helium into the ionic ground state $He^+(1s)$. A second FEL photon with the same energy, in combination with a collinear NIR beam with $\omega_{NIR} = 1.58\,eV$ ($\lambda = 784\,nm$), subsequently produced He^{2+}. The intensity of the XUV pulse irradiating the He atom was estimated as $I \approx 1.0 \times 10^{13}\,W/cm^2$, while the measurements were performed at two intensities of the NIR pulse, $I_{NIR} = 7.3 \times 10^{11}\,W/cm^2$ and $I_{NIR} = 1.4 \times 10^{12}\,W/cm^2$, respectively. The energy of 48.37 eV corresponds to the $He^+(1s)$–$He^+(3p)$ transition in the field-free case. Therefore, with the above intensities of the photon beams, corresponding to the multiphoton regime, it is expected that the second ionization proceeds predominantly via the $He^+(3p)$ state with subsequent ionization by absorption of a few NIR photons. The measured circular dichroism is defined as $CD = (P_+ - P_-)/(P_+ + P_-)$, where P_+ and P_- are the angle-integrated photoemission probabilities for circularly polarized pulses with the same (+) or opposite (−) helicity, respectively.

An unexpected result of the experiment was a sharp decrease of the CD of the main photoelectron line from $0.98^{+0.02}_{-0.11}$ to $0.169^{+0.06}_{-0.10}$ when the intensity of the NIR beam increased only twice, from 7.3×10^{11} to $1.4\times10^{12}\,W/cm^2$. The experimental result reported in [1] was reproduced by TDSE calculations, but possible mechanisms for the effect remained unclear. It was suspected that the reason for the sharp decrease in the CD is a different helicity-dependent shift of the $He^+(3p, m = \pm1)$ states due to the presence of the circularly polarized NIR field: with increasing NIR intensity a magnetic substate $He^+(3p, m)$ with $m \neq 0$ moves out of resonance faster for co-rotating XUV and NIR fields than for the counter-rotating case. A prerequisite for this phenomenon is the axial component of the polarizability, which has opposite signs for the $He^+(3p, m \neq 0)$ magnetic sublevels, depending on the chirality of the NIR field.

We further investigated this subject theoretically on the basis of the corresponding quasienergy curves and confirmed that the proposed mechanism is the principal source of the fast CD decrease with increasing NIR intensity. The mechanism is illustrated by further TDSE calculations. We also explain the main features of the photoelectron resonance structures as function of the NIR intensity. Detailed results will be presented at the conference.

References

[1] Ilchen M *et al* 2017 *Phys. Rev. Lett.* **118** 013002

*E-mail: grum@sinp.msu.ru

Coherent control schemes for the photoionization of helium in the extreme ultraviolet spectral region

C-N Liu[1]* and T Morishita[2]†

[1]Department of Physics, Fu-Jen Catholic University, Taipei 24205, Taiwan
[2]Institute for Advanced Science, The University of Electro-communications, 1-5-1 Chofu-ga-oka, Chofu-shi, Tokyo 182-8585, Japan

Synopsis We present a theoretical study for coherent control based on resonant excitation to doubly excited states of helium. Bichromatic schemes are designed, using XUV light from the seeded free-electron laser FERMI, where the phase of the third harmonic relative to the fundamental can be controlled. In addition to demonstrate the interference effects, we provide detailed theoretical analyses to identify the FEL parameters, such as intensities and pulse durations, needed for experimentally observable coherent control of the photoionization yields.

Coherent control has reached a high level of sophistication in experiments with optical lasers, enabling an initial quantum state to be steered to a desired final state, but has not yet been applied with FELs. However, FEL source at FERMI, the first source of short-wavelength light possessing the full coherence of optical lasers, has the unique characteristics that the phase of the harmonics relative to the fundamental can be controlled. In addition, FERMI has the fine tunability and wavelength stability required to excite very narrow resonances. As the high intensity of the FEL opens up possibilities in nonlinear process, these characteristics of FEL at FERMI create a unique opportunity to explore nonlinear coherent control in EUV regime.

We consider two fundamental plus third harmonic schemes, based on the resonant excitation of helium atoms to doubly excited states [1]. The first one is a singly resonant scheme: the fundamental photon energy is chosen to be resonant with the doubly excited ${}_2(0,1)_2^+$ states via a three-photon process, while the third harmonic excites this state directly. The second one is a doubly resonant scheme where the $1s2p\ {}^1P^o$ state is excited resonantly by the fundamental. With further two-photon absorption, the doubly excited ${}_2(0,1)_3^+\ {}^1P^o$ state is reached.

Calculations were performed to obtain the photoelectron yields by solving the TDSE. In addition to demonstrating the interference effects, we aim to identify the optimal FEL parameters related the experimental conditions available at FERMI. The dependence of the photoionization probability from TDSE calculation on the pulse duration and intensity for both the one- and three-photon absorption processes can be used to estimated by perturbation theory. Consequently, one can find the parameters for comparable transition amplitudes of these two processes in order to achieve experimentally observable coherent control of the photoionization yields.

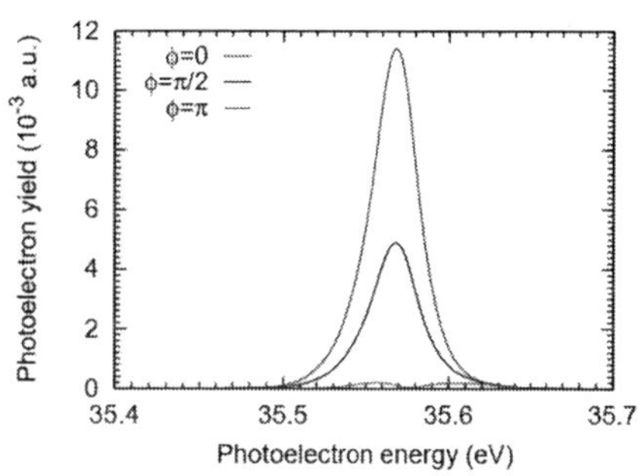

Figure 1. Calculated photoelectron spectra of He for fundamental wavelength irradiance of 10^{13} W/cm^2, third harmonic irradiance of 2×10^8 W/cm^2, pulse duration of 33.2 fs, and relative phase ϕ between the fundamental and third harmonic of 0, $\pi/2$ or π rad. Photon energies: fundamental, 20.05 eV; third harmonic, 60.15 eV.

References

[1] Giannessi L *et al* 2018 *Sci. Rep.* 8 7774

*E-mail: cnliu@mail.fju.edu.tw
†E-mail: toru@pc.uec.ac.jp

Polarization phenomena in bichromatic ionization of Ne and Ar induced by circularly polarized VUV femtosecond pulses

A N Grum-Grzhimailo[1] *, E V Gryzlova[1] and M M Popova[2]

[1]Skobeltsyn Institute of Nuclear Physics, Lomonosov Moscow State University, Moscow 119991, Russia
[2]Faculty of Physics, Lomonosov Moscow State University, Moscow 119991, Russia

Synopsis Interference effects resulting from direct single-photon and resonant two-photon ionization of neon and argon by bichromatic circularly polarized radiation were investigated in the framework of the perturbation theory. Analytical formulas are obtained for the differential cross section of the process in multielectron noble gas atoms. The possibility of coherent control of the photoelectron angular distribution and spin polarization will be discussed.

Interference phenomena resulting from two-pathway ionization of atoms and from photodissociation of molecules by photons of two different frequencies ω_1 and $\omega_2 = n\omega_1$ are well-known in optical and IR regions [1, 2]. Application of the highly coherent and stable pulses from seeded FEL FERMI [3] allows extending such studies to the VUV region [4]. Therefore there is a high demand on the corresponding theoretical considerations.

Interference phenomena induced by the fundamental frequency and its second harmonic, $\omega + 2\omega$, of linearly and circularly polarized light in the region of intermediate resonances were investigated for simple case of hydrogen in [5, 6]; the case of Ne was analyzed for linearly polarized light in [7]. The present study extends the approach to $\omega + 2\omega$ ionization of Ne and Ar by circularly polarized radiation.

The photoelectron angular distribution (PAD) induced by circularly polarized $\omega + 2\omega$ light from a noble gas can be parameterized in the form

$$\frac{dW}{d\Omega} = \frac{W_0}{4\pi}\Big(1 + \sum_{k=2,4} \beta_k P_k(\cos\theta) + |\beta_1| \sin\theta \cos(\varphi - \psi_1) + |\beta_3| \sin^3\theta \cos(\mu\varphi - \psi_3)\Big), \quad (1)$$

where $\mu = 1$ for equal helicities of ω and 2ω. For opposite helicities the β_1 parameter turns to zero and $\mu = 3$. The exact values of β_k ($k = 1, 2, 3, 4$) and $\psi_{1,3}$ depend on the particular atom, relative intensity of the harmonics and their energy. Typical PADs are shown in the figure for neon in the case of ω near the $2p^53s\ ^1P_1$ resonance (16.84 eV).

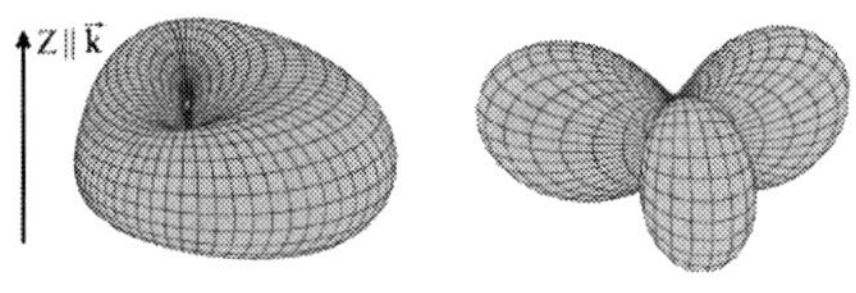

Figure 1. PADs for equal (left) and opposite (right) helicities calculated for the fundamental harmonic intensity $I = 10^{12}$ W/cm^2 and 1% admixture of the second harmonic

The general form of the PADs for different parameters of the FEL pulse will be presented at the conference, with illustrative calculations for neon and argon atoms. Furthermore, the possibility of coherent control of the photoelectron spin polarization will be discussed.

E.V.G. acknowledges support by the Foundation for the Advancement Development of Theoretical Physics and Mathematics ‘BASIS’.

References

[1] Yin Y *et al* 1992 *Phys. Rev. Lett.* **69** 2353
[2] Baranova N B *et al* 1990 *Sov. Phys. JETP* **71** 1043
[3] Allaria E *et al* 2012 *Nature Photonics* **6** 699
[4] Prince K C *et al* 2016 *Nature Photonics* **10** 176
[5] Grum-Grzhimailo A N *et al* 2015 *Phys. Rev. A* **91** 063418
[6] Douguet N *et al* 2016 *Phys. Rev. A* **91** 033402
[7] Gryzlova E V *et al* 2018 *Phys. Rev. A* **97** 013420

*E-mail: grum@sinp.msu.ru

The Small Quantum System (SQS) Instrument at European XFEL: Results of commissioning and first experiments

M Meyer[1,†], T M Baumann[1*], A Achner[1], R Boll[1], A De Fanis[1], P Grychtol[1], M Ilchen[1], T Mazza[1], J Montaño[1], V Music[1], Y Ovcharenko[1], N Rennhack[1], D E Rivas[1], R Wagner[1], P Ziołkowski[1], D Rolles[2], P Walter[6], B Erk[3], B Keitel[3], E Plönjes[3], K Mann[4], S Lange[4], B Schäfer[4], J Chalupský[5], T Burian[5], K Juranova[5], L Vyšín[5], V Hájková[5] and L Juha[5]

[1]European XFEL GmbH, Holzkoppel 4, 22869 Schenefeld, Germany
[2]J. R. Macdonald Laboratory, Department of Physics, Kansas State University, Manhattan, KS 66506, USA
[3]Deutsches Elektronen-Synchrotron (DESY), Notkestrasse 85, 22607 Hamburg, Germany
[4]Laser-Laboratorium Göttingen e.V. (LLG), Hans-Adolf-Krebs-Weg 1, 37077 Göttingen, Germany
[5]Institue of Physics, Academy of Science, Na Slovance 2, 18221 Prague 8, Czech Republic
[6]Linac Coherent Light Source, SLAC National Accelerator Laboratory, Menlo Park, CA 94025, USA

Synopsis The Small Quantum System (SQS) instrument on the soft X-ray undulator SASE3 of the European XFEL went into operation during 2018. We present first results from the instrument characterizing the focus properties and the time-of-flight as well as velocity map imaging spectrometers of the AQS chamber.

The SQS instrument is designed for experiments on atomic and molecular systems, as well as clusters, nano-particles and small bio-molecules. It is located at the SASE3 soft X-ray undulator of the European XFEL, which produces FEL radiation in the photon energy range between 250 eV and 3000 eV. The instrument enables studies of non-linear and multi-photon processes, as well as time-resolved experiments on fast dynamic processes.

The SASE3 undulator and the SQS instrument were taken into operation during 2018. The first commissioning activities were dedicated to characterize and optimize the focussing properties of the Kirkpatrick-Baez (KB) mirror system. A Hartmann-type wavefront sensor was utilized and the results compared to imprint measurements using PbI_2 coated CVD diamond substrates. The focus size was determined to be 1.3(3)x1.2(3) μm^2 at a photon energy of 1050 eV and consistently below 2x2 μm^2 for the energy range 500 - 1200 eV available at that time.

The first experiments were performed in the AQS (Atomic-like Quantum Systems) end-station of SQS, which is equipped with up to six high-resolution electron time-of-flight spectrometers as well as a Velocity Map Imaging (VMI) spectrometer. Sequential ionization of Neon and Xenon was investigated at photon energies between 900 eV and 1200 eV and pulse energies of up to 2 mJ. The measured charge states of up to Xe^{41+} are compared to similar studies at other FELs [1,2] demonstrating directly the feasibility of studies in the non-linear regime at SQS.

Furthermore, angle-resolved, high-resolution electron spectroscopy was used to study the sequential ionization of the Neon 1s core-electron and the corresponding Auger decay of the single- or double core hole (cf. [3]). During this first campaign at SQS, advantage was taken from the possibility to easily tune the photon energy of the SASE3 undulator. This enabled for the first time to study beside the direct also the resonant double core-hole Auger decay in Neon around a photon energy of 1000 eV [4].

During 2019, the other two endstations of SQS are taken into operation: the reaction microscope (SQS-REMI) for coincidence studies on electrons and ions and the NQS (Nano-sized Quantum Systems) for imaging and scattering experiments on clusters or nano-particles.

References

[1] Sorokin AA *et al* 2007 *Phys. Rev. Lett.* **99** 213002
[2] Rudek B *et al* 2012 *Nat. Phot.* **6** 858
[3] Young L *et al* 2010 *Nature* **466** 56
[4] Mazza T *et al* to be published

* E-mail: thomas.baumann@xfel.eu
† E-mail: michael.meyer@xfel.eu

Journal of Physics: Conference Series **1412** (2020) 112006 doi:10.1088/1742-6596/1412/11/112006

Attosecond delays in photoionization studied with coherent-controlled FEL

D You[1], K Ueda[1 *], O Tugs[2], Y Orimo[2], T Sato[2], K L Ishikawa[2], E V Gryzlova[3], E I Staroselskaya[3], A N Grum-Grzhimailo[3], P A Carpeggiani[4], T Csizmadia[5], M Füle[5], N G Harshitha[5], G Sansone[6], P K Maroju[6], M Meyer[7], T Mazza[7], A D'Elia[8], C Callegari[9], M Di Fraia[9], O Plekan[9], L Giannessi[9,10], E M Allaria[9], G De Ninno[9,11], M Trovò[9], L Badano[9], B Diviacco[9], D Gauthier[9], N S Mirian[9], G M Penco[9], P R Rebernik[9], S Spampinati[9], C Spezzani[9], G Gaio[9] and K C Prince[9,12]

[1] Institute of Multidisciplinary Research for Advanced Materials, Tohoku University, Sendai, 980-8577, Japan
[2] The University of Tokyo, Tokyo, 113-8656, Japan
[3] Skobeltsyn Institute of Nuclear Physics, Lomonosov Moscow State University, Moscow, 119991, Russia
[4] Technische Universität Wien, Wien, 1040, Austria
[5] ELI-ALPS, Szeged, 6728, Hungary
[6] The University of Freiburg, Freiburg, 79085, Germany
[7] European XFEL, Schenefeld, 22869, Germany
[8] The University of Trieste, Trieste, 34127, Italy
[9] Elettra Sincrotrone Trieste, Basovizza, 34149, Italy
[10] ENEA Centro Ricerche Frascati, Frascati, 00044, Italy
[11] Laboratory of Quantum Optics, University of Nova Gorica, Nova Gorica, 5001, Slovenia
[12] Molecular Model Discovery Laboratory, Swinburne University of Technology, Hawthorn, 3122, Australia

Synopsis When an electron is ejected from an atom after absorption of a photon, the photoelectron wave packet has an extremely short group delay between the photon absorption and the electron emission. This interval, called the Eisenbud-Wigner-Smith delay, is on the order of a few attoseconds. Here, we present a new method to measure the photoemission delay, using coherent-controlled free-electron laser pulses.

The age of attosecond physics was ushered in by the invention of methods for probing samples on a time scale less than femtoseconds [1], and currently, many such ultrafast phenomena are being investigated. Photoemission is one such process, which has attracted much interest, particularly regarding the photoemission delay between the absorption of a photon and the emission of a photoelectron, which is called the Eisenbud-Wigner-Smith (EWS) delay [2].

Here, we present a new method to measure the EWS delay. In this method, we use short wavelength EUV light, consisting of phase-locked fundamental (ω) and second harmonic (2ω) pulses. It requires extremely accurate phase control (few attoseconds). Such fine control is available from the Italian free-electron laser (FEL), FERMI [3]. We report the measurement of the EWS delay difference between one- and two-photon ionizations using atomic Ne, as a demonstration of the new method. We carried out the measurement at the LDM beam-line, FERMI. The photon energies were set to 14 eV, 16 eV, and 19 eV for ω, which are below the Ne ionization threshold, while the second harmonics 2ω are above the threshold. Phase-locked bichromatic light beams crossed the atomic gas jets of He-Ne mixtures (We used the He for a calibration). A velocity map imaging spectrometer measured ejected electrons. The target processes are the photoionizations of Ne by one or two photon(s): $\mathrm{Ne} + 2\omega \rightarrow \mathrm{Ne}^+ + e^-$,

$\mathrm{Ne} + \omega + \omega \rightarrow \mathrm{Ne}^+ + e^-$ (non-resonant).

Because these electrons emitted by the two different pathways interfere with each other, the electron angular distribution is correlated to the difference of the unique phase shifts. The phase shift difference can be extracted by scanning the optical phase difference between ω and 2ω FEL pulses. The EWS delay τ is defined as the derivative of phase shift η with respect to the photoelectron kinetic energy E: $\tau = \hbar \cdot \partial\eta/\partial E$ [2]. We have measured the phase shift differences at several photon energies and found their slope to estimate the delay difference, in order to demonstrate the feasibility of the method.

References

[1] Krausz F *et al* 2009 *Rev. Mod. Phys.* **81** 163
[2] Pazourek R *et al* 2015 *Rev. Mod. Phys.* **87** 765
[3] Prince K C *et al* 2016 *Nat. Photonics* **10** 176

* E-mail: kiyoshi.ueda@tohoku.ac.jp

Time-resolved dynamics of thiophene dication – probing parent molecule survival times and multi-step dissociation processes of cyclic molecules by free-electron-laser experiments combined with theoretical simulations

E Kukk[1]*, H Fukuzawa[2,7], K Nagaya[3,7], J Niskanen[1], Y Luo[2], J Peschel[4], K Kooser[1], E Itl[1], E Pelimanni[5], S Saito[2], D You[2], S Maclot[6], A Nizou[3,7], D T Ha[1], H Myllynen[1] and K Ueda[2,7]

[1]Department of Physics and Astronomy, University of Turku, FI-20014 Turku, Finland
[2]Institute of Multidisciplinary Research for Advanced Materials, Tohoku University, Sendai 980-8577, Japan
[3]Department of Physics, Kyoto University, Kyoto 606-8502, Japan
[4]Department of Physics, Division of Atomic Physics, Lund University, Box 118, SE-221 00 Lund, Sweden
[5]Nano and Molecular Systems Research Unit, Faculty of Science, 90014 University of Oulu, Finland
[6] Department of Applied Physics, KTH Royal Institute of Technology, SE-10691 Stockholm, Sweden
[7]RIKEN SPring-8 Center, Kouto 1-1-1, Sayo, Hyogo 679-5148, Japan

Synopsis Thiophene is a cyclic molecule that becomes unstable as a molecular dication. We followed the multi-step dissociation dynamics triggered by soft x-ray FEL pulses from SACLA, core-ionizing the S 2p orbital and creating parent dications by the Auger decay. The time-evolution was probed by optical laser pulsesthat reveal the survival lifetime of the parent dication and the build-up of fragmentation proceeding by ring opening and rupture, accompanied by hydrogen ejection and secondary dissociation. Experimental results are compared to statistical analysis of molecular dynamics simulations.

In a pump-probe experiment performed at the SACLA free-electron-laser facility in Japan, thiophene (C_4H_4S) molecules were core-ionized from the S $2p$ orbital, using 180 eV photons as a pump. The subsequent Auger decay produced molecular dications with internal energy up to about 50 eV. The molecular dications are, except in the lowest energy states, unstable. Molecular dynamics simulations using density functional tight-binding theory (DFTB) [1] have shown that the ensuing dynamics spans a time range from tens of femtoseconds to tens of picoseconds, depending on the internal energy available. The cyclic parent molecules can occasionally exhibit long survival times even at high energies, before the first step of fragmentation – the ring opening – occurs. The dissociation then proceeds quickly with a rupture of another bond, creating two charged fragments. In addition, hydrogen ejection and secondary dissociation take place over the course this complex dynamics. Near-infrared (NIR) 800 nm Ti-Sapphire laser pulses were used to probe the status of the molecular dynamics with pump-probe delays in the femto- and picosecond scale. The optical laser pulses served to alter the dissociation pathways. In particular, a strong time-dependent enhancement of secondary dissociation channels was seen with characteristic rise- and delay times. These, an other time-dependent effects are used to obtain insight into the kinematics and dynamics of multi-step dissociation processes of cyclic molecules.

We interpret the experimental results with the help of a statistical analysis of trajectories from *ab initio* molecular dynamics simulations, carried out over a broad range of internal temperatures (as atomic velocities).

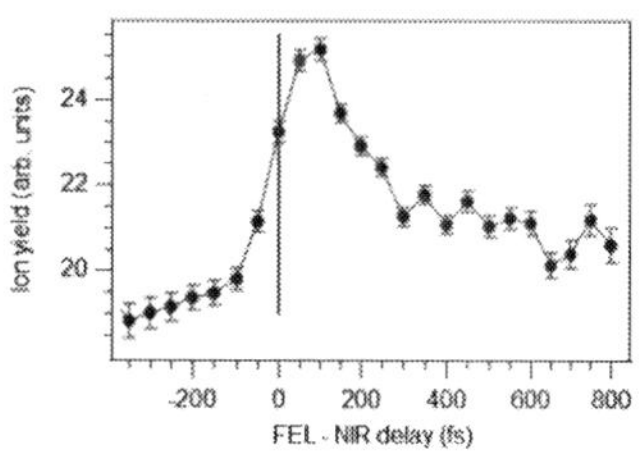

Figure 1. NIR-induced enhancement of the C^+ ion yield from secondary fragmentation.

References

[1] Kukk E *et al* 2015 *Phys. Rev. A* **91** 43417

*E-mail: edwin.kukk@utu.fi

Ultrafast ionization and fragmentation dynamics of polycyclic atomatic hydrocarbons by XUV radiation

B Manschwetus[1,*], P Chopra[1,2,8], D Tikhonov[1,2,8], H Köckert[5], S Gruet[1,2,8], J Lahl[3], S Maclot[13], J Peschel[3], A L Steber[1,2,8], J W L Lee[5], R Boll[10], S Bari[1], S Dörner[1], F Allum[5], B Erk[1], Lanhai He[6], C Passow[1], D Rompotis[1], A Lemmens[11,12], R Mason[5], D Heathcote[5], M Burt[5], J Wiese[6], M Johny[6], E Müller[1], F Ziaee[4], M M Kazemi[1], N Schirmel[1], P Olshin[7], X Cheng[1], D Ramm[1], S Trippel[6,8], T Mullins[6], A Rijs[11], J Küpper[6,8,9], C Vallence[5], M Brouard[5], D Rolles[4], P Johnsson[3] and M Schnell[1,2,8]

[1]Deutsches Elektronen Synchrotron (DESY), Hamburg, 22607, Germany
[2]Institute of Physical Chemistry, Christians-Albrecht-Universität zu Kiel, Kiel, 24118, Germany
[3]Department of Physics, Lund University, 22100 Sweden
[4]Kansas State University, Kansas, USA
[5]Chemistry Research Laboratory, University of Oxford, United Kingdom
[6]Center for Free-Electron Laser Science, Deutsches Elektronen-Synchrotron DESY, Hamburg, 22607, Germany
[7]Saint Petersburg State University, Saint Petersburg, Russia
[8]Center for Ultrafast Imaging, Universität Hamburg, Hamburg, Germany
[9]Department of Physics, Universität Hamburg, Hamburg, Germany
[10]European XFEL, Schenefeld, Germany
[11]Radboud University, FELIX Laboratory, the Netherlands;
[12]Van 't Hoff Institute for Molecular Sciences, University of Amsterdam, the Netherlands;
[13]KTH Royal Institute of Technology, Stockholm, Sweden

Synopsis In the interstellar medium polycyclic aromatic hydrocarbon molecules (PAH) are exposed to strong ionizing radation leading to complex organic photochemistry. We investigated these ultrafast fragmentation reactions after ionization of the PAHs phenanthrene, fluorene and pyrene at a wavelength of 30.3 nm using pump probe spectroscopy at a free electron laser. We observe double ionization and afterwards hydrogen abstraction and acetylene loss with characteristic time scales for the reaction processes below one hundred femtoseconds.

The interstellar medium (ISM) shows a surprisingly complex radiation driven organic chemistry. One important molecular species in the ISM are polycyclic aromatic hydrocarbons (PAHs). PAHs are very abundant in the galaxy with estimations from the characteristic IR emission bands showing up to 20% of carbon in the Milky Way exists in this form [1]. The harsh conditions in the ISM due to ionizing radiation cause complex fragmentation and isomerization reactions to take place in PAHs [2]. To explore these reaction processes on femtosecond timescales we performed pump probe experiments using the CAMP endstation [3] at the FLASH free electron laser facility. In our study, three PAHs fluorene, phenanthrene and pyrene were ionized by XUV radiation at 30.3 nm wavelength and the reaction dynamics was probed using laser pulses at 800 nm. Complex fragmentation patterns are observed for all three PAHs: multiple hydrogen and/or acetylene abstraction. Several fragmentation channels could be resolved by covariance mapping and characteristic timescales for the femtosecond decay process after ionization were retrieved.

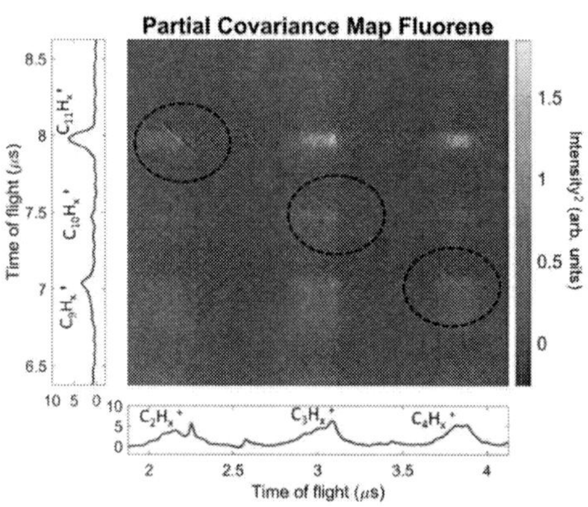

Figure 1. Partial covariance map of fluorene showing the two body fragmentation channels into C_xH_y and $C_{13-x}H_{10-y}$.

References

[1] Joblin C *et al* 2011 *EAS Public. Series* **46** 3
[2] Herbst E *et al* 2009 *Annu. Rev. Astron. Astrophys.* **47** 427
[3] Erk B *et al* 2018 *J. Synchrotron. Rad.* **25** 1529

* E-mail: Bastian.Manschwetus@desy.de

Ultrafast Structural Changes in Chiral Molecules Measured with Free-Electron Lasers

P Schmidt[1,2], V Music[1,2], G Hartmann[1], R Boll[2], B Erk[3], S Bari[3], F Allum[4], T M Baumann[2], G Brenner[3], M Brouard[4], M Burt[4], R Coffee[5], S Dörner[3], A Galler[2], P Grychtol[2], D Heathcote[4], L Inhester[6], M Kazemi[3], M Larsson[7], J Lee[4], Z Li[6], A Lutmann[5], B Manschwetus[3], L Marder[1], R Mason[4], S Moeller[5], T Osipov[5], H Otto[1], C Passow[3], D Rolles[8], P Rupprecht[9], K Schubert[3], L Schwob[3], R Thomas[7], C Vallance[4], C von Korff Schmising[10], R Wagner[2] P Walter[5], T J A Wolf[11], V Zhaunerchyk[12], M Meyer[2], A Ehresmann[1], A Knie[1], Ph V Demekhin[1] and M Ilchen[1,2*]

[1] University of Kassel, Institut für Physik, Heinrich-Plett-Straße 40, 34132 Kassel, Germany
[2] European XFEL GmbH, Holzkoppel 4, 22869 Schenefeld, Germany
[3] Deutsches Elektronen-Synchrotron (DESY), Notkestraße 85, 22607 Hamburg, Germany
[4] University of Oxford, The Department of Chemistry, Oxford OX1 2JD, UK
[5] SLAC National Accelerator Laboratory, 2575 Sand Hill Road, Menlo Park, California 94025, USA
[6] Center for Free-Electron Laser Science (CFEL), Notkestraße 85, 22607 Hamburg, Germany
[7] University of Stockholm, Alba Nova Institute, Roslagstullsbacken 21, 114 21 Stockholm, Sweden
[8] Kansas State University, 1228 N 17th St, KS 66506, United States
[9] Max-Planck-Institut für Kernphysik Heidelberg, Saupfercheckweg 1, 69117 Heidelberg, Germany
[10] Max Born Institute, Max-Born-Straße 2A, 12489 Berlin, Germany
[11] PULSE at Stanford, 2575 Sand Hill Road, Menlo Park, California 94025, USA
[12] University of Gothenburg, 405 30 Gothenburg, Sweden

Synopsis (X-ray) free-electron lasers are employed to site specifically interrogate atomic fragments during ultrafast photolysis of chiral molecules via time-resolved photoelectron circular dichroism.

The method of photoelectron circular dichroism (PECD) has been shown to be a powerful tool for chiral recognition over the last two decades[1, 2]. It provides up to orders of magnitude stronger effects than normal CD. The observable forward-backward asymmetry in the angular distribution of photoelectrons emitted from a chiral system is very sensitive to the electron energy and also to (ultrafast) changes of the underlying charge distribution. This has been demonstrated for vibrational excitations[3], multiphoton interactions[4] and ultrafast structural changes[5]. Such exceptional sensitivity of PECD opens unique perspectives for addressing nonlinear and ultrafast phenomena with VUV and X-ray free-electron lasers that can uniquely enable site selective interrogation of an observing (e.g. dissociating) atomic site. However, such pump-probe experiments on chiral systems are challenging in terms of their technical, physical and (stereo-)chemical complexity.

In the presented work, we show first approaches from the free-electron lasers LCLS (USA) and FLASH (Germany) to measure the time-resolved (TR-)PECD in chiral model systems, i.e. trifluoromethyloxirane ($C_3H_3F_3O$)[6] and iodomethylbutane($C_5H_{11}I$), respectively.

To measure the TR-PECD of these prototypical chiral molecules during (UV) laser or X-ray triggered fragmentation (pump), highly intense, circularly polarized XUV free-electron laser pulses were used as probe. The presentation will primarily focus on a recent experiment at FLASH (10/2018), where we employed a two-sided velocity map imaging spectrometer in order to obtain electron-ion correlations from a dissociating chiral system. In this case, atomic iodine is ejected from enantiomeric iodomethylbutane, serving as dynamic observer site for monitoring the evolving chirality of the residual molecule on a femtosecond timescale.

*E-mail: ilchen@physik.uni-kassel.de

References

[1] Böwering N 2001 *Phys. Rev. Lett.* **86** 1187
[2] Powis I 2000 *J. Chem. Phys.* **112** 301
[3] Garcia G *et al* 2013 *Nature Commun.* **4** 2132
[4] Lux C *et al.* 2012 *Angew. Chem. Int. Ed.* **51** 5001
[5] Beaulieu S *et al* 2016 *Faraday Discuss.* **194** 325
[6] Ilchen M *et al* 2017*Phys. Rev. A* **95** 053423

Collisions between cold molecules in a superconducting magnetic trap

M Pitzer[1 *], Y Segev[1], M Karpov[1], N Akerman[1], J Narevicius[1] and E Narevicius[1 †]

[1]Department of Chemical and Biological Physics, Weizmann Institute of Science, Rehovot 7610001, Israel

Synopsis We decelerate and trap molecular oxygen using time-dependent magnetic fields and superconducting magnets. The density-dependent, non-exponential decay in particle number provides a clear proof of molecule-molecule collisions within the trapped ensemble. The spatial distribution of molecules in the trap is found to change over time, allowing to set limits on the ratio between the elastic and inelastic collision cross sections. Our experimental scheme opens up new possibilities for studying quantum effects in chemistry as well as for evaporative cooling of molecules.

Cold dense ensembles of molecules allow studying a large variety of topics, from quantum phenomena in chemical reactions to the possible achievement of quantum degeneracy with complex species.

Following the success of atom laser cooling, Feshbach molecules are now routinely created by magneto-association of cold atomic species. Directly cooling and trapping neutral molecules has turned out to be much more complicated, mainly because of the large number of vibrational and rotational states that makes laser cooling a challenge. Deceleration and subsequent trapping of dense molecular ensembles has emerged as a promising alternative.

We present here first measurements of collisions between magnetically decelerated and trapped molecules[1].

A cloud of O_2 molecules is created by a pulsed supersonic expansion and decelerated in a moving magnetic trap decelerator[2]. At the end of the deceleration, a superconducting magnetic trap, made of high-T_c superconducting tape, is switched on, allowing constant or time-dependent traps to be realized. The detection is carried out by a 2+1 REMPI (resonance enhanced multi-photon ionization) scheme, yielding the particle density integrated over the laser propagation direction.

We observe a clear non-exponential decay that depends on the O_2 density (Figure 1).

We provide an independent proof of molecular collisions by co-loading lithium atoms together with the molecular oxygen. We observe that the lifetime of lithium is reduced from background collision limited lifetime of 14 s to 1.9 s due to the collisions with oxygen.

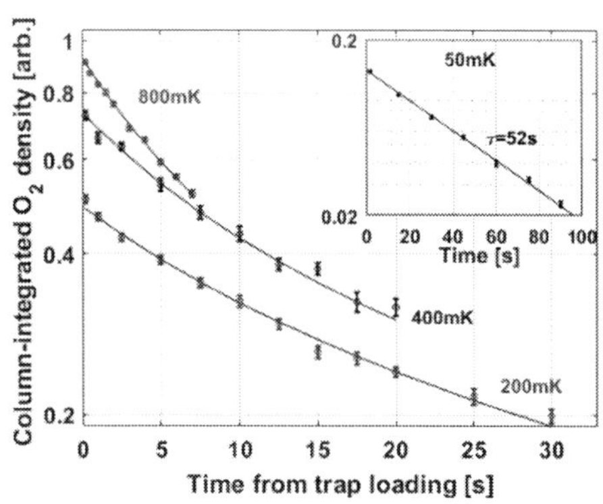

Figure 1. Time evolution of the column-integrated O_2 density for different trap depths, given in E_{max}/k_B. While the decay in the shallow trap (50 mK) is exponential due to pure background collisions, higher densities show a non-exponential slope due to collisions between trapped molecules.

Our detection scheme allows probing the density at different locations in the trap. We observe a faster decay at the edges of the trap, indicating a redistribution due to elastic collisions. With the help of a classical simulation of collision dynamics, we are able to estimate the ratio of elastic to inelastic cross sections. Although for $^{16}O_2$, this ratio is not sufficient to apply direct evaporative cooling, the versatility of our method enables us to study now if different oxygen isotopes or other paramagnetic species have more favorable collisional properties.

References

[1] Segev Y, Pitzer M, Karpov M *et al* 2019 arxiv 1902.04549
[2] Akerman N *et al* 2017 *Phys. Rev. Lett.* **119** 073204

* E-mail: martin.pitzer@weizmann.ac.il
† E-mail: edn@weizmann.ac.il

Rydberg polyatomic molecules: Electronic structure and experimental proposal for their creation

R González-Férez[1] *, S T Rittenhouse[2], P Schmelcher[3,4] and H R Sadeghpour[5]

[1] Instituto Carlos I de Física Teórica y Computacional and Departamento de Física Atómica, Molecular y Nuclear, Universidad de Granada, 18071 Granada, Spain
[2] Department of Physics, The Naval Academy, Annapolis, Maryland 21402, USA
[3] Zentrum für optische Quantentechnologien, Universität Hamburg, Germany
[4] The Hamburg Centre for Ultrafast Imaging, Universität Hamburg, Luruper Chaussee 149, 22761 Hamburg, Germany
[5] ITAMP, Harvard-Smithsonian Center for Astrophysics 60 Garden St., Cambridge, Massachusetts 02138, USA

Synopsis We describe the electronic structure and main features of ultralong range Rydberg molecules formed by a Rydberg atom, either Rb or K, and the heteronuclear diatomic molecule KRb. By analyzing their electronic structure, we propose a protocol to create these molecules experimentally from a mixture of ultracold atoms and ultracold molecules.

The traditional Rydberg molecules are formed by a tightly bound positively charged core and an excited electron circulating around it. A more exotic type of Rydberg molecules were theoretically predicted to exist when a ground-state atom locally probes the Rydberg electronic wavefunction [1]. These ultralong-range exotic species were experimentally observed in 2010 [2], showing that, analogously to Rydberg atoms, they have a large sensitivity to external field, which allows their control. Another type of ultralong-range Rydberg molecules are theoretically predicted to exist if a heteronuclear diatomic molecule, a Λ-doublet or a rotating polar molecule, is immersed into the wave function of a Rydberg atom [3, 4, 5, 6]. The anisotropic scattering of the Rydberg electron from the permanent electric dipole moment of the polar molecule is responsible for the binding mechanism in these Rydberg molecules.

In this work, we describe the main properties of these exotic ultralong range Rydberg molecules formed by a Rydberg atom and a heteronuclear diatomic molecule. We explore the electronic structure of triatomic Rydberg molecules formed from a potassium or rubidium Rydberg atom and the KRb molecule. Our focus is on the interaction of Rydberg states, having s angular momentum, with ground and rotationally excited KRb diatomic molecules. These Rydberg molecules could be created by exciting Rydberg atoms in a mixture of ultracold atoms and ultracold molecules, i.e., molecules such as K-KRb or Rb-KRb would be created experimentally. We have found adiabatic electronic states evolving from the Rydberg degenerate manifold with a significant contribution from a low angular momentum Rydberg state. These mixed electronic states are a few GHz deep and have several vibrational bound states [7]. We propose a protocol to create these molecules experimentally in these electronic states from a mixture of ultracold atoms and ultracold molecules [7].

References

[1] Greene C H, Dickinson A S and Sadeghpour H R 2000 *Phys. Rev. Lett.* **85** 2458
[2] Bendkowsky V *et al* 2010 *Phys. Rev. Lett.* **105** 163201
[3] Rittenhouse S T and Sadeghpour H R 2010 *Phys. Rev. Lett.* **104** 243002
[4] Rittenhouse S T, Mayle M, Schmelcher P and Sadeghpour H R 2011 *J. Phys. B* **44** 184005
[5] Mayle M, Rittenhouse S T, Schmelcher P and Sadeghpour H R 2012 *Phys. Rev. A* **85** 052511
[6] González-Férez R, Sadeghpour H R and Schmelcher P 2015 *New J. Phys.* **17** 013021
[7] González-Férez R, Rittenhouse S T, Schmelcher P and Sadeghpour H R 2019, *submitted*

*E-mail: rogonzal@ugr.es

Ion-ion collision-induced evaporative cooling of atomic and molecular ions by autoresonance in an electrostatic ion beam trap

O Heber[1,*], R K Gangwar[1,3], K Saha[1,4], M L Rappaport[2] and D Zajfman[1]

[1]Department of Particle Physics and Astrophysics, Weizmann Institute of Science, Rehovot 7610001, Israel
[2]Department of Physics Core Facilities, Weizmann Institute of Science, Rehovot 7610001, Israel
[3]Department of Physics, Indian Institute of Technology, Tirupati 517506, India
[4]Department of Physics, Indian Institute of Technology, Dharwad 580011, India

Synopsis A new method for cooling beams of any ion is introduced. The method is autoresonance acceleration of ions inside an Electrostatic Ion Beam Trap. Intrabunch collisions of trapped (molecular) ions transfer energy from the "cold" part of the population of the ion distribution to the "hot" part, which, in turn, evaporates from the bunch. As a result, bunches of ions were cooled from about 45 K to well below 1 K.

Translational cooling of bunches of atomic and molecular ions is a requisite in several research areas. An Electrostatic Ion Beam Trap (EIBT), which can trap any ion with any mass or charge using the same tuning conditions, is therefore ideal for ion beam cooling.

A chirped sinusoidal voltage is applied to one of the EIBT mirror electrodes. In this procedure, called autoresonance[1] (AR), a bunch of ions is accelerated out of the rest of the ion beam population. The bunch acceleration efficiency is governed by the strength of the chirped electric field. Depending upon the chirped field intensity and rate, one can control the velocity distribution in the bunch.

A cooling process has been demonstrated in the EIBT that, by using autoresonance, reduced the temperature of ions from an initial value of ~45 K down to about 0.15 K in 80 ms [2]. The EIBT is tuned so that ion density is maximized at the turning point of the trap to achieve thermalization and effective evaporation.

Figure 1 shows the calculated bunch internal temperature as a function of the AR voltage using the measured ion bunch velocity distributions. The AR threshold potential for ion bunch acceleration is about 0.052 V in this data set. The arrow in the figure indicates the initial temperature of the ions in the trap before the AR process.

During AR, it has been shown [2] that the ion-ion collisions transfer kinetic energy from the cold population to the hotter population, which in turn is evaporated from the ion bunch, hence reducing the bunch's temperature and increasing the phase-space density. Further experiments and theoretical models are ongoing to improve the cooling efficiency and to achieve lower temperatures.

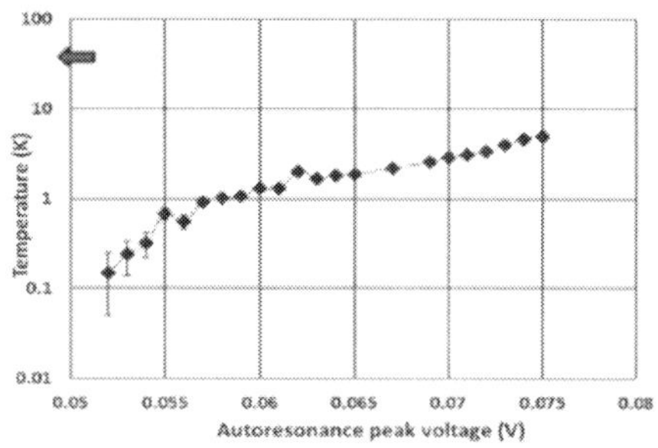

Figure 1. Temperature of the ion bunch after autoresonance acceleration. The arrow indicates the intial temperature of the ions in the EIBT.

Since this process is applicable for any ion and any charge, it has the potential to be useful for reaction studies. Moreover, even without a sufficient number of ions in the trap for intrabeam collisions to produce significant cooling, the AR process by itself is a filtering process for cold population selection.

References

[1] Rajasekar S, Sanjuan M A F 2016 "Nonlinear Resonances – 2016", Springer Series in Synergetics. Springer, Cham 293

[2] Gangwar R K *et al* 2017 *Phys, Rev. Lett.* **119** 103202

* E-mail: oded.heber@weizmann.ac.il

Measurements of spatial correlations in quantum gases using ultralong-range strontium Rydberg molecules

J D Whalen[1] *, R Ding[1], S K Kanungo[1], F B Dunning[1], T C Killian[1], S Yoshida[2] and J Burgdörfer[2]

[1]Department of Physics & Astronomy, Rice university, Houston, Texas, USA
[2]Institute for Theoretical Physics, Vienna University of Technology, Vienna, Austria, EU

Synopsis The photoexcitation of ultralong-range Rydberg molecules is demonstrated as a probe of spatial correlations in a quantum gas on previously inaccessible length scales of 1400-3200 a_0. Measurements using cold (~1 μK) (fermionic) ^{87}Sr and (bosonic) ^{84}Sr reveal the effects of quantum statistics on the pair correlation function $g^{(2)}(R)$.

The understanding of quantum gases can be enhanced by *in situ* measurements of spatial correlations. Here we demonstrate that photoexcitation of ultralong-range Rydberg molecules can provide a novel probe of nonlocal pair correlations. Rydberg dimer molecules [1] comprise a ground-state atom bound to a Rydberg atom by scattering of the Rydberg electron. The outer regions of the resulting molecular potential are shown in Fig. 1 for different n values together with the wavefunctions for the ground $\nu = 0$ molecular dimer states which are strongly localized near the electron outer classical turning point at $R_n \sim 2n^2$. Thus, the molecular formation rate is governed by the probability of finding atom pairs with separation R_n.

Here we use non-degenerate quantum gases of (spin-polarized) fermionic ^{87}Sr ($I = 9/2$) and bosonic ^{84}Sr ($I = 0$) to measure the effects of quantum statistics on the molecular excitation rate. To extract the pair-correlation function $g^{(2)}(R)$ we compare excitation rates in the (polarized) Fermi- or Bose-gas to an unpolarized sample of ^{87}Sr where the tenfold-degenerate ground state provides a good approximation to a gas of uncorrelated particles. Figure 1 shows the dependence of the measured molecular production rate, normalized to that seen for an unpolarized sample, on R_n. As n, i.e., R_n, decreases, suppression of the excitation rate in the spin-polarized ^{87}Sr gas due to Pauli exclusion (antibunching), and the enhancement for the ^{84}Sr gas due to bunching are clearly evident. Figure 1 also includes the expected behavior of the pair correlation function $g^{(2)}(R)$ for non-degenerate gases of non-interacting bosons and fermions. The measurements mirror well this behavior demonstrating that Rydberg molecule formation can provide a novel probe of pair correlations in cold gases that is suitable for use as a diagnostic in many other studies.

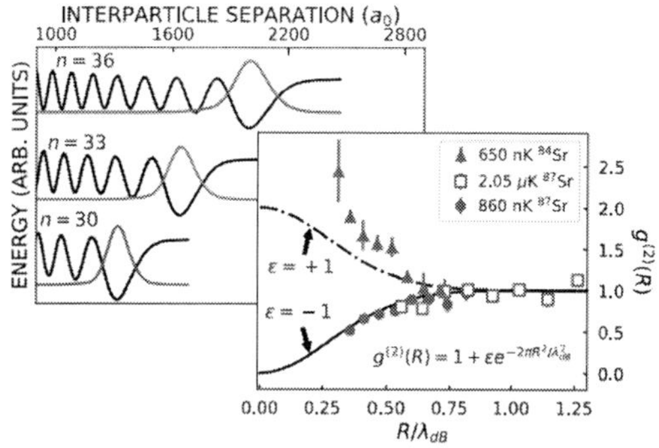

Figure 1 - Left: Black lines, molecular potentials for $5sns\ ^3S_1 + 5s^2\ ^1S_0$ pairs (not to scale); colored lines, the radial wavefunction of the corresponding highly localized ground $\nu = 0$ molecular state. Right: Extracted values of the pair-correlation function from comparing boson/spin-polarized fermion molecular excitation rates to unpolarized excitation rates plotted versus $R_n \sim 2n^2$ scaled by the thermal de Broglie wavelength $\lambda_{dB} = h/\sqrt{2\pi m k_B T}$ (see text). The lines show the calculated values of $g^{(2)}(R)$.

Research supported by the AFOSR, NSF, Robert A. Welch Foundation and FWF (Austria).

References

[1] DeSalvo B J *et al* 2015 *Phys. Rev. A* **92** 031403(R)

* E-mail: jwhalen@rice.edu

Excitation of vibrationally-excited ultralong-range strontium Rydberg molecules in quantum gases

R Ding[1], J D Whalen[1], S K Kanungo[1], T C Killian[1], F B Dunning[1*], S Yoshida[2] and J Burgdörfer[2]

[1]Department of Physics & Astronomy, Rice University, Houston, Texas, USA
[2]Institute for Theoretical Physics, Vienna University of Technology, Vienna, Austria, EU

Synopsis The photoexcitation rate for ultralong-range Rydberg molecules is sensitive to the initial spatial separation of the atoms from which they are formed. Here we explore the photoexcitation of different molecular vibrational states as a probe of spatial correlations in a cold atomic gas.

Rydberg dimer molecules comprise a ground-state atom bound to a Rydberg atom by scattering of the Rydberg electron [1]. The resulting molecular potential for a $5s31s\,^3S_1$-$5s^2\,^1S_0$ atom pair is shown in Fig. 1(a) together with the wavefunctions for the v=0, 1, 2 and 3 molecular vibrational levels. The v=0 wavefunction is strongly localized near the electron outer classical turning point whereas those for the v=1, 2, and 3 states extend to smaller internuclear separations. Since photoexcitation of a particular molecular level requires an atom pair with the appropriate initial separation, *R*, this suggests that measurements of the formation rates for different vibrational levels can be used to probe spatial correlations in a cold gas over a broad length scale.

Here we compare results for molecule formation in a cold (T~1 μK) gas of spin-polarized ^{87}Sr, for which quantum statistics are expected to reduce the pair correlation function $g^{(2)}(R)$ at small *R* due to Pauli exclusion (antibunching), with results for an unpolarized sample which provides a good approximation to a gas of uncorrelated particles. Figure 1(b) shows the results of theoretical predictions for the excitation of the v=0, 1, and 2 states in both polarized and unpolarized samples. As seen in Fig. 1(b), the calculations predict a reduction in the production of the v=1 and, to a lesser extent, the v=2 excited states, relative to the ground v=0 state, in the polarized sample, a manifestation of Pauli exclusion affecting the Franck-Condon factor for vibrational states that extend to small *R*. Figure 1(b) also includes the results of experimental measurements which are seen to be in good accord with the theoretical predictions. The data thus demonstrate that measurements of Rydberg molecule formation can provide, even for a single value of *n*, a probe of spatial correlations over a sizable range of internuclear separations.

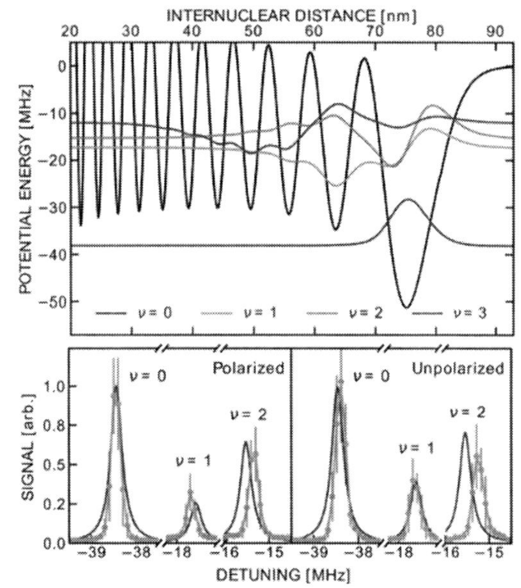

Figure 1. (top) Molecular potential for a $5s31s\,^3S_0$-$5s^2\,^1S_0$ atom pair together with the the v=0, 1, 2 and 3 vibrational wavefunctions. (bottom) Calculated (---) and measured (o) v=0, 1, and 2 excitation features for polarized, and unpolarized, T~1 μK ^{87}Sr gases (see text). The experimental and theoretical v=0 features are normalized to each other.

Research supported by the AFOSR, NSF, Robert A. Welch Foundation, and FWF (Austria).

References

[1] DeSalvo B J *et al* 2015 *Phys. Rev. A* **92** 031403

* E-mail: fbd@rice.edu

Photodetachment of negative ion beams for the ion-neutral merged-beam experiments at RICE

M Iizawa[1,3]*, S Iida[1,2], S Kuma[3], T Azuma[2,3] and Y Nakano[1,3]

[1]Department of Physics, Rikkyo University, Tokyo, 171-8501, Japan
[2]Department of Physics, Tokyo Metropolitan University, Tokyo, 192-0397, Japan
[3] AMO Physics Laboratory, RIKEN, Saitama, 351-0198, Japan

Synopsis Generation of neutral atomic beam is under development for the ion-neutral merged-beam experiment at the RIKEN Cryogenic Electrostatic ion storage ring (RICE). The beam is produced by a photodetachment of a negative ion beam using a high power laser diode array with a wavelength of 808 nm. In order to increase the detachment efficiency, we have developed an optical cavity built with a planar mirror. In this presentation we will present the current status of the neutral atom beam source.

The low-energy collision of molecular ions with neutral species is one of the most important processes in the production of complex organic molecules (COMs) in interstellar spaces. Arrhenius-type models are reasonable approximations to predict the reaction rate at different temperatures, however, under cryogenic temperatures, some reactions are known to show characteristic behaviors beyond the model. Experimental studies are needed to investigate the dynamics of such low energy collisions and eventually to give a better understanding of interstellar chemistry.

To measure the reaction cross section between hydrocarbon ions and neutral atoms at low collision velocity corresponding to the temperature of interstellar cloud (10–100 K), the ion and neutral beams will be merged in the RIKEN Cryogenic Electrostatic ion storage ring (RICE). The storage ring provides a cryogenic environment down to 4 K and a high collision luminosity [1]. For this experiment, a neutral atomic beam has to be prepared with a well defined relative velocity with respect to the ion beam. To prepare the neutral beam in the atomic ground state with a small velocity and angular spread, we adopted a method of electron detachment of negative ion beam by a high-power laser diode [2].

When the laser crosses with the negative ion beam at 90 degree, the photodetachment probability η is obtained by

$$\eta = 1 - \exp\left(-\frac{P\sigma}{vwE}\right) \qquad (1)$$

where P is the laser power, E is the photon energy, σ is the photodetachment cross section, v is the beam velocity and w is the laser spot width along the beam. The photodetachment cross section σ of C^- is $1.4 \times 10^{-17}\,\mathrm{cm}^2$ at the photon energy of 1.5 eV [3]. We have tested the photodetachment of 10 keV C^- ion beam using an 808 nm CW laser diode array with a maximum output power of 5 kW. The measured probability of detachment was on the order of 10^{-3}, which well agreed with the estimation using Eq. (1). In order to increase the intensity of a neutral beam, a multi-pass optical cavity was built with three planar mirrors. The mirror and lens configurations were optimized by an optical lay tracing simulation using Zemax Optic Studio, which estimates an enhancement of the laser photon flux by nearly an order of magnitude.

References

[1] Nakano Y *et al* 2017 *Rev. Sci Instr* **88** 033110
[2] O'Connor A P *et al* 2015 *Rev. Sci. Instr.* **86** 113306
[3] Seman M L *et al* 1962 *Phys. Rev.* **125** 1602

*E-mail: iizawamasatomi@rikkyo.ac.jp

Reaction of dication C^{++} with molecular hydrogen at temperature 20 K

R Plašil [*], A Kovalenko, T D Tran, S Rednyk, P Dohnal, Š Roučka and J Glosík

Department of Surface and Plasma Science, Faculty of Mathematics and Physics, Charles University, Prague, 180 00, Czech Republic

Synopsis The reaction of dication C^{++} with H_2 was studied at temperature 20 K using a linear 22-pole radiofrequency ion trap. The overall reaction rate coefficient was measured as $1.4 \pm 0.3 \times 10^{-10}$ cm^3 s^{-1}. The upper limit for formation of CH^+ in this reaction was evaluated as 4×10^{-12} cm^3 s^{-1}.

The methylidyne cation, CH^+, was the first molecular ion identified in interstellar space. Despite its abundance in a number of different interstellar environments, there are still queries concerning the formation of this ion. While models of chemical kinetics predict the observed abundances of neutral CH quite well, the observed column densities of CH^+ are often orders of magnitude higher than calculated ones, which are taking into account only singly charged ions.

An alternative way of formation of CH^+ was proposed by Langer [1] in a reaction of doubly charged C^{++} with a hydrogen molecule. The theoretical calculation [2] predicts the possibility of CH^+ formation, however, without quantitative information.

$$C^{++} + H_2 \rightarrow CH^+ + H^+ \quad (1)$$
$$\rightarrow C^+ + H_2^+ \text{ or } (H^+ + H) \quad (2)$$
$$\rightarrow C + H^+ + H^+ \quad (3)$$

To study the title reaction, we used a temperature variable linear 22-pole radiofrequency ion trap [3]. The primary C^{++} ions are produced in a separate source and injected into the ion trap. The ions are cooled in collisions with helium buffer gas and react with added H_2. After various trapping times, the ions are extracted from the trap and analyzed. An example of the evolution of the ion composition is shown in Fig. 1.

The overall $C^{++} + H_2$ reaction rate coefficient is $1.4 \pm 0.3 \times 10^{-10}$ cm^3 s^{-1} at 20 K and it is close to the value measured by Smith and Adams [4] at 300 K. However, there is a significant difference in product distribution. In our experiment, more than 50% of C^{++} is converted to C^+ in the reaction channel (2). It is in contradiction to the result of Smith and Adams as they observed especially channel (3) with more than 90% dominance. In our current setup of the trap we cannot confine efficiently light and fast products of the reaction, H^+ or H_2^+, so the number of the charges in the trap is not conserved.

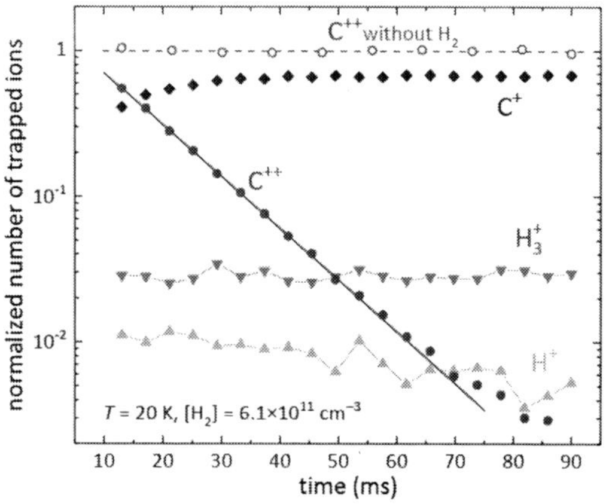

Figure 1. Time evolution of normalized numbers of ions in the trap. Open circles show traping of C^{++} without H_2 reactant.

The derived upper limit of the reaction rate coefficient of the formation of methylidyne cation CH^+ may be evaluated at 20 K as 4×10^{-12} cm^3 s^{-1}. This reaction rate coefficient is three orders of magnitude lower than Langevin value, thus comparing it to other loss processes, it is unlikely that channel (1) is important for the formation of CH^+ in cold interstellar media.

This work was supported by the Czech Science Foundation project GACR 17-18067S.

References

[1] Langer W D 1978 *Astrophys. J.* **228** 860
[2] Bacchus-Montabonel M-C and Wiesenfeld L 2013 *Chem. Phys. Lett.* **583** 23
[3] Zymak I *et al* 2013 *Astrophys. J.* **768** 86
[4] Smith D and Adams N G 1981 *Mon. Notices Royal Astron. Soc.* **197** 377

[*] E-mail: Radek.Plasil@mff.cuni.cz

Laser-Assisted Self Induced Feshbach Resonance : a new tool for controlling ultracold atomic collisions

A Devolder[1 *], E Luc-Koenig[1], O Atabek[2], M Desouter-Lecomte[3] and O Dulieu[1]

[1] Lab. Aimé Cotton, CNRS, Univ. Paris-Sud, ENS Paris-Saclay, Univ. Paris-Saclay, Bât. 505, 91405 Orsay, France
[2]Institut des Sciences Moléculaires d'Orsay, CNRS, Univ. Paris-Sud, Univ. Paris-Saclay, Bât. 520, 91405 Orsay, France
[3]Laboratoire de Chimie Physique, CNRS, Univ. Paris-Sud, Univ. Paris-Saclay, Bât. 349, 91405 Orsay France

Synopsis : We propose a new kind of Feshbach Resonance (FR) : the Laser-Assisted Self-Induced Feshbach Resonance (LASIFR), applicable for all mixtures of ultracold atoms. The interspecies scattering length characterizing elastic collisions, a key property for many applications of ultracold quantum gases, can be controlled by a LASIFR through the laser intensity and the frequency. It can be also used for the formation of ultracold polar molecules.

Ultracold gases composed of two atomic species offer numerous possibilities for studying novel few- and many-body phenomena like Efimov states, or crystalline quantum phases. All these applications require a fine control of the scattering properties, in particular of the interspecies scattering length.

Magnetic Feshbach Resonance (MFR) is the main tool for tuning the interspecies scattering length of two different alkali-metal atoms[1]. For a mixture of alkali-metal atoms with closed-shell atoms like alkaline-earth or Ytterbium atoms, MFR cannot be easily implemented [2]. To circumvent this pitfall, we propose a new concept of FR: the Laser-Assisted Self-Induced Feshbach Resonance (LASIFR). A continuous-wave laser couples the state of the colliding atoms to a weakly-bound rovibrational level of the related molecular ground state (Fig. 1a). In the field-dressed picture, this coupling can be interpreted as a Feshbach Resonance between channels sharing the same electronic molecular state but with different number of photons (Fig.1). The coupling relies on the variation of the intrinsic permanent dipole moment of the molecular state with the interatomic distance. As for MFR, the two channels have a long lifetime.

We focus on the example of ^{87}Rb and ^{84}Sr mixture [2]. We show how the interspecies scattering length can be controlled by a LASIFR. We also demonstrate that the resonance width can be tuned by the laser intensity, a remarkable property of the LASIFR.

Ultracold quantum gas mixtures are currently a convenient platform for the formation of ultracold polar molecules, *inter alia* promising systems for quantum simulation [3]. We prove that the proposed LASIFR represents a powerful tool for ultracold molecule formation in two respects. First, the optical control of the LASIFR induces a strong increase of the pair probability density at short distances, enabling a huge increase of the population transfered by STimulated Rapid Adiabatic Passage (STIRAP) via a vibrational level of an electronic excited state. Secondly, in analogy with magnetoassociation using magnetic field ramping, chirping of the laser frequency accross the LASIFR induces an adiabatic passage between the initial state of the atom pair and a vibrational level of the electronic ground state.

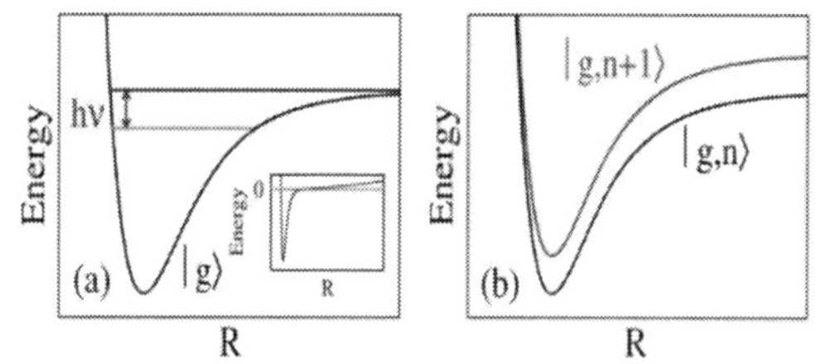

Figure 1. Scheme for the LASIFR in the bare-state (a) and dressed-state (b) pictures.

Work supported by CEFIPRA under project #5404-1

References

[1] Tiesinga E *et al* 1993 *Phys. Rev. A* **82** 4114
[2] Barbé V *et al* 2018 *Nat. Phys.* **14** 881
[3] Micheli A *et al* 2012 *Phys. Rev. Lett* **109** 115302

* E-mail: adrien.devolder@u-psud.fr

Optical shielding for ultracold ^{39}K-Cs binary collisions

T Xie[1*], A Orbán[2], O Dulieu[1] and N Bouloufa-Maafa[1†]

[1] Lab. Aimé Cotton, CNRS, Univ. Paris-Sud, ENS Paris-Saclay, Univ. Paris-Saclay, Bât. 505, 91405 Orsay, France
[2]Hungarian Academy of Science, Institute for Nuclear Research, Debrecen, 4026, Hungary

Synopsis The optical shielding efficiency in ultracold collisions is investigated for the first time with atoms of different species (^{39}K-Cs) with a close-coupling approach. The spontaneous emission is taken into account through a complex optical potential. The results show that an efficient shielding can be achieved in the presence of a blue-detuned laser. The role of the light polarization on the shielding efficiency is carefully investigated.

Research focusing on the formation of ultracold atomic and molecular quantum gases is a continuously expanding field due to its envisioned applications such as quantum-controlled chemistry or quantum simulation. One important limitation for such developments is the occurrence of inelastic or reactive collisions between ultracold particles. In this theoretical work, we investigate ways to suppress such processes using suitable laser light. We propose to couple the initial colliding particle state to a repulsive excited molecular state using a laser field detuned to the blue of a relevant transition, thus preventing the particles to come close to each other. This "blue-shielding" technique, described in detail in [1, 2], is applied here for the first time to ultracold collisions between atoms of different species (^{39}K-Cs) at temperatures close to quantum degeneracy (microkelvin range).

We adopt the field-dressed-state approach to describe our system (Fig. 1). The semiclassical Landau-Zener model has been used previously to successfully model the blue-shielding of atomic collisions in the millikelvin range [1]. It fails in the ultracold limit as the de Broglie wavelength becomes larger than the interaction region. Therefore we developed a fully quantum close coupling numerical code for ^{39}K-Cs collisions in the dressed-state framework. Molecular excited states correlated to the second excited asymptote K(4^2P) + Cs(6^2S) are repulsive at large interatomic distance, and are thus suitable for optical shielding [3]. Among them the $(3)^3\Sigma^+$ state possesses a high barrier at short distances [4]. It can be coupled to lowest $a^3\Sigma^+$ initial state of the colliding atoms by polarized light. We demonstrate that a strong laser field with a small blue detuning (Δ~100 MHz) from the K(4^2P)- K(4^2S) atomic transition is indeed possible. The spontaneous emission is taken into account by a complex optical potential through the modified formula of shielding rate [5].

We have investigated the influence of several parameters on the efficiency of the shielding: the detuning Δ, the laser coupling, the relative slopes of initial and excited potential curves, the temperature (fixing the number of involved partial waves), the spontaneous emission rate, and the light polarization [6,7].

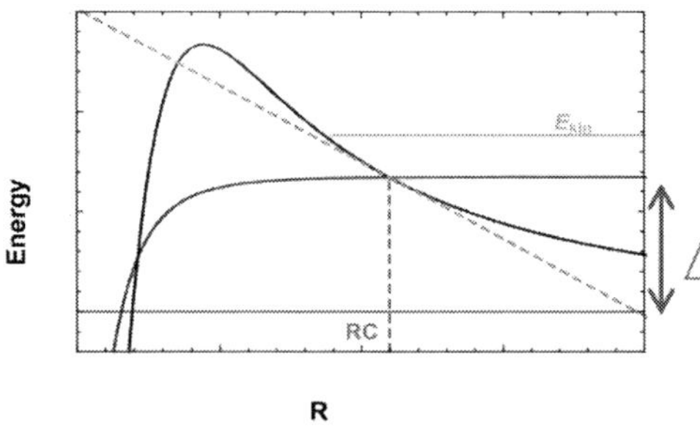

Figure 1. Scheme of optical shielding in the dressed-state picture. The entrance channel for colliding atoms with energy E_{kin} dressed with a photon detuned by Δ crosses a repulsive channel in R_c.

* E-mail: ting.xie@u-psud.fr
† E-mail: nadia.bouloufa@u-psud.fr

References

[1] Suominen K-A *et al* 1996 *Phys. Rev. A* **53** 1678
[2] Weiner J *et al* 1999 *Rev. Mod. Phys.* **71** 1
[3] Orbán A *et al* 2015 *Phys. Rev. A* **92** 032510
[4] Marinescu M *et al* 1999 *Phys. Rev. A* **59** 390
[5] Nicholson T L *et al* 2015 *Phys. Rev. A* **92** 022709
[6] Zilio S C *et al* 1996 *Phys. Rev. Lett.* **76** 2033
[7] Napolitano R *et al* 1997 *Phys. Rev. A* **55** 1191

Modeling ultracold lithium ion-atom collision

A Pandey[1] *, M Niranjan[1], N Joshi[1], S A Rangwala[1] and O Dulieu[2]

[1]Raman Research Institute, C. V. Raman Avenue, Sadashivanagar, Bangalore, 560080, India
[2] Lab. Aimé Cotton, CNRS, Univ. Paris-Sud, ENS Paris-Saclay, Univ. Paris-Saclay, Bât. 505, 91405 Orsay, France

Synopsis Accurate estimation of the scattering properties at the ultracold temperatures is challenging due to the demand for the precise knowledge of interactions between the colliding partners. We identify the aspects of interactions that play the defining roles in the scattering calculations. We compute potential energy curves for the $^7Li^+$-^7Li, $X^2\Sigma_g^+$ and $A^2\Sigma_u^+$, and determine their collisional properties.

In the study of the ion-atom interactions at ultracold temperatures, Li^+-Li is one of the most advantageous candidates for attaining the *s*-wave and few partial wave regime for their mutual scattering. This is because of the combination of its reduced mass and polarizability [1,2].

Scattering processes in the $^7Li^+$-^7Li ion-atom system for temperatures starting from a few K to sub-μK are investigated. Accurate potential energy curves (PECs) of the ground and first excited state of the Li_2^+ molecular ion, $X^2\Sigma_g^+$ and $A^2\Sigma_u^+$, are calculated, (Fig. 1), in order to estimate the low energy scattering phase shifts, which are employed to compute the ion-atom total scattering cross section for the system. The PECs are computed using the multi-reference configuration interaction (MRCI) method with Dunning correlation-consistent polarized, core-valence, augmented, 5-zeta, (aug-cc-pCV5Z), basis sets in the range of internuclear separations 2-50 atomic units. Numerical extrapolations to generate the PECs outside this range are performed.

We discuss the contributions arising from the $X^2\Sigma_g^+$ and $A^2\Sigma_u^+$ states for both scenarios when the identities of the participating entities, Li^+ and Li, are retained or interchanged in the collision events. These two collision channels can be described only in the high energy limit as direct elastic and resonant charge exchange (RCE) collisions [3]. At very low energies, it is impossible to assign the outcome of the collision to one or other of these channels. Results of the cross section calculations will be presented at the conference.

In addition, we examine the effects of minor alterations in small-R regions of the calculated PECs, where the ab-initio molecular calculations can have significant uncertainties, on the scattering parameters. We find that in the *s*-wave limit, the cross section terms involving $X^2\Sigma_g^+$ state is sensitive to the minor alterations in the PEC; the effect is amplified as the low energy scattering is close to a pole. We provide a comparison of the properties of PECs and scattering parameters with previous experimental and theoretical results and review the essential elements for reliable estimation of the collision cross sections [4].

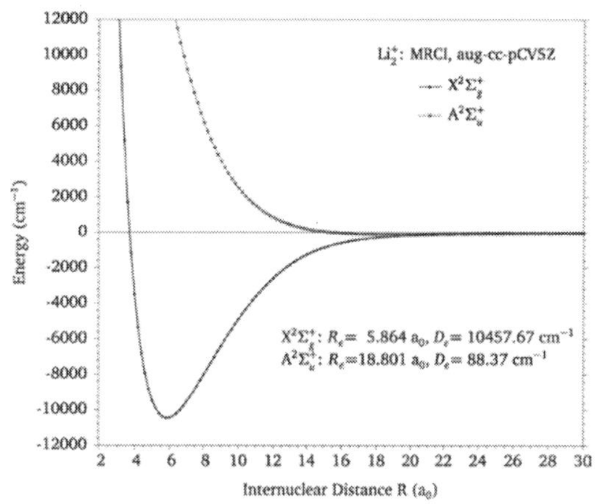

Figure 1. The first two states of Li_2^+, $X^2\Sigma_g^+$ and $A^2\Sigma_u^+$, computed using the MRCI method with aug-cc-pCV5Z basis set are shown.

References

[1] Côté R and Dalgarno A 2000 *Phys. Rev.* A **62** 012709
[2] Massey H and Smith R A 1933 *Proc. R. Soc. Lond.* A **142** 142
[3] Dutta S and Rangwala S 2018 *Phys. Rev.* A **97** 041401
[4] Zhang P, Bodo E and Dalgarno A 2009 *J. Phys. Chem.* A **113** 15085

* E-mail: amrendra@rri.res.in

Chiral Bloch oscillations of a biased bosonic ladder in an artificial gauge field

X Qiao[1], X B Zhang[1], A X Zhang[1], Z F Yu[1] and J K Xue[1]*

[1]College of Physics and Electronics Engineering, Northwest Normal University, Lanzhou 730070, China

Synopsis Rich chiral Bloch oscillations are predicted to distinguish the Meissner phase, Vortex phase and two new Plane Wave phases of the biased non-interacting bosonic ladder in an artificial gauge field.

Ladder systems as simple models exhibiting behavior of many artificially manufactured structures and naturally found compounds are gaining growing interest in strongly correlated electron systems [1]. In recent years, experimental realization and theoretical investigation of bosonic ladder system have been tremendous activity in topological physics, quantum Hall effects and novel phases of many-body system [2]. Most studies are interested in the ground state, phase transition processes and other equilibrium properties of this ladder system in an artificial gauge field. But the research of the dynamical features of the system is still an undeveloped issue. Bloch oscillation (BO) is one of the fundamental and important dynamical phenomena of the quantum mechanics. When considering a longitudinal dc gravitational field along the legs, the BOs in this ladder system will be predicted.

In this paper, we investigate the ground states and phase transitions of the biased non-interacting bosonic two leg flux ladder, and distinguish different phases intuitively via the features of the chiral BO in the presence of a longitudinal dc gravitational field along the legs. The full phase diagrams and the chiral BOs corresponding to different phases are summarized in Figure 1. When there is no energy bias $\tilde{\omega}_\perp = 0$, the ground states of the ladder system have a phase transition from the Vortex phase to the Meissner phase with increasing rung-to-leg coupling ratio $\tilde{K}$. The waveforms of the chiral BOs in one Bloch period have double breathing structure in the Vortex phase [Fig. 1 (f)] and single breathing structure in the Meissner phase [Fig. 1 (d)]. Particularly, the adding energy bias results in the emergence of two new Plane Wave phase, that break both time-reversal and Z_2 symmetry, identified by asymmetry double breathing structure BOs in the Plane Wave I (PWI) phase [Fig. 1 (e)] and asymmetry single breathing structure BOs in Plane Wave II (PWII) phase [Fig. 1 (c)].

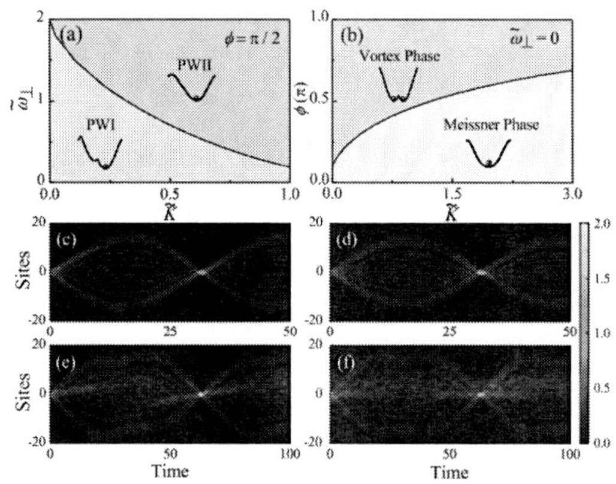

Figure 1. (color online) Phase diagrams in $(\tilde{K}, \tilde{\omega}_\perp)$ plane (a) with the magnetic flux $\phi = \pi/2$ and $(\tilde{K}, \phi)$ plane (b) with $\tilde{\omega}_\perp = 0$. Chiral Bloch oscillations for breathing mode of a single site in the PWII phase (c), the Meissner phase (d), the PWI phase (e) and the Vortex phase (f).

This work is supported by the National Natural Science Foundation of China under Grants No. 11764039, 11847304, 11865014, 11475027, 11305132, and 11274255, by Natural Science Foundation of Gansu province under Grant No. 17JR5RA076, and by Scientific research project of Gansu higher education under Grand No. 2016A-005.

References

[1] Atala M *et al* 2014 *Nat. Phys.* **10** 588
[2] Tai M E *et al* 2017 *Nature* **546** 519

*E-mail: xuejk@nwnu.edu.cn

Quantum phases of biased bosonic ladders subject to a magnetic field

X Qiao[1], X B Zhang[1], A X Zhang[1], Z F Yu[1] and J K Xue[1]*

[1]College of Physics and Electronics Engineering, Northwest Normal University, Lanzhou 730070, China

Synopsis The quantum phase transition of the Plane Wave I phase, the Plane Wave II phase, and interestingly, a new phase named Moving Biased Vortex phase are revealed in the biased bosonic two-leg ladder system.

The realization of the optical lattices and artificial gauge fields provide an ideal playground to explore intriguing phenomena in condensed physics [1]. Due to the one-dimensional system has no response to orbital magnetic magnetic field effects, the two-leg ladder system occupies a decisive position in the investigations of the quantum Hall effects. Recently, the experiment realized the quantum phase transition from the Vortex phase to the Meissner phase in the ladder system, which also open a new path to implement spin-orbit (SO) coupling in one dimensional quantum gases [2]. The adding Zeeman field leads to rich phase transition processes in spin-orbit-coupled Bose-Einstein condensate [1]. However, it is still an opening subject to explore the ground state properties in the ladder system with the similar Zeeman field.

In analogy to the one-dimensional SO coupling system with the Zeeman field, the adding controlled energy bias between the two legs of the ladder can realize a biased bosonic ladder, which has been shown to possess the essence of such SO coupling system. The ground states and the phase transition conditions are obtained theoretically in this biased ladder system subject to a magnetic field. As shown in Figure 1, rich phases are predicted in the presence of the energy bias, including the Plane Wave I (PWI) phase, the Plane Wave II (PWII) phase, and interestingly, a new phase named Moving Biased Vortex phase (MBVP). For the fixed interaction strength [Fig. 1 (a)], the increasing energy bias ω and rung-to-leg coupling ratio $\tilde{K}$ lead to the phase transition from the MBVP to the PWI phase. Then with further increase of the energy bias and the coupling ratio strength, the ladder system enters PWII phase. For the fixed energy bias [Fig. 1 (b)], the direct phase transition between the MBVP and PWII phase appears with the increase of the interaction strength.

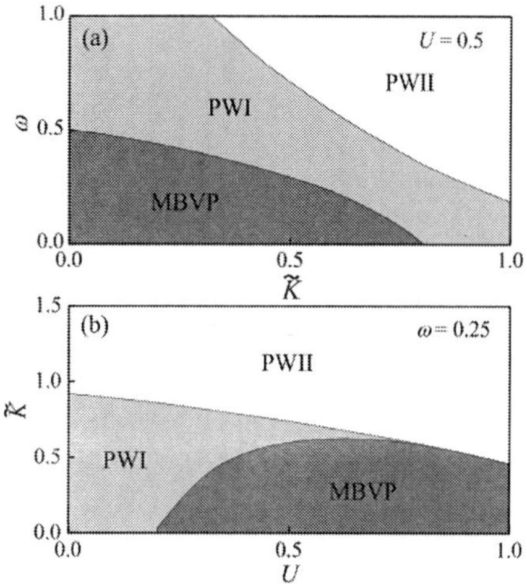

Figure 1. (color online) Phase diagrams of the two-leg biased bosonic ladder with the magnetic flux $\phi = \pi/2$ in $(\tilde{K}, \omega)$ plane (a) with the interaction strength $U = 0.5$ and $(\tilde{K}, U)$ plane (b) with the energy bias $\omega = 0.25$.

This work is supported by the National Natural Science Foundation of China under Grants No. 11764039, 11847304, 11865014, 11475027, 11305132, and 11274255, by Natural Science Foundation of Gansu province under Grant No. 17JR5RA076, and by Scientific research project of Gansu higher education under Grand No. 2016A-005.

References

[1] Li J R *et al* 2017 *Nature* **543** 91
[2] Atala M *et al* 2014 *Nat. Phys.* **10** 588

*E-mail: xuejk@nwnu.edu.cn

Atomic and molecular Auger decay in $CHCl_3$ molecule

D N Vasconcelos[1†], A C F Santos[1*], V Morcelle[4], M A MacDonald[2], M M Sant'Anna[1], B N C Tenório[3], A B Rocha[3], V B Bonfin[1], N Appathurai[2] and L Zuin[2]

[1]Instituto de Física, Universidade Federal do Rio de Janeiro - 21941-972 Rio de Janeiro, RJ, Brazil
[2]Canadian Light Source Inc., Saskatoon, SK S7N 2V3, Canada
[3]Instituto de Química, Universidade Federal do Rio de Janeiro- 21941-909 Rio de Janeiro, RJ, Brazil
[4]Departamento de Física, Universidade Federal Rural do Rio de Janeiro, CEP 23890-000, RJ, Brasil

Synopsis The dissociation following Cl 2p core–shell excitation in $CHCl_3$ molecule is studied. It is observed the presence of atomic Auger lines associated with the decay of dissociated Cl^* atoms. Calculations were compared with the corresponding experimental photoexcited and photoionized spectra to study the competition between the atomic and molecular Auger decays.

The Auger electron spectra after the chlorine 2p excitation [1] and ionization in the gas phase $CHCl_3$ molecule were investigated both theoretically and experimentally. Experimentally, in this study the molecule was photoionized by soft X-rays from the monochromatic synchrotron radiation in a beamline of the variable line space plane, grating monochromator (VLS-PGM) energy range from 5.5 eV to 250 eV, of the Canadian Light Source Inc. synchrotron radiation laboratory (Saskatoon, Canada). The molecular and atomic auger transitions are examined and assigned.

The atomic decay regime is expressed by sharp features in energy, suggesting well-defined initial and final states [1]. The photoexcited or photoionized molecule gives rise to the transition of an electron with a vacancy in a core electron shell that is strongly unstable and short-lived.

The molecular decay regime, on the other hand, represented as broad and structureless peaks and the widths of the broad structures reflect the slope of the dissociative molecular potential.

A series of high-level ab initio quantum mechanical calculations have been performed at multiconfigurational self-consistent field (MCSCF) and multireference configuration interaction (MRCI) levels of theory to compute the Cl L2,3VV Auger lines [2]. Calculations were compared with the corresponding experimental photoexcited and photoionized spectra.

Acknowledgments

The work is financially supported in part CNPQ. One of us, A.C.F Santos, acknowledges support from CAPES. We express our gratitude to staff of CLS for their valuable help during the course of the experiments. Research described in this paper was performed at the Canadian Light Source, which is supported by the Canada Foundation for Innovation, the Natural Sciences and Engineering Research Council of Canada, the University of Saskatchewan.

References

[1] Menzel A, Langer B, Viefhaus J, Whitfield S B and Becker U 1996 Chem. Phys. Lett. **258** 265
[2] Santos A C F, Vasconcelos D N B, MacDonald M A, Sant'Anna M M, Tenório B N C, Rocha A B, Morcelle V, Appathurai N and Zuin L 2018 *J. Chem. Phys.* **149** 054303

† E-mail: deboranb@ufrj.br
* E-mail: toni@ufrj.br

A traveling wave Zeeman decelerator

T Damjanovic[1], N Vanhaecke[2,3,†], J Cromiéres[3], H Haak[2], G Meijer[2], S willitsch[1] and D Zhang[1,‡]*

[1]Department of Chemistry, University of Basel, Klingelbergstrasse 80, Basel, 4056, Switzerland
[2]Fritz-Haber-Institute der Max-Planck-Gesellschaft, Garadayweg 4-6, 14195 Berlin, Germany
[3]Laboratoire Aimé Cotton, CNRS, Université Paris-Sud, ENS Cachan, Université Paris-Saclay, 91405 Orsay, France
[†]Present address: European Patent Office, Patentlaan 2, 2288 EE Rijswijk, The Netherlands
[‡]Present address: Institute of Atomic and Molecular Physics, Jilin University, Qianjin Avenue 2699, Changchun, 130012, Peoples Republic of China

Synopsis We developed a new-concept Zeeman decelerator which produces a traveling magnetic trap. Atoms and molecules possessing a magnetic dipole moment, in so-called low field seeking states, are trapped around a node of a propagating wave provided that the initial velocity of the wave matches a velocity populated in the supersonic beam. In addition, three dimensional confinement is achieved by controlling the radial orientation of the trap, which can be done fully independently from its longitudinal motion.

The development of new methods to decelerate and trap molecules paves the way for a range of studies, such as high-resolution spectroscopy, collision experiments or chemical reaction studies. While Stark decelerator allows one to manipulate the longitudinal motion of polar molecules, Zeeman deceleration uses time-dependent, inhomogeneous magnetic fields to control the motion of paramagnetic molecules [1, 2]. Zeeman deceleration constitutes an alternative approach to decelerating molecules, especially for non-polar molecules, provided that they possess a magnetic dipole moment.

We developed a traveling-wave Zeeman decelerator based on a novel double helix coil geometry. It produces a traveling-wave magnetic field with controllable velocity and orientation. Atoms and molecules possessing a magnetic dipole moment, in so-called low field seeking (LFS) quantum states, are confined around the trap minimum, provided that the initial velocity of the trap matches a velocity class populated in the supersonic beam. In addition, the magnetic fields provide real-time tri-dimensional confinement of the particles in LFS states, in analogy with the traveling-wave Stark decelerator [3].

We tested the decelerator with OH radical molecules. The OH molecules are generated using a pulsed discharge valve by dissociating water molecules seeded in Kr. The mean initial velocity of the OH molecules is around 445 m/s. The OH molecules in the ground state X $^2\Pi_{3/2}, J = 3/2, M_J = 3/2$ are guided through the decelerator with the moving trap velocity matching the mean velocity of the OH beam as shown in Fig.1. The trajectory simulation can reproduce the experimental results with no free parameters involved.

*E-mail: dongdongzhang@jlu.edu.cn

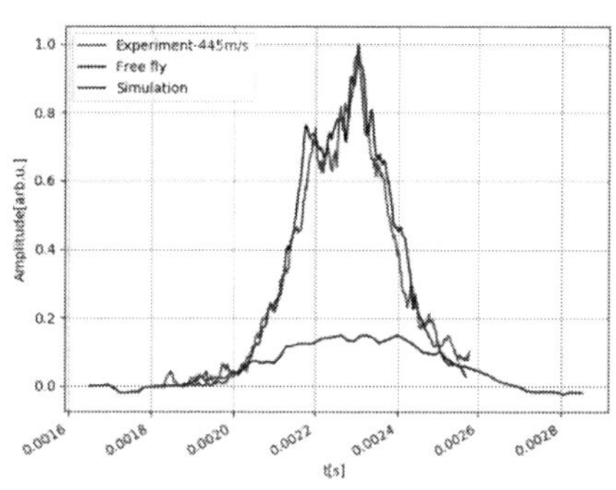

Figure 1. Comparison of the TOF (time of flight) profiles with decelerator on (red) and off (black). In the experiment, the decelerator is used to guide (without any deceleration or acceleration) OH molecules with the mean velocity of 445 m/s through the decelerator.

References

[1] van de Meerakker S Y T *et al* 2012 *Chem. Rev.* **112** 4828
[2] Vanhaecke N *et al* 2007 *Phys. Rev. A* **75** 031402
[3] Osterwalder A *et al* 2010 *Phys. Rev. A* **81** 051401

Calculations of long-range three-body interactions for Li($2\,^2S$)-Li($2\,^2S$)-Li$^+$($1\,^1S$)

P G Yan[1,2], L Y Tang[2] *, Z C Yan[1,2,3] and J F Babb[4]

[1]Department of Physics, University of New Brunswick, Fredericton, New Brunswick, E3B5A3, Canada
[2]State Key Laboratory of Magnetic Resonance and Atomic and Molecular Physics, Wuhan Institute of Physics and Mathematics, Chinese Academy of Sciences, Wuhan 430071, People's Republic of China
[3]Center for Cold Atom Physics, Chinese Academy of Sciences, Wuhan 430071, People's Republic of China
[4]ITAMP, Harvard-Smithsonian Center for Astrophysics, Cambridge, Massachusetts 02138, USA

Synopsis Using perturbation theory for energies, we evaluate the additive and nonadditive interaction coefficients C_4, C_6, C_7, C_8 and C_9 for the Li($2\,^2S$)-Li($2\,^2S$)-Li$^+$($1\,^1S$) system. The obtained coefficients C_n are evaluated with highly accurate variationally-generated nonrelativistic wave functions in Hylleraas coordinates. The nonadditive interaction coefficients depend on the geometrical configurations of this three-body system and on the different position of the ion for each configuration. Our calculations may be of interest for the study of three-body recombination and for constructing precise potential energy surfaces.

In the present work, using perturbation theory up to the fourth order, we derive the formulas for the long-range interaction coefficients for the Li($2\,^2S$)-Li($2\,^2S$)-Li$^+$($1\,^1S$) system, and evaluate these coefficients with highly accurate, variationally determined wave functions in Hylleraas coordinates.

Table 1. The additive interaction coefficients $C_n^{(23)}$ of the $^\infty$Li($2\,^2S$)-$^\infty$Li($2\,^2S$)-$^\infty$Li$^+$($1\,^1S$) system in atomic units.

Coefficient	Value
$C_4^{(23)} = C_{4,ind}^{(23)}$	82.0563(6)
$C_{6,ind}^{(23)}$	711.631(1)
$C_{6,disp}^{(23)}$	3.3208(5)
$C_6^{(23)} = C_{6,ind}^{(23)} + C_{6,disp}^{(23)}$	714.951(1)
$C_{8,ind}^{(23)}$	19824.64(1)
$C_{8,disp}^{(23)}$	152.884(3)
$C_{8,hyp}^{(23)}$	127.5(9)
$C_8^{(23)} = C_{8,ind}^{(23)} + C_{8,disp}^{(23)} + C_{8,hyp}^{(23)}$	20105.0(9)
$C_7^{(23)} = C_{7,DDQ}^{(23)}$	27143.0(2)
$C_9^{(23)}$	1080753.2(1)

We present the additive interaction coefficients $C_4^{(IJ)}$, $C_6^{(IJ)}$, $C_7^{(23)}$, $C_7^{(31)}$, $C_8^{(IJ)}$, $C_9^{(23)}$, $C_9^{(31)}$ and their contributions to each from the induced parts; $C_{4,ind}^{(IJ)}$, $C_{6,ind}^{(IJ)}$, and $C_{8,ind}^{(IJ)}$; the dispersion parts $C_{6,dis}^{(IJ)}$ and $C_{8,dis}^{(IJ)}$ and the hyperpolarizability part $C_{8,hyp}^{(IJ)}$, which enters in the fourth order. Here "additive" means pairwise amongst the three particles. An example of $C_n^{(23)}$ for the infinite nuclear masses is listed in Table 1. We also present the third-order nonadditive interaction coefficients $C_9^{(12,23)}$, $C_9^{(31,12)}$, $C_7^{(12,23,31)}$, $C_9^{(12,23,31)}$. These nonadditive coefficients depend on the geometrical structure of these three particles. In addition, for some configurations, these nonadditive interaction coefficients also depend on the different positions of the ion. Here 'nonadditive" means collectively amongst the three particles [1, 2] An example of $C_{7,DD}^{(12,23,31)}$, $C_{9,DDD}^{(12,23,31)}$ for the infinite nuclear masses with different ion positions (Figure 1) is listed in Table 2.

*E-mail: lytang@wipm.ac.cn

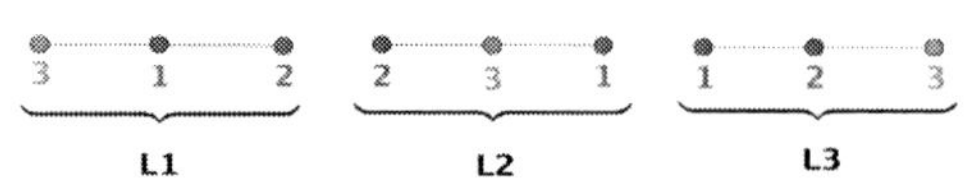

Figure 1. Three different types of straight lines L1, L2 and L3 with different ion positions, where the ion is placed on the position 3.

Table 2. The nonadditive interaction coefficients $C_{7,DD}^{(12,23,31)}$ and $C_{9,DDD}^{(12,23,31)}$ of the $^\infty$Li($2\,^2S$)-$^\infty$Li($2\,^2S$)-$^\infty$Li$^+$($1\,^1S$) system in atomic units.

Coefficient	L1	L2	L3
$C_{7,DD}^{(12,23,31)}$	53866(1)	−53866(1)	53866(1)
$C_{9,DDD}^{(12,23,31)}$	535.92(2)	535.92(2)	535.92(2)

References

[1] Yan P G *et al* 2016 *Phys. Rev. A* **94** 022705
[2] Yan P G *et al* 2018 *Phys. Rev. A* **97** 042710

Non-adiabatic quantum dynamics of the ultracold Li+LiNa→ Li_2+Na chemical reaction

B K Kendrick[1]*, M Li[2], H Li[2], S Kotochigova[2], J F E Croft[3] and N Balakrishnan[4]†

[1] Theoretical Division (T-1, MS B221), Los Alamos National Laboratory, Los Alamos, NM 87545, USA
[2] Department of Physics, Temple University, Philadelphia, Pennsylvania 19122, USA
[3] Department of Physics, University of Otago, Dunedin, New Zealand
[3] The Dodd Walls Centre for Photonic and Quantum Technologies, New Zealand
[4] Department of Chemistry and Biochemistry, University of Nevada, Las Vegas, NV 89154, USA

Synopsis We report non-adiabatic dynamics of the Li+LiNa→Li_2+Na chemical reaction at cold and ultracold temperatures employing accurate ab initio electronic potential energy surfaces in a quantum dynamics formulation employing a diabatic representation. Results are compared against those from a single adiabatic ground state potential energy surface and a universal model based on the long-range interaction potential. We discuss signatures of non-universal behavior in the total rate coefficients as well as strong non-adiabatic effects in the state-to-state rotationally resolved rate coefficients.

Ultracold molecules, polar molecules in particular, are now at the forefront of precision spectroscopy, sensing, controlled studies of chemical reactions, quantum many-body physics, and quantum computing. Polar molecules comprised of heteronuclear alkali metal dimers such as KRb, NaK and LiNa have attracted considerable attention in recent years in controlled studies of chemical reactions. Electronically non-adiabatic effects are expected to play an important role in atom-dimer reactions involving these systems, e.g., K+KRb or Li+LiNa. These systems exhibit a conical intersection (CI) between the ground electronic state and the first excited doublet electronic state. The CI becomes accessible even for collision energies in the ultracold limit for ground state reactants. Explicit quantum calculations for these reactions remain a formidable challenge even for dynamics on a single adiabatic electronic potential surface.

Here, we report a first principles quantum dynamics study of non-adiabatic effects in Li+LiNa(v = 0, j = 0) →$Li_2(v', j')$+Na reaction and compare results against a calculation that neglects non-adiabatic and geometric phase effects as well as that obtained using a universal model based on long-range forces. Details of the quantum dynamics method are given elsewhere [1]. Figure 1 compares non-thermal rate coefficients for total angular momentum quantum number $J = 0$ (s-wave scattering in the incident channel) from the two-state calculations within a diabatic representation against a single surface calculation that excludes any geometric phase and non-adiabatic effects. The results correspond to even exchange symmetry (exchange of the two identical Li atoms). It is seen that the limiting values of the total and vibrationally resolved rate coefficients differ by factors of about 1.2 and 1.6. For rotational resolved rates, the results differ by an order of magnitude or more between 2-state and single surface calculations.

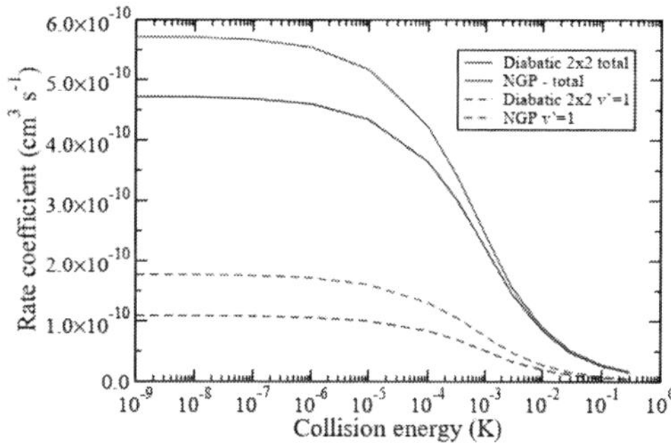

Figure 1. Rate coefficients for the Li+LiNa(v = 0, j = 0) → $Li_2(v')$+Na reaction as a function of the collision energy. The diabatic 2×2 results correspond to the two-state calculation that implicitly accounts for the geometric phase effect. The NGP (no-geometric phase) result corresponds to a single ground state adiabatic potential. The results are for even exchange symmetry of the identical Li atoms and do not include nuclear spin-statistics factors.

References

[1] Kendrick B K 2018 *J. Phys. Chem.* **148** 044116

*E-mail: bkendric@lanl.gov
†E-mail: naduvala@unlv.nevada.edu

Journal of Physics: Conference Series **1412** (2020) 122017 doi:10.1088/1742-6596/1412/12/122017

Mutual Neutralization in Cl^- + C^+ Collisions

P Hedvall[1] *, Larson[1] † and A E Orel[2]

[1]Department of Physics, Stockholm University, Albanova University Center, S-106 91 Stockholm, Sweden
[2]Department of Chemical Engineering, University of California, Davis, CA 95616, USA

Synopsis

The mutual neutralization reaction of Cl^-+ C^+ $\rightarrow$ Cl + C is studied from first principles. The cross section and the thermal rate coefficient are calculated for low collision energies, and the effect of spin-orbit coupling is estimated using a semi-empirical method. Recent experiment has measured a thermal rate coefficient [1], which allows for a comparison between theoretical and experimental studies of the reaction.

Mutual neutralization of oppositely charged ions is a reaction driven by non-adiabatic couplings. The cross section and the thermal rate coefficient for mutual neutralization of Cl^-+ C^+ is calculated from first principles. The 5 lowest molecular electronic states of $^2\Sigma^+$ symmetry and 7 lowest electronic states of $^2\Pi$ symmetry, as displayed in Figure 1, together with the associated non-adiabatic couplings have been calculated at internuclear distance between 1.5 to 10 a_0 using the multi reference configuration interaction method. The potential energy curves describing the electronic states have been transformed into a strictly diabatic basis, resulting in a set of coupled differential equations describing the motion of the nuclei. These were solved numerically and the cross section for neutralization into C+Cl fragments was obtained [2].

The importance of the spin-orbit coupling was investigated using a semi-empirical method [3], where the spin-orbit Hamiltonian was approximated using the asymptotic atomic spin-orbit coupling parameter of Cl(^{2}P). For this, all doublet and quintet molecular states associated with the relevant asymptotic limits where also calculated and included in the model.

Without the spin-orbit coupling the thermal rate coefficient for mutual neutralization in Cl^- + C^+ at 300 K was calculated to be 3.7×10^{-10} cm^3 s^{-1}, which is roughly an order of magnitude smaller than the experimentally obtained value of $4.2 \pm 1.1 \times 10^{-9}$ cm^3 s^{-1}. The effect of spin-orbit coupling was found to be minor, resulting in a thermal rate coefficient of 3.9×10^{-10} cm^3 s^{-1}. A subject for future study is the effect of associative ionization, which may be a possible source for the discrepancy between the measured and calculated results.

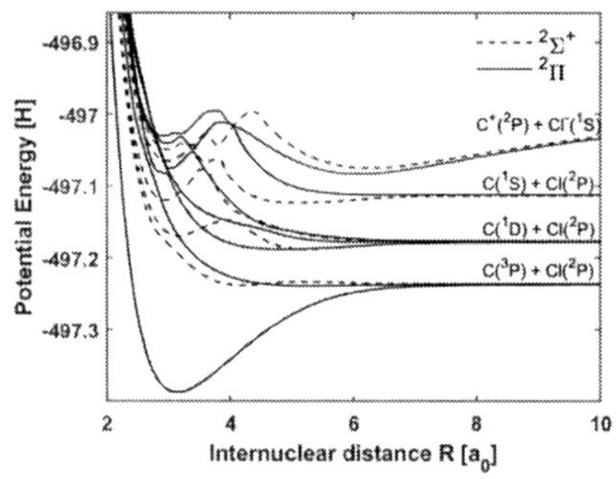

Figure 1. Potential energy curves of CCl showing the 5 lowest molecular electronic states of $^2\Sigma^+$ symmetry and 7 lowest electronic states of $^2\Pi$ symmetry with corresponding asymptotic atomic limits.

References

[1] Sawyer J C *et al* 2019 *to be published*
[2] Stenrup M *et al* 2009 *Phys. Rev. A.* **79** 012713
[3] Cohen J S *et al* 1974 *J. Chem. Phys* **61** 3230

*E-mail: patrik.hedvall@fysik.su.se
†E-mail: aasal@fysik.su.se

Reaction Microscope experiments with shaped laser pulses

D Martínez Farfán[1]*, P Fross[1], S Kerbstadt[2], Y Mi[1], N Camus[1], X Sun[1], M Wollenhaupt[2], T Pfeifer[1] and R Moshammer[1]

[1]Max-Planck-Institut für Kernphysik, Heidelberg, 69117, Germany
[2]Carl von Ossietzky Universität Oldenburg, Institut für Physik, Oldenburg, D-26129, Germany

Synopsis Using a 4f pulse shaper in combination with a Reaction Microscope we performed 800 nm pump-probe experiments with orthogonally polarized pulses on H_2.

Laser control by pulse shaping [1] is a well known technique that allows to customize the optical properties of laser pulses, like polarization, relative phases and time delays. In recent experiments it was used to control the electron emission direction and symmetry in strong field ionization of atoms or molecules. Therefore, combining the potential of the Reaction Microscope to record all particles in coincidence and to obtain their 3D momentum distributions, with the versatility of the pulse shaper technology holds promise for interesting experiments.

In this work, using a pulse shaper with 4f-geometry and a dual-layer Liquid-Crystal Spatial Light-Modulator (LC-SLM), two time-delayed pulses have been generated in order to perform pump-probe experiments with H_2. In addition, the results are compared with a previous measurement where a conventional Mach–Zender interferometer was used to generate the pulse pairs.

In the present experiment, linearly polarized light at 800 nm with a pulse duration of ~35 fs has been used to feed the pulse shaper. The achieved intensity is in the order of 10^{14} W/cm^2. The pump-probe measurement covered a time-range from -600 to 600 fs.

In order to access the dynamics, the transversal and longitudinal momenta of the charged particles created in the interaction have been reconstructed. For H_2, three well known main ionization channels are active: Single ionization (SI) leading to an H_2^+ ion, the dissociative ionization (DISS) leading to an H^+ and an H fragment, and the coulomb explosion (CE) leading to two H^+ ions.

* E-mail: denhi.martinez@mpi-hd.mpg.de

Figure 1 shows preliminary results of our experiment.

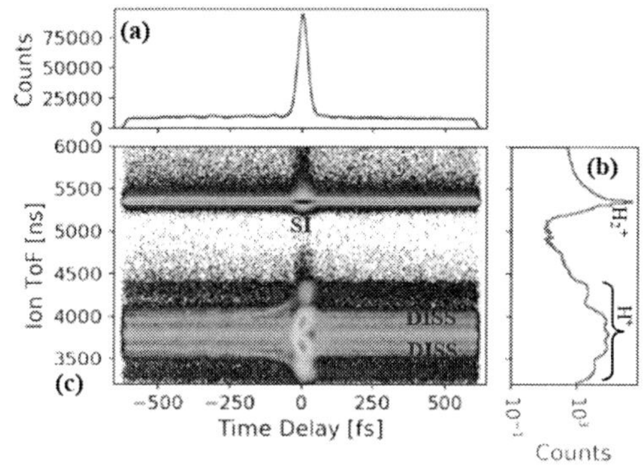

Figure 1. (a) Projection of (b) onto TOF-axis. (b) Ion time-of-flight (TOF). (c) TOF versus pump-probe delay.

In Fig. 1 the ion time-of-flight versus the time delay is shown with their respective projections. In Fig.1 (a) the typical enhancement of ionization near time zero is shown. The ion time-of-flight distribution is shown in Fig. 1 (b) where the fragment ions are indicated. And in Fig. 1 (c) the three main channels can be distinguished. Due to the orthogonal polarization of pump and probe a central peak in between the dissociation channels is clearly visible. The Coulomb-explosion channel appears weaker in comparison with the dissociation and also it is possible to observe dynamics between those channels.

Reference

[1] Brixner T and Gerber G 2001 *Opt. Lett.* 26 557-559

Ionization of large helium nanodroplets

L Lundberg[1 *], F Laimer[1], P Martini[1], L Kranabetter[1], L Tiefenthaler[1], S Albertini[1], F Zappa[1], M Gatchell[1,2] and P Scheier[1 †]

[1] Institute for Ion Physics and Applied Physics, University of Innsbruck, Austria
[2] Department of Physics, Stockholm University, Sweden

Synopsis We have produced and mass-selected large, highly charged helium droplets containing up to 10^{10} He atoms and several tens of charges. They are stable with both multiple positive and negative charge carriers and are stabilized by losing small fragments of excess charge instead of disintegrating (highly asymmetric fission).

Helium droplets containing between 10^5 and 10^{10} He atoms [1] were formed and subjected to electron beams of defined energy and current [2]. Mass per charge distributions were determined by electrostatic energy analyzers for positively and negatively charged droplets [3]. Utilizing two ionization sources, each followed by a spherical sector field analyzer, it was possible to unambiguously determine the charge state as well as the mass of charged He droplets. Charge states higher than 65 for positively charge droplets and 5 for negatively charge droplets could be assigned, and neutralization as well as increasing of the charge state could be achieved with the second ion source. Perfect fractional numbers of the mass per charge ratios of product ions with respect to the charged precursor droplets (see panel b) of Figure 1) indicate that fragmentation exclusively happens via very asymmetric Coulomb explosion and evaporation of He atoms, with the mass of the parent droplets decreasing only very slightly. These low-mass ions have been investigated by mass spectrometer systems in past experiments [2]. Coulomb repulsion between charges of same polarity in highly-charged He droplets will lead to minimum energy configurations in the form of Coulomb crystals with the charges near the surface of the droplets. Dopants will be more readily polarized and attracted by the charged centers than the surrounding He which thereby act as seeds for cluster growth.

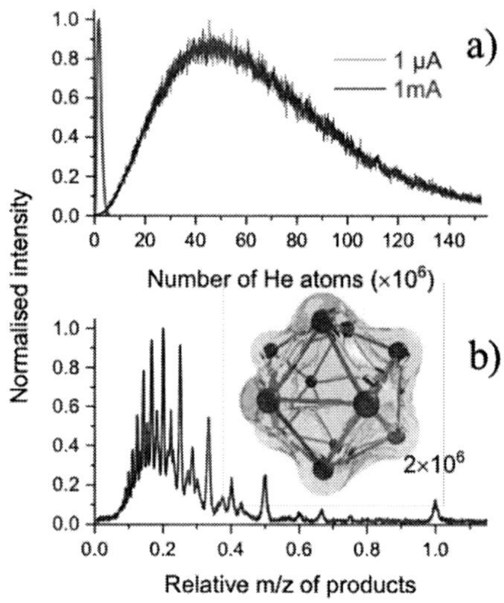

Figure 1. Panel a) shows size distributions measured for soft (1 μA) and hard (1 mA) ionization conditions. The panel designated with b) shows the mass per charge distribution of product ions formed upon ionization of a size-selected precursor droplets consisting of $2 \times 10^6 \times z$ He atoms. The inset is an artists view of a Coulomb crystal in a large helium droplet.

This work was supported by the EU commission, EFRE K-Regio FAENOMENAL EFRE 2016-4

References

[1] Gomez L F *et al* 2011 *J. Chem. Phys.* **135** 154201
[2] Mauracher A *et al* 2018 *Phys. Rep.* **751** 1
[3] Henne U and Toennies J P *et al* 1998 *J. Chem. Phys.* **108** 9327

*E-mail: linnea.lundberg@uibk.ac.at
†E-mail: paul.scheier@uibk.ac.at

Solitons in two-leg bosonic ladder subject to an artificial magnetic field

L X Cai[1], **A X Zhang**[1]* **and J K Xue**[1]†

[1]College of Physics and Electronics Engineering, Northwest Normal University, Lanzhou 730070, China

Synopsis The nonlinear dynamics in two-leg ladder system subject to an artificial magnetic field are discussed in detail. The critical conditions for the formation of soliton and for the transition from diffusion to self-trapping are presented, and confirmed by the numerical simulations of the full discrete nonlinear Schrödinger equation.

Two-leg bosonic ladder in the presence of an artificial magnetic field created by photon-assisted tunneling attracts considerable attention, which has been widely studied and discussed both theoretically and experimentally [1, 2]. Since its applicability and simplicity for study the response of neutral atom to magnetic field. So far, the topic which have been discussed most in this system are Meissner phase and vortex phase characterized by the behavior of the chiral edge current, and the quantum phase transition between the phases of Meissner and vortex [3, 4]. However, rich nonlinear dynamics, e.g., dynamical localization phenomena, such as soliton and self-trapping, are not discussed in the system.

In this paper, we pay our attention to the nonlinear dynamics of BECs trapped into two-leg ladder subject to an artificial magnetic field. We have shown that such a system really possesses rich nonlinear phenomena in the regions of negative and positive effective mass. The self-trapping and diffusion can be observed when effective mass is negative, and there are soliton and breather besides self-trapping and diffusion in the region of positive effective mass. Furthermore, the coupled effects of the repulsive atomic interaction g, the initial population difference of ultracold atoms between the two legs s_0, the artificial magnetic flux ϕ and the hopping matrix elements along the rungs of the ladder K on the dynamics have been discussed in detail. The solitonic and self-trapped phenomena are depicted in Fig. 1. The study of nonlinear dynamics for two-leg bosonic system serve as a suitable tool for understanding the properties of atomic gases coupled to artificial magnetic field, and paved the way for realizing quantum transport.

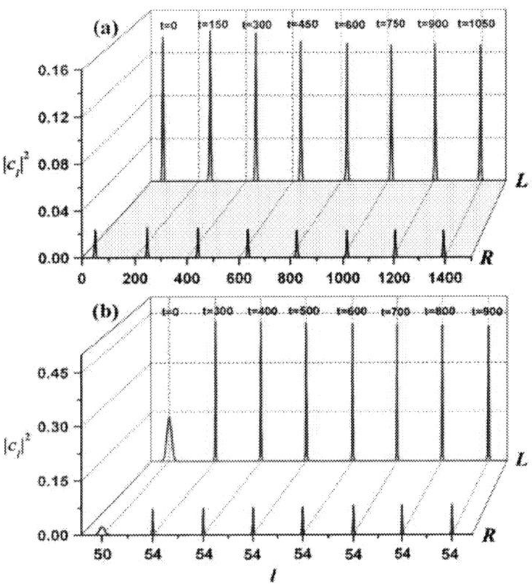

Figure 1. (color online) The soliton (a) and self-trapped (b) solutions against time t and lattice site l. (a) with $\theta_0 = \pi$, $s_0 = 0.7$, $p_0 = 0.5\pi$, $g = 0.95$, $\phi = 5.47$ and $K = 1.48742$, (b) is the same as (a) but $g = 10$, $K = 1.85206$. The blue and red lines represent the evolution of Gaussian wave packets with time in right (R) and left (L) leg, respectively.

This work is supported by the National Natural Science Foundation of China under Grants No. 11764039, 11847304, 11865014, 11475027, 11305132, and 11274255.

References

[1] Atala M *et al* 2014 *Nat. Phys.* **10** 588
[2] Natu S 2015 *Phys. Rev. A* **92** 053623
[3] Piraud M *et al* 2015 *Phys. Rev. B* **91** 140406(R)
[4] Greschner S *et al* 2018 *Phys. Rev. A* **97** 033619

*E-mail: zhangax@nwnu.edu.cn
†E-mail: xuejk@nwnu.edu.cn

Development of a cryogenic ion mobility spectrometer for an isomer-selected photoexcitation and reaction studies

Y Shiina[1], **K Takaya**[2], **H Ueta**[3], **T Hirayama**[1] **and Y Nakano**[1]*

[1]Department of Physics, Rikkyo University, Tokyo 171-8501, Japan
[2]Center for Preventive Medical Science, Chiba University, Chiba 277-0882, Japan
[3]Advanced Science Research Center, Japan Atomic Energy Agency, Ibaraki 319-1195, Japan

Synopsis A new development aiming experimental studies of isomer-selected gas-phase reactions and photoexcitation processes is being conducted. The experiment is based on isomer separation by an ion mobility spectrometry and ion accumulation in a temperature-variable radiofrequency ion trap. The laser plasma light source at Rikkyo University is currently under an upgrade for providing ultraviolet light at the wavelength region from 90 to 200 nm. We report on the details of the experimental setup and current status of the development.

Ion mobility spectrometry (IMS) is a powerful technique to separate isomers, the molecules having the same chemical composition but different structures, which measures the difference of their drift velocity in a buffer gas. While IMS technique is rather established in analyzing large biomolecules in particular drugs and explosives, a recent experiment showed that the method still has a high resolving power for smaller molecules when the buffer gas is cooled down [1]. Combining the cryogenic IMS devices with ion trapping techniques, one is provided the opportunity to carry out an isomer-selected studies of gas-phase reactions and photoexcitation processes. In the context of astrochemistry, Cuadrado et al. observed using the IRAM 30 m telecsope that the abundance ratios of tran-cis formic acid (HCOOH) is strongly depending on the position in the Orion Bar photodissociation region. This may suggest that the photoisomerization of formic acid molecules is induced by the far UV radiation from neighboring objects.

FUV light is available using a laser plasma light source at Rikkyo University [2], which is based on plasma emission from a metal target irradiated by a 532 nm Q-switched Nd:YAG laser at a repetition rate of 50 Hz and a pulse energy of 500 mJ. In order fo choose an optimum metal target for the wavelength region from 90 to 200 nm, the photon energies lower than the hydrogen ionization limit, we have carried out a spectral simulation using the Flexible Atomic Code (FAC) for various metals, *i.e.*, tungsten, copper, molybdenum etc. A dedicated FUV beamline is under development currently with a simple monochrometer (ACTON VM-502) to test the first operation of the laser plasma source at the FUV range.

The existing IMS source apparatus has been reconstructed to match the beamline of the laser plasma light source, and tested under the room temperature with an electron impact ionization source. Beside this, an electrospray ion source and a temperature-variable ion trap are planned to be integrated into the apparatus. The status of the experimental developments will be presented.

References

[1] Takaya K *et al* (in preparation)
[2] Iwabuchi A *et al* 2015 *J. Vac. Soc. Jpn* **58** 165

*E-mail: nakano@rikkyo.ac.jp

Coincident detection of recoil ions and photoelectrons after multiphoton-ionisation of cold ^{6}Li atoms trapped in a dipole trap

N Kurz[1]*, **D T M Mootheril**[1], **S Mezinska**[1], **T Pfeifer**[1] **and A Dorn**[1]

[1]Max Planck Institute for Nuclear Physics, Heidelberg, 69117, Germany

Synopsis We report on performing high resolution recoil ion momentum spectroscopy in a reaction microscope (ReMi) on a target of dipole-force trapped cold ^{6}Li atoms in a non-interacting $|1\rangle$-$|2\rangle$ spin mixture and coincident detection of recoil ions and photoelectrons. Photo-ionization occurs from the electronic ground state of the ^{6}Li atoms and IR photons significantly contribute to the ion yield via 3-photon resonances. High resolution time-of-flight spectra are obtained by modulating the trap depth synced with the arrival of the fs laser pulses.

The implementation of a magneto-optically trapped target in a reaction microscope has led to several studies centered on strong-field ionization of atomic lithium [1, 2, 3]. However while laser-cooled targets are limited to atomic physics, dipole traps also allow to store molecules due to their dynamical polarizability. In order to investigate molecules created by different association techniques from cold atomic gases, we follow the approach of loading atoms from a ^{6}Li MOT into a dipole trap realized by a strongly focused IR beam at 1070 nm ($2w_0 = 96\,\mu m$). This allows for field-free kinematically complete measurement of the momentum distribution of a target with high phase space density [4]. Here we report on energy resolution benchmarking experiments with an atomic sample of ^{6}Li trapped in their electronic ground state $2^2S_{\frac{1}{2}}$ and a target temperature of $\approx 350\,\mu$K, in which ionisation has been performed by a focused beam of fs laser pulses. The MOT is loaded via a Zeeman-slowed atomic beam with equilibrium atom numbers of $\approx 2 \cdot 10^7$ atoms and around 2% of the atoms are transferred into the dipole-force trap. The target was ionised using ≈ 32 fs laser pulses with a photon energy of 1.55 ± 0.1eV. Prior to ionization, the atoms resided in their electronic ground state $2^2S_{\frac{1}{2}}$. Due to the high bandwidth of the fs laser, apart from direct 4-photon ionisation, also 3-photon resonances represent a significant ionization mechanism. By absorption of 3 fs photons, intermediate highly excited states are populated (4p, 5p) from which atoms can get ionized either via absorption of another fs laser photon or by a dipole laser photon ($E_\nu = 1.16\,$eV). This IR assisted ionization via 3-photon resonances leads to a significantly higher ion yield, but also smears out the energy resolution along the spectrometer axis since a time-of-birth time stamp is not available. In order to obtain high resolution time-of-flight spectra, we modulate the trap depth synced with the fs laser, so that the expected arrival of the fs pulse is embedded in a turn-off time-window of about 1 μs. Coincident detection of photoelectrons and ions has been achieved (see Fig. 1). These experiments serve as a benchmark for future experiments, in which homonuclear ^{6}Li dimers are created by photo- or magneto-association, and stored in the optical dipole force trap in order to study their properties using e.g. Coulomb explosion imaging.

*E-mail: kurzn@mpi-hd.mpg.de

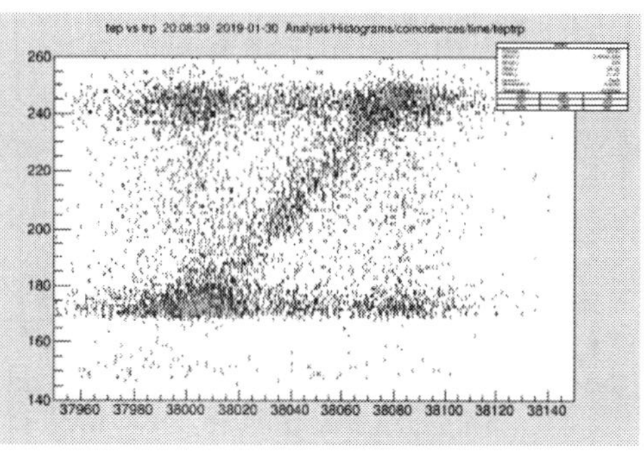

Figure 1. Photoelectron TOF vs recoil ion TOF for ionisation out of dipole trap.

References

[1] Schuricke M *et al* 2011 *Phys. Rev. A* **83** 023413
[2] Zhu G *et al* 2009 *Phys. Rev. Lett.* **103** 103008
[3] Sharma S *et al* 2018 *Phys. Rev. A* **97** 043427
[4] Wessels P *et al* 2018 *Comm. Physics* **1** 32

Low-energy scattering properties of ground-state and excited-state positronium collisions

M D Higgins[1] *, K M Daily[1] and C H Greene[1,2] †

[1]Department of Physics and Astronomy, Purdue University, West Lafayette, Indiana 47907, USA
[2]Purdue Quantum Center, Purdue University, West Lafayette, Indiana 47907, USA

Synopsis Low-energy elastic and inelastic scattering in the Ps(1s)-Ps(2s) channel is treated in a four-body hyperspherical coordinate calculation. Adiabatic potentials are calculated for triplet-triplet, singlet-singlet, and singlet-triplet spin symmetries of coupled electrons and coupled positrons, with total angular momentum L = 0 and positive parity. Preliminary estimates of the s-wave scattering length in the elastic Ps(1s)-Ps(2s) channel were obtained. Spin recoupling is implemented to obtain calculations of experimentally relevant scattering properties for different Ps spin states.

Ground-state Ps collisions have been extensively studied due to growing interests in the creation of a spin-polarized Ps Bose-Einstein condensate (BEC) [?, ?] and references therein. Experimental efforts have been made to probe BEC densities of atomic hydrogen by measuring the cold-collision frequency shift in the spectrum for ground-state and excited-state atomic collisions [?, ?]. For experimentalists to probe Ps BEC densities using the same techniques, knowledge of s-wave scattering lengths for ground-state excited-state atomic collisions should be well characterized. This recognition forms the motivation for this study.

The four-body Hamiltonian is investigated to study the interaction of two electrons and two positrons. The interactions considered in this study include only two-body Coulombic interactions; spin-spin, spin-orbit and relativistic interactions are neglected. This problem is solved using hyperspherical coordinates, which treats fragmentation pathways on an equal footing. The Hamiltonian is solved in a coupled electron and coupled positron spin basis in order to simplify the anti-symmeterization requirements of the total wavefunction. States with total orbital angular momentum $L = 0$ and positive parity are considered in this work in representing s-wave collisions.

From the adiabatic/diabatic potential curves, s-wave scattering properties were investigated in the Ps($1s$)-Ps($2s$) elastic and inelastic channels. Spin recoupling was implemented to represent experimentally relevant spin states of each Ps atom in a two-body collision, both in spin states coupled to total spin S_{tot} and uncoupled spin states specifying spin and spin projection quantum numbers for both Ps atoms. Scattering cross-sections are provided for the spin-flip conversion from the triplet to singlet spin states to give preliminary estimates of the spin-conversion rate.

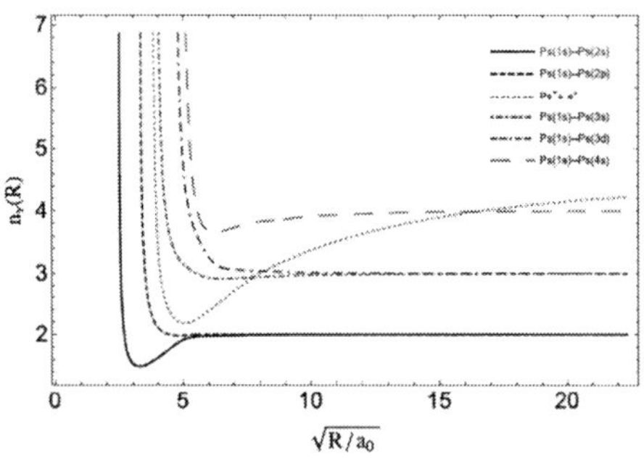

Figure 1. Adiabatic and diabatic potential curves for the singlet-triplet spin configuration shown on an effective principle quantum number scale for the first 6 channels.

This work was supported in part by the U.S. Department of Energy, Office of Science.

References

[1] Ivanov I A *et al* 2001 *Phys. Rev. Lett.* **87** 063201
[2] Daily K M *et al* 2015 *Phys. Rev. A* **91** 012512
[3] Killian T C *et al* 1998 *Phys. Rev. Lett.* **81** 3807
[4] Killian T C 2000 *Phys. Rev. A* **61** 033611

*E-mail: higgin45@purdue.edu
†E-mail: chgreene@purdue.edu

A linear polar molecule in a two-color laser field: a symmetry analysis

D Mellado-Alcedo[1], N R Quintero[2] and R González-Férez[1*]

[1] Instituto Carlos I de Física Teórica y Computacional and Departamento de Física Atómica, Molecular y Nuclear, Universidad de Granada, 18071 Granada, Spain
[2] Departamento de Física Aplicada I, Universidad de Sevilla, 41011 Sevilla, Spain

Synopsis We describe the field-dressed rotational dynamics of a linear polar molecule in a a non-resonant two-color cw laser field, and analyze the orientation and alignment as the field parameters are varied.

An experimental technique to induce molecular orientation is based on the use of non-resonant two-color laser fields [1], which couples pendular states with different parity due to the interaction with the molecular hyperpolarizability. In this work, we investigate the rotational dynamics of a linear polar molecule in a a non-resonant two-color cw laser field. We consider non-resonant two-color laser fields having parallel polarizations in the regime where the time-average approximation does not hold [2]. To do so, we increase the frequencies of the laser fields till the time-average over the rapid oscillations of the laser field becomes non-negligible. Note that we still assume that the laser field does not drive any electric, vibrational or rotational transitions in the molecule, which is described within the rigid-rotor approximation. Thus, the field-dressed rigid-rotor Hamiltonian includes the interaction of the two-color electric field with the permanent electric dipole moment, the polarizability, and the hyperpolarizability of the molecule.

In the regime beyond the time-average approximation, we explore the symmetries of the two-color electric field given by the biharmonic function [3] and analyze their impact on the field-dressed rotational dynamics. We investigate the rotational dynamics of a rigid-rotor molecule in detail as the parameters of the two-color electric field, i.e., the frequencies of the two electric field components, their relative phase, and their relative strength, are varied. We show that the alignment and orientation are symmetric and antisymmetric, respectively, as a function of the relative phase of the two components of the electric field. We also provide the field parameters needed to reach the largest orientation.

References

[1] Kraus P M, Rupenyan A and Wörner H J 2012 *Phys. Rev. Lett.* **109** 233903
[2] Pershan P S, van der Ziel J P and Malmstrom L D 1966 *Phys. Rev.* **143** 574
[3] Cuesta J A, Quintero N R and Álvarez-Nodarse R 2013 *Phys. Rev. X.* **3** 041014

*E-mail: rogonzal@ugr.es

Tunnelling of the Bose-Einstein condensates in a tilted shallow trap

Y Jian[1] and J K Xue[1]*

[1]College of Physics and Electronics Engineering, Northwest Normal University, Lanzhou 730070, China

Synopsis Using variational method and numerical simulation, the tunnelling of Bose-Einstein condensates (BECs) in a tilted shallow trap is studied. The stable bound state, the quasi-bound state and the diffusion state are predicted. For a driven shallow trap, the resonance tunnelling phenomenon is observed.

In recent years, the tunnelling of BECs has been studied in various experimental and theoretical work[1]. Recent experiments show that the tunnelling rate is an experimentally measurable parameter, which contains important information about the atomic dynamics in the well and the initial state of the quantum system[2]. So far, the studying of the tunnelling dynamics is still an active research field. Usually, the tunnelling phenomenon is studied in a symmetric trap. However, it is more general to study the tunnelling in an asymmetrical shallow well. In addition, the modulation of trap plays an important role in the tunnelling dynamics of the system.

Using variational method and numerical simulation, the tunnelling of BECs in a tilted shallow trap is discussed in detail. Two critical points for the change of system state are obtained: the critical atomic interaction g_{crit} and the maximum atomic interaction g_{max}. The average tunnelling rate $\overline{\gamma}$ for the tilted shallow trap is concretely shown in Fig. 1 (a). When $g < g_{crit}$, $\overline{\gamma} \to 0$, the system is in the stable bound state and the condensates are completely trapped; When $g_{crit} < g < g_{max}$, the system is in the quasi-bound state and the tunnelling takes place; When $g > g_{max}$, the system is in an unstable state and particles quickly overflow from the potential well. The tunnelling of the system strongly depends on the atomic interaction g. The dynamics of the condensates in the parametrically modulated trap is shown in Fig. 1 (b), where the barrier waist is modulated as $\sigma(t) = \sigma_0(1 + \epsilon\cos(\Omega t))$, σ_0 corresponds to the constant part and ϵ to the amplitude of the modulating part, and Ω is the modulation frequency. The instability of the system behaves obvious resonance character, i.e., as $\Omega \to \Omega_0$ (Ω_0 is inherent frequency of the system), the average tunnelling rate $\overline{\gamma}$ and the maximum energy difference $\Delta E_{max} = E - E_0$ increase rapidly, where E and E_0 are the energies of the modulation system and the non-modulation system respectively. These predictions can be verified experimentally.

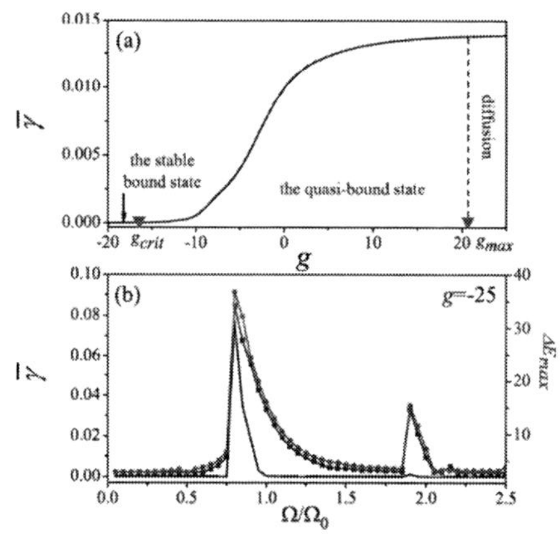

Figure 1. the average tunnelling rate $\overline{\gamma}$ for the tilted shallow trap (a) and the modulated trap (b). Solid line for $\overline{\gamma}$; dot line for ΔE_{max}.

This work is supported by the National Natural Science Foundation of China under Grants No. 11764039, 11847304, 11865014, 11475027, 11305132, and 11274255, by Natural Science Foundation of Gansu province under Grant No. 17JR5RA076, and by Scientific research project of Gansu higher education under Grand No. 2016A-005.

References

[1] Alcala D A *et al* 2018 *Phys. Rev. A.* **98** 023619
[2] Potnis S *et al* 2017 *Phys. Rev. Lett.* **118** 060402

*E-mail: xuejk@nwnu.edu.cn

Formation of H_3^- molecular ions by Three-Body Recombination

H da Silva Jr[1], **O Dulieu**[1*] **and M Raoult**[1†]

[1]Lab. Aimé Cotton, CNRS, Univ. Paris-Sud, ENS Paris-Saclay, Univ. Paris Saclay, Orsay, France

Synopsis The simplest triatomic molecular ion H_3^-, is predicted to be stable, but has not yet been directly observed experimentally. He we model its formation via Three-Body Recombination of cold and trapped H^- ions colliding with H_2 molecules in the presence of a buffer gas of He atoms.

The simplest triatomic molecular ion H_3^-, is predicted to be stable, but has not yet been directly observed experimentally. Cold ion traps currently developed in Innsbruck [1] should allow its formation by Three-Body Recombination (TBR) by inserting a dense H_2 beam seeded in He buffer gas pulse in a 22-pole ion trap containing H^-. We evaluate the corresponding TBR rate constant in order to validate the possibility to observe H_3^- under such experimental conditions. TBR has been widely studied theoretically since twenty years [2, 3], based on a formalism written in hyperspherical collective coordinates, leading to the following TBR cross section:

$$\sigma = \frac{32\pi^2}{k^5}\sum_J (2J+1)\sum_{i,f}|S^J_{f,i}|^2 \qquad (1)$$

where S is the scattering matrix, i and f label the entrance (three-body continuum) and exit (two-body recombination) channels, $k = (2\mu E)^{\frac{1}{2}}$ is the hyperradial momentum in the entrance channel. The rate constant is then given by $K = \frac{k}{\mu}\sigma$. The specificity of this work is that the TBR probability in Eq.1 is rather determined with Jacobi coordinates, which usually describe the atom-diatom continuum, the diatom being in a bound state. However, as implemented in [4] in another context, the diatom dissociation continuum can be accounted for by adding box-states in the rovibrational basis set used to build the set of closed-coupled differential equations which is solved to obtain the S matrix. The discrete probability to form a diatom in a bound state i_o is given by

$$P_{i_o}(\epsilon) = \sum_{b\in box} P_{i_o,b}(\epsilon_b) = \sum_{b\in box} |S_{i_o,b}|^2 \qquad \epsilon_b < \epsilon \qquad (2)$$

where ϵ is the difference between total energy E and the three-body break-up threshold, ϵ_b is the difference between E and the diatom vibrational energy. $P_{i_o}(\epsilon)$ is expressed in terms of a cumulative continuum density of probability

$$P_{i_o}(\epsilon) = \int_0^\epsilon \frac{dP_{i_o}}{d\epsilon'}(\epsilon')d\epsilon', \qquad (3)$$

$$\frac{dP_{i_o}}{d\epsilon}\left(\frac{\epsilon_b + \epsilon_{b+1}}{2}\right) = \frac{P_{i_o,b}(\epsilon_b) + P_{i_o,b+1}(\epsilon_{b+1})}{2(\epsilon_b - \epsilon_{b+1})} \qquad (4)$$

We report in Fig.1 the total TBR density of probability obtained by summing over all the accessible diatom bound states i_o. One can see that it is independent of the box size, as it should be.

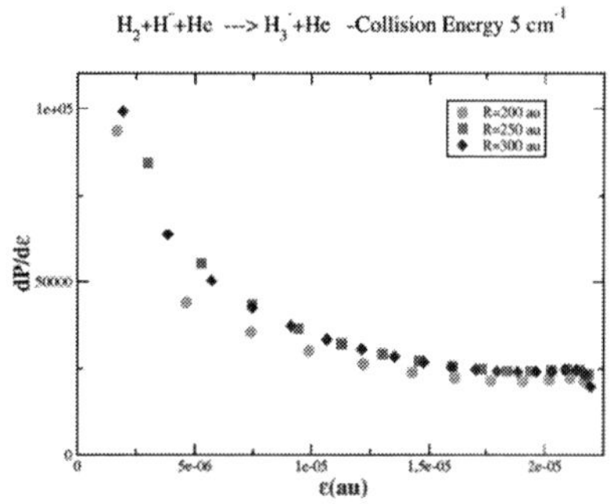

Figure 1. Total density of probability for different box sizes.

From this result, a TBR rate constant $K = 10^{-32}\text{cm}^6/\text{s}$ is obtained for a total angular momentum $J = 0$. This value is already larger than the lowest limit which can be measured in the Innsbruck set-up. The evaluation of the higher partial wave contributions is in progress.

References

[1] Yuen C H *et al* 2018 *Phys. Rev. A* **97** 022705
[2] Esry B D *et al* 1999 *Phys. Rev. Lett.* **83** 1751
[3] Suno H *et al* 2009 *Phys. Rev. A* **80** 062702
[4] Gauyacq J P 1986 *J. Phys. B: At. Mol. Phys.* **19** L477

*E-mail: olivier.dulieu@u-psud.fr
†E-mail: maurice.raoult@u-psud.fr

Journal of Physics: Conference Series **1412** (2020) 122027 doi:10.1088/1742-6596/1412/12/122027

Design and simulation of a novel magneto-optic trap

D M Mootheril*, N Kurz, S Mezinska, T Pfeifer and A Dorn

Max Planck Institut für Kernphysik, Heidelberg, 69117, Germany

Synopsis In order to study ultracold gases by photoionization within a Reaction Microscope (ReMi), we investigate a new magneto-optical trap (MOT) configuration for cooling and trapping the atoms. Here we present the simulation results. In comparison with the conventional MOT, the new MOT has four laser beam axes. Two radial laser beam axes along x and y directions and two laser beam axes inclined $\pm 45^0$ to the MOT/Feshbach coil axis (z axis).

Ultra-cold quantum gases are ideal systems to study few- and many-body phenomena, since they can be prepared with well-defined temperature, density, interaction strength and dimensionality. The final goal of our experiment is to map the mean square of the wave function of an ultracold target in the position space by photoionization. The experiment combines a multiparticle imaging spectrometer, 'Reaction Microscope' (ReMi) [1] and an ultracold fermionic target ^{6}Li. Lithium atoms from an oven are initially cooled by the Zeeman slower and trapped in a MOT. They are transferred to an optical dipole trap and evaporatively cooled further at the BEC side of the Feshbach resonance to create ultracold ^{6}Li dimers. Atoms of these dimers will be ionized by femto second laser pulses instantaneously and the ions gain momentum from the Coulomb repulsion which depends on the initial inter-particle separation. The fragments will be detected by the time and position sensitive detectors of the ReMi and the initial position distribution and thereby the spatial wavefunction of the molecules can be calculated.

To enable projection of ions to the ReMi detector within a strong magnetic field of Feshbach coils, the presently used MOT coils with axis perpendicular to ReMi axis have to be replaced by a pair of MOT/Feshbach coils with axis coinciding with the ReMi axis. As a consequence, laser beams for the MOT cannot be aligned as in the conventional MOT configuration, since the optics for the axial beam of the MOT coil, which is along the ReMi axis, will obstruct the detector. We propose another new configuration for the MOT, in which, instead of one beam along the ReMi axis, two beams inclined $\pm 45^0$ to this axis can be used, while the radial orthogonal beam directions (x and y) remain the same. This give an effective inward force towards the MOT centre along the axial direction.

Due to different magnetic field vectors along the axial and the radial directions of the MOT, the 45^0 beams alone create an effective outward force along the radial direction. This can be compensated by aligning the radial beams 45^0 to the plane of the axial beams and choosing the intensity of each of the axial beams to be half the intensities in one of the radial beams.

We have calculated the 3D trajectories of the atoms moving in the direction of the MOT center in the new laser beam arrangement. The simulation is done using the rate equation model described as in [2].

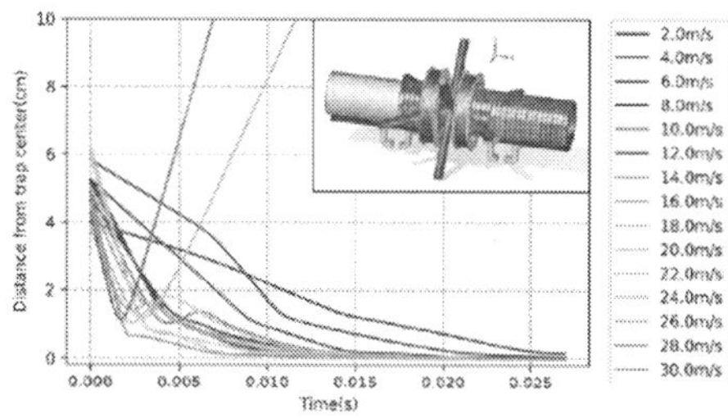

Figure 1. Distance of atoms from the MOT center for different initial velocity groups with time. Laser beam detuning -18.7Γ. Legend shows the initial velocities. Inset shows the new MOT arrangement with ReMi electrodes.

At a higher detuning of -18.7Γ compared to that used in a conventional MOT configuration (-5Γ detuning) the simulation shows the atoms slow down and get trapped at the MOT center. The MOT will be changed to the new configuration and tested later on. This experimental setup is an ideal and novel tool to study the ultracold samples by photoionization.

References

[1] R. Moshammer *et al* 1994 *Phys. Rev. Lett* **73** 3371
[2] M R Tarbutt 2015 *New Journal of Physics* **17** 015007

*E-mail: deepthy.mootheril@mpi-hd.mpg.de

Hybrid Electrostatic Ion Beam Trap (HEIBT): design and simulation of ion-ion and ion-neutral low-energy collisions and ion-laser photoreactions

A Shahi[1], R Singh[2 †], Y Ossia[1], M L Rappaport[2], O Heber[2], D Zajfman[2] and D Strasser[1 *]

[1]The Hebrew University of Jerusalem, Institute of Chemistry, Jerusalem 91904, Israel
[2]Weizmann Institute of Science, Department of Particle Physics and Astrophysics, Rehovot 7610001, Israel

Synopsis A new hybrid electrostatic ion beam trap (HEIBT) is designed and simulated for low-energy ion-ion, ion-neutral merged beam collision studies as well as ion-laser interaction experiments. The HEIBT is made possible by the design of dichroic electrostatic mirrors that reflect or transmit ion beams of different charge and energy. The experimental setup allows fragment imaging detection of the investigated reaction products ejected outside the trap.

Ion reactions with neutral and other ionic species and their interaction with light are at the foundation of chemical sciences. Furthermore, gaseous ion chemistry plays an important role in many interstellar, planetary and other partially ionized environments [1,2]. While typical ion-neutral and ion-ion collision studies are performed using flowing afterglow type measurements, state-of-the-art merged beam experiments allow modeling-free measurement of reaction rates and product branching ratios [3-5]. Several large scale experimental storage ring setups have been designed and constructed to allow merged beam experiments on mass-selected trapped ion beams [6,7].

The electrostatic ion beam trap (EIBT) allows trapping an ion beam between two electrostatic mirrors. Thus it has been proved to be extremely valuable research tool in many research areas including molecular dynamics, spectroscopy, mass spectrometry, lifetime measurement of metastable species and ion collective dynamics [8]. Here we present a hybrid EIBT (HEIBT) design that relies on the development of a "dichroic" electrostatic mirror – allowing for example reflecting a fast anionic beam, while transmitting a cation beam with minimal focusing.

Figure 1 shows the schematic of HEIBT, consists two simultaneously trapped, which are merged in a central field free region. The inner HEIBT mirrors trap the 2.1 keV F^- beam, while transmitting the 4.2 keV F_2^+ beam, which is trapped between the outer HEIBT mirrors. In the central field free region both ion beams are merged at the same lab frame velocity allowing for low-energy collision studies. Preliminary simulation support as low as ~4 meV resolution of the collision energy. Neutral products of such ion-ion interaction as well as cationic products of ion-laser interaction can be readily imaged behind the HEIBT mirrors [9].

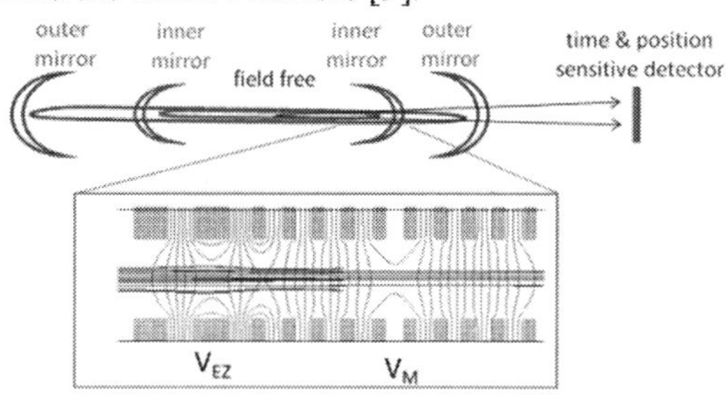

Figure 1. Schematic of the HEIBT showing simultaneous trapping of a merged cation and anion beam (top) as well as simulated cation and anion trajectories through the dichroic HEIBT mirror (bottom).

References

[1] Larsson M *et al* 2012 *Rep. Prog. Phys.* **75** 066901
[2] Geppert W D *et al* 2103 *Chem. Rev.* **113** 8872
[3] Shuman N S *et al* 2014 *J. Chem. Phys.* **140** 224309
[4] Peart B *et al* 1992 *J. Phys. B: At. Mol. Opt. Phys.* **25** 5109
[5] Reutte N d *et al* 2018 *Phys. Rev. Lett.* **121** 083401
[6] Thomas R D *et al* 2011 *Rev. Sci. Instrum.* **85** 065112
[7] Gangwar R K *et al* 2017 *Phys. Rev. Lett.* **119** 103202
[8] Zajfman D *et al* 1997 *Phy. Rev. A* **55** R1577
[9] Shahi A *et al* 2017 *J. Phys. Chem. A* **121** 3037

* E-mail: strasser@huji.ac.il
† E-mail: raj.sing@weizmann.ac.il

 1412 (2020) 122029 doi:10.1088/1742-6596/1412/12/122029

Entropic uncertainty as a predictor of avoided crossing for confined system

S Saha[1*], J Jose[1†] and P C Deshmukh[2]

[1] Department of Physics, IIT Patna, Bihta-801106, Bihar, India
[2] Department of Physics, IIT Tirupati, -517506, Andhra Pradesh, India

Synopsis Avoided Crossings are one of the unique spectroscopic features of confined atomic system. Shannon's information entropy as a predictor of these phenomena by varying the perturbation (confinement strength) on ground state and some of excited states of H atom is studied in this work. In addition to the energy level repulsion at avoided crossing, it is observed that Shannon's entropy also assures the exchange of information of involved states, and also validates entopic uncertainty.

The electronic structure properties of H endohedrally confined by fullerene C_{60} is investigated employing two model potentials 1) an annular square well potential (ASW) [1] and 2) a Gaussian annular square well potential (GASW). Electronic properties such as the radial wavefunctions, probability density distribution, energy spectrum, entropy etc. are studied for these two model cases using Numerov's technique. The momentum-space wave functions are generated from Fourier transforms. A systematic study of the evolution of electronic properties of these confined systems as a function of the strength of the model potential is carried out. As the strength of the confined potential is increased, the energy levels display "avoided crossing" between two different levels of like symmetry. As a consequence of avoided crossing, mirroring of behavior of these states also happens; known as mirror collapse [2, 3]. Sensitivity of avoided crossing and mirror collapse to different model potentials in the case of H are studied in this work.

The information entropy, or Shannon entropy, in atomic system reveals deeper knowledge of the electron correlations, density function, and distinct spectroscopic phenomena in investigating the structure of atomic system. The Shannon entropy of the normalized one electron density $\rho(\vec{q})$ in position / momentum space is defined as [4]

$$S = -\int \rho(\vec{q}) \ \ln \rho(\vec{q}) \ d\vec{q}. \qquad (1)$$

Shannon entropy gives quantitative information about the delocalization of the electronic cloud. Variation of the Shannon entropy (Fig.1) of H under external confinement using different model potentials, with an emphasis in the region of avoided crossing is also studied in this work.

Dramatic change in the entropy is observed at the avoided crossing. Information regarding delocalization of electrons as the confinement varies is studied for the first time, to the best of our knowledge, in endohedral atomic systems. The present work demonstrates through the radial and momentum entropy that the confined electron is more delocalized in the GASW model. In addition, avoided crossing occurs at different strengths of confinement for different model potentials. It is clear from the present work that entropy is an excellent spectator of avoided crossing.

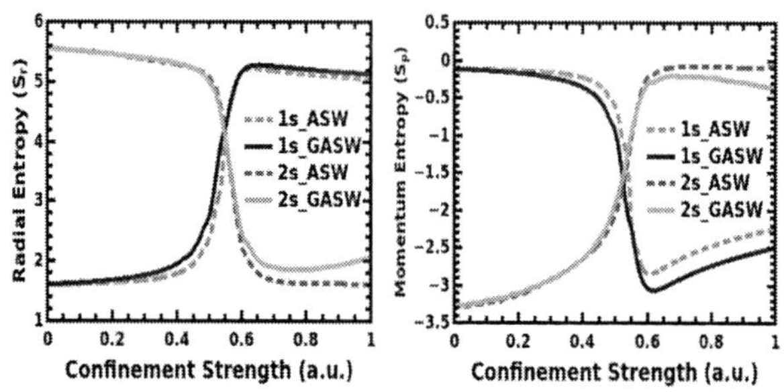

Figure 1: Radial Entropy (S_r) and Momentum Entropy (S_p) as a function of confinement strength.

References

[1] Varma H R, Deshmukh P C, Dolmatov V K and Manson S T 2007 *Phys. Rev. A* **76** 012711
[2] Connerade J P, Dolmatov V K, Lakshmi P A and Manson S T 1999 *J. Phys. B* **32** L239
[3] Lin C Y and Ho Y K 2012 *J. Phys. B* **45** 145001
[4] Nascimento W S and Prudente F V 2018 *Chem. Phys. Lett.* **691** 40

* E-mail: subhasish.pph16@iitp.ac.in
† E-mail: jobin.jose@iitp.ac.in

On how classical uncertainties might affect the coherence properties of collisions processes, and how to control them

R O Barrachina[1,2], F Navarrete[3] and M F Ciappina[4]*

[1]Centro Atómico Bariloche and Instituto Balseiro, Comisión Nacional de Energía Atómica (CNEA) and Universidad Nacional de Cuyo, Av. Bustillo 9500, 8400 Bariloche, Argentina
[2]Consejo Nacional de Investigaciones Científicas y Técnicas (CONICET), Argentina
[3]Department of Physics, Kansas State University, Manhattan, Kansas 66506, USA
[4]Institute of Physics of the ASCR, ELI-Beamlines, Na Slovance 2, 182 21 Prague, Czech Republic

Synopsis In principle, the quantum coherence properties of any collision experiment should be affected by unavoidable classical uncertainties. However, they seem to produce no observable effect. This mystery can be solved by introducing the concept of "coherence length" which, in addition, provides a way to control and modify the aforementioned quantum properties. Finally, we discuss different methods to achieve a transition between coherent and incoherent collisions, evaluating their strengths and weaknesses.

No matter how much we struggle to reduce the classical uncertainties in a collision experiment, we could never suppose them null. In other words, it would never be possible to describe all the different scattering events by a single and unique wave function. However, this is what we actually do in the standard scattering theory, and even assume that this wave function can be represented by a plane wave [1].

A key element for unravelling this conundrum is the coherence length ℓ [2]. The standard "fully coherent" scattering theory applies whenever ℓ is much larger than any characteristic size of the scattering event. Otherwise, a partially coherent (or even incoherent) calculation is mandatory.

In recent years, a series of experiments have explored this transition between coherent and incoherent collisions (see, e.g. [3]). Lacking a proper quantum mechanical definition of ℓ, it has been a common practice to borrow it from Optics. Thus, some authors [4] have modified the angle $\theta_L = \Delta R/L$ subtended by the collimator of aperture ΔR as seen from the target at a distance L , i.e. by assuming $\ell_L \propto \lambda/\theta_L$, with $\lambda = h/P$ the wavelength of the projectile of momentum P; while others [5, 6] have tried to set the focusing of the projectile's beam, as characterized by its angular dispersion θ_P, i.e. by defining $\ell_P \propto \lambda/\theta_P$. However, it is legitimate to question the validity of each approach, and the appropriateness to use them indistinctly. To answer these questions, we developed in a previous work a full quantum mechanical definition of ℓ [2, 7, 8], proving that $\ell \propto \lambda/\theta_L$ whenever $\theta_P \approx 0$. This result validates the first strategy, but a complete analysis incorporating both effects was still lacking. In the present communication we present this full and comprehensive study for the first time, which leads to the following expression:

$$\ell \propto \lambda \sqrt{\frac{1}{(\Delta p/P)^2 + \theta_P^2} + \frac{1}{(\Delta r/L)^2 + \theta_L^2}} ,$$

where $(\Delta r, \Delta p)$ are the quantum uncertainties in position and momentum of the projectile, respectively. This result shows that both strategies, i.e. modifying θ_L or θ_P, do not lead to similar results. In particular, it can be demonstrated that while modifying θ_L is an effective method for exploring the aforementioned coherent–incoherent transition, the variation of θ_P might become inconsequential in most experimental set-ups.

References

[1] Taylor J R 1972 *Scattering Theory: The Quantum Theory of Nonrelativistic Collisions* (Hoboken, NJ: Wiley)
[2] Fabre I *et al* 2018 *Eur. J. Phys.* **39**, 015401
[3] Schulz M 2017 *Adv. Atom. Mol. Opt. Phys.* **66**, 507
[4] Egodapitiya K N *et al* 2011 *Phys. Rev. Lett.* **106**, 153202
[5] Gassert H *et al* 2016 *Phys. Rev. Lett* **116**, 073201
[6] Gatzke J *et al* 2017 *J. Phys.: Conf. Ser.* **875**, 092006
[7] Sarkadi L *et al* 2016 *Phys. Rev. A* **93**, 032702
[8] Navarrete F *et al* 2017 *Nucl. Instr. Meth. B* **408**, 165

*E-mail: marcelo.ciappina@eli-beams.eu

Molecular Auger Interferometry

M Khokhlova[1 *], **B Cooper**[2], **K Ueda**[3], **K C Prince**[4], **P Kolorenč**[5], **M Ivanov**[1,6] **and V Averbukh**[1]

[1]Blackett Laboratory, Imperial College London, London SW7 2AZ, UK
[2]Department of Physics and Astronomy, University College London, London WC1E 6BT, UK
[3]Institute of Multidisciplinary Research for Advanced Materials, Tohoku University, Sendai 980-8577, Japan
[4]Elettra-Sincrotrone Trieste SCpA, Basovizza-Trieste 34149, Italy
[5]Charles University, Institute of Theoretical Physics, V Holešovičkách 2, Prague 180 00, Czech Republic
[6]Max Born Institute for Nonlinear Optics and Short Pulse Spectroscopy, Max-Born-Strasse 2A, Berlin D-12489, Germany

Synopsis We propose a theory of interferometric measurement of a normal Auger decay width in molecules. Molecular Auger interferometry is based on the coherent phase control of Auger dynamics in a two-colour ($\omega/2\omega$) laser field. We show that, in contrast to atoms, in oriented molecules of certain point groups (e.g. CH_3F) the relative $\omega/2\omega$ phase modulates the total ionisation yield. A simple analytical formula is derived for the extraction of the widths of Auger-active states from a molecular Auger interferogram, avoiding the need of either attosecond or high-resolution spectroscopy.

We present Auger interferometry, i.e. a theory of coherent control of Auger decay and single-photon laser-enabled Auger decay (spLEAD) [1] in ionised molecules belonging to the molecular point groups C_s, C_n, C_{nv}, $C_{\infty v}$, C_{3h}, D_3, D_{3h}, and T_d. It is introduced as a measurement of the total yield of a normal Auger decay (or spLEAD) in oriented molecules as a function of the relative $\omega/2\omega$ phase. Our analytical theory predicts that the Auger decay width can be reconstructed from the relative phase scan of the Auger yield modulation, shown in Fig. 1. The interference contrast $\mathcal{M}$ onto which the decay width is mapped can be maximised at any decay width by controlling the ratio of the ω- and 2ω-field intensities. As a result, the interferometric measurement proposed here is free of the limitations of both high-resolution Auger electron spectroscopy [2] (struggling to characterise decay widths smaller than the instrumental energy resolution) and attosecond time-resolved spectroscopy [3] (struggling to characterise decay rates faster than the available pulse durations). We suggest a method of extracting this information from the Auger interferograms by applying a simple analytical formula

$$\Gamma_1 = \frac{1-\sqrt{1-\mathcal{M}^2}}{\mathcal{M}}\frac{d_{01}d_{1E}\mathcal{E}_1^2}{d_{0E}\mathcal{E}_2}.$$

An illustrative example dealing with inner valence hole decay in CH_3F shows that molecular Auger interferometry is well within the present-day experimental capabilities of the modern FEL facilities, such as FERMI@Elettra. More generally, the same coherent control scheme can be used to study any type of ultrafast hole dynamics, e.g. hole migration [4].

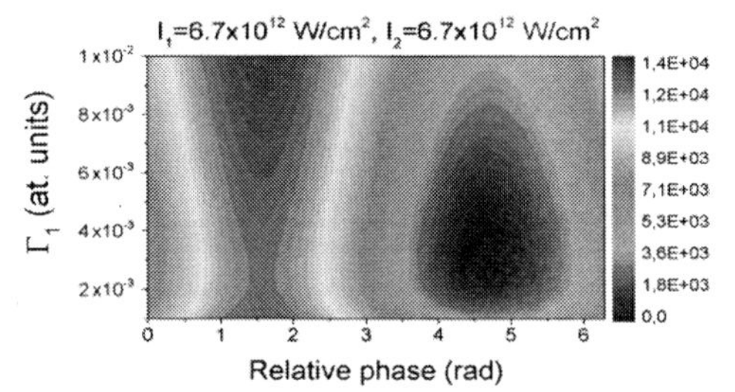

Figure 1. Auger interferograms. The total yield of electrons emitted by a bichromatic laser field for the case of CH_3F^+ as a function of the relative phase ϕ and the AAS width Γ_1 (y-axis).

*E-mail: m.khokhlova@imperial.ac.uk

References

[1] Cooper B and Averbukh V 2013 *Phys. Rev. Lett.* **111** 083004; Iablonskyi D *et al* 2017 *Phys. Rev. Lett.* **119** 073203
[2] Jurvansuu M, Kivimäki A and Aksela S 2001 *Phys. Rev. A* **64** 012502
[3] Drescher M*et al* 2002 *Nature* **419** 803
[4] Calegari F *et al* 2014 *Science* **346(6207)** 336-339; Kraus P M *et al* 2015 *Science* **350(6262)** 790-795

Understanding the formation of metastable furan dication in collisions with ions

E Erdmann[1*], S Indrajith[2], J Chiarinelli[3,4], S Díaz-Tendero[5], N F Aguirre[5,6], M Alcamí[5], A Domaracka[2], P Rousseau[2], B A Huber[2], P Bolognesi[3], L Avaldi[3] and M Łabuda[1]

[1]Gdańsk University of Technology, Narutowicza 11/12, 80-233 Gdańsk, Poland
[2]Normandie Université, ENSICAEN, UNICAEN, CEA, CNRS, CIMAP, 14000 Caen, France
[3]CNR - Istituto di Struttura della Materia, Area della Ricerca di Roma 1, Monterotondo Scalo, Italy
[4]Dipartimento di Scienze, Università di Roma Tre, 00146 Roma, Italy
[5]Departamento de Química, Módulo 13, Universidad Autónoma de Madrid, 28049, Madrid, Spain
[6]Theoretical Division, Los Alamos National Laboratory, Los Alamos, New Mexico 87545, USA

Synopsis This work relies on complementary theoretical and experimental studies of the processes induced by ion-furan collisions. Results of the Molecular Dynamics simulations and exploration of the energy profiles combined with coincidence mass spectrometry provide complete picture of the fragmentation of furan dication.

The main aim of this work is to investigate and understand the fragmentation processes of a furan dication induced by ion collisions with complementary theoretical methodology and experimental technique. The furan molecule has been chosen as a target due to its model structure that can be seen as an elementary analogue of deoxyribose, fundamental component of the DNA backbone chain. Depending on the charge state, the furan molecule is expected to dissociate in various ways [1].

Our theoretical approach consists of *ab initio* Molecular Dynamics describing the evolution of the system following deposition of the energy, and exploration of the Potential Energy Surface (PES) studied at high level of theory. Therefore, the dynamical and energetical considerations of the studied process are expected to create a complete picture of the fragmentation mechanism.

Results of the dynamics simulations lead to the observation of four different processes: isomerization, skeleton fragmentation, H^+/H_2^+ loss and H/H_2 loss. Total occurrence of the most abundant channels varies depending on the energy deposited to the system (from 5 to 30 eV). More insight into three- and four-body fragmentation processes is obtained by investigation of the order of events occurring as the trajectory progresses. Therefore, average time scales of sequential events leading to the major channels are also estimated.

Calculated fragmentation pathways, presenting minima and transition states, indicate stability of intermediate structures leading to channels with many fragments. Investigation of the PES and MD suggests that fragmentation of the furan dication follows a dissociation path with one channel clearly distinguished as most probable and producing fragments of formyl cation (CHO^+) and cyclopropenyl cation ($C_3H_3^+$).

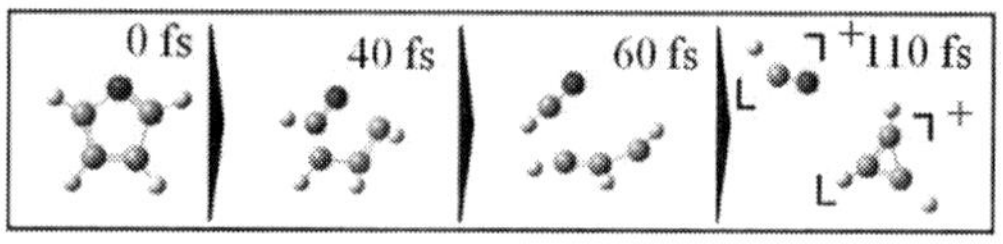

Figure 1. Snapshots of the dynamical simulation.

In order to support theoretical results, coincidence time-of-flight (TOF) mass spectrometry has been employed to study the interaction of 46 keV O^{6+} ions with neutral furan. Measured relative intensities of the coincident ion pairs are in good agreement with calculated energy barriers of the lowest-energy pathways. Geometrical parameters of the measured correlation islands in the fragment-fragment map indicate the mechanism of a specific dissociation channel. Long tail in the TOF mass spectrum points out to slow charge separation associated with the production of a metastable furan dication. This feature can be explained by the properties of the calculated PES.

Acknowledgments: This work has been done within the XLIC COST Action and CNRS-PICS BIFACE project.

References

[1] Erdmann E *et al* 2018 *J. Phys. Chem. A.* **122** **4153**

* E-mail: ewa.erdmann@pg.edu.pl

Oscillator strengths and integral cross sections of the valence-shell excitations of HCl studied by fast electron scattering

Y-C Xu[1,2], X-J Du[1,2], Y-W Liu[1,2] and L-F Zhu[1,2]*

[1]Hefei National Laboratory for Physical Sciences at Microscale and Department of Modern Physics, University of Science and Technology of China, Hefei, Anhui 230026, People's Republic of China
[2]Synergetic Innovation Center of Quantum Information and Quantum Physics, University of Science and Technology of China, Hefei, Anhui 230026, People's Republic of China

Synopsis Oscillator strengths and integral cross sections (ICSs) of HCl are of great importance for the understanding of many different aspects of atmospheric physics. Based on the generalized oscillator strengths (GOSs) of the valence-shell excitations of HCl measured by the high-energy electron scattering, the optical oscillator strengths (OOSs) have been obtained by extrapolating the GOSs to the limit of the squared momentum transfer $K^2 \to 0$. Furthermore, the ICSs of the valence-shell excitations from the threshold to 5000 eV have been obtained systematically with the aid of the BE-scaling method.

Hydrogen chloride (HCl) and its dynamic parameters are important in a number of practical applications, including etching and plasma processing [1], interstellar and atmospheric photochemistry [2, 3, 4]. For example, HCl was detected on Venus for the first time by Connes *et al.* [5] in 1967. Although HCl is a minor constituent of the Venusian atmosphere, it plays an important role in the photochemistry and atmospheric evolution. The OOSs of HCl have been used to explain the exceptionally large $[D]/[H]$ isotopic ratio in the Venusian atmosphere [6]. Furthermore, from the viewpoint of fundamental research, as a typical heteronuclear diatomic molecule, HCl is used to test the theoretical methods and calculational codes strictly [7]. So the determination of the oscillator strengths and ICSs of HCl has been a subject of continuously increasing interests both experimentally and theoretically.

In this work, the GOSs of the valence-shell excitations of HCl have been determined at an incident electron energy of 1500eV and an energy resolution of 70meV. Besides, the OOSs have been obtained by extrapolating the GOSs to the limit of the squared momentum transfer $K^2 \to 0$ and the ICSs of the valence-shell excitations from the threshold to 5000 eV have been obtained systematically with the aid of the BE-scaling method [8, 9]. Present GOSs and ICSs of the $C^1\Pi_1\,(\nu' = 0)$ excitation of HCl are shown in the Fig.1.

In summary, the GOSs, OOSs and ICSs of the valence-shell excitations of HCl have been determined by fast electron impact, and the GOSs and ICSs are the only reported data, to the best of our knowledge. The present reported dynamic parameters of HCl can serve as the basics input parameters of the theoretical models to simulate the processes in planetary atmosphere and explain the astronomical observations.

*E-mail: lfzhu@ustc.edu.cn

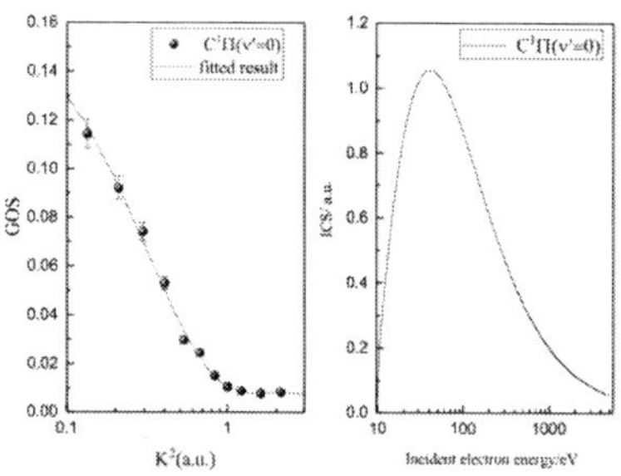

Figure 1. The present GOSs and ICSs of the $C^1\Pi_1\,(\nu' = 0)$ excitation of HCl.

References

[1] Ogawa T *et al* 1994 *J Chem Phys.* **101** 3969
[2] Prinn R G 1971 *J Atmos Sci.* **28** 1058
[3] McElroy M B *et al* 1973 *J Atmos Sci.* **30** 1437
[4] Yung Y L *et al* 1982 *ICARUS.* **51** 199
[5] Connes P *et al* 1967 *Astrophys J.* **147** 1230
[6] Mohammed B *et al* 2001 *Astrophys J.* **559** 179
[7] Brion C E 2005 *J Electron Spectrosc* **144** 127
[8] Kim Y K 2001 *Phys. Rev. A.* **64** 032713
[9] Kim Y K 2007 *Phys. Rev. A.* **126** 064305

Observation of indirect ionization of W^{7+} in EBIT plasma

Q Lu[1,2], J He[1,2], H Tian[1,2], M Li[1,2], Y Yang[1,2], K Yao[1,2], C Chen[1,2], J Xiao[1,2 *], J G Li[3 †], B Tu[4] and Y Zou[1,2]

[1]Institute of Modern Physics, Department of Nuclear Science and Technology, Fudan University, Shanghai 200433, China
[2]Key Laboratory of Nuclear Physics and Ion-beam Application (MOE), Fudan University, Shanghai 200433, China
[3]Institute of Applied Physics and Computational Mathematics, Beijing 100088, China
[4]Max-Planck-Institute für Kernphysik, Saupfercheckweg 1, 69117 Heidelberg, Germany

Synopsis In this work, visible and EUV spectra of W^{7+} have been measured using the high-temperature super-conducting electron-beam ion trap at the Shanghai EBIT laboratory under extremely low-energy conditions. The relevant atomic structure has been calculated by using the flexible atomic code package and the GRASP2K code. A hypothesis for the charge-state evolution of W^{7+} is proposed based on our results. The occurrence of W^{7+} ions results from indirect ionization caused by stepwise excitation between some metastable states of lower-charge-state W ions, at the nominal electron beam energy of 59 eV.

Based on the previous study by Mita *et al.*[1], the spectra of W^{7+} are measured in the visible and EUV range at SH-HtscEBIT[2] under extremely low electron beam energy conditions. The 574.49(3) nm M1 line of W^{7+} is observed at the nominal electron beam energy of 59 eV which is below the ionization energy of W^{5+}. The multi-configuration Dirac-Hartree-Fock calculation further confirms the identification of this line. A hypothesis of charge-state evolution from W^{5+} to W^{7+} is proposed, based on our theoretical studies on the energy levels of these charge states, in order to explain the appearance of W^{7+} spectra. Indirect ionization via cascade excitations from the long-lived metastable states of lower charge W ions play a key role in occurrence of W^{7+}. In addition, the EUV spectra at 75 eV as well as the FAC calculations also prove that W^{7+} can be and is created via indirect ionization out of W^{5+}.

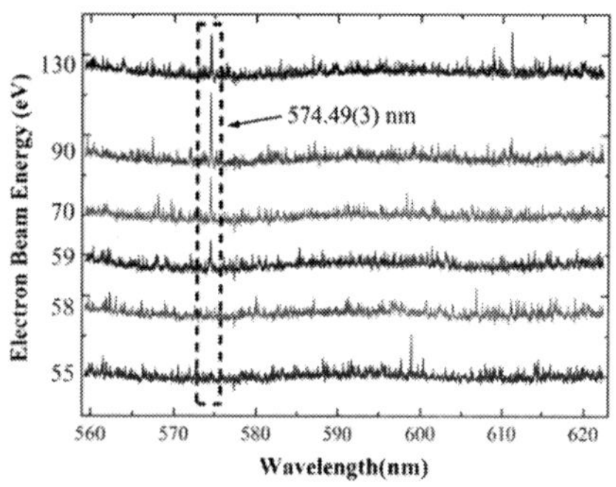

Figure 1. Spectra of tungsten obtained by SH-HtscEBIT at the nominal electron beam energy of 55, 58, 59, 70, 90 and 130 eV in the range of 559-623 nm. The accumulation time of each spectrum is 2 hours. The line at 574.49(3) nm is the M1 transition between the fine structure splitting in the $4f^{13}5s^25p^6\ ^2F$ ground term of W^{7+}.

References

[1] Mita M *et al* 2017 *Atoms* **5** 13
[2] Xiao J *et al* 2013 *Proceedings of IPAC2013* MOPFI066

* E-mail: xiao_jun@fudan.edu.cn
† E-mail: Li_Jiguang@iapcm.ac.cn

A new F-wave resonant state in the e^+-He^+ system

Z-C Yan[1] * and Y K Ho[2] †

[1] Department of Physics, University of New Brunswick, Fredericton, New Brunswick, Canada
[2] Institute of Atomic and Molecular Sciences, Academia Sinica, Taipei 106, Taiwan

Synopsis Using the complex scaling method, together with employing highly correlated Hylleraas wave functions in which all the interparticle coordinates are included, we have calculated for the first time a F-wave resonance state in the positron-helium positive ion system near the postronium (ground state) formation threshold.

Bhatia and Drachman reported two S-wave resonances in positron scattering by a helium positive ion [1]. A subsequent paper by Ho [2] confirmed such results, and in the same paper P-wave resonances were also reported. These findings have since attracted considerable interests to investigate resonances in the e^+-He^+ system [3-7]. There have been continuous interests to investigate atomic resonances involving positron, with the latest references on this topic that can be found in a recent paper on resonances in positron-helium scattering [8]. Here, we extend our earlier works [4,7] to investigate the resonances in e^+-He^+ scattering. The method of complex coordinate rotation [9] is used, and the basis set is constructed using Hylleraas coordinates

$$\chi_{ijk}(\alpha,\beta)=r_1^i r_2^j r_{12}^k e^{-\alpha r_1-\beta r_2}\, \mathcal{Y}_{l_1 l_2}^{LM}\left(\hat{r}_1,\hat{r}_2\right) ,$$

where $\mathcal{Y}_{l_1 l_2}^{LM}(\hat{r}_1,\hat{r}_2)$ is the vector coupled product of solid spherical harmonics for the electron (1) and the positron (2). The explicit form for the wave function is

$$\Psi(\vec{r}_1,\vec{r}_2)=\sum_{ijk} a_{ijk}\chi_{ijk}(\alpha,\beta),$$

where i, j, k are all integers. For F-states (L=3), four blocks of basis sets with individual angular momenta (l_1, l_2) ≡ (3,0), (2,1), (1,2) and (0,3), coupled to a total of 4979 terms, are used. A new stablized complex eigenvalue is found to be located around (E_r, E_i) = (-0.4834, -0.614) Ryd. We show this newly-found F-wave resonance state in Fig.1, together with the earlier reported S-, P-, and D-states [2,4]. These rotor-like stabilizied complex eigenvalues exhibit a regular pattern, indicating that they might belong to a family of resonances.

* E-mail: zyan@unb.ca
† E-mail: ykho@pub.iams.sinica.edu.tw

Figure 1. Complex poles in e^+-He^+ scattering.

We will show detailed calculations for this F-wave resonance state at the meeting, together with some updated results for resonance states with other angular momenta.

Acknowledgement

ZCY was supported by NSERC of Canada. YKH was supported by MOST in Taiwan.

References

[1] Bhatia A K and Drachman R J 1990 *Phys. Rev. A* **42** 5117
[2] Ho Y K 1996 *Phys. Rev. A* **53** 3165
[3] Igarashi A and Shimamura I 1997 *Phys. Rev. A* **56** 4733
[4] Ho Y K and Yan Z-C 2002 *Phys. Rev. A* **66** 062705
[5] Igarashi A and Shimamura I 2004 *Phys. Rev. A* **70** 012706
[6] Yamanaka N *et al* 2004 *Phys. Rev. A* **70** 062701
[7] Ho Y K and Yan Z-C 2009 *J. Phys.B* **42** 044006
[8] Yan Z-C and Ho Y K 2018 *Phys. Rev. A* **98** 062702
[9] Ho Y K 1983 *Phys. Rep.* **99**, 1

A resonant ionization spectroscopy scheme for photodetachment studies

J Welander[1], J E Navarro Navarrete[2], M Kodambattil[3], J Sundberg[1], D Lu[1*] and D Hanstorp[1†]

[1]Department of Physics, University of Gothenburg, Gothenburg, 41296, Sweden
[2]Instituto de Ciencias Nucleares, Universidad Nacional Autónoma de México, México City, 04510, México
[3]International School of Photonics, Cochin University of Science and Technology, Kerala, 682022, India

Synopsis A refined Resonant Ionization Spectroscopy (RIS) scheme for high resolution Laser Photodetachment Threshold Spectroscopy (LPTS). In this work the scheme is used to measure the electron affinity of cesium.

Negative ions are fragile quantum systems where electron correlation effects are enhanced due to the nature of their loosely-bound outermost electrons. The binding energy of negative ions is referred to as the Electron Affinity (EA) of the corresponding neutral atoms. A common technique widely applied to measure EAs is Laser Photodetachment Threshold Spectroscopy (LPTS) [1], where the detachment cross sections can be measured by counting the number of generated neutral atoms as a function of photon energy. However, according to the Wigner's threshold law [2], photodetachment cross sections associated with an outgoing p-wave electron have a less pronounced onset, and can hence not be determined very precisely. In this work, we present a refined Resonant Ionization Spectroscopy (RIS) [3] scheme utilizing photon energies sufficiently high to leave the residual atom in an excited state. This opens for the possibility of selecting a single channel of photodetachment that can circumvent the issue of a slow onset of the photodeatchment at the threshold.

We will present a recently updated RIS scheme that employs a new detector named Rydberg Atom Detector for Anion Research (RADAR), which contains three segments (Fig. 1). Negative ions interact with two lasers along the interaction region in a collinear overlap. The first laser photodetach the negative ions and the second is resontly exciting the residual atom to a Rydberg state. The Rydberg atoms are ionized in the Field Ionizer (FI). Positive ions produced by field ionization leave the FI with an energy that differs from the energy of positive ions produced through collisions in the interaction region.The kinetic energy tagged positive ions follow respective trajectories in the electrostatic analyzer and are finally detected by a Position Sensitive Detector (PSD).

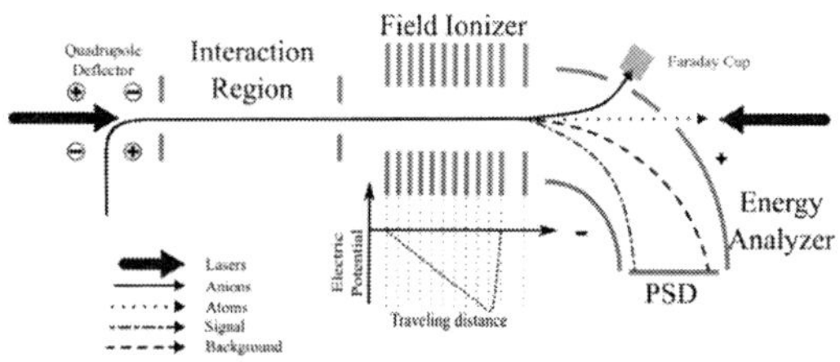

Figure 1. A schematic dragram of RADAR.

In this work, a high resolution experiment where RADAR is used to determine the EA of Cs^- will be presented. The experiment was conducted at the Gothenburg University Negatvie Ion Laser LAboratory (GUNILLA) [4]. By tuning the laser responsible for detachment over the opening of $Cs(6p_{3/2})$ channel, state selective LPT with a sharp onset at the threshold was realized. Both co- and counter propagating alignments of the ion/laser beams was applied in order to correct for the Doppler shift to all orders [5]. The EA could thus be determined limited only by the bandwidth of the laser.

In the future, RADAR will be used to measure the branching ratios of photodetachment that leaves the residual atom in Rydberg states all the way up to the double detachment limit and, eventually, used to investigate the double photodetachment process of atomic anions.

References

[1] Pegg D J 2004 *Rep. Prog. Phys.* **67** 857
[2] Wigner E P 1948 *Phys. Rev.* **73** 1002
[3] Haeffler G *et al* 1996 *Phys. Rev. A* **53** 4127
[4] Diehl C *et al* 2011 *Rev. Sci. Instrum.* **82** 053302
[5] Juncar P *et al* 1985 *Phys. Rev. Lett.* **54** 1

* E-mail: di.lu@physics.gu.se
† E-mail: dag.hanstorp@physics.gu.se

Model potential parameters for alkali metal atomic lithium

S F Zhao[1]*, Z F Tang[1], Z H Jiao[1], W Li[2], G L Wang[1] and X X Zhou[1]

[1]College of Physics and Electronic Engineering, Northwest Normal University, Lanzhou, 730070, China
[2]College of Electronic Information Engineering, Lanzhou Institute of Technology, Lanzhou, 730050, China

Synopsis The model potential proposed by Laughlin (Phys.Rev.A 64,013406 (2001)) can give accurate bound-state energies of atoms, which is very useful in atomic physics. We determine model potential parameters for alkali metal atomic lithium (Li) by comparing the present calculated bound-state energies with both the experimental singlet energy values and quantum defect calculations.

Model potential method has been widely used to study structures and dynamics of alkali metal atoms [1, 2, 3, 4, 5]. Recently this method was extended to investigate strong-field phenomena by solving the time-dependent Schrödinger equation (TDSE) of alkali metal atoms in an intense laser field [6, 7]. Chu *et al.* [6] proposed an accurate model potential for studying the spectral and temporal structures of high-order harmonic generation (HHG) of Na exposed to intense mid-ir laser field. In Ref. [6], the fine structures and significant enhancement of below-threshold harmonics can be well understood by the strong coupling of the 3s-np states and the 3s-3p multiphoton resonance. Based on an accurate model potential and a synchrosqueezing transform (SST) technique, Li *et al.* [7] explored the dynamic origin of near- and below-threshold harmonic generation of Cs in an intense mid-infrared laser field. Milošević *et al.* [8] determined numerically the static ionization rate by the complex rotation method and compared with those from the Stark-corrected Ammosov-Delone-Krainov (ADK) model.

In this work, we present model potential parameters for Li in Table 1 and the corresponding model potential has the following form [6]:

$$v_l = -\frac{\alpha}{2r^4}W_6(\frac{r}{r_c}) - \frac{1}{r} - (\frac{N-S}{r} + A_1)e^{-B_1 r} - (\frac{S}{r} + A_2)e^{-B_2 r}, \quad (1)$$

where $N = 2$ and $\alpha = 0.1894$ is the Li^+ core dipole polarizability. Here, W_6 is a core cut-off function and $r_c = 1.426$ is an effective Li^+ core radius. We obtained bound-state energies by solving numerically the time-independent Schrödinger equation of Li and found that all the calculated bound-state energies using the present potential parameters agree very well with the experimental singlet energy values and quantum defect calculations.

Table 1. Model potential parameters for Li.

l	S	A_1	A_2	B_1	B_2
0	3.30637491	-2.6898612	-0.73157041	2.45	1.44
≥ 1	1.34201612	-1.3439454	-0.44155158	3.05	1.53

Project supported by the National Natural Science Foundation of China (Nos. 11664035, 11764038, 11864037, 11765018) and the Foundation of Northwest Normal University (No. NWNU-LKQN-17-1).

References

[1] Hellmann H 1935 *J. Chem. Phys.* **3** 61
[2] Bardsley J N 1970 *Chem. Phys. Lett.* **7** 517
[3] Philippe M *et al* 1979 *J. Phys. B* **12** 2493
[4] Marinescu M *et al* 1994 *Phys. Rev. A* **49** 982
[5] Schweizer W *et al* 1999 *Atomic Data and Nuclear Data Tables* **72** 33
[6] Chu X *et al* 2001 *Phys. Rev. A* **64** 013406
[7] Li P C *et al* 2015 *Nat. Commun.* **6** 7178
[8] Milošević M Z *et al* 2015 *Phys. Rev. A* **91** 023424

*E-mail: zhaosf@nwnu.edu.cn

1412 (2020) 132008 doi:10.1088/1742-6596/1412/13/132008

Energy levels structure of Pr II - Gd II ions

L Radžiūtė[1] *, G Gaigalas[1], P Rynkun[1], D Kato[2,3] and M Tanaka[4]

[1]Institute of Theoretical Physics and Astronomy, Vilnius University, Vilnius, LT-01513, Lithuania
[2]National Institute for Fusion Science, Toki, 509-5292, Japan
[3]Department of Advanced Energy Engineering Science, Kyushu University, Fukuoka, 816-8580, Japan
[4]Astronomical Institute, Tohoku University, Sendai, 980-8578, Japan

Synopsis Observations of a kilonova associated with GW170817 provide a unique opportunity to study heavy element synthesis in the universe. However, the atomic data of r-process elements are not yet complete enough to decipher the light curves and spectral features of kilonovae. We performed extended atomic calculations for singly ionized elements with $Z = 59 - 64$, by employing GRASP2K package. These computations are based on strategies of Nd II published in [1]. Accuracy of data is evaluated by comparing computed energy levels with the NIST database and other authors.

The calculations were done using multi-configuration Dirac-Hartree-Fock (MCDHF) and configuration interaction (RCI) methods [2, 3], which are implemented in the general-purpose relativistic atomic structure package GRASP2K [4]. The transverse-photon (Breit) interaction, the vacuum polarization, and the self-energy corrections were included in the RCI calculations. Energy levels were computed for states of the configurations: $[\mathrm{Xe}]4f^{N-1}5d6s$, $[\mathrm{Xe}]4f^{N-1}5d6p$, $[\mathrm{Xe}]4f^{N-1}6s6p$, $[\mathrm{Xe}]4f^{N-1}5d^2$, $[\mathrm{Xe}]4f^{N}6s$, $[\mathrm{Xe}]4f^{N}6p$ and $[\mathrm{Xe}]4f^{N}5d$, where $N = 3 - 8$ for Pr-Gd ions respectively. For Sm II and Eu II calculations were performed also for states of $[\mathrm{Xe}]4f^{N+1}$ configuration, for Gd II were added states of $[\mathrm{Xe}]4f^{N-1}6s^2$ configuration. Energy levels were computed up to 10 eV limit for opacity computation of neutron stars merger one day after event.

The configuration space was increased step by step with increasing the number of layers. The orbitals of previous layers were held fixed and only the orbitals of the newest layer were allowed to vary. Sets of virtual orbital were generated by single (S), double (SD) or triple (SDT) substitutions. SDT substitutions were used to generate active space for configuration $[\mathrm{Xe}]4f^{N+1}$, SD for $[\mathrm{Xe}]4f^{N}\{6s, 5d, 6p\}$ and S for $[\mathrm{Xe}]4f^{N-1}\{5d6s, 5d6p, 6s6p\}$ and $[\mathrm{Xe}]4f^{N-1}\{6s^2, 5d^2\}$. The largest active space used in the computations was {8s, 8p, 7d, 6f, 5g}. Substitutions were done only from valence orbitals. Each configuration were computed separately.

*E-mail: laima.radziute@tfai.vu.lt

Figure 1 shows the distribution of the energy level number over the relative difference compared to NIST for Sm II, as example. Data for other ions and it's accuracy will be also presented in the abstract.

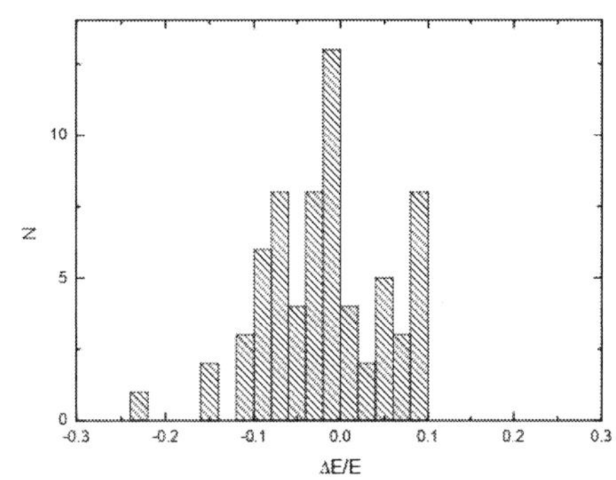

Figure 1. Distribution of the energy levels (N) according the disagreement with NIST database for Sm II.

The analysis of the results shows, that energy levels is suitable for computation opacity of neutron stars merger.

Acknowledgement: *This research was funded by a grant (No. S-LJB-18-1) from the Research Council of Lithuania.*

References

[1] Gaigalas G *et al* 2019 *ApJS* **240** 29
[2] Grant I P 2007 *Relativistic Quantum Theory of Atoms and Molecules* (New York: Springer)
[3] Froese Fischer C *et al* 2016 *J. Phys. B* **49** 182004
[4] Froese Fischer C *et al* 2019 *Comput. Phys. Comm.* **237** 184

Theoretical study of the dielectronic recombination process of Mg-like W^{62+} ions

L Y Xie[1]*, G W Nai[1], J Jiang[1], D H Zhang[1] and C Z Dong[1]

[1]Key laboratory of Atomic and Molecular Physics & Functional Materials of Gansu Province, College of Physics and Electronic Engineering, Northwest Normal University, Lanzhou, 730070, P. R. China

Synopsis The dielectronic recombination of Mg-like W^{62+} ($3s^2$) ions was studied using the flexible atomic code based on the relativistic configuration interaction method with considering Breit and QED corrections. Theoretical results are compared with the experimental results from electron beam ion trap, good consistency is obtained.

Tungsten (Z=74) is one of the most important materials used for plasma facing divertor plates in thermonuclear fusion devices [1]. For divertor plasma modeling and radiative cooling studies, accurate atomic data of highly ionized tungsten ions are essential and important. Dielectronic recombination (DR) is the most important recombination process in highly ionized plasmas, it plays an important role in the determination of radiation losses and radiative spectrum of plasmas [2-4].

In this work, DR cross sections and rate coefficients have been calculated for 2s/2p-core excitation of Mg-like tungsten (W^{62+}) using the flexible atomic code [5] based on the relativistic configuration interaction method with considering Breit and QED corrections. In the calculations, the dominant DR, and trielectronic recombination (TR) channels were included as described as following

$$W^{62+}(3s^2)+e^- \rightarrow W^{61+}\begin{cases}(2s2p)^7 3s^2 3lnl'(DR) & (1a)\\ (2s2p)^7 3s3l^2nl'(TR) & (1b)\\ (2s2p)^7 3s^2n'l'n''l''(DR) & (1c)\\ (2s2p)^7 3s3ln'l'n''l''(TR) & (1d)\end{cases} \quad (1)$$

$$\rightarrow W^{61+} + h\nu$$

where $l=p,d$, $n=3-12$ in (1a) and (1b), $n'=4-5$, $n''=4-6$ in (1c) and (1d). It is found that DR contributions from the channel (1a) dominate total cross sections, and TR contributions from (1b) cannot be neglected. The contributions from (1c) and (1d) serve the higher energy resonances above 7.5 KeV (see Figure 1). The plasma rate coefficients were deduced from the calculated recombination spectra, and the contributions from the higher excitated resonance groups $(n>12)$ were evaluated by an extrapolation method, which is about 5.6% to total rate coefficient. Our theoretical results of cross section were compared with the experimental measurements of electron beam ion trap by Watanabe *et al* [6], good consistency is obtained.

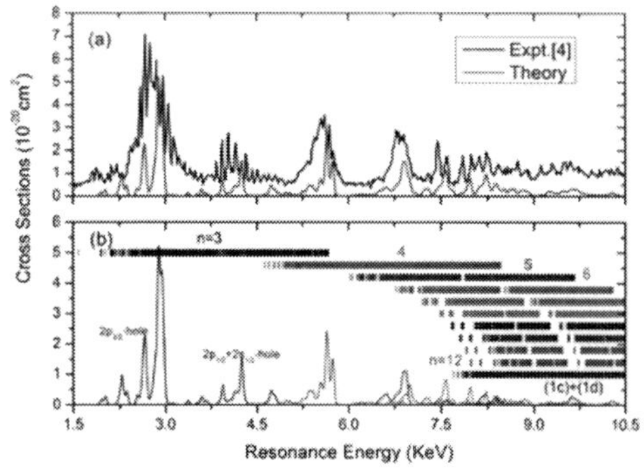

Figure 1. (a) Comparison of the theoretical (red line) and EBIT experimental (black line) cross sections [6] for W^{62+} DR spectrum in the resonance energy range 1.5-10.5 KeV. (b) Detailed theoretical cross sections are show for the different double excited configuration complexes, namely $(2s2p)^7 3s^2 3lnl'$+$(2s2p)^7 3s3l^2nl'$ (n=3-12) in Eq.(1). The corresponding resonance positions were indicated by short bars in different colors.

The work was supported by the National Key Research and Development Program of China under Grant No.2017YFA0402300), the NSCF (Grant Nos: 11564036, 11774292, U1530142, 11464042, 11874051) and the Fund of NWNU (NWNU-LKQN-15-3).

References

[1] Naujoks D *et al* 1996 *Nucl. Fusion*. **36** 671
[2] Behar E *et al* 1999 *Phys. Rev. A* **59** 278
[3] Preval P S *et al* 2016 *Phys.Rev. A* **93** 042703
[4] Dou L J *et al* 2018 *J Quant Spectrosc Radiat Transf* **152** 171
[5] Gu M F *et al* 2003 *Astrophysical Journal* **582** 1241
[6] Watanabe H *et al* 2007 *Plasma and Fusion Research* **2** 027

* E-mail: xiely@nwnu.edu.cn

Theoretical studies of energy spectra and E1 transitions of Ni II ion

P Rynkun[1] *, G Gaigalas[1] and P Jönsson[2]

[1]Institute of Theoretical Physics and Astronomy, Vilnius University, Vilnius, LT-10222, Lithuania
[2]Group for Materials Science and Applied Mathematics, Malmö University, Malmö, SE-20506, Sweden

Synopsis Accurate atomic data of the iron group elements are of major importance in astrophysics. In the present work energy spectrum calculations are performed for 332 lowest states for Ni II ion. Energy levels are compared with NIST database recommended values. All computations were done using the general-purpose relativistic atomic structure package GRASP2018.

Nickel belongs to the iron group elements, and it is one of the most abundant element from this group in cosmic objects. In the determination of abundances the accurate transition characteristics are needed. In the NIST [1] database there are not much electric dipole transition data for Ni II ion (only transitions between [Ar]$3d^9$ (ground), $3d^8 4s$, and $3d^8 4p$ configurations).

In this work the multiconfiguration Dirac-Hartree-Fock and relativistic configuration interaction (RCI) methods, which are implemented in the general-purpose relativistic atomic structure package GRASP2018 [2], were used for energy spectra computations. The transverse-photon (Breit) interaction, the vacuum polarization, and the self-energy corrections were included in the RCI calculations. Atomic state functions (ASFs) were obtained as expansions over jj-coupled configuration state functions. To provide the LSJ labeling system the ASFs were transformed using the method provided by Gaigalas *et al.* [3].

The energy spectra for 332 lowest states for Ni II ion were computed. The calculations are in progress. So here are presented the preliminary results where just valence-valence electron correlations are included. The inactive core used in the present calculations is [Ar]. For the construction of the ASFs single and double substitutions from $3d^9$, $3d^8\{4s, 4p, 4d, 4f, 5s, 5p, 5d, 6s, 6p\}$, $3d^7 4s^2$, and $3d^7 4s4p$ configurations to active orbitals set $\{7s, 7p, 6d, 5f, 5g\}$ were allowed.

In Figure 1 excitation energies of the lowest states for each configuration are compared with data from NIST [1] database. As it seen from the figure, the largest disagreements are for energy levels of $3d^8 4s$ and $3d^7 4s^2$ configurations. For other lowest states of studied configurations the difference with NIST is less than 1%, except $3d^7 4s4p$ configuration.

*E-mail: pavel.rynkun@tfai.vu.lt

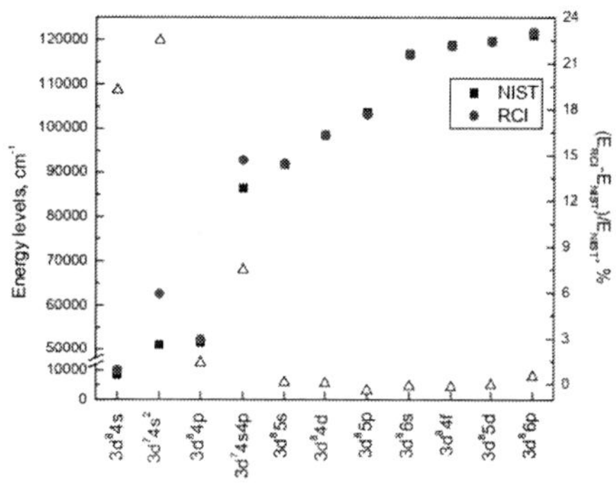

Figure 1. Excitation energy of the lowest state for each configuration. Black squares show the data from the NIST [1] while red circles show GRASP2018 results (RCI). The right Y axis and blue open triangles show the difference (in %) between RCI and NIST results for plotted states.

Acknowledgments: This research is funded by the European Social Fund under the No 09.3.3-LMT-K-712 "Development of Competences of Scientists, other Researchers and Students through Practical Research Activities" measure.

References

[1] Kramida A *et al* NIST Atomic Spectra Database (ver. 5.6.1), [Online]. Available: `https://physics.nist.gov/asd` [2019, February 21]. National Institute of Standards and Technology, Gaithersburg, MD.
[2] Fischer C F *et al* 2019 *Comp. Phys. Commun.* **237** 184
[3] Gaigalas G *et al* 2017 *Atoms* **5** 6

Near-saddle-point-energy single- and two-photon Stark spectra of Sr

D Spasopoulos[1], **P Kalaitzis**[1 *] **and S Cohen**[1]

[1]Atomic and Molecular Physics Laboratory, Physics Department, University of Ioannina, 45110 Ioannina, Greece

Synopsis We present single-photon ionization Stark spectra of Sr as a function of static electric field strength and just above the saddle-point-energy. The spectra exhibit a significant number of avoided crossings between Stark resonances, superimposed on an ionization background due to the continuum excitation. The latter may be avoided by employing single- and two-photon ionization in conjunction with phase sensitive coherent control techniques. This scheme will lead to the isolation of the resonances in differential ionization cross section measurements.

When an atom is placed in a uniform static electric field of strength *F*, its ionization threshold is lowered from its zero-field value *E*=0, to the so-called saddle point energy, classically given by $E_{sp}=-2F^{1/2}$ a.u. Within the $E_{sp}\leq E\leq 0$ energy range quasi-bound Stark states (resonances) are degenerate with the continuum ones. In non-hydrogenic atoms these states are coupled and the resonance-continuum coupling leads to autoionization [1]. Resonance-resonance coupling is more interesting and leads to avoided crossings of the resonance energies as a function of *F* [1]. In the vicinity of an avoided crossing one of the participating resonances is partially decoupled from the degenerate continua and its spectral width exhibits a minimum. Then, the outgoing photoelectron escapes to infinity solely via tunneling. This fact is very attractive for Photoionization Microscopy (PM) studies, the term denoting the measurement of the probability density of slow (meV) photoionized electrons in the presence of *F* [2]. It turns out that the image of the probability density is proportional to the squared modulus of the electronic wave function. Decoupled PM images have been recorded so far only for He atom [2]. As a first step towards the extension of similar observations to heavier atoms, we recorded (using a typical atomic beam/time-of flight ion spectrometer setup) Stark spectra of strontium for a number of field strengths and just above the respective saddle point energies.

Figure 1(a) shows a number of recorded spectra where several avoided crossings between pairs of resonances may be identified. Figure 1(b) shows the aforementioned spectral width narrowing as a function of *F* observed for the specific resonance indicated in Fig. 1(a). We have also employed two-photon excitation and the corresponding spectra are generally found to be similar to the spectra of Fig. 1 but with a worse spectral resolution.

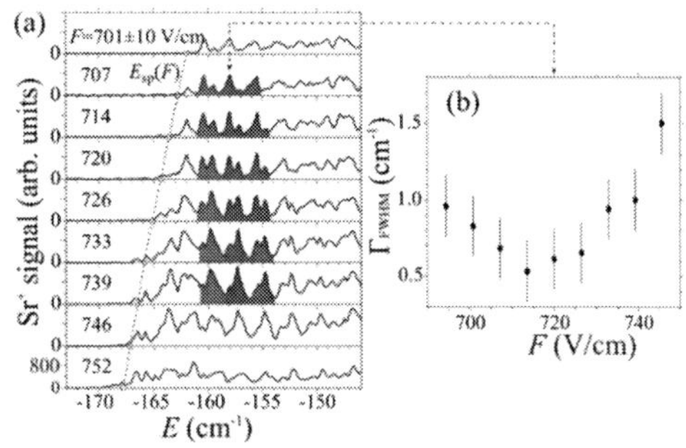

Figure 1. (a) Single-photon Stark spectra of Sr out of its ground state for several field values. The open circles mark the respective saddle point energies which are connected with a dashed line. Three avoided crossings between pairs of resonances are shown with three different colours. (b) Measured spectral width of the resonance indicated by an arrow in (a) as a function of *F*.

Finally, we have performed simultaneous single- and two-photon ionization studies in conjunction with phase sensitive coherent control techniques [3] and the latter techniques are currently applied to PM.

The research (MIS: 5005247) is implemented through the Operational Program "Human Resources Development, Education and Lifelong Learning" and is co-financed by the European Union (European Social Fund) and Greek national funds.

References

[1] Liu J Y *et al* 1985 *Phys. Rev. Lett.* **55** 189
[2] Stodolna A S *et al* 2014 Phys. Rev. Lett. **113** 103002
[3] Shapiro M and Brumer P 2000 *Adv. At. Mol. Opt. Phys.* **42** 287

* E-mail: pkalaitz@cc.uoi.gr

Electron-impact excitation of the $(4p^5 5s5p)^4S_{3/2}$ quasimetastable state in Rb

V Roman[1 *] **and S Demes**[2]

[1]Institute of Electron Physics, National Academy of Sciences of Ukraine, Uzhgorod, 88017, Ukraine
[2]Institute for Nuclear Research, Hungarian Academy of Sciences, Debrecen, 4026, Hungary

Synopsis The ejected-electron excitation function of the $(4p^5 5s5p)^4S_{3/2}$ quasimetastable state in rubidium was experimentally investigated by electron-impact in the energy range from the excitation threshold at 16.64 eV to 100 eV. The obtained data at 0.12 eV energy resolution indicate the presence of negative-ion resonances just above excitation threshold. A comparison is made with the optical excitation function of the $(4p^5 5s5p)^4S_{3/2} \rightarrow (4p^6 5p)^2P_{3/2}$ transition (λ = 82.37 nm), measured earlier with a lower energy resolution of $\approx$1.5 eV.

It is known that in alkali atoms (K, Rb, Cs) there is a subclass of p^6-core excited quartet states which is due to the mixing with doublet states of the same total electronic angular momentum J possess both radiative and electronic decay channels [1]. As a rule, the radiative decay dominates for such quasimetastable states, while autoionization probability remains small. The latter makes difficult the experimental observation of quasimetastable states in ejected-electron spectra [2]. The optical excitation function for the $(4p^5 5s5p)^4S_{3/2} \rightarrow (4p^6 5p)^2P_{3/2}$ transition (λ = 82.37 nm) in Rb was measured earlier with an incident-electron energy resolution of $\approx$1.5 eV [3]. In the present work we obtained the ejected-electron excitation function of the $(4p^5 5s5p)^4S_{3/2}$ state with an improved 0.12 eV energy resolution by using energy-selective electron beam and by accurate measurement of the ejected-electron spectra in a broad impact energy range.

The apparatus and the experimental procedure were described in details earlier [4]. The ejected-electron spectra were measured at 54.7° observation angle. Due to the overlapping of the $^4S_{3/2}$ line with the high intensity neighboring line $(4p^5 4d5s)^4P_{1/2}$ with 16.69 eV excitation threshold, the total relative uncertainty is approximately 40% for the most of our data.

Figure 1 shows the measured ejected-electron excitation function of the $(4p^5 5s5p)^4S_{3/2}$ state and the optical excitation function for λ = 82.37 nm [3] in an impact-energy range from the 16.64 eV excitation threshold up to 100 eV. Our comparison shows that there is a fairly good agreement between the two types of data. Due to the better energy resolution of the present measurements, the "fine" structure *a-b* is clearly seen in the ejected-electron function just above the excitation threshold. The broad maximum *c* in our data [3] corresponds to the spin-exchange character of the electronic excitation of the $^4S_{3/2}$ state. The notable scatter of experimental data, which is observed between 20 and 25 eV in the ejected-electron excitation function could reflect the combined effect of several overlapping resonances.

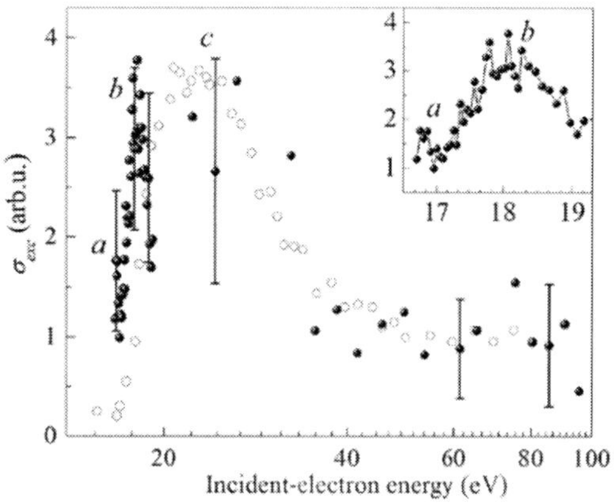

Figure 1. The electron-impact excitation functions of the $(4p^5 5s5p)^4S_{3/2}$ state in rubidium: • - present data; ○ - optical data [3]. The inset shows the threshold region of the present data. The data [3] and our data are normalized with incident-electron energy of 65 eV.

In order to obtain the absolute values of the excitation cross sections for the $(4p^5 5s5p)^4S_{3/2}$ and other p^6-core excited states in rubidium, we have performed relativistic distorted-wave calculations.

References

[1] Mendelson A *et al* 1987 *Phys. Rev. A.* **35** 2095
[2] Borovik A *et al* 2006 *Phys. Rev. A.* **73** 062701
[3] Bogachev G 1974 *Ukr. Phys. J.* **12** 2061
[4] Borovik A *et al* 2005 *J. Phys. B* **38** 1081

* E-mail: viktoriyaroman11@gmail.com

Approaches for constraining uncertainty and degeneracy in geometry reconstruction of molecules from Coulomb Explosion Imaging data

K Tian*, A Ramadhan, M D Singh and J Sanderson†

University of Waterloo, Waterloo, ON N2L 3G1, Canada

Synopsis Starting with simulated OCS molecular data we use two methods to propose a framework for reconstructing geometries, using Coulomb explosion imaging (CEI), for use in the production of molecular movies. The first method is a lookup table, which helps us investigate the existence and nature of multiple solutions to the geometry reconstruction problem and define regions of degeneracy in molecular structure. The second is a new method using nonlinear constrained optimization. We also use this method to investigate the uncertainty in geometry reconstructions as a function of measurement error.

Coulomb explosion imaging (CEI) is a technique for studying the structure and ultrafast dynamics of small molecules in the gas phase. It is possible to reconstruct the molecular structure with knowledge of the momentum vectors of every ionic fragment. We perform geometry reconstruction for simulated carbonyl sulfide (OCS) in the $OCS \rightarrow O^{2+} + C^{2+} + S^{2+}$ concerted fragmentation channel. In all cases we use a simple point like ion approach calculating final momentum vectors, for initial geometry by integrating the Coulomb repulsion to infinity. We then test the usefulness of two approaches, a look up table and a nonlinear constrained optimization approach, in terms of their speed and the uncertainty of their results.

A look up table [1] is a very simple approach for reconstructing geometries. Simulating Coulomb explosions is computationally fast, so we can simulate the explosion of many geometries and create a large mapping of geometries to asymptotic momentum vectors. Then reconstructing a geometry from measured momentum vectors is simply a matter of looking up the geometry that produces the most similar momentum vector arrangement. We test its feasibility for triatomic and larger molecules.

The nonlinear constrained optimization approach is faster and more sophisticated. To set up the statistical model, we use the interior-point method to find the geometries whose post-explosion momentum vectors most precisely match the measurements within desired constraints. This method also allows us to investigate degeneracy geometry and quantify the uncertainty on the geometry reconstructions based on the uncertainty in the momentum vectors, resulting from experimental parameters. We will therefore be able to present a two-step approach in which the look up table reveals degeneracy regions the r_{CO}, r_{CS}, θ_{OCS} space for OCS and uncertainty is dealt with by convex hulls, which are three dimensional regions representing probable values for r_{CO}, r_{CS}, θ_{OCS}, calculated using the nonlinear constrained optimization method.

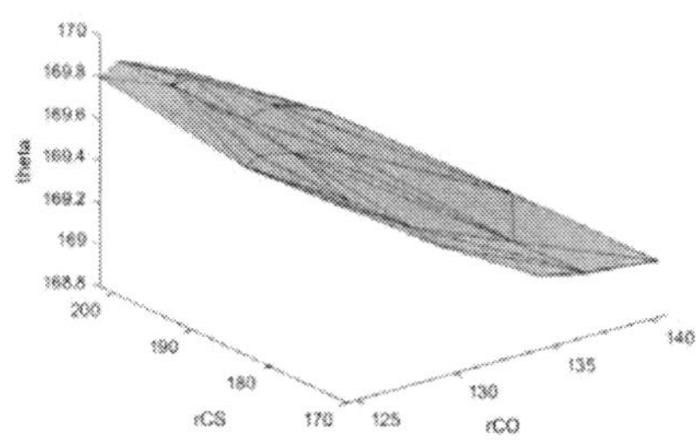

Figure1. 3D convex hull for the reconstruction of OCS (2,2,2) geometries from momentum vectors with 2% uncertainty, from the starting geometry (r_{CO}, r_{CS}, θ_{OCS}) = (130pm,190pm, 169°)

References

[1] Kunitski M 2015 *et al Science* **348** 551

*E-mail: k3tian@uwaterloo.ca
†E-mail: j3sander@uwaterloo.ca

Infrared laser spectroscopy measurement of the electron affinity of thallium

C W Walter*, N D Gibson†, K R Patel and S E Spielman

Denison University, Granville, Ohio 43023, USA

Synopsis The electron affinity of thallium, Tl, has been measured using tunable infrared laser photodetachment threshold spectroscopy. The relative cross section for neutral atom production following photodetachment from Tl^- was measured with a crossed laser-ion beam apparatus over the photon energy range 300 - 900 meV. A single *s*-wave threshold was observed near 320 meV due to the opening of the ground state phodetachment channel, yielding a preliminary value for the electron affinity of thallium. The present results are compared to previous experimental and theoretical results.

The extra electron in a negative ion is not bound by a net Coulomb field, therefore electron correlations are responsible for their stability. Thus studies of negative ions yield key insights into the dynamics of multi-electron interactions and serve as important tests of detailed atomic structure calculations.

Tremendous progress in understanding negative ions has been made through both experiment and theory over the past two decades, and there is now good agreement for the electron affinities of most atoms in the periodic table. However, one of the largest remaining discrepancies in electron affinities occurs for the element thallium (Tl, Z=81). The electron affinity of Tl has previously been measured only in one study, which used laser photoelectron spectroscopy to determine a value of 377(13) meV [1]. A number of theoretical calculations of the electron affinity of Tl have been reported using different methods; the calculations scatter widely, ranging from 270 – 2415 meV (see for example [2,3] and references therein).

In the present experiments, the relative cross section for neutral atom production following photodetachment from Tl^- was measured with a crossed laser–ion beam apparatus over the photon energy range 300 - 900 meV. A single *s*-wave threshold was observed near 320 meV, with essentially zero photodetachment signal below this threshold (see Fig. 1). Because no thresholds were observed at higher energy, we conclude that this threshold is due to the opening of the transition from the negative ion ground state Tl^- ($6p^2\ {}^3P_0$) to the neutral atom ground state Tl ($6p\ {}^2P_{1/2}$), which defines the electron affinity of thallium. Furthermore, the absence of photodetachment signal below this threshold indicates that the excited fine structure levels of Tl^- (${}^3P_1$ and ${}^3P_2$) are either unbound or very weakly bound such that the ion beam has no significant remaining population of these excited states that would photodetach at lower photon energies.

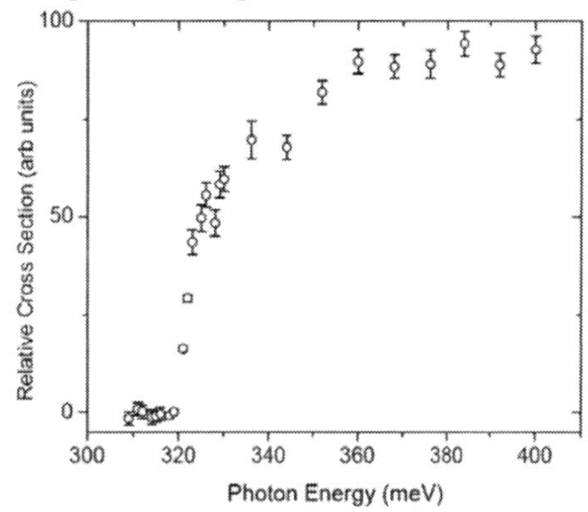

Figure 1. Measured photodetachment spectrum from Tl^- showing a single threshold near 320 meV.

The present results yield a definitive measurement of the electron affinity of Tl, which will help resolve the existing discrepancies and evaluate theoretical approaches used to analyze this complex, heavy ion.

References

[1] Carpenter D L, Covington A M and Thompson J S 2000 *Phys. Rev. A* **61** 042501

[2] Li J *et al* 2012 *J. Phys. B* **45** 165004

[3] Felfli Z, Msezane A Z and Sokolovski D 2012 *J. Phys.* **45** 189501

* E-mail: walter@denison.edu

† E-mail: gibson@denison.edu

Generalized oscillator strengths of the valence-shell excitations of sulphur dioxide studied by fast electron impact

X -J Du[1,2], Y -C Xu[1,2], Y -W Liu[1,2] and L -F Zhu[1,2 *]

[1]Hefei National Laboratory for Physical Sciences at Microscale and Department of Modern Physics, University of Science and Technology of China, Hefei, Anhui 230026, People's Republic of China

[2]Synergetic Innovation Center of Quantum Information and Quantum Physics, University of Science and Technology of China, Hefei, Anhui 230026, People's Republic of China

Synopsis Generalized oscillator strengths (GOSs) of the valence-shell excitation of sulphur dioxide are of great importance for modeling the planet's atmospheric phenomena and astronomical observations, which are determined at an incident electron energy of 1.5keV in this work. The present GOSs are helpful to assign the experimental spectra and explore excitation dynamics.

Sulphur dioxide is an important constituent of the interstellar medium and the atmospheres of Venus and Jupiter's satellite Io. Sulfur cycle is one of the main chemical processes in the atmosphere of Venus and Mars [1, 2]. Besides, sulphur dioxide is a key component of tropospheric and stratospheric aerosols and acid rain. Therefore, the valence-shell excitations of sulfur dioxide have attract extensive interests spanning more than eight decades. However, the dynamic parameters of the valence-shell excitations of SO_2 have not been paid enough attention, and this is what we want to do in this work. Using the fast electron energy loss spectroscopy, the GOSs of the valence-shell excitations of SO_2 have been measured in this work at an incident electron energy of 1.5keV and an energy resolution of 70meV. The present GOS for the $\tilde{C} + \tilde{D}$ excitation of SO_2 is illustrated in Fig.1. It is obvious that the GOS of $\tilde{C} + \tilde{D}$ states shows a typical momentum transfer dependence behavior of a dipole-allowed transition, i.e., the GOS decrease with the increasing of the squared momentum transfer. It is also noticed that the GOS has a minimum at about 2 a.u.. Furthermore, the GOSs for other valence-shell excitations have also been obtained.

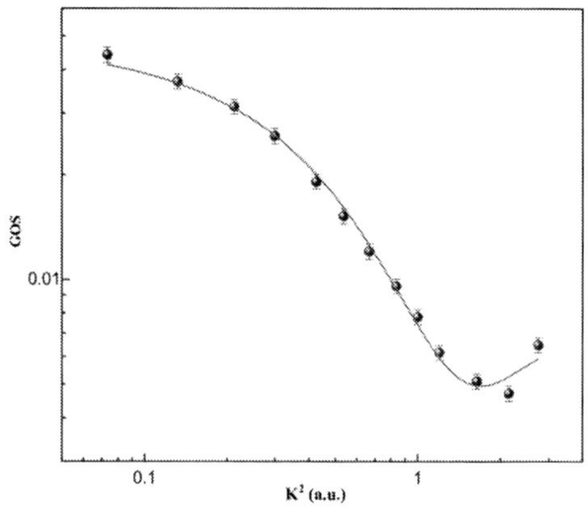

Figure 1. The GOSs for the $\tilde{C} + \tilde{D}$ excitations of SO_2. The blue dots are the present GOSs. The Megenta line are fitting curve.

In summary, the GOSs of the valence-shell excitations of sulphur dioxide have been determined by a fast electron impact method, and the present GOSs can serve as basic data to modeling planetary atmospheric phenomena and evolution of the atmosphere of the early Earth.

References

[1] Farquhar J *et al* 2000 *Nature* **404** 50
[2] Mills F P *et al* 2007 *Geophys. Monogr. Ser.* **176** 73

*E-mail: lfzhu@ustc.edu.cn

Coexistence of cubic and tetragonal phase with colored excitonic emissions in mixed halide Perovskite $CH_3NH_3PbI_{3-x}Cl_x$ thin film

A M M Tanveer Karim[1], M S Hossain[1] and M K R Khan[2]*

[1] Department of Physics, Rajshahi University of Engineering & Technology, Rajshahi-6204, Bangladesh
[2] Department of Physics, University of Rajshahi, Rajshahi-6205, Bangladesh

Synopsis $CH_3NH_3PbI_{3-x}Cl_x$ crystalline film composed of cuboids and polyhedral structure was prepared by dipping and withdrawing method at ambient. X-ray diffraction shows the presence of tetragonal and cubic phases having high absorption coefficient of 10^6 m^{-1}. Photoluminescence study confirms shallow and deep level emissions. Temperature dependent-cathodoluminescence shows a green luminescence at 2.51 eV accompanied by weak red and ultraviolet luminescence at 80 and 300 K.

The ability of organic-inorganic perovskites $CH_3NH_3PbX_3$ (X = Br, I and Cl) to convert visible light to electricity was first discovered in 2006 using $CH_3NH_3PbBr_3$ as a sensitizer on nano-porous TiO_2 in a liquid electrolyte-based dye-sensitized solar cell but the device power conversion efficiency (PCE) was found very low Researchers try to boost up the PCE of solar cells using perovskite nano-particle or thin film over conventional dyes replacing the halogen site partially but still PCE is not satisfactory. In this paper we show the coexistence of tetragonal and cubic phases in $CH_3NH_3PbI_{3-x}Cl_x$ thin film deposited on glass substrate by dipping and withdrawing method at ambient atmosphere. The coexistence of tetragonal and cubic phases and the physical properties have been explained on the basis of the result of the X-ray diffraction (XRD), scanning electron microscope (SEM), Raman spectroscopy, ultraviolet (UV)-visible spectroscopy, photoluminescence (PL) and temperature dependent cathodoluminescence (CL). Perovskite solution synthesis and $CH_3NH_3PbI_{3-x}Cl_x$ film preparation have been given elsewhere [1]. The SEM micrograph shows cuboids and polyhedral structure with grain size lies 100 nm to 2 μm. XRD patterns show the coexistence of cubic ($CH_3NH_3PbCl_3$) and tetragonal $CH_3NH_3PbI_{3-x}Cl_x$ phases accompanied by a PbI_2 phase due to decomposition of $CH_3NH_3PbI_{3-x}Cl_x$ in air. From Raman spectrum, vibrational modes centered at 110, 174 cm^{-1} for $CH_3NH_3PbI_{3-x}Cl_x$ phase corresponds to hetero-polar ionic/covalent interactions of Pb-X bonds in the inorganic framework while vibrational mode at 1419 cm^{-1} for $CH_3NH_3PbCl_3$ phase corresponds to CH_3 stretching or bending of the C-H bonds. While vibrational mode at 1419 cm^{-1} for $CH_3NH_3PbCl_3$ phase corresponds to CH_3 stretching or bending of the C-H bonds.

Furthermore, vibration at 1032 cm^{-1} seems to arise from the organic part ($CH_3NH_3^+$).The PL emission at 1.64 eV may be due to band edge (BE) emission attributed free carrier excition. Besides, BE emission peak at 1.64 eV, there are two other peaks and PL at 2.01 eV and 2.14 eV refers to shallow level defects of vacancies or interstitials of organic part, otherwise deep level defects of halogen interstitials or antisites of lead and halogen as suggested by DFT calculations [2].

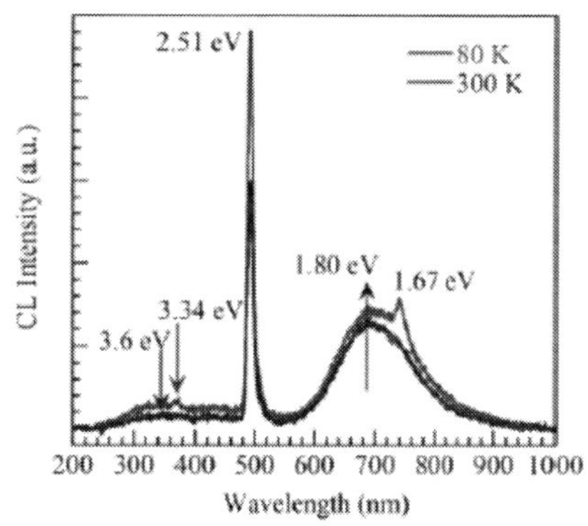

Figure 1. CL intensity vs. wavelength at 80 and 300 K

References

[1] Tanveer Karim A M M *et al* 2018 *ChemistrySelect* **3** 6525
[2] Buin A *et al* 2014 *Nano Lett.* **14** 6281

* E-mail: fkrkhan@yahoo.co.uk

Resonance contribution to electron-impact excitation rate coefficients of helium-like S^{14+} ions

Z W Wu[1] *, C Ren[1], J Jiang[1], L Y Xie[1], D H Zhang[1] and C Z Dong[1] †

[1]Key laboratory of Atomic and Molecular Physics & Functional Materials of Gansu Province, College of Physics and Electronic Engineering, Northwest Normal University, Lanzhou, 730070, P. R. China

Synopsis Electron-impact excitation rate coefficients of S^{14+} ions have been studied by using a relativistic flexible atomic code that can account systematically for configuration interactions. Moreover, we also considered the resonance excitation contributions to the rate coefficients. It is found that the contributions of resonance excitation on rate coefficients is very important.

Electron-impact excitation of positive ions is an important atomic process in high-temperature plasmas. It can proceed through direct excitation (DE) channels or indirect channels involving autoionizing resonances that are termed as resonance excitation (RE) [1, 2]. Due to its abundance in most of astrophysical plasmas, the structure and collision dynamics of sulfur have been widely studied in the last years. The parametric studies on the atomic process of sulfur and its ions are not only beneficial to plasma diagnosis, but also can provide effective data support for astrophysical and laboratory plasma modeling[3].

In the present work, electron-impact excitation rate coefficients of S^{14+} ions have been studied by using a relativistic flexible atomic code that can account systematically for configuration interactions. Moreover, we also considered resonance excitation contributions to the rate coefficients. In table 1, we list the excitation energies of S^{14+} ions at the collision electron energies 2.4~3.0 keV.

Table 1. Excitation energies (in units of keV) of S^{14+} ions.

He-like S^{14+}			
Final states	Excitation energies	Final states	Excitation energies
$(1s_{1/2}2s_{1/2})_1$	2.4292	$(1s_{1/2}3p_{1/2})_0$	2.8789
$(1s_{1/2}2s_{1/2})_0$	2.4477	$(1s_{1/2}3p_{1/2})_1$	2.8790
$(1s_{1/2}2p_{1/2})_0$	2.4458	$(1s_{1/2}3p_{3/2})_2$	2.8795
$(1s_{1/2}2p_{1/2})_1$	2.4462	$(1s_{1/2}3p_{3/2})_1$	2.8829
$(1s_{1/2}2p_{3/2})_2$	2.4478	$(1s_{1/2}3d_{3/2})_1$	2.8819
$(1s_{1/2}2p_{3/2})_1$	2.4603	$(1s_{1/2}3d_{3/2})_2$	2.8819
$(1s_{1/2}3s_{1/2})_1$	2.8744	$(1s_{1/2}3d_{5/2})_3$	2.8821
$(1s_{1/2}3s_{1/2})_0$	2.8792	$(1s_{1/2}3d_{5/2})_2$	2.8823

* E-mail: zhongwen.wu@nwnu.edu.cn
† E-mail: dongcz@nwnu.edu.cn

Figure 1 shows the electron impact excitation rate coefficients of S^{14+} ions in the energy range of 2.4-3.0 keV. The solid red line represents the result of the direct excitation, while the gray area denotes the results with the resonance excitation included. Configurations $1s3l3l$ (l=s, p, d) and $1s3l4l'$ (l'=s, p, d, f) correspond to the resonance excitation processes.

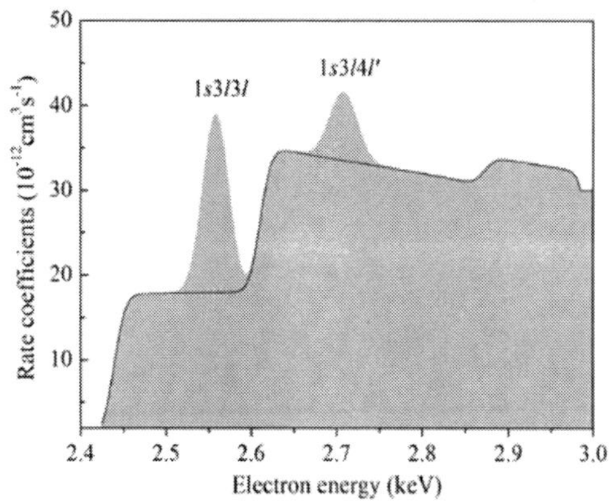

Figure 1. A synthesized theoretical electron-impact excitation rate coefficients of S^{14+} ions with the DE and RE channels included.

This work has been supported by the National Key Research and Development Program of China (2017YFA0402300) and the National Natural Science Foundation of China under Grant Nos. 11804280, 11864036, 11874051. Z. W. Wu acknowledges the support of the Scientific Research Program of the Higher Education Institutions of Gansu Province, China (Grant No. 2018A-002).

References

[1] Dong C Z *et al* 1998 *Chin. Phys. Soc.* **7** 4
[2] Jiang J *et al* 2008 *Phys. Rev. A* **78** 022709
[3] Wu Z W *et al* 2012 *Phys. Rev. A* **86** 022712

Development of the numerical Hartree-Fock method for screened atoms with Debye-Hückel potential

L G Jiao[1] *, J Ma[2] † and Y K Ho[3]

[1]College of Physics, Jilin University, Changchun, 130012, P.R. China
[2]College of Science, Shenyang Aerospace University, Shenyang, 110136, P.R. China
[3]Institute of Atomic and Molecular Sciences, Academia Sinica, Taipei 10617, Taiwan

Synopsis The difficulties in developing the numerical Hartree-Fock programs for atoms interacting with screened Coulomb potential have been overcome. The binding energies of atoms for various screening parameters are obtained in high accuracy. C^{4+} ion is investigated to demonstrate the validity of our program.

The screened Coulomb potential (SCP) has been extensively used to model different physical systems, such as atoms embedded in weakly-coupled plasmas and the impurity ions in quantum dot. In these systems, the Coulomb interaction between charged particles is replaced by SCP in the Debye-Hückel form $V(r) = \frac{Z_1 Z_2}{r} e^{-\lambda r}$, where λ is the screening parameter related to physical quantities of the system.

For few-electron atoms interacting with SCP, both their bound and continuum properties have been investigated and many physical quantities have been determined in very high accuracy, e.g. the application of Hylleraas basis function in calculating the bound and resonant states of screened two-electron systems [1]. However, accurate calculation of many-electron screened atoms is still a formidable task. The Hartree-Fock method serves as the foundation of computational atomic physics consistently in nowadays and its extension to screened atomic systems would be useful in many aspects.

Based on the pioneering work of Fischer *et al* [2] on numerical HF method, we have extended the programs to screened atomic systems. The most intricate difficulties in calculating the two-electron Slater integrals F^k and G^k under SCP have been overcome by developing a numerical integration and a differentiation method. We demonstrate the capabilities of our programs by calculating the screened C^{4+} ions in three different models (screened V_{en}, V_{ee}, and $V_{en} + V_{ee}$) and compare the results with Čertík's Roothaan HF calculations [3] and our Hylleraas-basis variational predictions [1]. The ground state energies are shown in Fig. 1. It is clear that the HF predictions of atomic properties are feasible in a variety of screening parameters. Calculations on more complex atoms are in progress.

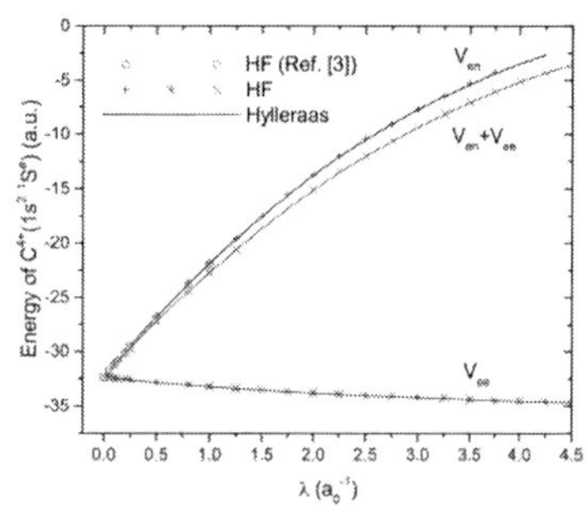

Figure 1. Energies of screened C^{4+} ion in its ground state ($1s^2\ ^1S^e$) at various screening parameters. Scatters: Hartree-Fock method, lines: Hylleraas-basis variational method.

References

[1] Jiao L G and Ho Y K 2015 *Comput. Phys. Comm.* **188** 140
[2] Fischer C F, Brage T and Jönsson P 1997 *Computational Atomic Structure*, Institute of Physics Publishing, Bristol and Philadelphia
[3] Čertík O and Winkler P 2013 *Int. J. Quantum Chem.* **113** 2012

*E-mail: lgjiao@jlu.edu.cn
†E-mail: majia@sau.edu.cn

K-shell electron-impact excitation and polarization of x-ray emission of Fe^{20+} ion

J M Zhang[1], L Y Xie[1 *], D H Zhang[1], J Jiang[1], Z W Wu[1] and C Z Dong[1 †]

[1]Key Laboratory of Atomic and Molecular Physics and Functional Materials of Gansu Province, College of Physics and Electronic Engineering, Northwest Normal University, Lanzhou, 730070, P. R. China.

Synopsis The total and magnetic cross sections for *K*-shell electron impact excitation (EIE) of Fe^{20+} have been calculated using the fully relativistic distorted-wave approach. Degrees of linear polarization of x-ray emission following EIE are presented for several stronger $2p\rightarrow1s$ transitions, and compared with the values for the same transition lines but from dielectronic recombination process of Fe^{21+} in the ground state, obvious differences are found that can be employed in diagnosing the formation mechanism of corresponding x-ray lines.

Electron-ion collisions, such as electron impact excitation (EIE), dielectronic recombination (DR) etc, are fundamental atomic processes in astrophysical and laboratory plasmas. Accurate data of cross sections, rate coefficients, and degrees of linear polarization of radiations following EIE and DR processes have been widely applied for testing atomic theory, revealing detailed information of collision dynamics, and diagnosing plasmas [1]. Iron is a very important element in astrophysics, high-quality atomic data of iron ions are much interested [2].

Table 1. The calculated transition energy ($E_{x\text{-ray}}$ / eV) and linear polarization (P) of x-ray lines from *KLL* DR process of Fe^{21+} ion. Label represent x-ray transition lines.

Excited states	Final states	Label	$E_{x\text{-ray}}$	P
$[1s2p^2_{1/2}2p_{3/2}]_2$	$[2p_{1/2}2p_{3/2}]_1$	A_1	6508.36	0.47
	$[2p_{1/2}2p_{3/2}]_2$	A_2	6502.90	-0.69
$[1s2p^2_{1/2}2p_{3/2}]_1$	$[2p^2_{1/2}]_0$	B_0	6543.35	0.58
	$[2p_{1/2}2p_{3/2}]_1$	B_1	6534.27	-0.41
	$[2p_{1/2}2p_{3/2}]_2$	B_2	6528.80	0.07
$[(1s2p_{1/2})_0 2p^2_{3/2}]_2$	$[2p_{1/2}2p_{3/2}]_1$	C_1	6530.85	0.51
	$[2p_{1/2}2p_{3/2}]_2$	C_2	6327.72	-0.77
$[(1s2p_{1/2})_1(2p^2_{3/2})_2]_1$	$[2p^2_{1/2}]_0$	D_0	6546.61	0.54
	$[2p_{1/2}2p_{3/2}]_1$	D_1	6541.15	-0.37
	$[2p_{1/2}2p_{3/2}]_2$	D_2	6525.21	0.07

In the present work, we mainly pay our attentions on *K*-shell EIE process of Fe^{20+} ion. Using the fully relativistic distorted-wave approach [3], total and magnetic cross sections, and degrees of linear polarization of x-ray emissions were systematically calculated for *K*-shell EIE process of Fe^{20+} ion with incident electron energies in the range of 1-6 times thresholds. In the calculations, effects of electron correlations, quantum electrodynamics, and Breit interactions were well considered. The linear polarizations for the same x-ray lines but from the *KLL* DR process of Fe^{21+} ion were also calculated. Obvious differences are found between the linear polarizations of the same x-ray lines from EIE and DR processes, respectively.

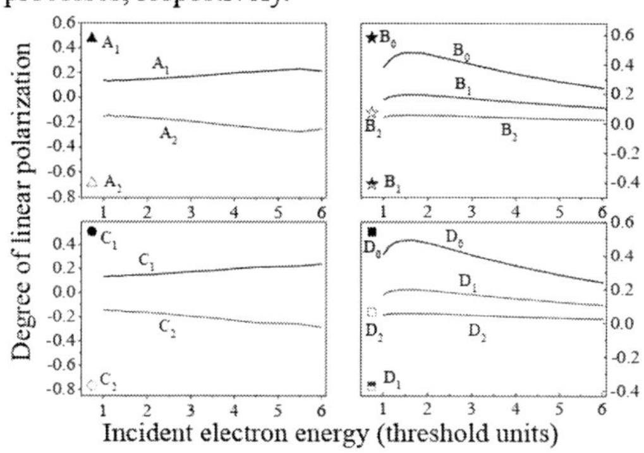

Figure 1. The degrees of linear polarization of x-ray lines following *K*-shell EIE process of Fe^{20+} ion as functions of incident electron energy (solid lines), compare to the polarizations of the same x-ray emissions but from *KLL* DR process (see table 1) of Fe^{21+} ion in the ground state (scattered plots).

This work was supported by the National Key Research and Development Program of China under Grant No.2017YFA0402300, and the Natural Science Foundation of China (Grant Nos. U1530142, 11564036, 11774292, 11464042, 11874051) and the Fund of NWNU (NWNU-LKQN-15-3).

References

[1] Ren C *et al* 2018 *Phys. Rev. A* **98** 012711
[2] Chen H *et al* 2005 *Astrophys. J.* **618** 1086
[3] Jiang J *et al 2008 Phys. Rev. A* **78** 022709

* E-mail: xiely@nwnu.edu.cn
† E-mail: dongcz@nwnu.edu.cn

Rayleigh scattering formalism for the tune-out wavelength: Application to the $2\,^3S$ state of helium

G W F Drake[1] *, J G Manalo[2], P-P Zhang[3], B M Henson[4] and K G H Baldwin[4]

[1]Department of Physics, University of Windsor, Windsor, ON N9B 3P4 Canada
[2]Department of Physics, University of Ottawa, Ottawa ON K1N 6N5 Canada
[3]Wuhan Institute of Physics and Mathematics, Chinese Academy of Sciences, Wuhan, Peoples Republic of China
[4]Laser Physics Centre, Research School of Physics and Engineering, The Australian National University, Canberra, ACT 2601, Australia

Synopsis The tune-out wavelength for an atom interacting with an incident laser beam is reformulated as a zero in the Rayleigh scattering cross section instead of the dynamic polarizability (or AC Stark shift). The two pictures are equivalent in lowest order, but not when higher-order retardation effects are taken into account. Detailed high-precision calculations in Hylleraas coordinates will be presented for the tune-out wavelength of helium at 413 nm near the $2\,^3S - 3\,^3P$ optical transition, including D-wave retardation corrections, and compared with recent high-precision measurements.

The tune-out wavelength is the wavelength at which the interaction between an atom and an incident laser beam vanishes. It has traditionally been viewed as a zero in the frequency-dependent polarizability, or AC Stark shift [1]. This view is appropriate for an atom situated in an optical lattice that is fixed in space and oscillates in time. However, this does not take into account the photonic nature of the interaction. For recent high-precision experiments involving the interaction of a helium atom in the $2\,^3S$-state with a laser beam in the form of a travelling plane-wave [2], it is more appropriate to think of the tune-out wavelength near 413 nm as a zero in the Rayleigh scattering cross section rather than the dynamic plarizability. It can be measured to very high precision and it provides a novel test of QED for an atomic property other than the energy. The 413 nm transition is the one closest to the $1s2s\,^3S - 1s3p\,^3P$ transition of ^{4}He.

The purpose of this paper is to formulate the tune-out wavelength as a zero of the Rayleigh scattering cross section instead of the dynamic polarizability. The two pictures are equivalent in lowest order, but not when higher-order retardation corrections are taken into account. These are of the same α^2 order as the relativistic corrections, and so must be included before a meaningful test of QED can be made. Detailed calculations in Hylleraas coordinates will be presented in both the length and velocity forms of the interaction. The results for the tune-out wavelength λ_t in the velocity (V) and length (L) forms in the nonrelativistic limit for the case of infinite nuclear mass are [3, 4]

$$\lambda_t^{(V)} = 413.038\,304\,3858(1) \text{ nm}$$

$$\lambda_t^{(L)} = 413.038\,304\,3893(35) \text{ nm}$$

As indicated, the velocity form is indeed more rapidly convergent. The above lays a firm foundation for the calculation of finite nuclear mass, retardation, relativistic and QED corrections, as will be discussed at the conference.

Research support by NSERC and by SHARCNET are gratefully acknowledged.

References

[1] Zhang Y H *et al* 2016 *Phys. Rev. A* **93** 052516
[2] Henson B M *et al* 2015 *Phys. Rev. Lett.* **115** 043004
[3] Manalo J G 2018 *M.S. Thesis* University of Windsor
[4] Drake G W F *et al* 2019 *Hyperfine Int.* in press

*E-mail: gdrake@uwindsor.ca

Combined MCDHF-CI and MBPT calculations for n=4 to n=3 x-ray transitions in Ni-like tungsten

K Kozioł* and J Rzadkiewicz†

National Centre for Nuclear Research (NCBJ), Andrzeja Sołtana 7, 05-400 Otwock-Świerk, Poland

Synopsis The $4d \to 3p$, $4p \to 3s$, and $4f \to 3p$ x-ray transitions in Ni-like tungsten ions have been studied theoretically. The Multiconfiguration Dirac–Hartree–Fock method and the large-scale relativistic Configuration Interaction and Many Body Perturbation Theory methods have been employed in order to take into account electron correlation effects on the lines wavelengths.

Tungsten is chosen as a plasma facing material in modern large tokamaks, such as JET and ITER. Therefore, spectroscopic studies of tungsten ions are a tool for diagnostics relevant for a wide range of electron temperatures [1].

Table 1. Wavelengths of Ni1, Ni2, and Ni3 transitions (Å) for various theoretical approaches.

	Ni1	Ni2	Ni3
AMBIT:			
MCDHF	5.1986	5.2528	4.6394
+CI	5.2007	5.2537	4.6420
+CI+MBPT	5.2011	5.2543	4.6424
GRASP2K (cited from Ref. [2]):			
MCDHF	5.1947	5.2486	4.6367
+CI(FCI)	5.1994	5.2517	4.6519
+CI(FCI*)	5.1988	5.2510	4.6413
+CI(CV)	5.2010	5.2541	4.6318
+CI(CV*)	5.2004	5.2532	4.6417
Experiment:			
Ref. [3]	5.2008(3)	5.2540(3)	
Ref. [4]	5.2002(9)	5.2520(16)	4.6372(10)

In our previous paper [3] the x-ray transitions in Ni- and Cu-like tungsten ions in the 5.19–5.26 Å wavelength range that are relevant as a high-temperature tokamak diagnostic have been studied. In our next paper [2] various Multiconfiguration Dirac–Hartree–Fock (MCDHF) plus Configuration Interaction (CI) approaches (named FCI, FCI*, CV, and CV* in Table 1; see [2] for detailed description), employing GRASP2K [5] code, have been tested in order to reproduce and predict experimental wavelengths of $4d \to 3p$ transitions in Ni- and Cu-like tungsten ions.

*E-mail: Karol.Koziol@ncbj.gov.pl
†E-mail: Jacek.Rzadkiewicz@ncbj.gov.pl

In the present work the calculations of the wavelengths of the $4d \to 3p$, $4p \to 3s$, and $4f \to 3p$ x-ray transitions in Ni-like tungsten ions have been carried out by means of the AMBIT [6] code, based on MCDHF method with CI and Many Body Perturbation Theory (MBPT) approaches in order to take into account electron correlation effects on the lines wavelengths. Applying both the MCDHF-CI and the MCDHF-CI + MBPT approaches instead the "pure" MCDHF approach allowed to reduce substantially the experiment–theory gap in wavelengths. Results of this study provide an important benchmark for x-ray measurements in tokamaks.

Acknowledgments: The work was partly supported by the Polish Ministry of Science and Higher Education within the framework of the scientific financial resources in the years 2016–2019 allocated for the realization of the international co-financed project. This work has been carried out within the framework of the EUROfusion Consortium and has received funding from the Euratom Research and Training Programme 2014–2019 under Grant Agreement No. 633053.

References

[1] Pütterich T, Neu R, Dux R *et al* 2008 *Plasma Phys. Controlled Fusion* **50** 085016
[2] Kozioł K and Rzadkiewicz J 2018 *Phys. Rev. A* **98** 062504
[3] Rzadkiewicz J, Yang Y, Kozioł K *et al* 2018 *Phys. Rev. A* **97** 052501
[4] Clementson J, Beiersdorfer P, Brown G V, and Gu M F 2010 *Phys. Scr.* **81** 015301
[5] Jönsson P, Gaigalas G, Bieroń J, Froese Fischer C, and Grant I P 2013 *Comput. Phys. Commun.* **184** 2197
[6] Kahl E V and Berengut J C 2018 *Comput. Phys. Commun.* **238** 232

Photodetachment Studies of Ir^- Ions at DESIREE

M K Kristiansson[1] *, J Sundberg[2], D Hanstorp[2], G Eklund[1], H Cederquist[1], H Zettergren[1] and H T Schmidt[1]

[1] Department of Physics, Stockholm University, Albanova, 10691 Stockholm, Sweden
[2] Department of Physics, University of Gothenburg, 41296 Gothenburg, Sweden

Synopsis Anions of iridium were stored in the cryogenic storage ring, DESIREE [1, 2, 3] and investigated using laser photodetachment. The ions were stored for up to 2000 seconds and the relaxation of several metastable states were studied by selecting different photon energies. The threshold of the previously not observed $^3P_2$ level is located and the lifetime of this sate is found to be longer than the storage time of the ion beam. We also find that a state exists higher in energy than the previously known level with a lifetime in the range of hundreds of milliseconds.

We have used laser photodetachment to investigate anions of iridium produced in a cesium sputter ion source and stored in the cryogenic storage ring, DESIREE [1, 2, 3]. We applied a widely tunable OPO, optical parametric oscillator to be able to address a range of excited state energies. Due to the excellent DESIREE vacuum conditions, we could store the ions for many hundreds of seconds and thereby study any relaxation of longlived metastable levels. Iridium has an electron affinity of 1.56 eV [4] and the ground state of the anion is [Xe]$4f^{14}5s^25p^65d^86s^2$ $^3F_4$, where the term scheme is determined by the partly filled 5d shell. Predicted bound excited states are metastable as they are of the same configuration and therefore the same parity as the ground state. The $^3F_3$ level can decay to the ground state by a magnetic-dipole transition while the $^3P_2$ level can only decay by an electric-quadrupole transition. The higher lying fine structure components, $^3F_2$ and $^3P_1$ have been assumed to be unbound [4, 5]. We performed systematic studies of the photodetachment signal as function of photon energy and time after ion injection (milliseconds to 2000 seconds).

Preliminarily, we reached the following conclusions: We locate the threshold for detachment from the $^3P_2$ level and find that this level is bound more strongly than predicted by an amount of the order of 100 meV. The threshold obtained is shown in Figure 1. The OPO is scanned over a range of photon energies and the signal is normalized to the pulse energy of the laser. The binding energy of $^3P_2$ level is obtained by fitting the data to the Wigner law [6].

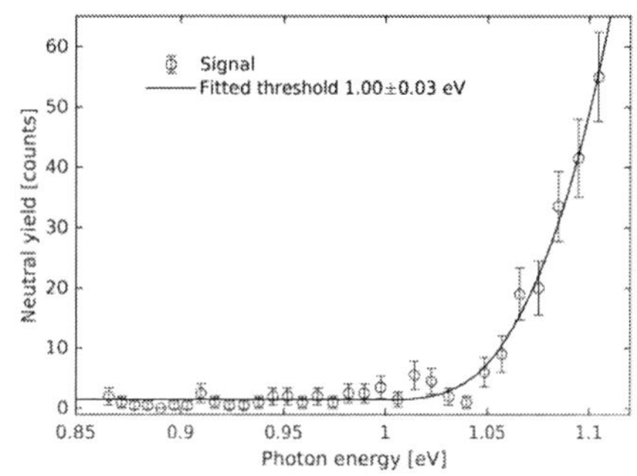

Figure 1. Photodetachment signal of Ir^-. The solid line is a Wigner law fit resulting in a preliminary threshld value of 1.00 ± 0.03 eV.

We find that the lifetime of this level is longer than the storage time of the beam. We find that a state exists higher in energy than the previously known levels with a lifetime in the range of hundreds of milliseconds. A final analyzis of the experimental results will be presented at the conference.

References

[1] Schmidt H T *et al Rev. Scient. Instr.* 2013 **84** 055115
[2] Bäckström E *et al Phys. Rev. Lett.* 2015 **114** 143003
[3] Chartkunchand K C *et al Phys. Rev. A* 2016 **94** 032501
[4] Bilodeau R C *et al Phys. Rev. A* 1999 **61** 012505
[5] Thogersen J *et al Phys. Rev. Lett.* 1996 **76** 2870
[6] Wigner E P *Phys. Rev.* 1948 **73** 1002

*E-mail: moa.kristiansson@fysik.su.se

Theoretical study on *KLL* dielectronic and higher-order recombination of boron-like isoelectronic sequence

L Y Xie[1*], S M Lu[1], D H Zhang[1], J Jiang[1], Z W Wu[1] and C Z Dong[1]

[1]Key Laboratory of Atomic and Molecular Physics and Functional Materials of Gansu Province, College of Physics and Electronic Engineering, Northwest Normal University, Lanzhou, 730070, P. R. China.

Synopsis The *K*-shell dielectronic (DR), trielectronic (TR), and quadruelectronic (QR) recombination have been systematically studied for highly charged B-like ions with Z=14-50 by using the FAC code. Strong configuration mixing are found that significant influence the Auger rate, and TR strengths, and leads to irregular changes of TR/DR strength ratio along B-like sequence.

DR and higher-order processes with the active excitation of more than one electron may contribute considerably to recombination of highly charged ions and affect plasma parameters noticeably [1]. Lots of experimental and theoretical studies have focused on the DR and higher-order resonant recombination process. For example, Beilmann *et al* [1] recently studied the DR and the higher-order TR, and QR processes for Li- to O-like Ar, Fe, and Kr ions at high electron energy resolution in electron beam ion traps, and found that the TR process contributes about 10% to the total recombination cross section in most cases, and its relative strength decrease with increasing Z.

In the present work, *KLL* inter-shell resonant recombination processes are systematically calculated for B-like ions with Z=14-50 by using the flexible atomic code (FAC) [2] based on the relativistic configuration interaction approach. The detailed resonant energy, resonant strength, Auger rate, cross section is obtained and compared with the previous results. Strong configuration interaction is found between the resonance states with the same parity and total angular momentum. Such strong configuration interactions is responsible for energy-level crossing, and the great enhancement of the TR strength, and result in an irregular change of the TR/DR strength ratio at Z=22-25 region along the B-like sequence. In Fig.1, the total resonant strength (10^{-19} cm^2.eV) and the TR/DR strength ratio are plotted for B-like ions. Our calculated TR/DR strength ratio is in good agreement with experimental results [1] for Ar^{13+} and Fe^{21+} ions. Larger differences are found between the present results and previous calculations and measurement [4] for Si^{9+} ion. In the case of Si^{9+}, the DR peaks are strongly mixed with TR peaks, experimental measurements are much difficult, and the high possible energy resolution is required to resolve TR within the DR forest [3].

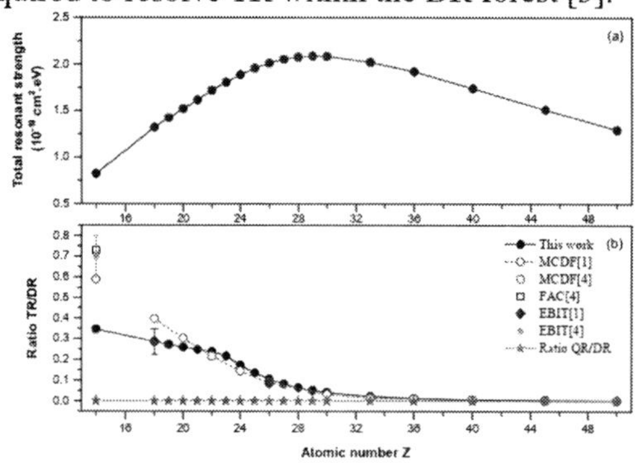

Figure 1. (a) The total resonant strength (10^{-19} cm^2.eV) (DR+TR+QR) for B-like isoelectronic sequence ions. (b) The calculated TR/DR strength ratio are compared with experimental and other theoretical results [1,4].

The work was supported by the National Key Research and Development Program of China under Grant No.2017YFA0402300, and the Natural Science Foundation of China (Grant Nos. U1530142, 11564036, 11774292, 11464042, 1874051) and the Fund of NWNU (NWNU-LKQN-15-3).

References

[1] Beilmann C *et al* 2013 *Phys. Rev. A* **88** 062706
[2] Gu M F 2008 *Can. J. Phys.* **86** *675*
[3] Schnell M *et al* 2003 *Phys. Rev. Lett.* **91** 043001
[4] Baumann T M *et al* 2014 *Phys. Rev. A* **90** 052704

* E-mail: xiely@nwnu.edu.cn

Determination of the electron affinity of astatine

D Leimbach[1,2,3 *], S Rothe[2 †], J Sundberg[3] and D Hanstorp[3] for the IS615 Collaboration

[1]Institut für Physik, Johannes Gutenberg-Universität, 55122 Mainz, Germany
[2]CERN, 1211 Geneva, Switzerland
[3]University of Gothenburg, 40530 Gothenburg, Sweden

Synopsis Astatine is a promising candidate for targeted alpha therapy in nuclear medicine. However, due to its scarce nature, some of its fundamental properties that determine its chemical characteristics such as the electron affinity, the binding erergy of the additional electron in a negative ion, are still unknown. Here we present the method and results of the first measurement of the electron affinity of astatine, performed with the Gothenburg ANion Detector for Affinity measurements by Laser PHotodetachment (GANDALPH) at CERN-ISOLDE.

Astatine, the purely radioactive element 85, is the rarest naturally occurring element on earth [1]. One of its longer lived isotopes, ^{211}At is a promising candidate for targeted alpha therapy (TAT), a method of treating cancer directly at the location of the tumor cell level using alpha particle emitting components [2].

However, some of the fundamental properties to predict the chemistry of astatine are not known. One of these missing parts is the electron affinity (EA), the binding energy of the additional electron in its negative ion. The EA, combined with the previously measured ionization potential (IP) [3], determines the electronegativity, which serves as valuable benchmark for quantum chemical calculations.

Astatine atoms were produced at the CERN-ISOLDE radioactive ion beam facility by impinging a 1.4 GeV proton beam onto a thick Th/Ta-foil target. The atoms were ionized with an LaB_6 surface ionizer and accelerated to an energy of 20 keV .

In order to determine the EA of astatine, the dedicated Gothenburg Anion Detector for Laser PHotodetachement (GANDALPH) [4] was connected to a port of the ISOLDE low-energy beamline. In GANDALPH, the negative ion beam was overlapped with a frequency tuneable laser beam in either co- or counterpropagating geometry.

The resulting neutral particles were detected by allowing them to impinge on a graphene coated glass plate, thereby creating secondary electrons which were guided to an electron multiplier. By scanning the laser frequency around the photodetachment threshold, one can obtain the EA by fitting the Wigner law to the resulting spectrum.

Following the first measurement of the electron affinity of the EA of a radiogenic iodine isotope in 2016 [4], the GANDALPH beamline received several upgrades for this measurement campaign resulting in the first reported measurement of the EA of astatine.

We will present the GANDALPH beam-line as well as the value of the EA of astatine and give an outlook for future photodetachment experiments using radioactive isotopes.

References

[1] Asimov I *et al* 1953 *J. Chem. Educ.* **30** 616
[2] Mulford D A *et al* 2005 *J. Nucl. Medicine* **46** 199S
[3] Rothe S *et al* 2013 *Nature Communications* **4** 1835
[4] Rothe S *et al* 2017 *J. Phys. G* **44** 104003

* E-mail: david.leimbach@cern.ch
† E-mail: sebastian.rothe@cern.ch

Ultrafast laser induced structural dynamics probed by time-resolved photoelectron spectroscopy

M de Anda Villa[1]*, A Lévy[1†], J Gaudin[2], R Bouillaud[2], D Descamps[2], B Dorado[3], N Fedorov[2], R Grisenti[4], M Hatifi[1], H Jouin[2], E Lamour[1], S Macé[1], P Martin[2], S Petit[2], C Prigent[1], R Sobierajski[5], S Steydli[1], M Trassinelli[1], A Wawro[5] and D Vernhet[1]

[1]Institut des Nanosciences de Paris, CNRS UMR 7588, Sorbonne Université Campus UPMC, 75005 Paris, France.
[2]Université de Bordeaux, CEA, CNRS, CELIA, 33400 Talence, France
[3]CEA, DAM, DIF, F-91297, Arpajon, France
[4]Institut für Kernphysik J.W. Goethe Universität Max-von-Laue-Str. 1, 60438 Frankfurt (M), Germany
[5]Institute of Physics of the Polish Academy of Sciences, PL-02-668 Warsaw, Poland

Synopsis This work aims to study laser-induced lattice dynamics when there is strong electron-lattice temperature non-equilibrium for which the material properties are poorly known. The proposed experimental technique is time-resolved photoelectron spectroscopy (Tr-PES). The main challenges of such technique are the superposition of pump and probe spectra and the distortion of the probe spectra due to the interaction with pump-induced electrons, called space charge effect (SCE). We strongly reduced this effect by tuning the 100eV XUV beamline we have developed, enabling us to investigate Copper samples.

When ultra-short laser pulses induce a non equilibrium regine of the electron-lattice temperature, theoretical description is difficult due to the limited knowledge of material properties in these conditions [1], including the electron-phonon coupling factor, the electronic and ionic thermal conductivities, and the electron heat capacity. In this context, time-resolved photoelectron spectroscopy (Tr-PES) experiments may help to improve our understanding of the physical processes and material properties in this regime.

The excitation and relaxation of matter can be diagnosed by studying the time evolution of their PES spectra, which is influenced by the lattice state/temperature. However, such a pump/probe experiment, based on the detection of probe-induced photoelectron spectra, is technically challenging due to the presence of a pump-induced high flux of photoelectrons that can cause a loss of information by overlapping and/or interacting with the probe electrons, distorting the detected spectrum (called space charge effect, SCE) [2].

To prevent the overlap of photoelectron spectra, we have developed a dedicated XUV beamline based on High-order Harmonic Generation [3] that delivers femtosecond pulses with the required photon energy (80 – 100 eV). This beamline will be fully described during the conference.

Thanks to the complete experimental and theoretical investigations on SCE, we have been able to optimize the experimental parameters in order to perform Tr-PES pump/probe experiments on Copper samples where the SCE has been strongly reduced. It allowed us to separate the influence of lattice dynamics on the PES from the SCE, as illustrated Figure 1.

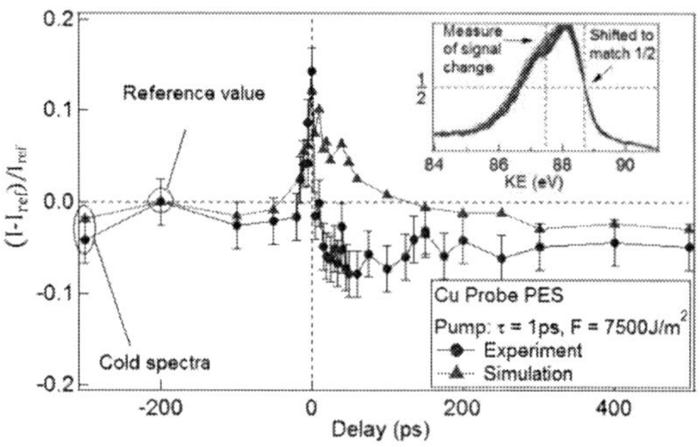

Figure 1. Pump/probe measurements and their corresponding SCE simulations showing the Cu PES evolution. The difference in positive delays is attributed to the PES modification due to lattice dynamics. Inset: spectra normalized and superimposed to match the signal value of $I = 1/2$. The signal I is then measured at $KE = 87.5eV$.

References

[1] Lin Z *et al* 2008 *Phys. Rev. B.* 77 075133
[2] Oloff L *et al* 2016 *J. Appl. Phys. B.* **119** 225106
[3] L'Huillier A *et al* 2017 *Phys. Rev. Lett.* **70** 774

* E-mail: manuel.andavilla@insp.jussieu.fr
† E-mail: levy@insp.jussieu.fr

Accurate electron affinity of lanthanum atom

Y Lu, R Tang, X Fu and C Ning *

Department of Physics, State Key Laboratory of Low Dimensional Quantum Physics,Tsinghua University, Beijing 10084, China

Synopsis Atomic lanthanum anion La^- is the best candidate for laser cooling of any atomic anions by far. The previous theoretical studies showed that La^- has very a complicated electronic structure, which cannot be well resolved by the traditional photoelectron spectroscopy. In the present work, we used the slow-electron velocity-map imaging method in combination with an ion trap to measure the electronic structure of La^-.The electron affinity of La is accurately determined to be 4496.97(20) cm^{-1} or 0.557553(25) eV.

Only a few atomic anions have bound excited states with an opposite parity to the ground state, which is required for the electric dipole transition used for laser cooling. The previous experimental and theoretical studies have shown that La^- is the most promising candidate for laser cooling of any atomic anions. Although several electric dipole transitions have been identified via the tunable infrared spectroscopy and the resonant frequency of the laser cooling transition have been accurately determined to be 96.592 713(91) THz [1]. However, the electron affinity (EA) of La still remains poorly measured.

In this work we report the high resolution photoelectron spectroscopy of lanthanum anion La^- via the slow-electron velocity-map imaging method in combination with an ion trap. This method has very high energy resolution, typically a few cm^{-1} near the photodetachment threshold [2]. This is an essential ability for resolving the dense electronic states of lanthanides. Moreover, our spectrometer can work on the trap-on mode and the trap-off mode. In the trap-off mode, ions took about 0.4 ms to fly from the ion source to the photodetachment zone. Therefore, the short-lived excited states have more chance to survive in comparison with the trap-on mode with a typical trap period 45 ms. As shown in Figure 1, more than 20 sharp peaks are resolved. Among them, peak *a*, which is from the groud state $^3F^e{}_2$ of La^- to the state $^2F^e{}_{5/2}$ of La, is chosen to measure the EA value of La. To determine the binding energy of peak *a* as accurate as possible, a series of spectra were obtained near the threshold of the transition. The electron affinity of La is accurately determined to be 4496.97(20) cm^{-1} or 0.557553(25) eV.

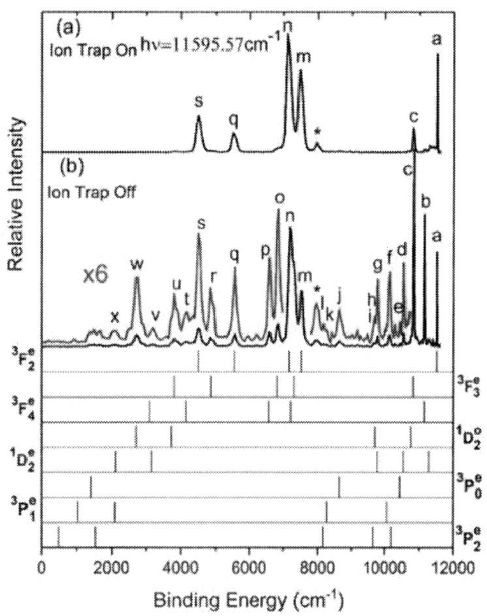

Figure 1. Photoelectron spectra of La^- obtained with the trap-on mode (a) and the trap-off mode (b). The red curve shows the weak peaks multiplied by a factor 6 for a clear view. The sticks below the spectra indicate the binding energy of photodetachment channels from the states labeled on the left or right sides.

References

[1] Cerchiari G *et al* 2018 *Phys. Rev. Lett.* **120** 133205

[2] Tang R *et al* 2018 *J. Chem. Phys.* **149** 134304

* E-mail: ningcg@tsinghua.edu.cn

Bayesian atomic structure calculations for collisional problems

A M P Mendez[1]*, J I Di Filippo[2], S D Lopez[1] and D M Mitnik[1,2]

[1]Instituto de Astronomía y Física del Espacio, Consejo Nacional de Investigaciones Científicas y Técnicas and Universidad de Buenos Aires, Buenos Aires, Argentina
[2]Departamento de Física, Universidad de Buenos Aires, Buenos Aires, Argentina

Synopsis The calculations of collisional processes require an accurate description of the target. In general, the atomic structure is obtained through tedious iterations in which a variety of configurations and parameters are chosen to minimize the differences between the numerical and experimental values of the energies and the oscillator strengths. Using a Bayesian machine learning analysis through a Tree–structured Parzen Estimator, we can reproduce the experimental atomic structure with high accuracy. Results for neutral beryllium are presented.

The computation of collisional rates demands an accurate description of the ions involved in the processes [1]. However, the determination of an adequate atomic structure can signify lots of time and computational resources. The optimization of the target wavefunctions relies on considering a configuration interaction (CI) expansion, in which additional levels are added to improve the accuracy of the calculations. Often, the radial wavefunctions are obtained with model potentials containing scaling parameters, which are then varied. The design of automatic optimization is a challenging task. The number of configurations included in the CI, and the variation of the scaling parameters, lead to erratic oscillations in the results. This behavior implies the lack of a systematic and logic prescription for this procedure.

In this contribution, we implemented the Bayesian method via a Tree–structured Parzen Estimator from HYPEROPT [3] to optimize the atomic structure of the ions. The sequential model–base optimization turns to be an excellent machine learning methodology to minimize the scalar–valued error functions. To illustrate the procedure, we considered the neutral beryllium atom. The correct description of the target has proven to be necessary to obtain accurate electron impact excitation and electron impact ionization cross sections [4, 5].

The atomic structure of Be is calculated with the AUTOSTRUCTURE code [2], and the orbital radial wavefunctions are obtained with a Slater-type–orbitals containing λ_{nl} scaling parameters. These parameters are then optimized within a Bayesian model. The target expansion involved the $2snl$ configurations up to $nl = 5g$, and $2p^2$. We included the first six orbitals in our *search space*, and they were varied around $\lambda_{nl} = 1$. The results obtained after 10000 evaluations are given in Fig. 1. The agreement for the $2s^2\,{}^1S$, $2s2p\,{}^3P$ and $2s2p\,{}^1P$ energies with the benchmark data [6, 7] is better than 0.1%.

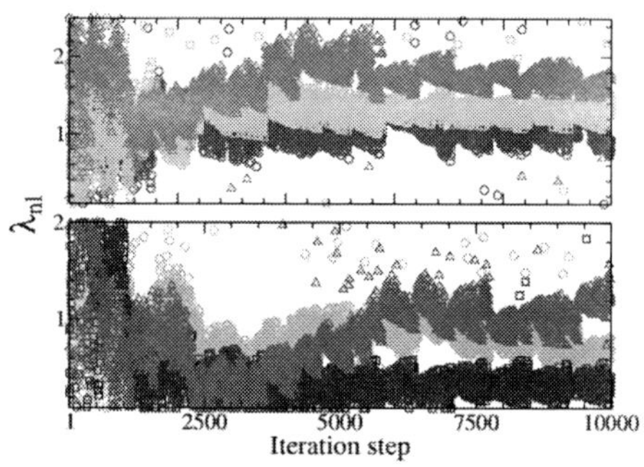

Figure 1. Bayesian optimization for the scaling parameters. Up: λ_{1s} and λ_{2l}. Down: λ_{3l}.

References

[1] Bartschat K *et al* 2004 *J. Phys. B* **37** 2617
[2] Badnell N R 2011 *Comput. Phys. Commun.* **7** 1528
[3] Bergstra J *et al* 2015 *Comput. Sci. Discov.* **1** 014008
[4] Zatsarinny O *et al* 2016 *J. Phys. B* **49** 235701
[5] Ballance C P *et al* 2003 *Phys. Rev. A* **68** 062705
[6] Jönsson P *et al* 1999 *J. Phys. B* **32** 1233
[7] Kramida A *et al* NIST Atomic Spectra Database (version 5.6.1)

*E-mail: alemendez@iafe.uba.ar

The radiative decay process of 2s2p configuration of He-like ions *

M X Cao[1], X B Ding[1 †], D H Zhang[1], M W Zhang[2], Y L Xue[2], D Y Yu[2] and C Z Dong[1 ‡]

[1]Key Laboratory of Atomic and Molecular Physics and Functional Materials of Gansu Province, College of Physics and Electronic Engineering, Northwest Normal University, Lanzhou 730070, China
[2]Institute of Modern Physics, Chinese Academy of Sciences, Lanzhou 730000, China

Synopsis The electron correlation effects are crucial to the two-electron one-photon transition of double K hole excited state. Appropriate correlation model was constructed to investigate this process of 11 selected He-like ions of Z from 10 to 47. Good agreement to the previous work and some new transition data were obtained.

The inner-shell hole state is a special excited state which refers to the inner-shell orbital of an atom or ion is unoccupied, while the outer shell orbital is occupied by electrons. These exotic atoms are extremely unstable and mainly decay through non-radiative Auger and radiative process. The former is usually faster than the later. With development of the weak signal detection, it is possible to investigate the decay process from the photon perspective with higher accuracy. Such investigations are helpful to explain the effects of electron correlation, coupling scheme change and would be useful to provide diagnostic information for astrophysical and laboratory plasma.

The radiative decay process of inner-shell excited states with two K vacancies can be fulfilled either by one-electron one-photo(OEOP) or by two-electron one-photon(TEOP) transitions. The OEOP transition from $2s2p$ to $1s2s$ configuration while the TEOP transition from $2s2p$ to $1s^2$ configuration. The present calculations[1] are carried out for 11 selected He-like ions of Z from 10 to 47 by using GRASP2K code[2] based on the multi-configuration Dirac-Fock (MCDF) method.

The correlation model used in the present work similar to the model used by Kadrekar and Natarajan[3]. The minimum basis (MB) was generated by considered limited expansion allowing single and double (SD) substitutions of electrons from the reference configurations, the study on the correlation effects were carried out with respect to MB. The active space was expanded and all the newly added orbital functions were optimized while previous orbitals were kept fixed. Good convergence and agreement with the previous work were obtained.

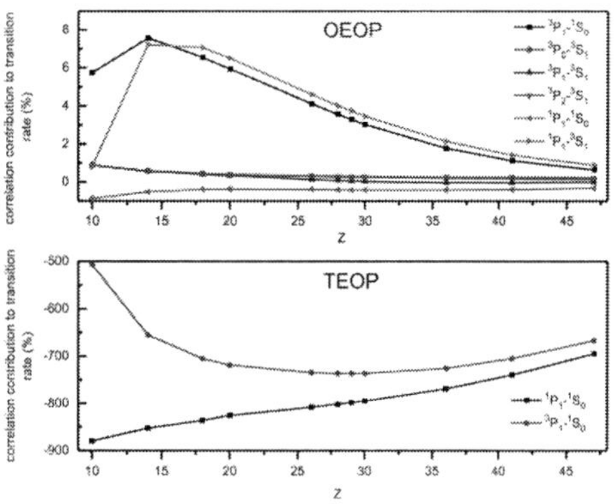

Figure 1. Electron correlation effect on the transition rates for OEOP and TEOP transitions in He-like ions.

The electron correlation contribution on the OEOP and TEOP transition rates are shown in Fig.1. Competition between the Coulomb interaction of the nucleus and electrons and the electron-electron interaction was found for the low Z elements in OEOP transition rate. While this competition will dominated by the Coulomb interaction for th high-Z ions. The correlation contribution for TEOP is much larger than that for OEOP which indicates the TEOP is more sensitive to the correlation effects. The calculated data will be helpful to the future investigations on OEOP and TEOP transition of He-like ions.

References

[1] Ding X *et al* 2019 arXiv: 1901.05689
[2] Jönsson P *et al* 2007 *Comput. Phys. Commun* **177** 597
[3] Kadrekar R *et al* 2011 *Phys. Rev. A* **84** 062506

*Supported by NSFC (Grant NO.U1832126).
†E-mail: dingxb@nwnu.edu.cn
‡E-mail: dongcz@nwnu.edu.cn

B-spline basis implementation of the Hartree-Fock method for arbitrary central potentials

D T Waide[1]*, D G Green[1] and G F Gribakin[1]

[1]School of Mathematics & Physics, Queen's University Belfast, Belfast BT7 1NN, United Kingdom

Synopsis The Hartree-Fock equations are solved using B-splines. Convergence of self-consistency iterations is achieved by turning on the electron interaction gradually. We test the ground and excited-state sets for atoms and an electron gas confined in a harmonic potential and the jellium model of alkali-metal clusters. Comparisons are made with the Thomas-Fermi approximation and dipole polarizabilities are found and compared with experiment.

The Hartree-Fock (HF) method provides a reliable and effective first step in the calculation of atomic structure [1]. The electronic states it generates can also be used a basis for higher-order calculations, e.g., using many-body theory [2].

For electrons confined in a central potential, it is convenient to expand the radial wavefunctions in a B-spline basis and convert the integro-differential HF equations into a generalised eigenvalue problem. We have developed a new code that solves it iteratively for the ground-state orbitals and then constructs sets of excited electron or positron states. Convergence of the self-consistency iterations is aided by setting the elementary charge in the electron Coulomb interaction to a value less than unity (in atomic units) and gradually increasing it to unity as the solution converges. The code generates compact and effectively complete basis sets, covering both bound states and the positive-energy continuum, that can be used for many-body theory calculations of atomic processes, e.g., positron scattering and annihilation [3].

We have tested the code for atoms and atomic negative ions, and performed calculations for N electrons confined by either a harmonic potential (modelling a uniform electron gas) or by that of a uniformly charged sphere (the jellium model of alkali metal clusters). For the harmonically-trapped electrons, the electron radial charge density has been compared with that of the Thomas-Fermi model. Static dipole polarizabilites α were calculated in the HF and Random-Phase Approximation with Exhange (RPAE), which takes into account electron screening effects that are crucial for the electron gas. The results of modelling of sodium clusters are shown in Fig. 1. The static dipole polarizability of electrons trapped in a harmonic potential, corresponding to an infinite positive background charge density, is also shown. We plan to use the new code for studying positron interactions with an electron gas and systems such as clusters.

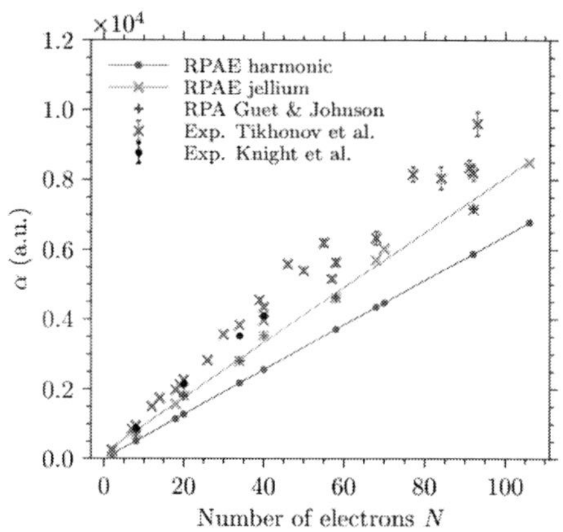

Figure 1. Static dipole polarizability for N electrons in the harmonic and jellium model, compared with the jellium calculations by Guet & Johnson [4] and experimental data for N-atom sodium clusters from Knight *et al.* [5] and Tikhonov *et al.* [6].

References

[1] Froese Fischer C 2011 *Comp. Phys. Comm.* **182** 1315
[2] Amusia M Ya and Chernysheva L V 1997 *Computation of Atomic Processes* (IoP Publ., Bristol)
[3] Green D G, Ludlow J A and Gribakin G F 2014 *Phys. Rev. A* **90** 032712
[4] Guet C and Johnson W R 1992 *Phys. Rev. B* **45** 11283
[5] Knight W D, de Heer W A, Clemenger K and Saunders W A 1985 Solid State Comm. **53** 445
[6] Tikhonov G, Kasperovich V, Wong K and Kresin V V 2001 *Phys. Rev. A* **64** 063202

*E-mail: dwaide01@qub.ac.uk

Three-body fragmentation of CF_4^{3+} induced by 1 keV electron collision

L Chen[1], Z J Shen[1], E L Wang[2], X Zhao[1], X Shan[1] and X J Chen[1] *

[1] Hefei National Laboratory of Physical Sciences at the Microscale and Department of Modern Physics, University of Science and Technology of China, Hefei, 230026, China

[2] Max Plank Institute for Nuclear Physics, Heidelberg, 69117, Germany

Synopsis We report the investigation on three-body fragmentation of CF_4^{3+} induced by 1 keV electron collision utilizing an ion momentum spectrometer. With the help of coincidence measurement for three fragment ions, the channel $CF_4^{3+}\rightarrow F^++CF2++F+$ is identified. By employing Dalitz plot and Newton diagram, a sequential path-way and a concerted pathway are uncovered, which are further separated by momentum correlation of two F^+ ions. We find that the CF_4^{3+} fragmentation prefers a sequential pathway $CF_4^{3+}\rightarrow F^++CF_3^{2+}\rightarrow F^++F^++CF_2^+$.

Investigations on dissociative ionization of polyatomic molecules are of extensive interest in physics, chemistry as well as biology, and of great significance in applications like radiation damage, planet atmospheres and plasma industry. One of the key task in this field is to figure out which parameter control the selective breakup of molecular bonds. In past several decades, great progress has been made in the development of coincidence momentum imaging techniques [1, 2], making it feasible to reveal multi-body fragmentation dynamics for small molecules.

The CF_4 molecule, in its ground states, has a tetrahedral structure with high T_d symmetry. It is an important plasma etching gas in the semiconductor industry and also a potent greenhouse gas in the earth atmosphere. In the past decades, the ionization and dissociation of CF_4 has been extensively studied.

In this work, three-body fragmentation dynamics of CF_4^{3+} is studied by utilizing a newly built ion momentum spectrometer. The precursor ion CF_4^{3+} is created by the collisions of CF_4 with 1 keV electron emitted from a photoelectron emission gun. With the help of ion-ion coincidence TOF map, the channel $CF_4^{3+}\rightarrow F^+ +F^+ +CF_2^+$ is clearly identified. Dalitz plot [3] and Newton diagram are used to visualize of the three-body processes for this channel.

As shown in Fig. 1(a), the obtained Dalitz plot demonstrates a butterfly structure. Momentum correlation of two F^+ ions is displayed in Fig. 1(b), two symmetric structures along a diagonal where the same momenta of two F^+ ions are labeled as A region, and the diagonal regioin is labeled as B. In Fig. 1(c), the Newton diagram for region A shows a circular structrure, which is characteristic of a sequential pathway (i.e., $CF_4^{3+}\rightarrow F_a^++CF_3^{2+}\rightarrow F_a^++F_b^++CF_2^+$). In region B, the momentum values of two F^+ fragment ions are almost the same. Newton diagram for region B is shown in Fig. 1(d), where the momenta of the first F^+ ion (F_f^+) and the second F^+ ion (F_s^+) distribute symmetrically with respect to the momentum of CF_2^+, a clear evidence of the existence of a concerted pathway (i.e., $CF_4^{3+}\rightarrow F_f^++F^{s+}+CF_2^+$).

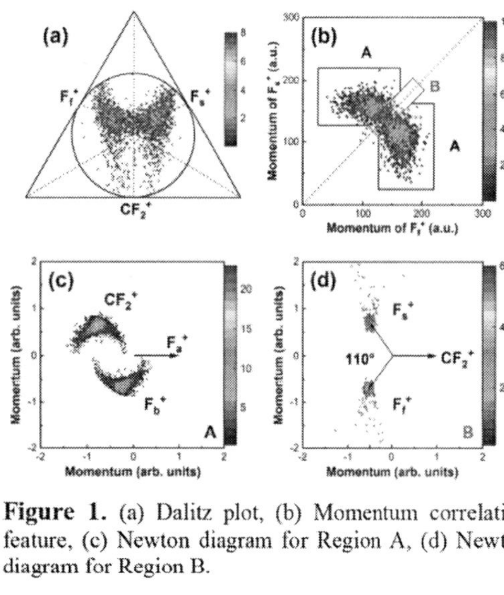

Figure 1. (a) Dalitz plot, (b) Momentum correlation feature, (c) Newton diagram for Region A, (d) Newton diagram for Region B.

References

[1] Dörner R *et al* 2000 *Phys. Rep.* **330** 95
[2] Ullrich J *et al* 2003 *Rep. Prog. Phys.* **66** 1463
[3] Dalitz R H *et al* 1953 *Mag. J. Sci.* **44** 1068

* E-mail: xjun@ustc.edu.cn

Concept and simulations of a high-resolution asymmetric von Hamos X-ray spectrometer for CRYRING@ESR electron cooler

P Jagodziński[1], D Banaś[2]*, M Pajek[2], A Warczak[3], H F Beyer[4], A Gumberidze[4], G Weber[5], Th Stöhlker[4,5] and M Trassinelli[6]

[1]Department of Mathematics and Physics, Kielce University of Technology, 25-314 Kielce, Poland
[2]Institute of Physics, Jan Kochanowski University, 25-406 Kielce, Poland
[3]Institute of Physics, Jagiellonian University, 30-348 Cracow, Poland
[4]Helmholtzzentrum für Schwerionenforschung, D-64291 Darmstadt, Germany
[5]GSI Helmholtz-Institut Jena, D-07743 Jena, Germany
[6]Institutdes NanoSciences de Paris, CNRS, Sorbonne Université, F-75005 Paris, France

Synopsis A concept of a high resolution asymmetric von Hamos X-ray spectrometer for the CRYRING@ESR electron cooler is described and its characteristics obtained by ray-tracing Monte-Carlo simulations are presented. The spectrometer will be used to study the QED effects in H-like medium-Z ions by measuring the energies of X-rays from radiative recombination of highly charged ions with cooling electrons, with a ppm precision of energy determination.

The design of a high resolution asymmetric von Hamos spectrometer (AvH) for low-energy X-ray spectroscopy experiments at the electron cooler of CRYRING@ESR [1] in the international Facility for Antiproton and Ion Research (FAIR) in Darmstadt is presented. The spectrometer will allow to measure, with a high resolution down to 70 meV, the low-energy X-rays (5-10 keV) from radiative recombination (RR) of stored bare or few-electron heavy ions interacting with cooling electrons.

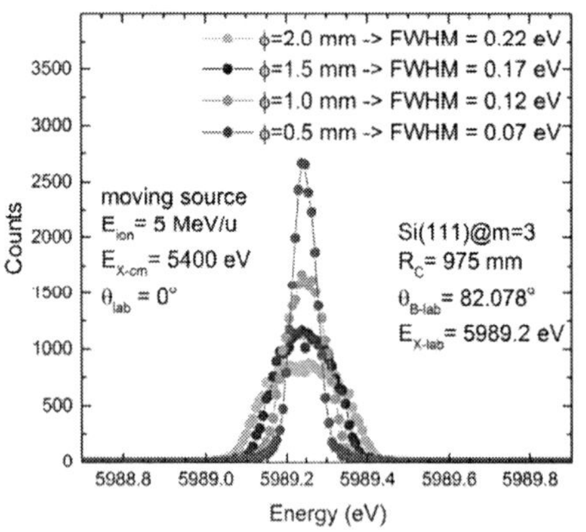

Figure 1. Simulated X-ray line profiles expected for asymmetric von Hamos spectrometer for different diameter of ion beam.

Performed X-ray-tracing simulations [2] show that the energies of the X-ray transitions can be measured for medium-Z bare ions with relative precision of a few ppm, which gives access to study the QED effects. For medium $Z = 20-30$ H-like ions the nuclear size effect is much smaller than one-loop QED corrections [3] and thus such measurements are less sensitive to uncertainties of determination of nuclear charge distribution.

The proposed AvH spectrometer benefits from the unique features of RR X-ray emission in the electron cooler of CRYRING, namely, a extremely long and narrow (1 m × 1 mm) X-ray source accepted by von Hamos geometry and very cold transverse electron beam temperature of about meV achieved by application of adiabatic magnetic expansion [4] of the electron beam. This effect increases the intensities of RR X-rays and, consequently, a precision of determination of X-ray energies. In order to control the Doppler effect two copies of the AvH spectrometers will be installed on both sides of the electron cooler to detect blue/red shifted RR X-rays, what allows to eliminate completely the influence of Doppler effect.

References

[1] Lestinsky M *et al* 2016 *Eur. Phys. J. Special Topics* **225** 797
[2] Jagodziński P *et al* 2014 *Nucl. Instrum. Methods* **A753** 121
[3] Lindgren I *et al* 1999 *Phys. Scr.* **T80A** 133
[4] Danared H *et al* 1994 *Phys. Rev. Lett.* **72** 3775

*E-mail: dariusz.banas@ujk.edu.pl

Theoretical investigation of electron-ion recombination process of Fe^{14+} ion

L Y Xie[1*], J P Pan[1], D H Zhang[1], J Jiang[1] and C Z Dong[1]

[1]Key Laboratory of Atomic and Molecular Physics and Functional Materials of Gansu Province, College of Physics and Electronic Engineering, Northwest Normal University, Lanzhou 730070, P. R. China

Synopsis The $\Delta n=1$ dielectronic, trielectronic, and quadruelectronic recombinations of Fe^{14+} ion are systematically studied using the flexible atomic code based on the relativistic configuration interaction method. The theoretical rate coefficients are identified and compared with the experimental measurements at heavy-ion storage ring (TSR), good agreement is obtained.

Dielectronic recombination (DR) is the dominant electron-ion recombination process in both laboratory and astrophysical plasmas. Accurate DR rate coefficients are important for interpreting and modelling the level populations, ionization balance, and radiative spectra of these plasmas [1]. Iron is the most abundant element in the cosmos. Recently, Bernhardt *et al* [2] measured the absolute DR rate coefficients of Fe^{14+} by employing electron-ion merged-beams technique at a heavy-ion storage ring for electron-ion collision energies in the range 0-2600 eV, which covered both the $\Delta n=0$ and $\Delta n=1$ DR processes. For the former, theoretical calculations have been performed by Lukić *et al* [3], good consistent is acquired with the experimental measurements [2, 3]. However, for the later, few theoretical calculation has been carried out.

In this work, the $\Delta n=1$ DR process was systematically studied by using the flexible atomic (FAC) code [4]. The $2s/2p \rightarrow 3l$ transitions are considered, that include the dominant DR, trielectronic recombination (TR), and quadruelectronic recombination (QR) channels described as following

$$Fe^{14+}\left(2s^2 2p^6 3s^2\right)+e^- \begin{cases} Fe^{13+**}\left(2s2p\right)^7 3s^2 3l nl' & (DR) \\ Fe^{13+***}\left(2s2p\right)^7 3s3l3l'nl'' \rightarrow Fe^{13+} + h\nu & (TR) \\ Fe^{13+****}\left(2s2p\right)^7 \left(3l3l'\right)^3 nl'' & (QR) \end{cases} \quad (1)$$

Where l, $l'=p$, d, and n=3-10. We calculated the resonance energies, resonance strengths, and cross sections for $\Delta n=1$ transition. The corresponding electron-ion recombination rate coefficient are obtained by convoluting the calculated cross sections with the experimental electron energy distribution $kT_{\perp}=0.0002$ eV, $kT_{\parallel}=1.6$ eV at heavy-ion storage ring [2], and compared with the experimental results [3], good agreement is obtained (see Fig.1 (a)). We identified the experiment observed resonance peaks, and found that TR and QR contributions can't be neglected (see Fig.1 (b)).

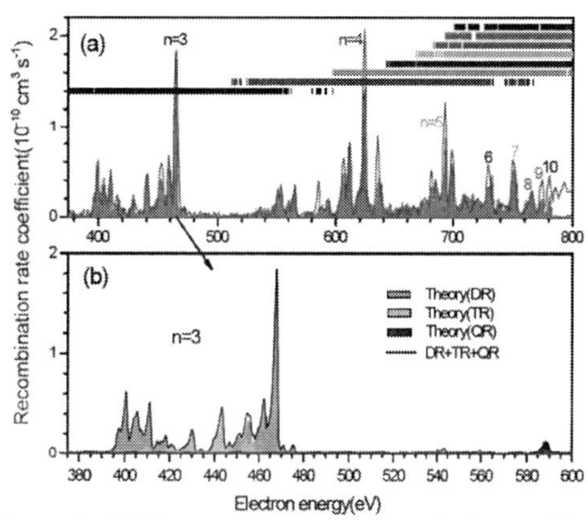

Figure 1. (a) Comparison of the theoretical recombination rate coefficients (shaded gray area) and the experimental results (red solid line) [2]. The vertical bars in different colors denote calculated resonance positions. (b) The contributions of DR, TR and QR processes to total rate coefficients for the n=3 resonance groups of Fe^{14+}.

The work was supported by the National Key Research and Development Program of China under Grant No.2017YFA0402300, and the Natural Science Foundation of China (Grant Nos. U1530142, 11564036, 11774292, 11464042, 1874051) and the Fund of NWNU (NWNU-LKQN-15-3).

References

[1] Altun Z *et al* 2007 *Astron.&Astrophys.* **474** 1051
[2] Bernhardt D *et al* 2014 *Phys. Rev. A* **90** 012702
[3] Lukić D V *et al* 2007 *Astrophys. J.* **664** 1244
[4] Gu M F 2008 *Can. J. Phys.* **86** 675

* E-mail:xiely@nwnu.edu.cn

Ab initio investigation of the line-shape parameters for atmosphere-relevant molecular systems

H Jóźwiak[1]*, F Thibault[2], H Cybulski[1], N Stolarczyk[1], M Gancewski[1] and P Wcisło[1]

[1]Institute of Physics, Faculty of Physics, Astronomy and Informatics, Nicolaus Copernicus University in Toruń, Grudziadzka 5, 87-100 Toruń, Poland
[2]Institut de Physique de Rennes, UMR CNRS 6251, Université de Rennes 1, Campus de Beaulieu, Bât.11B, F-35042 Rennes, France

Synopsis We demonstrate the results of the first *ab initio* investigation of the line-shape parameters for two molecular systems important for atmospheric studies: CO-N_2 and O_2-N_2. We provide the pressure broadening and shift coefficients with their speed dependencies for purely rotational lines, calculated from highly accurate potential energy surfaces with the close-coupling scheme. This is the first, fully quantum approach to the problem of determination of the spectral line shapes for the systems important for terrestrial atmospheric measurements.

The reduction of systematic errors in the atmospheric measurements of the Earth requires careful treatment of effects perturbing the shapes of the spectral lines. The accuracy of the classical or semi-classical calculations of the line-shape parameters is not always satisfactory. However, fully quantum calculations of the pressure broadening and shift coefficients for the helium-perturbed H_2 system resulted in a very good agreement with experimental data [1].

We present the results of the investigation of the line-shape parameters for two atmospheric systems based on *ab initio* quantum-scattering calculations. The first one is the nitrogen-perturbed CO molecule, for which we investigated the R(0) purely rotational line. We use the most accurate potential energy surfaces (PESs), calculated by means of the state-of-the-art quantum chemistry methods [2, 3, 4]. The PESs are expanded over bispherical harmonics leading to 205 radial coupling terms. The close-coupling equations are solved for a wide range of kinetic energies. The generalized spectroscopic cross sections, obtained from the scattering matrices, lead to the first-ever pressure broadening and shift coefficients for the CO-N_2 system, calculated using the fully quantum approach.

The second system is the nitrogen-perturbed oxygen. We use the four-dimensional PES [5], which is also expanded over the bispherical harmonics, leading to 85 radial coupling terms. The same methodology is used to obtain the generalized spectroscopic cross sections from the close-coupling equations and the scattering matrix. This is the first *ab initio* investigation of the line-shape parameters in the O_2-N_2 system. The data provided through the investigation of both systems can be used for populating the HITRAN database [6].

References

[1] Słowiński M *et al in preparation*
[2] Cybulski H *et al* 2018 *Phys. Chem. Chem. Phys* **20** 12624
[3] Surin L A *et al* 2018 *J. Chem. Phys.* **148** 044313
[4] Liu J M *et al* 2018 *Phys. Chem. Chem. Phys* **20** 2036
[5] Bartolomei M *et al* 2014 *J. Phys. Chem. A* **118** 6584
[6] Gordon I E *et al* 2017 *J. Quant. Spectrosc. Radiat. Transf.* **203** 3

*E-mail: hubert.jozwiak@stud.umk.pl

Relativistic configuration interaction with core polarization method for the divalent systems

X Wang[1], J Jiang[1]*, Z W Wu[1], L Y Xie[1], D H Zhang[1] and C Z Dong[1]

[1]Key Laboratory of Atomic and Molecular Physics and Functional Materials of Gansu Province,College of Physics and Electronic Engineering, Northwest Normal University, Lanzhou, 730070, P. R. China

Synopsis The relativistic configuration interaction with core polarization method and program for the divalent systems have been developed. Based on this new program, the energy levels, transition matrix elements, static and dynamic dipole polarizabilities of He, Be, and Mg atoms have been calculated systematically.

The divalent atoms have been widely used in high-precision physics, such as, atomic clock, quantum information. The divalent systems are of the advantageous candidates in atomic clocks because its narrow spin-forbidden transition is less susceptible to effect of the external field. The main uncertainties of these high-precision measurements are from the blackbody radiation shifts and ac stark shifts. The static and dynamic polarizabilities of the divalent atoms are very important to evaluate these uncertainties.

The relativistic semiempirical core polarization method has already succeeded in computing the atomic structures of monovalent systems previously [1]. Recently, the relativistic configuration interaction with core polarization(RCICP) method has been developed by us to deal with the divalent systems. The basic strategy is in a frozen core model in which the atoms are partitioned into valence and core electrons. The wave functions of core electrons are represented by the Dirac-Fock calculations. The single electron orbitals are expanded in a large basis of L-spinors and S-spinors, which can be regarded as relativistic generalizations of the familiar Laguerre- and Slater-type orbitals respectively. The total number single electron orbitals usually can be extended to 1000 in the present program. The wave functions of atomic states are built as a linear combination of two valence electron configurations of the same J-value and parity. Core-valence correlations are represented by introducing semiempirical one- and two-body core polarization potentials, which are tuned to ensure that the energies for the valence electrons agree with experiments.

Table 1. The static dipole polarizabilities (in a.u.) of the ground states of He, Be, and Mg atoms.

Method	He $1s^2$	Be $2s^2$	Mg $3s^2$
RCICP	1.38306	37.61	70.32
CICP [2]		37.73	71.35
MBPT [3]		37.76	71.33
Expt. [4]	1.383759(13)		

As a example, the energy levels and transition matrix elements of He, Be, and Mg atoms are calculated systematically. Then the static and dynamic dipole polarizabilities are further determined by the sum-over states approach. Table 1 lists the static dipole polarizabilities of the ground states of He, Be, and Mg atoms. The present RCICP results agree well with all other results [2, 3, 4]. This method and program can also be applied in the calculation of the energy levels, transition matrix elements, static and dynamic dipole polarizabilities of more complicated atoms, such as Yb, Ba, and Sr.

This work has been supported by National Key Research and Development Program of China (2017YFA0402300) and National Natural Science Foundation of China (Grant Nos. 11774292, 11564036, 11874051, 11864036, 11804280).

References

[1] Jiang J *et al* 2016 *Phys. Rev. A* **94** 062514
[2] Cheng Y J *et al* 2013 *Phys. Rev. A* **88** 022511
[3] Porsev S G and Derevianko A 2006 *J. Exp. Theor. Phys.* **102** 195
[4] Schmidt J W *et al* 2007 *Phys. Rev. Lett.* **98** 254504

*E-mail: phyjiang@yeah.net

Molecular-frame angular distributions of spectator resonant interatomic coulombic decay electrons in neon dimers

A Mhamdi[1] *, J Rist[2], D Aslitürk[2], M Weller[2], N Melzer[2], D Trabert[2], M Kircher[2], I Vela-Pérez[2], J Siebert[2], S Eckart[2], S Grundmann[2], G Kastirke[2], M Waitz[2], A Khan[2], M S Schöffler[2], F Trinter[3,4], R Dörner[2], T Jahnke[2] and Ph V Demekhin[1] †

[1]Institut für Physik und CINSaT, Universität Kassel, Kassel, 34132, Germany
[2]Institut für Kernphysik, J. W. Goethe-Universität, Frankfurt am Main, 60438, Germany
[3]Deutsches Elektronen-Synchrotron (DESY), FS-PETRA-S, Hamburg, 22607, Germany
[4]Fritz-Haber-Institut der Max-Planck-Gesellschaft, Molecular Physics, Berlin, 14195, Germany

Synopsis The spectator resonant interatomic Coulombic decay (SRICD) following the $2s \longrightarrow 5p$ excitation of Ne dimers has been examined experimentally and theoretically. The molecular frame angular distributions (MFADs) of the SRICD electrons depend strongly on the direction of the polarization vector of the exciting photon with respect to the molecular axis. We demonstrate that this effect is caused by the scattering of the low-energy SRICD electron on the density of the spectator electron.

It is widely accepted that the creation of a core hole by photoionization and its subsequent Auger decay can be described, in a good approximation, independently. This approximation is known as the two-step model [1], in which the electron emitted by the Auger decay is not influenced by the initial excitation or ionization process [2]. For instance, within the two-step model, the MFADs of the Auger electrons are independent of the polarization properties of the ionizing/exciting photons. For a resonant Auger decay, the two-step model holds if the excited electron only witnesses the decay process as a spectator.

Here, we demonstrate that, for low-energy electrons, the generally-admitted two-step approximation breaks down. For this purpose, we performed a joint experimental and theoretical study of SRICD after inner-valence excitation of the neon dimer into the $5p$ state. Experimentally, the momenta of all charged particles created after the absorption of the photon and subsequent SRICD were coincidently measured using the COLTRIMS technique [3]. We observe a strong dependency of the SRICD electron angular emission distribution on the polarization direction of the exciting photons with respect to the dimer axis (cf. Fig. 1b and 1c), which is in contradiction with the two-step model. For a deeper understanding of these observations, we performed electronic structure and dynamics calculations by employing the Single Center method [4]. The present observations are explained by the substantial scattering of the escaping low-energy SRICD electron on the density of the excited spectator electron [5] (see Fig. 1a).

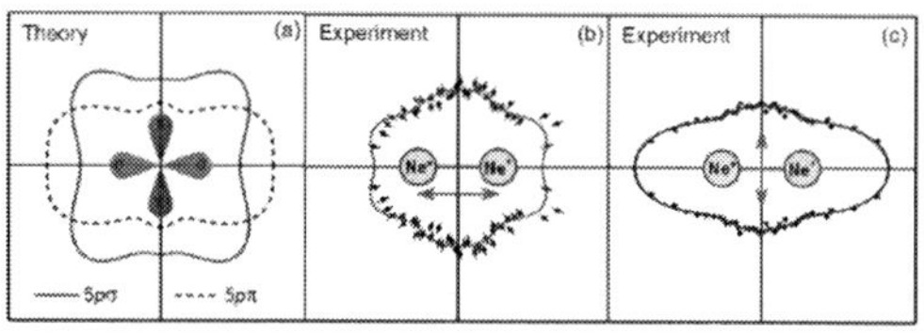

Figure 1. MFADs of SRICD electrons after $2s \longrightarrow 5p$ excitation. Panel (a): Theoretical MFADs for $5p\sigma$ and $5p\pi$ excitations of Ne_2. Panels (b) and (c): Corresponding experimental results, where the polarization direction of the exciting light is oriented in parallel or perpendicularly to the dimer axis, as indicated by the double arrows.

References

[1] Armen G B *et al* 2000 *J. Phys. B* **33** R49
[2] Kuznetsov V V and Cherepkov N A 1996 *J. Electron Spectrosc. Relat. Phenom.* **79** 437
[3] Dörner R *et al* 2000 *Phys. Rep.* **330** 95
[4] Demekhin Ph V *et al* 2011 *J. Chem. Phys.* **143** 024113
[5] Mhamdi A *et al* 2018 *Phys. Rev. Lett.* **121** 243002

*E-mail: abir.mhamdi@uni-kassel.de
†E-mail: demekhin@physik.uni-kassel.de

Integral cross sections of the valence-shell excitations of carbonyl sulphide studied by the high-energy electron scattering

Y-W Liu[1,2]*, X-J Du[1,2], Y-C Xu[1,2] and L-F Zhu[1,2]†

[1]Hefei National Laboratory for Physical Sciences at Microscale and Department of Modern Physics, University of Science and Technology of China, Hefei, Anhui 230026, People's Republic of China

[2]Synergetic Innovation Center of Quantum Information and Quantum Physics, University of Science and Technology of China, Hefei, Anhui 230026, People's Republic of China

Synopsis The integral cross sections (ICSs) of the valence-shell excitations of carbonyl sulphide have significant applications in the astronomical observation. Based on the generalized oscillator strengths (GOSs) of these excitations measured by the high-energy electron scattering, the ICSs of the valence-shell excitations of OCS have been determined in a wide energy region from the threshold to 5 keV with the aid of the *BE*-scaling method.

The ICSs of the valence-shell excitations of carbonyl sulphide (OCS) can be used as the important and basic input parameters of the theoretical models for understanding the observation spectroscopy of sulphur-containing species, and modeling the chemistry and physics processes in planetary and interstellar space. However, the ICSs are only reported by the low- or moderate-energy electron scattering at 100, 200 and 300 eV [1] to our knowledge.

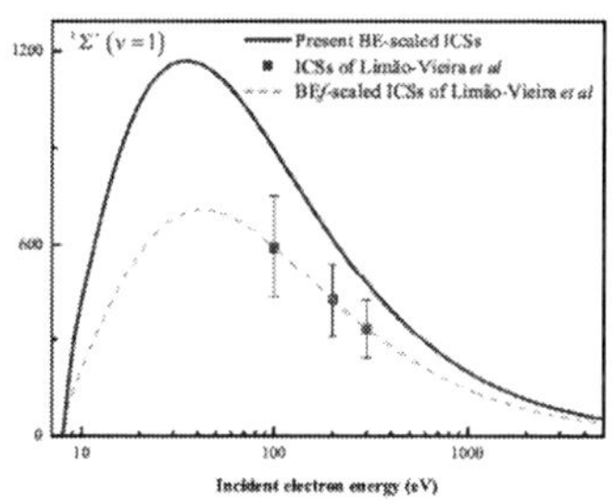

Figure 1. The present *BE*-scaled ICSs and the previous ones [1] of the $^1\Sigma^+$ $(v = 1)$ excitation of OCS molecule.

In the present work, the GOSs of the valence-shell excitations of OCS have been determined by the high-energy electron scattering at an incident electron energy of 1500 eV and an energy resolution of about 70 meV. The BE-scaling method is widely used to calculate the ICSs of the dipole-allowed transitions of atoms and molecules, which is developed by Kim [2, 3] and extensively used by Tanaka et al [4]. With the aid of the BE-scaling method, the present *BE*-scaled ICSs of the valence-shell excitations of OCS have been obtained in a wide energy region from the threshold to 5 keV.

The present *BE*-scaled ICSs of the $^1\Sigma^+$ $(v = 1)$ excitation are shown in Figure 1 along with the previous ICSs of Limão-Vieira et al [1] by the *BEf*-scaling method and by integrating the differential cross sections (DCSs) from 0° to 180°. It is clear in Figure 1 that, the present *BE*-scaled ICSs are obviously higher than the *BEf*-scaled ICSs of Limão-Vieira et al [1] in the whole region from the threshold to 5 keV and the experimental ICSs of Limão-Vieira et al [1] at 100, 200 and 300 eV.

In summary, the present ICSs of OCS molecule are accurate, and can serve as the basics input parameters of the theoretical models to simulate the processes in planetary atmosphere and explain the astronomical observations.

References

[1] Limão-Vieira P *et al* 2015 *J. Chem. Phys.* **142** 064303
[2] Kim Y K 2001 *Phys. Rev. A.* **64** 032713
[3] Kim Y K 2007 *J. Chem. Phys.* **126** 064305
[4] Tanaka H *et al* 2016 *Rev. Mod. Phys.* **88** 025004

*E-mail: liuyawei@ustc.edu.cn
†E-mail: lfzhu@ustc.edu.cn

Polarizabilities and tune-out wavelengths of the hyperfine ground states of ^{133}Cs atom

J Jiang*, **X J Li**†, **Z W Wu, D H Zhang, L Y Xie and C Z Dong**

Key Laboratory of Atomic and Molecular Physics and Functional Materials of Gansu Province, College of Physics and Electronic Engineering, Northwest Normal University, Lanzhou 730070, China

Synopsis The static and dynamic electric dipole polarizabilities of the ground state and hyperfine ground states of the ^{133}Cs have been calculated by using relativistic configuration interaction plus core polarization (RCICP) approach. The tune-out wavelengths of ground and hyperfine ground states are also determined.

Cs atoms have been widely used in high-precision measurements, such as, atomic clock, electric-dipole moment of the electron (eEDM). The accurate atomic parameters of Cs are very important for these measurements.

The polarizabilities of the ground states of ^{133}Cs atom are calculated by using relativistic configuration interaction plus core polarization (RCICP) approach. After considering the hyperfine splittings, the energies and transition matrix elements of the hyperfine structure of ^{133}Cs are computed. Then the scalar (α_1^S) and tensor (α_1^T) polarizabilities of the hyperfine ground states of ^{133}Cs are determined.

Table 1. lists present calculations of scalar and tensor dipole polarizabilities of the hyperfine ground states of ^{133}Cs. The difference between the scalar polarizabilities of hyperfine states, which is related with the hyperfine stark shift, agree with other calculations , −0.01825[1], and experiments, −0.01809(6)[2], very well.

Table 1. The dipole α_1^S and tensor α_1^T polarizabilities (in a.u.) of the hyperfine ground states $5s_{1/2}$ F=3,4 of ^{133}Cs. Δ : difference of polarizabilities for F = 3 and F=4 states. The notation a[b] means a×10^b.

	F=3	F=4	Δ
α_1^S	399.0807	399.0989	−0.0182
α_1^T	1.08[-4]	−2.02[−4]	3.10[−4]

Table 2. gives the tune-out wavelengths of the two hyperfine ground states of the $6s_{1/2}$ state of ^{133}Cs. The longest tune-out wavelengths near 894.58 nm occur in the hyperfine splitting of the $6p_{1/2}$ state. These tune-out wavelengths are hard to detect due to the very small energy splittings of the hyperfine states. The second tune-out wavelengths (near 880 nm) lie between the excitation thresholds of the $6p_{1/2}$ and $6p_{3/2}$ states. These tune-out wavelength are defined as primary tune-out wavelengths. The measurement of the primary tune-out wavelength can be used to determine $6s_{1/2} \to 6p_{1/2}$ matrix elements with very high-accuracy. The other tune-out wavelengths occur in the hyperfine splitting of the $6p_{3/2}$ state.

Table 2. Tune-out wavelengths λ_{zero} (in nm) of the $6s_{1/2}$, F=4 and $6s_{1/2}$ F=3 states of ^{133}Cs. $\Delta\lambda$ is the shift of the primary tune-out wavelength compared to the tune-out wavelengths of the 6s state.

F=3		F=4	
λ_{zero}	$\Delta\lambda(10^{-2})$	λ_{zero}	$\Delta\lambda(10^{-2})$
894.580130		894.603630	
880.238941	-1.30	880.263118	1.12
852.335407		852.357404	
852.334875		852.356781	

This work was supported by National Key Research and Development Program of China (2017YFA0402300) and National Natural Science Foundation of China (Grant Nos. 11774292, 11564036, 11874051, 11864036, 11804280).

References

[1] Angstmann E J *et al* 2006 *Phys. Rev. A* **74** 023405
[2] Mowat J R *et al* 1972 *Phys. Rev. A* **5** 1059

*E-mail: phyjiang@yeah.net
†E-mail: 710746305@qq.com

Resolving and imaging the stereo-configuration of OCS dimers in gas phase

X Yu [1,2], X Li[1,2], Z Yin[1,2], C Wang[1,2], S Luo[1,2]* and D Ding [1,2]†

[1] Institute of Atomic and Molecular Physics, Jilin University, Changchun, P R China
[2] Jilin Provincial Key Laboratory of Applied Atomic and Molecular Spectroscopy, Jilin University, Changchun, P R China

Synopsis Polar, non-polar and cross-shaped isomers of OCS dimer are observed and the corresponding structures are resolved using *fs* laser-induced Coulomb explosion. These three structures are confirmed in the simulations, and demonstrate that the dimer structures can be reconstructed from the measured momenta distributions. Our study will promote the real time imaging the evolution of molecular dimer at *fs* time scale after excitation and ionization.

Imaging three-dimensional (3D) structures of molecules and clusters is an important task for physicists and chemists. The femtosecond laser triggered Coulomb-explosion (CE) can be used to imaging the dynamics structure evolution of molecules and clusters. The cross-shaped structure of $(CS_2)_2$ and the planar structures of $(OCS)_2$ in the helium nanodroplets have been observed by Coulomb-explosion imaging (CEI) method combing with molecular alignment technology [1,2]. It is a challenge to reconstruct the structures of clusters especially for some extremely weakly bond systems and clusters with multiple isomers.

We experimentally investigate CE of OCS dimer in strong laser fields using COLTRIMS setup. A 40 *fs*, 800 nm near circularly polarized laser (its ellipticity ~ 0.95) is used to ionize OCS dimer in the molecular beam. The relative momenta of quadruple coincident fragment ions (CO^++S^++CO^+ +S^+) are measured, and the critical bond angles of the dimers are determined from the obtained relative momenta of four ions.

Here, we show that the 3D structures of dimer isomers can be resolved and reconstructed from the CEI combining with the classical-trajectory-Monte-Carlo molecular dynamics method. The measured and calculated linear momenta distributions of three different isomers of $(OCS)_2$ are shown in Figure 1(a)(b)(c). This measurement confirm the structures of polar and non-polar dimers, more importantly, the measured angular distributions corresponding to the cross-shaped structure of dimer is the first experimental observation after predicted from theoretical calculations [3,4]. The current results will enable tracking the ultrafast evolution of cluster structures in *fs* time scale by take advantage of ultrafast laser induced CEI, and researchers can observe a cluster writhing, bonds vibrating, breaking and deformation in real time.

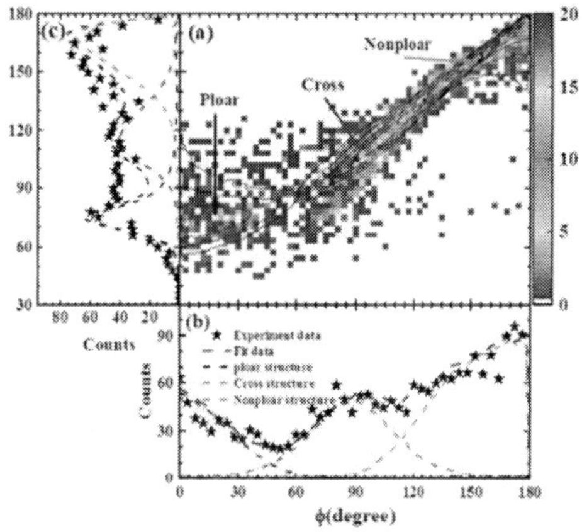

Figure 1. (a) the correlation map distribution of $\theta_{Pco\text{-}Pco}$ and ϕ. (b) the distribution as function of dihedral angle ϕ integrated over $\theta_{Pco\text{-}Pco}$. (c) the distribution as function of intersection angle $\theta_{Pco\text{-}Pco}$ integrated over ϕ.

References

[1] Pickering J *et al* 2018 *Phys. Rev. Lett.* **120** 113202
[2] Pickering J *et al* 2018 *J. Chem. Phys.* **149** 154306
[1] Brown J *et al* 2012 *J. Chem. Phys.* **136** 134306
[2] Sahu N *et al* 2013 *J. Phys. Chem. A* **117** 10964

* E-mail: luosz@jlu.edu.cn
† E-mail: dajund@jlu.edu.cn

Photodissociation processes of $O_2^+(H_2O)$ studied by ion imaging experiments

Y Nakashima, M Nakano, K Okutsu, Y Ito and F Misaizu*

Department of Chemistry, Graduate School of Science , Tohoku University, Sendai, 980-8578, Japan

Synopsis Photodissociation processes in the visible and ultraviolet region of water-oxygen molecular complex ion, $O_2^+(H_2O)$, were studied by ion imaging experiments and theoretical calculations. Two fragment ions, H_2O^+ and O_2^+, were observed in the experiment; the former fragment ion was produced by direct dissociation after excitation, whereas the latter was generated by nonadiabatic transitions from the initially excited states to the lower electronic states.

In the mesosphere, the atmospheric region below the altitude of ~100 km, there exist NO^+ and O_2^+ ions predominantly, and also the clusters including those ions [1]. Among such clusters, water-oxygen molecular complex ions, $O_2^+(H_2O)$, probably play a role in generating proton hydrates, $H_3O^+(H_2O)_n$, which become important below 85 km in the atmosphere. Photochemistry of $O_2^+(H_2O)$ was studied against this background [2]. In this study, visible and ultraviolet photodissociation processes of $O_2^+(H_2O)$ have been investigated more in detail by ion imaging experiments and theoretical calculations.

In the experiments, a linear-type double reflectron time-of-flight mass spectrometer with a position sensitive detector [3] was used to measure velocity and angular distributions of photofragment ions produced from the $O_2^+(H_2O)$ ions. Cluster ions including precursor ions were produced by supersonic expansion of gas mixture of $H_2O/O_2/He$ and electron impact ionization. Mass-selected target $O_2^+(H_2O)$ ions were irradiated with a linearly-polarized photodissociation laser at 473 or 355 nm. The O_2^+ and H_2O^+ fragment ions were again mass-analyzed and detected. In the theoretical calculations, potential energy curves (PECs) of the ground and excited states were calculated.

As a result of the experiments, the product ratios of the fragment ions, O_2^+:H_2O^+, were 7:3 at 473 nm and 2:8 at 355-nm excitations. Both fragment ions were found to have angular distributions parallel to the laser polarization direction. The kinetic energy release (KER) distribution of the O_2^+ fragment ions at 473 nm was determined by analyzing the observed O_2^+ image. The determined average KER was 0.41 ± 0.02 eV, which was comparable to a calculated value based on the prior distribution, 0.31 eV.

From the excited-state calculation at state-averaged CASSCF(13,11)/aug-cc-pVTZ level, transitions of $O_2^+(H_2O)$ at 473 and 355 nm photoexcitation were assigned to $\tilde{B}$ and/or $\tilde{D} \leftarrow \tilde{X}$. The PECs of the $\tilde{B}$ and $\tilde{D}$ states were repulsive along the O_2-H_2O distance as shown in Fig.1. The adiabatic dissociation limits were $O_2(X\ ^3\Sigma_g^-) + H_2O^+$ for $\tilde{B}$ state and $O_2(a\ ^1\Delta_g) + H_2O^+$ for $\tilde{D}$ state. Therefore, H_2O^+ ion was produced in the adiabatic dissociation processes. On the other hand, O_2^+ ion was produced with a statistical energy partitioning in the dissociation of $O_2^+(H_2O)$ after nonadiabatic transitions from the excited states to $\tilde{A}$ or $\tilde{X}$ states, both of which correlate to O_2^+ $(X\ ^2\Pi_g)$ + H_2O at the dissociation limit.

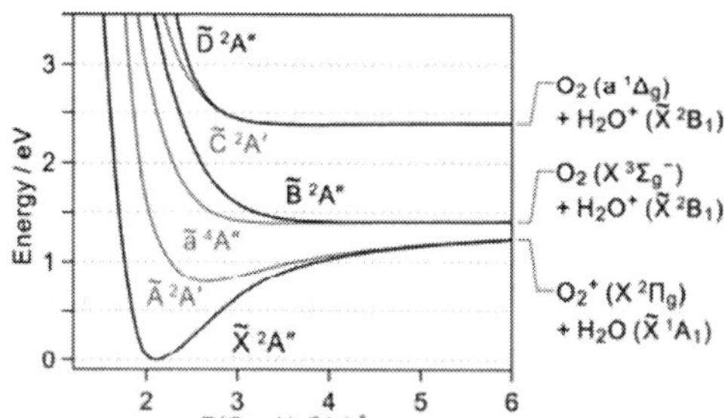

Figure 1. Potential energy curves of $O_2^+(H_2O)$ as a function of O_2-H_2O distance and adiabatic dissociation limits calculated at SA-CASSCF(13,11)/aug-cc-pVTZ.

References

[1] Shuman N S *et al* 2015 *Chem. Rev.* **115** 4542
[2] Angel L and Stace A J 1999 *J. Phys. Chem. A* **103** 2999
[3] Okutsu K *et al* 2017 *Rev. Sci. Instrum.* **88** 053105

* E-mail: misaizu@m.tohoku.ac.jp

Ab initio quantum calculations of collisional effects in molecular spectra

H Jóźwiak[1]*, F Thibault[2], N Stolarczyk[1] and P Wcisło[1]

[1]Institute of Physics, Faculty of Physics, Astronomy and Informatics, Nicolaus Copernicus University in Toruń, Grudziadzka 5, 87-100 Toruń, Poland
[2]Institut de Physique de Rennes, UMR CNRS 6251, Université de Rennes 1, Campus de Beaulieu, Bât.11B, F-35042 Rennes, France

Synopsis We present the methodology of obtaining the dataset of the beyond-Voigt line-shape parameters for spectroscopic databases from fully *ab initio* quantum-scattering calculations. It provides not only the pressure broadening and shift coefficients, but also their speed dependencies and the complex Dicke parameter. This approach is based on the calculations of the generalized spectroscopic cross section, resulting from the close-coupling equations solved for a given scattering system. The calculations were validated on accurate experimental spectra of the helium-perturbed hydrogen molecule.

In molecular spectroscopy, the collisions manifest themselves as a perturbation of the spectral line shapes. Proper treatment of these effects is crucial in the interpretation of highly accurate molecular spectra [1], analyses of the planetary atmospheres [2, 3] or studies of *ab initio* molecular interactions [4, 5, 6, 7]. Recently, a new structure of the most commonly-used spectroscopic database HITRAN [1, 8] has been proposed, enabling the beyond-Voigt line-shape effects to be represented. However, populating the entire database (all the bands, branches or temperature ranges) with purely experimental parameters is very difficult.

We present a new methodology of obtaining the line-shape parameters beyond the Voigt profile, based on *ab initio* quantum-scattering calculations. The process begins with the ultra-accurate potential energy surface (PES), calculated by means of the state-of-art quantum chemistry methods [5, 7]. Dynamical calculations on the PES are used to solve the close-coupling equations for a wide range of kinetic energies of the colliding pair. From the solutions of the close-coupling equations, the scattering matrices are constructed. Afterwards, the generalized spectroscopic cross sections are calculated for a given spectroscopic transition. Finally, the line-shape parameters are determined, taking into account various effects, such as the speed-dependence or the influence of the velocity-changing collisions. We provide the temperature dependence of the pressure broadening and shift coefficients, of their speed dependencies, as well as the complex Dicke parameters.

The methodology was successfully tested on the H_2-He system [7, 9], for which the calculations were validated by the ultra-accurate experimental data [10]. This approach was also used for several transitions of the helium- or self-perturbed D_2 [11, 12], and is currently being implemented on the systems of atmospheric interest, like CO-N_2 or O_2-N_2.

References

[1] Wcisło P *et al* 2016 *J. Quant. Spectrosc. Radiat. Transf.* **177** 75
[2] Miller C E *et al* 2005 *C. R. Phys.* **6** 876
[3] Smith W H *et al* 1989 *Icarus* **81** 429
[4] Tran H *et al* 2011 *J. Chem. Phys.* **131** 154303
[5] Bakr B W *et al* 2013 *J. Chem. Phys.* **139** 144305
[6] Wcisło P *et al* 2015 *Phys. Rev. A* **91** 052505
[7] Thibault F *et al* 2017 *J. Quant. Spectrosc. Radiat. Transf.* **202** 308
[8] Gordon I E *et al* 2017 *J. Quant. Spectrosc. Radiat. Transf.* **203** 3
[9] Jóźwiak H *et al* 2018 *J. Quant. Spectrosc. Radiat. Transf.* **219** 313
[10] Słowiński M *et al in preparation*
[11] Martinez R Z *et al* 2018 *J. Raman Spectrosc.* **49** 1339
[12] Wcisło P *et al* 2018 *J. Quant. Spectrosc. Radiat. Transf.* **213** 41

*E-mail: hubert.jozwiak@stud.umk.pl

Statistical process in nitrogen containing naphthalene derivatives in collision with fast electrons and protons

P K Najeeb[†] *, M V Vinitha[†], A Kala[†], P Bhatt[†], C P Safvan[‡], S Vig[†] and U Kadhane[†]

[†]Indian Institute of Space Science and Technology, Thiruvananthapuram, 695547, India
[‡]Inter University Accelerator Centre, Aruna Asaf Ali Marg, New Delhi, 110067, India

Synopsis The effect of nitrogen substitution and isomerization in naphthalene under swift charged particle collision is investigated. The ratios of single ionization to capture cross-section are measured for the case of naphthalene and two of its nitrogen-containing derivatives. The propensity of HCN loss in quinoline and isoquinoline is shown to alter systematically from evaporation dominated to randomization dominated process in the case of monocation. The influence of plasmon excitation process in polycyclic aromatic nitrogen heterocycles (PANHs) is identified for the first time.

Nitrogen-containing polycyclic aromatic hydrocarbon (PANH) systems are essential ingredients in bio-molecules and their spectroscopy is interesting because of their possible role in prebiotic chemistry [1, 2]. Addition of nitrogen into PAH structure leads to considerable weakening of the structure and faster dissociation for a given amount of excitation energy [3, 4].

The present work explores the structural and dynamical implications of nitrogen and their positional identity in naphthalene. We have separated the statistical (evaporative) and non-statistical (knock-out) decay channels with proton (velocity ranges ~ 1.42 - 2.4 a.u.) as well as electron (velocity ranges ~ 4.3 - 8.6 a.u.) as the projectiles. The location of nitrogen have no observable effect as far as the ratio of ionization to capture is concerned. With electron as well as proton as the projectiles, all the three target molecules shows a decreasing yield of non-statistical channels except the statistical channels like H_2, C_2H_2 or HCN loss as a function of projectile velocity [Figure 1]. H_2 loss process is found to be completely suppressed by the inclusion of N atom in the ring.

At lower velocities (< 4 a.u.), a combination of plasmon excitation and electronic stopping is observed to deposit the higher amount of internal energy, which in turn increases the decay rate as well as the rate of randomization. But we observe that the rate constant of randomization is higher than that of HCN evaporation. This is accompanied by additional energy loss due to plasmon excitation. It is known that energetically quinoline is more stable than isoquinoline but it is observed that sufficient internal energy leads to rapid rearrangement and the evaporative decay contribution from both isomers would become equally significant.

*E-mail: najeeb.kty@gmail.com

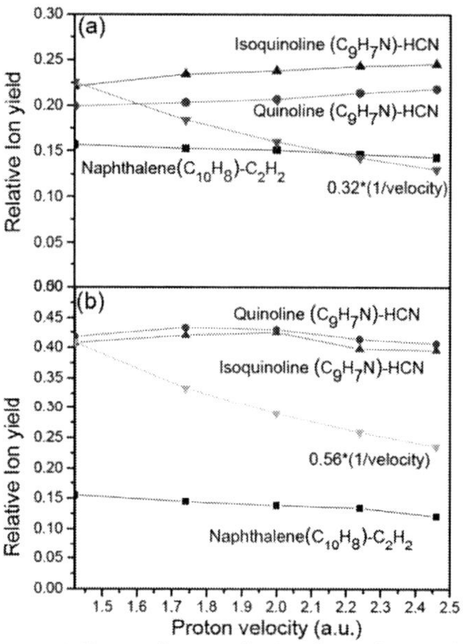

Figure 1. Proton beam velocity dependence of statistical and non-statistical channels in the case of a) electron emission b) electron capture.

References

[1] Mattioda A L *et al* 2003 *J. Phys. Chem. A* **107** 1486
[2] Galué H A *et al* 2010 *Astron. Astrophys.* **517** A15
[3] Stockett M H *et al* 2014 *Phys. Chem. Chem. Phys.* **16** 21980
[4] Najeeb P K *et al* 2017 *Int. J. Mass Spectrom.* **414** 23

Pressure manipulation of exciton dynamics in few-layered MoS_2

Q Li[1]*, L Sui[2], Y Jiang[1], A Chen[1], W Jin[1], H Liu[1], K Yuan[2] and M Jin[1]†

[1]Institute of Atomic and Molecular Physics of Jilin University, Changchun, 130012, China
[2]Dalian Institute of Chemical Physics, Chinese Academy of Science, Dalian, 116023, China

Synopsis Transition metal dichalcogenides have recently attracted great interest as an important class of two-dimensional semiconductors because of their distinctive electronic, optical and catalytic properties. Here we demonstrate the first experimental study on ultrafast excited state dynamics of few-layered molybdenum disulfide under high pressure by means of broadband transient absorption spectroscopy combined with diamond anvil cell devices.

Molybdenum disulfide (MoS_2), which is a representive semiconductor in the class of transition metal dichalcogenides (TMDCs), has attracted great attentions for its distinctive electronical, optical properties and potential applications in electronic and optoelectronic devices. MoS_2 has a layered structure and different layers are combined by weak van der waals' forces. Bulk MoS_2 is an indirect-bandgap semiconductor. When thinned down from bulk size to monolayer, it transforms from indirect-bandgap to direct-bandgap semiconductor and emerges strong photoluminescence[1]. Meanwhile, the band structure of MoS_2 shows obvious dependence on number of layers. Theoretical and experimental studies on structure and bandgap changes of atomically thin few-layered MoS_2 under pressure show that when pressure is applied, the distance between different layers will decrease, which will enhance the interlayer coupling effect and thus significantly changes the structure and bandgap.

Here we demonstrate the first experimental study on pressure-induced ultrafast dynamic changes of MoS_2 conducted by using diamond anvil cells combining with femtosecond transient absorption spectroscopy. Dispersion of few-layered MoS_2 in DMF solvent was obtained through liquid phase mechanical exfoliation with a tip sonicator. The dispersion was then put into our symmetric DAC apparatus with a culet size of 600um for in-situ high-pressure experiments. Femtosecond transient absorption pump-probe measurements were performed with a Ti:sapphire laser system with central wavelength of 800nm and a duration of 35fs.

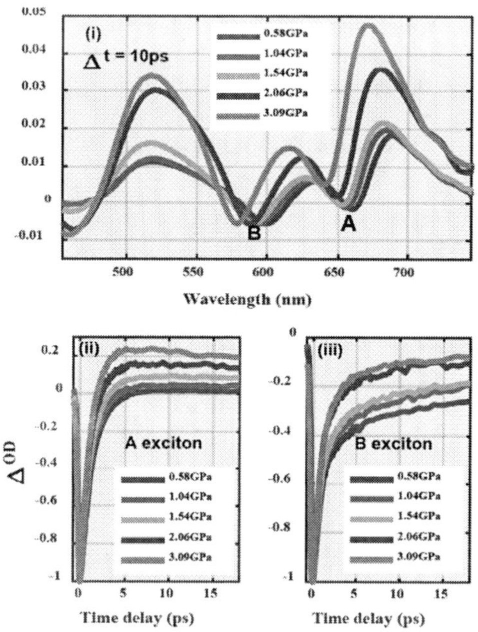

Figure 1. (i) Transient absorption(TA) spectra at time delay of 10ps under various pressures as indicated, normalized by the B exciton peaks. (ii)(iii) Normalized TA kinetics of A and B excitons at various pressures.

As shown in Fig 1, with the pressure increases, the A and B exciton peaks are dramatically blue shifted, and both of their decay dynamics get faster accordingly. This shows the ability of pressure manipulation of exciton dynamics in few-layered TMDCs semicondors.

References

[1] Kin F M *et al* 2010 *Phys. Rev. Lett.* **105** 136805

* E-mail: liqy17@mails.jlu.edu.cn
† E-mail: mxjin@jlu.edu.cn

Bidimensional Sturmian basis for bound states and collision problems

J M Randazzo[1]* **and L U Ancarani**[2]†

[1]Instituto Balseiro and CONICET, 8400 S. C. de Bariloche, Río Negro, Argentina.
[2]Université de Lorraine, CNRS, LPCT, F-57000 Metz, France.

Synopsis We present a novel spectral approach to deal both with bound states and with stationary continuum waves associated to collision problems. It is based on the definition of a two dimensional radial basis which replaces the combination of single particle basis sets as employed in most of the usual spectral methods. The new basis elements are externally defined as solutions of a generalized Sturmian eigenvalue equation containing the main properties of the physical problem under consideration.

Solving the Schrödinger equation for systems involving more than two particles still constitutes a quite formidable task in the continuum regime. For the simplest three body Coulomb systems, 'numerically exact' bound and continuum solutions have been obtained, but require extensive computational resources.

Most methods use two independent relative vectors (*e.g.*, the electron-nucleus distances r_1 and r_2 in the case of Helium) and consider a partial wave decompositions with respect to the third coordinate (*e.g.*, the electron-electron distance $r_{12} = |\mathbf{r}_1 - \mathbf{r}_2|$). When the searched three-body wave function is expanded in terms of the eigenfunctions of the total angular momentum, a coupled set of bidimensional differential equation is obtained; one way to solve it is by writing the two coordinates radial wave functions in terms of products of some one particle basis functions, one for each coordinate. The size of the related linear system of equations grows, roughly speaking, as the square of the one dimensional basis size, multiplied by the number of considered partial waves. For atomic scattering problems, extended spatial domains must be span; this requires large basis sets and computation costs can become prohibitive. The situation is even worse for polyatomic molecules, for which the additional issue of molecular orientation further increases the number of required calculations.

With the aim of tackling always more computationally demanding scattering problems, it is important to gain in efficiency by looking for the best basis functions. With this purpose in mind we explore here a new strategy that considers a two dimensional basis instead of products of one dimensional ones.

In the last decade, Generalized Sturmian Functions have been developed [?] for bound states (e.g., [?]) and scattering (e.g., [?]) problems. The basis elements (1DSF) are solutions of a 1D radial eigenvalue problem. In the present spectral approach we introduce a 2D Sturmian eigenvalue problem. As an example, we show in Fig. 1 the bidimensional structure of four bound type 2DSF.

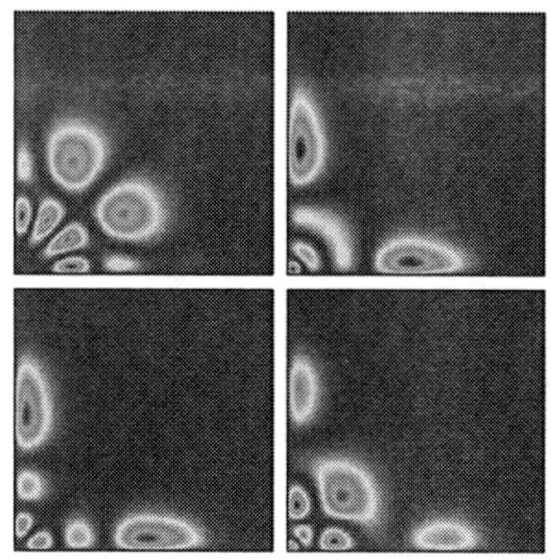

Figure 1. Radial structure of four 2DSF.

At the conference, we will compare the two basis sets by exploring their performance in describing the solution of the Helium atom bound states, and the continuum wave functions arising in the single photon double ionization of Helium.

References

[1] Gasaneo G *et al* 2013 *Adv. Quant. Chem.* **67** 153
[2] Randazzo J M *et al* 2010 *Phys. Rev. A* **81** 042520
[3] Randazzo J M *et al* 2015 *Eur. Phys. J. D* **69** 189

*E-mail: randazzo@cab.cnea.gov.ar
†E-mail: ugo.ancarani@univ-lorraine.fr

Jahn-Teller effect on the $1t_2$ orbital electron momentum distribution of Methane

S S Niu, X Shan *, Y G Tang, Z H Liu and X J Chen †

Hefei National Laboratory for Physical Sciences at the Microscale and Department of Modern Physics, University of Science and Technology of China, Hefei, Anhui, 230026, China

Synopsis We report an experimental study about the Jahn-Teller effect resulted from the ionization of the highest occupied molecular orbital (HOMO) $1t_2$ of methane. The difference of the electron momentum distributions for the Jahn-Teller splitting components reveals the strong interaction between electronic and nuclear vibrational motion.

The Jahn-Teller (JT) effect essentially means that a nonlinear molecule with a spatially degenerate electronic ground state will undergo a geometrical distortion that removes the degeneracy, because the distortion lowers the overall energy of the species [1]. The JT effect may enable the splitting of ionization band and the change of orbital electron distribution assicaited with the distortion of molecular geometry [2-6], which plays a crucial role in the structure and dynamics of symmetric nonlinear molecular systems.

Methane is a prototype of tetrahedral molecules wiht a T_d symmetry in the ground state and its electronic configuration is $(1a)^2(2a)^2(1t_2)^6$. The highest occupied orbital (HOMO) is a triply degenerate $1t_2$ orbital. If one electron is removed from the HOMO, the CH_4^+ ion is formed in the $^2T_2$ state (T_d) which is unstable by the Jahn-Teller effect and distorts the molecule to a lower symmetry (D_{2d}, C_{3v} or C_{2v} three possibilities) [3-6].

In this work, we report the measurement on the JT effect that results from the ionization of methane using a high-resultion (e, 2e) electron momentum spectrometer. The measured binding energy spectrum in Figure 1 showed a very broad envelope for the ionization band of $1t_2$ orbital that is the result of the JT effect, which agreed well with the high resultion photoelectron spectrum [5], corresponding to three JT splitting components. The electron momentum distributions (EMDs) for these three components have been extracted and compared with the calculated ones for the relevant states with the T_d , D_{2d}, C_{3v} and C_{2v} symmetry of molecular geometry. The distinct difference between the EMDs for the JT splitting components has been observed, as shown in Figure 1, indicating the strong vibronic coupling in the ionization of the triply degenerate $1t_2$ orbital. The detailed results and discussion will be presented in the poster.

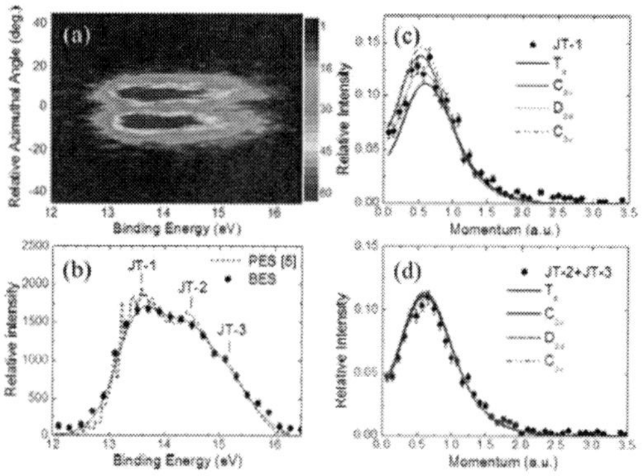

Figure 1. (a) Two dimensional electron density map, (b) binding energy spectrum, (c) and (d) electron momentum distributions for the JT splitting components together with the calculated ones for the states with T_d C_{2v}, D_{2d} and C_{3v} symmetry of of molecular geometry.

References

[1] Jahn H A and Teller E 1937 *Proc. R. Soc. Lond. A* **161** 220.
[2] Li Z J *et al* 2009 *J. Chem. Phys.* **130** 054302
[3] Rabalais J W *et al* 1971 *Phys. Scr.* **3** 13
[4] Mondal T and Varandas A J C 2011 *J. Chem. Phys.* **135** 174304
[5] Potts A W *et al* 1972 *Proc. R. Soc. Lond. A* **326** 165
[6] Takeshita K, 1987 *J. Chem. Phys.* **86** 329

* E-mail: xshan@ustc.edu.cn
† E-mail: xjun@ustc.edu.cn

Momentum transfer dependence behavior of the $3s^{-1}nl$ autoionization states of argon studied by high-resolution fast electron impact

L-F Zhu[1,2]*, X-J Du[1,2] and Y.-W. Liu[1,2]

[1]Hefei National Laboratory for Physical Sciences at Microscale and Department of Modern Physics, University of Science and Technology of China, Hefei, Anhui 230026, People's Republic of China

[2]Synergetic Innovation Center of Quantum Information and Quantum Physics, University of Science and Technology of China, Hefei, Anhui 230026, People's Republic of China

Synopsis Momentum transfer dependence behavior of the autoionization states plays an unique role in understanding the electronic correlation effect. In the present work, the $3s^{-1}nl$ autoionization states and their peak profiles of argon have been measured at different scattering angles by fast electron impact. The momentum transfer dependence behaviors of the dipole-allowed and dipole-forbidden autoionization states of argon have been elaborated.

The in-depth study of the electron-electron correlation has always been one of the most important contents in atomic and molecular physics. The effects of configuration interaction between the atomic discrete energy level series and the continuum give arise to a phenomenon of autoionization [1]. Fano profile has been used to describe the autoionization states, so the evolution of the peak profile parameters with the momentum transfer gives insight to these processes.

Traditionally, the autoionization states have been widely studied by the optical methods [2, 3]. However, only the dipole-allowed autoionization states can be achieved by this experimental technique. Herein we report the investigations of both dipole-allowed and dipole-forbidden autoionization states of argon by the fast electron energy loss spectroscopy operated at different momentum transfers.

The electron energy loss spectra in 11-13 eV and 23-31 eV were measured at an incident electron energy of 1.5 keV and an energy resolution of 85 meV, which correspond to the valence-shell and $3s^{-1}nl$ excitation regions. By fitting the measured spectra and normalizing to the $3p^{-1}4s$, the obtained Fano profile parameters of the transition of $3s^{-1}4p$ state are shown Fig.1. It can be seen that the experimental q factor of this work agrees well with the calculations in the whole momentum transfer region within the experimental uncertainties. Other peak profile parameters and their momentum transfer dependence behaviors of the $3s^{-1}4p$ are reported for the first time, which are also shown in Fig.1.

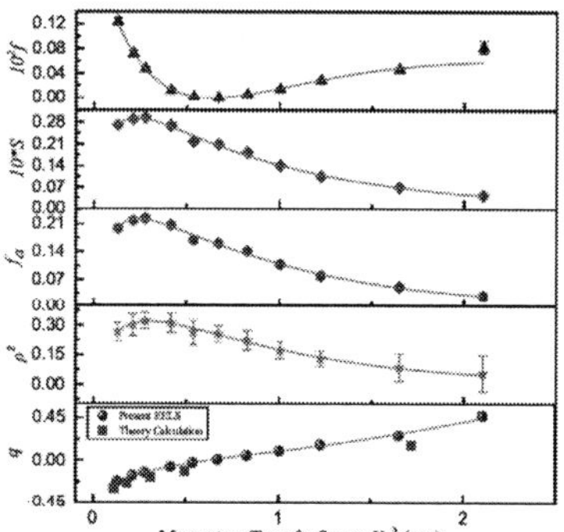

Figure 1. Momentum transfer dependent behaviours of peak profile parameters of the $3s^23p^6 \rightarrow 3s3p^64p$. The dark yellow lines are the fitting result.

In summary, the absolute generalized oscillator strength density in the auto-ionization region of argon has been determined by the high-energy electron scattering. The Fano profile parameters and their momentum transfer dependence behaviors of both the dipole-allowed and dipole-forbidden autoionization states have been reported.

References

[1] Fano U *et al* 1961 *Phys. Rev.* **124** 1866
[2] Liu X J *et al* 2003 *Phys. Rev. Lett.* **91** 193203-1
[3] Ström M *et al* 2006 *Phys. Rev. Lett.* **97** 253002

*E-mail: lfzhu@ustc.edu.cn

Probing functional group dependence in dissociative electron attachment using negative ion momentum imaging

S Gokhale[1,2], K Gope[1,3], V Tadsare[1], M Bain[4], A Raghav[2], M N R Ashfold[4], E Krishnakumar[1,5] and V S Prabhudesai[1 *]

[1]Tata Institute of Fundamental Research, Mumbai, 400005, India
[2]University of Mumbai, Mumbai, 400098, India
[3]The Institute of Chemistry, The Hebrew University of Jerusalem, 9190401, Israel
[4]University of Bristol, Bristol BS8 1TS, United Kingdom
[5]Raman Research Institute, Bangalore, Karnataka, 560080, India

Synopsis Dissociative electron attachment (DEA) is the most efficient way to explore the structure and dynamics of excited states of molecular negative ions. More importantly, DEA has been shown to be a tool for chemical control. One crucial aspect of the chemical control using DEA is the functional group dependence present in this process. Here we present cases on the effect of delocalized electrons on DEA patterns observed in simple aromatic compounds.

It has been shown that H^- formation from DEA to aliphatic compounds follows a pattern that depends on the functional group present and roughly mimics that shown by the prototype hydride - e.g. H^- from DEA to alcohols and amines show parallels with that from DEA to water and ammonia, respectively [1]. This functional group dependence of DEA process leads to site selective fragmentation of N-H, C-H and O-H bonds using electron energy as a control parameter. Beyond 4eV of electron energy, this site selectivity has been found to stem from core excited resonances. In the case of N-H and O-H bonds, the localization of excitation is understood to be due to lone pair excitation of N or O atom. As noted above, prior studies of functional group dependent site selectivity have been largely restricted to aliphatic compounds [1]. However, in aromatic organic molecules like aniline, benzyl amine, pyrrole, delocalized electrons influence this site selectivity due to mixing of lone pair electrons.

In pyridine, where lone pair electrons of N do not become part of delocalzed pi electron cloud, H^- from C-H site shows exactly same DEA beahviour as that from benzene [2]. On the other hand, in aniline, H^- from its N-H site shows deviation in DEA dynamics as compared to that from ammonia because of influence lone pair electrons. Wheras in benzyl amine, as lone pair elctrons are shifting away from delocalized pi electron cloud, H^- from N-H site shows similar DEA behavior as that from ammonia at same electron energy [3].

Pyrrole is a five membered heteroatomic aromatic compound in which lone pair electrons of N become part of delocalized pi electron cloud. H^- from its N-H site (Figure 1(a)) shows similar angular distribution as that from aniline. H^- from its C-H site (Figure 1(b)) shows intriguing dynamics unlike that from benzene. We believe that this distinguishing feature might be a signature of mixing of lone pair electrons with delocalized pi electron cloud. We also compare pyrrole results with N-methyl pyrrole under similar conditions.

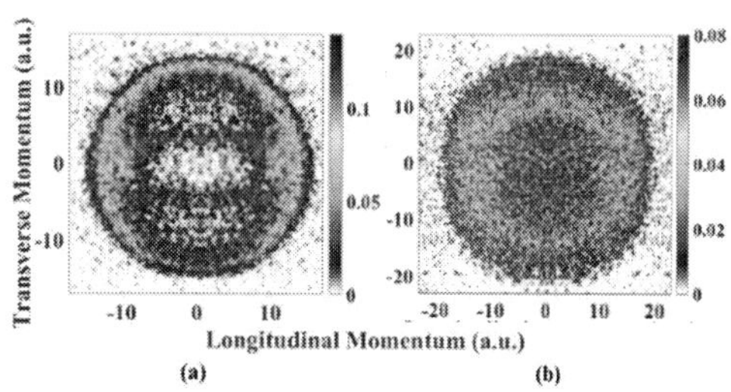

Figure 1. H^- momentum images at (a) 5.5 eV and (b) 9.5 eV from DEA to pyrrole.

References
[1] Prabhudesai V S *et al* 2005 *Phys. Rev. Lett.* **95** 143202
[2] V Tadsare 2018, *Ph. D. Thesis*, Tata Institute of Fundamental Research, India
[3] Ram N B *et al* 2012 *J. Chem. Phys.* **136** 164308

* E-mail: vaibhav@tifr.res.in

Investigating the CO trimer geometry using Coulomb imaging technique

X Fléchard[1 *], A Méry[2], B Gervais[2], S Guillous[2], V Kumar[2], J Matsumoto[3], J Rangamma[2], F Ropars[2], H Shiromaru[3] and A Cassimi[2]

[1] Normandie Univ, ENSICAEN, UNICAEN, CNRS/IN2P3, LPC Caen, 14000 Caen, France
[2] CIMAP, CEA-CNRS-ENSICAEN-UNICAEN, Normandie Université, BP5133, F-14050 Caen Cedex 04, France
[3] Department of Chemistry, Tokyo Metropolitan University, 1-1 Minamiosawa, Hachiouji-shi, Tokyo 192-0397, Japan

Synopsis The Coulombic explosion of multi-charged $(CO)_3^{q+}$ molecular ions produced in low energy collisions with Ar^{9+} projectiles was investigated using recoil ion momentum spectrometry (RIMS). A preliminary analysis of the data clearly shows a dominant triangular cyclic structure. Events that may correspond to a linear chain are also observed, but the latter could as well originate from false coincidences, associated to the fragmentation of dimers. A careful analysis of this source of background is proposed to disentangle trimer fragmentation events from false coincidences.

Direct observation of weakly bound structures, such as small molecular clusters, remains an experimental challenge. Infrared (IR) spectroscopy, combined with calculations of intermolecular potential surfaces, has for long been the only way to access such structures. In the last decade, a new technique providing more direct measurements has been developed: by Coulomb exploding the cluster, all momentum vectors of the charged ions or molecular ions can be measured in coincidence, giving access to the initial 3D relative position prior dissociation. A nice illustration of this technique, called "Coulomb imaging", can be found in [1], where such measurements were performed to determine the geometry of argon and neon dimer, trimer, and tetramer.

Our study focuses on the structure of molecular CO trimers. Few experimental results from IR spectroscopy were made available for this structure, as the $(CO)_3$ transitions were found partly overlapped and obscured by those of $(CO)_2$. The work from Rezaei and collaborators [2] has previously shown a strong evidence for a planar, cyclic and C-bonded structure of the trimer. But satellite peaks and other spectroscopic features remained unexplained. This shortcoming called for a new investigation of the structure of $(CO)_3$ using a completely different and independent technique.

In the present experiment, Coulomb explosion is induced by collisions between a supersonic gas jet, providing the $(CO)_3$ molecular trimers, and Ar^{9+} ions delivered by the low energy beam line of the GANIL/ARIBE facility. The fragments momenta are measured using a COLTRIMS setup. A clean data selection is first achieved by imposing momentum conservation laws to the fragments resulting from Coulomb explosion. The momentum vectors of the fragments are then further analyzed to infer the geometry of the trimer.

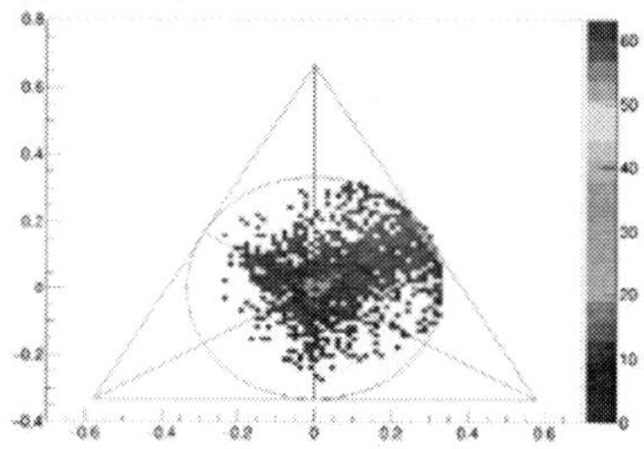

Figure 1. Dalitz plot obtained for the fragmentation of $(CO)_3^{3+}$ in three singly charged $(CO)^+$ fragments.

The plot of Fig. 1 clearly shows two distinct contributions. The dominant one, in the center, can be associated to a regular triangular structure. The weaker contribution, visible on the right edge of the Dalitz plot, could either indicate a linear structure or originate from false coincidences between a dissociating dimer and a monomer. A careful analysis of the background has been performed to quantify this possible contribution.

References

[1] Ulrich B *et al* 2011 *J. Phys. Chem. A* **115** 6936
[2] Rezaei M *et al* 2013 *J. Chem. Phys.* **138** 071102

* E-mail: flechard@lpccaen.in2p3.fr

Quenching of interstellar carbenes: Interaction of C_3H_2 with He and H_2

M Ben Khalifa[1,2], E Sahnoun[1,2], L Wiesenfeld[1]*, F Khadri[2], K Hammami[2] and O Dulieu[1]

[1] Lab. Aimé-Cotton, CNRS, Univ. Paris-Sud, ENS Paris-Saclay, Univ. Paris-Saclay, 91405 Orsay, France
[2] LSAMA, Université de Tunis-El Manar, 2092 El Manar Tunis, Tunisia

Synopsis C_3H_2 molecules are ubiquitous in the Galaxy and present important challenges in their chemistry, whether cyclic or linear, and possibly substituted with D. Obtaining theoretical rotational quenching rates by collisions with He and H_2 as projectiles is of great importance in order to get quantitative abundance estimates. It is all the more challenging that many rotational lines are excited during the collision, resulting in computationnally very heavy quantum dynamics. The quality of various approximations will be discussed, as well as the quenching rates for astrophysical applications.

In order to characterize the chemical composition, isotopic enrichment and possibly, history of the interstellar medium, it is necessary to get an accurate estimate of the molecular densities of the polar molecules observed by rotational spectra. Since the intensity of the spectral lines depends on both photon and molecular collisions, knowing the quenching/excitation rates is an essential ingredient, as soon as the spontaneous photon emission rates becomes comparable to the collisional quenching rates.

Many complex molecules are observed, and especially so, in pre-stellar and proto-stellar environments. Among those, hydrocarbons are particularly challenging, because of the absence of permanent electric dipole moment for the ones, CH_4 and C_2H_n, n=2,4,6. Carbenes $c-C_3H_2$ and $l-C_3H_2$ are the lightest hydrocarbons to display a permanent electric dipole moment and hence, to be detectable by rotational spectroscopy. The cyclic form, cyclopropenylidene, is ubiquitous in the Milky Way and external galaxies. Its abundance is high enough to warrant observations of ^2H and ^{13}C substituted species. The linear form, propadienylidene, is less abundant but widely observed.

We present here a precise *ab initio* Potential Energy Surface for the interaction C_3H_2 with helium [1] and H_2 [2], by means of a CCSD(T)-F12a/aug-cc-pVTZ formalism. We conduct quantum dynamical scattering in order to get precise inelastic cross sections using a coupled-channel and/or coupled-states approach for solving the nuclear motion. By averaging over a Maxwell distribution if speeds, we get the relevant rates for the astrophysical needs. For the low temperature scattering of $c-C_3H_2$ with H_2, we perform extensive tests on the quantum dynamics formalism, to validate the various methodological choices and to balance precision and acceptable computer load.

*E-mail: laurent.wiesenfeld@u-psud.fr

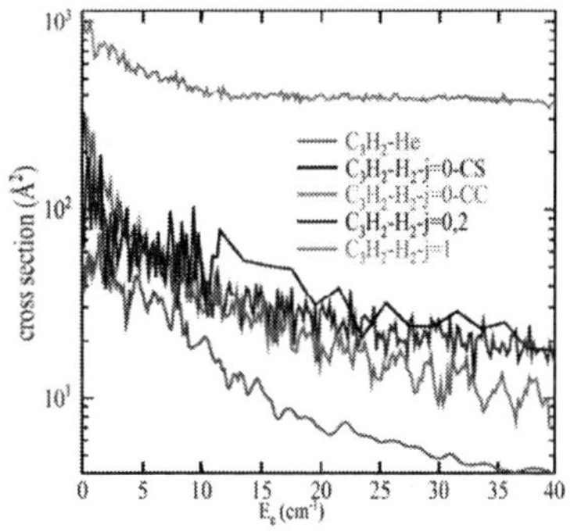

Figure 1. Comparison between different projectiles and approximations of quantum dynamics, for the rotational quenching of the $1_{10} \rightarrow 1_{01}$ transition of $c-C_3H_2$. E_C, collision energy. Note the many resonances for quenching by H_2, $j=0$.

References

[1] Ben Khalifa M, Sahnoun E, Wiesenfeld L, Khadri F, Hammami K, Dulieu O, Spezzano S and Caselli P 2019 *Phys. Chem. Chem. Phys.* **21** 1443
[2] Ben Khalifa M, Wiesenfeld L and Hammami K 2019 *Phys. Chem. Chem. Phys.* submitted

Autodetachment spectroscopy of the metastable HD^- and D_2^- anions

R Marion[1] *, M Čížek[2] and X Urbain[1]

[1]Institute of Condensed Matter and Nanosciences, Université Catholique de Louvain, Chemin du Cyclotron 2, B-1348 Louvain-la-Neuve, Belgium

[2]Charles University Prague, Faculty of Mathematics and Physics, Institute of Theoretical Physics, 180 00 Praha 8, Czech Republic

Synopsis The autodetachment spectrum of decaying HD^- and D_2^- has been recorded with a velocity-map imaging spectrometer. Coincident detection of the neutral molecule permits the selection of events occuring in the center of the spectrometer. The discrete features of the spectrum are assigned to transitions between well identified ro-vibrational levels. Energies, lifetimes and branching ratios predicted with the non-local potential framework developed for associative detachment satisfactorily reproduce the experimental results.

The existence of long-lived molecular hydrogen anions was established by accelerator mass spectrometry [1]. Their metastability is a result of fast molecular rotation causing the potential energy minimum of the H_2^- ground state to lie outside the H_2 potential well. A single rovibrational level of D_2^- was found to decay significantly by tunneling to $D + D^-$ [2]. All other levels decay via autodetachment, with lifetimes ranging from fs to ms [3, 4]. Coulomb explosion imaging of H_2^- and D_2^- has provided the average internuclear distance distribution [5, 6].

In the present study, we directly record the ejected electron energy spectrum by means of a velocity-map imaging (VMI) spectrometer (see Fig. 1). The coincident detection of the electron and its parent molecule permits to locate the emission point, hence to retrieve regular momentum images. A rich spectrum emerges, that allows for a unique determination of the initial and final ro-vibrational levels. The detachment occurs in the exponential tail of the vibrational wavefunctions, with branching among all accessible vibrational levels of the neutral molecule.

The energies and lifetimes of the relevant HD^- and D_2^- levels were computed earlier in the nonlocal potential scattering framework [4]. This work has been expanded here to take explicitly into account the change in angular momentum imposed by parity conservation. The vibrational branching ratios are satisfactorily reproduced by analyzing the partial cross sections for the associative detachment reaction $H(D) + D^- \to HD(D_2) + e^-$.

This work was supported by the Fonds de la Recherche Scientifique-FNRS through IISN grant No. 4.4504.10.

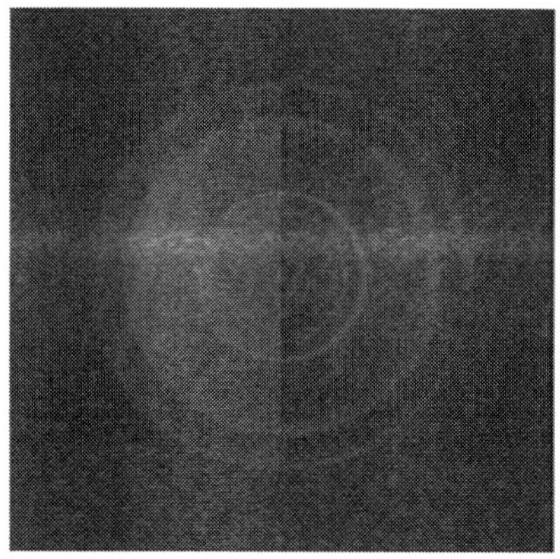

Figure 1. Velocity map of ejected electrons from D_2^-. Left: histogram of hit positions. Right: map after Abel inversion.

References

[1] Golser R *et al* 2005 *Phys. Rev. Lett.* **94** 223003
[2] Mezdari F, de Ruette N and Urbain X 2017 *Phys. Rev. A* **96** 032512
[3] Heber O *et al* 2006 *Phys. Rev. A* **73** 060501
[4] Čížek M, Horáček J and Domcke W 2007 *Phys. Rev. A* **75** 012507
[5] Jordon-Thaden B *et al* 2011 *Phys. Rev. Lett.* **107** 193003
[6] Herwig P *et al* 2013 *Phys. Rev. A* **87** 062513

*E-mail: raphael.marion@uclouvain.be

Spectroscopy and kinetics of IR laser interaction with atmospheric molecules: Effects of cooling and chaos

E Ternovsky[1,2]*, A Glushkov[1,2], A Svinarenko[1,2] and Yu Bunyakova[1]

[1]Dept. of Applied mathematics, Odessa State Environmental University, Odessa, 65016, Ukraine
[2]Intern. Res. Centre of Quant. Optics and Laser Phys., Odessa State Environmental University, Odessa, 65016, Ukraine

Synopsis Spectroscopy and kinetics of energy and heat exchange in the mixture CO_2-N_2-H_2O under passing the CO_2 laser pulses is studied. To describe the energy exchange and relaxation processes in the CO_2.– N_2 – H_2O mixture, which interacts with a laser pulse , we use the modified 3-mode model of the kinetic processes New data on the resonant absorption relative coefficient $\alpha(CO_2)$ for the European mid-latitude atmospheric model are presented. It is also studied a chaotic behaviour of the carbon and nitrogen dioxide concentration time series (Munich & Gdansk reg.) and the low-D chaos effect is found.

We study spectroscopy and kinetics of the energy and heat exchange in the mixture CO_2-N_2-H_2O under passing the CO_2 laser pulses.To describe energy exchange and relaxation processes in the CO_2-N_2-H_2O mixture, which interacts with laser radiation, we start from the modified 3-mode model of kinetic processes [1,2]. It is studied a kinetics of three vibrational levels: 100, 001 (CO_2) and v = 1 (N_2). The balance equation system for relative populations is [3]:

$$\frac{dx_1}{dt} = -\beta(\omega + 2gP_{10})x_1 + \beta\omega x_2 + 2\beta g P_{10})x_1^0,$$

$$\frac{dx_2}{dt} = \omega x_1 - (\omega + Q + P_{20})x_2 + Qx_3 + P_{20}x_2^0,$$

$$\frac{dx_3}{dt} = \delta Q x_2 - (\delta Q + P_{30})x_3 + P_{30}x_3^0. \qquad (1)$$

$x_1=N_{100}/N_{CO_2}$, $x_2=N_{001}/N_{CO_2}$, $x_3=\delta N_{N_2}/N_{CO_2}$; N_{100}, N_{001} are populations of the levels 100, 001(CO_2); N_{N_2} is level population v=1 (N_2); N_{CO_2} is concentration of the CO_2 molecules; δ is a ratio of the total concentrations of CO_2 and N_2 in an atmosphere ($\delta=3.85\cdot10^{-4}$); x_1^0, x_2^0, x_3^0 are the equilibrium relative populations for temperature T: $x_i^0 = \exp(-E_i/T)$; E_1,E_2 in (1) are the energies (K) of the levels 100, 001 P_{10}, P_{20} and P_{30} are the probabilities (s^{-1}) of the collisional deactivation of levels 100, 001 (CO_2) and v = 1 (N_2), Q is a probability (s^{-1}) of the resonant transfer in the reaction $CO_2 \rightarrow N_2$,ω is the probability (s^{-1}) of excitation, $g=3$, $\beta = (1+g)^{-1}$ [1]. A total dynamics includes processes of vibration-transitional relaxation (VT-relaxation), intermodal vibration-vibration relaxation (VV'-relaxation). A renewal process of thermodynamic equilibrium is linked with a fast exchange of level 100 with a ground state and relatively slow relaxation of high level 001 [1]. It results in an effect of the environment cooling. New data on a temporary dependence of the resonant absorption relative coefficient α_{CO2} (cm^{-1}) for gaussian laser pulse ($\lambda=10.6\mu m$, $I=10^5$ W/cm^2) on a height (H, km; the European mid-latitude atmospheric model) are listed in Table 1.

Table 1. Temporary dependence of the α_{CO2}: our data.

t, μs	H=0	H=10km
0	1.0	1.0
1	0.57	0.13
2	0.46	0.05
3	0.59	0.19
4	0.64	0.28

The chaotic behaviour of the carbon and nitrogen dioxide concentration time series (Munich & Gdansk reg.) is studied too with using a group of the chaos theory methods: the autocorrelation function,mutual information, correlation dimension, false nearest neighbours methods (in version [4]). A spectrum of the fractal dimensions, Lyapunov exponents, Kaplan-Yorke dimension, Kolmogorov entropy etc are also calculated for studied system.

References

[1] Zuev V *et al* 1991 *Nonlinear optics of atmosphere* (Leningrad)
[2] Khetselius O 2015 *Frontiers in Quantum Methods and Applications in Chemistry and Physics (Springer)* **29** 55
[3] Bunyakova Yu *et al* 2016 *Photoelectr.* **25** 68
[4] Buyadzhi V *et al* 2019 *Adv. Quant. Chem.* **78** 171

* E-mail: ternovskyev@gmail.com

Uncertainty assessment of Rydberg and doubly-excited molecular states ab initio calculations using Quantum Defect Theory scaling laws

M Telmini[1] * and Ch Jungen[2] †

[1]LSAMA, Department of Physics, Faculty of Science of Tunis, University of Tunis El Manar, 2092 Tunis, Tunisia
[2]Department of Physics and Astronomy, University College London, London WC1E 6BT, United Kingdom

Synopsis We discuss the accuracy of various published *ab initio* calculations of highly excited Born-Oppenheimer bound and doubly excited states in diatomic molecules. We show that some of these results appear questionable and in a few cases even clearly unreliable, a fact which may also affect the prediction of collision cross sections based on such data. Our analysis uses Quantum Defect Theory scaling laws and is supported by variational R-matrix calculations. Some simple recipes are discussed, which may help users of *ab initio* quantum chemistry packages to assess the quality of their results.

Uncertainty estimates in theoretical calculations and modeling are becoming an important issue in present-day Atomic and Molecular Physics, both in spectroscopic and collisional applications [1]. Until recently, numerical theoretical data were typically published either in graphical form or as extensive data tables, usually without much indication as to their expected uncertainty. New theoretical data were compared to theoretical or experimental results available in the literature, and the agreement between the new results and existing data was then taken as an indicator of the accuracy of the results obtained.

To some extent the comparison of new data with previous ones may indeed serve as a rough measure of the uncertainty. However, when new calculations are presented for which no previous results exist to compare with, it is desirable to provide estimated uncertainties or error bars along with the data. Such estimates are important when experimental work based on high quality theoretical predictions is carried out, and they may become crucial, for instance when theoretical and experimental data are combined in metrology.

The present contribution deals with first principles calculations of highly excited molecular electronic states in instances where spectroscopic level accuracy is needed to interpret levels and transitions observed in high-resolution spectroscopy. Specifically we present a critical review of selected published quantum-chemical calculations of Rydberg and doubly-excited Born-Oppenheimer electronic states of diatomic hydrogen H_2 and the helium hydride ion HeH^+, which serve here as benchmark systems for homonuclear and heteronuclear diatomic molecules in general.

On the basis of Rydberg scaling laws we are able to show that certain published calculations are not converged and doubtless deviate substantially from the exact values. We further demonstrate that the alternative approach that we have been developing for a number of years now, based on variational R-matrix theory[3], may help to avoid these drawbacks and is able to provide a comprehensive insight into the structure and dynamics of highly-excited states.

References

[1] Chung H-K *et al* 2016 *J. Phys. D: Appl. Phys.* 49 363002
[2] Telmini M and Jungen Ch 2019 to be submitted
[3] Telmini M and Jungen Ch 2003 *Phys. Rev. A* 68 062704

*E-mail: mourad.telmini@fst.utm.tn
†E-mail: c.jungen@ucl.ac.uk

Virtual Laboratory Astrophysics and the STARK-B database VAMDC node: a resource for electron and ion impact widths and shifts of isolated lines

S Sahal-Bréchot [1*], M S Dimitrijević [2,1 †] and N Moreau [1]

[1]Sorbonne Université, Observatoire de Paris, Université PSL, CNRS, LERMA, F-92190 Meudon, France
[2]Astronomical Observatory, Volgina 7, 11060 Belgrade, Serbia

Synopsis Stark broadening theories and calculations can now be considered as mature. The fast computers allow to calculate a great number of atomic data needed for accurate spectroscopic diagnostics and modeling, such as for stellar atmospheres and envelopes, laboratory and technological plasmas, and laser equipments. Thus the access to these data via an on line database is essential. The STARK-B database is intended to meet this need. It contains calculated widths and shifts of isolated atomic lines due to electron and ion impacts. It is still under development and will be presented at the Conference.

Stark broadening theories and calculations have been very much developed for about 50 years. The theory can now be considered as mature for many applications, especially for accurate spectroscopic diagnostics and modelling. This requires the knowledge of numerous collisional line profiles, in particular for low abundant atoms and ions which are now also used as probes for current spectroscopic diagnostics in astrophysics. Thus, the access to such data via an on line database becomes essential. STARK-B [1] was created for fulfilling this need. It is a collaborative project between the Astronomical Observatory of Belgrade (AOB) and the LERMA. This database is maintained and developed at the Paris Observatory and contains our calculated widths and shifts of isolated lines of neutral and ionized atoms due to electron and ion impacts. Our calculations are based on the impact Semi-Classical-Perturbation method (SCP) [2] and earlier papers cited therein, and on the Modified Semi Empirical method (MSE) [3]. Our quantum calculations based on the Superstructure and Distorted Wave method [4] are not yet implemented, this will be for a future stage. STARK-B is devoted to modeling and spectroscopic diagnostics of stellar atmospheres and envelopes, laboratory plasmas, laser equipments and technological plasmas. Thus the domain of temperatures and densities covered by the tables is wide and depends on the ionization degree of the considered ion. STARK-B is a node of VAMDC (Virtual Atomic and Molecular Data Centre) [5], which is an European Union funded collaboration between groups involved in the generation and use of atomic and molecular data. The aim of VAMDC is to build a secure, documented, flexible and interoperable e-science environment-based interface to existing atomic and molecular data. It complies with the standards of the Virtual Observatories. One can access to the 33 current VAMDC databases via the portal of its home page [6].

STARK-B opened at the end of 2008. Today, the database contains our calculated data for a varied number of transitions of more than 130 neutral or ionized atoms, published in international refereed journals (more than150 papers). We continue to implement our already published data, and the new calculated ones, as our work progresses and published. We do not implement unpublished data.

At the Conference, the key points of the methods and assumptions used in the calculations will be highlighted, The database, its current developments and an outlook of our ongoing work will be presented.

References

[1] http://stark-b.obspm.fr
[2] Sahal-Bréchot S Dimitrijević M S and Ben Nessib N 2014 *Atoms* **2** 225
[3] Dimitrijević M S and Konjevic J 1980 *JQSRT* **24** 451
[4] Elabidi H Ben Nessib N and Sahal-Bréchot S 2004 *J Phys B : At Mol Opt Phys* **37** 63
[5] Dubernet M L et al. 2016 *J. Phys. B: At. Mol. Opt. Phys.* **49** 074003
[6] http://www.vamdc.org/

* E-mail: sylvie.sahal-brechot@obspm.fr
† E-mail: mdimitrijevic@aob.rs

Double-continuum R-matrix theory and codes for many-electron atoms: ionizing electron collisions and time-dependent laser-induced double-ionization

M Plummer[1 *] and C J Noble[1]

[1]Scientific Computing Department, STFC Daresbury Laboratory, Sci-Tech Daresbury, Cheshire WA4 4AD, United Kingdom

Synopsis An extension of ab initio many-electron R-matrix theory is described which is relevant both to the time-dependent description of excitation and single and double ionization of atoms in laser pulses, and to time-independent calculation of intermediate energy electron collisions where ionized states of the target atom can be excited. Corresponding general many-electron 'inner region' code modules have been developed for use in combination with existing R-matrix method codes for diverse applications with up to two electrons explicitly described by a continuum orbital basis.

R-matrix theory in atomic physics allows ab initio theoretical study of many-electron systems with continuum states of one or more electrons. [1] extensively reviews the theory and applications. We extend this theory in which the 1-electron orbital functions used are orthogonal ([2] reviews the alternative 'B-spline R-matrix' or BSR method using non-orthogonal orbitals).

R-matrix theory for N+1 electrons divides configuration space into two regions: a sphere centred on the 'target' atom in which the all-electron Hamiltonian is diagonalized using configuration interaction (CI) functions built from 1-electron spin-orbitals (a spinor, a spherical harmonic and a radial function), and an outer region where one electron moves in multichannel potentials generated by the target electrons. The antisymmetric CI functions include radial continuum orbitals, finite at the sphere boundary, for one electron with all other radial orbitals decaying to zero. The new work, for N+2 electrons, allows two continuum orbitals in each inner region CI function and links to two types of outer region, the first with one electron outside the sphere, for which corresponding (N+1)-electron functions may include one continuum orbital, and the second with N electrons in the sphere and two interacting electrons outside it.

The theory leads to complications at the region boundaries if the wavefunction and its derivatives are to be continuous. Also, explicit antisymmetrization of necessary recouplings may lead to linear dependence. A singular value decomposition method introduced by Gorczyca *et al* [3] in another context is applied in a new way to overcome these difficulties.

Coding builds on earlier work [4, 5] for efficient many-electron angular integral calculation, with systematic progression for matrices with N, N+1 and N+2 electrons. High performance computing methods cope with large bases (for numerical and B-spline radial orbitals) and high values of continuum orbital angular momentum needed in multiphoton calculations.

The codes are designed both for electron collisions and to be used by the 'R-Matrix with time-dependence' (RMT) laser-atom code with applications in multiphoton ionization, rescattering, harmonic generation and coherent control of electronic transitions [6]. RMT has already been independently extended to explicit treatment of He double ionization and collisional ionization of 2-electron atoms [7, 8].

References

[1] Burke P G 2011 *R-matrix Theory of Atomic Collisions* (London New York: Springer Series on Atomic, Optical and Plasma Physics)
[2] Zatsarinny O and Bartschat K, 2013 *J. Phys. B* **46** 112001
[3] Gorczyca T W, Robicheaux F, Pindzola M S, Griffin D C and Badnell N R 1995 *Phys. Rev.* A **52** 3877
[4] Burke P G, Burke V M and Dunseath K M 1994 *J. Phys. B* **27** 5341
[5] Burke V M 1998 *Comput. Phys. Commun.* **114** 210
[6] Moore L R, Lysaght M A, Nikolopoulos L A A, Parker J S, van der Hart H W and Taylor K T 2011 *J. Mod. Optics* **58** 1132
[7] Wragg J, Parker J S and van der Hart H W 2015 *Phys. Rev.* A **92** 022504
[8] Wragg J and van der Hart H W 2016 *Phys. Rev.* A **94** 032706

* E-mail: Martin.Plummer@stfc.ac.uk

Formation of positive and negative ions in the gas phase of D-ribose by low-energy electrons

I Chernyshova[1] *, J Kontros[1], O Shpenik[1] and S Demes[2] †

[1]Institute of Electron Physics, Nat. Acad. Sci. of Ukraine, 21, Universitetska str., 88017 Uzhgorod, Ukraine
[2]Institute for Nuclear Research, Hungarian Acad. Sci., 18/c, Bem square, 4026 Debrecen, Hungary

Synopsis Using a hypocycloidal electron spectrometer the total dissociative electron attachment cross section and the ionization cross section for D-ribose molecule were studied in the E<30 eV energy range. A clear structure near zero energy and in the 6.0–9.5 eV range is observed in the dissociative electron attachment cross section. These features appear due to the formation of negative ion fragments of the molecule. The ionization potential of the D-ribose molecule was determined at 9.74 ± 0.02 eV. In the total ionization cross section weak features are observed, which most likely due to the formation of positive ion fragments of the initial molecule.

D-ribose ($C_5H_{10}O_5$) belongs to monosaccharides, the most abundant group of biomolecules, with five C atoms. This molecule is a part of the building blocks that form DNA and RNA molecules. D-ribose is also contained in ATP and NADH, which are critical for the metabolism. Here we report on the results of our studies on the total cross sections for the production of both positive and negative ions of the D-ribose molecule by slow (<30 eV) electrons.

In our experiments, a hypocycloidal electron monochromator [1] was used to produce an electron beam with at least 0.2 eV energy resolution. The ribose powder (Sigma-Aldrich, 99% purity) placed into a quartz ampoule was heated up to ~80°C in a stainless-steel reservoir, filling thus the gas cell (at $P≈10^{-3}$Torr). The produced ions were extracted to the collector mounted perpendicularly to the electron beam direction. A low negative potential was applied to the collector for detecting the positive ions, while for the negative-ion detection its polarity was changed to positive.

Figure 1 shows the total cross section of dissociative electron attachment to D-ribose molecule measured at 80°C in the 0.00–11.00 eV electron energy region. An intense peak near zero energy and two lower-intensity maxima in the range of 6.00–9.50 eV can be observed. Based on the results of Ref. [2] we assume that the observed features appear due to the formation of $C_5H_8O_4^-$, $C_5H_6O_3^-$, $C_4H_5O_3^-$, $C_3H_4O_2^-$, $C_3H_3O_2^-$, $C_2H_3O_2^-$ and OH^- ion fragments.

We also measured the total ionization cross section for the ribose molecule within the energy range from the threshold up to 28 eV. The ionization potential was determined at E_i = 9.74 ± 0.02 eV. The thorough measurements confirm that the observed weak features in the ionization curve are to be related to the formation of positive ion fragments of the initial molecule.

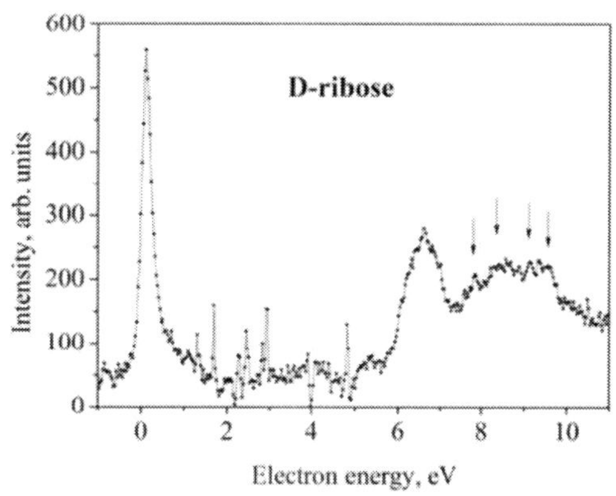

Figure 1. The total dissociative electron attachment cross section for the D-ribose molecule measured by slow electrons.

References

[1] Kontros J E *et al* 2002 *J. Phys.B.* **35** 301
[2] Ptasinska S *et al* 2004 *J.Chem.Phys.* **120** 8505

* E-mail: irinav.chernyshova@gmail.com
† E-mail: demes.sandor@atomki.mta.hu

Role of methylation in electron scattering on $X(CH_3)_4$ (X = C, Si, Ge) molecules

S Stefanowska-Tur[1]*, Cz Szmytkowski[1], E Ptasińska-Denga[1] and P Możejko[1]

[1]Department of Atomic, Molecular and Optical Physics, Faculty of Applied Physics and Mathematics, Gdańsk University of Technology, Gabriela Narutowicza 11/12, Gdańsk, 80-233, Poland

Synopsis Total cross sections (TCS) for low-energy electron scattering from $X(CH_3)_4$ (where X = C, Si, Ge) molecules have been measured using linear transmission method. Present results and those obtained previously for XH_4 molecules were used to determine how methylation of the target is reflected in TCS energy dependence.

In recent years, modern techniques of nanostructures production, like focused electron beam induced deposition (FEBID) method, are of great interest. The precursors used in FEBID method often contain methyl groups, therefore it is important to understand how the replacement of H atoms with CH_3 groups affects processes occuring during electron collisions. In this work, we present our investigation on the role of methylation based on TCS results for $X(CH_3)_4$ and XH_4 molecules.

The absolute total cross sections (TCS) for electron scattering from studied molecules have been measured using the electrostatic 127° electron spectrometer working within the linear transmission mode [1] for impact energies from 0.6 to 300 eV. The TCS at given electron energy has been obtained from intensities of transmitted electron currents taken in the absence and presence of target in the scattering chamber and the target vapour pressure.

To investigate how methylation affects TCS energy dependence, we carried out detailed analysis of experimental TCS for tetrahedral compounds of carbon, silicon and germanium XH_4 and for their fully methylated derivatives $X(CH_3)_4$ [2]. Preliminary results for Si compounds are shown in fig. 1. TCS curves are similar in the shape, but methylated molecule has additional enhancement extending between 12 and 20 eV. Simple estimation of TCS for 1,2,3 and 4 times methylated SiH_4 shows that high-energy structure becomes more visible with increasing number of CH_3 group in target. Analogous studies show the same dependence for C and Ge compounds. The main maxima of TCS are also shifted toward high energy for methylated molecules.

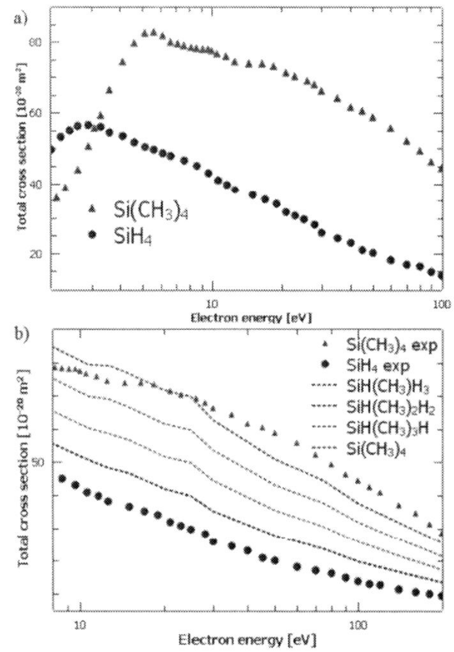

Figure 1. a) TCS for SiH_4 [3] and $Si(CH_3)_4$. b) Dot-dot lines mean estimated TCS for once, twice, three times and four times methylated SiH_4.

This work has been partially supported by the Polish Ministry of Science and Higher Education and the Diamond Grant program (Project No. DI2015 018945).

References

[1] Szmytkowski Cz *et al* 2001 *Vacuum* **63** 549
[2] Stefanowska-Tur S *et al* 2019 *J. Chem. Phys* **150** 094303
[3] Szmytkowski Cz *et al* 1997 *J. Phys. B: At. Mol. Opt. Phys.* **19** 4363

*E-mail: s141549@student.pg.edu.pl

$K\alpha_{1,2}\alpha_3$ and $K^h\alpha_{1,2}$ satellite transition energies and rates for Ar and Ca

K Kozioł* and J Rzadkiewicz†

National Centre for Nuclear Research (NCBJ), Andrzeja Soltana 7, 05-400 Otwock-Świerk, Poland

Synopsis The calculations of radiative energies and transition rates for $K\alpha_{1,2}\alpha_3$ and $K^h\alpha_{1,2}$ satellite (M_1, $M_{2,3}$, and N_1 spectator holes) lines for Ar and Ca have been carried out by using the GRASP2K and AMBIT codes. The influence of electron correlation on transition rates has been studied.

The radiative transitions in K-shell hollow (i.e., having empty K shell) atoms can occur via one-electron one-photon (OEOP: $1s^{-2} \rightarrow 1s^{-1}2p^{-1}$, named $K^h\alpha_{1,2}$) or via much less probable two-electron one-photon (TEOP: $1s^{-2} \rightarrow 2s^{-1}2p^{-1}$, named $K\alpha_{1,2}\alpha_3$) transitions. It has been found that TEOP transition rates are very sensitive to the electron correlations [1], while OEOP transition rates are only weakly sensitive.

Table 1. Transition rates for $K^h\alpha_{1,2}$ and $K^h\alpha_{1,2}M_1^{-1}$ lines of Ar, calculated in length (A_L) and velocity (A_V) gauge. 'CI' means extensive Configuration-Interaction calculations.

Code	Transition rate (s^{-1})	
	A_L	A_V
$K^h\alpha_2$ ($1s^{-2} \rightarrow 1s^{-1}2p_{1/2}^{-1}$)		
GRASP2K	2.66E+14	2.57E+14
GRASP2K (CI)	2.69E+14	2.57E+14
AMBIT	2.56E+14	2.45E+14
$K^h\alpha_1$ ($1s^{-2} \rightarrow 1s^{-1}2p_{3/2}^{-1}$)		
GRASP2K	3.85E+12	3.44E+12
GRASP2K (CI)	3.54E+12	3.39E+12
AMBIT	3.43E+12	3.29E+12
$K^h\alpha_{1,2}M_1^{-1}$ ($1s^{-2}3s^{-1} \rightarrow 1s^{-1}2p^{-1}3s^{-1}$)		
GRASP2K	2.79E+14	2.68E+14
GRASP2K (CI)	2.64E+14	2.55E+14
AMBIT	2.58E+14	2.49E+14

We reported recently that incorporation of $K\alpha_{1,2}\alpha_3$ and $K^h\alpha_{1,2}$ satellite transitions is crucial to proper estimate the OEOP-to-TEOP branching ratio and $K\alpha_{1,2}\alpha_3$ linewidth [1]. Then the calculations of radiative energies and transition rates for satellite $K\alpha_{1,2}\alpha_3$ and $K^h\alpha_{1,2}$ lines for Ar and Ca atoms have been carried out by means of the GRASP2K [2] and AMBIT [3] codes, based on the Multi-Configuration Dirac–Hartree–Fock method. The influence of electron correlation on transition rates has been studied. The results of our studies set new theoretical limits for the TEOP transitions in the low-Z atomic range. We also hope that this work will guide future experiments with a high accuracy.

*E-mail: Karol.Koziol@ncbj.gov.pl
†E-mail: Jacek.Rzadkiewicz@ncbj.gov.pl

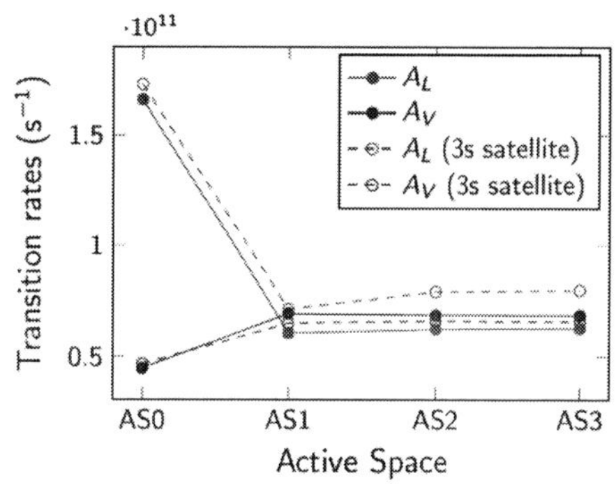

Figure 1. Convergence of calculations of $K\alpha_{1,2}\alpha_3$ and $K\alpha_{1,2}\alpha_3M_1^{-1}$ transition rates for Ar for various configuration-interaction active spaces.

Acknowledgments: The work was supported by the Polish Ministry of Science and Higher Education within the framework of the National Centre for Nuclear Research Grant for Young Scientists and PhD Students 2018–2019.

References

[1] Kozioł K and Rzadkiewicz J 2017 *Phys. Rev. A* **96** 031402
[2] Jöönsson P, Gaigalas G, Bieroń J, Froese Fischer C and Grant I P 2013 *Comput. Phys. Commun.* **184** 2197
[3] Kahl E V and Berengut J C 2018 *Comput. Phys. Commun.* In Press, DOI: 10.1016/j.cpc.2018.12.014

Investigation of TEOP transitions in dielectronic recombination with H-like ions

K N Lyashchenko[1,2]*, **O Yu Andreev**[2,3] **and D Yu**[1]†

[1]Institute of Modern Physics, Chinese Academy of Sciences, Lanzhou 730000, China
[2]St. Petersburg State University, St. Petersburg, 199034, Russia
[3]Peter the Great St. Petersburg Polytechnic University, St. Petersburg 195251, Russia

Synopsis We consider theoretically the process of dielectronic recombination (DR) with H-like ions where two-electron one-photon (TEOP) transitions take place. We focus our attention on the channel with transitions to the ground state of the He-like ions and emission of $K_{\alpha\alpha}$ photon. Total and differential cross sections of DR with TEOP are calculated for wide range of H-like ions: from B^{4+} to Xe^{53+}. The resonant Fano-like structure of the cross section is observed. The resonant energies, widths and TEOP transition probabilities are calculated.

The autoionizing states are responsible for many resonant processes in collisions of highly charged ions with electrons and other atomic particles. Such resonant processes as dielectronic and trielectronic recombination, excitation-autoionization and many others actively proceed in highly ionized laboratory and astrophysical plasmas. In particular, the autoionizing states participate in two-electron one-photon (TEOP) transitions, which are single photon transitions where two electrons change their quantum numbers [1]. Such transitions occur only due to the interelectron interaction and represent a process which is very sensitive to the description of the interelectron correlation.

In the present theoretical investigation we focus our attention on TEOP transitions occurring during the DR with H-like ions. Schematically, this process can be described as

$$\begin{aligned} e^- + A^{(Z-1)+}(1s) \quad &\to \quad A^{(Z-2)+}(2l_j, 2l'_{j'})_J \\ &\qquad\downarrow \\ &\quad A^{(Z-2)+}(1s^2) + \hbar\omega\,. \end{aligned}$$

This process was considered within QED theory. The Line-Profile Approach was employed [2]. The total and differential cross sections for various H-like ions as well as resonant energies, widths and transition rates were calculated. In figure below we present the total cross section of DR with TEOP transitions to the ground state for initially H-like B^{4+}, Ar^{17+} and Xe^{53+} ions. The cross section clearly shows the resonant Fano-like profiles due to the participation of intermediate autoionizing $(2s, 2p_{3/2})_1$ state.

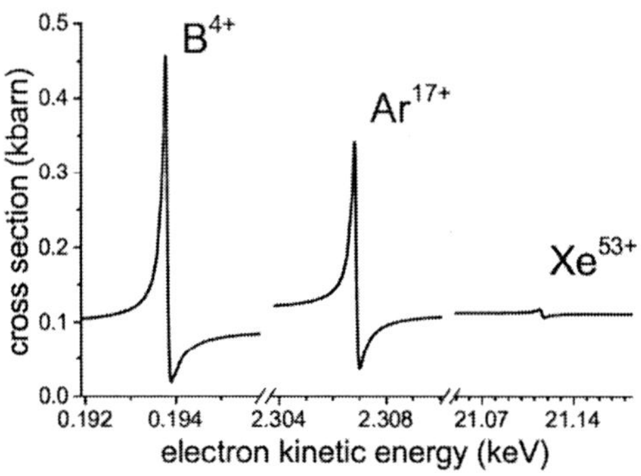

Figure 1. The total cross section for the DR with TEOP transitions (in kbarn) as a function of kinetic energy of the incident electron for various ions. The resonant structure is determined by the autoionizing $(2s, 2p_{3/2})_1$ state.

References

[1] Zou Y *et al* 2003 *Phys. Rev. A* **67** 042703
[2] Andreev O Yu *et al* 2008 *Phys. Rep.* **455** 135

*E-mail: k.n.lyashchenko@yandex.ru
†E-mail: d.yu@impcas.ac.cn

Journal of Physics: Conference Series **1412** (2020) 142004 doi:10.1088/1742-6596/1412/14/142004

Cascade and direct processes in multiple Auger decay of core-excited states of N_2

T Odagiri[1] *, T Taniguchi[1], T Kaneyasu[2], H Tanaka[3], J Adachi[3], P Lablanquie[4] and Y Hikosaka[5]

[1]Department of Materials and Life Sciences, Sophia University, Tokyo 102-8554, Japan
[2]SAGA Light Source, Tosu 841-0005, Japan
[3]Photon Factory, Institute of Materials Structure Science, Tsukuba 305-0801, Japan
[4]Sorbonne Universités, UPMC Univ Paris 06, CNRS, LCP-MR (UMR 7614), 11 rue Pierre et Marie Curie, 75231 Paris Cedex 05, France
[5]Institute of Liberal arts and Sciences, University of Toyama, Toyama 930-0194, Japan

Synopsis We report here a detailed experimental study on multiple Auger decay after core excitation into the π^* valence and Rydberg orbitals in N_2 by an electron-electron coincidence technique.

Relaxations of core-hole states by emission of several electrons are known as multiple Auger transitions. They are specially interesting in molecules because their dynamics involve interplay between electronic and nuclear motions, as demonstrated in CO by Kaneyasu et al [1]. They remain however little studied because efficient coincidence techniques are required. In the present study, electrons emitted after core excitation of N_2 were energy-analyzed and detected in coincidence to obtain electron-energy correlations for understanding detailed dynamics of multiple Auger decay in molecules.

The experiments were performed at the soft X-ray undulator beamline BL-16A of the Photon Factory, KEK. In order to obtain longer time interval between the light pulses, we used a mechanical chopper of which rotation synchronizes with the RF signal of the synchrotron ring and a repetition rate of 146 kHz was obtained [2]. Electrons emitted by photoabsorption were collected by an inhomogeneous magnetic field, energy-analyzed by a 2.5 m long TOF tube and detected in coincidence by a time-to-digital converter.

Figure 1 represents the energy correlations between the two electrons in the resonant double Auger processes after the 1s $\rightarrow \pi^*$ core excitation at 401.08 eV in N_2. The kinetic energies of fast electron, KE_{fast}, were converted to the binding energies of singly charged states by subtracting them from the photon energy, i.e. $h\nu - KE_{fast}$. The bottom and left panels in figure 1 are the projections on each axis. The peaks in the kinetic energy spectrum of slow electrons on the left are attributed to electrons emitted in autoionizations of superexcited neutral nitrogen atoms which are produced in cascade double Auger processes: $N_2^* \rightarrow N_2^{+*} + e^-_{fast}$ followed by $N_2^{+*} \rightarrow N^* + N^+$ and then by $N^* \rightarrow N^+ + e^-_{slow}$. The island structures on the energy correlation indicate that the intermediate singly charged states for the cascade double Auger processes lie at around 60 eV, which is formed by a spectator Auger transition.

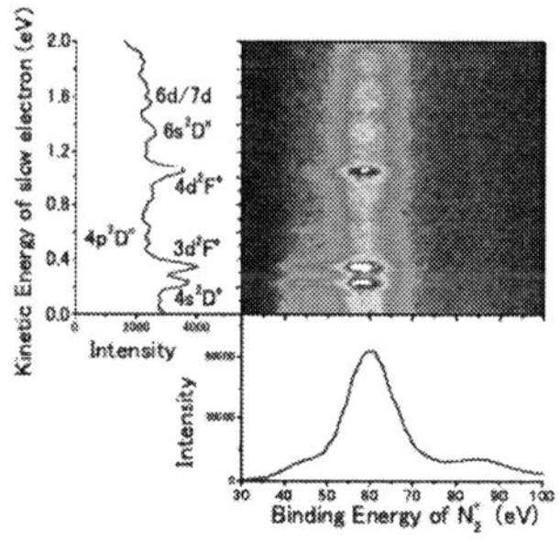

Figure 1. Energy correlations between two electrons in the resonant double Auger processes and projections on each axis (bottom and left) at 1s $\rightarrow \pi^*$ core excitation in N_2 (see text for details).

References

[1] Kaneyasu T 2008 *Phys. Rev. Lett.* **101** 183003
[2] Kaneyasu T 2017 *J. Chem. Phys.* **147** 104304

*E-mail: odagiri.t@sophia.ac.jp

On unified contribution of pair Coulomb excitations in three-body reactions

S Levin*, A Budylin† and Y Koptelov‡

[1]St-Petersburg State University, St-Petersburg, 198504, Russia

Synopsis In the present work we obtain the unified contribution of pair excited states to the asymptotics (at infinity in configuration space) of the three-body absolutely continuous spectrum eigenfunctions. Note that it is the case that is most interesting and rich from the physical point of view, as it is connected in particular with re-clustering of charged quantum complexes.

The unified contribution of the pair Coulomb excited states to the scattering state asymptotics in the scattering problem of three three-dimensional charged quantum particles for large values of $x, y,\ y \gg x,\ R = \sqrt{|\alpha| x}$ ($\mathbf{z} = (\mathbf{x}, \mathbf{y})^t$ is the standard vector of Jacobi coordinates connected with the pair of particles interacted by the Coulomb pair potential of attraction, $\mathbf{q} = (\mathbf{k}, \mathbf{p})^t$ is the corresponding vector of momenta, α is the constant proportional to the multiplication of charges in a separated pair of particles) is the following:

$$\Psi_c(\mathbf{z}, \mathbf{q}) \sim U(\mathbf{y}, \mathbf{q}, -\hat{x}) \frac{1}{4\sqrt{\pi} R^{5/2}}$$

$$\times \left(\hat{\Upsilon}(R) + \hat{\Upsilon}^*(R)\right) \left((1 + O\left(\frac{1}{R}\right)\right).$$

Here

$$\hat{\Upsilon}(R) \equiv -ie^{2iR} \int_0^\infty \frac{dte^{-t/3}}{C(R) - D \ln t},$$

$B = -0.310 \pm 0.001, D = -1.26 \pm 0.001$ are the real values defined in terms of the solution of some model problem, while the function $C(R)$ is defined below:

$$C(R) = B + D\left(3 \ln R - 2 \ln 2 + i\frac{\pi}{2}\right).$$

The function $U(\mathbf{y}, \mathbf{q}, -\hat{x})$ is a certain known smooth function of its arguments.

The main term of this expression can be presented as

$$\Psi_c(\mathbf{z}, \mathbf{q}) \sim \frac{3}{2\sqrt{\pi}} U(\mathbf{y}, \mathbf{q}, -\hat{x}) \frac{\sin(2R)}{C(R) R^{5/2}}.$$

At that the contributions of such a type generated by different pair subsystems (with the Coulomb pair potentials of attraction) should be constructed independently. These contributions become essentially important in different asymptotic domains of configuration space.

References

[1] Budylin A M, Koptelov Ya Yu and Levin S B 2017 *Zap. Nauch. Sem. POMI* **461** 65-94

*E-mail: s.levin@spbu.ru
†E-mail: a.budylin@spbu.ru
‡E-mail: kopya239@yandex.ru

Electron-impact excitation of forbidden and allowed transitions in Cr II

S S Tayal[1] **and O Zatsarinny** [2*]

[1]Department of Physics, Clark Atlanta University, Atlanta, GA 30314, USA
[2]Department of Physics and Astronomy, Drake University, Des Moines, Iowa 50311, USA

Synopsis The B-spline R-matrix method is used to investigate the electron-impact excitation of Cr II. The close-coupling expansion contains 512 fine-structure levels of Cr II and includes all levels of the $3d^5$, $3d^44s$, $3d^34s^2$, $3d^44p$ and $3d^34s4p$ configurations. The present results considerably expand the existing data sets for Cr II, allowing a more detailed treatment of the available measured spectra from different space observatories.

The radiative and collision rates of Cr II are of significant importance for the analysis of stellar and nebular spectra. Until recently there was an apparent lack of the collision data available in the literature for the electron-impact excitation of Cr II. Extensive R-matrix calculations were carried out by Wasson et al. [1,2] who reported data for selected transitions between levels of the $3d^5$, $3d^44s$, and $3d^44p$ configurations. The purpose of the present work is to perform more elaborate and more extensive calculations for the electron scattering from Cr II to provide independent test of the existing data sets and to obtain more realistic estimations for accuracy of collision parameters. We extend the scattering model by including all fine-structure transitions among the 512 levels arising from the principal $3d^44s$, $3d^5$, $3d^44p$, $3d^34s^2$, and $3d^34s4p$ configurations.

The present calculations were performed with the BSR codes [3] which are based on the R-matrix method and employ B-splines as a basis for the representation of the continuum orbitals. The BSR codes were considerably extended and modified in last two years in order to deal with such extremely complicated atomic systems as iron-peak elements. The use of non-orthogonal orbital sets allowed us to generate a more accurate description of the target states than those used in previous collision calculations. In particular, we use extensive multi-configuration expansions with carefully chosen configurations and, as a distinguish feature of the present calculations, we employ the term-dependent one-electron orbitals in configurations with different occupation in the 3d-subshells. To further improve the accuracy of our final collision rates, experimental energies through the fine-tuning process have been used, not only for target level energies, but also to enhance the accuracy of the term-mixing coefficients in the wave functions.

The comparison of the present effective collision strengths for the electric dipole-allowed transitions with R-matrix calculation [2] is given in Fig.1 The overall agreement is very reasonable considering the complexity of the system. Some discrepancies can be explained on the basis of more accurate target wave functions used in our calculations. Comparison between these two extensive independent calculations carried out with different computational methods allows us to make conclusive assessment of the accuracy of the effective collision strengths in these data sets. Overall, our results are estimated to be accurate to 30% or better for most of the transitions, with lower accuracy for transitions between closely lying levels.

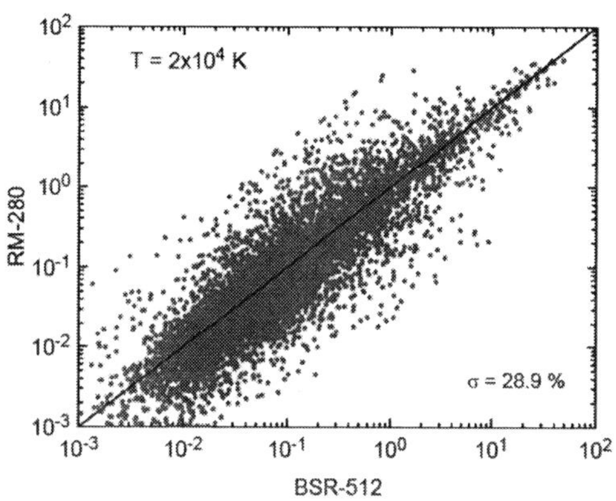

Figure 1. Global comparison of effective collision strengths obtained in the present BSR-512 model with the RM-280 calculations of Wasson et al. [1,2].

This work was supported by the US National Science Foundation under grants AST-1714159 and PHY-1520970, and the XSEDE allocation PHY-090031.

References

[1] Wasson I R *et al* 2010 *A&A* **524** A35
[2] Wasson I R *et al* 2011 *ApJS* **196** 24
[3] Zatsarinny O 2006 *Comp. Phys. Commun.* **174** 273

* E-mail: oleg.zatsarinny@drake.edu

K-shell Photoionization of the N^+, NH^+ and NH_2^+ ions

B M McLaughlin[1] *, J P Mosnier[2] †, E T Kennedy[2], E Sokell[3], J M Bizau[4], D Cubaynes[4], S Guilbaud[4] and S Carniato[5]

[1]Centre for Theoretical Atomic, Molecular and Optical Physics, School of Mathematics and Physics, Queen's University of Belfast, Belfast, BT7 1NN, United Kingdom
[2]School of Physical Sciences and National Centre for Plasma Science and Technology, Dublin City University, Glasnevin, Dublin 9, Ireland
[3]School of Physics, University College Dublin, Belfield, Dublin 4, Ireland
[4] Institut des Sciences Molculaires dOrsay, CNRS, Universit ParisSud, and Universit Paris-Saclay, F-91405 Orsay, France and Synchrotron SOLEIL, LOrme des Merisiers, Saint-Aubin, BP 48, F-91192 Gif-sur-Yvette Cedex, France
[5]Sorbonne-Universit, Laboratoire de Chimie Physique-Matire et Rayonnement (LCPMR) - UMR 7614, 4 Place Jussieu, 75005 Paris, France

Synopsis Single and double photoionization cross-sections in the photon region straddling the nitrogen K-edge and up to photon energies of ~450 eV, for the atomic N^+ and molecular NH^+ and NH_2^+ species were measured at the SOLEIL radiation facility in Orsay, France. The measurements are compared with theoretical estimates.

Inner-shell photoionization of positive ions is a key process in various laboratory and astrophysical plasmas and interstellar clouds near star-forming regions. Ionized fragments created by the photoionization processes influence the dynamics and species abundance of these reactive gaseous media. Modelling of these processes, thus, requires the experimental knowledge of the absolute values of the photon cross-sections involved. Measurements are important to benchmark state-of-the-art theoretical calculations.

The MAIA photon-ion merged-beam apparatus at the SOLEIL synchrotron in France was used to measure single and double photoionization cross-sections in the photon region straddling the nitrogen K-edge and up to photon energies ~450 eV, for the atomic N^+ and molecular NH^+ and NH_2^+ species. This extends prior work [1] on N^+ into the Rydberg region ($1s \rightarrow np$, $n > 2$ excitations), manifesting as sharp structures between 420 and 435 eV, up to and over the K-ionization threshold.

Photoionization of the NH^+ and NH_2^+ molecular ions were measured for the first time. The main fragmentation routes following $1s$ excitation in NH^+ and NH_2^+ produced mostly the N^{2+} species. The spectral patterns are similar to N^+: Strong discrete structures near 400 eV ($1s \rightarrow 2p$ excitations), followed by a Rydberg structure in the 410 - 430 eV regions. Significant broadening effects are apparent, testimony to the influence of the molecular vibrational degrees of freedom on the inner-shell photoionization process. The N^+ experimental spectra are compared with results of ab-initio RMPS and MCDF calculations, and the molecular spectra with the configuration-interaction single (CIS) method. Figure 1 illustrates our experimental and theoretical work for the atomic N^+ ion in the region of the K-edge.

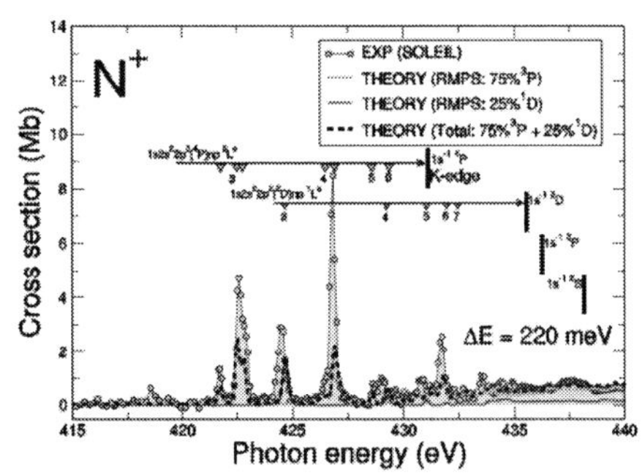

Figure 1. N^+ ions, theory and experiment, in the K-edge photon energy region.

References

[1] Ghariabeh M F *et al* 2011 *J. Phys.B.: At. Mol. Opt. Phys.* **44** 175208

*E-mail: bmcl104@gmail.com
†E-mail: jean-paul.mosnier@dcu.ie

Study of electron impact fine-structure excitation and polarization of highly charged tungsten ions

N Shukla[1*] and R Shrivastava[1†]

[1]Indian Institute of Technology Roorkee (IITR), Roorkee, India-247667

Synopsis Study of electron-impact excitation of few highly charged tungsten ions viz. Tm-, Er- and Ho-like has been done using relativistic distorted wave theory (RDW). The dipole allowed transitions of these ions are considered and electron-impact excitation cross section are reported in the wide range of energy. These calculated cross section are also fitted with suitable analytical expressions for the direct use in plasma modelling. Furthermore, the polarization of the emitted photons on consequent decay of the excited tungsten ions are calculated by utilizing density matrix theory.

Spectroscopic studies of tungsten and its ions have become hot topic for researcher as the divertor region of the world's largest tokamak i.e. International Thermonuclear Experimental Reactor (ITER) [1] is planned to be made up of tungsten. Because of its exceptional chemical and physical properties it is preferred over the other materials. The ITER device will have a very high temperature gradient from divertor to core region and due to this temperature gradient the tungsten can be stripped to its distinct ionic states and can get further excited to several fine structure states and there after decay by photon emissions. To analyze spectroscopically the emitted lines and know about the details of the ITER plasma, the accurate and complete set of electron-impact excitation cross section data are required in the wide range of incident electron energy. These results are the prime ingredient for the diagnostics of ITER plasma.Therefore, in the present work our aim is to study the electron-impact excitations of few highly charged tungsten ions and report their excitation cross sections as well as polarization of their subsequent decay photon emissions.

Recently Clementson *et al.* [2] conducted spectroscopic measurements of Tm-, Er- and Ho-like tungsten ions at Sustained Spheromak Physics Experiment (SSPX) in Livermore. They have identified few transitions which give intense lines from W^{5+}, W^{6+} and W^{7+} tungsten ions. We have performed the calculations of the electron-impact excitation cross sections for such transitions as there are no previous theoretical or experimental reports of the results of the cross section and polarization available for their identified lines. We have used fully relativistic distorted wave theory (RDW) [3] to describe the electron-impact excitation process. The bound state wave functions of tungsten ions in their initial and final states are obtained in the multi-configuration Dirac-Fock approximation by using GRASP2k code [4]. To ensure the accuracy of the evaluated wave functions of the different tungsten ions, we have calculated the excitation energies and oscillator strengths for different transitions considered through GRASP2k code and compared with the available results of Clemenston *et al.* [2]. On comparing our results with their values we find good agreement and thus fixed our bound state wave functions of the tungsten ions. Further, through these obtained bound state wave functions the distortion potential is calculated and thereafter projectile distorted wave functions in initial/final channel are obtained by numerically solving the Dirac equations. Finally, we evaluated the T- matrix for each transition and calculated the magnetic sublevel as well as total cross section of the different transitions of W^{5+} to W^{7+} tungsten ions. Furthermore, we have employed the density matrix theory [3,5] and magnetic sublevel cross sections to calculate the linear polarizations of the emitted photons due to decay of the excited tungsten ions. To facilitate the plasma modelling calculations the obtained data of cross sections are fitted with multi parameter analytical functions.

References

[1] Ikeda K 2010 *Nucl.Fusion* **50** 014002
[2] Clementson *et al* 2010 *J. Phys. B.* **43** 144009
[3] Sharma L *et al* 2011 *Phys.Rev.A* Phys. **83** 1
[4] Jönsson P *et al* 2013 *Comput.Phys Commun.* **184** 2197
[5] Sharma L *et al* 2010 *Phys.Rev.A* **81** 1

* E-mail: phyneelam@gmail.com
† E-mail: rajsrfph@iitr.ac.in

On the effect of cascade transitions after electron-impact excitation of Zn by spin-polarized electrons

K Bartschat[1]*, O Zatsarinny[1], C J Bostock[2], D V Fursa[2], I Bray[2] and A N Grum-Grzhimailo[3]

[1]Department of Physics and Astronomy, Drake University, Des Moines, Iowa 50311, USA
[2]Institute for Computation and Department of Physics, Curtin University, Perth, 6102 WA, Australia
[3]Skobeltsyn Institute of Nuclear Physics, Lomonosov Moscow State University, Moscow 119991, Russia

Synopsis We investigate the possible effect of cascade transitions from the $(4s5p)^3P_{0,1,2}$ states to the $(4s5s)^3S_1$ state of Zn. Our calculations suggest that the excitation cross sections and the polarization of the cascade radiation cannot explain the serious controversy between two existing experimental datasets below the cascade threshold and raise additional questions regarding the measurements at higher energies.

In a recent paper, Clayburn and Gay [1] reported their measurements of the angle-integrated linear polarization P_2 in the $(4s5s)^3S_1 \to (4s4p)^3P_0$ transition in Zn after impact excitation by a transversally spin-polarized electron beam. With the light detector placed at right angle to the incident beam direction along the direction of the electron polarization P_e, P_1 and P_2 are the linear polarizations for $(0°, 90°)$ and $(45°, 135°)$ transmission, respectively, while P_3 is the circular polarization. P_1 is independent of P_e, while P_2 and P_3 are proportional to P_e.

Clayburn and Gay noted significant disagreement between their $P_2/P_e \approx 0$ data and the $P_2/P_e \approx -10\%$ measurements of Pravica *et al.* [2] in the cascade-free region of incident electron energies below ≈ 7.6 eV, while relatively good agreement was found above that threshold. According to Bartschat and Blum [3], one would expect $P_1 \approx 0$, $P_2/P_e \approx 0$, and $P_3/P_e \approx -1$ in the cascade-free region. Above the cascade threshold, significant deviations are only possible for sufficiently large excitation cross sections and strong polarization of the cascade radiation.

We performed fully relativistic convergent close-coupling (RCCC-94) and semi-relativistic B-spline R-matrix (BP-BSR-43) calculations with 94 and 43 states, respectively. As seen in Fig. 1, accounting for cascades indeed reduces the magnitude of the circular polarization, in agreement with the measurements. However, we find $P_2/P_e \approx 0$ at all incident energies considered. Our results agree with general theory [3] but are in strong contradiction with both experimental datasets above the cascade threshold.

*E-mail: klaus.bartschat@drake.edu

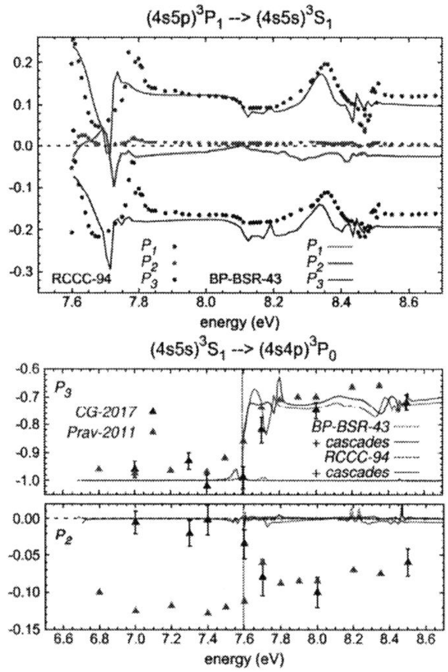

Figure 1. P_1, P_2/P_e, and P_3/P_e for the $(4s5p)^3P_1 \to (4s5s)^3S_1$ cascade transition (top), as well as P_3 (center) and P_2 (bottom) for $(4s5s)^3S_1 \to (4s4p)^3P_0$ in Zn. CG-2017 [1]; Prav-2011 [2].

Work supported by the United States National Science Foundation and the Australian Research Council.

References

[1] Clayburn N B and Gay T J 2017 *Phys. Rev. Lett.* **119** 093401
[2] Pravica L *et al* 2011 *Phys. Rev. A* **83** 040701
[3] Bartschat K and Blum K 1982 *Z. Phys. A* **304** 85

R-matrix electron-impact excitation data for the C-like isoelectronic sequence

J Mao[1,2]* and N R Badnell[1]

[1]Department of Physics, University of Strathclyde, Glasgow G4 0NG, UK
[2]SRON Netherlands Institute for Space Research, Sorbonnelaan 2, 3584 CA Utrecht, The Netherlands

Synopsis Emission and absorption features from C-like ions are served as temperature and density diagnostics of astrophysical plasmas. R-matrix electron-impact excitation data for C-like ions in literature cover merely a handful of ions. Here we present a systematic intermediate coupling frame transformation R-matrix calculations for C-like ions from N II to Kr XXXI. The effective collision strengths cover a wide range of temperature. Our results are compared with archival databases and literature results.

Emission and absorption features from C-like ions are served as temperature and density diagnostics of astrophysical plasmas [1]. R-matrix electron-impact excitation data for C-like ions in literature cover merely a handful of ions.

Here we present a systematic intermediate coupling frame transformation R-matrix calculation [2] for C-like ions from N II to Kr XXXI (i.e., N^+ to Kr^{30+}). The effective collision strengths over a wide temperature range $z^2(10^1,\ 10^7)$ K, where z is the ionic charge plus one.

The systematic calculation includes a total of 590 fine-structure levels in both the configuration interaction target and close-coupling collision expansion, following the case study of Fe XXI [3]. These levels arise from 24 configurations with $n \leq 4$ and 3 configurations $2s^2\ 2p\ 5l$ with $l = 0 - 2$. We used the AUTOSTRUCTURE code [4] to calculate the target structure.

The R-matrix calculation consists of an inner and outer region calculation. The inner region calculation is further split into an exchange and non-exchange calculation. The exchange calculation included angular momenta up to $2J = 23$, while the non-exchange calculation extended to $2J = 77$. For higher angular momenta up to infinity, we use the top-up formula of the Burgess sum rule [5] for dipole allowed transitions, and a geometric series for the remaining non-forbidden transitions [6].

The outer region calculation has three parts. A fine energy mesh (3600–30000 points in total, increasing with the ionic charge) is used between the first and last thresholds for the exchange calculation in the resonance region. A coarse energy mesh (about 1000 points) is used from the last threshold up to $3I_p/z^2$, where I_p is the ionization potential, to determine a smooth background of the exchange calculation. A coarse mesh (about 1400 points) is used from the first threshold up to $3I_p/z^2$ for the non-exchange calculation.

Dipole and Born limits are used from the last calculated energy point up to infinite to complete the Maxwellian-averaged effective collision strengths [7].

We compare present results of a few ions with archival database or results in literature [8, 9]. The comparison covers energy levels, transition rates, and effective collision strengths.

The electron-impact excitation data will be archived according to the Atomic Data and Analysis Structure (ADAS) data class adf04. The data will also be incorporated into the spectral codes like CHIANTI [10] for plamsa diagnostics.

References

[1] Mao J 2017 *A&A* **607** 100
[2] Griffin D C 1998 *et al J. Phys. B* **31** 3713
[3] Fernndez-Menchero L 2016 *J. Phys. B* **49**, 085203
[4] Badnell N R 2011 *Comp. Phys. Comm.* **182** 1528
[5] Burgess A 1974 *J. Phys. B* **7** L364
[6] Badnell N R 2001 *et al J. Phys. B* **34** 5071
[7] Burgess A 1992 *Astron. Astrophys.* **254** 436
[8] Liang G 2011 *et al Astron. Astrophys.* **533** 87
[9] Tayal K P 2017 *et al Astrophys. J.* **850** 147
[10] Dere K P 1997 *et al Astron. Astrophys.* **125** 149

*E-mail: junjie.mao@strath.ac.uk

Electron impact excitation of highly charged Ge-Like Xe ion

S Bharti[1] *, L Sharma[1] † and R Srivastava[1]

[1]Department of Physics, Indian Institute of Technology Roorkee, Roorkee, 247667, India

Synopsis A systematic study of electron impact excitation of Ge-like highly charged xenon ions have been carried out in Extreme Ultraviolet (EUV) range within fully relativistic multiconfiguration Dirac-Fock framework. The cross section calculation reported here are performed for dipole allowed transitions for Xe^{22+} ions from ground state $4s^24p^2$ to $4s^24p^3$ and $4s^24p4d$ excited fine structure levels. The analytic fitting for the cross section data for incident electron energy range within 50 to 1500 eV have been provided for further applications in various plasma diagnostic techniques.

Highly charged ions (HCI) have emerging applications in various field of science and technologies such as plasma physics to diagnose and model the plasma properties for fusion devices [1, 2]. The extreme ultraviolet (EUV) radiation emitted by the highly charged ions carry information about plasma parameters and impurities in plasmas. Moreover, EUV radiation is expected to be important in the development of the next generation lithography, which is extremely important for further improvement of semiconductor devices. Since in plasmas, electron impact excitation of ions is one of dominant processes, it would be essential to use excitation cross sections for plasma diagnosis.

In the present work, we have considered electron collisions with highly charged Ge-like Xe ions. These ions can be used in the International thermonuclear Experimental Reactor (ITER) for cooling the edge plasma. The available Experimental data for HCI is very limited, therefore, role of reliable theoretical methods is very crucial. We have used fully relativistic distorted wave method to calculate cross sections for all the 69 dipole allowed fine-structure transitions arising from excitation of $4s4p^3$ and $4s^24p4d$ levels from $4s^24p^2$ in Ge-like Xe ions in the incident electron energy range from excitation threshold to 1.5 keV. Such a study is challenging as Xe^{22+}, being open-shell and heavy ion, has many fine-structure levels with appreciable splitting.

In order to obtain the cross sections, the bound state wavefunctions are obtained within the multi-configuration Dirac-Fock approach. Since the accuracy of the atomic wavefunctions determines the validity of the cross sections, we have compared our calculated excitation energies and oscillator strengths for the above fine-structure transitions with the available other theoretical results [3, 4]. The projectile and scattered electron wavefunctions are obtained using distortion potential as spherically averaged static potential of the ion in its initial and final state, respectively. As an example, we have plotted in Fig-1 the cross sections for excitation of Xe ions from $4s^24p^2$ $^3P_0$ to excited $4s^24p^3$level.

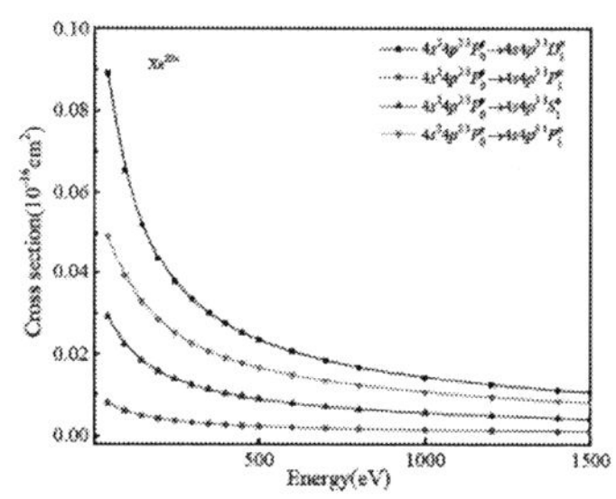

Figure 1. Integrated cross sections for electron excitation of Xe^{22+} ion: for transition from ground state $4s^24p^2$ $^3P_0$ to excited $4s^24p^3$level.

Analytic fitting of the calculated cross sections data are also done for providing cross-sections at any desired energy in plasma applications. The details of the work will be presented at the conference.

References

[1] Gilaspy J D 2001 *J. Phys. B* **34** R93
[2] Liedahl D A 2001 *Spectroscopic Challenges of Photoionized Plasmas, ASP Conference Series 247* 417
[3] Nagy O *et al* 2012 *At. Data Nucl. Data Tables* **98** 373
[4] Hao L H *et al* 2017 *J Appl Spectros* **84** 337

* E-mail: sbharti@ph.iitr.ac.in
† E-mail: lalitfph@iitr.ac.in

The generalized two-photon excitation cross-section of Xe revisited

C Blondel*, J Alkhoury, T Jannaud and C Drag†

Laboratoire de Physique des plasmas, École polytechnique, Centre national de la recherche scientifique, université Paris-sud, Sorbonne Université, Observatoire de Paris, route de Saclay, F-91128 Palaiseau, France

Synopsis Two-photon absorption laser-induced fluorescence (TALIF) is a technique commonly used for density and/or temperature diagnostics, mainly of oxygen atoms, in flames and plasmas. Usually, density measurements are based on the comparison with the signal obtained, at a similar wavelength, in pure xenon. The cross-section of xenon thus appears a key parameter, the measure of which has not been consolidated by independent measurements. The present experiment is a direct measurement of this two-photon cross-section.

Laser diagnostics in plasmas or flames frequently resorts to the so-called TALIF technique, which consists in recording the fluorescence that follows two-photon excitation of an atomic species. The density of oxygen atoms in plasmas can for instance be monitored via the 3p $^3P \rightarrow 3s\ ^3S^0$ 844 nm fluorescence that follows two-photon excitation of the 3p ^3P level from the $2p^4\ ^3P$ ground term, at the wavelength λ=226 nm. A widespread protocol is to compare the fluorescence yield to that measured in pressure-controlled Xe vapor, when illuminated by the same optical system. Xe atoms can be excited at the nearby wavelength of 224 nm to a $6p'[3/2]_2$ state, then de-excited by spontaneous emission to the $6s'[1/2]_1$ level at the wavelength λ=835 nm, in an atomic scheme quite similar to the one implemented in oxygen, which makes dealing with calibration issues easier, both for the exciting light and for fluorescence detection.

Quantitative analysis, however, relies on an assumed ratio of the two two-photon cross-sections $\sigma^{(2)}$ involved, namely $\sigma^{(2)}(Xe)/\sigma^{(2)}(O)$. Yet this ratio was measured only once [1] and only the cross-section of oxygen $\sigma^{(2)}(O)$ has been measured [2] or calculated [3] directly.

A direct measurement of the cross-section $\sigma^{(2)}(Xe)$ of Xenon has thus appeared desirable. We made it using an injected single-mode Ti:Sapphire laser [4], which gives perfect control of the photon statistics and assures a good reproducibility of the results. An example of a spectrum obtained in the vicinity of the $6p'[3/2]_2$ resonance is given by figure 1. The corresponding two-photon excitation cross-section of the $6p'[3/2]_2$ level has been found roughly twice smaller than what had been admitted previously. Results on other resonances will be shown at the conference.

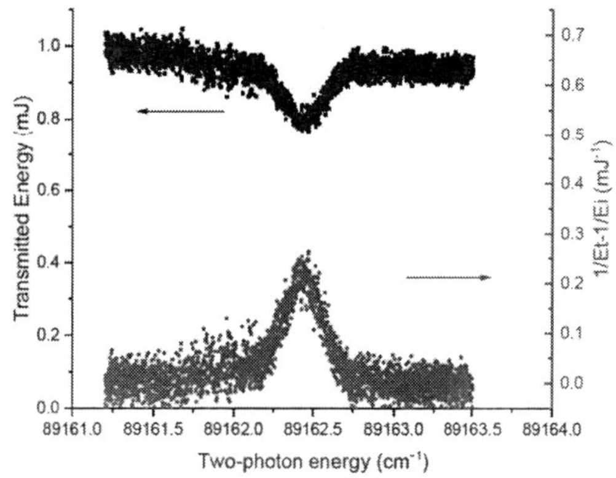

Figure 1. Two-photon absorption spectrum around the $6p'[3/2]_2$ resonance of Xe. The cell length is 51 cm and the pressure 8.61 kPa. The laser pulse duration is about 6 ns and the beam waist about 350 μm. Left axis: transmitted energy. Right axis: 1/(transmitted energy) - 1/(incident energy).

References

[1] Niemi K, Schulz-von der Gathen V and Döbele H F 2005 *Plasma Sources Sci. Technol.* **14** 375

[2] Bamford D J, Jusinski L E and Bischel W K 1986 *Phys. Rev. A* **34** 185

[3] Saxon R P and Eichler J 1986 *Phys. Rev. A* **34** 199

[4] Lottigier P, Jucha A, Cabaret L, Blondel C and Drag C 2019 *Appl. Phys. B* **125** 14

* E-mail: christophe.blondel@lpp.polytechnique.fr
† E-mail: cyril.drag@lpp.polytechnique.fr

Development of accurate anharmonic vibrational state theory using a novel vibrational coordinate based on backflow transformation

K Ishii[1 *], M Tashikawa[1,2] and Y Kita[1†]

[1]Graduate School of Nanobioscience, Yokohama City University, Yokohama, 236-0027, Japan
[2]Shool of Data Science, Yokohama City University, Yokohama, 236-0027, Japan

Synopsis The anharmonic vibrational state theory based on quantum Monte Carlo method is one of the most accurate methods to solve a vibrational Schrödinger equation of molecular systems. In anharmonic vibrational state analyses, one should take account of the inter-mode couplings, which cannot be included by the conventional normal mode coordinates only. In this study, thus, we have developed a novel vibrational coordinate based on backflow transformation which enable us to take correlation effects in quantum many-body systems directly.

The vibrational analysis using a first principle calculation is necessary to assign peaks in IR spectra, and to determine molecular geometries. The anharmonic vibrational state theory based on quantum Monte Carlo (VQMC) method is one of the most accurate methods to solve a vibrational Schrödinger equation of molecular systems. In this method, an accurate trial wave function being able to describe molecular vibrations precisely is indispensable to obtain reliable theoretical predictions for vibrational frequency, molecular geometry, *etc.* Vibrational configuration interaction (VCI) method is known to generate an accurate trial wave function, however, its applicable scope is restricted to small molecular species.

In this study, thus, we have developed a novel vibrational coordinate based on backflow transformation (BF) [1], which enable us to take correlation effects in quantum many-body systems directly, in order to generate an accurate vibrational trial wave function. The theoretical accuracy of BF in quantum vibrational problems has been tested for the vibrational ground states of H_2O, HCN, and H_2CO molecules. The variational Monte Carlo method [2], which is one of the QMC methods, was used to analyze the molecular properties of the ground state. The backflow transformation was applied to vibrational self-consistent field (VSCF) wave function. For comparing, we also calculated using Reptation Monte Carlo (RMC) method [3], which is able to calculate the exact eigenvalue and eigenfunction at the vibrational ground state numerically.

Figure 1 shows the zero-point energy (ZPE) of H_2O molecule with several vibrational methods. The ZPE decreases by the introduction of BF, and ZPE with BF is much lower than that with VCI. Since ZPE is a variational energy, these results mean that the introduction of BF is effective for the improvement of vibrational trial wave function, and the BF calculation is more accurate than the conventional VCI calculation. Furthermore, the ZPE result with BF method is quite consistent with the exact ZPE value with RMC method. We also got similar results for HCN and H_2CO molecules. The other details will be discussed on poster.

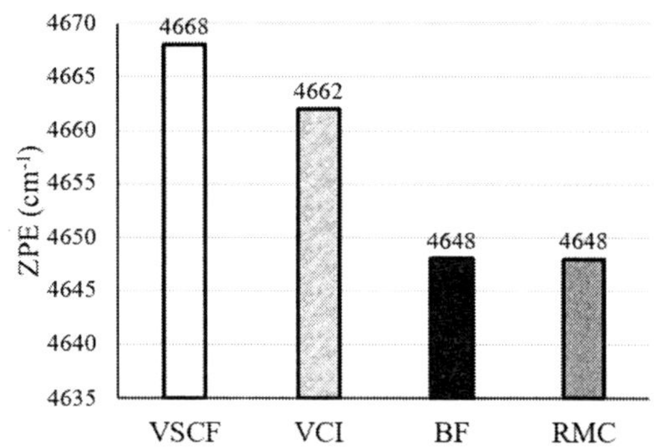

Figure 1. The zero-point energy (ZPE) of H_2O with several wave functions and methods.

References

[1] Lopéz Ríos P *et al.* 2006 *Phys. Rev. E* **74** 066701
[2] Needs R J *et al* 2010 *J. Phys. Condens. Matter* **22** 023201
[3] Toulouse J *et al* 2007 *J. Chem. Phys.* **126** 084102

* E-mail: n175202g@yokohama-cu.ac.jp
† E-mail: ykita@yokohama-cu.ac.jp

Electron emission from CH_4 molecule induced by 5.5 MeV/u bare C ions

A Mandal[1], C Bagdia[1], M Roychowdhury[1], J M Monti[2], R D Rivarola[2] and L C Tribedi[1*]

[1]Department of Nuclear and Atomic Physics, Tata Institute of Fundamental Research, Colaba, Mumbai, 400005, India
[2]Institute de Fisica Rosario (CONICET-UNR), Universidad Nacional de Rosario, 2000 Rosario, Argentina

Synopsis We present the experimental and theoretical study of electron emission from methane (CH_4) in collisions with 5.5 MeV/u bare carbon projectiles. We have measured double differential cross section of electron emission and compared the data with the prior form of the continuum distorted wave – eikonal initial state (CDW-EIS) model calculation. The CDW-EIS calculations are in very good agreement with the data.

The measurement of double differential cross section (DDCS) of electron emission have importance in understanding the basic collision processes and as well as in areas such as, astrophysics and radiation biology. Moreover, such studies provide a test to the validity of the available theoretical frameworks. Methane (CH_4), the simplest hydrocarbon, has significant presence in both the interstellar medium and in the solar system. It is one of the most abundant species in the atmospheres of many planets [1]. Although there are few experimental investigations on the ionization of CH_4 upon electron and proton impact reported in the literature [2, 3], studies on interaction of CH_4 with fast heavy ions is scarce. The case of heavy ions provides the stringent test to the theoretical models as the perturbation strength is large.

In the present study, we measure the DDCS of electron emission from CH_4 molecule upon 5.5 MeV/u bare C ion impact. The C ions collide with the CH_4 targets in a scattering chamber that is flooded with the CH_4 gas at a constant pressure. The ejected electrons are detected using an electrostatic hemispherical analyzer setup with a channel electron multiplier (CEM) detector [4]. Electron yields are measured in the range from 1 eV to 400 eV at 12 different emission angles.

Energy and angular distribution of electron emission is compared with the prior form of the continuum distorted wave-eikonal initial state (CDW-EIS-prior) model calculations [5]. The values of DDCS at emission angles of 45° and 135° are plotted in Fig. 1. The KLL Auger emission peak of C appearing in the experimental spectrum at ~240 eV is marked in the Figure. The CDW-EIS calculations are in excellent agreement with the measured DDCS values (Fig. 1). Similar agreement is also found for the DDCS data at other angles. The detailed results of comparisons of measured and calculated cross sections and discussions will be presented.

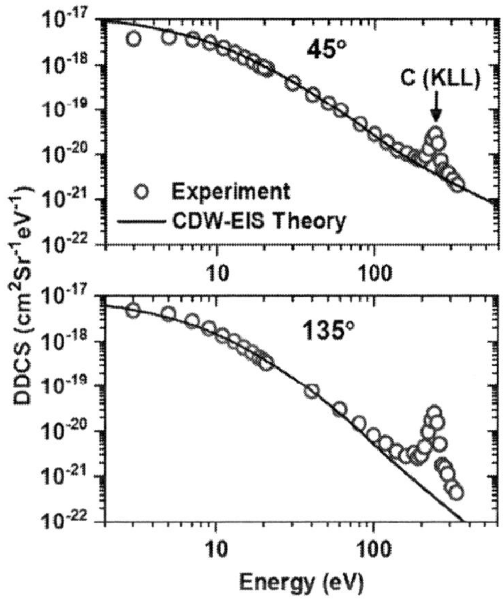

Figure 1. DDCS spectrum of electron emission from CH_4 at angles of 45° and 135°. Solid line: the theoretical values from CDW-EIS prior calculations.

References

[1] Formisanoet V *et al* 2004 *Science* **306** 1758
[2] Dogan M *et al* 2013 *J. Spec.* **2013** 192917
[3] Tachino C A *et al* 2013 *J. Phys. Conf. Ser.* **2014** 102029
[4] Roy Chowdhury M *et al* 2017 *J. Phys. B* **50** 155201
[5] Fainstein P D *et al* 1996 *Nucl. Instr. Meth. Phys. Res. B.* **107** 19

* E-mail: lokesh@tifr.res.in

Impact parameter sensitive study of inner-shell atomic processes in Xe^{54+}, $Xe^{52+} \rightarrow Xe$ collisions

A Gumberidze[1]*, C Kozhuharov[1], R T Zhang[1,2], S Trotsenko[1], Y S Kozhedub[3], R D DuBois[4], H F. Beyer[1], K-H Blumenhagen[5,6], C Brandau[1,7], A Bräuning-Demian[1], W Chen[1,8], O Forstner[5,6], B Gao[1,9], T Gassner[1], R E Grisenti[1,10], S Hagmann[1,10], P-M Hillenbrand[1], P Indelicato[11], A Kumar[12], M Lestinsky[1], Yu A Litvinov[1], N Petridis[1], D Schury[1,13], U Spillmann[1], C Trageser[1,7], M Trassinelli[13], X Tu[1,9] and Th Stöhlker[1,5,6]

[1]GSI Helmholtzzentrum für Schwerionenforschung, 64291 Darmstadt, Germany
[2]Columbia Astrophysics Laboratory, Columbia University, New York, NY 10027, USA
[3]Department of Physics, St. Petersburg State University, St. Petersburg 198504, Russia
[4]Department of Physics, Missouri University of Science and Technology, Rolla, Missouri 65409, USA
[5]Helmholtz-Institut Jena, 07743 Jena, Germany
[6]IOQ, Friedrich-Schiller-Universität Jena, 07743 Jena, Germany
[7]Physikalisches Institut, Justus-Liebig-Universitt Gießen, 35392 Gießen, Germany
[8]Chinese Academy of Sciences, Institute of High Energy Physics, Dongguan 523803, China
[9]Institute of Modern Physics, Chinese Academy of Sciences, 730000, Lanzhou, China
[10]Institut für Kernphysik, Universität Frankfurt, 60486 Frankfurt am Main, Germany
[11]LKB, Sorbonne Université, CNRS, ENS-PSL Research University, Collège de France, F-75005 Paris, France
[12]Nuclear Physics Division, Bhabha Atomic Research Centre, Trombay, Mumbai 400085, India
[13]Institut des NanoSciences de Paris, CNRS, Sorbonne Université, F-75005, Paris, France

Synopsis In this work, we present a pilot experiment in the experimental storage ring (ESR) at GSI devoted to impact parameter sensitive studies of inner shell atomic processes for bare and He-like xenon ions (Xe^{54+}, Xe^{52+}) colliding with neutral xenon gas atoms. The projectile and target x-rays have been measured at different observation angles for all impact parameters as well as for the impact parameter range of ~35 - 70 fm.

In this contribution, we present an impact parameter sensitive study of inner shell atomic processes in symmetric ion-atom collisions. The measurement was performed at the Experimental Storage Ring (ESR) at GSI Darmstadt with bare and He-like xenon ions (Xe^{54+}, Xe^{52+}) colliding with neutral xenon gas atoms. This choice of the projectile charge states was made in order to compare the effect of a filled K-shell with the empty one. The beam energy (for both charge states) was 50 MeV/u. This value of the beam energy was chosen as a compromise between the adiabaticity of the collision and the reasonable beam lifetime/intensity in the ESR after deceleration. In order to obtain information concerning the impact parameter and, in particular to pick out close collisions which are especially important for observing two-center (or quasi-molecular) effects, the scattered projectile ions which had undergone close collisions with the target atoms were detected by a particle detector (plastic scintillator) mounted in a specially constructed movable pocket at ~3.5 m downstream from the target. In addition to the detector for the scattered projectiles, the x-rays emitted from the interaction zone were observed by an array of semiconductor and scintillator detectors mounted at different angles with respect to the ion beam direction. The physical processes leading to the x-ray emission are: excitation of the projectile/target electrons and electron capture from the neutral target into the highly-charged xenon ions. By looking at the coincidences between the x-rays and the scattered projectiles, we were able to clearly demonstrate the possibility of picking out the characteristic x-rays stemming from the close collisions only. This offers us a new observable for the atomic collision experiments at the ESR storage ring. The experimental results will be presented together with the state-of-the-art relativistic calculations [1].

*E-mail: a.gumberidze@gsi.de

References

[1] Kozhedub Y S *et al* 2014 *Phys. Rev. A* **90** 042709

 1412 (2020) 142016 doi:10.1088/1742-6596/1412/14/142016

Electron impact excitation of the 4 $^1S_0$-4 $^1P_1$ state of Zn atom

L Y Xie [1] †, X S Cheng [1] *, D H Zhang[1], J Jiang[1], Z W Wu[1] and C Z Dong[1]

[1]Key laboratory of Atomic and Molecular Physics & Functional Materials of Gansu Province, College of Physics and Electronic Engineering, Northwest Normal University, Lanzhou, 730070, P. R. China

Synopsis The electron impact excitation total and differential cross sections, Stokes parameters, and coherence parameters of Zn atom from the ground state 4^1S_0 to the excited state 4^1P_1 were systematically calculated using the fully relativistic distorted wave method. The calculated results are compared with the experimental data and other theoretical results, good agreement is obtained.

Electron impact excitation (EIE) is fundamental atomic process in astrophysical and laboratory plasmas, and relevant cross sections is needed for modelling and diagnosing of different types of plasmas. There has been a growing interest in studying of EIE process of Zn atom due to its possible application as a replacement for mercury in high-pressure discharge lamps. Recently years, though many attentions have been focused on the scattering parameters of Zn atom both experimentally and theoretically, some larger differences are still existed between theoretical and experimental results [1-2].

In this work, the EIE total and differential cross sections, Stokes parameters (P_1, P_2 and P_3), and electron impact coherence parameters (EICP) of Zn atom from the ground state 4^1S_0 to the excited state 4^1P_1 were calculated using the fully relativistic distorted wave method (RDW) [3]. Valence-valence (VV) and core-valence (CV) electron correlations are well considered for calculating the target states. There are total 90 $J^\pi = 0^+$ and 192 $J^\pi=1^-$ relativistic configuration state wave functions are included for initial state 4^1S_0 and final state 4^1P_1 with considering single, double, and triplet electron excitation of 3*d*-core, and 4s, 4p valence electron up to *nl*=6*s*, 6*p* orbitals, and EOL calculations are performed for the EIE initial and final state respectively using GRASP2K code [2]. It is found that VV and CV electron correlation have significant influence on both cross sections, and Stokes and EICP parameters. In figure 1, the Stokes and EICP (P_L, L, P^+, γ) parameters are shown for Zn atom with incident electron energy 40eV as an example. Our results are in good agreement with the experimental data and better than previous RDW calculations [1].

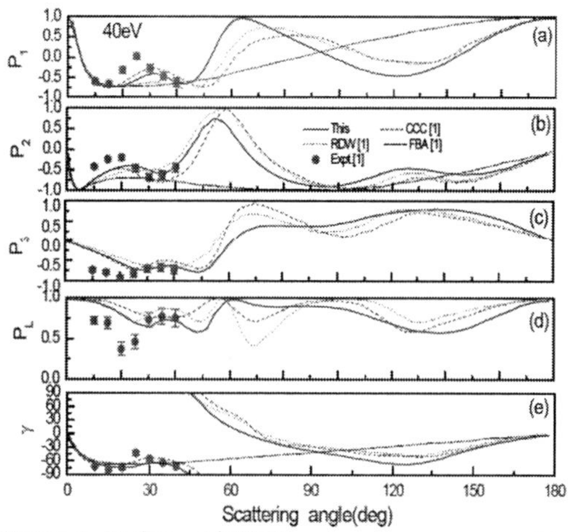

Figure 1. Stokes and EPIC parameters correspond to electron impact excitation process of Zn from the ground state 4^1S_0 to 4^1P_1 state excitation state with incident electron energy 40eV.

The work was supported by the National Key Research and Development Program of China under Grant No.2017YFA0402300, and the Natural Science Foundation of China (Grant Nos. U1530142, 11564036, 11774292, 11464042, 1874051) and the Fund of NWNU (NWNU-LKQN-15-3)

References

[1] Piwiński M *et al* 2018 *J. Phys. B* **51** 085002
[2] Clayburn N B *et al* 2017 *Phys. Rev. Lett.* **119** 03401
[3] Jiang J *et al 2008 Phys. Rev. A* **78** 022709
[4] Jönsson P *et al* 2007 *Comput. Phys. Commun* **117** 597

* E-mail: xiely@nwnu.edu.cn

Relation between the spatial shape of excited molecular orbital and molecular-frame electron scattering cross section

N Watanabe*, **S Yamada and M Takahashi**

IMRAM, Tohoku University, Katahira 2-1-1 Aoba-ku Sendai 980-8577, Japan

Synopsis We report the molecular-frame electron scattering cross sections of the $2\sigma_g \rightarrow 1\pi_g$ excitation in O_2 and N_2, obtained using an electron-ion coincidence technique. The vector correlation between the scattered electron and the fragment ion has been measured at incident electron energy of 1.4 keV. The results exhibit clear molecular-orientation dependence reflecting the nodal structure of the $1\pi_g$ orbital. It strongly suggests that the electron-ion coincidence technique can provide a powerful tool to investigate spatial shapes of unoccupied molecular orbitals.

The spatial distribution of a particular molecular orbital plays a crucial role in determining molecular properties. The high-energy electron scattering cross section may bring valuable information about molecular orbitals participating in the target excitation process, since it is related to the Fourier transform of the overlap between the electronic wavefunctions of the initial and final states. In this study we have carried out molecular-frame electron scattering experiments for single electron excitations of N_2 and O_2 [1-3] to elucidate how the scattering cross section reflects the spatial shape of the excited orbital.

For transition to a short-lived excited state followed by Auger decay, the orientation of the target molecule can be determined if the residual ion dissociates much faster than it rotates. Since the recoil direction of the fragment ion coincides with the molecular orientation, the scattering cross section can be measured in the molecular-frame by detecting not only the electron but also the ion. On this basis, electron-ion coincidence experiments were conducted using an apparatus developed in our laboratory [4].

Figure 1 shows the molecular-frame electron scattering cross section for the excitation from the $2\sigma_g$ orbital to the $1\pi_g$ orbital in O_2. It was constructed by plotting the number of the electron-O^+ coincidence events at $\theta = 8.2°$ as a function of the angle ϕ_K made between the recoil direction of O^+ and the momentum transfer vector $\boldsymbol{K}$. The angular distribution exhibits minima at $\phi_K = 0°, 90°$, and 180°. A similar angular distribution has been observed also for N_2 [1].

To get an understanding of the result we examine the transition matrix of the excitation, $\langle 2\sigma_g|\exp(i\boldsymbol{K}\cdot\boldsymbol{r})|1\pi_g\rangle$. The $2\sigma_g$ orbital is totally symmetric and its shape is rather isotropic. On the other hand, the $1\pi_g$ orbital has an anisotropic shape with a nodal plane through two O atoms. The reflection in the plane inverts the sign of the $1\pi_g$ orbital, while it leaves the $\exp(i\boldsymbol{K}\cdot\boldsymbol{r})$ term unchanged when $\boldsymbol{K}$ is parallel to the molecular axis. Thereby, the integrand in the transition matrix is odd under the reflection and the scattering cross section vanishes at $\phi_K = 0°$ and 180°. A similar consideration shows that the transition provability becomes zero for 90°. The observed minima therefore reflect the nodal structure of the $1\pi_g$ orbital, indicating that the scattering cross section exhibits a molecular-orientation dependence reflecting the shape of the orbital to which the target electron is excited.

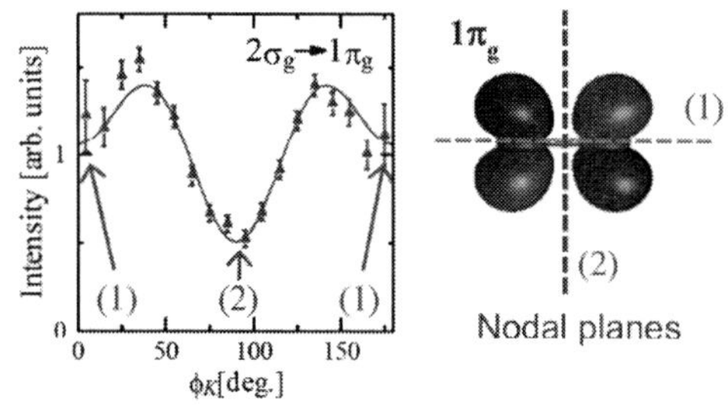

Figure 1. Molecular-frame electron scattering cross section of the $2\sigma_g \rightarrow 1\pi_g$ excitation in O_2.

References

[1] Watanabe N, Yamada S and Takahashi M 2017 *Phys. Rev. A* **95** 060702(R)

[2] Watanabe N, Yamada S and Takahashi M 2018 *Phys. Chem. Chem. Phys.* **20** 1063

[3] Watanabe N, Yamada S and Takahashi M 2019 *Phys. Rev. A* **99** 022704

[4] Watanabe N *et al* 2018 *Rev. Sci. Instrum.* **89** 043105

* E-mail: noboru.watanabe.e2@tohoku.ac.jp

Energetic processing of hydrocarbons molecular species by ionizing radiation towards the formation of cyclic species

S Indrajith[1] *, J Kočišek[2] †, A Domaracka[1], P Rousseau[1] and B A Huber[1]

[1]Normandie Université, ENSICAEN, UNICAEN, CEA, CNRS, CIMAP, Caen, 14000, France
[2]J. Heyrovský Institute of Physical Chemistry v.v.i., the Czech Academy of Sciences, Prague, 18223, Czech Republic

Synopsis Here we report on experimental studies of the collisions between butadiene clusters and three different ionizing radiation sources: 70 eV electrons, 270 nm photons and 3 keV Ar^+ ions. New molecular species are produced by growth mechanism inside of clusters.

Cassini-Huygens mission brought interesting results concerning one of Saturns biggest satellites, Titan. In fact, it has been shown that this moons typical orange-brownish haze layers could be due to the presence of aerosols. These aerosols are essential for Titans climate, radiation balance and atmosphere chemistry. The formation of such aerosols is a complex process ranging from the molecular scale to large particles scale. Polycyclic Aromatic Hydrocarbons (PAHs) and their clusters are important components in terrestrial soot formation. Given this, their presence in Titans atmosphere can be considered as a nucleation agent, which triggers aerosol formation in Saturn moons low temperature atmosphere. However, since it is mainly composed of nitrogen, methane and a small amount of more complex hydrocarbons, understanding the way such PAHs are formed is mandatory[1].

Ionizing radiation sources (photons, electrons and ions) can trigger gas-phase polymerization, ion-molecule reactions and nanoparticle grain surfaces catalysis. These processes contribute to the formation of complex organic synthesis pathways in atmosphere and space. Simple ionization may induce chemical reactions and isomerization processes.

To understand them, simple linear hydrocarbons like butadiene (a non-aromatic molecule) have been chosen to study the formation of more complex cyclic species (aromatic). The irradiation of hydrocarbons clusters has been performed using different radiation sources (electrons, UV photons and ions)(Fig.1). This allowed the determination of the balance between ionization induced reactivity and collision induced one. The latter is the typical collision occurring in Titans atmosphere involving slow and heavy ions (few keV O cations). It has been shown that this triggers reactivity, as observed in C_{60} clusters [2] or PAH ones [3]. These experiments were perfomed at ARIBE, the low energy ion beam facility of GANIL in Caen and in J.Heyrovský Institute of Physical Chemistry in Prague.

*E-mail: suvasthika.indrajith@ganil.fr
†E-mail: jaroslav.kocisek@jh-inst.cas.cz

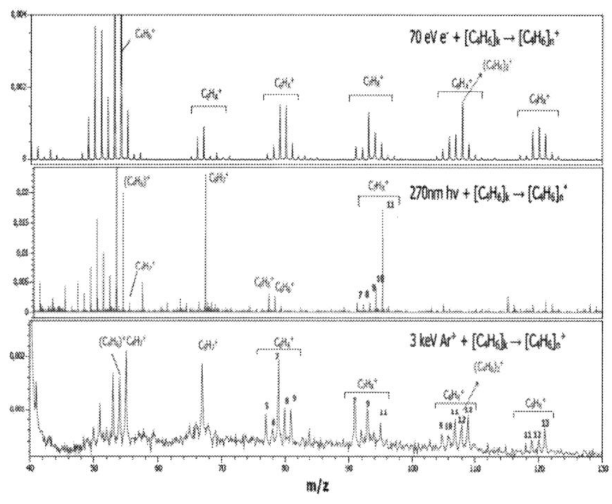

Figure 1. Mass spectra of butadiene clusters in collision with electrons, photons and ions showing growth products between the monomer and the dimer.

References

[1] Long Z *et al* 2018 *Nature Astronomy* **2** 973
[2] Delaunay R *et al* 2018 *Carbon* **129** 766
[3] Delaunay R *et al* 2015 *J. Phys. Chem. Lett.* **6** 1536

Nonlocal treatment of resonant vibrational excitation in electron-dioxygen collision

V Alt* **and K Houfek**†

Charles University, Faculty of Mathematics and Physics, Institute of Theoretical Physics, V Holešovičkách 2, Prague, 180 00, Czech Republic

Synopsis The cross sections for vibrational excitations of O_2 through collisions with electron are calculated using the nonlocal discrete-state-in-continuum model and its local approximation and compared to experimental results.

Our interest lies in investigating nuclear dynamics of diatomic molecules in resonant collisions with electrons. This work specifically aims at calculating the cross sections for vibrational excitation of O_2 by an electron impact via the $^2\Pi_g$ resonance.

The process starts with obtaining the potential energy curves of O_2 and its negative ion. This was done by standard ab-initio quantum chemistry methods (CASSCF, MRCI – implemented in MOLPRO package [1]) under different conditions. Quality of the resulting curves was controlled by comparing calculated electron affinities with experimental values. The O_2 potential energy curve was additionally split due to the spin-orbit interaction which turned out to be essential for comparison with experiment.

The resonant part of the O_2^- potential energy curve was found by fixed-nuclei scattering calculation using the R-matrix method within the UKRmol suite of codes [2]. These calculations provide eigenphase sums from which we extract the resonance energy and width.

We treat nuclear dynamics using the nonlocal discrete-state-in-continuum model [3] based on the projection-operator approach. We then compare the resulting cross sections to experimental data [4] along with results obtained within the local complex potential approximation (LCP) [5].

Acknowledgement: The study was supported by the Charles University, project GA UK No. 643216, and by the Grant Agency of Czech Republic under contract number GACR 19-20524S.

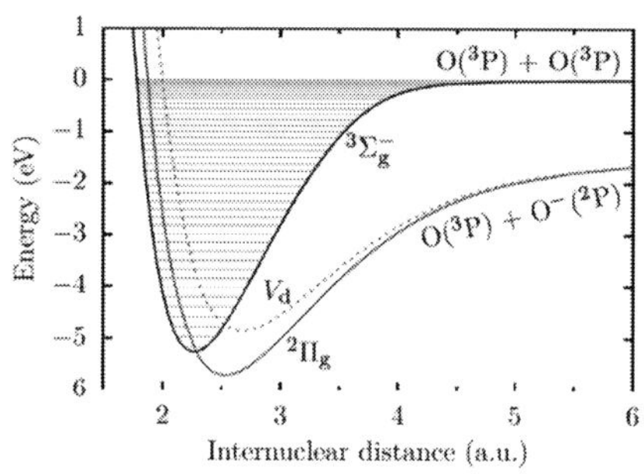

Figure 1. Potential energy curves of O_2 and O_2^- and discrete state potential V_d

References

[1] Werner H-J, Knowles P J, Knizia G and others, see http://www.molpro.net
[2] Carr J M, Galiatsatos P G, Gorfinkiel J D and others 2012 *Eur. Phys. J. D* **66**
[3] Domcke W 1991 *Phys. Rep.* **208** 97
[4] Allan M 1995*J. Phys. B* **28** 5163
[5] Dubé L and Herzenberg A 1979 *Phys. Rev. A* **20** 194

*E-mail: vaclav.alt@utf.mff.cuni.cz
†E-mail: houfek@mbox.troja.mff.cuni.cz

Distribution of dipole oscillator strengths for the formation of excited hydrogen atom from C_2H_2

Y Kumagai[1,2]*, T Odagiri[3]†, M Nakano[1,4], I H Suzuki[3,5], K Hosaka[1], M Kitajima[1] and N Kouchi[1]

[1]Department of Chemistry, Tokyo Institute of Technology, Tokyo 152-8551, Japan
[2]Present address: Department of Applied Physics, Tokyo University of Agriculture and Technology, Koganei-shi, Tokyo 184-8588, Japan
[3]Department of Materials and Life Sciences, Sophia University, Tokyo 102-8554, Japan
[4]Present address: Institute for Excellence in Higher Education Division of Developmental Research in Education Programs, Tohoku University, 41 Kawauchi, Aoba-ku, Sendai, 980-8576, Japan
[5]Photon Factory, Institute of Materials Structure Science, KEK, Tsukuba 305-0801, Japan

Synopsis Dipole oscillator strengths for the formation of excited hydrogen atom from C_2H_2 were measured to investigate formation and decay of molecular superexcited states. It seems that non-adiabatic transition plays an important role in the formations of the hot hydrogen atom.

In the interaction of ionizing radiation with matters, molecules are not necessarily ionized directly but are excited to superexcited states which are doorway states into various channels [1]. Among the decay channels dissociations are known to be of particular importance in radiolysis since the fragments can work as hot species in the chemical stage of radiation interaction. In the present study, distributions of dipole oscillator strengths for dissociations of acetylene to form excited hydrogen atom were measured as a function of incident photon energy to reveal the role of superexcited states in the fragment formations and accumulate quantitative data for understanding radiation interactions.

Experiments were conducted by using linearly polarized synchrotron light derived from the 3-m normal incidence monochromator at beamline 20A of KEK-PF. Yields of the H(2p), H(n=4) and H(n=6) atoms were monitored by detecting the Lyman-α, Balmer-β and Balmer-δ photons, respectively. The H(2s) atoms were detected by a combination of an external electric field and a photon detector. By the angle resolved detection of the H(2s) atom the the electronic symmetries of the superexcited states ($^1\Sigma_u^+$ or $^1\Pi_u$) were resolved. Deteils of the apparatus and procedures to obtain the oscillator strengths are described elsewhere [2, 3].

In figure 1, distributions of the dipole oscillator strengths for the formation of excited hydrogen atoms are shown. The resontant peaks in the distributions are due to the superexcited states of C_2H_2. The peak structures are similar with each other, which seems to substantiate the significant role of nonadiabatic transitions in the dynamics of the superexcited states involved.

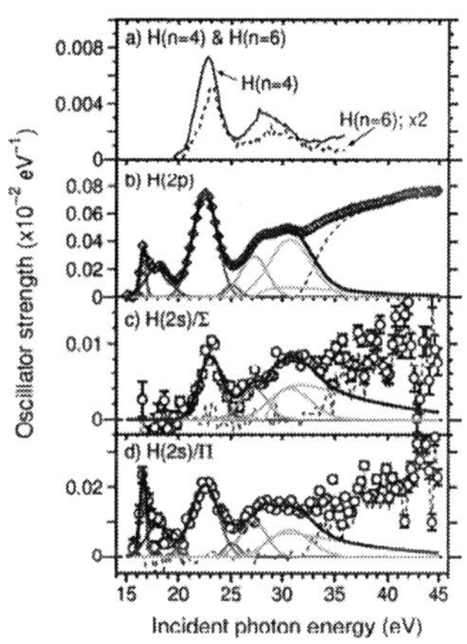

Figure 1. The dipole oscillator strengths for the formations of (a) the H(n=4) or H(n=6) atom, (b) the H(2p) atom, (c) the H(2s) atom via the $^1\Sigma_u^+$ states and (d) the H(2s) atom via the $^1\Pi_u$ states, respectively. The solid curves are fits to the experimental data.

References

[1] Hatano Y 1999 *Phys. Rep.* **313** 109
[2] Kato M 2002 *J. Phys. B* **35** 4383
[3] Kumagai Y 2018 *J. Chem. Phys.* **149** 244302

*E-mail: kumagai@go.tuat.ac.jp
†E-mail: odagiri.t@sophia.ac.jp

Complex scaling in relativistic calculations of cross sections for electron-ion recombination processes

A I Bondarev[1,2]*, V A Zaytsev[1], V M Shabaev[1] and X Ma[3]

[1]Department of Physics, St. Petersburg State University, 199034 St. Petersburg, Russia
[2]Center for Advanced Studies, Peter the Great St. Petersburg Polytechnic University, 195251 St. Petersburg, Russia
[3]Institute of Modern Physics, Chinese Academy of Sciences, 730000 Lanzhou, China

Synopsis We present an approach to calculate cross sections for electron-ion recombination processes. It is based on the combination of relativistic many-body perturbation theory and complex scaling method. A preliminary comparison between measured and calculated spectra for the electron-ion recombination rate coefficients for Be-like Ca will be given.

In electron-ion recombination, along with a process of radiation recombination, which can happen at any collision energy, there are resonant processes, where the capture of an incident electron simultaneously involves the target excitation, and photon emission completes the recombination process. Dielectronic recombination (DR) is a fundamental process which occurs through a doubly excited autoionizing state. Analysis of the DR spectra is required for diagnostics of hot laboratory plasmas, cosmological and fusion researches (see, e.g., Refs. [1, 2, 3, 4]). The DR experiments can also be used for studying such subtle effects as isotope shifts [5] and nuclear excitation [6]. All these and many other applications demand an accurate theoretical investigation of the recombination process.

Precise calculation of the DR process is quite a challenging task. Indeed, in this process the incident electron is captured into an autoionizing state which is embedded into the positive energy continuum. Such a state can not be described by square-integrable functions which form the basis of conventional high-precision many-electron methods. This problem can naturally be solved with the usage of the complex scaling approach (see recent review [7]). In this approach, the transformation of the radial coordinate, $r \to re^{i\theta}$, leads to the detachment of the autoionizing levels from the continuum. Thus, wave functions of these states become square-integrable and can be used in usual many-electron techniques.

The present approach is based on relativistic many-body perturbation theory combined with complex scaling. The complex scaled many-electron wave functions are constructed from antisymmetrized products of the scaled one-electron Dirac orbitals. These orbitals are the eigenfunctions of the dilated Dirac equation in a spherical cavity of a finite radius which we solve numerically by means of the finite-basis-set approach employing B-splines [8]. In our method, the DR spectrum is calculated through the photoionization cross section which, in turn, is obtained from the amplitude of the elastic photon scattering in accordance to the optical theorem.

As a first application, we plan to use the approach for the calculation of the electron-ion recombination rate coefficients in B-like Ca. This process was recently studied experimentally at the CSRm (Lanzhou, China) [9]. At incident electron energies close to the threshold there exists sizable disagreement between the experimental data and available theoretical predictions.

This work was supported by RFBR-NSFC (Grants No. 17-52-53136 and No. 11611530684). A.I.B. acknowledges support from the Ministry of Education and Science of the Russian Federation (Grant No. 3.1463.2017/4.6).

*E-mail: a.bondarev@spbu.ru

References

[1] Beiersdorfer P *et al* 1993 *Astrophys. J.* **409** 846
[2] Widmann K *et al* 1995 *Rev. Sci. Instr.* **66** 761
[3] Porquet D *et al* 2010 *Space Sci. Rev.* **157** 103
[4] Schippers S 2012 *J. Phys.: Conf. Ser.* **388** 012010
[5] Brandau C *et al* 2008 *Phys. Rev. Lett.* **100** 073201
[6] Borisyuk P V *et al* arXiv:1804.00299v1 [nucl-th]
[7] Lindroth E and Argenti L 2012 *Adv. Quantum Chem.* **63** 247
[8] Johnson W R *et al* 1988 *Phys. Rev. A* **37** 307
[9] Wang S X *et al* 2018 *Astrophys. J.* **862** 134

Coincidence time-of-flight mass spectrometry of electron impact on anthracene using a field programmable gate array

P J M van der Burgt[1] * and M L Gradziel[1]

[1]Department of Experimental Physics, National University of Ireland Maynooth, Maynooth, Co. Kildare, Ireland

Synopsis A new data acquisition system for electron-impact time-of-flight mass spectrometry has been set up, in which a field programmable gate array is used for the timing and the recording of mass spectra on an event by event basis. Fragmentation studies of anthracene have shown clear evidence for double ionization above 21 eV electron impact, which provides the motivation to look for coincident fragments. The first results for coincidence mass spectrometry of anthracene will be presented at the conference.

Mass spectra for positive ions produced by low-energy electron impact on anthracene and phenanthrene [1, 2] have produced clear evidence for double ionization above 21 eV. Groups of fragments with 6 and 7 carbon atoms clearly show the presence of doubly-charged fragments at half-integer masses. The smaller fragments with 1 to 4 carbon atoms all show broadened peaks, and these fragments may be partly or mostly due to energetic charge-separation fragmentations of doubly-charged anthracene and phenanthrene. For this reason it is of interest to perform coincident time-of-flight mass spectrometry using electron impact. In ion-molecule collisions this technique has been applied regularly, including to doubly-ionized anthracene molecules produced by double electron transfer to 5 keV protons [3].

In our earlier work [1, 2] we measured mass spectra for positive ions using a refectron time-of-fight mass spectrometer, where electron impact energy was varied from 0 to 100 eV in steps of 0.5 eV. Ion yield curves and appearance energies of most of the fragment ions of anthracene and phenanthrene were determined. Two key components in our time-of-flight mass spectrometry experiment were a FastComtec 7886S multichannel scaler and a digital delay generator. The multichannel scaler was used for recording mass spectra, and the delay generator was used to synchronise the pulsing of the electron gun, the ion extraction voltage, and the start of the multichannel scaler. Data acquisition was automated with a PC using LabView control.

We have implemented a new data acquisition system in which a field programmable gate array (FPGA, National Instruments cRIO9075) is used for the timing of the pulsing of the electron gun and the ion extraction voltage, and for the recording of mass spectra on an event by event basis. We have developed LabView code for both the communication between the FPGA and a PC, running on the FPGA chassis, and the control of the experiment and the data acquisition, running on the PC.

Initial tests with the FPGA have shown that we can accurately measure mass spectra, and first results of the coincidence tests will be presented at the conference.

References

[1] van der Burgt P J M, Dunne M and Gradziel M L 2018 *Eur. Phys. J. D* **72** 31
[2] van der Burgt P J M, Dunne M and Gradziel M L 2019 *J. Phys. Conf. Ser.* submitted
[3] Reitsma G, Zettergren H, Martin S, Brédy R, Chen L, Bernard J, Hoekstra R and Schlathölter T 2012 *J. Phys. B: At. Mol. Opt. Phys.* **45** 215201

* E-mail: peter.vanderburgt@mu.ie

To disentangle strong resonant Auger decay lines after the O1s $\rightarrow 3\sigma_u^*$ excitation in O_2 molecules

B Ding[1], W Q Xu[1], C Nicolas[2], S Song[1,3], Y Jiang[1], M Patanen[2,4], S Nandi[2], Z Xiao[3], C Miron[2,5] and X J Liu[1,3]*

[1]ShanghaiTech University, Shanghai, 201210, China
[2]Synchrotron SOLEIL, Gif-sur-Yvette Cedex, 91192, France
[3]Beihang University, Beijing, 100191, China
[4]University of Oulu, Oulu, 90014, Finland
[5]Université Paris-Saclay, Gif-sur-Yvette Cedex, 91191, France

Synopsis Atomic features may appear in the Auger electron spectra after an electron is excited from a core orbital to an antibonding orbital in molecules, as a result of complex competitions among electron and nuclear motions. Relatively weak atomic features were widely discussed in spectator Auger decay processes on top of O1s $\rightarrow 3\sigma_u^*$ excitation in O_2 molecules. Thee strong Auger decay processes are heavily-overlapped in the electron spectrum and not discussed up to now. Here, using multi-particle coincidence setup, we separate these Auger decay processes for the first time.

Morin and Nenner discovered atomic features in the resonant Auger electron spectrum following the core excitation of HBr molecule [1]. Atomic features were also observed for O_2 [2, 3, 4]. However, for O_2, the strongest atomic peaks $(2s^22p^3)\,^2P$ and ^2D are two or three times stronger than $(2s^22p^3)^2D$ [5]. Only the latter was widely discussed.

With the EPICEA setup at PLEIADES beamline at synchrotron SOLEIL, we measured in coincidence the kinetic energy of resonant Auger electron and the kinetic energy release of fragments in Fig. 1. So we can identify the dissociation processes and the dissociation limit energies E_{diss} are listed in Table. 1.

The sharp peaks due to the decays from Rydberg states, while the energy position corresponding to strong Atomic auger decay are expanded due to fast nuclear motion caused by antibonding orbital $3\sigma_u^*$.

Table 1. The dissociation limit energy E_{diss} after the resonant Auger decay in O_2 molecule.

	Energy (eV)	Assignment O + O$^+$
I	1.96	$(2p^4)^1D + (2p^3)^4S$
II	3.41	$(2p^4)^3P + (2p^3)^2D$
III	5.04	$(2p^4)^3P + (2p^3)^2P$
IV	6.79	$(2p^4)^1D + (2p^3)^2P$
V	10.75	$[(2p^3)^4S]3p + (2p^3)^4S$
VI	12.54	$[(2p^3)^4S]3s + (2p^3)^2D$
VII	14.00	

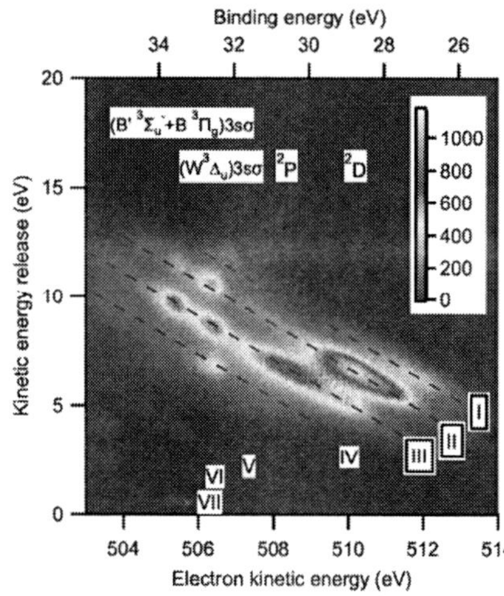

Figure 1. The correlation between the electron kinetic energy and the kinetic energy release of fragments.

References

[1] Morin P and Nenner I 1986 *Phys. Rev. Lett.* **56** 1913
[2] Bjorneholm O *et al* 2000 *Phys. Rev. Lett.* **84** 2826
[3] Travnikova O *et al,* 2010 *Phys. Rev. Lett.* **105** 233001
[4] Liu X J *et al* 2014 *Nat. Photon.* **9** 120
[5] Caldwell C D *et al* 1994 *J. Elec. Spec. Relat. Phenom.* **67** 243

*E-mail: liuxj@shanghaitech.edu.cn

Electron induced excitation of Furfural ($C_5H_4O_2$)

P Modak*, **A Singh and B Antony**†

AMPL, IIT(ISM) Dhanbad-826004, India

Synopsis The present study reports excitation cross section for three low-lying excited states of furfural ($C_5H_4O_2$). These states have a significant contribution to the elastic channel as well as dissociation processes. R-matrix method is used to study the e-$C_5H_4O_2$ problem under configuration interaction (CI) approximation. The ground state geometry is considered for the construction of Hamiltonian of the problem. Results show resonances for these sates.

In our recent study into electronic excitation, we have chosen the case of furfural. This is identified as a key element in the commercial realization of bio-refineries [1]. It is produced through atmospheric-plasma treatments of biomass [2]. Hence, cross sections data for the e-$C_5H_4O_2$ system are very crucial input for modeling and simulation purposes. The use of electron-induced excitation data in the plasma model provides insight into the inelastic processes inside a plasma. It helps in controlling the chemical mechanism occurring in plasma and eliminating a particular undesired channel. Here, cross-section data for three lower excited states of furfural are reported since these states are very dominant in the absorption of incident flux.

R-matrix method [3] was used to study the scattering problem of the multi-electron system. We used complete active space configuration interaction (CASCI) approximation and all excited states below 25 eV having the same symmetry as that of the ground state are included to obtain a fair description of the present target. The short-range interactions viz. static, exchange and correlation are considered inside a sphere of radius r=10 a_0 and long-range forces are taken outside this sphere. Thus, the configuration space is divided into two regions and Hamiltonian is constructed for them separately. The present target has a very high value of dipole moment. Hence, long-range forces are taken care of using a large value to r for the outer region to fulfill the asymptotic boundary condition. The K-matrices are then obtained and used to calculate S and T-matrices. Scattering observables are also found from K matrices. Born correction is added to the results to include higher angular momentum contributions.

The present calculation is an attempt to study three attractive excited states of furfural.The excitation cross-section for the present target is represented in Fig.1. The integrated cross section for the $^3A'$ state increases smoothly above the threshold and form a bound state at 6.5 eV. This state has the largest contribution towards the inelastic channel as shown in Fig.1. This is quite expectable since this state has the same symmetry as that of the ground state. The second bound state is observed at 9 eV. The position of shape resonances for the other two states are at 8.5 and 7 eV respectively.

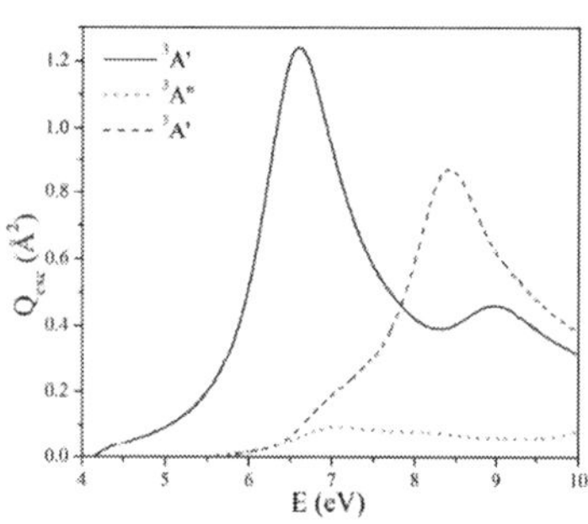

Figure 1. Excitation cross-section for the present target.

References

[1] Ragauskas A J *et al* 2006 *Science* **311** 484
[2] Amorim J *et al* 2013 *Plasma Processes Polym.* **10** 670
[3] Tennyson J *et al* 2010 *Physics Reports* **491** 29-76

*E-mail: paresh@ap.ism.ac.in
†E-mail: bobby@iitism.ac.in

Suppression of the AC-Stark shift by vortex light beams

S A-L Schulz[1,2]*, R A Muller[1,2] and A Surzhykov[1,2]

[1]Physikalisch-Technische Bundesanstalt, D-38116 Braunschweig, Germany
[2]Technische Universität Braunschweig, D-38106 Braunschweig, Germany

Synopsis In this work, we present how the AC- Stark shift can be suppressed by using twisted light beams for photoexcitations. Twisted light beams carry a non-zero projection of orbital angular momentum (OAM) onto their propagation direction, have a helical phase front and an intensity minimum in the beam center. Atoms located in this minimum experience a reduced AC-Stark shift and, therefore, twisted beams may be considered as valuable tool to drive atomic clock transitions. In this contribution, particular emphasis is paid to the electric-octupole (E3) transition $^2S_{1/2} \to {}^2F_{7/2}$ in the $^{171}Yb^+$ ion.

Since the groundbreaking work of Allen et al. in 1992 [1], the interest in twisted light has grown steadily both in experiment and theory. The phasefront of vortex light beams is twisted around the propagation direction of light. Therefore, they have a phase singularity and an intensity minimum in the beam centre. A beam is called "twisted", if the projection of orbital angular momentum (OAM) onto their propagation direction is non-zero [2]. Theses beam properties can be used to suppress the AC-Stark shift.

The electric-octupole (E3) hyperfine clock transition $^2S_{1/2}(F=0) \to {}^2F_{7/2}(F=3)$ in the $^{171}Yb^+$ ion is with 10^{-9} Hz extremely narrow, hence, the lifetime of the excited state is considerable ~ 6 years long [3]. To achieve the required population in the excited state a high spectral power density of the laser is needed. This leads to a significant AC-Stark effect, also known as light shift. The AC-Stark shift for hyperfine transitions is given by

$$\Delta W_{AC} = -\frac{\alpha_S}{2}E^2 \qquad (1)$$
$$-\frac{\alpha_T}{2}\left(\frac{3m_F - F(F+1)}{F(2F-1)}\right)\frac{3\cos^2\theta - 1}{2}E^2$$

where α_S is the scalar and α_T the tensorial polarizability, θ the angle between the E-field and the quantization axis (B-field), F is the hyperfine quantum number and m_F its projection. From Eq. (1) we see, that either the scalar (α_S) or the tensorial polarizability (α_T) or the electric field can be reduced to decrease the light shift.

In the following, we present how it is possible to excited a single cooled atom or ion with a low electric field strength. There are many different types of vortex light beams. In this work the calculations were done for twisted Bessel beams.

Fig. 1 illustrates E^2 (a), the AC-Stark shift (b) and the transition probability for the E3 transition (c) for a twisted beam with an OAM of 1 as functions of the impact parameter, which indicates the distance from the beam centre to the position to the atom.

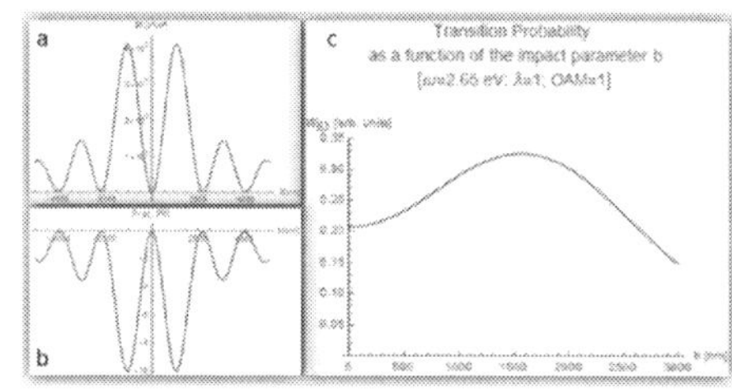

Figure 1. Function of E^2 (a), the light shift (b) and the transition probability for the E3 hyperfine transition (c) for a beam with an OAM of 1.

Fig. 1(a) shows that the electric field has a minimum in the beam centre (b=0), hence, as we know from Eq. (1), also the light shift (b). From (c) we see that the intensity for this beam with an OAM of 1 is non-zero. Therefore atoms should be located in this minimum, because then they undergo a suppressed AC-Stark shift.

Based on this calculations these results may be implemented in the $^{171}Yb^+$ single-ion clock experiments at PTB [3].

References

[1] Allen L *et al* 1992 *Phys. Rev.* A **45** 8185
[2] Scholz-Marggraf H M *et al* 2014 *Phys. Rev.* A **90** 013425
[3] Huntemann N *et al* 2016 *Phys. Rev. Lett.* **116** 063001

*E-mail: sabrina.schulz@ptb.de

Energy deposit by electron excitation in C_nN^+ projectiles (n=1-3) colliding at intermediate velocity with He atoms : semi-empirical estimates and calculations

T Mahajan[1], **K Béroff**[1 *], **B Pons**[2], **C Illescas**[3], **M Chabot**[4], **T IdBarkach**[4], **A Jorge**[3 †], **N F Aguirre**[5] **and S Diaz-Tendero**[3]

[1]Institut des Sciences Moléculaires d'Orsay, CNRS and Paris Sud University, 91405 Orsay, France
[2]Université de Bordeaux-CNRS-CEA, CELIA UMR5107, F-33405 Talence, France
[3]Universidad Autonoma de Madrid, Cantoblanco, E-28049 Madrid, Spain
[4]Institut de Physique Nucléaire d'Orsay, CNRS and Paris Sud University, 91406 Orsay, France
[5]Theoretical division, Los Alamos National Laboratory, Los Alamos, 87545 New Mexico, city, USA
[†] present address: Department of Physics and Astronomy, York University, Toronto, Ontario, Canada

Synopsis Using measured dissociation branching ratios and theoretical dissociation energies, we extracted semi-empirical estimates of the energy deposit due to electron excitation in C_nN^+-He collisions at v=2.2 a.u. We compared these estimates to calculations based on the IAE (Independent Atom and Electron) model where the atom(ion)-atom excitation is treated within the Classical Trajectory Monte Carlo (CTMC) approach. A good agreement has been found. This atomic approach is also able to reproduce the experimental Kinetic Energy Release (KER) in CN^+ dissociation measured in 3.5 a.u CN^+-He collision.

Experiments have been performed at the Tandem accelerator in Orsay (France). We used the AGAT setup for excitation of the molecule and detection of its fragmentation. The way the internal energy distribution (IED) of the C_nN^+ was derived from the fragmentation pattern has been explicited elsewhere [1]. KER measurements were performed using a dedicated position sensitive detector [2].

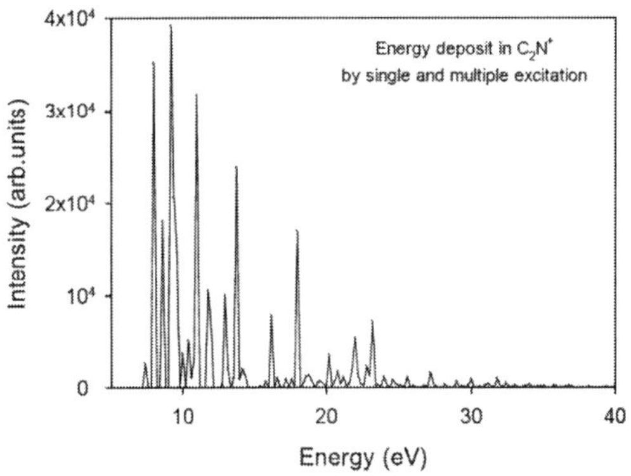

Figure 1. Calculated internal energy deposit in C_2N^+ following electron excitation

On the other hand, a IAE/CTMC calculation of the energy deposit resulting from atomic assumed was performed. The C_nN^+ molecule was supposed to be made of independent atomic constituents colliding independently with He atoms. Excitation of atoms in final (n,l) levels were calculated with CTMC. Energies of final atomic excited states were obtained from NIST database. Figure 1 shows the internal energy deposit in C_2N^+ calculated with the IAE/CTMC model. As will be shown at the conference, this distribution is compatible with the one obtained from fragmentation branching ratios [1].

In the case of CN^+ dissociation at v=3.5 a.u we measured the KER for both C^+/N and C/N^+ channels [2]. By assuming dissociation towards fragments in their electronic ground states to occur, we could reconstruct, using calculated dissociation energies, the internal energy of the molecule CN^+ following electronic excitation. Results will be presented and compared to the IAE/CTMC calculation at the conference.

References

[1] IdBarkach T *et al* 2018 *Mol. Astrophysics* **12** 25
[2] Jallat A 2015, *PhD Thesis University Paris Sud*, unpublished

* E-mail: karine.beroff@u-psud.fr

Femtosecond nonadiabatic cascade and subsequent photofragmentation of XUV excited caffeine molecule

K Yamazaki[1]*, A Marciniak[2] , S Maeda[3], M Reduzzi[4], V Despré[2], M Hervé[2], M Meziane[2], T Niehaus[2] , V Loriot[2], A Kuleff[5,6], B Schindler[2] , I Compagnon[2,7] , G Sansone[4,8] and F Lépine[2†]

[1]Tohoku University, Sendai, 980-8577, Japan
[2]Université Claude Bernard Lyon 1, Villeurbanne, F-69622, France
[3]Hokkaido University, Sapporo, 060-0810, Japan
[4]Politecnico Piazza, Milan, I-20133, Italy
[5]Universität Heidelberg, Heidelberg, D-69120, Germany
[6]ELI-ALPS, Szeged, H-6728, Hungary
[7]Institut Universitaire de France IUF, Paris, F-75005, France
[8] Albert-Ludwigs-Universität Freiburg, Freiburg, D-79104, Germany

Synopsis We found that the XUV excited caffeine undergoes nondadabatic cascade *via* more than 100 shake-up states within the time constant of 40 fs by using femtosecond XUV-NIR pump–probe photo-ion spectroscopy and the SAC-CI level of *ab-initio* many-body theory. This nonadiabatic cascade results in the photofragmentation to specific large fragments. Furether ionization by NIR probe during the cascade increases the small fragments CH_3^+, CO^+ *etc*. This results show that both vibronic coupling and electron correlation are the keys for the femtochemistry in the highly-correlated electronic excited states.

Electron wave packet generated by an extreme ultraviolet (XUV) pulse with femtosecond or attosecond duration easily modulated by nonadiabatic coupling between electron and nuclear degrees of freedom. This results in the nonradiative decay and photofragmentation with specific fragmentation pattern. In this report, We will discuss the femtosecond nondaidabatic cascade (FNC) and subsequent photofragmentation of highly-excited caffeine molecule (CAFF) by femtosecond XUV-NIR pump–probe photoion spectroscopy and the SAC-CI level of many-body *ab-initio* theory.

We found that the XUV excited monocation CAFF+ undergoes the FNC among more than 100 Rydberg 3*p* shake-up states in the elerctron binding enrgy of 21-25 eV shown in Figure 1 with the mean time constant of 40 fs. This FNC results in the photofragmentation to specific large fragments. Furether ionization by NIR probe during the cascade increases the small fragments CH_3^+, CO^+ *etc*. These results show that nonadiabatic coupling and electron correlation are the keys for the ultrafast reaction dynamics in the highly-correlated electronic excited states.

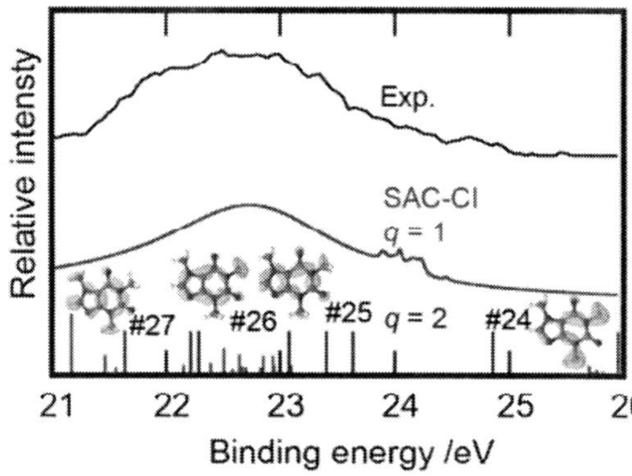

Figure 1. Comparison between experimental spectrum measured of caffeine [2] and simulated photoelectron spectrum at the SAC-CI level of theory. [1]

References

[1] Marciniak A *et al* 2018 *J. Phys. Chem. Lett.* **9** 6927
[2] Feyer V *et al.* 2009 *Chem. Phys.* **358** 33

* E-mail: kaoru.yamazaki@imr.tohoku.ac.jp
† E-mail: franck.lepine@univ-lyon1.fr

Photoionization of Ti^{3+}: β parameter and Momentum Transfer

W Eissner** and L Vo Ky‡

*Universität Stuttgart, Institut für Theoretische Physik I, Pfaffenwaldring 57, 70550 Stuttgart, Germany
‡retired from LERMA, Observatoire de Paris, 92195 Meudon Cedex, France

Synopsis Key issue is the presentation of the *unpublished* 4 angular factors arising from the 4 paths when momentum transfer induces asymmetry β in photoionization calculations. They require functional expansions such as the frozen cores approach which are not bedevilled by the prior-post dilemma. It is essential to consider *all* magnetic 2-body terms in the Breit-Pauli hamiltonian that contribute to finestructure.

With the concept of angular *Momentum Transfer* (MT) Fano and Dill [1] offer a transparent alternative to the conventional approach of Jacobs [2] in computing asymmetry parameters β, part of stage bound-free (**stgbf**) of the Belfast Breit-Pauli-R-matrix code (BPRM) [3] shaped to the needs of the Opacity Project.

In the photoionization cross section up to second harmonic,

$$\sigma(\vartheta) = \frac{\sigma}{4\pi}\left[1 + \beta\left(\frac{3\cos^2\vartheta - 1}{2}\right)\right],$$

the spatial direction ϑ of the photoelectron is observed against the photon propagation axis, which enters the MT balance by its spin $\hbar$.

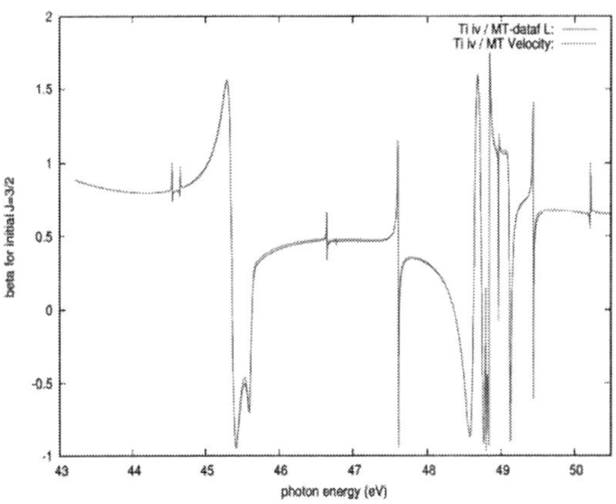

Figure 1. Calculated parameter β for photoionizing the Ti IV ground state $^2D_{3/2}$ in dipole length and velocity formulation.

The twin peak at 44.5 eV in the Figure near the first ionization threshold of 43.267 eV has been measured in high resolution on the right shoulder of a giant resonance 1.5 eV wide by Schippers et al. [4], verified as a Rydberg doublet over a PEC [5] resonance by Eissner et al. [6]. Here we employ their expansion with 7 target or frozen core terms from 2 spectroscpic configurations of Ti V:

	E/eV		BPRM0	+TEC[7]	BPRM	NIST
1	$3s^23p^6$	$^1S_0$	0	0	0	0
2	$3s^23p^53d$	$^3P^o_0$	33.81	33.967	33.774	34.026
3		$_1$	33.91	34.091	34.094	34.142
4		$_2$	34.19	34.345	34.342	34.382
⋮		$^3F^o$, $^1D^o$, $^3D^o$				
12		$^1F^o_3$	38.56	38.623	38.609	38.613
13		$^1P^o_1$	50.21	49.019	49.020	49.014

The MT version of routine **beta** involves four paths for the pair j, j' of transfer momenta related to collision channel orbital ℓ:

$$\begin{array}{lll} l_j = l_{j'} = \ell - 1 & \text{only } \ell \geq 2 & \text{(d1)} \\ l_j = \ell - 1; l_{j'} = \ell + 1 & \ell \geq 1 & \text{(d2)} \\ l_j = l_{j'} = \ell & \ell \geq 1 & \text{(d3)} \\ l_j = l_{j'} = \ell + 1 & \ell \geq 0 & \text{(d4)} \end{array}$$

leading to 4 types of angular integrals $A(\ell; j, j')$ in the dipole transition amplitude such as

$$A(\ell; \ell-1, \ell-1) = (-1)^\ell \langle \ell-1\ 0\ \ell-1\ 0|2\ 0\rangle \times\sqrt{2/3}(2\ell-1)(2\ell+1)\begin{Bmatrix} 1 & \ell-1 & \ell \\ \ell-1 & 1 & 2 \end{Bmatrix}$$

After some unwieldy evaluation one arrives at

$$\begin{array}{lrcl} \text{d1}: & A(\ell; \ell-1, \ell-1) & = & (\ell-1)/3 \\ \text{d2}: & A(\ell; l_j, \ell-1, \ell+1) & = & -[\ell(\ell+1)]^{1/2} \\ \text{d3}: & A(\ell; \ell, \ell) & = & -(2\ell+1)/3 \\ \text{d4}: & A(\ell; \ell+1, \ell+1) & = & (\ell+2)/3. \end{array}$$

References

[1] Fano U and Dill D 1972 *Phys. Rev. A* **6** 185-92
[2] Jacobs V L 1972 *J. Phys.* B **5** 2257–71
[3] Berrington K A *et al* 1995, *CPC* **92** 290-420
[4] Schippers S *et al* 2004 *J. Phys.* B **37** L209-16
[5] Yu Y and Seaton M J 1987 *J. Phys.* B **20** 6409-29
[6] Eissner W *et al* 2009 ICPEAC *Abstracts*
[7] Zeippen C J *et al* 1977 *MNRAS* **181** 527-40

*E-mail: we@itp1.uni-stuttgart.de

Heavy metal spectroscopy in extremely metal rich subdwarfs

L Fernández-Menchero*, C P Ballance and C A Ramsbottom

Queen's University Belfast, BT7 1NN, United Kingdom

Synopsis Helium-rich subdwarf B stars represent a small group of low-mass hot stars. A number of significant but unfamiliar absorption lines have been identified of heavy-metal ions, such as germanium, strontium, yttrium and zirconium. The abundances in the photosphere of theses element was estimated between 1 000 and 10 000 times the solar. We calculate the collision-radiative data for the ions Sr^{+}, Y^{2+} and Zr^{3+}. Comparing with measured spectra we can determine these abundances. We compare the obtained results with the RMATRX and DARC methods.

Subdwarf stars with helium-rich atmospheres (He-sdB and He-sdO) compromise about the 5% of the total subdwarf population. An example of extremely metal-rich He-sdO is LSIV–14°116 [1]. Using high-resolution spectroscopy in the optical, in this star they were spotted lines of Sr II, Y III and Zr IV.

We calculate level energies, radiative transition coefficients and effective collision strenghts of these observed ions. With these data the abundances can be determined accurately, with the possibility of using these spectral lines as diagnostics for relevant plasma parameters. We compare our results obtained with the semi-relativistic Breit-Pauli R-Matrix (BPRM) formalism and the fully relativistic Dirac Atomic R-Matrix Code (DARC) to limit the uncertainties.

The atomic structure of the target was calculated using AUTOSTRUCTURE (AS) for the semi-relativistic calculation and GRASP for the fully relativistic. For a proper comparison, we use exactly the same set of configurations in the Configuration Interaction (CI) expansions with both methods. For all the systems, we expand atomic wave functions of the Rb-like target in a set of 16 non-relativistic configurations, we include the ground configuration $4p^6\,4d$, all the one-electron excitations $4p^6\,nl$, with $n = 4-8$ and all possible values of l for $n \leq 5$, $l = 0-2$ for $n = 6, 7$, and the 8s; and finally the two-electron excitations $4p^4\,4d\,5s^2$ and $4p^4\,4f\,5s^2$ to include continuum states in the expansion. In total we get 82 relativistic levels, from which 24 are spectroscopic and the rest are immersed in the autoionization continuum.

We achieve a good agreement, below the 5%, in the excitation energies between both methods and with the observed values tabulated in the NIST data base [2].

We have also compared the oscillator strengths gf and the Einstein spontaneous emission coefficients A for the transitions present in the optical spectrum, between 4 000 and 5 000 Å. We obtain a good agreement, approximately 20%, between AS and GRASP and the preliminary calculations of [1] with CIV3.

Finally, we use the plasma diagnostics code COLRAD to calculate the collision radiative model of the stellar atmosphere of LSIV–14°116 and produce synthetic spectra, which can be compared with the observed ones of [1]

References

[1] Naslim N, Jeffery C S, Behara N T and Hibbert A 2011 *Mon. Not. R. Astron. Soc.* **412** 363
[2] Kramida A and Ralchenko Y, Reader J and NIST ASD Team 2018 NIST Atomic Spectra Database (ver. 5.5.6), Online Available: `https://physics.nist.gov/asd`. National Institute of Standards and Technology, Gaithersburg, MD.

*E-mail: l.fernandez-menchero@qub.ac.uk

L-changing through very-long-range interactions in high-*n*, *n*~300, Rydberg-Rydberg collisions

S Yoshida[1], J Burgdörfer[1], G Fields[2], R Brienza[2] and F B Dunning[2*]

[1]Institute for Theoretical Physics, Vienna University of Technology, Vienna, Austria, EU
[2]Department of Physics & Astronomy, Rice University, Houston, Texas 77005-1892, USA

Synopsis State-changing in thermal-energy collisions between strontium atoms in high-*n*, *n*~300, n^1F_3 Rydberg states is studied in an atomic beam and analyzed using classical trajectory Monte Carlo simulations. The *L*-changing cross section is large, ~10^{-4} cm^2, much greater than the corresponding "hard-sphere" cross section.

Because of their large dipole moments, Rydberg atoms provide a valuable tool for exploring long-range interactions. Such measurements, however, are complicated by collisions which can lead to rapid state changing (and to ionization). Here we examine collisional state changing in a strontium atom beam by exploiting dipole blockade to create a string of high-*n*, *n*~300, n^1F_3 Rydberg atoms through (pulsed) multiphoton excitation using focused laser beams that define a small excitation volume (~50 μm on a side). Excitation of a Rydberg atom suppresses further excitation until the atom has traveled one blockade radius (~100 μm) in the beam direction. If, once this occurs, excitation of an additional Rydberg atom occurs quickly, a string of Rydberg atoms is created with approximately equal initial separations but a distribution of initial velocities. The state-changing due to their subsequent collisions is monitored by observing the damping of quantum beats induced by sudden application of a "pump" field with rise time $t_R << T_n$, the classical orbital period.[1]

Figure 1 shows the time dependence of the quantum beats seen following application of the pump field. Two very-different beat frequencies are evident and correspond to the electron orbital period (~4 ns) and to Stark precession of the electron orbit in the pump field. Comparison of quantum beat data recorded with less than one Rydberg atom excited per laser pulse (i.e., no collisions expected) and with ~15 Rydberg atoms created per laser pulse shows that collisions lead to a marked reduction in the quantum beat amplitudes. In particular, the fast beats are sensitive to *n*-changing collisions, the slow beats to L-changing processes.

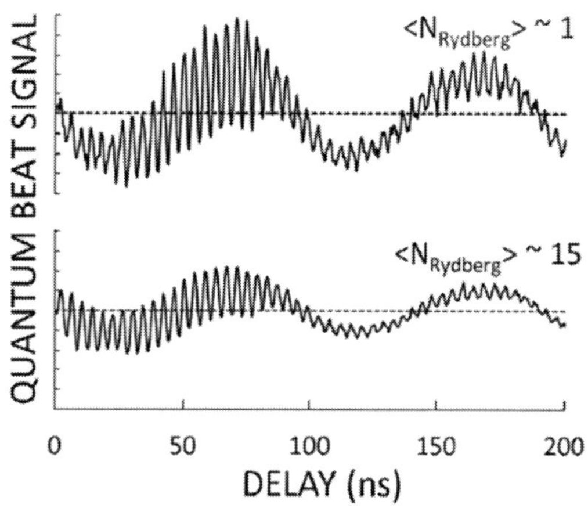

Figure 1. Time dependence of quantum beat signals recorded when creating ~1 and ~15 Rydberg atoms per (~6 μs) laser pulse.

The experimental data are analyzed using classical trajectory Monte Carlo simulations that use an effective four-body Hamiltonian. These simulations show that while collisions lead to only small changes in *n*, they cause dramatic changes in the *L* distributions, which results in a large reduction in the predicted quantum beat amplitudes. Theory and experiment show that, even for impact parameters *b*~50μm, collisions lead to strong *L*-changing. The cross-sections, ~10^{-4} cm^2, are large, much greater than those typically associated with neutral-neutral collisions.

Research supported by the NSF, Robert A. Welch Foundation, and FWF (Austria)

References

[1] Yoshida S *et al* 2007 *Phys. Rev. A* **75** 013414

* E-mail: fbd@rice.edu

Model for vibronic excitation of CH^+ by resonant scattering of electron

X Jiang[1], **C Yuen**[2], **P Cortona**[1], **V Kokoouline**[2] **and M Ayouz**[3] *

[1] SPMS, CentraleSupélec, Université Paris-Saclay, 91190, Gif-sur-Yvette, France
[2] Department of Physics, University of Central Florida, 32816, Florida, USA
[3] LGPM, CentraleSupélec, Université Paris-Saclay, 91190, Gif-sur-Yvette, France

Synopsis A new model for theoretical study of vibronic (de-)excitation (including vibrational (de-)excitation) cross sections of CH^+ by electron impact was developed based on the multichannel quantum defect theory (MQDT) close-channel-elimination procedure and vibrational-frame-transformation theory of Chang and Fano. The successful application of this model suggested a promising calculation method for vibronic (de-)excitation of diatomic ion core by resonant scattering of electron.

Vibrational-frame-transformation theory of Chang and Fano [1] is extensively used to describe the vibrational (de-)excitation of nonresonant electron molecule (ions) collision system. Fixed-nuclei R-matrix method is perfectly combined with this theory to provide a smooth short-range scattering matrix (S-matrix) in the Body frame (BF) [2, 3]. However, difficulties emerged for this theory to understand the vibrational excitation for the e-CH^+ collision system of great astrophysical and technological plasma interest. This was ascribed to the strong energy dependence of the short-range S-matrix for the collisions. Meanwhile, the non-adiabatic method coupling with vibrational motion of the nuclei for this collision system would be extremely time-consuming due to the low-lying electronically excited states of CH^+. Accordingly, this process was far from being precisely modeled in theory.

In this study, we propose a new model based on the MQDT close-channel-elimination procedure and the vibrational-frame-transformation theory to compute vibronic (de-)excitation cross sections and thermally-averaged rate coefficients for e-CH^+ collisions. In this model, the R-matrix formalism is employed to evaluate the electron-ion S-matrix in BF frame for a fixed geometry of the ion core. The electron scattering energy was extended to a highly excited state level where the S-matrix with multiple open electronic excitation channels is smooth. The S-matrix describing the vibronic transition is obtained from the vibronic-frame-transformation and MQDT close-channel-elimination procedure. The electronic excitation cross sections obtained in this model was compared with that from UKRmol suite [4], which was shown in Figure 1. A well matched behavior suggested the validation of our model. Finally, fitting formulas with a few numerical parameters were derived for the obtained rate coefficients. The interval of applicability of the formulas is from 40 to 10,000 K.

This model that has been applied successfully to the e-CH^+ system provides a potentially unified theoretical treatment for the vibronic (de-)excitation of other diatomic molecular ions by resonant scattering of electron. Further attempts will be taken to generalize this model to those collisions of polyatomic molecular ions and electrons.

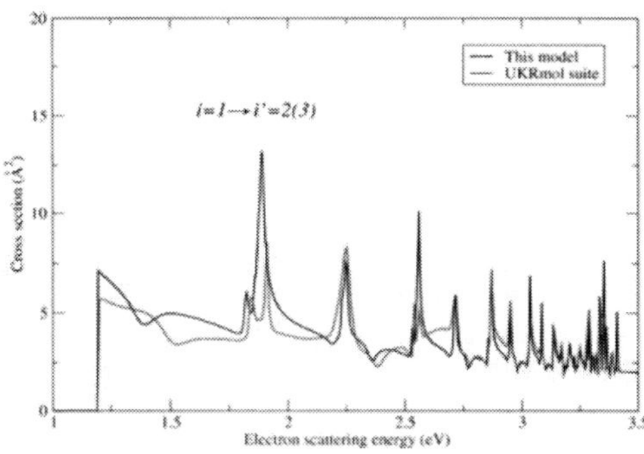

Figure 1. The comparison of the electronic excitation cross section for $X^1\Sigma^+$ (i=1)-$a^3\Pi$ (i=2(3)) at equilibrium nuclear distance (R=2.137 bohr).

References

[1] Chang E S, Fano U 1972 *Phys. Rev. A.* **6** 173
[2] Ayouz M, Kokoouline V 2016 *Atoms.* **4** 30
[3] Aymar M *et al* 1996 *Rev. Mod. Phys.* **68** 1015
[4] Tennyson J *et al* 2007 *J. Phys. : Conf. Ser.* **86** 012001

* E-mail: mehdi.ayouz@centralesupelec.fr

Atomic diagnostics of ultrahigh laser intensities

M F Ciappina[1]*, S V Popruzhenko[2], G Korn[1], T Ditmire[3] S V Bulanov[1] and S Weber[1]

[1]Institute of Physics of the ASCR, ELI-Beamlines, Na Slovance 2, 182 21 Prague, Czech Republic
[2]Prokhorov General Physics Institute of the Russian Academy of Sciences, Moscow, 119991, Russia
[3]Center for High Energy Density Science, C1510, Department of Physics, University of Texas at Austin, Austin, Texas 78712, USA

Synopsis We suggest and discuss a method for precise measurement of ultrahigh laser intensities in the $10^{20} - 10^{24}$W/cm^2 range of intensities. The method is based on the observation of multiple sequential tunneling ionization of heavy atoms in the focus of an intense laser beam. It can be of use for probing electromagnetic fields of record power expected to achieve at new multi-petawatt laser facilities.

The continuous development of optical and infrared high-power laser sources leads to a progressive growth in available intensities of electromagnetic radiation. Presently, several lasers of petawatt power [1] deliver pulses of intensity up to $I \approx 10^{21}$ W/cm^2. The forthcoming commissioning of new 10-PW class laser facilities worldwide opens a way to laboratory experiments with electromagnetic fields $I \simeq 10^{24}$ W/cm^2.

In view of these expectations, the problem of precise determination of the electromagnetic field intensity in the ultra-high power regime becomes of exceptional importance. Here we suggest [2] using multiple tunneling ionization of heavy atoms as a probe of laser intensity. We apply the setup of [3] in a way opposite to that employed in those experiments where the tunneling ionization theory has been validated assuming that the laser intensity is known. In contrast to other methods, this approach allows probing the intensity of an electromagnetic field directly and locally in time and space and with a considerable sensitivity which stems from the highly nonlinear dependence of the tunneling ionization probability on the electromagnetic field strength, so that a relatively small variation in intensity leads to orders of magnitude change in the ionization rate. The maximal ionic charge observed in a laser focus can be used as a sensitive measure of the peak intensity. With the peak intensity increasing, deeper electronic shells are stripped out by the laser: thus a target consisting of relatively heavy atoms is required.

We verify the validity of this idea by numerically modeling the sequential tunneling ionization of krypton and xenon in a tightly focused laser pulse of intensity $10^{20} - 10^{24}$W/cm^2. We demonstrate that the off-set in the charge distribution of high-Z ions can indeed be used as a measure and introduce a simple analytic theory which reproduces our numerical results with a reasonable accuracy and therefore can be used for estimates. Constraints imposed on possible experimental realizations are also discussed.

*E-mail: marcelo.ciappina@eli-beams.eu

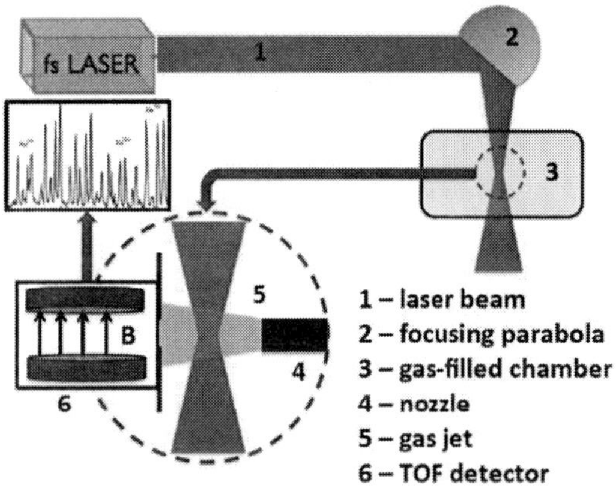

Figure 1. A schematic of the suggested experiment. A tightly focused laser pulse ionizes a gas or a mixture of several gases kept at a sufficiently low pressure to avoid plasma and space charge effects. Highly charged ions produced via sequential tunneling ionization are collected by a TOF detector.

References

[1] Danson C *et al* 2015 *High Power Laser Sci. Eng.* **3** e3
[2] Ciappina M *et al* 2019 *Phys. Rev. A* **99** 043405
[3] Chowdhury E A *et al* 2001 *Phys. Rev. A* **63** 042712; Yamakawa K *et al* 2003 *Phys. Rev. A* **68** 065403

Electron-impact ionization cross section calculations for selected ribonucleosides

P Możejko[1] *

[1]Department of Atomic, Molecular and Optical Physics, Faculty of Applied Physics and Mathematics, Gdańsk University of Technology, Gabriela Narutowicza 11/12, 80-233 Gdańsk, Poland

Synopsis Total cross sections for the single electron-impact ionization of selected ribonuclosides (guanosine, adenosine, cytidine and uridine) have been calculated for electron energies ranging from the ionization threshold up to 5 keV.

Accurate data concerning electron collisions with matter are of great importance in many fields of science including space sciences, astrobiology and radiation physics and chemistry. It is well established now, that low- and intermediate-energy electron interactions should be taken into account in the studies of the processes induced within biological material by primary ionizing radiation [1]. Thus qualitative and quantitative studies and description of the electron interaction with components of biomolecules are strongly desired.

In the present work we have studied positive ionization of selected ribonucleosides (guanosine, adenosine, cytidine and uridine) by electron impact. Ribonuleosides consists simply of a nucleobase and a five-carbon sugar-ribose.

Total cross sections for the single electron-impact ionization of the studied targets have been derived using the binary-encounter-Bethe (BEB) model [2] for energies between the ionization threshold up to 5 keV. Due to the (BEB) method the electron-impact ionization cross section of a given molecular orbital can be calculated according to:

$$\sigma = \frac{S}{t+u+1}\left[\frac{\ln t}{2} - \frac{\ln t}{2t^2} + \frac{t-1}{t} - \frac{\ln t}{t+1}\right],$$

where $u = U/B$, $t = T/B$, $S = 4\pi a_0^2 N R^2/B^2$, $a_0 = 0.5292$ Å, $R = 13.61$ eV, U is the kinetic energy of the given orbital, B is the electron binding energy, T is the energy of impinging electron and N is the orbital occupation number. The total cross section for electron-impact ionization can be obtained as a sum of ionization cross sections calculated for all molecular orbitals. The used theoretical approaches and computational procedures are described in detail in our previous works [3, 4]. The BEB method provides usually ionization cross sections which are in quite good agreement (within ±15%) with experimental data [5]. The resulting total ionization cross sections for the single-ionization of guanosine ($C_{10}H_{13}N_5O_5$), adenosine ($C_{10}H_{13}N_5O_4$), cytidine ($C_9H_{13}N_3O_5$) and uridine ($C_9H_{12}N_2O_6$) are shown in figure 1.

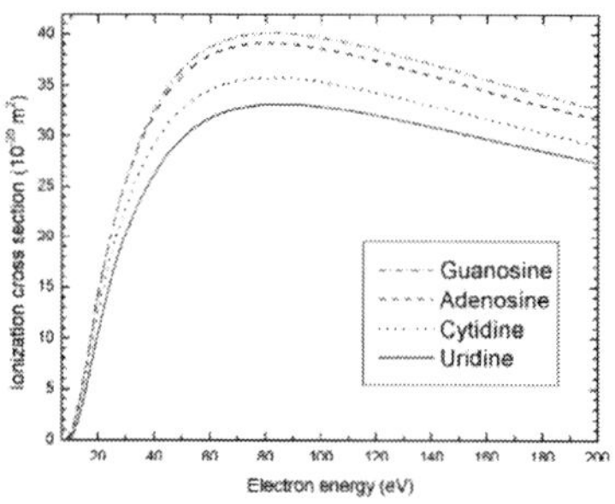

Figure 1. Comparison of ionization cross sections claculated in the present work for studied targets.

This work has been partially supported by the Polish Ministry of Science and Higher Education. Numerical computations have been performed at the Academic Computer Center (TASK) in Gdańsk.

References

[1] Sanche L 2005 *Eur. Phys. J. D* **35** 367
[2] Hwang W *et al* 1996 *J. Chem. Phys.* **104** 2956
[3] Możejko P *et al* 2003 *Radiat. Environ. Biophys.* **42** 201
[4] Możejko P *et al* 2005 *Radiat. Phys.Chem.* **73** 77
[5] Karwasz G P *et al* 2014 *Int. J. Mass Spectrom.* **365-366** 232

*E-mail: paw@pg.edu.pl

Post-collision interaction in sequential double-ionization of helium

A N Artemyev*, A I Streltsov and Ph V Demekhin

Institut für Physik und CINSaT, Universität Kassel, Kassel 34132, Germany

Synopsis We investigate the post-collision interaction between a slow electron released by a pump pulse from the neutral helium atom and a fast electron released by a time-delayed probe pulse from the created ion. To this end, we numerically solve the time-dependent Schrödinger equation in the basis of the essential eigenstates of the unperturbed Hamiltonian. We demonstrate that the energy exchange between the slow and the fast electrons and the recapture of slow photoelectrons can be controlled by the properties of the pulse-sequence.

The effect of the energy exchange between a slow primary electron released from the system and a fast secondary electron emitted from the remaining ion is called post-collision interaction (PCI, [1, 2]). Typically, the PCI is studied by the inner-shell ionization of a neutral atom which is followed by the Auger decay in the created ion. In the present work, we investigate PCI by the pump-probe sequental double-ionization scheme. Here, the neutral atom is ionized by a pump pulse with a carrier frequency slightly above its ionization potential (IP). Subsequently, a time-delayed probe pulse ionizes the remaining ion releasing thereby a fast photoelectron.

The present calculations are performed for a helium atom. To this end, we solve the time-dependent Schrödinger equation (TDSE) with the following total Hamiltonian:

$$H = \frac{\vec{p}_1^{\,2}}{2} + \frac{\vec{p}_2^{\,2}}{2} - \frac{2}{r_1} - \frac{2}{r_2} + \frac{1}{|\vec{r}_1 - \vec{r}_2|} + V_{int}(t)\,. \quad (1)$$

Here, $V_{int}(t)$ describes the interaction of the two electrons with the laser fields in the dipole length or velocity gauges. Calculations are performed for the sine-squared pulses. We select the carrier frequency ω_1 of the pump pulse to be 1.5 eV above IP of the neutral helium. For the probe pulse, we take the carrier frequency ω_2 to be 10 eV above the IP of the helium ion. The intensities of both pulses are chosen to be in the linear ionization regimes.

In order to solve the TDSE with the Hamiltonian (1), we use a basis of the essential stationary eigenstates of the unperturbed Hamiltonian constructed over the finite-element discrete-variable representation reported in Ref. [3] and references therein. Calculations are performed for different durations of the pump and probe pulses, as well as for different time delays. Our approach allows to follow evolution of the PCI effects in real time. As an example, Fig. 1 demonstrates time evolution of the radial electron density, calculated for 5 fs pump and 2 fs probe pulses with the time-delay of 3 fs. From this figure one can follow how the fast photoelectron overtakes the slow one.

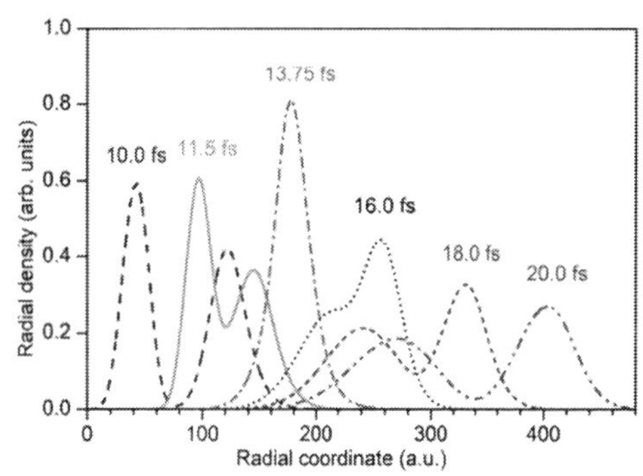

Figure 1. Time evolution of the electron density.

References

[1] Schmidt V 1982 *AIP Conf. Proc.* **94** 544
[2] Schmidt V 1992 *Rep. Prog. Phys.* **55** 1483
[3] Artemyev A N *et al* 2017 *Phys Rev A* **96** 033410

*E-mail: anton.artemyev@physik.uni-kassel.de

Time-dependent perturbation theory beyond the dipole approximation for two-photon ionization of atoms

M Wang[1], H Liang[1], X Xiao[1], S Chen[1] and L Peng[1,2,3]*

[1]State Key Laboratory for Mesoscopic Physics and Collaborative Innovation Center of Quantum Matter, School of Physics, Peking University, Beijing 100871, China
[2]Beijing Academy of Quantum Information Sciences, Beijing 100193, China
[3]Collaborative Innovation Center of Extreme Optics, Shanxi University, Taiyuan, Shanxi 030006, China

Synopsis We develop a time-dependent perturbation theory beyond the dipole approximation to analyze the two-photon ionization dynamics of atoms exposed to short laser pulses with an arbitrary polarization. Subsequent applications of the PT method in the nonresonant regime allow us to analyze one of the most important nondipole effects, i.e., the linear-momentum-transfer law in the two-photon ionization process and to unveil the important role of the interferences between the dipole and nondipole transition pathways to the photoelectron momentum shift along the laser-propagation direction.

We develop a time-dependent perturbation theory (PT) beyond the dipole approximation to treat the two-photon ionization of atoms. We investigate the two-photon ionization of the hydrogen atomic system exposed to a laser pulse with a central frequency ranging from 0.28 to 0.48 a.u. and with a pulse duration varying from 0.87 to 20 fs.

The general good performance of the PT method has been discussed in a number of physical quantities and excellent agreements with the accurate TDSE results have been achieved. However, deviations do appear when the laser spectra have significant components resonant with an intermediate state. The limitation comes from the inadequacy of the present PT model in dealing with the resonant case where the strong coupling between the two states must be properly accounted for.

We apply the PT method in the nonresonant regime to analyze one of the most important nondipole effects, i.e., the linear-momentum-transfer law in the two-photon ionization process. We identify the reasons for the minima in the momentum shift for the case of a linear polarization, i.e., they originate from the interferences among the dipole and nondipole ionization paths, and the oscillation behavior of the ratios of the ionization probabilities through different dipole paths to the total ionization probability.

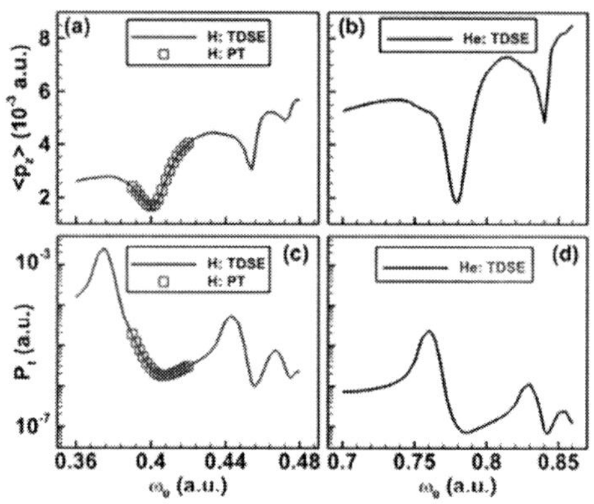

Figure 1. Expectation of the linear momentum transferred to the photoelectron ionized from (a) the hydrogen and (b) the helium atomic system. The two-photon ionization probability is shown as a function of the central frequency for (c) the hydrogen and (d) the helium atomic systems. The square symbols in (a) and (c) are results from the PT method.

References

[1] Pronin E *et al* 2009 *Phys. Rev. A* **80** 063403
[2] Wang M *et al* 2017 *Phys. Rev. A* **96** 043414

* E-mail: liangyou.peng@pku.edu.cn

Water molecule ionization by proton impact experiment and GTO-First Born theory

R Della Picca*, **D Cariatore, J Fiol, D Fregenal and G Bernardi**

CONICET and Centro Atómico Bariloche, CNEA, Bariloche, 8400, Argentina

Synopsis In this communication we present calculations and measurements of water ionization by protons at intermediate and high impact energies. The main goal of the present theoretical approach is the development of a method to routinely incorporate sophisticated wavefunctions for small and medium-sized molecules in the determination of ionization cross-sections.

The ionization of water molecules by charged particles is a fundamental process in describing the interaction of radiation with biological environments. The requirement of accurate ionization cross sections has stimulated the investigation of these molecular process from both theoretical and experimental points of view. A common characteristic in the theoretical models is the description of water ground state by means of the molecular wave function proposed by Moccia in 1964 [1]. Moccia used a basis set of Slater orbitals (STO) centered at the oxygen to construct the ground state. The great advantage of these orbitals is that the involved integrals with Coulomb wavefunctions could be analytically carried out employing the well know Nordsieck integral representation [2].

Nowadays, quantum chemistry packages (like GAMESS(US) [3]) routinely provide molecular orbitals based on Gaussian-type orbtials (GTO) basis functions virtually for every molecule. Conversely to STO case, the GTO-Coulomb integrals are not effortless to carry out.

In this work we rewrite the GTO water ground state obtained from density functional theory (same as in [4, 5]) as a sum of STOs centered in each nucleus $\gamma = \mathrm{O}, \mathrm{H}_1$ and H_2:

$$\phi_i(\mathbf{r}) = \sum_{j,\gamma} C_{j,\gamma} \underbrace{Y_{\ell_j}^{m_j}(\check{r}_\gamma)\, r_\gamma^{n_j-1}\, e^{-\alpha_j r_\gamma}}_{\mathrm{STO}(\gamma,j)} .$$

Then we perform the calculation of single ionization doubly differential cross section (DDCS) within the First Born approximation solving analytically the Coulomb space integrals [2], allowing us to easily extend the present procedure to other approximations. Despite this benefit, the molecular orientation average must be done carefully due the multicentre character of the wavefunctions.

Additionally, we have measured the DDCS of H_2O vapor by 1 MeV protons at the 1.7 MV Tandem accelerator at Centro Atmico Bariloche. As we show in figure 1, good agreement between theoretical and experimental data is found.

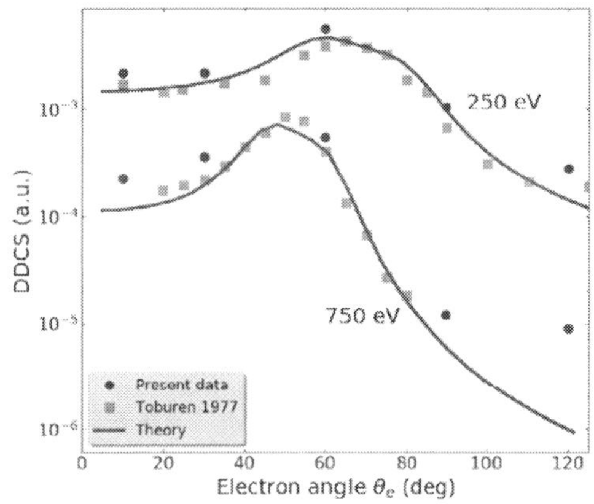

Figure 1. DDCS for single ionization of H_2O vapor by 1 MeV protons for 250 eV and 750 eV electron energies as a function of the electron angle. Theory: solid line. Present experiment: blue circles. Toburen and Wilson experiment [6]: orange squares.

References

[1] Moccia R 1964 *J. Chem. Phys.* **40** 2186
[2] Datta S 1985 *J. Phys. B* **18** 853
[3] Schmidt M W *et al* 1993 *J. Comput. Chem.* **14** 1347
[4] Dubois A *et al* 2011 *Phys. Rev. A* **84** 012708
[5] Della Picca R *et al* 2012 *J. Phys. B* **45** 194009
[6] Toburen L H *et al* 1977 *J. Chem. Phys.* **66** 5202

*E-mail: renata@cab.cnea.gov.ar

Target dependence of post-collision effects in ionization by proton impact

A Silvus[1], M Dhital[1], S Bastola[1], J Buxton[1], Z Klok[1], E Ali[1], M F Ciappina[2], D H Madison[1] and M Schulz[1*]

[1]Dept. of Physics and LAMOR, Missouri University of Science & Technology, Rolla, MO 65409, USA
[2]Institute of Physics of the ASCR, ELI-Beamlines, Na Slovance 2, 182 21 Prague, Czech Republic

Synopsis Double differential cross sections (DDCS) for ionization in p + Ne, Ar collisions were measured and calculated for a broad range of projectile energy losses as a function of scattering angle. PCI effects in the DDCS were found to be become increasingly prominent with increasing ionization potential *I* probably due to a greater importance of the nucleus-nucleus interaction when *I* becomes larger.

We have measured recoil ions produced in ionization of Ne and Ar by 75 keV p impact in coincidence with the momentum analyzed scattered projectiles. From the data we extracted double differential cross sections (DDCS) for a broad range of fixed energy losses ε as a function of projectile scattering angle θ_p.

In Fig. 1 the average scattering angle θ_{ave}, obtained by integrating the angular distribution of the DDCS, is shown for the Ar target as a function of the ejected electron to projectile speed ratio v_{el}/v_p. Up to about $v_{el}/v_p = 0.95$ θ_{ave} increases slowly, but at $v_{el}/v_p = 1.03$ a pronounced minimum is found. This structure can be explained by a focusing effect due to the post-collision interaction PCI. There, ionization involves at least 2 interactions between the projectile and the active electron. In the first, the electron is lifted to the continuum and in the second both particles attract each other towards the initial beam axis leading to the narrowing of the angular distribution of the DDCS. Between $v_{el}/v_p = 1.1$ and 1.2 θ_{ave} remains rather flat before increasing again around $v_{el}/v_p = 1.3$. This behavior is due to ionization from the initial 3s state for which the ionization potential *I* is significantly larger than for the 2p state. Since v_{el} was calculated using *I* for the 3p state, the matching speed for 3s ionization occurs at a value larger than $v_{el}/v_p = 1$, as indicated by the arrow in Fig. 1.

Along with data taken previously for He [1] and H and H_2 [2] θ_{ave} can now be compared for 5 different targets. The results show that the minimum close to the matching speed becomes increasingly pronounced with decreasing *I*. However, comparison to theory suggests that this does not necessarily mean that PCI becomes more important. Rather, the minimum gets increasingly washed out by an increasing role of the nucleus-nucleus interaction with increasing *I*.

This work was supported by NSF and by ADONIS.

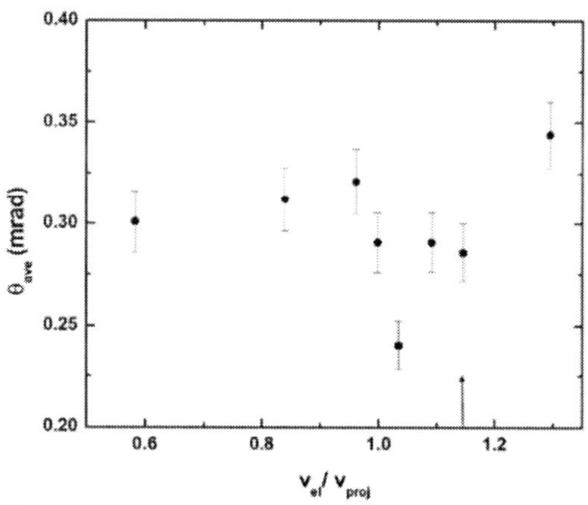

Figure 1. Average scattering angle θ_p of DDCS as a functiom of the electron to projectile speed ratio for the Ar target.

References

[1] Vajnai T *et al* 1995 *Phys. Rev. Lett.* **74** 3588
[2] Schulz M *et al* 2010 *Phys. Rev.* A **81** 052705

* E-mail: schulz@mst.edu

Kinetic energy release in dissociation of $NH_3{}^{q+}$ under ion impact

C P Safvan[1] *, T Sairam[2] and P Bhatt[1]

[1]Inter University Accelerator Centre, Aruna Asaf Ali Marg, New Delhi, 110067, India [2]Tata Institute of Fundamental Research, Hyderabad, 500107, India

Synopsis The dynamics of the dissociation of multiply charged ammonia ions is investigated. The pathways for sequential and concerted dissociation are separated, and the kinetic energy released along the different pathways is measured. The measured values are compared with *ab-initio* quantum chemical calculations to identify the repulsive states of the molecular ion which are responsible for the two pathways.

Dissociation of molecules under charged particle impact has been an active field of study. Concerted and sequential dissociation of multiply charged polyatomic molecular ions is being studied by several groups to understand the role of excited states in such processes. We have studied the dissociation of ammonia under energetic highly charged ion impact. The (multiply charged) projectile ion beam is obtained from an electron cyclotron resonance ion source. These ions are made to collide with an effusive jet of neutral NH_3 molecules at room temperature. The technique of recoil ion momentum spectrometry and multi-hit ion detection are used in unison [1] to measure the kinetic energy released in the dissociation of NH_3^{q+} (q=2,3). The measured data contains a mixed signature (black curve in Figure 1) for the events arising from the concerted and sequential dissociation dissociation of multiply ionized NH_3. We have used the technique of Dalitz plot [2] to disentangle the events arising from the symmetric concerted dissociation of NH_3^{3+} from the rest of the events.

The theoretical *ab initio* calculations using the quantum chemistry package GAMESS are performed to obtain the possible electronic states for the doubly and triply ionized ammonia molecule. The energy values for these states are found to be in reasonable agreement with our experimental results and are compared with earlier reported experimental result in literature [3]. These calculations further help in identifying the pathway of dissociation of doubly charged NH_3 when it dissociates sequentially.

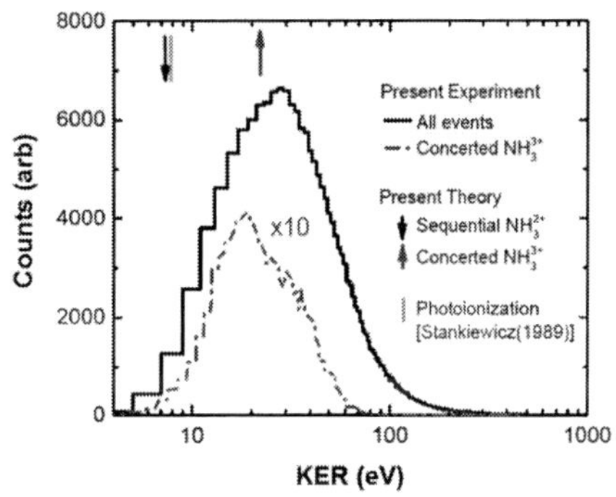

Figure 1. Kinetic energy release distribution for the dissociation of NH_3^{q+} (q=2,3)

References

[1] Bhatt P *et al* 2017 *Phys. Rev. A* **96** 22710
[2] Dalitz R 1953 *The London, Edinburgh and Dublin Philosophical Magazine Journal of Science* **44** 1068
[3] Stankiewicz M *et al* 1989 *J. Phys. B* **22** 21

*E-mail: cp.safvan@gmail.com

Journal of Physics: Conference Series **1412** (2020) 152008 doi:10.1088/1742-6596/1412/15/152008

Ionization cross-sections of L-shell and L-subshells by electron impact

M A R Patoary[1] *, M Shorifuddoza[1], A K F Haque[1] and M A Uddin[1]

[1]Department of Physics, University of Rajshahi, Rajshahi-6205, Bangladesh

Synopsis Vaisburd and Evdokimov proposed an empirical VE model to calculate the electron impact single ionization cross-sections of atoms and molecules. The extension of the modified VE model (MVE) has been proposed incorporating the relativistic effects. This model is applied to investigate the total L-shell as well as L1-, L2- and L3-subshells ionization cross-sections of various atomic targets with atomic numbers Z =1892 for incident energies up to 1 GeV. The results from the MVE model are compared with the available experimental data and theoretical findings.

The cross-sections of electron impact inner shell ionization (EIISI) are useful in various fields such as atomic physics, astrophysics, plasma physics and radiation science [?]. In literature, the EIISI cross-sections data are available for the K-shell. But it is limited to the L-shell as well as M-shell ionization. There are few quantum mechanical calculations are available namely plane wave Born approximation (PWBA) of Khare and Wadehra, the relativistic plane wave Born approximation (RPWBA) of Schofield, the distorted-wave Born approximations (DWBA) of Merlet et al., subconfiguration-average distorted-wave (SCADW) calculations of Pindzola and so on. Most of the quantum calculations are computationally difficult, time consuming, and produce results at some discrete incident energy. The analytical, semi-empirical or semi-classical models are available in literature, but these are limited in terms of species dependent parameters and energy ranges. Therefore it is necessary to develop an easy-to-use method to generate the cross-sections data of EIISI for a wide range of energies and atomic targets.

Vaisburd and Evdokimov [?] proposed an empirical VE model to calculate the electron impact single ionization cross-sections of atoms and molecules. The VE model combines the Lorentzian expression along with the Bethe formula. It is observed that, the VE model is limed to targets with $Z \leq 38$ and high-energy regime. In order to overcome the barrier of VE model, we proposed a new model namely MVE model [?, ?] which combines the Bethe term, the Lorentz expression, and Gryzinskis relativistic factor[?].

In order to verify our theory, we have performed the numerical calculations for K-shell as well as L-shell cases. Our model describes satisfactorily wide range of energies and atomic targets. In Fig 1, the result for L-shell ionization cross-sections of Au are compared with the experimental data as well as other theoretical findings and found good agreements.

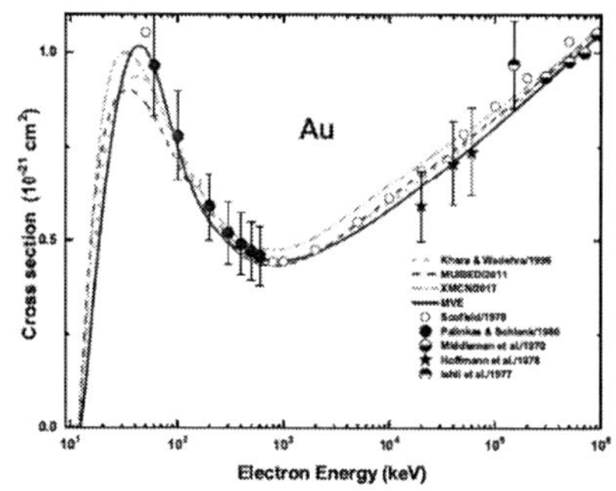

Figure 1. Electron impact total L-shell ionization cross-sections for Au. The sources of experimental EIISI cross-sections data are in solid symbols, present MVE result in solid blue line.

*E-mail: atiqur_physics@ru.ac.bd

References

[1] Haque A K F *et al* 2018 *Adv. Quantum Chem.* **77** 121

[2] Vaisburd D I *et al* 2003 *Russ. Phys. J.* **46** 1160

[3] Patoary M A R *et al* 2017 *Int. J. Mass Spectrom.* **415** 1

[4] Patoary M A R *et al* 2018 *Phys. Scr.* **93** 115401

[5] Gryzinski M 1965 *Phys. Rev.* **138** 336

First experimental results from a new high-power electron gun: Electron-impact ionisation of multiply charged xenon ions

B M Döhring[1,2]*, **B Ebinger**[1,2], **A Borovik Jr**[1], **K Huber**[1], **A Müller**[3] **and S Schippers**[1]

[1]I. Physikalisches Institut, Justus-Liebig-Universität Gießen, 35392 Giessen, Germany
[2]GSI Helmholtzzentrum für Schwerionenforschung GmbH, 64291 Darmstadt, Germany
[3]Institut für Atom- und Molekülphysik, Justus-Liebig-Universität Gießen, 35392 Giessen, Germany

Synopsis During recent years we have developed a new high-current electron gun for the Giessen electron-ion crossed-beams experiment aiming at an extension of the previously available collision-energy range of 1–1000 eV to 3500 eV. Here, we report on the first measurements of absolute cross sections for electron-impact ionisation of multiply charged xenon ions employing the new electron gun. Comparisons with available experimental and theoretical data from the literature provide convincing evidence that the new electron gun can be operated reliably over the entire design energy-range.

Electron-impact ionisation (EII) of ions is a fundamental atomic collision process that governs the charge balance in hot plasmas. Experimental measurements of the corresponding cross sections on an *absolute* scale are indispensable for testing the theoretical approaches that provide most of the data required for applications such as magnetically confined fusion [1] or light sources for EUV lithography [2]. In order to expand our experimental capabilities we have developed a new high-current electron gun [3] for the Giessen electron-ion crossed-beams experiment aiming at an extension of the previously available collision-energy range of 1–1000 eV to 3500 eV.

Figure 1 compares two sets of experimental cross sections for single ionisation of Xe^{12+} ions that were measured with our old and with our new electron gun. Both data sets agree with one another within the experimental uncertainties as well as with the result of a distorted wave calculation [1]. With the old gun, the cross sections could only be measured up to the region of the cross-section maximum. The new gun permits measurements at energies well beyond that limit. With increasing ion charge state, the cross-section maximum moves to higher energies. Therefore, measurements with more highly charged ions will greatly benefit from our expanded capabilities. On the poster we will present results for single and double electron-impact ionisation of Xe^{q+} ions with $q = 12 - 14$. Finally, we would also like to mention, that our new gun serves as a prototype for a high-intensity transverse electron target in the low-energy heavy-ion storage ring CRYRING [4] of the international FAIR accelerator facility in Darmstadt, Germany.

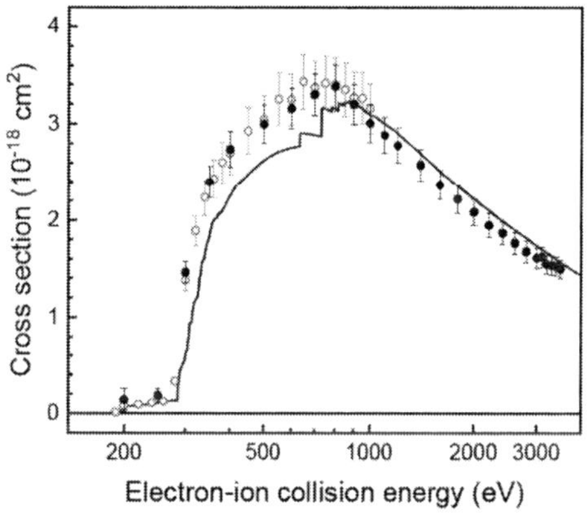

Figure 1. Experimental (open symbols: old gun; full symbols: new gun) and theoretical (solid line) cross sections for electron-impact single ionisation of Xe^{12+} ions.

References

[1] Borovik A *et al* 2015 *J. Phys. B* **48** 035203
[2] Borovik A *et al* 2013 *J. Phys. B* **46** 175201
[3] Ebinger B *et al* 2017 *Nucl. Instrum. Meth. B* **408** 317
[4] Lestinsky M *et al* 2016 *Eur. Phys. J. ST* **225** 797

*E-mail: michel.doehring@physik.uni-giessen.de

Role of the variable charge in (e,3e) processes on argon

A Herbadji[1], A Mansouri[1], C Dal Cappello[2], I Kada[1] and S Houamer[1*]

[1]LPQSD, Department of physics, Faculty of Science, University Sétif-1, 19000, Setif, Algeria
[2]SRSMC, UMR CNRS 7565, University of Lorraine, BP 70239, 54506 Vandoeuvre-les-Nancy, France

Synopsis : The differential fivefold cross section for the double ionization of argon by electron impact is calculated taking into account the variations of the effective charges describing each ejected electron during the reaction; the energy of the electron projectile is fixed in the intermediate range, and the ejected electrons have low energies. The initial state is described by the Clementi wavefunction, and the ejected electrons in the final state by Coulomb waves. A comparison is made with the measurements of Jia et al [1].

Collision systems studies are still a very powerful tool to probe the structural properties of atoms and molecules. The calculation of the cross sections for (e,2e) and (e,3e) reactions has been the subject of many works, one of the last studies takes into account the variations of the change of the target felt by the projectile during the reaction of ionization process [2]. In this work we calculate the differential fivefold cross sections in the reaction (e,3e) of the argon atom. Calculations are made in the first Born approximation. The initial state is described by the Hartree Fock wave functions, and the ejected electrons in the final state are described by Coulomb waves, the projectile in the initial and final channel being described by a plane wave. The variations of the charge of the target as a function of 'r' is calculated using the properties of the static potential felt by the projectile during the reaction. To compare our model with the experiment, we investigate the double ionization of argon atoms, by electron impact in the experimental conditions of Jia et al. [1] where the incident electron energy is fixed at 561.4 eV and the scattered angle θ_s = 6.5°. The ejected electron energies are chosen to be (E_1,E_2) = (9eV,9eV).

The correlations between ejected electrons are described by the Ward-Macek method [3]. Fivefold cross sections (FDCS) are calculated as a function of the angle of the first ejected electron, the second is fixed by the experiment study. As an example, we present in figure 1, the FDCS for one case of the angle of the second ejected electron (θ_2 = 110°), we compare our theoretical results : fixed charge (dashed line) and variable charge (solid line) with experimental data [1].

The model with the variable charge makes important corrections to the curves of the FDCS cross sections when compared to those calculated using the fixed charge model for all cases considered in reference[1], the maximums of the reaction foreseen by the experiment are better described in this model.

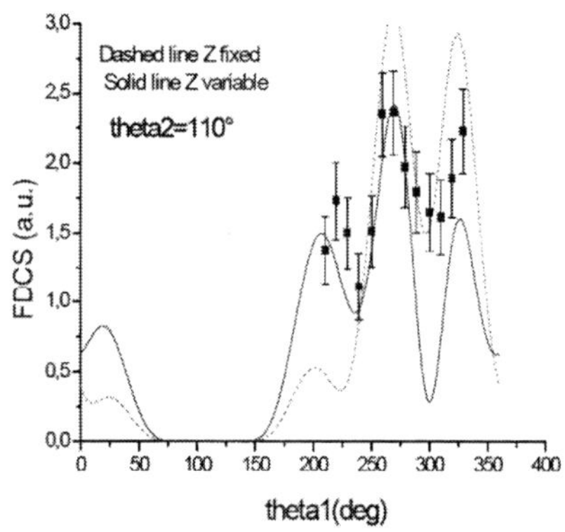

Figure1: FDCS for argon atoms, the incident electron energy is E_s = 561.4 eV and the ejected electrons energies are (E_1,E_2) = (9eV,9eV). The scattering angle is θ_s = 6.5° and θ_2 = 110°.

References

[1] Jia C C *et al* 2002 *J. Phys. B* **35** 1103
[2] Chinoune M *et al* 2016 *J. Phys. B* **49** 205201
[3] Ward S J and Macek J H 1994 *Phys. Rev. A* **49** 1049

* E-mail: s_houamer@univ-setif.dz

Electron-impact ionisation of W^{q+} ions with $1 \leq q \leq 19$

D Schury[1,2,3]*, A Becker[1], A Borovik Jr[4]†, B Ebinger[3,4], M F Gharaibeh[1,5,7], J Hellhund[1], K Huber[4], J Rausch[1], K Spruck[1,8], F Jin[4,6], S Schippers[4] and A Müller[1]

[1]Institut für Atom- und Molekülphysik, Justus-Liebig-Universität Gießen, 35392 Giessen, Germany
[2]Institut des Nanosciences de Paris, Sorbonne Université, 75005 Paris, France
[3]GSI Helmholtzzentrum für Schwerionenforschung GmbH, 64291 Darmstadt, Germany
[4]I. Physikalisches Institut, Justus-Liebig-Universität Gießen, 35392 Giessen, Germany
[5]Department of Physics, Jordan University of Science and Technology, Irbid 22110, Jordan
[6]Department of Physics, National University of Defense Technology, Hunan 410073, Changsha, China
[7]Department of Mathematics, Statistics and Physics, P.O.Box 2713, Qatar University, Doha, Qatar
[8]Institut für Medizinische Physik und Strahlenschutz, THM, 35390 Giessen, Germany

Synopsis Electron-impact ionisation cross sections of tungsten ions in the energy range from the ionisation threshold to 1000 eV were measured. The experimental cross sections are compared with results from sub-configuration averaged distorted wave calculations. Derived plasma rate coefficients are provided for convenient use in plasma modelling.

Tungsten is considered a suitable material for plasma-facing components in fusion reactors due to its good thermo-mechanical properties and its low sputtering yield, low chemical erosion, low tritium retention and high durability against neutron flux [1]. As these components will be subject to high particle fluxes, tungsten impurities in the fusion plasma cannot be avoided [2]. The tungsten ions are expected to absorb collisional energy of free electrons and emit strongly in the X-ray and EUV energy range [3] which leads to energy loss in the optically thin plasma which cannot reabsorb the radiation. Control and understanding of the conditions in the plasma requires extensive modelling and this, in turn, reliable data on atomic processes. This motivated our present study on electron-impact ionisation of tungsten ions which complements related work on photoionisation [4] and recombination [5] of tungsten ions. We measured absolute cross sections for electron-impact single ionisation of tungsten ions of the charge states 1 to 19 employing the animated-beam technique at the Giessen electron-ion crossed-beams experiment [6, 7]. In addition we performed accompanying sub-configuration averaged distorted wave calculations (SCADW) using the FAC code [8] in order to guide the interpretation of the measured cross sections. For easier access in plasma modelling, plasma rate coefficients are derived from the experimental and theoretical cross sections.

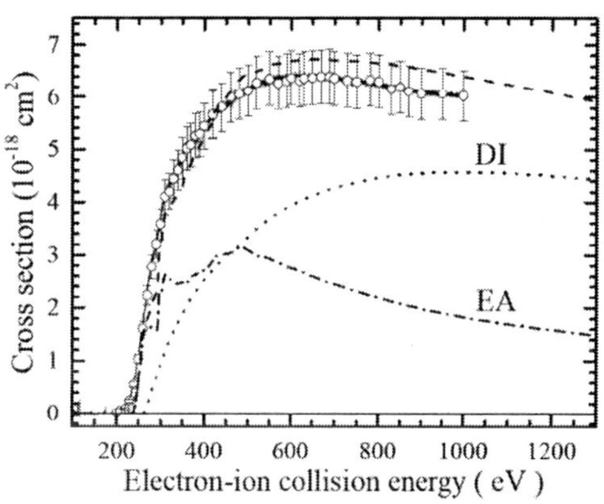

Figure 1. Cross sections for electron-impact single ionisation of W^{12+} ions: Experimental data (solid line, open symbols with quadrature sum of the statistical and systematic uncertainties as error bars) compared to our results of SCADW calculations (dashed line), comprising direct ionisation (DI, dotted line) and excitation autoionisation (EA, dash-dotted line).

References

[1] Philipps V 2011 *J. Nucl. Mater.* **415** S2
[2] Dux R *et al* 2011 *Nucl. Fusion* **51** 053002
[3] Pütterich T *et al* 2010 *Nucl. Fusion* **50** 025012
[4] Müller A *et al* 2017 *J. Phys. B* **50** 085007
[5] Krantz C *et al* 2017 *J. Phys. B* **50** 052001
[6] Rausch J *et al* 2011 *J. Phys. B* **44** 165202
[7] Borovik A *et al* 2016 *Phys. Rev. A* **93** 012708
[8] Gu M F 2008 *Can. J. Phys.* **86** 675

*E-mail: Daniel.Schury@physik.uni-giessen.de
†E-mail: Alexander.Borovik@physik.uni-giessen.de

H_2O single ionization by light particle-impact: fully differential cross sections

E Acebal[1] * and S Otranto[1]

[1]Instituto de Física del Sur (IFISUR), Departamento de Física, Universidad Nacional del Sur (UNS), CONICET, Av. L. N. Alem 1253, B8000CPB-Bahía Blanca, Argentina

Synopsis Fully differential cross sections for the single ionization of the $1b_1$ and $3a_1$ orbitals of H_2O by electron- and positron-impact are calculated by means of the CDW-EIS model at an impact energy of 81 eV. A proper average procedure over molecular orientations is performed. Main differences between positron- and electron-impact are observed for coplanar geometries in the binary structures. In addition, an enhancement of the binary region and a decrease of the recoil lobe are observed for positron-impact.

Over the last decades, radiation therapy is becoming one of the most common modes of cancer treatment. This increases the importance of studying the dynamics of ionization radiation with living tissue. In this sense, the ionization of the water molecule (H_2O) by particle-impact has become of great interest since this single compound represents up to sixty percent of the human body.

Fully differential cross sections (FDCS), in this sense, provide the most exhaustive test for state-of-the-art theoretical models and a clear challenge for current experimental setups.

Recently, a complete kinematically study of the single ionization of the two most outer valence orbital of H_2O by electron-impact at 81 eV has been reported [1]. The theory under use was based on the DWBA, and the authors concluded that higher-order terms in the projectile –target interaction might be relevant at this impact energy based on the discrepancies observed with the experimental data. Subsequently, in a previous work, we analysed those data with the CDW-EIS model [2] and found good overall agreement with the reported data.

In this work, we focus ourselves on the influence of the projectile charge sign in the single ionization of the $1b_1$ and $3a_1$ orbitals of H_2O. Fully differential cross sections were calculated by means of the CDW-EIS model and contrasted after a proper averaged procedure over molecular orientations.

In Figure 1, we compare positron- and electron-impact results for one of the four geometries reported for the scattering plane. It can be seen how the recoil lobe is decreased for positron-impact. We also observe a maximum among the direction of $\boldsymbol{k}_1$, which is slightly suggested for electron projectiles but no fully apparent. These features obey to the post-collisional interaction. The same situation stands for the four geometries reported.

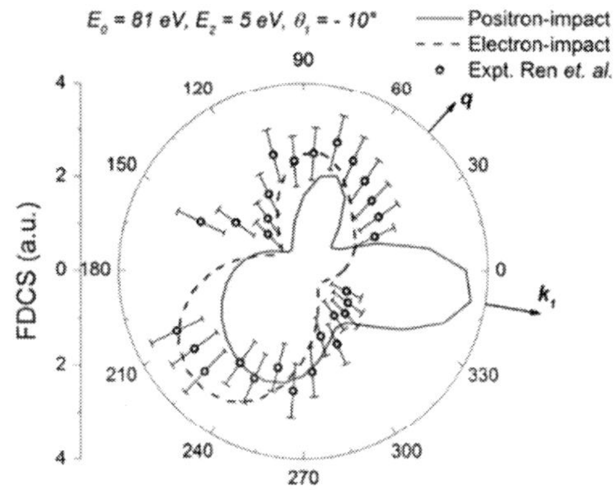

Figure 1. FDCS for positron- (solid line) and electron- (dashed line) impact ionization of H_2O, for the scattering plane xz. Cross sections represent the sum of $1b_1$ and $3a_1$ orbitals. Experimental data for electron-impact from Ref. [1] scaled to the present theoretical results.

In addition, we present FDCS for electron- and positron-impact ionization of the $1b_1$ and $3a_1$ orbitals of H_2O, for the denominated semi-perpendicular and full perpendicular plane. In these cases, similar structures are observed with an enhancement of the binary region and a decrease of the recoil lobe for positron-impact.

References

[1] Ren X *et al* 2017 *Phys. Rev. A* **95** 022701
[2] Acebal E and Otranto S 2018 *Phys. Rev. A* **98** 012703

* E-mail: emiliano.acebal@uns.edu.ar

Photoionization of superheavy elements

R Khademhosseini[1*], A Razavi[1], D A Keating[1], P C Deshmukh[2] and S T Manson[1†]

[1]Department of Physics and Astronomy, Georgia State University, Atlanta, Georgia 30303, USA
[2]Department of Physcis, Indian Institute of Technology-Tirupati, Tirupati 517506, India

Synopsis Calculations of the photoionization cross sections for the superheavy atoms of atomic numbers 102, 112 and 118 have been performed using the relativistic-random-phase approximation (RRPA) which is based on the Dirac equation. The results show that relativistic effects cause anomalies in level ordering for some inner shells, and that interchannel coupling is of great importance.

Superheavy elements offer a "laboratory" to study the effects of strong relativistic interactions within the context of atomic systems. Specifically, an investigation of the photoionization of the superheavy atoms, Z=102 (No), Z=112 (Cn) and Z=118 Og) is performed employing the relativistic-random-phase approximation (RRPA) [1] methodology which is based upon the Dirac equation, so that relativistic effects are included on an *ab initio* basis. In addition, significant amounts of initial-state correlation (two-particle two-hole excitations) and final-state correlation (interchannel coupling among all open and closed single-excitation channels) are included.

The structure of the higher-Z atoms is peculiar, owing to relativistic effects, in that the 5f thresholds lie between the $6p_{3/2}$ and $6p_{1/2}$ thresholds owing to their huge (relativistic) spin-orbit splitting. In addition, the outer-shell structure of Cn is $6d_{3/2}{}^4 7s^2 6d_{5/2}{}^6$ [2]. The atoms studied are all closed-shell, which simplifies the interpretation of the results, and lie below Yb, Hg and Rn in the periodic table.

As an example, the total photoionization cross section for Og is shown in Fig. 1. The cross section is dominated by the 7p subshells at the near-threshold peak, then, by the 6d sbnshells at the next peak. Starting at about 6 a.u., the cross section is dominated by the broad peaks in the 5f subshells which exhibit Cooper minima at about 15 a.u.; the total cross section is dominated by the second maxima in the 5f's from there to the opening of the 4f photoionization channels just below 50 a.u. In addition, it is found that the dominant channels exhibit very significant interchannel coupling on the weaker channels so that, for example, in the 10 a.u. photon energy region, all of the subshell cross sections look like mini-versions of the 5f cross sections. It is also found that this interchannel coupling could alter the uncoupled cross section by a factor of ten or more.

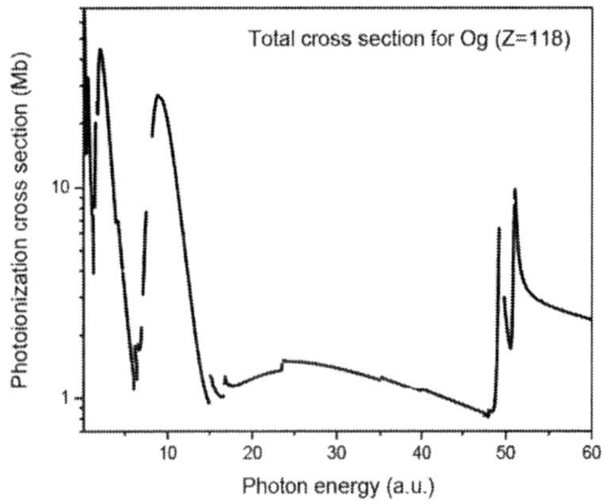

Figure 1. Total photoionization cross section for Og (Z=118) from threshold to 60 a.u. The gaps represent the energy regions of autoionizing resonances below each threshold which are omitted from the plot.

This work was supported by the US Department of Energy.

References

[1] Li J *et al* 2007 *Sci. China Ser. G* **50** 707
[2] Johnson W and Lin C D 1979 *Phys. Rev. A* **20** 964

* E-mail: rkhademhosseini1@student.gsu.edu
† E-mail: smanson@gsu.edu

Quantum vortex surfaces: a dizzying new effect in single ionization collisions

F Navarrete[1] and R O Barrachina[2,3] *

[1]Department of Physics, Kansas State University, Manhattan, Kansas 66506, USA.
[2]Centro Atmico Bariloche and Instituto Balseiro, Comisin Nacional de Energa Atmica (CNEA) and Universidad Nacional de Cuyo, Av. Bustillo 9500, 8400 Bariloche, Argentina.
[3]Consejo Nacional de Investigaciones Cientficas y Tcnicas (CONICET), Argentina.

Synopsis We study the appearance of quantum vortices in the transition matrix element of single ionization collisions. We relax all the usual geometrical restrictions on the momenta of the final-state, and demonstrate that previously observed vortex points and lines correspond to 2D and 3D intersections of a much more complex structure: a quantum vortex surface. Finally, we theoretically evaluate this complex structure for the single ionization of hydrogen by positron impact.

Back in the early 1990's, a deep minimum was experimentally observed in the differential cross section of the (e,2e) process in helium [1]. A similar minimum was theoretically found by Brauner and Briggs [2] for the ionization of hydrogen by the impact of positrons. Two decades later, these minima were shown to be isolated zeroes, and recognized as quantum vortices [3, 4, 5]

For certain projectile energies, more than one vortex point were observed in collinear cuts of the fully differential cross section (FDCS) and were supposed to be independent. However, when one of the locked dimension was released, they turned out to be the intersections of a single vortex line with the 2D cut of the corresponding restricted geometry [6], and later identified as a vortex ring [7]. A schematic plot of their morphology is given in Fig.1 on the right, where we can see their closed structure, and also a closed vortex field around them.

Now, since the FDCS depends of four independent variables, it is not unreasonable to assume that the previous argument can be applied again, and that the vortex ring would in fact be the cut of a vortex surface. In this communication we demonstrate that this is actually the case. In the Fig.1, on the left, we show the geometry of the problem. A positron of initial momentum K_i (not shown in the figure) along the z axis, is scattered with momentum K in the plane xz, removing an electron of momentum k from a target at the origin. Here we extend previous works by releasing the coplanar constraint, calculating vortex rings in all the momentum space of the electron at different final angles θ of the positron. This procedure allows us to take a glimpse to different cuts of a vortex surface of the FDCS in a 4D momentum space.

*E-mail: barra@cab.cnea.gov.ar

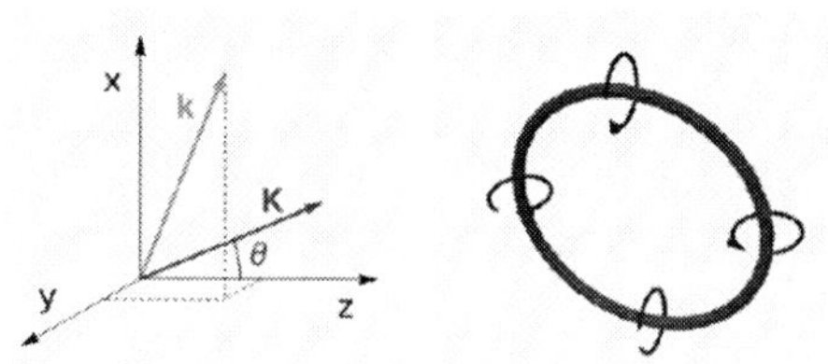

Figure 1. On the left, the momenta of the electron k (green) and the positron K (red). The positron is scattered in the plane xz, for which we calculate vortex rings in the momentum space of the electron for different positron angles θ. On the right, a scheme of a vortex ring (red) and the vorticity around it.

References

[1] Murray A J and Read F H 1993 *Phys. Rev. A* **47** 3724
[2] Brauner M and Briggs J S 1991 *J. Phys. B* **24** 2227
[3] Macek J H *et al* 2010 *Phys. Rev. Lett.* **104** 033201
[4] Navarrete F *et al* 2013 *J. Phys. B* **46** 115203
[5] Navarrete F and Barrachina R O 2015 *J. Phys. B* **48** 055201
[6] Navarrete F and Barrachina R O 2016 *Nucl. Instr. Meth. B* **369** 72
[7] Navarrete F and Barrachina R O 2016 *J. Phys. Conf. Ser.* **875** 012022

Spectroscopy and photoionization of C^+ ions by the relativistic eigenchannel R-matrix method

X F Yue[1*], R Jin[2], H R Feng[1], M C Jiao[1], X Gao[3,4†] and J M Li[2,5,6]

[1]Department of Physics and Information Engineering, Jining University, Qufu, 273155, China
[2]Department of Physics and Center for Atomic and Molecular Nanosciences, Tsinghua University, Beijing, 100084, China
[3]Center for Computational Sciences, University of Tsukuba, 1-1-1 Tennodai, Tsukuba, Ibaraki 305-8577, Japan
[4]Beijing Computational Science Research Center, Beijing, 100193, China
[5]Key Laboratory for Laser Plasmas (Ministry of Education) and Department of Physics and Astronomy, Shanghai Jiao Tong University, Shanghai, 200240, China
[6]Collaborative Innovation Center of Quantum Matter, Beijing, 100084, China

Synopsis The scattering matrices of C^+ are calculated by using our relativistic eigenchannel R-matrix method. With the merits of the analytical continuation properties of the scattering matrices, lots of atomic dynamical data with accuracy are obtained. Much physical quantities are acquired by applying the multichannel quantum defect theory (MQDT) procedure. The present calculated results are in good agreement with the available experimental data, which has been shown by the newly proposed projected high dimensional quantum defect graph (symmetrized), namely JHANGZ plot.

Carbon is one of the most abundant metals in the universe. C II is an important ion in astrophysical sources, such as the interstellar medium. Accurate features of C II are considerable helpful for accurate spectral analysis. Kjeldsen et al. [1] carried out the first measurements of the absolute photoionization cross sections of C II from the ionization threshold at 24 eV to 31 eV using the merged ion-photon beam setup. Their experimental results are generally agreement with nonrelativistic theoretical data in the near-threshold region, but differences in the magnitude of the continuum cross section of 15%-35% at higher energies and several deviations in the resonance structure. Nahar [2] reported the photoionization cross sections including relativistic effects in the Breit-Pauli R-matrix approximation for the fine-structure levels of the ground configuration $2s^22p$ of C II. Our motivation of the present work is to in further study relativistic effects in relation to the experimentally observed features with our relativistic eigenchannel R-matrix method (R-R-eigen) [3-4].

The R-R-eigen method has been presented in detail in Ref. [3-4]. Here we only describe the present calculation. We can readily obtain the scattering matrices which demonstrates good analytic continuation in discrete and continuum regions. The scattering matrices provides abundant of physical parameters needed in our MQDT calculations. Accuracy of short range scattering matrices can be tested by the available exactly experimental spectroscopic data of atomic energy levels. For the photoionization problem, because of selection rule, finite number of partial waves are needed. The $1/2^-$ partial wave is shown as an illustration example. The calculation photon energy is considered up to 1.0 Ry in order to cover important autoionization resonances associated with thresholds $[1s^22s^12p^1]$ $^3P_{0,1,2}$ and $^1P_1$ (with IPs 0.47 Ry and 0.93 Ry respectively). With the energy increases, the number of physical channels n_p will change. Five energy regions in the present calculation, i.e., 1, 3, 7, 13 and 16 physical channels are selected. The results demonstrate that all the quantum defects vary smoothly in different energy regions and connect with each other. The discrete energy levels and autoionization resonant levels without missing any lines are obtained, which have been compared with the available spectroscopic data through the newly proposed JHANGZ plot [4].

* E-mail: xfyuejnu@jnxy.edu.cn
† E-mail: xgao@csrc.ac.cn

References

[1] Kjeldsen H *et al* 1999 *Astrophys. J.* **524** L43
[2] Nahar N S 2002 *Phys. Rev. A* **65** 052702
[3] Zeng D L *et al* 2017 *Phys. Rev. A* **95** 042508
[4] Jin R *et al* 2017 *Sci. Rep.* **7** 11589

Experimental and theoretical study of singly ionizing 1-MeV p+He collisions at different energy and momentum transfer values

O Chuluunbaatar[1,2], K A Kouzakov[3], S A Zaytsev[4], A S Zaytsev[4], V L Shablov[5], Yu V Popov[6,1], H Gassert[7], M Waitz[7], H-K Kim[7], T Bauer[7], A Laucke[7], Ch Müller[7], J Voigtsberger[7], M Weller[7], J Rist[7], K Pahl[7], M Honig[7], M Pitzer[7], S Zeller[7], T Jahnke[7], L Ph H Schmidt[7], H Schmidt-Böcking[7], R Dörner[7] and M S Schöffler[7]*

[1]Joint Institute for Nuclear Research, Dubna, Moscow region 141980, Russia
[2]Institute of Mathematics, National University of Mongolia, UlaanBaatar, Mongolia
[3]Faculty of Physics, Lomonosov Moscow State University, Moscow 119991, Russia
[4]Pacific National University, Khabarovsk 680035, Russia
[5]Obninsk Institute for Nuclear Power Engineering of the National Research Nuclear University MEPhI, Kaluga Region, 249040 Russia
[6]Skobeltsyn Institute of Nuclear Physics, Lomonosov Moscow State University, Moscow 119991, Russia
[7]Institut für Kernphysik, Universität Frankfurt, 60438 Frankfurt, Germany

Synopsis We present ultra-high-resolution COLTRIMS data on single ionization of helium induced by 1 MeV proton impact and compare it with theoretical calculations. Our analysis clearly shows the limitations of the first Born approximation.

In our earlier study [1], we reported high-resolution COLTRIMS data for the emission of electrons with energy $E_e = 6.5 \pm 3.5$ eV in 1-MeV proton-helium collisions at momentum transfer $q = 0.75 \pm 0.25$ a.u. The data exhibited a well-pronounced nodal structure, which is in agreement with predictions of the first Born approximation (FBA) for the electron three-dimensional angular distribution. At the same time, some discrepancy between the FBA theory and experiment was found in the scattering plane: the binary and recoil peaks appeared to be shifted towards smaller emission angles with respect to the FBA calculations.

In this contribution we present the experimental data and numerical calculations for the fully differential cross section (FDCS) in the form

$$\mathrm{FDCS} = \frac{d^5\sigma}{dE_e d\theta_e d\phi_e dq d\phi_q}, \quad (1)$$

where θ_e and ϕ_e are the polar and azimuthal angles of the emitted electron, and ϕ_q is the azimuthal angle of the momentum transfer q. We study the cases of momentum-transfer values $q = 0.5, 1.0,$ and 1.75 a.u. and (for each momentum-transfer value) the ejected-electron energies $E_e = 2.5, 5, 10$ and 20 eV. Fig. 1 shows experimental and FBA values for FDCS when $E_e = 20$ eV. The FBA calculations are averaged within the experimental E_e- and q-gates.

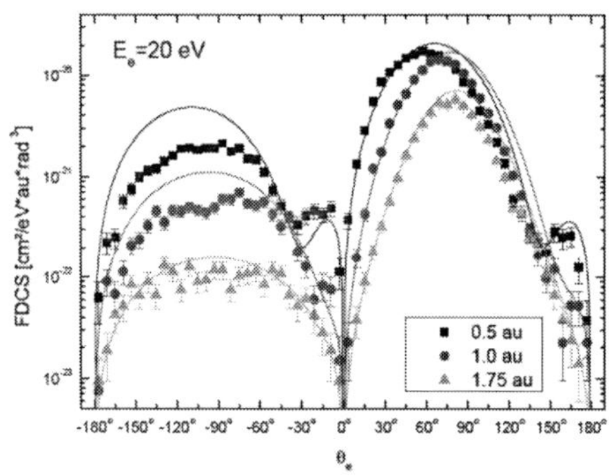

Figure 1. Experimental angular distribution in the scattering plane for $E_e = 20 \pm 5$ eV kinetic energy. The black squares correspond to a momentum transfer $q = 0.5 \pm 0.15$ a.u., the red circles to $q = 1.0 \pm 0.25$ a.u. and the blue triangles to $q = 1.75 \pm 0.4$ a.u. Solid lines correspond to the FBA using a Hartree-Fock wave function of helium.

We examine theoretical approaches beyond the FBA, such as the 3C model with effective charges [2], replacing potentials by T matrices, and semiclassical post-collision interaction.

References

[1] Gassert H *et al* 2016 *Phys. Rev. Lett.* **116** 073201
[2] Chuluunbaatar O *et al* 2017 *Phys. Rev. A* **96** 042716

*E-mail: schoeffler@atom.uni-frankfurt.de

Nonlinear Cooper minimum as a precise tool for understanding multiphoton photoionization

J Hofbrucker[1,2,3]*, **A V Volotka**[2,3] **and S Fritzsche**[1,2,3]

[1] Theoretisch-Physikalisches Institut, Friedrich-Schiller-Universität Jena, Jena, D-07743, Germany
[2] Helmholtz-Institut Jena, Jena, D-07743, Germany
[3] GSI Helmholtzzentrum für Schwerionenforschung GmbH, Darmstadt, D-64291, Germany

Synopsis A new approach to accurately assess multiphoton ionization is suggested. Vanishing of the dominant ionization channel in nonresonant (direct) multiphoton ionization is predicted for a specific incident photon energy. The exact energy position of such *nonlinear Cooper minimum* can be accurately measured and requires calculations of the complete electronic spectrum. Measurements of various observables at these photon energies are desirable for further evaluation of theoretical calculations at hitherto unreachable accuracy.

The advances and increasing number of free electron lasers (FELs) revolutionized the exploration of inner-shell electron dynamics and nonlinear light-matter interaction by delivering intense extreme-ultraviolet and x-ray beams. However, the accuracy of the extracted cross sections is bound by the systematic uncertainty in the beam parameters which determine the absolute intensity. This causes large experimental uncertainties which in return do not allow to resolve differences between various theoretical approaches [1].

We suggest alternative measurable quantities which promise more accurate comparison of experiment with theory. It is generally expected that the properties of multiphoton ionization are dominantly determined by ionization channel with the highest angular momentum. Here, we show that this rule is broken not only at photon energies matching intermediate electronic resonances, but also at a nonlinear equivalent of Cooper minimum, where the dominant ionization channel drops to zero. We further suggest that measuring observables such as fluorescence polarization, photoelectron or Auger distributions (see Figure 1 for a graphical example) will accurately determine the position of these *nonlinear Cooper minima* and will allow for comparison of experimental measurements with theory at hitherto unreachable precision. Measurements of fluorescence or angular distributions of escaping electrons as a characteristic of multiphoton ionization are well established, and hence, measurements of these properties at the nonlinear Cooper minimum could readily find applications beyond fundamental importance, as it could also serve for applied fields such as spectroscopy of atoms and molecules [2].

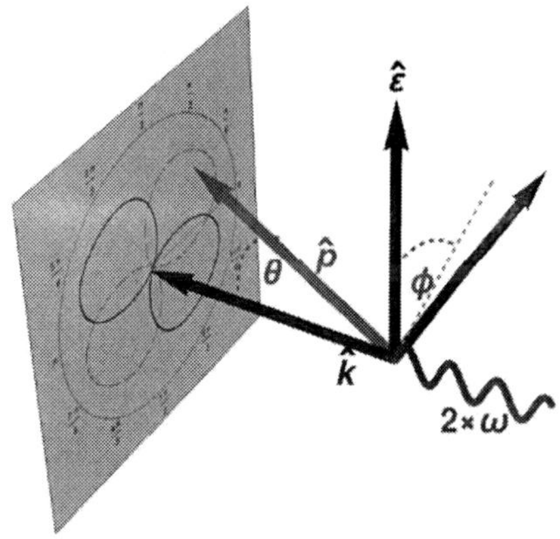

Figure 1. Schematic representation of possible determination of nonlinear Cooper minimum position with the detection of maximum elliptical dichroism in two-photon ionization of He. Photoelectron distribution are detected in dipole plane for incident left- (black) and right-(green) handed elliptically polarized light.

References

[1] Ghimire S *et al* 2016 *Phys. Rev. A* **94** 043418
[2] Hofbrucker J *et al* 2018 *Phys. Rev. Lett.* **121** 053401

*E-mail: j.hofbrucker@uni-jena.de

Auger-like correlations in the two-photon above threshold ionization of atoms and molecules in the vicinity of the giant resonance

B M Lagutin[1] *, **I D Petrov**[1], **N M Novikovskiy**[2], **V L Sukhorukov**[2,3], **Ph V Demekhin**[3] **and A Ehresmann**[3]

[1]Rostov State University of Transport Communications, Rostov-on-Don, 344038, Russia
[2]Institute of Physics, Southern Federal University, Rostov-on-Don, 344090, Russia
[3]Institute of Physics and Center for interdisciplinary Nanostructure Science and Technology (CINSaT), University of Kassel, Kassel, D-34132, Germany

Synopsis The two-photon ionization of the Ne–CH_4 and Ar–SH_4 isoelectronic systems was studied theoretically for the photon energies exceeding the ionization threshold of the outer shells. The calculations demonstrate that the two-photon ionization of the investigated objects is almost entirely a collective process if the one-photon energy corresponds to the energy of the $n\ell \dashrightarrow \varepsilon\ell'_R$ giant resonance. The largest contribution to the transition amplitude stems from the Auger-like correlation $n\ell^2 \overset{(2)}{\dashrightarrow} \varepsilon\ell'^2_R \Rightarrow n\ell\varepsilon''\ell''$.

The multiphoton ionization of atoms is a matter of theoretical studies since a long time [1, 2, 3, 4]. The renewed interest in the description of the multiphoton ionization processes is connected with the progress in the development of the advanced light sources.

Recently, an extended investigation of the two-photon ionization of Ar has been performed in the exciting-photon energy range below [5] and above [6] the $3p^5$ threshold with taking into account all single and double excitations in addition to the Lowest Order Perturbation Theory (LOPT) approximation.

It was revealed [6] that in the Above Threshold Ionization (ATI) of Ar the Auger-like correlation $3p^2 \overset{(2)}{\dashrightarrow} \varepsilon d^2_R \Rightarrow 3p\varepsilon' f$ plays a decisive role increasing the Generalized 2-Photon Ionization Cross Section (G2PICS) in several times (see Fig. 1 where the total 3p G2PICS computed in the Hartree-Fock (HF) and Hartree-Slater (HS) approaches are depicted). This correlation corresponds to the process when two photons excite two electrons in the continuum (dashed arrow) with the subsequent Auger-like decay and has an exceptionally large value when the exciting-photon energy corresponds to the energy of the single-electron giant resonance $3p \dashrightarrow \varepsilon d$.

In the present work we report the results of the G2PICS calculations for the isoelectronic pairs Ne–CH_4 and Ar–SH_4. In the case of Ne where the giant resonance is not formed yet the Auger-like correlation changes the LOPT results by less than 20%, whereas in CH_4 where the $\sigma_{1t_2(2p)}(\omega)$ is about 50 Mb at the threshold the Auger-like correlation leads to the enhancement of G2PICS by more than 100%.

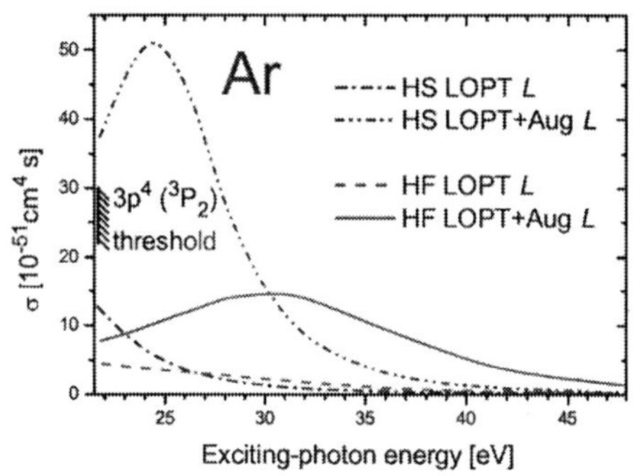

Figure 1. Total 3p G2PICSs of Ar computed in the HF and in the local HS approximations with and without taking into account Auger-like correlations. Our HS LOPT result coincides with calculation of Ref. [7].

*E-mail: bml58@mail.ru

References

[1] Lambropoulos P 1974 *Phys. Rev. A* **9** 1992
[2] Pindzola M S and Kelly H P 1975 *Phys. Rev. A* **11** 1543
[3] Aymar M and Crance M 1980 *J. Phys. B: At. Mol. Phys.* **13** L287
[4] Starace A F and Jiang T F 1987 *Phys. Rev. A* **36** 1705
[5] Petrov I D *et al* 2016 *Phys. Rev. A* **93** 033408
[6] Petrov I D *et al* 2019 *Phys. Rev. A* **99** 013408
[7] Pi L W and Starace A F 2010 *Phys. Rev. A* **82** 053414

Tracing inter-Coulombic decay of molecular dimers

W Iskandar[1]*, A S Gatton[1], B Gaire[1], F P Sturm[1], K A Larsen[1,2], E G Champenois[1,2], N Shivaram[1], A Moradmand[3], J B Williams[4], B Berry[5], T Severt[5], I Ben-Itzhak[5], D Metz[6], H Sann[6], M Weller[6], T Jahnke[6], R Dörner[6], D Slaughter[1] and Th Weber[1]

[1]Chemical Sciences Division, Lawrence Berkeley National Laboratory, Berkeley, CA-94720, USA
[2]Graduate Group in Applied Science and Technology, University of California, Berkeley, USA
[3]Department of Physics, Auburn University, Auburn, AL-36849, USA
[4]Department of Physics, University of Nevada, Reno, NV-89557, USA
[5]J. R. Macdonald Laboratory, Department of Physics, Kansas State University, Manhattan, KS-66506, USA
[6]Institut für Kernphysik, Goethe-Universitaet, Max-von-Laue-Str.1, D-60438 Frankfurt am Main, Germany

Synopsis We have conducted an experimental study on the photo double ionization (PDI) of carbon dioxide dimers and oxygen dimers, while focusing on the dissociation dynamics upon single photon absorption. The results in terms of the kinetic energy and angular distributions of the charged particles show unambiguous experimental evidence of intermolecular Coulombic decay (ICD) in carbon dioxide dimers. In the oxygen dimer, the results show that ICD is accompanied by knock-off ionization mechanisms.

Excited systems embedded in an environment can efficiently transfer their energy to neighboring species via an ultrafast de-excitation mechanism known as the Inter-Coulombic Decay (ICD). Theoretical and experimental investigations have shown that ICD is a very common decay route in nature as it occurs after exciting a loosely bound system (e.g. van der Waals or Hydrogen bonds) by ion, electron, or photon impact. A lot of studies have been carried out on atomic clusters, but significantly less progress is made for molecular clusters. Compared to atomic clusters, new questions arise in the investigation of molecular clusters. For example, how does the higher structural complexity of molecular clusters affect ICD? What other ultrafast relaxation processes can be present, and how do they compete with ICD?

For isolated molecules the pathways leading to the ejection of two electrons and a subsequent ionic fragmentation are direct photodissociation, predissociation, or autoionization. In the case of molecular dimers these processes ionize only one-site of the dimer directly, leaving the second site intact, unless a secondary reaction like ICD, charge transfer, or two-site electron knock-off process is triggered. In this work, we investigate the competition between one-site dissociation and two-site fragmentation of the molecular dimers via photoionization of valence electrons. The experiments were performed at beamline 10.0.1.3 at the Advanced Light Source in Berkeley. We have investigated the fragmentation dynamics of CO_2 dimer and O_2 dimer targets after single photon absorption. Specifically, we focused on the investigation of the symmetric fragmentation channels of the doubly charged dimers. A COLd Target Recoil Ion Momentum Spectroscopy (COLTRIMS) was employed to measure in coincidence the 3D momenta of the emitted ions and electrons.

We found that the direct dissociation or autoionization of CO_2^{+*} on a single site of a CO_2 dimer is suppressed due to the fast relaxation of the dimer via ICD. For both, CO_2 dimers and O_2 dimers, the kinetic energy release of the dissociating ions enabled us to pinpoint the internuclear distance at the time of fragmentation and to deduce the geometry of the dimers. Moreover, the kinetic energy range of the emitted electrons enabled us to coarsely identify the contributing electronic states. For CO_2 dimers the results show a photoelectron emission pattern preferentially aligned along the polarization direction and an isotropically emitted ICD electron. For O_2 dimers the relative emission angle between the two electrons showed contributions from ICD as well as knock-off processes [1].

References

[1] Iskandar W *et al* 2019 *Phys. Rev. A* **99** 043414

*E-mail: wiskandar@lbl.gov

General formula of shakeup and shakeoff probabilities for arbitrary shells

T Mukoyama[1 *]

[1]Institute for nuclear Research of the Hungarian Academy of Sciences (ATOMKI), H-4026, Hungary

Synopsis General expressions for the shakeup and shakeoff probabilities following sudden change in central potential have been obtained for arbitrary atomic shells by the use of nonrelativistic hydrogenic wave functions. The final formulae are given in terms of Gauss-type hypergeometric functions. The obtained results are used to calculate the shakeup and shakeoff probabilities for outer-shell electrons following β^- decay.

When there is sudden change in the central potential atomic electrons have small probability to be excited into a vacant state (shakeup) or to be ejected in the continuum (shakeoff). This shake process is the fundamental process in atomic physics and caused by nuclear decay, such as α decay, β decay electron capture, and internal conversion, or inner-shell vacancy creation following photoionization and charged-particle impact [1]. The theoretical calculations are usually performed in the framework of the sudden approximation and the probabilities are expressed in terms of imperfect overlap integral between the initial- and final-state wave functions.

In order to study various properties of the shake process in more detail and to obtain rough estimate of the shake probabilities it is sometimes useful to express shakeup and shakeoff probabilities in a simple analytic form. For K-, L-, and M-shell electrons such analytical expressions have been reported [2]. In the present work, general formulae for arbitrary shells are presented.

Using the nonrelativistic hydrogenic wave functions, overlap integrals between the initial bound-state wave function with nuclear charge Z and wave functions for final bound and continuum states with nuclear charge Z' are calculated.

The final results are expressed in terms of Gauss-type hypergeometric functions. It is interesting to note that the shakeup probability can be written in the same form as the shakeoff probability, except for the normalization constant and one parameter.

In Table 1, the shakeup probabilities per electron for N-subshell electrons in Xe following β^- decay are listed for dominant transitions of the subshells.

Table 1. Shakeup probabilities per electron for N-subshell electrons in Xe following β^- decay.

Transition	Probability (%)
4s → 6s	0.023
4p → 6p	0.025
4d → 5d	0.100

Figure 1 shows the shakeoff probabilities per electron for N-subshell electrons in Xe accompanying β^- decay as a function of the ejected electron energy. Within the same shell the shakeoff probability for the subshell with larger orbital angular momentum decreases more rapidly with the ejected electron energy.

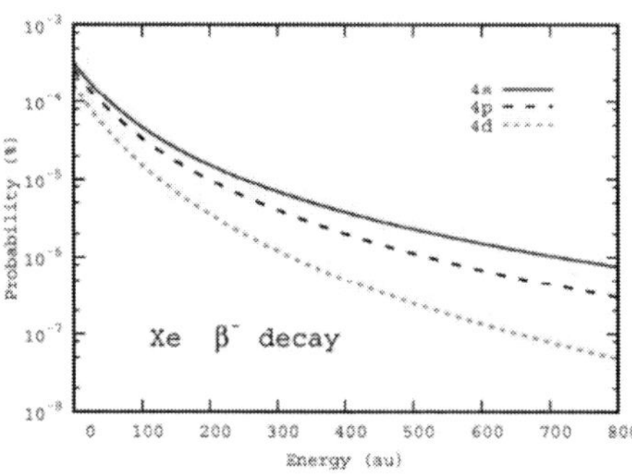

Figure 1. N-subshell shakeoff probabilities (%) per electron of Xe following β^- decay as a function of ejected electron energy.

Here only the case where the nuclear charge changes was considered. However, the present results can also be used to calculate the shake probabilities accompanying inner-shell vacancy creation by the use of appropriate screening constants.

References

[1] Freedman M 1974 *Ann. Rev. Nucl. Sci.* **22** 209
[2] Mukoyama T 2005 *X-Ray Spectrom.* **34** 64

* E-mail: hakutei@rhythm.ocn.ne.jp

Electron emission in collisions between dressed ions and multielectronic targets: the role of the projectile distortion

J M Monti[1] *, M A Quinto[1] and R D Rivarola[1]

[1]Instituto de Física Rosario, CONICET and Universidad Nacional de Rosario, Rosario, S2000EKF, Argentina

Synopsis A further improvement of the CDW and CDW-EIS distorted wave models for ionization by dressed-ion impact is made by taking into account the projectile form factor into the projectile distortions.

The electron emission mechanism on atomic collisions has been target of active research for many decades. It has implications on many areas as astrophysics, fusion reactors, and radiation damage among others. Many theoretical frames have been developed in order to investigate this reaction for low, intermediate-high, and high collision energy regimes. In the particular case of the intermediate-high impact energies and bare projectiles the distorted wave formalisms, such as the Continuum Distorted Wave (CDW) [1] and the Continuum Distorted Wave-Eikonal Initial State (CDW-EIS) [2] arose as reliable methods to calculate multiply differential and total cross section.

The CDW and CDW-EIS were extended to consider partially dressed projectiles [3,4] describing the non-Coulomb projectile potential with a parametric Green-Sellin-Zachor one [5] that can be separated in two terms: a pure Coulomb potential given by the net charge q of the projectile and a short range part determined by its electronic cloud. In such extensions, the initial channel distortion is considered as a hypergeometric function, for CDW, and an eikonal phase, for CDW-EIS, describing the interaction between the fully screened projectile and the target active electron. These functions give a correct asymptotic description of such interaction at large separations but they might differ from the exact function for smaller distances where the short-range part of the projectile potential is not negligible. This discrepancy is expected to be larger for smaller impact energies and projectiles with large number of electrons.

In this work we introduce a correction to the net charge q in the CDW and CDW-EIS projectile distortions taking into account the screening due to the projectile form factor.

Details of the calculation and preliminary results on electron emission in collisions between dressed ions and atoms will be shown during the conference.

* E-mail: monti@ifir-conicet.gov.ar

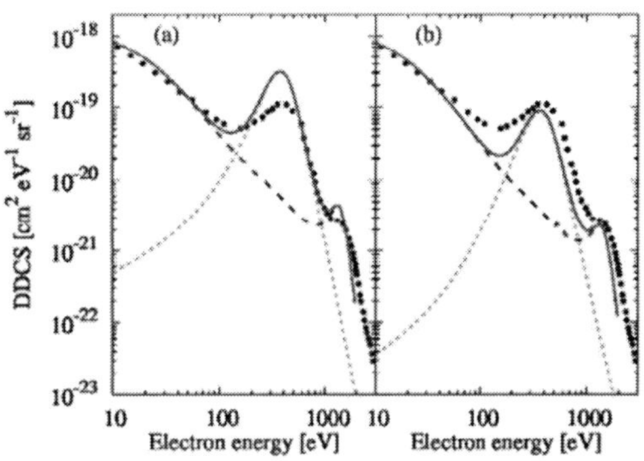

Figure 1. CDW-EIS double differential cross section for ionization of H_2O molecules by 750 kev/u He^+ impact as a function of the ejected electron energy, for a fixed emission angle of 20°. Panel (a): usual projectile distortion with projectile net charge; panel (b): projectile distortion taking into account the projectile form-factor. Theory: blue dashed-line, target ionization; green dotted-line, projectile electron loss; red solid-line, target ionization plus projectile electron loss. Experiments: solid circles [6].

References

[1] Belkić Dž 1978 *J. Phys. B: At. Mol. Phys.* **11** 3529
[2] Crothers D S F and McCann J F 1983 *J. Phys. B: At. Mol. Phys.* **16** 3229
[3] Monti J M, Rivarola R D and Fainstein P D 2008 *J. Phys. B: At. Mol. Opt. Phys.* **41** 201001
[4] Monti J M, Rivarola R D and Fainstein P D 2011 *J. Phys. B: At. Mol. Opt. Phys.* **44** 195206
[5] Green A E S, Sellin D L and Zachor A S 1969 *Phys. Rev.* **184** 1
[6] Dubois R D and Manson S T 1990 *Phys. Rev. A* **42** 1222

Improvement of the target continuum state in distorted wave theories: a dynamic-depending charge

M F Rojas[1]*, **J M Monti**[1], **M A Quinto**[1] **and R D Rivarola**[1]

[1]Instituto de Física Rosario, CONICET and Universidad Nacional de Rosario, Rosario, S2000EKF, Argentina

Synopsis In order to improve the representation of the residual-target final state whilst maintaining low computational times, a dynamic-depending charge defined by the emission energy and angle of the ionized electron is considered in the residual-target continuum state.

Distorted wave theories such as the Continuum Distorted Wave (CDW) [1] and the Continuum Distorted Wave–Eikonal Initial State (CDW-EIS) [2] were proven to be quite successful for studying single electron ionization on many collision systems. They were initially developed to consider bare projectiles and monoelectronic targets for which the initial bound and the final continuum states were analytical hydrogenic functions. Later, an extension was made by Fainstein *et al.* in order to consider multielectronic targets [3]. There, the active electron initial bound wavefunction is considered as a Roothaan–Hartree–Fock function expanded in a Slater–type basis [4], and the residual–target final continuum state is taken as an analytical hydrogenic one with an effective charge given by $Z_{eff} = \sqrt{-2n^2 \varepsilon_0}$ being n the principal quantum number of the initial bound orbital and ε_0 its ionization energy. This corresponds to approximating the non-Coulomb potential of the residual target with a pure Coulomb potential generated by a charge Z_{eff}. In this way analytical expressions of the transition amplitude can be obtained which can be numerically integrated to obtain the differential and total cross sections. Another approach is to numerically solve the Schrödinger equation within a CDW theory of ionization by ion impact. As a result the initial and final states of the target are calculated in equal ground, and no approximation is made regarding the non-Coulomb nature of the target potential. The drawback of this method is that for each electron emission energy and angle the continuum state of the target must be numerically calculated resulting in large computational times.

In order to improve the residual-target final state, and save calculation time, we maintain the analytical hydrogenic function but instead of a fixed effective charge, depending only with the orbital bound energy, we explore the inclusion of dynamic-depending charge defined by the emission energy and angle of the ionized electron.

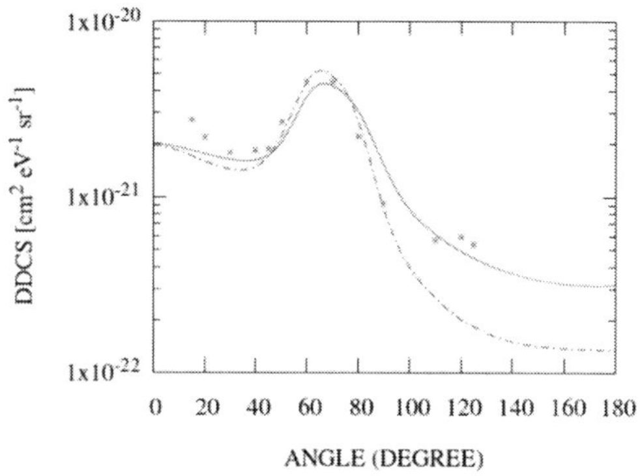

Figure 1. Double differential cross section for ionization of Ar atoms by 1 MeV H^+ impact as a function of the emission angle, for a fixed emission energy of 250 KeV. Theory: violet dashed-line, usual CDW-EIS; green solid-line, CDW-EIS with emission-angle depending charge. Experiments: blue asterisks, data of Toburen [5].

References

[1] Belkić Dž 1978 *J. Phys. B: At. Mol. Phys.* **11** 3529
[2] Crothers D S F and McCann J F 1983 *J. Phys. B: At. Mol. Phys.* **16** 3229
[3] Fainstein P D, Ponce V H and Rivarola R D 1988 *J. Phys. B: At. Mol. Opt. Phys.* **21** 287
[4] Clementi E and Roetti C 1974 *At. Data Nucl. Data Tables* **14** 17
[5] Toburen L H 1979 *At. Atomic Data and Nucl. Data Tables* **23** 43

* E-mail: rojas@ifir-conicet.gov.ar

Robust transverse structures in rescattered photoelectron wavepackets and their consequences

T Bredtmann[1]* and S Patchkovskii[1]†

[1]Max-Born-Institute, Max-Born-Straße 2A, Berlin 12489, Germany

Synopsis Initial-state symmetry has been under-appreciated in strong-field spectroscopies, where laser fields dominate the dynamics. We demonstrate numerically that the transverse photoelectron phase structure, arising from the initial-state symmetry, is robust in strong-field rescattering, and has pronounced effects on strong-field photoelectron spectra. Interpretation of rescattering experiments need to take these symmetry effects into account. In turn, robust transverse photoelectron phase structures may enable attosecond sub-Ångström super-resolution imaging with structured electron beams.

Symmetries, exact and approximate, as well as symmetry breaking underpin our understanding of perturbative spectroscopies, where field-free symmetries of initial and final states strongly constrain the range of possible outcomes. In contrast, in intense infrared laser fields, the overall dynamics is dominated by the field itself. The field symmetry determine many qualitative features of these processes, while the initial-state symmetry is often less important.

A recent study on molecular trans-butadiene[1] has shown that the rescattering probability in trans-butadiene is specific to the ionization channel and the molecular orientation, rather than a property of the driving field alone. These results may be interpreted as a signature of the transverse phase structure, imprinted on the recolliding electron wave packet (REWP) within each ionization channel.

In molecular systems, the initial-state symmetry effects are intertwined with contributions due to the asymmetry of the potential, orientation, and nuclear motion. This complexity makes the effects of the symmetry alone hard to isolate and model. Instead, we consider the simplest-possible example of the initial-state symmetry: a one-electron atom initially in an antisymmetric p state[2].

We demonstrate that the phase structure of the REWP due to the initial-state symmetry strongly affects strong-field rescattering (Fig 1). The effects are robust to the field misalignment with respect to the molecular symmetry element, and survive focal-spot intensity averaging. These effects need to be accounted for in the interpretation of strong-field recollision experiments both in the lower-energy (holographic) and the higher-energy (LIED) region.

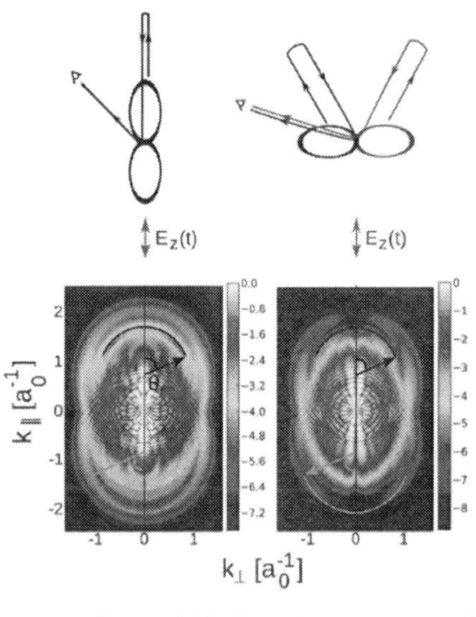

Figure 1. Strong-field photoelectron spectra (100 TW cm^{-2}, 800 nm, 13.5 fs) for the He^+ ion. Left column: symmetric initial state ($2p_z$). Right column: anti-symmetric initial state ($2p_x$). The $10U_p$ ($6U_p$) recollision-energy features shown with black (gold) lines.

References

[1] Schell F *et al* 2018 *Sci. Adv.* **4** 8148
[2] Bredtmann T and Patchkovskii S 2018 arxiv:1810.12672

*E-mail: Timm.Bredtmann@mbi-berlin.de
†E-mail: Serguei.Patchkovskii@mbi-berlin.de

Photoionization of low-charged silicon ions

T Buhr[1] *, S O Stock[2,3], A Perry-Sassmannshausen[1], S Reinwardt[4], M Martins[4], S Ricz[5], A Müller[6], S Fritzsche[2,3] and S Schippers[1]

[1]I. Physikalisches Institut, Justus-Liebig-Universität Gießen, 35392 Giessen, Germany
[2]Helmholtz-Institut Jena, 07743 Jena, Germany
[3]Theoretisch-Physikalisches Institut, Friedrich-Schiller-Universität Jena, 07743 Jena, Germany
[4]Institut für Experimentalphysik, Universität Hamburg, 22761 Hamburg, Germany
[5]Institute for Nuclear Research, Hungarian Academy of Sciences (MTA Atomki), 4001 Debrecen, Hungary
[6]Institut für Atom- und Molekülphysik, Justus-Liebig-Universität Gießen, 35392 Giessen, Germany

Synopsis Single and multiple photoionization of Si^{1+}, Si^{2+}, and Si^{3+} ions have been investigated near the silicon K-edge using the PIPE setup at beamline P04 of the synchrotron light source PETRA III operated by DESY in Hamburg, Germany. Pronounced resonance structures are observed for all ions which are associated with excitation or ionization of a K-shell electron. The experimental cross sections are compared with results from theoretical calculations.

Since silicon is one of the most abundant heavy elements in the Universe, the detailed study of the electronic structure of the silicon atom and silicon ions in various charge states is of fundamental astrophysical interest. Single and multiple photoionization of Si^{q+} ($q = 1, 2, 3$) ions have been experimentally investigated in the photon energy range 1830 eV to 2100 eV. The measurements were carried out at the PIPE setup [1] at beamline P04 of the third generation synchrotron light source PETRA III at DESY (Hamburg, Germany) employing the photon-ion merged-beams technique [2]. This facility enables us to record excitation functions and to determine absolute photoionization as well as photoexcitation cross sections. Several reaction channels were investigated. Distinct resonance structures are observed for all ions. The strongest resonances are associated with the photoexcitation of a $1s$ electron to an atomic np subshell (n=3,4,5,...) and the final ionic charge states are created by subsequent multiple autoionization of the associated $1s$ hole. As an example, figure 1 shows the measured cross sections for triple photoionization of Si^{2+} ions. Resonances with principal quantum numbers n ranging from 3 to 5 can be clearly seen. Furthermore the experimental cross sections are compared with theoretical values obtained from multiconfiguration Dirac-Fock calculations (MCDF). This theoretical description takes into account the initial excitation or ionization and the subsequent cascade of Auger processes.

This work was supported by the Bundesministerium für Bildung und Forschung (BMBF, Grant numbers: 05K10RG1, 05K10GUB, 05K16RG1, 05K16GUC and 05K16SJA), by the Deutsche Forschungsgemeinschaft (DFG, Grant number: Schi 378/12-1) and by the EU CALIPSOplus Transnational Access Programme DESY Photon Science. We thank the staff of the beamline P04 for their valued support.

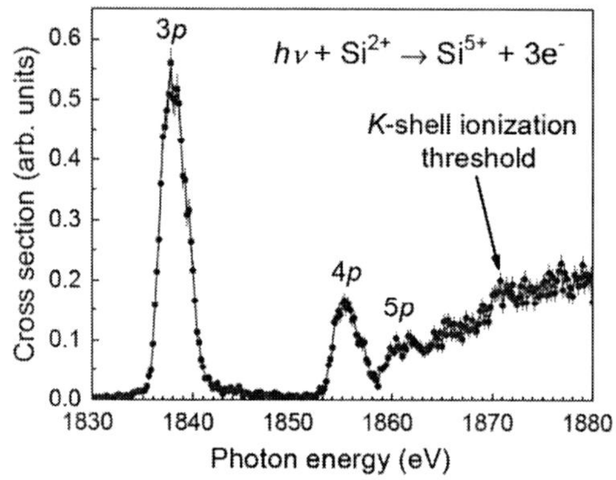

Figure 1. Experimental cross sections for triple photoionization of Si^{2+} ions as a function of the photon energy. Resonances are labelled by the np subshell to where the $1s$ electron is excited.

References

[1] Schippers S *et al* 2014 *J. Phys. B* **47** 115602
[2] Schippers S *et al* 2016 *Contemp. Phys.* **57** 215

*E-mail: ticia.buhr@physik.uni-giessen.de

 1412 (2020) 152025 doi:10.1088/1742-6596/1412/15/152025

Effect of SOIAIC on photoionization time delay of the Xe 3d subshell

S Banerjee[1] *, P C Deshmukh[2] and S T Manson[3]

[1] Department of Physics, Indian Institute of Technology Madras, Chennai 600036, India
[2] Department of Physics, Indian Institute of Technology Tirupati, Tirupati 517506, India
[3] Department of Physics and Astronomy, Georgia State University, Atlanta, Georgia 30303, USA

Synopsis The effect of spin-orbit activated interchannel coupling (SOIAIC) has been studied on the time delay spectrum of 3d photoionization in xenon. The relativistic-random-phase-approximation with relaxation (RRPA-R) has been used to calculate the Wigner-Eisenbud-Smith (WES) time delay in photoionization. SOIAIC has a strong influence on the WES time delay for all ejection angles of the photoelectrons with respect to the incident direction and the polarization vector of the photon absorbed.

The spin orbit interaction in atoms often produces strong effects on various photoionization parameters. The spin orbit interaction activated interchannel coupling (SOIAIC) is a special class of the many body interactions that is amenable to interpretation in terms of the interchannel coupling. The first theoretical report [1] on this effect, based on SPRPAE, explained the significant structure in the measured $3d_{5/2}$ cross section in Xe [2]. The relativistic approach [3], based on the RRPA-R methodology [4], resulted in excellent agreement with the previous results. SOIAIC effects on various photoionization parameters [5, 6] have been reported for several atoms. The study of intrinsic time delay, (also known as the Wigner-Eisenbud-Smith time delay) has emerged as a very hot topic of research in recent days, due to the advancement in ultrafast laser technology. The SOIAIC effect is very pronounced in Xe 3d photoionization. There are rather only a few studies [7, 8] on the effect of SOIAIC on photoionization time delay. The present study aims at investigating the WES time delay at different angles of the photoelectron ejection with respect to the incident direction of the electromagnetic radiation and its polarization direction. Our primary interest lies in understanding the importance of electron correlations. The RRPA-R method has been used, since its results agreed well with the experimental cross-sections.

The energy region above the $3d_{3/2}$ threshold has been investigated in Figure 1 to study the SOIAIC effect. With the inclusion of channels from $3d_{3/2}$ subshell, the time delay for the ionization from the $3d_{5/2}$ subshell shows an oscillatory pofile, with substantial decrement (~50 as) followed by a rise, for all ejection angles. In addition, the energy at which the time delays for different angles is the same shifts to a lower energy region on including SOIAIC.

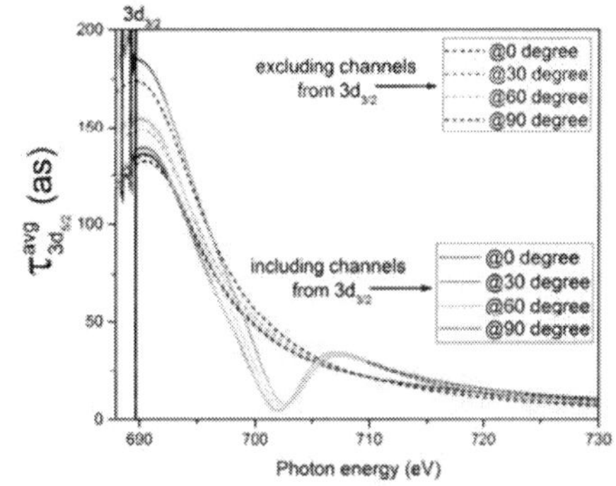

Figure 1. WES time delay from $3d_{5/2}$ subshell of Xe with and without the inclusion of interchannel coupling from $3d_{3/2}$ subshell.

References

[1] Amusia M Ya *et al* 2002 *Phys. Rev. Lett.* **88** 093002
[2] Kivimaki A *et al* 2000 *Phys. Rev. A* **63** 012716
[3] Radojević V *et al* 2003 *Phys. Rev. A* **67** 022719
[4] Radojević V *et al* 1989 *Phys. Rev. A* **40** 727
[5] Amusia M Ya *et al* 2004 *Phys. Rev. A* **70** 062709
[6] Kumar S S *et al* 2009 *Phys. Rev. A* **79** 043401
[7] Keating D A *et al* 2017 *J. Phys. B* **50** 175001
[8] Mandal A *et al* 2018 *Springer Proceedings* of ISAMP TC-7.

* E-mail: souravbanrj444@gmail.com

Journal of Physics: Conference Series **1412** (2020) 152026 doi:10.1088/1742-6596/1412/15/152026

Outer-shell photodetachment of Li^- near inner-shell thresholds

T W Gorczyca[1] * and S T Manson[2] †

[1]Western Michigan University, Kalamazoo, MI 49008-5252 USA
[2]Georgia State University, Atlanta, GA 30303, USA

Synopsis The effect of inner-shell excited channels on outer-shell photodetachment is investigated using a correlated and coupled-channel R-matrix method. The simplest multishell negative ion of Li^- is treated near the $1s$ ionization threshold to determine strong "hole-state" continuum coupling with the direct photodetachment channels.

Recent work has shown that the photoionization of outer shells of atoms are profoundly affected by correlation in the form of interchannel coupling in the vicinity of inner-shell thresholds [1, 2]. Since negative ions, by their very nature, are much more highly-correlated than atoms, it is expected that these effects should be even more pronounced in negative ion photodetachment.

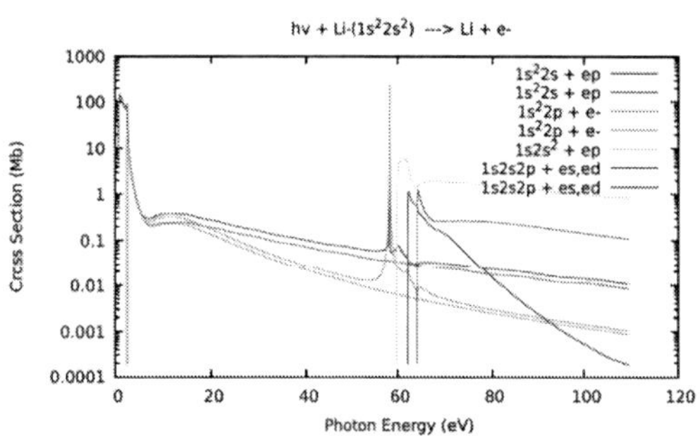

Figure 1. Outer-shell photodetachment from Li^- without inner-shell channels (green and orange curves, smooth through the $1s^{-1}$ threshold) and with the $1s^{-1}$ channels included (violet and cyan). The dominant inner-shell cross section onsets are seen above about 60 eV.

Thus, we have started a program to explore this effect, and we have looked at the photodetachment of the simplest closed-shell negative ion with more than one shell, Li^- in the ground $1s^2 2s^2$ state, using a version of the Belfast R-matrix methodology that has been modified to accommodate photodetachment [3]. The results, displayed in Fig. 1, show that the outer-shell photodetachment cross sections leading to the ground state $1s^2 2s$ state of Li, along with a number of excited $1s2nl$ states (detachment plus excitation), are indeed very strongly altered owing to the interchannel coupling with the inner-shell photodetachment channels; note the logarithmic scale.

Work supported partially by NASA and DOE.

References

[1] Drube W *et al* 2013 *J. Phys B* **46** 245006
[2] Keating D A *et al* 2018 *Phys. Rev. A* **98** 013420
[3] Gorczyca T W *et al* 2003 *Phys. Rev. A* **68** 050703

*E-mail: gorczyca@wmich.edu
†E-mail: smanson@gsu.edu

Photodouble ionization of water

J M Randazzo[1,*], G Turri[2], P. Bolognesi[3], H Sugimoto[2] , J Mathis[2], L U Ancarani[4,†] and L Avaldi[3]

[1]Division Colisiones Atomicas, CAB, Instituto Balseiro and CONICET, 8400 S. C. de Bariloche, Rio Negro, Argentina
[2]Embry-Riddle Aeronautical University, Physical Sciences Department, Daytona Beach, FL, (USA)
[3]CNR-Istituto di Struttura della Materia, Area della Ricerca di Roma 1,00015 Monterotondo Scalo (Italy)
[4]Université de Lorraine, CNRS, LPCT, 57000 Metz, France

Synopsis The photodouble ionization of water has been studied experimentally by photoelectron-photoelectron angular resolved coincidence experiments and theoretically by applying the Generalized Sturmian Functions approach. The measured differential cross sections present a rich angular distribution that can be, in part, reproduced by a first order treatment of the interaction considering a two active electrons target.

Double and multiple electron emissions from an atom or a molecule by the absorption of a single photon represent minor processes in the radiation-matter interaction. However, they are of extreme interest, because they are determined by electron correlation. The detection in coincidence of the two correlated electrons allows a complete investigation of the different active mechanisms depending on the total energy and energy sharing between the two electrons. A detailed understanding of the photodouble ionization (PDI) has been achieved in two-electron systems [1]. In diatomic and polyatomic molecules experimental and theoretical studies of the mechanisms of PDI are scarse.

In this combined experimental and theoretical work the PDI of water has been investigated. The measurements have been performed at the Gas Phase beamline of Elettra at about 20 eV above the double ionization threshold. Two types of measurements have been performed. In the first one the energy of each electron has been fixed to 10 eV and the photon energy has been scanned. In this way the binding energy spectrum of the water dication has been reconstructed. In the second type, at a fixed photon energy three independent coincidence angular distributions have been measured fixing the direction of one electron at 0°, 30° and 60° with respect to the direction of the linear polarization of the incoming radiation. Two photon energies (63 and 65 eV) have been chosen in order to explore dication states with different symmetries.

Describing accurately two electrons escaping from an atomic target is already a challenge [3]. Here, we further have the difficulties related to the multicenter nature of the target, and molecular orientation issues. As a first step we have considered the water molecule as a two active electrons target, with one-center expansions as in [3], and adapted the atomic Generalized Sturmian Function approach [4] to calculate the electron scattering states, and finally the differential cross sections.

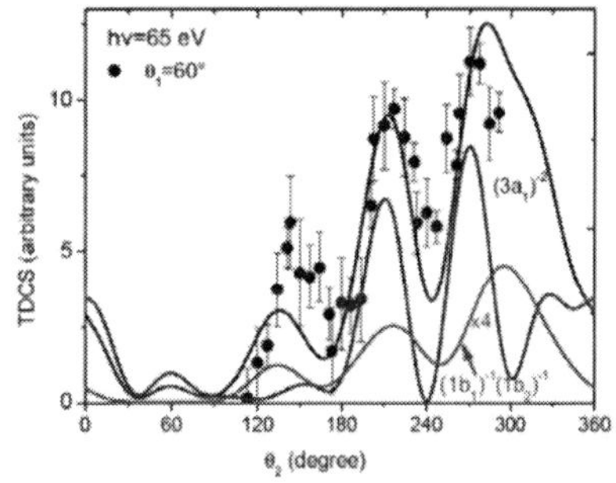

The experiment-theory comparison is fair, considering some of the theoretical approximations, and the mixture of states that are involved at 65 eV.

Figure 1. Angular distribution at 65 eV photon energy and θ_1=60°. The contribution of the $(1b_2)^{-1}(b_1)^{-1}\ {}^1A_2$ and $(3a_1)^{-2}\ {}^1A_1$ states of H_2O^{2+} as well as their combination (black full line) are shown.

References

[1] Avaldi L and Huetz A 2005 *J. Phys. B* **38** S861
[2] Turri G *et al* 2019 *J. Phys. B (accepted)*
[3] Granados-Castro C M and Ancarani LU 2017 *Eur. Phys. J. D* **71** 65
[4] Randazzo J M *et al* 2015 *Eur. Phys. J. D* **69** 189

* E-mail: randazzo@cab.cnea.gov.ar
† E-mail: ugo.ancarani@univ-lorraine.fr

Using 'dressed' convoluted quasi-Sturmian functions as a basis for Coulomb three-body continuum problems

A S Zaytsev[1], D S Zaytseva[1], L U Ancarani[2]* and S A Zaytsev[1]

[1]Pacific National University, Khabarovsk 680035, Russia
[2]Université de Lorraine, CNRS, LPCT, F-57000 Metz, France

Synopsis We present an *ab initio* approach to the two electron continuum problem based on an expansion on convoluted quasi Sturmian functions (CQS) equipped with an appropriately built electron-electron phase factor. Such 'dressed' CQS functions constitute a suitable basis to solve Coulomb three-body scattering problems as illustrated through numerical applications to double ionization processes on Helium. They put in evidence, once more, the importance of electron-electron correlation over the whole space.

In this contribution we consider the Coulomb three-body scattering problem which results from the double ionization of Helium by impact of fast projectiles or the double photoionization; to first order, such a process reduces to solving a Schrödinger equation

$$\left[E - \hat{H}\right] \Phi^{(+)}(\mathbf{r}_1, \mathbf{r}_2) = \hat{W}_{fi}\Phi^{(0)}(\mathbf{r}_1, \mathbf{r}_2) \qquad (1)$$

with a three-body Helium Hamiltonian $\hat{H}$. Assumed to be initially in the Helium ground state $\Phi^{(0)}$, after the double ionization process the two electrons escape from the nucleus with total energy $E = k^2/2$; $\hat{W}_{fi}$ is the interaction operator related to the studied process.

We suggest [1] solving equation (1) with a truncated expansions in CQS functions. To make the expansion converge, we dress the 'bare' CQS functions with a phase factor $e^{i\mathcal{W}}$, such that $\mathcal{W}$ approaches the Coulomb phase $-\frac{1}{k}\frac{\rho}{r_{12}}\ln(2k\rho)$ in the limit $\rho \to \infty$ ($\rho = \sqrt{r_1^2 + r_2^2}$); r_{12} is the interelectronic distance.

To test our approach and compare it with other *ab initio* methods, we calculate fully differential cross sections for $(\gamma, 2e)$ and $(e, 3e)$ processes on Helium. We demonstrate that our 'dressed' CQS functions constitute a suitable basis which allows us to provide converged and sensible partial ionization amplitudes, and thereafter cross sections. This is in contrast with 'bare' CQS which also yield cross sections but with amplitudes having a divergent phase.

In the $(\gamma, 2e)$ example shown in Fig. 2, our triple differential cross sections (TDCS) calculated with the 'bare' CQS match very well the ECS [5] and GSF [4] results, and also the experimental data [2]; our 'dressed' CQS cross section, on the other hand, is noticeably higher in magnitude, even more than the CCC [3] result. The comparison between such *ab initio* approaches solving exactly the same problem, illustrates the mathematical and numerical difficulties associated with the Coulomb three-body problem.

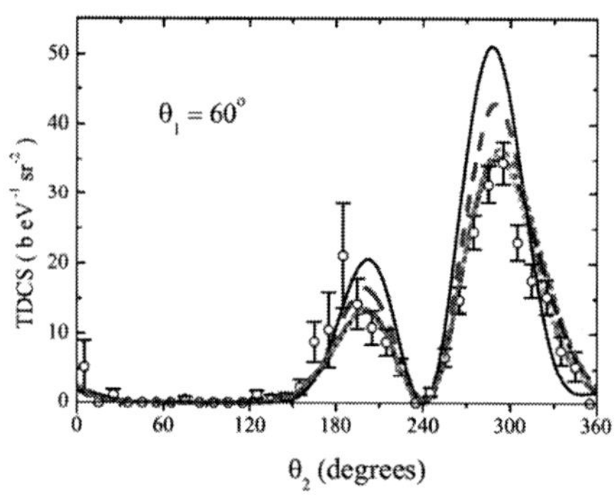

Figure 1. TDCS as a function of θ_2 for $(\gamma, 2e)$ on Helium for two electron escaping in the polarization plane with $E_1 = E_2 = 10$ eV and one electron detected at $\theta_1 = 60^0$. Experimental data [2]; present 'bare' CQS (dotted line); present 'dressed' CQS (solid line); CCC [3] (dashed line); GSF [4] (dot-dashed line) and ECS [5] (short dashed line).

References

[1] Zaytsev A S *et al* 2017 *Eur. Phys. J. D* **71** 177
[2] Bräuning H *et al* 1998 *J. Phys. B* **31** 5149
[3] Kheifets A S *et al* 2004 *Phys. Rev. A* **69** 050701(R)
[4] Randazzo J M *et al* 2015 *Eur. Phys. J. D* **69** 189
[5] McCurdy C W *et al* 2004 *Phys. Rev. A* **69** 032707

*E-mail: ugo.ancarani@univ-lorraine.fr

Ionization of rubidium by electron impact

K Tőkési[1,2*] and T Mukoyama[1]

[1]Institute for Nuclear Research, Hungarian Academy of Sciences (Atomki), 4026 Debrecen Bem tér 18/c, Hungary
[2]ELI-ALPS, ELI-HU Non-profit Ltd., Dugonics tér 13, H-6720 Szeged, Hungary

Synopsis The cross sections for ionization by electron impact of rubidium 5s shell have been calculated by the Classical Trajectory Monte Carlo method. The kinetic energy of the projectile is in the range between 5 and 1000 eV. We compared our results with other theoretical and experimental data..

We have many experimental investigations of the ionization process by electrons and other charged particles from the beginning of the 20th century. Rubidium is an important element in laser physics, but only a limited number of theoretical studies has been reported. It is interesting to calculate ionization cross section by electron impact for this element and to test theoretical models.

The classical treatment of various collision problems has been quite successful to obtain ionization cross sections. In particular the Classical Trajectory Monte Carlo (CTMC) method is widely used [1]. It is a non-perturbative method and hence all the interactions between colliding particles can be taken into account exactly within the framework of the classical dynamics.

We describe the collision system as a three-body model: the projectile, an atomic electron (e^-), and the remaining rubidium ion (Rb^+). The target atom is modeled like a tiny solar system, in which the electron is moving on Kepler orbit around the nucleus. We investigated the ionization cross sections for rubidium using two models. In the first model, the effective static target nucleus charge is used and Coulomb force acts between the colliding particles. In this case we calculate the effective charge of the valence electron by the Slater's rules. In the second approach we represent the rubidium atom with Garvey model potential. The two parameters of the potential, ξ and η, represent quantum mechanical effects by taking into account the independent particle model. The binding energy of the rubidium 5s shell valence electron is 0.1504 a.u. (4.1912 eV) based on the theoretical data of the ionization energy. The mass of Rubidium atom is 156918.88 a.u. For the first model, the corresponding effective nuclear charge is taken to be $Z_{eff} = 2.2$.

In the model 2, the parameters of the Garvey type potential are chosen as $\xi = 0.1686$ and $\eta = 0.1619$.

Figure 1 shows the ionization cross sections of Rb 5s shell by electron impact. While the data of CTMC with the static potential are in agreement with the calculations with Lotz's formula [2], the CTMC results using the Garvey model potential during the simulations are in agreement with the binary encounter approximation (BEA) and with the results in Ref. [3].

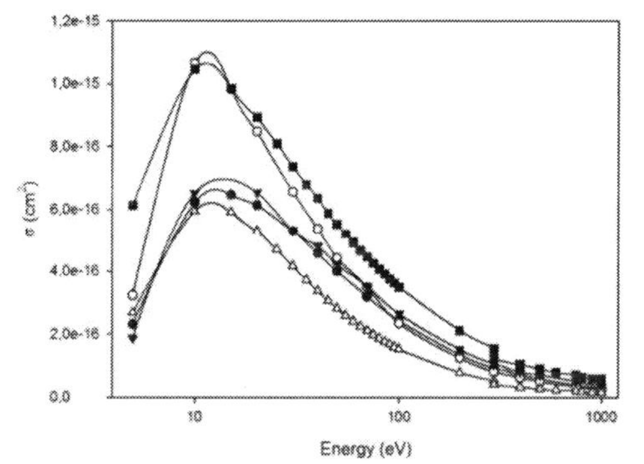

Figure 1. Electron-impact ionization cross sections of rubidium 5s shell. O: present conventional CTMC results using the effective static target nucleus charge, ▼: present CTMC results with Garvey-type model potential, △: present BEA simulation, ■: semi-empirical formula of Lotz [2], ●: simulation from Ref. [3].

The work was supported by the National Research, Development and Innovation Office (NKFIH) Grant KH126886

References

[1] Tőkési K and Hock G 1997 *Nucl. Instr. Meth. Phys. Res. B* **124** 398
[2] Lotz W 1967 *Z. Phys.* **206** 205
[3] Kim Y-K *et al* 1998 *Phys. Rev.* A**57** 246

* E-mail: tokesi@atomki.mta.hu

The importance of energy-differential cross sections highlighted in a benchmark for positronium formation in the Galaxy

D G Cocks[1] *, H Chaudhary[1] and J R Machacek[1]

[1]Research School of Physics and Engineering, Australian National University, Canberra, 2601, Australia

Synopsis We detail a benchmark model aimed at improving the implementation of Monte Carlo codes for simulations of positrons propagating through interstellar media. The benchmark highlights the key features of cross sections sets which lead to major differences in astronomical observables. One of these is the energy-differentiation of the ionisation cross section. This leads us to call for improved cross section sets for all constituents of interstellar media.

Many simulations (see [1, 2] and references therein) have been made to follow positrons released from a radioactive source in the Galaxy, to their eventual annihilation. The pathways of either positronium formation or direct annihilation with an electron can be interpreted from astronomical measurements. Hence, the "positronium formation fraction" is an important distinguishing feature that is used to help interpret astronomical measurements in a sense of the abundance of radioactive sources, amongst other properties of the Galaxy.

The majority of Monte Carlo simulations have predicted similar Ps formation fractions, resulting apparently from a choice of different input sets of cross sections. However, the implementation of these simulations contains some subtle differences in the choice of e.g. ionisation energy sharing. Furthermore, it is difficult to specify the exact implementation details in the confines of a single manuscript and so it is difficult to reproduce each simulation's results exactly.

Hence, we have proposed [3] a benchmark using an analytically defined cross section that is motivated by atomic hydrogen, which can allow for detailed comparison of the numerical results from our calculations.

As part of this process, we have also scanned a range of parameters for the cross sections, in order to determine which features the results are most sensitive to. While many perturbations of the cross sections produce easily explainable variations in the results, some, such as the location of the maximum in the cross section, as not as obvious.

We also identify a quantity to be extremely important – the energy sharing ratio between the positron and electron+ion in an ionisation event. Close to the Ps formation threshold, the energy sharing ratio has a strong influence on the Ps formation fraction, and at high energies, it has a strong influence on the propagation time. As the energy sharing is a distribution that varies significantly with energy, it is important to accurately estimate this quantity in a simulation.

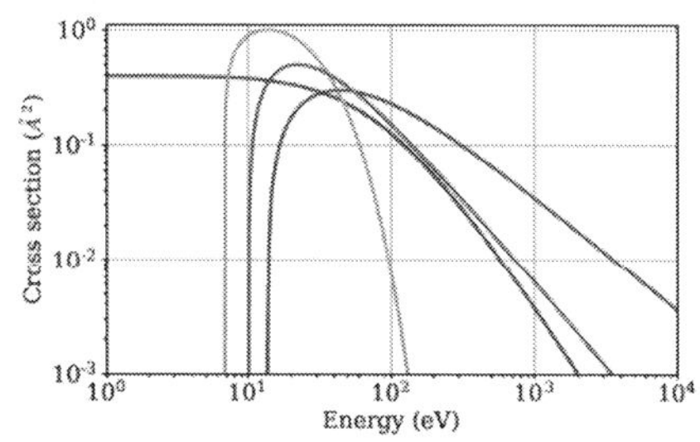

Figure 1. The set of cross sections, showing simple analytical forms for the elastic (blue), Ps formation (orange), excitation (red) and ionization (green) processes.

References

[1] Prantzos N *et al* 2011 *Rev. Mod. Phys.* **83** 1001
[2] Jean P *et al* 2009 *Astron. Astrophys.* **508** 1009
[3] Cocks D G, Chaudhary H and Machacek J R, 2019 submitted

*E-mail: daniel.cocks@anu.edu.au

Proton scattering from ground and excited states of atomic hydrogen

I B Abdurakhmanov[1]*, O Erkilic[1], Sh U. Alladustov[1], A S Kadyrov[1], and I Bray[1]

[1]Curtin Institute for Computation and Department of Physics and Astronomy, Curtin University, GPO Box U1987, Perth, WA 6845, Australia

Synopsis The semiclassical two-centre convergent close-coupling approach is applied to study total ionisation, Balmer emission as well as various excitation and electron-capture transitions in proton collisions with the ground and $n = 2$ states of atomic hydrogen, where n is the principal quantum number. The results for polarisation fraction of the Balmer-α emission induced in p-H(1s) collisions are in good agreement with experiment at intermediate energies, but surprisingly not at higher energies.

Apart from representing fundamental interest, proton-hydrogen collisions are also important in practical applications such as plasma modeling and astrophysics. For instance, collisional excitation of hydrogen is the one of the main processes in fast astrophysical shocks. Modeling these processes require an accurate knowledge on Balmer emission which can be obtained from cross sections for certain excitation transitions. In addition, accurate data for proton scattering from ground and excited states of hydrogen is crucial for plasma control and diagnostics in the proposed fusion experiment, ITER.

To provide extensive cross section data for proton-hydrogen scattering we applied our two-centre wave-packet convergent close-coupling (WP-CCC) approach [1]. The approach assumes a classical motion for the projectile and is based on the solution of the three-body Schrödinger equation using the two-center expansion of the total scattering wave function. The scattering wave function is expanded in an orthonormal basis set built from negative-energy eigenstates and wavepacket pseudostates representing the continuum of both the target atom and the atom formed by the projectile after capturing the electron. With a sufficiently large basis, due to the strong coupling between channels, the method produces converged cross sections for direct-scattering, ionisation and electron-capture processes simultaneously. Since all of the interaction matrix elements in the scattering equations are calculated independently from a particular initial state the computer code has been designed in such a way that it produces results for transitions from multiple initial states in one instance.

*E-mail: Ilkhom.Abdurakhmanov@curtin.edu.au

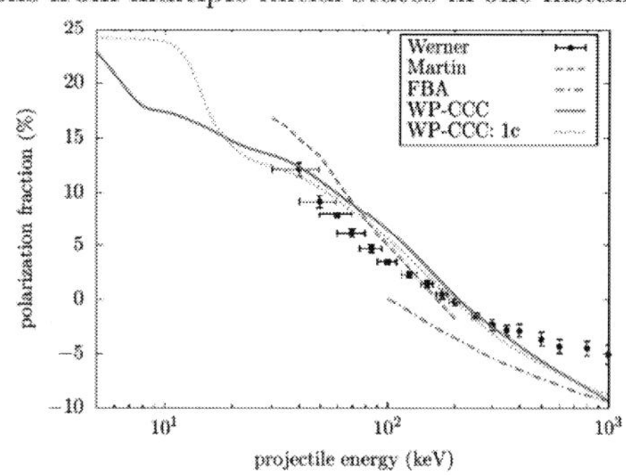

Figure 1. Polarisation fraction of the proton impact induced Balmer-α radiation of hydrogen. The present WP-CCC results are compared with the experiment of Werner *et al* [2] and theoretical calculations of Martin [3].

In Fig 1 we present the WP-CCC results in comparison with experiment [2] and theoretical calculations [3]. We see reasonably good agreement with the experimental data up to 350 keV. However, at higher energies our results significantly underestimate them. We also present the one-centre WP-CCC results (denoted as WP-CCC: 1c). At high energies the full and one-centre results coincide as the electron transfer channels are negligible and do not influence excitation of the target. This gives us confidence in the accuracy of the present calculations and the disagreement with experiment at energies above 400 keV remains a puzzle.

References

[1] Abdurakhmanov I B *et al* 2018 *Plasma Phys. Control. Fusion* **60** 095009
[2] Werner A *et al* 1996 *J. Phys. B* **29** 125
[3] Martin J 1999 *J. Phys. B* **32** 501

Projectile coherence effects analyzed using impact parameters determined by classical trajectory Monte Carlo calculations

L. Nagy[1] *, **F. Járai-Szabó**[1] **S. Borbély**[1] **and K. Tőkési**[2,3]

[1]Faculty of Physics, Babeş-Bolyai University, str. Kogălniceanu nr. 1, 400084 Cluj, Romania
[2]Institute for Nuclear Research, Hungarian Academy of Sciences, Hungary
[3]ELI-HU Non-profit Ltd., Szeged, Hungary

Synopsis We analyze the dependence of the fully differential ionization cross section on the coherence width of the wave packet associated to the projectile. The wave packet beside its size is characterized by the closest distance of its center relative to the target (the impact parameter). We determine the distribution of the possible impact parameters by classical trajectory Monte Carlo calculations.

The possible influence of projectile coherence on the fully differential cross sections (FDCS) in case of ionization by ion impact has attracted much interest in the last years [1–7]. Experimental and theoretical studies have shown [1, 2, 3, 6, 7], that the cross sections depend on the width of the wavepacket associated to the projectile. However, there are also arguments [4, 5] that this may be not the case.

Our group has performed calculations for the ionization of helium by 75 kev proton [3] and 100 MeV/u C^{6+} (or 3 MeV proton) [7] projectiles. We have associated to the projectile a wavepacket with variable width, and have shown, that the FDCS depend on the width of the wavepacket, in reasonable agreement with the experiments. In our method we have calculated the transition probability amplitude as a function of the impact parameter, and have multiplied it by a Gaussian, modeling the finite width of the wave packet. The FDCS for a certain momentum transfer can be obtained by a Fourier transform of this convoluted probability amplitude.

The shortcoming of this method was the assignment of the impact parameter to a certain momentum transfer on the basis of classical elastic scattering, while the studied process is inelastic. In order to correct this problem we have performed classical trajectory Monte Carlo (CTMC) calculations to obtain the cross section distribution for a given momentum transfer and ejected electron momentum.

As it may be observed on Figure 1, for a given momentum transfer and ejected electron momentum the distribution of possible impact parameters has a sharp maximum. This result confirm our previous assumption, that we assign a certain impact parameter to a momentum transfer. However, the impact parameter obtained by the CTMC is much smaller, than that obtained by classical scattering (~2 a.u.). Detailed results for the dependence of the FDCS on the coherence widths of the projectile will be presented at the conference.

*E-mail: lnagy@phys.ubbcluj.ro

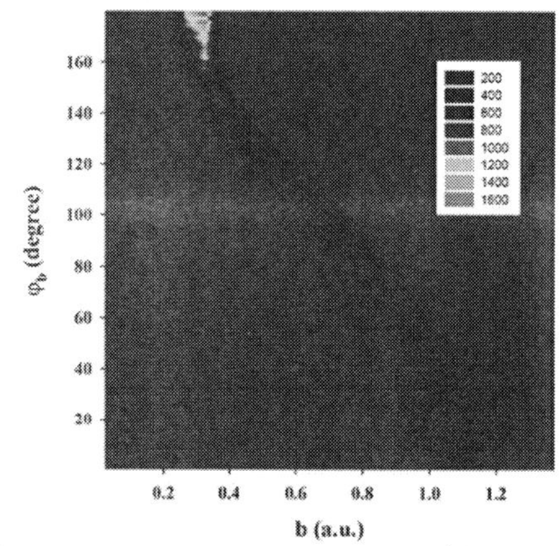

Figure 1. Impact parameter value (b) and angle (φ_b) distribution for the ionization of He by 100 MeV/u C^{6+} impact at the binary peak, if the momentum transfer is 0.75 a.u. and the ejected electron energy 6.5 eV.

References

[1] Wang X *et al* 2012 *J. Phys. B* **45** 211001
[2] Járai-Szabó F and Nagy L 2015 *Eur. Phys. J. D* **69** 4
[3] Arthanayaka T *et al* 2016 *J. Phys. B* **49** 13LT02
[4] Gassert H *et al* 2016 *Phys. Rev. Lett.* **116** 073201
[5] Kouzakov K A 2017 *Eur. Phys. J. D* **71** 63
[6] Navarrete F *et al* 2017 *Nucl. Instrum. and Meth. B* **408** 165
[7] Nagy L *et al* 2018 *J. Phys. B* **51** 144005

Coherence properties of the projectile's beam: The missing piece of the C^{6+} + He ionization puzzle

F Navarrete[1], M F Ciappina[2]* and R O Barrachina[3,4]

[1]Department of Physics, Kansas State University, Manhattan, Kansas 66506, USA
[2]Institute of Physics of the ASCR, ELI-Beamlines, Na Slovance 2, 182 21 Prague, Czech Republic
[3]Centro Atómico Bariloche and Instituto Balseiro, Comisión Nacional de Energía Atómica (CNEA) and Universidad Nacional de Cuyo, Av. Bustillo 9500, 8400 Bariloche, Argentina
[4]Consejo Nacional de Investigaciones Científicas y Técnicas (CONICET), Argentina

Synopsis We analyze the influence of the coherence of the projectile's beam in scattering phenomena. We focus our study in the ionization of He by C^{6+} projectiles at 100 MeV/amu. Our calculations, which had been previously performed for only the scattering and perpendicular collision planes and within the First Born approximation, were repeated for an ampler set of collision planes and with a more refined model. We obtain a very good agreement with the experimental results.

The ionization of atoms and molecules by the collision of energetic ions has been the subject of extensive studies for many decades (see e.g. [1]). Increasingly sophisticated theoretical models have being keeping the pace of new experimental achievements as, for instance, the development of the COLTRIMS technique and the measurement of Fully Differential Cross Sections (FDCS). Nevertheless, all theoretical models have failed to explain an experiment which has been coined as the C^{6+} – puzzle [2].

Our motivation in studying these intriguing results, is not only to improve our understanding of them, but to build a generalization of the standard Scattering Theory that would allow for the inclusion of coherence effects. In fact, we propose that the coherence properties of the projectile's beam should be considered in order to satisfactorily explain these experimental results. The so-called coherence length [3] allows for an operational definition of the partially coherent cross section in terms of the traditional fully coherent one, as described in [4].

Here we replace the First Born approximation of our previous calculation [4] by a Born initial state and continuum distorted wave final state (CDW-1B) model. Furthermore, we evaluate the FDCS for the full range of the ejected electron's azimuthal angle $0 \leq \phi_e \leq 90^o$. In the figure we show the $\phi_e = 20^o$ case, as representative of the much better agreement achieved by the partially coherent calculations with the experimental results over the entire range of ϕ_e.

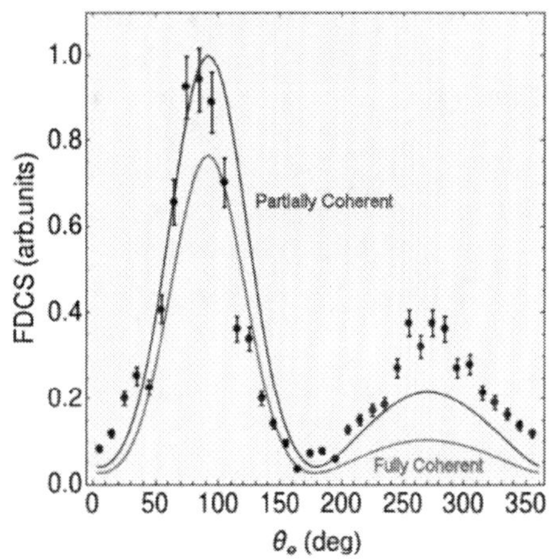

Figure 1. CDW-1B calculation of the FDCS for the ionization of He by 100 MeV/amu C^{6+} ions as a function of the polar angle θ_e of 6.5 eV ejected electrons, at an azimuthal angle $\phi_e = 20^o$. The curves have been normalized by the maximum at $\phi_e = 0^o$. The partially (blue line) and fully (red line) coherent calculations are compared with the experimental results (black dots) [2].

References

[1] Stolterfoht N, DuBois R D and Rivarola R D 1997 *Electron Emission in Heavy Ion-Atom Collisions* (Springer Verlag, Heidelberg)
[2] Schulz M *et al* 2003 *Nature* **422** 48
[3] Fabre I *et al* 2018 *Eur. J. Phys.* **39** 015401
[4] Navarrete F *et al* 2017 *Nucl. Instr. Meth. B* **408** 165

*E-mail: marcelo.ciappina@eli-beams.eu

Coulomb explosion of polycyclic aromatic hydrocarbons induced by heavy cosmic rays: carbon chains production rates

M Chabot[1] *, K Béroff[2], E Dartois[2] and T Pino[2]

[1]Institut de Physique Nucléaire d'Orsay (IPNO), CNRS-IN2P3, Univ. Paris Sud, Université Paris-Saclay, F-91406 Orsay, France
[2]Institut des Sciences Moléculaires d'Orsay (ISMO), CNRS, Univ. Paris Sud, Université Paris-Saclay, F-91405 Orsay, France

Synopsis The interstellar medium contains both polycyclic aromatic hydrocarbons and cosmic rays. The frontal impact of a single heavy cosmic ray strips out many electrons. The highly charged species then relax by multi-fragmentation, potentially feeding the interstellar medium with hydrocarbon chains. We model both ionization(s) and fragmentation processes and compute the fragments production rates of particular interest for astrophysical models.

Low energy CRs consist of projectiles from proton to nickel with energy from 10 keV to GeV. Astrophysical PAHs are carbons structures with cycles containing from ten's to few hundred's of carbon atoms. We report on the coulomb explosion of highly positively charged PAHs induced by the interstellar low energy CRs leading to the formation of long carbon chains.

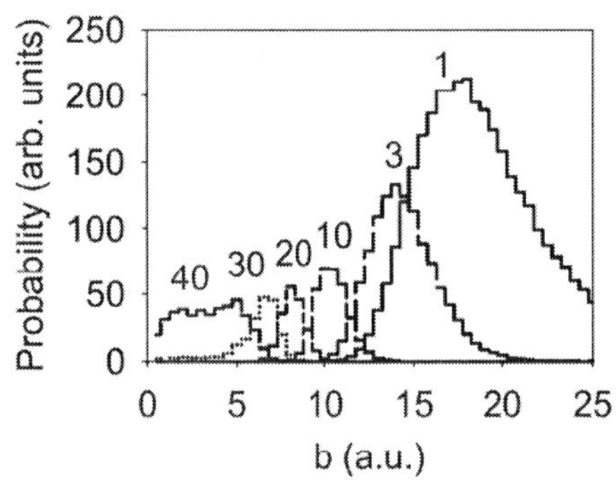

Figure 1. Relative ionization probabilities as function of the impact parameter for C_{60} and iron projectile at 1.1 MeV/u. The label on the curves corresponds to the ionization fold.

To calculate the ionization cross sections, we employ the so-called independent atom and electron (IAE) model [1] where all atoms and electrons are treated as independent. The IAE probabilities thus consist of the products of atomic probabilities. The impact parameter probabilities of Q fold ionization's for an iron projectile at 5MeV/u on a C_{60} is shown in Fig.1 as an example of such calculation.

To calculate the charge state above which the PAH multi-fragments, we used a statistical model in a microcanonical formalism based on the experimental work of S. Martin [3]. The internal energy was compute with the IAE model. To retrieve the figure of fragmentation in the multi fragmentation process of the multi-charged species, we extend a statistical model construct initially for small multi-charged carbon C_n(n=2 to 10)[2].

The size of the carbon chains resulting from the coulomb explosion extends up to 10 to 15(±3) carbon atoms, depending on the size, compactness and hydrogenation degree of the PAHs. Their production rates are weakly depending on their sizes. The carbon chains production rates range from a few to many tens of percents of the H_2 ionization rate, depending on the PAH structure and the adopted CR type (i.e. Galactic standard spectrum or closer to a local CR source with a steeper low energy flux distribution) [4].

References

[1] Wohrer K *et al* 2017 *J. Phys. B: At. Mol. Opt. Phys.* **29** L755
[2] Chabot M *et al* 2010 *Phys. Rev. Lett.* **104** 043401
[3] Martin S *et al* 2008 *Phys. Rev. A* **77** 043201
[4] Chabot M *et al* 2017 *arXiv* 1709.07803

*E-mail: chabot@ipno.in2p3.fr

Double photoionization of He isoelectronic sequence

J Koncevičiūtė[1]* and V Jonauskas[1]†

[1]Institute of Theoretical Physics and Astronomy, Vilnius University, Saultekio av. 3, LT-10257 Vilnius, Lithuania

Synopsis Double photoionization of He-like ions is investigated using a few-step approach, that treats direct double photoionization as a sequence of two- (photoionization-ionization) and three-step (photoionization-excitation-ionization) processes.

The emergence and advancement of attosecond science provide the fastest time resolutions, corresponding to the timescale for electron motion in atoms [1], thus enabling oservation of a rapid response of electron in atoms to light waves, and electronic correlations.

To interpret the results of such experiments theoretical methods capable to describe the dynamics of a few-electrons systems are needed. Direct double photoionization is of greatest interest among these processes. This process cannot occur without electron correlations therefore it is thought that it cannot be treated perturbatively.

The aim of this work is to apply a few-step approach [2] to study double photoionization of the ground states of He-like ions with Z up to 10. This method treats direct double photoionization as a sequence of two- (photoionization-ionization) and three-step (photoionization-excitation-ionization) processes, thus simplifying the solution of the complex three-body Coulomb problem. This method requires relatively low computational resources and it has already been proved successful in the study of direct double ionization process for light ions. Therefore it is of great practical importance.

Flexible atomic code, which implements the Dirac-Fock-Slater approach, has been applied to calculate energy levels, radiative transition probabilities, as well as electron-impact excitation and single ionization cross sections. Electron-impact excitation and single ionization processes were investigated using the distorted-wave approximation. Coulomb (CG) and Babushkin (BG) gauges were used for the radiative transition calculations.

For He it is shown that the better agreement with experimental results is achieved when Coulomb gauge is used for the radiative transition calculations. Though positions of the peaks of obtained theoretical crossections correspond to that of experimental results, both gauges give significantly lower results at the peak. Similar tendency has been observed for other studied He-like ions.

In order to achieve better agreement with experimental results, theoretical cross sections were scaled to experimental ones. Scaling factors were obtained by a least squares method. It has been shown that scaling factors for CG and BG converge with increasing ionisation stages of investigated He-like ions.

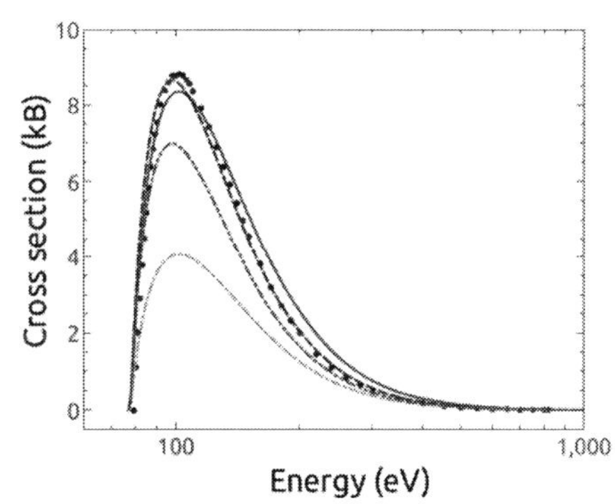

Figure 1. Double photoionization cross section of He. Solid line (red): scaled BG; long-dashed line (blue): scaled CG; pointed-dashed line (orange): BG; double-pointed-dashed line (green): CG; black points: experiment [3]

References

[1] Kim K T *et al* 2014 *Nat. Photon.* **8** 187
[2] Jonauskas V *et al* 2014 *Phys. Rev. A* **89** 052714
[3] Samson J A R *et al* 1998 *Phys. Rev. A* **57** 1906

*E-mail: jurgita.konceviciute@tfai.vu.lt
†E-mail: valdas.jonauskas@tfai.vu.lt

Experimental alternative to investigate the radiation-induced radical chemistry at the molecular level

A Beydoun[1] *, F Lecomte[1], N Nieuwjaer[1], C Desfrançois[2] and B Manil[1]

[1]Université Paris 13, Sorbonne Paris Cité, Laboratoire de Physique des Lasers, CNRS (UMR 7538), F-93430 Villetaneuse, France
[2]CNRS (UMR 7538), Laboratoire de Physique des Lasers, Université Paris 13, Université Sorbonne Paris Cité F-93430 Villetaneuse, France

Synopsis This project aims to develop an original approach to investigate the damage at the molecular level on biomolecules from the indirect irradiation effects. We will construct a new device to put biomolecules in the gas phase, based on the soft laser desorption from microdroplets directly into vacuum, and this source will be coupled with an irradiation platform.

Biologically relevant material is usually associated with considerable amounts of water; e.g. the living cell contains about 70% water. When ionizing radiation interacts with such material one must consider two main modes of energy deposition: the *direct effect* (ionizing radiation is directly absorbed by the biomolecule) and the *indirect effect* (ionizing radiation is absorbed by the surrounding molecules). In the *indirect effect*, radical cations, electrons and slow ions (essentially oxygen ions) are formed. These reactive intermediates interact also with the biomolecules.

The radiation damage, usually obtained from studies where the living cells are directly irradiated with ionizing particles, is consequently the sum of these different effects. In this case, the only contribution of the radical chemistry in the global damage is reasonably impossible to extract. However, the radical species playing an important role in the formation of lesion clustering at the end of the Bragg peak, the understanding of the radical chemistry at the molecular level becomes crucial with the advent of cancer treatments by ionizing radiation [1].

We propose here an alternative to study separately the part of damage due to the radial species. We are currently developing an original source producing biomolecules in gas phase, which is based on the ultra soft laser desorption from liquid microdroplets under vacuum [2]. The desorbed biomolecules, initially present in the liquid, will thus be studied, in the gas phase by mass spectrometry, close to their native forms in solution. Upon irradiation by home-made mid-IR broad band laser pulses (centered on an absorption band of the solvent), analyte ions are ejected directly into high vacuum, with a minimum of perturbations. The possibility to irradiate the droplet with low energy ions gives then a unique access to the damage mechanisms linked with the radical species (see Fig. 1).

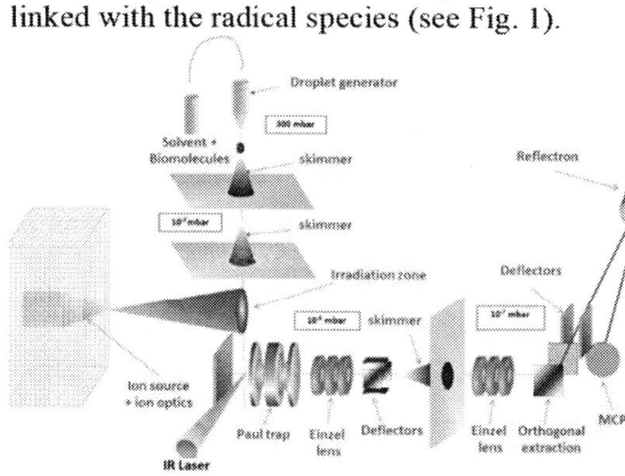

Figure 1. . Schematic view of the experimental set-up.

The low energy ions cannot penetrate the microdroplet and create only water free-radials (OH^- and solvated electrons) at the droplet surface, which interact later with the biomolecules included into the liquid droplet. This experimental method simulates, close to *real conditions*, the radical chemistry steps occurring during the irradiation.

References

[1] Nikjoo H *et al* 2001 *Radiat. Res. : Conf. Ser.* **156** 577
[2] Morgner N *et al* 2006 *Aust. J. Chem. : Conf. Ser.* **59** 109

* E-mail: ali.beydoun@univ-paris13.fr

Auger decay of Rubidum atom after 3d-shell ionization

J Palaudoux[1*], F Penent[1†], M Khalal[1], K Jänkälä[2], J Keskinen[2], M Huttula[2], J-M Bizau[3,4], D Cubaynes[3,4], S Guilbaud[3], M Zitnik[5], K Bucar[5], K Ito[4], L Andric[1] and P Lablanquie[1]

[1]Sorbonne Université, LCP-MR,CNRS UMR 7614,, 4 place Jussieu, 75005 Paris, France
[2]Nano and Molecular Systems Research Unit, University of Oulu, P.O. Box 3000, 90014 Oulu, Finland
[3]Univ. Paris-Sud, ISMO, CNRS UMR 8214, , Université Paris-Saclay, 91405 Orsay, France
[4]Synchrotron SOLEIL, l'Orme des Merisiers, Saint-Aubin, BP 48, 91192 Gif-sur-Yvette Cedex, France
[5]Jozef Stefan Institute, Jamova cesta 39, SI-1001 Ljubljana, Slovenia

Synopsis We present here the different Auger decay paths following 3d inner-shell ionization of Rubidium atom and we observe the peculiar behaviour and correlation effects due to the outer, unpaired 5s electron. This electron can be submitted to shake-up during 3d ionization. Also shake-up and shake-down of the outer electron are observed in the Auger decay. Cascade double Auger decay is the dominant process leading to Rb^{3+} ion.

We have studied Rb 3d inner-shell ionization by detecting in coincidence photo- and Auger electrons. The experiment, performed on PLEIADES and SEXTANTS beamlines of synchrotron SOLEIL, uses a resistively heated oven to produce atomic Rb vapour and a magnetic bottle spectrometer to analyze electrons [1]. This apparatus provides 4π detection for all the electrons with a resolution $\Delta E/E \sim 1.6\%$, enabling detection in coincidence of all the electrons emitted in multi-ionization processes.

Rb 3d inner-shell ionization is peculiar since the outer 5s electron can be easily simultaneously excited to nl orbitals (nl = 5s, 5p, 6s, 6p, 5d and 7s), giving rise to $3d_j^{-1}nl$ satellite photoelectrons lines. With the $3d_{5/2}$-$3d_{3/2}$ spin-orbit splitting, this produces a congestion of lines in the conventional Auger spectrum, escaping to easy interpretation until now [2].

In a first study, we have filtered out the total Auger spectrum to obtain the partial Auger spectra associated to each $3d_j^{-1}nl$ state. Using configuration interaction Dirac-Fock calculations, we could assign the $M_{4,5}N_{2,3}N_{2,3}$ and $M_{4,5}N_1N_{2,3}$ Auger transitions in the spectra [3].

Here, we pursue this study to obtain the complete Auger spectrum of the $3d_j^{-1}5s$ states and their $3d_j^{-1}nl$ satellites.

The first result concerns the Auger decay in which the outer nl electron participates to the decay, populating Rb^{2+} $4s^24p^{-1}5s^{-1}$ or $4s^{-1}4p^65s^{-1}$ states. This participator decay is very weak (<2%) with a probability decreasing with n, as predicted by our calculations.

To assign the Auger lines, we measured first the Auger decay of the core-valence doubly ionized states Rb^{2+} $3d_j^{-1}5s^{-1}$, isoelectronic with Kr^+ $3d_j^{-1}4s^24p^6$ states, the Auger spectra of which was previously studied in detail [4]. The Rb^{2+} $3d_j^{-1}5s^{-1}$ Auger spectra is obtained from with 3-electron events (2 photo-electrons + 1 Auger electron) in the reaction: Rb + hν → Rb^{2+} $3d_j^{-1}5s^{-1}$ + $2e^-_{Photelectron}$ (core-valence double ionization) → Rb^{3+} + e^-_{Auger} (Auger decay)

The Auger spectra from Rb^+ $3d^{-1}nl$ states is simply deduced from that of the core-valence $Rb^{2+}3d_j^{-1}5s^{-1}$ states by considering the outermost electron as spectator, only 4s and 4p electrons being involved in the Auger decay. However, weak shake-up (5s being promoted to 5p or 6s) and shake-down (6s being shaken down to 5p or 5s) processes are also observed that amount to less than a few % of the total decays.

Finally, double Auger decay of the Rb^+ $3d_j^{-1}nl$ states was characterized. This process is stronger than single Auger decay (76 vs 24%) The first step Auger decay, with 5s electron spectator, leads mostly to Rb^{2+} states with binding energies higher than the $4s^{-1}4p^{-1}(^3P)5s$ level above the Rb^{3+} ground state. These states further decay by emission of the 5s electron in a cascade Double Auger decay.

References

[1] Palaudoux J *et al* 2015 *Phys. Rev. A* **92** 012510
[2] Jänkälä K *et al* 2006 *Phys. Rev. A* **74** 062704
[3] Keskinen J *et al* 2017 *Phys. Rev. A* **95** 043402
[4] Palaudoux J *et al* 2010 *Phys. Rev. A* **82** 043419

* E-mail: jerome.palaudoux@upmc.fr
† E-mail: francis.penent@upmc.fr

 1412 (2020) 152038 doi:10.1088/1742-6596/1412/15/152038

Multi-electron/ ion coincidence spectroscopy in Xe triple ionization

M A Khalal[1], J Palaudoux[1], P Lablanquie[1], L Andric[1], M Zitnik[2], K Bucar[2], J-M Bizau[3, 4], D Cubaynes[3, 4] and F Penent[1*]

[1] Sorbonne Université, CNRS, LCP-MR, 75005 Paris, France
[2] J. Stefan Institute, Jamova 39, 1000 Ljubljana, Slovenia
[3] ISMO, CNRS UMR 8214, Université Paris-Sud, Bâtiment 350, F-91405 Orsay cedex, France
[4] Synchrotron SOLEIL, L'Orme des Merisiers, Saint Aubin, F-91192 Gif-sur-Yvette cedex, France

Synopsis The study of atomic multiple photoionization by a single photon can be performed by detecting in coincidence all the corresponding electrons with a magnetic bottle time of flight spectrometer. To reveal minor triple ionization channels of Xe atom that are hindered by radom coincidences, we have detected three electrons in coincidence with the corresponding Xe^{3+} ion, thanks to a new permanent magnet design using a drilled soft iron pole.

Single photon mutiple photoinization processes result directly from electron correlations. They can be studied most efficiently by multi-electron coincidence spectroscopy [1] with a magnetic bottle time-of-flight (MB-TOF) spectrometer that provides both very high detection efficiency (>70%) and good energy resolution ($\Delta E/E<2\%$). Nevertheless, minor ionization channels can be hindered by random coincidences if they appear in the same energy window than major processes.

Our experimental setup has been modified to add a ion time-of-flight in order to detect ions and electrons in coincidence. Electrons are flying in a 2 m long MB-TOF whereas, after 300 ns time delay after ionization, a ~200V electric pulse repels the ions, through a specially designed "drilled magnet", in the opposite direction into the ion TOF.

The experiment was performed on the SEXTANTS beamline at SOLEIL. Xenon multiple photoionization was studied at hν=220 eV (Fig 1) and 135 eV. Different charge states ions (Xe^{n+}, n=2-4) are detected in coincidence with the associated electrons. It allows us to filter and select electrons that come from double, triple or quadruple ionization.

Thanks to the detection of n electrons with the associated Xe^{n+} ion, random coincidences are reduced drastically. At 135 eV, this has allowed the observation of the minor $4d^{-1}5s^{-1}$ core/inner-valence double ionization channel leading to Xe^{3+} after Auger decay. Without Xe^{3+} ion detection, this process is masked by random coincidences related to the dominant 3d single Auger decay.

Another electron / ion coincidence apparatus was recently developed [2]. One advantage of our set-up is that ions are detected through the drilled magnet. This offers a better accessibility to the interaction region, for an oven and a cold trap, for further studies of metallic vapors.

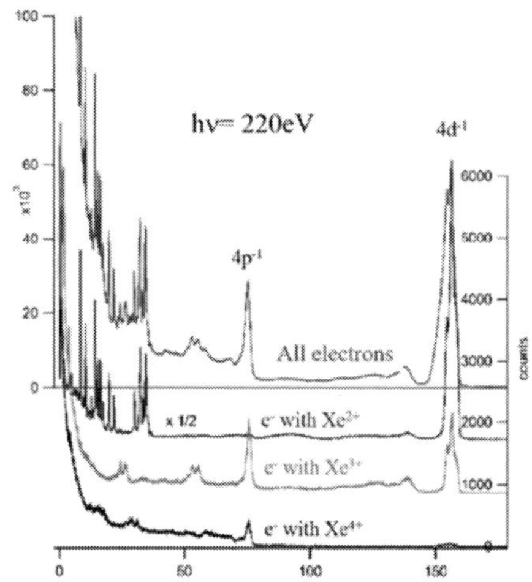

Figure 1. Coincidence between one electron and Xe^{n+} ion.

References

[1] Penent F *et al* 2005 *Phys. Rev. Lett.* **95** 083002

[2] Eland J H D *et al* 2015 J. *Phys. B.* **48** 205001

* E-mail: francis.penent@upmc.fr

Journal of Physics: Conference Series **1412** (2020) 152039 doi:10.1088/1742-6596/1412/15/152039

Inner-shell ionization of W^{25+}

S Pakalka[1] *, A Kynienė[1] and V Jonauskas[1]

[1]Vilnius University, Institute of Theoretical Physics and Astronomy, Saulėtekio av. 3, Vilnius, LT-10257, Lithuania

Synopsis Inner-shell ionization by electron impact is investigated for the W^{25+} ion. Decay of inner-shell vacancy through Auger transitions is analyzed. Contribution of direct ionization process to double and triple ionization is presented.

Inner-shell ionization by electron impact can lead to cascade of radiative and Auger transitions. Direct ionization of the valence shells is stronger compared to the ionization from the inner shells. On the other hand, the inner-shell ionization can be important for ionization balance in plasma. What is more, excitation with multiple autoionization can also contribute to higher ionization stages.

The aim of the current work is to analyze electron-impact ionization from the inner shells of tungsten ions with the open $4f$ shell in the ground configuration and to estimate contribution of inner-shell ionization to multiple ionization.

Energy levels and ionization cross sections are studied using Flexible Atomic Code (FAC) [1] which implements the Dirac-Fock-Slater approach. Electron-impact ionization cross sections are obtained in the distorted wave approximation.

Theoretical single ionization threshold for the W^{25+} ion is 782 eV and the double ionization threshold is equal to 1612 eV. Direct ionization from the $4s$, $4p$, $4d$, and $4f$ shells of the ground $4s^2 4p^6 4d^{10} 4f^3$ configuration leads to single ionization since the energies of the corresponding configurations are below the double ionization threshold. On the other hand, ionization from the deeper shells is the source of multiple ionization.

Electron-impact direct ionization cross sections from shells of the W^{25+} ion are presented in Fig. 1. It can be seen that contribution from the deeper shells that can contribute to multiple ionization reaches ~10%. Energy levels of the W^{26+} $3d^9 4f^3$ configuration are above the triple ionization threshold. Current data show that decay of this configuration ends in W^{27+}. The configuration with the largest probabilities decays to $4d^8 4f^3$ (46%), $4p^5 4d^9 4f^3$ (26%), and $4d^9 4f^2$ (12%). On the other hand, influence of correlation effects for these configurations can lead to triple ionization of W^{25+}. It was demonstrated that correlation effects are important in triple ionization of Se^{3+} [2]. What is more, decay of the W^{25+} $3p^5 4f^8$ configuration leads to double and triple ionization.

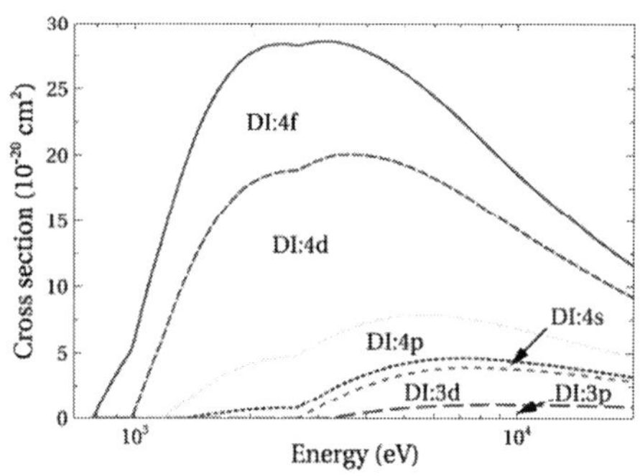

Figure 1. Direct ionization cross sections for the shells of W^{25+}.

References

[1] Gu M F 2008 *Can J Phys* **86** 675
[2] Koncevičiūtė J *et al* 2018 *Phys. Rev. A* **97** 012705

*E-mail: saulius.pakalka@tfai.vu.lt

 1412 (2020) 152040 doi:10.1088/1742-6596/1412/15/152040

Multiple ionization effect in Dy, Ho and Er at low proton energy

M Kaur[1,2*], H Mohan[1], A K Jain[1], P S Singh[2] and S Sharma[3]

[1]Department of Physics, M. L. N. College, Yamuna Nagar, 135 001, Haryana, India
[2]Department of Physics, Punjabi University, Patiala, 147 002, Punjab, India
[3]Department of Chemistry, M. L. N. College, Yamuna Nagar, 135 001, Haryana, India

Synopsis The L shell X-ray production cross sections for elements (Z = 66, 67 and 68) at three different low proton energies i.e. 260, 280 and 300 keV have been investigated. Calculations are based on the ECPSSR model and its modified form i.e. ECPSSR-UA model. During calculations with ECPSSR theory, a significant phenomenon known as multiple ionization (MI) has been included. Theoretically calculated values are compared with the experimental values. The calculations with ECPSSR-MI shows improved results.

The emission of X-rays is a fundamental process in the interaction of charged particles with matter. Ion moving into it never pierces directly rather it penetrates via several collisions. This effect is known as multiple ionization (MI). Its ignorance may leads to an erratic results at lower energy side [1]. Because it alters the parameters such as fluorescence and Coster-Kronig yields [2]. These varying parameters further affect X-ray production cross sections and their energies. A compilation of Verma [3] and Jaskola *et al* [4] shows MI effect for heavy projectile ions. However, the case of protons is still largely unexplored [5].

In the present paper, work has been performed for L_α cross sections of Dy, Ho and Er at three different proton energies i. e. 260, 280 and 300 keV (see Fig. 1). Calculations are made with ECPSSR model with the MI effect as well as with united atom (UA) correction using parameters recommended by Campbell [6, 7]. The calculated values are then compared with experimental points [8]. The theoretical predictions with MI effect are represented by the solid line while those with UA are by dashed line. Whereas, experimental values are denoted by the different symbols. From Figure 1, it is observed that the X-ray cross sections get improved by the aid of the MI effect. It brings theoretical values near the experimental points. The improvement is more at lower energy than that of higher one. For example, in case of Dy the difference between experimental and calculated values at 260 keV reduces from 16% to 6% while at 300 keV this difference gets reduced to 3.5% from 12.5%. All of this illustrates that the MI effect is mainly prominent in the lower energy region. The detailed results will be discussed during the conference.

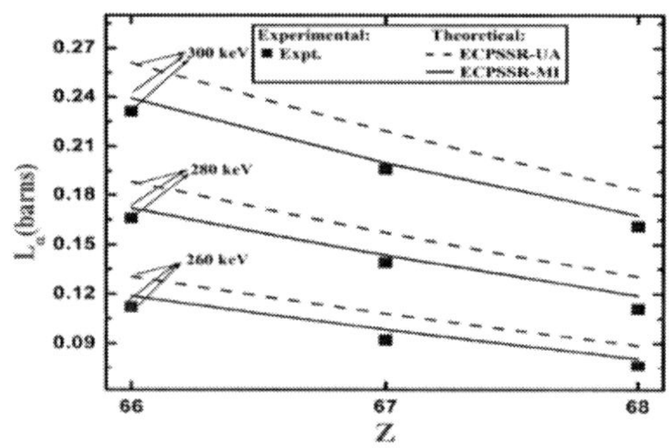

Figure 1. L_α cross sections of **low Z** elements at different proton energies.

References

[1] Jain A K *et al* 2010 *Nucl. Instr. Meth. B* **268** 1790
[2] Kaur M *et al* 2018 *Radiat. Phys. Chem.* **151** 120
[3] Verma H R 2018 *Radiat. Phys. Chem.* **150** 30
[4] Jaskola M *et al* 2017 *Nucl. Instr. Meth. B* **408** 146
[5] Miranda J *et al* 2004 *Radiat. Phys. Chem.* **69** 257
[6] Campbell J L 2003 *At. Data Nucl. Tables* **85** 291
[7] Campbell J L 2009 *At. Data Nucl. Tables* **95** 115
[8] Singh P S 1989 Ph. D. Thesis Punjabi University Patiala, Punjab

* E-mail: manpuneetkaur1007@gmail.com

Strong bond rearrangement and hydrogen-pair emission in linear geometry following water ionization by electron impact

N Ferreira[1], L Sigaud[2] and E C Montenegro[3*]

[1]Centro Federal de Educação Tecnológica Celso Suckow da Fonseca, Rio de Janeiro, 20271-110, Brazil
[2]Universidade Federal Fluminense, Niteroi, 24210-346, Brazil
[3]Universidade Federal do Rio de Janeiro, Rio de Janeiro, 21941-972, Brazil

Synopsis The $H^0 + O^+ + H^0$ fragmentation of water by electron impact is studied using the delayed extraction time-of-flight technique to uncover kinematic attributes of suprathermal O^+ ions. It is found that the H^0 pair ejections occur under a large angular rearrangement, in an approximate linear geometry, and leaving the O^+ near at rest, with ~50% of the O^+ produced in water fragmentation by swift electrons presenting this feature. It is shown that this behavior is closely associated to the excitation of satellites one-electron-two-vacancies states near the $2a_1$ orbital.

Collisions between electrons and water molecules have received great attention in the last decades due to their importance on radiobiology and in understanding the inventory of planetary atmospheres and magnetospheres. These motivations have greatly influenced the choice of making water a benchmark molecule to understand the collision and fragmentation dynamics of multi-electron molecular systems. Valence shell electrons are the major players in these processes and rule the main fragmentation pathways. Among all valence electrons, the removal of an inner valence shell (2a1 MO) electron results in a particularly rich and complex dynamics as it throws the system in a set of states that cannot be described within the independent particle model or the Born-Oppenheimer approximation.

An important question to be posed is if these mixed states can anyhow be linked to the final fragmentation products. We have addressed this issue using the DETOF technique [1] to show that the O^+ fragment is a clear identifier of the kinematic and dynamic signatures of three-body fragmentation associated to shake-up (satellite) states. We found that the kinetic energy distribution of O^+ ions from single-ionized water has a suprathermal narrow component with average kinetic energy, relative to the center of mass (CM), of ~0.038 eV. This component has a threshold below the vertical ionization energy of the orbital 2a1 thus corresponding to satellites one-electron-two-vacancies excited states. This very narrow contribution makes a strong constraint for the kinetic energy release of three-body kinematics, and restricts the neutral hydrogen-pair ejection to a large angular bond rearrangement, towards the fragmentation near the linear configuration of the water molecular ion [2]. This behavior accounts for ~50% of the O^+ produced by electron impact. The remaining O^+ are produced by a vertical transition followed by Auger decay of the evolving valence orbitals during the molecular break-up.

Satellite states are also found to have a main role in multiple-hydrogen ejection from singly-ionized CH_4 [3]. Differences and similarities between CH_4 and H_2O for full H^0 removal following single ionization by electron and proton impact will be explored at the conference.

* E-mail: montenegro@if.ufrj.br

References

[1] Sigaud L *et al* 2016 *Rev. Sci. Instrum.* **87** 083112
[2] Ferreira N *et al* 2017 *J. Phys. Chem.*. **121** 3234
[3] Luna H *et al* 2019 *Phys. Rev. A* **99** 012709

Fully differential study of ionization of H_2 by p impact near velocity matching

M Dhital[1], S Bastola[1], A Silvus[1], A Hasan[2], B R Lamichhane[1], E Ali[1], M F Ciappina[3], R A Lomsadze[4], D Cikota[1], B Boggs[1], D H Madison[1] and M Schulz[1*]

[1]Dept. of Physics and LAMOR, Missouri University of Science & Technology, Rolla, MO 65409, USA
[2]Dept. of Physics, UAE University, P.O. Box 15551, Al Ain, Abu Dhabi, UAE
[3]Institute of Physics of the ASCR, ELI-Beamlines, Na Slovance 2, 182 21 Prague, Czech Republic
[4]Tbilisi State University, Tbilisi 0179, Georgia

Synopsis Fully differential cross sections (FDCS) for ionization in p + H_2 collisions in the region of the electron-projectile velocity matching were measured and calculated. The FDCS were found to be very sensitive to the details of the few-body dynamics underlying the reaction.

We have measured fully momentum-analyzed scattered projectiles and recoil ions, produced in ionization of H_2 by 75 keV p impact, in coincidence. The ejected electron momentum was deduced from momentum conservation. From the data we extracted fully differential cross sections for fixed projectile energy losses of ε = 50,53,57, and 60 eV, for various fixed scattering angles θ_p, and for electrons ejected into the scattering plane as a function of the emission angle. Here, ε = 57 eV corresponds to an electron speed equal to the projectile speed. The data were compared to two conceptually very similar distorted wave calculations.

The FDCS for θ_p = 0.55 mrad are shown in Fig. 1. Two separate peak structures can be seen, one near the direction of the momentum transfer **q**, indicated by the vertical arrows, and one at θ_{el} = 0. The former, known as the binary peak, is a signature of the first-order mechanism. The latter, to which we refer as the forward peak, is a result of a higher-order mechanism, known as post-collision interaction PCI, involving at least 2 interactions between the projectile and the active electron. The first interaction lifts the electron to the continuum and in the second interaction the electron and the scattered projectile attract each other towards the initial beam axis. As expected, the forward to binary peak intensity ratio maximizes at the matching velocity (ε = 57 eV).

The dashed and solid curves show continuum distorted wave – eikonal initial state (CDW-EIS) and 3-body distorted wave (3DW) calculations, respectively. While at θ_p = 0.1 mrad both calculations are in reasonable agreement with experiment, increasing discrepancies and differences between both models are observed with increasing θ_p. At θ_p = 0.55 mrad there is not even qualitative agreement. PCI effects are most pronounced in the experimental data and much stronger in the 3DW model compared to the CDW-EIS calculations. Considering that both theoretical approaches are conceptually very similar these surprisingly large differences demonstrate the high sensitivity of the FDCS to the details of the few-body dynamics in this kinematic regime.

This work was supported by NSF and by ADONIS.

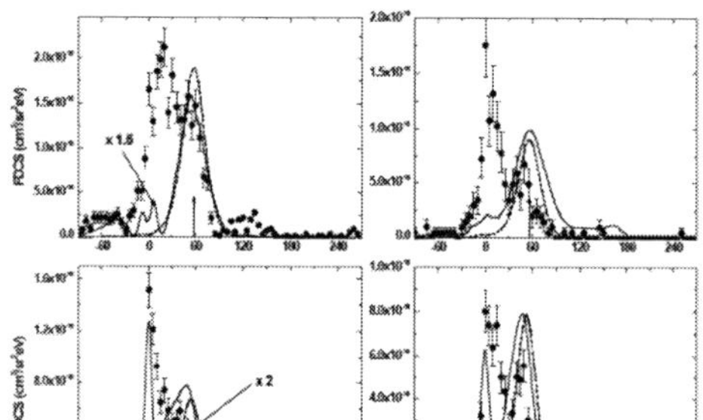

Figure 1. FDCS for electrons ejected into the scattering plane at θ_p = 0.55 mrad. Dashed curves, CDW-EIS; solid curves, 3DW.

* E-mail: schulz@mst.edu

Nitrogen K-shell photoionization of NO^+

M F Gharaibeh[1] *, J M Bizau[2], D Cubaynes[2], S Guilbaud[2], M O A El Ghazaly[3] and O A Al Hagan[3]

[1]Department of Mathematics, Statistics and Physics, P.O.Box 2713, Qatar University, Doha, Qatar
[2]Institut des Sciences Moléculaires d'Orsay (ISMO), CNRS UMR 8214, Université Paris-Saclay, Bât. 520, F-91405 Orsay cedex, France
[3]Department of Physics, Faculty of Science, King Khalid University, P. O. Box 9004, Abha 61413, Saudi Arabia

Synopsis Photoionization measurements of NO^+ at Nitrogen K-shell, performed at the SOLEIL Light Source will be presented . The main observed structure corresponds to the 1s to π^* resonant transition.

Our knowledge on the Universe is carried to us by photons, which are dispersed and detected by the orbiting Chandra and XMM-Newton satellites [1]. Molecular ions have major influence in the chemistry of planetary atmospheres, as well as in astrophysical and laboratory plasmas. The accurate knowledge of their electronic structure and of their photoexcitation cross sections is of crucial importance to evaluate their chemical reactivity. Nitrogen and Oxygen are from the most abundant gasses in the space as well as the NO molecules.

Combining a merged ion-photon beam end-station [Permanently installed on PLÉIADES Beam-line] and named by MAIA (Multi-Analysis Ion Apparatus) with an electron-cyclotron-resonance ion source and a third-generation synchrotron radiation source makes possible high quality quantitative studies of photo-ionization of singly and multiply charged ions. We are going to present our recent study of molecular ions (NO^+). The PLÉIADES BL covers the photon energy range required to investigate photon studies of K-Shell region of Nitrogen and Oxygen, in particular measurements for photoionization of NO^+ by detecting the N^{2+} fragment in the photon range of the nitrogen K-shell. We found the fragmentation channel of O^{2+} having similar intensity in this energy range. Up to our knowledge the only previous studies on photoionization of NO molecule were performed on neutral molecules and not in ionic state. MAIA setup has already shown its capability to study the inner-shell photoexcitations of molecular ions such as CH^+, OH^+, and SiH^+[2].

We managed to detect the N^{2+} fragments produced in the Nitrogen K-shell PI of NO^+ in the energy range 398-414 eV which covers the region of the $1s \rightarrow \pi^*$ excitation (See Figure 1).

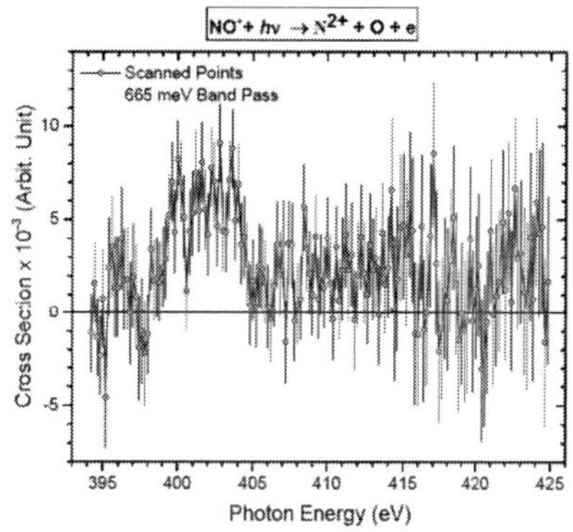

Figure 1. N^{2+} fragments produced in the Nitrogen K-shell photoionization of NO^+ in the energy range 398-414 eV with 670 meV band pass.

References

[1] Costantini E *et al* 2012 *Astron. and Astrophys.* **539** A32
[2] Mosnier J P *et al* 2016 *Phys. Rev. A.* **93** 061401(R)

* E-mail: mgharaibeh@qu.edu.qa

Triple differential cross sections for electron-impact ionization of atomic and molecular targets

S Houamer[1] *, T Khatir[1] and C Dal Cappello[2]

[1]*LPQSD, Department of physics, Faculty of Science, University Sétif-1, 19000, Setif, Algeria*
[2]*SRSMC, CNRS 7565, Univ. of Lorraine, BP 70239, 54506 Vandoeuvre-les-Nancy, France*

Synopsis Triple differential cross section for ionization of atomic and molecular targets by electron impact is calculated using approximate three body waves. The post-collisionnal interaction (PCI) between the outgoing electrons as well as the distortion effects of the ejected electron are included. The effect of the short range potential is evenly investigated. Results are compared with available data.

Electron impact ionization of matter is a fundamental process in collision physics where cross sections are calculated and measured since quite a while, the process has then gained a lot of interest during the past years. Much emphasis has been devoted to cross sections calculations because of basic importance in many practical applications such as astrophysics, plasma physics and irradiation of living matter. The ionization process has in fact been widely dedicated to atomic targets in contrast to molecules where further calculations are still to be developed because of the encountered problems in theory and experiments.

We consider here an approach which includes post-collisionnal interaction (PCI) between the outgoing electrons as well as the distortion effects induced by the residual ion to investigate the ionization mechanism for some atoms and small molecules. (i) firstly the BBSR model where the outgoing electrons are represented by Coulomb waves taking into account their interactions with the residual ion. (ii) secondly the BD model where the ejected electron is represented by a distorted wave while the scattered one is described by a Coulomb wave. (iii) and in a third step the BDSR model which includes of the short range potential in the of BD model . In all cases the impact electron is described by a plane wave and a Coulomb interaction is used to account for the PCI, a full description of the inclusion of distortion effects can be found in our previous works [1]. The method is applied to study the ionization process in the case of some atoms and molecules.

In figure 1 the TDCS is displayed in the case of the $2a_1$ molecular orbital of water molecule at 250 eV projectile energy, where theory and experiments are normalized to BD model in the binary region. It is seen that BD and BDSR models reproduce better the recoil region than BBSR which underestimates somewhat the peak for this inner orbital. It should be indicated that the distorted effects considered in our models are introduced approximately using a variable effective charge but enabling a precious economy of computation time. A full numerical shame of the distortion effects should likely improve the results but requires more computational time. We come to the conclusion that the post collision interactions in the final state as well as the distortion effects of the outgoing electrons are to be taken into account especially for inner orbitals.

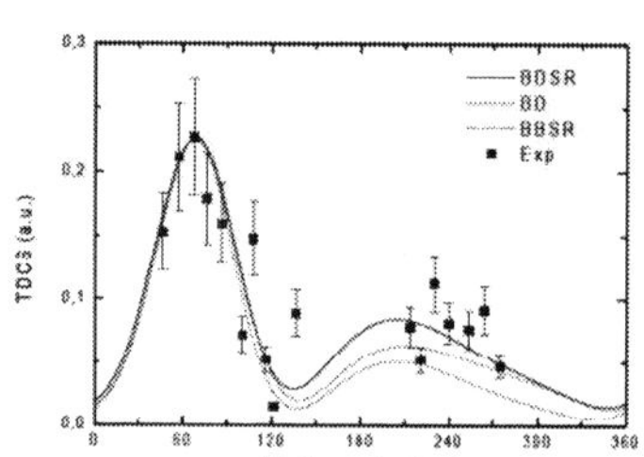

Figure 1. TDCS vs ejection angle for the ionization of the 2a1 orbital of H_2O . Kinematical parameters are: $\theta_S = 15°$, *Ei=250 eV* and *Ee=10 eV* [2].

References

[1] Houamer S *et al* 2017 *Eur. Phys. J. D* **71** 17

[2] Milne-Brownly D S *et al* 2004 *Phys. Rev. A* **69** 032701

* E-mail: s_houamer@univ-setif.dz

 1412 (2020) 152045 doi:10.1088/1742-6596/1412/15/152045

Differential cross sections for ionisation of helium by proton impact

S U Alladustov[1]*, I B Abdurakhmanov[1], A S Kadyrov[1], I Bray[1] and K Bartschat[2]

[1]Department of Physics and Astronomy, Curtin University, GPO Box U1987, Perth, WA 6845, Australia
[2]Department of Physics and Astronomy, Drake University, Des Moines, Iowa 50311, USA

Synopsis A wave-packet convergent close-coupling approach is applied to study differential ionisation in proton-helium collisions. The fully differential cross sections for electrons ejected with energy of 5.4 eV are investigated in 75 keV proton-helium collisions. We also study the doubly differential cross sections for electrons ejected with energies up to 65 eV at intermediate projectile energies (50-150 keV). The results are converged regarding the number of pseudostates included and are in fairly good agreement with the available experimental data.

In our recent work we developed a wave-packet convergent close-coupling (WP-CCC) approach to calculate the total cross sections for single ionisation and electron capture in proton-helium collisions [1]. The approach is based on expansion of the total scattering wave function in a basis made of target- and projectile-centered pseudostates and includes electron exchange in the rearrangement channel. The target is treated in the frozen-core approximation, where correlations between the electrons are taken into account. Direct ionisation of helium and electron capture to the continuum of the projectile are accounted for by including positive-energy states for both the target and projectile. The continuum states are constructed using the wave-packet method by discretising the continuum of the target and the hydrogen atom formed after electron capture by the projectile. This allows us to include pseudostates with arbitrary energies and distribution, making the method ideal for differential ionisation studies.

Here we extend our method to calculate differential cross sections for single ionisation of helium. As an example, Fig. 1 shows our results for the fully differential cross section for single ionisation of He by protons with incident energy 75 keV. The results are provided for different values of the transverse momentum transfer q_t. The experimental data of Schulz *et al* [2] are normalised to our results. We also compare our results with the first-order perturbative calculations of Schulz *et al* [2], Ma *et al* [3] and Niu *et al* [4]. Regarding the shape of the angular dependence, all calculations are in generally good agreement, but none of them can reproduce the experimental data perfectly.

*E-mail: shukhrat.alladustov@postgrad.curtin.edu.au

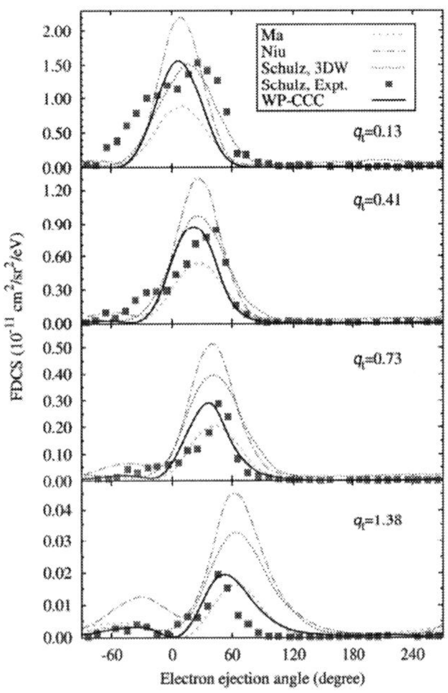

Figure 1. Fully differential cross sections for electrons with an energy of 5.4 eV ejected into the scattering plane in 75-keV p+He collisions. The WP-CCC results are compared with the experimental data of [2] (multiplied by 5.3 at q_t=0.13 au, 2.9 at q_t=0.41 au, 1.15 at q_t=0.73 au and 0.44 at q_t=1.38 au, where q_t is the transverse component of momentum transfer) and perturbative calculations [2, 3, 4].

Work supported by the Australian Research Council and the United States National Science Foundation.

References

[1] Alladustov S *et al* 2019 *Phys. Rev. A* **99** accepted
[2] Schulz M *et al* 2006 *Phys. Rev. A* **73** 062704
[3] Ma X Y *et al* 2012 *EPL* **98** 53001
[4] Niu X *et al* 2017 *Phys. Rev. A* **875** 022703

Classical calculations of differential electron emission in fast heavy-ion collisions with water molecules

A Jorge[1] *, M Horbatsch[1], C Illescas[2] and T Kirchner[1]

[1]Department of Physics and Astronomy, York University, Toronto, Ontario, M3J 1P3, Canada
[2]Departamento de Química, Universidad Autónoma de Madrid, Madrid, 28049, Spain

Synopsis We present a classical study of differential electron emission cross sections for highly-charged projectiles at $v_p \sim 10$ a.u. impinging on water molecules [1, 2]. Comparisons with Continuum Distorted Wave Eikonal Initial State theoretical results and with experimental data are presented. We implement a time-dependent screening effect in our model and account for the molecular target structure by means of a three-center potential.

In this work we study collisions of fast and highly-charged ions with the water molecule. This involves dealing with both the effect of a multicenter target as well as the effect of the so-called saturation behaviour [2]. The saturation behaviour happens when for $Z_p/v_p \sim 1$, Z_p and v_p being the charge and velocity of the projectile, the measured electron emission increases at a lesser rate than predicted by the Bethe-Born model.

In our theoretical model we consider one active electron in the water molecule target which interacts with the H_2O^+ core through a three-center model potential [3], which has screened charges for the oxygen and the two hydrogens. We apply the independent electron model (IEM) to obtain the multiple ionization probablities with a binomial analysis based on the single-ionization probabilities. We propose to make the screened charges in the model potential time-dependent, so that the screening decreases with increasing net ionization probability during the collision dynamics, in the spirit of time-dependent mean field theory [4].

In Figure 1 we show the doubly-differential cross sections for the electron emission energy 200 eV, which is sufficiently high so that the ionization mechanism can be considered classical. We find that, in order to properly represent the backward-forward asymmetry, the multi-center nature of the molecule has to be taken into account. We also find that the measured cross sections lie between the static and the time-dependent screening potential results for $Z_p/v_p = 0.73$, showing that for $Z_p/v_p \rightarrow 1$ the IEM with a static potential may become inaccurate.

*E-mail: albama@yorku.ca

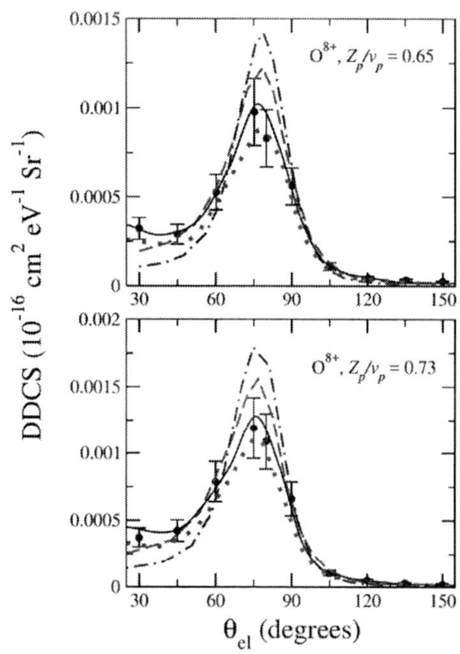

Figure 1. DDCS for net ionization as a function of electron emission angle in degrees, for the electron emission energy $E_{el} = 200$ eV. Present CTMC results with static (—) and with time-dependent screening potential (· · ·), experimental data (•), prior (— — —) and post (— · —) CDW-EIS data. [1, 2]

References

[1] Bhattacharjee S *et al* 2017 *Phys. Rev. A* **96** 052707
[2] Bhattacharjee S *et al* 2018 *Eur. Phys. J. D* **72** 15
[3] Illescas C *et al* 2011 *Phys. Rev. A* **83** 052704
[4] Kirchner T *et al* 2000 *Phys. Rev. A* **61** 052710

 1412 (2020) 152047 doi:10.1088/1742-6596/1412/15/152047

Laser-induced deformation of an inner valence molecular orbital in ethanol

H Akagi[1 *], T Otobe[1] and R Itakura[1]

[1] Kansai Photon Science Institute, QST, Kizugawa, Kyoto 619-0215, Japan

Synopsis We reveal that an intense laser field deforms an inner valence orbital (10a') in the ethanol molecule. Using photoelectron-photoion coincidence momentum imaging, we measure the recoil-frame photoelectron angular distribution (RFPAD) for the CD_2OH^+ production channel of CH_3CD_2OH induced by a circularly polarized intense ($I_0 \approx 8\times10^{13}$ W/cm^2) near-infrared ($\lambda \approx 795$ nm) laser field. *Ab initio* simulations show that the orbital deformation depends strongly on the laser field direction, and that the measured RFPAD cannot be reproduced without taking the orbital deformation into account.

In the last two decades, angular dependence of tunnel ionization probabilities has been studied mainly for diatomic and triatomic molecules [1–3]. These studies have shown that the shape of molecular orbital (MO) from which an electron tunnels strongly affects the angular dependence of the ionization probability. Recently, it has been suggested that laser electric fields deform MOs in molecules larger than triatomic ones, affecting the ionization probabilities [4,5].

In this study, we measured the recoil frame photoelectron angular distribution (RFPAD) of of ethanol (CH_3CD_2OH), using photoelectron-photoion coincidence momentum imaging [6]. An unaligned ethanol molecule was singly ionized in a circularly polarized intense laser field ($\lambda \sim 795$ nm, $I_0 \sim 8\times10^{13}$ W/cm^2), and three-dimensional momentum vectors of the electron and ion were measured in coincidence. Here, we focused on CD_2OH^+, which is produced by an electron hole formation in the 10a' MO (HOMO-1) [7]. In the circularly-polarized field, the freed electron drifts perpendicularly to the initial tunneling direction [2,3]. The recoil direction of CD_2OH^+ ion reflects the orientation of the parent molecule. Figure 1(A) shows the recoil-frame photoelectron momentum distribution. The electron in the 10a' tunnels preferentially from the CH_3 side of CH_3CD_2OH.

To explain the experimental result, we simulated angular dependence of the tunnel ionization probability from the MOs using density functional theory [8]. The simulated RFPAD from the 10a' shows the preferential electron tunneling from the CH_3 side; thus, it is somewhat similar to the measured RFPAD, as shown in Fig. 1(B).

The discrepancy between them is attributed to electron tunneling from the 3a" (HOMO), followed by the subsequent electronic excitation to the first excited state (with a hole in the 10a') of the parent ion. We evaluated the contribution of the stepwise process to be 0.48, using the measured ion yields. Taking this contribution into account, the simulation reproduces the experiment.

To evaluate the effect of MO deformation, we simulated RFPADs from the field-free MOs. The simulation cannot reproduce the measured RFPAD, suggesting that the MO deformation is needed to explain the experiment.

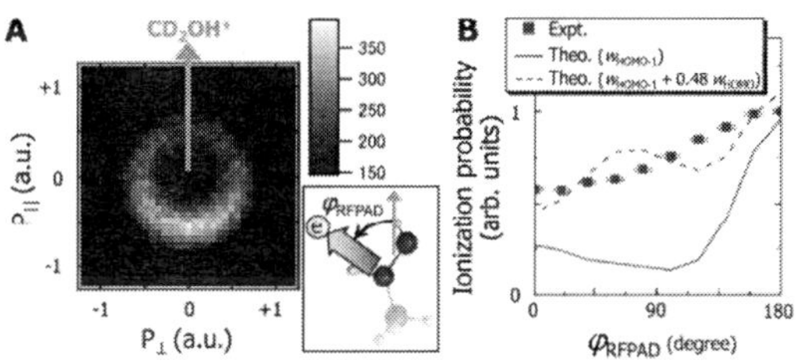

Figure 1. **(A)** Recoil-frame photoelectron momentum distribution for the CD_2OH^+ channel. **(B)** Comparison of the measured RFPAD with theory. Solid line, simulated RFPAD from the 10a'; dotted line, same as solid line but including 0.48 of the 3a" RFPAD.

References

[1] Alnaser AS *et al* 2004 *Phys. Rev. Lett.* **93** 113003
[2] Staudte A *et al* 2009 *Phys. Rev. Lett.* **102** 033004
[3] Akagi H *et al* 2009 *Science* **325** 1364
[4] Kraus P M *et al* 2015 *Nat. Commun.* **6** 7039
[5] Jochim B *et al* 2017 *Sci. Rep.* **7** 4441
[6] Hosaka K *et al* 2013 *J. Chem. Phys.* **138** 204301
[7] Niwa Y *et al* 1982 *Int. J. Mass. Spectrom. Ion. Phys.* **42** 91
[8] Otobe T *et al* 2004 *Phys. Rev. A* **69** 053404

* E-mail: akagi.hiroshi@qst.go.jp

Three body dissociation of SO_2^{q+}

J Rajput[1] *, H Kumar[1] †, P Bhatt[2] and C P Safvan[2]

[1]Department of Physics and Astrophysics, University of Delhi, Delhi-110007, India
[2]Inter University Accelerator Centre, Aruna Asaf Ali Marg, New Delhi-110067, India

Synopsis Multiply charged ion impact induced three body fragmentation pathways of SO_2^{q+} (q=3, 4) molecular ions have been studied using the technique of recoil ion momentum spectroscopy. An attempt has been made to distinguish the three-body breakup based on mechanism of dissociation, whether sequential or concerted.

In a collision of highly charged ions with neutral molecules one or more electrons can be removed from the target molecule by either ionization or electron capture process. Such an interaction causes production of multiply changed molecular ions which subsequently fragment into atomic and/or molecular ions. By measuring the momenta of the fragment recoil ions, information about the dissociation pathways can be deciphered.

We performed an experiment in which a beam of multiply charged ions, 450 keV Xe^{9+}, are made to collide with neutral SO_2 molecules. The experiment was performed at Low Energy Ion Beam Facility (LEIBF) of Inter University Accelerator Centre (IUAC), New Delhi, India. The technique of multi-hit recoil ion momentum spectroscopy employing a position sensitive microchannel plate (MCP) detector was used to measure the time and position information of the recoil ions generated upon fragmentation of SO_2^{q+}. Electrons emitted during the interaction are used as timing reference for starting the data acquisition. The three-dimensional momentum vectors are derived from the measured time of flight and position information of the detected fragment ions. The multi-hit capability of the setup aids in gaining knowledge about the parent ion from which these fragments originate. For more details of experimental setup see reference [1].

We have identified complete three body dissociation channels of SO_2^{q+} from the triple ion coincidence maps. Many three body dissociation channels are observed as O^{l+} + O^{m+} + S^{n+}, where l, m and n range from 1 to 3 with l+m+n=q.

In this presentation, the distinct dissociation pathways of SO_2^{q+}, (q=3-4) will be discussed. For example, Dalitz plot [2] for three-body breakup of SO_2^{3+} into (O^+ + O^+ + S^+) will be used to gain insight into dissociation mechanisms.

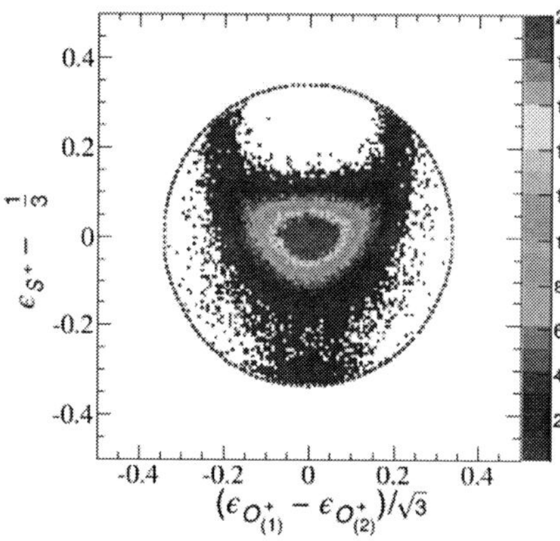

Figure 1. Dalitz plot of SO_2^{3+} ion dissociating into (O^+ + O^+ + S^+) channel, where $\epsilon_i = \frac{|\vec{p_i}|^2}{\sum|\vec{p_i}|^2}$, p_i is momentum of the ith fragment. Concerted process makes up the bulk of the image while sequential processes involving metastable SO^{2+} appear as diagonal structure.

References

[1] Kumar A *et al* 2014 *J. Mass Spect.* **374** 44-48.
[2] Dalitz R 1953 *The London, Edinburgh and Dublin Philosophical Magazine Journal of Science* **44** 1068

*E-mail: jrajput.du@gmail.com
†E-mail: harendraamu@gmail.com

A look into the photoionization of medium-sized molecular systems from an XCHEM perspective

P Fernández-Milán[1]*, M Klinker[1], J González-Vázquez[1] and F Martín[1,2,3]

[1]Departamento de Química, Universidad Autónoma de Madrid, Madrid, 28049, España
[2]Instituto Madrileño de Estudios Avanzados en Nanociencia (IMDEA-Nanociencia), Madrid, 28049, España
[3]Condensed Matter Physics Center (IFIMAC), Universidad Autónoma de Madrid, Madrid, 28049, España

Synopsis In recent years, the XCHEM methodology has been successfully applied to describe photoionization of many-electron atoms and diatomic molecules. In this communication we present the first application of XCHEM to describe photoionization of medium-sized halomethanes and pyrazine

The recent XCHEM [1] code allows the study of photoionization considering the effect of electron correlation in the configuration interaction picture and taking into account the coupling between several continua.

It is especially appropriate for the study of low energy ionization, when the electron still feels the ionic core and electron correlation still plays a role; as recently shown in our study of the Hopfield resonant series in the nitrogen molecule [2].

At the core of the XCHEM method lies the hybrid Gaussian-B-spline basis [3] (GABS), used to represent the bound and continuum states of hydrogen with essentially exact results. It consists of a set of monocentric gaussians (MC) centered at the center of mass (CoM) of the molecule and a box of B-splines that ranges from a distance to the CoM R_0 to a sufficiently far distance R_∞. The XCHEM model couples this basis with the polycentric gaussian (PC) basis that results from a standard quantum chemistry calculation (QCP). The R_0 distance is chosen so that the overlap between PC gaussians and B-splines is negligible. The MC gaussians serve as a bridge communicating B-splines and PC gaussians.

The XCHEM model wavefunction is constructed by augmenting with a single electron in the GABS basis as well as in the PC guassians of the active space of the parent ions.

$$\Psi = \sum_i c_i \Xi_i + \sum_{\alpha j} c_{\alpha j} \Phi_{\alpha j} \qquad (1)$$

where Ξ represent short-range states, that come from optimizing the CI vector while maintaining the same set of orbitals. The second term represents an augmentation in each of the considered α parent ions by using all the available orbitals j, be it PC gaussians, MC gaussians or B-splines.

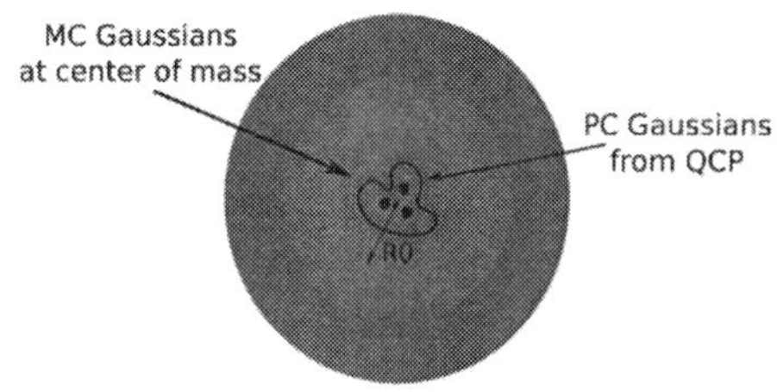

Figure 1. Schematic representation of the GABS hybrid basis. At the center lies the molecule represented as black dots. The black line surrounding the atoms represents the extension of the polycentric gaussians that come from a standard QCP calculation. The blue and purple areas correspond to the monocentric gaussians and B-spline functions respectively. The area in between the previous ones is where MC gaussians overlap with B-splines. R_0 is the distance from the center of mass of the molecule at which B-splines start.

This model has been applied up to now to small systems, atoms like hydrogen, helium and neon [4] and diatomics as hydrogen and nitrogen. In this communication we present results for the photoionization of medium-sized molecules, halomethanes and pyrazine.

References

[1] Marante C *et al* 2017 *J. Chem. Theory Comput.* **13** 499–514
[2] Klinker M *et al* 2018 *J. Phys. Chem. Lett.* **9** 756-762
[3] Marante C *et al* 2014 *Phys. Rev. A* **90** 012506
[4] Marante C *et al* 2017 *Phys. Rev. A* **96** 022507

*E-mail: pedro.fernandezm@uam.es

Energy dependent relative cross sections in carbon 1s photoionization

O Travnikova[1,2 *], M Patanen[3], J Söderström[4], A Lindblad[4], J J Kas[5], F D Vila[5], D Céolin[2], C Nicolas[2], T Marchenko[1,2], G Goldsztejn[1], L Journel[1,2], R Guillemin[1,2], T X Carroll[6], K J Børve[7], P Decleva[8], J J Rehr[5], N Mårtensson[4], M Simon[1,2], S Svensson[4] and L J Sæthre[7 †]

[1]LCPMR, CNRS, Sorbonne Université, UMR7614, Paris, France
[2]Synchrotron SOLEIL, Saint-Aubin, F-91192, France
[3]Nano and Molecular Systems Research Unit, 90014 University of Oulu, Finland
[4]Department of Physics and Astronomy, Uppsala University, 75120 Uppsala, Sweden
[5]Department of Physics, University of Washington, Seattle, Washington 98195-1560, USA
[6]Division of Natural Sciences, Keuka College, New York 14478, USA
[7]Department of Chemistry, University of Bergen, NO-5007 Bergen, Norway
[8]Dipartimento di Scienze Chimiche e Farmaceutiche, Universitá di Trieste and IOM-CNR, 34127 Trieste, Italy

Synopsis We demonstrate that the possibility of monitoring relative photoionization cross sections over a large photon energy range extending to hard x-rays allows studying and disentangling shake processes and intramolecular inelastic scattering effects. An "EXAFS"-type of oscillatory behaviour of the relative C 1s cross sections is observed in a wide photon energy range up to 1 keV due to multiple intramolecular elastic scattering of the photoelectron. These effects are also present in solids and constitute a significant part of all elastic and inelastic scattering effects occurring in solids.

Relative intensities of the C 1s photoelectron lines from chemically inequivalent carbon atoms in the same molecule have been measured as function of incident photon energy in the range of 300 – 6000 eV. We have used different beamlines at two synchrotron radiation facilities. We present relative cross sections for the chemically shifted C 1s lines in the photoelectron spectra of ethyl trifluoroacetate (the "ESCA" molecule), a series of chloro-substituted ethanes as well as 2-butyne [1]. In the lower energy range (photon energy below 1 keV), the cross sections show EXAFS-type of wiggles as was previously observed for a series of chloroethanes [2]. The current findings have been supported by theoretical calculations based on a multiple scattering model.

The use of soft and hard x-rays provides a more complete picture of the dominant processes accompanying photoionization. Such processes reduce the main line intensities by 20–60%, cf. Figure 1. The low energy results are dominated by elastic scattering of the outgoing photoelectron giving rise to EXAFS-like intensity modulations. Using the high energy results it has also been possible to discern the process of intramolecular inelastic scattering of the outgoing electron. This effect relates to the notion of the Inelastic Mean Free Path (IMFP) commonly used in photoemission studies of solids.

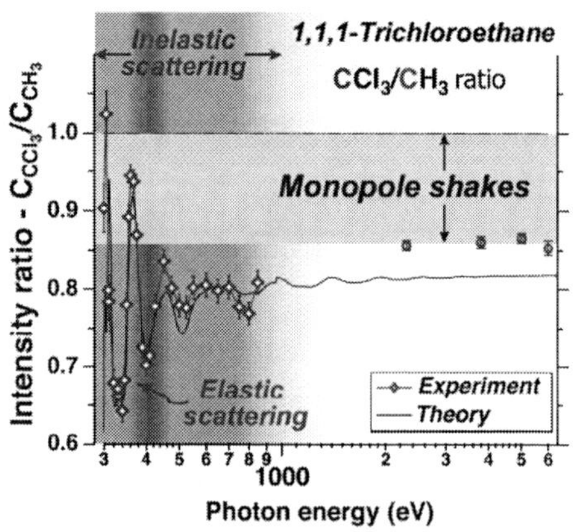

Figure 1. The relative cross section ratios up to 6 keV for C 1s ionization of the chemically different carbon atoms in 1,1,1-trichloroethane, CH_3CCl_3.

References

[1] Travnikova *et al submitted*
[2] Söderström *et al* 2012 *Phys. Rev. Lett.* **108** 193005

*E-mail: Oksana.Travnikova@upmc.fr
†E-mail: Leif.Saethre@uib.no

Enormous enhancement of molecular ionization at high x-ray intensity

L Inhester[1,2]*, Y Hao[3], S-K Son[1,2], and R Santra[1,2,4]

[1]Center or Free-Electron Laser Science, DESY, Hamburg, 22607, Germany [2]The Hamburg Centre for Ultrafast Imaging, 22761 Hamburg, Germany [3]Department of Physics, University of Science and Technology Beijing, Beijing 100083, P. R. China [4]Department of Physics, Universität Hamburg, 20355 Hamburg, Germany

Synopsis The ultra-intense and ultra-short x-ray pulses provided by X-ray Free-Electron Lasers (XFELs) sequentially ionize molecular samples many times. In this context, we have recently discovered that under intense x-ray radiation the total charge of a molecule is enhanced compared to the sum of charges of its constituent isolated atoms[2, 3]. We report here on new theoretical results[5] for iodobenzene (C_6H_5I) that show an even stronger ionization enhancement than previously observed for iodomethane (CH_3I)[3].

Molecules exposed to intense x-ray radiation undergo multiple ionization due to repeated core ionization and subsequent Auger decay cascades. In such a scenario the molecule can populate a huge number (trillions for CH_3I) of different highly excited electronic configurations. We employ a kinetic Monte Carlo technique[4] to solve rate equations for the electronic configurations. Within this scheme all the electronic transition rates are calculated on the fly[1, 2].

We have modeled the x-ray radiation damage in CH_3I and C_6H_5I when exposed to very intense x-ray light with photon energy of 8.3 keV and pulse duration of 30 fs, as produced in recent experiments at XFEL[3]. Because the photoabsorption cross section is much lower on carbon and hydrogen atoms, the photoabsorption happens almost exclusively on the iodine atom. In Figure 1 we show the average charge state obtained after such an x-ray pulse as a function of fluence. We observe that the average total charge in a CH_3I molecule is significantly enhanced compared to what one obtains for a simulation with isolated atoms. This effect is explained by recurrent charge rearrangement in the molecule upon multiple x-ray ionization. Electrons on the methyl part are pulled towards the strongly ionized iodine atom and thereby more electrons become available for further ionization that would not have been available in independent atoms.

We now show that this charge-rearrangement-enhanced x-ray ionization of molecules (CREXIM) is much more pronounced for C_6H_5I, where the iodine atom is attached to a more electron rich chemical environment[5]. The final charges on the carbon atoms depend on their relative position to the iodine. Strikingly, we obtain an enormous enhancement for this molecule of $\simeq +20$ charges (see Fig. 1) demonstrating the huge impact of intra molecular charge rearrangement processes for the radiation damage. The large magnitude of the enhancement manifests that intra-molecular electron rearrangment is of huge relevance for prospective XFEL applications at high fluence.

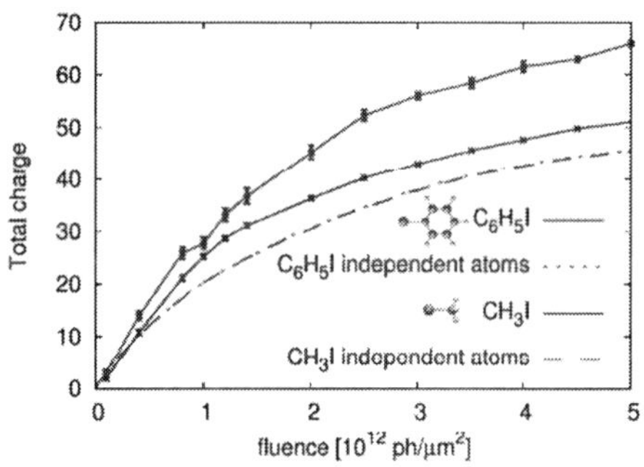

Figure 1. Average charge of CH_3I and C_6H_5I after exposure to the x-ray pulse. The solid line show results from the molecular calculation, the dotted/dashed line from the independent atom model.

References

[1] Hao Y *et al.* 2015 *Struct. Dyn.* **2** 041707
[2] Inhester L *et al.* 2017 *Phys. Rev. A* **94** 023422
[3] Rudenko A *et al.* 2017 *Nature* **546** 129
[4] Son S K *et al.* 2012 *Phys. Rev. A* **85**, 063415
[5] Hao Y *et al.* 2019 *submitted*

*E-mail: ludger.inhester@desy.de

N-NO & NN-O bond cleavage dynamics in the N_2O^{2+}

K Gope[1 *], **I Luzon**[1] **and D Strasser**[1 †]

[1]Institute of Chemistry, Hebrew University of Jerusalem, Jerusalem, city, 91904, Israel

Synopsis N_2O Coulomb explosion is initiated by single-photon double ionization with an ultrafast EUV pulse and is probed by delayed near-IR pulses. The competing two- and three-body dissociation dynamics are reflected in the time resolved branching ratios and in the three-body momentum correlation spectra. Channel resolved kinetic energy release spectra exhibit up to a 0.9 eV shift as a function of time delay – interpreted in terms of the long-bond distance difference between the potentials of a Rydberg ion state and its dication core.

Describing competing bond cleavage dynamics of excited polyatomic systems is one of the great challenges of experimental and theoretical quantum chemistry. Moreover, understanding the dissociation dynamics of multiply-ionized systems is essential to unravel multiply-ionized systems dynamics occur naturally in different planetary and inter-stellar environments due to the ionizing cosmic ray radiation. Accordingly, the simple triatomic benchmark systems of N_2O^{2+} has attracted significant experimental and theoretic attention.[1–3] We time-resolve the competing bond cleavage dynamics dynamics of the dication using ultrafast high-order harmonics generation (HHG) pulses and time-dealyed fs near-IR pulses in a pump-probe manner as shown in Fig. 1.[4] Using 3D coincidence fragment imaging of the Coulomb explosion products, we have measured time resolved branching ratios, kinetic energy release (KER) and in the three-body momentum correlation spectra of both $N^+ + N^+ + O$ as well as the $N^+ + N + O^+$ three-body breakup channels. These results are interpreted in terms of the competition between N-NO & NN-O bond cleavage in two- and three-body dissociation dynamics.

The transient enhancement of NN-O over the N-NO breakup (see Fig. 1) is assigned to excitations of the dication from the two low lying singlet states to the $^1\Pi$ potential. Furthermore, a ~0.9eV asymptotic KER shifts that develop on the hundreds of fs time scale, is assigned to the potential gap between Rydberg ion states and their dication core, developing due to screening of the long-range Coulomb repulsion at distances. The observed effect of the potential gap between a molecular Rydberg cation and its dication core provides valuable insights for interpreting time resolved and control studies in N_2O, as well as in other complex systems exhibiting Rydberg ion dynamics.

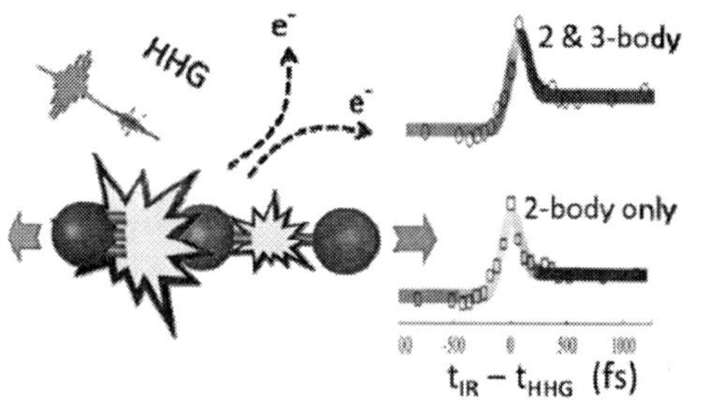

Figure 1. Schematic representation of the HHG pump and near-IR probe experiment. Also showing the Coulomb explosion 2- and 3-body channel yields as a function of the pump-probe delay.

References

[1] Zhou X *et al* 2012 *Nat. Phys.* **8** 232
[2] Ueyama M *et al* 2005 *J. Chem. Phys.* **123** 154305
[3] Taylor S *et al* 2006 *J. Chem. Phys.* **124** 204319
[4] Luzon I *et al* 2017 *Phys. Chem. Chem. Phys.* **8** 232

* E-mail: krishnen.gope@mail.huji.ac.il
† E-mail: strasser@huji.ac.il

Electron-impact ionization of Fe^{3+}

A Kynienė*, S Kučas, Š Masys, S Pakalka and V Jonauskas

[1]Vilnius University, Institute of Theoretical Physics and Astronomy, Saulėtekio av. 3, Vilnius, LT-10257, Lithuania

Synopsis Electron-impact ionization cross sections are studied for levels of the ground and first excited configurations of the Fe^{3+} ion. Convergence for the excitation-autoionization cross sections due to excitations to the high-nl shells is investigated. Single configuration and configuration interaction methods are used in the study. Scaling is applied to the distorted wave cross sections of the excitation-autoionization and collisional ionization processes.

Fundamental atomic data of ionization and recombination processes are necessary for identification of the composition of astrophysical plasmas, modeling the structure and dynamics of laboratory plasma and high-temperature plasmas in stars.

In this work, electron-impact ionization for the Fe^{3+} ion is investigated. The study of direct ionization (DI) and excitation-autoionization (EA) processes is carried out for levels of the ground [Ne]$3s^2 3p^6 3d^5$ and first excited [Ne]$3s^2 3p^6 3d^4 4s$ configurations. Convergence for the EA cross sections due to excitations to the high-nl shells is investigated.

Energy levels, radiative and Auger transition probabilities, electron-impact excitation and ionization cross sections for the Fe^{3+} ion are generated using the Flexible atomic code (FAC) [1]. The code implements the Dirac-Fock-Slater approach. Electron-impact ionization and excitation cross sections are obtained in the distorted wave (DW) approximation. Single configuration and configuration interaction (CI) methods are used in the study of ionization cross sections. In addition, scaling is applied to the DW cross sections of the EA and DI processes.

Previous calculations for the ionization cross sections of Fe^{3+} were performed using the configuration-average DW approach [2]. Our data shows that many configurations have energy levels that straddle the single ionization threshold. This demonstrates that study of energy levels is needed.

Comparison of theoretical cross sections with experimental values for Fe^{3+} is presented in Fig. 1. Theoretical data correspond to the $3d^2_{3/2}(2)3d^2_{5/2}(4)4s_{1/2}$ $J = 13/2$ level. It can be seen that scaling plays an important role in producing a better agreement with measurements. However, the scaled single-configuration results overestimate the experimental values at the lower energies of the incident electron. On the other hand, the scaled CI data are below the measurements near the peak of cross sections on the low energy side. The good agreement with experimental values is obtained at the higher energies for both scaled calculations.

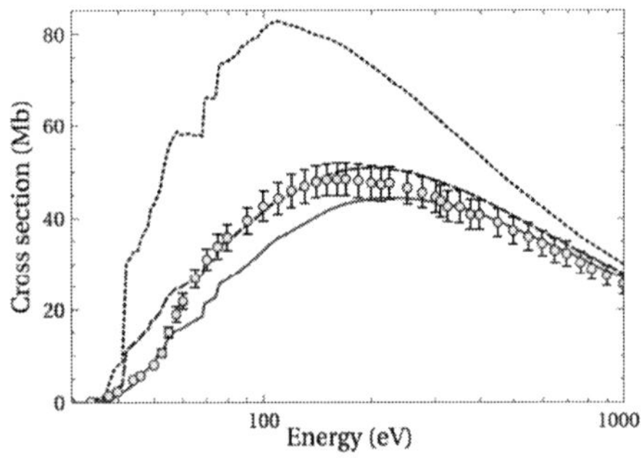

Figure 1. Comparison of theoretical cross sections with experimental values [3]. Dotted line (black) - single-configuration, dashed line (blue) - scaled single-configuration, and solid line (green) - scaled CI calculations.

References

[1] Gu M F 2008 *Can J Phys* **86** 675
[2] Pindzola M S *et al* 2018 *J. Phys. B: At. Mol. Opt. Phys.* **51** 015202
[3] Stenke M *et al* 1999 *J. Phys. B: At. Mol. Opt. Phys.* **32** 3627

*E-mail: ausra.kyniene@tfai.vu.lt

Radiative and Auger cascades in Fe^{3+}

V Jonauskas[1] *, P Drabužinskis[1], S Kučas[1] and J Koncevičiūtė[1]

[1]Vilnius University, Institute of Theoretical Physics and Astronomy, Saulėtekio av. 3, Vilnius, LT-10257, Lithuania

Synopsis Radiative and Auger cascades following creation of the $2s$ or $2p$ shell vacancy in the Fe^{2+} ion are studied by performing calculations for energy levels and subconfigurations. Radiative and Auger spectra and charge state distribution determined by cascade are investigated. Branching ratios of the cascade are analyzed and main decay mechanisms are identified for all levels of the studied configurations.

Electromagnetic emission from accretion disk of active galactic nucleus is the main source that determines higher ionization stages in the surrounding matter. Inner-shell vacancy created by photon leads to cascade of radiative and Auger transitions. Spectral lines of the Fe^{2+} and Fe^{3+} ions were observed in active galactic nuclei [1, 2]. Therefore, the higher ionization stages of iron can be also produced in such objects.

The aim of the current work is to study radiative and Auger cascade following the creation of L shell vacancy in Fe^{2+}. Energy levels, electric dipole and Auger transitions are calculated using Flexible Atomic Code (FAC) [3] which implements the Dirac-Fock-Slater approach. Atomic data are studied using single-configuration approximation. Previous calculations for cascade were performed by considering transitions among configurations [4].

The main decay channels from the lowest level of the $2s\,3d^6$ configuration following the $2s$ shell vacancy creation for Fe^{2+} are shown in Fig. 1. Population transfer from Fe^{4+} $2p^53d^5$ is not presented in this figure. The strongest branch corresponds to $Fe^{3+}\,2s\,3d^6 \rightarrow Fe^{4+}2p^53d^5 \rightarrow Fe^{5+}3p^43d^5$ (26%) sequence of transitions. Auger decay of $Fe^{5+}3p^43d^5$ transfers only 0.5% of population to $Fe^{6+}\,3p^53d^3$. The study demonstrates that the highest ion yield corresponds to Fe^{5+} ($\sim$ 80%). The highest ionization stage produced by cascade reaches Fe^{7+}, however, with very low population ($<$ 1%). It has to be noted that ion yield depends on the initial level of the $2s\,3d^6$ configuration. This configuration has 63 energy levels. Variation of population for Fe^{5+} is less than 10% for the different levels of the initial configuration. The ground $3d^3$ configuration of Fe^{5+} is mainly populated in the cascade. However, small part of population ($<$ 5%) appears in the long lived levels of the excited $3p^53d^4$ configuration.

Decay of the Fe^{3+} $2p^53d^6$ vacancy reaches Fe^{6+}. There are 180 energy levels of the $2p^53d^6$ configuration. Population of ions depends on the initial level the cascade starts. Variation of the population for the Fe^{6+} ion from 0.2% to 8.1% is obtained for the different initial levels. The main decay branches of cascade from the lowest level of Fe^{3+} $2p^53d^6$ are $Fe^{3+}\,2p^5\,3d^6 \rightarrow Fe^{4+}\,3p^43d^6 \rightarrow Fe^{5+}\,3p^53d^4$ (23%) and $Fe^{3+}\,2p^5\,3d^6 \rightarrow Fe^{4+}\,3s\,3p^53d^6 \rightarrow Fe^{5+}\,3p^53d^4$ (9%).

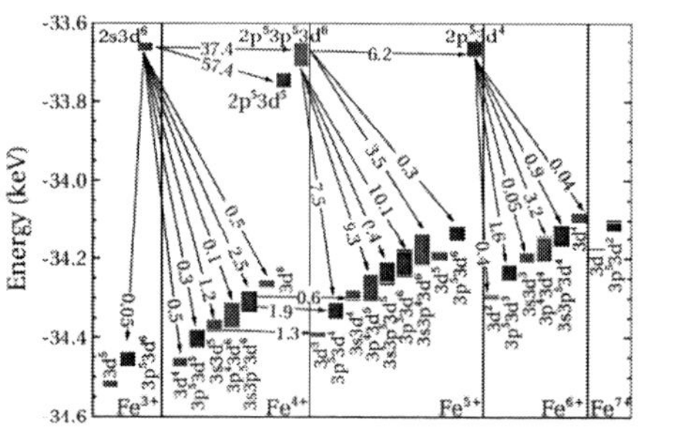

Figure 1. Radiative and Auger cascade corresponding to decay of the Fe^{3+} $2s\,3d^6$ configuration. Blue (dark) represents even parity configurations; red (light) represents odd parity configurations.

References

[1] Laor A *et al* 1997 *Astrophys J* **489** 656
[2] Vestergaard M *et al* 2001 *Astrophys J Suppl Ser* **134** 1
[3] Gu M F 2008 *Can J Phys* **86** 675
[4] Kaastra J S *et al* 1993 *Astron Astrophys Suppl Ser* **97** 443

*E-mail: valdas.jonauskas@tfai.vu.lt

Revisit of (e,2e) & (e,3e) processes on atoms with twisted electron impact

N Dhankhar[1] *, A Mandal[1], D Sebilleau[2] and R Choubisa[1] †

[1]Department of Physics, BITS Pilani, Pilani Campus Vidya Vihar, Pilani-333031, India
[2]Department of Material and Nanosciences, Institute Of Physics of Rennes, University of Rennes-1, 35042 Rennes-Cedex, France

Synopsis We compute Triple Differential Cross Section (TDCS) and Five-fold Differential Cross Section (FDCS) for (e,2e) & (e,3e) processes on atoms respectively to investigate the effects of non zero angular momentum and the apex angle of the twisted electron. We observed that the differential coincidence cross sections largely depend on these parameters.

Since the first measurement of (e,2e) process on atoms [1, 2], there have been extensive theoretical and experimental studies on (e,2e) process on atoms, molecules, and solid targets. Till now, most of the activities have been confined to an electron beam which moves in the straight line and hit the target. Recently, it has been possible to generate twisted electron beam that carries orbital angular, angular momentum apart from its linear momentum [3]. When we study scattering/ionization processes with a twisted electron beam, we can get more information as we have more control parameters to play, like Orbital Angular Momentum (OAM), the apex angle of the beam and the impact parameter. There have been few theoretical studies on the scattering of twisted electrons by microscopic targets(see e.g.[4]). This field is largely unexplored to electron impact ionization process on atoms and other microscopic targets. In this communication, we present the results of our calculations of (e,2e) & (e,3e) processes on atomic targets to elucidate the effects of OAM and apex angle of the twisted electron beam on the coincidence differential cross sections in various geometrical modes. We describe incident electron as twisted electron plane wave [4], scattered electron as plane wave, the ejected electron as Coulomb wave (or 2 Coulomb waves with Gamow factor(2CG) for the ejected electrons in the (e,3e) process)and a Hydrogen-like wave function for bound electron(or He like wave wave function for (e,3e) process). Through our study on (e,2e) process, we found that the TDCS depends largely on OAM and apex angle of the twisted electron beam. We also found that the spin asymmetry in TDCS also depends on these parameters. We are in process to compute the FDCS for the (e,3e) process on the atoms with twisted electrons. We will present our detailed finding on (e,2e) and (e,3e) processes in the conference.

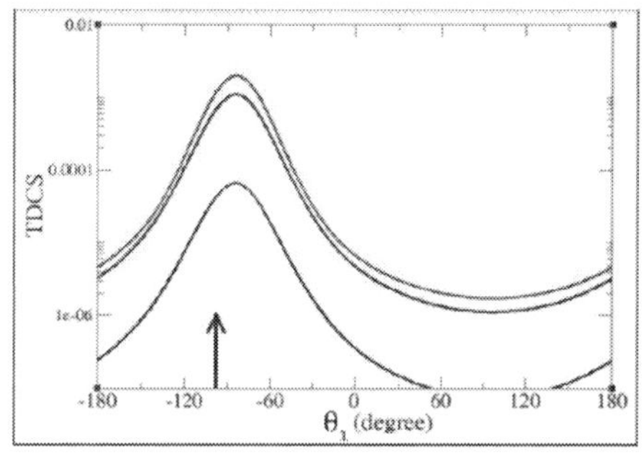

Figure 1. TDCS as a function of the angle of the ejected electron (θ_1) for He atom in coplanar asymmetric geometry. E_i(incident energy) = 100keV, E_1(ejected electron energy) = 125eV, scattering angle(θ_s) = 7^0, apex angle of twisted electron(θ_p) = 5^0. Black curve(OAM = 1/2), Blue curve(OAM = 3/2), Violet curve(OAM = 5/2). Arrow represents the direction of momentum transfer.

References

[1] Ehrhdart H *et al* 1969 *Phys. Rev. Lett.* **22** 89
[2] Amaldi U *et al* 1969 *Review of Scientific Instruments* **40** 1001
[3] Uchida M *et al* 2010 *Nature* **464(7289)** 737
[4] Serbo V *et al* 2015 *Phys. Rev A* **92** 012705

*E-mail: p20170427@pilani.bits-pilani.ac.in
†E-mail: rchoubisa@pilani.bits-pilani.ac.in

The role of multiple ionization in fast highly charged ions collisions on H_2O

N Bachi[1] *, G S Otero[1], S Otranto[1] and R E Olson[2]

[1]Instituto de Física del Sur (IFISUR), Departamento de Física, Universidad Nacional del Sur (UNS), CONICET, Av. L. N. Alem 1253, B8000CPB-Bahía Blanca, Argentina
[2]Physics Department, Missouri University of Science and Technology, Rolla MO 65409, USA

Synopsis An adaptive classical trajectory Monte Carlo model that explicitly considers the eight electrons corresponding to the valence molecular orbitals of the H_2O molecule is introduced. In this dynamical model, the electrons' binding energies change throughout the collision ensuring the proper energy deposition by the projectile for multiple electron removal. Present results are relevant to the ion-therapy program, specially at irradiation planning stages in which detailed information on the electronic emission spectra are needed as input data.

The last few years have seen a steep increase in the number of radiotherapy facilities based on ion beams [1]. This is due to the fact that ion-beam therapy delivers a localized dose of energetic electrons minimizing the unnecessary irradiation of healthy tissue surrounding a tumour. In this context, irradiation planning is an essential tool of the process and is based on numerical simulations which ultimately rely in electron emission cross sections regarding the interaction of the ions of the beam with biological molecular targets.

In this work, we introduce an adaptive classical trajectory Monte Carlo method in which the eight valence electrons of the H_2O molecule are explicitly considered. Although the electronic correlation is not included and screening is only accounted for, electrons' binding energies are changed throughout the collision as the electrons are being emitted. The present method has two clear advantages compared to the independent electrons and sequential electron schemes previously used in the literature [2,3]. First, in the entrance channel the projectile sees the correct electron density and energies. Second, the adaptive scheme ensures that the projectile invests the proper amount of energy in collisional events leading to multiple electron removal.

Singly differential cross sections in energy and angle and recoil charge distributions are studied at an impact energy of 4 MeV/amu for projectiles with charges in the range 1+ to 54+. The electron emission spectra fractions for single and multiple ionization are explicitly determined for the different projectiles under consideration. As a result, we observe that multiple ionization dominates the energetic emission for electron energies greater than about 50 eV.

In Fig. 1 we show the net singly differential cross section in electron emission energy for C^{6+} impact. Our results are compared to the recent experimental data of Bhattacharjee *et al.* [4].

The present methodology is of potential relevance as input data generator for the ion-therapy irradiation simulation codes under use.

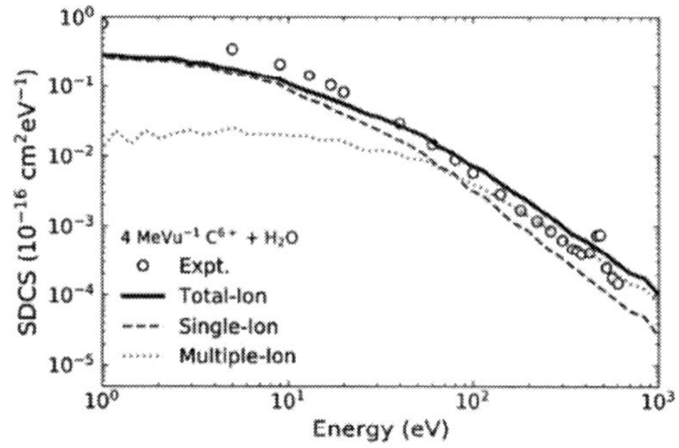

Figure 1. Net singly differential cross section in electron emission energy for 4 MeV/amu C^{6+} + H_2O collisions. The separate contributions from the single ionization and multiple ionization channels are explicitly indicated.

References

[1] Dosanjh M 2018 *Cern Courier* **58** 32
[2] Olson R E *et al* 1987 *J. Phys. B : At. Mol. Phys.* **20** L809
[3] Olson R E *et al* 1989 *Phys. Rev. A* **39** 5572
[4] Bhattacharjee S *et al* 2017 *Phys. Rev. A* **96** 052707

* E-mail: nicolas.bachi@uns.edu.ar

Journal of Physics: Conference Series **1412** (2020) 152057 doi:10.1088/1742-6596/1412/15/152057

Photoionization of the open-shell nitrogen atom confined in C_{60}

T W Gorczyca[1]* and S T Manson[2]†

[1]Western Michigan University, Kalamazoo, MI 49008-5252 USA
[2]Georgia State University, Atlanta, GA 30303, USA

Synopsis Endofullerene photoionization of open-shell N@C_{60} is investigated using a coupled-channel, correlated R-matrix approach with additional modifications for a finite spherical-shell fullerene potential. Particular focus is on the initial-state dependence of the photoionization cross section. Comparisons show significant differences between the $1s^22s^22p^3(^4S)$ initial state and the metastable ^2D and ^2P states, as well as strong confinement oscillatory structure.

Since almost all of the work on the response of confined atoms to ionizing radiation has considered closed-shell atoms [1], we have embarked on a program to investigate the response of open-shell atoms with confinement. To this end, calculations of the photoionization of each of the three states, ^4S, ^2D and ^2P, arising from the ground $1s^22s^22p^3$ configuration of the nitrogen atom confined in C_{60}, N@C_{60}, have been performed using a variation of the Belfast R-matrix code that we have modified to simulate the confinement [2], from the thresholds to a photon energy of 29 eV. For comparison, free atom calculations have also been done using the same methodology. The results show that the confinement has very significant effects on the cross sections, both in resonances regions and in the open continuum, both qualitatively and quantitatively (see Fig. 1). Low-energy confinement resonances are clearly seen. In addition, the ^4S cross sections (confined and free) shown in Fig. 1 are rather different from their ^2D and ^2P counterparts owing to the significant differences between the quartet and doublet final continuum states.

Work supported partially by the NASA and DOE.

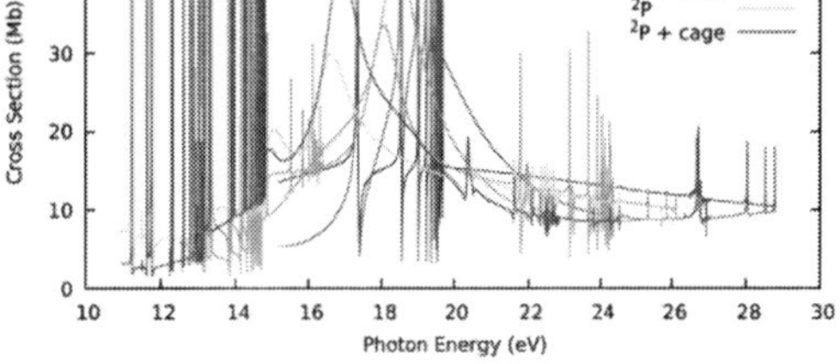

Figure 1. Photoionization of nitrogen ground and metastable $2p^3$ states with and without endofullerene confinement.

References

[1] Dolmatov V K 2009 *Adv. Quantum Chem.* **58** 13
[2] Gorczyca T W *et al* 2012 *Phys. Rev. A* **86** 033204

*E-mail: gorczyca@wmich.edu
†E-mail: smanson@gsu.edu

Cascade of elementary processes in Se^{1+}

V Jonauskas[1]*, S Kučas[1], Š Masys[1] and J Koncevičiūtė[1]

[1]Vilnius University, Institute of Theoretical Physics and Astronomy, Saulėtekio av. 3, Vilnius, LT-10257, Lithuania

Synopsis Analysis of radiative and Auger cascade following creation of the K shell vacancy in the Se atom is presented. Calculations of radiative and Auger transition probabilities are performed by using Dirac-Fock-Slater approximation. Radiative and Auger spectra as well as ion yield produced by cascade are studied.

Merger of two neutron stars or neutron star with black hole leads to production of heavier than iron elements through the rapid neutron capture (r-process) nucleosynthesis [1]. Electromagnetic radiation from the merged stars as the counter-part of the gravitational waves is observed by the onboard satellites (Chandra X-ray Observatory, XMM-Newton). Interaction of x-ray radiation with atoms and ions in the surroundings of the merged stars can lead to emission of electron from inner shell. The highly excited state decay through cascade of elementary processes. The cascades of elementary processes are the main source of formation for the higher ionization stages of ions in the surrounding of merged stars.

Cascade of radiative and Auger transitions following a vacancy creation in the K shell of the Se^{1+} ion is investigated by considering transitions among subconfigurations. Energies of subconfigurations, radiative and Auger transition probabilities are studied using Flexible Atomic Code (FAC) [2] which implements Dirac-Fock-Slater approximation.

Ion yields produced by cascades are presented in Fig. 1. The results correspond to decay of the $1s_{1/2}4p^2_{1/2}4p^2_{3/2}$ subconfiguration. Radiative transitions play a significant role in the first step of cascade. The Se^{1+} $1s\,4p^4$ configuration decays with the probability of $\sim$54% to Se^{1+} $2p^5 4p^4$ through radiative transitions. Population of the Se^{1+} $3p^5 4p^4$ and $4p^3$ configurations reached by radiative transitions corresponds to $\sim$7.2% and $\sim$0.4%, respectively. Transition to the next ionization stage from Se^{1+} $1s\,4p^4$ amounts to 38%. The Se^{1+} $2p^5 4p^4$ and $3p^5 4p^4$ configurations decay further mainly through Auger transitions and that leads to negligible contribution to Se^{1+}. Decay of other two subconfigurations ($1s_{1/2}4p^1_{1/2}4p^3_{3/2}$ and $1s_{1/2}4p^4_{3/2}$) of the $1s\,4p^4$ configuration agrees within 1% for the ion yield.

The data for Auger cascade are compared with the results that in addition to Auger transitions involve the radiative process. It can be seen that radiative transitions play a crucial role in the formation of ions in the various ionization stages. The higher ionization stages are reached when only Auger cascade is included in the study. The highest population of ions corresponds to Se^{7+} for the Auger cascade. The radiative and Auger cascade produces the Se^{4+} and Se^{5+} ions with the largest population. It has to be noted that the highest ionization stage reached by the cascades is larger than the number of shells in the Se^{1+} ion.

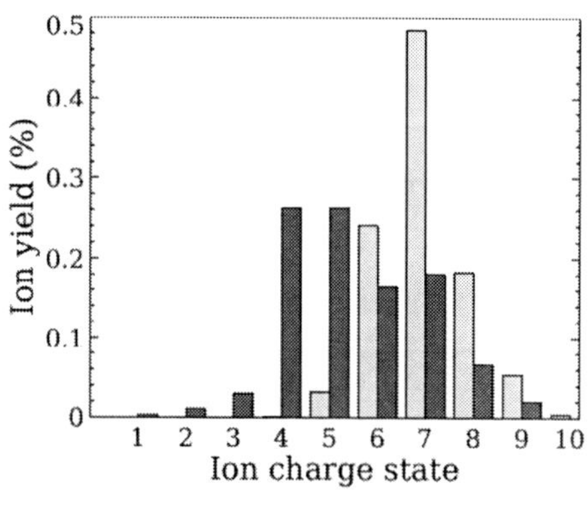

Figure 1. Ion yield produced in Se^{1+} by cascades: Auger cascade (yellow), radiative and Auger cascade (green).

References

[1] Lattimer J M *et al* 1974 *Astrophys. J.* **192** L145
[2] Gu M F 2008 *Can J Phys* **86** 675

*E-mail: valdas.jonauskas@tfai.vu.lt

Adamantane dication fragmentation dynamics following ion collisions

C Nicolafrancesco[1,2*], S Indrajith[1], S Maclot[3], A Domaracka[1], B Huber[1], S Díaz-Tendero[4] and P Rousseau[1]

[1]Normandie Université, ENSICAEN, UNICAEN, CEA, CNRS, CIMAP, 14000 Caen, France
[2]Synchrotron SOLEIL, L'Orme des Merisier, 91190 Saint Aubin, France
[3] Biomedical and X-Ray Physics, Department of Applied Physics, AlbaNova University Center, KTH Royal Institute of Technology, SE-10691 Stockholm, Sweden
[4] Departamento de Química, Módulo 13, Universidad Autónoma de Madrid, 28049 Madrid, Spain

Synopsis Being an important exponent of diamondoids molecular class, and abundantly found in the interstellar medium, getting a better insight on its behaviour when exposed to energetic particles is of great interest. In order to do that, we performed gas phase ion-collisions to study the fragmentation dynamics of adamantane analysing the charged molecular fragments by a multicoincident mass spectrometer. With the complementary approach of quantum chemistry calculations, we obtained a better understanding on adamantane fragmentation dynamics.

Diamondoids are a class of carbon nanomaterials based on carbon cages with well-defined structures formed by C(sp^3)-C(sp^3)-hybridised bonds and fully terminated by hydrogen atoms. All diamondoids are variations on the adamantane molecule model, the most stable among all the isomers, with formula $C_{16}H_{16}$. Adamantane and diamondoids have proven to be of great interest under various points of view: from biomedical applications to nanomaterials and optoelectronics [1]. Another very interesting aspect is the interstellar abundance of these kind of molecules. In fact, in the interstellar medium, molecules are exposed to energetic particles, such as electron, photons, or ions [2].

In order to get a better insight of the behaviour of adamantane in such an environment, we have performed studies using ion collisions in the gas phase, which better resembles the interstellar medium. Under these conditions, the molecule is typically ionised, excited, and undergoes fragmentation. The produced charged molecular fragments are recorded with multicoidencident mass spectrometry. This allows us to infer the energy and charge redistribution processes after the collisions. Quantum chemistry simulations provided insights into the underlying fragmentation mechanisms.The two complementary approaches provided us with a deeper understanding on the complete picture of the fragmentation dynamics.

In this communication, we will present results on the fragmentation dynamics of doubly charged adamantane cation produced in collision with 120 keV Ar^{8+} ions.

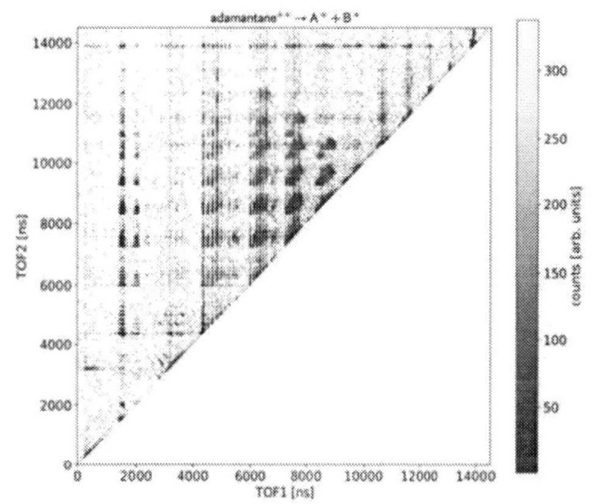

Figure 1. Coincidence map of the fragmentation of adamantane dication produced in collision with 120 keV Ar^{8+}.

References
[1] Gunawan M A *et al* 2014 *New J. Chem.* **38** 28
[2] Bouwman J *et al* 2018 *Chem. Phys. Chem.* **19** 3211

* E-mail: chiara.nicolafrancesco@ganil.fr

Electron-induced decomposition of the coenzyme Q_0

J Ameixa[1,2*], R Meißner[1,2], E Arthur-Baidoo[1], C Lochmann[1], M Ončák[1], M K Beyer[1], F Ferreira da Silva[2] and S Denifl[1†]

[1]Institut für Ionenphysik und Angewandte Physik, Leopold-Franzens Universität Innsbruck, Technikerstrasse 25, A-6020 Innsbruck, Austria

[2]Laboratório de Colisões Atómicas e Moleculares, CEFITEC, Departamento de Física, Faculdade de Ciências e Tecnologia, Universidade NOVA de Lisboa, Campus de Caparica, 2829–516 Caparica, Portugal

Synopsis We report on the formation of anions through electron attachment to the coenzyme Q_0 in the gas-phase. It represents the functional group of the electron carrier coenzyme Q_{10}, which plays a key role in the electron transport chain in the mitochondrial complex. The parent anion is most abundantly formed in a near-zero eV resonance, presumably correlating with the electronic ground state of the molecule. This observation accounts for the ability of coenzyme Q_{10} to capture and transport electrons in biological processes.

Coenzyme Q_{10} is a key component in the electron transport chain, which generates ATP by oxidative phosphorylation in the mitochondria [1]. It is comprised of a quinone derivative group, known as coenzyme Q_0 (CoQ_0), as shown in Figure 1, and a hydrophobic-tail formed by ten isopropenoid subunits [2].

In this contribution, we investigate the fundamental electron-induced reactions with CoQ_0 by performing dissociative electron attachment (DEA) studies on CoQ_0 in the gas-phase. Briefly, DEA occurs when a molecule captures a low-energy electron (<15 eV) leading to the formation of a temporary negative ion which may decay through bond dissociation yielding a negatively charged fragment ion and neutrals. The present study was carried out by using a hemispherical electron monochromator coupled with a quadrupole mass spectrometer.

We report the formation of six anions formed upon DEA to CoQ_0 (Table 1). Remarkably, the parent anion is not formed exclusively through a single resonance centered at ~0 eV, but also by further ones at 0.55 and 8.25 eV, as shown in figure 1. The main fragmentation channels correspond to the formation of anionic demethylated CoQ_0 and di-demethylated CoQ_0, alongside neutral methyl radicals. Moreover, anionic methyl was also observed. Anions with m/z 124 and m/z 108 are formed through several bond cleavages. Finally, the thermodynamic threshold for each DEA reaction was calculated theoretically at the B3LYP/aug-cc-pVDZ level.

The calculations on the TD-CAM-B3LYP/aug-cc-pVDZ level of theory reproduced the low-lying resonances, but do not provide evidence for the 8.25 eV resonance. This requires further investigation.

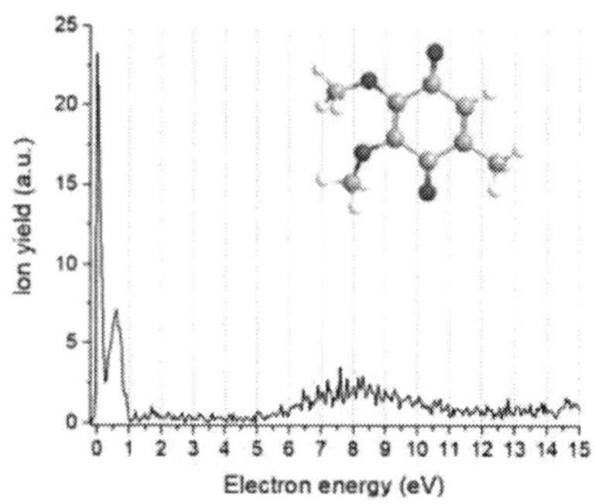

Figure 1. Ion yield of CoQ_0^-, m/z 182, formed upon electron attachment to CoQ_0 (2,3-dimethoxy-5-methyl-1,4-benzoquinone) along with the molecular structure.

Mass (m/z)	Anion	Resonance energies (eV)			
		1st	2nd	3rd	4th
182	CoQ_0^-	~0.0	0.55	8.25 (?)	
167	$(CoQ_0{-}CH_3)^-$	~0.0	1.92		
152	$(CoQ_0{-}2CH_3)^-$	~0.0	2.61	3.07	4.72
124	$C_7H_8O_2^-$ $C_6H_4O_3^-$	5.46	6.63		
108	$C_7H_8O^-$ $C_6H_4O_2^-$	5.30	7.03	9.30	
15	CH_3^-	8.01	9.70		

Table 1. Positions of the resonance maxima for the anions formed upon electron attachment to CoQ_0. The 3rd resonance of the parent anion (marked with "?"), is currently under investigation.

References

[1] Karp G 2010 *Cell and Molecular Biology: Concepts and Experiments* p 185 ed Wiley (New York)

[2] Lenaz G *et al* 2009 *Biochim. Biophys. Acta - Bioenerg.* **1787** 563

* E-mail: j.ameixa@campus.fct.unl.pt
† E-mail: stephan.denifl@uibk.ac.at

Influence of ionizing radiation on molecular recognition in the gas phase: VUV and X-ray photoabsorption of antibiotic-receptor non-covalent complexes

M Abdelmouleh[1*], M Lalande[1], A Espinosa Rodriguez[1,2], V Vizcaino[1], J Rangama[1], T Schlathölter[2] and J-C Poully[1]

[1]CIMAP, UMR 6252 (CEA/CNRS/ENSICAEN/Université de Caen Normandie), Caen, France
[2]Dpto. de Física Atómica, Molecular y Nuclear. Facultad de CC Físicas. Avda. Complutense, s/n - 28040 Madrid, Spain
[3]Zernike Institute for Advanced Materials, University of Groningen, Nijenborgh 4, 9747AG Groningen, The Netherlands

Synopsis The intrinsic effects of ionizing radiation on molecular recognition between an antibiotic and its receptor have been probed by coupling a synchrotron beamline to a tandem mass spectrometer. Protonated vancomycin-receptor non-covalent complexes in the gas phase undergo inter- as well as intramolecular bond cleavage due to internal energy transfer after photoabsorption in the VUV-X energy range. The deprotonated complex undergoes electron detachment, but the native interactions underlying molecular recognition are not systematically lost.

The structure of biological molecules plays an important role in their activity, especially in the molecular recognition between drugs and their receptors. This is illustrated in figure 1 for the glycopeptide antibiotic vancomycin bound to its peptidic receptor Ac_2-^{L}Lys-^{D}Ala-^{D}Ala in condensed phase, where the two structures tightly fit each other. When isolated in the gas phase, this native structure is lost for gas-phase protonated but retained for deprotonated complexes.[1,2] Moreover, in the case of a chemio-radiotherapy to cure cancer, drugs can interact with ionizing radiation. Therefore, it is crucial to know if molecular recognition is affected by irradiation. We have thus studied photoionization of isolated protonated as well as deprotonated vancomycin/receptor complexes by means of a synchrotron beamline coupled to a tandem mass spectrometer, which allowed identifying the irradiation products. After absorption of one VUV photon by doubly-protonated Vancomycin, we observe non-dissociative ionization (NDI), but also peptide backbone fragmentation as well as sugar bond cleavages due to electronic to vibrational energy transfer. The latter rises with photon energy, making NDI to vanish at X-ray energies. For the doubly protonated vancomycin-receptor complex, NDI is not observed following photoabsorption, probably because of the low binding energy of this non-covalent complex. Interestingly, when deprotonated, the intact complex is observed after electron detachment, together with fewer fragmentation channels, the main ones being loss of receptor and CO_2. The latter comes only from vancomycin, not from the receptor, which can be explained by the retained native structure of the deprotonated complex, where the receptor's CO_2 group is strongly bound to Vancomycin, even after electron detachment. Our results indicate that the geometrical structure of a drug-receptor complex has a large effect on the processes induced by photon irradiation in the VUV-X energy range. Furthermore, molecular recognition is not systematically lost after photoabsorption.

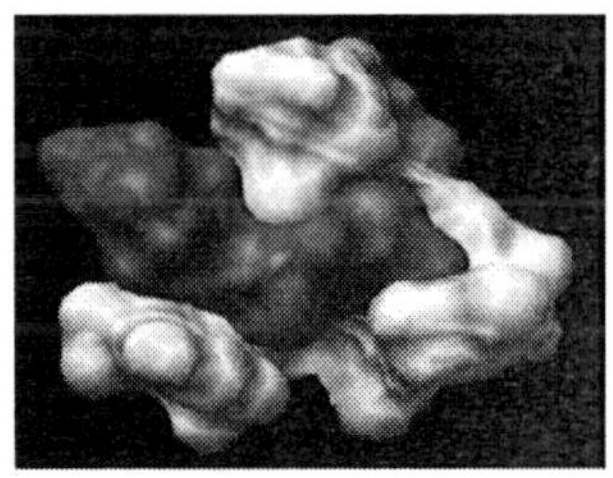

Figure 1. molecular surfaces of the native structure of vancomycin (white) and its receptor (purple) in a crystal (Protein Data Bank 1FVM).

References

[1] Poully J C *et al* 2010 *International Journal of Mass Spectrometry* **297** 28–35
[2] Poully J C *et al* 2010 *Phys. Chem. Chem. Phys.* **12** 3606

* E-mail: marwa.abdelmouleh@ganil.fr

Angular distribution in two-body dissociation of SO_2^{q+}

J Rajput[1] *, H Kumar[1], P Bhatt[2] and C P Safvan[2] †

[1] Department of Physics and Astrophysics, University of Delhi, Delhi - 110007, India
[2] Inter-University Accelerator Centre, Aruna Asaf Ali Marg, New Delhi - 110067, India

Synopsis The two-body dissociation of multiply charged SO_2 molecules is studied using a recoil ion momentum spectrometer. These molecular ions were generated in collision of the neutral parent with an energetic highly charged ion. From the measured momenta of the recoil ions, angular distributions are generated with respect to the incoming beam direction and are found to vary strongly with kinetic energy release.

The energetics and dynamics of the dissociation of multiply charged molecular ions is an active field of study. Such investigations provide us information about the unstable, excited molecular ionic states that are not amenable to study by conventional spectroscopic techniques.

We have conducted experiments to get insight into the dynamics of dissociation of multiply charged SO_2 molecules. These were generated from the neutral parent upon interaction with 450 keV Xe^{9+} ions. The multiply ionized molecular ions dissociate into molecular and/or atomic fragments which are extracted by a uniform electric field into a linear recoil ion momentum spectrometer and detected in coincidence with ejected electrons. The multi-hit capability of the data-acquisition setup allows us to detect all fragments from a specific molecular ion. The experiments were conducted at the Low Energy Ion Beam Facility at the Inter University Accelerator Centre, New Delhi India, and the details of the setup are described in [1].

From an analysis of the recorded events, we have determined the kinetic energy release distribution (KERD) for all two body dissociation channels observed for breakup of SO_2^{q+}. The KERD of each fragmentation channel carries information about the excited electronic states of the respective parent molecular ion. We have investigated the effect of the orientation of the molecule with reference to the projectile beam direction [2] on the energetics of the dissociation pathways. We find that the kinetic energy of the fragment ions is strongly dependent on this orientation. For example, the figure shows the angular distribution of O^+ fragment ions for two sets of kinetic energy release values but from the same breakup channel. The anisotropy and the orientation effects are observed to be a strong function of the kinetic energy release (and hence the excited states from which they originate).

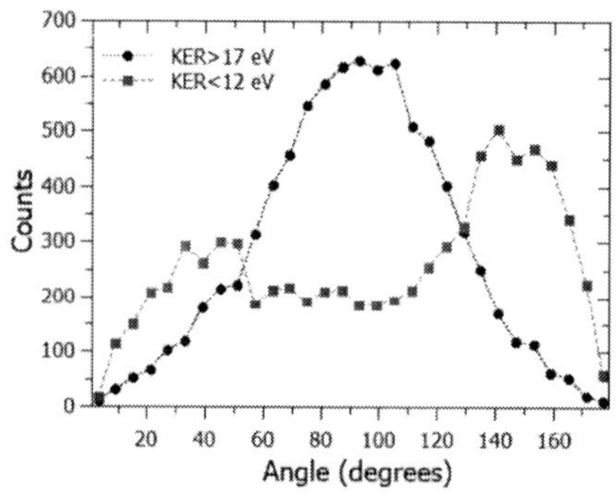

Figure 1. Angular distribution of O^+ fragment with respect to the direction of incoming projectile beam for the two-body dissociation channel $SO_2^{3+} \rightarrow O^+ + SO^{2+}$ for two different ranges of measured kinetic energy release. The kinetic energy release for this fragmentation channel ranges from 9 eV to 22.5 eV with the most probable value being 13.5 eV.

References

[1] Kumar A *et al* 2014 *J. Mass Spect.* **374** 44.
[2] Rajput J *et al* 2017 *J. Phys.: Conf. Ser.* **875** 102003

*E-mail: jrajput.du@gmail.com
†E-mail: cp.safvan@gmail.com

Plasmon assisted isomerization in naphthalene and azulene

M V Vinitha [1†], P K Najeeb[1], A Kala[1], P Bhatt[2], C P Safvan[2], S Vig[3] and U Kadhane[1]

[1]Department of Physics, Indian Institute of Space Science and Technology, Thiruvanathapuram-695 547, India
[2]Inter-University Accelerator Centre, Aruna Asaf Ali Marg, New Delhi-110 067, India
[4]Department of Earth and Space Science, Indian Institute of Space Science and Technology, Trivandrum-695 547, India

Synopsis Plasmon excitation in polycyclic aromatic hydrocarbons is a dominant excitation mode in fast charged particle collisions. Excitations of this nature lead to several secondary processes like ejection of neutral atomic or molecular fragments. The present work explores the influence of plasmon excitation on isomerization on two small size PAHs namely, naphthalene and azulene. Moreover, the existence of double plasmon excitation is demonstrated with the help of experimental observations of dication dissociation and plasmon excitation model calculations.

Polycyclic aromatic hydrocarbons (PAHs) and their cations are of significance in astronomical context for the last few decades. It is believed that molecular dynamics has an important role to play in their formation, existence and survival inspite of harsh interstellar conditions. Naphthalene and azulene are the simplest isomeric pairs in the family of PAHs. The major reaction channels of PAH cations are H,2H/H_2 and C_2H_2 elimination. But in case of naphthalene and azulene monocations, isomerization channel is seen to be competing with low energy dissociation channels [1,2].

The dominant energy loss mechanism in proton-PAH interaction is identified as single plasmon excitation at about 17eV [3]. The excited molecule relaxes via single ionization. Previous experiments using photoelectron spectroscopy [4] have demonstrated that collective excitation is coupled to the inner valence electron, which is ejected with minimal kinetic energy. The remainder of the energy is retained in the molecule as internal energy and utilized for various (statistical) processes. In the present proton velocity regime (1.41 and 2.4 a.u.), electron capture also offers constant energy deposition as in case of plasmon excitation.

It is confirmed that cross sections for electron capture (EC) and electron emission (EE) via plasmon excitation are identical in both the isomer targets studied here. Although the later dynamics of isomers is structure dependent, energetic fragmentation channels are found to be identical for the two targets. However, the low energy dissociation channels like H/C_2H_2 loss are not similar in the two molecules, which are quite sensitive to minor variation of available energy. The azulene fragment spectrum shows excess H/C_2H_2 yield compared to naphthalene which was explained using a plasmon excitation driven isomerization model.

In the same work, the origin of double ionization of $C_{10}H_8$ isomers is explained using multiplasmon resonance. By invoking the multiplasmon excitation model, double ionization cross section is found to be in good agreement with the double plasmon excitation cross section. The constant energy distribution expected from plasmon excitation is corroborated by the constant relative yields of dication products, in the investigated proton energy regime. Further investigation carried out on the energetics of different dication dissociation dynamics agree well with the energetics of double plasmon resonance.

References

[1] Dykov Y A *et al* 2006 *Phys. Chem. Chem. Phys.* **8** 1404
[2] Cui W *et al* 2000 *J. Phys. Chem. A.* **104** 6339
[3] Mishra P M *et al* 2013 *Phys. Rev. A* **88** 052707
[4] Mishra P M *et al* 2014 *J. Phys. Chem. A.* **118** 3128

†E-mail: vinithamvv@gmail.com

Absolute total cross sections of multiple electron processes for molecules of biological interest by fast proton impact

P N Terekhin[1,2 *], M A Quinto[3], J M Monti[3,4], O A Fojón[3,4] and R D Rivarola[3,4]

[1]Department of Physics and Research Center OPTIMAS, TU Kaiserslautern, Kaiserslautern, 67663, Germany
[2]National Research Center "Kurchatov Institute", Moscow, 123182, Russia
[3]Instituto de Física Rosario (CONICET-UNR), Bv 27 de Febrero 210 bis, 2000 Rosario, Argentina
[4]Laboratorio de Colisiones Atómicas, Facultad de Ciencias Exactas, Ingeniería y Agrimensura, Universidad Nacional de Rosario, Av. Pellegrini 250, 2000 Rosario, Argentina

Synopsis We present the study of multiple electron processes produced by impact of proton beams on molecular targets from intermediate to high projectile impact energies. Transition probabilities and absolute cross sections were determined by using the Continuum Distorted Wave-Eikonal Initial State approximation. Results were compared with previous calculations and also with recent experimental data.

Information about differential ionization cross sections and total multiple electron processes (single and multiple ionization, single capture, transfer-ionization) cross sections is of great interest in many areas of physics, for example, in atmospheric, plasma, medical and other branches of Physics. Nowadays, these data are relevant in the area of radiotherapy. Living matter consists mainly of liquid water and DNA-like structures. However, other simple molecules such as N_2, O_2, CO, CO_2, CH_4 are of importance in biological processes. Then, absolute total cross sections of these molecules by impact of ion beams are of fundamental interest as well.

In the present work we calculate total cross sections for single and multiple electron ionization, electron capture and transfer-ionization reactions of several molecular targets (N_2, CO, CO_2) by proton impact. The three-body Continuum Distorted Wave-Eikonal Initial State approximation (3B-CDW-EIS) [1,2] is used to calculate transition probabilities (capture and ionization) as a function of the impact parameter and absolute cross sections for the considered collisions. The initial wavefunctions of the active electrons bound to a particular molecular orbital are described employing the complete neglect of the differential overlap (CNDO) approximation [3]. A trinomial distribution analysis has been employed to compute exclusive probabilities using the independent electron (IEL) model, where electron correlation is neglected [4].

The results are compared with other theoretical calculations and available experimental data (see Figure 1).The developed method for calculation of one-electron ionization and capture probabilities allows to study multiple electron processes for charged particles passing through matter. It also can be used to investigate molecular fragmentation mechanisms.

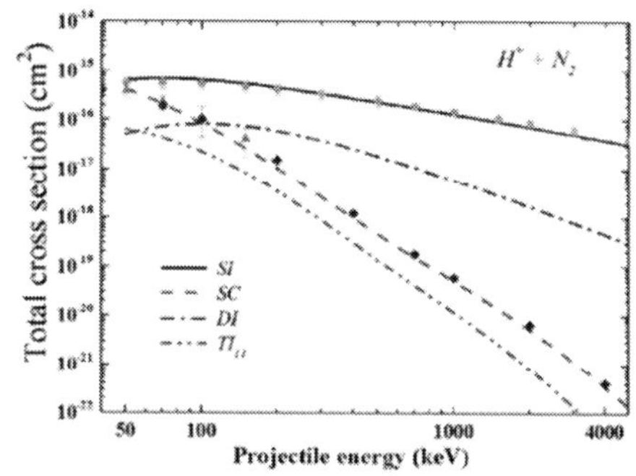

Figure 1. Total multiple electron processes cross sections as a function of the incident energy for H^+ impacting on N_2: single ionization (SI), single capture (SC), double ionization (DI), transfer-ionization (TI_{11}). Experiments: circles and triangles [5], diamonds [6].

References

[1] Monti J M *et al* 2013 *J. Phys. B: At. Mol. Opt. Phys.* **46** 145201
[2] Quinto M A *et al* 2018 *J. Phys. B: At. Mol. Opt. Phys.* **51** 165201
[3] Pople J *et al* 1965 *J. Chem. Phys.* **43** S136
[4] Bradley J *et al* 2004 *J. Phys. B: At. Mol. Opt. Phys.* **37** 3723
[5] Rudd M E *et al* 1985 *Rev. Mod. Phys.* **57** 965
[6] Barnett C F *et al* 1977 *Atomic Data for Controlled Fusion*

* E-mail: terekhin@physik.uni-kl.de

Ab initio study of energy characteristics of small polyatomic molecules in threshold electron-impact dissociative ionization processes

S Demes[1 *] **and E Remeta**[2 †]

[1]Institute for Nuclear Research, Hungarian Acad. Sci., 18/c, Bem square, 4026 Debrecen, Hungary
[2]Institute of Electron Physics, Nat. Acad. Sci. of Ukraine, 21, Universitetska str., 88017 Uzhgorod, Ukraine

Synopsis A joint theoretical study of electron-impact dissociative ionization (DI) processes of small polyatomic molecules is presented. Different organic (CH_4 and C_2H_6) and inorganic (SF_6 and S_n, n = 2-8) target molecules were examined. The investigated dissociative ionization processes take place at threshold energies, so our main focus is to study the appearance energies (E_{ap}) of different ion fragments as well as to analyse the corresponding DI reaction pathways.

The appearance energy characterizes that minimal collision energy in a DI process, which is necessary in order to observe the appearance of a concrete ion fragment. While the main ion products in electron impact measurements are usually well-known, it is challenging to measure by spectroscopy the different secondary products, which are mainly consists of neutral fragments. So, if one would like to characterize the complete reaction channels in such DI experiments, one should utilize different high-level *ab initio* methods [1].

In order to obtain the energy characteristics of the reaction products in threshold processes, i.e. the appearance energy (E_{ap}) or the more general ionization energy (I) and electron affinity (E_a), we performed total energy calculations by the GAMESS-US software package [2]. The corresponding molecular energy characteristics are derived from these data simply. They are considered as the reaction enthalpy change, and calculated as the total energy difference of the initial and final products [3]. So the E_{ap} value is valid for particular fragmentation channels.

The systematic theoretical study of the various neutral and ion fragments of CH_4, C_2H_6, SF_6 and S_n target molecules was performed by means of higher level *ab-initio* theories, such as CCSD(T), DFT or MP2. We used large basis sets in the calculations, e.g. aug-cc-pVTZ. The total energies of all molecular species are calculated as the sum of total electronic and zero-point vibrational energies. Some relevant energy characteristics of the observed fragments in a DI processes of ethane molecule are shown in Table 1.

* E-mail: demes.sandor@atomki.mta.hu
† E-mail: remetoveyu@gmail.com

Table 1. Calculated energy characteristics of the neutral and ion fragments of ethane molecule.

Energy char.		Calc.	Exper.
$E_a(C_2H)$		2.91	2.97
$I(C_2H_4)$		10.33	10.52
$I(C_2H_2)$		11.13	11.40
DI reaction channel: $C_2H_6 + e^- \rightarrow$			
E_{ap}	$\rightarrow C_2H_5^+ + H + 2e^-$	12.43	12.45
E_{ap}	$\rightarrow C_2H_3^+ + H_2 + H^- + e^-$	14.52	13.76
E_{ap}	$\rightarrow CH_4^+ + C + H_2 + 2e^-$	20.01	20.40
E_{ap}	$\rightarrow CH_3^+ + CH_3 + 2e^-$	13.54	13.65
E_{ap}	$\rightarrow CH_3^+ + CH_3^- + e^-$	13.64	13.56
E_{ap}	$\rightarrow C_2^+ + 6H + 2e^-$	34.88	31.50

The theoretical values in Table 1 are calculated by CCSD(T)/aug-cc-pVDZ theory [2]. The experimental values are taken from the NIST database [4]. Our results are in a fairly good agreement with these measurements, which indicates the validity of the proposed theoretical approach for the description of electron-impact DI processes. More detailed results will be presented at the conference.

References

[1] Gross J H 2011 *Principles of Ionization and Ion Dissociation in Mass Spectrometry*. Berlin, Springer-Verlag, pp. 21-66

[2] Schmidt M W, Baldridge K K *et al.* 1993 *J.Comp. Chem.* **14** 1347

[3] Demesh S and Remeta E 2015 *Eur. Phys. J. D* **69** 168

[4] https://webbook.nist.gov/chemistry/form-ser/

Photoionization of H_2 using the molecular R-matrix with time approach

J Benda[1], Z Mašín[2,*], G S J Armstrong[3], D D A Clarke [3], A C Brown[3 †], J D Gorfinkiel[1 ‡] and H van der Hart[3]

[1]School of Physical Sciences, The Open University, Walton Hall, MK7 6AA Milton Keynes, UK
[2]Institute of Theoretical Physics, Faculty of Mathematics and Physics, Charles University, V Holešovičkách 2, 180 00 Prague 8, Czech Republic
[3]Centre for Theoretical Atomic, Molecular and Optical Physics, School of Mathematics and Physics, Queen's University Belfast, Belfast BT7 1NN, United Kingdom

Synopsis We present results of the first calculations using the variational *ab initio* molecular R-matrix with time approach. We have calculated two and four-photon ionization cross sections for H_2 and studied the effects of electron correlation and choice of the Gaussian atomic basis sets. Our results are compared with earlier calculations.

The atomic R-matrix with time dependence (RMT) [1] approach has been applied successfully to the study of the interaction of atoms with ultrashort laser pulses. The approach is based on the R-matrix method that divides space into an inner and an outer region. RMT uses highly accurate wavefunctions in the inner region that describe multielectronic interactions precisely. In the outer region, the ionized electron becomes distinguishable; here the single-electron time-dependent Schrödinger equation is solved, with the electron subject to the long-range potential of the molecule and the laser field.

Recently, we have extended the original atomic RMT software [2] to treat interaction of laser fields with molecules in the fixed-nuclei approximation. The molecular, inner-region data required (transition dipole moments, etc.) are generated using the UKRmol+ suite [3], a highly accurate set of (time-independent) programs initially designed to treat electron and positron scattering from polyatomic molecules.

The RMT approach is very general and allows the study of a range of processes from the perturbative to highly non-linear regimes. Detailed understanding of the molecular RMT models for multiphoton ionization is a prerequisite for its application to processes such as strong-field ionization and high-harmonic generation.

In order to gain this understading and validate the molecular RMT calculations, we have studied two- and four-photon ionization of H_2 for which earlier results are available. Figure 1 shows our simplest results produced using a Gaussian basis set for the bound orbitals and B-splines for the description of the continuum. Our cross sections are in good agreement with those obtained with the diatomic R-matrix Floquet approach [4] but less so with other calculations. We have studied the effect of the modelling of electron correlation for multiphoton ionization and will compare the results of our close-coupling models with similar results obtained using other approaches.

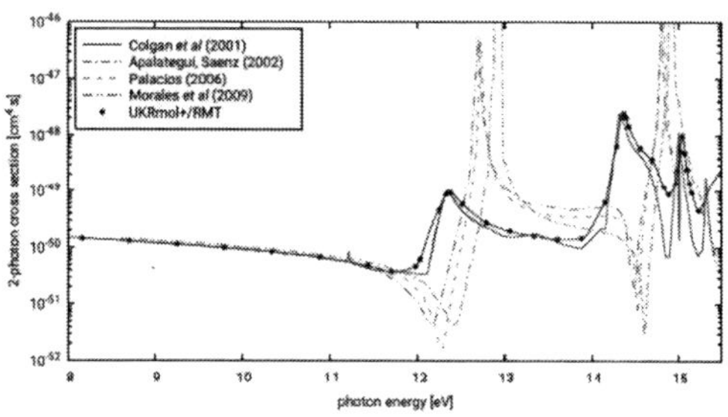

Figure 1. Cross section for two-photon ionization of H_2. Our initial results show good agreement with earlier Floquet calculations.

*E-mail: zdenek.masin@utf.mff.cuni.cz
†E-mail: andrew.brown@qub.ac.uk
‡E-mail: J.Gorfinkiel@open.ac.uk

References

[1] Moore L R *et al* 2011 *J. Mod. Optics* **58** 1132
[2] Brown A C *et al Computer Phys. Comm.*, submitted
[3] Benda J *et al Computer Phys. Comm.*, submitted
[4] Colgan J *et al* 2001 *J. Phys. B* **34** 2089

The interaction of 355nm laser light on Adenine and Uracil

L Hallado[1,2]* , E Prieto[1,2], I Álvarez[1], A Guerrero[1] and C Cisneros[1 †]

[1] Instituto de Ciencias Físicas, Universidad Nacional Autónoma de México, Cuernavaca, México
[2] Programa de Maestría y Doctorado en Ciencias Químicas, Universidad Nacional Autónoma de México, Ciudad de México, México

Synopsis Dissociation and ionization of adenine and uracil were investigated in the multiphoton absorption regime using laser pulse at 355nm in conjunction with TOF mass spectrometry in reflectron mode (R-TOF). In contrast of previous investigations where a notorious amount of low mass ions was detected arising from both molecules.

Different ions formation from Adenine (A) and Uracil (U) have been previously reported using electrons, synchrotron radiation or laser at different wavelengths and techniques O. Ghafur et al, M.Rahman and M Ryszka et al[1,2,3].

The present results show the multiphoton dissociation using laser radiation of 355nm in the range of 109-1010 W/cm2 intensity. Both molecular beams were produced from Sigma Aldrich samples with 99% purity and introduced into a reservoir with an inert gas (Ar), and heated to ≅220°C-230°C. Mass spectra were obtained using time of flight spectroscopy. An intense fragmentation of both molecules was observed, where the most intense signals correspond to fragments of low molecular weight. Among the identified ions, 19 coincidences were found, including the prebiotic precursors fragments HCN^+ and NH_3CN^+ that intervened in the generation of these molecules. For the fragments with greater signal intensity their formation energy was calculated, observing that ions whit the same m/z ratio require different energies, which can be an indication of the structural difference between these molecules.

The energy absorbed to form different ions was calculated from the relation $Y=\sigma^n$. Y is the ion yield, σ is the nth-order cross section, and I is the laser intensity.

We estimate the number photons absorbed using the mention equation. Then it was observed the differences on the number of photons needed to produce each ion, depending whether they are originated from (A) or (U). For the case of ions m/z 14 and m/z 28 from (U) the calculated energies correspond closely to the ones calculated [1]. To our knowledge, there is not reported similar information for the case of adenine. Some of the results for (A) are shown in the figure 1.

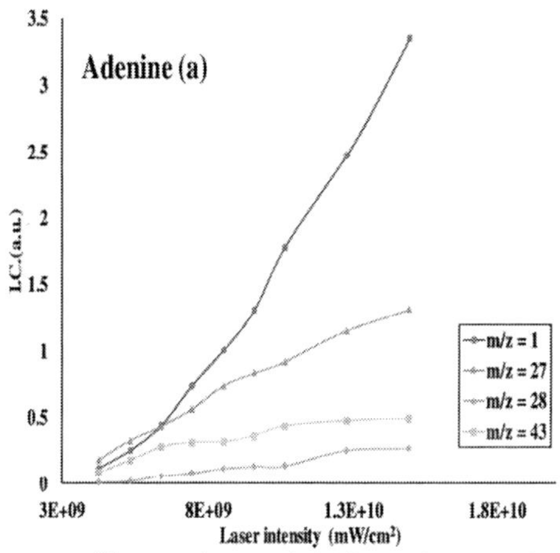

Figure 1. Ion yield in arbitrary units as a function of the laser intensity

This work was supported by UNAM-PAPIIT IN104318 and IN104019.

References

[1] Ghafur O *et al* 2018 *J. Chem. Phys.* 149 034301
[2] Rahmanand M 2016 *J.Chem. Phys.* 144 161102
[3] Ryszka M *et al* 2016 *Int. J. Mass Spectrom* 396 48

* E-mail: carmen@icf.unam.mx

Ion induced fragmentation of diatomic molecular clusters

A Méry[1] *, A N Agnihotri[1], J Douady[1], X Fléchard[2], B Gervais[1], S Guillous[1], W Iskandar[3], E Jacquet[1], V Kumar[1], J Matsumoto[4], J Rangama[1], F Ropars[1], C P Safvan[5], H Shiromaru[4], D Zanuttini[1] and A Cassimi[1]

[1] CIMAP, CEA-CNRS-ENSICAEN-UNICAEN, Normandie Université, BP5133, F-14050 Caen Cedex 04, France
[2] Normandie Univ, ENSICAEN, UNICAEN, CNRS/IN2P3, LPC Caen, 14000 Caen, France
[3] Chemical Science Division, Lawrence Berkeley National Laboratory, Berkeley, CA-94720, USA
[4] Department of Chemistry, Tokyo Metropolitan University, 1-1 Minami-Osawa, Hachioji, Tokyo, Japan 192-0397
[5] Inter University Accelerator Center, Aruna Asaf Ali Marg, New Delhi 110067, India

Synopsis Weakly bound atomic or molecular clusters are of growing interest in the fields of atmospheric science, hadron-therapy or astrophysics. We will present our recent experimental results on the fragmentation of diatomic molecular dimers after irradiation by highly charged ions. For $(N_2)_2$ dimers, the measured KER spectrum is very similar to the one of the N_2 monomers but is globally shifted towards higher energies. Moreover, sequential dissociation of the $(CO)_2$ dimers has been observed when metastable states of the CO^{2+} dications are populated.

Diatomic molecular dimers such as $(N_2)_2$ and $(CO)_2$ are model systems containing both covalent intramolecular bonds and van der Waals intermolecular bonds. Following multiple electron capture, both intramolecular and intermolecular fragmentation may occur. We will focus on some particular 3-body dissociation channels where only one of the constitutive molecule breaks while the other cation remains in a non-dissociative state. For such channels, the influence of a second spectator ion on the molecular dissociation has been investigated.

We use a COLTRIMS setup (COLd Target Recoil Ion Momentum Spectroscopy) to perform cross beam collisions between molecular dimers produced by a supersonic gas jet and highly charged ions delivered by the low energy beam line of the GANIL/ARIBE facility. The charged fragments produced in the collision are measured in coincidence and their three-dimension momentum vectors are computed thanks to their position and time-of-flight on a position sensitive detector. A clean data selection is then achieved by imposing momentum conservation laws. The kinetic energy of each fragment as well as the KER (Kinetic Energy Release) are deduced from the momentum vectors and m/q identification of the fragments. Finally, a comparison between events originating from monomer or dimer targets is performed from a single experiment.

In the case of $(N_2)_2$ dimers, the measured KER spectra are globally shifted towards higher energies in comparison to the fragmentation of the N_2 monomer. The value of the energy shift is fairly well reproduced by accounting for the additional Coulomb potential energy due to presence of the neighbor ion [1]. This indicates that the presence of the second molecule has a weak influence on both electron capture and molecular fragmentation dynamics. For $(CO)_2$ dimers, similar results are observed but concerted and sequential fragmentation of the dimer have also been identified. These two processes are respectively associated to the population of either short or long lived excited states of the CO dication. As a consequence, molecular fragmentation within such dimers can be used as a tool to distinguish between short lived and metastable states of the monomer cations.

References

[1] Méry A *et al* 2017 *Phys. Rev. Lett.* **118** 233402

*E-mail: alain.mery@ganil.fr

Absolute cross sections for electron capture and loss of O^+ in water molecule

V Oliveira[1], A H A Gomes[1], A C F Santos[1*] and K Tőkési[2,3†]

[1]Instituto de Física, Universidade Federal do Rio de Janeiro - 21941-972 Rio de Janeiro, RJ, Brazil
[2]Institute for Nuclear Research, Hungarian Academy of Sciences (Atomki), Debrecen, Hungary
[3]ELI-ALPS, ELI-HU Non-profit Ltd., Dugonics tér 13, H-6720 Szeged, Hungary

Synopsis Absolute cross sections for single and double electron loss and single and double electron capture for O^+ projectiles colliding with water vapor have been measured. In addition, we have performed many-body classical trajectory Monte Carlo calculations, where the O^+ projectile was either one or more particle, depending on the model used. The water target is modeled as two virtual particles. The calculated cross sections are in agreement with the experimental data.

The study of ionization of water molecules by charged particle impact plays an important role in many research fields such as radiobiology, radioprotection and medical physics, like cancer therapy. In particular, over the last few years, the application of light-ion beams to external beam radiotherapy has been spreading worldwide. Oxygen beams are thought to be a therapeutic complement to carbon beams because they provide a higher linear energy transfer with an even more reduced lateral scattering, which is advantageous for the treatment of radio-resistant tumours. Around the Bragg peak, oxygen beams consist mainly of neutral and singly-charged components.

In this work, we present experimental and theoretical data on the absolute cross sections for single and double electron loss and single and multiple electron capture by oxygen projectiles colliding with the water molecule.

Singly charged oxygen beams with energies ranging from 200 -1200 keV were produced from O^- beams, delivered by a cesium sputtering source. The beam were accelerated by the 1.7 MV tandem accelerator at the Federal University of Rio de Janeiro. The O^+ beam is then directed to a 12-cm-long differentially pumped gas cell, containing water vapor. We have also performed classical trajectory Monte Carlo calculations. This approximation seems to be useful in treating atomic collisions where the quantum mechanical ones become very complicated or unfeasible. This is the case usually when higher order perturbations should be applied or many particles take part in the processes.

* E-mail: toni@if.ufrj.br
† E-mail: tokesi@atomki.mta.hu

Figure 1 shows our results. We obtained a good agreement between our measured and calculated data.

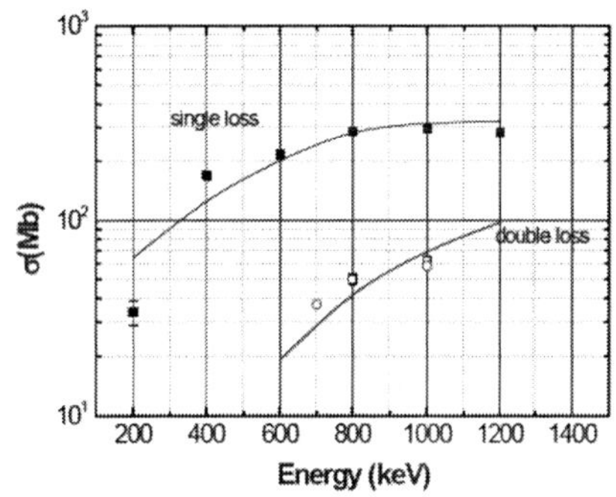

Figure 1. Single and double electron loss cross sections of the projectile in collision between O^+ and water. Solid squares: experiment: single loss (present data); open squares: double loss (present data); Full lines: Theory, present results of our classical simulations for single (black line – four-body model) and double loss (red line – five-body model) of O^+ on water.

Ackowledgments This work is supported in part by CNPq. One of us, A.C.F. Santos, acknowledges support from CAPES (Programa Abdias Nascimento). One of us (KT) was support by the National Research, Development and Innovation Office (NKFIH) Grant KH126886.

References

[1] Deng Z *et al* 2005 *Phys. Rev. Lett.* **95** 153201
[2] Kurz C *et al* 2012 *Phys. Med. Biol.* **57** 5017
[3] Sant'Anna M.M *et al* 2006 *Phys. Rev. A* **74** 022701

Electron–beam ion trap/source for ion–molecule collisions in the non-perturbative regime

B Bapat[1]*, **D Sharma**[1] **and S Srivastav**[1]

[1]Indian Institute of Science Education and Research Pune, Homi Bhabha Road, Pune, 411008, India

Synopsis An Electron Beam Ion Trap and Source has been installed at the Indian Institute of Science Education and Research Pune. The source, manufactured by Dreebit GmbH, will provide ions up to Ar^{16+} at energies up to 30 keV/q. These will be used for studying ion–molecule collisions in the non-perturbative regime.

Among the variety of ion sources, the electron beam ion trap/source, which works on the principle of magnetic compression of an intense electron beam, creating a significant space charge potential and enhancing the rate of collisional ionisation, has the advantage of producing ions with very high charge states at low energies [1].

Low energy, highly charged projectiles are of great importance in exploring the response of molecules to slow, but heavy perturbations. The typical 'interaction time' of a projectile with a molecule is taken to be a/v, where a is the nominal size of the molecule and v is the projectile speed. When this time is comparable to the vibrational time scales of molecules, we move significantly away from the perturbative regime, requiring the use of non-perturbative quantum mechanical treatments to predict the outcome of such collisions [2].

To investigate such collisions, one needs projectile energies of the order 10 keV/amu and variable charge states. These are well provided by an ion source such as EBIS/T, with suitable acceleration. The source at IISER Pune is based on a thermionic electron emitter and permanent magnets for compression of the electron beam [3]. The complete source is raised to a positive potential between 10–30 keV, allowing the extraction of 10–30 keV/q beams. Charge states up to Ar^{16+} can be produced and extracted. The distribution of charge states in the source/trap can be controlled by changing the conditions in the trap, viz. the electron flux, electron energy, trapping time and pressure. The trap can be operated in the pulsed mode or a continuous, 'leaky' mode. A Wien Filter with an aperture of 1.0 mm is employed to separate the charge states before the ions enter the beamline. The beamline consists of an einzel lens and an electrostatic quadrupole deflector, which couples downstream to the collision chamber.

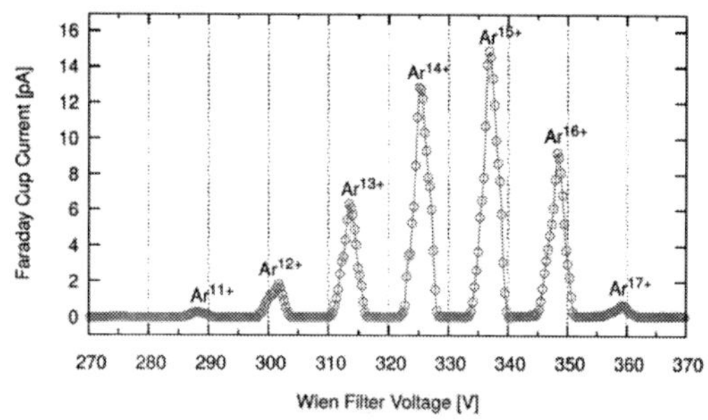

Figure: Charge state spectrum of Argon ions recorded after the Wien Filter. The trap was operated in the leaky mode.

The ion beam crosses an effusive gas jet in the collision chamber. Analysis of the collision fragments is done by a multi-coincidence of an ion momentum spectrometer. Projectile ions that have undergone charge change in the collision are separated by a cylindrical sector charge state analyser. We expect to present preliminary results at the conference.

References

[1] Donets E D 1998 *Rev. Sci. Instr.* **69** 614
[2] Lüdde H J, Spranger T, Horbatsch M and Kirchner T 2009 *Phys. Rev. A* **80** 060702R
[3] Schmidt M, Zschornack G, Kentsch U and Ritter E *Rev. Sci. Instr.* **85** 02B704

*E-mail: bhas.bapat@iiserpune.ac.in

Dissociative multiple ionization of the H_2O molecule induced by ion impact: experimental and theoretical results

S T S Kovács[1] *, P Herczku[1], Z Juhász[1], L Sarkadi[1], L Gulyás[1], J-Y Chesnel[2] and B Sulik[1]

[1]Institute for Nuclear Research, Hungarian Academy of Sciences (MTA Atomki), Debrecen, H-4026, Hungary
[2]Centre de Recherche sur les Ions, les Matériaux et la Photonique, Normandie Univ, ENSICAEN, UNICAEN, CEA, CNRS, CIMAP, Caen, 14000, France

Synopsis Fragment energy spectra of the water molecule impacted by singly charged (SCI) and highly charged (HCI) ions have been measured in crossed beam experiments. Double differential fragmentation cross sections (DDCS) have been determined. Similarities and differences of the measured DDCSs induced by the medium- and low-energy projectiles have been analyzed. Multiple target ionization cross sections have been deduced and they were further analyzed through the results of the CTMC and CDW-EIS models.

We present experimental results on the Coulomb explosion of H_2O molecules, bombarded by different types of projectile ions. The initial multiple vacancy production in the target molecule is further analyzed through the results of CTMC and CDW-EIS calculations.

Gas phase water molecules were impacted by medium-energy singly charged (560 keV C^+ and 650 keV N^+) and low-energy highly charged (60 keV N^{6+} and 70 keV O^{7+}) ions [1, 2]. The energy and angular distributions of the charged fragments were measured by single-stage electrostatic spectrometers. Double differential fragmentation cross sections have been deduced.

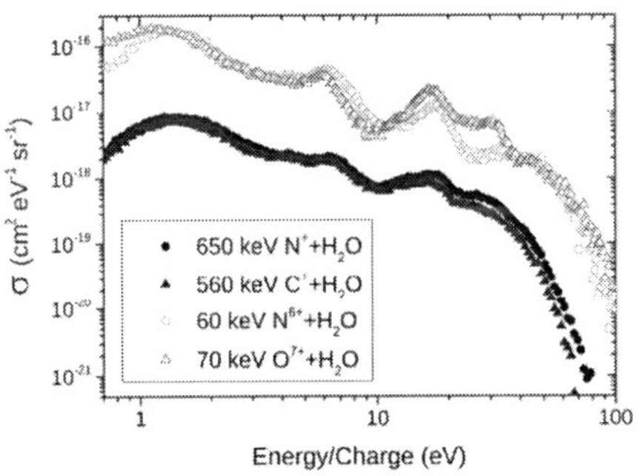

Figure 1. Measured fragmentation cross sections for the four different projectiles at 45 observation angle.

The shape of the spectra is found to be very similar for C^+, N^+ and N^{6+} impact (see Figure 1.). Though the dominant ionization mechanism and the average perturbation strength is different for SCIs and HCIs, the similarity indicates that the same fragmentation channels appear in the collisions. While the SCI induced spectra are almost identical, slight differences in the relative intensities of the structures were found between the SCI and HCI induced spectra, as well as between the spectra induced by the two HCIs. It was attributed to the selectivity of the electron capture process compared to the direct ionization.

From the fragment ion spectra multiple ionization (MI) cross sections have been deduced. We found that the highest degrees of ionizations for the two SCIs approach those for the HCIs with non-negligible yields. The MI cross sections were further analyzed through the results of classical (CTMC) and quantum mechanical (CDW-EIS) model calculations [1, 3]. The calculated MI cross sections almost coincide with the experimental data for C^+ and N^+ impact, and are only in qualitative agreement with the measured HCI results. Further analysis have been performed through the calculated impact parameter dependent MI probabilities.

Work was supported by the Hungarian OTKA (Grants No. K109440 and K128621), the CNRS-MTA Cooperation (Grant No. 26118) and by the PHC Balaton Program(Grant No. 38620NH/TéT 16-1-2016-2026).

References

[1] Kovacs S T S *et al* 2017 *Phys. Rev. A* **96** 032704
[2] Sobocinski P *et al* 2006 *J. Phys. B* **39** 927
[3] Kovacs S T S *et al* 2016 *Phys. Rev. A* **94** 012704

*E-mail: kovacs.sandor@atomki.mta.hu

Angle-resolved Auger spectroscopy of K-shell ionized molecules

S Fuchs[1,2*], D Paul[1,3], T Buhr[1], S Ricz[4] and S Schippers[1]

[1]I. Physikalisches Institut, Justus-Liebig-Universität Gießen, 35392 Giessen, Germany
[2]GSI Helmholtzzentrum für Schwerionenforschung GmbH, 64291 Darmstadt, Germany
[3]Max-Planck-Institut für Kernphysik, 69117 Heidelberg, Germany
[4]Institute for Nuclear Research of the Hungarian Academy of Sciences, 4001 Debrecen, Hungary

Synopsis We report on measured oxygen K-LL Auger spectra induced by 1.5-keV electron impact on O_2, CO_2, and N_2O gaseous targets. An ESA-22-type electron spectrometer was employed which is capable of simultaneous energy and angular analysis covering a polar angular range of 15° to 345°. The measurements were carried out in preparation for future experiments with more complex chiral molecular targets and a spin-polarized electron source. The present experimental data are compared with results from earlier experimental studies.

In order to study the angle-dependent interactions of polarized electrons as well as circular polarized photons with chiral molecules, an ESA-22-type electron spectrometer [1, 2] in combination with an electron gun and focusing optical elements was upgraded and tested. The electron spectrometer is capable of simultaneous energy and angular analysis covering almost the entire polar angular range of emitted electrons.

As test experiments, oxygen K-LL Auger spectra induced by 1.5-keV electron impact were measured using O_2, CO_2, and N_2O gas targets. Auger electrons were detected simultaneously in the polar angular range of 15° to 345° relative to the electron-beam direction. As an example, figure 1 shows an oxygen K-LL Auger spectrum of CO_2 at a polar angle of 195° with respect to the incoming electron-beam direction. The present experimental data are compared with results from earlier experimental studies [3, 0, 0]. Overall, a fairly good agreement is found. Moreover, the angular distribution of the Auger electrons was investigated for each Auger peak separately, and the related anisotropy coefficients were determined.

For a better understanding of the experimental uncertainties, we have carried out electron-optical simulations of our spectrometer using the SIMION program together with the script language LUA. Some of the aspects investigated were the influence of the position of the source volume on the spectrometer transmission, on the spectrometer constants and on the electron angular distributions. For future experiments with chiral molecular targets we plan to combine the ESA-22-type electron spectrometer with a spin-polarized electron source which was recently developed at GSI [0].

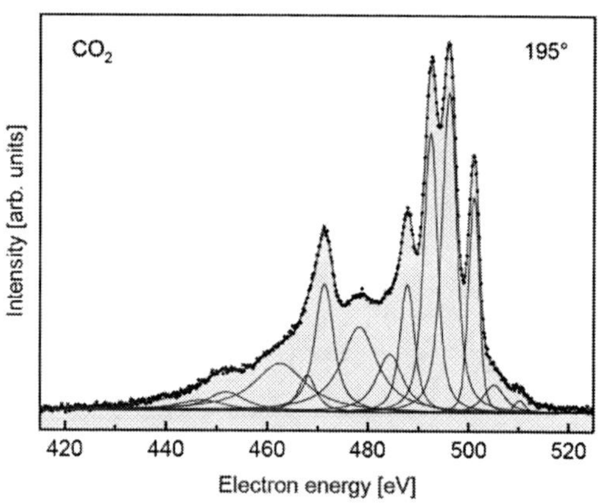

Figure 1. O K-LL Auger spectrum of CO_2 at a polar angle of 195° including all resolved peaks. The experimental data points are shown in black, the blue curves represent the fitting functions of the single peaks and the red line represents their sum.

References

[1] Ricz S *et al* 2002 *Phys. Rev. A* **65** 042707
[2] Ábrók L *et al* 2016 *Nucl. Instrum. Meth. B* **369** 24
[3] Moddeman W E *et al* 1971 *J. Chem. Phys.* **55** 2317
[4] Hiltunen A *et al* 1999 *Nucl. Instrum. Meth. B* **154** 267
[5] Griffiths W J *et al* 1991 *J. Phys. B* **24** 4187
[6] Schury D 2017 PhD Thesis, JLU Gießen

*E-mail: sebastian.fuchs@physik.uni-giessen.de

Electron impact ionization of pyrimidine: differential and total cross sections

L Mouawad[1] *, P -A Hervieux[1] †, C Dal Cappello[2], J Pansanel[3], V Robert[4] and Z El Bitar[3]

[1]Université de Strasbourg, CNRS, IPCMS UMR 7504, F-67000 Strasbourg, France
[2]Université de Lorraine, CNRS, LPCT UMR 7019 , F-57000 Metz, France
[3]Université de Strasbourg, CNRS, IPHC UMR 7178, F-67000 Strasbourg, France
[4]Université de Strasbourg, CNRS, Laboratoire de Chimie Quantique UMR 7177, F-67000 Strasbourg, France

Synopsis We provide theoretical cross sections for the ionization of pyrimidine by electron impact. The theoretical framework is based on a quantum approach making use of particular tools to overcome the computing challenges. The developed methodology allows to determine not only the total but also the triple, double and simple differential cross sections. This approach was previously used to calculate the triple differential cross sections of many molecules. Here, we demonstrate that it can be used to calculate more integrated cross sections of a molecule of biological interest: pyrimidine.

Providing cross sections for the electron impact ionization of biological molecules has become an urgent requirement for valid investigations of particle interactions within the biological medium. Due to the lack of cross section data for complex biological molecules, interaction cross sections with water are often used to model interactions in living tissue. However, recent studies show that using water to mimic biological systems is not quantitatively correct [1]. We propose a methodology to calculate differential and more integrated cross sections for the ionization of biologically relevant molecules by electron impact [2, 3]. It is based on a distorted wave model (DW) within the framework of the First Born Approximation. We show here the resultant cross sections for pyrimidine: a structural unit of DNA and RNA bases thymine, cytosine and uracil. In figure 1 we compare the triple differential cross sections (TDCS) obtained with the present model (solid line) to those computed using the orientation averaged molecular orbital (OAMO) approximation (dashed line) and the experimental data from [4]. The OAMO approximation was proposed as a solution to overcome the difficulty of calculating cross sections for an average molecular orientation by performing this average on the level of the molecular wave functions. It was used to determine the TDCSs for many molecules but was found to deplete important information about the molecular orbital structure. Figure 1 shows that the cross sections calculated with the DW model are in better agreement with the experimental data.

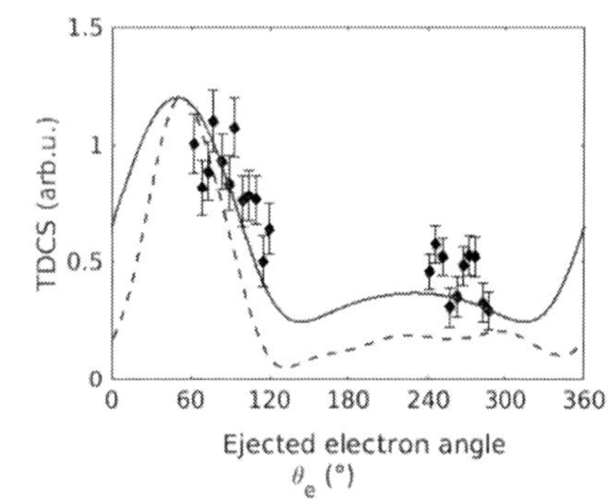

Figure 1. Summed TDCSs for the electron ionization of pyrimidine orbitals $10a_1$, $1b_1$ and $6b_2$ calculated with the DW (solid line) approach, compared to OAMO TDCSs (dashed line) and experimental data (solid circles) from [4].

References

[1] Fuss M C *et al* 2015 *J. Appl. Phys.* **117** 214701
[2] Mouawad L *et al* 2017 *J. Phys. B* **50** 215204
[3] Mouawad L *et al* 2018 *J. Phys. B* **51** 175201
[4] Builth-Williams JD *et al* 2012 *J. Chem. Phys.* **136** 024304

*E-mail: lena.mouawad@hotmail.com
†E-mail: paul-antoine.hervieux@ipcms.unistra.fr

A quantum approach to calculate differential and total cross sections for electron impact ionization of molecular targets

L Mouawad[1] *, P-A Hervieux[1] †, C Dal Cappello[2], J Pansanel[3], V Robert[4] and Z El Bitar[3]

[1]Université de Strasbourg, CNRS, IPCMS UMR 7504, F-67000 Strasbourg, France
[2]Université de Lorraine, CNRS, LPCT UMR 7019 , F-57000 Metz, France
[3]Université de Strasbourg, CNRS, IPHC UMR 7178, F-67000 Strasbourg, France
[4]Université de Strasbourg, CNRS, Laboratoire de Chimie Quantique UMR 7177, F-67000 Strasbourg, France

Synopsis We present a theoretical approach to calculate the cross sections for the ionization of molecules by single electron impact. It is based on the First Born approximation (FBA), describing the ejected electron by a distorted wave (DW). The cross sections are calculated for an average molecular orientation with the proper average method. Single-center molecular wave functions are generated using Gaussian making this method easily applicable to any molecule. We show the calculated cross sections of some molecular targets of varying complexity.

We propose a quantum approach to calculate total and differential cross sections for the electron impact ionization of molecules. The developed approach was detailed in [1, 2] where it was used to calculate triply differential cross sections (TDCS) for the electron impact ionization of formic acid and tetrahydrofuran. By integrating the TDCS, we can easily retrieve the doubly differential, simply differential and total cross sections. Another interesting feature of this methodology is that it can be applied to any molecular target. It is particularly adapted for complex molecules for which theoretical cross section data are very scarce but much needed. The theoretical framework is based on the First Born Approximation (FBA) and the ejected electron is described by a distorted wave. The molecular wave functions are generated with a quantum chemistry program and are converted to single center wave functions thereby reducing the complexity of multi-atom systems. All wave functions are written in partial wave expansion making it possible to translate the theory to a parallel program hence reducing the computing time. As an example, we show in the figure below the TDCSs (solid line) for the ionization of phenol. They are in better agreement with the experimental data than the ones calculated with another theoretical approach making an approximation for orientation averaging: the average orientation molecular orbital approximation OAMO (dash-dotted line) [3].

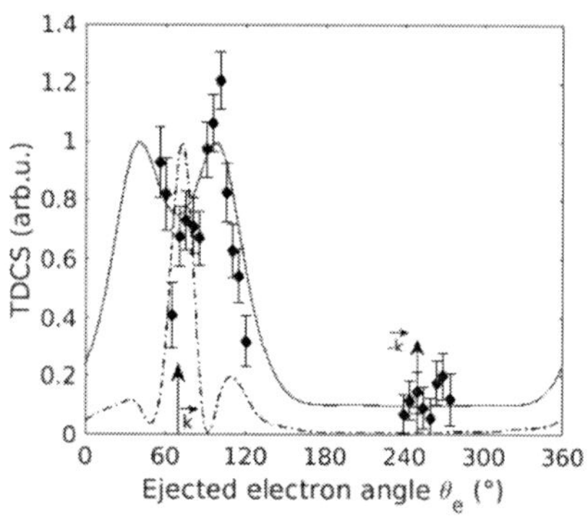

Figure 1. Summed TDCSs for the electron impact ionization of the highest occupied and next highest occupied molecular orbitals of phenol calculated using the present model (solid line), compared to theoretical OAMO TDCSs (dash-dotted) and experimental data (solid circles) from [3].

Hence, we demonstrate that it is possible to calculate the cross sections for complex molecular targets at low computational cost without compromising the validity of the theory.

References

[1] Mouawad L *et al* 2017 *J. Phys. B* **50** 215204
[2] Mouawad L *et al* 2018 *J. Phys. B* **51** 175201
[3] Da Silva G B *et al* 2014 *J. Chem. Phys.* **141** 214307

*E-mail: lena.mouawad@hotmail.com
†E-mail: paul-antoine.hervieux@ipcms.unistra.fr

Channel-resolved molecular Auger spectroscopy

U Ablikim[1,2], R Forbes[3,4], C Bomme[5], T Osipov[6], H Xiong[7], R Bilodeau[2,7], D Kilcoyne[2], A Rudenko[1] D M P Holland[8], N Berrah[7] and D Rolles[1*]

[1]J.R. Macdonald Laboratory, Department of Physics, Kansas State University, Manhattan, KS, 66506, USA
[2]Advanced Light Source, Lawrence Berkeley National Laboratory, Berkeley, CA, 94720, USA
[3]Department of Physics and Astronomy, University College London, Gower Street, London, WC1E 6BT, UK
[4]Department of Physics, University of Ottawa, 150 Louis Pasteur, Ottawa, ON, K1N 6N5, Canada
[5]Deutsches Elektronen-Synchrotron DESY, 22607 Hamburg, Germany
[6]SLAC National Accelerator Laboratory, Menlo Park, CA, 94025, USA
[7]Department of Physics, University of Connecticut, Storrs, CT, 06269, USA
[8]Daresbury Laboratory, Daresbury, Warrington, Cheshire WA4 4AD, UK

Synopsis Molecular Auger spectra typically consist of many overlapping lines that are hard or impossible to resolve experimentally even with highest-resolution electron spectroscopy techniques. Since molecular Auger energies depend on non-adibatatic dynamics, they can also be difficult to calculate precisely, such that the assignment of molecular Auger transitions is often ambiguous. Here we show that by measuring Auger-electron multi-ion coincidences, we can disentangle the spectra into the contributions from specific ionic final states, which helps significantly with their interpretation and assignment.

When an atom or molecule is inner-shell ionized by an soft X-ray photon, the inner-shell hole can relax via Auger decay, leading to a number of different multiply charged final ion states. In atoms, the resulting Auger electron spectrum therefore displays a series of typically well-resolvable peaks, which can be directly linked to the corresponding ionic final states. The Auger specta of (polyatomic) molecules are more complex due to the increased density of valence states as well as the effect of nuclear motion and molecular bonding. This results in a large number of possible decay channels and extremely convoluted Auger electron spectra, making assignment of the particular electronic and ionic final state challenging.

Here we demonstrate that Auger-electron multi-ion coincidence spectroscopy can be utilized to separate the contributions associated with specific dicationic final states and with multiple (cascaded) Auger processes. The ability to separate different final state contributions is based on the fact that population of a specific set of states will result in Coulomb explosion into a particular set of ionic fragments.

We have chosen methyl iodide (CH_3I) as a model system to test the validity of our coincidence channel-resolved Auger approach. This system [1-3] has emerged as an important model system to study X-ray processes in molecules since it can be compared to the closely related benchmark system of Xe. The availability of ultrafast light sources in the XUV and soft X-ray range has also resulted in recent renewed interest in the photodissociation of CH_3I [4,5].

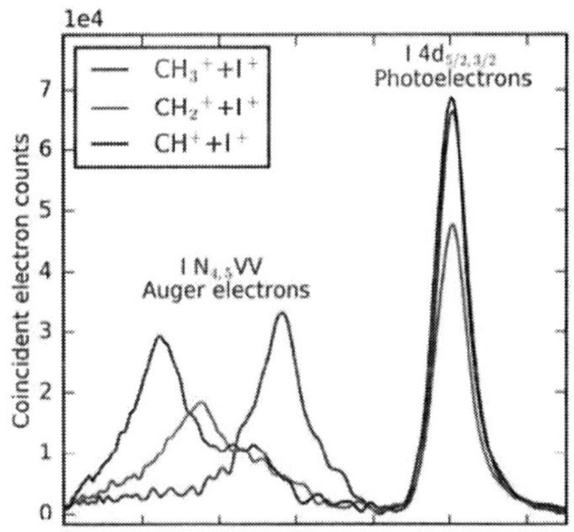

Figure 1. CH_3I electron spectra at 107 eV photon energy measured in coincidence with CH_3^+ and I^+ ions (blue), CH_2^+ and I^+ ions (green), and CH^+ and I^+ ions (black).

References

[1] Forbes R *et al* 2018 *J. Chem. Phys.* **149** 144302
[2] Forbes R *et al* 2018 *J. Chem. Phys.* **149** 094304
[3] Rudenko A *et al* 2017 *Nature* **546** 129
[4] Erk B *et al* 2014 *Science* **345** 288
[5] Brauße F *et al* 2018 *Phys. Rev. A* **97** 043429

* E-mail: rolles@phys.ksu.edu

Electron-impact ionization cross section calculations for 5-chloropyridine and 5-bromopyridine molecules

B Żywicka[1] * and P Możejko[2] †

[1]Nicolaus Copernicus High School No. I, ul. Wały Piastowskie 6, 80-855 Gdańsk, Poland
[2]Department of Atomic, Molecular and Optical Physics, Faculty of Applied Physics and Mathematics, Gdańsk University of Technology, Gabriela Narutowicza 11/12, 80-233 Gdańsk, Poland

Synopsis The total cross sections for the single electron-impact ionization of 5-chloropyridine (5-C_5H_4ClN) and 5-bromopyridine (5-C_5H_4BrN) molecules have been calculated using binary-encounter-Bethe method for electron energies ranging from the ionization threshold up to 5 keV.

In the present work we have studied positive ionization of 5-chloropyridine (5-C_5H_4ClN) and 5-bromopyridine (5-C_5H_4BrN) molecules by electron impact. The total cross sections (ICSs) for the single electron-impact ionization have been calculated using the binary-encounter-Bethe (BEB) model [1] for electron energies ranging from the ionization threshold up to 5 keV. All quantities necessary in the BEB calculations, like the electron binging energy and the orbital kinetic energy, have well defined physical meaning and have been evaluated with the Hartree-Fock method using the GAUSSIAN code and the Gaussian 6-311G++(2d,2p) basis set. Since obtained that way ionization energies usually differ from experimental ones, we have performed also outer valence Green function calculations of correlated electron affinities and ionization potentials [2]. ICS obtained with the BEB method usually are in good agreement (within ±15%) with experimental data [3].

In figure 1 the calculated ICSs are shown. The ionization threshold behavior of the obtained ionization cross sections is very similar. Both ICS functions almost merge for electron-impact energies up to 50 eV. For electron energies higher than 50 eV the ionization cross section for 5-bromopyridine molecule is slightly higher than that for 5-chloropyridine molecule.

The ionization thresholds for studied targets calculated in the present work are listed in table 1 together with respective values obtained recently for pyridine, 2-chloropyridine and 2-bromopyridine molecules [4].

Table 1. The ionization thresholds for pyridine molecule and its halogenated derivatives.

Molecule	Ionization Potential (eV)
Pyridine	9.638 [4]
2-chloropyridine	9.520 [4]
2-bromopyridine	9.372 [4]
5-chloropyridine	9.509
5-bromopyridine	9.372

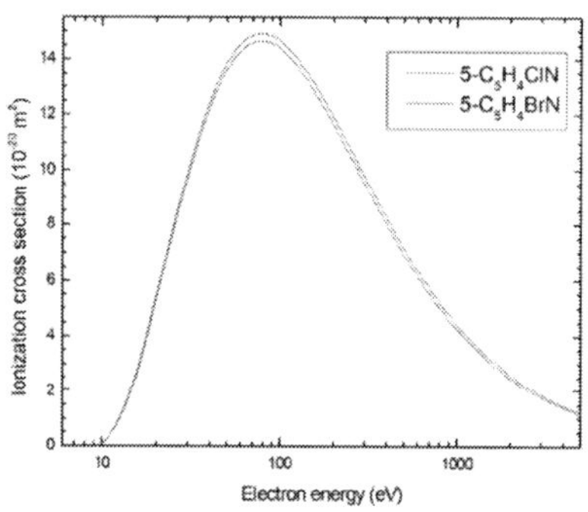

Figure 1. Comparison of the cross sections for electron-impact ionization of 5-chloropyridine and 5-bromopyridine molecules.

This work has been supported by the Polish Ministry of Science and Higher Education. Numerical computations have been performed at the Academic Computer Center in Gdańsk.

References

[1] Hwang W *et al* 1996 *J. Chem. Phys.* **104** 2956
[2] Zakrzewski V G *et al* 1993 *J. Comp. Chem.* **14** 13
[3] Karwasz G P *et al* 2014 *Int. J. Mass Spectrom.* **365-366** 232
[4] Szmytkowski Cz *et al* 2018 *Mol. Phys.* **117** 395

*E-mail: boshena@mif.pg.gda.pl
†E-mail: paw@pg.edu.pl

Unravelling the molecular phenomena of hydrogen peroxide (H_2O_2) through many-body fragmentation dynamics

M Nrisimhamurty *, **A Mandal, L C Tribedi and D Misra** †

Department of Nuclear and Atomic Physics, Tata Institute of Fundamental Research (TIFR), 1, Dr. Homi Bhabha Road, Navy Nagar, Colaba, Mumbai-400005, Maharashtra, India

Synopsis Ion-induced fragmentation dynamics of $H_2O_2^{3+}$ has been studied to distangle molecular phenomena like concerted and sequential fragmentation routes.

Upsurge in the development of the state-of-the-art recoil-ion momentum spectroscopy (RIMS) experimental technique [1, 2] opened doorways to address plethora of physics problems from a molecular perspective. One among those that has been the center-of-interest during tri- or poly-atomic molecule interaction with any probe is: *fragmentation*. The underlying question here in this case is 'how do the control parameters play their key role during the breakup of a molecule'? In 2010, N. Neumann *et al.* [3] and R. Guillemin *et al.* (2014) [4] in their studies on CO_2^{3+} and CS_2^{3+} molecular dissociation dynamics, respectively, had come up with an idea of Dalitz plot and Newton diagram to unveil various physical dissociation processes like sequential- or concerted-fragmentation dynamics. Indeed their work not only revolutionized the field of molecular physics but also promoted these physical methodologies as potential tools to identify the source of dissociation in the molecule under study. However, it still remains as an inevitable challenge during an ion-induced polyatomic molecular breakup process in intermediate energy range (projectile vel. (v_p) $\approx$ electron vel. (v_e)) because of their multi-electronic nature, structural-complexity and various physical processes that lead to dissociation. Hence, we report here about the three-body fragmentation studies of one such polyatomic molecule, namely, H_2O_2.

Generally, hydrogen peroxide (H_2O_2) is a non-planar molecule that plays a prominent role in comprehending the atmospheric chemistry of ozone and is which has been discovered recently in the interstellar medium [5]. To probe this system, 1 MeV Ar^{8+} ions produced by the electron cyclotron resonance ion accelerator (ECRIA) at TIFR, Mumbai has been used as projectile, in the present work. Electrons ejected during the interaction are considered as 'start' whereas recoil ions projected onto a position- and time-sensitive detector serve as 'stop'. From the coincidence observations, it is clear that $H_2O_2^{3+}$ fragments into only one-channel with H^+, H^+ and O_2^+ as its recoil ions. In the present results, two-step sequential dissociation is also likely to occur along with concerted decay as shown in the Fig. 1. Further more interesting details related to the molecular structure of this quantum system (H_2O_2) will be presented and discussed.

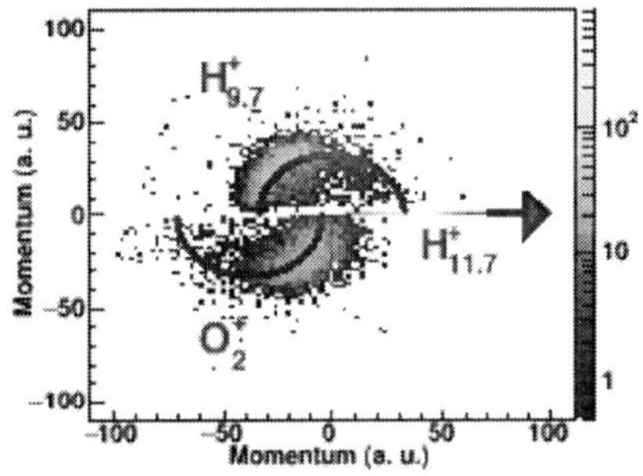

Figure 1. Three-body Newton diagram exhibiting the sequential and direct-dissociation nature with $H^+_{11.7}$ along the preferential axis.

References

[1] Ullrich J *et al* 1997 *J. Phys. B* **30** 2917
[2] Khan A *et al* 2015 *Rev. Sci. Instrum.* **86** 043105
[3] Neumann N *et al* 2010 *Phys. Rev. Lett.* **104** 103201
[4] Guillemin R *et al* 2015 *Nat. Commun.* **6** 6166
[5] Bergman P *et al* 2011 *A&A* **531** L8

* E-mail: nrisimhamurtytifr@gmail.com
† E-mail: dmisra@tifr.res.in

 1412 (2020) 152078 doi:10.1088/1742-6596/1412/15/152078

Energy-transfer in collisions of low velocity ions with CO molecules

M Nakamura[1] * and A Ichimura[2]

[1]College of Science and Technology, Nihon University name, Funabashi, 274-8501, Japan
[2]Faculty of Science, Tokyo University of Science, Shinjyuku-ku, Tokyo, 162-8601, Japan

Synopsis The energy transfer in collisions of low velocity ions with CO molecules is theoretically investigated. Energy loss spectrum of ions scattered by CO molecules is obtained with using the classical trajectory (CT) calculation and the sudden-limit models. Both of the calculations well reproduce an experimental spectrum with a considerable accuracy. Rotational excitation is found to be principal energy transfer process from translational to internal motions.

Sakon and coworkers have performed an exposure experiment outside the International Space Station for different examples such as carbonaceous chondrites [1]. To get insights into this experiment as an example, we attempt to investigate the interaction of CO molecules with inter-planetary ions. Collisional excitation of CO molecules with proton impacts has been extensively studied by Kumar *et al.*[2]. In the present work, we calculate the energy-loss spectra in collisions of ions (H^+, Li^+, and Na^+) with CO molecules. Both sudden-limit models [3, 4, 5] and classical trajectory (CT) calculation are applied to the spectrum. The interaction potentials are expressed by analytical forms to reproduce the results of the SCF calculations given by Grice *et al.*[6].

Figure 1 represents the calculated intensity of Li^+ ions scattered by CO molecule as a function of the normalized energy loss $\epsilon = \Delta E/E$ at the collision energy $E_{\rm lab} = 19$ eV and scattering angle $\theta_{\rm lab}= 60$ deg. in the laboratory frame. An experimental spectrum [7] is shown together. As is shown in the figure, the experimental spectrum has double peaks and a plateau. A peak at $\epsilon \sim 0$ corresponds to the elastic scattering between an ion and a molecule. The other peak and the plateau correspond to the scattering of ions by O and C atoms, respectively. The CT calculation shows triple peaked structure. The shape and position of two peaks are well reproduced. However, the plateau structure observed in the experiment is not well reproduced. The hard-shell model for a rigid-rotor molecule gives similar spectrum to the spectrum obtained with the CT calculation. This result means that rotational excitations are principal energy transfer processes from translational to internal degrees of freedoms.

This work is financially supported by Institute for Quantum Chemical Exploration.

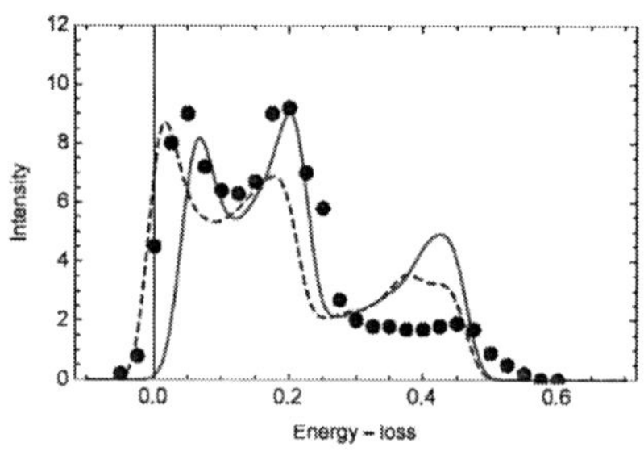

Figure 1. The intensity of Li^+ scattered by CO as a function of the normalized energy loss ϵ. Solid and dashed curves represent the spectra with the classical trajectory and that with a sudden limit model[3, 4], respectively. The dotted curve represents an experimental spectrum by Gierz *et al.* [7].

References

[1] Sakon I *et al.* 2018 *Journal of Japan Society for Aeronautical and Space Science* **66** 381 (in Japanese)
[2] Kumar T J D *et al* 2006 *J. Chem. Phys* **124** 034314
[3] Ichimura A and Nakamura M 2004 *Phys. Rev. A* **69** 022716
[4] Nakamura M and Ichimura A 2005 *Phys. Rev. A* **71** 062701
[5] Nakamura M and Ichimura A 2016 *Phys. Rev. A*, **94** 022717
[6] Grice S T *et al* 1993 *J. Chem. Phys.* **99** 7619
[7] Gierz U *et al* 1984 *Chem. Phys. Lett.* **110** 115

*E-mail: nakamura.masato@nihon-u.ac.jp

Fitting continuum wave functions with complex Gaussians

A Ammar[1] *, A Leclerc[1] and L U Ancarani[1] †

[1]Université de Lorraine - CNRS, UMR 7019, LPCT, Metz, F-57000, France

Synopsis One particle atomic and molecular bound functions can be easily fitted with a set of real Gaussian type orbitals (r-GTOs). Here we explore how continuum wave functions such as Bessel, Coulomb or generalized Sturmian functions can be represented by a set of real or complex Gaussian type orbitals (c-GTOs) that can, ultimately, enter a molecular scattering calculation.

The well known r-GTOs are largely used in quantum chemistry and molecular physics due to remarkable mathematical properties, including the Gaussian product theorem that greatly facilitates a number of integral calculations. A set of N r-GTOs can also be used to represent atomic and molecular one particle bound functions like, for example, Slater orbitals [1]. Representing scattering states is much more delicate because of their oscillating nature. Some studies have shown, though, that it is feasible [2, 3], but with a large number N of nodeless r-GTOs (logically, N increases as the energy of the continuum state increases). An alternative approach is to consider c-GTOs (Gaussians with complex exponents) that have an intrinsic oscillating behavior making them, a priori, more suitable to represent continuum functions.

In this contribution, we wish to explore the representation of spherical Bessel, Coulomb and generalized Sturmian functions [4] by a set of c-GTOs, and compare the efficiency with respect to using r-GTOs. As an example, here, we consider a set of 9 sample Coulomb functions $F_l(Z, k, r)$ for $l = 0, 1, ..., 8$, with energy $\frac{k^2}{2} = 2$ a.u. and a charge $Z = 1$. We have fitted them with either N r-GTOs or N c-GTOs

$$F_l(Z, k, r) = \sum_{i=1}^{N} c_{i,l} \exp(-\alpha_i r^2) \qquad (1)$$

by applying the approach of Nestmann and Peyerimhoff [2] in a range $r \in [0, 12]$ a.u. and using a quadratic approximation method [5]. Figure 1 shows the fitting of the $l = 0$ Coulomb function taking $N = 12$ in both expansions. It is easy to clearly appreciate the superiority of c-GTOs.

We are now in the process of applying c-GTOs representations of continuum states in scattering problems such as photoionization of small molecules.

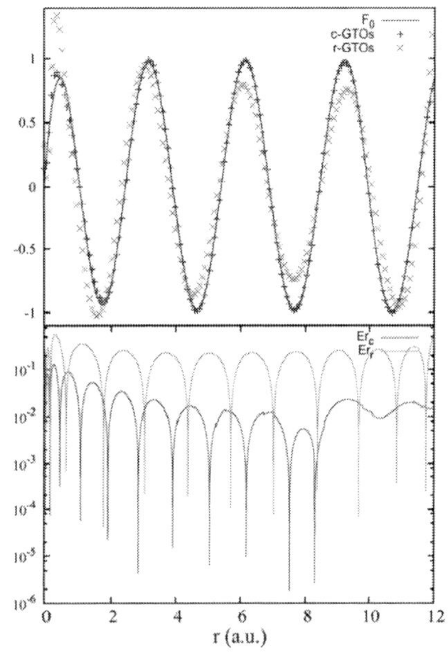

Figure 1. Upper panel: the coulomb function for $l = 0$ (blue), with its fitting by 12 c-GTOs (red dots) and by 12 r-GTOs (green dots). Bottom panel: the respective absolute error.

References

[1] O-ohata K *et al* 1966 *J. Phys. Soc. Jpn.* **21** 2306
[2] Nestmann B *et al* 1999 *J. Phys. B: At. Mol. Opt. Phys.* **23** L773
[3] Faure A *et al* 1999 *Comput. Phys. Comun.* **144** 224
[4] Gasaneo G *et al* 2013 *Adv. Quantum Chem.* **67** 153
[5] Powell M G D 2009 *Univ. Cambridge* **DAMTP** NA06

*E-mail: abdallah.ammar@univ-lorraine.fr
†E-mail: ugo.ancarani@univ-lorraine.fr

Experimental investigation of dissociative ionization of NeAr dimer by electron impact

S Yan[1*], X Ma[1†], X L Zhu[1], S F Zhang[1] and D M Zhao[1]

[1] Institute of Modern Physics, Chinese Academy of Sciences, Lanzhou 730000, China

Synopsis A two body fragmentation experiment of NeAr dimer by 480 eV electron impact was performed. By reconstructing the momenta of Ne^+ ion, Ar^+ ion and one electron, the 2D coincidence map between Kinetic energy release and electron energy is obtained, from which two islands are observed. It is found that, Interatomic Coulombic Decays induced by Ne^+Ar ion and Ar^+Ne ion can both take place in our experiment, as well as Radiative Charge Transfer. This is different from X ray in which only one decay process is dominant in one incident energy.

The atoms (molecules) in the cluster are related to each other through weak interactions, such as Van der Waals bond, hydrogen bond. It can be considered that each atom (molecules) exists in the environment formed by other atoms (molecules). This is similar with the weak bonding system-biological tissue. So we can use the fragmentation experiment of cluster to simulate the radiation damage process [1-5].

In this work, the NeAr fragmentation experiment is performed in the Reaction microscope in Institute of modern physics, Chinese Academy of Sciences. We use 480 eV electron as the projectile, and detect the Ne^+/Ar^+ ion pair and one emitted electron in coincidence. By reconstructing the 3D momenta of each fragment, the coincident map between kinetic energy release (KER) and electron energy are obtained.

As shown in figure 1, two significant peaks are observed, one (peak A) lies at the low energy area of electron, and the other one (peak B) locates at the high energy area (about 7 eV). It can be concluded that peak B arise from the Interatomic Coulombic Decay (ICD) of Ne^+Ar ion, and parts of peak A arise from Ar^+Ne ion. The corresponding positions are similar with that of X ray experiment [5], but the relative contribution is significantly different. In the X ray experiment, the amplitude of peak B is at least four times larger than that of peak A.

In order to explain this difference, we focus on the low energy electron events, and found that the characteristic of corresponding KER distribution is same with Radiative Charge Transfer (RCT). This means, part events of peak B come from RCT process. The origin is that the continuous energy deposition of the electron experiment will populate many ionization states; but in X ray experiment, only several states will be populated predominantly through resonance absorption.

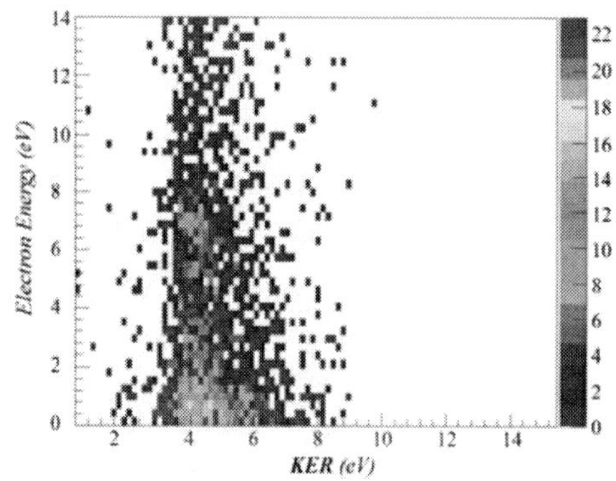

Figure 1. Correlation map between electron energy and KER.

The work was partly supported by the National Key R&D Program of China under Grant No. 2017YFA0402300 and the National Nature Science Foundation of China under Grant No. U1532129.

References

[1] Cederbaum L S *et al* 1997 *Phys. Rev. Lett.* **79** 4778
[2] Jahnke T *et al* 2004 *Phys. Rev. Lett.* **93** 163401
[3] Ren X *et al* 2018 *Nat. Comm.* **6** 139
[4] Yan S *et al* 2018 *Phys. Rev. A* **97** 010701(R)
[5] Ouchi T *et al* 2017 *Chem. Phys.* **482** 17

* E-mail: yanshuncheng@impcas.ac.cn
† E-mail: x.ma@impcas.ac.cn

Photoemission delays in 2D model molecules

B Wurmser[1] *, M Labeye[1], R Taïeb[1] and J Caillat[1]

[1]Laboratoire de Chimie Physique - Matière et Rayonnement, Sorbonne Université, Paris, France

Synopsis Photoelectron spectroscopy is one of the main techniques used to probe matter since the understanding of the photoelectric effect. Within the last decade, a new aspect has been given to spectroscopy, revisiting it in the time domain. It allows one to resolve the electron dynamics at the attosecond (10^{-18} s) time scale. Achieving such resolution experimentally brought one to establish innovative set up, and theoretical background is needed to validate, predict and interpret measurements. Much progress has been made in the attoscience field, this work is a step forward in understanding the electron dynamics for molecular continuum states at the attosecond timescale.

The recent progress in ultra-short laser pulses generation allows one to resolve temporally phenomena that was thought to occur instantly: photoionization processes. Photoemission delays can be observed through pump-probe experiments (RABBIT, streaking) [1–4]. In diluted systems, these delays are expressing the detailed influence of the parent ion on the electron. Therefore, they are highly depending on the species, but also on the channel the electron is ionized through for given species. They will also rely on the chirality for a molecule, the latter being involved in manifold natural phenomena [5].

In quantum physics, the dynamics of an electron is encoded in the phase of its wave function, and the dynamics are expressed in terms of group delay [6]. The ionization continuum being degenerated, there are multiple ways to describe it. The Selected Continuum Wave Function (SCWF) model [7] embeds all the information needed to describe the dynamics of photoemission, and provides an unambiguous and straightforward interpretation of the delays that have been recently discussed in the framework of attosecond science.

For a single-photon transition from a bound state $|\psi_{ini}\rangle$ with energy E_{ini}, the SCWF is defined in the framework of the first-order perturbation theory at a given energy E as $|\phi_{E,\mathrm{SCWF}}\rangle = \sum_\nu \langle\phi_{E,\nu}|\hat{\mathrm{d}}|\psi_{ini}\rangle|\phi_{E,\nu}\rangle$ where $\hat{\mathrm{d}}$ is the dipole operator coupling the initial and final states, and $\{|\phi_{E,\nu}\rangle\}$ an arbitrary orthonormal eigenbasis degenerated for the energy E. The SCWF model has already been verified with a time-dependent wave packet propagation in one-dimensional systems in [7]. Here, the two-dimensional approach allows to retrieve the angular dependent delays. Time-independent computations have been run to investigate ionization in an H_2^+ model molecule for a single-photon transition (Figure 1). The photoemission delays obtained from the time-independent approach are to be compared with a time-dependent propagation of a state at the same energy in order to validate the formalism of the SCWF. This work is the first step towards the extension to more realistic molecules.

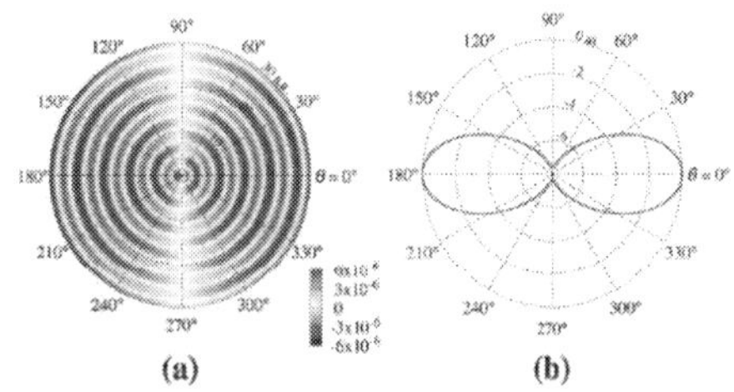

Figure 1. **(a)** The Selected Continuum Wave Function of a 2D model H_2^+ electron ionized in a single-photon transition at $\omega = 1$ a.u. and polarized linearly along the molecular axis. **(b)** Angular photoemission delays relative to the delay computed along the molecular axis ($\theta = 0°$) for the same transition.

References

[1] Cavalieri A L *et al* 2007 *Nature* **449** 1029-1032
[2] Haessler S *et al* 2009 *Phys. Rev. A* **80** 011404
[3] Schultz M *et al* 2010 *Science* **328** 1658-1662
[4] Klünder K *et al* 2011 *Phys. Rev. Lett.* **106** 143002
[5] Beaulieu S *et al* 2017 *Science* **358** 1288-1294
[6] Wigner E P 1955 *Phys. Rev.* **95** 145-147
[7] Gaillac R *et al* 2016 *Phys. Rev. A* **93** 013410

*E-mail: basile.wurmser@sorbonne-universite.fr

Photodissociation dynamics of the diamondoid adamantane induced by attosecond XUV pulses

S Maclot[1,2]*, J Peschel[2], J Lahl[2], H Wikmark[2], P Rudwaski[2], H Coudert-Alteirac[2], P Rousseau[3], S Indrajith[3], B A Huber[3], S Díaz-Tendero[4] and P Johnsson[2] †

[1]Biomedical and X-Ray Physics, Department of Applied Physics, AlbaNova University Center, KTH Royal Institute of Technology, SE-10691 Stockholm, Sweden
[2]Department of Physics, Lund University, P.O. Box 118, 22100 Lund, Sweden
[3]Normandie Université, ENSICAEN, UNICAEN, CEA, CNRS, CIMAP, 14000 Caen, France
[4]Departamento de Química, Módulo 13, Universidad Autónoma de Madrid, 28049 Madrid, Spain

Synopsis Adamantane is the simplest of the diamondoid molecules, which due to their high stability are of high interest both in astrophysics and nanotechnology. This work investigates the molecular photodissociation after ionization by attosecond XUV pulses. The fragmentation dynamics is inferred by means of velocity map imaging spectrometry, covariance analysis techniques and quantum chemistry calculations.

Carbon is, by mass, the fourth most abundant element in the universe. In combination with its ability to form complex organic compounds, this makes it a major player for the evolution of the universe as well as all life. Diamondoids are a class of carbon nanomaterials based on carbon cages with well-defined structures formed by C(sp^3)-C(sp^3)-hybridised bonds and fully terminated by hydrogen atoms. All diamondoids are variants of the adamantane molecule, the most stable among all the isomers with formula $C_{10}H_{16}$. Today, diamondoids are attracting increasing interest for use as an applied nanomaterial [1]. In space, diamondoids have been found to be the most abundant component of presolar grains [2], and due to their high stability, they are expected to be abundant in the interstellar medium [3].

A recent experiment has been performed at Lund University using attosecond XUV pulses produced via high-order harmonic generation (Fig. 1a) to study the photodissociation of adamantane. The photoions were detected by a velocity map imaging spectrometer [4] recording the mass spectrum and the ion momentum distribution on a single shot basis, allowing for the use of covariance analysis [5] to disentangle the dissociation dynamics. The mass spectrum (Fig. 1b) shows the production of a wide distribution of $C_xH_y^+$ fragments resulting from dissociation of singly and doubly charged adamantane molecules. The experimental data is currently being evaluated in detail and compared to the results of potential energy surface calculations and molecular dynamics simulations performed by the theory group in Madrid. These show that, in the dication, the available photodissociation channels have a common first step of cage-opening and hydrogen migration occurring within 500 fs via a barrier-less transition (Fig. 1a). In Figure 1c,d and e, one of the possible fragmentation channels following the cage-opening, $C_{10}H_{16}^{2+} \rightarrow C_2H_5^+ + C_8H_{11}^+$, is highlighted as an example.

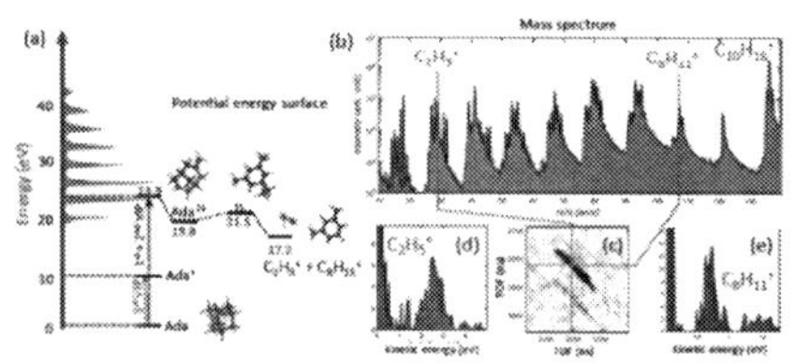

Figure 1. Example of experimental and theoretical results.

References

[1] Stauss S 2017 *Pan Stanford* "Diamondoids: Synthesis, Properties, and Applications"
[2] Anders E *et al* 1993 *Meteoritics* **28** 490
[3] Henning T *et al* 1998 *Science* **282** 2204
[4] Rading L *et al* 2018 *Appl. Sci.* **8** 998
[5] Frasinski L J 2016 *J. Phys. B: At. Mol. Opt. Phys.* **49** 152004

*E-mail: maclot@kth.se
†E-mail: per.johnsson@fysik.lth.se

Target sensitivity of photoelectron holograms

S Borbély[1]*, A Tóth[2], G Zs Kiss[1,3] and L Nagy[1]

[1]Faculty of Physics, Babeş-Bolyai University, Kogălniceanu Street 1, 400084 Cluj-Napoca, Romania
[2]ELI-ALPS, ELI-HU Non-Profit Ltd, H-6720 Szeged, Dugonics tér 13, Hungary
[3]Department of Molecular and Biomolecular Physics, National Institute for Research and Development of Isotopic and Molecular Technologies, Donath Street 61-103, 400293, Cluj-Napoca, Romania

Synopsis Photoelectron holograms appear as a modulation in the momentum distribution of the photoelectrons created during the ionization of atomic and molecular targets by few-cycle laser pulses, and it is the direct result of the continuum electonic wave packet scattering on the target. Using first principle calculations we show that the pattern of the formed photoelectron hologram is strongly influenced by the shape of the target's binding potential.

Previous experimental and theoretical investigations [1, 2, 3, 4, 5] showed that the photoelectron hologram observed in the momentum distribution of the continuum electrons created during the ionization of atomic targets by ultrashort laser pulses is the result of wave packet interference. The interference occurs between continuum electronic wave packet created during the same half cycle of the driving field, but which follow different spatial paths under the combined action of the laser field and the binding potential. In this interference scenario, also called spatial interence [2, 3], two characteristic wave packet trajectories exist [1, 3]: along the first trajectory the returning electron does not approach its parent ion too closely, hence it is only weakly scattered, while the second trajectory passes nearer, leading to a strongly scattered photoelectron. Considering the wave packet following the first path as a reference wave, while the wave packet following the second path as a signal wave, their interference can be interpreted as the hologram of the target. The shape of the resulting hologram is determined by the phase difference accumulated by the electron along the two distinct paths. This is directly influenced by the length of those paths (mainly defined by the ionization energy and the laser pulse parameters), and by the depth of the encountered binding potential [6].

In order to investigate the influence of the target's binding potential on the hologram, we have performed first principle calculations for different targets (H, He, Ar, Ne, and H_2^+), where by keeping the same direct and scattered paths, the differences in the holograms obtained for different targets can be attributed to the different binding potentials.

*E-mail: sandor.borbely@phys.ubbcluj.ro

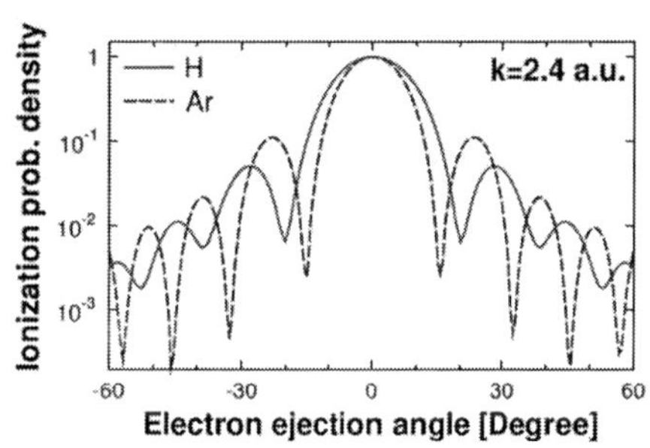

Figure 1. The angular distribution of photoelectrons emitted from the H and Ar targets with the constant momentum of $k = 2.4$ a.u. The parameters of the driving laser pulse are: $\omega = 0.4445$ a.u. (photon energy), $\tau = 28.17$ a.u. (pulse duration), $E_0 = 1$ a.u. (pulse strength).

In Fig. 1 the angular distribution of the photoelectrons with fixed momentum 2.4 a.u. is shown for the H and Ar targets. One can observe that the location of the deep interference minima is strongly influenced by the binding potential of the target.

References

[1] Y. Huismans *et al* 2011 *Science* **331** 61
[2] X.-B Bian *et al* 2011 *Phys. Rev. A* **84** 043420
[3] S. Borbély *et al* 2013 *Phys. Rev. A* **87** 013405
[4] H. Agueny *et al* 2018 *Phys. Rev. A* **98** 023414
[5] S. D. López *et al* 2019 *Eur. Phys. J. D* **73** 28
[6] S. Borbély *et al* 2019 *Phys. Rev. A* **99** 013413

 1412 (2020) 152084 doi:10.1088/1742-6596/1412/15/152084

Geometrically reconstructing laser Coulomb-exploded molecules

M D Singh[1]*, A Ramadhan[1], K Tian[1], J Sanderson[1], B Wales[1], R Karimi[1,3], S Beaulieu[2], W K Liu[1], J C Kieffer[2] and F Légaré[2]

[1]University of Waterloo, Waterloo, N2L 3G1, Canada
[2]Institut National de la Recherche Scientifique, Varennes, J3X 1S2, Canada
[3]Amirkabir University of Technology, Tehran, 15875-4413, Iran

Synopsis Geometrical reconstructions are performed for Carbonyl Sulfide (OCS) through Coulomb Explosion Imaging (CEI) employing intense femtosecond laser ionization with pulse durations 7 fs, 30 fs, 60 fs, 100 fs, and 200 fs, using two methods that significantly differ in approach. Firstly, a look up table is employed to reconstruct OCS (2, 2, 2) channel and highlight the presence of degenerate regions. Following this a more sophisticated constrained nonlinear optimization approach improves on precision and gives an indication of uncertainty. Results are compared with equilibrium geometry.

The focus of an ultrafast laser pulse can carry with it an immensely strong electric field comparable to that of the electronic bonds of a molecule. These focused pulses can highly ionize a molecule inducing a Coulomb explosion through which atomic fragment ions receive substantial momentum. These momentum vectors measured in coincidence carry information about the molecule's initial geometry. Since no unique solutions exist for molecules bigger than diatomic, a process of simulation must take place before geometry information can be returned.

The lookup table is a simple approach whereby Coulomb explosions are simulated with varying initial structures and the resulting momentum vectors are stored. Initial geometries can then be mapped to momentum vectors. The (2,2,2) channel of Carbonyl Sulfide OCS $\rightarrow O^{2+} + C^{2+} + S^{2+}$ was produced via pulses 30 fs ($I = 3 \times 10^{15}$ W/cm^2) and the momentum vectors were extracted and searched for in the lookup table. Initial geometries were mapped and returned with particular attention to their uniqueness. The reconstruction provides a modal (most probable) and average geometry consisting of a C-O (rCO) and C-S (rCS) bond length, and bond angle. These results are compared to the equilibrium state of OCS. The same analysis is repeated for 7 fs, 60 fs, 100 fs, and 200 fs.

Molecular reconstruction can be treated as a constrained nonlinear optimization problem. The interior-point method is employed to find geometrical solutions that best match with the measurements within box constraints of 1.00 Å $\leq$ rCO, rCS $\leq$ 5.00 Å and bond angle between 140 deg and 180 deg. OCS (2, 2, 2) is again used for this reconstruction attempt. Figure 1 shows the returned geometry for 30 fs pulse duration and compared to the equilibrium geometry. Of the total 1126 optimal geometries found, 1072 corresponded to a single geometry, 18 mapped to 2 distinct degenerate geometries, and 6 mapped to 3 distinct degenerate geometries. The results show that bending increases with pulse duration and stretching of the bonds, particularly the CO bond, has occurred [1]. This result will be compared with the few cycle (7fs) and longer pulses.

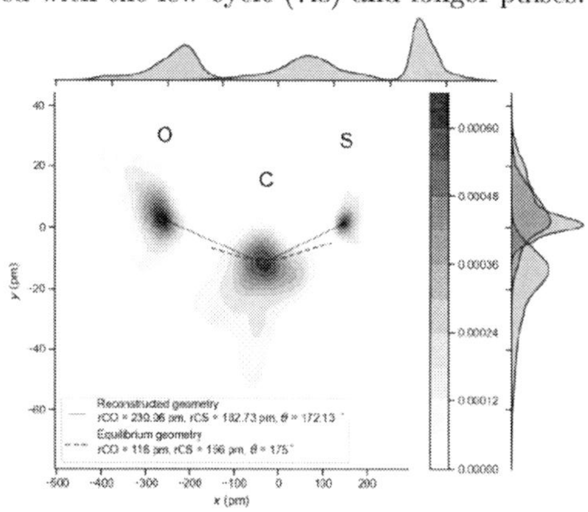

Figure 1. Probability density of each atomic fragment's position. The equilibrium geometry is shown for comparison.

*E-mail: Michael93Singh@gmail.com

References

[1] Bocharova I *et al* 2011 *Phys. Rev. Lett.* **107** 063201

Determination of energy-transfer distributions in ionizing ion-molecule collisions

S Indrajith[1] *, E Erdmann[2], J Chiarinelli[3,4], A Domaracka[1], M Labuda[2], L Avaldi[3], M Alcami[5], N F Aguirre[5,6], S Díaz-Tendero[5], P Bolognesi[3], P Rousseau[1] and B A Huber[1]

[1]Normandie Université, ENSICAEN, UNICAEN, CEA, CNRS, CIMAP, Caen, 14000, France
[2]Gdańsk University of Technology, ul. Narutowicza 11/12, Gdańsk, 80-233, Poland
[3]CNR—Instituto di Struttura della Materia, Area della Ricerca di Roma 1, Monterotondo Scalo, Italy
[4]Dipartimento di Scienze, Universit di Roma Tre, Roma, Italy
[5]Departamento de Química, Módulo 13, Universidad Autónoma de Madrid, Madrid, 28049, Spain
[6]Theoretical Division, Los Alamos National Laboratory, Los Alamos, New Mexico, 87545, USA

Synopsis The main objective of this study is to determine the energy transfer occuring in ion-molecule collisions. In order to solve this problem, we followed two approaches; the first one by validating a purely experimental method and the second one by testing a new theoretical model M3C (Microcanonical Metropolis Monte Carlo).

Nowadays, several studies regarding medical applications on radiation damage are being carried out, in particular, concerning hadrontherapy where ions are used as ionizing particles. In order to better understand the fundamental processes causing these damages, such as ionization and fragmentation, it is necessary to study the system at the molecular level, thus in the gas phase. The comparison between theoretical models and experimental results requires the knowledge of the energy-transfer distribution during ion-molecule collisions. The aim of this study is to determine this energy distribution. We have chosen to investigate the furan molecule, a very reactive organic species, composed of an aromatic ring with five atoms including one oxygen atom. This molecule can be considered as a model or analogue of the nucleotides sugar, a part of the DNA [1]. In order to determine the energy-transfer distribution during the collision between an ion and a furan molecule, we have applied two approaches. The first one is purely experimental; the distribution is obtained by fitting ion collision mass spectra with a combination of all PEPICO (Photon-Electron Photon-Ion COincidence) spectra after photon interaction [2]. The second method relies on a new theoretical model: M_3C (Microcanonical Metropolis Monte Carlo) which provides fragmentation probabilities as a function of the excitation energy of the molecule [3]. Once again, the energy distribution is obtained by fitting ion collision mass spectra with the branching ratios obtained by this model. Figure 1 shows good agreement between the energy distribution of furan obtained by collisions of Ar^+ and Xe^{25+} ions compared with PEPICO data and M_3C calculations. The result is considered as an accurate prediction of the excitation energy distribution of complex molecular ions produced in collision with fast ions. It should be noted that this method does not require the knowledge of the initial and final states of the projectile nor the ionization potential of the target.

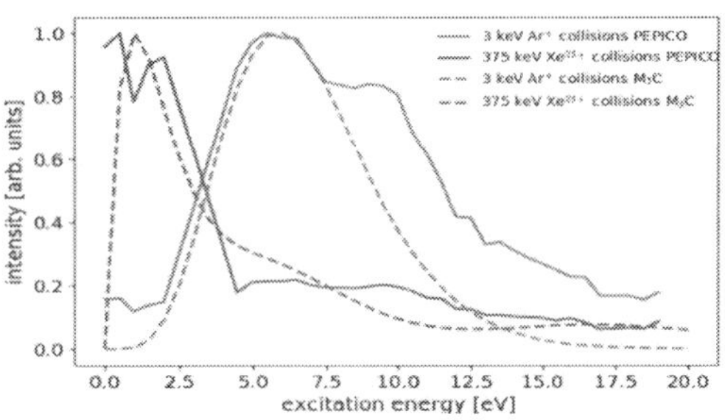

Figure 1. Distributions of the energy-transfer in collisions of Ar^+ and Xe^{25+} with the furan molecule obtained with the two approaches proposed here.

References

[1] Hervé du Penhoat M A *et al* 2015 *Phys. Chem. Chem. Phys.* **17** 32375
[2] Maclot S *et al* 2016 *Phys. Rev. Lett.* **117** 073201
[3] Aguirre N F *et al* 2017 *J. Chem. Theory Comput.* **13** 992

*E-mail: suvasthika.indrajith@ganil.fr

Triple Differential Cross Sections for electron impact ionization of methane at intermediate energy

C Granados[1*], E Ali[2], A Sakaamini[3], M Harvey[3], L U Ancarani[4], A J Murray[3] M Dogan[5], C Ning[6], J Colgan[7] and D Madison[2†]

[1]Institut für Physik, Martin-Luther Universität Halle-Wittenberg, 06099 Halle(Saale), Germany
[2]Department of Physics, Missouri University of Science and Technology, Rolla, 65401, USA
[3]Photon Science Institute, University of Manchester, Manchester M13 9PL, UK
[4]Université de Lorraine, CNRS, LPCT, F-57000 Metz, France
[5]ecol Academy, Afyon Kocatepe Konutları, Afyon, CK1, C3-030916, Turkey
[6]Department of Physics, Tsinghua University, Beijing, China
[7]Theoretical Division, Los Alamos National Laboratory, Los Alamos, 87545, USA

Synopsis In a joint experimental and theoretical collaboration, triple differential cross sections are presented for electron-impact single ionization of the *1t₂* and *2a₁* states of CH_4 at an incident electron energy of 250 eV. Experimental data from two different laboratories are compared with molecular 3-body distorted wave (M3DW) and generalized Sturmian function (GSF) calculations. Overall, we find a good experiment-experiment, and also a relatively good experiment-theories, agreement. We observe that for the $1t_2$ state, similarly to ionization of atomic p-states, the peak is sometimes doubled.

In this work we compare measurements taken at Manchester, UK and Afyon, Turkey with theoretical M3DW and GSF calculations of triple differential cross sections (TDCS) for 250 eV electron-impact single ionization of the HOMO $1t_2$ and NHOMO $2a_1$ states of CH_4.

Measurements were made for coplanar asymmetric scattering, for several different projectile scattering angles and for ejected electron energies of either 30 eV or 50 eV.

For 50 eV data were collected for five different projectile scattering angles. An example, at 25°, is shown in Fig. 1. For each theory, results calculated at the centroid angle and convoluted over the experimental uncertainty are presented; for this case, this operation clearly makes very little difference. As can be seen, the agreement between the two measurements is very good, and the overall agreement between the two theoretical calculations and the experiment data is also very good.

For 30 eV ejected electrons, the agreement between experiment and theory is reasonable, but not as satisfactory as found for the 50 eV case. Overall the M3DW results are in a little better shape agreement with experiment than the GSF calculations.

The full set of measurements for both the HOMO and NHOMO states will be presented at the conference, together with details of the theoretical models and of the experiments. For ionization of the $2a_1$ state a single maximum is observed in the binary region while for the $1t_2$ state, similarly to ionization of atomic p-states, this peak is sometimes doubled with a minimum close to the momentum transfer direction. The evolution of the single or double binary peak with varying scattering angle will be explored in detail using both M3DW and GSF calculations, and conclusions drawn as to the origins of rapidly changing features. More details are given in Ref. [1] and in references therein.

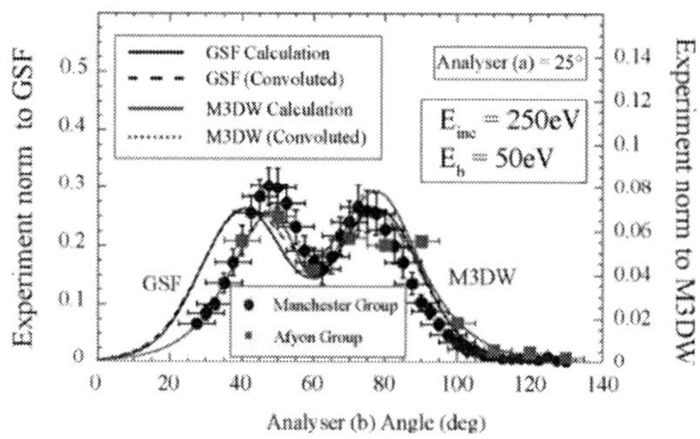

Figure 1. Cross section data from the Manchester (black circles) and Afyon (red squares) groups, compared to the GSF and M3DW theories for ionization from the $1t_2$ orbital of CH_4. The M3DW and GSF calculations yield TDCS with a factor of ~4 difference in magnitude, so that two vertical scales are used to represent them. The experimental data are normalized to theory.

References

[1] Ali E *et al* 2019 *J. Chem. Phys.* Accepted for publication

* E-mail: carlos.granados@physik.uni-halle.de
† E-mail: madison@mst.edu

Absolute cross section for electron emission from uracil upon ion collision using velocity map imaging technique

N Sens[1], M Ryszka[1], A Méry[1], J-C Poully[1], J-Y Chesnel[1] and V Vizcaino[1 *]

[1]Centre de Recherche sur les Ions, les Matériaux et la Photonique (CIMAP), Normandie Univ, ENSICAEN, UNI-CAEN, CEA, CNRS, 14000 Caen, France

Synopsis Ion-induced electron emission from RNA base uracil has been measured using a newly developed experimental set-up based on velocity map imaging (VMI) technique. Electrons emitted in a solid angle of 4π steradians are extracted and analyzed in energy and angle by the multi-electrodes VMI spectrometer. By monitoring the target density, the projectile ion beam intensity and the beams overlap with the use of a quartz crystal micro balance and an ion beam profiler, we were able to estimate absolute singly- and doubly- differential cross sections for electron emission.

In recent years, ion collision with biologically relevant molecules has received increasing interest due to applications in radiation biology. Absolute total or differential cross sections for electron emission have been measured for small biomolecules such as nucleobases [1-5] or amino acids [6]. Experiments were mainly performed with bare ions such as protons [1,3,4,7], C^{6+} or O^{8+} [2] with a kinetic energy of a few MeV. Absolute cross sections were measured using electrostatic analyzers (parallel plate or hemispherical) mounted on a turntable. Measured electron yields were usually put on an absolute scale by normalization to other collisional processes (for which cross sections are known) such as Rutherford forward scattering [1,4] or projectile elastic scattering at right angle [3].

We have recently developed a new experimental set-up to measure absolute singly or doubly differential cross section for electron emission from atoms or molecules using a velocity map imaging (VMI) spectrometer. This technique allows detecting all electrons emitted in a solid angle of 4π steradians at once, while with hemispherical or parallel plate electrostatic analyzer typical accessible angular ranges are between 20° and 160° due to mechanical constraints. Moreover, with the VMI spectrometer, very low energy electrons (below 1eV) can be detected. To the best of our knowledge, this technique has never been employed to measure ion-induced electron emission and had to be adapted to allow the projectile ion beam to enter the interaction region despite a collection electric field of about 100V/cm.

In our set-up, the target beam crosses orthogonally the projectile ion beam with typical kinetic energies from a few keV to a few MeV. Emitted electrons with energies up to 200 eV are extracted by the multi-electrodes VMI spectrometer and focused onto a position sensitive detector (micro-channel plates coupled to a phosphor screen). After reconstruction of the 2D image recorded by the camera, one can extract the electron yield as a function of energy or emission angle or both. The target density is measured with the use of a quartz crystal micro balance. A density of about 2.10^9 molecules/cm^3 was found for the uracil beam produced by the effusion cell.

In this contribution, we will present the experimental set-up and our methodology to determine absolute cross section. Moreover recent measurements of absolute singly- and doubly- differential cross sections for electron emission from uracil will be discussed and compared to previous data.

The authors acknowledge the support of the Agence Nationale de la Recherche (ANR), under grant ANR-16-CE30-0006. N Sens acknowledges the Normandy Regional Council for his PhD grant.

References

[1] Itoh A *et al* 2013. *Phys. Rev. A* **88** 052711
[2] Tribedi LC *et al* 2012. *Eur. Phys. J. D* **66**, 1
[3] Moretto-Capelle P *et al* 2006 *Phys. Rev. A* **74** 062705
[4] Iriki Y *et al* 2011. *Phys. Rev. A* **84** 052719
[5] Champion C *et al* 2013 *Nucl. Instrum. Methods Phys. Res. Sect. B Beam Interact. Mater. At.* **314** 66
[6] Sato, S. *et al.* 2007 *Nucl. Instrum. Methods Phys. Res. Sect. B-Beam Interact. Mater. At.* **256** 506
[7] Champion C *et al.* 2014 *J. Phys. Conf. Ser.* **488** 012038

* E-mail: violaine.vizcaino@ganil.fr

A new experimental setup to measure absolute cross section for electron emission from atoms, molecules or nanoparticles upon ion collisions

N Sens[1], M Ryszka[1], A Mery[1], J-C Poully[1], J-Y Chesnel[1] and V Vizcaino[1 *]

[1]Centre de Recherche sur les Ions, les Matériaux et la Photonique (CIMAP), Normandie Univ, ENSICAEN, UNICAEN, CEA, CNRS, 14000 Caen, France

Synopsis A newly developed crossed beam experimental setup that utilizes a velocity map imaging technique to simultaneously measure both emission energy and angle of electrons ejected from atoms, molecules or nanoparticles upon ion collision will be presented. The incident ion beam is produced at the GANIL facility with incident energies from a few keV to a few MeV and the target beam density is monitored by a quartz crystal microbalance allowing for estimating absolute single or double differential cross sections.

Electron emission following ion collisions with atoms or molecules has been studied for several decades to better understand fundamental ionization mechanisms (soft collisions, binary encounter…) [1] but also for its wide range of applications including radiation cancer therapy. In particular, electron emission might explain the efficiency of nanoparticles (NP) when used as radiosensitizers, to enhance tumor destruction [2]. However, the abundance and energy of these electrons have never been measured. In this project, we aim to quantify electron emission from various targets (atoms, biomolecules and NP) following ion beam irradiation.

To do so, the target beam crosses orthogonally the projectile ion beam produced by the GANIL beamlines with typical kinetic energies from a few qkeV to a few MeV. Emitted electrons with energies up to 200 eV are extracted and analyzed by a multi-electrode velocity-map-imaging (VMI) spectrometer. By monitoring the target density, the projectile ion beam intensity and the beam overlap with the use of a quartz crystal microbalance and an ion beam profiler, we are able to estimate absolute double (vs. energy and angle) or single (vs. energy) differential cross sections for electron emission.

The setup allows us to produce a wide range of target beams. A supersonic jet source allows for measuring electron emission from noble gas atoms, e.g. the well-defined electrons from autoionizing states of Ar^+ emitted in the range from 1 eV to 20 eV are used for calibration purposes. With the use of an effusive cell, we were also able to produce in the gas phase a steady effusive beam of biomolecules (e.g. uracil) as well as metallic atoms (e.g. silver). Moreover, we are developing a nebulizer – aerodynamic lens system dedicated to experiments on metallic nanoparticles of various sizes (from nm to few tenths of nm).

In this contribution, we will present the development and characterization of the set-ip in particular the design of the new NP source.

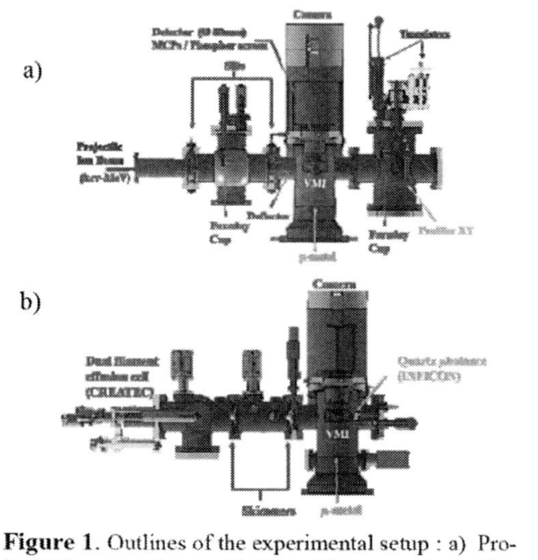

Figure 1. Outlines of the experimental setup : a) Projectile beam section, and b) Target beam section with an effusive cell installed.

The authors acknowledge the support of both the ANR (grant ANR-16-CE30-0006) and the Normandy Regional Council.

References

[1] Stolterfohlt N, DuBois RD, Rivarola RD, 1997 *Springer-Verlag, Berlin Heidelberg*

[2] Porcel E *et al.* 2014 *Nanomedicine Nanotechnol. Biol. Med.* **10** 1601–1608

* E-mail: violaine.vizcaino@ganil.fr

New structures in fully differential cross sections for ionization of He by proton impact

M Dhital[1], S Bastola[1], A Silvus[1], A Hasan[2], B R Lamichhane[1], E Ali[1], M F Ciappina[3], R A Lomsadze[4], D H Madison[1] and M Schulz[1*]

[1]Dept. of Physics and LAMOR, Missouri University of Science & Technology, Rolla, MO 65409, USA
[2]Dept. of Physics, UAE University, P.O. Box 15551, Al Ain, Abu Dhabi, UAE
[3]Institute of Physics of the ASCR, ELI-Beamlines, Na Slovance 2, 182 21 Prague, Czech Republic
[4]Tbilisi State University, Tbilisi 0179, Georgia

Synopsis Fully differential cross sections (FDCS) for ionization in p + He collisions were measured and calculated for several projectile energy losses. The electron angular dependence of the FDCS reveals structures not seen in previous studies.

We have measured fully momentum-analyzed scattered projectiles and recoil ions, produced in ionization of He by 75 keV proton impact, in coincidence. The ejected electron momentum was deduced from momentum conservation. From the data we extracted fully differential cross sections for various fixed projectile energy losses in the range ε = 62.5 to 85 eV, for various fixed proton scattering angles θ_p, and for electrons ejected into the scattering plane as a function of the polar emission angle θ_e. The data were compared to two conceptually very similar distorted wave calculations.

The FDCS for ε = 62.5 eV are shown in Fig. 1 for proton scattering angles (from top to bottom) of 0.1, 0.2, 0.3 and 0.5 mrad as a function of θ_e. The data are dominated by a pronounced peak structure at $\theta_e = 0$. This forward peak is due to the post-collision interaction (PCI) between the scattered projectile and the electron lifted to the continuum in a preceding primary interaction. Since at ε = 62.5 eV the electron speed is close to the projectile speed, PCI effects nearly maximize at this energy loss. The binary peak, which usually dominates the FDCS for much smaller ε, is resolved from the forward peak only at the largest θ_p. For the smaller θ_p it is manifested as a "shoulder" on the large angle wing of the forward peak.

Apart from the forward and binary peak additional oscillating structures are observed especially at the smallest θ_p. Some oscillations, which have not been observed in previous experiments, are also present in the 3DW calculations. However, they occur at different θ_e.

A possible explanation will be discussed. This work was supported by NSF and by ADONIS.

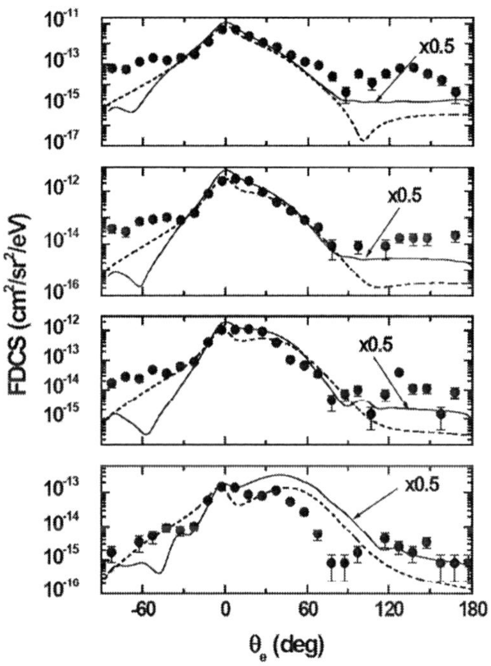

Figure 1. FDCS for ε = 62.5 eV and for electrons ejected into the scattering plane at θ_p (from top to bottom) = 0.1, 0.2, 0.3, and 0.55 mrad. Dashed curves, CDW-EIS; solid curves, 3DW.

* E-mail: schulz@mst.edu

Photoionization of water: influence of the initial state representations

L Martini[1,2]*, D I R Boll[1] and O A Fojón[1,2,3]

[1]Grupo de Colisiones Atómicas, Instituto de Física Rosario (CONICET-UNR), Rosario, 2000, Argentina
[2]Facultad de Ciencias Exactas, Ingeniería y Agrimensura (FCEIA-UNR), Rosario, 2000, Argentina
[3]Escuela de Ciencias Exactas y Naturales (FCEIA-UNR), Rosario, 2000, Argentina

Synopsis We study theoretically the photoionization of water molecules by monochromatic linearly polarized photons. We elaborated a model in which the final state wavefunctions are given by Coulomb continuum wavefunctions with distorted charges and the bound states are represented by using different wavefunctions. We contrast our calculations for photon energies between 20 and 300 eV with more elaborated theories and experimental results. We achieve a very good agreement with experiments, especially at enough high energies where elaborated results are not available.

We analyze the effects of using different representations of the initial bound states for several water molecule orbitals in the photoionization by high energy photon impact. We employ a model in which the final state wavefunction is given by a Coulomb continuum wavefunction with distorted charges Z_{ef} to approximately take into account the screening provoked by the passive electrons on the reaction. Then, the initial state wavefunctions are represented using linear combinations of Slater type orbitals (STO). On one hand, we use the Moccia's monocentric wavefunctions [1, 2]. This representation considers the electronic structure of the molecule bound states in a basis of STO centered on the oxygen atom. On the other hand, we include the multicenter geometry of the water molecule by describing the ground state with a linear combination of STO centered on each atom of the molecule. We use four different basis sets. The first one is given by the minimal basis compounds of the seven atomic occupied orbitals of the hydrogen and oxygen atoms. In the other ones, additional STO orbitals centered on the oxygen or the hydrogen atoms are included.

We get analytical expressions for the transition matrix elements that enable the obtention of cross sections by numerical quadratures with low computational cost. We calculate cross sections within the dipole approximation for several water orbitals in the energy range [20,300] eV (for instance, see figure 1). We compare our predictions with more elaborated theories and experimental results. We obtain a very good agreement with experimental data, in particular, above 70 eV where there is a lack of elaborated results.

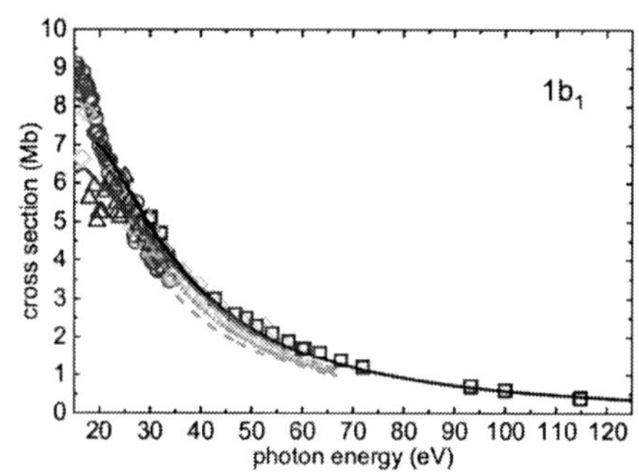

Figure 1. Total cross sections as a function of the photon energy for photoionization of water. Our results (▬▬) for the $1b_1$ orbital with Moccia's wavefunctions and $Z_{ef} = 1$. Calculations from [3] (▬▬) and from [4] for VWN GS (- - -), VWN TS (· · · ·) and LB94 GS (▬▬). Experiments: [5] (○), [6] ( ), [7] (△), [8] (□).

References

[1] Moccia R *et al* 1964 *J. Chem. Phys.* **40** 2186
[2] Martini L *et al* 2017 *Papers in Physics* **9** 090006
[3] Gozem S *et al* 2015 *J. Phys. Chem. Lett.* **6** 4532
[4] Stener M *et al* 2000 *J. Phys. B: At. Mol. Opt. Phys.* **33** 1081
[5] Brion C E *et al* 1985 *Chem. Phys.* **100** 291
[6] Tan K H *et al* 1978 *Chem. Phys.* **29** 299
[7] Truesdale C M *et al* 1982 *J. Chem. Phys.* **76** 860
[8] Banna M S *et al* 1986 *J. Chem. Phys.* **84** 4739

*E-mail: martini@ifir-conicet.gov.ar

Two-body dissociation of C_3H_4 isomers investigated by 50 keV/u Ne^{8+} impact

Y Li[1], S Xu[1*], D L Guo[2], S Jia[1], X Jiang[1], X L Zhu[2] and X Ma[2†]

[1]School of Science, Xi'an Jiaotong University, Xi'an 710049, China
[2]Institute of Modern Physics, Chinese Academy of Sciences, Lanzhou 730000, China

Synopsis Fragmentation dynamics of two isomers of C_3H_4, propyne (CH_3CCH) and allene (CH_2CCH_2), is investigated by 50 keV/u Ne^{8+} impact. Isomerization processes are identified in two-body fragmentation channels of $C_3H_4^{2+}$ for both isomers. Besides, we observe two new dissociation channels of $C_3H_4^{3+}$ trications, $H^+ + C_3H_3^{2+}$ and $H_2^+ + C_3H_2^{2+}$. For $CH_2CCH_2^{3+}$, the peak and shoulder structures in the kinetic energy release spectrum of $H^+ + C_3H_3^{2+}$ channel are attributed to different geometries of the $C_3H_3^{2+}$ products by comparing with theoretical calculation.

Fragmentation dynamics of hydrocarbon cations has drawn great interest due to its significance in many research fields, such as interstellar molecular evolution and radiation damage of biological tissues. Once a hydrocarbon molecule is ionized, the protons could response rapidly to the disruption of electronic structures and migrate fast along the carbon skeleton due to its small mass compared with C atom. In other words, isomerization occurs. C_3H_4 is the smallest hydrocarbon molecule that possesses stable isomers in the nature. Two isomers of it, propyne (CH_3CCH) and allene (CH_2CCH_2) with linear carbon skeleton, have been chosen as the model system to investigate the isomerization in the fragmentation processes [1, 2].

In the present work, we investigate the two-body dissociation of propyne and allene by using 50 keV/u Ne^{8+}. Isomerization processes are identified in the fragmentation of both isomers. The observation of channel $CH_2^+ + C_2H_2^+$ for CH_3CCH^{2+} as well as channels $CH^+ + C_2H_3^+$ and $CH_3^+ + C_2H^+$ for $CH_2CCH_2^{2+}$ exhibit the existence of indirect processes in which proton migration occurs prior to decomposition of the parent dication.

It is noteworthy that we observe two new two-body dissociation channels of $C_3H_4^{3+}$ trications, $H^+ + C_3H_3^{2+}$ and $H_2^+ + C_3H_2^{2+}$. While the former is observed in the fragmentation of both isomers, the latter is only observed in the fragmentation of CH_3CCH^{3+}. Figure 1 presents the kinetic energy release spectrum of $H^+ + C_3H_3^{2+}$ channel for $CH_2CCH_2^{3+}$. The peak and shoulder structures in this figure indicate that more than one pathways may contribute to this channel. By comparing with calculations in Reference [3], these structures are attributed to different geometries of the $C_3H_3^{2+}$ products.

More details of this study will be presented at the conference.

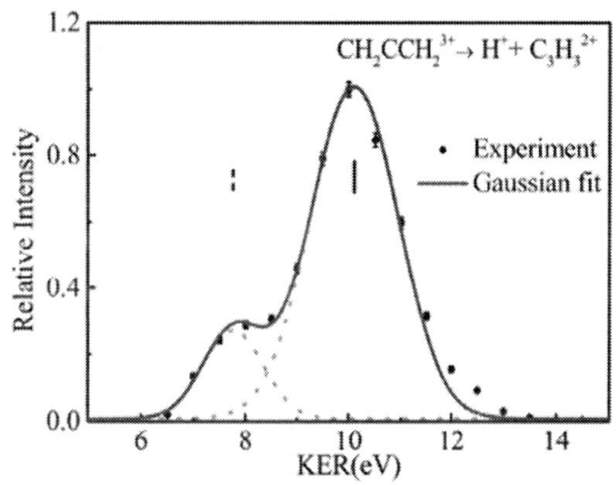

Figure 1. KER distribution of H^++$C_3H_3^{2+}$ dissociation channel of $CH_2CCH_2^{3+}$. The black dots with error bars present the experimental data, and the two green dotted curves are Gaussian fits of the peak and shoulder structures while the red solid line is the sum of these two fits. Blue dotted and solid vertical lines are centering values of the two Gaussian fits.

References
[1] Okino T *et al* 2012 *Phys. Chem. Chem. Phys.* **14** 4230
[2] Xu H *et al* 2009 *Chem. Phys. Lett.* **469** 255
[3] Mebel A *et al* 2008 *J. Chem. Phys.* **129** 224311

* E-mail: xushenyue@xjtu.edu.cn
† E-mail: x.ma@impcas.ac.cn

Development of pulse selectors for the synchrotron radiation pulses from the Photon Factory 2.5 GeV ring to study multiple photoionization

J Adachi[1] *, H Tanaka[1], T Kosuge[1], H Ishii[1], I H Suzuki[1], T Kaneyasu[2] †, T Taniguchi[3], T Odagiri[3], S Ohtaki[4], Y Tsuji[4], K Soejima[4], P Lablanquie[5] and Y Hikosaka[6]

[1] Photon Factory, Institute of Materials Structure Science, KEK, Tsukuba 305-0801, Japan
[2] SAGA Light Source, Tosu 841-0005, Japan
[3] Department of Materials and Life Sciences, Sophia University, Tokyo 102-8554, Japan
[4] Faculty of Science, Niigata University, Niigata 950-2182, Japan
[5] Sorbonne Universités, UPMC Univ Paris 06, CNRS, LCP-MR (UMR 7614), 11 rue Pierre et Marie Curie, 75231 Paris Cedex 05, France
[6] Institute of Liberal arts and Sciences, University of Toyama, Toyama 930-0194, Japan

Synopsis We have developed three pulse selectors to obtain soft x-ray pulses with repetition rates suitable for electron coincidence measurements. The devices enable us to pick out the light pulse from the solo bunch located in the center of the 364-ns dark gap of the hybrid-filling operation. We report the development status and show some typical results of the coincidence measurements with the pulse selectors.

To investigate resonant and normal Auger processes in atoms and molecules, the coincidence measurements with synchrotron radiation (SR) pulses are very useful. The coincidence measurements need adequately isolated SR pulses, when we utilize an advantage of a magnetic bottle electron spectrometer.

We have developed three pulse selectors (Table 1), which are mechanical choppers synchronous with the revolution of the SR ring, to pick out the soft x-ray pulses with the fixed repetition rates of 40 k-228 kHz in the center of the dark gap (~360 ns) of the hybrid-filling mode [1] at the Photon Factory (PF) 2.5 GeV ring. Each pulse selector comprises a bearing, a phase-locked-loop-controlled motor system, and a many-slit disk fixed at the rotation axis.

Table 1. Characteristics and specifications of the three pulse selectors.

	#1	#2	#3
Repetition	40 kHz	146 kHz	228 kHz
Slit height	60±10 μm	*60±10 μm	70±10 μm
Slit width	0.7 mm	1.5 mm	3 mm
Vacuum	3×10^{-3} Pa	2×10^{-3} Pa	4×10^{-6} Pa
Bearing	Air	Air	Magnetic
Geometry†	⊥	//	⊥

*An additional pre-slit is needed to restrict the SR spot size to be about 10 μm in operation. †It means that the axis of the rotating body is placed in perpendicular (⊥) or in parallel (//) to the SR propagation.

The pulse selectors were used at undulator beam lines, BL-2B, 13A, and 16A of the PF to perform multi-electron coincidence experiments [2,3] and to evaluate the performances (also at 28B and old 19B). Figure 1 shows one of the best results with the pulse selector #3. The bunch-train part slightly contributes and its integrated amount is estimated to be 1.5% of the solo part.

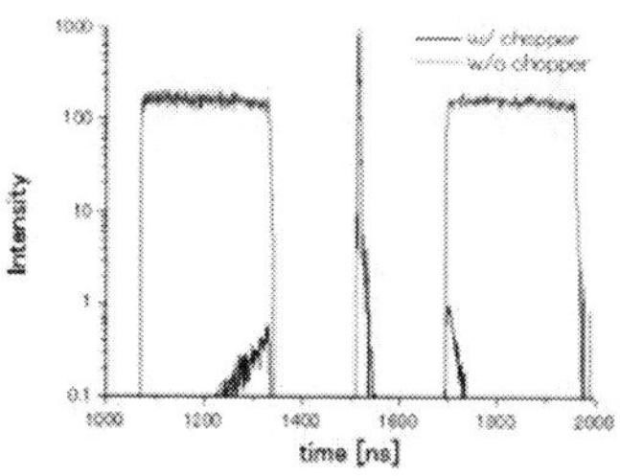

Figure 1. Intensities (semi-log scale) of the soft x-ray pulses at BL-13A of Photon Factory with (black line) or without (red line) the pulse selector #3.

We report the development status as well as some unsettled issues and show the examples for the results of the coincidence measurement.

References

[1] Takai R *et al*, 2010 *Proc. IPAC* **10** 2564
[2] Kaneyasu T *et al* 2017 *J. Chem. Phys.* **147** 104304.
[3] Hikosaka Y *et al* 2016 *Phys. Rev.* **A 93** 063412.

* E-mail: adachij@post.kek.jp
† E-mail: kaneyasu@saga-ls@jp

Electron emission from fluorene molecule in collisions with 3.5 MeV/u Si^{8+} ions

C Bagdia[1], M Roychowdhury[1], A Mandal[1], C Champion[2] and L C Tribedi[1]*

[1]Department of Nuclear and Atomic Physics, Tata Institute of Fundamental Research, Colaba, Mumbai, 400005, India
[2]Université Bordeaux 1, CNRS/IN2P3, CENBG, Chemin du Solarium, BP120, 33175 Gradignan, France

Synopsis We present measured absolute double differential cross section (DDCS) of electron emission from fluorene ($C_{13}H_{10}$) molecule upon 3.5 MeV/u Si^{8+} ion impact. We have compared the energy and angular distributions of the DDCS with CB1 (1st Born with correct boundary conditions) calculations. The e-DDCS shows an indication of collective excitation peak in low energy part of the spectrum at a few backward angles. The shape of the distribution has also been compared with that for a smaller molecule i.e. CH_4.

Polycyclic aromatic hydrocarbon (PAH) molecules are of recent interest because of their relevance in astrophysics and astrochemistry. These molecules are considered to be abundant in intersteller medium (ISM) [1]. According to a class of researchers, some of the unidentified mid-infrared bands in emission spectra from the ISM can be explained from various electronic excitations of these molecules [2,3].

The PAH molecules have a plannar geometry with a delocalized π-electron cloud around covalently bonded carbon cage providing extra stability. It has been predicted that these delocalized electrons can oscillate collectively about C cage upon external perturbation [4]. Recently, signatures of such collective excitation for coronene ($C_{24}H_{12}$) molecule, upon ion impact has been reported [5,6]. To explore the effect of the number of benzoidal rings and the number of delocalized π-electrons, ion impact ionization of fluorene ($C_{13}H_{10}$) molecule, another member of PAH family, has been investigated. Fluorene molecule has two benzoidal rings connected by a pentagon ring whereas coronene molecule has seven benzoidal rings.

We have measured the electron absolute double differential cross section (e-DDCS) in collisions of 3.5 MeV/u Si^{8+} ions with $C_{13}H_{10}$ molecule. A 14 MV Pelletron accelerator in TIFR, was used. The experiments have been performed using effusive vapor jet of fluorene molecules. A hemispherical electrostatic nalyzer was used. The measurements have been performed in the electron energy range of 1 to 600 eV and in the angular range of 30^0 to 160^0. The measured DDCS is compared with the CB1 calculations [7] for e- energy above 11 eV. The CB1 calculation underestimate the data by factor of 2 (see Fig. 1) and more incase of exreme forward and backward angles. The peak at 240 eV correspond to C-KLL auger electron emission. The indication of a peak in the low energy part of the spectrum in a backward angles could possibly be due to plasmon excitation. The detailed analysis and comparison with CH_4 data will be presented.

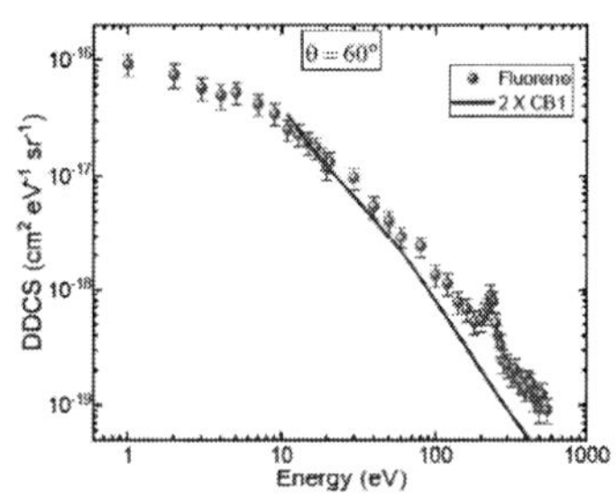

Figure 1. e-DDCS spectrum versus ejected electron energy. Line: CB1 scaled up by factor of 2.

References

[1] Tielens A G G M 2008 *Annu. Rev. Astron. Astrophys.* **46** 289
[2] Leger A *et al* 1985 *Astron. Astrophys.* **146** 81
[3] Xiang F Y *et al* 2011 *Astrophys. J.* **733** 91
[4] Jochims H W *et al* 1997 *Int. J. Mass Spectrom. Ion Process.* **167/168** 35
[5] Biswas S *et al* 2015 *Phys. Rev. A* **92** 060701
[6] Biswas S *et al* 2017 *Sci. Rep.* **7** 5560
[7] Champion C *et al* 2010 *Phys. Med. Biol.* **55** 6053

* E-mail: lokesh@tifr.res.in

 1412 (2020) 152094 doi:10.1088/1742-6596/1412/15/152094

Ionization and fragmentation of fluorene in collisions with 50-300 keV proton

C Bagdia[1] and L C Tribedi[1] *

[1]Department of Nuclear and Atomic Physics, Tata Institute of Fundamental Research, Mumbai, 400005, India

Synopsis In this report, we present the study of ionization and fragmentation of fluorene ($C_{13}H_{10}$) molecule upon intermediate energy proton impact. We have measured the TOF spectra of fluorene molecule to identify various fragmentation channels. We have also measured the projectile energy dependence of single and double ionization of fluorene molecule.

The ionization and fragmentation studies of polycyclic aromatic hydrocarbon (PAHs) molecules have attracted special attention recently, owing to their abundance in interstellar medium (ISM) [1,2]. These molecules have large number of delocalized π-electron cloud over honecomb like C-skeleton showing collective electron behavior [3-5]. Apart from their astrophysical relevance, these molecules are important from fundamental molecular physics and they also have applications in UV plasmonic devices. In ISM these molecules are exposed to various radiation such as photons, electrons and ions. Many of the photon impact studies involving PAH molecules are reported in literature whereas very few studies with ion impact are available.

In this work we present the study of ionization and fragmentation of fluorene ($C_{13}H_{10}$) molecule upon proton impact using the double field time-of-flight (TOF) spectrometer. The proton beam for the experiment has been obtained from 400kV TIFR-ECRIA, Mumbai [6]. The time of flight spectra of recoil ions are recorded for projectile in the energy range of 50 keV to 300 keV. The effusive jet of fluorene molecule has been obtained by heating the molecule around 50^{0}C.

The figure 1 shows the typical TOF spectrum of fluorene molecule upon 250 keV proton impact. The various fragmentation channels and products have been identified from the TOF spectrum. We have also measured the projectile energy dependence of single and double ionization of fluorene molecule. We have also compared the energy dependence of ratio of double ionization to single ionization with that for He and Ne. The ratio is substantially higher for fluorene as compared to He and Ne. In case of fluorene the ratio is about 48% at 50 keV which reduces gradually to 24% at 300 keV. In contrast, for the Ne-atom this ratio decreases from 20% to about 12% and for He atom its only about 6%-2%. The apparently large ratios in case of Fluorene can be qualitatively understood indicating the effect of strong e^--e^- correlation among the collectively excited plasmon in such PAH molecules.

The detailed results of various fragmentation channels and energy dependence of ionization will be presented.

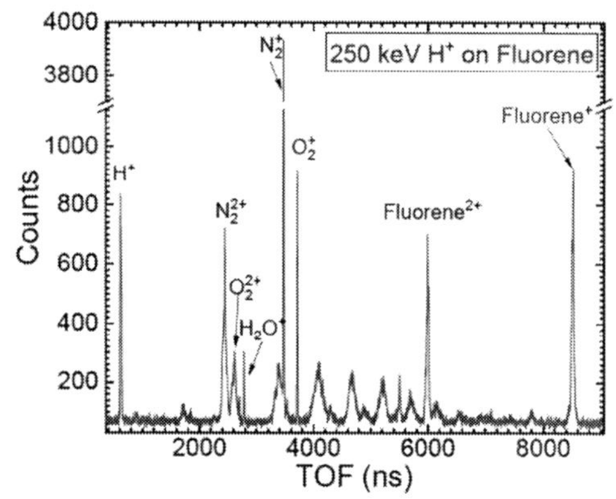

Figure 1. Time of flight spectrum of fluorene molecule upon 250 keV proton impact.

References

[1] Tielens A G G M 2008 *Annu. Rev. Astron. Astrophys.* **46** 289
[2] Lopez-Puertas M *et al* 2013 *Astrophys. J.* **770** 132
[3] Jochims H W *et al* 1997 *Int. J. Mass Spectrom. Ion Process.* **167/168** 35
[4] Biswas S *et al* 2015 *Phys. Rev. A* **92** 060701
[5] Biswas S *et al* 2017 *Sci. Rep.* **7** 5560
[6] Agnihotri A *et al* 2011 *Phys. Scr.* **144** 014038

* E-mail: lokesh@tifr.res.in

Photoionization microscopy: linear polarization rotation effects

P Kalaitzis[1] *, **S Danakas**[1], **F Lépine**[2], **C Bordas**[2] **and S Cohen**[1] †

[1]Atomic and Molecular Physics Laboratory, Physics Department, University of Ioannina, 45110 Ioannina, Greece
[2]Institut Lumière Matière, Université Lyon 1, CNRS, UMR 5306, 10 rue Ada Byron 69622 Villeurbanne Cedex, France

Synopsis Photoionization microscopy (PM) aims at imaging the flux of slow (meV) electrons released by atomic photoionization in the presence of a static electric field. We discuss the rotation of the linear light polarization with respect to the electric field direction. Two-photon ionization out of an m=0 initial state allows for the population of $|m|$=0,1,2 Stark states. The resulting $|m|$-beating effects in the PM images are demonstrated theoretically (in hydrogen atom) and experimentally (in Mg atom). Implications on the applicability of tomographic techniques for slow electrons are also discussed.

The purpose of photoionization microscopy (PM) is the measurement of the electron current probability density j in the case of photoionization of an atom in the presence of a static electric field [1]. The low (meV) energy of the outgoing electrons and the high spatial resolution of PM allows for the observation of the radial quantum interference patterns exhibited by j. For simplifying the analysis of these patterns PM experiments were so far restricted to the excitation of final Stark states of a single magnetic quantum number $|m|$. Here we study the simultaneous excitation of multiple-$|m|$ final Stark states. We examine a two-photon excitation scheme out of an m=0 ground state. The employed geometry is shown in Fig. 1(a). The field **F** is directed towards the positive z-axis, the laser beam propagates along x-axis and the linear laser polarization vector **ε** lies in the yz plane. The angle between **ε** and **F** is Θ. The 2D position sensitive detector is placed at the macroscopic distance z_{det}<0.

Figure 1(b) shows theoretical images (computed along the lines described in [2]) for several values of Θ and fixed strength F and electron excitation energy E (with respect to the zero-field limit, E=0). For Θ=0° only m=0 states are excited, while for Θ=90° excitation of both $|m|$=0 and 2 is allowed, but the latter dominates the image. The more interesting Θ=45° image shows a quasi-horizontal nodal line signaling the additional excitation of $|m|$=1 states. Experimental PM images obtained by two-photon ionization of Mg atom reproduce qualitatively these effects. In addition, multiple-$|m|$ excitation is also evident in other observables, such as the total ionization cross section.

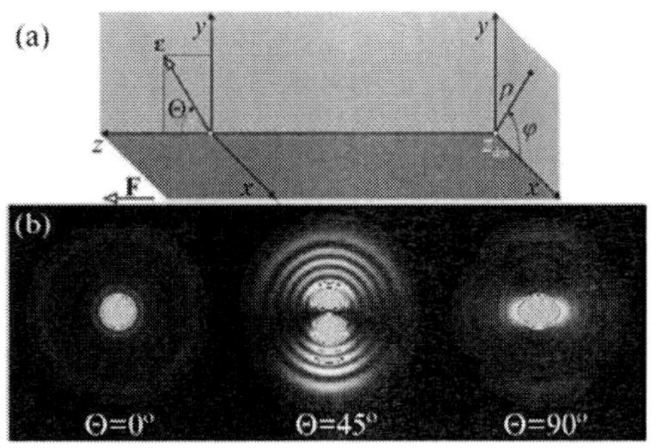

Figure 1. (a) Geometry of the present calculations and experiment. (b) Calculated hydrogenic PM images for F=680 V/cm, electron excitation energy E≈−15.3 meV and different angles Θ. Linear colorscale, common for all images. Intensity cuts allow for better visualization of the Θ-dependent radial quantum interferences.

Note finally that the Θ-dependence of the relative contributions of different-$|m|$ states poses new challenges on the applicability of tomographic reconstruction techniques [3] when meV electrons are involved, because these techniques assume the reconstruction of a *single* 3D object, rotated (by rotating **ε**) and projected on the 2D detector by a field **F** assumed to be weak enough that its influence can be ignored.

References

[1] Kondratovich V D and Ostrovsky V N 1990 *J. Phys. B* **23** 3785
[2] Kalaitzis P *et al* 2018 *Phys. Rev. A* **97** 053412
[3] Smeenk C *et al* 2009 *J. Phys. B* **42** 185402

* E-mail: pkalaitz@cc.uoi.gr
† E-mail: scohen@uoi.gr

Cross section scaling for the ionization of H_2O by highly charged ions

N Bachi[1] *, S Otranto[1] and R E Olson[2]

[1]Instituto de Física del Sur (IFISUR), Departamento de Física, Universidad Nacional del Sur (UNS), CONICET, Av. L. N. Alem 1253, B8000CPB-Bahía Blanca, Argentina
[2]Physics Department, Missouri University of Science and Technology, Rolla MO 65409, USA

Synopsis A classical trajectory Monte Carlo model that explicitly considers the ten electrons of the H_2O molecule is used to study electron production in collisions involving bare ions. Present results are contrasted to the reported experimental data and to other theoretical results obtained by means of distorted wave models. Focus is made on the energy range 100 (keV/amu) /Z_P-1000 (keV/amu)/Z_P along which a simple scaling for the net cross section is introduced.

Recent projections for 2030 suggest that the number of Cancer diagnosis and deaths in a period of almost two decades will be increased by 74% and 59%. In addition to the strong efforts devoted to prevention campaigns, design and improvement of treatment techniques and hospital-held facilities that could allow ample patient treatment conforms a fast developing field. This is particularly the case of ion-therapy, which in contrast to photon-beam and electron-beam therapy, allows the irradiation of deep seated tumors with minimal irradiation of surrounding organs and healthy tissue. It is not surprising then that ion-therapy facilities are expected to double by 2021 from those that were available or in construction by 2016 [1].

In ion-therapy, most of the biological damage to the tumoral cells is due to secondary electrons and free radicals emiited following the ion impact. Hence, an accurate description of the main physical mechanisms leading to electron production is desirable, specially during the planning stage of the irradiation procedure which strongly relies on a numerical simulation that uses collisional cross sections as input data.

In this work, a classical trajectory Monte Carlo model in which the ten electrons of the H_2O molecule are explicitly considered in an independent atom picture for the study of H^+, He^{2+}, C^{6+} and Si^{13+} collisions at impact energies in the range 100-10,000 keV/amu. Coulomb potentials with effective charges are used to represent the interaction of the respective electrons with the molecular centers. Since multielectronic targets are classically unstable, the electron-electron correlation is not explicitly considered.

Results are compared to experimental data reported by several laboratories throughout the past 4 decades and to recent CDW-EIS calculations of Rivarola's group [2].

In Fig. 1 we show the obtained scaling for the total net cross section $\sigma_{net} = \sigma_{1\text{-ion}} + 2\,\sigma_{2\text{-ion}} + 3\,\sigma_{3\text{-ion}} \ldots$ as a function of the impact energy. Present results are compared to the scaling reported for $\sigma_{loss} = \sigma_{1\text{-ion}} + \sigma_{2\text{-ion}} + \sigma_{3\text{-ion}} + \ldots$ for H(1s) target wich has a similar ionization potential [3].

Figure 1. Linear Fit of the CTMC data for

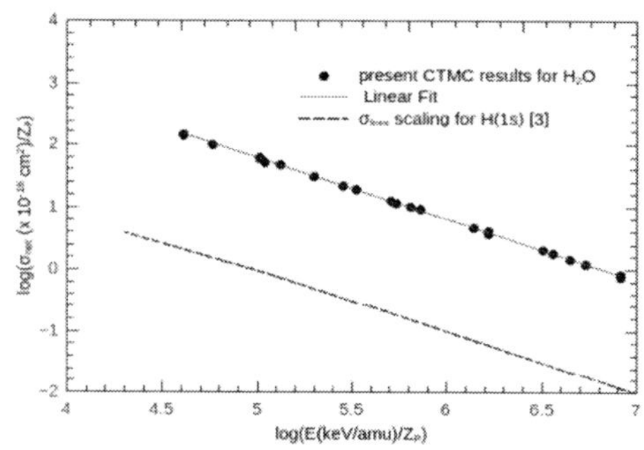

log(σ_{net}/Z_P) vs log(E/Z_P).

References

[1] Dosanjh M 2018 *Cern Courier* 58 32
[2] Bhattacharjee S *et al* 2017 *Phys. Rev.* A **96** 052707
[3] Olson R E *et al* 1978 *Phys. Rev. Lett.* **41** 163

* E-mail: nicolas.bachi@uns.edu.ar

Study of chiral asymmetries in the inner-shell photoionization of partially oriented trifluoro-methyloxirane

G Nalin[1], S Grundmann[1], G Kastirke[1], K Fehre[1], D Trabert[1], J Rist[1], M Weller[1], M Waitz[1], J Siebert[1], I Vela-Perez[1], A Khan[1], N Anders[1], M Kircher[1], R Tomar[1], M Hofmann[1], F Trinter[1,2,3], C Küstner-Wetekam[4], L Marder[4], J Viehmann[4], H Fukuzawa[5], K Ueda[5], J B Williams[6], T Jahnke[1], R Dörner[1] and M S Schöffler[1*]

[1] Institut für Kernphysik Goethe-Universität Frankfurt Max-von-Laue-Str. 1, 60438 Germany
[2] Photon Science, Deutsches Elektronen-Synchrotron DESY, Hamburg 22607, Germany
[3] Molecular Physics, Fritz-Haber-Institut der Max-Planck-Gesellschaft, Berlin 14195, Germay
[4] Institute of Physics and CINSaT, University of Kassel, Heinrich-Plett-Str. 40, 34132 Kassel, Germany
[5] Department of Physics, University of Nevada, Reno, Nevada 89557, USA
[6] Institute of Multidisciplinary Research for Advanced Materials, Tohoku University, Sendai 980-8577, Japan

Synopsis Photoelectron circular dichroism (PECD) was determined for O(1s) photoionization of methyloxirane (C_3H_6O) and trifluoro-methyloxirane ($C_3H_3F_3O$). Several electron energies above the threshold were chosen to compare the partially fixed-in-space electron distributions.

When exceeding a certain size, most molecules become chiral in their structure: they exist as two enantiomers, which are mirror images of each other. Therefore they cannot be superimposed without breaking/rearranging of bonds. The common properties - as melting point, density etc. - of these enantiomeres are the same, and even the rovibronic sublevels are identical (except the tiny effect of weak interaction). After ionization with a single photon the photo electron angular distribution is symmetric in forward/backward direction with respect to the light's propagation direction. It turns out that the combination of circularly polarized light and chiral molecules break this symmetry. This effect is called photoelectron circular dichroism (PECD) [1]. This rather strong orbital-dependent chiroptical effect reaches up to a few percent and shows rich photon energy dependence. In previous experiments we observed a strong enhancement of the PECD for uniaxially oriented methyloxirane (C_3H_6O) molecules [2]. In the current study, trifluoro-methyloxirane ($C_3H_3F_3O$) was core photo-ionized, creating photoelectrons with either 4, 6, 8 and 11.5 eV, launched from the O(1s). The energies were chosen to give an overall maximum positive, negative or 0 PECD signal [3]. The experiments were conducted on SEXTANTS beamline at the Synchrotron SOLEIL using the well-established COLTRIMS (Cold Target Recoil Ion Momentum Spectroscopy) technique [4]. High-efficiency-MCPs combined with a mesh-free spectrometer were used to enhance the multiple particle detection efficiency [5]. This allows post-orienting the molecule not only by one-axis, but also by two or even in 3d and even determining the absolute handedness [6,7]. Alternating between right- and left-handed circularly polarized radiation the PECD in the angular distribution was extracted from the photoelectron images obtained in coincidence with the molecular break-ups and compared for the 11.5 eV case with the results for C_3H_6O.

* E-mail: schoeffler@atom.uni-frankfurt.de

References

[1] Ritchie B 1976 *Phys. Rev A* **14** 359
[2] Tia M *et al* 2017 *Journal of phys. chem. lett.* **8** 2780
[3] Ilchen M *et al* 2017 *Phys. Rev. A* **95** 053423
[4] Dörner R *et al* 2000 *Phys. Rep.* **330** 95
[5] Fehre K *et al* 2018 Rev. Sci. Instr. **89** 045112
[6] Pitzer M *et al* 2013 *Science* **341** 1096
[7] Fehre K *et al* 2019 *Science Advances* **5** 7923

Single electron capture cross sections for collisions of multi-charged ions with helium-like atomic system

S Samaddar[1], K Purkait[1] and M Purkait[1]*

[1]Department of Physics, Ramakrishna Mission Residential College, Narendrapur, Kolkata-700103, India

Synopsis A four-body distorted wave method (DW-4B) is applied to calculate the total cross section for single electron capture in the collisions of multi-charged ions with helium-like atomic system. The proper long range behavior of the Coulomb potential is introduced. The present theoretical method is satisfactory in the intermediate and high-energy regime and is quite reliable compared to other theoretical methods.

It is well known that ion-atom collisions involve long-range Coulomb potential that persist even when the scattering particles are at infinite separations from each other. This causes distortion of the unperturbed channel states and modification of the corresponding perturbation potentials. The long range nature of the Coulomb interaction yields considerable mathematical difficulties in treating collision processes and requires special attention for unambiguous and proper applications. The theoretical description of four-body process is very complicated. The influence of electron-electron interaction is important to study the dynamics in the collision phenomena. Such correlations are one of the causes of the transition from the initial state to the final state. Different theories [1-3] have been developed to study the single-electron capture (SEC) by a fast bare ion with helium. In the present work, we have studied the SEC from He-like atom by bare ions. The prior form of the scattering amplitude may be written in the form

$$T_{if}^{(-)} = \langle \psi_f^- | V_i | \psi_i \rangle,$$

with

$$\psi_f^- = \phi_f(\vec{s}_1) e^{i\vec{k}_f \cdot \vec{R}_T} D(Z_T, \vec{v}_f, \vec{r}),$$

$$D^+(Z_T, \vec{v}_f, \vec{r}) = e^{\frac{\pi}{2}} \Gamma(1 + i\alpha)\, {}_1F_1(i\alpha; 1; iv_f r - i\vec{v}_f.\vec{r}),$$

$$D^-(Z_T, \vec{v}_f, \vec{r}) = e^{\frac{\pi}{2}\alpha} \Gamma(1 - i\alpha)\, {}_1F_1(-i\alpha; 1; -iv_f r - i\vec{v}_f.\vec{r}),$$

and

$$\alpha = \frac{Z_T}{v_f}.$$

For the initial-state eigenfunction we write

$$\psi_i = e^{i\vec{k}_i \cdot \vec{R}_T} \phi_i(\vec{r}_1) e^{\frac{\pi}{2}\gamma_P} \Gamma(1 - i\gamma_P)\, {}_1F_1(i\gamma_P; 1; ik_i R_P - i\vec{k}_i.\vec{R}_P),$$

where

$$\gamma_P = \frac{2Z_P}{v_i}.$$

The initial residual potential is

$$V_i = \frac{2Z_P}{R_P} - \frac{Z_P}{s_1} - \frac{Z_P}{s_2}.$$

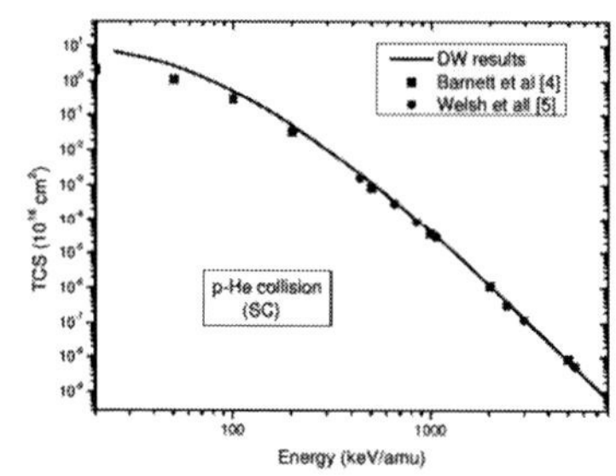

Figure 1. $p - He$ collision for single charge (SC) transfer.

The authors gratefully acknowledge the financial support from the Council of Scientific & Industrial Research (CSIR), New Delhi, India through Project No. 03/1366/16/EMR-II.

References

[1] Samanta R, Purkait M and Mondal C R 2011 *Phys. Rev. A* **83** 032706
[2] Mancev I, Milojevic N and Belkic Dz 2015 *Phys. Rev. A* **91** 062705
[3] Guo D L *et all* 2012 *Phys. Rev. A* **86** 052707
[4] Barnett C F *et all* 1977 *Atomic Data for Controlled Fusion Research*
[5] Welsh L M *et all 1967 Phys. Rev. A* **158** 85

*E-mail: mpurkait_2007@rediffmail.com

Distorted wave model for sub-shell electron capture cross sections of noble gases by bare ion impact

M Purkait[1] *

[1]Department of Physics, Ramakrishna Mission Residential College, Narendrapur, Kolkata-700103, India

Synopsis Electron capture from noble gases (He, Ne and Ar) are investigated in the framework of the quantum-mechanical distorted wave model. In this model, we use an independent electron model taking particular care of the representation of the bound and continuum target states. The results for partial and total capture cross sections are compared with other theoretical calculations and experimental data. The present results are found to be in very good agreement with the available experimental findings.

Atomic data for single charge transfer between different combination of bare ions and noble gases are in very great demand to study the behaviours of impurity ions in all types of plasma. The dynamics of many-particle collisions is a very complex problem which involves the description of bound and continuum states of atoms and the possibility of many open channels. Different theoretical model have been developed to study such process by a model one-active electron system. According to this model, there is only one-active electron whereas the other non-captured electron(s) are passive. The net results of such a 'frozen core' approximation is a reduction of the many particle problem to a three-body problem. In the case of asymmetric collisions, the long-range Coulomb interactions in both the channels are very important [1] and should be considered properly. In the present work we propose a distorted wave (DW) formalism with the independent electron model (IEM), taking account of the Coulomb interactions.

In the DW formalism, we choose the final state eigenfunction ψ_f which may be written as

$$\psi_f = e^{i\vec{k}_f \cdot \vec{r}_f}\phi_f(\vec{r}_P)e^{\frac{\pi}{2}\alpha_T}\Gamma(1 + i\alpha_T)_1F_1(-i\alpha_T; 1; -iv_f r_T - i\vec{v}_f.\vec{r}_T),$$

and for the initial state eigenfunction we write,

$$\chi_i^+ = e^{i\vec{k}_i.\vec{r}_i}\phi_i(\vec{r}_T)e^{\frac{\pi}{2}\alpha_{PT}}\Gamma(1 - i\alpha_{PT})_1F_1(i\alpha_{PT}; 1; ik_i r_i - i\vec{k}_i.\vec{r}_i),$$

where $\alpha_T = \frac{Z_T^{eff}}{v_f}$ and $\alpha_{PT} = \frac{Z_P}{v_i}$.
The residual potential of the initial channel is

$$V_i = \frac{Z_P}{r_i} - \frac{Z_P}{r_P}.$$

*E-mail: mpurkait_2007@rediffmail.com

Finally, the total cross sections in the DW method are given by

$$\sigma = \frac{\mu_i \mu_f}{4\pi^2}\frac{k_f}{k_i}\int |T_{if}^{(-)}|^2 d\Omega,$$

where

$$T_{if}^- = \langle\psi_f|V_i|\chi_i^+\rangle.$$

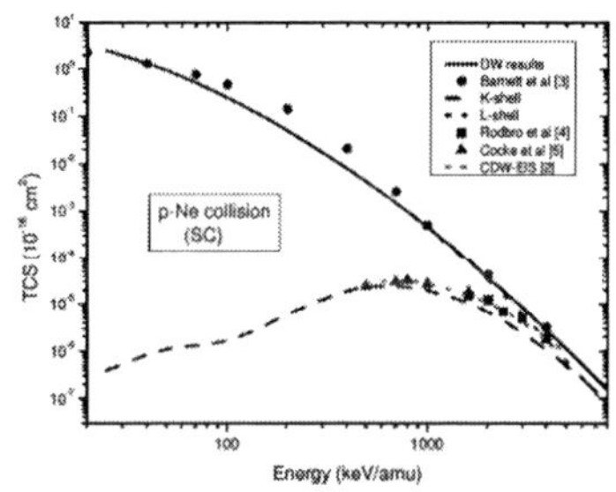

Figure 1. $p - Ne$ collision for single charge (SC) transfer.

The authors gratefully acknowledge the financial support from the Council of Scientific & Industrial Research (CSIR), New Delhi, India through Project No. 03/1366/16/EMR-II.

References

[1] Belkic Dz, Gayet R and Salin A 1979 *Phys. Rep.* **56C** 290
[2] Abufager P N *et al* 2004 *J. Phys. B: At. Mol. Opt. Phys.* **37** 817
[3] Barnett C F *et al* 1977 *Atomic Data for Controlled Fusion Research*
[4] Rodbro M *et al* 1979 *Phys. Rev. A* **19** 1936
[5] Cocke C L *et al* 1977 *Phys. Rev. A* **16** 2248

Li-like carbon Auger KLL measurements of mixed-state $C^{4+}(1s^2\,{}^1S, 1s2s\,{}^3S)$ ions in 6-15 MeV collisions with He and comparison to 3eAOCC calculations

I Madesis[1,2], A Laoutaris[1,2], E P Benis[3], T J M Zouros[1,2*] and A Dubois[4]

[1] Department of Physics, University of Crete, P.O. Box 2208, GR 71003 Heraklion, Greece
[2] Tandem Accelerator Laboratory, INPP, NCSR Demokritos, GR 15310 Ag. Paraskevi, Greece
[3] Department of Physics, University of Ioannina, GR 45110 Ioannina, Greece
[4] Sorbonne Université, CNRS, Laboratoire de Chimie Physique-Matière et Rayonnement, Paris, France

Synopsis New state-resolved Li-like carbon Auger KLL measurements of mixed-state $C^{4+}(1s^2\,{}^1S, 1s2s\,{}^3S)$ ions in 6-15 MeV collisions with He are presented. Results of three-electron semiclassical atomic orbital AOCC coupled channel (3eAOCC) calculations are also presented and compared.

State-resolved cross sections for the production of $1s2s^2\,{}^1S$, $1s2s2p\,{}^4P$, $1s2s2p\,{}^2P_\pm$, and $1s2p^2\,{}^2D$ were determined using our zero-degree Auger projectile spectroscopy (ZAPS) setup at the 5.5 MV tandem accelerator at Demokritos [1]. These states are populated predominantly by transfer-excitation processes from the $1s^2$ and/or by single electron transfer processes from the $1s2s\,{}^3S$ metastable components.

Using our recently reported dual measurement technique [2], we have separated ground state from metastable state KLL contributions in collisions of mixed-state $C^{4+}(1s^2\,{}^1S, 1s2s\,{}^3S)$ ions in 6-15 MeV collisions with He.

For the interpretation and understanding of the experimental results we also performed calculations over the 0.5-18 MeV range using a semiclassical atomic orbital close-coupling approach, based on an asymptotic (atomic) description of the neutral and charged collision partners [3,4]. The electronic dynamics is then treated quantum mechanically solving the three-electron time-dependent Schrödinger equation with full configuration interaction (CI).

To describe electron excitation and transfer to doubly excited states on the carbon center, three active electrons are taken into account: two of them are reserved for the He-like C^{4+} projectile, while the third is initially on the He target, accounted for by a He^+ model potential binding a unique electron which can be transferred in the collision to C^{4+} to describe 1-, 2- and 3-open-shell electronic configurations of the C^{3+} ions. An illustration of our results is shown in Fig. 1 for the production of the $C^{3+}(1s2p^2\,{}^2D)$ state. Results for the other KLL states will also be presented.

* E-mail: tzouros@physics.uoc.gr

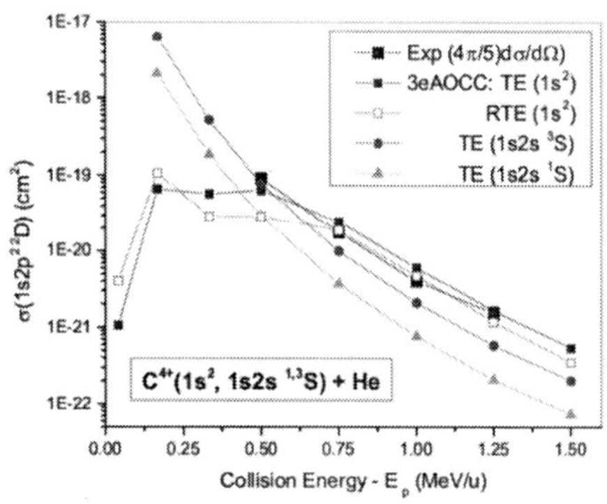

Figure 1. Cross sections for the production of the $C^{3+}(1s2p^2\,{}^2D)$ state in collisions of $C^{4+}(1s^2\,{}^1S, 1s2s\,{}^{1,3}S)$ ions with He. ZAPS measurements (black squares), 3eAOCC results: (squares) from $1s^2$, (circles) from $1s2s\,{}^3S$ and (triangles) from $1s2s\,{}^1S$ metastable states, respectively. RTE assessment (open squares) is also indicated in good overall agreement with experiment.

References

[1] Madesis I *et al* 2015 *J. Phys: Conf. Ser.* **583** 012014.
[2] Benis E P and Zouros T J M 2016 *J. Phys. B* **21** L671
[3] Sisourat N *et al* 2011 *Phys. Rev. A* **84** 052722
[4] Gao J W *et al* 2018 *Phys. Rev. A* **96** 052709

We acknowledge support of this work by the project "Cluster of Accelerator Laboratories for Ion-Beam Research and Applications – CALIBRA" (MIS 5002799) implemented under the Action "Reinforcement of the Research and Innovation Infrastructure", funded by the Operational Programme "Competitiveness, Entrepreneurship and Innovation" (NSRF 2014-2020) and co-financed by Greece and the European Union (European Regional Development Fund).

Shrinking diffraction "fringes" in positronium formation from C_{60} to C_{240}

P –A Hervieux[1] *, A R Chakraborty[2] @ and H S Chakraborty[2] †

[1]Université de Strasbourg, CNRS, Institut de Physique et Chimie des Matériaux de Strasbourg, 67000 Strasbourg, France

[2]Department of Natural Sciences, D L Hubbard Center for Innovation, Northwest Missouri State University, Maryville, Missouri 64468, USA

Synopsis Due to the bulk electron capture from the molecular shell of fullerenes by incoming positron projectiles and a dephasing effect across the shell width, the positronium formation signal yields diffraction fringes in the target recoil momentum space. Our results show the variance in the separation and shape of the fringe pattern as a function of the fullerene size from C_{60} to larger C_{240}.

Varieties of matter have been probed by the positronium (Ps) formation upon bombarding them with positron beams. They range from atoms, molecules, polymers to solids, liquids, surfaces/films, metal-organic-frameworks and embedded nanostructures. However, using nanoparticles as targets in Ps formation can yield novel effects, since solid-like delocalized electrons in nanoparticles are contained in atom-like finite spaces. Fullerenes can be excellent prototypes for such studies. We recently took pilot strides in probing the Ps formation from gas phase C_{60} and predicted a novel diffraction phenomenone that underpins the Ps formation at energies above the C_{60} plasmon excitation [1, 2]. The effect induces fringes in C_{60} recoil momentum which can be called the diffraction resonances. In the present study, we use C_{240} as the diffractor and show the critical dependence of these resonances on the fullerene size.

The structure of a C_{240} molecule is described by the local-density approximation augmented by the LB94 exchange-correlation functional [3]. The positron impact on C_{240} leading to the Ps formation is treated by the continuum distorted-wave-final-state approximation [4].

We compare results of C_{240} in Figure 1(a) as a function of the recoile momentum with those in 1(b) of C_{60}. Results are presented as the ratios of Ps(1*s*) formation cross sections captured from selected states of the target. This facilitates a better representation of resonances by roughly neutralizing the decay of the cross sections. As seen, the fringe separation is almost halved going from C_{60} to C_{240}, in reciprocal proportion of their radii, in analogy to the shrinking fringe pattern with increasing slit size in classical single-slit experiments. Also, the home/homo-1 shape is very different in C_{240} compared to C_{60}.

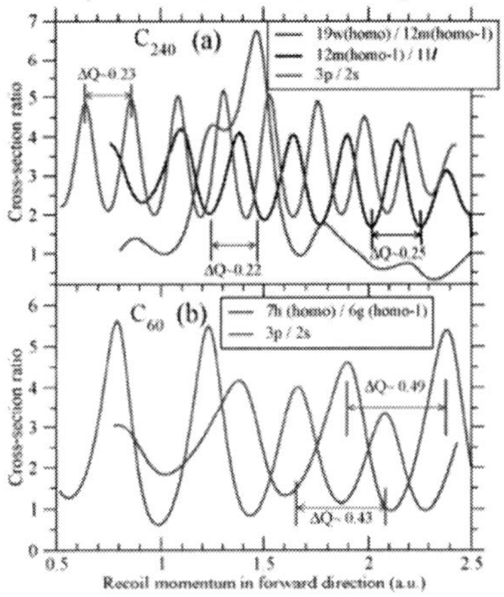

Figure 1. (a) Ps(1*s*) cross section ratios of C_{240}. (b) Same for C_{60}.

The work may encourage experimental applications of Ps formation spectroscopy to gas-phase nanosystems. Supported by the National Science Foundation grant PHY-1806206.

References

[1] Hervieux P-A *et al* 2017 *Phys. Rev. A* **95** 020701(R)
[2] Chakraborty *et al* 2017 *J. Phys. Conf. Ser,* **875** 072002
[3] Choi J *et al* 2017 *Phys. Rev. A* **95** 023404
[4] Fojón O A *et al.* 2001 *J. Phys. B* **34** 4279

* E-mail: paul-antoine.hervieux@ipcms.unistra.fr
@ E-mail: arc84c@mst.edu
† E-mail: himadri@nwmissouri.edu

Radiative double electron capture for 2.11 MeV/u $F^{9+,8+}$ + N_2, Ne

D S La Mantia*, P N S Kumara, S L Buglione, C P McCoy, C J Taylor, J S White, A Kayani and J A Tanis

Department of Physics, Western Michigan University, Kalamazoo, MI 49008 USA

Synopsis The first successful observation of radiative double electron capture for gas targets is reported. RDEC has been observed for 2.11 MeV/u $F^{9+,8+}$+N_2, Ne collisions. Cross sections for both projectile charge states and targets were determined and compared with previous results and theory. This abstract gives an overview of the results found for both gas targets.

Radiative double electron capture is a fundamental atomic process where two electrons are captured from a target to bound states in a projectile simultaneous with emission of a single photon [1]. RDEC can be considered the inverse of double photoionization for ion-atom collision systems. RDEC is related to the well-known process of radiative electron capture [2], in which a single electron is captured to a bound state with the simultaneous emission of a photon. Cross sections for RDEC for 2.11 MeV/u $F^{9+,8+}$+N_2, Ne have been determined [3].

This work was performed using the tandem Van de Graaff accelerator facility at Western Michigan University (WMU). A beam of F^{9+} or F^{8+} ions was directed toward a differentially-pumped gas cell with target pressures set to the single-collision regime (8 mTorr for N_2 and 15 mTorr for Ne). A Si(Li) x-ray detector mounted at 90° to the beamline collected photons and separate silicon surface-barrier detectors collected the charge-changed projectiles. Appropriate electronics and an event-mode data acquisition system were used to assign the measured x rays to the corresponding single (Q-1) or double (Q-2) charge-changed particles, or vice versa. Typical spectra obtained are shown in Fig. 1.

The first successful observations of RDEC were performed at WMU using 2.38 MeV/u O^{8+} [4] and 2.11 MeV/u F^{9+} [5] projectiles incident on thin carbon foils. Unavoidable multiple-collision effects were present in these measurements causing RDEC events to be observed in both the Q-2 and Q-1 channels. These previous measurements provided the motivation for the present single-collision gas target work. Contaminants were avoided with high-purity gases. From the data obtained, RDEC cross sections for the fully-stripped fluorine beam on N_2 and Ne were 2.5-4 times smaller than the corresponding cross sections for O^{8+} and F^{9+} incident on carbon.

*E-mail: david.s.lamantia@wmich.edu

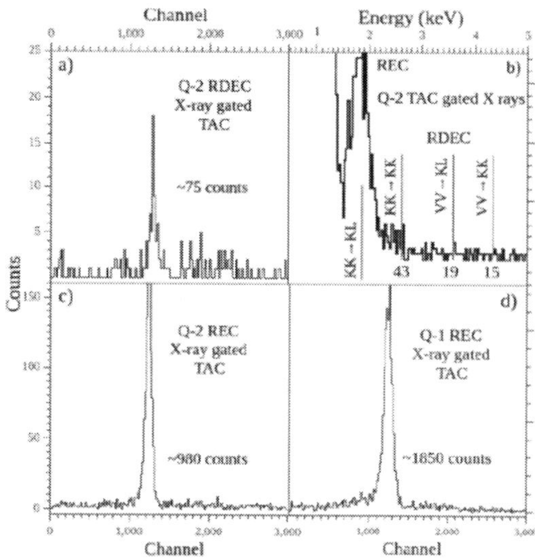

Figure 1. Spectra for 2.21 MeV/u F^{9+}+N_2: (a) Q-2 TAC spectrum associated with RDEC energy photons, (b) x-ray spectrum associated with Q-2 particles, (c) Q-2 and (d) Q-1 TAC spectra associated with REC energy photons.

This work was supported in part by National Science Foundation Grant PHY-1707467.

References

[1] Miraglia J and Gravielle M S 1987 *ICPEAC XV: Book of Abstracts* p. 517 Brighton, U.K.
[2] Stöhlker T *et al* 1995 *Phys. Rev. A* **51** 2098
[3] La Mantia D *et al* 2019 Submitted to *Phys. Rev. Lett.*
[4] Simon A *et al* 2010 *Phys. Rev. Lett.* **104** 123001
[5] Elkafrawy T *et al* 2016 *Phys. Rev. A* **94** 012705

Ionization and electron capture in $Li^{3+} + H_2O$ collisions

C Illescas[1], **M A Lombana** [1], **L Méndez** [1,†,*], **I Rabadán** [1] and **J Suárez** [2]

[1]Departamento de Química, Universidad Autónoma de Madrid, 28049 Madrid, Spain
[2]Dipartimento di Chimica, Università degli Studi di Milano, 20133-Milano, Italy

Synopsis Total cross sections for ionization and electron capture in $Li^{3+} + H_2O$ collisions are calculated at energies $20 \leq E \leq 750$ keV/u. Three different models are applied: the classical trajectory Monte Carlo method, an expansion in terms of asymptotic-frozen-molecular-orbitals and the numerical solution of the time dependent Schrödinger equation. One-electron transition probabilities from the three methods show general good agreement and with those from the calculation of Luna *et al Phys Rev. A* **93**, 052705.

We carry out a theoretical study of ionization and electron-capture reactions in Li^{3+} collisions with H_2O at energies $20 \leq E \leq 750$ keV/u. The processes considered are:

$$Li^{3+} + H_2O \to Li^{(3-p)+} + H_2O^{(q+p)+} + qe^-,$$

where p electrons are exchanged and q electrons are released. Luna *et al* [1] have studied these processes in a combined theoretical and experimental work.

At the energies of the present work one can assume that the projectile follows straight-line trajectories and that the nuclei of the molecular target remain fixed during the collision. Our calculations are based on the independent electron approximation, where the active electron moves in an effective potential. The solution of the semiclassical equation for the one-electron problem provides one-electron probabilities for capture and ionization, which are then combined to yield the corresponding many-electron probabilities. We have solved the one-electron dynamical problem by applying three methods: the Classical Trajectory Monte Carlo (CTMC) method [2], the asymptotic-frozen-molecular-orbitals expansion (AFMO)[3] and the numerical solution of the time dependent Schrödinger equation (GridTDSE method) [4].

The agreement between the different models is illustrated in Fig. 1, where we plot the electron production cross sections obtained (see [3]) as $\sigma^{\rm EP} = 2\sum_{k=1}^{4}\sigma_k^{\rm ion}$, with $\sigma_k^{\rm ion}$ the one-electron ionization cross sections from the water molecular orbitals $2a_1$, $1b_2$, $3a_1$ and $1b_1$. We also compare in the figure our results with the calculation of Ref. [1] for electron production with $q + p \leq 4$. We find general good agreement between the calculations for $E > 100$ keV/u although the AFMO calculation overestimates the cross section at low E because ionization and capture are not correctly separated [3]. As explained in [1], some differences with the experimental values might be due to the fact that a specific ion can be formed from the fragmentation of different molecular cations.

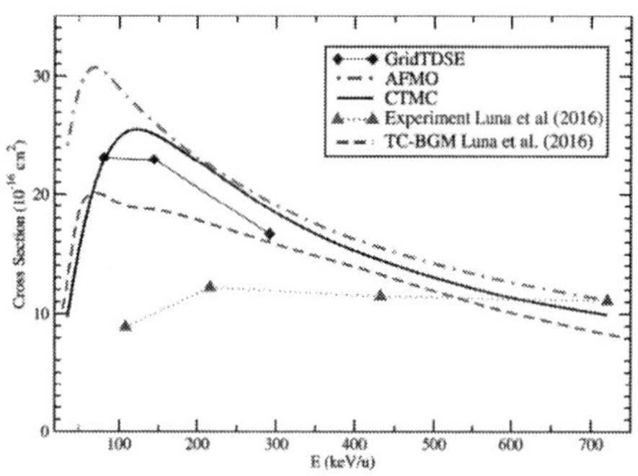

Figure 1. Total cross sections for electron production from CTMC, AFMO and GridTDSE calculations compared to TC-BGM calculations and experiments [1].

This work has been partially supported by Ministerio de Ciencia, Innovación y Universidades (Spain) (project FIS2017-84684-R).

References

[1] Luna H *et al* 2016 *Phys. Rev. A* **93** 052705
[2] Illescas C *et al* 2011 *Phys. Rev. A* **83** 052704
[3] Errea L F *et al* 2013 *Phys. Rev. A* **87** 032709
[4] Errea L F *et al* 2015 *Chem. Phys.* **462** 17

*E-mail: l.mendez@uam.es

 IOP Publishing
1412 (2020) 162007 doi:10.1088/1742-6596/1412/16/162007

State-selective single electron capture in slow O^{6+}-He collisions

J W Xu[1,2], X L Zhu[1,2*], W T Feng[1], D M Zhao[1], S F Zhang[1,2], D L Guo[1], Y Gao[1] and X Ma[1,2†]

[1]Institute of Modern Physics, Chinese Academy of Sciences, Lanzhou 730000, Peoples Republic of China
[2]University of Chinese Academy of Sciences, Beijing 100049, Peoples Republic of China

Synopsis The state-selective single-electron capture processes were measured by cold-target recoil-ion momentum spectroscopy in O^{6+}-He collisions between 18keV and 96keV. From the longitudinal momentum spectrum of recoil ions, the single electron captured into n=3 states of projectile ion is predominant and n=4 states have a small contribution. The oscillation structure of angular differential cross sections was observed. The present experimental results are compared with other experiments at low energy and the predictions of the molecular Columbic barrier model.

The charge transfer processes play an important role in different fields for keV collisions of highly charged ions with atoms and molecules. The solar wind contains a large number of minor/heavy ion species with a range of charge states, such as O^{6+}, C^{5+}, N^{5+}, and Si^{10+}. These ions will readily charge transfer with cometary neutrals, producing ions which can be highly excited and consequently emit photons in the extreme ultraviolet and x-ray part of the spectrum, such as comet Hyakutake [1].

In the present work, the recoil ion momentum of single-electron capture (SEC) in collisions of O^{6+} ions with He was systematically measured at impact energies range from 18 to 96 keV. Our experiments were conducted at EBIS platform combined COLTRIMS [2, 3] at the Institute of Modern Physics in Lanzhou.The fully differential cross sections have been obtained.

From the longitudinal momentum spectrum of recoil ions, the SEC into n=3 states of projectile ion is predominant and n=4 states is very small in research energy range. With the increasing of projectile energy, the relative contribution of SEC n=4 states increases, while the relative contribution of SEC n=3 states decrease slightly. The present results agree well with the measurement Gordeev et al [4]. The experimental results were qualitatively explained by the prediction of the molecular Columbic barrier model (MCBM) [5]. Also the obviously oscillation structure was observed at the angular differential cross sections. Fig. 1 shows the experimental angular differential cross sections for SEC into n=3 states in O^{6+}-He collisions at an impact energy of 36 keV. Two peaks can be clearly observed in figure 1. The calculated angular distribution (red solid line) based on MCBM model is also shown in the figure. Comparison of the experimental results and the calculation, the discrepancy is evident. In order to understand this discrepancy, the new theoretical work is needed.

*E-mail: zhuxiaolong@impcas.ac.cn
†E-mail: x.ma@impcas.ac.cn

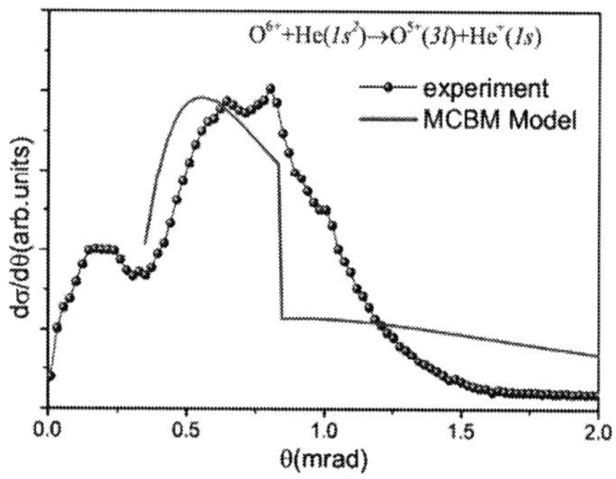

Figure 1. Angular differential cross sections for electron captured to the $O^{5+}(1s^2 3l)$ states at an impact energy of 36 keV.

This work is supported by the National Key *R&D* Program of China under Grant No. 2017YFA0402400.

References

[1] Cravens TE *et al* 1997 *Geophys Res Lett.* **24** 1
[2] Zhu X L *et al* 2019 *NIMB* **in press**
[3] Ma X *et al* 2011 *Phys Rev A.* **83** 052707
[4] Gordeev Y S *et al* 1983 *Phys Rev Lett.* **50** 1842
[5] Niehaus A *et al* 1986 *J. Phys. B: At. Mol. Opt. Phys.* **19** 2925

Coupled-channel calculations of ionisation of atomic hydrogen by multiply-charged bare ions

I B Abdurakhmanov*, K Massen-Hane, J Faulkner, Sh U Alladustov, A S Kadyrov and I Bray

Curtin Institute for Computation and Department of Physics and Astronomy, Curtin University, GPO Box U1987, Perth, WA 6845, Australia

Synopsis The wave-packet convergent close-coupling method has been applied to calculate the total and fully differential cross sections for electron capture, excitation and ionisation processes in collisions between multiply-charged bare ions and atomic hydrogen.

Scattering of multiply-charged bare ions on atomic hydrogen is a three-body problem where the target electron is exposed to the asymmetric two-center Coulomb field. For highly-charged incident ions, due to the long-range attraction of the projectile, the multiple-scattering effects are important in a wide energy range, making the first-order approaches problematic even at considerably high impact energies. Apart from fundamental interest these collisions are also of considerable importance due to their relevance in many applied areas of physics. For instance, an accurate knowledge of the differential and total ionisation cross sections for these collisions are important in the diagnostics of fusion plasma.

We applied our two-centre wave-packet convergent close-coupling (WP-CCC) approach [1] to calculate total and fully differential cross sections for electron capture, excitation and ionisation processes occurring in these collisions. The usage of the two-centre basis constructed from the wave-packet pseudostates allows one to discretize the continuum arbitrarily which is ideal for differential ionisation studies.

In Fig. 1 we show the energy dependence of the total cross section for ionisation of atomic hydrogen by (a) He^{2+} and (b) C^{6+} ions. For both considered projectiles the current WP-CCC results slightly overestimate experimental measurements at intermediate energies. This appears to be a common feature of the other two displayed coupled-channel approaches [2, 3] as they also overestimate the measured points roughly by a similar amount. The only exception is the CDW-EIS calculations of Rivarola *et al* [4] which overlap with the lower limit of the experimental point for C^{6+}-impact ionisation.

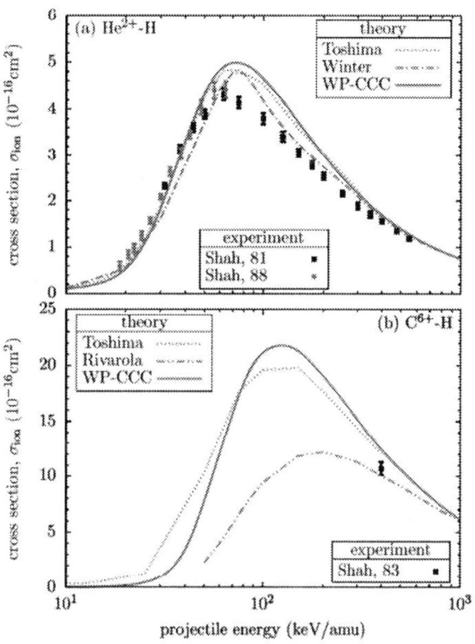

Figure 1. The total cross section for ionisation of atomic hydrogen by (a) He^{2+} and (b) C^{6+} ions: the present WP-CCC results are compared with the experiments of Shah *et al* [7, 6, 5] and theoretical calculations of Winter [2], Toshima [3] and Rivarola *et al* [4].

*E-mail: Ilkhom.Abdurakhmanov@curtin.edu.au

References

[1] Abdurakhmanov I B *et al* 2018 *Phys. Rev. A* **98** 062710
[2] Winter T G 2007 *Phys. Rev. A* **76** 062702
[3] Toshima N 1994 *Phys. Rev. A* **50** 3940
[4] Rivarola R D *et al* 1989 *Phys. Scr.* **1989** 101
[5] Shah M B *et al* 1988 *J. Phys. B* **21** 2455
[6] Shah M B *et al* 1983 *J. Phys. B* **16** L449
[7] Shah M B *et al* 1981 *J. Phys. B* **14** 2361

Cross sections for electron capture process of biological molecules by proton and α-particle impact

K Purkait[1] **and M Purkait**[1] *

[1] Department of Physics, Ramakrishna Mission Residential College, Narendrapur, Kolkata-700103, India

Synopsis Distorted wave model has been used to predict total cross sections (TCSs) of single electron capture process for several ion-biological molecule collisional systems in the intermediate and high energy regime. TCSs are calculated in the Independent Electron Model (IEM) In this work, we studied with water, nitrogen, oxygen, methane, adenine or cytosine targets impacted by protons and α-particles with kinetic energy ranging from 25 keV/amu to 10,000 keV/amu. It is observed that the present investigation produces results which are in good agreement with experimental observations for all cases.

Ion beam cancer therapy is based on the irradiation of the tumor by a beam of fast ions that causes the cell damage colliding with the bio-molecules. The damage can be either direct, where the DNA strands are broken by ion impact, or indirect, when the breaking of DNA strands provided by secondary particles formed in ion collision of other species. Therefore, the cross sections for different reactions in ion collisions with bio-molecules are required in the simulation of travelling ion in the biological medium, which is needed to understand the mechanism of the biological damage and to evaluate the optical dose. In this work we focus on the electron capture process in proton-biological molecule collisions in the framework of boundary correlated continuum intermediate state (BCCIS) approximation [1]. The prior form of the transition matrix element is

$$T_{if} = \langle \psi_f^- | (V_i) | \chi_i \rangle.$$

We write the final wavefunction as

$$\psi_f^- = e^{\frac{\pi}{2}(\alpha_1 - \alpha_2)} \Gamma(1+i\alpha_1)\Gamma(1-i\alpha_2) e^{i\vec{k}_f \cdot \vec{R}_P} \phi_f(\vec{r}_P) {}_1F_1\{-i\alpha_1; 1; -i(v_f r_T + \vec{v}_f . \vec{r}_T\}{}_1F_1\{-i\alpha_2; 1; -i(k_f R_T + \vec{k}_f . \vec{R}_T)\},$$

The wavefunction in the initial channel is given by

$$\chi_i = \phi_i(\vec{r}_T)\xi_i^+(\vec{R}_T),$$

where $\phi_i(\vec{r}_T)$ is the initial bound state wavefunction and $\xi_i^+(\vec{R}_T)$ is the Coulomb distorted wave in the entrance channel. The initial bound wavefunction, $\phi_i(\vec{r}_T)$ of each molecular orbital is considered as a linear combination of atomic orbitals (LCAOs). The complete neglect of differential overlap method (CNDO) is used for the effective occupation electron analysis of the different molecular orbitals [2,3].

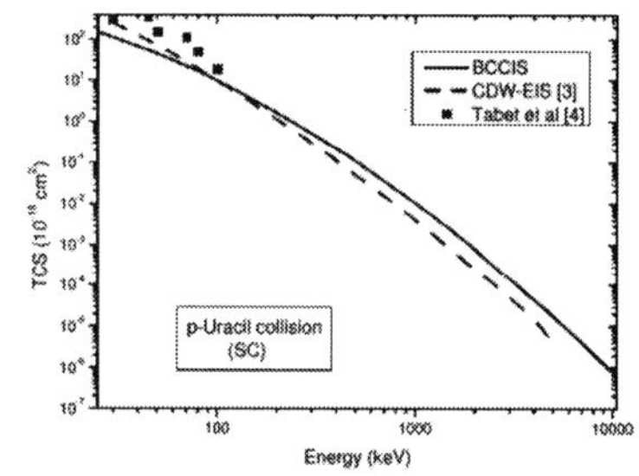

Figure 1. $p-Uracil$ collision for single charge (SC) transfer

The authors gratefully acknowledge the financial support from the Higher Education, Science & Technology and Biotechnology, Govt. of West Bengal, India under Grant No. 239(sanc.)/ST/P/S & T/16G-48/2017.

References

[1] Das M, Purkait M and Mondal C R 1998 *Phys. Rev. A* **57** 3573
[2] Galassi M E *et al* 2012 *Phys. Med. Biol.* **57** 2081
[3] Champion C *et al* 2012 *Phys. Med. Biol.* **57** 3039
[4] Tabet J *et al* 2010 *Phys. Rev. A* **82** 022703

*E-mail: mpurkait_2007@rediffmail.com

Fragmentation of hydrocarbons by collision. AGAT@ANDROMEDE.

T IdBarkach[1] *, M Chabot[1], K Béroff[2], S Della-Negra[1], J Lesrel[1] and S Diaz-Tendero[3]

[1]Institut de Physique Nucléaire d'Orsay (IPNO), CNRS-IN2P3, Univ. Paris Sud, Université Paris-Saclay, F-91406 Orsay, France

[2]Institut des Sciences Moléculaires d'Orsay (ISMO), CNRS, Univ. Paris Sud, Université Paris-Saclay, F-91405 Orsay, France

[3]Departemento de Quimica Modulo 13, Universidad Autonoma de Madird, 28409, Madrid, Spain

Synopsis We are studying small hydrocarbons produced in high velocity (3.4 a.u.) collision between CH_y^+ cations and He atoms. During the collision those hydrocarbons gain some energy and release this energy doing fragmentation. Thanks to the detector AGAT we have been able to measure fragmentation branching ratios for CH_y^{q+} (y=2-4, q=0-3). We also constructed semi-empirical breakdown curves for CH_y^{q+} using these experimental branching ratios.

The classical scenario of the main hydrocarbon gas phase synthesis in astrochemistry, starts with the reaction of C^+ ions with hydrogen atom or molecule. Then, the formed $CH_y^{(+)}$ species can react with another C atom or ion, and give heavier hydrocarbon species. Since $CH_y^{(+)}$ molecules are the mother of all the other hydrocarbons, reliable simulations in this field need precise both total reaction rates and branching ratio (BR). To document all the BRs, whatever the chemical or physical process at play, we construct, on both experimental and theoretical grounds, semi empirical Break Down Curve (BDC) [1]. These curves are internal energy dependent branching ratio as illustrated in Fig. 1 in case of CH_2 molecule.

The experiment was done using the AGAT silicon multidetector and the ANDROMEDE accelerator. CH_y^+ molecules produced at high velocity are collided with He atom at rest in the lab. Thanks to the experimental developments, all fragments, neutral or charged, are separately identified, allowing to resolve all fragmentation channels. Therefore we have been able to measure fragmentation branching ratio for CH_y^{q+} (n=2-4, q=0-3). In addition, using a CCD camera to detect neutral fragments [2], kinetic energy distribution (KED) of neutral fragments in each fragmentation channel is also measured.

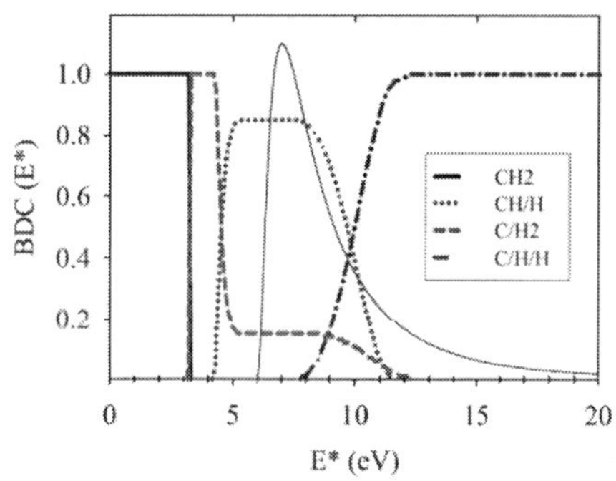

Figure 1. Semi empirical break down curve of the CH_2 molecule.

The semi empirical BDCs construct using these BRs and KEDs and with theoretical dissociation energies of [3], are found in good agreement with existing photodissociation, dissociative recombination and electron impact experimental BRs.

References

[1] Chabot M *et al* 2013 *Astrophys. J.* **771** 90
[2] Chabot M *et al* 2011 *Rev. Sci. Instrum.* **82** 10 103301
[3] Sanchez J P *et al* 2016 *J. Phys.Chem.* **120** 588

*E-mail: idbarkach@ipno.in2p3.fr

The low energy beamline of the FISIC experiment: current status of construction and performance

D Schury[1]*, A Méry[2], J M Ramillon[2], L Adoui[2], J-Y Chesnel[2], A Lévy[1], S Macé[1], C Prigent[1], J Rangama[2], P Rousseau[2], S Steydli[1], M Trassinelli[1], D Vernhet[1], A Gumberidze[3], Th Stöhlker[3,4,5], A Bräuning-Demian[3], C Hahn[4], U Spillmann[3] and E Lamour[1]

[1]Institut des Nanosciences de Paris, Sorbonne Université, CNRS UMR 7588, Paris, 75005, France
[2]CIMAP, CEA/CNRS/ENSICAEN/Université de Caen Normandie, Caen, 14050, France
[3]GSI Helmholtzzentrum für Schwerionenforschung, Darmstadt, 64291, Germany
[4]Friedrich-Schiller-Universität Jena, Jena, 07743, Germany
[5]Helmholtz-Institut Jena, Jena, 07743, Germany

Synopsis Ion-ion collisions between slow (kev/u) and fast (MeV/u) ions play an important role in for example astrophysical or inertial fusion plasmas as well as in ion-matter interaction. In this regime the energy transfer is maximum, as all primary electronic processes reach their maximum. At the same time up to today no reliable experimental data exists while being difficult to treat accurately by theory. We present the current status and performance of the low energy beam-line of the FISIC experiment which aims at filling in the blanks in this regime.

So far, ion-ion collisions in atomic physics were performed mainly in the context of magnetically confined plasmas using crossed beam devices in the low-energy domain where electron capture is the dominant process [1, 2] or fast-ion/plasma-discharge experiments with a charge state distribution as target [3, 4]. Up to today no reliable experimental data exists for fast (MeV/u) and slow (keV/u) ion collisions in the regime where the ion stopping power (energy transfer) is maximum. There all the primary electronic processes (electron capture, -loss and -excitation) reach their maximum and therefore the role of the individual processes on the collision is almost impossible to disentangle, while being of high importance in ion-matter interaction, including biological materials. To fill in the blanks we are designing a new ion-ion collision experiment, the FISIC project (**F**ast **I**on **S**low **I**on **C**ollisions) [5], a mobile experimental set-up, although being quite complex, being able to conduct crossed beam ion-ion collisions at different high energy ion beam facilities. Today we present the current status of the set-up and performance of the low-energy branch of FISIC, which will deliver a medially- to highly-charged pure low-energy ion beam which will then be crossed with high energy ions. Test results, mainly on the charge state purification [7], carried out at the ARIBE beam-line at the GANIL facility in Caen and at Sorbonne Université in Paris, utilizing the SIMPA ion source, will be presented. The obtained experimental data can be favourably compared with numerical simulation. Proposed experiments at different facilities such as CRYRING@ESR [8] or S3/Spiral2/GANIL [9] and the status of the integration of our beam-line into the existing infrastructure will be discussed.

*E-mail: daniel.schury@insp.upmc.fr

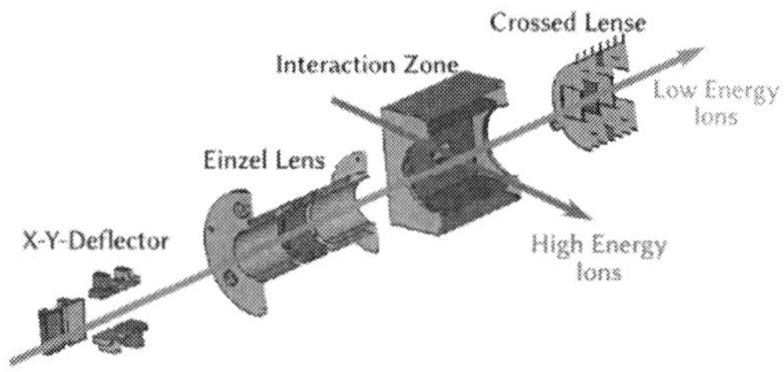

Figure 1. 3D model of the experimental area with indicated ion beams.

References

[1] Meuser S *et al* 1996 *Rev. Sci. Instrum.* **67** 2752
[2] Bräuning H *et al* 2005 *J. Phys. B* **38** 2311
[3] Fite WL *et al* 1960 *Phys. Rev.* **119** 663
[4] Cayzac W *et al* 2017 *Nat. Commun.* **8** 15693
[5] Lamour E *et al* 2013 *online* ANR-13-IS04-0007
[6] Gumberidze A *et al* 2010 *Rev. Sci. Instrum.* **81** 033303
[7] Schury D *et al* 2019 *Rev. Sci. Instrum.* In Review
[8] Lestinsky M *et al* 2016 *Eur. Phys. J* **225** 797-882
[9] Dechery F *et al* 2015 *Eur. Phys. J* **A 51** 66

Low energy fragments from O^{2+} + H_2 collisions following single and double electron removal from the target

Z Juhász[1*], V Vizcaino[2], J-Y Chesnel[2], STS Kovács[1], P Herczku[1], S Demes[1], R Rácz[1], S Biri[1] and B Sulik[1]

[1] Institute for Nuclear Research, Hungarian Academy of Sciences (Atomki), Debrecen, H-4026, Hungary
[2] Centre de Recherche sur les Ions, les Matériaux et la Photonique, Normandie Univ, ENSICAEN, UNICAEN, CEA, CNRS, CIMAP, Caen, 14000, France

Synopsis Fragmentation of H_2 by 20-keV O^{2+} was studied. A time of flight setup was developed to measure the low-energy fragment ions originating from singly ionized target molecules. High-energy fragments stemming from double electron removal were measured by an electrostatic spectrometer. The contribution of the low-energy fragments is significant but smaller than that of the energetic ones. Though single electron removal is more probable it less likely leads to dissociation. A statistical type model shows a good agreement with the experimental results.

Ion-impact-induced fragmentation of molecules in the gas phase is a key process in astrophysical environments and may provide important information for radiolysis studies. Emission of low-energy fragment ions is a strong output channel following the excitation and moderate ionization of the target molecules. However, its experimental investigation is difficult since low energy ions are sensitive to disturbing electric fields.

A time-of-flight measuring device was developed at Atomki for studying fragment ions in the 0.1-2 eV energy range to extend the results for 20-keV O^{2+} + H_2 collisions obtained by an electrostatic spectrometer at Ganil [1]. The emitted ions fly through a well-shielded field-free region before their detection on a 2D position sensitive detector. Their energy can be determined from their flight times. The detector can be rotated around the center of the chamber, so the angular distribution of the ions can be determined as well in a wide range.

We measured the energy and angular distribution of the emitted protons. No significant anisotropy was observed except for increased background-like contribution around 90° observation angle due to binary knock out by the projectile.

The measured cross sections show a maximum in the 0.1-0.3 eV range, which is most likely due to dissociation of H_2^+ ions created by single electron capture from the target. This is in accordance with the results of a recent thermodynamic model for molecular collisions [2]. Single capture does not always lead to dissociation of H_2^+. Its ground state is stable, dissociation occurs if it was created with internuclear separation far from equilibrium. Our ab initio quasi-molecular calculations for single capture show an order of magnitude higher cross sections than for the observed total low-energy proton production indicating that dissociation is indeed only partial.

At higher energies, a peak appears at 9.7 eV due to Coulomb explosion of the doubly ionized hydrogen molecules. It corresponds to about 90 % of the emitted protons. The thermodynamic model reproduces the observed cross section. It shows that transfer ionization has a similar contribution to double electron removal from the target as double capture. The energy of the peak indicates that Coulomb explosion is initiated at nearly equilibrium internuclear separation of H_2. In experiments by O^{5+} projectiles, it was found that the internuclear separation changes during the collision process and it increases with decreasing impact energy [3]. The thermodynamic model shows peakshift of -0.9 eV compared to the experiment, indicating an overestimation of this effect. But its existence suggests that fragmentation efficiency in single electron removal, which is dependent on the internuclear separation, may be sensitive on projectile energy.

This work was supported by Hungarian NRDIO-OTKA Grants No. K109440 and K128621, the French-Hungarian Cooperation Program PHC Balaton (No. 38620NH / TÉT_16-1-2016-0126) and the International Scientific Cooperation Project (PICS-CNRS No. 7739 / NKM-115/2017), and by the Hungarian NIIF Institution.

References

[1] Frémont F et al 2006 *Phys. Rev. A* **74** 052707
[2] Juhász Z 2016 *Phys. Rev. A* **94** 022707
[3] Sobocinski P et al 2002 *J. Phys. B* **35** 1353

* E-mail: zjuhasz@atomki.hu

Quasi-equilibrium in charge-state evolution for 1.0 and 2.0 MeV/u carbon ions after carbon foil penetration

A M Imai[1 *], M Sataka[2], M Matsuda[3], S Okayasu[3], K Kawatsura[4], K Takahiro[5], K Komaki[6], H Shibata[7] and K Nishio[3]

[1]Department of Nuclear Engineering, Kyoto University, Kyoto 615-8540, Japan
[2]Research Facility Center for Science and Technology, University of Tsukuba, Ibaraki 305-8573, Japan
[3]Japan Atomic Energy Agency (JAEA), Tokai, Ibaraki 319-1195, Japan
[4]Theoretical Radiation Research Laboratory, Kyoto 606-0966, Japan
[5]Kyoto Institute of Technology, Kyoto 606-8585, Japan
[6]RIKEN, Wako, Saitama 351-0198, Japan
[7]The Institute of Scientific and Industrial Research, Osaka University, Osaka 567-0047, Japan

Synopsis After our observation of "quasi-equilibrium" in charge-state evolution for 2.0 MeV/u sulfur ions after carbon foil penetration, we have extended the initial ions to 1.0 and 2.0 MeV/u carbon ions. A significant quasi-equilibrium was observed for 1.0 MeV/u C+C collisions, where the equilibrium mean charge-state was 4.9, while for 2.0 MeV/u C+C collisions, where the equilibrium mean charge-states was 5.6, was a weak quasi-equilibrium.

Equilibrium and pre-equilibrium charge-state distributions for 1.0 and 2.0 MeV/u C^{q+} (q = 1–6) ions after penetrating carbon foils have been investigated experimentally after similar investigation of ours using 2.0 MeV/u S^{q+} (q = 6–16) initial ions[1]. In the previous measurements, charge-state distributions, mean charge-states, and distribution widths for projectile ions without K-shell holes, S^{q+} (q = 6–14), once coincided at a target thickness of 6.9 μg/cm^2 (12.3 in mean charge-state), showing a "quasi-equilibrium," and simultaneously evolved to establish the real equilibrium (12.68 in mean charge-state) when the foil thickness was further increased, whereas those for projectile ions with K-shell hole(s), $S^{15,16+}$, evolved directly to the real equilibrium, established at a target thickness of around 100 μg/cm^2 or greater. Simulations using the ETACHA code[2] as well as a solution of simple rate equations showed the quasi-equilibrium was brought by a difference between the reaction-rates for K- and L-shell processes.

In the present measurements, we have observed a significant quasi-equilibrium for 1.0 MeV/u C+C collisions, where the mean charge-states for $C^{1,3,4+}$ initial ions coincided even at the thinnest measured foil thickness and evolved simultaneously until the real equilibrium of 4.9 established at around 3.0 μg/cm^2 in the target thickness. As for 2.0 MeV/u collisions, whose equilibrium mean charge-state was 5.6, a weak quasi-equilibrium was observed.[3]

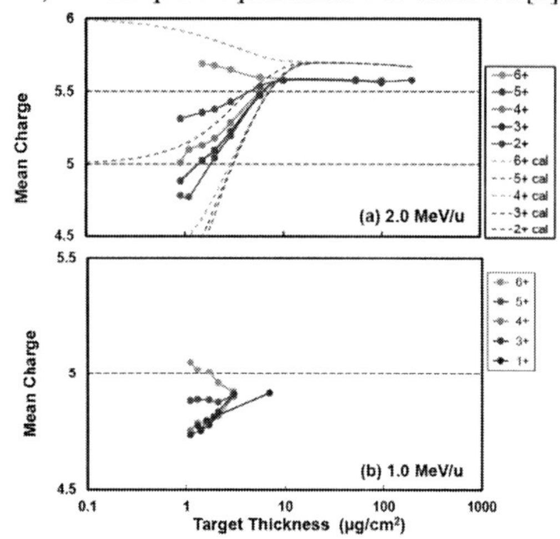

Figure 1. Mean charge-state evolution for (a) 2.0 and (b) 1.0 MeV/u C^{q+} (q = 1–6) initial ions after carbon foil penetration.

References

[1] Imai M *et al* 2009 *Nucl. Instr. and Meth. B* **267** 2675
[2] Lamour E *et al* 2015 *Phys. Rev. A* **92** 042703
[3] Imai M *et al* 2015 *Nucl. Instr. and Meth. B* **354** 172

* E-mail: imai@nucleng.kyoto-u.ac.jp

Journal of Physics: Conference Series **1412** (2020) 162014 doi:10.1088/1742-6596/1412/16/162014

Single and double electron capture associated with characteristic K-shell X-ray emission for $F^{9+,8+,7+}$+Ar

D S La Mantia*, **P N S Kumar and J A Tanis**

Department of Physics, Western Michigan University, Kalamazoo, MI 49008, USA

Synopsis Cross sections for target Ar K-shell ionization associated with single and double electron capture, as well as the corresponding cross sections for total electron capture and the projectile F K x rays, were determined for 1.8-2.2 MeV/u $F^{9+,8+,7+}$ projectiles. The Ar K x-ray coincidence cross section ratios for double to single capture are found to vary strongly with the incident charge state, with the ratios for F^{9+} well exceeding unity.

Ion-atom collisions in experiments provide a tool for the investigation of fundamental physics. Charge-changing cross sections and K-shell ionization are of particular importance in testing various theories. Cross sections for target Ar K-shell ionization associated with single and double electron capture, as well as the corresponding cross sections for total electron capture and the projectile F K x rays, were determined for 1.8-2.2 MeV/u $F^{9+,8+,7+}$ projectiles. The present investigation follows work done for fully-stripped fluorine [1] and oxygen [2] projectile ions on Ar.

This work was performed using the tandem Van de Graaff accelerator facility at Western Michigan University (WMU). A beam of F^{9+}, F^{8+}, or F^{7+} ions was directed toward a differentially-pumped gas cell with the target Ar pressure set to the single-collision regime (8 mTorr). A Si(Li) x-ray detector mounted at 90° to the beamline collected photons and separate silicon surface-barrier detectors collected the charge-changed projectiles. An event-mode data acquisition system was employed to assign the measured x rays to the corresponding single (Q-1) or double (Q-2) charge-changed particles.

The total capture cross sections behave as expected, decreasing strongly with increasing incident projectile energy and decreasing charge state. The cross sections for electron capture associated with characteristic projectile F K x-ray production behave essentially as anticipated, decreasing with projectile energy and being several orders of magnitude smaller than the total capture cross sections. However, the double to single electron capture cross section ratios associated with target Ar K-shell ionization are found to remain relatively constant with projectile energy. Most surprising, the ratios of double to single capture for incident F^{9+} projectiles are found to well exceed unity for all energies, with values of ~2.5 as shown in Fig. 1.

At these projectile energies, when a target K-shell electron is liberated the likelihood of capture is relatively small. The emission of a target Ar K x ray followed by the capture of one or two electrons implies a closer collision and it may be more likely for double capture to occur for F^{9+} incident ions than single capture.

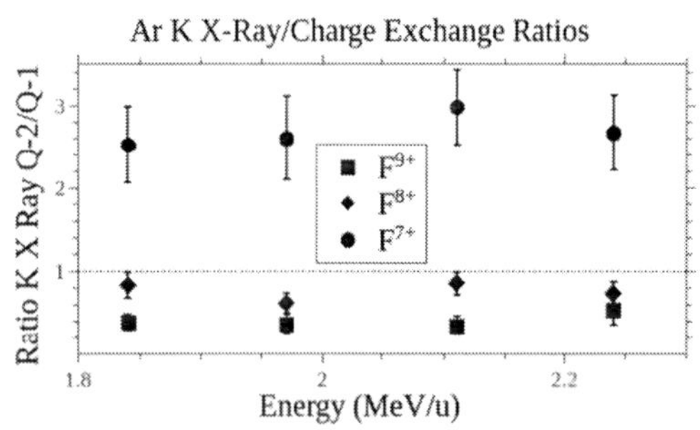

Figure 1. Ratios of double to single electron capture (Q-2/Q-1) associated with target Ar K-shell ionization cross sections for F^{7+}, F^{8+}, and F^{9+} projectile ions. The average relative uncertainties for the ratios for these projectiles are 30%, 20%, and 20%, respectively, in going from 7+ to 9+.

This work was supported in part by National Science Foundation Grant PHY-1707467.

References

[1] La Mantia D *et al* 2017 *Nucl. Inst. Meth. B* **408** 107

[2] Houck J *et al* 1997 *Phys. Rev. A* **56** 1954

*E-mail: david.s.lamantia@wmich.edu

Convergent close-coupling approach to proton collisions with helium

S U Alladustov[1]*, I B Abdurakhmanov[1], A S Kadyrov[1], I Bray[1] and K Bartschat[2]

[1]Department of Physics and Astronomy, Curtin University, GPO Box U1987, Perth, WA 6845, Australia
[2]Department of Physics and Astronomy, Drake University, Des Moines, Iowa 50311, USA

Synopsis We develop a two-center wave-packet convergent close-coupling approach to proton collisions with helium including electron exchange. The convergence of the final results is studied in terms of the included projectile- and target-centered pseudostates. We present electron-capture and single- and double-ionisation cross sections in the projectile energy range from 15 keV to 1 MeV. We also provide partial cross sections for electron capture and direct excitation into the $n = 2$ shell states of hydrogen and helium, where n is the principal quantum number of the final state. Our results are in good agreement with available experimental data and other calculations.

Proton collisions with the helium atom are described by the four-body Schrödinger equation, which cannot be solved analytically. We have developed a two-center wave-packet convergent close-coupling (WP-CCC) approach to the problem including electron exchange in the rearrangement channel. This is achieved by expanding the total scattering wave function in a basis of target- and projectile-centered pseudostates. The Schrödinger equation is transformed into a set of coupled differential equations for the unknown expansion coefficients. By solving the resulting set of coupled equations we obtain transition amplitudes into all included channels. The latter are used to calculate the necessary cross sections. We also study the convergence of the results by increasing the basis size.

In our approach the target is treated in the frozen-core approximation. Correlations between the electrons are taken into account. The He wave functions and energy levels are obtained using a numerical approach.

To account for direct ionisation of helium and electron capture to the continuum, positive-energy states for both the target and projectile are included. The continuum states are constructed with the wave-packet method by discretizing the continuum of the target and the hydrogen atom formed after electron capture by the projectile. This allows us to include pseudostates with arbitrary energies and distribution.

We calculated cross sections for electron capture and single ionisation of helium in the projectile energy range from 15 keV to 1 MeV. A comparison of the WP-CCC results for electron capture with the experimental data of [1]–[4] and the time-dependent density-functional-theory (TDDFT) calculations of Baxter and Kirchner [5] is shown in Fig. 1.

*E-mail: shukhrat.alladustov@postgrad.curtin.edu.au

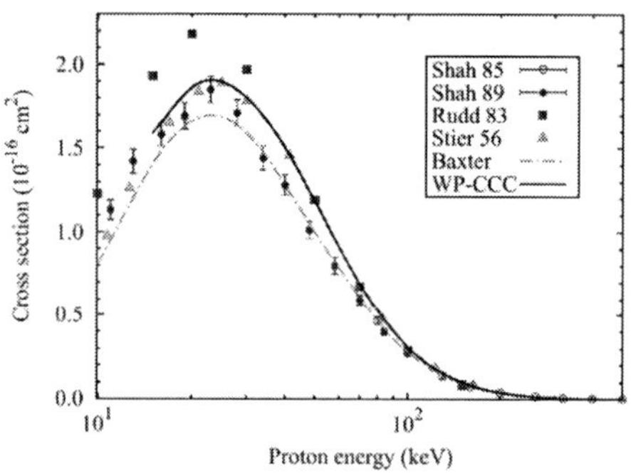

Figure 1. Total cross section for electron capture in p+He($1s^2$) collisions as a function of the incident proton energy. The present WP-CCC results are represented as a black solid line. Also shown are the experimental data from [1]–[4] and the calculations of Baxter and Kirchner [5].

Work supported by the Australian Research Council and the United States National Science Foundation.

References

[1] Shah M B and Gilbody H B 1985 *J. Phys. B: At. Mol. Phys.* **18** 899
[2] Shah M B *et al.* 1989 *J. Phys. B: At. Mol. Opt. Phys.* **22** 3037
[3] Rudd M E *et al.* 1983 *Phys. Rev. A* **28** 3244
[4] Stier P M and Barnett C F 1956 *Phys. Rev.* **103** 896
[5] Baxter M and Kirchner T 2016 *Phys. Rev. A* **93** 012502

Charge exchange UV-Visible and EUV spectroscopy of multiply charged heavy ions

N Numadate[1,2]*, Y Yamada[1], T Ohna[1]† and H Tanuma[1]‡

[1]Department of Physics, Tokyo Metropolitan University, Tokyo 192-0397, Japan
[2]Institute for Laser Science, The University of Electro-Communications, Tokyo 182-8585, Japan

Synopsis To provide spectroscopic data of multiply charged heavy ions related not only the development of EUV light and soft X-ray sources and also optical measurements of neutron star mergers, charge exchange spectroscopy was performed using an ECR ion source with two spectrometers in EUV and UV-Visible regions.

Spectroscopic data for heavy element atoms are needed for various fields in fundamental and applied physics and technology. However, their database is very far from complete sets of data.

For example, the development of 13.5 nm extreme ultra-violet (EUV) light sources for the next generation semi-conductor lithography had needed spectroscopic data of multiply charged Xe and Sn ions. To provide these data, we have performed charge exchange spectroscopic (CXS) measurements using a 14.25 GHz electron cyclotron resonance (ECR) ion source [1]. Usually CXS is used for measurements of charge-transfer cross-sections. But, this method is very useful for measurements of transtion wavelengths of charged-selected multiply charged atomic ions. Now new light sources for more shorter wavelengths should be developed for another EUV lithography technology and soft X-ray microscope for observation of cells using the water window [2].

The origin of heavier elements than iron in the universe has still remained an open question. Supernova explosion has been one strong candidate of the heavy element synthesis. But, recent works have suggested that the contribution of r-process in neutron star mergers must be comparable or dominant. While supernova can be investigated in observation of SNR (supernova remnants), neutron star mergers were difficult to be observed. However, in 17 August, 2017, the first gravitation wave from a neutron star merger was detected and various kinds of electromagnetic waves from gamma-ray to micro-wave were observed after this event GW170817 [3]. By optical telescopes, a promising optical and near-infrared counterpart was discovered and named as SSS17a [4]. The optical spectra and brightness of SSS17a show modestly good agreement with prediction by theoretical radiation transport calculation [5]. However, more accurate simulation needs huge amount of spectroscopic date on multiply charged heavy elements. Therefore, theoretical calculations for several heavy element have performed by our collaborators [6, 7]. To confirm the accuracy of theoretical calculations, the corresponding experimental measurements must be carried out by ourselves.

The multiply charged ions of Er and Bi were produced with the ECR ion source, and the extracted ion beam was fed into a collision gas cell after the charge-selection by a dipole magnet. The photo-emissions from the collision region have been observed with two different spectrometers, namely a grazing incident spectrometer equipped with a electric cooled CCD camera for EUV region and a Czerny-Turner mounting spectrometer equipped with a liquid nitrogen cooled CCD camera for ultraviolet-visible region.

References

[1] Ohashi H *et al* 2010 *J. Phys. B* **43** 065204
[2] Myneni *et al* 1999 *Science* **286** 1335
[3] Abbott B P *et al* 2017 *ApJL* **848** L12.
[4] Utsumi Y *et al* 2017 *Publ. Astron. Soc. Japan* **69** 101
[5] Tanaka M *et al* 2017 *Publ. Astron. Soc. Japan* **69** 102
[6] Tanaka M *et al* 2018 *ApJ* **852** 109
[7] Gaigalas G *et al* 2019 *ApJS* **240** 29

*E-mail: numadate-naoki@ed.tmu.ac.jp
†E-mail: ohna-takumi@ed.tmu.ac.jp
‡E-mail: tanuma-hajime@tmu.ac.jp

Electron correlation in simultaneous electron transfer and excitation: results of new 2eAOCC calculations and comparison to experiment in 50-500 keV $He^+ + H/H_2$ collisions

A Dubois[1*] **and T J M Zouros**[2†]

[1] Sorbonne Université, CNRS, Laboratoire de Chimie Physique-Matière et Rayonnement, Paris, France
[2] Department of Physics, University of Crete, P.O. Box 2208, GR 71003 Heraklion, Greece

Synopsis New two-electron semiclassical atomic orbital coupled channel calculations (2eAOCC) are compared to experiment and to older coupled channel calculations. Progress in these type of calculations is assessed in an effort to investigate the effect of electron correlations in this fundamental collision system.

The study of multi-electron processes occurring in ion-atom collisions is still a theoretical challenge due to the importance of static and dynamic correlation in these transitions. It is therefore interesting to investiage the dynamics of true two-electron systems, where the transfer of an electron and the excitation of another is likely. From this point of view, He^+- H collisions, inducing transfer-excitation (TE) processes to autoionizing states of He, can be considered as a benchmark system.

Here, we present cross sections for such TE processes, especially

$$He^+ + H \rightarrow He(2l2l'\ {}^1L) + H^+$$

for impact energies ranging from 50 to 500 keV. Results are obtained with a full configuration interaction (CI) close coupling semiclassical approach [1]: the two electrons are taken into account and the time-dependent Schrödinger equation is solved with an unprecedented large basis set to allow for convergence of the non perturbative scheme. These results are compared with three-decade old theoretical results [2], as well as with experimental data obtained with zero-degree Auger projectile spectroscopy for dihydrogen molecular target [3].

The discussion and interpretation of the results in terms of one- and two-step mechanisms, so called resonant (correlated) transfer-excitation (RTE) [4] and non resonant transfer-excitation (NTE) [5], respectively, are particularly interesting when comparing with converged two-electron ab initio calculations, in which all mechanisms are described coherently. We shall present cross sections for production of $2p^2\ {}^1D$ (see Fig. 1), $2p^2\ {}^1S$ and $2s2p\ {}^1P$ autoionising states.

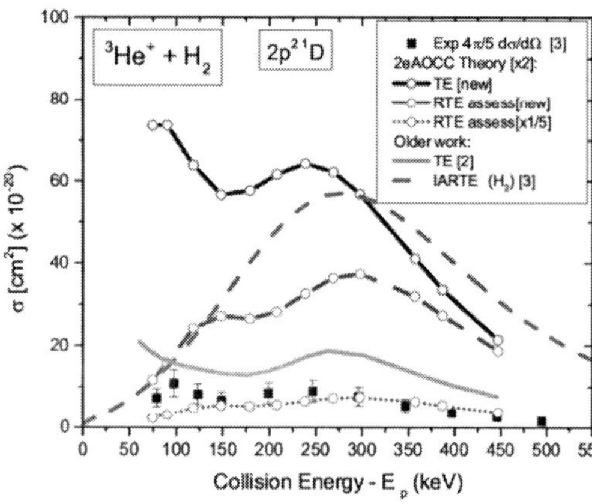

Figure 1. Comparison of cross sections for the production of the He($2p^2\ {}^1D$) state in collisions with hydrogen. Theory: He^+ + H, 2eTE cross sections multiplied by two to account for H_2 target in experiment (squares) [3]. Impulse Approximation RTE (IARTE) (long dashes violet).

* E-mail: alain.dubois@sorbonne-universite.fr
† E-mail: tzouros@physics.uoc.gr

References

[1] Gao J W, Wu Y, Sisourat N, Wang J G and Dubois A 2017 *Phys. Rev. A* **96** 052703
[2] Fritsch W and Lin C D 1988 *Phys. Rev. Lett.* **61** 690
[3] Zouros T J M, Schneider D and Stolterfoht N 1988 *J. Phys. B* **21** L671
[4] Tanis J A *et al* 1982 *Phys. Rev. Lett.* **49** 1325
[5] Brandt D 1983 *Nucl. Instrum. Meth. B* **214** 93

Energy deposition of highly charged ions transmitted through single layer MoS_2

S Creutzburg [1,2*], J Schwestka [3], H Inani [4], M K Tripathi [4], P L Grande [5], R Heller [1], N Klingner [1], A Niggas [3], R Kozubek [6], L Madauß [6], S Facsko [1], J Kotakoski [4], M Schleberger [6], F Aumayr [3] and R A Wilhelm [1,3]

[1] Helmholtz-Zentrum Dresden-Rossendorf, Ion Beam Center, Dresden, Germany
[2] Technische Universität Dresden, Dresden, Germany
[3] TU Wien, Institute of Applied Physics, Vienna, Austria
[4] University of Vienna, Faculty of Physics, Vienna, Austria
[5] Federal University of Rio Grande do Sul, Ion Implantation Laboratory, Porto Alegre, Brasil
[6] University Duisburg-Essen, Faculty of Physics and CENIDE, Duisburg, Germany

Synopsis Highly charged ions (HCIs) are an efficient tool for the perforation of suspended 2D materials. Only a fraction of their potential energy is transferred to the atomically thin target during the very short interaction time and is available for pore formation. Charge exchange spectra were measured for highly charged xenon ions transmitted through suspended, single layer MoS_2 in order to determine the deposited potential energy available for pore formation. Additionally, charge exchange dependent ion stopping responsible for kinetic sputtering was measured simultaneously.

2D materials have come into the spotlight because their intriguing features make them promising candidates for future electronic and optoelectronic applications. In order to utilize 2D materials in specific applications surface-sensitive modification techniques have to be applied. Highly charged ions have been demonstrated as a promising tool for nanostructuring of 2D materials [1, 2]. They store potential energy in the keV range, which is deposited in close vicinity to a surface upon neutralization. The impact leads to strong electronic excitations, which may subsequently result in nanostructure formation. Even in freestanding 2D layers pore formation [1, 2] induced by HCI impact was observed despite their ultra-small thickness limiting the fraction of deposited energy. For the present study, charge exchange of highly charged xenon ions passing freestanding monolayers of MoS_2 was measured in coincidence with the time-of-flight (TOF) of the projectiles. Angle and charge state resolved spectra were recorded. A bimodial exit charge state distribution is observed as shown in Fig. 1. The distribution at high outgoing charge states is characterized by small scattering angles. This points to collisions occurring in a narrow range of larger impact parameters with the target atoms. The distribution at low charge states accompanied by larger scattering angles originates from close collisions with the target atoms. A model description for the two distinctly separated charge state distributions is discussed based on a spatially confined Interatomic Coulombic Decay process. Our experimental data is put into context of recently observed pore formation on freestanding MoS_2 by HCIs [2].

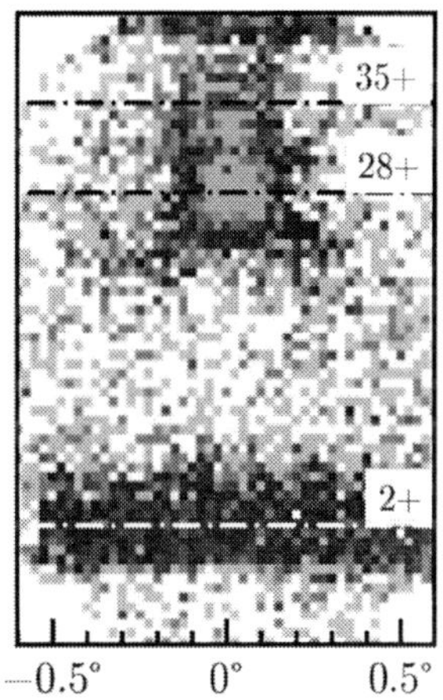

Figure 1. Charge state and scattering angle resolved spectrum for incident xenon ions of charge state 35+ and kinetic energy of 100 keV transmitted through single layer MoS_2.

*E-mail: s.creutzburg@hzdr.de

References

[1] Wilhelm R A *et al* 2015 *2D Mater.* **2** 035009
[2] Kozubek R *et al* 2019 *J. Phys. Chem. Lett.* **10** 904-910

Electron capture by low velocity highly charged ions A tribute to Michel Barat

P Roncin[1] * **and L Guillemot**[1]

[1]Institut des Sciences Moléculaires d'Orsay, CNRS, Univ. Paris-Sud, Université Paris-Saclay, Orsay, France

Synopsis Michel Barat passed away last November at the age of 80. We briefly review the contributions under his supervision where we revisited atomic collisions in the blooming context of highly charged ions.

One of us started his PHD by constructing the setup that Michel had imagined. A 2D position sensitive detector in the focal plane of an electrostatic analyser [1]. Such detector was not available but taking a proposal from D. de Bruijn, FOM Amsterdam, we built it and were able to record, at once, the energy and angular scattering profile in coincidence with recoil ions and emitted photons (UV and Xrays). We were ready to do physics with these weird ions *e.g.* O^{8+}. **In what sense is the captured electron rotating?** With NO Andersen we could show that the captured electron tends to rotate along the intuitive direction [2]; A runner trying to jump into a merry go round will find which way to go. However this should not hold in the molecular regime where the electron motion is much faster that of the nuclei. V. N. Ostrovsky designed an elegant text-book quasi-molecular interpretation, that we calculated with A. Dubois *et al.* **Why do we observe higher** n **value?** The selective electron capture into a given n shell was well explained by the coulombic barrier model but evidence for higher n capture were reported. Michel could interpret the population of higher n values by transitions at inner crossings [3] predicted by the Barat-Lichten-Fano electron promotion model. **Are the electrons captured simultaneously or one after the other?** We observed both situations [4] and we reformulated the question differently. What is the driving mechanism in simultaneous double electron capture ? With V Sidis, we showed that double electron capture can be explained [5] by a second order effect without the need of di-electronic interaction. The same mechanism was found responsible for other two electron process such as Transfer-Excitation.**How do two highly excited electrons manage to avoid autoionisation?** The data indicated an initial population of quasi symmetric doubly excited states *e.g.* O^{6+}*(4l,4l')* [6] way above several ionisation limits and predicted to be auto-ionising. We showed that a significant part of the *(4l,4l')* population transfers to the straddling *Rydberg* by di-electronic interaction and calculate the effect with H. Bachau.

Figure . Michel Barat played a key role to promote the physics of atomic and molecular collisions.

With A. K. Kazansky [7] we showed that the electric field of the receding target transforms these linear Rydberg into circular Rydberg states where both electron orbits do not overlap anymore. With S. Martin we identified the population of long-lived triply-excited double Rydberg.In less than ten years Michel *et al* published more than 40 papers, always with great enthusiasm. Upon turning towards atomic collisions at surfaces we of course started by constructing an ambitious 2π multi coincidence setup [8].

*E-mail: philippe.roncin@u-psud.fr

References

[1] Roncin P *et al* 1986 *J. Phys. E.* **19** 37
[2] Roncin P *et al* 1990 *Phys. Rev. Lett.* **65** 3261
[3] Roncin P *et al* 1990 *J. Phys. B.* **23** 1215
[4] Guillemot L *et al* 1990 *J. Phys. B.* **23** 4293
[5] Barat M *et al* 1990 *J. Phys. B.* **16** 2811
[6] Roncin P *et al* 1991 *EuroPhys. Lett* **16** 551
[7] Kazanski AK *et al* 1994 *J. Phys. B.* **27** 5537
[8] Morosov VA *et al* 1996 *Rev. Sci. Inst.* **67** 2163

Kinetic energy release (KER) of vibrationally cold HeH^+: experiments at the reaction microscope in the Frankfurt low-energy storage ring (FLSR)

J C Müller*[1], R Dörner[1], L Ph H Schmidt[1], M S Schöffler[1] and K E Stiebing[1]

[1] Institute for Nuclear Physics, Goethe University, Max von Laue Str. 1, D-60438 Frankfurt am Main, Germany

Synopsis The installation of a reaction microscope (COLTRIMS [1]) in the electrostatic Frankfurt Low Energy Storage Ring (FLSR) [2] has been completed. A first experiment, exploring the KER for dissociative electron transfer to HeH^+ ions at 20 keV from a cold supersonic He gas target have been carried. Even at the moderate vacuum of 10^{-8} mbar a change of the KER and the Q value distribution due to cooling of the molecular ions during the storage time can be observed. In this contribution the final experiment, performed after heating the ring will be discussed.

The HeH^+ molecular ion instantly decays into Helium and Hydrogen after capturing an electron. The kinetic energy release (KER) changes with the electronic state of the HeH molecule after capturing the electron, while the population of the excited states depends on the vibrational level of the HeH^+ ion [3].

An exploratory measurement of a 20 keV HeH^+ ion beam, stored in FLSR, had been carried out, where the fragmentation was measured in coincidence with recoil ions from the residual gas (Hydrogen). These experiments have shown that the KER distribution changes with storage time [4].

To identify the electronic state of the HeH^+ molecule, the Q value of the reaction has to be measured with sufficient accuracy as well. For this purpose, a supersonic gas-jet target and a reaction microscope were installed in the storage ring, allowing for a measurement of both the KER and the Q value of the reaction:

$$HeH^+ + He \rightarrow He + H + He^+ + KER$$

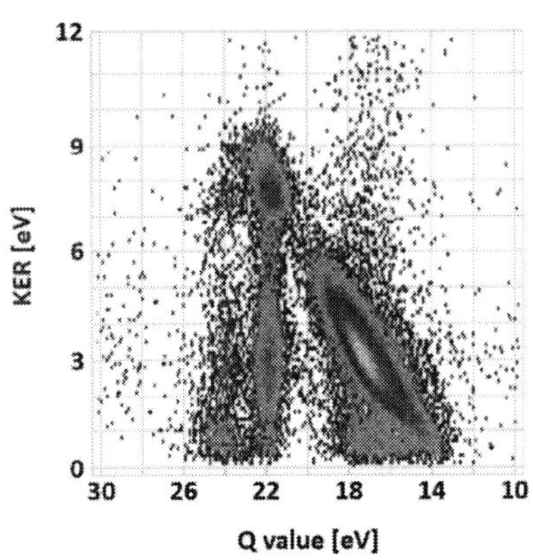

Figure 1: Dissociative electron capture of vibrationaly hot HeH^+. Horizontal axis KER, vertical axis Q value.

A first measurement of this reaction had been performed at a still moderate residual gas pressure of $1x10^{-8}$ mbar. The KER-Q-value distributions at the start of the storage time and after 100 ms, given in Figure 1 and 2 respectively, show a distinctly sharper KER- and Q-value distribution after cooling.

The results from the final measurement at a residual gas pressure of better than $1x10^{-10}$ mbar will be presented at the conference. and especially the contributions to the KER at about 1 eV will be discussed.

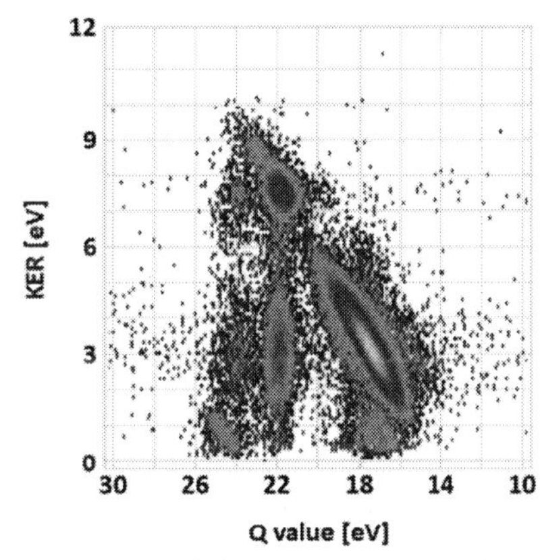

Figure 2: Dissociative electron capture of vibrationally cold HeH^+ Horizontal axis: KER, vertical axis Q value.

References

[1] Dörner R *et al* 2000 *Phys Rep.* **330** 95
[2] Stiebing K E *et al* 2010 *Nucl. Instr. Meth.* **A614** 10
[3] Theodorakopoulos G *et al* 1987 *J. Phys. At. Mol. Phys.* **20** 2339
[4] Müller J C *et al* 2017 *J. Phys. Conf. Series* **875** 092002

E-mail: janmueller@atom.uni-frankfurt.de

Production of molecular dications by electron impact investigated by (e, e+ion) measurement

S Xu[1*], Y Wang[1], C Ma[1], Y Li[1], J Zhou[1], L Ding[1], Z F Xu[1], A Dorn[2] and X Ren[1,2†]

[1]School of Science, Xi'an Jiaotong University, Xi'an 710049, China
[2]Max Planck Institut für Kernphysik, 69117 Heidelberg, Germany

Synopsis We present (e, e+ion) measurements for double ionization of nitrogen and oxygen molecules employing the reaction microscope in which the momentum vectors for final-state charged reaction products are determined experimentally. By measuring the final state electrons in coincidence with the molecular dication, the related electronic states and, thus, the production mechanisms of these dications could be identified and studied in detail.

When a molecules is populated to doubly charged states, it is usually unstable and may undergo fragmentation. However it is also possible that the molecular dication is metastable and could survive for a quite long time. The mechanisms for formation of metastable dications are of great interest in many applied fields, such as plasma devices, radiation chemistry and biology, planetary atmospheres and the physics of cometary tails. In this respect, studies of electron-molecule collisions have made tremendous contributions to our understanding of the reaction mechanisms [1].

In an electron-impact ionization process, the molecular dications can be produced by two different pathways, (1) the direct double ionization process with both electrons being removed in the collision with the projectile and (2) the Auger process following single ionization of an inner shell electron. In Refs. [2,3] these two pathways are identified according to the ratio between doubly charged and singly charged parent molecules as function of the incident electron energy. If both the parent dication and the scattered/emitted electron from one collision event could be measured in coincidence, then the related electronic state information could be obtained via the energy spectra of scattered projectile and emitted electron. This will provide direct information to determine the pathways for formation of metastable dications.

We have obtained such a measurement using the reaction microscope technique in which the momentum vectors and, consequently, the kinetic energies for all final-state charged reaction products are determined in coincidence. In our measurements the molecular nitrogen (N_2) or oxygen (O_2) dication produced in collisions with electrons of 120 eV impact-energy is detected in coincidence with either the scatted electron or one of the emitted electrons, in a so-called (e, e+ion) measurement. During the offline data analysis, the related electronic states could be determined through the energy loss spectrum while the energy distribution of the emitted electron may present the typical features of an Auger process if the dication is produced through pathway (2). Figure 1 presents the time-of-flight (TOF) vs. position 2D plot of the doubly charged and singly charged parent cations of N_2. The background signal around the N_2^{2+} dication is quite weak, indicating that N^+ fragments are well eliminated in our measurement. Details of this study will be presented at the conference.

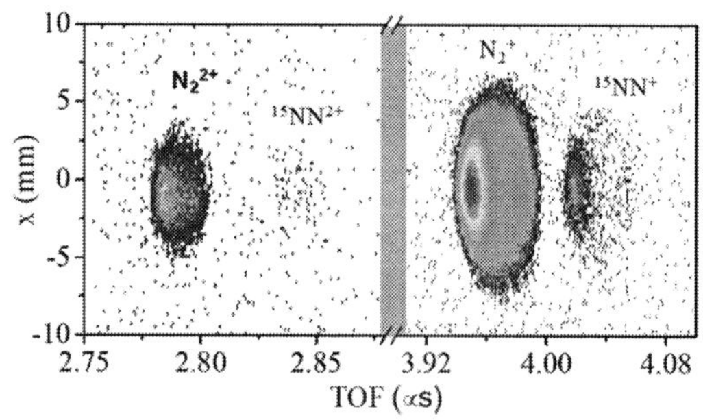

Figure 1. TOF vs. x-coordinate of the impact position for doubly and singly charged parent cations of N_2.

References

[1] Bartschat K *et al* 2016 *PNAS* **113** 7026
[2] Sigaud L *et al* 2018 *Phys. Rev. A* **98** 052701
[3] Sigaud L *et al* 2017 *Phys. Rev. A* **95** 012704

* E-mail: xushenyue@xjtu.edu.cn
† E-mail: ren@mpi-hd.mpg.de

Three-body fragmentation dynamics of $CH_2CCH_2^{3+}$ investigated by Ne^{8+} impact

C Ma[1], S Xu[1*], D M Zhao[2], D L Guo[2], S C Yan[2], W T Feng[2], X L Zhu[2] and X Ma[2†]

[1]School of Science, Xi'an Jiaotong University, Xi'an 710049, China
[2]Institute of Modern Physics, Chinese Academy of Sciences, Lanzhou 730000, China

Synopsis The three-body fragmentation dynamics of triply charged allene ($CH_2CCH_2^{3+}$) induced by 50-keV/u Ne^{8+} ion impact is studied by measuring the charged fragments in triple coincidence using reaction microscope. We focus on the fragmentation dynamics of two completely measured fragmentation channels, $H^++CH^++C_2H_2^+$ and $H^++CH_2^++C_2H^+$. Both concerted and sequential dissociation mechanisms are observed for these two channels. The estimation of relative contributions of different fragmentation pathways is presented.

Hydrocarbon molecules widely exist in the nature. The investigation of the fragmentation mechanisms of hydrocarbons is not only of fundamental interest but also important in many application fields such as plasma physics, chemistry of planetary atmospheres, and evolution of the interstellar mediums. Nevertheless, the three-body fragmentation of hydrocarbons has obtained much less attention, due to the fact that a hydrocarbon molecule usually consists of more than three nuclei leading to a much more complicated fragmentation mechanisms.

In this work, we chose allene, CH_2CCH_2, as the target and studied experimentally the three-body fragmentation dynamics of allene induced by Ne^{8+} ion. The projectile energy is chosen as 50-keV/u corresponding to the velocity of 1.4 atomic unit. With this energy the Ne^{8+} ion passes through the allene molecule within time scale of ~100 as which is much shorter than that of the nuclei motion. Consequentially the structure of the precursor $CH_2CCH_2^{3+}$ is the same as neutral molecule without any deformation. This is very different from ionization and dissociation induced by intense laser fields for which the interaction usually sustains for tens of femtoseconds [1, 2].

For analyzing the time of flight (TOF) correlation spectra, three-body fragmentation channels $H^+ + CH^+ + C_2H_2^+$ and $H^+ + CH_2^+ + C_2H^+$ are investigated in detail. By analysis with Newton diagram, Dalitz plot and kinetic energy correlation spectra, concerted and sequential dissociation mechanisms are identified for these two channels. As shown in Figure 1, in the Newton diagram of channel $H^+ + CH_2^+ + C_2H^+$ two most intense areas are originated from concerted fragmentation process. In addition, two different kinds of semicircular structures are observed. For inner semicircular structures (marked as blue solid semicircules), it is a typical feature of the fragmentation process for which one CH bond and one CC bond in $CH_2CCH_2^{3+}$ break sequentially. Furthermore, the outer semicircular structures (marked as grey dashed semicirculars) indicate that the sequential fragmentation process in which breakage of CC bond prior to CH breakage also exists in this channel.

More results will be discussed at the conference.

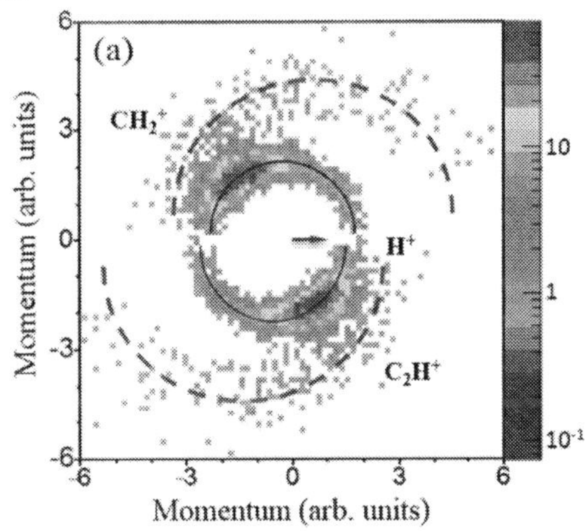

Figure 1. Newton diagram for $H^++CH_2^++C_2H^+$ channle. Blue solid semicirculars and grey dashed semicirculars are meant to guide the eye.

References

[1] Okino T *et al* 2012 *Phys. Chem. Chem. Phys.* **14** 10640
[2] Xu H *et al* 2009 *J. Chem. Phys.* **131** 151102

* e-mail: xushenyue@xjtu.edu.cn
† e-mail: x.ma@impcas.ac.cn

Non-local complex potential theory of dissociative electron attachment: inclusion of two vibrational modes

H B Ambalampitiya[1]* **and I I Fabrikant**[1]†

[1]Department of Physics and Astronomy, University of Nebraska-Lincoln, Lincoln, NE 68588, USA

Synopsis The nonlocal complex potential theory is developed to treat the dissociative electron attachment to molecules when more than one vibrational mode is allowed in the neutral molecule. Temperature-averaged cross sections for electron attachment to CF_3Cl are obtained assuming two vibrational degrees of freedom.

Dissociative electron attachment (DEA) to polyatomic molecules plays an important role in many applications [1]. DEA cross sections are significantly affected by the initial vibrational state of the target molecule. Previous DEA calculations involving several nuclear degrees of freedom have been done in the framework of the local approximation for the nuclear motion in the intermediate resonant state. However, the actual potential describing this motion is a nonlocal energy dependent operator [2].

Previous nonlocal DEA calculations have so far assumed only one vibratioinal degree of freedom in the neutral molecule. In the present paper we extend the formalism of nonlocal complex potential (NLCP) theory to incorporate two vibrational degrees of freedom in the target molecule. We then apply the theory to the process [3]

$$e^- + CF_3Cl(\nu_2, \nu_3) \to CF_3Cl^- \to CF_3(\nu_2') + Cl^-,$$

where ν_2 and ν_3 stand for symmetric deformation ("umbrella") and C-Cl symmetric stretch vibrations, respectively. To obtain an expression for the cross sections of the above process, we calculate the exact matrix elements of the two-dimensional Green's function by solving a set of homogeneous coupled-channel equations with a real potential matrix. The 2D *ab initio* potential energy surfaces of Ref. [3] were used in our DEA calculations. The adjacent figure shows our results for temperature-averaged cross sections which are compared with the previous 1D results [4] and experimental data [5].

Inclusion of the second vibrational mode does not lead to a better description of the low-energy peak at the higher vibrational temperature. The long-standing disagreement with the experiment [5] remains unexplained.

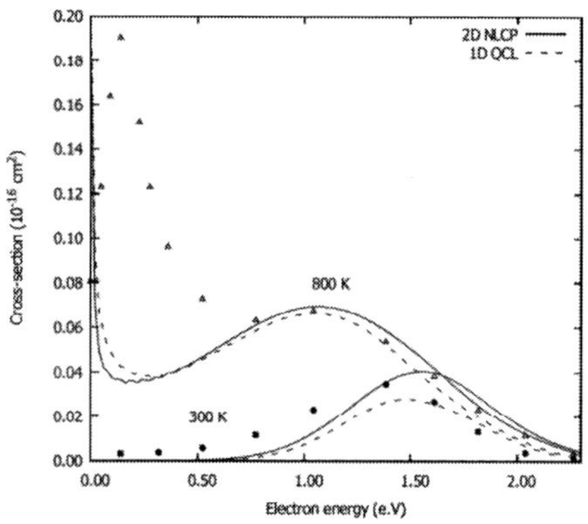

Figure 1. Temperature-averaged cross sections at 300 and 800 K. Solid lines are present 2D NLCP calculations. Dashed curves are 1D quasiclassical calculations [4]. Experimental data points [5] are taken from Ref. [6]

References

[1] Fabrikant I I *et al* 2017 *Adv. At. Mol. Opt. Phys.* **66** 545
[2] Bardsley J N 1968 *J. Phys. B: At. Mol. Phys.* **1** 349
[3] Tarana M *et al* 2009 *Phys. Rev. A.* **79** 052712
[4] Wilde R S *et al* 1999 *J. Phys. B: At. Mol. Opt. Phys.* **32** 663
[5] Hahndorf I *et al* 1994 *Chem. Phys. Lett.* **231** 460
[6] Lehr L *et al* 1996 *Chem. Phys. Lett.* **250** 515

*E-mail: harindranath@huskers.unl.edu
†E-mail: ifabrikant@unl.edu

Theoretical study of vibrational excitation and dissociative electron attachment of NO_2 by an electron impact

H Liu[1], S F Santos[2], C Yuen[3], P Cortona[1], V Kokoouline[3] and M Ayouz[4] *

[1] SPMS, CentraleSupélec, Université Paris-Saclay, 91190, Gif-sur-Yvette, France
[2] Department of Physics, Rollins College, 32789, Florida, USA
[3] Department of Physics, University of Central Florida, 32816, Florida, USA
[4] LGPM, CentraleSupélec, Université Paris-Saclay, 91190, Gif-sur-Yvette, France

Synopsis Cross sections for vibrational excitation and dissociative electron attachment (DEA) of the nitrogen dioxide (NO_2) molecule in collisions with low-energy electrons are determined theoretically. The thermally-averaged rate coefficients of vibrational excitation are presented for the temperatures from 10 K to 10000 K. The obtained DEA cross section for NO_2 agrees well with experimental data.

NO_2 is a common atmospheric pollutant which can cause health hazards. Electron scattering from this molecule has attracted great attention due to its increasing concentration. Notably, the processes of vibrational excitation and dissociative electron attachment play a crucial role in depollution. However, the theoretical description of these two processes for this open-shell molecule is still an extremely challenging task to date.

In this study, electron-impact vibrational excitation cross sections of NO_2 are calculated by a theoretical approach that combines the normal modes approximation for the vibrational states of the target molecule, fixed-nuclei R-matrix formalism [1] to determine the e-NO_2 scattering matrices, and the vibrational frame transformation [2] employed to evaluate the scattering matrix for vibrational transition. Corresponding thermally-averaged rate coefficients for vibrational excitation in ground electronic state of NO_2 are presented for the temperatures region 10 K-10000 K for each normal mode.

The cross sections for DEA to NO_2 are estimated through a simplified approach that following the Bardsley [3] and O'Malley [4] theory proposed for diatomic targets. We extends the formalism of resonant scattering which has already satisfactorily applied to diatomic molecules to polyatomic molecules relies on a generalized dissociation coordinate. The calculated DEA cross section is compared with the experimental data of Rangwala S A *et al.* [5]. An excellent agreement could be seen in figure 1, especially for the first peak at about 1.46 eV.

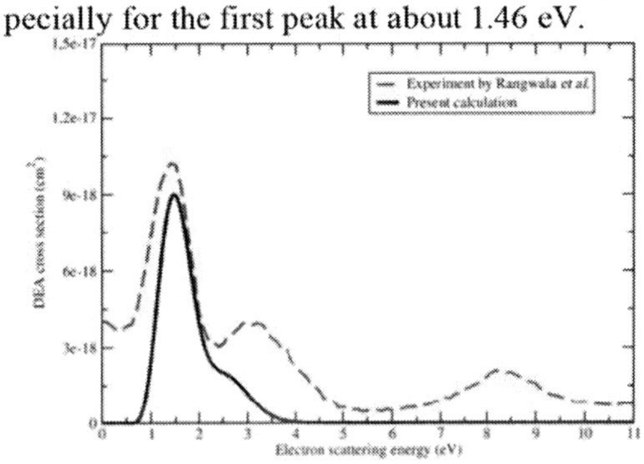

Figure 1. Calculated DEA cross section (black solid line) compared with experimental data (red dashed line) of Rangwala S A *et al.* [1].

We expect that the currently obtained e-NO_2 collisions deduced vibrational excitation and DEA data could be used as the fundamental input quantities for the kinetic modeling of the diffused low-temperature NO_2-containing plasma. Moreover, investigation of applying our model for other triatomic molecules by electron impact will be carried out in the further work.

References

[1] Tennyson J 2010 *Phys. Rep.* **491** 29.
[2] Chang E S, Fano U 1972 *Phys. Rev. A.* **6** 173
[3] Bardsley J N 1968 *J. Phys. B.* **1** 365
[4] O'Malley T F 1966 *Phy. Rev.* **150** 14
[5] Rangwala S A *et al* 2003 *Phys. Rev. A* **68** 052710

* E-mail: mehdi.ayouz@centralesupelec.fr

Electronic reactive collisions in cold ionised media: from mechanisms to new state-to-state cross sections and rate coefficients

J Zs Mezei[1] *, A Orbán[1], K Chakrabarti[2], Y Moulane[3], M D Epée Epée[4], O Motapon[4], D Talbi[5] and I F Schneider[6,7] †

[1]Institute for Nuclear Research, Hungarian Academy of Sciences, Debrecen, H-4001, Hungary
[2]Department of Mathematics, Scottish Church College, University of Calcutta, Calcutta, 700006, India
[3]High Energy Physics and Astrophysics Laboratory, Cadi Ayyad University, Marrakech, 40000, Morocco
[4]Dept. of Physics, Faculty of Sciences, University of Douala, Douala, 24157, Cameroun
[5]LUPM CNRS-UMR5299, Univ. Montpellier, Montpellier,34095, France
[6]LOMC CNRS-UMR6294, Univ. Normandie, Univ. du Havre, Le Havre, 76058, France
[7]LAC CNRS-UMR9188, Univ. Paris Sud, ENS Cachan, Univ. Paris-Saclay, Orsay, 91405, France

Synopsis The major mechanisms governing the dynamics of electron-driven reactions of molecular cations will be illustrated.

Electron-impact dissociative recombination, ro-vibrational (de)excitation and dissociative excitation of molecular cations

$$AB^+ + e^- \to AB^{*,**} \to \begin{cases} A + B \\ AB'^+ + e^- \\ A + B^+ + e^- \end{cases} , \quad (1)$$

are at the heart of molecular reactivity in the cold ionised media [1], being major molecular ion destruction reactions, producing often atomic species in metastable states, inaccessible through optical excitations. They involve super-excited molecular states undergoing predissociation and autoionization, having thus strong resonant character. We use the Multichannel Quantum Defect Theory [2], capable to account the strong mixing between ionization and dissociative channels, open - direct mechanism - and closed - indirect mechanism, via capture into prominent Rydberg resonances [3] correlating to the ground and excited ionic states, and the rotational effects. These features will be illustrated for several cations of high astrophysical and planetary relevance such as CO^+ [4], SH^+ [5], and CH^+ [6, 7], comparisons with other existing theoretical and experimental results being performed. A representative example is shown in figure 1. Advancement in the theoretical treatment - addressing the effect of spin-orbit coupling for HCl^+, polyatomic systems and pre-

dic 1.

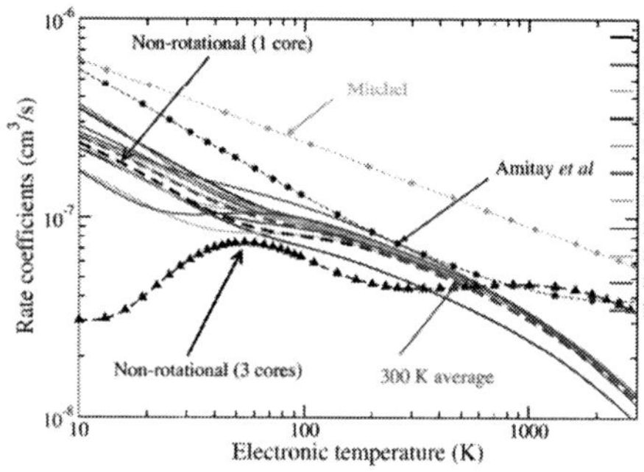

Figure 1. Maxwellian rate coefficients for dissociative recombination of vibrationally relaxed $CH^+(N_i^+)$ with electrons as functions of the kinetic temperature. Our results are compared with the experimental results of Amitay *et al.* [8] and of Mitchell [9].

References

[1] I. F. Schneider, O. Dulieu, and J. Robert (editors) 2015 *Eur. Phys. J. Web of Conf.* **84**
[2] Jungen Ch (editor), 1996 *Molecular Applications of Quantum Defect Theory*, (IoP Publish. Bristol)
[3] Schneider I F *et al* 1991 *J. Phys. B* **24** L289
[4] Moulane Y *et al* 2018 *A&A* **615** A53
[5] Kashinski D O *et al* 2017 *J. Chem. Phys.* **146** 204109
[6] Faure A *et al* 2017 *MNRAS* **469** 612
[7] Chakrabarti K *et al* 2018 *J. Phys. B* **51** 104002
[8] Amitay Z *et al* 1996 *Phys. Rev. A* **54** 4032
[9] Mitchell J B A 1990 *Phys. Rep.* **186** 215

*E-mail: mezei.zsolt@atomki.mta.hu
†E-mail: ioan.schneider@univ-lehavre.fr

Dissociative recombination and vibrational excitation of molecular cations with electrons: application to H_2^+, BeH^+ and their isotopomers

N Pop[1*], F Iacob[2†], J Zs Mezei[3], S Niyonzima[4], O Motapon[5] and I F Schneider[6,7]

[1]Dept. of Fundamental Physics for Engineers, Politehnica University, Timisoara, 300006, Romania
[2]Physics Faculty, West University of Timişoara, Timişoara, 300223, Romania
[3]Inst. of Nuclear Research of the Hungarian Academy of Sciences, Debrecen, H-4001, Hungary
[4]Department of Physics, Faculty of Sciences, University of Burundi, Bujumbura, 1550, Burundi
[5]Department of Physics, Faculty of Sciences, University of Douala, 24157, Cameroon
[6]Laboratoire Ondes et Milieux Complexes, CNRS, Univ. du Havre, Le Havre, 76058, France
[7]Laboratoire Aimé Cotton, CNRS, ENS Cachan and Univ. Paris-Sud, Orsay, 91405, France

Synopsis Cross sections and Maxwell rate coefficients for the low-energy reactive collisions of electrons with Hydrogen and Beryllium hydride cations will be presented, as well as a discussion on the major relevant mechanisms and features – direct/indirect process, ro-vibronic couplings, ro-vibrational dependence.

Using a stepwise method based on the Multichannel Quantum Defect Theory (MQDT) [1], cross sections and Maxwell rate coefficients have been obtained for dissociative recombination (DR), elastic collisions (EC), vibrational excitation (VE, inelastic collisions), vibrational de-excitation (VdE, super-elastic collisions) of $\mathbf{H_2^+}$ and $\mathbf{HD^+}$ for numerous ro-vibrational states of the ion [2, 3].

A very good agreement is found between our results and other computations, as well as with experiment. These results will complete the existing data-bases for electron-impact collision processes.

We have expanded our studies on $\mathbf{BeH^+}$ [4] to $\mathbf{BeD^+}$ [5] and $\mathbf{BeT^+}$ [6] cations. A complete set of vibrationally resolved rate coefficients for BeT^+ cation reactive collisions with electrons below the ion dissociation threshold will be provided.

The resulting data are useful for the modeling of the kinetics of the Early Universe and of the magnetic-confinement-fusion-edge plasma in JET and, later, in ITER.

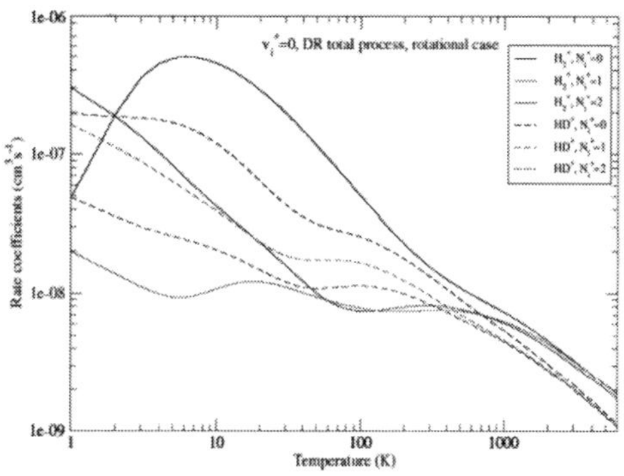

Figure 1. Dissociative recombination Maxwell rate coefficients of H_2^+ and HD^+ in ro-vibrational states N_i^+=0-2, v_i^+= 0.

References

[1] Jungen Ch 2011 *Handbook of High Resolution Spectroscopy*, Wiley & Sons, New York, 471
[2] Motapon O *et al* 2014 *Phys. Rev. A* **90** 012706
[3] Epée Epée MD *et al* 2015 *MNRAS* **455** 276–281
[4] Niyonzima S *et al* 2017 *Atomic Data and Nuclear Data Tables* **115-116** 287
[5] Niyonzima S *et al* 2018 *Plasma Sources Sci. Technol.* **27** 025015
[6] Iacob F *et al* 2019 *AIP* **2071**(1) 020007

* E-mail: nicolina.pop@upt.ro
† E-mail: felix.iacob@e-uvt.ro

Electron dissociative attachement to ArH^+, HD^+, N_2^+ and CO_2

A Abdoulanziz[1*], E Djuissi[1], C Argentin[1], Y Moulane[2], A Bultel[3], J Zs Mezei[4], J Tennyson[5], K Chakrabarti[6], V Laporta[1,7] and I Schneider[1,8†]

[1]LOMC UMR CNRS 6294, Université du Havre, Le Havre, 76058, France
[2]Space Sciences, Technologies & Astrophysics Research Institute, University of Liege, Liege, Belgium
[3]CORIA, UMR CNRS 6614, Université de Rouen, Saint Etienne du Rouvray, 76801, France
[4]Institute for Nuclear Research, Hungarian Academy of Sciences, Debrecen, H-4001, Hungary
[5]Dept. of Physics and Astronomy, University College London, London WC1E 6BT, UK
[6]Dept. Of Mathematics, Scotish Church College 1&3 Urquhart Sq., Kolkata 700 006, India
[7]P.Las.M.I. lab, Nanotec, CNR, 70126 Bari, Italy
[8]Laboratoire Aimé-Cotton CNRS-UPR-3321, Université Paris-Sud, 91405 Orsay, France

Synopsis We will present our recent results in the study of electron-induced reactivity of ionized and neutral molecules. The Configuration Interaction method and the Multichannel Quantum Defect Theory have been employed in order to obtain cross sections for collisions of electrons with ArH^+, HD^+, N_2 and CO_2.

Using the Multichannel Quantum Defect Theory (MQDT) [1-6], we computed the cross sections and the rate coefficients for the dissociative recombination (DR) (1.1) and the related competitive processes - vibrational excitation or de-sexcitation (VE or VdE) (1.2) ($v_f^+ > v_i^+$ or $v_f^+ < v_i^+$) and dissociative excitation (DE) (1.3):

$$AB^+(v_i^+) + e^- \longrightarrow A + B \quad (1.1)$$
$$\longrightarrow AB^+(v_f^+) + e^- \quad (1.2)$$
$$\longrightarrow A^+ + B + e^- \quad (1.3)$$

AB standing for ArH[7], HD and N_2[6]. Here v_i^+ and v_f^+ stand for the initial and final vibrational quantum number of the target ion.

These processes occur in the interstellar media, in (re)entry, fusion edge and laboratory plasmas.

The configuration interaction method [8] has been alternatively used for exploring the dissociative electron attachement (1.4) and dissociative excitation (1.5) of CO_2 whose removal is crucial for depollution:

$$CO_2(X) + e^- \longrightarrow CO(X) + O^-(^2P), \quad (1.4)$$
$$\longrightarrow CO(X) + O(^3P) + e^- \quad (1.5)$$

The study of the reactional dynamics relies on the calculation of the relevant potential energy curves and of the corresponding mutual interactions. In these calculations one bond of the target (OC-O) has been stretched only, in order to obtain CO in its electronic ground state.

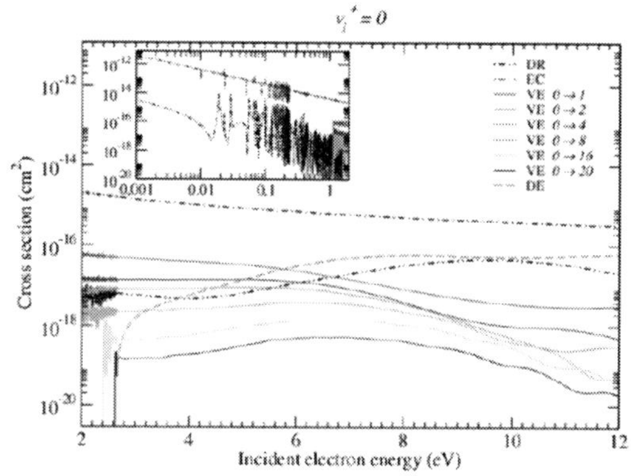

Figure 1. Reactive collisions of HD^+ in its ground vibrational state with electrons.

References

[1] Giusti A 1980 *J. Phys. B* **13** 3867
[2] Guberman S L *et al* 1991 *The Journal of Chemical Physics* **95** 2602
[3] Chakrabarti K *et al* 2013 *Phys. Rev. A*, **87** 022702
[4] Motapon O *et al* 2014 *Phys. Rev. A* **90** 012706
[5] Epée Epée M *et al* 2015 *MNRAS* **455** 276
[6] Little D A *et al* 2014, *J. Phys. B* **47** 105204
[7] Abdoulanziz A *et al* 2018 *MNRAS* **479** 2415
[8] Bardsley J N 1968 *J. Phys. B: At. Mol. Phys.* **1** 349

* E-mail: abdillah.abdoulanziz@univ-lehavre.fr
† E-mail: ioan.schneider@univ-lehavre.fr

Jahn-Teller effect among electronic resonant states of H_3

P Hedvall[1] * and Larson[1] †

[1]Department of Physics, Stockholm University, Albanova University Center, S-106 91 Stockholm, Sweden

Synopsis The Jahn-Teller conical intersection among the electronic resonant states of H_3 is studied. We have performed electron scattering calculations and obtained complex adiabatic potential energy surfaces that we describe using a generalized Jahn-Teller Hamiltonian. The non-adiabatic coupling and the geometric phase were also investigated.

The electronic bound states of H_3 are known to exhibit a Jahn-Teller conical intersection. A phenomena which has been studied extensively for bound states. Here we study the electronic resonant states of H_3 with energies above the potential energy surface of the H_3^+ ground state. These resonant states are important for dissociative recombination of H_3^+ at higher collision energies, and previous studies have indicated that these resonant states also exhibit a Jahn-Teller conical intersection [1].

We have performed electron scattering calculations on the lowest lying resonant states of ^2A' symmetry, using the complex Kohn variational method [2] to compute the potential energies and the autoionization widths of the resonant states. A local complex approximation [3] is applied where the adiabatic resonant states are described by adding an imaginary term to the potential energy. Following the ideas of Feuerbacher *et al.* [4], a complex Jahn-Teller Hamiltonian is introduced for which the parameters are extracted by a least square fit. By diagonalising the complex Jahn-Teller Hamiltonian we obtain complex adiabatic potential energy surfaces describing the system. In Figure 1 the imaginary part corresponding the autoionization width is displayed. We also calculate the non-adiabatic coupling and study the geometric phase for these resonant states.

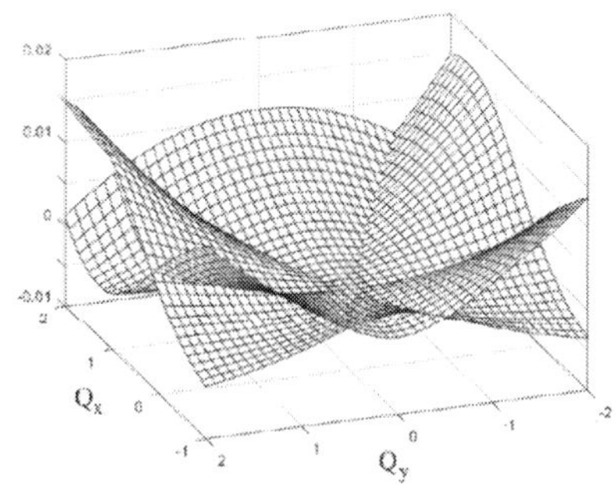

Figure 1. The imaginary part of the complex adiabatic potential energy surfaces. These terms represents the autoionization widths of the resonant states of H_3, as a function of the vibrational normal modes. At the totally symmetric configuration of the H_3 system, the surfaces cross indicating a complex conical intersection with seams of intersections expanding outwards.

References

[1] Orel A E *et al* 1993 *Phys. Rev. Lett.* **71** 4315
[2] Rescigno T N *et al* 1995 *Moden Electronic structure theory vol 1*, Singapore: World Scientific, p. 501.
[3] McCurdy C W *et al* 1983 *J. Chem. Phys.* **78** 6773
[4] Feuerbacher S *et al* 2004 *J. Chem. Phys.* **121** 5

*E-mail: patrik.hedvall@fysik.su.se
†E-mail: aasal@fysik.su.se

Theoretical investigation of dissociative electron attachment (DEA) study for halogen targets

M Vinodkumar[1 *], H Yadav[1,2 †], C Limbachiya[3] and P C Vinodkumar[2]

[1]V. P. & R. P. T. P. Science College, Vallabh Vidyanagar, Gujarat, 388120, India
[2]Department of Physics, Sardar Patel University, Vallabh Vidyanagar, Gujarat, 388120, India
[3]Department of Applied Physics, The M. S. University of Baroda, Vadodara, 390002, India

Synopsis Recently dissociative electron attachment (DEA) study has geared up due to its many fold applications in various fields. Here we have presented some of the results of DEA cross sections of few halogen targets of interest.

Scattering study is important since decades, still it is a developing field with many investigations at low energy interactions. Dissociative electron attachment (DEA) is one such significant phenomena.

DEA finds direct relevance with resonance phenomenon and is closely associated with bond breaking of the molecule on electron impact. DEA gives insight into the fundamental aspects of local chemistry induced by the incident electron. DEA occurs at very low energy below the ionization threshold of the target and sustains for 10^{-12}s to 10^{-15}s. DEA is one of the important phenomena responsible for the breaking of DNA strands due to secondary electrons generated by X-rays or high energy radiations.

To calculate the DEA cross sections, we have used DEA estimator via Quantemol-N [1] based on UK R-matrix code which is widely used for low energy calculations. The important inputs extracted from this codes are resonance width and resonance position using Breit Wigner profile.

The total DEA cross section is in its simplest form can be viewed as a product of resonance cross-section $\sigma_r(E_i)$ due to formation of transient negative ion (TNI) that consequently gives rise to different resonance states and the survival probability 'S' of these resonances that occur during the electron interaction [2]. Thus we can represent DEA cross-section $\sigma_T(Ei)$ as:

$$\sigma_T(E_i) = \Sigma_j\, C_j\, S_j\, \sigma_r(E_i)$$

where E_i represents the incident electron energy, the summation runs over the different resonances which are accounted. The coefficient C_j relates to energy dependent widths of these resonances. Important inputs are the dissociation energy (D_d) calculated as the difference between the target dissociation energy (D_b) and the electron affinity (E_a) of the outgoing fragment, $D_d = D_b - E_a$.

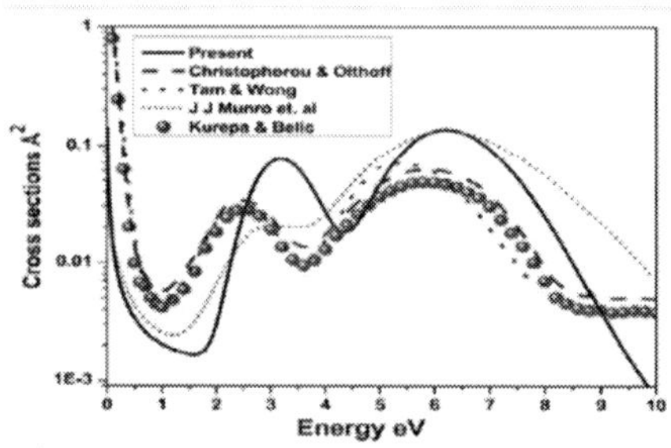

Figure 1.DEA cross sections, where the solid line represents the present calculated data.

Figure 1, presents the sample result of DEA cross sections for Cl^- anion. Qualitatively our results resemble the data predicted by Christophorou & Olthoff [3], Tam & Wong [4] and Kurepa & Belic [5]. But the difference between present and other reported data may be attributed to accuracy in predicting the resonances at low energies.

Dr. Minaxi Vinodkumar acknowledges DST-SERB, New Delhi for Major research project [EMR/2016/000470] for financial support.

References

[1] Tennyson J 2010 *Phys. Rep.* **491** 29-76
[2] Yadav H *et al* 2018 *J. Phys. B: At. Mol. Opt. Phys.* **51** 045201
[3] Christophorou L G and Olthoff J K 1999 *J. Phys. Chem. Ref. Data* **28** 131
[4] Tam W C and Wong S F 1978 *J. Chem. Phys.* **68** 5626
[5] Kurepa M V and Belic D S 1978 *J. Phys. B: At. Mol. Opt. Phys.* **11** 3719

* E-mail: minaxivinod@yahoo.co.in
† E-mail: hitesh0507@gmail.com

 1412 (2020) 172008 doi:10.1088/1742-6596/1412/17/172008

Dissociative Electron Attachment to MgCN

A Orel[1] * and Å Larson[2] †

[1]Department of Chemical Engineering, University of California, Davis, CA 95616 USA
[2]Department of Physics, Stockholm University, AlbaNova University Center, SE-106 91 Stockholm, Sweden

Synopsis Dissociative electron attachment (DEA) to the molecule MgCN and its isomer MgNC have been proposed as possible sources of CN^- in the interstellar media. We have carried out electron scattering calculations using the Complex Kohn Variational Method as a function of the internal degrees of freedom to obtain the resonance energy surfaces and autoionization widths. We use this data as input to form the Hamiltonian relevant to the nuclear dynamics. The multidimensional wave equation is solved using the MultiConfiguration Time-Dependent Hartree (MCTDH) approach. We compute the DEA cross sections.

The mechanism for the formation of CN^- in circumstellar envelopes such as IRC +10216 is unclear [1]. It has been proposed that dissociative electron attachment to MgCN or MgNC, the reactions:

$$e^- + \text{MgCN} \rightarrow \text{Mg} + CN^-$$
$$e^- + \text{MgNC} \rightarrow \text{Mg} + CN^-$$

could be involved in CN^- production [2] . Both MgCN and MgNC have been observed in IRC + 10216 [3, 4, 5].

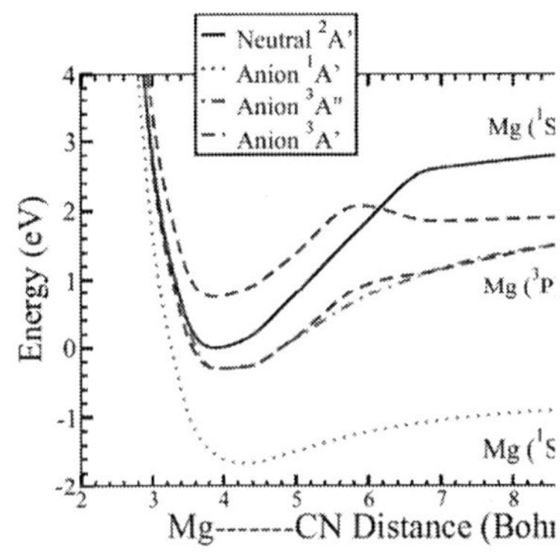

Figure 1. Potential energy curves for MgCN and $MgCN^-$ as a function of Mg to CN distance

We have carried out a series of structure and electron scattering calculations using the Complex Kohn variational method [6] to study the positions and widths of the electronic resonant states in this system. It was found that the ground state of $MgCN^-$ for all geometries (including $MgNC^-$) lies below the neutral. There are anion states that cross the neutral, but these dissociate into the excited states of Mg.

The resonance positions and autoionization widths were calculated on a grid of Jacobi coordinates and used as input to a time-dependent treatment of the nuclear dynamics, in one-, two- and three-dimensions. The multidimensional wave equation is solved using the MultiConfiguration Time-Dependent Hartree (MCTDH) approach [7]. The cross sections for dissociative electron attachment were determined. We will discuss the results and their implications for the formation of CN^- in the circumstellar media.

References

[1] Agundez M *et al* 2010 *A&A* **517** L2
[2] Petrie S 1996 *Mon. Not. Astron. Soc.* **281** 137
[3] Kawaguchi K *et al* 1993 *Astrophys. J.* **406** L29
[4] Guelin M *et al* 1993 *A&A* **280** L19
[5] Ziurys L *et al* 1995 *Astrophys. J.* **445** L47
[6] Rescigno, T N *et al* 1995 *Modern Electronic Structure Theory* **1 501**
[7] Beck M H *et al* 2000 *Physics Reports* **324 1**

*E-mail: aeorel@ucdavis.edu
†E-mail: aasal@fysik.su.se

Journal of Physics: Conference Series **1412** (2020) 172009 doi:10.1088/1742-6596/1412/17/172009

Dissociative electron attachment to supersonically cooled oxygen

D Nandi[1] *, I Jana[1] † and V Ramaprasad[1]

[1]Department of Physical Sciences, Indian Institute of Science Education and Research Kolkata, Mohanpur 741246, India

Synopsis Formation of O^- and O_2^- ions via DEA to a supersonic jet of O_2 having weakly bound van der Waals clusters is reported. The ion-yields show three peaks near 7, 11 and 16 eV electron energies where the 7 eV peak arises from the $^2\Pi_u$ state of O_2^-. The 11 and 16 eV peaks are ascribed to two resonance states $^2\Sigma_g^+$ and $^2\Sigma_u^+$ respectively, through a violation of σ^- selection rule. Ar was used as a refrigerant carrier gas and cross-section dependence of the two new peaks on the proportion of carrier gas used is experimentally observed, eventually leading to an optimum $O_2 : Ar$ ratio.

Electron attachment to clusters of O_2 is known to produce two homologous series $(O_2)_n^-$ and $(O_2)_nO^-$, [1] through

$$(O_2)_x + e^- \rightarrow (O_2)_{x-1}O^- + O. \quad (1)$$

The dimer O_2^- ion is known to produce a peak near 7 eV along with the high cross-section 0 eV peak.

Electron attachment process to isolated O_2 molecules and small van der Waals clusters formed in a supersonic jet is reported here. Along with the previously reported 7 eV resonance, two new resonant peaks near 11 and 16 eV can be observed, indicating violation of the selection rule for electron attachment to isolated molecules. Cross-sections of these new symmetry-forbidden peaks also indicate a dependence on the proportion of the carrier gas Ar, acting as a refrigerant for O_2 molecules in the supersonic jet.

The ion-yields for O^- and O_2^- ions are shown in Fig. 1 for O_2:Ar ratio 1:10. At stagnation pressures 5, 5.5 and 6 bar for O_2:Ar ratios 1:4, 1:10 and 1:12 respectively, two new features near 11 and 16 eV incident electron energies for both ions can be observed, along with the 7 eV peak. These new features are much more pronounced for O^- than for O_2^- for O_2:Ar ratios 1:4 and 1:10. On the contrary, the two new peaks become more pronounced for O_2^- ions as we increase the Ar proportion (1:12). The O_2:Ar ratio 1:10 gives the optimum peak ratio in both cases with the highest number of counts.

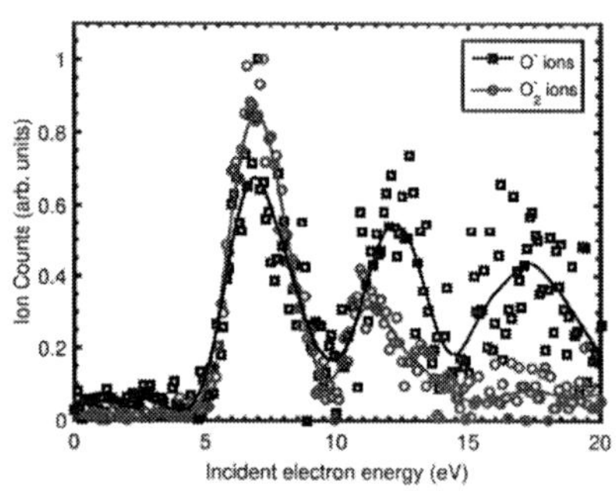

Figure 1. Ion-yields for O^- and O_2^- ions with O_2:Ar ratio 1:10 at a stagnant reservoir pressure of 5.5 bar.

Electron attachment to the neutral molecule at 6.5 eV occurs due to a transition of the molecule to a $^2\Pi_u$ state from the ground state $^3\Sigma_g^-$. According to the σ^- selection rule, transitions from σ^- to σ^+ states and vice-versa are forbidden. But the two new peaks support the selection-rule violating transitions from $^3\Sigma_g^-$ to $^2\Sigma_g^+$ and $^3\Sigma_g^-$ to $^2\Sigma_u^+$ respectively [2]. This points to an important observation that the σ^- selection rule is clearly violated in the condensed phase.

We thus show the existence of symmetry-forbidden transitions in molecular aggregates and provide experimental evidence for variation of relative cross-sections with carrier gas proportion.

References

[1] Märk T D *et al* 1985 *Phys. Rev. Lett.* **55** 2559
[2] Illenberger E 1992 *Chem. Rev.* **92** 1589-1609

*E-mail: dhananjay@iiserkol.ac.in
†E-mail: irinajana@gmail.com

On-the-fly dynamics simulation of dissociative electron attachment to chloroethane

F Kossoski[1] * and **M Barbatti**[1]

[1] Aix Marseille Univ, CNRS, ICR, Marseille, France

Synopsis We present a novel theoretical approach for describing dissociative electron attachment. It is is based on classically propagating the nuclei on the potential surface of the resonant anion, which is described by bound state methods, while electron autodetachment is taken into account with an ad-hoc model built from scattering calculation results. As a first application, we considered the electron induced dissociation of chloroethane. Comparison to experimental data is very encouraging. The simulations reveal further details on the underlying dynamics of the dissociation.

Electron capture by a molecular target can give rise to a resonant anion, which decays either by detachment of the excess electron or by dissociation. The later process, dissociative electron attachment (DEA), plays key roles in radiation-induced biological damage, current and nascent technological applications, interstellar medium, among others [1].

Existing theoretical descriptions of DEA usually rely on propagating the nuclear wavepacket on complex potential energy surfaces, which are built from computationally demanding scattering calculations [2]. Despite being successful, most applications are limited to small molecules [3] or to models of reduced dimensionality [4]. Molecular dynamics simulations for the later stages of DEA are sometimes employed [5], after the anion becomes stable against autoionization. However, we are not aware of any current and feasible method capable of describing DEA to systems with more than a few vibrational degrees of freedom while considering for the finite lifetimes of the resonances.

We propose an alternative approach to the problem, where the resonant anions are described with bound state methods, nuclei are propagated classically, and electron autodetachment is accounted for by a model that requires few scattering calculations. This approach is benchmarked against quantum wavepacket propagation on model potential energy curves. Different curve steepness, magnitudes of the imaginary component and reduced mass were considered. We found a very similar overall behavior for the cases considered.

The very first application of the proposed approach concerns DEA to chloroethane, where the chloride ion is abstracted via formation of a σ^* resonance [6]. An ensemble of classical trajectories were propagated on potential surfaces computed on-the-fly at the Multireference configuration interaction with singles and doubles level of theory. The σ^* character of the anion state was confirmed with electron scattering calculations performed with the Schwinger multichannel method [7]. These later calculations further provided resonance widths, which were employed for modeling the autodetachment probability during the dynamics propagation.

The computed DEA cross section showed the same experimentally observed [6] displacement from the vertical resonance energy ($\sim$2.5 eV) to the maximum ion yield energy ($\sim$1.5 eV), as well as comparable magnitudes. This supports the adopted autodetachment model and the overall proposed strategy to the problem. We have also addressed the dependence of the DEA cross section on several components of the simulation: the accuracy of the autodetachment model, of the electronic structure and scattering calculations. Finally, the dynamics simulation provides a detailed time-dependent picture on the relaxation of the resonant anion.

*E-mail: fabris.kossoski@univ-amu.fr

References

[1] Arumainayagam C *et al* 2010 *Surf. Sci. Rep.* **65** 1
[2] Gertitschke P L *et al* 1993 *Phys. Rev. A* **47** 1003
[3] Haxton D *et al* 2007 *Phys. Rev. A* **75** 12711
[4] Gallup G A *et al* 2011 *Phys. Rev. A* **83** 12706
[5] Feng W *et al* 2015 *J. Phys. Chem. A* **119** 1838
[6] Aflatooni K *et al* 2000 *J. Chem. Phys.* **113** 1455
[7] da Costa R *et al* 2015 *Eur. Phys. J. D* **69** 159

Methylation effect in e⁻– scattering on methyl-substituted ethylenes

S Stefanowska-Tur[1] *, Cz Szmytkowski[1], E Ptasińska-Denga[1] and P Możejko[1]

[1]Department of Atomic, Molecular and Optical Physics, Faculty of Applied Physics and Mathematics, Gdańsk University of Technology, Gabriela Narutowicza 11/12, Gdańsk, 80-233, Poland

Synopsis Methylation effect has been observed and studied in electron-scattering from selected hydrocarbon molecules. In measured total cross section (TCS) functions we have noticed energy shifts and changes in the intensity of observed structures.

Electron-scattering cross sections data for hydrocarbons are of great importance in many fields of science and technology. Systematic studies of TCS for electron scattering from molecules allow also to notice relationships between shape and magnitude of TCS energy dependence and some physical properties of targets.

In this work, we present how methylation can affect TCS for ethylene and its methylated derivatives. All data have been obtained in our laboratory using the electrostatic 127° electron spectrometer working in the linear transmission mode [1].

Table 1. Location of the first maximum in TCS for studied targets.

Molecule	E_{1max} [eV]
Ethylene	1.9
Propene	2.2
2-Methylpropene	2.4
2-Methyl-2-butene	2.6
2,3-Dimethyl-2-butene	2.7

To investigate how the replacement of H atoms in target molecule with CH_3 affects TCS energy dependence, we compared results for ethylene [2], propene [3], 2-methylpropene [4], 2-methyl-2-butene [5] and 2,3-dimethyl-2-butene [5] molecules (fig. 1). Table 1 gives the location of low-energy TCS maxima: increase in the number of methyl groups in the molecule causes a shift of this structures toward higher energies. This effect can be associated with different redistribution of electric charge in the methyl-substituted molecules. A weak high-energy structure is also observed, near 25 eV, which becomes more pronounced as the number of CH_3 groups in a target molecule increases.

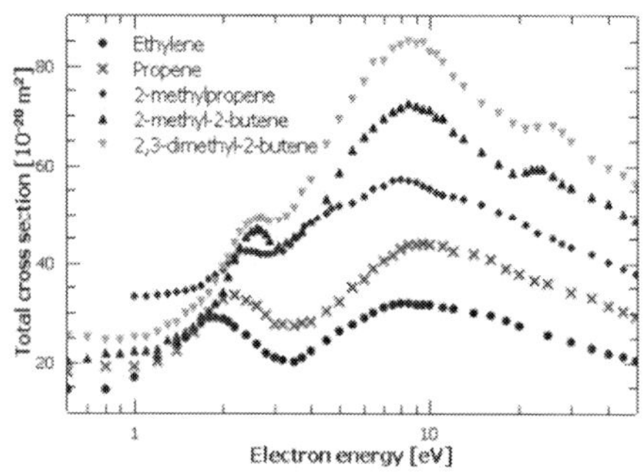

Figure 1. TCS for methyl-substituted ethylenes: ethylene [2], propene [3], 2-methylpropene [4], 2-methyl-2-butene [5] and 2,3-dimethyl-2-butene [5].

This work has been supported in part by the Polish Ministry of Science and Higher Education (MNiSzW Project 2018-2019). S. Stefanowska-Tur kindly acknowledge the support of the Polish Ministry of Science and Higher Education within the Diamond Grant program (Project No. DI2015 018945).

References

[1] Szmytkowski Cz *et al* 2001 *Vacuum* **63** 549
[2] Szmytkowski Cz *et al* 2003 *Phys. Rev. A* **68** 032715
[3] Szmytkowski Cz *et al* 2002 *J. Phys. B: At. Mol. Opt. Phys.* **35** 3781
[4] Możejko P *et al* 2012 *J. Phys. B: At. Mol. Opt. Phys.* **45** 145203
[5] Szmytkowski Cz *et al* 2015 *J. Chem. Phys.* **143** 064306

*E-mail: s141549@student.pg.edu.pl

Electron elastic collisions with Pu, Am and Lr actinide atoms

A Z Msezane[1*] and Z Felfli[1]

[1] Center for Theoretical Studies of Physical Systems, Clark Atlanta University, Atlanta, Georgia 30314, USA

Synopsis The robust Regge-pole methodology which embeds the crucial electron correlations and the vital core-polarization interaction is used to explore negative-ion formation in the actinide atoms Pu, Am and Lr through the electron elastic total cross sections (TCSs) calculation. The TCSs are found to be characterized by ground, metastable and excited anionic formation, whence we extract the anionic binding energies and compare them with existing electron affinities. The TCSs exhibit both atomic and molecular behavior in addition to polarization-induced metastable TCSs.

Here we probe the response of atomic Pu, Am and Lr to low-energy electron collision through the elastic TCSs calculation. For complex heavy systems (atoms and fullerene molecules) these TCSs are characterized generally by ground, metastable and excited states negative ion formation [1-3]. The TCS with account of essential electron-electron correlation effects [4] is calculated from (atomic units are used)

$$\sigma_{tot}(E) = 4\pi k^{-2}\int_0^{\infty} \mathrm{Re}[1 - S(\lambda)]\lambda d\lambda - 8\pi^2 k^{-2}\sum_n \mathrm{Im}\frac{\lambda_n \rho_n}{1+\exp(-2\pi i\lambda_n)} + I(E) \quad (1)$$

In Eq. (1) S(λ) is the S-matrix, $k = \sqrt{2mE}$, m being the mass and E the impact energy, ρ_n is the residue of the S-matrix at the nth pole, λ_n and I(E) contains the contributions from the integrals along the imaginary λ-axis. The TCS calculation uses the potential which accounts for the vital core-polarization interaction

$$U(r) = -\frac{Z}{r(1+\alpha Z^{1/3} r)(1+\beta Z^{2/3} r^2)} \quad (2)$$

In Eq. (2) Z is the nuclear charge, α and β are variation parameters. The potential has the appropriate asymptotic behavior and accounts properly at low E for the polarization interactions. The powers of Z as coefficients of r and r^2 in Eq. (2) ensures that spherical and non-spherical as well as small and large systems are correctly treated. Fig. 1 shows the TCSs for Am; they typify those for Pu and Lr as well. Three observations are made: The ground state TCS is typical of those for Au and Pt; the highest excited state TCS exhibits fullerene behavior [1]; the metastable TCS (brown curve) is the polarization-induced TCS due to size effect. Indeed, any very sharp peak could easily be misidentified with the electron affinity of Am.

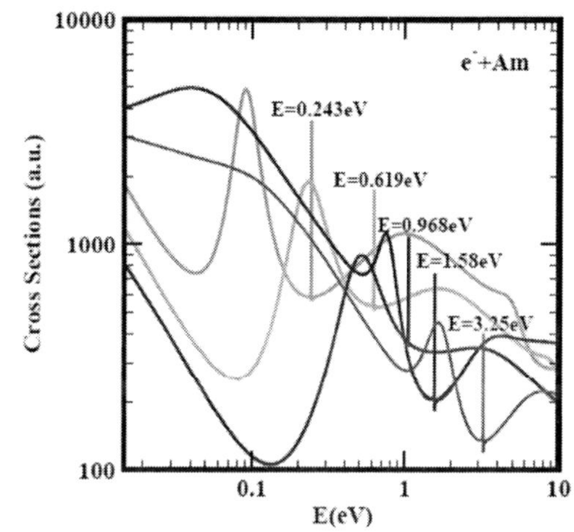

Figure 1. Electron elastic TCSs for Am, red (ground); blue and brown (metastable) and orange and green (excited) states curves. The brown curve with a deep minimum is the polarization-induced TCS due to size.

Research is supported by US DOE Office of Basic Energy Sciences, Office of Energy Research

References

[1] Msezane A Z and Felfli Z 2018 *Chem. Phys.* **503** 50
[2] Felfli Z and Msezane A Z 2018 *Euro Phys. J. D* **72** 78
[3] Felfli Z *et al* 2019 *Applied Physics Research* **11** 52
[4] Sokolovski D *et al* 2007 *Phys. Rev. A* **76** 026707

* E-mail: amsezane@cau.edu

A study of the critical minima and spin polarization in the e-Sr elastic scattering

M N A Abdullah[1*], A K F Haque[2], A Kumar[3] and M A Uddin[2]

[1]Department of Physics, Jagannath University, Dhaka 1100, Bangladesh
[2]Department of Physics, University of Rajshahi, Rajshahi 6205, Bangladesh
[3]Department of Physics, IFTM University, Moradabad 244102, India

Synopsis A complex optical potential (OP) is employed within the framework of the Dirac partial wave method to study the critical minima (CM) in the differential cross sections and the spin polarization for the elastic scattering of electrons by the Sr atom in the energy range E_i = 1-1000 eV. The complex optical potential is composed of the static, exchange, polarization and absorption potentials. The number of CM points has been found to be 7 in the DCSs of e-Sr scattering, while the number of maximum spin polarization points observed is 13.

The critical minima (CM) [1] in differential cross section (DCS) are the points, on the plane constituted by the scattering angle and projectile energy axes, where the DCS attains its smallest values. The role of CM is very important in electron-atom scattering as they are a sensitive test for the validity of experimental procedures and theoretical predictions. The Sherman function, which is a measure of the spin asymmetry in the elastic scattering of transversely polarized electrons from targets, also provides detailed information of the electron-atom collision process and the spin-dependent interactions in the scattering process [2].

The strontium (Sr) is one of the five alkaline earth metals with electron configuration $5s^2$. The scattering of electrons from Sr atom has drawn considerable interest due to the presence of low-lying shape resonance (SR) structures and the existence of its negative ion Sr^-. The DCS minima in the elastic electron-atom scattering are sensitive to the theoretical formalism used for prediction. The exchange potential along-with the target polarization potential is crucial in this case. The complex OP is a combination of the components of OP with use of analytical electron density function of the target atom. Here the electron charge density for neutral atom, generated numerically by the Hartree-Fock method, is represented by a parametrized analytical form of Toshikatsu Koga [3]. The correlation polarization potential is a combination of the long-range polarization potential of Sun et al [4] and a short range local density approximation (LDA) correlation potential. The imaginary part, responsible for absorption, is expressed in terms of a parameter-free absorption potential derived from the Linhard dielectric theory [5] applied to the free electron gas (FEG) considering electron-hole and plasmon interactions.

A three-dimensional plot of the CM is shown in figure 1, which clearly indicates its positions in DCSs for e-Sr scattering. A total of seven CM points have been observed in the DCSs of e-Sr scattering. The critical energy E_c, in this case, has been found to vary between 3.715 eV corresponding to θ_c=54.5° and 483.0 eV corresponding to θ_c=126.5°. The critical angle θ_c varies from 54.5° at E_c=3.715 eV to 150.0° at E_c=126.0 eV.

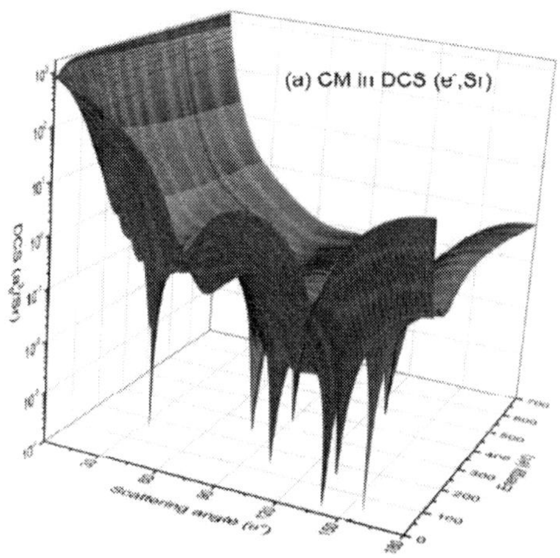

Figure 1. Critical minima in the DCS of e-Sr scattering.

[1] Sienkiewicz J E et al 2003 *Radiation Physics and Chemistry* **68** 285.
[2] Yuan J and Zhang Z 1993 *Phys. Rev. A* **48** 2018.
[3] Koga T 1997 *Theor. Chim. Acta* **95** 113.
[4] Sun J *et al* 1998 *Eur. Phys. J. D* **4** 83.
[5] Salvat F 2003 *Phys. Rev. A* **8** 012708.

* E-mail: maaa05@gmail.com

 1412 (2020) 182004 doi:10.1088/1742-6596/1412/18/182004

Scattering of $e^{\mp}$ from H_2O

A K F Haque[1] *, M Shorifuddoza[1], M A R Patoary[1] and M A Uddin[1]

[1]Department of Physics, University of Rajshahi, Rajshahi, 6205, Bangladesh

Synopsis Theoretical differential, elastic, total, momentum transfer, viscosity cross sections and spin-polarization for the elastic scattering of $e^{\mp}$ by H_2O molecule in the energy range 10 eV-10 keV have been calculated. These calculations have been carried out within the framework the Dirac partial wave analysis. We have employed a complex optical potential and additivity rule (AR) to compute the cross sections. Present results reasonably agree with available experimental data and other theoretical findings.

Water is the second most abundant substance in the universe and a fundamental biological molecule that makes up 75% of our body. Hence, a database on cross section for the collision of $e^{\mp}$ with H_2O are growing in different branches of physics such as astrophysics and atmospheric physics, radiation biology and plasma physics [1]. Therefore, e^{-}-H_2O scattering have been studied extensively both, theoretically and experimentally, since the early part of the twentieth century. Different cross sections for present system have been computed and measured by several authors for different energy range. Among them, Muñoz et al [2] carried out experimental measurements along with theoretical calculations over a broad energy range.

In this work, the net cross section for water molecule is obtained by AR, where the molecular cross section is given as

$$\sigma_{H_2O} = \sum_i \sigma_i = 2\sigma_H + \sigma_O$$

Here, σ_i is the cross section for i'th atom.

Actually, the projectile-atom interaction is approximated by an effective local optical potential in order to obtain the scattering characteristics easily although it is a non-local many-body problem. In this work we have used a complex optical potential (COP) under the framework of Dirac partial wave analysis. The real part of the COP includes static, exchange and correlation-polarization potentials. On the other hand, the imaginary component represents the absorption of particles into various inelastic channels during the collision.

The total cross section (TCS) of electron scattering in the energy range 10 eV-10 keV are depicted in Figure 1. A good agreement is found with available experimental data and comparable with other theoretical calculations.

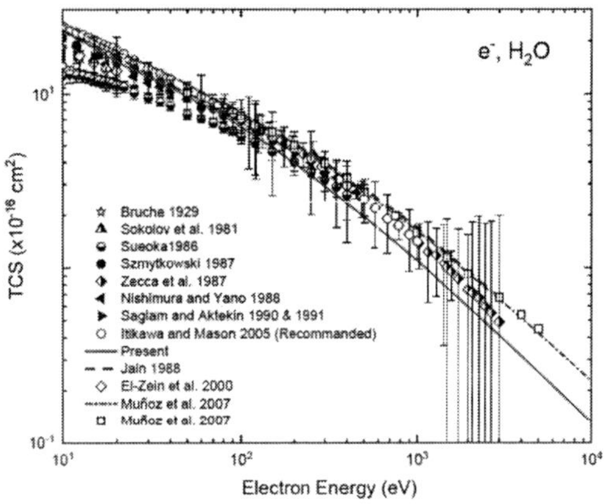

Figure 1. TCS for the scattering of e^{-}-H_2O. Present result is thick solid line (red). Broken curves are other theoretical calculations and various symbols are experimental measurements.

As we have ignored the geometry of the molecule and bonding effects, a small deviation is observed at low energy from experimental data. However, with the increase of electron energy, the contributions of these effects are become less important.

Therefore, this method may be used to calculate the scattering characteristics with reasonable accuracy over a wide energy range.

References

[1] Faure A *et al* 2004 *J. Phys. B: At. Mol. Opt. Phys.* **37** 801
[2] Muñoz A *et al* 2007 *Phys. Rev. A* **76** 052707

*E-mail: fhaque2001@yahoo.com

Separation of electronic states for small molecular ions by the mobility measurements in cooled He gas

T Yamazaki[1], **K Mori**[1] **and H Tanuma**[1]*

[1]Department of Physics, Tokyo Metropolitan University, Tokyo 192-0397, Japan

Synopsis In some cases of the ion mobility measurements in cooled He gas, we have observed the separation of electronic states for small molecular ions. Both NH^+ and ND^+ show the separation of $X^2\Pi$ and $a^4\Sigma$ states. The meta-stable ND_2^+ and NH_3^+ were observed, but these of NH_2^+ and ND_3^+ do not appear in the measurements. The reason of the isotope dependence in mobility measurement is discussed using the calculated potential energy curves for NH_2^+ and ND_2^+.

We have measured the mobilities of atomic and small molecular ions in cooled helium gas at 77 and 4.3 K using a selected-ion-injection drift tube mass spectrometer [1]. As an electron impact ion source has been used to produce ions with gas or vapor from liquid substances in our apparatus, not only the ions in their electronic ground states, but also those in the electronic excited states could be produced, even though the fraction of the excited state is difficult to predict in a simple way. Most of the electronic excited states of atomic and molecular ions have very short life times.

However, some excited states have no allowed transition to lower states, and they could be regarded as meta-stable states. In the drift tube filled with a buffer gas, each ion collide with gas molecules a thousand times. In the slow collisions of meta-stable ions with buffer gas molecules, collisional de-excitation must be considered [2]. However, some ions in the meta-stable state can survive and be detected after passing through the drift tube. Because the interaction potentials between the ions and the gas molecules depend on the electronic states of the ions, the mobility of the meta-stable state might be deferent from that of the ground state. Therefore, we expect that the meta-stable states could be observed as extra peaks in arrival time spectra in mobility measurements.

We have already reported the mobility of the ground (^{2}P) and meta-stable (^{4}P) states for C^+ ions in He gas as a function of effective temperature at gas temperature of 77 and 4.3 K [3]. Recently, we have performed the ion mobility measurements for various small molecular ions produced from NH_3 and ND_3 gases. In the series of the experiments, we have observed the meta-stable NH^+, ND^+, ND_2^+, and NH_3^+ ions. For NH^+ and ND^+, we consider that the separation of $X^2\Pi$ and $a^4\Sigma$ states are observed.

On the other hand, it is very interesting that the meta-stable NH_2^+ and ND_3^+ states were not shown in the mobility measurements. This fact means that the strong isotope effects have been observed in two kinds of ion species. In the Born-Oppenheimer approximation, the interaction potentials should not depend on the isotope. Therefore, an intramolecular vibrational mode can be considered as one of the reasons for these isotope effects. For NH_2^+, the bending potential curves have been calculated, and the intramolecular potentials of the X^3B_1 and a^1A_1 states show a crossing around 90 degree of H-N-H angle [4]. Therefore, the vibrational energy levels of the bending motion must be related to the isotope effect in NH_2^+ and ND_2^+. The detail mechanism will be discussed in the conference. On the other hand, NH_3^+ and ND_3^+ have six vibrational modes, and they are much more complex to consider the origin of the isotope effect. At present, we cannot understand the reason of this isotope dependence between NH_3^+ and ND_3^+.

References

[1] Tanuma H *et al* 2000 *Rev. Sci. Instr.* **71** 2019
[2] Koizumi T 1992 *J. Phys. B* **25** L335
[3] Matoba S *et al* 2008 *J. Phys. B* **41** 145205
[4] Osman G *et al* 1997 *J. Mol. Spectrosc.* **186** 319

*E-mail: tanuma-hajime@tmu.ac.jp

Low energy electron mass stopping power in H_2

M Zawadzki[1,2]* **and M A Khakoo**[1]†

[1]Department of Physics, California State University, Fullerton, California 92831, USA
[2]Atomic Physics Division, Department of Atomic, Molecular and Optical Physics, Faculty of Applied Physics and Mathematics, Gdańsk University of Technology, ul. Gabriela Narutowicza 11/12, 80-233 Gdańsk, Poland

Synopsis We present experimental mass stopping powers of electrons in gaseous H_2 obtained with a newly developed electron time-of-flight spectrometer, for the incident electron energy range of 10 eV to 25 eV. In our procedure the average energy loss is derived from our conversion of measured electron time-of-flight spectra into equivalent electron energy loss spectra so as to obtain the values of mass stopping power for electron scattering from H_2.

Considerable effort has been made in the past to investigate the electron collisions with molecular hydrogen. Before considering more complex molecular systems it is fundamentally important to investigate this simplest molecule, to provide insights into quantitative studies of electron-molecule collision processes.

The mass stopping power (MSP) - a useful parameter obtained from collision studies - is of fundamental importance in biomedical dosimetry, radiation physics, chemistry, medicine and biology involving neutrons, protons, X-rays and electrons [1]. The MSP is defined by the well-known formula:

$$\mathrm{MSP} = \frac{1}{\rho}\frac{dE}{dx} = \frac{N_\mathrm{a}}{M}\langle E_\mathrm{L}\rangle\,\sigma_\mathrm{inel}\,, \qquad (1)$$

where N_a is Avogadro's number, M is the molar mass of the molecule, in this case for H_2 = 2.016 g/mol. $\langle E_\mathrm{L}\rangle$ is the mean inelastic energy loss, and σ_inel is the integral inelastic cross section in a.u. and ρ, is the density of gaseous H_2.

Although, MSPs are widely used in many fields they are not easy to obtain experimentally. This is because in these measurements the complete electron energy loss spectrum should be measured, and the spectrum should be corrected for instrumental transmission effects.

Our measurements were obtained from a recently constructed time-of-flight (TOF) spectrometer, consisting of an intense energy unselected pulsed electron gun and a TOF analyzer [2]. In our studies we collide low-energy electrons with target H_2 gas molecules in the crossed beam configuration, and we measure transmission-effects-free TOF electron scattering spectra and used them to provide experimental MSP of electrons in H_2 in the low electron energy range, as presented in Fig. 1.

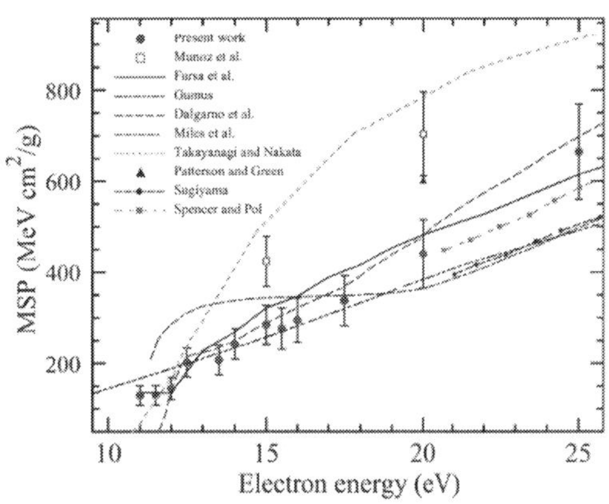

Figure 1. MSP for electron scattering from the ground state of H_2 compared with (many) available theoretical and (one other) experimental data.

Acknowledgments: M.A.K. and M.Z. acknowledge support from a National Science Foundation research Grant Number NSF-RUI-PHY 1606905. M.Z. acknowledges the Fulbright Program to conduct this work at California State University Fullerton.

References

[1] Mayles P, Nahum A and Rosenwald J C 2007 *Handbook of Radiotherapy Physics: Theory and Practice* (CRC Press, Boca Raton)
[2] Zawadzki M *et al* 2018 *Phys. Rev. A* **98** 062704

*E-mail: mateusz.zawadzki@pg.edu.pl
†E-mail: mkhakoo@fullerton.edu

Low energy elastic electron scattering from benzonitrile (C_6H_5CN)

G Tatreau [1], B Diaz[1], F Alharthi[1], M Zawadzki[1,2*] and M A Khakoo[1†]

[1]Department of Physics, California State University, Fullerton, California 92831, USA
[2]Atomic Physics Division, Department of Atomic, Molecular and Optical Physics, Faculty of Applied Physics and Mathematics, Gdańsk University of Technology, ul. Gabriela Narutowicza 11/12, 80-233 Gdańsk, Poland

Synopsis We present experimental differential elastic scattering cross sections (DCSs) for low energy electron scattering from benzonitrile along with integral and momentum-transfer cross sections that are determined from these DCSs. The measurements of DCSs are obtained using the relative flow method with helium as the standard gas, in a crossed electron-molecular gas beam arrangement. Our measurements are made at incident electron energies in the range of $1-30$ eV and scattering angles in the range of (10°-130°).

Recently, in many previous theoretical and experimental electron collision studies the molecular targets containing a ring structure were of great interest due to its relevance in many fields, i.e. medicine, biology and bio-physics [1]. In present work we focus on providing the experimental insight into integral cross sections (ICSs) and momentum transfer cross sections (MTCSs) for benzonitrile.

This aromatic organic compound is a phenyl group bonded to a cyanide group (see figure). Such configuration, where a triple CN bond is present is very interesting from electron-molecule interaction point of view in terms of electron attachment to a system different from a C=C bond commonly found in organic compounds.

Our apparatus consists of a high energy-resolution electron gun/analyzer system, both are equipped with double hemispherical energy selectors to provide well-defined electron beam energy profiles and finesse. Typical electron currents ranged around 19-26 nA, with a corresponding energy resolution of between 40 and 55 meV, full width at half maximum. The experimental setup is located in a magnetically shielded vacuum chamber with a base pressure of 10^{-8} torr.

In order to evaluate ICSs and MTCSs for benzonitrile we performed series of measurements of elastic differential cross section (DCS) for a range of scattering angles (10°-130°) and energies (1 eV to 30 eV). ICSs and MTCSs were calculated from the measured DCS by extrapolating the DCS to 0° and 180° and applying standard numerical integration methods to them. Each DCS set was extrapolated to 0° and 180° by a polynomial curve [2]. The present results are presented in Fig. 1; both the behavior of the elastic ICSs and MTCSs, indicate a possible resonance located around 3.5 eV.

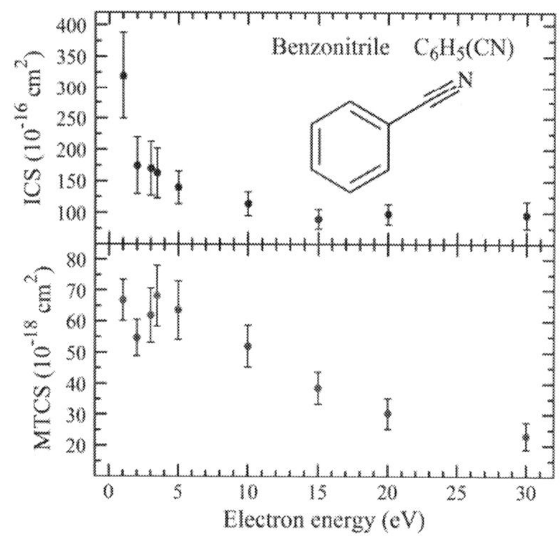

Figure 1. ICSs and MTCSs for elastic electron scattering by benzonitrile.

Acknowledgments: The work carried out at California State University, Fullerton was funded by the National Science Foundation under grant NSF-RUI-PHY 1606905 and by a Fulbright Fellowship to M Zawadzki.

References

[1] Gorfinkiel J D and Ptasinska S 2017 *J. Phys. B: At. Mol. Opt. Phys.* **50** 182001
[2] Fedus K, Navarro C, Hargreaves L R, Khakoo M A, Silva F M, Bettega M H F, Winstead C and McKoy V 2014 *Phys. Rev. A* **90** 032708

*E-mail: mateusz.zawadzki@pg.edu.pl
†E-mail: mkhakoo@fullerton.edu

Negative ion formation in fullerene molecules C_{44}, C_{74}, C_{100}, C_{124} and C_{136}: determination of their electron affinities

A Z Msezane[1*] and Felfli[1†]

[1] Center for Theoretical Studies of Physical Systems, Clark Atlanta University, Atlanta, Georgia 30314, USA

Synopsis Here we investigate through the electron elastic total cross sections (TCSs) calculations the variation of the electron affinity (EA) with the fullerene size from C_{44} to C_{136} and contrast their EAs with that of C_{60}. The TCSs are found to be characterized generally by ground, metastable and excited anionic formation during the collisions. The extracted ground state anionic binding energies for C_{44}, C_{74}, C_{100}, C_{124} and C_{136} are 3.25eV, 4.03eV, 3.67eV, 3.06 eV and 3.75eV, respectively. These corresponding to the fullerenes EAs demonstrate their wide variation.

The EAs are extracted from our Regge-pole calculated ground-state electron elastic TCSs. Recently, it has been demonstrated for the first time that the ground state anionic binding energies (BEs) extracted from our Regge-pole calculated TCSs for C20 through C92 matched excellently their measured EAs [1, 2]. The Regge-pole method requires no assistance whatsoever from either experiment or other theory.

The TCS which embeds fully the essential electron-electron correlation effects [3] is calculated from (atomic units are used)

$$\sigma_{tot}(E) = 4\pi k^{-2}\int_0^\infty \mathrm{Re}[1 - S(\lambda)]\lambda d\lambda - 8\pi^2 k^{-2}\sum_n \mathrm{Im}\frac{\lambda_n \rho_n}{1+\exp(-2\pi i\lambda_n)} + I(E) \quad (1)$$

In Eq. (1) S(λ) is the S-matrix, $k = \sqrt{2mE}$, m being the mass and E the impact energy, ρ_n is the residue of the S-matrix at the nth pole, λ_n and $I(E)$ contains the contributions from the integrals along the imaginary λ-axis. The TCS calculation uses the ABF potential which accounts for the vital core-polarization interaction

$$U(r) = -\frac{Z}{r(1+\alpha Z^{1/3} r)(1+\beta Z^{2/3} r^2)} \quad (2)$$

In Eq. (2) Z is the nuclear charge, α and β are variation parameters. The potential has the appropriate asymptotic behavior and accounts correctly at low E for the polarization interaction. The powers of Z in front of r and r^2 in Eq. (2) ensures that spherical and non-spherical as well as small and large systems are correctly treated. Fig. 1 presents the typical TCSs for the relatively large C100. Indeed, they are characterized by ground, metastable and excited negative ion formation and Ramsauer-Townsend minima. The ground state anionic BE at the minimum is identified with the EA of C100.

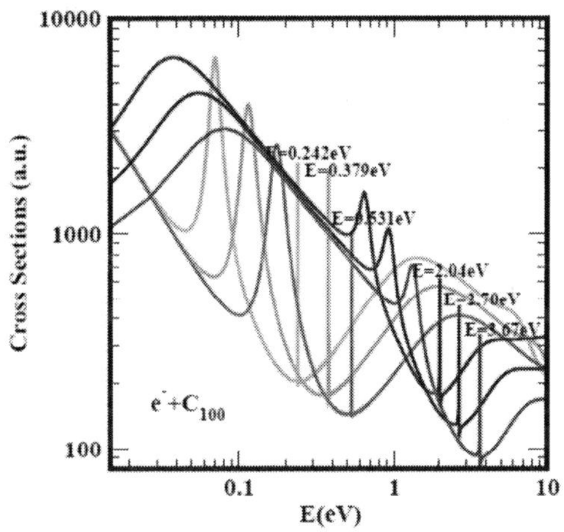

Figure 1. Electron TCSs for C100; the red, blue and brown curves are respectively for ground and metastable TCSs. The pink, green and orange curves represent TCSs for excited states. Very sharp peaks denote anionic BEs.

Research is supported by US DOE Office of Basic Energy Sciences, Office of Energy Research.

References

[1] Msezane A Z and Felfli Z 2018 *Chem. Phys.* **503** 50
[2] Felfli Z and Msezane A Z 2018 *Euro Phys. J. D* **72** 78
[3] Sokolovski D *et al* 2007 *Phys. Rev. A* **76** 026707

* E-mail: amsezane@cau.edu

Low-energy electron scattering from methylene radicals: multichannel-coupling effects

K D Wang* **and H X Zhang**
College of Physics and Materials Science, Henan Normal University, Xinxiang, 453007, P. R. China

Synopsis We reported elastic and excitation cross sections for electron collisions with methylene radical using R-matrix method for electron energies ranging from 0.01 to 15 eV in close-coupling (CC) approximation . We discuss the multi-channel-coupling effects on the calculated cross sections and resonances, and how the number of excited states included in the target state impacts the convergence of the elastic and the $X^3B_1 \rightarrow 1^1A_1$ and $X^3B_1 \rightarrow 1^1B_1$ excitation cross sections, especially at higher impact energy. Finally, we estimate the accuracy of the $1^1A_1$and 1^1B_1 excitation cross sections.

At low energies only the elastic channel is electronically open. With increasing the impact energy, other electronic states can be excited giving rise to the important questions: (1) how many of these states must be included in a calculation in order to predict an accurate cross section? (2) how do these multichannel effects affect the elastic cross sections and excitation cross sections? and (3) how sophisticated must the electronic states description be in order to accurately calculate the cross sections? For the purpose, we performed five different calculations within 1-state, 3-state, 15-state, 20-state, and 24-state close-coupling (CC) approximations for electron collision with methylene radical using R-matrix method [1].

Table 1. Parameters of identified e-CH_2 resonances in 24-state CC model. (in eV)

State	Type	Position	Parent state
$^2B_1$	shape	0.05	X^3B_1
$^2A_1$	shape	0.64	X^3B_1
$^2B_2$	Feshbach	7.35	1^3A_2
$^2A_1$	Feshbach	7.48	1^3A_1
$^2B_1$	Feshbach	8.63	2^3B_1

We compare elastic (integrated and differential), momentum-transfer, and excitation cross sections for e-CH_2 collision in five different CC models [2]. The results indicated that multichannel coupling has a substantial effect on the elastic and excitation cross sections and more target states are needed to include in the CC calculations in order to obtain the convergent cross section. This effect becomes even more important at higher impact energy. The obtained resonance parameters in 24-state CC model are listed in Table 1 along with the tentative assignments.

There are threshold energy differences of 0.77 eV for 1^1A_1 excited state and 0.31 eV for 1^1B_1 excited state between our calculations and experiment values from Ref. [3]. We estimate that these energy differences may introduce a 17% uncertainty in the $X^3B_1 \rightarrow 1^1A_1$ (see Fig. 1) and a 13% uncertainty for the $X^3B_1 \rightarrow 1^1B_1$ excitation cross section.

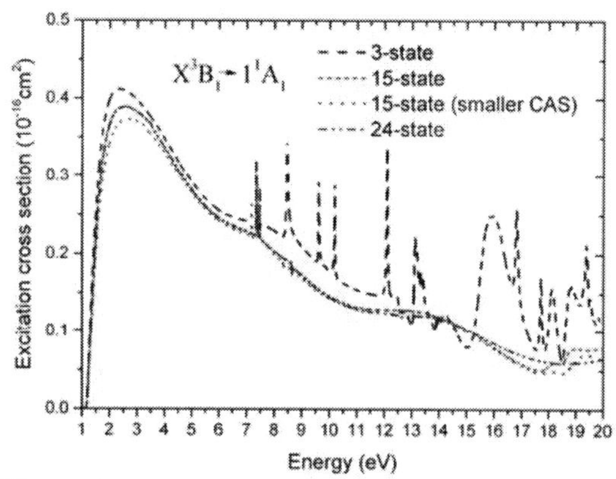

Figure 1. Excitation cross sections in 3-state, 15-state, 15-state (with a smaller CAS), 20-state, and 24-state CC models. The 15-state excitation threshold has been shifted lower by 0.28 eV.

This work was supported by National Natural Science Foundation of China under Grants No. U1504109 and 11604085.

References

[1] Tennyson J 2010 *Phys. Rep.* **491** 29
[2] Wang K *et al* 2018 *Phys. Rev. A* **97** 012703
[3] Jensen P *et al* 1988 *J. Chem. Phys.* **89** 1327

* E-mail: wangkd@htu.cn

Time delay in slow electron elastic scattering by potentials with arising levels

M Ya Amusia[1,2*] **and A S Baltenkov**[3†]

[1] Racah Institute of Physics, the Hebrew University, Jerusalem, 91904, Israel
[2] Ioffe Physical-Technical Institute, St. Petersburg, 194021, Russia
[3] Arifov Institute of Ion-Plasma and Laser Technologies, Tashkent, 100125, Uzbekistan

Synopsis We study Wigner time delay of slow electrons by spherically symmetric square potential well as a function of the well parameters. We show that the electron interaction with scattering center qualitatively depends on presence of discrete levels in the well. Electron retention mode by a potential well with no discrete level is replaced by the mode at which after level appears the incident electron hardly enters the scatterer. This behavior of time delay is universal since it is typical not only for the first but also for following arising s-levels.

1. In a recent paper [1] we consider the partial Wigner-times delay in slow electron elastic scattering by the fullerene C_{60} shell. We show that for small electron energies the sign of a partial time delay depends on the presence of the discrete level with corresponding orbital momentum l in the C_{60} potential well. The lth time delay is positive when there are no discrete l-levels in the well and negative, when these levels exist.

2. We investigated this specific feature of Wigner time behavior calculating the s-scattering phases and times delay for slow electron elastic scattering by spherically symmetric square potential well, the small changes in the parameters of which (potential well depth U and its radius R) lead to arising of the first (as in case [1]) and of arbitrary number of s-bound states. We derived the general formula for the derivative with respect to electron wave number k of the s-phase shift. This formula includes two types of terms: the first of them depend on k and R only while the second one is the function of the well parameters R and U. For small electron energy the first term plays a decisive role as was shown in [2], and reflects the causality principle. The second one is determined by the shape of the potential well and its parameters.

3. We focused here on studiing time delays as function of the scattering potential parameters. We show here that the Wigner time delay changes sign each time a discrete s-level appear in the well. Time delay as a function of U experiences instant jumps from positive to negative value. Amplitudes of jumps increase with decrease of k. Times delay for potential well, where variation of R leads to appearence of discrete levels, also change sign at respective critical radii values. So, the U- and R-dependences of the time delay are alternating functions. The periods of their oscillations increase as n^2 in the fist case and as n in the second one, where n is the level number. Simple resonance behavior of these functions for small k changes with its growth, becoming much more complex. Electron retention or its pushing out by the scatterer strongly depends on the closeness of potential parameters to their critical values, see Figure 1.

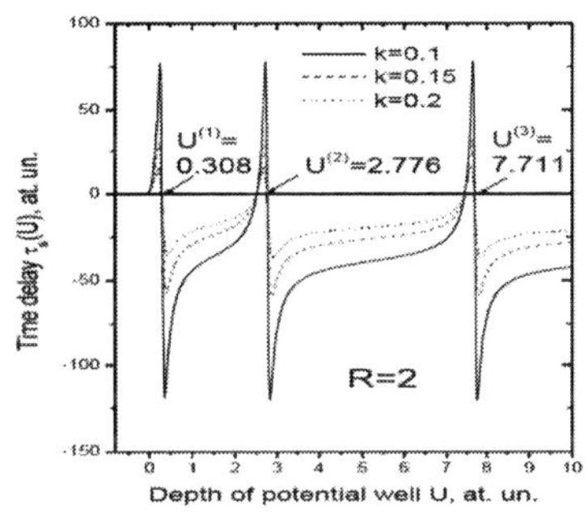

Figure 1. Time delay, function of U; k=0.1, 0.15, 0.2.

4. In spite of simplicity, the performed analysis makes it possible to draw some peculiarity in the behavior of the Wigner time delay and, con-se-quently, in the character of the electron interaction with the scattering potential.

References

[1] Amusia M and Baltenkov A 2019 *J. Phys.B : At. Mol. Opt. Phys*: **52** 015101
[2] Wigner E 1955 *Phys Rev* **98** 145

* E-mail: amusia@012.net.il
† E-mail: arkbalt@mail.ru

On the $e^- - C_{60}$ elastic angle-differential scattering cross section

V K Dolmatov [1]*, M Ya Amusia[2,3]† and L V Chernysheva[3]

[1] Department of Physics and Earth Science, University of North Alabama, Florence, AL 35632, U.S.A.
[2] Racah Institute of Physics, Hebrew University, 91904 Jerusalem, Israel
[3] A. F. Ioffe Physical-Technical Institute, 194021 St. Petersburg, Russia

Synopsis We study theoretically the experimental angle-differential elastic electron scattering off a C_{60} fullerene with the aim to elucidate whether its features depend primarily on the details of the full molecular potential of C_{60} or they can be understood without reference to the full nature of the potential. We demonstrate the success of the latter option.

The molecular potential of C_{60} is essentially multi-centred and, thereby, nonspherical. Then, the scattering process can be thought to depend entirely on the fine details associated with the molecular potential, such as, e.g., bonds between the carbon atoms which make the cage, the molecular resonances due to the ℓ-mixing between different angular momentum of scattering channels, *etc.* We study in the present work the extent to which this perception is correct. To meet the goal, we attempt to interpret the experimentally measured $e^- - C_{60}$ elastic angle-differential scattering cross section, $\frac{d\sigma^{\rm exp}}{d\Omega}$ [1], without regard for the full nature of the molecular potential of C_{60}. We limit the discussion to low electron energies, E. Then, the electron wavelength exceeds noticeably the bond length between the carbon atoms of the cage. This washes out the multi-centred details of the C_{60} potential and turns it into a homogenous attractive spherical potential. We tie up the latter to the geometry of C_{60}, such as its average radius, R_c, and the thickness of the cage, Δ, whereas the depth, U_0, of the potential is determined by making the potential to predict the known electron affinity of C_{60}; see details in [2, 3] and references therein. The polarization potential of C_{60} is approximated by $V_{\rm pol} = -\frac{\alpha}{2(r^2+R_c^2)^2}$, where $\alpha \sim R_c^3$ is the dipole static polarizability of C_{60}. Thus calculated $\frac{d\sigma}{d\Omega}$, along with $\frac{d\sigma^{\rm exp}}{d\Omega}$ and $\frac{d\sigma}{d\Omega}$ calculated with a representation of the full nature of the molecular potential [4], are plotted in figure 1. Our $\frac{d\sigma}{d\Omega}$ matches $\frac{d\sigma^{\rm exp}}{d\Omega}$ reasonably well. Furthermore, it predicts the same narrow resonances at $E \approx 2$ eV at $\theta = 80$ and $90^{\rm o}$ as in work [4], as well as the emergence of a 3.5 eV-resonance at $80^{\rm o}$ reminiscent of that in $\frac{d\sigma^{\rm exp}}{d\Omega}$. Thus, the features of $\frac{d\sigma^{\rm exp}}{d\Omega}$ are largely independent of the full nature of the C_{60} molecular potential. Rather, they correlate with the inherent geometry of C_{60}.

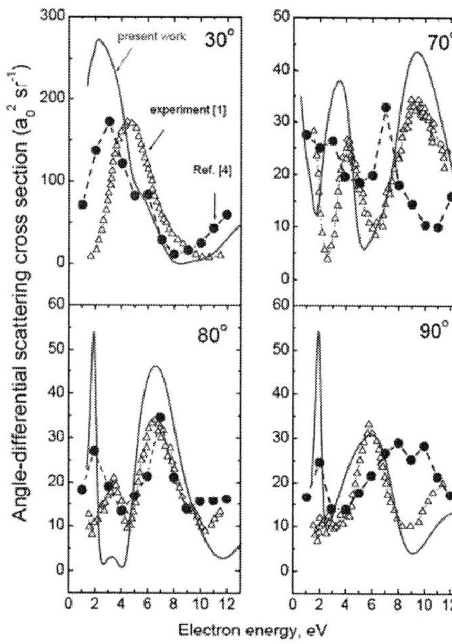

Figure 1. $e^- - C_{60}$ elastic angle-differential scattering cross section, $\frac{d\sigma}{d\Omega}$. Solid line, present data. Solid circles-dashed lines, results of Ref. [4]. Open triangles, experiment [1]. Parameters: $R_c \approx 7$, $\Delta \approx 1.9$, $U_0 \approx 0.302$ and $\alpha = 850$ *au*.

*E-mail: vkdolmatov@una.edu
†E-mail: amusia@012.net.il

References

[1] Tanaka H *et al* 1994 *J. Phys. Soc. Japan* **63** 485
[2] Dolmatov V K *et al* 2017 *Phys. Rev. A* **95** 012709
[3] Amusia M Ya *et al* 2018 *arXiv:* 1810.03700
[4] Gianturco F A *et al* 1999 *J. Phys. B* **32** 2181

Integral cross sections for elastic electron scattering by methane molecule

J B Maljković [1] *, J Vuković[2], K Tőkési[3,4], B Predojević[2] and B P Marinković[1] †

[1]Institute of Physics Belgrade, University of Belgrade, Pregrevica 118, 11080 Belgrade, Serbia
[2]Faculty of Science, University of Banja Luka, Mladena Stojanovića 2, 78000 Banja Luka, Republic of Srpska, Bosnia and Herzegovina
[3]Institute for Nuclear Research, Hungarian Academy of Sciences (ATOMKI), Debrecen, Hungary
[4]ELI-ALPS, ELI-HU Non-profit Kft., Szeged, Hungary

Synopsis We have performed a combined experimental and theoretical study of electron elastic differential and integral cross sections from methane molecule (CH4) in the intermediate energy range. Data have been compared with available data sets in the literature as well as with recommended data set [Song *et al J. Phys. Chem. Ref.Data* **44** 023101 (2015)].

Methane gas has been investigated widely in past years by electron collisions [1,2]. We measured electron elastic differential cross sections (DCSs) for methane molecule (CH_4) in the energy range from 50 eV to 300 eV and angular range from 20° to 130°. DCSs were then extrapolated to zero and 180° and integrated in order to obtain the integral cross sections. The experimental setup based on a crossed beam technique comprising of an electron gun, a single capillary gas needle and a detection system with a channeltron was used to measure differential cross sections.

The absolute scale for the cross sections is obtained by relative-flow method using argon gas as a reference [3]. For the interpretation of the measured data we applied the partial expansion method to calculate the elastic cross sections for electron scattering from methane.

Here we present our daa point at 50 eV and we find it in an excellent agreement with the recommended set of data [1]. At the conference we will present our data points in a full scale of electron impact energies, from 50 to 300 eV in steps of 50 eV and calculated ICS curves obtained based on IAM method and screening correction factors.

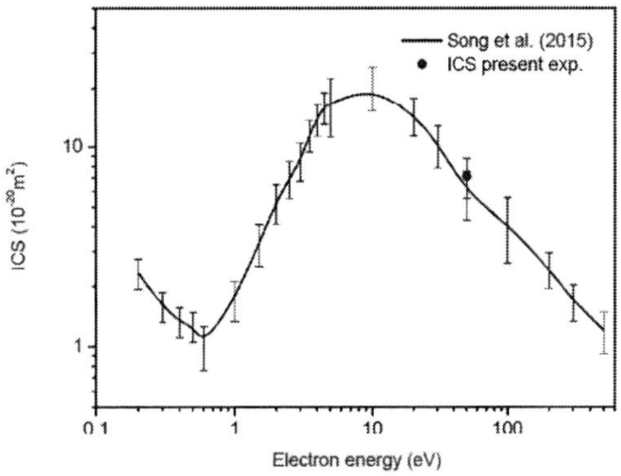

Figure 1. Energy dependence od integral cross sections for elastic electron scattering by methane molecule.

This work has been supported by the MESTD Grant #OI 171020. BPM acknowledges the grant of VAMDC consortium under Support Actions" Funding Scheme of The Seventh Framework Program. One of us KT was also supported by the National Research, Development and Innovation Office (NKFIH) Grant KH126886, and by the European COST Action CA15107 (MultiComp).

References

[1] Song M-Y *et al* 2015 *J. Phys. Chem. Ref. Data* **44** 023101
[2] Marinković B P *et al* 2019 *Atoms* **7** 11
[3] Ranković M Lj *et al* 2018 *Eur. Phys. J D* **72** 30

* E-mail: jelena.maljkovic@ipb.ac.rs
† E-mail: bratislav.marinkovic@ipb.ac.rs

Dissociation of furan-2-carboxylic acid by low energy electrons

M Zawadzki[1,2*], J Kočišek[1] and J Fedor[1]

[1]J. Heyrovský Institute of Physical Chemistry vvi, Academy of Sciences of the Czech Republic, Dolejškova 3, 182 23 Praha, Czech Republic
[2]Atomic Physics Division, Department of Atomic, Molecular and Optical Physics, Faculty of Applied Physics and Mathematics, Gdańsk University of Technology, ul. Gabriela Narutowicza 11/12, 80-233 Gdańsk, Poland

Synopsis Electron attachment to molecular target can lead to a variety of negative fragment ions. We experimentally probe the main dissociative electron attachment channels in furan-2-carboxylic acid ($C_3H_3O_2$ – COOH) - a molecular target composed of two important structural units: a furan ring and a carboxylic group. In our studies we utilise nearly mono-energetic electrons at electron energies from 0 eV up to 15 eV. Our detection setup is based on extraction lens system and quadrupole mass spectrometer, that enables measurement of the energy dependent mass-selected yields of newly formed stable anions.

Many investigations have been focused on the understanding of physico-chemical processes of molecular complexes containing different heterocyclic rings [?]. Such compounds have important applications in many disciplines including medicine, biology, chemistry, pharmacology and material science, which place these molecular systems in a very interesting point of view. On the other hand molecular complexes containing carboxylic group are interesting due to the ease of forming double-hydrogen-bonded complexes [?].

The replacement of the adjacent to single C–O bond hydrogen atom in the furan ring by a carboxylic group results in formation of the furan-2-carboxylic acid molecule (F2CA) - see Fig. 1. Effectively, the molecular structure consists of a five-membered aromatic ring and a carboxylic group connected via single C–C bond.

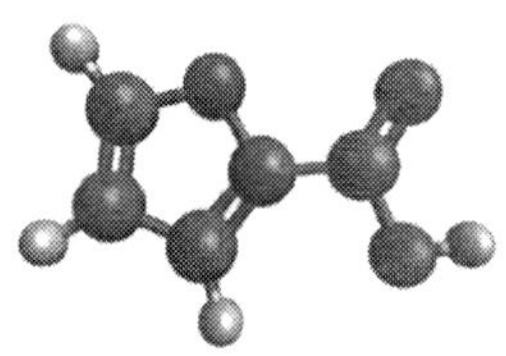

Figure 1. Ball-and-stick model of the structure of F2CA.

In our studies we collide low-energy electrons with the target gas molecules in the collision chamber, and our detection system is used for analysing negative ions created in the dissociative electron attachment (DEA) process [?]. We experimentally probe the main DEA channels in F2CA. Electron attachment to F2CA leads to a variety of negative fragment ions. Detailed energetics of formation of the DEA channels will be discussed. The example of such molecular fragmentation is shown in Eq. ??:

$$e^- + \underbrace{\overbrace{C_4H_3O_2}^{\text{ring}}\overbrace{COOH}^{\text{carboxyl}}}_{M} \xrightarrow{M^{\#-}} \underbrace{(C_4H_3O_2\,COO)^-}_{(M-H)^-} + H\,, \quad (1)$$

where $M^{\#-}$ denotes a transient negative anion.

Dominant fragments arise from the cleavage of a single bond. Other fragments can be described using more complex fragmentation reactions. The DEA processes show one prominent resonance at 1.2 eV and three distinct energy bands in the region of 5–10 eV. The dissociation channel via C_2H^- anion is especially interesting as the theoretical and experimental results reveal the possible hydrogen migration between a ring and carboxylic structure prior to dissociation.

Acknowledgments: The work was supported by the Czech Science Foundation Grant no. 16-10995Y. MZ acknowledges the Academic Computer Centre in Gdańsk (TASK) for the Computational Grant.

References

[1] Gorfinkiel J D and Ptasinska S 2017 *J. Phys. B: At. Mol. Opt. Phys.* **50** 182001
[2] Fabrikant I I, Eden S, Mason N J and Fedor J 2017 *Adv. At. Mol. Opt. Phys.* **66** 545
[3] Stepanović M, Pariat Y and Allan M 1999 *J. Chem. Phys.* **110** 1137

*E-mail: mateusz.zawadzki@pg.edu.pl

New magnetized electron beam apparatus and measurement of total electron scattering cross section of N_2O at intermediate energy region

H Pyun[1,2], D C Kim[1*], Y Kim[1], Y R Choi[1], Y Park[1], M-Y Song[1], Y-W Kim[1], J-S Yoon[1], H Cho[2], J P Sullivan[3] and S J Buckman[3]

[1] Plasma Technology Research Center, National Fusion Research Institute, Gunsan, 54004, South Korea
[2]Department of Physics, Chungnam National University, Daejeon, 34134, South Korea
[3]Plasma Research Laboratories, RSPE, Australian National University, Acton ACT, Australia

Synopsis A new electron scattering apparatus is developed based on concepts from the positron system built by Surko *et al.* In an electron-version of the system, the electron beam is provided by a cathode and the rest of the system is correspondingly modified to a simpler system. The system was tested with argon to verify its operation. N_2O is important in atmosphere and plasma, and in the intermediate energy region, there are two previous results which are in disagreement. We present here a result for the total electron cross section of N_2O.

The principle of our new apparatus is based on concepts from the positron system built by Surko *et al* [1,2]. Later, Sullivan *et al* [3] have developed a similar system. In an electron-version of the system, electron beam is provided by a cathode and the rest of the system is correspondingly modified to a much simpler and smaller system. The Garcia group developed the first electron version, also motivated by concepts for the positron system [4]. In our new apparatus, the idea is similar to the positron and Garcia systems, but the details are different in design and in data reduction procedures. The schematic diagram of our apparatus is given in Fig. 1.

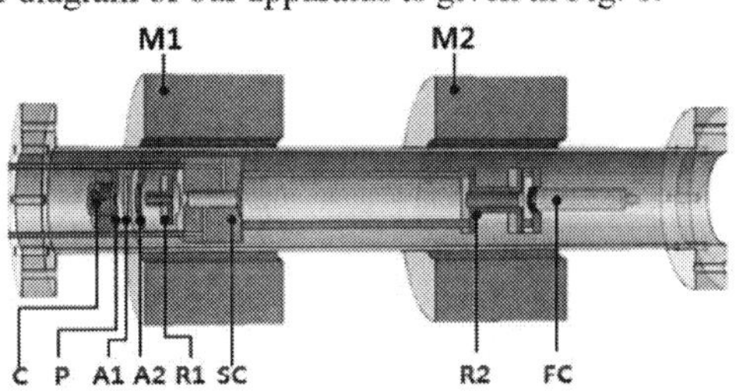

Figure 1. Schematic diagram of the system. C, cathode; P, pierce; A, anode; SC, scattering cell; R, retarding potential analyzer (RPA). FC, Faraday cup.

The core part of the system is housed in a cylindrical, differentially pumped vacuum chamber. M1 and M2 are the magnets, normally operating from 450 to 500 gauss. In principle, this system can be operated to measure not only total scattering cross sections (TCS) but also differential and excitation cross sections with proper magnetic field intensity ratio of M1 and M2. For the TCS measurements presented here, both magnets were operated at the same magnetic field. RPA 1 is designed to act as a crude monochromator and RPA2 acts as an energy analyzer.

First, to demonstrate that our new system is working properly, it was tested with argon as a reference to measure the TCS in the intermediate energy region from 5 eV to 50 eV. Our measurement agrees quite well with other previous measurements [5, 6].

Nitrous oxide (N_2O) is a potential greenhouse gas and important in the atmosphere and in various plasma processes. However, in the intermediate energy region, there are only two previous results and they are in disagreement. This situation motivated us to measure e-N_2O TCS and our measurement results will be presented at this meeting.

References

[1] Surko C M *et al* 1988 *Phys. Rev. Lett.* **61** 1831
[2] Sullivan J P *et al* 2002 *Phys. Rev. A* **66** 042708
[3] Sullivan J P *et al* 2008 *Rev. Sci. Instrum.* **79** 113105
[4] Lozano A I *et al* 2018 *Rev. Sci. Instrum.* **89** 063105
[5] Szmytkowski Cz *et al* 1996 *Phys. Scr.* **54** 271
[6] Baek W Y *et al* 2003 *J. Phys. B* **36** 731

* E-mail: dchcharm@nfri.re.kr

Halogen loss induced by electron collisions in halouracils at low energies

L M Cornetta[1]* and M T do N Varella[1]†

[1]Instituto de Física - University of São Paulo, São Paulo, ZIP 05508-090, Brazil

Synopsis We employed both eletronic scattering techniques and *ab initio* quantum dynamics to study electron-driven reactions in halouracils at low energies. Particularly, it was addressed halogen dissociation and H-loss probabilities in anionic 5-BrU and 5-IU. Besides, we also propose a model of how to include auto-ionization probabilities during the dynamics.

Dynamics of transient negative ions (TNI's) is a subject of different complexities. The formation of such systems begina when an electron attaches to a closed-shell neutral molecule in its ground electronic state, though being related to the formation of resonances. The dynamics triggered by the formation of resonant states could decay to different channels, but in either case the auto-detachment process is one of the possible outcomes.

In this context, radiosensitizers like 5-halouracils constitute an interesting group for studying dissociative electron attachment (DEA), so it has been shown they might decay to different dissociative channels under low energy (0 $\sim$ 3 eV) electron collisions[1]. From previous theoretical works[2], both anionic 5-BrU and 5-IU present a set of low-lying anionic states. In the case of 5-BrU we observe one valence π_1^* state, a σ_{CBr}^* antibonding resonance and a π_2^* resonance at energies -0.30 eV, 0.78 eV and 1.50 eV. Theoretical predictions rely on the most significant dissociation mechanism to be the electron attachment on π_2^* state followed by a π^*/σ^* coupling/crossing.

Combining Schwinger multichannel (SMC) method for describing resonances with *ab initio* multiple spawning (AIMS) methodology[3] we have approached TNI's dynamics as a photochemical reaction of the anion species (Fig. 1). We also have implemented the auto-ionization probability during the dynamics *a posteriori*.

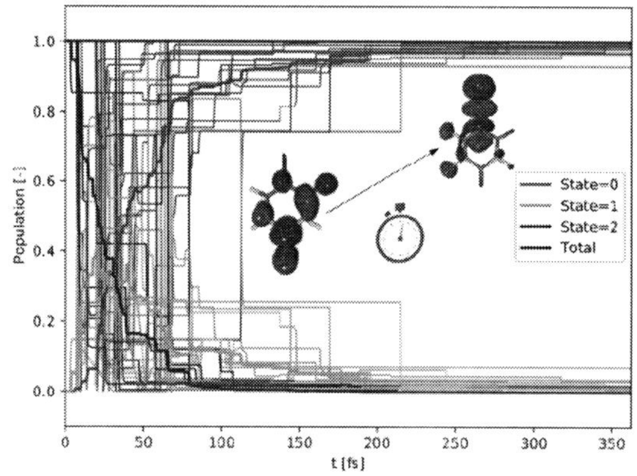

Figure 1: Population of the involved electronic states for 122 trajectories obtained for anionic 5-BrU. Thinner lines represent distinct initial conditions while thicker lines represent the average over all trajectories at each time.

The present work has been leading us to a clearer comprehension of different features, like the dissociative mechanisms induced by low-energy electron attachments, the presence and absence of distinct signals in DEA experiments and the role of the auto-ionization probabily.

References

[1] Abouaf R and Dunet H 2005 *European Physical Journal D.* **35** 405–410

[2] Kossoski F and Varella M T do N 2015 *Phys. Chem. Chem. Phys.* **17** 17271

[3] Martinez T J 2006 *Accounts of Chemical Research.* **39(2)** 119–126

*E-mail: lucas.cornetta@usp.br

†E-mail: mvarella@if.usp.br

Low energy differential elastic electron scattering from trichloromethane

B Diaz[1], G Tatreau[1], B Hlousek[1], M Zawadzki[1,2*] and M A Khakoo[1†]

[1]Department of Physics, California State University, Fullerton, California 92831, USA
[2]Atomic Physics Division, Department of Atomic, Molecular and Optical Physics, Faculty of Applied Physics and Mathematics, Gdańsk University of Technology, ul. Gabriela Narutowicza 11/12, 80-233 Gdańsk, Poland

Synopsis Experimental differential cross sections for low energy electron scattering from trichloromethane is measured utilizing a crossed electron-molecular beam experiment via the relative flow method, for the incident electron energies in the range of $E_0 = 0.5\,\text{eV}$-$30\,\text{eV}$ and the scattering angles in the range of $\theta = 10° - 130°$.

Low-energy electron collision processes play an important role in several areas including astrophysics, radiation biology, atmospheric chemistry and technology. When considering the inelastic and elastic collisions, the former constitutes a dominant process of low-energy electron transport through gaseous media, plasmas and condensed matter. In our present work we provide the elastic differential cross sections (DCSs) for low energy electron scattering from trichloromethane - the organic compound with formula $CHCl_3$.

The experimental setup employs a crossed electron/target beam arrangement. The energy separation of electrons is made at high resolution (40-50 meV FWHM) using electrostatic lenses combined with hemispherical analyzers [1]. In order to accurately determine background scattered electrons we employ a movable collimated gas beam source [2] developed in our laboratory.

The DCSs for trichloromethane, $\mathcal{Q}_A(E_0,\theta)$ are obtained from the relative-flow formula:

$$\mathcal{Q}_A(E_0,\theta) = \mathcal{Q}_{He}(E_0,\theta)\frac{\mathcal{R}_{He}\, I_{S,A}}{\mathcal{R}_A\, I_{S,He}}\sqrt{\frac{M_{He}}{M_A}}, \quad (1)$$

where $\mathcal{R}$ stands for the relative flow rates, I_S represents the scattering rates and M represents the molar masses where the subscripts indicate the gas species, with A being the unknown gas (trichloromethane) whose DCS is to be determined and He the standard gas (helium) whose DCSs are known; thus our measured trichloromethane DCSs were normalized to elastic DCSs for helium.

*E-mail: mateusz.zawadzki@pg.edu.pl
†E-mail: mkhakoo@fullerton.edu

We will present the experimental data on differential cross sections for elastic electron scattering on $CHCl_3$ for range of energies (0.5, 1, 2, 3, 4, 5, 7, 10, 15, 20, 30 eV) and the scattering angles $(10° - 130°)$. An example of the data for four incident electron energies is shown in Fig. 1.

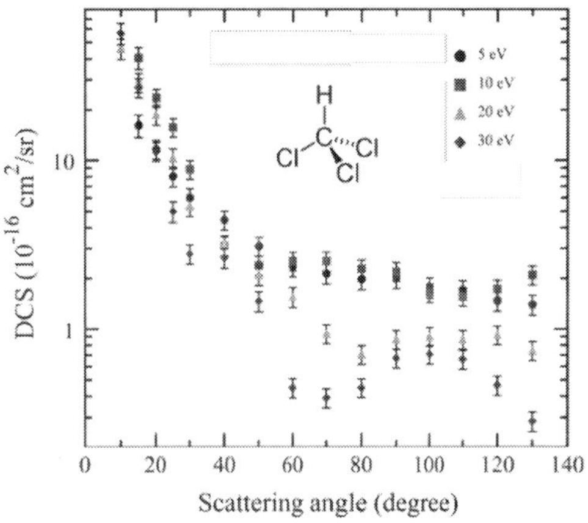

Figure 1. Differential cross sections for elastic electron scattering by trichloromethane.

Acknowledgments: The work carried out at California State University, Fullerton was funded by the National Science Foundation under grant NSF-RUI-PHY 1606905. M.Z. acknowledges the Fulbright Program to conduct this work at California State University Fullerton.

References

[1] Zawadzki M and Khakoo M A 2018 *J. Chem. Phys.* **149** 124304
[2] Hughes M, James Jr K E, Childers J G and Khakoo M A 2003 *Meas. Sci. Technol.* **14** 1

Total cross section measurements for electron scattering from tin(IV) chloride ($SnCl_4$)

P Możejko[1] *, S Stafanowska-Tur[1], E Ptasińska-Denga[1] and Cz Szmytkowski[1] †

[1]Department of Atomic, Molecular and Optical Physics, Faculty of Applied Physics and Mathematics, Gdańsk University of Technology, Gabriela Narutowicza 11/12,80-233 Gdańsk, Poland

Synopsis Total cross section for electron scattering from $SnCl_4$ molecules has been measured for energies from 0.6 to 300 eV. Obtained results have been compared with total cross sections for electron scattering from other tetrachoride molecules: XCl_4 where X=C, Si, Ge.

Precise experimental data concerning electron interactions with matter, including molecules in the gas phase, are very important for understanding and modelling many physical and technological processes. Tin tetrachloride is one of the simplest molecular compound which can be used as a precursor in nanostructure production with focused electron beam deposition (FEBID).

In the present experiment we have measured absolute total cross section (TCS) for low- and intermediate-energy electron scattering from $SnCl_4$ molecules. Measurements have been carried out with the electrostatic 127° electron spectrometer [1] working in the linear transmission mode for electron energies ranging from 0.6 to 300 eV. The TCS $\sigma(E)$ at given electron impact energy E has been obtained according to the attenuation formula:

$$I_n(E) = I_0(E) \exp\left[-nL\sigma(E)\right].$$

Here, $I_n(E)$ and $I_0(E)$ are the measured intensities of the electron beam passing the distance L through the reaction volume in the presence and absence of the target vapour, respectively; n is the number density of the target molecules in the scattering cell. Obtained TCS is shown in figure 1a for impinging electron energies ranging from 20 eV up to 300 eV. In figure 1b TCS for electron scattering from $SnCl_4$ molecules is compared with TCSs for CCl_4 [2], $SiCl_4$ [3] and $GeCl_4$ [4] molecules. All data have been taken with the same experimental method and the same apparatus. It is worth to note that the TCS value in the presented energy region is correlated with the static electric polarizability of the target.

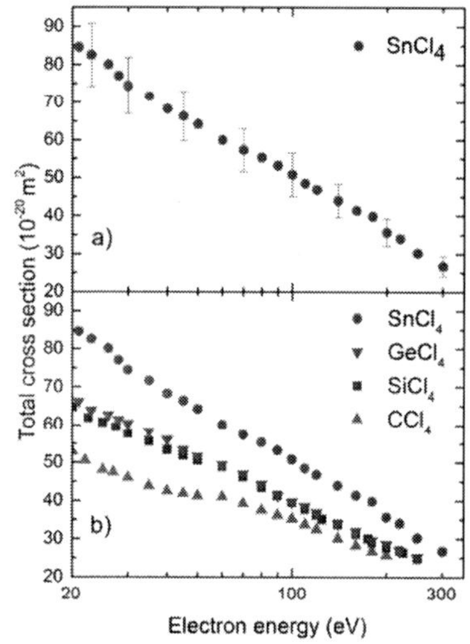

Figure 1. a) Present experimental total cross section (TCS) for the electron scattering from $SnCl_4$ and b) its comparison with TCSs for CCl_4 [2], $SiCl_4$ [3] and $GeCl_4$ [4] molecules. Error bars correspond to overall experimental uncertainties.

References

[1] Szmytkowski Cz *et al* 2001 *Vacuum* **63** 549
[2] Szmytkowski Cz *et al* 1992 *Chem. Phys. Lett.* **199** 191
[3] Możejko P *et al* 1999 *Eur. Phys. J. D* **6** 481
[4] Szmytkowski Cz *et al* 1997 *J. Phys. B* **30** 4363

*E-mail: paw@pg.edu.pl
†E-mail: czsz@mif.pg.gda.pl

Investigation of electron scattering asymmetries in halocamphors

J C Ruivo[1] *, F Kossoski[2] and M T do N Varella[1] †

[1]Instituto de Física da Universidade de São Paulo, Rua do Matão 1731, 05508-090, São Paulo, SP, Brazil
[2]Aix Marseille Univ, CNRS, ICR, Marseille, France

Synopsis We present a study on the electron transmission and dissociative electron attachment asymmetries in halocamphor species. Our results consist in their anion state spectra, applied in a model for the spin-polarized electron scattering problem under the influence of a low energy resonance.

Motivated by the Vester-Ulbrich hypothesis for biological homochirality [1], chiral sensitivity in longitudinally-polarized spin electron scattering has been investigated in the last decades. Electron transmission asymmetry experiments reported by Mayer *et al.* show that the magnitude of scattering asymmetries for Bromocamphors are enhanced by the formation of resonances, in a phenomenon called electron circular dichroism (ECD) [2].

More recently, the dissociative electron attachment (DEA) asymmetries for 3-Bromocamphor (3BrC), 3-Iodocamphor (3IC) and 10-Iodocamphor (10IC) were reported by Dreiling *et al.* [3]. The results obtained do not seem consistent with theoretical predictions, and points out the need for the development of a new model to provide an understanding of the magnitudes and energy dependence of the asymmetries.

We investigated the temporary electron attachment in the halocamphor molecules by elastic scattering calculations using the Schwinger multichannel method with pseudopotentials (SMCPP) [4]. The anionic spectra were analyzed and their resonances were characterized, from which provided essential information to understanding the expected dissociation mechanisms and the related chiral asymmetries.

A model, proposed elsewhere, is being developed for the spin-dependent electron scattering problem under the influence of a low energy resonance. Provided that the anion state spectra are known, we intend to explain the experimental results of asymmetries.

The electron transmission asymmetry for 3-Bromocamphor was computed in a first, preliminary calculation, which considered its anion spectra (Table 1). Figure 1 shows the obtained results and indicates a qualitative agreement with the experimental data measured by Mayer *et al.*[2].

Table 1. Energies of anionic states and widths (in parenthesis), in eV, obtained from SMCPP calculations. States are labeled according to orbital character.

	σ_1^*	σ_2^*
3-Bromocamphor	0.32 (0.192)	2.34 (0.49)
	σ^*	π^*
3-Iodocamphor	-0.01	2.00 (0.50)
10-Iodocamphor	0.25 (0.03)	1.77 (0.19)

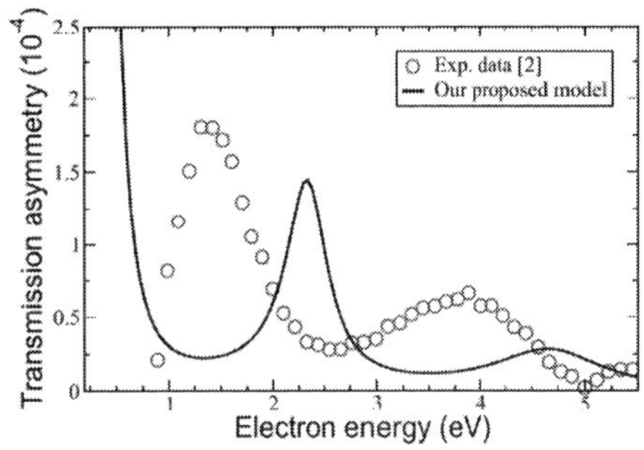

Figure 1. Asymmetry obtained by our model against experimental data [2], for 3BrC.

The authors acknowledge financial support from Brazilian agencies CNPq and FAPESP.

References

[1] Ulbrich T *et al* 1962 *Tetrahedon* **18** 629
[2] Mayer S *et al* 1996 *J. Phys. B: At. Mol. Opt. Phys. B* **29** 3497
[3] Dreiling J *et al* 2016 *Phys. Rev. Let.* **116** 093201
[4] da Costa R *et al* 2015 *Eur. Phys. J. D* **69** 159

*E-mail: ruivo@if.usp.br
†E-mail: mvarella@if.usp.br

1412 (2020) 182019 doi:10.1088/1742-6596/1412/18/182019

Electron impact scattering study on formyl fluoride (HFCO)

T Jani[1] *, H Yadav[1,2] †, D Prajapati[1,3], M Vinodkumar[2] and P C Vinodkumar[1]

1 Department of Physics, Sardar Patel University, Vallabh Vidyanagar - 388 120, India
2 V. P. & R. P. T. P. Science College, Vallabh Vidyanagar - 388 120, India
3 Shree M R Arts & Science College, Rajpipla - 393 145, India

Synopsis Electron impact scattering studies with organic compounds play a significant role in studies related to astrophysics and radiation physics. We present here electron impact scattering studies on formyl fluoride as no cross-sectional data is available to the best of our knowledge. Total scattering cross sections (TCS) on comparing with iso-electronic molecule reveals that at low energy molecular dipole moment plays a significant role in cross-section deviation and at high energy the tendency seems to be same in all the formyl series.

A descriptive and detailed theoretical analysis is carried out on electron impact scattering of formyl fluoride (HFCO) over a wide energy range, ranging from 0.1 eV to 5000 eV. Such a wide energy range enables us to calculate variety of processes e.g. eigen phase study to identify the electron impact resonances, vertical electron excitations, differential cross sections, momentum transfer cross sections, ionization cross sections and total scattering cross sections.

To compute cross section up to such a wide energy range, we have used an amalgamation of two theoretical methodologies. For low energy, R-matrix [1] using the Quantemol-N package from 0.1 eV to 20 eV. And for intermediate to high energy, we have adopted Spherical Complex Optical Potential (SCOP) from ionization threshold to 5000 eV. There is smooth cross-over of total cross sections at 20 eV, using R-matrix and SCOP formalism.

Table 1. Calculated target property for HFCO

	Present	Other
Ground state Energy (Hartee)	-212.58	-212.62 [2]
Dipole momentum (Debye)	2.02	2.02 [3]
Molecular polarizability (Å^3)	1.23	1.96 [3]

We have noticed that there is a good agreement of the TCS for formyl fluoride and formic acid [4]. This may be due to the fact that they are isoelectronic (24 electrons). At higher energies beyond 1000 eV, the TCS from all these targets tend to merge implying that at high energies cross sections do not depend on the structure of target but depends on the geometry and effective interaction time between the target and the electron. As the energy increases, the effective time the electron spends with the target decreases and hence the cross sections decreases.

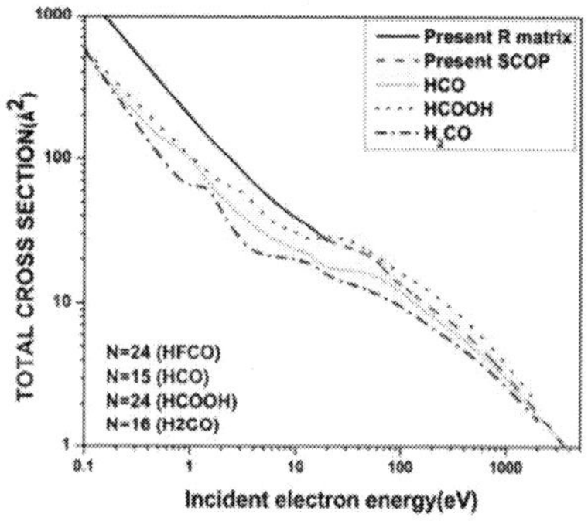

Figure 1. TCS curve for HFCO, in comparison with HCO^5, $HCOOH^4$ and H_2CO^4.

The present study is important due to detection of such organic molecules in astrophysical environment as they are key constituents for formation of larger amino acids and play vital role in evolution of life.

Minaxi Vinodkumar acknowledges DST-SERB, New Delhi for Major research project [EMR/2016/000470].

References

[1] Tennyson J 2010 *Phys. Rep.* **491** 29
[2] Ha T K *et al* 1975 *J. Mol. Struct.* **27** 225
[3] *https://cccbdb.nist.gov*
[4] Vinodkumar M *et al* 2011 *Phys. Rev. A* **84** 052701
[5] Yadav H *et al* 2017 *Mol. Phys.* **115** 952

* E-mail: tejasjani4596@gmail.com
† E-mail: hitesh0507@gmail.com

Electron driven scattering studies involving diborane

H Yadav[1,2]*, D Prajapati[1,3], M Vinodkumar[2] and P C Vinodkumar[1]

[1]Department of Physics, Sardar Patel University, Vallabh Vidynagar, Gujarat, 388120, India
[2]V. P. & R. P. T. P. Science College, Vallabh Vidynagar, Gujarat, 388120, India
[3]Shree M R Arts & Science College, Rajpipla, Gujarat, 393145, India

Synopsis In the present case we study the electron impact scattering cross sections on diborane (B_2H_6) for the energy range 0.1 eV to 5000 eV. The present study provides differential cross sections (DCS), momentum transfer cross sections (MTCS), ionization cross sections and total scattering cross sections (TCS). The experimental determinations of these cross sections are rather difficult dur to its complex structure and stability. These cross sections are important in many applied fields.

The development of ab-initio methods [1] to study low energy electron-molecule collisions has made the possible calculation of reliable cross sections. Several studies involving polyatomic and linear targets were tested at different levels of approximations. This has allowed us to compute the cross section for important molecular targets, for which the experimental data are very rare or difficult to obtain.

Diborane (B_2H_6) is one such peculiar molecule in the series of X_2H_6. But diborane just does not follow the generality in the geometry of the X_2H_6, it has a bent shape bond between B-H-B that is known as 3-centre-2-electron bonding i.e. banana bond. This particular bending of the bond or the electron sharing bond can also be considered as due to the resonances of the molecular electrons.

The diborane molecule has many applications in the applied fields. It is used as a rocket propellent in space science, in reactions it works as reducing agent and it is also used as a doping agent in the manufacturing of semiconductor devices.

The only work reported on electron impact scattering on diborane is the theoretical estimation by M. H. F. Bettega et al [2], where they have used IAM model to compute the cross section using H_3B-BH_3 system.

For the present energy range we have used two different methodologies to calculate the cross section. For low energy range (0.1 eV to 20 eV) we have used UK-Rmol based R-matrix via Quantemol-N [1], while for intermediate to high energy (ionization threshold to 5000 eV) we have employed Spherical Complex Optical Potential (SCOP) [3,4].

* E-mail: hitesh0507@gmail.com

In figure 1, we present the sample result of diborane differential cross section for 7.5 eV. The present result shows a better agreement with Bettega et al [2]. In the forward angle scattering for the DCS, while in the backward scattering beyond 90 degrees, the present result shows some feature which is absent in the case of Bettega et al [2]. This may be accounted due to the limitations of IAM model.

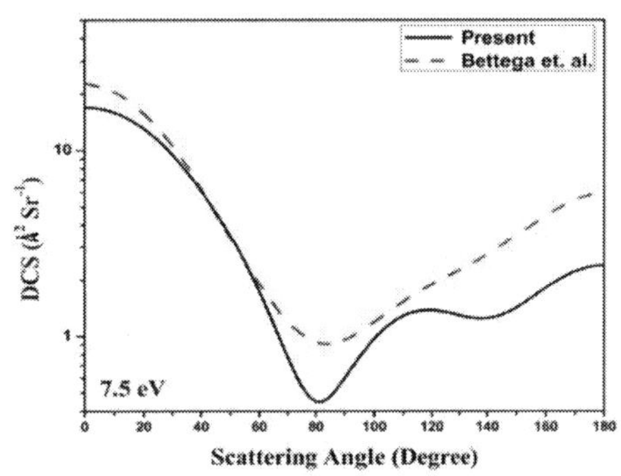

Figure 1. Differential Cross section (DCS), where the solid line represents the present calculated data and the dash line represents the reported data of Bettega et al [2]

Minaxi Vinodkumar acknowledges DST-SERB, New Delhi for the partial funding through a major research project [EMR/2016/000470].

References

[1] Tennyson J 2010 *Phys. Rep.* **491** 29
[2] Bettega M H F *et al* 1998 *Eur. phys. J. D* **1** 291
[3] Prajapat D *et al* 2018 *Molecular Physics* doi: 10.1080/00268976.2018.1554191
[4] Vinodkumar M *et al* 2006 *Eur. Phys. J. D* **37** 67

 1412 (2020) 182021 doi:10.1088/1742-6596/1412/18/182021

Electron impact cross sections for H_2, HD and D_2

D Prajapati[1,2]*, H Yadav[2,3], M Vinodkumar[3] and P C Vinodkumar[2]

[1]Shree M R Arts & Science College, Rajpipla - 393 145, India
[2]Department of Physics, Sardar Patel University, Vallabh Vidyanagar - 388 120, India
[3]V P & R P T P Science College, Vallabh Vidyanagar - 388 120, India

Synopsis The demand for electron-impact cross sections has increased tremendously in recent years. There is, however, a special interest in such cross sections for hydrogen molecules and its isotopomers, HD and D_2. We have computed target properties & electron scattering cross sections for H_2, HD and D_2 and discuss different phenomena and their importance.

The deuterated hydrogen (HD) molecule is the simplest two-electron system after hydrogen and deuterium molecules (H_2, D_2). These molecules are important in tokamak edge plasmas, planetary atmospheres and at different astrophysical sites. There are enough experimental as well as theoretical cross section data available for H_2, but limited cross sectional data are available for HD and D_2. Here we provide electron impact cross sectional data for HD and D_2 in comparison with H_2 molecule.

A theoretical study is carried out for electron interactions with H_2, HD and D_2 with impact energies ranging from 0.01 to 5000 eV. As single theoretical method for this wide energy ranges is not suitable, for the present study here we use two different theoretical formalism. For low incident energy 0.01 eV to 20 eV, we have performed R-matrix calculations using Complete Active Space - Configuration Integration (CAS-CI)[1, 2]. The R-matrix calculations were performed using close coupling approximation. Low energy computations using cc-pVTZ basis set provide us electronic excitation, eigenphase sum, Integral cross section (ICS), excitation (ECS), differential (DCS), momentum transfer (MTCS), ionization and total cross sections (TCS), i.e. the summation of total elastic (Q_{el}) and total inelastic (Q_{inel}) cross sections. For intermediate to high energy (~16 eV to 5000 eV), we have adopted Spherical Complex Optical Potential (SCOP) formalism [3]. In the SCOP method with single center approximation we compute ionization cross section and total cross section. We found a good matching of the total cross sections by R-matrix and by SCOP at 16 eV. The target properties reported using quantum chemistry codes are shown in Table 1.

*E-mail: dineshphy13@gmail.com

Table 1. Target properties for H_2, HD and D_2.

Target property (unit)	H_2	HD	D_2
Ground State Energy (Hatree)	-1.172	-1.172	-1.172
Ionization Potential (eV)	16.17	16.17	16.17
Polarizability($\AA^3$)	0.870	0.874	0.866
Rotational Constants (cm^{-1})	60.853	45.655	30.444

As a sample result we report here TCS data for e - H_2, HD and D_2 in Figure 1. The present results from 1 eV to 16 eV, are in good agreement with available experimental data of Golden et al.[4].

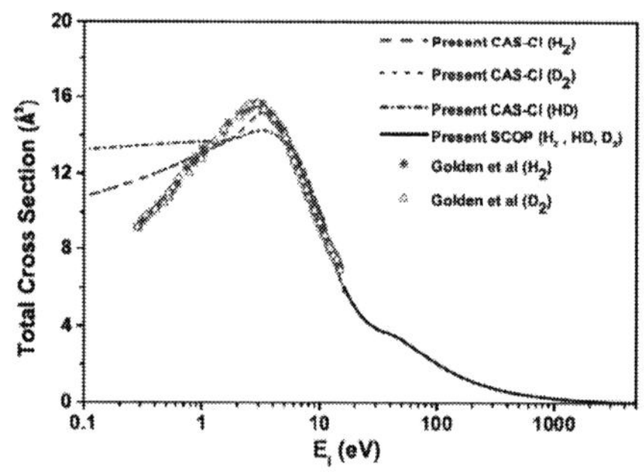

Figure 1. Total Scattering Cross Sections (TCS) for e - H_2, HD and D_2

Minaxi Vinodkumar acknowledges DST-SERB, New Delhi for the Major research project [EMR/2016/000470] for financial support under which part of this work is carried out.

References

[1] Vinodkumar M *et al* 2014 *Phys. Rev. A.* **89** 062715
[2] Prajapati D *et al* 2018 *Eur. Phys. J. D.* **72** 210
[3] Yadav H *et al* 2017 *J. Phys. B.* **4** 51
[4] Golden D *et al* 1966 *Phys. Rev. A.* **146** 40

Energy loss of laser-accelerated ions in dense ionized matter

Y Zhao[1 *], J Ren[1], Z Deng[2], W Qi[2], R Cheng[3], X Wang[1], W Liu[1], B Ma[1], S Yin[1], B Chen[1], J Feng[1], S Wang[2], Q Fan[2], B Cui[2], X Zhou[3], W Zhou[2] and L Cao[2]

[1]Xi'an Jiaotong University, Xi'an, 710049, China
[2] Laser Fusion Research Center, Chinese Academy of Engineering Physics, Mianyang, 621000, China
[2] Institute of Modern Physics, Chinese Academy of Sciences, Lanzhou, 710049, China

Synopsis Knowledge about the energy loss of heavy ion in dense ionized matter is of great importance in fields such as inertial confinement fusion, fast ignition as well as intense heavy ion beam driven high energy density physics. The energy loss of 3.4 MeV proton in TAC foam plasma was measured on the XGIII laser facility. It was found that the measured energy loss is far above the Beth prediction.

Ion stopping in dense ionized matter is the basic process in inertial confinement fusion and fast ignition. The precise knowledge about the stopping power is of great importance, since it determines the range of ions in fusion pellet. To fully understand the energy deposition process of ion in plasma, large amount of experimental and theoretical work has been carried out during the past years, but mostly restricted to high energy region. Recently, Cayzac and co-workers [1] performed an experiment to measure the energy deposition of ions in laser-generated plasma near Bragg peak and discriminate the various models. In this report, we investigated the energy deposition of laser-accelerated ions in dense C-H-O plasmas.

The experiment was carried out on XGIII laser facility in Mianyang. The ion beam was generated through TNSA process with ps laser. Through selecting energy with a magnet, short pulsed quasi mono-energetic ions were produced. The dense ionized matter was generated through heating the TAC foam with ns laser-driven gold hohlraum radiation. With such kind of scheme, large scale($\sim mm$), uniform and long-lived($\sim ns$) plasma, which has the temperature of 19eV and free electron density of $4 * 10^{20} cm^{-3}$, was produced in laboratory[2]. Since the time-scale of ion beam is several magnitude lower than the plasma living time, it can be thought that the energy loss measurement was carried out in case of stable isolated plasma.

The energy spectra of ions were measured with a Thomson Parabola, and the result is shown in Fig. 1. The solid lines indicate the theoretical tracks of different ion species. The black dots are the signal of neutral particles, the blue and red dots are protons with and without plasma target. Evident energy shift was observed. Through analysing the relative deflection distance, the energy of ions was obtained. It was found that the measured energy loss is far above the Beth prediction, and the beam intensity may play important roles in the stopping process. Details will be reported in the conference.

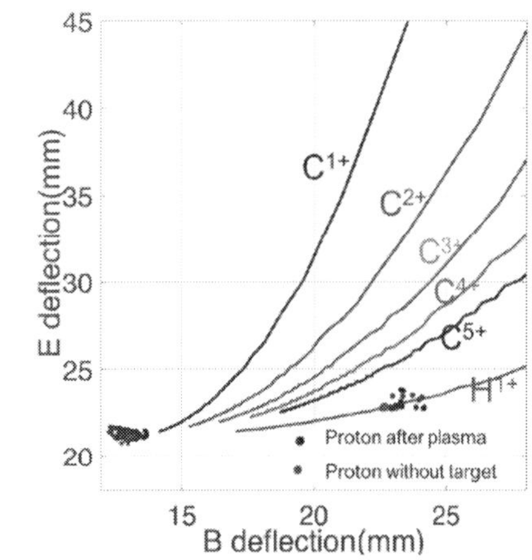

Figure 1. Thomson parabola spectrometry of protons with(blue dots) and without plasma(red dots), respectively. The solid lines are simulation lines of various species.

References

[1] Cayzac W *et al* 2017 *Nature Comm.* **8** 15693
[2] Rosmej O *et al Nucl. Instr. Meth. A* **653** 52

*E-mail: zhaoyongtao@xjtu.edu.cn

EUV spectroscopy of Kr, Y and Mo ions using CoBITs

E Sokell[1] *, S Ali[2] , T Kondo[2], M Monobe[2], S Ohishi[2], H A Sakaue[3] and N Nakamura[2]

[1] School of Physics, University College Dublin, Belfield, Dublin 4, Ireland
[2] Institute for Laser Science, The University of Electro-communications, Chofu, Tokyo 182-8585, Japan
[3] National Institute for Fusion Science, Toki, Gifu 509-5292, Japan

Synopsis CoBITs have been used to produce Kr, Y and Mo ions and EUV spectra have been recorded. The data indicate that the CoBIT spectra are significantly different to laser-produced plasma emission spectra for similar ion stages. This difference is attributed to the different number density of electrons in the two kinds of plasmas. The evolution of the line profiles with electron beam energy for a number of different ions has been studied.

CoBIT (Compact electron Beam Ion Trap) spectra are far simpler than those obtained from laser produced plasmas (LPPs) [1], the differing number density of electrons strongly influences the emission lines that appear in the EUV spectra from the different plasmas. In LPPs the number density of electrons is sufficiently high that metastable levels are generally collisionally depopulated before they can radiate. In contrast in CoBIT plasmas, such metastable levels live long enough to radiate and may give rise to optically forbidden lines, such as the MoXV $3d^{10}$ (J=0) - $3d^9 4s$ (J=2) lines [2]. The prominent appearance of these lines in the CoBIT spectra, indicates that the population of the $3d^9 4s$ state is enhanced within the electron beam ion trap in comparison to LPPs.

The study of dipole forbidden lines is of relevance to a number of fieds including astrophysics and fusion science. Indeed a comprehensive theoretical study of the E1, M1 and E2 transitions of the Ca-like tungsten ion has been recently completed and the data compared reasonably well with the available experiments [3]. In the present work, the evolution of the intensity of E1 and E2 transitions with CoBIT beam energy has been studied for the Ni-like ions Kr^{8+} and Mo^{14+}. The equivalent lines have also been observed in Y^{11+}. The data were taken at the Tokyo and NIFS CoBITs [4], devices designed for spectroscopic studies of moderately charged ions which utilize a superconducting wire with a high critical temperature to produce a 0.2T magnetic field that compresses the electron beam. This configuration means that the devices can be operated at liquid nitrogen, rather than liquid helium as for a convential EBIT, temperatures.

Figure 1 shows EUV spectra recorded at CoBIT energies between 290 and 430 eV. At the lowest beam energy lines associated with Mo^{13+} are present [1], which die out and are replaced with lines associated with Mo^{14+} as the beam energy is increased. The energy required to produced Mo^{13+} is 280 eV and for Mo^{14+} it is 303 eV [5]. In the case of Mo^{15+}, lines appear at energies below the ionisation energy reuired to produce Mo^{15+} of 544 eV [5] and this indicates that ionisation from metastable levels may be occurring.

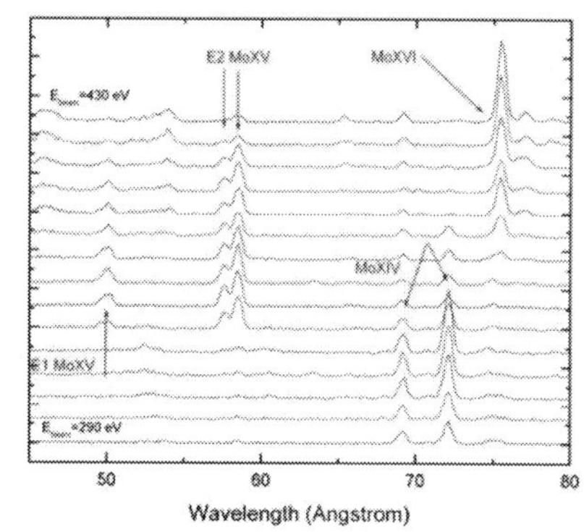

Figure 1. EUV spectra for Mo, for CoBIT beam energies ranging from 290 to 430 eV in 10 eV increments. The calibration is preliminary but considered to be good to ~ 1 Angstrom. To show the spectra more clearly, each spectra has been offset from the one at the next lowest beam energy.

References

[1] Sokell E *et al* 2017 *J. Phys. : Conf. Ser.* **875** 052045
[2] Klapisch M *et al* 1978 *Phys. Rev. Lett.* **41** 403
[3] Ding X *et al* 2017 *J. Phys. B:* **50** 045004
[4] Nakamura N *et al* 2008 Rev. Sci. Inst. **79** 063104
[5] https://physics.nist.gov/PhysRefData/ASD/ionEnergy.html

* E-mail: emma.sokell@ucd.ie

Measurement and analysis of emission spectrum of laser-produced Ni plasma in the range of 7-14 nm

S Q He[1], M G Su[1]*, S Q Cao[1], L Wu[1], H D Lu[1], Q Min[1], D X Sun[1] and C Z Dong[1]

[1]Key laboratory of Atomic and Molecular Physics & Functional Materials of Gansu Province, College of Physics and Electronic Engineering, Northwest Normal University, Lanzhou, 730070, China

Synopsis We measured and analyzed the emission spectrum of laser-produced Ni plasma in the range of 7-14 nm. The dominant transition arrays were identified as $3p^63d^n$-$3p^63d^{n-1}f$ and $3p^53d^{n+1}$-$3p^53d^n4f$ from Ni^{6+} to Ni^{9+} ions with aid of the Hartree-Fock calculations with configuration interaction effects. The plasma parameters have been obtained by comparison of experimental and simulated spectrum based on a steady-state collisional-radiative (CR) model.

Highly charged ions of middle- and high-Z elements exist widely in astrophysical, fusion plasmas and laboratory plasmas. Spectral structure analysis of highly-charged ions ablated by high power laser pulse can reveal abundant information on plasmas, such as electron temperature, electron/ion density, particle and energy transport, and the evolution of these parameters.

In this work, the EUV spectra of laser produced Ni plasmas in the 7-14 nm range were measured using spatio-temporally resolved emission spectroscopy, as is indicated in Fig. 1. The characteristic features of experimental spectra are dominated by three intense, superimposed, quasi-continuous bands, which mainly located around 9.5 nm, 10.5 nm and 11.9 nm.

In order to identify the structures of the experimental spectrum, a series of calculations were performed with the Hartree-Fock method by Cowan codes [1], and detailed comparative analysis between experimental and theoretical simulated profiles based on the assumption of a normalized Boltzmann distribution among excited states and a steady-state collisional-radiative model [2].

The results show that the dominant spectral profiles arose from the $3p^63d^n$-$3p^63d^{n-1}f$ and $3p^53d^{n+1}$-$3p^53d^n4f$ transition arrays of Ni^{6+} to Ni^{9+} ions, and there is good agreegement between experimental and theoretical spectral profiles at a electron temperature of 30 eV and a electron density of 6.8×10^{20} cm^{-3}. The inserted figure shows the calculated ion fraction.

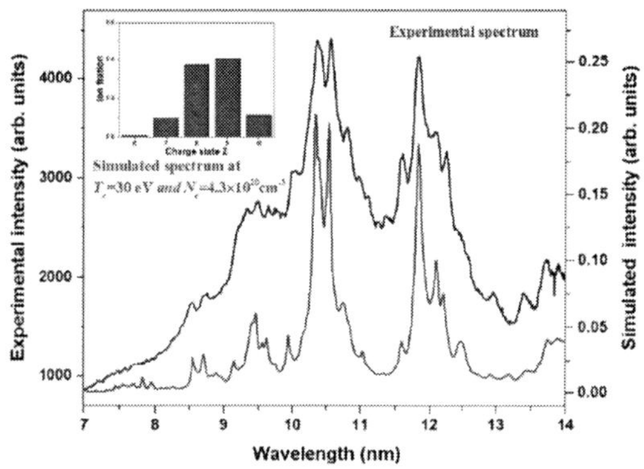

Figure 1. Comparison between the experimental spectrum and simulated spectrum.

This work is supported by the National Key Research and Development Program of China (Grant no. 2017YFA0402300), the Natural Science Foundation of China (NSFC) (Grant nos. 11874051, 11564037, 61741513).

References

[1] Cowan R D 1991 *The Theory of Atomic Structure and Spectra* (Berkeley, CA: University of California Press)

[2] Colombant D *et al* 1973 *J. Appl. Phys.* **44** 3524.

* E-mail: nwnu_sumg @163.com

1412 (2020) 192004 doi:10.1088/1742-6596/1412/19/192004

Diagnosing the lightning plasma temperature based on the collisional-radiative model

W Zhang, J Jiang* and P Yuan

Key Laboratory of Atomic and Molecular Physics and Functional Materials of Gansu Province, College of Physics and Electronic Engineering, Northwest Normal University, Lanzhou 730070, China

Synopsis The atomic parameters of the NII ion, including the energy levels, transition probabilities, collision sections, and collision excitation rate coefficients, have been calculated based on the multi-configuration Dirac-Fock method by implementing Flexible Atomic Code. The ratios of line intensity, $I_{567.96nm}/I_{500.52nm}$, have been calculated by using the collisional-radiative model. The temperatures of six lightning plasma channels were then obtained, and they are consistent with temperatures obtained by using the multi-spectral lines method.

The temperature of lightning plasma is one of the basic parameters directly reflecting the physical characteristics of the lightning discharge process[1, 2]. Analysis of the lightning spectrum is an effective way to diagnose the lightning plasma temperature at present. It would be significant to analyze the physical characteristics and mechanism of lightning if the temperature can be acquired by combining theoretical calculations with the characteristic lines in lightning spectra.

In this work, the energy levels, transition probabilities, and collision excitation rate coefficients of the NII ion have been studied by using the Flexible Atomic Code (FAC) program, which is a fully relativistic program for computing both atomic structure and scattering parameters [3]. The relative intensity of the spectral lines for every transition near the NII 500.52 and 567.96 nm lines were obtained by using collisional-radiative model (CRM). The temperature of the lightning plasma was then obtained further by matching the ratio of total intensities of the characteristic lines in the experimental spectrum with the ratios of total relative intensities of NII lines at 567.96 and 500.52 nm.

Six lightning flashes (labeled as A, B, C, D, E, and F) had been recorded by a slitless spectrograph with a high-speed camera as a recording system on the Qinghai plateau of China[4], their temperatures were obtained by using the CRM and were listed in Table 1 respectively. The results are consistent with those obtained by diagnosing the actual spectrum of these lightning using the multi-spectral line method[4]. It can be seen that their differences were only within 2500 K.

Table 1. Temperature of the lightning plasma

lightning plasma	Temperature (K)	
	CR model	Multi-spectral line method
flash A	21800	20000
flash B	22200	20500
flash C	20500	19500
flash D	23500	24400
flash E	26300	23900
flash F	30000	29600

This work is supported by National Natural Science Foundation of China under Grant Nos. 11564036, 11774292 and 11475139. The authors thank all participants of the field experiments and especially grateful to the help of Datong County Education Bureau for the experiment site.

References

[1] Nakiimana A *et al* 2013 *Chin. Phys. B* **22** 014209
[2] Mu Y *et al* 2015 *J. Atomos. Sol. Terr. Phys* **145** 98
[3] Gu M 2004 *Phys. Rev. A* **70** 062704
[4] Cen J *et al* 2015 *Appl. Phys. Lett.* **106** 054104

*E-mail: phyjiang@yeah.net

Plasma dynamic simulations for reducing the erosion of discharge channel walls of Hall Effect Thruster (HET)

B Priyanka[1 *], **M M Asif**[1] **and R K Umesh**[1 †]

[1]Indian Institute of Space Science & Technology, Thiruvananthapuram, 695547, India

Synopsis To assess and improve the design aspects during the design and development of HETs, it is essential to perform plasma dynamic simulations on these thruster designs. A 2D-3V (two dimensional-three velocity) Particle-in-cell Monte Carlo (PIC-MC) model has been used to simulate the standard SPT-100, and the design was optimized. Later, a completely new HET geometry has been simulated with an aim to optimize the thruster design with reduced plasma wall interaction.

Of the two primary means of propulsion available today for spacecraft mobility, namely chemical and electric, electric propulsion (EP) is an area of high interest, as the key attraction of EP systems has been the requirement of lower amount of fuel and highly efficient utilization of propellant mass. HETs are electrostatic thrusters which utilize a cross-field described by the Hall Effect to generate the plasma discharge with magnetized electrons and unmagnetized ions in the axial electric and radial magnetic fields applied in an annular ceramic channel.

The shape of the magnetic field which entraps the energetic electrons inside the channel, and the erosion of the discharge channel walls are the most crucial design elements that directly impacts the thruster lifetime and performance. Hence theoretical modeling of the plasma dynamics is of great importance, and with a right emphasis on theoretical modeling, the iteration cycles required in developing an EPS can be reduced. The present work has been done with an aim of developing an erosion less HET using a theoretical design and microscopic predictions on the plasma discharge parameters and magnetic parameters, and thus to enable an optimization and characterization of the modeled design.

The discharge channel plasma and the ion beam outside the discharge channel of both SPTs have been mathematically modelled using the full PIC method of simulation [1] incorporating the Monte-Carlo Collisions (MCC). The geometry of the new HET along with the magnetic flux lines has been designed in such a way to reduce the plume interaction with the walls, and an electrostatic solver has been used for the simulation.

The Xe neutrals are treated as background gas and is predefined to give a steady state profile, with maximum density of $10^{21}/cm^3$ just at the exit of the AGD, with a linear density drop from anode to exit of the channel.

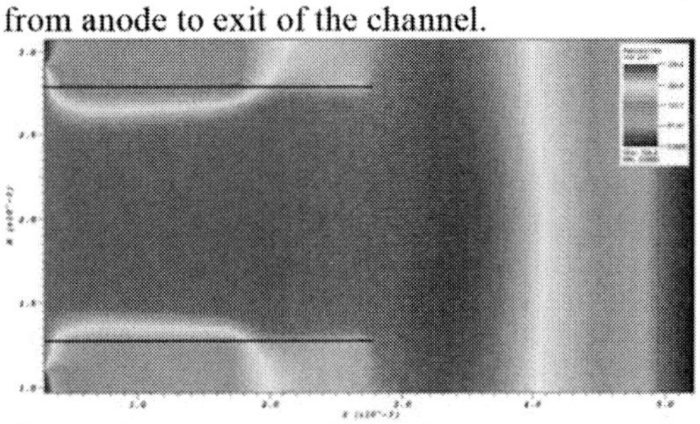

Figure 1. Plasma potential gradiant (at 0.5μs) in the discharge channel.

The electrons are assumed to have an initial electron energy of 25eV, and the cathode is modelled as electron emitter close to exit with V=0. The particle collision is handled by MCC, and both the electron-neutral collisions and ion-neutral collisions have been incorporated.

The simulation was carried out for about 0.5μs, and it was observed that the plasma potential acquires a steady state within this time. The measurements on the exhaust velocities of Xe^+ from the HET gives a value of above 20,000m/s. The simulation could also give interesting results on the plasma potential, Xe ion density and electron density.

References

[1] Birdsall C K and Langdon A B 1988 *Plasma Physics Via Computer Simulation*, Institute Of Physics series in plasma physics

* E-mail: priyankab@iist.ac.in
† E-mail: umeshrk@iist.ac.in

Spectroscopic investigations of YAG-laser-driven microdroplet-tin plasma sources of extreme ultraviolet radiation for nanolithography

O Versolato[1] *, A Bayerle[1], M Bayraktar[2], L Behnke[1], H Bekker[3], Z Bouza[1,4], J Colgan[5], J R Crespo López-Urrutia[3], M J Deuzeman[1,6], R Hoekstra[1,6], D Kurilovich[1,4], B Liu[1,4], R Meijer[1,4], A Neukirch[5], L Poirier[1], S Rai[1,6], A Ryabtsev[7], J Scheers[1,4], R Schupp[1,4], J Sheil[1], F Torretti[1,4], W Ubachs[1,4] and S Witte[1,4]

[1]Advanced Research Center for Nanolithography (ARCNL), Amsterdam, 1098 XG, The Netherlands
[2]Industrial Focus Group XUV Optics, University of Twente, Enschede, 7522 NB, The Netherlands
[3]Max-Planck-Institut für Kernphysik, Saupfercheckweg 1, Heidelberg, 69117, Germany
[4]Department of Physics and Astronomy, and LaserLaB, Vrije Universiteit, Amsterdam, 1081 HV, The Netherlands
[5]Los Alamos National Laboratory, Los Alamos, NM 87545, USA
[6]Zernike Institute for Advanced Materials, University of Groningen, Groningen, 9747 AG, The Netherlands
[7]Institute of Spectroscopy, Russian Academy of Sciences, Troitsk, Moscow, 108840, Russia

Synopsis Highly charged tin ions are the sources of extreme ultraviolet (EUV) light at 13.5-nm wavelength in laser-produced transient plasmas for next-generation nanolithography. Generating this EUV light at the required power, reliability, and stability however presents a formidable task that combines industrial innovations with challenging scientific questions. We will present work on the spectroscopy of tin ions in and out of YAG-laser-driven plasma and present a surprising answer to the key question: what makes that light?

Extreme ultraviolet (EUV) at 13.5-nm wavelength light will soon be used for state-of-the-art nanolithography. Current lithography technology uses 193-nm light; EUV wavelengths enable higher resolution. This step, towards using EUV light, is crucial for continuing the miniaturization of the features on chips as represented by Moore's law. To produce the required EUV light, highly charged tin ions are bred in laser-driven transient, dense plasmas. Near the 13.5 nm wavelength that can efficiently be reflected from available multilayer optics, the EUV spectrum of highly charged Sn ions is dominated by intense unresolved transition arrays (UTAs) from the resonance transitions $4p^6\ 4d^m$ - $4p^5\ 4d^{m+1}$ + $4d^{m-1}\ 4f$ in Sn^{8+}-Sn^{14+} (see figure) although other, more highly excited states also significantly contribute. UTAs from serendipitously aligned, strongly interacting configurations from several Sn charge states contribute to a remarkably efficient production of EUV. A detailed understanding of the complex atomic structure of tin ions is of a prerequisite to establish and understand the fundamental atomic physics limitations to the conversion efficiency of drive laser light into useful EUV radiation. In this contribution, we will present work on the spectroscopy of tin ions both in [1] and out [2,3] of the laser-driven plasma. Detailed comparisons are made with ongoing large-scale simulation efforts on the YAG-laser-produced plasma.

These studies enable finding a surprising answer to the key question: what makes that light?

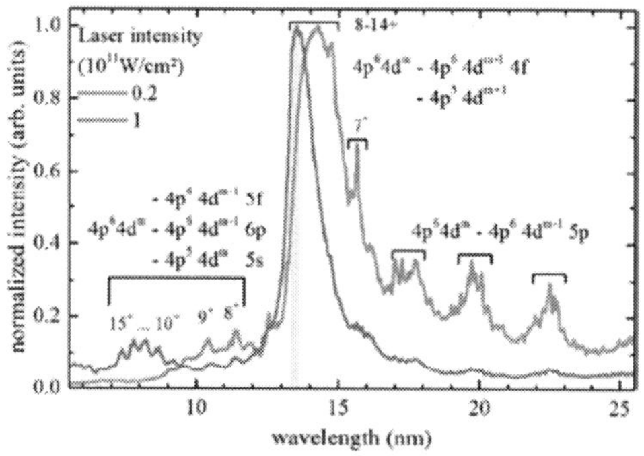

Figure 1. Emission spectra resulting from a tin droplet illuminated with a Nd:YAG laser beam at two intensities. Emission features attributed to the various Sn charge states are labeled (m is an integer between 0 and 6).

References

[1] Torretti F *et al* 2018 *J. Phys. B* **51** 045005
[2] Torretti F *et al* 2017 *Phys. Rev. A* **95** 042503
[3] Windberger A *et al* 2016 *Phys. Rev. A* **94** 012506

* E-mail: o.versolato@arcnl.nl

Electron capture to the Rydberg atomic states in ternary collisions with neutral particles

A A Narits[1] *, K S Kislov[1,2] † and V S Lebedev[1]

[1]P.N. Lebedev Physical Institute, Leninskiy prospect 53, Moscow 119991, Russian Federation
[2]Moscow Institute of Physics and Technology, Institutskiy per. 9, Dolgoprudny, 141701, Russian Federation

Synopsis We report on the results of theoretical comparative studies of different physical mechanisms of free electron capture to the Rydberg atomic states and bound-bound transitions in discharge plasmas of Ne/Xe and Ar/Xe mixtures containing atomic and molecular ions. Particular attention is paid to the role to the resonant non-adiabatic free-bound and bound-bound transitions due to collisions with neutral particles. We show that in a wide range of low-temperature plasma parameters the contribution of resonant processes to the population of low-lying Rydberg states is predominant.

Comparative theoretical studies of different physical mechanisms of free electron capture to the Rydberg atomic states and bound-bound $n \to n'$ transitions between the highly excited states of atom are carried out. The studied processes include the collisions with free electrons of plasma and resonant free-bound and bound-bound transitions in the collisions with neutral atoms of buffer gas. Particular attention is paid to the influence of these resonant processes caused by the non-adiabatic energy exchange between the outer and inner electrons of a quasi-molecule. We consider the low-temperature plasmas formed in the afterglows of pulse discharges in noble gas mixtures. For such conditions it is important to take into account the contributions from large number of rotational and vibrational states of the weakly bound heteronuclear molecular ion to the total electron capture cross section. To this end we develop a semi-analytic technique, based on [1], for the self-consistent evaluation of the integral contribution of all rovibrational states of a molecular ion to the Boltzmann-averaged cross sections of the dissociative recombination and ternary electron capture.

Our main goal is to study the influence of resonant non-adiabatic free-bound and bound-bound inelastic transitions on the population of different atomic levels, and to compare the contributions of the resonant processes to that of the widely known mechanism of collisions with free electrons. Specific calculations are performed for two systems of inert gas mixtures, Ne/Xe and Ar/Xe, in conditions typical for experimental studies. We show that unlike the case of electron capture due to collisions with free electrons resonant processes of capture and discrete transitions result in populating mostly low-lying Rydberg states of atoms. As can be seen from fig. 1, the resonant non-adiabatic capture can bring the predominant contribution to the population of low-lying Rydberg levels even in case of very high ionization degrees κ of plasma, up to $\kappa = 10^{-4}$.

*E-mail: narits@sci.lebedev.ru
†E-mail: kislov93@mail.ru

This work was supported by the Russian Science Foundation (grant 19-79-30086).

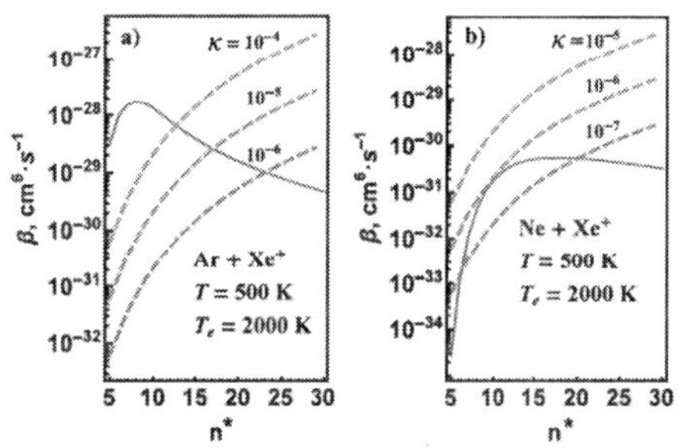

Figure 1. Rate constants of electron capture to Xe(n) levels via the collisions with free electrons of plasma and due to resonant capture by collisions with buffer-gas atoms Ar (panel a) and Ne (panel b) for gas temperature T = 500 K and electron temperature T_e = 2000 K. Solid curves, resonant electron capture; dashed curves, electron capture by collisions with free electrons $\beta_{ee}^{(n)} \cdot \kappa$.

References

[1] Lebedev V S and Presnyakov L P 2002 *J. Phys. B: At. Mol. Opt. Phys.* **35** 4347

Strong enhancement of electron-ion recombination induced by free-bound and bound-bound resonance transitions

K S Kislov[1,2]*, **A A Narits**[1]† **and V S Lebedev**[1]

[1]P.N. Lebedev Physical Institute, Leninskiy prospect 53, Moscow 119991, Russian Federation
[2]Moscow Institute of Physics and Technology, Institutskiy per. 9, Dolgoprudny, 141701, Russian Federation

Synopsis The role of free-bound and bound-bound resonant nonadiabatic transitions on electron-ion recombination in plasma of noble gas mixtures is studied. A kinetic model of recombination is proposed which includes the energy relaxation in collisions with electrons, resonant and non-resonant electron capture to Rydberg states through collisions of atomic ions with neutral particles, and resonant $n \to n'$ transitions. It is shown that the resonant processes considered may strongly enhance total recombination coefficient even at relatively high degrees of ionization of plasma.

We report the results of theoretical studies of the resonant and non-resonant capture of electrons to the Rydberg atomic states in ternary collisions with neutral particles of buffer atoms and bound-bound $n \to n'$ transitions between the highly excited states. The resonant processes are due to the nonadiabatic energy exchange between the outer and inner electrons of a quasi-molecule temporarily formed during a collision of atomic ion (or ionic core of the Rydberg atom) with a neutral projectile. The main goal of the present study is the self-consistent description of the influence of the resonant processes on the kinetics of electron-ion recombination of electrons with atomic and molecular ions in afterglow plasmas of gas mixtures created by electron beam or pulse gas discharges. We are particularly interested in situations when the plasmas contain a small concentration of heteronuclear molecular cations. An analytic kinetic model based on the modified diffusion approximation [1] is proposed, which treats in the uniform manner the aforementioned processes, as well as the standard energy relaxation mechanism caused by collisions with free electrons of the plasmas.

We have carried out the calculations for systems of inert gas mixtures in conditions typical for low-temperature pulse discharge plasmas. It is shown that the presence of effective resonant processes in quasimolecular systems leads to a strong enhancement of the total recombination coefficient as compared to that associated with standard mechanism of ternary recombination of atomic ions in collisions with free electrons in a wide range of the electron (T_e) and gas temperatures (T) and the ionization degrees κ, up to $\kappa = 10^{-5}$, as shown on fig. 1. The role of alternative non-resonant mechanisms of electron capture, as well as of other competing processes is discussed.

This work was supported by the Russian Science Foundation (grant 19-79-30086).

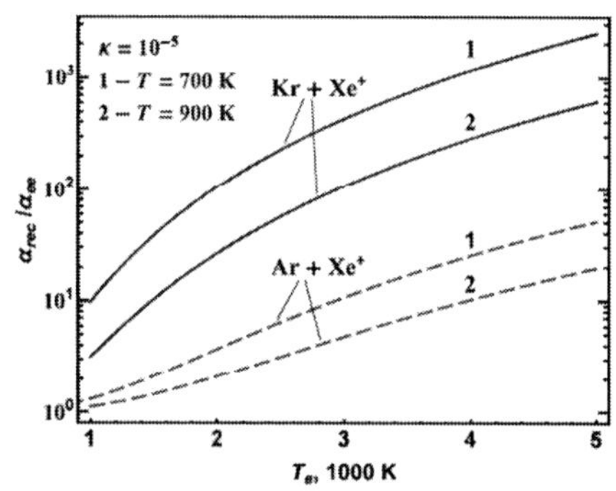

Figure 1. T_e-dependencies of the ratio of the total recombination coefficient α_{rec}, which includes the resonant and non-resonant processes considered, to the three-body recombination coefficient α_{ee} via collisions with free electrons of the plasma for Ar+Xe$^+$ + e and Kr+Xe$^+$ + e systems at gas temperatures equal to 500 K and 700 K and ionization degree $\kappa = 10^{-5}$.

References

[1] Biberman L M, Vorob'ev V S and Yakubov I T 1987 *Kinetics of Nonequilibrium Low-Temperature Plasmas* (New York: Consultants Bureau)

*E-mail: kislov93@mail.ru
†E-mail: narits@sci.lebedev.ru

Identification of EUV spectral profiles of laser-produced Cr plasmas

L Wu[1], M G Su[1*], Q Min[1], S Q Cao[1], S Q He[1], D X Sun[1] and C Z Dong[1†]

[1]Key laboratory of Atomic and Molecular Physics & Functional Materials of Gansu Province, College of Physics and Electronic Engineering, Northwest Normal University, Lanzhou, 730070, China

Synopsis Laser-produced chromium (Cr) plasmas have been measured by using a temporal-spatially resolved laser-produced plasma emission spectroscopy technique. The emission peaks in the wavelength range of 6.5-15 nm were identified from the 3p-4d, 5d, and 3p-4s transition lines from Cr^{5+} to Cr^{10+} ions with aid of Cowan and Flexible Atomic Code (FAC) atomic structure calculations, and through comparative analysis between experimental and simulated spectra.

Laser-produced plasma (LPP) is a kind of high density and temperature plasma, which is formed using a high energy Q-switched laser focused onto a solid target, and consists of a large number of highly charged ions and emits intense radiation from visible light to X-ray. Especially, for the highly charged ions of some low Z elements, their spectral features in the EUV and soft-X-ray region exist some isolated lines with good resolution and have typical spectral characteristics, such as C, Si, Al, S and O. At present, such individual lines usually were used as calibrated light source of EUV spectrograph. Recently, the Cr element was noticed since the spectral features of highly charged Cr ions also show the isolated peak structures in the special spectral region.

In this work, EUV spectra of laser-produced Cr plasmas were measured by using a temporal-spatially resolved laser-produced plasma emission spectroscopy technique. Figure 1(a) shows an experimental spectrum in the 6.5-15.0 nm range with a laser power density of 2.0×10^{11} W/cm^2. The spectrum was recorded at a distance of 1.5 mm from target surface and a time delay of 55 ns.

In order to explain this spectrum, the systematic theoretical calculations using the Cowan[1] and FAC code[2] were carried out, and a steady-state collisional-radiative model was used to calculate the ion fractions of different ionization stages as a function of electron temperature and electron density. Figure 1(b) and (c) show the simulated spectra according to the Cowan and FAC results. Good agreement were obtained at an electron temperature of 28.2 eV and an electron density of 6.4×10^{20} cm^{-3}. The results show that the dominant experimental peaks were identified as the 3p-4d, 5d and 3p-4s transitions of Cr^{5+}-Cu^{10+} ions. All of 21 new lines from highly charged Cr ions were identified, and the corresponding atomic parameters were given. We hope these data can be useful in the future plasma diagnose and wavelength calibration.

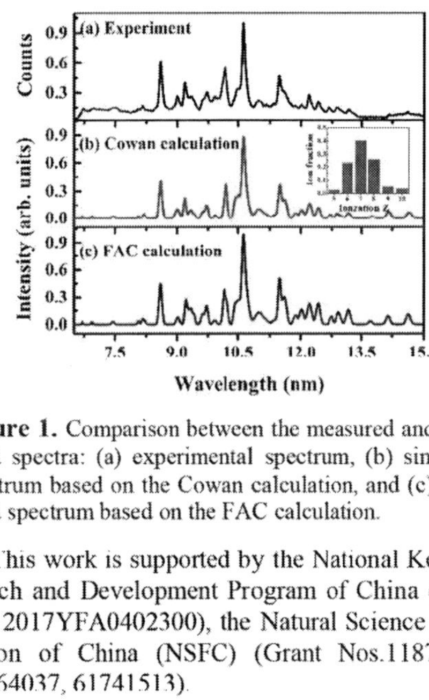

Figure 1. Comparison between the measured and simulated spectra: (a) experimental spectrum, (b) simulated spectrum based on the Cowan calculation, and (c) simulated spectrum based on the FAC calculation.

This work is supported by the National Key Research and Development Program of China (Grant No. 2017YFA0402300), the Natural Science Foundation of China (NSFC) (Grant Nos.11874051, 11564037, 61741513).

References

[1] Cowan R D 1981 *The Theory of Atomic Structure and Spectra* (Berkeley :University of California Press)
[2] Gu M F 2008 *Can. J. Phys.* **86** 675

* E-mail: nwnu_sumg@163.com
† E-mail: dongcz@nwnu.edu.cn

Ladder climbing in the spherical plasma

P F Geng, R A Tang, A X Zhang and J K Xue*

College of Physics and Electronics Engineering, Northwest Normal University, Lanzhou 730070, China

Synopsis We report the ladder climbing (LC) of the electron plasma density wave in the spherical plasma.

The quantum mechanics is closely linked to the classical mechanics. The researches about the ladder climbing (LC) and the autoresonance (AR) maybe help bridge between the quantum and classical physics [1]. AR is a nonlinear phase-locking phenomenon and LC is its quantum counterpart. Recently, the research in Ref. [2] reports first that the classical system also exists LC, which is very useful for the acceleration and manipulation of the electrons.

However, the theory given in Ref. [2] is limited to the one-dimensional plasma. Apparently, the one-dimensional model is too simplified to describe the general and real problems, we should consider the case of the higher dimension, particularly, for the spherical plasma, which is significant for the injection and acceleration of electrons, and the majority of the experimental devices related to the inertial confinement fusion are spherical due to its spherical symmetry. Whether the spherical plasma also can exhibit LC and exist the difference with the one-dimensional case are still an unresolved problem.

Our work suggests that the spherical plasma also can achieve the LC, and which exists the obvious differences with the one-dimensional plasma case. Figure 1 shows the evolutions and distributions of the spherical and one-dimensional plasma density wave in the LC regime. In the energy level space (the first column of Fig. 1), the behave of the spherical plasma wave is extremely analogous to the one-dimensional case. However, compared with the one-dimensional case, the spherical plasma needs larger driving strength (the driving strength for the spherical plasma is 1.66 times that of the one-dimensional case) to achieve this LC phenomenon due to the level spacing in the spherical plasma is broadened. In the coordinate space (the second column of Fig. 1), compared with the one-dimensional plasma where the perturbed density wave behaves as a sine wave, the change of the spherical plasma density wave is nonuniform, and the closer to the sphere center, the more violent change is, i.e., the spherical plasma wave shows a clear compression character, this can be related to the dynamics of the spherically convergent plasma fusion (SCPF) and can offer some directions to the related experiments.

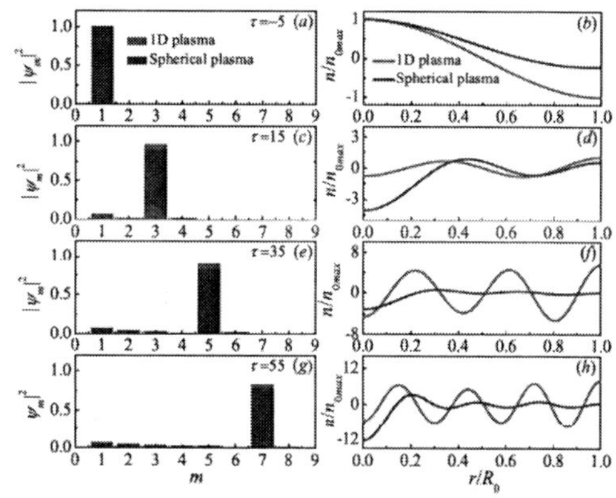

Figure 1. The evolutions of the plasma wave in the LC regime at different time τ. The first column: the occupation numbers $|\psi_m|^2$ against energy level m. The second column: the perturbed density n against the position of the plasma r.

This work is supported by the National Natural Science Foundation of China under Grants No. 11865014, 11765017, 11847304, 11764039, 11475027, 11305132, and 11274255.

References

[1] Barth I *et al* 2011 *Phys. Rev. A* **84** 013837
[2] Barth I *et al* 2015 *Phys. Rev. Lett.* **115** 075001

*E-mail: xuejk@nwnu.edu.cn

Chirped laser acceleration of electron in plasma channel

**Y N Hu[1], X B Zhang[1], A X Zhang[1] and J K Xue[1]*

[1] College of Physics and Electronics Engineering, Northwest Normal University, Lanzhou, 730070, China

Synopsis Considering the asymmetry of chirped laser pulses, the dynamics of direct electron acceleration by chirped laser in a homogeneous cylindrical plasma channel is given.

In recent years, laser-plasma interaction has attracted widespread attention because of its wide application, such as X ray generation and particle accelerator. How to enhance the energy of electrons by the laser pulse and generate high-energy electrons is an important subject of laser-plasma interaction. The laser wakefield accelerator and laser beat wave accelerator are developed. The result shows that dozens of GeV energetic electrons can be obtained by accelerating electrons directly in vacuum with chirped laser pulses [1, 2]. However, laser can be scattered easily in a vacuum, while it can be solved exactly in plasma [3].

In order to obtain higher energy electron, we investigate the dynamics of single electron in homogeneous cylindrical plasma channel during the direct acceleration of linearly and Gaussian chirped laser pulses. Figure 1 presents the numerical results of electron dynamics, where a_0 is the intensity of laser pulse, ω_{p0} is the density of plasma channel, c is the chirped parameter, γ is the relativistic factor.

Figure 1 shows γ_{max} as a function of intensity of chirped laser pulse a_0, density of plasma channel ω_{p0} and chirped parameter c. The linearly(Fig. 1 (a), (b) and (c)) and Gaussian(Fig. 1 (d), (e) and (f)) chirped laser pulse are used. We can easily see that the threshold of the electron oscillation instability is reduced significantly and the region of electron oscillation instability is expanded with chirped parameters. Especially for $c = -0.5$, the maximum value of electron energy can reach to 66.2 for Gaussian chirped laser(Fig. 1 (f)). We set $\omega_{p0} = 0.05$, $a_0 = 6$ and compare with un-chirped case(black solid line in Fig. 1 (g) with $c = 0$), the maximum value of electron energy can reach to 14.3 for Gaussian chirped laser(red dashed line in Fig. 1 (g) with $c = -0.4$). In conclusion, in a homogeneous plasma channel, the threshold of electron instability oscillation can be reduced by adjusting the chirped parameter, and electron can gain more energy from the chirped laser pulse.

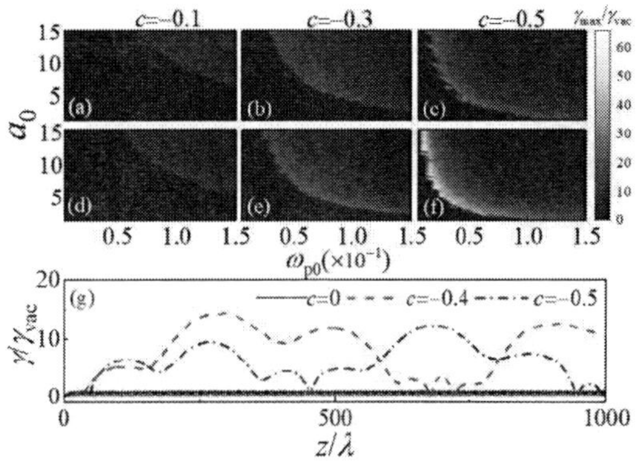

Figure 1. (color online) The electron energy gain against chirped parameters c, laser intensity a_0 and density of plasma channel ω_{p0}.

This work is supported by the National Natural Science Foundation of China under Grants No. 11865014, 11765017, 11847304, 11764039, 11475027, 11305132, and 11274255, by Natural Science Foundation of Gansu province under Grant No. 17JR5RA076, and by Scientific research project of Gansu higher education under Grand No. 2016A-005.

References

[1] Sohbatzadeh F *et al* 2007 *Phys. Plasmas.* **13** 123108
[2] Malka G *et al* 1997 *Phys. Rev. Lett.* **78** 3314
[3] Arefiev A V *et al* 2012 *Phys. Rev. Lett.* **108** 145004.

*E-mail: xuejk@nwnu.edu.cn

 1412 (2020) 192012 doi:10.1088/1742-6596/1412/19/192012

Investigation of temporal-resolved spectra of laser-produced titanium plasmas

D H Zhang* and J N Zhao

Key laboratory of Atomic and Molecular Physics & Functional Materials of Gansu Province, College of Physics and Electronic Engineering, Northwest Normal University, Lanzhou, 730070, China

Synopsis Laser-produced titanium (Ti) plasmas have been measured using temporal-spatially resolved laser-produced plasma emission spectroscopy technique. The emission peaks in the wavelength range of 7.5-15 nm were identified from the 3s-(4,5)p, 3p-4d, and 3d-4f transition lines from Ti^{5+} to Ti^{11+} ions with aid of NIST database and Cowan atomic structure calculations, as well as comparative analysis between experimental and simulated spectra based on a simplified radiation hydrodynamic model.

In recent decades, the laser-produced plasma has gained universal acceptance as standard laboratory pulsed short wavelength light source and ion source, as well as good research carriers of basic atomic physics and plasma physics [1]. Due to the inherent transient and inhomogeneous nature of laser produced plasma, a temporal-spatially resolved laser produced plasmas spectroscopy technique is useful to investigate the spectral properties and dynamics of transient species in plasmas, especially for the production of highly charged ions when the laser pulse with higher power density was employed.

In this work, the temporal behaviors of laser produced Ti plasmas have been investigated with a laser power density of 1.8×10^{11} W/cm^2 at the distances of 1.5 mm from target surface. The emission spectra in the wavelength range of 7.5–15 nm are shown in Figure 1.

The emission peaks were identified from the 3s-(4,5)p, 3p-4d and 3d-4f transition lines from Ti^{5+} to Ti^{11+} ions with aid of NIST database and Cowan atomic structure calculations. It can be seen that the spectra of highly charged Ti ions rapidly change with increasing time delay, which have a maximum of the emission intensity around the time delay of 40 ns, and then disappear after 70 ns.

In order to deeply understand the time features of ionized stages of highly charged ions in plasma, a simplified radiation hydrodynamic model was employed to simulate the spectral profiles at the different delay times, and reconstruct the temporal and spatial evolution of the plasma.

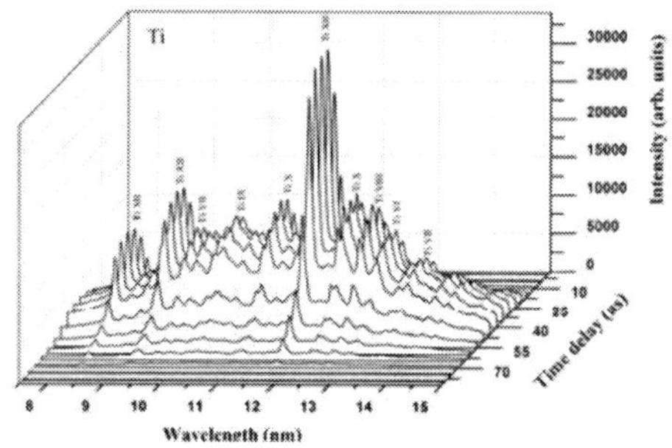

Figure 1. Temporal evolution of EUV emission spectra from laser-produced titanium plasmas.

This work is supported by the National Key Research and Development Program of China (Grant No. 2017YFA0402300), the Natural Science Foundation of China (NSFC) (Grant Nos. 11864036).

References

[1] Amoruso S *et al* 2012 *J. Phys. B* **32** R131

* E-mail: Zhangdh@nwnu.edu.cn

Response of nonequilibrium nonstationary plasma created by intense femtosecond UV laser pulse in rare gases: emission in the THz frequency band

A V Bogatskaya[1,2*] and A M Popov[2,3†]

[1]D. V. Skobeltsyn Institute of Nuclear Physics, Moscow State University, Moscow, Russia
[2]P. N. Lebedev Physical Institute, RAS, Moscow Russia
[3]Department of Physics, Moscow State University, Moscow, Russia

Synopsis We analyse the response of the current in plasma created in the process of gas ionization by an intense UV laser pulse in the terahertz frequency band. Our consideration allows us to take into account both temporal retardation of the EVDF evolution with respect to the external electric field and the effect of the relaxation of the strongly nonequilibrium EVDF due to elastic electron - atomic collisions The suggested theory enables to properly describe propagation of short and ultra-short THz pulses in laser plasma waveguides.

Plasma formed in gaseous media by powerful femtosecond laser pulses is widely studied nowadays as an effective source of radiation in different frequency band from THz to EUV. Recently, nonequilibrium plasma channels formed in gases characterized by an energy interval with an increasing transport cross section $\sigma_{tr}(\varepsilon)$ (for example, with a Ramsauer minimum) by a powerful femtosecond UV laser pulse were proposed for amplification of (sub)THz radiation [1,2]. Such amplification can be realized if the position of the peak in photoelectron spectrum is located in the range of positive derivative of the transport cross section over energy $d\sigma_{tr}/d\varepsilon$. It was shown that for xenon plasma at atmospheric pressure it is possible to reach amplification of radiation with $\omega \leq \nu \approx 10^{12}$ c^{-1} at the nanosecond time scale.

From practical point of view it is interesting first of all to analyze the process of amplification of extremely short THz pulses of few (up to one or two) cycles duration, as pulses of such duration are typically generated experimentally in laser - plasma interaction. Such pulses have wide spectrum while till present the theoretical analysis of amplification has been performed for quasimonochromatic radiation only. Hence, for such short pulses the approach typically used for the analysis of the kinetic Boltzmann equation should be modified. Also, it is assumed typically that anisotropic part of the electron velocity distribution function (EVDF) corresponds to the instantaneous electric field strength value. This assumption is also violated when considering EVDF evolution in process of extremely short THz pulses propagation. It should be also mentioned that temporal relaxation of EVDF will lead to some difficulties in using the well-known Fourier methods of describing of plasma dielectric properties due to the appearance of the parametric frequency conversion in the propagated/amplified pulse spectrum.

Our research is aimed at developing the analytical model for analysis of plasma electromagnetic features in rare gases which is beyond limitations on the duration of amplified THz pulses and their spectral width. The model is applied to study the possibility of amplification of THz pulses of ultrashort duration in relaxing xenon plasma formed by powerful femtosecond KrF pulse.

This research was supported by the Russian Science Foundation (project no. 18-72- 00125).

References

[1] Bogatskaya A *et al* 2013 *JETP Letters* **97** 388
[2] Bogatskaya A *et al* 2018 *Laser Physics* **28** 115301

* E-mail:annabogatskaya@gmail.com
† E-mail:alexander.m.popov@gmail.com

Electron density diagnosis of laser-produced plasmas based on spectral line broadening of highly charged ions

K P Wang, M G Su*, S Q Cao, P P Ma, D X Sun, Q Min and C Z Dong

Key Laboratory of Atomic and Molecular Physics & Functional Material of Gansu Province, College of Physics and Electronic Engineering, Northwest Normal University, Lanzhou, 730070, China

Synopsis Spectral line profile of Ti XII 11.64 nm was measured using spatio-temporally resolved laser-produced plasma technique. The electron density of laser-produced Ti plasmas was calculated based on the spectral line Stark broadening of highly charged ions.

The state parameters of laser-produced plasmas (LPP) can be diagnosed according to the broadening of spectral lines. For neutral atoms or singly-ionied spectral lines, can get the relatively accurate diagnosis results using Girem's method [1]. However, expansion from tens to hundreds of nanoseconds of LPP usually involved the EUV/soft x-ray radiation of highly charged ions, there is evident deviation from Girem's method. In order to realize the diagnosis of plasma based on highly charged ions spectral line, Dimitrijevi *et al* presented a method for calculating Stark broadening of highly charged ions spectral line, which can obtain more accurate results [2]. In this work, the electron density of laser-produced Ti plasmas was calculated based on spectral line broadening using Dimitrijevi's method, this give an extra method for the electron density diagnoses of LPP during the EUV/soft x-ray radiation stage.

Stark broadening of Ti XII 11.64 nm as a function of electron density at different electron temperature were shown in figure 1. It can be found that the broadening increases with the increase of electron density, but no evident changes for different electron temperature. Here, an estimate of of the electron temperature at 30 eV was used to calculate the electron density. Spectral line profile of Ti XII 11.64 nm was measured using spatio-temporally resolved LPP technique with the laser power density of 2×10^{11} W/cm^2, the spatial detection position of 1.0 mm and at a delay time of 40 ns. Full width at half maximum (FWHM) is 0.039 nm by a Lorentz fitting (see inset, figure 1) and the corresponding electron density is about 3.6×10^{21} cm^{-3} under this experimental conditions.

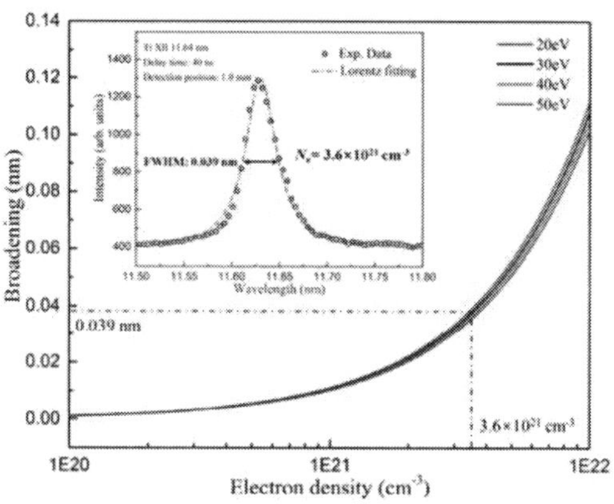

Figure 1. Stark broadening of Ti XII 11.64 nm as a function of electron density at different electron temperature (subfigure shows the diagnostic result of electron density by base on Stark broadening).

This work is supported by the National Key Research and Development Program of China (Grant No. 2017YFA0402300), the Natural Science Foundation of China (NSFC) (Grant Nos. 11874051, 11564037, 61741513).

References

[1] Griem H R 1968 *Phys. Rev.* **165** 258
[2] Dimitrijevic M S *et al* 1980 *Astron. & Astrophys.* **24** 451

* E-mail: nwnu_sumg@163.com

Evidence of trielectronic recombination for Ar ions in an EBIT

D S La Mantia[1,2]*, W Biela[1], J A Tanis[2] and A Warczak[1]

[1]Institute of Physics, Jagiellonian University, Krakow 30-348 PL
[2]Department of Physics, Western Michigan University, Kalamazoo, MI 49008 USA

Synopsis Trielectronic recombination has been investigated for Ar ions in a compact, room-temperature electron beam ion trap. A significant increase in hollow-core emission has been observed at the expected beam energy (~ 6.4 keV) for the KK→LMM resonance transition.

The recombination of an electron and an ion can result in nonresonant capture with the emission of a photon, called radiative recombination (RR), or resonant capture with the excitation of one or more bound-state electrons. In the latter case, the ion can then decay to the ground state through the emission of an electron (Auger decay) or a photon. Trielectronic recombination (TR) is a process where the resonant capture of a free electron to an ion bound state causes two core electrons to be excited to a higher atomic shell. In highly-charged ions, photon emission is the dominant subsequent relaxation channel.

Unlike dielectronic recombination, TR has not been well-studied. TR has previously been investigated for LL resonances in Be-like Cl in a heavy-ion storage ring [1] and also for KL resonances in Kr^{30+} in a superconducting EBIT [2]. The TR process for KK resonances has not been studied to our knowledge. The present work follows observation of enhanced Ar K x-ray emission in an EBIT at electron energies ~ 6.5 keV, suggesting the presence of TR [3].

Here, TR has been investigated in the region of the KK→LMM transition for a dilute Ar plasma using the compact, room-temperature EBIT at the Jagiellonian University by scanning the electron beam energy. A Si(Li) x-ray detector with a 25 μm window positioned at 90° to the beam axis at the trap center collected emitted x rays. The trap was pulsed over 250 ms to generate significantly high Ar charge states. The presence of both K-shell electrons is necessary for TR to occur in KK→LMM transitions. The trap time was chosen to minimize K-shell RR, an indicator of a K-shell vacancy.

An electron falling to the K shell with one or two vacancies will result in the emission of a satellite K^s_α or hypersatellite K^h_α photon, respectively. At sufficiently high electron beam energy, the K^h_α line becomes significantly observable and one would expect the K^h_α/K^s_α ratio to increase smoothly. If, however, the beam energy is resonant to for KK transition in the ion, K^h_α photon production will subsequently increase and the ratio will rise more sharply. For KK→LMM, the TR transition is expected to occur at ~ 6.4 keV beam energy. A significant resonance has been observed at this energy, as shown in Fig. 1.

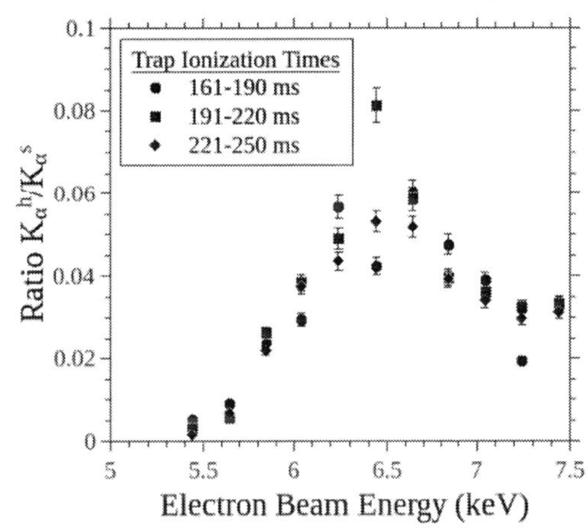

Figure 1. Ratio of Ar hypersatellite to satellite peaks (K^h_α/K^s_α) scanned in the expected KK→LMM TR electron beam energy (E_e) range.

This work was funded in part by National Science Foundation Grant PHY-1707467, the Kościuszko Foundation, and Grant DSC:004969-UJ.

References

[1] Schnell M *et al* 2003 *Phys. Rev. Lett.* **91** 043001
[2] Beilmann C *et al* 2009 *Phys. Rev. A* **80** 050702(R)
[3] Biela W *et al* 2018 *19th Int. HCI Conf.: Book of Abstracts* p. 162 Lisbon, PT

*E-mail: david.s.lamantia@wmich.edu

 1412 (2020) 192016 doi:10.1088/1742-6596/1412/19/192016

EUV spectroscopy of optically thin germanium plasmas in the 9-18 nm region

O Maguire[1], D Kos[1], G O'Sullivan[1] and E Sokell[1] *

[1] School of Physics, University College Dublin, Belfield, Dublin 4, Ireland

Synopsis EUV spectra have been recorded from optically-thin germanium laser produced plasmas created using a high-powered TEA CO_2 laser. The Cowan suite of atomic codes has been used to identify previously unidentified lines in Ge VI-XI. The absence in the spectra of $\Delta n=0$ transitions for ion stages lower than Ge^{9+} has been interpreted by calculating line strength ratios for of $\Delta n=0$ to $\Delta n=1$ transitions for the relevant ion stages.

Accurate atomic data is necessary for many technological developments including confinement for fusion [1], water window [2] and EUVL light sources [3]. Here the Cowan suite of atomic codes [4] has been used to identify previously unidentified lines in Ge VI-XI. These lines were observed in emission spectra obtained from optically-thin laser produced plasmas. The spectra were recorded in the 9-18 nm range. The high temperature plasmas were formed using a high-powered TEA CO_2 laser which delivered 50 mJ in 50 ns. The pulse was shortened in time using a plasma shutter [5]. The laser was focussed to a variety of power densities up to a maximum of 7 x 10^8 W/cm^2. The spectra were recorded at 45° to the target normal, incident laser direction, on an absolutely calibrated EUV Jenoptik ¼ m flat-field, grazing incidence spectrometer with a resolving power $\lambda/\Delta\lambda$ ~ 1700. The uncertainty on the wavelength calibration was ~0.005 nm.

The vast majority of the ~60 newly identified, experimental lines in Ge VI-XI were assigned to transitions of the type $3p^63d^n$-$3p^63d^{n-1}4p$ and $3p^63d^n$-$3p^63d^{n-1}4f$. These arrays were observed to move towards shorter wavelengths as the charge state increased. This behaviour for similar $\Delta n=1$ transitions is well known [6]. A feature associated with $3p^63d^n$-$3p^53d^{n+1}$ transitions in GeX was observed in the spectrum recorded at the highest power density. Such $\Delta n=0$ transitions have wavelengths that are essentially independent of charge state. The absence of similar features at lower plasma temperatures for lower charge states may be explained using the concept of total line strength [7].

Figure 1 shows the ratio of the total calculated line strength (S_{tot}) for $3p^63d^n$-$3p^53d^{n+1}$ to $3d^n$-$3d^{n-1}4p$ transitions. It is evident from the figure that once the 3d occupancy falls below five (germanium X) the ratio becomes greater than one. This explains why the $3p^63d^n$-$3p^53d^{n+1}$, $\Delta n=0$ transitions only appear in the spectra recorded from laser produced plasmas in which there is a significant contribution from Ge^{9+} ions to the emission spectrum. At lower power densities, where the spectra are dominated by ion stages for which the line strength ratio is lower, only contributions from $\Delta n=1$ transitions of the type $3p^63d^n$-$3p^63d^{n-1}4p$ and $3p^63d^n$-$3p^63d^{n-1}4f$ are evident.

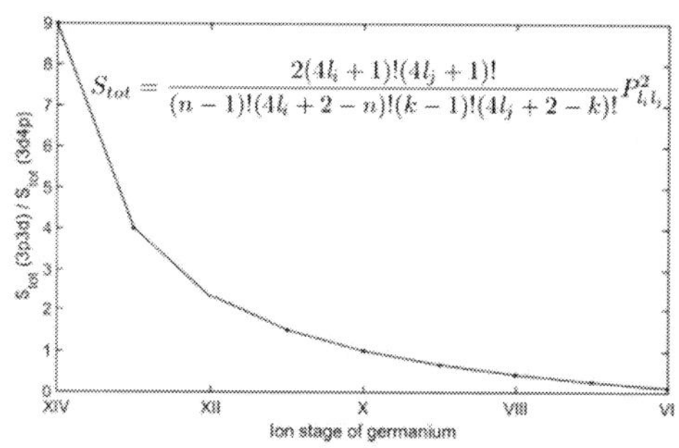

Figure 1. Ratio of the total calculated line strength (S_{tot}) for $3p^63d^n$-$3p^53d^{n+1}$ to $3d^n$-$3d^{n-1}4p$ transitions.

References

[1] Clementson J *et al* 2011 *Can. J. Phys.* 89 571
[2] Wachulak PW *et al* 2013 *Appl. Phys. B* 111 239
[3] Wagner C and Harned N 2010 *Nat. Photonics* 4 24
[4] Cowan R D 1968 *J. Opt. Soc. Am.* **58** 808
[5] Hurst N and Harilal S S 2009 *Rev. Sci. Instrum.* **80** 035101
[6] Ohashi H *et al* 2014 *Appl. Phys. Lett.* **104** 234107
[7] Cowan R D 1981 The Theory of Atomic Structure and Spectra

* E-mail: emma.sokell@ucd.ie

Effective modulation of the terahertz radiation force

X B Zhang[1], **X Qiao**[1], **A X Zhang**[1] **and J K Xue**[1] *

[1]College of Physics and Electronics Engineering, Northwest Normal University, Lanzhou 730070, China

Synopsis The radiation force of the terahertz field is realized and highly modulated by changing the spatial distribution of the THz field, which is controlled by the coupling effects of the laser field, external magnetic field, and self-generated plasma channel. Moreover, the stability of the terahertz radiation is also discussed.

Teraherta (THz) radiation has attracted much attentions for a variety of promising applications [1]. At present, the optical force are ideally studied to realize the optical trapping and manipulation of single cells [2]. In particular, the trapping and manipulation technology based on the infrared light has significantly less optical damage to a variety of living cells with more better biocompatibility [3]. But the realization of the teraherta radiation force (TRF) is still an undeveloped issue. That's, to develop the THz technology applications in optical trapping and manipulation, high performance modulators and active filters to control and manipulate the THz waves are in high demand. Unfortunately, many technology for controlling the THz wave strongly rely on the expensive THz transmitted system and the pump lights.

In this paper, we present an efficient method to control the intensity of the electric field, the spatial structure and the radiation force of the THz field with an external magnetic field, which is confirmed by the two dimensional particle-in-cell (PIC) simulations [Fig. 1]. When the magnetic field equal to zero, the twin-peak structure of the THz field is symmetry about the y = 0 axis [Fig. 1 (a) and (b)]. When the intensity of the magnetic field increases, the symmetry of the twin-peak structure (the intensity and the width of the each peak) will be controlled [Fig. 1 (a) and (b)]. Note that, the twin-peak structure of the THz field can generate the four-peak structure of TRF ($\mathbf{F}_{TRF} \propto \nabla E^2_{THz}$) which results in the formation of the double quasi-stable trapping region [I and II region of the Fig. 1 (c)]. Interestingly, the double quasi-stable trapping region can be modulated, and even be broken by changing the structure of the THz field with the strong magnetic field [Fig. 1 (d)]. In our scheme, the instabilities of the terahertz radiation is discussed in the strong laser field ($a \geq 4$) and super-strong magnetic field ($B \geq 60T$) [Fig. 1 (e) and (f)].

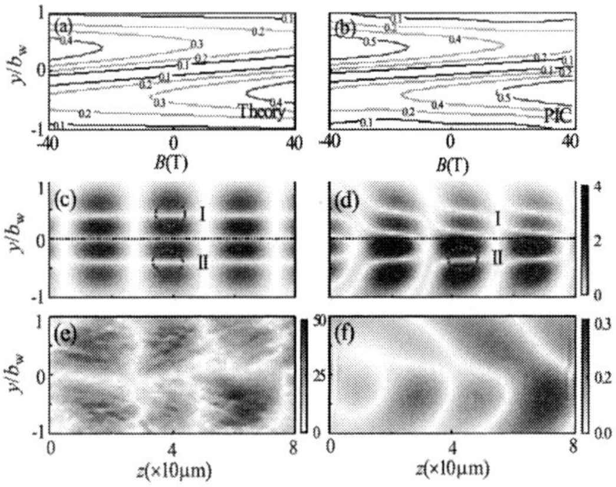

Figure 1. (color online) Variation of E_{THz} (MV/cm, the first row) against the transverse coordinate and the intensity of the magnetic field. The F_{TRF} (the second row) across the transverse coordinate and the longitudinal coordinate. The instabilities phenomenon of E_{THz} with the strong laser field (e) and super-strong magnetic field (f).

This work is supported by the National Natural Science Foundation of China under Grants No. 11865014, 11765017, 11847304, 11764039, 11475027, 11305132, and 11274255, by Natural Science Foundation of Gansu province under Grant No. 17JR5RA076, and by Scientific research project of Gansu higher education under Grand No. 2016A-005.

References

[1] Zhang Z 2018 *Nat. Photonics* **12** 554
[2] Wang L G *et al* 2013 *Optics Express* **19** 15
[3] Ashkin A *et al* 1987 *Nature (London)* **330** 769

*E-mail: xuejk@nwnu.edu.cn

Programmable plasma structure generated by the laser-plasma interaction

X B Zhang[1], X Qiao[1], A X Zhang[1] and J K Xue[1]*

[1]College of Physics and Electronics Engineering, Northwest Normal University, Lanzhou 730070, China

Synopsis The programmable plasma structure generated by the laser-plasma interaction is discussed theoretically. The multiple plasma density distribution patterns which includes single channel, double channel and slow-wave plasma structure is achieved by modulating the ponderomotive force controlled by the laser's beam profile and pulse width.

Extending laser-plasma interaction with modulated plasma density distribution patterns is crucial to many applications [1]. To further improve the quality of electron acceleration and radiation sources, the capability to tailor and modulate the transverse and longitudinal profile of plasma density is essential [2]. Generally, the transverse electrostatic field in the channel enhances the positive effect of the group velocity, thus superponderomotive electrons is produced in the plasma channel. The slow-wave plasma structure in longitudinal direction can be used for direct laser acceleration of charged particles and the generation of the quasi phase matched electromagnetic radiation [2, 3]. Although several techniques were performed for producing the useful plasma density distribution in the theories and experiments. There are still an opening subject to realize the programmable plasma structure in transverse and longitudinal profile.

In this paper, we discuss the generation of the multiple plasma density distribution patterns (single channel, double channel and slow-wave plasma structure) by an adjustable laser beam in a uniform plasma [Fig. 1]. When laser's beam profile is Gauss profile, the phase diagram of the ponderomotive force and electron density contains one region corresponding to the single channel generation [Fig.1 (a) and (c)]. Interestingly, when the uniform plasma is irradiated by the super-Gauss laser, the phase diagram of the ponderomotive force and electron density contains two region corresponding to the double channel generation [Fig.1 (b) and (d)]. When the pulse width of the laser is shorter enough, the slow-wave plasma structure will be achieved by the Gauss laser or super-Gauss laser [Fig. 1 (e) and (f)]. The programmable plasma structure will have the positive effects for electron acceleration and electromagnetic radiation.

*E-mail: xuejk@nwnu.edu.cn

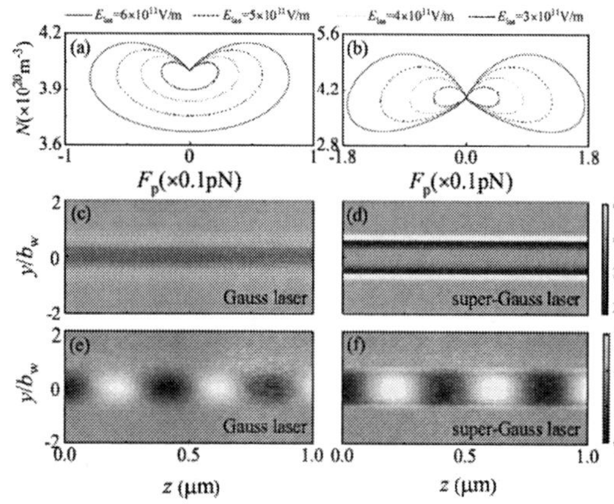

Figure 1. (color online) The phase diagram of the the ponderomotive force and electron density with the Gauss laser (a) and super-Gauss laser (b). The electron density generated across the transverse coordinate and the longitudinal coordinate with the longer pulse (c), (d) and shorter pulse (e), (f).

This work is supported by the National Natural Science Foundation of China under Grants No. 11865014, 11765017, 11847304, 11764039, 11475027, 11305132, and 11274255, by Natural Science Foundation of Gansu province under Grant No. 17JR5RA076, and by Scientific research project of Gansu higher education under Grand No. 2016A-005.

References

[1] Layer B D *et al* 2007 *Phys. Rev. Lett.* **99** 035001
[2] Hung T S *et al* 2012 *Phys. Plasmas* **19** 063109
[3] Malik A K *et al* 2011 *Appl. Phys. Lett.* **99** 071107

Polarization of terahertz radiation driven by incommensurate two-color laser pulses

Y N Li[1], G L Wang[1]*, L Zhang[2], Z H Jiao[1], S F Zhao[1] and X X Zhou[1]

[1]College of Physics and Electronic Engineering, Northwest Normal University, Lanzhou, 730070, China
[2]School of Mathematics and Physics, Lanzhou Jiaotong University, Lanzhou, 730070, China

Synopsis We investigate the polarization characteristics of terahertz (THz) emission from gaseous plasma induced by the two-color incommensurate femtosecond laser fields, which contains a circularly polarized laser field and a linear pulse. Our simulations show that strong linearly, elliptically, and circularly polarized THz radiation can be generated, by choosing appropriate wavelength, carrier-envelope phase and intensity for linear pump pulse.

To date, two-color commensurate-frequency laser fields, which are composed of a fundamental pulse (ω) and its second harmonic (2ω), have been widely used to generate elliptically polarized THz waves. Here we investigate the controlling of the polarization of THz radiation by using the *incommensurate* two-color fields, which has the following form

$$\begin{aligned} E_x(t) &= \tfrac{\sqrt{2}}{2}\sqrt{I_1}f(t)\cos(\omega_1 t+\varphi_1) \\ &\quad +\sqrt{I_2}f(t)\cos(\omega_2 t+\varphi_2), \qquad (1) \\ E_y(t) &= \tfrac{\sqrt{2}}{2}\sqrt{I_1}f(t)\cos(\omega_1 t+\varphi_1+\pi/2), \end{aligned}$$

where I_1, I_2, and φ_1, φ_2 are peak intensity and carrier-envelope phase for circularly (ω_1) and linearly (ω_2) polarized laser fields, respectively, $f(t)$ is the pulse envelope with the Gaussian shape $f(t) = e^{-2\ln 2t^2/\tau^2}$, τ is the pulse duration. The THz emissions are simulated by well-known photocurrent model [1].

Figure 1 shows one of our main simulation results. Here we take $I_2/I_1 = 0.5$ with total laser peak intensity of 2.0×10^{14} W/cm^2, $\varphi_{1,2} = 0$, λ_1=800 nm and τ = 8.6 fs, while λ_2 increases from 400 nm to 2800 nm. Two important observations can be made. First, the addition of a linear polarized laser pulse can effectively enhance radiated THz field from single-color circular laser fields (THz amplitude generated from monochromatic laser field has been enlarged by a factor of 100 in Fig. 1), which is crucial for the THz wave applications. Second, the ellipticity (ε) of THz wave depends evidentally on the wavelength of pump linear laser field. For example, for 28-THz radiation, we can control its polarizations by choosing proper wavelength λ_2. The THz wave changes from nearly linearly polarized state (800 nm+400 nm, ε=0.08), to elliptically polarized one (800 nm+1600 nm, ε=0.5) and to circular one (800 nm+2200 nm, ε=0.94). Such controlling can work over a very wide bandwidth. Furthermore, the phase delay and intensity ratio between two laser components can also be used to control the polarization degree and direction of emitted THz waves.

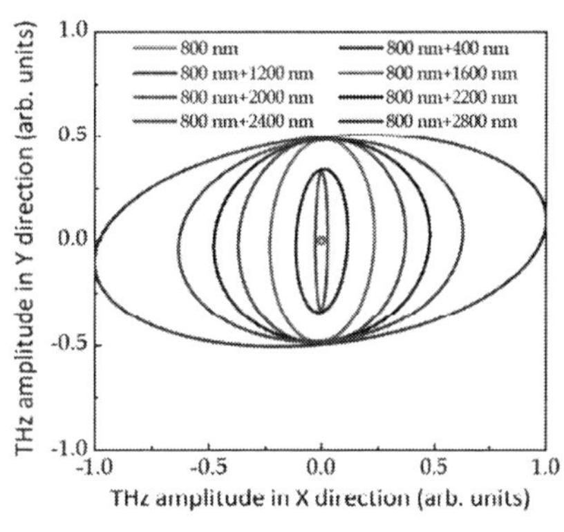

Figure 1. Polarization evolution of 28-THz radiation with the wavelength of pump linear polarized laser pulse from Ar gas. For the case of single-color 800-nm laser pulse, the THz amplitudes have been enlarged 100 times.

Project supported by the National Natural Science Foundation of China (Nos. 11864037, 11465016, 11765018, 11664035), and the Foundation of Northwest Normal University [NWNU-LKQN-17-1].

References

[1] Kim K Y *et al* 2007 *Opt. Express* **15** 4577

*E-mail: wanggl@nwnu.edu.cn

Spectroscopic diagnostic of isolated dense plasma generated through laser driven gold hohlraum radiation

J Ren[1]*, Y Zhao[1], Z Deng[2], W Qi[2], R Cheng[3], X Wang[1], W Liu[1], B Ma[1], S Yin[1], B Chen[1], J Feng[1], S Wang[2], Q Fang[2], B Cui[2], X Zhou[3], W Zhou[2] and L Cao[2]

[1]Xi'an Jiaotong University, Xi'an, 710049, China [2] Laser Fusion Research Center, Chinese Academy of Engineering Physics, Mianyang, 621000, China [2] Institute of Modern Physics, Chinese Academy of Sciences, Lanzhou, 710049, China

Synopsis Through indirectly heating a TAC foam with ns laser driven gold hohlraum radiation, isolated, highly homogeneous and long-living dense ionized matter was generated in laboratory. This provide a fantastic sample for investigation of dense plasma science and technology. The emission spectra of the plasma with time, spatial and spectral resolution were measured with various techniques. The temperature and electron density are precisely diagnosed on the basis of the spectra.

Dense plasma attracted a lot of attentions these years since it widely exists in the supernova, stellar interior, accration disks, and is the important state of matter in inertial confinement fusion. The routine way to generate dense plasma in laboratory is to heat solid samples with laser. In laser-solid interaction, the laser cannot penetrate into solid, but is reflected on the critical surface. The target bulk are heated and ionized through the propagation of these energetic electrons. In this way, the temperature as well as the free electron density shows great gradient, and the hydrodynamic behavior of the target cannot be neglected. This invokes great challenge for the fast diagnostics techniques and detailed understanding of target evolution. In order to generate uniform, long-living dense plasma in laboratory, we used TAC foam ($C_{12}H_{16}O_8$) as the sample material, which was heated by hohlraum radiation. With this method, the hydrodynamic effects of the target could be neglected in timescale of 10 ns [1, 2].

The measurement was carried out at the XGIII facility. Combined target consisting of a gold hohlraum converter and a 2 mg/cm^{-3} TAC foam layer was used. Within this scheme, the ns laser (2ω) with energy of 150 J interacts with the inner gold wall of the cylindrical hohlraum, which will induce blackbody-like radiation. The emitted X rays then volumetrically heat the foam to plasma state. The emission spectra of both gold hohlraum and CHO plasma were measured. Pinhole cameras which is sensitive to X rays are used to image the plasmas. XRDs were used for the time-resolved measurement of X-ray flux. Transmission Grating Spectrometer and Flat field Grating Spectrometer are used for the measurement of X ray emission with spectral resolution.

Fig.1 shows the emission lines of the CHO plasma in range of 10-28 nm. Through analysing the broadening and intensity of helium-like carbon lines with Saha-Boltzmann curve slope method and Stark line broadening method, it was obtained the electron temperature is around 19eV and the free electron density is about $1.0 \times 10^{21} cm^{-3}$.

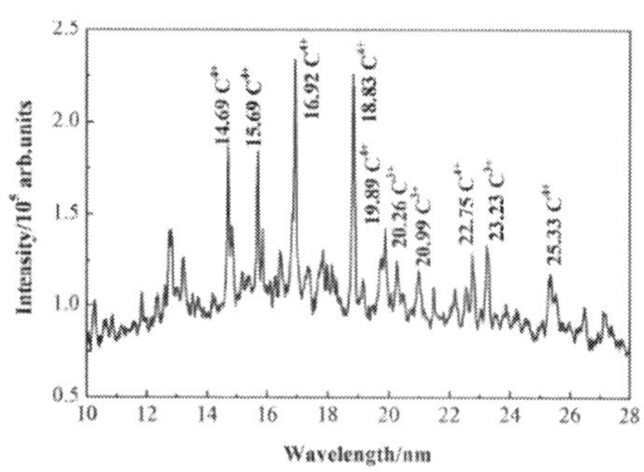

Figure 1. Self-emission spectra of the CHO plasma.

References

[1] Rosmej O *et al Nuclear Instruments and Methods in Physics Research Section A* 2011 **653** 52
[2] Vasina E*et al GSI* (2000-2002) Scientific Report

*E-mail: renjieru@xjtu.edu.cn

Absolute vibrational and electronic cross sections for low-energy electron scattering from condensed Thymidine

V Lemelin[1]*, A D Bass[1] and L Sanche[1]

[1]Département de médecine nucléaire et radiobiologie, Faculté de Médecine et Sciences de la santé, Université de Sherbrooke, Sherbrooke, J1H 5H3, Québec, Canada

Synopsis Absolute vibrational and electronic excitation cross sections (CSs) for low-energy electrons (1-18 eV) scattering from condensed thymidine (dT) are measured with High-Resolution Electron Energy Loss Spectroscopy (HREELS). EEL spectra are recorded and the vibrational and electronic modes of dT identified. Absolute CSs for these modes are extracted from each spectrum by a fitting procedure. Resonances are observed in the incident energy dependence of the CSs, below 3 eV and near 4, 8 and 10 eV. The CSs are compared to those of its constituents: thymine and tetrahydrofuran.

The primary goal of radiotherapy is to deliver lethal doses of radiation exclusively to cancer cells, while preserving healthy tissue. One novel approach is Targeted Radionuclide Therapy (TRT), in which Auger-electron-emitting radioisotopes are molecularly targeted towards cancer cells [1]. Biologically, significant damage is produced by the complex sequence of reactions initiated by the secondary species, which include the numerous low-energy electrons (LEEs). *Nanodosimetric* calculations are required to describe new modalities such as TRT. Modelling such treatments requires multiple parameters to describe the transport and reactions of secondary species in biological matter. Consequently, absolute cross sections (CSs) for the interactions of LEEs with DNA and/or its component sub-units in the condensed phase are required for cellular nanodosimetry [1]. Here we present absolute CSs for vibrational excitations of condensed Thymidine (dT), which is a nucleoeside, composed of Thymine (DNA base) and deoxyribose. Electron energy loss spectra (Fig. 1) for 1.7 monolayers (ML) of dT deposited on 3 ML of an Ar spacer on a Pt substrate at 19 K, were recorded in vacuum, for electron incident energies between 1 and 18 eV using a high resolution electron energy loss spectrometer. As in our earlier studies [2, 3], the CSs were determined at fixed energies, by fitting each energy loss spectrum with multiple Gaussian functions. These latter were centered at the expected energy loss of each mode. The energy dependence of the CSs for each excitation mode exhibits multiple features that can be related to transient anion states located at energies below 3 eV, and near 4, 8 and 10 eV. These absolute CSs values for dT are quantitatively compared to those previously measured for thymine [4] and tetrahydrofuran (a model for deoxyribose) [2]. Hence, we can find out, if the CSs for a complex molecule, such as dT, can be represented by the sum of CSs of its constituent subunits.

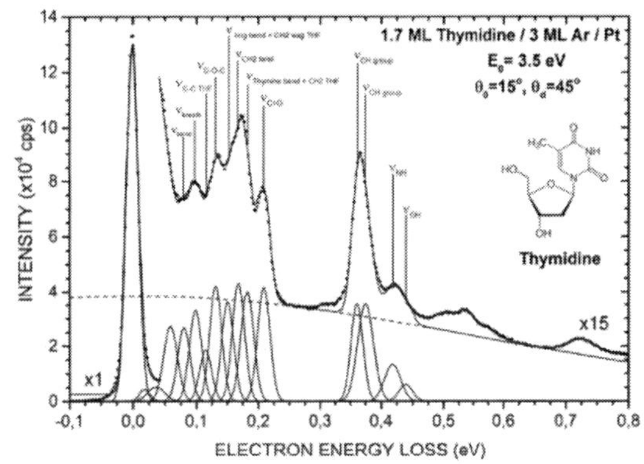

Figure 1. Representative EEL spectrum for incident electron of 3.5 eV scattering from 1.7 monolayer of thymidine.

References

[1] Rezaee M *et al* 2017 *Rad. Res.* **188** 123
[2] Lemelin V *et al* 2016 *J. Chem. Phys.* **144** 074701
[3] Lemelin V *et al* 2017 *J. Chem. Phys.* **147** 234305
[4] Lévesque P L *et al* 2003 *Nucl. Inst. and Meth. in Phys. Res. B* **208** 225

* E-mail: Vincent.Lemelin@USherbrooke.ca

Journal of Physics: Conference Series **1412** (2020) 202002 doi:10.1088/1742-6596/1412/20/202002

Observation of two-electron one-photon X-ray transitions in collisions of slow Xe^{26+} ions with beryllium surface

Ł Jabłoński[1*], D Banaś[1], P Jagodziński[2], A Kubala-Kukuś[1], D Sobota[1], I Stabrawa[1], K Szary[1] and M Pajek[1]

[1]Institute of Physics, Jan Kochanowski University, 25-406 Kielce, Poland
[2]Department of Mathematics and Physics, Kielce University of Technology, 25-314 Kielce, Poland

Synopsis Two-Electron One-Photon (TEOP) X-ray transitions were observed for the first time in collisions of highly charged Xe ions with metallic Be surface. The intensities of TEOP X-rays are well described by performed MCDF for one-electron one-photon (OEOP) dipole-allowed transition of Paschen and Brackett series assuming constant TEOP/OEOP branching ratio. The importance of the Internal Dielectronic Excitation (IDE) in relaxation of hollow Xe ions/atoms has been evidenced and discussed.

Slow highly charged ions (HCI) colliding with metal surface are rapidly neutralized forming highly excited Rydberg states, "hollow" atoms, which subsequently relax in a series of fast processes including Interatomic Coulombic Decay (ICD) [1] and X-ray andAuger electron emission cascade. It was evidenced the Internal Dielectronic Excitation (IDE) [2] contributes to relaxation of "hollow" atoms. We report here on the first observation of the TEOP X-ray transitions measured in interaction of slow Xe^{26+} ions with metallic beryllium surface. We show that TEOP process, predicted by Heisenberg in 1925 [3], contributes to X-ray emission from "hollow" atoms. In the present experiment the X-rays were emitted from 209 keV Xe^{26+} ions neutralized at metallic Be foil by capturing resonantly the electrons into Rydberg states with n≈25. The ion beam was produced by the EBIS-A facility [4]. For Xe^{26+} ions having filled M-shell the observed M-X-rays, which were excited by the IDE process, were measured with XFlash silicon drift detector (SDD). The measured M-X-rays were dominated by OEOP electric-dipole transitions (nf-3d) and reasonably well interpreted (Fig. 1)in terms of the MCDF calculations performed using GRASP2K code [5].

We have found that the high energy part of X-ray spectrum, extending roughly over doubled energy range of the OEOP transitions, can be explained by the TEOP transitions, having photon energies slightly bigger than doubled energy of the diagram M-X-ray transitions. Contribution of the TEOP transitions to the measured X-ray spectrum was calculated by assuming constant TEOP/OEOP ratio for these processes.

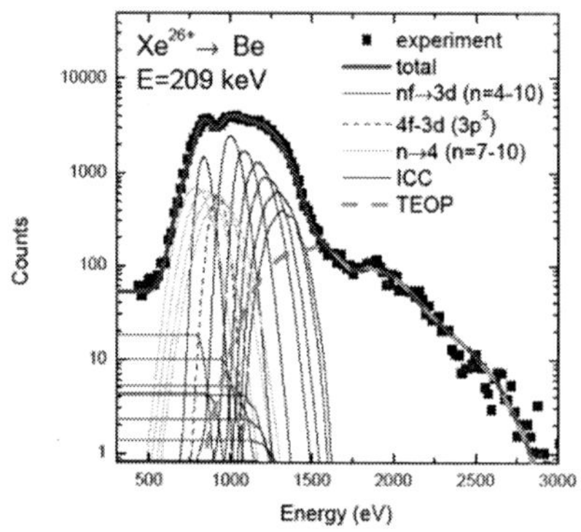

Figure 1. Measured spectrum of X-rays for 209 keV Xe^{26+} ions colliding with Be foil.

Consequently, the shape of TEOP was the same as for OEOP with stretched energy scale by a factor of about two.Despite of simplicity of the proposed approach, discussed OEOP and TEOP transitions reproduce reasonably well the measured X-rays. The results evidence importance of electron-electron correlations in relaxation of hollow atoms.

Acknowledgments

The financial support of the EU within the Polish Innovative Economy Operational Program (Contact No. WNP-POIG.02.02.00-26-023/08) is acknowledged.

References

[1] Wilhelm R *et al* 2017 *Phys. Rev. Lett.* **119** 103401
[2] Schuch R *et al* 1993 *Phys. Rev. Lett.* **70** 1073
[3] Heisenberg W 1925 *Z. Phys.* **32** 841
[4] Banaś D *et al* 2010 *J. Instrum.* **5** C09005
[5] Jönsson P *et al* 2013 *Comp. Phys. Comm.* **184** 2197

* E-mail: l.jablonski@ujk.edu.pl

Ammonia ices under low-energy electron irradiation - Towards quantitative desorption rates and effective cross sections

L Sala[1], L Amiaud[1], C Dablemont[1] and A Lafosse[1*]

[1] Institut des Sciences Moléculaires d'Orsay (ISMO), CNRS, Univ Paris Sud, Université Paris-Saclay, Orsay, F-91405, France

Synopsis Low-energy (< 20 eV) electron processing of NH_3 multilayer films, systems of interest in astrochemistry, leads to significant NH_3 desorption, fragmentation, and to a complex chemistry arising from recombination and secondary processes induced by reactive products. Quantitative desorption rates of neutral species and effective cross sections for selected processes were obtained by combining three surface-sensitive techniques: High Resolution Electron-Energy Loss Spectroscopy (HREELS), Temperature Programmed Desorption (TPD) and Electron Stimulated Desorption (ESD) of neutral species.

NH_3 is known to be a major component in ice mantles in molecular clouds with up to ~15 % of H_2O in certain regions of the interstellar medium [1]. Model predictions cannot account for the abundance of NH_3 [2], suggesting that desorption or conversion to other compounds should be considered. Among the various thermal and non-thermal mechanisms responsible [3], this study is dedicated to processes induced by low energy electrons (LEE, energy < 20 eV) [4]. LEE interactions with matter are considered to be the primary driving force in many radiation-induced desorption and chemical reactions in ices [5].

As schematized in figure 1, multilayers of NH_3 isotopologues ($^{14}NH_3$, $^{15}NH_3$, and $^{14}ND_3$) (~10 ML) were deposited on copper at 28 K and crystallized under annealing at 60 K. High-Resolution Electron Energy Loss Spectroscopy (HREELS) and Temperature Programmed Desorption (TPD) were used to control the film morphology and estimate film thickness, respectively. LEE irradiation (1-20 eV) was performed at 28 and 60 K. Electron Stimulated Desorption (ESD) was used to detect neutral fragments/molecules desorbing during irradiation while HREELS and TPD were used to characterize processed residues.

The desorption of intact NH_3 was observed which resulted in the direct erosion of the film proceeding through a mechanism consistent with desorption processes induced by electronic transition (DIET). Different fragmentation and recombination processes were also observed as evidenced by the detection of neutral radicals NH_x (x = 1,2), and stable molecules N_2 and H_2. Aside from desorption, a wealth of chemical processes was also observed at 13 eV. Temporal ESD at this energy allowed for the estimation of the effective cross section of NH_3 desorption and observing the delayed desorption of N_2 and H_2. TPD analysis of the residues also provided evidence of diazene N_2H_2 and hydrazine N_2H_4 synthesis in the film.

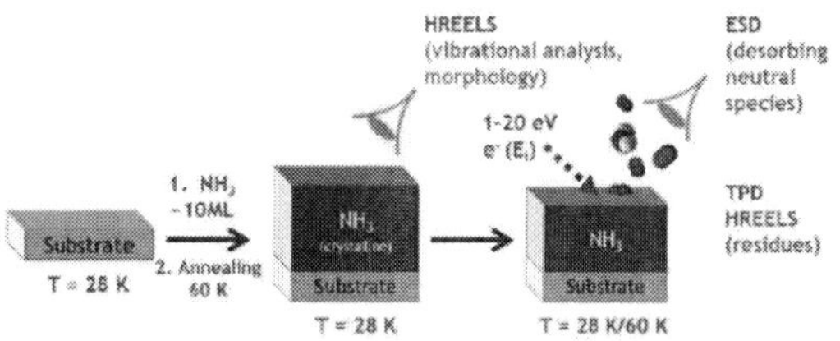

Figure 1. Scheme of the procedure used to study electron-induced processes in molecular layers of ammonia and its isotopologues.

References

[1] Jones A P and Williams D A 1984 *Mon. Not. R. Astron. Soc.* **209** 955
[2] Brown P D and Charnley S B 1990 *Mon. Not. R. Astron. Soc.* **244** 432
[3] Bertin M *et al* 2013 *The Astrophysical Journal* **779** 120
[4] Sala L A 2018 *PhD Thesis*, Université Paris-Sud 11: Orsay, France
[5] Boamah M D *et al* 2012 *Faraday Discussion* **168** 249

* E-mail: anne.lafosse@u-psud.fr

Base damages induced in DNA by 10 eV electrons: contribution to the radiosensitization mechanism of platinum chemotherapeutic drugs

W Liu[1], Y Gao[1], V Lemelin[2], Y Zheng[1]* and L Sanche[2]

[1]Research Institute of Photocatalysis, State Key Laboratory of Photocatalysis on Energy and Environment, Fuzhou University, Fuzhou 350116, P. R. China
[2]Group in the Radiation Sciences, Faculty of Medicine, Université de Sherbrooke, Sherbrooke, QC J1H 5N4, Canada

Synopsis Binding of Pt-drugs to DNA significantly enhaces base damages, crosslinks, double strand breaks (DSBs) and non-DSB cluster lesions induced by 10 eV electrons. This enhancement of potentially lethal lesions is due to a modification of electron-DNA interactions, which contributes to the synergy between chemo- and radio-therapy, when these modalities are administered in concomitance during cancer treatments.

DNA is the most critical target in Radiotherapy. The measurement of clustered damages in Pt-chemotherapeutic-drugs-DNA complexes induced by low-energy (0-30 eV) electrons (LEEs) may provide further understanding of the synergy mechanism in cancer treatment with concomitant chemoradiation therapy (CRT). In the present experiments, 5-monolayer films of Pt-DNA complexes (i.e., cisplatin, carboplatin and oxaliplatin) with Pt/DNA ratio 5:1 were irradiated with mono-energetic electrons of 10 eV, the most prominent energy for bond dissociation by LEE impact.[1] Damages to plasmid DNA including crosslinks, SSBs, DSBs and the loss of the supercoiled configuration were analyzed by the electrophoresis. Base damages (BDs) occurring within two turns of DNA helix were detected by the treatment of E. coli base excision repair endonuclease (Nth and Fpg).[2]

The total DNA damages to the three Pt-DNA complexes was 350 ± 42, 296 ± 45 and 434 ± 46 $\times 10^{-15}$ electron^{-1}molecule^{-1}, respectively, while the percentages of BDs in this total were 56%, 54% and 60%. Compared to unmodified DNA, binding of Pt-drugs to DNA increased BDs by factors of 1.9, 1.6 and 2.6, respectively, which partly resulted in significant enhancements of potentially lethal lesions (i.e., crosslinks, double strand breaks (DSBs) and non-DSB cluster damages). Specifically, compared to pure DNA, non-DSB cluster damage increases by 144%, 112%, and 141 % upon platination, respectively. The general mechanism of formation of clustered DNA damages related to BDs induced by LEEs is proposed in Fig. 1.

* E-mail: Yizheng@fzu.edu.cn

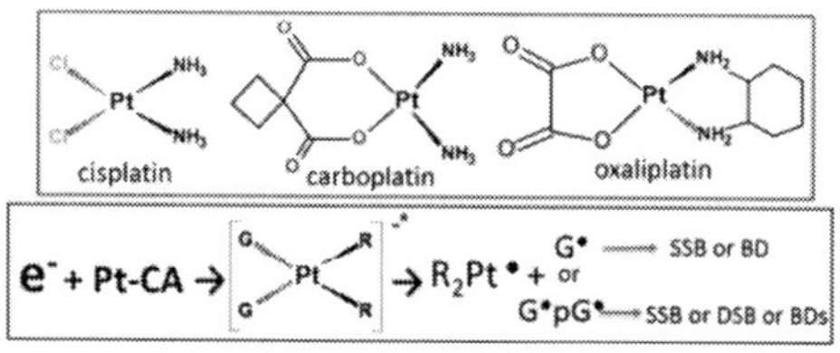

Figure 1. A) Chemical structure of the chemotherapeutic agents (CAs) cisplatin, carboplatin and oxaliplatin. B) A general pathway for LEE reactions with a Pt-CA bound to two Guanine (G) bases as in chemotherapy. R is the remaining component of the CA. A transient anion is first formed by electron capture into sigma orbitals between Pt and G. Subsequently, dissociation of one G-Pt bond or simultaneously G-Pt-G bonds occurs leading to various DNA damages [3].

The analysis of BDs and cluster lesions suggests that, along with previously measured DSBs, multiple lesions associated with BDs may constitute to a substantial proportion of LEE-induced clustered damages in cells. The large yields of BDs and non-DSB clustered damages measured in the present study demonstrate that the molecular radiosensitization mechanism, postulated from investigations with cisplatin [3] (Fig. 1) is persistent for other Pt-drugs, which further illustrates the important role of LEEs in CRT [1,3]. This study should contribute to the development of more effective CRT protocols for Pt-drugs in the clinic, via optimization of initial DNA damages in cancer cells.

References

[1] Sanche L 2016 *Radiat. Phys. Chem.* **128** 36
[2] Sahbani S *et al* 2014 *J. Phys. Chem. B* **118** 13123
[3] Dong Y *et al* 2017 *J. Phys. Chem. C* **121** 175

Low energy electron induced chemistry in CO_2:NH_3:CH_4 ices

S Esmaili[1], A D Bass[1]*, P Cloutier[1], V Lemelin[1], L Sanche[1] and M A Huels[1]

[1]Département de médecine nucléaire et radiobiologie, Faculté de médecine et des sciences de la santé, Université de Sherbrooke, Sherbrooke, J1J 1R7, Canada

Synopsis We report measurements of the electron stimulated desorption of anions from thin (nanometer scale) films of CO_2, NH_3, CH_4, and mixtures thereof.

More than 200 different, often complex, molecules are seen in the space environments [1]. It is thought that they form by the radiation and thermal processing of molecular ices found either on dust grains in dark cloud regions of the interstellar medium [2], or on the surfaces of planets and satellites [3]. Efforts to simulate in the laboratory the relevant radiation chemistry have focused on the effects of UV radiation and of fast ions and electrons [4]. More recently attention has been directed towards the role of secondary low energy electrons (LEEs) [5]. Using temperature programmed desorption (TPD) we recently reported the production of the simple amino acid glycine in multilayer films of a CH_4/CO_2/NH_3 mixture upon irradiation with 0-70 eV electrons at 22 K [6]. To better understand these results, we present complimentary data on LEE-induced chemistry occuring in similar films. Mass resolved electron stimulated desorption yields of anions are recorded as a function of electron fluence (e.g., Fig. 1). The increasing signals of new negative ion products desorbing during prolonged irradiation of CO_2/NH_3/CH_4 films included C_2^-, C_2H^-, $C_2H_2^-$, as well as CN^-, HCN^- and H_2CN^-. The gradual appearance of the desorption signals of these three latter anions suggests that they result from the dissociation of much larger molecules that contain new nitrile groups. The identification of these product ions was aided by use of $^{13}CO_2$ and $^{15}NH_3$ isotopes. The chemistry induced by electrons in pure films of CH_4, CO_2 and NH_3, as well as mixture films of CO_2/NH_3/CH_4 with composition ratios (1:1:1) were also studied by X-ray photoelectron spectroscopy (XPS). The XPS results show that electron irradiation of CO_2/NH_3 mixed films produces new chemical species containing C=O, O-H, C-O, C=N and N=O bonds.

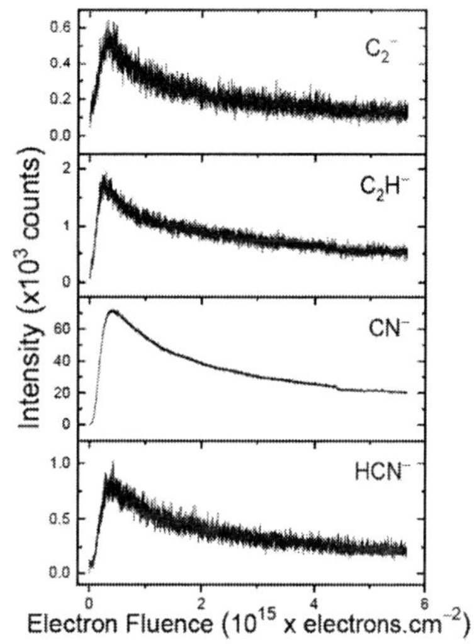

Figure 1. Variation with electron fluence at 70 eV, of the ESD signals of the indicated anions measured from a 10 ML film of CO_2/CH_4/NH_3 (1:1:1) mixture.

References

[1] Endres C P, Schlemmer S, Schilke P, Stutzki J and Müller H S P 2016 *J. Mol. Spectrosc.* **327** 95
[2] Moore M H and Hudson R L 2005 *Proc. Int. Astron. Union* **1** 247
[3] Bennett C J, Pirim C and Orlando T M 2013 *Chem. Rev.* **113** 9086
[4] Öberg K I 2016 *Chem. Rev.* **116** 9631
[5] Boyer M C, Rivas N, Tran A A, Verish C A and Arumainayagam C R 2016 *Surf. Sci.* **652** 26
[6] Esmaili S, Bass A D, Cloutier P, Sanche L and Huels M A 2018 *J. Chem. Phys.* **148** 164702

* E-mail: andrew.bass@USherbrooke.ca

Solar wind sputtering of lunar analogue material

P S Szabo[1], H Biber[1], D Weichselbaum[1], M Brenner[1], R Stadlmayr[1], D Mayer[1], A Mutzke[2], A Nenning[3], M Doppler[3], M Sauer[4], N Jäggi[5], A Galli[5], J Fleig[3], A Foelske-Schmitz[4], H Lammer[6], K Mezger[7], P Wurz[5] and F Aumayr[1,*]

[1] Institute of Applied Physics, TU Wien, Vienna, 1040, Austria
[2] Max-Planck-Institut für Plasmaphysik, Greifswald, 17491, Germany
[3] Institute of Chemical Technologies and Analytics, TU Wien, Vienna, 1060, Austria
[4] Analytical Instrumentation Center, TU Wien, Vienna, 1060, Austria
[5] Physics Institute, University of Bern, Bern, 3012, Switzerland
[6] Space Research Institute, Austrian Academy of Sciences, Graz, 8042, Austria
[7] Institute of Geological Sciences, University of Bern, Bern, 3012, Switzerland

Synopsis Sputtering by solar wind ions is investigated on thin films of the mineral wollastonite ($CaSiO_3$) as an analogue material for the lunar surface. Sputtering yields are measured with a Quartz Crystal Microbalance (QCM) using H and He ions as well as multiply charged Ar ions at different incidence angles. The results of these irradiations show that kinetic sputtering can be very well reproduced using SDTrimSP, while potential sputtering significantly increases the expected erosion of airless bodies in the solar system.

Solar wind sputtering represents an important contribution to surface erosions of bodies in the solar system that do not posses a protecting atmosphere. Sputtered atoms are emitted and form a thin exosphere around the body [1]. Analysis of this exosphere can give remote information on the surface compositions of moons and planets. This approach will also be important for the recently launched ESA mission BepiColombo to Mercury [2].

As experiments for analogue materials are scarce, we investigate the sputtering of the mineral wollastonite ($CaSiO_3$) by solar wind-relevant ions [3] on a quartz crystal microbalance (QCM). H and He ions, which make up most of the solar wind, as well as Ar ions are used as projectiles. Besides obtaining angular dependent yields, a special emphasis is put on potential sputtering by multiply charged He and Ar ions, as wollastonite is an insulator [4]. Measured yields are compared to SRIM [5] and SDTrimSP [6] simulations, which only include the kinetic sputtering contribution.

The measurements show that kinetic sputtering can be precisely described by SDTrimSP. Potential sputtering significantly increases the sputtering yield, but shows a strong fluence dependent behavior. Based on these findings we propose a model for the total erosion of $CaSiO_3$ by the solar wind. We find that potential sputtering significantly increases the erosion, especially under normal incidence. As this effect can not yet be simulated, our results prove the necessity of experimental research to understand these important space weathering processes.

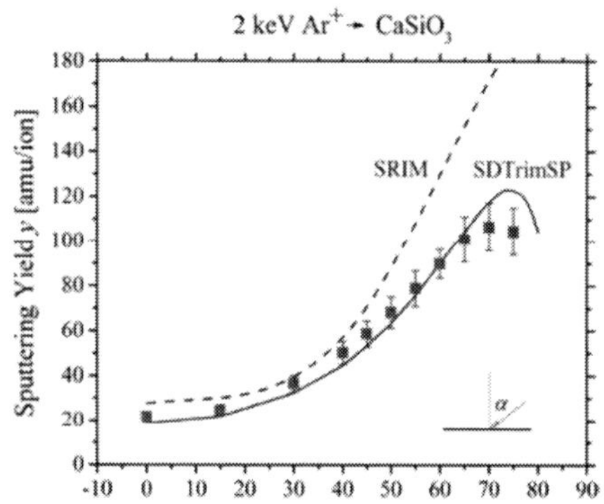

Figure 1. Experimental sputtering yields for singly charged Ar ion impact on wollastonite (red) compared to SDTrimSP (blue) and SRIM (black) simulation results.

References

[1] Hapke B *et al* 2001 *J. of Geophys. Res.* **106** 10039
[2] Wurz P *et al* 2010 *Planet. Space Sci.* **58** 1599
[3] Szabo P S *et al* 2018 *Icarus* **314** 98
[4] Aumayr F and Winter HP 2004 *Philos. Trans. Royal Soc. A* **362** 77
[5] Ziegler J *et al* 2010 *J. of Geophys. Res.* **106** 10039
[6] Mutzke A *et al* 2011 *Report IPP* **12/8**

* E-mail: aumayr@iap.tuwien.ac.at

Particle emission from two-dimensional MoS_2 induced by highly charged ion impact

L Skopinski[1]*, P Ernst[1], M Herder[2] and M Schleberger[1]†

[1]Fakultät für Physik and CENIDE, AG Schleberger, Universität Duisburg-Essen, 47057 Duisburg, Germany
[2]Fakultät für Physik and CENIDE, AG Wucher, Universität Duisburg-Essen, 47057 Duisburg, Germany

Synopsis For many attractive applications of single layer MoS_2 such as in optoelectronics e.g., the sample is supported by a substrate. Its importance for the modification through ion irradiation is here experimentally investigated by the analysis of sputtered particle of MoS_2 on SiO_2 and Au substrates under highly charged ion irradiation. The velocity distribution of the sputtered atoms is less affected by the substrate using highly charged projectiles than using slightly charged ones. Furthermore, we can show that potential sputtering causes additional emission of particles with lower kinetic energy.

Two-dimensional materials such as graphene or transition metal dichalcogenides have attracted much attention lately because of their huge potential for novel applications. They are however also very interesting from a basic science point of view because they represent ideal samples for studies in the field of ion-solid interaction. While graphene has been shown to be quite resistant to the effects of irradiation with highly charged ions (HCI), MoS_2 seems to be much more sensitive. It was recently shown that irradiation with HCIs can be used to fabricate pores in free-standing single layers of MoS_2, the size of which can be tuned via the potential energy of the primary ion [1]. This is important for applications such as DNA sequencing and desalination but many applications, e.g. in optoelectronics, are based on supported samples. The substrate plays an important role for the damage formation due to indirect sputtering as has been recently shown by Kretschmer et al. [2]. Experimentally, the interaction of (highly charged) ions with supported MoS_2 has not yet been investigated in detail.

Here we report on HCI induced damage of supported single layer MoS_2. Complementary to our earlier study [3] we use SiO_2 and Au as substrates and analyze the the emission of secondary ions and atoms under irradiation. Using laser-assisted time of flight spectrometry the velocity distribution of sputtered particels can be determined. The experiments show that the impact of the substrate is less significant for highly charged projectils than for slightly charged ones with the same kinetic energy (see Fig.1).

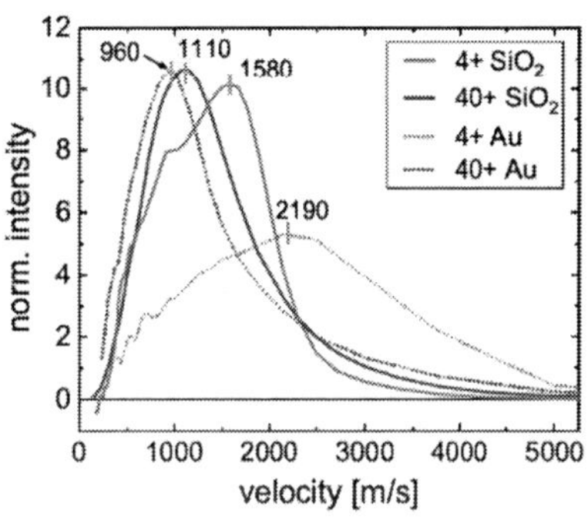

Figure 1. Velocity distribution of sputtered ^{98}Mo particles by Xe^{4+} (grey) and Xe^{40+} (red) from Au (dotted) and SiO_2 (solid) substrate.

Further, we analyzed MoS_2 on Au substrates to minimize potential sputtering from the substrate. We identified two contributions of the velocity distribution, one due to kinetic and one due to the potential energy of the projectile. The potential energy of the HCIs causes the emission of additional sputtered particles with lower kinetic energy in comparison to particles sputtered due to nuclear stopping.

*E-mail: lucia.skopinski@uni-due.de
†E-mail: marika.schleberger@uni-due.de

References

[1] Kozubek R *et al* 2019 *J. Phys. Chem. Lett.* **10** 904-910
[2] Kretschmer S *et al* 2018 *ACS Appl. Mater. Interfaces* **10** 30827-30836
[3] Hopster J *et al* 2013 *Nucl. Instrum. Methods Phys. Res* **317** 165-169

On the probability of close collisions of charged particles with atoms in a crystal

I V Kyryllin[1,2*] **and N F Shul'ga**[1,2]

[1]NSC "Kharkov Institute of Physics and Technology", Kharkiv, 61108, Ukraine
[2]Karazin Kharkiv National University, Kharkiv, 61022, Ukraine

Synopsis An investigation on the probability of close collisions of fast charged particles with atoms in a straight and bent crystal was carried out. On the basis of analytical calculation and numerical simulation we analyzed the dependence of this probability from the orientation of the crystal with respect to the direction of fast charged particles motion.

When a fast charged particle impinges on a crystal with a small angle with respect to one of the main crystallographic axes, correlations between consecutive collisions of the particle with lattice atoms appear. As a result of these correlations, the particle motion in the oriented crystal is basically defined by the continuous potential of atomic strings [1]. Due to strong intra-crystalline fields, the direction of a fast charged particle motion can be changed within quite small distances. The passage of fast charged particles through a bent crystal is of particular interest because, in this case, it is possible to deflect the particle beam direction with a small-sized crystal. There are several mechanisms of deflection of fast charged particles by a bent crystal connected with finite (channeling) and infinite (above barrier) motion in relation to bent atomic strings or bent crystal atomic planes. These mechanisms are realized in the planar and axial channeling in a bent crystal [2], in the stochastic mechanism of beam deflection (connected with multiple scattering by bent crystal atomic strings) [3] and volume reflection from bent crystal atomic planes [4].

The efficiency of the listed mechanisms of a beam deflection is defined by the stability of particle motion in the field of atomic strings or atomic planes. Close collisions of particle with atoms may lead to the escape of particle from channeling regime of motion. In the present work an investigation on the probability of close collisions of fast charged particles with atoms in a straight and bent crystal was carried out. On the basis of analytical calculation and numerical simulation we analyzed the dependence of this probability from the orientation of the crystal with respect to the direction of fast charged particles motion. Comparison of the probability of close collisions was done for three main mechanisms of beam deflection in a bent crystal: planar channeling, volume reflection and stochastic deflection. Peculiarities of the dependence of the probability of close collisions from the orientation of the crystal were explained with a help of random string approximation.

References

[1] Lindhard J 1965 *Danske Vid. Selsk. Mat. Fys. Medd.* **34** 14
[2] Tsyganov E N 1976 *Preprint* FermilabTM-682 TM-684
[3] Grinenko A A, Shul'ga N F 1991 *Pis'ma Zh. Eksp. Teor. Fiz.* **54** 520
[4] Taratin A M, Vorobiev S A 1987 *Phys. Lett.* A **119** 425

*E-mail: kirillin@kipt.kharkov.ua

Phonon contribution to grazing incidence fast atom diffraction

M S Gravielle* and J E Miraglia

Instituto de Astronomía y Física del Espacio (IAFE, UBA-CONICET), Buenos Aires, C1428EGA, Argentina

Synopsis We address the effect of lattice vibrations of the crystal on grazing incidence fast atom diffraction (GIFAD) from insulator surfaces. To describe the phonon contribution to GIFAD we introduce a semi-quantum method, named *Phonon- Surface Initial Value Representation* (P-SIVR), which represents the surface with a harmonic crystal model, while the scattering process is given by the *Surface Initial Value Representation* approach including phonon excitations. P-SIVR results for Ne/LiF(001) are compared with other theoretical methods and available experiments.

The observation of interference structures in grazing atom-surface scattering from ordered surfaces relies strongly on the coherence conditions. However, articles dealing with the decoherence introduced by thermal lattice vibrations or inelastic processes, like phonon excitations, are scarce [1]. Even though thermal vibrations of surface atoms were often expected to deteriorate the quantum coherence of the process, most of the theoretical descriptions of GIFAD assume an ideal and frozen crystal surface, with surface atoms at rest at their equilibrium positions.

In this work we study the contribution of both thermal vibrations and phonon excitations to GIFAD distributions. In order to describe the scattering from a realistic crystal, including phonon excitations, we develop a semi-quantum method - P-SIVR approach - which is based on the SIVR approximation [2]. Within this formalism the surface is represented with a harmonic crystal model, where a Maxwell-Boltzmann distribution of the initial crystal states is assumed. Contributions coming from the different final crystal states are incoherently added and the corresponding probability is then expanded in terms of the number n of excited phonons.

The P_n-SIVR approach for $n=0,1$ excited phonons is applied to study the influence of phonons in GIFAD patterns for Ne atoms grazingly impinging on a LiF(001) surface. In Fig. 1 a) the P_0-SIVR distribution for zero-phonon scattering (i.e., elastic collision) is compared with results derived with and without the inclusion of independent thermal vibrations of the lattice atoms. It is shown that, even at room temperature, the phonon contribution affects the relative intensity of the interference peaks. In addition, P_0-SIVR values are found in good agreement with available experimental data (Fig. 1 b)).

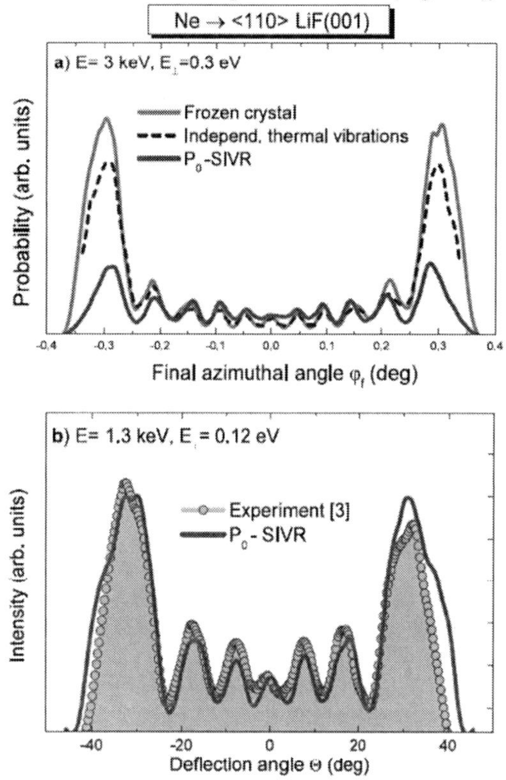

Figure 1. Angular distributions for Ne scattering from LiF(001) at room temperature, with different total, E, and normal, $E_{\perp}$, energies.

References

[1] Roncin P *et al* 2017 *Phys. Rev. B* **96** 035415
[2] Gravielle M S *et al* 2014 *Phys. Rev. A* **90** 052718
[3] Gravielle M S *et al* 2011 *Nucl. Instrum. Meth. Phys. Res. B* **269** 1208

* E-mail: msilvia@iafe.uba.ar

Fast atom interaction with surfaces at grazing incidence: classical and quantum scattering applied to thin film growth

E M Staicu Casagrande[1], A Momeni[1] and H Khemliche[1*]

[1]Institut des Sciences Moléculaires d'Orsay, CNRS, Univ. Paris-Sud, Université Paris-Saclay, Bât.520, Université Paris-Sud, F-91405 Orsay cedex, France

Synopsis Characterization of solid surfaces with heavy particles has mainly relied on the use of ions at energies from keV to MeV. Although atomic diffraction at thermal energies has been known since 1930 and became popular in the 1970's as a powerful surface characterization technique, it is nowadays rather confidential. Unexpectedly, diffraction with atoms in the keV range has also been observed despite the tininess of the de Broglie wavelength. Grazing Incidence Fast Atom Diffraction (GIFAD) has now become mature and allows the real-time analysis, during the growth, of the most fragile layers.

Particle-surface interactions play a key role in nanoscience and nanotechnology. One of the current challenge consists in engineering, with atomic level control, nanostructures with specific electronic, chemical or magnetic properties for applications in optoelectronics, sensing, or spintronics. Due to their large interaction cross-section, ions in the keV energy range have proven efficient for fine etching and characterizing solid surfaces. Yet, these charged particles are not well suited for a large class of promising but extremely fragile materials, such as organic layers.

One alternative solution for probing the outermost surface layer of these materials, is to use atoms instead of ions. In particular, when He atoms in the energy range 0.2-1 keV are scattered at grazing incidence, they provide a wealth of information on both structural and electronic properties without any damage. In addition, this analysis can be performed in real-time during film growth, thus revealing organization modes, phase transitions and dynamical structural properties.

In this context, we have developed a new technique, called GIFAD for Grazing Incidence Fast Atom Diffraction [1]. GIFAD relies on scattering of ~1 keV He atoms at grazing incidence angles near 1°. The scattered particles are collected on a position sensitive detector. For well-organized surfaces, coherent scattering yields diffraction patterns that bear sensitive information on crystallographic properties as well on the electronic corrugation.

Among the many examples that demonstrate the GIFAD capabilities, a recent work on the growth of a monolayer perylene ($C_{20}H_{12}$) on Ag(110) reveal an unexpected and complex organization dynamics [2]. In particular, crystallization of the first layer into a very dense packing occurs quasi-instantaneously at a critical coverage reached by molecule incorporation from the second layer. Following this transition, the diffraction pattern is clearly resolved, indicating a long rang order in the crystalline monolayer.

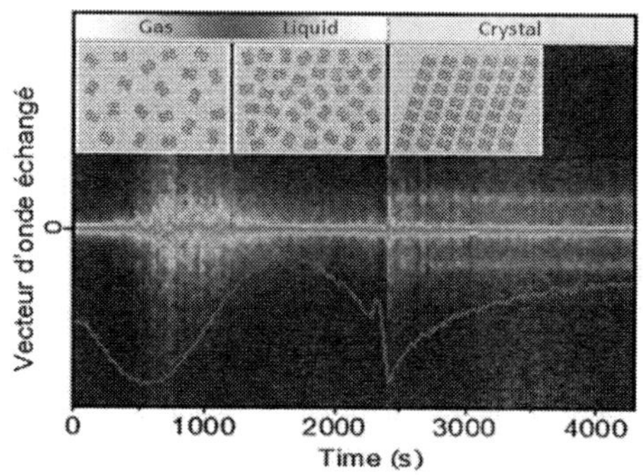

Figure 1. Evolution of the GIFAD reflectivity during deposition of perylene on Ag(110), with the evolution of the diffraction pattern.

References

[1] Khemliche H *et al* 2009 *Appl. Phys. Lett.* **95** 151901
[2] Momeni A *et al* 2018 *J. Phys. Chem. Lett.* **9** 908

* E-mail: hocine.khemliche@u-psud.fr

The role of contaminations on the interaction of highly charged ions with 2D materials

A Niggas[1]*, J Schwestka[1], S Creutzburg[2,3], F Aumayr[1] and R A Wilhelm[1,2]

[1]TU Wien, Institute of Applied Physics, Vienna, 1040, Austria
[2]Helmholtz-Zentrum Dresden-Rossendorf, Institute of Ion Beam Physics and Materials Research, Dresden, 01328, Germany
[3]TU Dresden, Dresden, 01062, Germany

Synopsis Contaminations on two dimensional materials can influence electronic, optical or mechanical material properties or even dominate them. We perform ion beam spectroscopy with highly charged ions as projectiles transmitted through freestanding 2D materials to study fundamental charge exchange processes between the ion and the samples under extreme conditions. Various *in situ* cleaning techniques to remove contaminations such as laser annealing and sample heating are employed. Their influence on our spectroscopy results is discussed here.

The rise of two dimensional materials such as free standing single layer graphene (SLG), MoS_2, and hexagonal BN (hBN) offers new possibilities to investigate the neutralisation of highly charged ions (HCI). Due to the complexity of the involved charge transfer processes, HCI neutralisation at a surface remains an unsolved puzzle.

Gruber et al. [1] reported on highly charged Xe ions capturing and stabilising a large amount of electrons during a very short interaction time with SLG (only a few fs), concluding ultra-fast HCI neutralisation and de-excitation.

However, contaminations can have a crucial influence on various measurements with 2D materials. For instance, Zan et al. showed that in presence of hydrocarbons, graphene shows self-healing effects as holes can be filled resulting in either perfect hexagonal graphene structure or a near-amorphous carbon layer [2]. The latter can also grow as an additional layer on top of the graphene under high energy electron irradiation as it is the case during transmission electron microscopy [2].

In our experiment, we use an electron beam ion source to produce ions with charge states in the range of Xe^{1+} to Xe^{44+} which can be extracted from the source with tunable kinetic energies (1-400 keV). The projectiles are transmitted through the freestanding 2D materials and capture as well as stabilise electrons due to the interaction. We measure the exit charge state of the primary ions after transmission in coincidence with emitted electrons from the sample due to ion impact. Additionally, we record the time of flight of Xe ions from the target position to the detector to gain information on their energy loss in the sample [3].

Both, exit charge state and energy loss critically depend on the sample thickness and for sure will be influenced by contaminations on either side of the two dimensional materials. As experimentalists we are therefore highly interested in confirming that our 2D samples are clean and the projectile exit charge state and the measured electron emission yields are dominated by the 2D layer rather than by contaminations. Hence, we compare measurements done with and without cleaning *in situ* and discuss various possible contaminations (e.g. water molecules, hydrocarbons). Different ways to achieve a clean sample are examined such as sample heating ($\leq$1000°C), low energy electron flooding triggering molecular desorption, and laser annealing/molecular bond breaking (450 nm).

References

[1] Gruber E *et al* 2016 *Nat. Commun.* **7** 13948
[2] Zan R *et al* 2012 *Nano Lett.* **12** 3936-3940
[3] Schwestka J *et al* 2018 *Rev. Sci. Instrum.* **89** 085101

*E-mail: niggas@iap.tuwien.ac.at

 1412 (2020) 202012 doi:10.1088/1742-6596/1412/20/202012

Highly charged ion impact on graphene leading to the emission of low energy electrons

J Schwestka[1*], A Niggas [1], S Creutzburg[2], R Kozubek[3], L Madauß[3], R Heller[2], M Schleberger[3], S Facsko[2], R A Wilhelm[1,2] and F Aumayr[1]

[1]TU Wien, Institute of Applied Physics, Vienna, 1040, Austria
[2]Helmholtz-Zentrum Dresden-Rossendorf, Institute of Ion Beam Physics and Materials Research, Dresden, 01328, Germany
[3]University Duisburg-Essen, Faculty of Physics and CENIDE, Duisburg, 47057, Germany

Synopsis Recent experiments found that the neutralisation of highly charged ions interacting with a freestanding single layer of graphene proceeds on a femtosecond time scale. This ultra-fast deexcitation was attributed to Interatomic Coulombic Decay (ICD), a process in which core holes in the projectile are filled by previously captured outer electrons and the energy is transferred to electrons of the surrounding carbon atoms. ICD therefore predicts the emission of many low energy electrons. We now present experimental evidence that e.g. Xe^{40+} indeed emits up to 85 electrons with energies below 20 eV.

For more than 20 years, there has been a model for the hollow atom (HA) formation and its deexitation in the vicinity of a solid surface via an auto-ionisation cascade [1]. Latest studies of highly charged ion (HCI) transmission through a freestanding two-dimensional graphene-membrane show that ion neutralisation and deexcitation take places within a few fs only and can no longer be explained by the comparatively slow Auger deexcitation cascades [2,3]. In a refined model Interatomic Coulombic Decay has now been proposed as the responsible mechanism for the observed fast deexcitation of the initially highly charged ion. Above the graphene sheet a HA is formed and upon entering the surface, the captured electrons are quenched into inner shells while the deexcitaion energy is released by excitation and ionisation of surrounding carbon electrons [3]. The rate of ICD depends on both the interatomic distance, which is small in case of an ion-atom collision, as well as the number of nearest neighbours, i.e. atoms participating in the process. This leads to a significantly higher efficiency of ICD in an embedded environment (solid surface) than in case of isolated gas atoms.

In our experiment we transmit highly charged projectiles through a freestanding single layer of graphene. We probe the rapid deexcitation of a HA via ICD by measuring the emitted electrons resulting from one single ion impact on the graphene sheet.

* E-mail: schwestka@iap.tuwien.ac.at

We find that the amount of emitted electrons is indeed as predicted by ICD (2-3 times the ion's initial charge state).

Further, we analyse the energy of these electrons (see Fig. 1) and find additional evidence for ICD by measuring that the vast majority of emitted electrons have energies well below 20 eV.

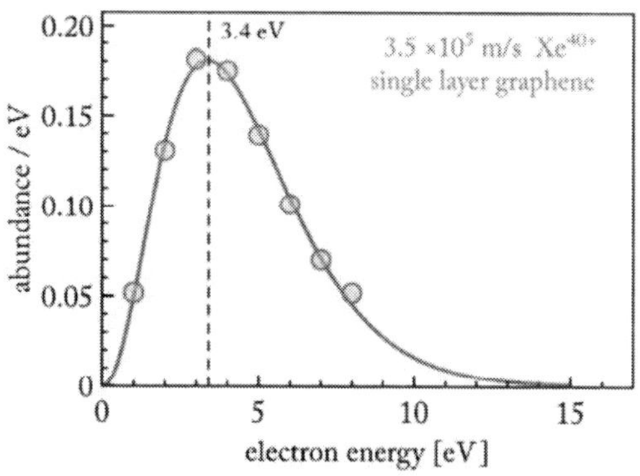

Figure 1. Measured energy distribution of emitted electrons due to 80 keV Xe^{40+} impact on single layer graphene.

References

[1] Arnau A *et al* 1997 *Surf. Sci. Rep.* **27** 113
[2] Gruber E *et al* 2016 *Nat. Commun.* **7** 13948
[3] Wilhelm R A *et al* 2017 *Phys. Rev. Lett.* **119** 103401

Investigation of the potential sputtering process on the solid Ne surfaces induced by grazing incidence of slow Ar^{q+}

H Sawa[1]*, H Ueta[1,2] and T Hirayama[1,2]

[1]Department of Physics, Rikkyo University, Tokyo 171-8501, Japan.
[2]Research Center for Measurements in Advanced Science (RCMAS), Rikkyo University, Tokyo 171-8501, Japan.

Synopsis For the quantitative investigation of the potential sputtering process on the surface of noble gas solid, we have performed the coincidence measurement of the ions desorbed and projectiles scattered from solid Ne surfaces, due to 3 keV Ar^{6+} impacts at grazing incidence angles. The dependence of Ne_n^+ desorption on the charge state of the scattered Ar projectiles is found.

Potential sputtering is a phenomenon caused by slow highly-charged ions impacting the surface of solid. It originates from the ionization and excitation of many surface atoms when electrons are captured by incident highly-charged ions, and has been observed on the surfaces of insulators such as alkali halides and oxides.

We have investigated the potential sputtering process on surfaces of noble gas solids. In our previous work [1, 2], we measured the mass distribution of ions desorbed from a surface of solid Ne under 1 keV Ar^{q+} ($q \leq 7$) impact, and reported that the potential sputtering of Ne_n^+ cluster ions was observed in the desorption yield of small size ($n \leq 3$) clusters.

To acquire more detailed information on Ne_n^+ desorption, we need to measure the potential and kinetic energies transferred from the incident ions to the target. For this purpose, we have constructed a new device for coincidence measurement of the ions desorbed and projectiles scattered from noble gas solid surfaces, due to slow multiply-charged ion impacts at grazing incidence angles [3]. This device is designed to measure flight times of the desorbed ions in coincidence with the scattered projectiles of specific charge states and scattering angle.

We have conducted the experiment on 3 keV Ar^{6+} impacts onto a solid Ne surface using the device. Charge states q of the scattered projectile were found to be distributed over $0 \leq q \leq 3$, and time-of-flight spectra of the desorbed ions in coincidence with the scattered $Ar^{0,1+,2+}$ were measured as shown in Fig. 1. While desorbed $Ne_n^+ (n = 1, 2, 3)$ are found in all spectra shown in Fig. 1., a ratio of the desorption yield of Ne^+ to Ne_2^+ increases as the charge state of the scattered projectiles decreases. Considering dissipated potential energy of incident Ar^{6+}, the change of the ratio is consistent with the dependence of Ne_n^+ desorption on the charge state q of the incident 1 keV $Ar^{q+} (q \leq 7)$ [1].

*E-mail: hiro.sawa@rikkyo.ac.jp

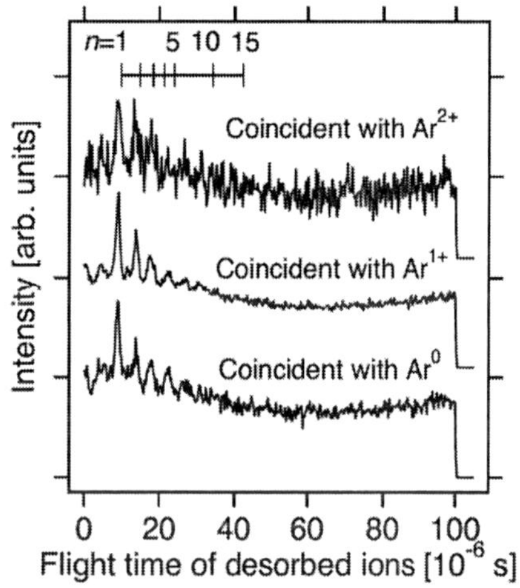

Figure 1. TOF spectra of the desorbed ions measured in coincidence with the scattered Ar ions from a surface of solid Ne (200 monolayers) by 3 keV Ar^{6+} impact. The n denotes the size of desorbed Ne_n^+.

References

[1] Fukai K, Fujita S, Tachibana T, Koizumi T and THirayama T 2010 *J. Phys. Cond. Matt.* **22** 084007
[2] Tachibana T, Fukai K, Koizumi T and Hirayama T 2010 *J. Phys. Cond. Matt.* **22** 475002
[3] Sawa H, Uchida S, Ueta H and Hirayama T, submitted to *X-Ray Spectrom.*

1412 (2020) 202014 doi:10.1088/1742-6596/1412/20/202014

Effect of secondary electron emission on the guiding properties of insulating capillaries

E Giglio[1]*, K Tőkési[2] and R D Dubois[3]

[1]Centre de Recherche sur les Ions, les Matriaux et la Photonique (CIMAP), Normandie Univ, ENSICAEN, UNICAEN, CEA, CNRS, 14000 Caen, France, EU
[2]Institute for Nuclear Research, Hungarian Academy of Sciences (ATOMKI), Hungary, EU
[3]Missouri University of Science and Technology, Rolla MO 65409 USA

Synopsis The transmission of 1 keV Ar^+ ions through a macroscopic glass capillary is simulated by taking explicitly into account the secondary electron (SE) emission generated by ion impacts inside the capillary. The SE emission re-distributes the injected charge in the capillary wall and thus modifies the guiding power of insulating capillaries. Especially for high beam intensities, SE are found to be crucial for explaining the observed time-evolution of the transmitted intensities.

In the present work, we simulate the transmission of 1 keV Ar^+ ions through a macroscopic borosilicate glass tube. The beam is tilted by 5° so that no geometrical transmission is possible. Our numerical code assumes that each Ar^+ ion that is not transmitted but hits the inner surface, injects holes at the impact point, which are immediately trapped by hole-centers at the inner surface [1]. Of the secondary electrons (SE) that are emitted from the impact point, one is picked up by the impacting Ar^+ projectile, while the remaining SE are emitted with a cosine angular distribution and propagated through the capillary until they hit the inner surface and are absorbed at the new impact point. The result is a re-distribution of the injected charge. We show that the secondary electron channel is crucial for explaining the observed time-resolved transmission currents in the case of relatively high intense ion beams [2]. Indeed, without the SE channel, we find that the transmission through our glass capillary is not stable (chaotic) in the high intensity regime, in conflict with the stable experimentally observed one.

Acknowledgements: This work was supported by the Hungarian National Research, Develop. and Innov. Off. (NKFIH) Grant No. KH126886 and by the bilateral relationships between France and Hungary in Science and Technology (S&T) under the N° 2017-2.2.5-TET-FR-2017-00008. This work was also supported by PICS N° 245 358 Hongrie 2018 founding and by Hubert Curien (PHC) Balaton 2018 program, N°40301VK and NKM-116/2018.

*E-mail: giglio@ganil.fr

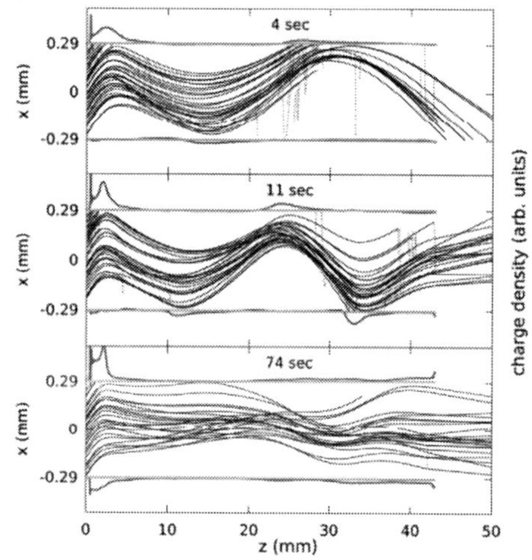

Figure 1. Snapshots of Ar^+ trajectories (black lines), projected on the x-axis, through the glass capillary at 4, 11 and 74 seconds after the beam is injected. Beam was tilted by 5° with respect to the capillary axis. The horizontal black lines at $x = \pm 0.29$ mm show the inner capillary surfaces. The red curves show the intensity (arb. units) of the positive charge patches at the inner surfaces. Note that for $t = 74$ seconds, the amount of charge on the lower surface at the entrance is due to surface charge relaxation along the angular direction. Blue dashed lines show the trajectories of secondary electrons, emitted from the ion impact point.

References

[1] Giglio E, Tőkési K and Dubois R D 2019 *Nuc. Inst. Meth. B* in press
[2] Dubois R D, Tőkési K and Giglio E submitted to *Phys. Rev. A*

Resonant anionic states of organic molecules adsorbed on metal surfaces

F Aguilar-Galindo[1] *, S Díaz-Tendero[1,2,3] and A G Borisov[4]

[1]Departamento de Química, Universidad Autónoma de Madrid, Madrid, 28049, Spain
[2]Instituto de Física de la Materia Condensada, Universidad Autónoma de Madrid, Madrid, 28049, Spain
[3]Institut for Advanced Research in Chemical Sciences, Universidad Autónoma de Madrid, Madrid, 28049, Spain
[4]Institut des Sciences Moléculaires d'Orsay, UMR 8214 CNRS-Université Paris Sud, Orsay, 91405, France

Synopsis The interaction of organic molecules with metal surfaces induces drastic changes in the electronic structure of the adsorbate. In particular, properties such as the energy and the lifetime of anionic states are strongly modified. Decay into the metal continuum must be considered in the electron dynamics, since in many cases it dominates over the decay through the vacuum. We present a generalized methodology to study this kind of systems in order to obtain the energy of the states, as well as their lifetime and decay paths.

Scanning Tunnel Microscopy (STM) is a powerful technique widely used in the study of surfaces and nanostructures. It can be used to inject electrons, populating different molecular empty states. After this process, in the case of molecules deposited on surfaces, the extra electron can decay through the substrate, emitting a photon. This fluorescence has been measured recently in a single molecule [1]. In these systems, the lifetime of the electron on the molecule is a key factor, since these anionic states play an important role in chemical reactivity. They have also high interest for further applications such as nanodevices and optoelectronics.

In this work, we present a deep study of the anionic resonant states of molecules adsorbed on metal surfaces. We first obtain the adsorption geometry using Density Functional Theory, which allows the inclusion of the periodic boundary conditions that are required in order to have a correct description of the electronic structure of the metal. With these calculations, complementary studies are carried out and further information is obtained [2]. Then, through the use of Wave Packet Propagation (WPP) calculations we study the quantum dynamics of the active electron and thus, we are able to obtain information of the resonant states. This methodology has been used to study metal states [3] and nanostructures supported on surfaces [4]. In this communication we present results for three organic molecules (acrylamide, acrylonitrile and acrolein) adsorbed on a pristine Cu(100) surface, although our procedure can be generalized to more complex systems.

Figure 1. Density (square of the resonant wavefunction) associated to the coupling of the π^* state of the acrolein with a Cu(100) surface. Spheres represent atoms which were explicitly used to obtain the adsorption geometry.

Our results show how the symmetry of the molecular state which couples with the surface affects the shape of the decay channel towards the bulk.

References

[1] Doppagne B *et al* 2018 *Science* **361** 251
[2] Aguilar-Galindo F and Díaz-Tendero S 2018 *J. Phys. Chem. C* **122** 27301
[3] Chulkov E V *et al* 2006 *Chem. Rev.* **106** 4160
[4] Díaz-Tendero S *et al* 2008 *Nano Lett.* **8** 2712

* E-mail: fernando.aguilar-galindo@uam.es

Influence of surface plasmon polaritons on heating of gold

P N Terekhin[1,2 *], P D Ndione[1], S T Weber[1] and B Rethfeld[1]

[1]Department of Physics and Research Center OPTIMAS, TU Kaiserslautern, Kaiserslautern, 67663, Germany
[2]National Research Center "Kurchatov Institute", Moscow, 123182, Russia

Synopsis We present a detailed investigation of periodic heating of gold and periodic structures formation on a surface induced by surface plasmon polaritons (SPP) after irradiation with ultrashort laser pulses. Conditions of the SPP excitation and influence of the angle of incidence and wavelength of the laser field on the resulting structured heating are analysed in detail.

The use of ultrashort laser pulses for surface nanostructuring is of great importance in technological applications. Surface plasmon polaritons, i.e. surface plasmons coupled to a laser-electromagnetic wave, can be excited on a rough material surface [1, 2]. The interference of the plasmons wave and the incoming pulse leads to a periodic modulation of the deposited laser energy along the surface of a sample.

The spatial and temporal evolution of the periodically modulated absorbed laser energy is studied after irradiation of gold in the framework of the two-temperature model (TTM) [3]. We present a new analytical source term for the TTM, which takes into account the excited plasmon subsystem and therefore spatial periodicity. This source term describes the energy absorbed by the electrons of a metal and consists of three terms originated from the interference of the SPP and laser fields, namely: laser-laser, laser-SPP and SPP-SPP.

Figure 1 shows the resulting rate of the increace of the electrons' energy density in dependence on time and lateral distance from a step edge at the interface between gold and air for a laser wavelength of 700 nm, pulse duration of 40 fs (full width at half maximum) and the radius of the Gaussian focus of 40 μm centered at the step edge. The periodic pattern of the absorbed energy distribution and the SPP propagating along the sample after the laser pulse can be observed in this plot.

Applying the developed energy source in the TTM model, we show that the electronic temperature follows the spatial profile of the source term including the oscillation of energy flux caused by the interference pattern. We investigate how these oscillations persist on large timescales when thermalization between electron and lattice subsystems is approached. Assuming that the final material modification is caused by the oscillating source term, we present results on the periodicity of the experimentally observed ripples for pulses incident at different angles relative to the surface normal. Future work will focus on implementation of the SPP waves into the combined atomistic-continuum molecular dynamic-TTM model [4].

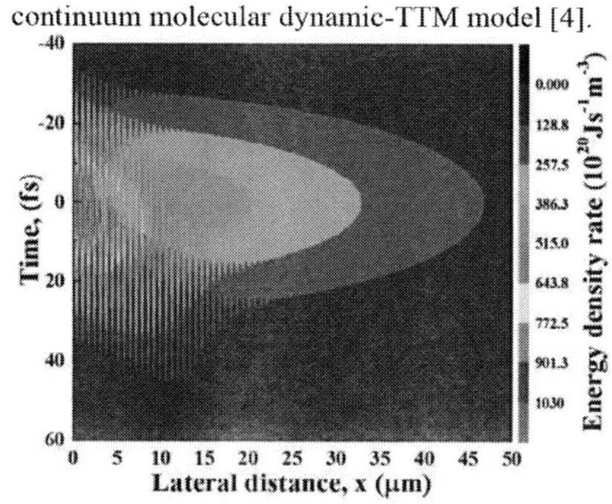

Figure 1. The energy absorbed by gold electrons at laser wavelength 700 nm and pulse duration 40 fs at the surface irradiated at normal incidence.

The developed method can be applied for different samples and arbitrary parameters of the laser pulses. It can be used to study the features of the excited SPP and the mechanisms of laser energy absorption under controlled conditions and the formation of morphological effects. Such periodic surface structures could be of interest for medical applications, tribology, wetting, and others.

References

[1] Maier S A 2007 *Plasmonics, Fundamentals and Applications* (Springer, Berlin)

[2] Raether H 1988 *Surface Plasmons on Smooth and Rough Surfaces and on Gratings* (Springer-Verlag, Berlin)

[3] Anisimov S I *et al* 1974 *Sov. Phys.-JETP* **39** 375

[4] Ivanov D S *et al* 2017 *Appl. Phys. A* **123** 744

* E-mail: terekhin@physik.uni-kl.de

Charge-transfer dynamics in anion collisions with vicinal nanostepped surfaces

J Shaw[1] *, D Monismith[2], Y Zhang[1], D Doerr[1] and H S Chakraborty[1] †

[1]Department of Natural Sciences, D L Hubbard Center for Innovation, Northwest Missouri State University, Maryville, Missouri 64468, USA
[2]Software Maintenance Group, Tinker AFB, Oklahoma, USA

Synopsis We investigate the electron dynamics at various monocrystalline metal surfaces with stepped vicinal nanostructures. The unoccupied bands of the surface superlattice are resonantly excited *via* the charge transfer interaction of the surface with a H^- ion which flies by at grazing angles. The ion survival exhibits modulations as a function of the vicinal-terrace size and shows peaks at energies that access the image state subbands.

Vicinal surfaces of regular arrays of linear steps, prepared by cutting and polishing a single crystal by ultra-high vacuum methods, are the simplest models of lateral nanostructures on surface. These systems closely mimic rough industrial surfaces and are critical for their catalytic properties, without losing the lattice periodicity [1]. Thus, understanding electronic motions in these surfaces are valuable.

The charge transfer interaction dynamics of an ion with a surface is very sensitive to the surface band structure. The energy conserving transfer of an electron, the resonant charge transfer (RCT), occurs when the shift of the ion level enables the transfer to (from) an unoccupied (occupied) resonant state of the substrate. While studies on flat surfaces are aplenty [2], the first study on vicinally stepped Pd(111) is reported by us [3]. Calculations are now extended for other surfaces, such as, Cu(111) and Au(100) with vicinal steps, to probe important effects of superlattice dispersions.

We model the vicinal structure in a Kronig-Penny scheme. The RCT interactions of the H^- ion grazing the surface is simulated in a quantum mechanical wave packet propagation approach utilizing the split-operator Crank-Nicholson propagation method [3]. Thread based parallel computing using OpenMP was employed. Visualizations of the evolving wave packet density from the ion to the surface and back in the close interaction zone as well as its decay through the surface into the bulk *via* subbands of various surface specific states are made. Further, the survival probability of the ion is calculated as a function of ion's parallel velocity for varied sizes of the vicinal terrace.

The H^- survival probability are presented in Figure 1 showing strong modulations for stepped surfaces compared to the steady result of flat precursor surface. When the ion's parallel energy parabolic dispersion matches an image subband state, a resonance-type condition is reached, and, consequently, the recapture rate by the ion increases. This effect produces peaks in the survival probability as seen which appear largely insentive to the ion's strike location.

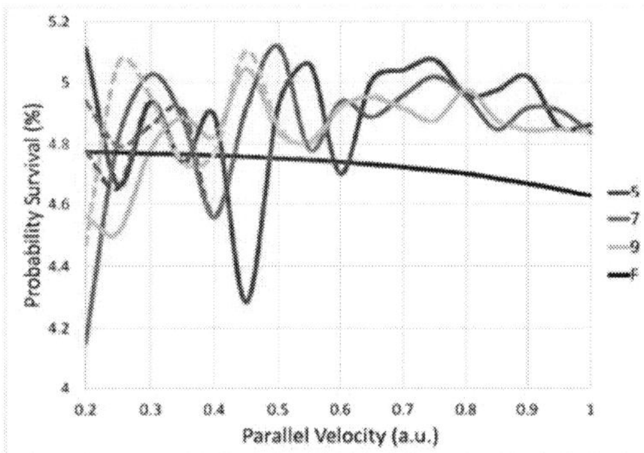

Figure 1. Ion survival for flat (F) and stepped Cu(111) as the ion strikes the center (solid) and step (dashed) of the terrace of sizes in units of lattice spacing.

The work serves as the proof-of-principle in the utility of our computational method to address surfaces with nanometric patterns.

Supported via XSEDE allocation NSF ACI-1548562, Stampede Supercomp TACC allocation TG-PHY150001, and NSF PHY-1806206.

References

[1] Jenkins S J *et al* 2007 *Surf. Sci. Rep.* **62** 373
[2] Schmitz A *et al* 2010 *Phys. Rev. A* **81** 042901
[3] Shaw J *et al* 2018 *Phys. Rev. A* **98** 052705

* E-mail: jshaw1969@gmail.com
† E-mail: himadri@nwmissouri.edu

Coherent effect in ionization loss of relativistic electron ensembles in ultrathin targets

S V Trofymenko[1,2*] **and N F Shul'ga**[1,2]

[1]National Science Center 'Kharkiv Institute of Physics and Technology', Kharkiv, 61108, Ukraine
[2]Karazin Kharkiv National University, Kharkiv, 61022, Ukraine

Synopsis Ionization loss of relativistic electron bunches in ultrathin targets is considered. It is shown that the effect of coherent amplification of ionization loss comparing to Bethe-Bloch result, analogous to such effect in radiation, can take place for sufficiently small bunches with high particle densities. It is demonstrated that such effect can manifest itself for bunch parameters achievable on modern accelerator facilities.

Ionization loss of a particle ensemble, moving in substance, is usually just a sum of independent losses of separate particles of the ensemble. In the present work it is shown that the value of a particle ensemble (bunch) ionization loss can exceed the mentioned sum by several orders of magnitude for ensembles of sufficiently high density and small spatial size. Such effect is analogous to the effect of coherent amplification of radiation intensity by electron bunches. The point is that the process of ionization loss, like radiation, is characterized by its typical frequencies ω_{0i} (the characteristic atomic frequencies of the substance) and corresponding wavelengths $\lambda_{0i} \sim 1/\omega_{0i}$ Therefore, in the case when the longitudinal size (along the direction of particles motion) of a particle ensemble is confined within a wavelength λ_{0i}, it is natural to expect strong interference effects in ionization loss originating from the interference of the proper fields of the particles. For this also the transversal size of the ensemble should not exceed the characteristic transversal size ρ_0 of a particle's proper field in substance. Due to polarization of the substance (causing the so-called density effect in ionization loss) this size is estimated as $\rho_0 \sim 1/\omega_p$, where ω_p – is the plasma frequency. This value is rather small and extremely high particle densities of the ensembles are needed for manifestation of the interference effects in ionization loss. Due to this fact fact that up to now the discussed effects we considered just for the simplest cases of naturally created ensembles of several particles situated very close to each other. This, particularly, includes Cudakov effect [1] of a high-energy electron-positron pair ionization loss suppression in the vicinity of its creation point and a related effect in ionization loss of charged particles created in the result of a low-energy molecule splitting in substance [2, 3].

We show that the condition on the particle density in the bunch is significantly weakened if consider ionization loss in sufficiently thin targets, where the density effect is absent. The same holds for thin boundary layers of thick targets, where the discussed effects can lead to a considerable amplification of a bunch energy deposition in the vicinity of the target surface. In such layers the value of ρ_0 is defined by the typical size of the Coulomb field in vacuum and reads $\rho_0 \sim \gamma/\omega_{0i}$, where γ is the particle Lorenz-factor. For ultrarelativistic particles this value significantly exceeds $1/\omega_p$. In this case the discussed effects can be manifested at bunch parameters achievable on modern free-electron lasers (e. g., European XFEL), as well as on a series of accelerators presently under construction (e. g., SINBAD). In the present work we estimate the value of ionization loss for the bunch parameters typical for the mentioned facilities and investigate its dependence on the bunch size and shape.

References

[1] Chudakov A E, 1955 *Izv. Akad. Nauk SSSR* **19** 589

[2] Brandt W, Ratkowski A, Richte R H, 1974 *Phys. Rev. Lett.* **33** 1325

[3] Kagan Y M, Kononets Y V, Dzhamankyzov N K, 1978 *Sov. Phys. JETP* **47** 148

* E-mail: trofymenko@kipt.kharkov.ua

The missing single-scattering peak in tin ion collisions on Mo and Ru targets

S Rai[1,2]*, M J Deuzemann[1,2], W Noordam[1], E Holleman[1], O O Versolato[2] and R Hoekstra[1,2]

[1] Zernike Institute for Advanced Materials, University of Groningen, 9747 AG, The Netherlands
[2] Advanced Research Centre for Nanolithography, Amsterdam 1098 XG, The Netherlands

Synopsis We report the absence of single-scattering peaks in collisions of tin ions on Mo and Ru targets. For these systems standard simulation packages as e.g. SRIM predict prominent single-scattering peaks. To identify the potential cause of the single-scattering peak missing, we changed charge state, energy, and species of the projectiles, detected ions and neutrals separately and studied collisions of Kr on Cu which has a similar mass ratio of projectie and target particles.

The study of the interaction of energetic tin ions with materials facing laser-produced, EUV generating Sn plasma is of importance to next generation EUV light based nanolithography tools. Energetic keV tin ions ejected from the plasma may damage the multilayered mirrors used to collect 13.5-nm EUV light from tin plasma source[1].

At the ZERNIKELEIF facility at the Zernike Institute for Advanced Materials, mass and charge-state selected Sn^{q+} ions can be produced and transported through a central beam and finally collided upon a solid target of choice in the Surface Physics (SirΦ) setup. Kinetic energy spectra of scattered tin particles are detected by an Electrostatic Analyser (ions) or by a Time Of Flight system (ions and neutrals).

Figure 1 shows a typical kinetic energy spectrum of tin ions scattered off Mo. The initial beam of 14 keV Sn^{2+} ions impinges on the target at grazing incidence of 15° and the ions are detected at a scattering angle of 30°. The standard software package SRIM has been used to produce a synthetic energy spectrum for comparison. The single-scattering peak of Sn on Mo is clearly visible in the simulated spectrum at an energy of around 8keV but is clearly absent in the experimental data.

To further investigate this remarkable observation, experiments have been conducted with tin ions in charge states of 1+ to 4+, energies of 5 – 25 keV, Xe and Kr ions instead Sn ions, and Ru instead of Mo as target. For Kr and Xe as projectiles and Ru instead of Mo the differences are even larger. All this hints at SRIM having diffulties in simulating slow heavy particle and impact on a heavy target, which might have different reasons including the neglect of prescattering in front of the surface, inefficiencies in the model potentials.

As a first further test we performed experiments with 7 keV Kr^{1+} impinging on Cu. The Kr-Cu system was choosen as their mass ratio is similar to Sn-Mo/Ru. Unlike in the case with Sn ions scattering off Mo or Ru, the single scattering peak shows up not only in the simulations but also in the experimental data. Experiments are ongoing to further narrow down the causes for the absence of a single-scattering peak in keV heavy particles colliding on heavy, metallic targets.

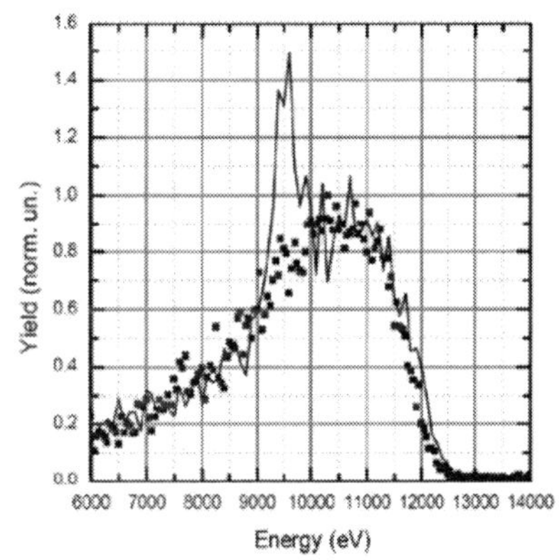

Figure 1. Experimentally obtained (squares) and simulated energy spectra(solid line) for 14 keV Sn^{2+} ions scattered from Mo. The incident angle is 15° with respect to the target surface and the scattering angle is 30°.

References

[1] Versolato O O 2019 *Plasma Sources Sci. Technology*, in press

* E-mail: s.rai@rug.nl

Journal of Physics: Conference Series **1412** (2020) 202020 doi:10.1088/1742-6596/1412/20/202020

Theoretical models of stopping and ionization of metallic targets and oxydes bombarded by H_2^+ molecules

C D Archubi[1] *, D Arbó[1] and N R Arista[2]

[1] Comisión Nacional de Investigaciones Científicas y Técnicas-Instituto de Astronomía y Física del Espacio (CONICET-IAFE), Universidad de Buenos Aires, Argentina
[2] Comisión Nacional de Energía Atómica (CONEA), Centro Atómico Bariloche, Argentina

Synopsis We study the vicinage effects in the interaction of H_2^+ projectiles with C, Al, Si, Al_2O_3 and SiO_2 targets considering the excitation of inner shells. On one side we extend the use of the semiclassical impact-parameter model for the excitation of atomic shells, considering quantum corrections and the role of target screening in the vicinage effects. On the other hand, we adapt our extended wave packet model (EWPM) [3] to the calculation of stopping ratios and ionization cross sections for correlated ions. Lindhard model is used to evaluate valence electrons contributions.

Most of the theoretical descriptions of the vicinage effect have been made using the dielectric formulation. This is a natural way to represent the electrodynamical interactions in a dense medium and is particularly useful to consider the interaction with conduction or valence band electrons. An interesting dielectric model developed in the nineties by Kaneko [1] extends the possibilities of the dielectric approach to evaluate the contributions of inner shells, opening in this way the possibility of more comprehensive stopping power calculations. The so called wave-packet model (WPM) of Kaneko, based on the use of gaussian distributions of electrons in momentum space, provides a significant advantage in terms of analyticity as it yields closed analytical expressions for the dielectric function, with similar properties to those of the Lindhard dielectric model (LDM) corresponding to a free-electron gas, but using more appropriate velocity distributions for each atomic shell.

Our EWPM introduces modifications to the WPM originally proposed by Kaneko, using the Levine and Louie technique [2,3] to take into account the energy gaps corresponding to the different atomic levels of the target.

On the other hand we extend the classical electrodynamics atomic-oscillator model of Bohr to a semiclassical impact parameter model (SIPM) for the case of correlated ions. To perform this extension we consider short-distance corrections and screening effects. This approach permits us to evaluate the rol of close and distant collisions on the vicinage effect, showing the growing importance of distant collisions with increasing energies.

We have performed specific calculations of vicinage effects in energy losses and inner-shell ionization for various accessible targets (C, Al, Si, Al_2O_3 and SiO_2), and compared the calculations with existing experimental results.

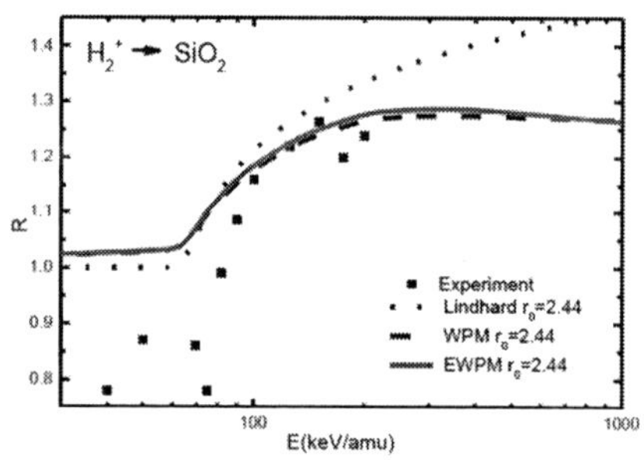

Figure 1. Lindhard electron gas and total contribution to the stopping ratio of H_2^+ projectiles traversing a SiO_2 target as a function of the projectile energy for a typical internuclear distance r_0.

Inner-shell excitations produce an additional decrease in the vicinage effect. The present work strongly suggests new experiments at intermediate and higher energies to test our new theoretical predictions and to reveal differences between models.

References

[1] Kaneko T 1995 *Phys Rev A* **51** 535
[2] Levine Z H and Louie S G 1982 *Phys Rev B* **25** 6310
[3] Archubi C D and Arista N R 2017 *Phys Rev A* **96** 062701

* E-mail: archubi@iafe.uba.ar

Many body phenomenon within multiple scattering approach in PEELS

A Mandal[1] *, V Kochetov[2], N Dhankhar[1], D Sébilleau[3] † and R Choubisa[1]

[1]Department of Physics, BITS Pilani Pilani Campus Vidya Vihar, Pilani-333031, India
[2]Institute for Physics, Universität Rostock, Albert-Einstein-Straße 23 – 24, 18059 Rostock, Germany
[3]Department of Material and Nanosciences, Institute Of Physics of Rennes, University of Rennes-1, 35042 Rennes-Cedex, France

Synopsis Plasmon excitation within multiple scattering approach is being explored. The plasmon loss is described within the so-called quasi-boson approximation, originally proposed by Hedin and further extended by Fujikawa. We compare the different fluctuation potentials i.e. Plasmon-pole model, Bechstedt model and other available potentials based on beyond-RPA dielectric functions and use either the exact core state wave function or more accurate model than a simple delta function used previously for total energy loss in Aluminium targets.

We consider only plasmon excitation which are excited by electrons. Although all the intrinsic properties of the plasmons remain valid in this case as we study them in PEELS. Photo emission energy loss spectroscopy (PEELS) aims at providing information on the dielectric function near the surface. It consists the monitoring of a plasmon satellite in a photoemission spectrum as a function of the direction of the outgoing electron that has suffered a plasmon loss. In order to be able to derive quantitative information on the (surface) dielectric constant, a suitable theoretical model is necessary upon which we are working presently. We intend to incorporate the present model into the MsSpec computer code MsSpec [1]. We use electron-boson Hamiltonian where plasmon loss is described within the quasi-boson approximation, originally proposed by Hedin [2] and further extended by Fujikawa and co-workers [3]. We have calculated the cross-section for the one-plasmon loss and compared it using various fluctuation potentials and calculated total energy loss for different core wavefunction for Aluminium 2p (Fig.1). The computed cross-section of plasmon peak in terms of core peak (extracted from MsSpec) for one plasmon loss is

$$I^1(\overrightarrow{\mathbf{k}}, \hbar\omega_{\mathbf{q}}, \hbar\omega_i) = I^0(\overrightarrow{\mathbf{k}}, \hbar\omega_{\mathbf{q}}) \frac{\alpha(\hbar\omega_i)}{\hbar\omega_i} \qquad (1)$$

We are computing the loss function using dielectric function. Overall we explore the importance of taking into account multiple scattering (MS) when the outgoing electron has lost energy through the excitation of a plasmon [4]. Presently the model is for one plasmon and and we plan to extend it to two,three plasmons. We also discuss our future plans to extend PEELS study to (e,2e) field on solid targets.

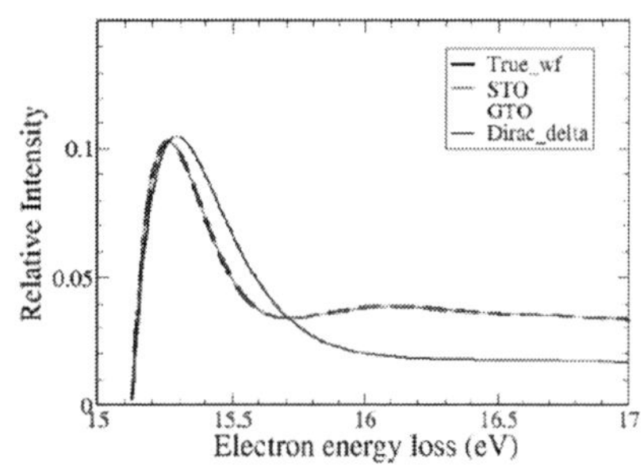

Figure 1. Total energy loss for Aluminium using various core-state wavefunction

References

[1] Sébilleau D *et al* 2011 *Computer physics Communications* **182** 2567-2579
[2] Hedin L *et al* 1998 *Phys. Rev. B* **58** 15565
[3] Kazama M *et al* 2014 *Phys. Rev. B* **89** 045110
[4] Sébilleau D *et al* 2018 *Multiple Scattering for Spectroscopies, Springer Proceedings in Physics* Vol 204

*E-mail: p2016009@pilani.bits-pilani.ac.in
†E-mail: didier.sebilleau@univ-rennes1.fr

Surface chemistry of colloidal surfactant-free gold nanoparticles generated by laser ablation

A Lévy[1,*], M De Anda Villa[1,†], J Gaudin[2], D Amans[3], V Blanchet[2], F Boudjada[3], J D Bozek[4], R E Grisenti[5,6], E Lamour[1], G Laurens[3], S Macé[1], A R Milosavljević[4], C Nicolas[4], I Papagiannouli[2], M Patanen[7], C Prigent[1], E Robert[4], S Steydli[1], M Trassinelli[1] and D Vernhet[1]

[1] Institut des Nanosciences de Paris, Sorbonne Université - Pierre et Marie Curie, CNRS, 75005 Paris, France
[2] Centre Lasers Intenses et Applications, Univ. Bordeaux, CNRS, CEA, 33405 Talence, France
[3] Institut Lumière Matière, Université Claude Bernard Lyon 1, CNRS, 69622, Villeurbanne, France.
[4] Synchrotron SOLEIL, L'Orme des Merisiers, Saint-Aubin, 91192 Gif-sur-Yvette Cedex, France
[5] GSI Helmholtzzentrum für Schwerionenforschung GmbH, 64291 Darmstadt, Germany
[6] Institut für Kernphysik, J. W. Goethe-Universität, 60438 Frankfurt am Main, Germany
[7] University of Oulu, Department of Physics, 90014 Oulu, Finland

Synopsis Surface chemistry of gold nanoparticles produced by laser ablation in liquids has been investigated using synchrotron radiation and the X-ray Photoelectron Spectroscopy technique. Thanks to the recorded core-level and valence spectra, combined with a detailed analysis based on Bayesian statistical methods, the oxide presence and its proportion on the nanoparticle surface have been determined.

A promising NanoParticle (NP) synthesis technique widely developed in the past few years is based on laser irradiation of a solid target in a liquid environment. The Pulsed Laser Ablation in Liquids (PLAL) method results in the formation of ligand-free nanoparticles. These bare nanoparticles in organic or non-organic solvents are well-suited for the development of different technological applications [1] usually requiring subsequent surface functionalization. In addition, the surface chemical composition of these nanoparticles is expected to take part in the colloidal stability of the PLAL product. The resulting surface charge could be responsible of the electrostatic repulsion impeding their aggregation. However, no consensus [2] has been drawn so far on these questions and thus a precise knowledge of their surface properties and composition is mandatory.

We performed an experimental investigation of the surface chemistry occurring at the PLAL nanoparticle surface at SOLEIL synchrotron facility on the PLEIADES beamline (Figure 1). The X-ray photoelectron spectroscopy measurements peformed on free-standing [3, 4] gold nanoparticles address the question of (i) their surface oxidation state and (ii) the chemical composition of their first's surface atomic layers. Signatures of halide-ions has been evidenced and the probability of finding gold oxidized atoms has been assessed, demonstrating that this technique provides a promising new way to study bare gold surfaces and a complementary insight to colloidal stability.

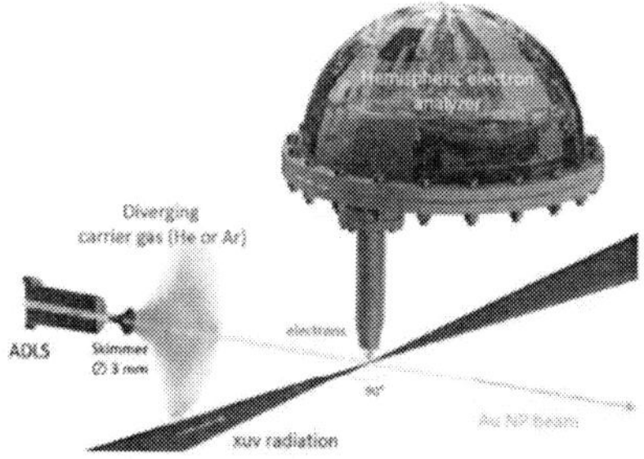

Figure 1. An aerosol of AuNPs is created using an atomizer. An aerodynamic lens focuses the AuNP beam to the interaction region with soft X-rays of PLEIADES beamline (Synchrotron SOLEIL).

References

[1] Daniel M-C and Astruc D 2004 *Catalysis and Nanotechnology. Chem. Rev.* 104 293

[2] Merk V *et al* 2014 *Langmuir* **30** 4213 ; Muto H *et al* 2007 J. *Phys. Chem. C* **111** 17221 ; Fong Y-Y *et al* 2013 *Langmuir* **29** 12452 ; Palazzo G *et al* 2017 *J. Colloid Interface Sci.* **489** 47

[3] Liu P *et al* 1995 Aerosol Sci. Technol. **22** 293

[4] Lindblad A *et al* 2013 Rev. Sci. Instrum. **84** 113105

* E-mail: anna.levy@insp.jussieu.fr
† E-mail: manuel.andavilla@insp.jussieu.fr

A quantitative study of ion guiding between two glass plates

R D DuBois[1*], E Giglio[2] and K Tőkési[3,4]

[1]Missouri University of Science and Technology, Rolla, MO USA
[2]Centre de Recherche sur les Ions, les Matériaux et la Photonique (CIMAP), F-14000, Caen, France
[3]Institute for Nuclear Research, Hungarian Academy of Sciences (Atomki), 4026 Debrecen Bem tér 18/c, Hungary
[4]ELI-ALPS, ELI-HU Non-profit Ltd., Dugonics tér 13, H-6720 Szeged, Hungary

Synopsis The transmission of a 1 keV Ar^{+} beam between two glass plates was studied. We present experimental data with the goal to provide quantitative information about the buildup and decay of charge on the glass plates and how this correlates with the observed guiding.

Since the original discovery of ion beam guiding by insulating capillaries, numerous studies have been performed. The vast majority of these provide qualitative information about the transmitted intensity as a function of time and/or angle between the capillary axis and the original beam direction. Models of the stochastic buildup of charge on the capillary surfaces have been used to simulate these properties [1]. However, quantitative information about the charge buildup, i.e., the deposition and redistribution, is needed for testing and improving these models.

In an earlier study of low energy ion beam guiding by a cylindrical glass capillary, we provided ~~some~~ quantitative experimental and theoretical information about how rapidly the charge was deposited and decayed away and how this influenced the beam transmission [2]. Here, Using our setup several studies were performed to provide similar quantitative information for a parallel plate geometry.

In the work presented here the bypass beams were prohibited from entering the guiding assembly. By using a beam intensity of ~100 pA, the relative amounts of current hitting the glass plates and being transmitted were measured as a function of time (see Fig.1). But the higher beam current also means guiding begins quite quickly. Data obtained after allowing the glass plates to discharge for 50 minutes showed that the transmitted current immediately increases while the current hitting the plates decreases. Ultimately, approximately ½ to ¾ of the injected charge is deposited in the plates and the transmitted beam intensity plateaus. During the entire process, the position of the bypass beams is also monitored in order to obtain information about the deposited charge. Simulations of this geometry are in progress.

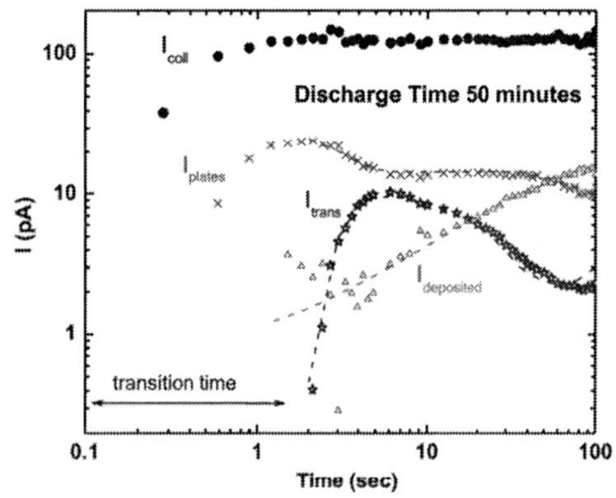

Figure 1. The amount of beam current on the collimating aperture, I_{coll}, impacting the plates, I_{plates}, and transmitted, I_{trans}, as a function of time. The decrease in $I_{plates} + I_{trans}$ is attributed to the amount of beam that is deposited in the plates, $I_{deposited}$.

Acknowledgements This work was supported by the Hungarian National Research, Development and Innovation Office (NKFIH) Grant No. KH126886 and by the Bilateral relationships between France and Hungary in Science and Technology (S&T) under the project number 2017-2.2.5-TET-FR-2017-00008. This work was also supported by PICS No 245 358 Hongrie 2018 founding and by Hubert Curien (PHC) Balaton 2018 program, project numbers 40301VK and NKM-116/2018.

References

[1] Schiessl K *et al* 2005 *Phys. Rev. A* **72** 062902.
[2] Giglio E *et al* 2015 *Nucl. Instr. and Meth. B* **354** 82

*E-mail: dubois@mst.edu

Formation of nanocraters on the surface of gold nanolayer by an impact of highly charged xenon ions

I Stabrawa[1], D Banaś[1]*, A Kubala-Kukuś[1], Ł Jabłoński[1], P Jagodziński[2], D Sobota[1], K Szary[1], M Pajek[1], E Mendyk[3], K Skrzypiec[3] and M Borysiewicz[4]

[1]Institute of Physics, Jan Kochanowski University, 25-406 Kielce, Poland
[2]Department of Mathematics and Physics, University of Technology, 25-314 Kielce, Poland
[3]Faculty of Chemistry, M. Curie-Skłodowska University, 20-031 Lublin, Poland
[4]Institute of Electron Technology, 02-668 Warsaw, Poland

Synopsis In this work we have systematically studied process of energy deposition and formation of nanostructures on metallic (Au nanolayer) surfaces as a result of the irradiation with slow highly charged xenon ions (HCI). Well pronounced modifications of the surfaces in form of craters were observed. We found that, in the systems considered, the kinetic energy influences the crater depth, while the potential energy affects the crater size at the surface. This finding showed that for gold nanolayers the potential energy plays important role in the nanostructures creation.

Modification of metal, semiconductor and insulator surfaces by an impact of single (i.e. each ion creates nanostructure) low-energy highly charged ions (HCI) is of great importance for developing new technologies and has the potential to introduce novel nanostructures and material properties not achievable by any other material processing methods [1].

When HCI approaches to the surface its kinetic and potential energy (the sum of binding energies of the removed electrons) is dissipated at the surface which leads to structural modification of the outermost surface layers and a nanostructure creation [2].

In this work, modifications of Au nanolayers deposited on a crystalline silicon Si (100) wafers caused by HCI xenon were analyzed. The nanolayers were irradiated with Xe^{q+} (q=15-40) ions in the energy range hundreds of keV (nuclear stopping power regime) at a fluence of about 10^{10} ions/cm^2 [3, 4]. For the first time, for such systems, dependence of the size of nanostructures on potential energy, kinetic energy, and nanolayer thickness were measured systematically. After irradiation the nanolayers were investigated using atomic force microscope. Well pronounced modifications of the nanoleyers surfaces, due to impact of the HCI ions, in the form of craters have been observed (Figure 1).

The results are consistent with previous experimental data and MD simulations for single ionized Xe and crystalline gold surface. We confirmed that the crater formation is due to HCI kinetic energy deposition, but we also found that role of HCI potential energy is very important. The results showed that kinetic energy has mainly influence on the crater depth, while potential energy of the HCI change the diameter of the molten region and thus the crater size at the surface. The results were qualitatively confirmed by preliminary 3D i-TS calculations [5].

*E-mail: d.banas@ujk.edu.pl

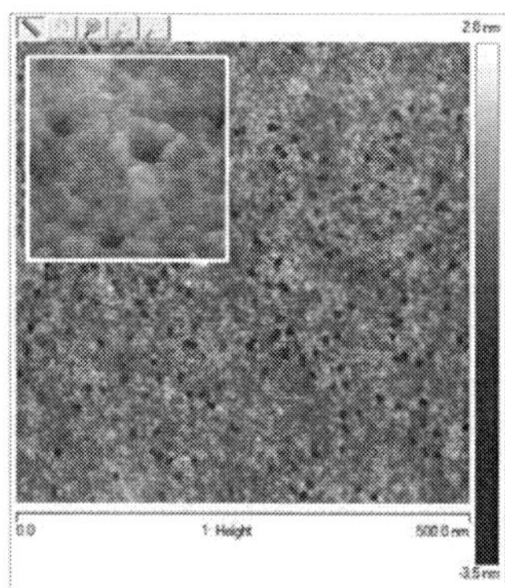

Figure 1. AFM images of Au nanolayer surface after irradiation by 360 keV Xe^{35+}. The inset shows enlarged picture of the nanostructures observed.

References

[1] Barth J V *et al* 2005 *Nature* **437** 671
[2] Aumayr F *et al* 2011 *J. Phys.: Condens. Matter* **23** 393001
[3] Banaś D *et al Nucl. Instr. Meth. B* **354** 125
[4] Stabrawa I *et al* 2017 *Nucl. Instr. Meth. B* **408** 235
[5] Dufour C *et al* 2017 *J. Phys.: Condens. Matter* **29** 095001

Journal of Physics: Conference Series **1412** (2020) 202025 doi:10.1088/1742-6596/1412/20/202025

Photoionization of endohedrals: complete semi-phenomenological approach

M Ya Amusia[1*] and L V Chernysheva[1]

[1]Racah Institute of Physics, the Hebrew University, Jerusalem 91904, Israel

Synopsis We developed a semi-phenomenological approach that permits to calculate the photoionization cross section of an endohedral A@C_N. We take, along with the static pseudopotential of atom A and fullerene C_N also the static and dynamic polarization of A and C_N by the incoming photons and outgoing electrons. The approach is general and applicable to any spherically-symmetric endohedrals. Concrete results are obtained for Ar@C_{60} and Xe@C_{60}. We take into account the multi-electron correlations in the frame of the random phase approximation with exchange (RPAE).

1. We have calculated photoionization cross sections of endohedral atoms A@C_N, for which developed an approach that is relatively simple and complete enough. We took into account the polarizability of the fullerene electron shell C_N that modifies the incoming photon beam and the alteration of the one-electron wave functions of the caged atom A under the action of C_N and atom A polarization potentials. Properly modified one-electron wave functions are initial in the account of multi-electron correlations in the frame of the random phase approximation with exchange (RPAE). As concrete objects, we have considered Ar@C_{60} and Xe@C_{60}.

2. The polarizability of the fullerenes C_N electron shell strongly affects the caged atom A photoionization cross section due to dynamic C_N polarization by the incoming photon beam. To study this effect, we assume the validity of a strong inequality $r_A \ll R$, instead of real simple one $r_A < R$, where r_A and R are the atomic and C_N radiuses, respectively. In the lowest order in $r_A / R \ll 1$ the effect of dynamic C_N polarization describes the photon frequency dependent factor $G(\omega) = \left[1 - \alpha_C(\omega)/2R^3\right]$, where $\alpha_C(\omega)$ is the C_N dynamic polarizability.

3. We took also into account the static polarization of C_N, using the following expression of the polarization potential U_C of fullerene C_N via its static dipole polarizability $\alpha_C \equiv \alpha_C(0)$: $W_C = -\alpha_C / 2(r^2 + R^2)^2$ for $r > R$ and for $r \leq R$ - $W_C = -\alpha_C / 8R^4$. We also took into account the polarization potential of the atom A $W_A = -\alpha_A / 2(r^2 + r_A^2)^2$, where α_A is the static atomic dipole polarizability. We describe the fullerenes shell by a Lorenzian potential $U_C(r) = -U_0 Rd / \left[(r-R)^2 + d^2\right]$, where d is its width, while $U_0 d / R$ is its depth. This potential represents the charge density distribution in C_N qualitatively better than the square well potential previously used by us and most of researchers in this area. Figure 1 depicts the RPAE cross-sections for $3p^6$ and $3s^2$ subshells of Ar and Ar@C_{60}. Subscripts WG, LG, LCAG denote the results that take into account along with factor G, the square well WG, Lorentz LG and polarization C(s) potentials of C_{60}, and the atomic potential A, entirely LCAG.

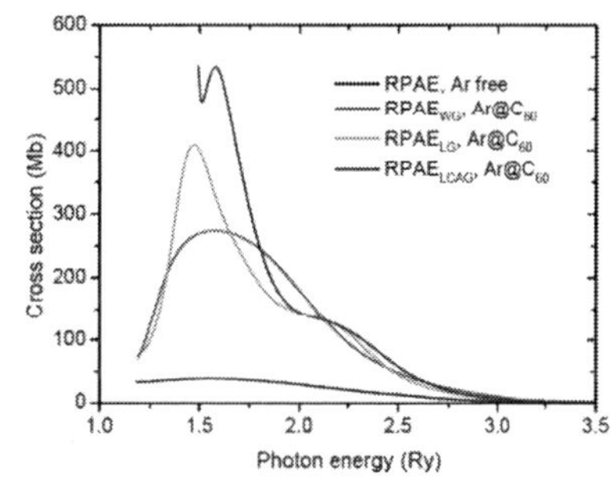

Figure 1. The RPAE photoionization cross-section of $3p^6$ and $3s^2$ subshells of Ar and Ar@C_{60}.

4. The results obtained demonstrate profound influence of static and dynamic polarizabilities at low photon energies. Obviously, the influence of the fullerene shell dies out, as it should be, with the photon energy growth. For more details and concrete results see [1]

References

[1] Amusia M and Chernysheva L 2019 *JETP Letters* **109** (in print)

* E-mail: amusia@vms.huji.ac.il

Revision of optical property of silicon by a reverse Monte Carlo analysis of reflection electron energy loss spectroscopy spectra

L H Yang[1], K Tőkési[2,3], J Tóth[2], B Da[4] and Z J Ding[1*]

[1]Hefei National Laboratory for Physical Sciences at Microscale and Department of Physics, University of Science and Technology of China, Hefei, Anhui 230026, P.R. China
[2]Institute for Nuclear Research, Hungarian Academy of Sciences (ATOMKI), Debrecen, Hungary, EU
[3]ELI-ALPS, ELI-HU Non-profit Ltd., Dugonics tér 13, H-6720 Szeged, Hungary
[4]Research and Services Division of Materials Data and Integrated System (MaDIS), National Institute for Materials Science (NIMS), 1-2-1 Sengen, Tsukuba, Ibaraki 305-0047, Japan

Synopsis The energy loss function (ELF) of silicon in a wide photon energy region (0-200 eV) was derived from reflection electron energy loss spectroscopy spectra with a theoretical analysis of the measured data. The accuracy of our result was justified by using the f- and ps-sum rules. Based on the new ELF, individual contributions of surface excitation and the bulk excitation to the REELS spectrum have been separated, and multi-scattering effect in the reflection electron energy loss spectroscopy spectrum has been studied in detail.

In recent years a well-established technique based on the reflection electron energy loss spectroscopy (REELS) has been developed to obtain optical constants in a wide range of electron energy loss. The REELS method does not require a complicated process for preparation of samples and the incident electron energy is usually around a few keV. However, the REELS spectrum usually contains not only bulk excitation but also surface excitation. To remove the surface excitation effect and multiple scattering effect from the REELS spectrum in data analysis, Da et al. [1] developed a reverse Monte Carlo (RMC) method for the derivation of the energy loss function (ELF) and thereby the dielectric function and optical constants of solids in a much wider photon energy range than that of the usual optical measurements [2,3]. The RMC method combines a Monte Carlo modelling of electron transportation for REELS spectrum simulation with a Markov chain Monte Carlo calculation of parameterized ELF. During the Monte Carlo simulations we used the Mott cross section with Thomas–Fermi–Dirac atomic potential to describe the electron elastic scattering, and the dielectric function theory for the description of the electron inelastic scattering processes [2]. The REELS spectra of Si were measured with primary electron energies at 3000, 4000 and 5000 eV in an energy-loss range of 0-200 eV. Then the RMC technique was applied to extract the ELF of Si. The reliability of the obtained data has been confirmed by applying the Thomas-Ritchie-Kuhn and the perfect-screening sum rules. Figure 1 shows that the agreement between the final simulated REELS spectra (blue line) and experimental REELS spectra (red dots) on the absolute intensity scale is excellent, covering the energy loss range from the elastic peak down to an energy loss of 100 eV, for the primary energy of 4000 eV. The inset of the figure shows the ELF obtained by the RMC method (blue line) in comparison with the results in databases of Palik (red circles) and Henke (black squares), which provides the revision of optical constants derived from optical measurements particularly around 10 eV.

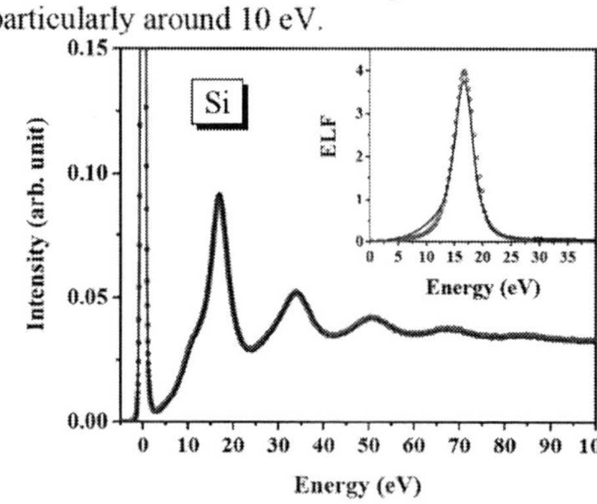

Figure 1. The REELS spectra and ELF of Si.

This work was support by the National Natural Science Foundation of China (No. 11574289) and by the National Research, Development and Innovation Office (NKFIH) Grant KH126886.

References

[1] B. Da *et al* 2013 *J. Appl. Phys.* **113** 21430
[2] H. Xu *et al* 2017 *Phys. Rev. B* **95** 195417
[3] L.H. Yang *et al* 2018 *Appl. Surf. Sci.* **456** 999

* E-mail: zjding@ustc.edu.cn

Photoionization and photofragmentation of singly charged positive and negative $Sc_3N@C_{80}$ endohedral fullerene ions

A Müller[1]*, M Martins[2], A L D Kilcoyne[3], R A Phaneuf[4], J Hellhund[1], A Borovik Jr[5], K Holste[5], S Bari[6], T Buhr[5], S Klumpp[6], A Perry-Sassmannshausen[5], S Reinwardt[2], S Ricz[7], K Schubert[6] and S Schippers[5]

[1]Institut für Atom- und Molekülphysik, Justus-Liebig-Universität Gießen, 35392 Giessen, Germany
[2]Institut für Experimentalphysik, Universität Hamburg, 22761 Hamburg, Germany
[3]Advanced Light Source, Lawrence Berkeley National Laboratory, Berkeley, CA 94720-8229, USA
[4]Department of Physics, University of Nevada, Reno, NV 89557, USA
[5]I. Physikalisches Institut, Justus-Liebig-Universität Gießen, 35392 Giessen, Germany
[6]Deutsches Elektronen-Synchrotron DESY, 22607 Hamburg, Germany
[7]Institute for Nuclear Research, Hungarian Academy of Sciences, 4001 Debrecen, Hungary

Synopsis Photoprocesses of the endohedral fullerene ions $Sc_3N@C_{80}^+$ and $Sc_3N@C_{80}^-$ in the gas phase have been experimentally investigated with emphasis on the specific effects of the encapsulated trimetallic nitride cluster Sc_3N on the observed reaction yields. Clear evidence of photoexcitation near the Sc L edge is obtained with the dominating contributions visible in the one- and two-electron-removal channels. The experiments are accompanied by model calculations employing density functional theory (DFT).

Employing the photon-ion merged-beams technique [1] at the ALS in Berkeley and at PETRA III in Hamburg, single-photon single and double ionization as well as single ionization accompanied by the release of a C_2 dimer were observed as a function of the photon energy for the positive $Sc_3N@C_{80}^+$ ion [2] and double detachment was measured for the negative $Sc_3N@C_{80}^-$ ion. Signatures of photoabsorption by atoms of the encapsulated Sc_3N cluster were expected and found near the absorption edges of scandium and nitrogen in the energy ranges 30 - 50 eV and 390 - 420 eV [3]. The main goal of this investigation was to study how the Sc_3N inner-shell photoabsorption features change under variation of the charge state q of the parent $Sc_3N@C_{80}^{q+}$ endohedral molecule. For comparison, results of previous measurements on photoabsorption by neutral $Sc_3N@C_{80}$ were included in the analysis [4].

The measured yield spectra of product ions arising from photoabsorption by the encapsulated Sc_3N cluster within $Sc_3N@C_{80}^{q+}$ at $q = -1, 0, +1$ exhibit very similar dependences on the photon energy. In particular, the resonance peaks found at different charge states q line up with one another. This finding is supported by DFT model calculations which show that the absorption cross sections of the encapsulated Sc and N atoms remain practically unchanged when the charge state q of the $Sc_3N@C_{80}^{q+}$ molecule is varied from $q = -1$ to $+4$ (see figure 1).

*E-mail: Alfred.Mueller@physik.jlug.de

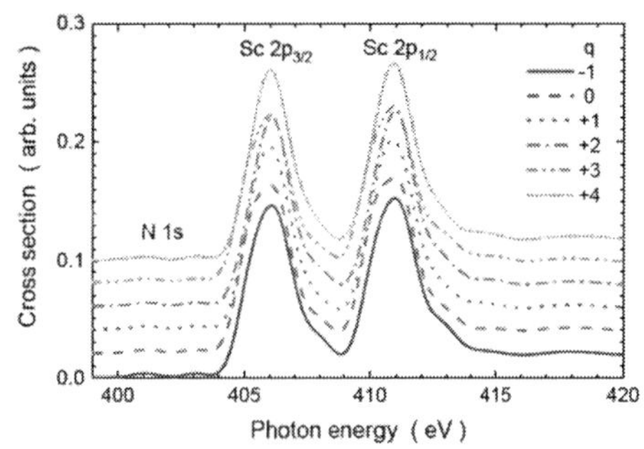

Figure 1. Relative photoabsorption cross sections of the $Sc_3N@C_{80}^{q+}$ molecule in different charge states q resulting from our model calculations. The results for charge states $q+1$ are vertically offset from those of q by 0.02 units ($q = -1, 0, +1, +2, +3$).

Spectra measured within this project across the C K edge at energies 280 - 390 eV [3] are remarkably similar to respective spectra obtained previously with $Lu_3N@C_{80}^{q+}$ ions [5].

References

[1] Schippers S *et al* 2016 *Contemp. Phys.* **57** 215
[2] Müller A *et al* 2017 *J. Phys.: Conf. Ser.* **875** 032033
[3] Müller A *et al* 2019, submitted to Phys. Rev. A
[4] Müller A *et al* 2007 *J. Phys.: Conf. Ser.* **88** 012038
[5] Hellhund J *et al* 2015 *Phys. Rev. A* **92** 013413

Characterizing crystalline defects in single Xe nanoparticles from angular correlations of single-shot diffracted X-rays

A Niozu[1,2]*, Y Kumagai[3], T Nishiyama[1,2], H Fukuzawa[2,3], K Motomura[3], M Bucher[4], Y Ito[3], T Takanashi[3], K Asa[1,2], Y Sato[1,2], D You[3], Y Li[3], T Ono[3], E Kukk[5], C Miron[6,7], L Neagu[7,8], C Callegari[9], M Di Fraia[9], G Rossi[10], D E Galli[10], T Pincelli[10], A Colombo[10], T Kameshima[11], Y Joti[11], T Hatsui[2], S Owada[2], T Katayama[11], T Togashi[11], K Tono[11], M Yabashi[2], K Matsuda[1], C Bostedt[4,12,13] K Nagaya[1,2] and K Ueda[2,3]

[1]Division of Physics and Astronomy, Kyoto University, Kyoto 606-8501, Japan
[2]RIKEN SPring-8 Center, Sayo, Hyogo 679-5148, Japan
[3]Institute of Multidisciplinary Research for Advanced Materials, Tohoku University, Sendai 980-8577, Japan
[4]Chemical Sciences and Engineering Division, Argonne National Laboratory, 9700 S. Cass Avenue, Argonne, IL 60439, USA
[5]Department of Physics and Astronomy, University of Turku 20014 Turku, Finland
[6]LIDYL, CEA, CNRS, Université Paris-Saclay, CEA Saclay, 91191 Gif-sur-Yvette, France
[7]Extreme Light Infrastructure - Nuclear Physiscs (ELI-NP), "Horia Hulubei" National Institute for Physics and Nuclear Engineering, 30 Reactorului Street, RO-077125 Măgurele, Jud. Ilfov, Romania
[8]National Institute for Laser, Plasma and Radiation Physics, 409 Atomistilor PO Box MG-36, 077125 Magurele, Jud. Ilfov, Romania
[9]Elettra - Sinchrotrone Trieste, 34149 Basovizza, Trieste, Italy
[10]Department of Physics, Universitá degli Studi di Milano, Via Celoria 16 - 20133 Milano, Italy
[11]Japan Synchrotron Radiation Research Institute (JASRI), Sayo, Hyogo 679-5198, Japan
[12]Laboratory for Femtochemistry, Paul-Scherrer Institute, CH-5232 Villigen PSI, Switzerland
[13]Institute of Chemical Sciences and Engineering, École Polytechnique Fédérale de Lausanne (EPFL), CH-1015 Lausanne, Switzerland

Synopsis We performed a wide-angle X-ray scattering experiment of single Xe nanoparticles using an X-ray free electron laser. We developed a novel analysis method that focuses on the angular correlation between plural Bragg spots in single-shot diffraction patterns. The angular correlations of the Bragg spots encode rich structural information and offer an evidence of twinning and stacking faults in Xe nanoparticles.

Characterizing crystalline defects in nanoparticles is of crucial importance for their applications in science and technology, since the defects potentially alter their desired properties. Single-shot X-ray diffraction with X-ray free electron lasers (XFELs) is a promising method to investigate the structure of nanoparticles, in which diffraction patterns of single nanoparticles are obtained with exposures to single XFEL pulses.

In this study, we carried out a wide-angle X-ray scattering experiment on Xe nanoparticles at an XFEL facility SACLA [1] in Japan. Single Xe nanoparticles (average radius $r \sim 65$ nm) were irradiated by 11.2-keV XFEL pulses, and the diffracted X-rays were recorded on a shot-by-shot basis with a multiport CCD sensor detector. The diffraction patterns contain Bragg spots correspoining to fcc{111}, {200}, {220} and hcp{101} reflections.

*E-mail: niozu.akinobu.85s@st.kyoto-u.ac.jp

In order to extract further structural information from the single-shot diffraction patterns, we developed an analysis technique that focuses on angular correlations [2] between plural Bragg spots in single-shot diffraction patterns. The analysis has revealed clear angular correlations between the Bragg spots, which provide rich structural information on the nanoparticles. We carried out a simulation to quantitatively evaluate the angular correlations and obtained a solid evidence of twinning and stacking faults in the Xe nanoparticles.

This study was supported by MEXT, JSPS KAKENHI, and IMRAM project.

References

[1] Ishikawa T *et al* 2012 *Nat. Photon.* **6** 540
[2] Mendez D *et al* 2016 *IUCrJ* **3** 420

Profile of energy-loss spectrum in collisions of ions with heteronuclear diatomic molecules

M Nakamura[1] * and A Ichimura[2]

[1]College of Science and Technology, Nihon University, Funabashi, 274-8501, Japan
[2]Faculty of Sciences, Tokyo University of Science, Shinjyuku-ku, Tokyo, 162-8601, Japan

Synopsis Profile of energy-loss spectrum in collisions of low velocity ions with heteronuclear diatomic molecules is theoretically investigated. With changing mass asymmetry in a target diatomic molecule, energy-loss spectrum is calculated with the classical trajectory calculation. Rotational excitation plays an essential role for energy-loss. The shape of the spectrum is found to show a drastic variation with mass asymmetry.

In large angle scattering of ions by molecules at the incident energy $E = 1$ eV – 1 keV, the energy transfer from translational to internal (rotational and vibrational) degrees of freedoms is caused by short-range repulsive forces. We have investigated the mechanism of energy transfer in collisions of ions with homonuclear diatomic molecules. Rotational excitations are found to be essential for the energy-loss processes [1, 2, 3].

We extend our study to the collision of ions with heteronuclear diatomic molecules. With changing mass asymmetry in the target molecule, we calculate a convoluted spectrum using classical trajectory method with applying a common interaction potential [2]. Calculation is performed for both rigid- and vibrating-rotor molecules. The asymmetry parameter α is given as $\alpha = (m_c - m_b)/(m_c + m_b)$, where m_b and m_c is the mass of two atoms constituting a molecule.

Figure 1 represents the intensity of Li^+ scattered by N_2 (α=0), CO (0.14), and BeF(0.36), as a function of a normalized energy-loss $\epsilon = \Delta E/E$ at $E = 27$ eV. As shown the figure, the difference in the spectrum between rigid-rotor and vibrating-rotor is small for all the cases. This finding indicates that the contribution of rotational excitation is essential for the energy-loss. When $\alpha = 0$, the spectra have two peaks; one corresponds to the elastic scattering and the other is due to impulsive scattering of ions by atoms in the target molecule. When $\alpha \neq 0$ an inelastic peak split into two peaks due to the difference in masses of atoms in the target. As the asymmetry is increases, one of the inelastic peaks overlaps with the elastic peak; the nearly elastic peek dominates the spectrum. Such a spectrum has been observed in He + HF collision [4].

*E-mail: nakamura.masato@nihon-u.ac.jp

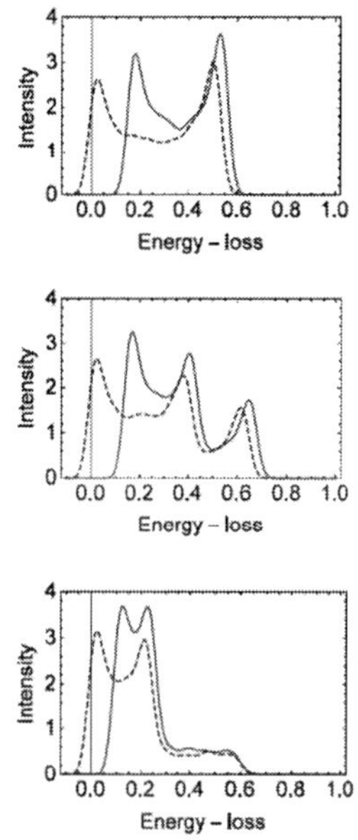

Figure 1. Energy loss spectra in the backward scattering of Li^+ by N_2 (upper), CO (middle) and BeF (lower) with the trajectory calculation at the collision energy E =27 eV. Solid and dashed curves represent the spectra for a vibrating rotor molecule and those for a rigid rotor molecule, respectively.

References

[1] Ichimura A and Nakamura M 2004 *Phys. Rev. A* **69** 022716
[2] Nakamura M and Ichimura A 2005 *Phys. Rev. A* **71** 062701
[3] Nakamura M and Ichimura A 2016 *Phys. Rev. A*, **94** 022717
[4] Gianturco F *et al.* 1985 *J. Chem. Phys.* **83** 1049

Deepening into the nucleation and fission processes of nano-hydrated ammonia clusters - a combined theoretical and experimental study

D Barreiro[1] *, B Oostenrijk[2], N Walsh[2], A Sankari[2], E P Månsson[3], S Maclot[4], S Sorensen[2], S Díaz-Tendero[1,5] and M Gisselbrecht[2]

[1]Department of Chemistry, Universidad Autonoma de Madrid, 28049, Madrid, Spain
[2]Department of Physics, Lund University, 22100, Lund, Sweden
[3]Attosecond Science group, DESY Photon Science Division, Schenefeld, Germany.
[4]KTH Royal Institute of Technology, Sweden
[5]IFIMAC, Universidad Autonoma de Madrid, 28049, Madrid, Spain

Synopsis While largely studied in the macroscopic scale, the dynamics leading to the nucleation and fission of atmospheric aerosols are very poorly understood at the nano or molecular scale. A model system consisting on ionized hydrogen-bonded ammonia and water molecules have been studied experimentally using mass- and 3D momentum spectroscopy and theoretically using ab initio molecular dynamics simulations.

The detailed mechanisms occurring at the molecular level in nucleation and fission processes have been reported to be relevant for the formation of atmospheric aerosols [1]. In the troposphere, cosmic radiation background is responsible for molecular ionization and have shown to be also crucial on the formation of these particles. As the concentration of anthropogenic ammonia in earth's atmosphere has increased in last years, our study model has consisted of multiply charged nano-sized clusters of hydrated ammonia, which may act as nucleation enhancers and therefore explain the particle formation rates of atmospheric aerosols containing, for example, sulfuric acid [2].

Advanced 3D momentum imaging mass spectrometry techniques have been implemented to uncover fundamental mechanisms related to charge and proton dynamics [3].The aggregates absorb a single soft X-Ray photon, resulting in the formation of multiple charged species. Thanks to our experimental set-up and selection of energies, we can focus on doubly charged clusters, the ones we are going to study in depth afterwards through ab initio molecular dynamics calculations. In our calculations, different cluster sizes of the type $(NH_3)_m(H_2O)_n$ have been taken into consideration. In order to find the most stable neutral conformers for these clusters, semi-classical calculations using Self-Consistent Charge Density-Functional Tight Binding based calculations (DFTB-SCC) were carried out [4] applying the Parallel Tempering Molecular Dynamics (PTMD) formalism [5]. DFTB in combination with Parallel Tempering Molecular Dynamics has demonstrated to contribute as an excellent tool to successfully explore the Potential Energy Surface (PES). These calculations were performed using specific water-ammonia interaction potentials developed by Cuny et al.[6]. To replicate the experimental conditions, a certain excitation energy, have been randomly redistributed among the nuclear degrees of freedom. Then, ab initio molecular dynamics calculations have been perform using the atom-centered Density Matrix Propagation (ADMP) method at a DFT/M062X level of theory.

Our results give an insight into the structure and charge location before and after ionization and also the mechanisms of the most probable fragmentation pathways. They support the role of ammonium as a charge carrier in the solution, preferentially bound to surrounding ammonia molecules, which could influence the atmospheric nucleation process.

*E-mail: dario.barreiro@uam.es

References

[1] Kulmala M *et al* 2013 *Science* **339** 943
[2] Almeida J *et al* 2013 *Nature* **502** 359
[3] Oostenrijk B *et al* 2018 *Phys. Chem. Chem. Phys.* **20** 932
[4] Elstner M *et al* 1998 *Phys. Rev. B* **58** 7260
[5] Oliveira L *et al* 2015 *Phys. Chem. Chem. Phys.* **17** 17079
[6] Cuny J *et al* 2018 *Mol. Simul.* **45** 249

Plasmons in graphene traversed by relativistic electronic beam with arbitrary trajectories

S Segui [1], J L Gervasoni[1,2,*], K Akbari[3], Z Mišković [3] and N Arista[2]

[1]Consejo Nacional de Investigaciones Científicas y Técnicas (CONICET), Argentina
[2]Comisión Nacional de Energía Atómica, Centro Atómico Bariloche, Argentina
[3]Department of Applied Mathematics, and Waterloo Institute for Nanotechnology, University of Waterloo, Waterloo, Ontario, Canada

Synopsis In this work we study the energy loss of a beam of relativistic electrons impinging on a graphene sheet under oblique incidence. Using several models for the electric conductivity of graphene, we cover a wide range of frequencies of the energy loss spectrum, from the terahertz to the ultraviolet. To describe this spectrum, we consider the contribution of electronic excitations as well as that of transition radiation.

In recent papers we have developed a fully relativistic description of the energy losses suffered by an energetic electron traversing perpendicularly a single layer of graphene (SLG) [1] as well as multilayered graphene stack (MLG) [2]. An extension of this work also considers the stopping of electrons travelling on parallel trajectories aloof a graphene layer [3].

In this presentation we develop the formalism for a generic oblique trajectory, and accordingly we consider a tensorial in-plane conductivity with different components along the longitudinal and the transverse directions with respect to the wave vector *k*.

The formulation is made for a general conductivity and we apply different models to describe both the low energy range (THz to mid-IR) and the UV range. In each case, we assess the contributions coming from radiative as well as non-radiative (ohmic) losses, identifying the longitudinal and transverse contributions [4]. In this way, we analyze their dependence of relevant parameters, like the electron velocity and the incidence angle.

Among other results, we find that transition radiation (TR) spectra in the THz range of frquencies exhibit strongly anisotropic patterns that may be readily distinghishable in catholuminiscence measurements in a STEM (see figure 1). We also show that the nonlocal effect in the graphene's conductivity is not important in the retarded regime. In the visible to the UV frequency range, our results demonstrate that strongly directional features arise in the in-plane distribution of the transverse contribution to the ohmic energy.

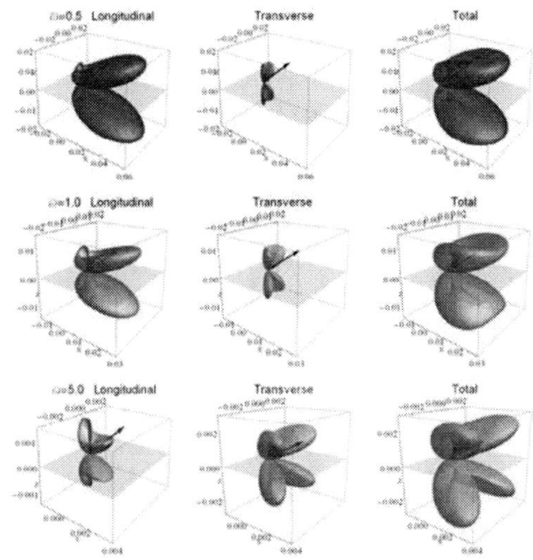

Figure 1. Longitudinal, transverse and total angular distributions of the spectral density for TR for three values of the reduced frequency $\omega=\omega/\omega_c$ with $\omega_c=\frac{1}{4}k_{TF}v_F^2/c$. Angle of incidence $\theta=60°$; velocity of the external particle $v=0.7c$.

References

[1] Mišković Z L, Segui S, Gervasoni J L and Arista N R 2016 *Phys. Rev. B* **94** 125414
[2] Akbari K, Mišković Z L, Segui S, Gervasoni J L and Arista N R 2017 *ACS Photonics* **4** 1980
[3] Mišković Z L, Akbari K, Segui S, Gervasoni J L and Arista N R 2018 *Nucl. Inst. Meth. B* **422** 18
[4] Akbari K, Mišković Z L, Segui S, Gervasoni J L and Arista N R 2018 *Phys. Rev. B* **98** 195410

* E-mail: gervason@cab.cnea.gov.ar
* E-mail: gervason@cab.cnea.gov.ar

Non-statistical fragmentation of C_{60} and the formation of endohedral defect fullerenes

M H Stockett[1]*, M Wolf[1], M Gatchell[1,2], H T Schmidt[1], H Zettergren[1] and H Cederquist[1]

[1]Department of Physics, Stockholm University, Roslagstullsbacken 21, SE-106 91 Stockholm, Sweden
[2]Institut für Ionenphysik und Angewandte Physik, Universität Innsbruck, Technikerstr. 25, A-6020 Innsbruck, Austria

Synopsis We report the first observation of the endohedral defect fullerene complex He@C_{59}^- and the first measurement of the threshold displacement energy of a fullerene.

The observation of fullerenes in various astronomical environments, and the identification of C_{60}^+ as the carrier of several Diffuse Interstellar Bands, has lead to a surge in interest in the formation and destruction pathways of fullerenes under different astrophysical conditions. Non-statistical fragmentation (*i.e.* prompt single-atom knockout) of fullerenes and other large molecules following collisions with energetic H and He atoms in supernova shocks is considered a key destruction mechanism [1].

We have investigated this process through laboratory experiments and classical Molecular Dynamics (MD) simulations of C_{60}^-+He collisions at center-of-mass energies of 20–80 eV. As the electron affinity of C_{60} is much lower than any of its dissociation energies, collisions depositing enough energy to induce statistical unimolecular dissociation most likely lead to electron loss and thus do not contribute to the measured negative ion product mass spectrum, eliminating the major source of background.

Non-statistical fragmentation is observed directly by detecting C_{59}^-, see Figure 1. We observe for the first time the endohedral defect fullerene complex He@C_{59}^-. Secondary decay products of these ions, C_{58}^- and He@C_{58}^-, are also observed. The edohedral character is confirmed by comparison to C_{60}^-+D_2 collisions (Figure 1). We measure the threshold center-of-mass energy for knocking out a single carbon atom from C_{60} in collisions with He. Combining this with our MD simulations, we determine a semi-empirical value for the (projectile-independent) threshold displacement energy, the minimum energy needed to remove a single carbon atom from the C_{60} cage [2]. Our value, 24.1±0.5 eV, is much higher than generally assumed previously for fullerenes (around 15 eV), and is similar to that determined for graphene and Polycyclic Aromatic Hydrocarbons [3].

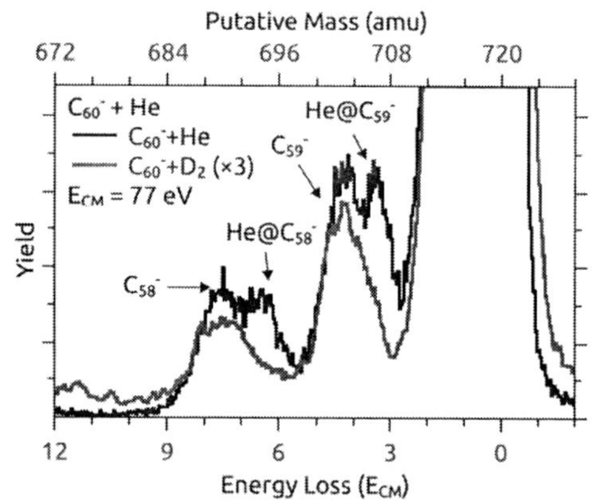

Figure 1. Negatively charged products of C_{60}^-+He/D_2 collisions.

References

[1] Micelotta E R, Jones A P and Tielens A G G M 2010 *Astron. Astrophys.* **510** A36
[2] Stockett M H *et al* 2018 *Carbon* **139** 906
[3] Stockett M H *et al* 2015 *J. Phys. Chem. Lett.* **6** 4504

*E-mail: Mark.Stockett@fysik.su.se

 1412 (2020) 202033 doi:10.1088/1742-6596/1412/20/202033

Modelling interactions of alkali-cation dimers in He clusters

N Alharzali[1,2], H Berriche[3], P Villarreal[1] and R Prosmiti[1]*

[1]Institute of Fundamental Physics (IFF-CSIC), CSIC Serrano 123, 28006 Madrid, Spain
[2]Laboratory of Interfaces and Advanced Materials, Faculty of Science, University of Monastir, 5019 Monastir, Tunisia
[3] Department of Mathematics and Natural Sciences, School of Arts and Sciences, American University of Ras Al Khaimah, RAK, P.O. Box 10021, UAE

Synopsis Doped He droplets constitute an important environment as ultracold homogeneous matrices for spectroscopic studies. Alkali metal dopants have a series of interesting properties due to their unusual bonding behavior. Neutral atoms attached to the surface of the droplet, forming eventually cold molecules via collisions, while charged dopants form solvation shells, related with the low mobility of the ions in experiments. So, investigations on the underlying intermolecular, markedly orientational ionic, forces, and on the miscoscopic structures of alkali-cation-He complexes is essential.

We present a detailed theoretical investigation on intermolecular interactions of cationic alkali-dimers solvated in small He clusters [1, 2]. The structural characteristics and stabilities of such cations are determined from high level *ab initio* electronic structure calculations. In particular, interaction energies are obtained from CCSD(T)/CBS and MRCI+Q calculations, depending on the single- or multi-reference character of the system under study. In turn, the potential energy surfaces are constructed employing an interpolation technique within the Reproducing Kernel Hilbert Space (RKHS) method, and were used to evaluate the trimers' nuclear quantum effects. By analyzing the potential anisotropy and the spatial arrangement of the ground and excited vibrational states, we found that energetically the most stable conformer corresponds to linear arrangements for all complexes studied.

Further, we have looked into the behavior of higher-order species with more He atoms surrounding the cationic dopant. A simplified sum of the three-body *ab initio*-based potential approach is employed to describe the microsolvation of the $A_2^+(^2\Sigma)$ molecular ions (A=Li, Na, K, Cs, and Rb) in He clusters by analyzing the structural stability of clusters with up to six He atoms in comparison with interactions energies obtained from quantum chemistry methods [1]. The optimized structures are analyzed in terms of their geometric features, confirming the dominance of ionic forces in driving the shape of the cluster, and their crucial role by changing the competitive strength of the cation-He interaction with respect to the He-He (pure solvent) network of interactions. Such interplay creates collocations, like the equilateral He_3 motifs (see figure) at the two ends of each ionic dimer, that provides some initial information on solvation process at miscroscopic level.

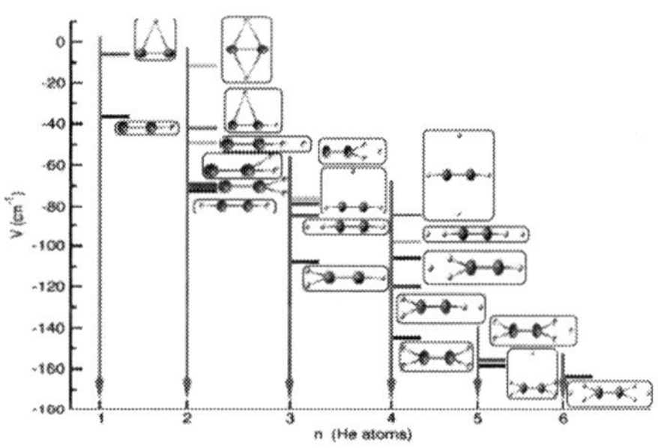

Figure 1. Optimal structures for the indicated He_n-A_2^+ clusters, where A=Li, Na, K, Cs, or Rb alkali-metal.

References

[1] Alharzali N *et al* 2019 *J. Comput. Chem.* submitted

[2] Alharzali N *et al* 2019 *Inter. J. Quant. Chem.* submitted

*E-mail: rita@ff.csic.es

Simulations of proton beam propagation in matter using entropic moment method

E Olivier[1]*, T H Nguyen-Bui[1]†, C Champion[1] and B Dubroca[2]

[1]CELIA, Université de Bordeaux – CNRS – CEA, 33405 Talence, France
[2]LCTS, CNRS – CEA – Safran, 33600 Pessac, France

Synopsis Atomic interactions of ions with the penetrated medium lead to a strong evolution of ion charge-state fractions as a function of the target thickness. In this paper, we present a fast and accurate method in order to simulate the slowing-down of ions in matter. Our model relies on the averaging of the transport equation leading to fast simulations of the deposited dose by particles beam. Simulations have been performed for the case of proton beams propagating through water. Results have been compared with Monte Carlo simulations.

Quantitative information on the penetration of charged ions through matter, in particular the energy loss, is of considerable interest in basic science, in medicine and in technology, and therefore has been studied through various methods [1]. Until the middle of the past century, studies of charged-particle penetration were stimulated almost exclusively by the needs of fundamental physics research, but applications in other areas gradually became important. Simulations of slowing-down of ion beams in thick targets aims at characterizing the spatial and energy distributions of incident particles and secondary fragments in matter. In particular, biological effects are due to energy deposition, which may be accurately calculated with Monte Carlo codes, but these are limited by the amount of needed statistics. Moreover, solving directly the system of linear transport equations for the motion of particles is computational time consuming, and thus nowadays incompatible with practical applications.

To overcome this limitation, a new code has been developed at CELIA [2]. This code resolves the linear Boltzmann equation using moment method. It consists in averaging the distribution function over the angles to obtain these moments. An entropy minimization theorem is used as closure relation for superior moments (M_2). Also, the M_1 model has been established. The equation system also obtained has been discretized by using a relaxation scheme [3]. This code is validated and exploited for electron and photon transport [2].

*E-mail: erwan.olivier@u-bordeaux.fr
†E-mail: thanh-ha.nguyen-bui@u-bordeaux.fr

Our objective is to extend it for heavy ion transport. For this paper, we focus on proton beams in matter. One of the difficulties of this study is to determine the stopping power of ions in any matter. By using theoretically-calculated cross sections for proton-matter interaction processes, we determine stopping power and dose, as well as properties such Bragg peak. This one is in good agreement with Monte Carlo simulations. Many test-case results will be presented in comparison with FLUKA [4], for 1D to 3D cases (see *e.g.* Fig. 1).

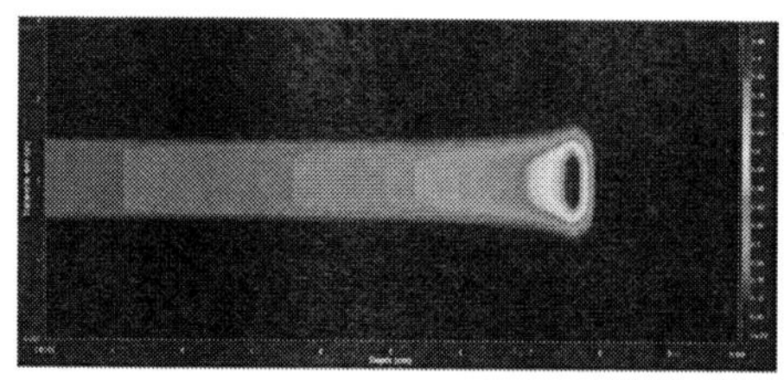

Figure 1. Deposited dose in water by 100 MeV protons, for a 2D simulation. Results are normalized to the maximum value.

The treatment of the physical processes required in the case of heavy ions (helium and carbon ions) is underway.

References

[1] Närmann A and Sigmund P 1994 *Phys. Rev. A* **49** 4709
[2] Caron J *et al* 2015 *Physica Medica* **31** 8 912-921
[3] Pichard T *et al* 2015 *Commun. Comput. Phys.* **19** 1 168-191
[4] Böhlen T *et al* 2014 *Nucl. Data Sheets* **120** 211-214

IRMPD spectroscopy and QC calculations on functionalized gold nanocluster ions

C Desfrançois[1], A Beydoun[1*], F Lecomte[1], N Nieuwjaer[1], B Manil[1] and D Scuderi[2]
[1]CNRS, Laboratoire de Physique des Lasers, Université Paris 13, F-93430 Villetaneuse, France
[2]CNRS, Laboratoire de Chimie Physique d'Orsay, Université Paris Sud, 91400 Orsay, France

Synopsis Many studies are devoted to functionalized metallic nanoparticles, for many applications: in Physics, due to their localized surface plasmon resonances; in Chemistry, due to their specific catalytic properties that depend on their surface nature; in Biology, due to their optical or magnetic properties coupled to their potential for targeting and vectorization of bioactive molecules within living cells. In this last domain, gold and silver nanoparticles are especially of interest and, for the control of their biological effects, it is very important to have a fine knowledge of the structural properties and the chemical stabilities of their functional ligands.

Indeed, many studies are devoted to the characterization of the nanoparticle – ligand interface, especially with NMR or IR methods [1]. However, data interpretations are somewhat difficult because of the relative lack of control in solution of the nanoparticle stoichiometry and of their ligand coverage.

With a bottom-up approach, we here report on a work in progress to study sub-nanometer metallic nanoclusters (few to few tens of metal atoms, few tens to few hundreds of ligand atoms) that allow for a size selection with mass spectrometry and hopefully for a precise characterization in the gas phase. Even if these functionalized nanoclusters do not display the same structural characteristics as the larger nanoparticles, their specific chemical-physical properties are more and more studied because they are very dependent from their size and structure and they are intermediate between those of the isolated metal – ligand and those of the nanoparticles in solution [2][3][4].
Moreover, the recently developed production methods for the synthesis of size-controlled functionalized nanoclusters in solution allow for efficient mass spectrometric experiments with rather high ion currents on a selected mass, either using FAB, MALDI or ESI techniques [5][6]. It is thus possible to consider now IRMPD experiments directly on these mass-selected ion species and to use these nanoclusters as model systems for understanding the effects of structures and of metal – ligand interactions. To our knowledge, this kind of experiments has been only very limited to few organometallic compounds containing only one metal atom [7][8][9] but, up to now, nothing has been published on IR spectroscopy studies of mass-selected functionalized gold or silver nanocluster ions in the gas phase.

First IRMPD spectroscopy experiments and Quantum Chemistry calculations have deal with gold clusters protected by phosphine ligands (Nanoprobes) and gold-sulfur hybrid clusters [10] (Dugourd's team). Preliminary results with their successes and difficulties will be presented in this contribution.

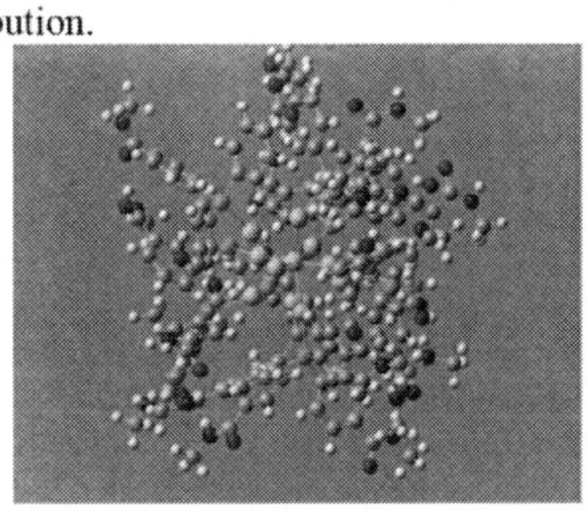

Figure 1. Gold – glutathion cluster anions synthesized in Philippe Dugourd's group.

References

[1] Zhang B and Yan B 2010 *Anal. Bioanal.Chem.* **396** 973
[2] Shang L *et al* 2011 *Nanotoday* **6** 401
[3] Lu Y and Chen W 2012 *Chem. Soc. Rev.* **41** 3594
[4] Russier-Antoine I *et al* 2016 *Nanoscale* **8** 2892
[5] Lu Y and Chen W 2015 *Anal. Chem.* **87** 10659
[6] Daly S *et al* 2015 *Phys. Chem. Chem. Phys.* **17** 25772
[7] Simon A *et al* 2004 *J. Am. Chem. Soc.* **126** 11666
[8] Butler M *et al* 2014 *J. Phys. Chem.* A **118** 4942
[9] Hopkins W S 2015 *Mol. Phys.* **113** 3151
[10] Bertorelle F *et al* 2017 *J. Phys. Chem. Lett.* **8** 1979

* E-mail: ali.beydoun@univ-paris13.fr

Diffraction of 80 eV hydrogen through suspended graphene

C Brand[1], **T Susi**[2], **J Kotakoski**[2] **and M Arndt**[1]
M Debiossac[1,3], **F Aguillon**[3] **and P Roncin**[3]*

[1]University of Vienna, Faculty of Physics, VCQ, Boltzmanngasse 5, A-1090 Vienna, Austria.
[2]University of Vienna, Faculty of Physics, PNM, Boltzmanngasse 5, A-1090 Vienna, Austria
[3]Institut des Sciences Moléculaires d'Orsay, CNRS, Univ. Paris-Sud, Université Paris-Saclay, Orsay, France

Synopsis We have investigated theoretically the possibility to diffract hydrogen atoms through a suspended graphene single layer. Using quantum and semi classical approaches we evaluate the momentum and energy exchange to the electronic and vibrational system and estimate their influence on the coherence and spot size.

Classical trajectories of H-graphene deflection have been determined using molecular dynamics (MD) based either on TDDFT or, more extensively, to DFT potential [1]. Both calculations yield generally similar results : a barrier of $3.8 \pm 1.3eV$ for atomic hydrogen to penetrate graphene at the center of the hexagonal ring, and a sputterization threshold of $82.5 \pm 2.5eV$.

The H-graphene potential energy landscape is approximated as being proportional to the electrostatic potential of the graphene sheet with a coefficient fitted to explicit DFT potential along the axes through the center of the hexagon.

Diffraction pattern has been determined within the eikonal approximation, whose validity has been shown to be excellent at the considered energy ($80eV$) by comparison with an exact 3D wavepacket calculation. When no decoherence effect is taken into account, a perfectly periodic hexagonal graphene layer gives rise to point like diffraction spots with an intensity distribution dominated by the Airy disk associated with the size of the central ring and elongated branches associated with the reduced repulsion perpendicular to the $C-C$ bound.

The main source of decoherence is expected to arise from the momentum spread induced by thermal movement of the lattice (Debye Waller factor). The diffraction pattern is calculated using a super cell of 256 unit cells distorted by the thermal displacements and convoluted by the beam parameter. The diffraction spot now acquires an additional broader contribution with a width increasing with the distance to the specular peak. The decoherence due to electronic excitation was estimated from TDDFT. For central trajectory, it indicates an electronic energy of $\sim 3.5eV$, significant but lower than the $\sim 4.7eV$ of the lowest in-plane plasmon. This later has a wavelength of $\sim 7nm$, encompassing hundreds of lattice cells and is expected to have a limited contribution to decoherence.

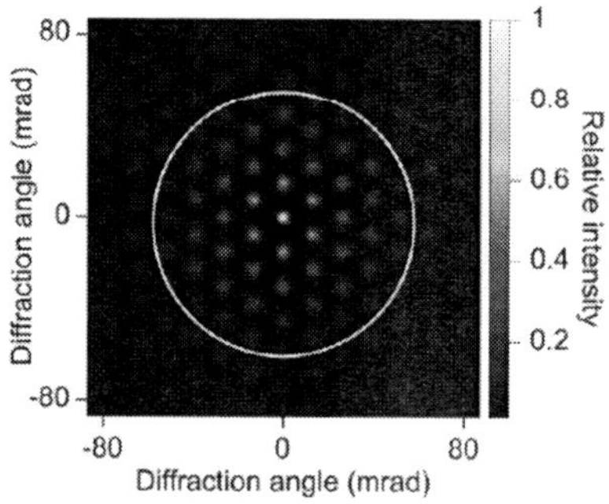

Figure 1. Diffraction pattern including decoherence effects. Around 8% of the atoms are diffracted by less than 52 mrad (white circle). The predicted signal-to-background ratio reaches up to 140.

The study suggests the possibility of constructing an ultra compact interferometer. Graphene technology is well advanced and slit with mono atomic thickness have already been demonstrated [2]. If space budget is less stringent, diffraction at crystal surfaces under grazing incidence can also provide a beam splitter with comparable deflection power and very limited decoherence [3] as the projectile is reflected above the surface with a reduce Debye-Waller factor and electronic excitation can be quenched by using wide band-gap insulator.

References

[1] Brand C *et al* 2019 *New J. Phys.* **accepted**
[2] Brand C *et al* 2015 *Nat. Nano.* **10** 845
[3] Debiossac M *et al* 2014 *Phys Rev lett* **112** 023203

*E-mail: philippe.roncin@u-psud.fr

Journal of Physics: Conference Series **1412** (2020) 212001 doi:10.1088/1742-6596/1412/21/212001

Energy selective time of flight spectrometer to study low energy dissociation channels of PAHs

M V Vinitha [1*] **and U Kadhane**[1]

[1]*Department of Physics, Indian Institute of Space Science and Technology, Thiruvanathapuram-695 547, India*

Synopsis Dissociation channels induced by sequential UV multiphoton absorption of vapour phase fluorene cation is investigated using energy selective ToF mass spectrometer. Neutral fluorene molecule is ionized by resonantly enhanced multiphoton absorption and dissociated in the same photon field. The main goal of this experiment is to determine the rate constants of statistical dissociation channels of PAH cations for a wide range of internal energies.

The polycyclic aromatic hydrocarbons (PAHs), their radical cations are possible carriers of diffuse interstellar bands. The infrared spectrum of most stellar objects is characterized by broad IR emission features that resembles the spectroscopic signatures of PAHs [1].These molecules are efficient absorbers of UV radiation. They excite to higher electronic states and undergo rapid internal conversion to the ground state, but stays vibrationally hot. Two major possible channels for the de-excitation of these highly vibrationally excited molecules are IR emission or fragmentation. However, in connection with the stability of these PAHs in the interstellar medium,fragmentation studies must be considered. It is important to study photodissociation rate constant and branching ratio as a function of internal energy. Usually Vacuum Ultraviolet source is employed for dissociation rate measurments, This technique generally offers dissociation rate constant values located at well above the dissociation thresolds. But UV multiphoton absorption technique is well suited for dissociation at lower energies

.

We have developed an in-house energy selective Time of Flight (ToF) mass spectrometer at molecular research lab at IIST. This instrument is built inspiring from ion-neutral coincidence spectrometer at laboratory of Atomic and Molecular Collisions (CNRS), University Paris-Sud XI [2]. Internally heated PAH ions are produced by sequential absorption of two or three UV photon (225-275nm). Using an electrostatic energy analyser accompanied by a position sensitive detector, the kinetic energy of the incoming fragment ions is measured along with their time of flight. If neutral product carry sufficient energy, it can be detected at the end of the ToF tube using an alternate detector and hence a complete detection of fragmentation products is possible. This can be used to asses the details of fragmentation mecanism.

Using a complex effect of ToF and residual energy of fragement ion, decay time constants are evaluated ranging across several orders of magnitudes in a single experiment. The 2D correlation diagram of the ToF and energy can only be analysed with the help of a very intricate Monte Carlo simulation code. It can simulate the molecular decay for the electrical and mechanical configuration of the spectrometer for various decay constants. The experiemntal data for H loss chnnel of smaller PAH ions is used to match with the simulation for a certain decay rate, hence those values are extracted out as H loss decay rate. This is repeated for a range of photon energies to study photodissociation dynamics and stability of PAH ions. Very consistant decay rate constants are measured for small PAHs at multiple wavelengths

References

[1] Tielens A G G M 2008 *Annu. Rev. Astron. Astrophys.* **46** 289

[2] Lepere V *et al* 2017 *J. Chem. Phys.* **8** 134313

* E-mail: vinithamvv@gmail.com

 1412 (2020) 212002 doi:10.1088/1742-6596/1412/21/212002

Ionization of pyrimidine molecule induced by proton

M A Quinto[1] *, J M Monti[1] and R D Rivarola[1]

[1]Instituto de Física Rosario, CONICET and Universidad Nacional de Rosario, Rosario, S2000EKF, Argentina

Synopsis In the present study will be investigated the ionization of isolated pyrimidine molecules induced by ions. The total and differential cross section have been computed with a quantum mechanical method within the continuum distorted wave-eikonal initial state (CDW-EIS) approximation. A comparison with the available data is presented.

The studies of the ionization processes induce by ion on molecules of biological interest has a great importance in radiobiology and medicine. In this context, the electron emission processes induced by charged particles on DNA components – including the nucleobases as well as the sugar-phosphate group – have been extensively studied on both experimental and theoretical sides [1,2]. However, some secondary biological molecules such as purine and pyrimidine have recived a particular interest of study since they are considered as a precursor of some DNA/RNA nucleobases (namely cytosine, thymine and uracil). Thus, pyrimidine has become a target of interest for radiobiology and it has been the subject of numerous electron and proton studies [3,4].

In the present work, we study the ionization of a isolated pyrimidine molecule induced by ions with a quantum mechanical framework within the CDW-EIS approximation [5]. A full set of total and differential cross sections will be present during the conference. A comparison with available experimental results will be made in every possible case (see Figure 1). The orbitals of the pyrimide molecular targer are described by linear combinations of their atomic compound orbitals (LCAOs). The corresponding binding energies and effective occupation electron analysis of each molecular orbital are obtained by using the GAUSSIAN09 software at the RHF/3-21G.

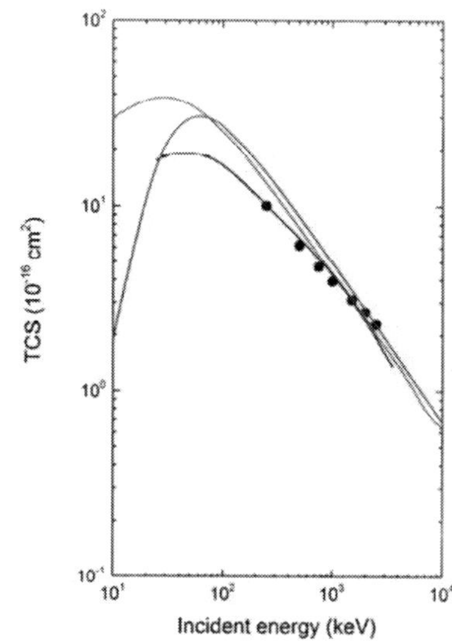

Figure 1. Total cross sections for single ionization of pyrimidine induced by proton. Theroretical calculations solid lines: CDW-EIS (red) and FBA (magenta) [3] . Semi-empirical results (blue) and experimental data are token from [4].

References

[1] Galassi M E *et al* 2012 *Phys. Med. Biol* **57** 2081
[2] Champion C *et al* 2012 *Phys. Med. Biol* **57** 3039
[3] Champion C *et al* 2015 *Eur. Phys. J. D* **69** 127
[4] Wolff W *et al* 2014 *J. Chem. Phys.* **140** 064309
[5] Monti J M *et al* 2014 *Appl Radiat Isot.* **83** 105–108

* E-mail: quinto@ifir-conicet.gov.ar

Resonance electron interaction with heterocyclic compounds: vibrational Feshbach resonances and hydrogen atom stripping

S A Pshenichnyuk[1 *], **I I Fabrikant**[2†], **A Modelli**[3,4], **S Ptasińska**[5,6] **and A S Komolov**[7]

[1]Institute of Molecule and Crystal Physics, Russian Academy of Sciences, Ufa, 450075 Russia
[2]Department of Physics and Astronomy, University of Nebraska-Lincoln, Lincoln, NE 68588, USA
[3]Università di Bologna, Dipartimento di Chimica "G. Ciamician", 40126 Bologna, Italy
[4]Università di Bologna, Centro Interdipartimentale di Ricerca in Scienze Ambientali, 48123 Ravenna, Italy
[5]Radiation Laboratory, University of Notre Dame, Notre Dame, IN 46556, USA
[6]Department of Physics, University of Notre Dame, Notre Dame, IN 46556, USA
[7]St. Petersburg State University, St. Petersburg 199034, Russia

Synopsis We present studies of low-energy resonance electron attachment in a series of 5-membered heterocyclic rings under gas-phase conditions by means of electron transmission spectroscopy and dissociative electron attachment spectroscopy.

Studies of dissociative electron attachment (DEA) to biorelevant molecules representing building blocks of living matter allows us to disclose the basic mechanisms of interaction of high energy radiation with living tissues [1-3].

In the present paper we study low-energy (0-15 eV) resonance electron attachment in a series of 5-membered heterocyclic rings (isoxazole, imidazole, pyrazole, pyrrole, 1-methyl- and 2-methylimidazole) under gas-phase conditions by means of electron transmission and DEA spectroscopies.

Sharp features, with fwhm less than 0.1 eV, observed in imidazole, pyrazole and pyrrole close to 0.45 eV, i.e. quite below the energy of their lowest-lying $\pi*$ shape resonances detected at 1.90, 1.87 and 2.33 eV (Figure 1), respectively, are associated with formation of diffuse negative ion states bound by long-range forces. The effective range theory (ERT) calculations support this interpretation. Since the permanent dipole moments of pyrazole and pyrrole are below the supercritical (2.5 D), the polarization interaction plays an important role in formation of these states.

In addition to the general observation of cleavage of the N-H bond at incident electron energy close to 2 eV, elimination of as much as three hydrogen atoms from the molecular negative ions are detected at higher energies with the only exception of methylated imidazoles. This complex process is associated with ring opening and formation of diatomic hydrogen as one of the neutral fragments that is the only way to satisfy the energetic conditions.

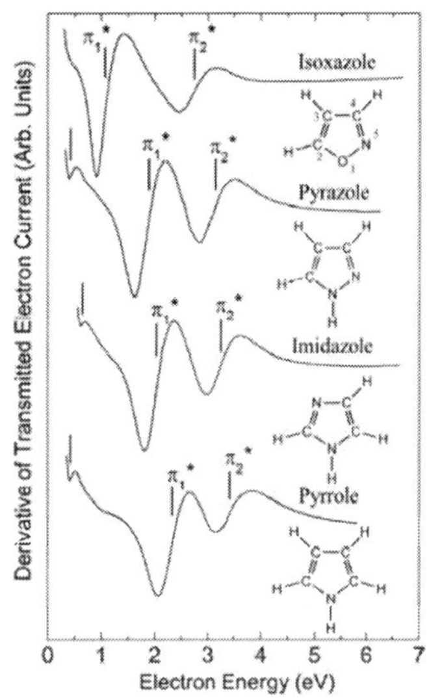

Figure 1. The first derivative of electron current transmitted through the gas-phase samples as a function of incident electron energy.

References

[1] Boudaiffa B *et al* 2000 *Science* **287** 1658
[2] Fabrikant I I *et al* 2017 *Adv. At. Mol. Opt. Phys.* **66** 545
[3] Gorfinkiel J D and Ptasinska S 2017 *J. Phys. B: At. Mol. Opt. Phys.* **50** 182001

* E-mail: sapsh@anrb.ru
† E-mail: ifabrikant@unl.edu

Simulation of single cell irradiation by proton beams using *TILDA-V*

M E Alcocer-Ávila[1] *, M A Quinto[2], J M Monti[2], P F Weck[3], O A Fojón[2], R D Rivarola[2] and C Champion[1] †

[1]Centre Lasers Intenses et Applications, Université de Bordeaux – CNRS – CEA, F-33400 Talence, France
[2]Instituto de Física Rosario, CONICET and Universidad Nacional de Rosario, Rosario, S2000EKF, Argentina
[3]Sandia National Laboratories, Albuquerque, NM 87185, USA

Synopsis The purpose of the present work is to emphasize the need for realistic modeling of the biological medium when studying the transport of protons in living matter at the very small scale. In this context, we have used the *TILDA-V* Monte Carlo track-structure code to simulate the irradiation of single cells with proton beams over the energy range of 10 keV – 100 MeV.

Radiation transport simulations are of great importance in various fields, such as radiobiology, medical physics and radiation protection. In this context, Monte Carlo codes are extremely useful tools to predict the absorbed dose in matter. However, when the dimensions of the system under study become smaller than the incident radiation track length, an event-by-event (or track-structure) approach is warranted to correctly determine the energy deposition in the target.

TILDA-V[1] is a Monte Carlo track-structure code based on a set of multiple differential and total cross sections, calculated within the quantum mechanical framework, for describing the main inelastic processes during the slowing-down of protons in water and DNA.

We report in this work the application of *TILDA-V* to a radiation microdosimetry case-study in which an individual cell was irradiated by protons arriving from random directions. The target was modeled as three concentric spheres representing the basic regions of a cell, i.e. the nucleus, the cytoplasm and the cell membrane. We assumed that the cytoplasm and the cell membrane consisted only of water with a density of 1 g cm^{-3}. On the other hand, to study the variation in the radiation dose deposited in the cell nucleus depending on its composition, three different simulations were carried out for each incident proton energy, assuming the nucleus contained: 1) water, 2) water with the density rescaled to 1.29 g cm^{-3} and 3) hydrated DNA.

Figure 1 shows the dose deposited in the cell nucleus depending on the proton incident energy and the composition of the nucleus.

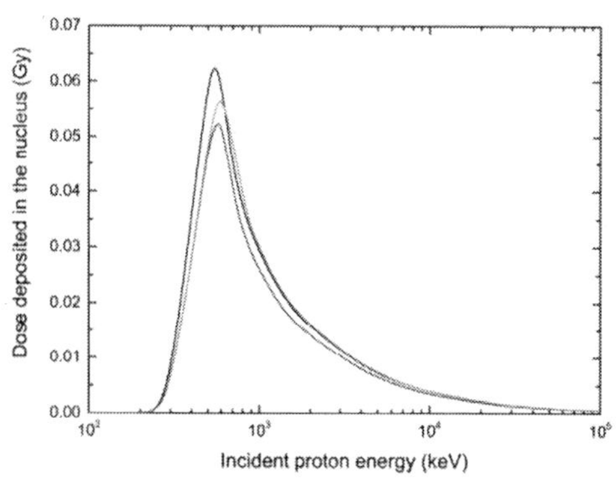

Figure 1. Dose deposited in the nucleus of a single cell as a function of the incident proton energy. The medium was modeled as water (blue line), water with a density rescaled to 1.29 g cm^{-3} (green line) and hydrated DNA (red line).

SNL is managed and operated by NTESS under DOE NNSA contract DE-NA0003525.

References

[1] Quinto M A *et al* 2017 *Eur. Phys. J. D* **71** **130**

*E-mail: mario-enrique.alcocer-avila@u-bordeaux.fr
†E-mail: christophe.champion@u-bordeaux.fr

Screened independent atom model calculations for electron emission from biologically relevant molecules by proton impact

H J Lüdde[1] *, **M Horbatsch**[2] **and T Kirchner**[2] †

[1]Frankfurt Institute for Advanced Studies (FIAS), D-60438 Frankfurt, Germany
[2]Department of Physics and Astronomy, York University, Toronto, Ontario, M3J 1P3, Canada

Synopsis Screened independent atom model calculations for the net ionization of complex biomolecules bombarded by energetic protons are presented. Scaling relations are examined and a simple parametrization of the cross section results in terms of an analytical formula is suggested.

The growing interest in collisions involving large biomolecules, coupled with the complexity of the problem, calls for the development of simplified theoretical models for the calculation of charge-changing cross sections. In recent work, we have introduced a variant of an independent atom model (IAM) which is easily applicable to a wide range of systems [1, 2]. It is based on a geometrical interpretation of an ionization or electron capture cross section as an effective area composed of overlapping circular disks. The latter represent the cross sections of the atomic constituents of the molecule of interest and are calculated using a well-tested density-functional-theory framework. The effective area calculation is carried out using a pixel counting method (PCM).

If the atomic cross sections are small and do not overlap the IAM-PCM results coincide with standard independent-atom-model additivity rule (IAM-AR) cross sections. However, at not too high impact energies atomic cross sections for ionization (and electron transfer) tend to be large and the overlap reduces the IAM-PCM cross section significantly compared to the IAM-AR result.

This effect is demonstrated in Fig. 1 for proton collisions from the two amino acids alanine and leucine. In addition to the IAM-AR and IAM-PCM cross sections the figure displays AR results from Ref. [3], in which experimental data for small molecular building blocks were used to assemble the cross sections of the amino acids. Not surprisingly, our two IAM calculations bracket these independent-molecule-model (IMM)-AR data and merge with them at high impact energies.

At the conference we will present IAM-PCM net ionization cross sections for a large class of systems including amino acids and nucleotides. We will discuss scaling relations with respect to the number of (active) electrons and suggest a parametrization of our results in terms of a simple analytical formula.

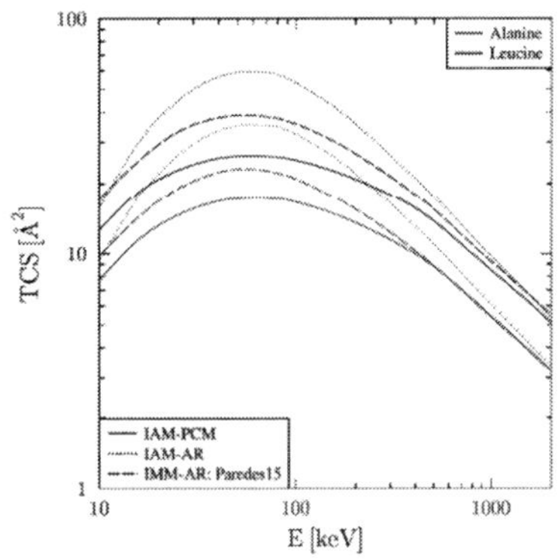

Figure 1. Total net ionization cross sections for proton collisions with the amino acids alanine ($C_3H_7NO_2$) and leucine ($C_6H_{13}NO_2$) as functions of impact energy. IAM-PCM and IAM-AR: present calculations; IMM-AR from [3].

References

[1] Lüdde H J *et al* 2016 *Eur. Phys. J. D* **70** 82
[2] Lüdde H J *et al* 2018 *Eur. Phys. J. B* **91** 99
[3] Paredes S *et al* 2015 *Eur. Phys J. D* **69** 178

*E-mail: luedde@itp.uni-frankfurt.de
†E-mail: tomk@yorku.ca

1412 (2020) 212006 doi:10.1088/1742-6596/1412/21/212006

Unexpected role of water on the fragmentation of hydrated biomolecule clusters

E Wang[1], **X Ren**[1,2]* **and A Dorn**[1]

[1]Max Planck Institute for Nuclear Physics, 69117 Heidelberg, Germany
[2]School of Science, Xi'an Jiaotong University, 710049 Xi'an, China

Synopsis We study the ionization and fragmentation of tetrahydrofuran (THF) and hydrated THF clusters induced by low-energy (65 eV) electron-collision using an (e, 2e + ion) triple coincidence method. The momentum vectors and the kinetic energies for all three final-state charged particles are determined. The ionic fragments are identified from the measured ion time-of-flight spectra. The influence of the water environment on the biomolecule is investigated with the ion fragment correlated binding energy spectra and triple-differential cross sections as well as *ab-initio* dynamical calculations.

Electron-impact ionization of atoms and molecules is a fundamental process which is relevant to understand and interpret a wide range of scientific phenomenon and technological applications, including radiation damage in biological tissue [1]. In this respect hydrated biomolecule clusters are interesting targets which allow to study the behavior of biomolecules in an aqueous environment, see e.g. [2].

In the present work, we report an experimental study on the electron-collision induced ionization and fragmentation processes in clusters consisting of water and bio-relevant molecules like DNA constituents. We use the multi-particle coincidence technique (reaction microscope) [3, 4] in which the momentum vectors and, consequently, the kinetic energies of all final state particles (electrons and ions) are measured in coincidence. The bio-molecule employed here is tetrahydrofuran (THF, C_4H_8O) which is often regarded as being the simplest molecular analog of deoxyribose, part of the DNA backbone linking the phosphate groups and the DNA bases. Figure 1 presents the measured fragment ion time-of-flight (TOF) spectra for both THF and hydrated THF clusters induced by electron ionization (65 eV). Compared to the fragmentation of isolated THF molecules where the most abundant fragment ion is identified as $C_3H_6^+$ [5], the observed major species for pure THF cluster fragmentation are $(C_4H_8O)H^+$, $(C_4H_8O)C_2H_4O^+$ and $(C_4H_8O)_2H^+$. For the hydrated THF clusters we find $(C_2H_4O)H_2O^+$, $(C_4H_8O)(H_2O)H^+$, $(C_4H_8O)(H_2O)_2H^+$ and $(C_4H_8O)(H_2O)_3H^+$ species. From the binding energy spectra and triple-differential cross sections correlated with specific ionic fragment as well as *ab-initio* dynamical calculations, we can investigate in detail the role of the environment for the modification of the biomolecular fragmentation. We observed that the water environment or a second THF acts as a catalyst for the break-up of THF ring structure. Detailed results will be presented at the conference.

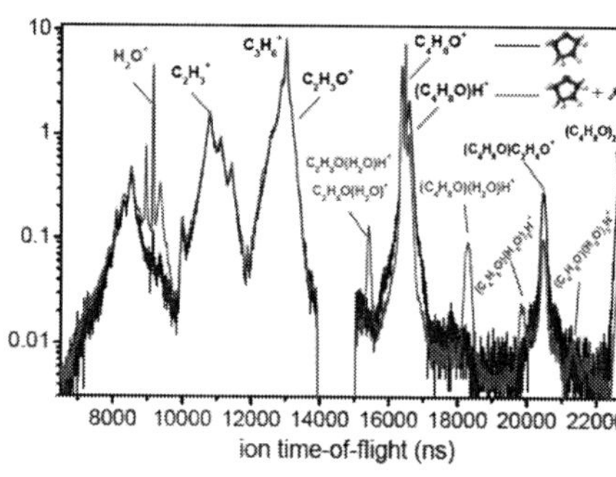

Figure 1. Measured fragment ion time-of-flight spectra of THF and hydrated THF clusters.

References

[1] Alizadeh E *et al* 2015 *Annu. Rev. Phys. Chem.* **66** 379
[2] Markush P *et al* 2016 *Phys. Chem. Chem. Phys.* **18** 16721
[3] Ren X *et al* 2018 *Nat. Phys.* **14** 1062
[4] Ren X *et al* 2016 *Nat. Commun.* **7** 11093
[5] Ren X *et al* 2014 *J. Chem. Phys.* **141** 134314

*E-mail: ren@mpi-hd.mpg.de

On the mechanisms of formation and decomposition of peptide bonds

P Bolognesi [1]*, J Chiarinelli[1,2], R Richter[3], M H Stockett[4], S Indrajith[5], C Nicolafrancesco[5,6], G Eklund[4], H Zettergren[4], H Sa'adeh[7], A Domaracka[5], P Rousseau[5] and L Avaldi[1]

[1]Istituto di Struttura della Materia, ISM-CNR, 00016 Monterotondo Scalo, Roma, Italy
[2]Dipartimento di Scienze, Università di Roma Tre, Roma, (Italy)
[3]Elettra-Sincrotrone Trieste, Strada Statale 14-km 163.5, Basovizza, Trieste, Italy
[4]Department of Physics, University of Stockholm, S-106 91 Stockholm, Sweden
[5]Normandie Universitè, ENSICAEN, UNICAEN, CEA, CNRS, CIMAP, Caen, France
[6]Synchrotron SOLEIL, L'Orme des merisiers,91190 Saint Aubin, France
[7]Department of Physics, The University of Jordan, Amman, Jordan

Synopsis The electronic structure and the fragmentation dynamics of a series of isolated cyclic-dipeptides (cyclo-Ala-Ala; cyclo-Gly-Ala; cyclo-Gly-Gly) have been studied by mass spectrometry, photoemission and photoelectron-photoion coincidence spectroscopy.

In the search of pathways to the evolution of life on the early Earth, dynamic chemical processes involving peptide bond formation and cleavage mechanisms are thought to be of central importance. The proposed coupling of the condensation of α-aminoacids into peptides with the peptide degradation 'returning' individual aminoacids might have provided an effective mechanism for reshuffling and building up the complexity of aminoacid sequences preceding the emergence of life. These processes of condensation and degradation rely on peptide bond formation and cleavage mechanisms. In this context, it has been proposed [1] that the cyclic dipeptides (also known as 2,5-diketopiperazines DKPs), the simplest peptide derivatives found in nature, may have been precursors in the formation of longer peptide chains.

The goal of this work is to provide information on the electronic structure and fragmentation dynamics of a series of isolated cyclic-dipeptides (cyclo-Ala-Ala, cyclo-Gly-Ala and cyclo-Gly-Gly). The basic mass spectrometric information on three cyclic dipeptides has been obtained in the CNR-ISM lab in Rome (fig.1), where a rare gas discharge light source is coupled with a time-of-flight spectrometer. The mass spectra have been collected at few photon energies from 10.3 to 21.22 eV .Then at the CIPO beamline at Elettra (Trieste) the photoemission spectra have been measured at 60 eV photon energy to identify the different valence electronic states. This information represents the basis for the energy selected fragmentation study achieved by detecting in coincidence the photoelectron and the photoions (PEPICO measurements).

The influence of the cyclic structure on the chemical reactivity and the tendency to form fragments,which can become precursors to peptidic chains, is addressed by comparing the present PES measurements and state-selected PEPICO fragmentation experiments with the results obtained on the isolated precursor aminoacids.

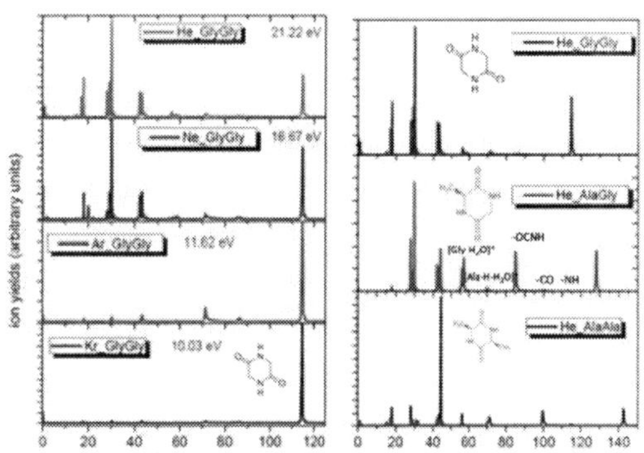

Figure 1. The mass spectra of cyclo-Gly-Gly at different photon energy (left panel) and of the three studied cyclic dipeptide at 21.2 eV (right panel).

Acknowledgment: MAECI Italy-Sweden Project 'Novel molecular tools for the exploration of the nanoworld'; CNRS PICS program (project BI-FACE)

References

[1] Danger G *et al* 2012 *Chem. Soc. Rev.* **41** 5416

* E-mail: paola.bolognesi@cnr.it

Radiodosimetry endpoints obtained with the *TILDA-V* Monte Carlo tracking structure code

M A Quinto[1] *, M E Alcocer-Avila[2], J M Monti[1], P F Weck[3], O A Fojón[1], R D Rivarola[1] and C Champion[2]

[1]Instituto de Física Rosario, CONICET and Universidad Nacional de Rosario, Rosario, S2000EKF, Argentina
[2]CELIA, Université de Bordeaux, CNRS, CEA, UMR 5107, F-33405, Talence, France
[3]Sandia National Laboratories, Albuquerque, NM 87185, USA

Synopsis The present study aims at scrutinizing at the nanometer scale the radiodosimetry endpoints induced by ion beams in a realistic biological medium. In this context, self-consistent quantum mechanical modeling of the ionization and the electron capture processes are reported within the continuum distorted wave-eikonal initial state (CDW-EIS) framework for both isolated water molecules and DNA components impacted by ions.

In radiobiology, predicting the evolution of irradiated biological samples is still nowadays an active field of research to identify DNA lesions or to adapt the radiotherapeutic protocols in radiation oncology. In this context, the numerical methods, based on Monte Carlo track-structure simulations, represent the most suitable and powerful tools for understanding the radiobiological damage induced by ionizing particles. However, most existing Monte Carlo track-structure codes require a large set of input data (cross sections) to describe at the atomic scale the various particle-induced interactions in the medium of interest (water in liquid/vapor phase and DNA). Also, the reliability of such numerical methods heavily depends on the accuracy of the input data used in the simulations, namely, the interaction cross sections needed for describing the various collisional processes involved in the slowing-down of the charged particles in the medium of interest [1].

In the present work, we report microdosimetry results obtained with the track-structure code *TILDA-V* [2]. *TILDA-V* (a French acronym for Transport d'Ions Lourds Dans l'Aqua & Vivo) is a code for modeling heavy ion and secondary electron histories in vapor water for impact energies ranging from10 keV/u to 100 MeV/u. This current version is based on a complete set of quantum-mechanically calculated multiple differential and total cross sections to describe all the inelastic processes (ionization and electron capture) occurring throughout the slowing-down of ions in water as well in DNA components (adenine (A), thymine (T), cytosine (C), guanine (G) and sugar phosphate (SP) backbone). Figure 1 reports the radial dose profiles, in the Bragg's peak region, induced by a proton beam with energy of 100 keV, in vapor water (1 g.cm^{-3}) and DNA (1.29 g.cm^{-3}).

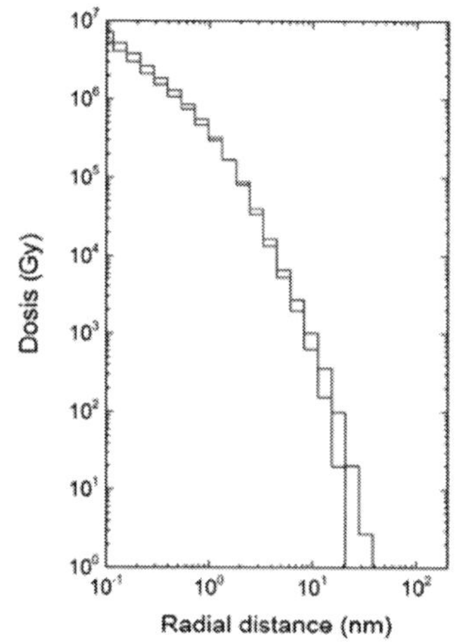

Figure 1. Comparison of radial dose profiles given by an incident proton energy of 100 keV, in vapor water (red line) and in hydrated DNA (green line).

SNL is managed and operated by NTESS under DOE NNSA contract DE-NA0003525.

References

[1] Champion C *et al* 2015 *Phys. Med. Biol.* **60** 7805–7828
[2] Quinto M A *et al* 2017 *Eur. Phys. J. D* **71** 130

* E-mail: quinto@ifir-conicet.gov.ar

Electron emission induced by Helium ions on water molecule

M A Quinto[1] *, J M Monti[1], C Champion[2] and R D Rivarola[1]
[1]Instituto de Física Rosario, CONICET and Universidad Nacional de Rosario, Rosario, S2000EKF, Argentina
[2]CELIA, Université de Bordeaux, CNRS, CEA, UMR 5107, F-33405, Talence, France

Synopsis This study focuses on the electron emission induced by He ions on water molecule. In this context, the ionization and the electron capture processes will be investigated using a quantum mechanical description within the continuum distorted wave-eikonal initial state (CDW-EIS) approximation.

Electron emission from water molecules by multi-charged fast ions has been vastly studied, in both the theoretical area as in the experimental one, over the last years (e.g. [1-3]). It is not the same case for electron emission, ionization and electron capture, from water molecules induced by dressed or neutral projectiles. Only a few experimental results are found in the bibliography, mainly for total cross sections [4-6]. Even less are the experiments that were performed in a way that electron emission from the water molecule target as well as from the ionic or atomic projectile can be distinguished. It is important to highlight that the study of electron emission from molecular targets by bare, partially-dressed, or neutral target is of main interest in many atomic and molecular physics as well as in plasma physics, and radiobiology.

With the intention of encouraging the scientific community to perform experiments where the target and projectile ionization can be distinguished, we present theoretical results regarding the ionization of vapour water molecules by He^{q+} projectile impact (wirh q=0, 1 and 2) . A set of total, simply, and doubly differential cross sections is calculated within the CDW-EIS appriximation.

Comparison with the existing experimental results will be made in every possible case (see Figure 1). Details on the calculations, results, and the theoretical model will be given during the conference.

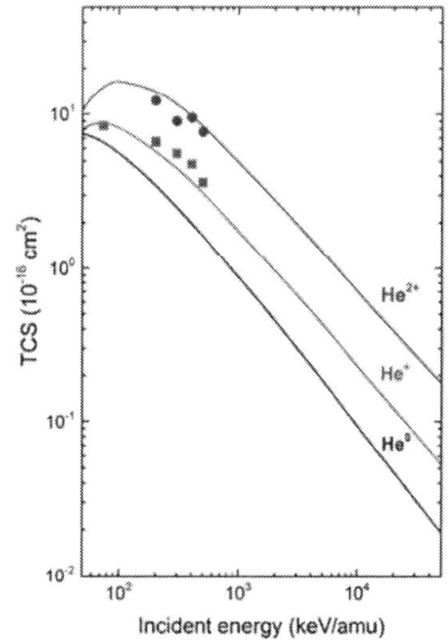

Figure 1. Total cross sections for single ionization of water by He ions. Solid lines present CDW-EIS theoretical calculation. Experimental data are token from [7].

References

[1] Nandi S *et al* 2013 *Phys. Rev. A* **87** 052710
[2] Tachino C *et al* 2014 *J. Phys. B: At. Mol. Opt. Phys.* **47** 035203
[3] Bhattacharjee S *et al* 2016 *J. Phys. B: At. Mol. Opt. Phys.* **49** 065202
[4] Gobet F *et al* 2006 *Chem. Phys. Lett.* **421** 68-71
[5] Bolorizadeh M A and Rudd M E 1986 *Phys. Rev. A* **33** 893
[6] Dingfelder M *et al* 2000 *Radiat. Phys. Chem.* **59** 255-275
[7] Toburen L H *et al* 1980 *Radiat. Res.* **82** 27-44

* E-mail: quinto@ifir-conicet.gov.ar

State-dependent fragmentation of protonated uracil and uridine

M Pitzer[1,2,*], **Ch Ozga**[2], **C Küstner-Wetekam**[2], **Ph Reiß**[2], **A Knie**[2], **A Ehresmann**[2], **T Jahnke**[3], **A Giuliani**[4] **and L Nahon**[4]

[1]Department of Chemical and Biological Physics, Weizmann Institute of Science, 76100 Rehovot, Israel
[2] Institute of Physics and Center for Interdisciplinary Nanostructure Science and Technology (CINSaT), University of Kassel, 34132 Kassel, Germany
[3] Institute for Nuclear Physics, Goethe University Frankfurt, 60438 Frankfurt, Germany
[4] Synchrotron SOLEIL, L'Orme des Merisiers, Saint Aubin, 91192 Gif-sur-Yvette, France

Synopsis Protonated uracil (m/z 113) and uridine (m/z 245) were exposed to synchrotron radiation in the photon energy range of 4.5 to 9 eV. The yield of various fragments as a function of photon energy shows maxima that can be attributed to excited electronic states and compared to calculated vertical excitation energies. For the nucleoside uridine, much less fragments were observed, indicating a stabilization of the nucleobase through the attached sugar molecule.

Nucleobases and nucleosides have large absorption cross sections for ultraviolet light. It is still not fully understood how the molecules relax from these photo-excited states and how radiation-induced damage is efficiently prevented. Recording the fragment mass spectra of gas phase molecules after irradiation can give important information to test theoretical models of de-excitation and fragmentation.

Many experiments have been performed on neutral species using synchrotron radiation, which provides high intensity and can be tuned over a wide spectral range [1]. Combining synchrotron radiation with Electrospray Ionization sources (ESI) and a tandem mass spectrometer [2] opens additional possibilities: On the one hand, (de)protonated or hydrated clusters can be investigated as first steps towards solvation. On the other hand, the pre-existing charge allows a mass-spectrometric approach to excited states that are below the ionization threshold of the neutral species.

Our contribution shows recent results for the protonated nucleobase uracil that exhibits a rich fragmentation behavior in the region of 4-9 eV photon energy [3]. Several distinct maxima in the fragment yields as a function of energy are visible (see Fig. 1).

Comparing these peaks with calculated excitation energies of protonated uracil allows investigating how excited states relate to the various fragments.

Photoexcitation of uridine, the nucleoside of uracil, in the same energy range shows almost exclusively protonated uracil as a charged fragment. This indicates that electronic excitation is efficiently driven into the weakest bond, the link between the sugar and the nucleobase and thus the nucleobase itself is stabilized.

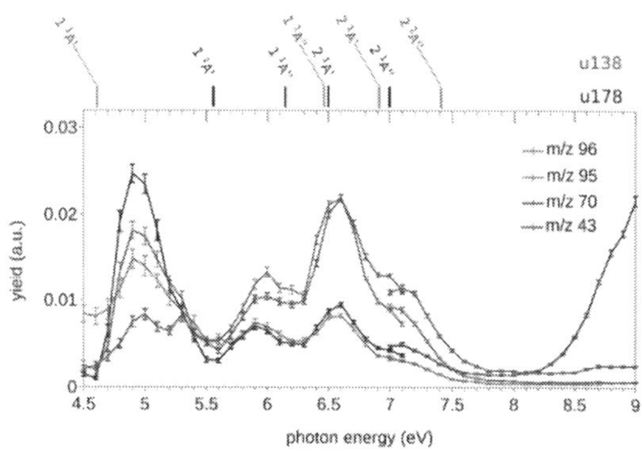

Figure 1. Fragment yields of protonated uracil as a function of exciting photon energy. The bars on top of the figure indicate vertical excitation energies calculated using EOM-CCSD for the two most abundant tautomers.

References

[1] Schwell A and Hochlaf M 2015 *Top. Curr. Chem.* **355** 155

[2] Milosavljević A R *et al* 2012 *J. Synchr. Rad.* **19** 174

[3] Pitzer M *et al* 2019 *J. Phys. Chem. A* **123** 3551

* E-mail: martin.pitzer@weizmann.ac.il

Towards prediction of the rates of antihydrogen positive ion production in antihydrogen-excited positronium reaction

T Yamashita[1,3*], Y Kino[1], E Hiyama[2,3], K Piszczatowski[4], S Jonsell[5] and P Froelich[4]

[1]Department of Chemistry, Tohoku University, Sendai, 980-8578, Japan
[2]Department of Physics, Kyushu University, Fukuoka, 819-0395, Japan
[3]Nishina Center, RIKEN, Wako, 351-0198, Japan
[4]Department of Chemistry, Uppsala University, Uppsala, Box 518 751-20, Sweden
[5]Department of Physics, Stockholm University, Stockholm, SE-10691, Sweden

Synopsis We present the 4-body calculation of the antihydrogen-positronium scattering aiming at the prediction of cross sections for the production of antihydrogen positive ions. The latter are expected to be a useful source of ultra-cold atoms for the test of matter-antimatter gravity. We convert the Schrödinger equation to a set of coupled integro-differential equations that involve intermediate states and are solved using the compact finite difference method. We will present the investigation of the rearrangement reaction between the ground-state antihydrogen atom and the excited positronium.

Antihydrogen positive ion ($\bar{\text{H}}^+$) consists of an antiproton ($\bar{\text{p}}$) and two positrons (e^+). $\bar{\text{H}}^+$ may be manipulated by electric field and be used in studies of particle physics and atomic physics. $\bar{\text{H}}^+$ ions are expected to be useful intermediates in the production of ultra-cold antihydrogen atoms which have been expected to be a good probe for the gravity between matter and antimatter [1,2]. It would be prepared by sympathetic cooling of $\bar{\text{H}}^+$ with Be^+ ions and the subsequent photodetatchment. $\bar{\text{H}}^+$ can also be utilized to develop an energy-tunable antihydrogen beam that will be used in atomic collision experiments.

Reaction between antihydrogen atom ($\bar{\text{H}} = \bar{\text{p}}\text{e}^+$) and positronium ($\text{Ps} = \text{e}^+\text{e}^-$) has been featured recently as one of the promising production schemes of H^+. A channel of H^+ production,

$$\bar{\text{H}}\,(1\text{s}) + \text{Ps}\,(1\text{s}) \rightarrow \bar{\text{H}}^+ + \text{e}^-, \tag{1}$$

opens at the collision energy 6.05 eV. A reaction between $\bar{\text{H}}\,(1\text{s})$ and the second excited Ps ($n = 3$) is also promising because the positron transfer to $\bar{\text{H}}$ from Ps ($n = 3$),

$$\bar{\text{H}}\,(1\text{s}) + \text{Ps}\,(n = 3) \rightarrow \bar{\text{H}}^+ + \text{e}^-, \tag{2}$$

can occur at the collision energy 1.7 meV. The rearrangement reactions (1) and (2) compete with several inelastic reactions of Ps (de-)excitation and require rigorous theoretical treatment.

Aiming at the prediction of cross sections in the reaction (2), the present work extends a coupled rearrangement channel scattering calculation method which has been adopted to elastic scattering between $\bar{\text{H}}$ and H [3]. A total scattering wavefuntion Ψ is written as

$$\Psi = \sum_{\upsilon} a_{\upsilon}\Phi_{\upsilon} + \sum_{c} \psi_c, \tag{3}$$

where the first summation with constants $\{a_{\upsilon}\}$ describes an intermediate state of the scattering state and ψ_c describes the initial and final states c. A set of $\{\Phi_{\upsilon}\}$ is constructed by a Gaussian expansion method (GEM) [4] to satisfy $\langle\Phi_{\upsilon'}|\Phi_{\upsilon}\rangle = \delta_{\upsilon'\upsilon}$. The Schrödinger equation with total 4-body Hamiltonian H, $H\Psi = E\Psi$, is converted to a set of coupled integro-differential equations under proper boundary conditions on asymptotic behavior of scattering wavefunction. We have solved the set of coupled equations by the compact finite difference method and have obtained the multi-channel S-matrix as a function of collision energy. Cross sections of the reactions (1) and (2) are extracted from the S-matrices as well as the cross sections of the competing inelastic reactions of Ps (de-)excitation.

We have found that the cross section of reaction (1) is consistent in magnitude with the previous calculation [5] by close-coupling method. We will present the formulation, benchmark calculations, and investigation on the reaction (2).

References

[1] Perez P and Sacquin Y 2012 *Class. Quantum Grav.* **29** 184008
[2] Sacquin Y 2014 *Eur. Phys. J. D* **68** 31
[3] Piszczatowski K *et al* 2014 *Phys. Rev. A* **89** 062703
[4] Hiyama E *et al* 2003 *Prog. Part. Nucl. Phys.* **51** 223
[5] Blackwood J E *et al* 2002 *Phys. Rev. A* **65** 030502

*E-mail: takuma.yamashita@riken.jp

Positronium formation in positron-helium collisions in Debye plasma

Y C Wang [1*], J Ma [2†], L G Jiao[3] and Y J Zhou[4]

[1]College of Physics Science and Technology, Shenyang Normal University, Shenyang, 110034, China
[2]College of Science, Shenyang Aerospace University, Shenyang, 110136, China
[3]College of Physics, Jilin University, Changchun, 130012, China
[4]Institute of Atomic and Molecular Physics, Jilin University, Changchun, 130012, China

Synopsis Positronium formation processes in positron-helium collisions with Debye potentials are studied using the screening approximation model in incident energy range from the threshold to 100 eV for screening parameters μ=0.00-0.25.

Positronium (Ps) formation processes in positron-helium collisions have now been studied by using the screening approximation model [1]. In the screening approximation model, a complex polarization potential is used to describe the formation of the Ps bound states in positron-atom systems. The electron binding energies and wave functions in Debye plasma environment are calculated by using single-active electron (SAE) approximation [2]. The binding energies of the different states in helium at different screening parameter are shown in table 1.

Table 1. The electron binding energies of ground state in a helium atom at different screening parameter μ.

μ	0.00	0.05	0.10
Energy	-0.90380	-0.84416	-0.78713
μ	0.20	0.25	
Energy	-0.68056	-0.63085	

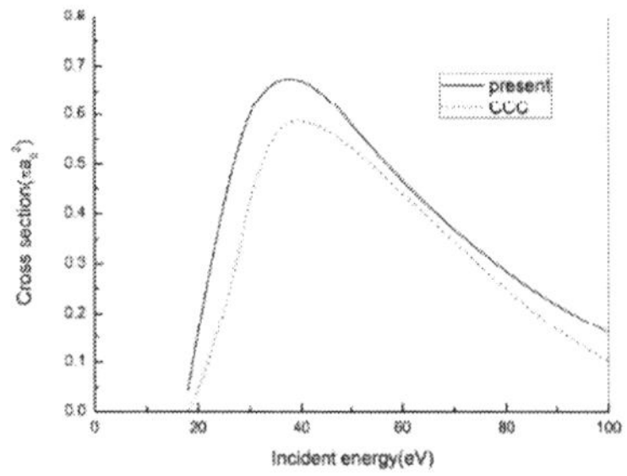

Figure 1. Ps formation cross sections with no screening. The present results are compared with the CCC calculations.

* E-mail: rickywangyc@aliyun.com
† E-mail: majia@sau.edu.cn

The present Ps formation cross section for unscreened case is displayed in figure 1, along with the CCC result [3]. The present result is higher in magnitude than cross sections of the CCC. In figure 2, Ps (1s and 2s) formation cross sections for incident positron energies in the range from the threshold to 100 eV are presented, for different values of the screening parameter μ=0.00-0.25. The Ps formation cross sections increase as the screening parameter μ increases.

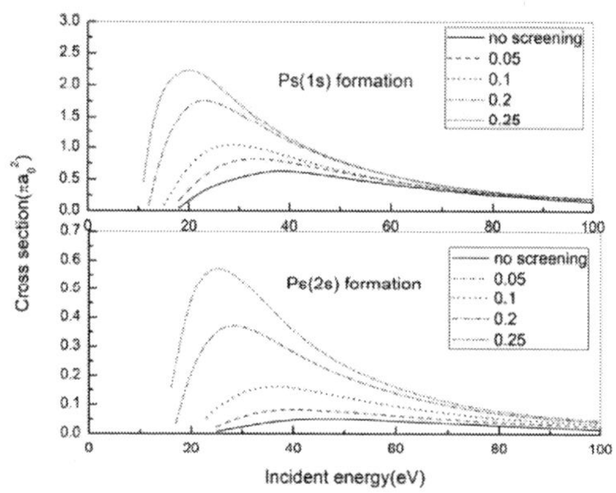

Figure 2. Ps (1s and 2s) formation cross sections under different screening parameters. solid lines, μ=0.00; dash lines, μ=0.05; dot lines, μ=0.10; dash dot lines, μ=0.20; dash dot dot lines, μ=0.25.

References

[1] Ma J *et al* 2012 *Phys. Plasmas* **19** 063303
[2] Tong X M and Lin C D 2005 *J. Phys. B: At. Mol.Opt. Phys.* **38** 2593
[3] Utamuratov R *et al* 2010 *J. Phys. B: At. Mol.Opt. Phys.* **43** 031001

Journal of Physics: Conference Series **1412** (2020) 222003 doi:10.1088/1742-6596/1412/22/222003

Elastic and inelastic scattering of positrons from neon

T Babij[1], Z Cheong[1], J Machacek[1], R P McEachran[1], S J Buckman[1] and J P Sullivan[1 *]

[1]Plasma Research Laboratory, Research School of Physics and Engineering, Australian National University, Acton ACT 2617, Australia

Synopsis Data will be presented for positron scattering from neon. As a "simple" target, neon provides a useful target for the testing and exploration of our fundamental understanding of scattering processes. The low-energy, high-resolution positron beamline based at the Australian National Universitywas used to measure elastic differential cross sections and total electronic and ionization cross sections for low energy scattering of positrons from neon. The experimental data presented will be compared to several recent theoretical calculations

Positron scattering from the noble gases has been a rich field for the development of both poitron scattering experimental techniques, as well as providing reasonably tractable targets for theoretical calculations. In particular, the development of the Surko trap and beam system [1] provided an opportunity to significantly improve the quality of experimental measurements, which has in turn spurred the development of more sophisticated theory [2].

This paper presents a study of positron scattering from neon, focusing on elastic scattering (through the elastic differential scattering cross section) and inelastic processes, in particular electronic excitation. Scattering measurements were performed using the positron beamline at the Australian National University [3]. This beamline provides a low-energy, high-resolution positron beam, and takes advantage of the measurement and analysis techniques developed for a magnetized beam by the UCSD group [4]. In this case, the pulsed positron beam was used to make measurements at energies from 1 to 30 eV, with an energy spread of 60-80 meV. The differential cross section was measured at a variety of different energies, and above the first electronic excitation threshold, total inelastic scattering measurements were also made. This is the first time that absolute measurements have been made for these processes and comparison ahs been made to a range of recent theoretical calculations.

Theoretical calculations for these processes have included the relativistic optical potential (ROP) and convergent close couplioing (CCC) techniques [5], as well as the many body calculations presented by Green et al. [6]. Some discrepancies between experiment and theory remain, as demonstrated in figure 1, which shows a comparison between theory and experiment for elastic scattering at 1eV. In this presentation, we will discuss the implications for the future of both experimental and theoretical progress, in particular with regards to excitation processes, which have largely remained unexplored by theory.

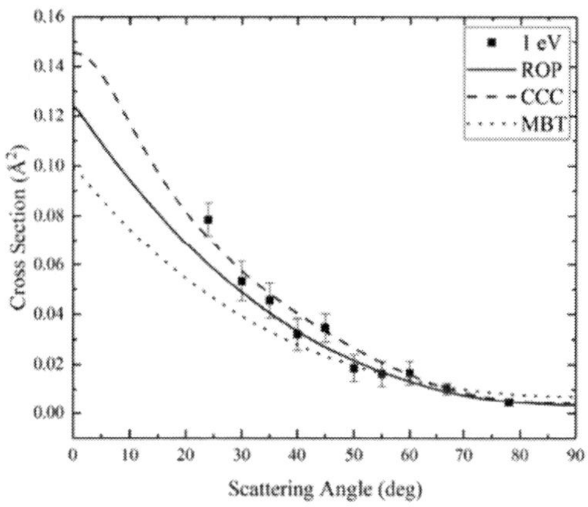

Figure 1. Comparison of experiments (blue circles) and theoretical data (solid line ROP [5], dashed line CCC [5], dotted line many body theory [6] for elastic scattering of 1 eV positrons from neon.

References

[1] Gilbert S J *et al* 1997 *Appl. Phys. Lett.* **70** 1944
[2] e.g.: Utamuratov R *et al* 2012 *Phys. Rev.* A **86** 062702
[3] Sullivan J P *et al* 2008 *Rev. Sci. Instr.* **79** 113105
[4] Sullivan J P *et al* 2002 *Phys. Rev.* A **66** 042708
[5] Jones A C L *et al* 2011 *Phys. Rev.* A **83** 032701
[6] Green D G *et al* 2014 *Phys. Rev.* A **90** 032712

* E-mail: james.sullivan@anu.edu.au

Configuration space method to calculate rearrangement matrix elements

D V Fursa*, R Utamuratov, A S Kadyrov and I Bray

Department of Physics, Curtin University, Perth, Australia

Synopsis A new method based on partial-wave expansion of the Ps radial wave functions has been developed for calculating the Ps-formation matrix elements. The method utilises the Euler-Wijngaarden tansformation to accelerate slowly converging series. The integrals in the method are free of the Coulomb singularity that appears in previously used momentum-space methods and therefore more straightforward to apply to complex and charged targets. The results are compared with the existing benchmark calculations for positron scattering on H and He^+ targets.

Studies of positron scattering are important for understanding some of the fundamental science problems and practical applications. One of the the main challenges for theoretical studies of positron scattering is the two-centre nature of the problem since a positron can capture an electron from the target and form Ps. Calculations of Ps-formation matrix elements require deriving and evaluating complicated expressions that are computationally costly. Previous state of the art methods [1, 2] applied to the positron scattering problem were based on momentum-space transformations and involved the Coulomb singularity that required special numerical techniques. As a result, application of the momentum-space based method to more complex and/or charged targets was quite challenging [3].

In this report we present a new configuration-space method [4] of calculating the Ps-formation matrix elements. The method is based on expansion of the Ps wave functions to separate electron and positron coordinates that allows one to obtain the rearrangement matrix elements in coordinate space. The expansion results in slow converging sums which, however, can be accelerated with the Euler-Wijngaarden tansformation [5].

As an example, figure 1 shows the absolute values of the off-shell rearrangement matrix elements for the H($2p$) to Ps($1s$) transition as a function of initial channel momentum q_α when q_β = 3.490 a.u. (as indicated by the vertical dashed line). Solid lines indicate the exact values [2]. Open triangles are the results of modified sum using the Euler-Wijngaarden acceleration technique [5]. Black circles represent the unmodified sum. Each panel shows results with a given value of $l_0^{\max}$, which is the maximum number of terms in the sum. This illustrates that sum convergence is dramatically improved with application of the series acceleration technique.

*E-mail: d.fursa@curtin.edu.au

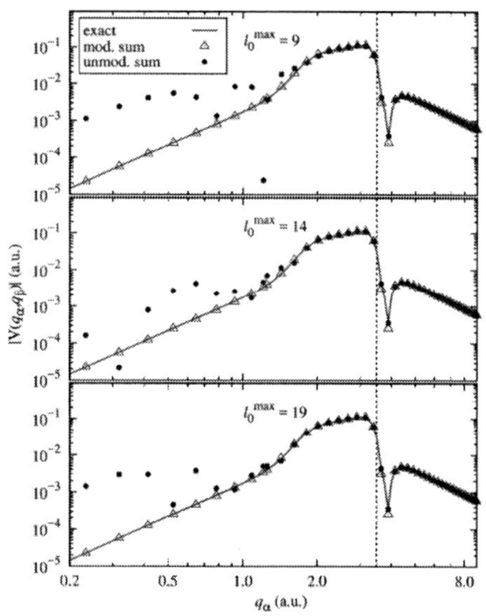

Figure 1. The absolute values of the off-shell matrix elements for the H($2p$) to Ps($1s$) transition.

References

[1] Mitroy J and Ratnavelu K 1994 *Aust. J. Phys.* **47** 721
[2] Kadyrov A S and Bray I 2002 *Phys. Rev. A* **66** 012710
[3] Rawlins C M *et al* 2018 *Eur. Phys. J. D* **72** 77
[4] Utamuratov R *et al* 2019 *Comput. Phys. Commun.*, in press
[5] Lauwerier H 1975 *SIAM J. Math. Anal.* **6** 96

Convergent close-coupling calculations of positron collisions with the hydrogen negative ion

D V Fursa*, **R Utamuratov, A S Kadyrov and I Bray**

Department of Physics, Curtin University, Perth, Australia

Synopsis Convergent close-coupling method has been applied to positron scattering by the hydrogen negative ion. Internal consistency of the method and convergence of the cross sections are achieved with use of both the single- and two-centre approaches. Calculations were performed using accurate target wave functions obtained with the multi-core approach.

Positron (e^+) scattering on the hydrogen negative ion (H^-) is an exotic collision problem representing a fundamental interest. Positron is the most accessible antimatter particle with a wide range of practical applications. The H^- ion is a unique system with just one bound state and plays an important role in astrophysics [1].

Positron scattering by H^- has been studied by various theoretical methods over the last few decades. Rudge [2] calculated electron detachment from H^- by electron and positron impact within the first Born approximation. McAlinden *et al* [3] applied a close-coupling method to calculate Ps formation in the ground and excited states. Ghosh and Sinha [4] calculated single- and double-ionization of H^- by positron impact using the Coulomb-distorted eikonal approximation. All these calculations were performed with simplified wavefunctions to represent the bound state of H^-.

We have recently reported a new method [5] of calculating Ps-formation matrix elements within the convergent close-coupling (CCC) approach. Here we report application of the method to e^+-H^- scattering. The CCC can be applied to this problem within the single- and two-centre approaches allowing for internal consistency checks. The single-centre approach is applicable to this system above the single-electron detachment threshold and can produce electron-loss cross sections. The two-centre approach allows calculations of the Ps-formation and electron-detachment cross sections explicitly.

H^- is a two-electron system with the electron detachment ($H^- \to e^-$-H) energy of 0.7542 eV [6]. Our calculations within the multi-core approximation gives the detachment energy 0.7326 eV.

Figure 1 shows results of calculations with three different basis sizes for the total scattering cross section after subtraction of the the Rutherford cross section. Basis size notation is $(N_L, N'_{L'})$, where N and L indicate the highest principal and orbital quantum numbers of the target states. Similarly, N' and L' indicate the Ps-basis size. All electronic states are pseudostates obtained with the Laguerre basis, the lowest states closely representing the true eigenstates. The single-centre calculations, denoted as CCC(25_7, 0), are in agreement with other two-centre calculations above 1 eV impact energies. Close agreement between the results of different bases indicate the convergence of the cross sections and internal consistency of the method.

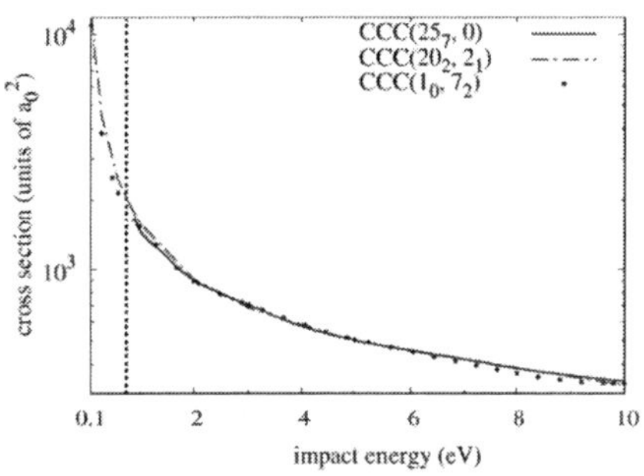

Figure 1. Total scattering cross section after subtraction of the Rutherford cross section.

*E-mail: d.fursa@curtin.edu.au

References

[1] Chandrasekhar S and Elbert D D 1958 *Astrophys. J* **128** 633
[2] Rudge M R H 1963 *Proc. Phys. Soc.* **83** 419
[3] McAlinden M T *et al* 2002 *Phys. Rev. A* **65** 032715
[4] Ghosh D and Sinha C 2004 *J. Phys. B* **38** 61
[5] Utamuratov R *et al* 2019 *Comput. Phys. Commun.* **239** 64-71
[6] Lykke K R *et al* 1991 *Phys. Rev. A* **43** 6104

Kinematically complete scattering cross sections in collisions between positron and hydrogen atom

K Tőkési[1,2*]

[1]Institute for Nuclear Research, Hungarian Academy of Sciences (Atomki), 4026 Debrecen Bem tér 18/c, Hungary
[2]ELI-ALPS, ELI-HU Non-profit Ltd., Dugonics tér 13, H-6720 Szeged, Hungary

Synopsis A three-body classical trajectory Monte Carlo (CTMC) method is used to describe positron and hydrogen atom collisions. The total and multi-differential ionization and charge exchange cross sections are calculated and compared with experimental data and other theoretical results.

Understanding the elementary scattering processes during atomic collisions is of fundamental importance both from the experimental and theoretical points of view. Ionization by positron impact has been extensively studied in recent decades. In most cases noble gas atoms were used as the target. For designing new experiments, such as production of antimatter, ionization cross sections for other atoms are also necessary. During the last two decades more and more studies also became available for positronium impact.

The classical treatment of various collision problems has been quite successful for the calculation of the ionization and charge exchange cross sections when the collision energy is sufficiently high. In particular the classical trajectory Monte Carlo (CTMC) method is widely used. It is a non-perturbative method and hence all the interactions between colliding particles can be taken into account exactly within the framework of the classical dynamics.

In this work a three-body CTMC technique is used to calculate the total and multi-differential scattering cross sections in collision between positron and hydrogen atom [1]. Our system consists of three particles, the positron as projectile, and the electron and proton as the target atom. The classical equations of motion for three-body system are solved numerically for large number of trajectories. The initial conditions are the impact parameter, the position and the velocity of the target electron. These are selected randomly for each trajectory. The initial conditions of the individual collision system were selected at a relatively large internuclear separation from the collision center in such a way that an initial binding energy of the H(1s) level (E_b=0.5 a.u.) was constrained.

For given initial parameters, the Newton's equations of motion were integrated with respect to time as independent variable by the standard Runge-Kutta method.

In this work we present singly differential as well as kinematically complete triply differential scattering cross sections for ionization and charge exchange processes at 50 and 100 eV positron impact energies. Fig. 1 shows the present CTMC results for angular differential ionization cross sections of hydrogen atom at 50 eV primary positron energy.

We find that the classical treatment of the scattering problem for light projectile impact describes reasonable well the various cross sections. Our results are compared with experimental data and other theoretical values.

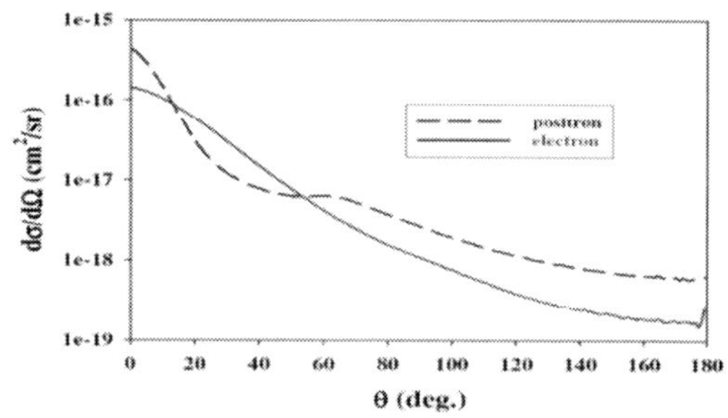

Figure 1. Present CTMC results for angular differential ionization cross sections of hydrogen atom at 50 eV primary positron energy.

The work was supported by the National Research, Development and Innovation Office (NKFIH) Grant KH126886

References

[1] Tőkési K and Hock G 1997 *Nucl. Instr. Meth. Phys. Res. B* **124** 398

* E-mail: tokesi@atomki.mta.hu

Charge exchange three- and four-body reactions in the presence of a laser field

K Lévêque-Simon[1] **and P -A Hervieux**[1]*

[1]Université de Strasbourg, CNRS, IPCMS UMR 7504, F-67000 Strasbourg, France

Synopsis Cross-sections computation of the formation of antihydrogen' atoms or ions resulting from two consecutive collisions between positronium atoms and antiprotons is a central issue of GBAR's experiment. In [2] cross-section calculations using a perturbative approach, have been undertaken, for the charge exchange three- and four-body reactions (1) and (2). In the present study, we investigate the influence of a laser field on these two reactions. We show that under special irradiation conditions, antihydrogen' atom or ion formation cross-sections may be increased by the presence of the laser field.

In order to make a direct observation of the effect of gravitation on antimatter, the GBAR [1] experiment aims at measuring the influence of Earth's gravity in the trajectory of antihydrogen atoms. The first step of the experiment involves two successive low energy collisions. Indeed, the antihydrogen ions will be produced in two steps. The first one is the charge exchange three-body reaction (1): $\bar{\mathrm{p}} + \mathrm{Ps}(n_p) \to \bar{\mathrm{H}}(n_h) + \mathrm{e}^-$ where $\bar{\mathrm{p}}$ stands for antiproton, $\mathrm{Ps}(n_p)$ for positronium (bound positron-electron pair) either in ground state ($n_p = 1$) or in an excited state ($n_p = 2$), $\bar{\mathrm{H}}(n_h)$ for antihydrogen (which can be also produced in an excited state $n_h \neq 1$) and e^- for electron. The second one is a more complex four-body reaction (2): $\bar{\mathrm{H}}(n_h) + \mathrm{Ps}(n_p) \to \bar{\mathrm{H}}^+ + \mathrm{e}^-$, in which the antihydrogen produced in the first reaction interacts with another positronium to create an antihydrogen ion.

In this contribution we explore the possibility to increase *significantly* the antihydrogen positive ion production rate by assisting the capture processes involved in the three- and four-body reactions (1) and (2) using a laser field having standard specifications (e.g. ion or excimer lasers). By using a formalism adapted from [2-4], we present an extensive study of the influence of the laser parameters (laser field strength and photon energy) on the charge exchange cross-sections in the energy range of interest for the GBAR's experiment. Under special irradiation conditions antihydrogen' atom and ion formation cross-sections may be increased by the presence of the laser field. Indeed, for reaction (1) we have chosen a laser frequency in such a way that the $1s$ ground state of positronium is dominantly coupled to the $2p$ state via the dipole interaction (cf. Figure 1). It corresponds to a krypton fluoride laser (KrF) having a frequency $\omega_l = 0.1837$ au which is very closed to the Bohr frequency $\omega_{1s-2p} = 0.1875$ au of Ps.

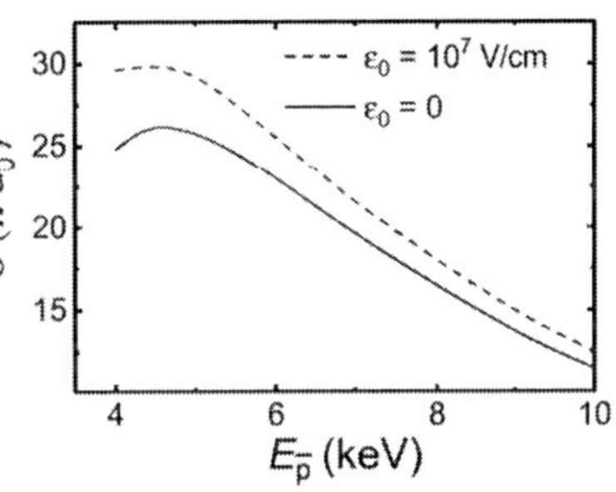

Figure 1. Cross-section of the three-body reaction $\bar{\mathrm{p}} + \mathrm{Ps}(1s) \to \bar{\mathrm{H}}(n_h = 2) + \mathrm{e}^-$ as a function of the antiproton energy. Solid line: laser off; Dashed line: KrF laser on with a frequency $\omega_l = 0.1837$ au and an electric field strength of 10^7 V.cm^{-1} corresponding to $I = 7.2 \times 10^{11}$ W.cm^{-2}.

References

[1] CERN report *et al* 2011 *SPSC-P-342*
[2] Comini P and Hervieux P -A 2013 *New J. Phys.* 15 09022
[3] Martin P *et al* 1989 *Phys. Rev. A* **39** 6178
[4] Joachain C J *et al* 2012 *Atoms in intense laser fields, Cambridge University Press*

*E-mail: hervieux@unistra.fr

Positronium formation in collisions between positrons and Rb, Cs atoms with screened Coulomb interaction

J Ma[1] *, Y C Wang[2] †, B Li[2] and Y B Wang[2]

[1] College of Science, Shenyang Aerospace University, Shenyang, 110136, China
[2] College of Physics Science and Technology, Shenyang Normal University, Shenyang, 110034, China

Synopsis Positronium formation processes in collisions between positrons and Rb, Cs atoms with Debye potentials are studied using the screening approximation model for screening parameters μ=0.00-0.25 at incident energy range 2-60 eV.

Positronium (Ps) formation processes in positron collision with rubidium and cesium in Debye plasmas are studied. The interaction among the charged particles in plasmas has been incorporated by using the Debye-Hückel potential, $V(r) = -\frac{Ze^2}{r}e^{-\mu r}$, where μ is the screening parameter. The method we used in this study is the screening approximation model, which has been used in study of Ps formation in positron-hydrogen collision [1]. In this model, an equivalent-local complex potential is used to describe the positronium formation process. The energies and wave functions of the targets in Debye plasma environment are calculated by using a model potential [2].

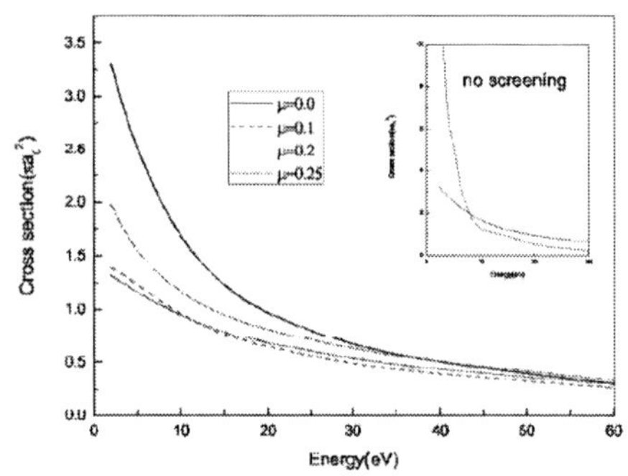

Figure 1. Ps(1s) formation cross sections in positron-rubidium collisions under different screening parameters. The no screening results compared with the calculations of coupled-state approximations(shot dash dot line) are given in the inset.

Ps(1s) formation cross sections from collisions between positrons and atoms(Rb, Cs) for different values of screening parameter μ=0.00-0.25 in the incident positron energy range of 2-60 eV are presented in figures 1 and 2, respectively. The available results of rubidium and cesium are limited. The present results for unscreened cases are compared with calculations of coupled-state approximations [3]. The only available results of Ps formation in positron collisions with Rb and Cs atoms under Debye plasma environments are carried out in the energy range of above 1 keV [4].

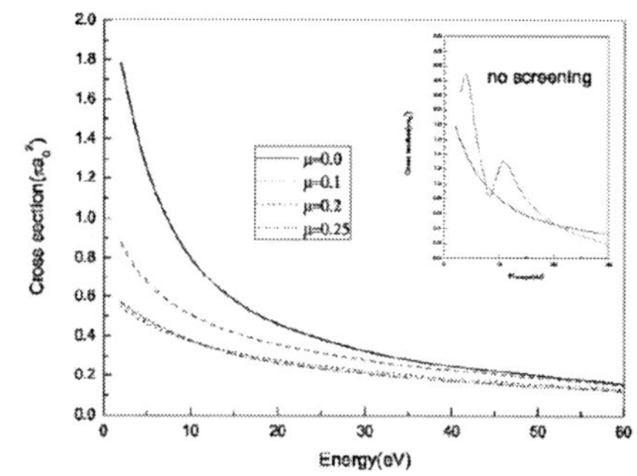

Figure 2. Ps(1s) formation cross sections in positron-cesium collisions under different screening parameters. The no screening results compared with the calculations of coupled-state approximations(shot dash dot line) are given in the inset.

This work was supported by the Natural Science Foundation of Liaoning Province of China (Grant No. 20180550772).

References

[1] Ma J *et al* 2012 *Phys. Plasmas* **19** 063303
[2] Schweizer W *et al* 1999 *Atomic Data and Nuclear Data Table* **72** 33
[3] Kernoghan A A *et al* 1996 *J. Phys. B: At. Mol. Opt. Phys.* **29** 3971
[4] Pandey M K *et al* 2016 *J. Phys. B: At. Mol. Opt. Phys.* **49** 034007

*E-mail: majia@sau.edu.cn
†E-mail: rickywangyc@aliyun.com

Electron and positron scattering from pyridine

N Sinha[1] *, A K Sahoo[1] and B Antony [1] †

[1]Atomic and Molecular Physics Laboratory, Indian Institute of Technology (Indian School of Mines) Dhanbad, Jharkhand-826004, India

Synopsis Advancement in medical technology is a much needed demand in the present scenario. To stem a robust technology, one should be aware of the knowledge behind such sciences. In view of this, we investigate here the interaction of biomolecules with the most elementary particle, electron along with its anti-particle positron. The cross sections computed here using SCOP formalism for pyridine molecule will be useful for simulations of bio molecular systems. We further compare the cross sections for e^- and e^+ scattering extracting some rudimentary information about these basic interactions.

Over the last one and half decades, positrons have served as an important tool to diagnose human cell disorders through positron emission therapy (PET) and positherapy [1]. Consequently the investigation of interaction of positrons with various biomolecules is a requisite for its further investigation. Further, biomolecules are highly sensitive to radiation, causing damage arising from different reactive processes. These impairments may lead to cell death or carcinogenesis [1]. Hence, a precise study on the interaction of electron is vital to model these bio-molecular systems. The present study aims to calculate the interaction probabilities, more commonly known as the scattering cross sections, for both electrons and positrons. In addition a comparative study of the various cross sections for electron and positron would enhance our basic understanding of matter-antimatter physics. We intend to provide a comprehensive set of cross sections for pyridine (C_5H_5N), which is indispensable for Monte Carlo simulations of these bio-molecular systems. Note that these models require knowing such parameters over a broad energy range, in principle from the high incident energy of primary particles down to their final thermalisation in the medium. Besides, it allows estimating the transport properties in biologically relevant media.

Spherical complex optical potential (SCOP) [2] and complex scattering potential-ionization contribution (CSP-ic) [2] methods are employed for the computations of the cross sections for pyridine molecule. Since pyridine is a complex molecule, more than one scattering centre is identified. For each centre, a group is formed containing certain number of atoms. Based on the bond length and molecular structure, the charge density and static potential of each atom is expanded from this central atom. After obtaining the total charge density of the concerned group, we normalize the expression so that total number of electrons in the group is reproduced. Thus we obtain a spherical density and potential for each group, which is approximate, but it has a dependance on molecular properties as well. The cross sections obtained from each group is then added to give cross sections for the molecule. In case of positron scattering, positronium formation process is adequately included in the computations through a modified absorption threshold [2]. In the literature sufficient studies are available for electron scattering; however, for positrons the condition is poor. There is only one previous study [3] found for positron scattering cross sections, that too limited to 20 eV. The present work is aimed to produce cross sections for a wide energy range from 1 eV to 5000 eV for positrons and from 5 eV to 5000 eV for electrons. Hence, it is presumed that the present set of cross sections will contribute to the paucity of literature available and will be vital for the biomedical field.

*E-mail: sinhanidhi00@gmail.com
†E-mail: bobby@iitism.ac.in

References

[1] Moadel R M *et al* 2005 *Can. Res.* **65 (3)** 698-702
[2] Sinha N *et al* 2018 *J. Phys. Chem. A* **122** 2513-2522
[3] Stevens D *et al* 2018 *J. Chem. Phys.* **148** 144308

Calculations of positron scattering, binding and annihilation for atoms and nonpolar molecules with a Gaussian basis

A R Swann[1] * and G F Gribakin[1] †

[1]School of Mathematics & Physics, Queen's University Belfast, Belfast BT7 1NN, United Kingdom

Synopsis Calculations of positron binding, scattering and annihilation using a model correlation potential and a basis of Gaussian states are carried out for a range of atoms and small molecules.

The possibility that neutral atoms could support a bound state for a positron was suggested by many-body-theory calculations in 1995 [1] and confirmed two years later by variational calculations [2]. Despite several detection schemes being proposed [3], such bound states have not yet been observed experimentally. In contrast, measurements using a trap-based positron beam have yielded measured values of the positron binding energies for over 70 *molecules* [4], most of which are nonpolar. *Ab initio* calculations of positron-molecule binding are very difficult. Exisiting calculations deal with strongly polar species [5], and only 6 molecules have been studied both experimentally and theoretically. To address this disparity, we have developed a model-correlation-potential approach [6], which we test here by doing calculations for atoms and small molecules.

We write the positron interaction with the target as $V(\mathbf{r}) = V_{\text{st}}(\mathbf{r}) + V_{\text{cor}}(\mathbf{r})$, where V_{st} is the electrostatic potential of the target, and V_{cor} is a model correlation potential, viz., $V_{\text{cor}}(\mathbf{r}) = -\sum_A(\alpha_A/2|\mathbf{r} - \mathbf{r}_A|^4)g_A(\mathbf{r})$, where the sum is over the target's constituent atoms A (if there are more than one), with the nucleus at $\mathbf{r}_A$ and the hybrid polarizability α_A [7]. The short-range cutoff function $g_A(\mathbf{r}) = 1 - \exp[-(|\mathbf{r} - \mathbf{r}_A|/\rho_A)^6]$ depends on the parameter ρ_A. It values are calibrated by comparison with existing high-quality calculations or experimental data, allowing predictions for other systems. We use a Gaussian basis set to solve the Schrödinger equation for the positron using a modified version of GAMESS [8], and obtain the negative-energy bound states (when they exist), as well as a set of discrete positive-energy positron pseudostates. The latter can be used to obtain the asymptotic s-wave scattering phase shifts (for atoms, spherical-top molecules, and other small nonpolar molecules) and the dimensionless energy-dependent normalized annihilation rate Z_{eff}. We can also calculate the rate of annihilation from bound states. Figure 1 shows Z_{eff} for atoms Be, Mg, He and Ar. The strong enhancement of Z_{eff} at low momenta for Be is due to the presence of a weakly bound state. We will also present the scattering phase shifts, Z_{eff} and binding energies (where they exist), for He, Be, Mg, Ar, H_2, N_2, Cl_2 (binding energy ~10 meV) and CH_4.

*E-mail: a.swann@qub.ac.uk
†E-mail: g.gribakin@qub.ac.uk

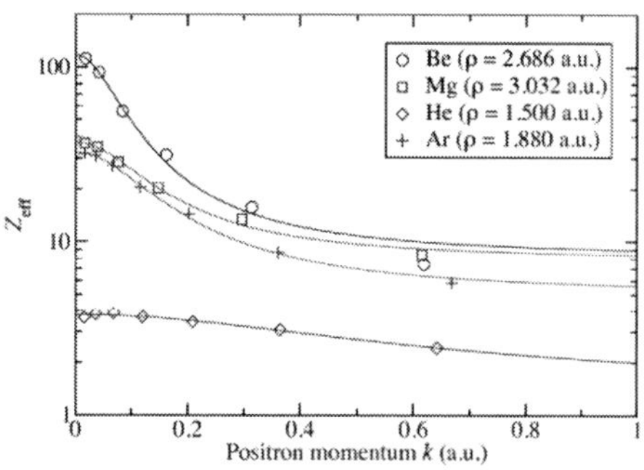

Figure 1. s-wave Z_{eff} for Be, Mg, He and Ar. Symbols, calculated values; curves, fits of the form $Z_{\text{eff}} = A/(B + k^2) + C$, with parameters A, B, C.

References

[1] Dzuba V A *et al* 1995 *Phys. Rev. A* **52** 4541
[2] Ryzhikh G G and Mitroy J 1997 *Phys. Rev. Lett.* **79** 4124; Strasburger K and Chojnacki H 1998 *J. Chem. Phys.* **108** 3218
[3] Swann A R *et al* 2016 *Phys. Rev. A* **93** 052712
[4] Danielson J R *et al* 2010 *Phys. Rev. Lett.* **104** 233201; 2012 *Phys. Rev. A* **85** 022709
[5] Gribakin G F, Young J A and Surko C M 2010 *Rev. Mod. Phys.* **82** 2557
[6] Swann A R and Gribakin G F 2018 *J. Chem. Phys.* **149** 244305
[7] Miller K J 1990 *J. Am. Chem. Soc.* **112** 8533
[8] Schmidt M W *et al* 1993 *J. Comput. Chem.* **14** 1347

Path integral molecular dynamics simulations for muoniated and hydrogenated thioacetone radicals

S Okano[1] *, Y Oba[1] and M Tachikawa[1,2] †

[1]Graduate School of Nanobioscience, Yokohama City University, Yokohama, 236-0027, Japan
[2]School of Data Science, Yokohama City University, Yokohama, 236-0027, Japan

Synopsis Theoretical understanding of hyperfine coupling constant (HFCC) is important to analyze muon spin rotation/relaxation/resonance (μSR) spectrum. We performed *ab initio* path integral molecular dynamics simulations to predict and analyze the reduced HFCCs of muoniated thioacetone radical (Mu-TACE) and hydrogenated thioacetone radical (H-TACE). Our predicted HFCC value of Mu in Mu-TACE was larger than that of H in H-TACE, because of the larger nuclear quantum effect of positive muon.

Muonium (Mu) is consisted of a positive muon (μ^+) and an electron (e^-) [1]. The mass of μ^+ is about 1/9 times smaller than proton. Accordingly, Mu atom can be regarded as an ultra-light isotope of hydrogen atom and is highly responsive to magnetic field. Such features have led Mu to the application for magnetic measurement technique such as muon spin resonance (μSR). This provides the informations of hyperfine structures of muoniated radicals, characterized by hyperfine coupling constant (HFCC): magnitude of the interaction between radical electron and nuclear spins. Although the HFCC values for various muoniated radicals have been measured by μSR, theoretical understanding of HFCC is needed to assign μSR spectra to their molecular structures. However, in some cases, it is difficult to estimate HFCCs precisely by using conventional *ab initio* molecular orbital method [2]. Systematic analysis for small molecules is essential to understand μSR spectra, even though they have no experimental data, e.g. muoniated thioacetone radical (Mu-TACE) as shown in **Figure 1**(a) [3]. Therefore, in this study, we calculated reduced HFCC values of Mu-TACE and hydrogenated thioacetone radical (H-TACE). To discuss the nuclear quantum effect of μ^+, we applied path integral molecular dynamics (PIMD) simulation for these species.

Table 1 shows our predicted HFCC values of Mu-TACE and H-TACE. For Mu-TACE, predicted HFCC value was 52.6 MHz, which is larger than that of H-TACE, 43.5 MHz. We have found that the C-Mu bond length and the electron density on Mu have positive correlations, as shown in **Figure 1**(b), since electron density on each Mu increases according to the dissociation from thioacetone molecule. We can conclude that the large nuclear quantum effect of μ^+ made the large HFCC of Mu-TACE.

Table 1. Predicted HFCC values [MHz] of Mu-TACE and H-TACE.

	Mu-TACE	H-TACE
HFCC	52.6	43.5

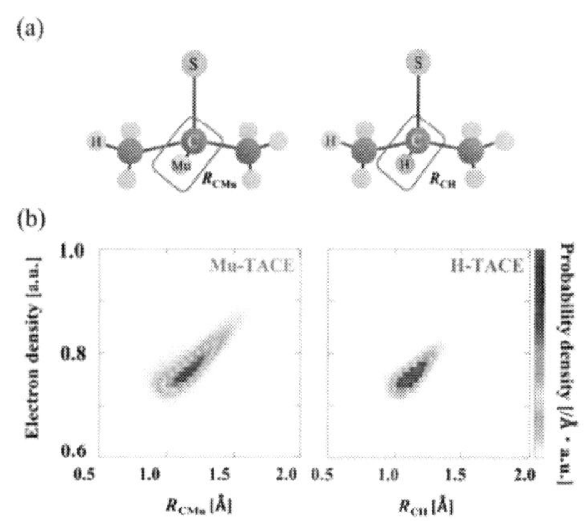

Figure 1. (a) Molecular structures of Mu-TACE and H-TACE, (b) two dimentional distributions for the bond length and electron density in Mu-TACE and H-TACE

References

[1] Percival P W 1979 *Radiochemica Acta,* **26** 1
[2] Oba Y, Kawatsu T, and Tachikawa M 2016 *J. Chem. Phys.* **145** 064301
[3] Kobayashi T *et al.* 2011 *Comp. Theor. Chem.* **963** 256

* E-mail: i150142g@yokohama-cu.ac.jp
† E-mail: tachi@yokohama-cu.ac.jp

Theoretical analysis of the binding of a positron to acetaldehyde molecule

K Dohi[1 *], M Tachikawa[1,2] and Y Kita[1 †]

[1]Graduate School of Nanobioscience, Yokohama City University, Yokohama, 236-0027, Japan
[2]School of Data Science, Yokohama City University, Yokohama, 236-0027, Japan

Synopsis We theoretically analyzed the positron affinity (PA) of acetaldehyde molecule at several vibrational excited states toward the elucidation of physical and chemical origin of H/D isotope shift in the molecular PA value. We found that (i) the acetaldehyde molecule has different PA values and its H/D isotope shift for each vibrational state, and (ii) H/D isotope shift in the PA values at vibrational excited states are mainly due to the change in the dipole moment by the vibrational excitations.

The positron (e^+) is now used in various scientific and technological areas. Positrons injected into a liquid or solid induce physical and chemical processes such as the formation of positron-molecular complexes which are bound states of a positron and atom/molecule. The positron affinity (PA), which is the binding energy of a positron, has been experimentally measured by Gribakin *et al* for many molecular species [1]. They have also reported that the PA values of acetaldehyde, acetone, and acetonitrile molecules become decreased by deuterium substitutions [2]. The mechanism of such H/D isotope effect, however, has not been elucidated yet. In this study, thus, we theoretically analysed the PA value of acetaldehyde molecule at several vibrational excited states, in order to elucidate the physical and chemical origin of H/D isotope shifts in the PA value.

In order to analyze the PA value including the effect of molecular vibrations, we introduced vibrational averaged PA (PA_v) defined by the following equation:

$$\mathrm{PA}_v \equiv \frac{\int \mathrm{PA}^{[X]}(\boldsymbol{Q})|\Psi_v(\boldsymbol{Q})|^2 \mathrm{d}\boldsymbol{Q}}{\int |\Psi_v(\boldsymbol{Q})|^2 \mathrm{d}\boldsymbol{Q}},$$

where $\boldsymbol{Q}$ is a set of vibrational normal coordinates and Ψ_v is the vibrational wave function of the v-th vibrational excited state. $\mathrm{PA}^{[X]}(\boldsymbol{Q})$ is the vertical PA value at the molecular geometry $\boldsymbol{Q}$, defined by the total energy difference of the parent molecule (X) and its positron attached system ($[X; \mathrm{e}^+]$) as $\mathrm{PA}^{[X]}(\boldsymbol{Q}) \equiv E^{[X]}(\boldsymbol{Q}) - E^{[X;\mathrm{e}^+]}(\boldsymbol{Q})$. In this study, $E^{[X;\mathrm{e}^+]}(\boldsymbol{Q})$ and $\Psi_v(\boldsymbol{Q})$ were calculated with configuration interaction level of multi-component molecular orbital theory [3] and vibrational quantum Monte Carlo method [4], respectively.

Figure 1 shows the H/D isotope shifts of PA_v values at the several fundamental tones and over tones of the acetaldehyde molecule (for excited states, vibration modes having the H/D shift of more than 1 meV were chosen). In Fig. 1, the H/D isotope shift is defined as $\Delta\mathrm{PA}_v \equiv \mathrm{PA}_v^{\mathrm{CD_3CDO}} - \mathrm{PA}_v^{\mathrm{CH_3CHO}}$. We found that the acetaldehyde molecule has different PA values and its H/D isotope shift for each vibrational mode. Figure 1 clearly shows that the H/D isotope shift which is qualitatively the same as the experimental value (-3 meV) was obtained only in the "Asymmetric vibration of CH_3". The other details will be discussed in the presentation.

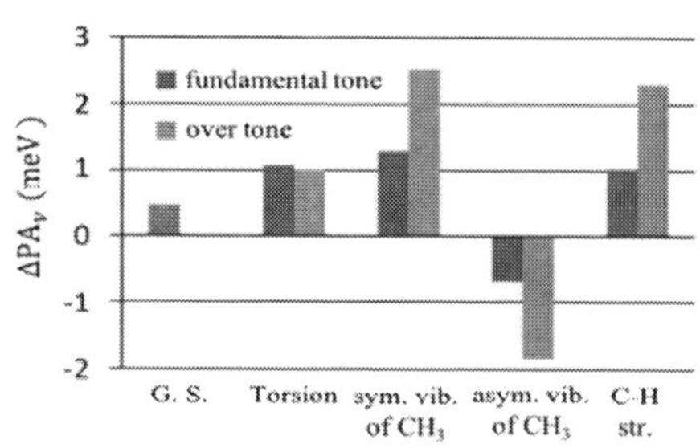

Figure 1. H/D isotope shift in vibrational averaged PA values of acetaldehyde molecule. The ground state (G.S.) and the vibrational modes having the shift more than 1 meV are shown.

References

[1] Gribakin G F *et al* 2010 *Rev. Mod. Phys.* **82** 2557
[2] Danielson J R *et al* 2012 *Phys. Rev. A* **85** 022709
[3] Tachikawa M 2002 *Chem. Phys. Lett.* 360 494
[4] Kita Y and Tachikawa M 2014 *Eur. Phys. J. D* **68** 116

* E-mail: n185217e@yokohama-cu.ac.jp
† E-mail: ykita@yokohama-cu.ac.jp

 1412 (2020) 222013 doi:10.1088/1742-6596/1412/22/222013

Four-body calculation of energy levels of muonic molecule dtμe in muon catalyzed fusion

M Niiyama[1], T Yamashita[1] and Y Kino[1] *

[1] Department of Chemistry, Tohoku University, Sendai 980-8578, Japan

Synopsis We investigate dtμe (= d + t + μ + e) four-body system which includes the most important bound states $(dt\mu)_{11}$ in muon catalyzed fusion (μCF). We directly estimate the energy shift more accurately than the previous perturbation calculation. The shift found to be one order of magnitude larger than previous calculation and to be more significant in the calculation of the μCF reaction rate.

A negatively charged muon (μ) has a lifetime of 2.2 μs and its mass is 207 times heavier than an electron. The muon can bind hydrogen nuclei in molecule which is less than several hundred fm in size. In the case of a deuterium (D_2) and tritium (T_2) mixture, the muon catalyzes a nuclear fusion reaction ($d + t \rightarrow \alpha + n + 17.6$ MeV) in the muonic hydrogen molecular ion dtμ. This phenomenon is so-called muon catalyzed fusion [1,2]. It has been observed that a muon can catalyze the d-t fusion reaction about 150 times during its lifetime. Though μCF researches had been stagnant due to low energy productivity, they become reevaluated because of recent progress in intense [3] and low cost [4] muon source.

Formation of $(dt\mu)_{v=1,J=1}$, which is the first excited state ($v = 1$) with total angular momentum of one unit ($J = 1$), in collision of a tμ atom and D_2 molecule is a resonance process and is one of the rate limiting processes. The muonic molecule $(dt\mu)_{11}$ forms inside the electronic molecule D_2, and is small enough to be regarded as a pseudo nucleus in the host electronic molecule. Since the excess energy due to the formation is used to the ro-vibrational excitation of the host molecule, the calculated rate of the resonance process is very sensitive to the binding energy of the $(dt\mu)_{11}$. In order to elucidate this μCF process and promote the fusion reaction, it is necessary to calculate the binding energy (2.7 keV) of the $(dt\mu)_{11}$ with an accuracy of 0.1 meV. The binding energy has been accurately calculated by many authors as non-relativistic Coulombic three-body problem. Moreover, some corrections to the three-body energy including relativistic and QED effects, particle spin (hyperfine) effects, nuclear electromagnetic structure effects, and energy shifts caused by the host molecule (finite size) have been calculated. The energy shift due to the sum of these effects is several tens meV, which is much larger than the required accuracy. In this study, we calculate $(dt\mu)_{11}e$ system to estimate the finite size effect. Previous calculation for the energy shift [5] was done by the 2nd order perturbation method, but the 1st and 2nd terms were 18.253 eV and -17.752 eV, respectively. In order to improve the accurate of the calculation, we perform four-body calculation [6]. Since the deexcitation energy of the $(dt\mu)_{11}$ is larger than the binding energy of the electron, the $(dt\mu)_{11}e$ system is a resonance state (Figure 1). We calculate the four-body system with a complex scaling method [7], and obtain both position and width of the resonance energy.

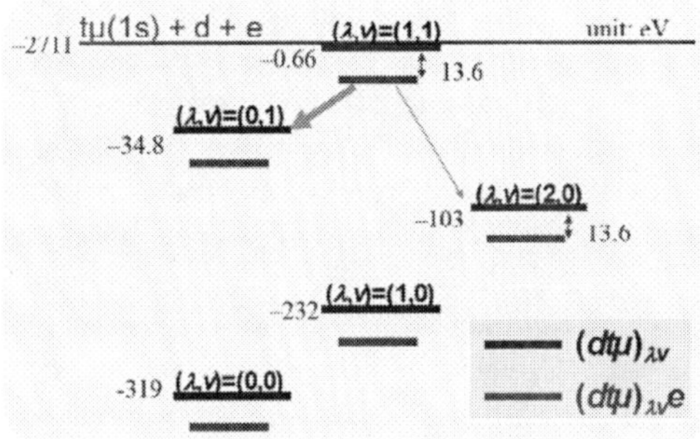

Figure 1. Energy levels of the dtμe system.

References

[1] Cohen J S and Leon M 1989 *Annu. Rev. Nucl. Part. Sci.* **39** 311
[2] Froelich P 1992 *Adv. Phys.* **41** 405
[3] Miyake Y 2014 *J. Phys. Conf. Ser.* **551** 012061
[4] Mori Y *unpublished*
[5] Harston M R et al 1997 *Phys. Rev. A* **56** 2685
[6] Hiyama E et al 2003 *Prog. Part. Nucl. Phys.* **51** 223
[7] Ho Y K 1983 *Phys. Rep.* **99** 1

* E-mail: y.k@m.tohoku.ac.jp

Dielectronic recombination rate coefficients for F-like nickel

S X Wang[1], Z K Huang[2], W Q Wen[2], H B Wang[2], X Xu[1], S Schippers[3], N Khan[2], L J Dou[2], S Mahmood[2,4], L J Mao[2], X M Ma[2], J Li[2], M T Tang[2], D Y Yin[2], Y J Yuan[2], J C Yang[2], C Y Chen[5], Y L Shi[6], C Z Dong[7], X Ma[2]* and L F Zhu[1]†

[1]Hefei National Laboratory for Physical Sciences at Microscale and Department of Modern Physics, University of Science and Technology of China, Hefei, Anhui 230026, PR China
[2]Institute of Modern Physics, Chinese Academy of Sciences, 730000 Lanzhou, PR China
[3]I. Physikalisches Institut, Justus-Liebig-Universität Gießen, 35392 Giessen, Germany
[4]Physics Division, PINSTECH, Nilore, Islamabad, 45650, Pakistan
[5]Shanghai EBIT Laboratory, Institute of Modern Physics, Fudan University, and the Key Laboratory of Applied Ion Beam Physics, Chinese Ministry of Education, Shanghai 200433, PR China
[6]Institute for Fundamental Physics, Tianshui Normal University, Tianshui 741000, China
[7]Key Laboratory of Atomic and Molecular Physics and Functional Materials of Gansu Province, College of Physics and Electronic Engineering, Northwest Normal University, Lanzhou 730070, China

Synopsis At the main cooler storage ring (CSRm) of IMP, Lanzhou, we have measured merged-beams rate coefficients for dielectronic recombination (DR) of F-like Ni in the energy range 0–160 eV which comprises all DR resonances associated with $\Delta N = 0$ core excitations. We compare our results with calculations by the flexible atomic code (FAC). Good agreement is found between calculated and measured resonance parameters except for the low-energy $2s\,2p^6\,(^2S_{1/2})\,6l$ resonances. In addition, we have also derived plasma rate-coefficients from our data for applications in astrophysics.

Accurate dielectronic recombination (DR) rate coefficients are essential for deducing the charge state balance and for interpreting the X-ray emission of astrophysical and fusion plasmas. However, the presently available atomic codes still cannot produce DR rates with sufficient accuracy, in particular, at low collision energies [1]. Therefore, experiments are vitally needed to benchmark the theoretical calculations. Here we present new results of a joint experimental and theoretical study of DR of F-like Ni^{19+} ions. The experimental data were obtained at CSRm (see, e.g., [2] and [3]), and the FAC code [4] was used for our calculations. The overall agreement between experiment and theory is rather satisfying (Figure 1), with the exception of the low-energy resonances, in particular, the recombination via the $2s\,2p^6\,(^2S_{1/2})\,6l$ intermediate states. Consequently, the experimentally derived and the theoretical plasma rate coefficients agree well with one another at temperatures where Ni^{19+} ions form in collisionally-ionized plasmas. At lower temperatures typical for photoionized plasmas, however, there are significant discrepancies due to the mismatch for the low-lying DR resonances.

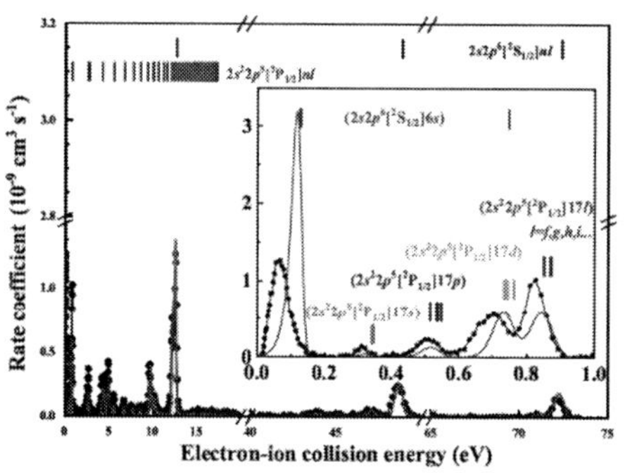

Figure 1. Measured (symbols) and calculated (solid line) recombination rate coefficients of F-like $Ni^{19+}(2s^2\,2p^5\,^2P_{3/2})$. The vertical bars denote positions of two DR series, $2s^2\,2p^5\,^2P_{1/2}\,nl$ (green) and $2s\,2p^6\,2S_{1/2}\,nl$ (blue), as estimated by the Rydberg formula. The inset shows the detailed comparison between the measurement and the theoretical prediction. The colorized vertical bars are the calculated resonance positions at low energies.

References

[1] Schippers S *et al* 2004 *Astron. Astrophys.* **421** 1158
[2] Huang Z K 2018 *et al* 2018 *Astrophys. J. Supp* **235** 2
[3] Wang S X *et al* 2018 *Astrophys. J.* **862** 134
[4] Gu M F 2008 *Can. J. Phys.* **86** 675

*E-mail: x.ma@impcas.ac.cn
†E-mail: lfzhu@ustc.edu.cn

Vibrational and rotational cooling dynamics of the triatomic molecular ion N_2O^+ stored in RICE

R Igosawa[1,2], A Hirota[1,2]*, T Yamaguchi[2], M Lindley[1], K Chartkunchand[1], Y Nakano[3], S Kuma[1] and T Azuma[1,4]†

[1]AMO Physics Laboratory, RIKEN, Saitama, 351-0198, Japan
[2]Department of Physics, Saitama University, Saitama, 338-8570, Japan
[3]Department of Physics, Rikkyo University, Tokyo, 171-8501, Japan
[4]Department of Physics, Tokyo Metropolitan University, Tokyo, 192-0397, Japan

Synopsis We studied radiative cooling dynamics of the triatomic molecular ion N_2O^+ stored in RICE. We observed the evolution of the populations at vibrational and rotational levels up to several 100 s by measuring neutral fragments from the laser-induced predissociation process. We found clear evidence of radiative cooling of the vibrational level populations with a time constant of several seconds. In contrast, absence of significant changes in the rotational levels revealed that the time scale of the rotational cooling is far longer, which is consistent to our simulation.

The RIKEN Cryogenic Electrostatic Ion Storage ring (RICE) has been operated at cryogenic temperatures down to 4.2 K [1], which enables the storage of atomic and molecular ions for long periods on the order of 1000 s or longer. RICE has opened a new opportunity to study excitation and de-excitation dynamics of these ions through observation of the time evolution of the populations at specific vibrational and rotational levels. Such behaviors of diatomic molecular ions have been recently reported at CSR, Heidelberg and DESIREE, Stockholm.

Here we report our first observation of these dynamics in the triatomic molecular ion of N_2O^+. This molecule has been extensively studied both experimentally and theoretically over the past decades, leading to availability of detailed spectroscopic information. Neutral fragments are produced via photo-induced predissociation processes after bound-bound state transitions in the ion, thus making it an excellent thermometer molecule to be probed by action spectroscopy of its vibrational and rotational levels.

After a specific period of ion storage, we introduced UV laser light provided by a narrow-band pulsed (10 Hz or 100 Hz) dye laser to the straight section of RICE for merging with the stored ion beam. Laser-induced dissociation of molecular ions was identified by detecting the neutral fragments. For spectroscopy, we monitored the neutral fragment yields as a function of the laser wavelength, and its storage time dependence was measured to study the cooling dynamics.

*E-mail: a.hirota.279@ms.saitama-u.ac.jp
†E-mail: toshiyuki-azuma@riken.jp

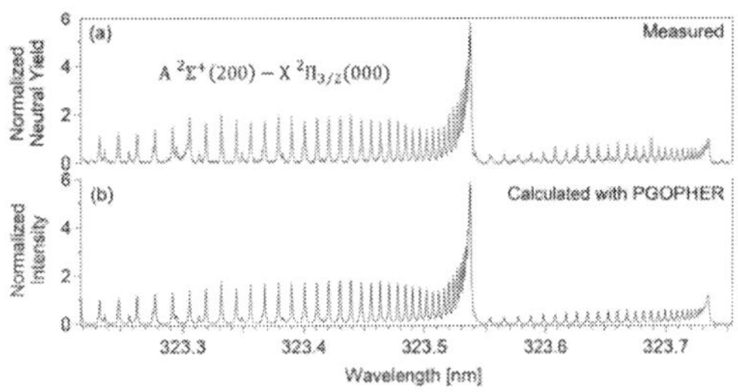

Figure 1. (a) measured spectrum at 9 ms storage. (b) simulated spectrum using the PGOPHER software at the rotational temperature of 300 K.

By comparing the high-resolution spectra composed of detailed vibrational and rotational level populations with a simulation by the PGOPHER software as shown in Fig. 1, we obtained information of the rotational temperature of isolated molecules. We found clear evidence of radiative cooling of the vibrational populations with a time constant of several seconds. In contrast, absence of significant changes in the rotational levels revealed that the time scale for the rotational cooling is far longer. Our observation is naturally explained by our simulation on the cooling rates.

References

[1] Nakano Y *et al* 2017 *Rev. Sci. Instrum.* **88** 33110

Towards investigations of pre-cooled ions stored in the RICE storage ring via action spectroscopy

KC Chartkunchand[1]*, S Menk[1], A Hirota[1,2], R Igosawa[1,2], S Kuma[1], Y Nakano[1,3] and T Azuma[1,4]†

[1]AMO Physics Laboratory, Cluster for Pioneering Research, RIKEN, Wakō-shi, 351-0198, Japan
[2]Department of Physics, Saitama University, Saitama, 338-8570, Japan
[3]Departmend of Physics, Rikkyo University, Tokyo, 171-8501, Japan
[4]Department of Physics, Tokyo Metropolitan University, Tokyo, 192-0397, Japan

Synopsis A new cryogenic linear ion trap beamline has been commissioned for use with the RIKEN Cryogenic Electrostatic (RICE) storage ring. Testing of the beamline at room-temperature and cryogenic conditions was conducted using methylene blue cations MB^+ and carbon cluster anions C_n^-. Action spectroscopy measurements of pre-cooled C_n^- anions injected into RICE will help determine the effectiveness of the so-called Pre-Trap Beamline, as well as provide structural and dynamical information on these cluster anions at cryogenic temperatures.

The recent introduction of cryogenic electrostic ion storage rings, with the RIKEN Cryogenic Electrostatic (RICE) ring being the newest [1], has opened up a new realm for the study of atomic and molecular ion cooling dynamics with their ability to store ions for hundreds to thousands of seconds or longer. Even at these low-temperature conditions ($\lesssim$10 K), however, the intrinisic relaxiation times of ion rovibronic levels can be comparable to or longer than the allowable ring storage time when, e.g., transitions from these levels are dipole-forbidden. It is thus of interest to be able to inject ions which have been pre-cooled to the rovibronic ground state into a storage ring for further study.

Recently, a new beamline capable of providing pre-cooled, keV-kinetic energy ion bunches has been commissioned at the RICE ring facility [2]. This so-called Pre-Trap Beamline utilizes a cryogenically cooled (5.7 K) RF octupole ion trap with He buffer gas for effective collisional cooling of ions to their ground rovibroinic state. Various types of ion sources can be installed on this beamline, such as electrospray ionization (ESI) sources and laser ablation ion sources (LAIS). Initial room-temperature testing and characterization of the Pre-Trap Beamline was carried out using methylene blue cations MB^+ produced from an ESI source. Measurements of the total number of ions stored in the trap and of the shape of ion bunchs extracted from the trap were found to be in good agreement with ion trajectory simulations.

Testing of the trap at cryogenic conditions, as well as of injection of ions into RICE, will be carried out using carbon cluster anions C_n^- (n = 3, 4, 5, ...) produced from a custom-build LAIS. Preliminary action spectroscopy experiments will be conducted using a tunable photon source to photodetach the C_n^- anions at varying storage times within the ring, providng information about the electronic, vibrational, and rotational structure of these anions. Comparisons between "hot" ions injected into RICE and those pre-cooled by the RF trap will help determine the effectiveness of the Pre-Trap Beamline. Additionally, the temporal behavior of the yield of neutral C_n clusters after photodetachment will help elucidate the various cooling mechanisms (e.g., radiative cooling, inverse internal conversion, etc.) that these anions undergo, as well as lifetimes of the rovibronic states involved. The main goal of this and related future studies will be to help shed light on the behavior of these ions in astrophysical environemnts, as well as on the approach towards ultracold conditions where quantum effects become manifest.

References

[1] Nakano Y *et al* 2017 *Rev. Sci. Instrum.* **88** 033110
[2] Menk S *et al* 2018 *Rev. Sci. Instrum.* **89** 113110

* E-mail: kiattichart.chartkunchand@riken.jp
† E-mail: toshiyuki-azuma@riken.jp

 1412 (2020) 232004 doi:10.1088/1742-6596/1412/23/232004

Coincident mapping of e$^+$ and e$^-$ from free-free pair production in a magnetic toroidal lepton spectrometer

S Hagmann[1,2*], P M Hillenbrand[1,3], Yu A Litvinov[1], U Spillmann[1] and Th Stöhlker[1,4,5]

[1]GSI Helmholtzzentrum Darmstadt, 64291 Darmstadt, Germany
[2]Inst. für Kernphysik, Universität Frankfurt, 60438 Frankfurt, Germany
[3]Universität Giessen, 35392 Giessen, Germany
[4]Helmholtz Institut Jena, 07743 Jena, Germany
[5]Fakultät für Physik, Universität Jena, 07743 Jena, Germany

Synopsis We discuss the electron-optical properties of a toroidal magnetic sector spectrometer and its suitablilty for electron-positron pair spectroscopy in relativistic ion-atom collisions in the future HESR storage ring at FAIR. With the simultaneous mapping of electrons and positrons and geometric invariants in the lepton trajectorties this instrument offers a very high efficiency for studies of vector momentum correlation in free-free pair production.

Electron–positron pair production, the excitation of an electron with negative energy from the completely filled Dirac sea into an unoccupied state with either a discrete positive energy (resulting in bound-free pairs) or positive energy in a continuum (resulting in free-free pairs) has evolved into a central topic of QED in extreme fields as the coupling between the lepton field and the electromagnetic field is close to one [1-2]. At the future relativistic storage-ring HESR at FAIR [3] such extreme fields may be produced and thus proposed kinematically complete experiments are focussing on momentum spectroscopy of emitted electrons and positrons and corresponding recoil ions.

The kinematic complete study of free-free pair creation $X^{Z+} + A \rightarrow X^{Z+} + \{A^*\} + e^+ + e^-$ demands for the intended spectrometer design to strongly emphasize a large combined detection efficiency (and solid angle) for simultaneous detection of electron and positron of a free-free pair. The electron-optical properties of a magnetic toroidal sector configuration for electrons and positrons were found to be favourable for implementation into a storage ring with a supersonic jet target while covering a wide range of lepton emission into the forward hemisphere. The calculations for the coil configuration adapted to the HESR ring structure, the magnetic B-field and the leptons trajectories in the toroidal B-field are performed using the TOSCA subpackage of the OPERA-3d 17-64 code [4].

Electrons and positrons are simultaneously mapped from the target zone onto large 2D position sensitive detectors in the focal plane in the toroidal magnetic field. We find a geometric invariance of lepton trajectories and describe how these considerably facilitate determination of coincident doubly differential emission cross sections and momentum correlations for electron-positron pairs.

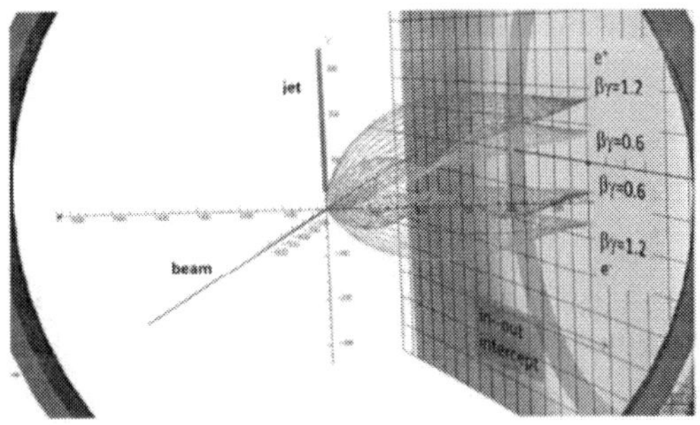

Figure 1. Simultaneous focusing of electrons and positrons of different momenta $\beta\gamma$=1.2/n onto the detector plane, only lepton trajectories for n=1, 2 are shown.

References

[1] Ruffini R *et al* 2010 *Phys. Rep. 487* 1-140
[2] Voitkiv A and Ullrich J 2008 *Relativistic Collisions (Springer Series AOPP vol 49)* (Berlin: Springer)
[3] Stoehlker T *et al* 2015 *Nucl. Instrum. Methods Phys. Res. B* **365** 680
[4] OPERA-3D User Guide, Vector Fields Limited, 24 Bankside, Kidington, Oxford OX5 1JE, UK

* E-mail: s.hagmann@gsi.de

1412 (2020) 232005 doi:10.1088/1742-6596/1412/23/232005

Atomic physics research at HIAF and future perspective

X Ma[1,2 *], W Q Wen[1,2], S F Zhan[1,2], Z K Huang[1], H B Wang[1], J Yang[1,2], X L Zhu[1,2], D Y Yu[1], R Cheng[1], L J Mao[1], J C Yang[1], L N Sheng[1], X H Zhou[1], H S Xu[1], Y J Yuan[1], J W Xia[1], H W Zhao[1], G Q Xiao[1] and W L Zhan[1]

[1]Institute of Modern Physics, Chinese Academy of Sciences, 730000, Lanzhou, China
[2]School of Nuclear Science and Technology, University of Chinese Academy of Sciences, 100049, Beijing, China

Synopsis A brief introduction to the undergoing HIAF projects in China is given, and the scientific goals as well as the atomic physics programs are discussed. A new science center in the Gulf area near Hongkong will provide unprecedented opportunities for novel scientific researches based on large scale facilities.

The High Intensity heavy ion Accelerator Facility (HIAF) [1] is an approved project for nuclear physics, atomic physics, high energy density physics, and applications. HIAF is located at the Huizhou area of Guangdong Province of South China and is under construction. An international facility - FAIR/GSI for ambitious physics programs with anti-proton and heavy ions is being constructed in Darmstadt Germany [2].

The HIAF consists of an super-conducting ECR source, a SC ion-Linac (iLinac) with a length of 100m, a Booster Ring (BRing), a radioactive beam line of fragmentation type, a spectrometer ring (SRing), and several experimental terminals at low- and high-energy ends as well as setups for in-ring experiments as shown in Figure 1. There is a possibility to build a MRing in future, which will provide the opportunity for the HCI beams to collide in the merging region with the SRing. The two beams are designed to cross with each other with a small angle of about 6 degrees. Therefore, the collision energy in the center of mass system can be detuned with changing the primary beam energies.

The BRing and SRing will be equipped with electron coolers, and a stochastic cooling system will be installed at the SRing. HIAF will provide ion beams of stable and unstable nuclei with high intensities and high quality. Highly charged ions can be injected in the SRing with $\sim 10^{11-12}$ particle per pulse (ppp).

The scientific programs for atomic physics research include the electron-positron pair creation in the super-critical strong field generated by merging two bare or highly charged heavy ion beams, electron-ion collision spectroscopy with an independent electron-target, ultra-fast collision dynamics in the relativistic energy regimes, X-ray spectroscopy at the internal-gas-target, laser cooling and laser spectroscopy of relativistic heavy ion beams, and the studies of Warm Dense Matter (WDM) driven by intense heavy ion beams. [3]

In addition, investigations of the properties of radioactive ions using the atomic physics methods are also under consideration at HIAF. Furthermore, a mega science center in the gulf area with cities of Guangzhou, Shenzhen, Hongkong, and Macao is on the horizon, and it will invest a huge budget into science and technologies. This will make it possible to combine the CiADS facility with HIAF, which will provide even more exotic beams for future atomic, nuclear, and nuclear astrophysics experiments. (CiADS: China initiative Accelerator Driven subcritical System).

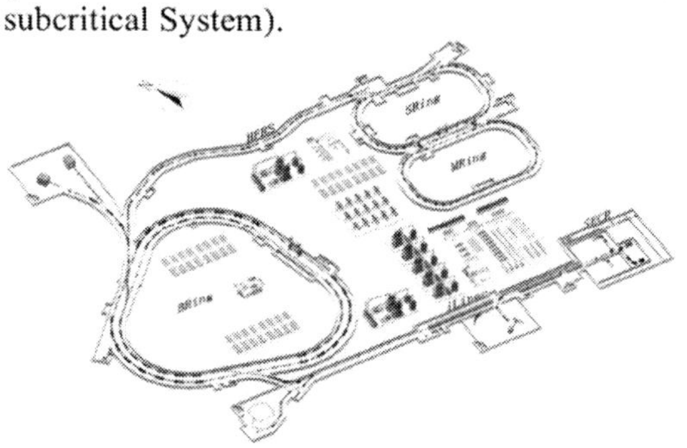

Figure 1. A schematic view of the HIAF complex.

References

[1] Yang J C *et al* 2013 *Nucl. Instrum. Methods Phys. Res.* B **317** 263
[2] www.fair-center.eu
[3] Ma X *et al* 2017 *Nucl. Instrum. Methods Phys. Res.* B **408** 169

* E-mail: x.ma@impcas.ac.cn

Performance of a scintillator-based ion detector for CRYRING@ESR

E B Menz[1,2,3 *], C Hahn[1,2,3], P Pfäfflein[1,2,3], G Weber[1,2] and Th Stöhlker[1,2,3]

[1] Helmholtz Institute Jena, Jena, 07743, Germany
[2] GSI Helmholtzzentrum für Schwerionenforschung, Darmstadt, 64291, Germany
[3] Friedrich Schiller University Jena, Jena, 07743, Germany

Synopsis A detector based on the scintillator material YAP:Ce and capable of counting single ions is presented. The detector consists of a YAP:Ce crystal and a light guide operating in ultra high vacuum and a conventional photomultiplier outside the vauum. The crystal demonstrated the necessary radiation hardness against heavy ion irradiation. The detector has been commissioned at CRYRING@ESR and its detection capabilities have been confirmed with beam from the local source.

The novel FAIR accelerator and storage ring complex, currently under construction at the site of the GSI Helmholtz Center for Heavy Ion Research near Darmstadt, Germany, achieved a major milestone with the first stored beam at CRYRING in 2017. Since then both beam acceleration and electron cooling have been achieved and the first user experiments are planned for 2019. To fully exploit the multifaceted field of research thus made accessible, the availability of robust and reliable ion detectors is of fundamental importance [1]. These sensors will need to provide single-hit detection efficiency at MHz count rates of ions with energies ranging from sub-MeV/u to 15 MeV/u, while withstanding the radiation damage imparted by the energetic ions.

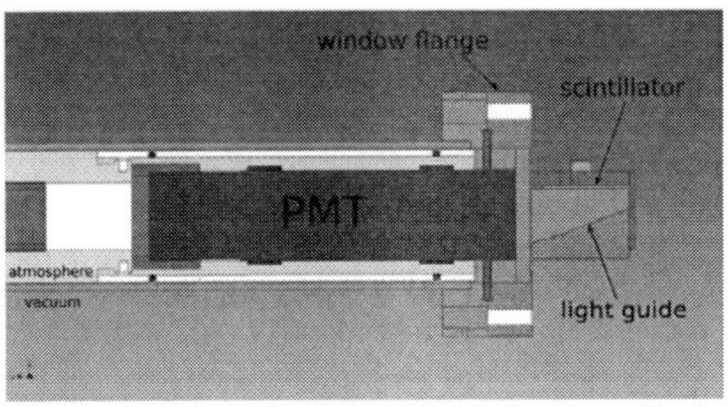

Figure 1. Schematic design of the detector with the scintillator crystal inside and the photomultiplier (PMT) outside the vacuum.

Given these restrictions, a detector system based on the YAP:Ce crystal scintillator provides an attractive approach, utilizing a material that is endowed with a significant degree of radiation hardness while remaining comparatively affordable [2]. A detector system was designed and constructed as part of the CRYRING@ESR instrumentation and has been installed at the ring.

The contribution details the detector design and outlines the results of characterization measurements before and after installation. A proof-of-concept measurement was conducted with various ion species at the 3 MV tandem accelerator JULIA operated by the University of Jena's Institute of Solid State Physics. This experiment confirmed the feasibility of the setup and found a destruction threshold fluence (50% signal amplitude loss) of approximately 10^{13} cm^{-2} across the range of ion species and energies investigated.

At CRYRING the detector is located downstream from the electron cooler and will be used for the observation of electron-ion recombination products. The system has already demonstrated its viability in commissioning measurements with H_2^+ and D^+ beams from the local ion source. These light ions could be detected even at the injection energy of 300 keV/u and from the measured count rate the lifetime of the H_2^+ beam was determined to be ~180 ms.

The setup will be used for e.g. dielectronic recombination and 1s Lamb shift measurements on slow, heavy, highly-charged ions of all species that the GSI accelerator complex is able to produce and transport through ESR.

References

[1] Lestinsky M *et al* 2016 *Eur. Phys. J. Special Topics* **225** 797
[2] Tokman M *et al* 2001 *Phys. Scripta* **2001** 406

* E-mail: e.menz@gsi.de

Journal of Physics: Conference Series **1412** (2020) 232007 doi:10.1088/1742-6596/1412/23/232007

Status of CRYRING@ESR and preparations for first experiments

M Lestinsky[1] *, Z Andelkovic[1], M Bai[1], A Bräuning-Demian[1], V Chetvertkova[1], S Fedotova[1], O Geithner[1], W Geithner[1], O Gorda[1], F Herfurth[1], A Kalinin[1], A Källberg[2], S Litvinov[1], E B Menz[1,3], K Mohr[1], R Sanchez[1], T Stöhlker[1,3,4], G Vorobjev[1] and U Weinrich[1]

[1]GSI Helmholtzzentrum für Schwerionenforschung GmbH, 64291 Darmstadt, Germany
[2]Fysikum, Stockholm University, SE-106 91 Stockholm, Sweden
[3]Helmholtz-Institut Jena, 07743 Jena, Germany
[4]Institut für Optik und Quantenelektronik, Friedrich Schiller Universität, 07743 Jena, Germany

Synopsis CRYRING was moved from Stockholm to Darmstadt, modernized and integrated into the GSI/FAIR beamline topology behind ESR. As CRYRING@ESR, it will receive and store heavy, highly charged ions from all species the present accelerator chain is capable of producing. An extensive research program on low-energy atomic collisions, spectroscopy and nuclear reactions was proposed. The facility is gradually completing commissioning, ion beams from the local injector branch have already been stored and prototype experiments performed. We present the machine status and highlight some planned experiments.

CRYRING@ESR[1] is a low-energy heavy-ion storage ring and the first beam installation realized within the FAIR project. Its full potential will be unleashed, once it receives ion beams from ESR, in all ion species that the GSI accelerator complex is able to produce. The storage of low-energy, heavy, highly charged ions allows for precision spectroscopy of atomic processes and dynamics in strong fields, deriving nuclear parameters like charge radii and nuclear magnetic moments through methods of atomic physics, or probing relevant reactions of nucleosynthesis networks in the Gamow window and many others [2]. Establishing the particle transfer from ESR is presently planned for late 2019.

With its additional ion source and local injector, the facility already runs on its own and provides ion beams e.g. for testing ring subsystems, experimental installations and establishing procedures for data handling and analysis. In the last engineering runs, CRYRING has stably demonstrated storage of ions from the local source, with beam currents close to the space-charge limit and with an ion beam lifetime defined only by residual gas induced loss channels. Moreover, it proved its capability for energy-ramping of the stored ions and for electron cooling, so that well-behaved beams of ions with a narrow beam profile and a sharply defined energy can be delivered to the experimenters' demands. The CRYRING@ESR project is gradually approaching the end of its commissioning period[3] and first experimental installations are ready for beam. Particle detectors for charge exchange reactions have produced beam-induced signal and the new CRYRING laser laboratory has prepared all systems for a prototype beamtime on fluorescence spectroscopy using Mg^+ beams from the local injector. The run shall commence already in March 2019. Three first experiment proposals were approved and necessary preparations are tackled in collaboration with partners from HI Jena, U Giessen, U Heidelberg and many others. This contribution shall discuss the present status and future plans for the machine, as well as lay out the immediate and long-term research program.

*E-mail: m.lestinsky@gsi.de

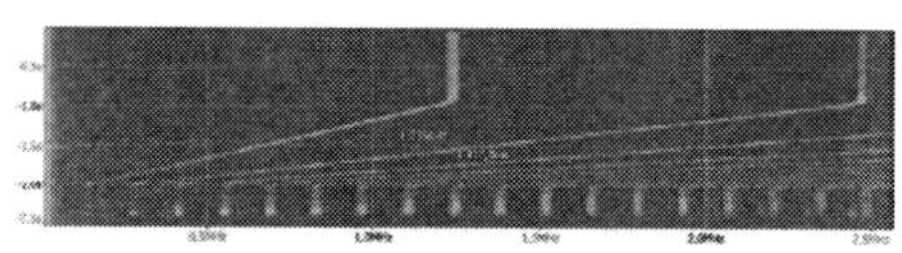

Figure 1. CRYRING@ESR Schottky noise spectrum of H_2^+ injected at 300 keV/u, accelerated to 24 MeV/u and stored for about one second. The fundamental revolution frequency is seen leftmost with several adjacent higher harmonics.

References

[1] Lestinsky M *et al* 2012 *GSI Tech Design Rep* CRYRING@ESR: A study group report
[2] Lestinsky M *et al* 2016 *Eur. Phys. J. ST* **225** 797
[3] Herfurth F *et al* 2017 *in proc. Cool'17* THM13

Final state resolved mutual neutralization of Li^+ and D^-

G Eklund[1]*, J Grumer[2], S Rosén[1], P K Najeeb[1], H Zettergren[1], H Cederquist[1], P S Barklem[2] and H T Schmidt[1]

[1] Department of Physics, Stockholm University, Stockholm, 10691, Sweden
[2] Department of Physics and Astronomy, Uppsala University, Uppsala, 75237, Sweden

Synopsis We have performed a merged beam experiment on mutual neutralization of Li^+ and D^-, at center of mass energies close to 0 eV using the DESIREE storage rings. The final state channels Li(3s) and Li(3p+3d) are resolved and the branching ratio is found to be more equal than predicted from theoretical models.

The measurement of Li abundance in stars is an important problem in astrophysics for understanding the Big Bang, stellar evolution and microscopic processes in stars. Observations of Li abundances in old, metal-poor stars are likely to be sensitive to collision processes between lithium and hydrogen atoms [1]. More recent calculations have shown that the most important process is the mutual neutralization (MN) reaction $Li^+ + H^- \rightarrow Li(3s) + H$ [2]. The aim of the present study is to provide experimental input to the theoretical models, as well as to investigate the importance of different final state channels of the MN process. Previous experimental studies have measured the total cross section of the MN process, at center of mass energies ranging from 0.6 to 300 eV [3]. The previous results were not final state resolved, and for astrophysical purposes it is desirable to perform the experiment at lower center of mass energies. We choose to study D^- rather than H^- as it provides more favorable experimental conditions.

The experiment is performed at the DESIREE facility [4, 5], which consists of two storage rings with a common straight section designed for studies of MN at center of mass energies close to 0 eV. The storage rings are enclosed in a single cryogenically cooled vacuum chamber operating at 13 K, which provides excellent vacuum conditions. The two ion beams are merged and a set of drift tubes is biased in order to control the collision energy of the ions. The products from the MN reaction are detected using an imaging detector, recording the positions and difference in arrival times between the products. The most important final channels of Li are the 3s, 3p and 3d states, with kinetic energy releases of 1.27, 0.81 and 0.77 eV respectively. The experimental resolution is sufficient to resolve the 3s from the 3p+3d states which is shown in Figure 1. Calculations of the cross section have predicted that the 3s final state is approximately a factor of 3 larger than 3p, and a factor of 10 larger than 3d [2]. The experimental results indicate that the branching ratio between the 3s and the 3p+3d channels are more equal than predicted.

*E-mail: gustav.eklund@fysik.su.se

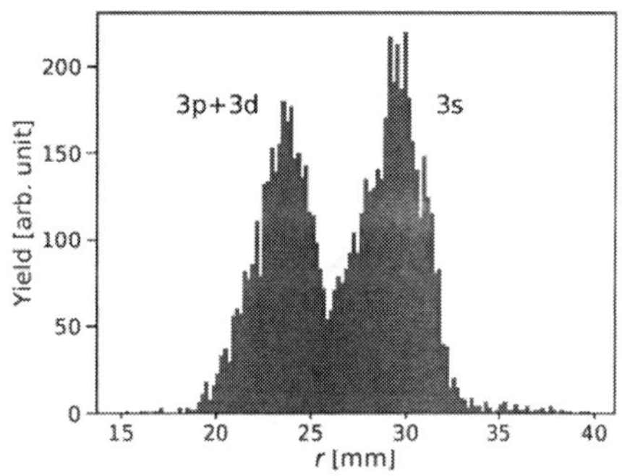

Figure 1. Histogram of the distances between the MN products at the time of arrival on the detector. The 3s and 3p+3d final states are clearly distinguishable.

References

[1] Steenbock W and Holweger H 1984 *Astron. Astrophys.* **130** 319323
[2] Barklem P S *et al* 2003 *Astron. Astrophys.* **409** L1-4
[3] Peart B and Hayton D A 1994 *J. Phys. B* **27** 2551-56
[4] Thomas R D *et al* 2011 *Rev. Sci. Instrum.* **82** 065112
[5] Schmidt H T *et al* 2013 *Rev. Sci. Instrum.* **84** 055115

State-selected ion-molecule reactions with VUV synchrotron radiation : the $O_2^+ + C_3H_6$ case

R Thissen[1,2*], C Romanzin[1,2], J Lemaire[1], S Thomas[3] and C Alcaraz[1,2]

[1] LCP, UMR8000 CNRS, Univ. Paris Sud and Univ. Paris-Saclay, Orsay, 91405, France
[2] Synchrotron SOLEIL, L'Orme des Merisiers, Saint Aubin, 91190, France
[3] CSNSM, UMR8609 IN2P3 - CNRS - Univ. Paris Sud, Orsay – France

Synopsis O_2^+ molecular ion is produced by photoionization close to threshold at SOLEIL, DESIRS beam line. By means of coincidence methods, ions are prepared in selected vibrational and electronic states (X, v=0,3,15 and a, v=3). Their reactivity with the propene in single collision regime is studied in the guided ion beam instrument CERISES, adding ability to vary the collision energy. Absolute reaction cross section and branching ratio are recorded for the 5 most intense reactive channels, described as dissociative charge transfer. Contrasted effects of internal and collision energy are observed.

In this contribution, we will present recent results on state-selected ion-molecule reactions obtained with the CERISES set-up: a quad/oct/oct/quad guided ion-beam experiment [1]. In these experiments, the photoionization of a neutral precursor by VUV synchrotron radiation and the TPEPICO technique are used to prepare state-selected cations before reaction, and absolute reaction cross sections are measured as a function of the cation excitation and collision energy.

Such measurements are needed to refine the description of reactive astrophysical or industrial plasmas. The present results were recorded at SOLEIL synchrotron in spring 2018 in the framework of a collaborative project aiming at a better description and optimization of industrial plasmas used for depollution of factory exhausts. To this end, the chemistry of ionic form of the major atmospheric species with main industrial pollutants has to be described.

Here, molecular oxygen, O_2 has been used to produce molecular ions by photoionization. By means of coincidence methods, they were prepared in selected, long-lived vibrational and electronic states (X, v=0-15 and a, v=0-4).

Their reactivity with the propene molecule (CH_2=CH-CH_3) has been studied by varying the excitation energy and the collision energy, and recording the absolute cross sections of all ionic products, as well as branching ratios.

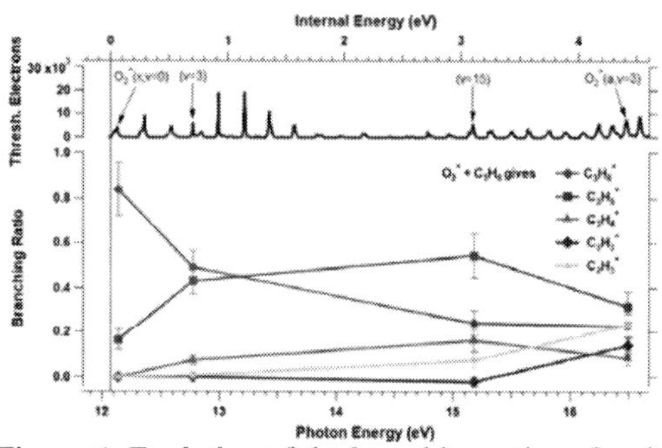

Figure 1. Evolution of the branching ratios of major products of the reaction of O_2^+ with propene, when the reactive ion is prepared in specific states (X, v=0,3,15, and a, v=3)

Contrasted results have been obtained, and though they can be described in terms of charge transfer and dissociative charge transfer, specific sensitivity to the ionic state as well as to the collision energy have been observed, and will be discussed.

References

[1] Cunha de Miranda B K *et al* 2015 *J. Phys. Chem. A* **119** 6082-6098.

* E-mail: roland.thissen@u-psud.fr

Production of cold molecular ions in superfluid helium droplets

A Iguchi[1,2]*, H Otani[1,2], S Kuma[1†], H Tanuma[3] and T Azuma[1,3]

[1]AMO Physics Laboratory, RIKEN Cluster for Pioneering Research, Wako, 351-0198, Japan
[2]School of Science, Kitasato University, Sagamihara, 252-0373, Japan
[3]Department of Physics, Tokyo Metropolitan University, Tokyo, 192-0397, Japan

Synopsis The capability of helium droplets to isolate cryogenic molecular ions at a temperature of 0.4 K is exploited to observe ultracold molecular processes in the RIKEN Cryogenic Electrostatic (RICE) storage ring. The development of a pulsed droplet beam consisting of extremely large helium droplets up to 10^{11} atoms is a promising step to apply this unique and robust cooling method to a wide variety of molecular ions. Production of ions by photoionization of neutral species inside helium droplets will provide internally cold ions below the ring temperature of 4 K.

The robust advantage of helium droplets to cool captured molecules instantaneously to the droplet temperature of 0.4 K is a potential experimental step to realize internally cooled molecular ions. This method is in a striking contrast to the cryogenic ion trap and storage ring techniques where the internal temperature of molecular ions, especially the rotational temperature, is in general difficult to manage to reach equilibrium with the environment. Weak intermolecular interactions with the captured species and the superfluidity of the droplets preserve the nature of the molecules such as their free-rotation as in the gas phase. In addition, the transparency of helium from the microwave to vacuum ultraviolet electromagnetic waves allows us to investigate them via various spectroscopic approaches. These unique properties of helium droplets are of benefit to the study of ultracold molecular dynamics being performed with the RIKEN Cryogenic Electrostatic (RICE) ring [1].

We recently developed a helium droplet source at the RICE facility as a new injection beamline of ultracold molecular ions. Helium droplets were produced in a pulsed beam by expanding a cold (< 20 K), high-pressure (2 MPa) He gas though a solenoid valve [2]. Improvement of the valve operation resulted in the controllable droplet size ranging from 10^5 to 10^{11} atoms per droplet by reducing the valve temperature down to 7 K. The extremely large helium droplets thus achieved enable us to study a wide variety of molecular ions. A spectroscopic setup was also constructed for detection of captured molecules. The droplets pass through a pick-up cell containing a sample vapor, where the collisional capturing efficiently produced cold molecules inside the droplets via evaporative cooling by surface helium atoms. Visible laser excitation of the molecules was monitored by detecting laser-induced fluorescence.

Production of molecular ions in helium droplets is being implemented in the photoionization setup constructed on the beamline. This method has advantages as (1) the initial state of the neutrals are cold to give better efficiency for resonant multiphoton photoionization and (2) the molecular ions are produced in the interior of the droplets at 0.4 K. The photoionization process is monitored as the yields of photoelectrons and ion-containing droplets detected by a channel electron multiplier equipped downstream of the beamline. The combination of the helium droplet beamline with the RICE storage ring will provide a new opportunity to explore slow dynamics of rovibrationally ultracold molecular ions at 0.4 K. Also, the innovative experimental approach to superfluid helium droplets will be established by controlling the droplet beam velocity by electrostatic fields.

References

[1] Nakano Y *et al* 2017 *Rev. Sci. Instrum.* **88** 033110
[2] Kuma S and Azuma A 2017 *Cryogenics* **88** 78

* E-mail: sc15307y@st.kitasato-u.ac.jp
† E-mail: susumu.kuma@riken.jp

Towards background-free studies of capture reactions in a heavy-ion storage ring

L Varga[1] *, K Blaum[2], T Davinson[3], J Glorius[1], B Jurado[4], C Langer[5], C Lederer-Woods[3], Yu A Litvinov[1], R Reifarth[5], Z Slavkovská[5], T Stöhlker[1,6], P J Woods[3] and Y M Xing[1]

[1]GSI, Darmstadt, 64291, Germany
[2]Max-Planck-Institut für Kernphysik, Heidelberg, 69117, Germany
[3]University of Edinburgh, Edinburgh, EH9 3FD, UK
[4]CENBG, Gradignan, 33170, France
[5]Goethe Universität, Frankfurt am Main, 60323, Germany
[6]Helmholtz Institute Jena, Jena, 07743, Germany

Synopsis Stored and cooled highly-charged ions offer unprecedented capabilities for precision studies in realm of atomic-, nuclear-structure and astrophysics. In context of the latter, after the successful investigation of the cross section of ^{96}Ru(p,γ) in 2009, in 2016 the first measurement of the ^{124}Xe(p,γ)^{125}Cs reaction was performed at the Experimental Storage Ring (ESR) at GSI.

Highly-charged stable or radioactive ions can be stored and cooled in a heavy-ion storage ring. Their charge state and their momentum can precisely be chosen and controlled over extended periods of time [1]. We have employed the unique feature of the ESR facility at GSI to address astrophysically relevant reactions. In a first, proof-of-concept, experiment ^{96}Ru(p,γ) has been measured [2]. The new experiment has been performed with decelerated fully-ionized ^{124}Xe ions [3]. Using a Double Sided Silicon Strip Detector (DSSSD), introduced directly into the Ultra High Vacuum environment of the storage ring, the ^{125}Cs proton-capture reaction products of interest have been successfully detected. The cross sections are measured at 5 different energies between 5.5 AMeV and 8 AMeV, on the high energy tail of the Gamow-window for hot, explosive scenarios such as supernovae and X-ray binaries. The well-understood atomic REC cross section for ^{124}Xe+H_2 reaction is used for luminosity calibration.

Elastic scattering on the internal H_2 gas jet target is the major source of background. Monte Carlo simulations show that an additional slit system in the ESR in combination with the energy information of the Si detector will make background free measurements of the proton-capture products possible, see Figure 1. This improvement is about to be implemented and will increase the sensitivity of the method tremendously.

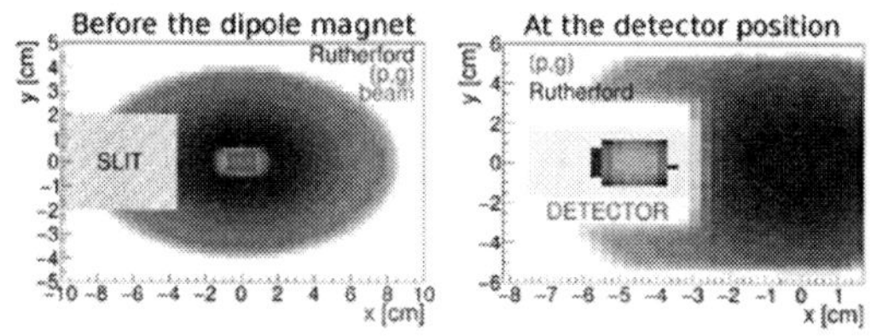

Figure 1. Simulated particle distributions before the dipole magnet (left) and at the detector position (right). Combining a scraping system with an offline energy separation of the measured ions can lead to background free measurement of the p-capture products.

In this contribution, the details of the ^{124}Xe(p,γ)^{125}Cs experiment and the data analysis will be introduced. As an outlook, future perspectives at ESR and CRYRING are discussed [4] focusing on precision (p,γ) reaction studies in the Gamow-window using stored and cooled, highly charged, radioactive ions.

References

[1] Bosch F *et al* 2013 *Prog. Part. Nucl. Phys.* **73** 84
[2] Mei B *et al* 2015 *Phys. Rev. C* **92** 035803
[3] Glorius J *et al* 2019 *Phys. Rev. Lett.* **122** 092701
[4] Woods P J *et al* 2015 *Phys. Scripta T* **166** 014002

* E-mail: L.Varga@gsi.de

Journal of Physics: Conference Series **1412** (2020) 232012 doi:10.1088/1742-6596/1412/23/232012

Vibrational autodetachment from hot copper dimer anions: breakdown of the Born-Oppenheimer approximation

E K Anderson[1*], A F Schmidt-May[2], P K Najeeb[1], G Eklund[1], S Rosén[1], M H Stockett[1], K Chartkunchand[1], M Kaminska[1], K Hansen[3], H Cederquist[1], H Zettergren[1] and H T Schmidt[1†]

[1]Department of Physics, Stockholm University, S-10691, Stockholm Sweden
[2]Institut für Ionenphysik und Angewandte Physik, Universität Innsbruck, A-6020 Innsbruck, Austria
[3]Center for Joint Quantum Studies and Department of Physics, Tianjin University, Tianjin 300072, China

Synopsis The decay of hot copper dimer anions is studied in a cryogenic storage ring. We detect atomic anion and neutral products and determine the fragmentation and vibrational autodetachment (VAD) rates separately. For short storage times the dimer anions predominantly decay by fragmentation of highly rotationally excited ions by tunneling through the rotational barrier. For storage times exceeding 100 ms, however, VAD is dominating. This process is driven by the very weak direct coupling between nuclear and electronic degrees of freedom.

We store Cu_2^- ion beams produced in a cesium sputter ion source in the cryogenic electrostatic ion-storage ring DESIREE [1]. After a straight section we record the rate of neutral decay products by means of a microchannelplate detector, while the rate of Cu^- atomic fragments is recorded after the other straight section by a similar detector. The count rates after detector background subtraction are shown in the upper panel of Fig. 1.

The similar rates of neutrals and ions shortly after injection are evidence that fragmentation dominates initially. This supports the conclusion of an earlier investigation [2] where only the neutral particles were detected and the signal was explained by highly rotationally excited ions decaying by tunneling through the centrifugal barrier.

Due to the excellent vacuum of DESIREE, we follow the decay to much longer times than previously possible. This ability, and the detection of ionic fragments, reveal the VAD process, which becomes dominant after 100 ms storage. The lower panel of Fig. 1 shows the ratio of the ionic to neutral products detected. The right side y-axis is corrected for the different detector efficiencies and gives the fraction of all decays which are due to fragmentation. For t>100 ms the VAD process dominates the decay. As the electron affinity of Cu_2 is smaller than the dissociation energy of Cu_2^- there exist a range of high vibrational excitation and low to moderate rotational excitation where VAD but not fragmentation is energetically possible. The absence of a crossing of the anion and neutral ground state potential energy curves [3] imply that the VAD process is driven by a direct coupling of electronic and nuclear degrees of freedom: i.e. a breakdown of the Born-Oppenheimer approximation.

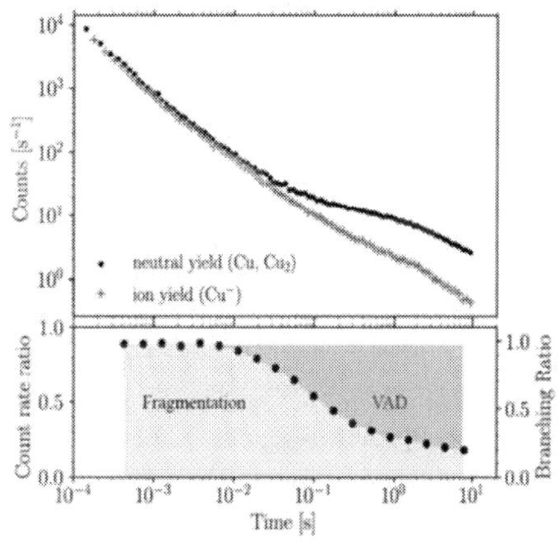

Figure 1. Upper panel: Count rate on the two detectors neutral products (•) and for atomic anion products (+). Lower panel: The ratio of the count rates of the ion and the neutral detectors. The right y-axis gives the fraction of all decays due to fragmentation.

References

[1] Schmidt H T *et al* 2013 *Rev. Sci. Instrum.* **84** 055115
[2] Fedor J, Hansen K, Andersen J U and Hvelplund P 2005 *Phys. Rev. Lett.* **94** 113201
[3] Ho J, Ervin K M, Lineberger W C 1990 *J. Chem. Phys.* **93** 6987

* E-mail: emma.anderson@fysik.su.se
† E-mail: schmidt@fysik.su.se

Radiative cooling dynamics of anthracene cations stored in DESIREE studied via the time evolution of 2-photon-absorption induced dissociation rate

J Bernard[1*], M C Ji[2†], S Martin[1], G Wenzel[3], A Al-Mogeeth[1], M H Stockett[2], H T Schmidt[2], H Zettergren[2] and C Joblin[3]

[1]Institut lumière matière (ILM), UMR5306 Université Lyon 1-CNRS, Université de Lyon 69622 Villeurbanne, France
[2]Department of Physics, Stockholm University, Roslagstullsbacken 21, SE-106 91, Stockholm, Sweden
[3]Institut de Recherche en Astrophysique et Planétologie (IRAP), Université de Toulouse (UPS), CNRS, CNES, 9 Av. du Colonel Roche, 31028 Toulouse Cedex 4, France

Synopsis Anthracene cations ($C_{14}H_{10}^+$) have been stored in the cryogenic (13 K) electrostatic storage ring DESIREE in Stockholm to study their radiative cooling dynamics on a long time scale up to 2 s. By monitoring the laser-induced delayed dissociation curves during the storage, the evolution of the dissociation rate was obtained. Its decrease as a function of time is closely related to the time evolution of the internal energy distribution of the stored anthracene cations and consequently to the energy loss due to radiative cooling, including both Poincaré and infrared fluorescences.

Measurements of radiative cooling rates of Polycyclic Aromatic Hydrocarbons (PAHs) in isolated conditions are essential to characterize the photostability of interstellar PAHs[1]. Previous studies using a compact electrostatic ion storage ring (Mini-Ring) have demonstrated that recurrent fluorescence, also denoted as Poincaré fluorescence, is an efficient radiative cooling mechanism in the time range up to 10 ms[2]. In the present work, we have extended these previous studies to a much longer time range (up to 2 s) by storing anthracene cations in the cryogenic (13 K) electrostatic storage ring DESIREE in Stockholm. On this long time scale, we could investigate the contribution to the radiative cooling due to slow infrared emission via vibrational transitions.

As in the previous study, the cations were produced in the ion source with a broad distribution of internal energies and injected in the ring. Delayed dissociation decay curves were then recorded after laser irradiation of the stored ion bunch every 100 ms with a pulsed nanosecond tuneable laser set at 350 nm. Dissociation rates were derived from the exponential fits to these decay curves as a function of storage time (Figure 1). In the time range 0.1 to 1 s, this rate is found to decrease almost linearly with storage time; whereas, after 1 s, the dissociation rate remains rather constant with time. Since dissociation rates are known to vary rapidly with the internal energy, Figure 1 can be used to further estimate the time evolution of the internal energy (or temperature) of the anthracene cations due to radiative emission. A Monte Carlo kinetics code [3] was used to simulate the radiative cooling of the ions and rationalise these results.

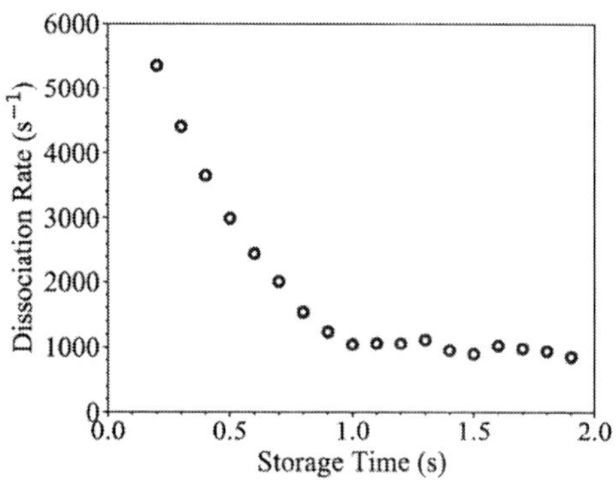

Figure 1. Time evolution of the dissociation rate of anthracene cations induced by two-photon absorption at 350 nm.

References

[1] Tielens A G G M 2013 *Rev. Mod. Phys.* **85** 1021
[2] Martin S *et al* 2013 *Phys. Rev. Lett.* **110** 063003
[3] Joblin C *et al* 2002 *Molecular Physics* **100** 3595

* E-mail: jerome.bernard@univ-lyon1.fr
† E-mail: mingchao.ji@fysik.su.se

Spontaneous decay of small carbon cluster dianions C_n^{2-} (n=7-11)

P K Najeeb[1]*, M H Stockett[1], E K Anderson[1], G Eklund[1], M K Kristiansson[1], S Rosén[1], K C Chartkunchand[2], H Gnaser[3], R Golser[4], D Hanstorp[5], H Cederquist[1], H T Schmidt[1] and H Zettergren[1]

[1]Department of Physics, Stockholm University, AlbaNova, Stockholm, SE-106 91, Sweden
[2]AMO Physics Laboratory, RIKEN Advanced Science Institute, Saitama, 351-0198, Japan
[3]Institut für Oberflächen-und Schichtanalytik, Technische Universität Kaiserslautern, D-67663, Germany
[4]Institute for Isotope Physics, University of Vienna, Vienna, A-1090, Austria
[5]Department of Physics, Chalmers University of Technology and Göteborg University, S-412 96, Sweden

Synopsis We have studied the stabilities of small carbon dianion clusters, C_n^{2-} (n = 7 − 11) in the cryogenic electrostatic ion storage ring, DESIREE. We observe spontaneous electron emission for tens of milliseconds, where dianions containing an even number of carbons decay slower in comparison to their neighbouring cluster sizes. For all the systems considered here, we find that there is a component which is stable at least on the time scales of hundreds of milliseconds.

Pioneering mass spectrometry experiments have shown that small carbon dianion clusters C_n^{2-} (n > 6) survive on microsecond timescales [1, 2, 3]. These results agree with theoretical calculations suggesting that they are stable or metastable with lifetimes exceeding microseconds [4, 5]. In this work, we take advantage of the extreme high vacuum ($\sim 10^{-14}$ mbar) and cryogenic conditions (13 K) of the electrostatic ion-storage ring DESIREE, to study the stabilities of these systems on much longer time scales.

Here, the doubly charged small carbon cluster anions are produced from a cesium-sputter negative ion source platform (at 10 kV) using a solid graphite target. The so formed 20 keV C_n^{2-} ion beam is mass to charge selected by a magnet and injected into one of the DESIREE storage rings [6]. We have monitored the decay of carbon cluster dianions due to electron emission by detecting the resulting monoanions on a position sensitive microchannelplate detector. This allows us to unambiguously separate such processes from the production of neutrals due to any contamination of monoanions present in the ion beam.

We observe spontaneous electron emission on time scales ranging from 10 - 50 ms, which follow power law dependencies as functions of time, $R(t) \propto t^{-\delta}$ (an example is shown for the case of C_7^{2-} in Fig.1). The power law exponents for C_7^{2-}, C_8^{2-}, C_9^{2-}, C_{10}^{2-}, C_{11}^{2-} are $\delta_7 = -1.66 \pm 0.02$, $\delta_8 = -1.47 \pm 0.07$, $\delta_9 = -1.76 \pm 0.02$, $\delta_{10} = -1.33 \pm 0.06$, $\delta_{11} = -1.79 \pm 0.02$ respectively. This result shows a clear even-odd effect, where the clusters containing an even number of carbon atoms decay slower compared to those with an odd number of carbons. On longer time scales we observe a low constant decay rate due to collisions with the residual gas. This shows that these systems are stable or metastable on time scales exceeding hundreds of milliseconds.

We are currently modeling the electron autodetachment rates using a detailed balance approach [7]. Comparisons with the experimental results will be presented at the conference.

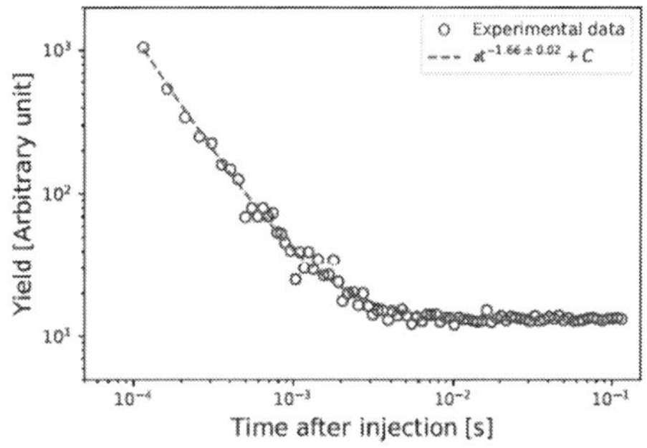

Figure 1. C_7^- yield (blue dots) from $C_7^{2-} \rightarrow C_7^- + e^-$ processes as a function of time. The dashed red line is a fit with a power law and a constant due to the collisions with residual gas

*E-mail: najeeb.punnakayathil@fysik.su.se

References

[1] Schauer S N *et al* 1990 *Phys. Rev. Lett.* **65** 625
[2] Gnaser H and Oechsner H 1993 *Nucl. Instrum. Methods Phys. Res. B* **82** 518-521
[3] Calabrese D *et al* 1996 *J. Phys. Chem.* **105** 2936
[4] Dreuw A *et al* 2002 *Chem. Rev.* **102** 181-200
[5] Sommerfeld T *et al* 1994 *J. Phys. Chem.* **98** 8914
[6] Chartkunchand K C *et al* 2018 *Rev. Sci. Instrum.* **89** 033112
[7] Hansen K 2013 *Statistical physics of nanoparticles in the gas phase* The Netherlands: Springer

An electrospray ionisation apparatus for gas phase study of biomolecules

J Chiarinelli[1,2], P Bolognesi [1 *], J M Ramillon[3], P Rousseau[3], A Domaracka[3], M C Castrovilli[1] and L Avaldi[1]

[1]Istituto di Struttura della Materia, ISM-CNR, 00016 Monterotondo Scalo, Roma, Italy
[2]Dipartimento di Scienze, Università di Roma Tre, Roma, Italy
[3]Normandie Univ, ENSICAEN, UNICAEN, CEA, CNRS, CIMAP, Caen, France

Synopsis A set-up to study large biomolecules in the gas phase based on an ElectroSpray Ionisation source has been installed at CNR-ISM in Rome, and it is currently under commissioning. Accurate ion optics simulation provide crucial feedback in the understanding and optimisation of its performance.

The ElectroSpray Ionization (ESI) [1] allows to bring large organic molecules (proteins, enzymes etc) as intact and isolated units in the gas phase. The technique is based on the use of a low-concentration solution of the molecule of interest flowing in a small capillary held at high voltage with respect to a grounded counter electrode. Driven by the electric field from the Taylor cone formed at the tip of the noozle, the charged droplets undergo a fast decrease in their size as the solvent evaporates, and a gas of molecular ions is formed. Transported towards the high vacuum stage of the set-up, after mass-over-charge filtering, the selected molecular ions can finally be used for soft-landing applications as well as for gas phase experiments.

The ElectroSpray Ionisation (ESI) set-up recently installed at the CNR-ISM laboratory in Rome, presently consists of i) a heated capillary to transport the charged species in vacuum, ii) a first differential pumping stage equipped with an octupole ion guide, iii) a quadrupole mass filter and quadrupolar deflector to direct the ion beam towards either a diagnostic or application stages (see Figure 1).

Figure 1. The ESI set-up installed at CNR-ISM, Rome.

The newly developed apparatus is currently under commissioning. Computer simulations by SIMION software [2] are being performed to optimise ion transmission through the numerous optical elements in the set-up. The comparison of the simulated results with the ion current and ion kinetic energy distribution measured at different check-points in the apparatus provides a valuable feedback for the simulations and deep insights on the behaviour of the apparatus. These are crucial for the optimisation of its performances as well as for further improvements of the set-up.

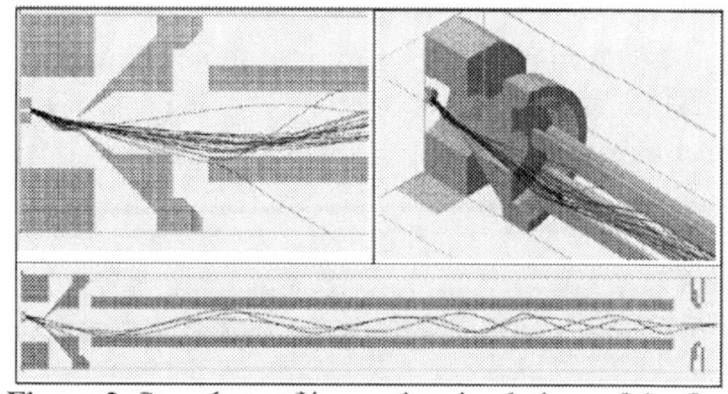

Figure 2. Snapshots of ion optics simulations of the first stage of the apparatus (capillary, skimmer and octupole) performed with Simion software.

Acknowledgment : MAECI Italy-Sweden Project 'Novel molecular tools for the exploration of the nanoworld', Regione Lazio DESIR project and CNRS-PICS BIFACE project. We acklowedge technical support from Aarhus University in the development of the first stage of the apparatus.

References

[1] Fenn J B *et al* 1989 *Science* **4926** 64
[2] https://simion.com/

* E-mail: paola.bolognesi@cnr.it

Journal of Physics: Conference Series **1412** (2020) 242002 doi:10.1088/1742-6596/1412/24/242002

Coupling of a velocity-map imaging spectrometer with an electron cyclotron resonance ion source

B Concina[1 *], F Lépine[1], S Martin[1] and C Bordas[1]

[1]Univ Lyon, Université Claude Bernard Lyon 1, CNRS, Institut Lumière Matière, F-69622 Villeurbanne, France

Synopsis A velocity-map imaging spectrometer has been coupled with an electron cyclotron resonance ion source for the first time. This new set-up aims at the study of the photodetachment of benchmark molecular anions, especially thermionic emission. First measurements have been taken on SF_6^-.

Electron cyclotron resonance (ECR) ion source can deliver intense current of atomic and molecular anions. Photoelectron imaging carried out with a velocity-map imaging (VMI) spectrometer requires relatively high ion current. As a consequence coupling these two devices is of particular interest.

A continuous beam of SF_6^- is produced by an ECR Nanogan ion source [1] and is mass selected by a magnet. A beam chopper is used to make ion bunches of 3-μs duration at 1-kHz repetition rate. The VMI spectrometer is mounted orthogonally to the ion drift tube [2]. The light pulses of nanosecond lasers operating at 1-kHz repetition rate are synchronized with the ion bunches. We have used diode pumped Nd:YAG and Nd:YLF lasers of 349-nm and 532-nm wavelength. The voltages applied to the position-sensitive detector (MCP and phosphor screen) are gated in order to reduce the background (and to allow accurate selection of the delayed thermionic signal of photoexcited anions). Photoelectron impacts are recorded by a CCD camera. Photoelectron images of the atomic anions S^- and F^- are used for the calibration of the VMI spectrometer.

Sulfur hexafluoride (SF_6) is widely used as a gaseous dielectric and in plasma etching. Electron attachment to this molecule at low energy is characterized by a very high cross section. It proceeds by s-wave electron capture and forms mostly long-lived SF_6^- anions by non-dissociative attachment. Thermionic emission from SF_6^- is closely related to the electron attachment to SF_6 through the detailed-balance principle. SF_6^- lifetime has been investigated by the measurement of neutralization rates over times up to 100 ms in a cryogenic ion-beam trap [3].

With the new set-up presented here, we have measured electron kinetic energy release distributions (KERDs) for SF_6^- photoexcited at 349 nm and 532 nm. These results will be discussed at the conference with particular attention to the angular distributions. First measurements on thermionic emission of this benchmark anion will be presented and discussed.

* E-mail: bruno.concina@univ-lyon1.fr

References

[1] Martin S *et al* 2012 *J. Phys.: Conf. Ser.* **388** 102011

[2] Concina B *et al* 2016 *Rev. Sci. Instrum.* **87** 033103

[3] Menk S *et al* 2014 *Phys. Rev. A* **89** 022502

Development of a detector to register low-energy, charge-changed ions from ionization experiments at CRYRING@ESR

A Borovik Jr[1]*, G Weber[2,3], V Hilbert[2,3], H Lin[2,3], P Pfäfflein[2,3,4], B Zhu[2,3], C Hahn[2,3,4], M Lestinsky[4], S Schippers[1], Th Stöhlker[2,3,4] and J Rothhardt[2,3]

[1]I. Physikalisches Institut, Justus-Liebig-Universität Gießen, 35392 Giessen, Germany
[2]Helmholtz-Institut Jena, 07743 Jena, Germany
[3]Institute of Applied Physics, Friedrich Schiller University, 07743 Jena, Germany
[4]GSI Helmholtzzentrum für Schwerionenforschung, 64291 Darmstadt, Germany

Synopsis A detector setup for registering ion species between the poles of a dipole magnet at CRYRING@ESR has been developed. It is based on a scintillator delivering light via a quartz light guide onto a semiconductor photomultiplier. The detector is capable of operating in a strong magnetic field. It can be swiftly retracted from the exposition area during the beam injection into the ring and repositioned back for the measurement cycle to avoid unnecessary exposition and, thus, to increase the scintillator life time.

In the frame of the upcoming proof-of-principle experiment, where C^+ ions stored in a storage ring shall be photoionized by XUV laser pulses [1], a specially tailored detector setup has been developed for counting product ions. Both photon beam and stored ion beam will be merged in the experimental section YR09 of the CRYRING. The product ions will be separated from the primary ions in the field of the dipole magnet. The trajectory of the product C^{2+} ions, according to ion-tracking simulations, allows for their detection exclusively inside the dipole-magnet chamber. This requires the detector to be positioned and operated in a strong magnetic field, which takes well-established detectors based on secondary-electron emission, as, e.g., [2], out of consideration. On the other hand, moderate ion energies, required <100 ns rise and fall times to be able to detect in coincidence with the XUV laser pulses, as well as the limited space inside the dipole-magnet vacuum chamber also limit the list of possible solutions. The CAD model of the detector setup is shown in Fig. 1. Product ions will hit the YAP:Ce scintillator plate producing light pulses which are to be registered by a 3×3 array of silicon photomultipliers (SiPMs), with an active area of 6.07 x 6.07 mm^2 each, placed on a custom-made interface board in a pocket tube outside of the vacuum. Due to the orientation of the available ports in the magnet dipole chamber, the photomultiplier cannot be mounted directly behind the scintillator. The produced photons must therefore be guided by a specially-shaped quartz crystal. The scintillator plate and the quartz light guide will be mounted onto a DN 16CF window, which is at the end of a pocket tube. To avoid unnecessary irradiation during the ion-beam preparation procedure, the entire construction can be swiftly retracted from the product-ion trajectory area and "hidden" behind a dedicated shielding. The whole construction can be fine-positioned along the pocket axis to account for possible displacements of the product-ion trajectory. This detector should enable a variety of photoionization experiments with stored highly-charged ions.

*E-mail: Alexander.Borovik@physik.uni-giessen.de

Figure 1. A cut view of the CAD model of the present detector setup. The solid (green) bow represents the simulated product-ion-beam trajectory.

References

[1] Lestinsky M *et al* 2016 *Eur. Phys. J. Special Topics* **225** 797882
[2] Spruck K *et al* 2015 *Rev. Sci. Instrum.* **86** 023303

Journal of Physics: Conference Series **1412** (2020) 242004 doi:10.1088/1742-6596/1412/24/242004

Characterisation of ion bunches by a single-pass non-destructive charge counter

M Kiffer[1] *, S Ringleb[1], N Stallkamp[1,2], S Kumar[3], B Arndt[4], M Vogel[2], W Quint[5] and Th Stöhlker[1,2,6]

[1]Friedrich-Schiller-Universität Jena, 07743 Jena, Germany
[2]GSI Helmholtzzentrum für Schwerionenforschung, 64291 Darmstadt, Germany
[3]Inter-University Accelerator Centre, 110067 New Delhi, India
[4]Goethe Universität Frankfurt, 60323 Frankfurt, Germany
[5]Physikalisches Institut, Ruprecht Karls-Universität Heidelberg, 69120 Heidelberg, Germany
[6]Helmholtz-Institut Jena, 07743 Jena, Germany

Synopsis We present non-destructive single-pass ion bunch detection and characterisation by measuring the induced image charge in a detection electrode. The presented technique allows direct determination of ion kinetic energy, absolute ion number and spatial ion bunch length. We will show the results of corresponding measurements with bunches of low-energy highly charged ions and discuss the minimum detectable number of charges.

The HILITE [1] Penning trap will provide well-defined ion targets to investigate laser-matter interactions in both the high-intensity and high-energy regime with highly charged ions. The experimental setup includes an Electron Beam Ion Trap (EBIT) and a velocity filter for access to highly-charged-ion bunches of a selected charge state.

For ion bunch detection and analysis we have implemented two distinct single-pass non-destructive charge counting devices [2]. They measure the time-dependent image charge induced by an ion bunch. From such a single-pass signal we can directly extract the kinetic energy, the number of ions and the bunch length. The non-destructive nature allows measurements of products and educts in laser-ion interactions.

We will show a brief overview of the HILITE experiment and present the implemented ion counting technique. This includes the theoretical background of induced image charges, the signal analysis method and calibration results for both kinetic energy and absolute ion numbers. Additionally, we will discuss the sensitivity of the setup and the minimum number of detectable ions. Also, we will outline approaches to further increase the signal-to-noise ratio to be even sensitive to less than ten ions per bunch.

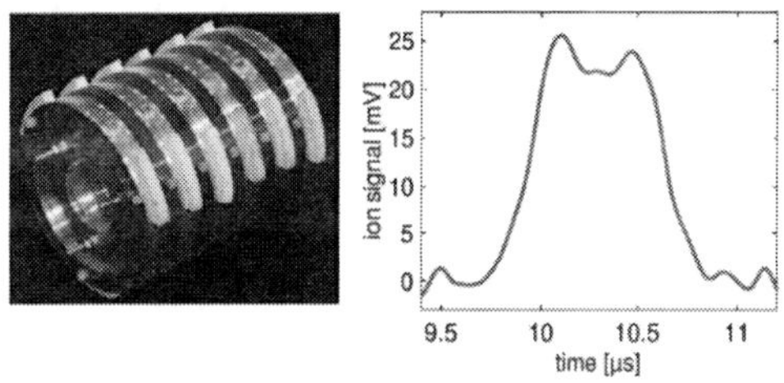

Figure 1. Picture of the detection electrode (left). Single-pass image charge signal induced by about 3000 Ar^{13+} ions (right).

References

[1] Vogel M, Quint W, Paulus G G and Stöhlker Th 2012 *Nucl. Inst. Meth.B* **285** 65
[2] Schmidt S *et al* 2015 *Rev. Sci. Inst.* **86** 113302

*E-mail: markus.kiffer@uni-jena.de

DAVIS: a direct algorithm for velocity-map imaging system

G R Harrison[1], J C Vaughan[1], B Hidle[1] and G M Laurent[1,*]

[1] Physics Department, Auburn University, Auburn, 36849, USA

Synopsis Recently, we reported a direct (non-iterative) algorithm to reconstruct the three-dimensional momentum-space picture of any charged particles collected with a velocity-map imaging system from the two-dimensional projected image captured by a position-sensitive detector. The method consists of fitting the measured image with an analytical 2D projection of a model 3D velocity distribution defined by the physics of the light-matter interaction. In this work, we present a systematic comparison of the level of performance (accuracy and speed) of our method with others well-known techniques for inverting experimental images.

Among the various techniques available to study the physics of the light-matter interaction, Velocity Map Imaging (VMI) systems emerged as a *first-choice* option as they simultaneously give access to the angular and energy distributions of the particles (electrons or ions) emerging from an interaction while retaining a high angular and energy resolution. The technique consists of projecting the particles onto a planar position sensitive detector. The full three-dimensional velocity distribution of the emerging particles can then be retrieved from the 2D projection by means of an inversion algorithm. Despite the wealth of algorithms available, the inversion problem remains an ongoing subject of study as algorithms with improved accuracy, speed, and versatility are still highly desirable.

Recently, we reported a direct (non-iterative) algorithm that bears a similarity to the popular pBASEX method [1]. Our method consists of fitting the experimental data with the two-dimensional projection of a model 3D velocity distribution defined by the physics of the light-matter interaction [2]. It is based on a proper analytical representation of the projection, which is expanded as a sum of Legendre polynomials. Both the velocity and velocity-dependent angular distributions are directly retrieved from the resulting expansion coefficients by using an analytical inversion procedure. Our method offers two main advantages over pBASEX. First, it is less sensitive to noise as the meaningful angle-correlated information is extracted from the raw data by the Legendre polynomial expansion before inversion. Second, it is computationally much cheaper as the 2D projection function is evaluated analytically rather than numerically.

*E-mail: glaurent@auburn.edu

In this work, we present a systematic comparison of the level of performance (accuracy and speed) of our method with others well-known techniques for inverting experimental images. This study aims to aid potential VMI users in choosing the inversion method that fits their experimental needs.

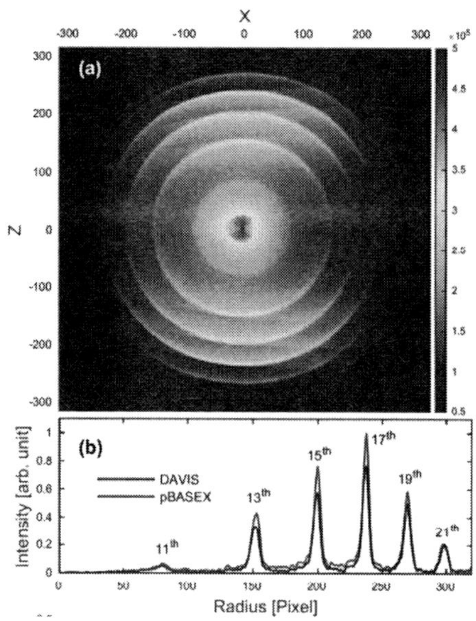

Figure 1. (a) Two-dimensional experimental image obtained for the one-photon ionization of argon by a linearly polarized attosecond pulse train made of odd harmonics. (b) Reconstructed velocity (radial) distributions obtained with DAVIS and pBASEX.

References

[1] Garcia G A *et al* 2004 *RSI* **75** 4989
[2] Harrison G R *et al* 2018 *J. Chem. Phys.* **148** 194101

DNA damage in Thymidyl(3'-5')thymidine (TpT) induced by very low energy electrons

G Khorsandgolchin[1], V Lemelin [1], P Cloutier [1], L Sanche[1*] and J R Wagner[1†]

[1] Département de médecine nucléaire et radiobiologie, Faculté de médecine et des sciences de la santé, Université de Sherbrooke, Sherbrooke, Québec CANADA

Synopsis The interaction of high-energy radiation with the biological matter produces copious amounts of non-thermal secondary low energy electrons (LEEs, 0-30 eV). Several studies have shown that 5-15 eV electrons induce different types of DNA damage by the formation of transient negative ions (TNIs), decaying into the dissociative electron attachment (DEA) process [1,2]. Here, our main focus is to understand the chemical mechanism of damage by very low energy electrons (vLEEs~1.8 eV) by using TpT as a simple model of DNA, solid-state LEE irradiation methods and LC-MS/MS analysis of the degradation products.

Our results show that there are three major types of damage, which are produced upon irradiation of TpT with vLEEs (~1.8 eV): 1) C-O bond cleavage (similar to a strand break); 2) release of non-modified thymine and 3) reduction of thymine to 5,6-dihydrothymine. C-O bond cleavage can occur in TpT at either the 3' or 5' phosphodiester termini. When the reaction occurs at the 3' position, the fragments are thymidine 5'-monophosphate (TMP5') and 3',2'-dideoxythymidine (ddT3') in a 2-fold greater yield compare to the reaction occurs at the 5' position, which the fragments are thymidine 3'-monophosphate (TMP3') and 5',2'-dideoxythymidine (ddT5'). Remarkably, the yield of thymidine monophosphates was higher than that of the corresponding dideoxythymidine product, implying that other reactions contributed to phosphodiester bond cleavage. As for the other types of damage, we observed that very low energy electrons (vLEEs) do not efficiently create N-glycosidic bond cleavage and induce similar yields between abasic sites (TpT containing an abasic site). Finally, through vLEE irradiation of TpT, a large amount of thymine converts to 5,6-dihydrothymine. This reaction can be explained by the electron attachment to the 5,6-double bond of thymine to produce the thymine anion radical. To conclude, this work confirms that vLEEs induce C-O bond cleavage of the phosphodiester bond as a principal reaction pathway.

References

[1] Park *et al* 2011 *Radiat. Res.* **175** 240
[2] Park *et al* 2013 *J. Phys Chem. B* **117** 10122

* E-mail: Leon.sanche@usherbrooke.ca
† E-mail: richard.j.wagner@usherbrooke.ca

Measurement of proton microbeam profiles in atmosphere produced by single tapered glass capillary optics with an end window

T Ikeda[1,2 *], M Ikekame[2,1], Y Hikima[2,1], M Mori[2,1], S Kawamura[2,1], T Minowa[2] and W-G Jin[2]

[1]RIKEN Nishina Center for Accelerator-Based Science, 2-1 Hirosawa, Wako, Saitama 351-0198 Japan
[2]Department of Physics, Toho University, 2-2-1 Miyama, Funabashi, Chiba 274-8510 Japan

Synopsis Profiles of 2.8 MeV H^+ microbeam extracted from a single tapered glass capillary optics were measured by two methods for a future irradiation to an outer organ of insect target. The halo component was analyzed to be 575 μm in full width at half maximum (FWHM) at a plastic tracking detector CR-39 located 1 mm downstream of the capillary. The core component was examined by knife-edge method according to the ion energies. The profiles consisting of more energy-loss events have wider FWHM, which can be reduced by Al thin layers.

MeV ion microbeam irradiation to biological target, *e.g.*, a part of a single outer organ of an insect (Fig.1 (a)) , has been developed, employing tapered glass capillary optics with an end window. The end window contributes in the close approach to the target even in air / liquid so that the broadening of the microbeam due to multiple-scattering is minimized. The beam size should be examined to be less than 100 μmϕ for 1 mm-irradiation distance or beam divergence of 1°, especially for the insect irradiation.

A 2.8 MeV H^+ beam from the RIKEN Pelletron accelerator was transmitted through a capillary, whose outlet and inlet sizes were 9.9 μmϕ and 1.8 mmϕ, respectively, (Fig.1 (b)), and then created a beam spot on a plastic tracking detector CR-39 piece at the irradiation distance of 1 mm (upper panel of Fig.1(c)). The green histogram in the lower panel expresses the sum of pit-darkness along with a horizontal thin zone shown as slice area in the upper panel. The beam structure produced by tapered glass capillary optics is known to consist of core and halo components [1]. The pit-darkness was used for the halo part as a density of the pits assuming that any pit is separated each other and all pits have roughly the same size. The histogram was analyzed to have two Gaussian peaks resulting in 575 μm in FWHM for the halo part.

The core structure was measured by knife-edge method with a step of 75 μm assuming a beam of Gaussian shape. The ions were counted by a PIN photodiode. Figure 1 (d) shows beam profiles at 11-mm downstream of the capillary outlet according to the pulse height. Each line number 1-5 corresponds to a set of lower- and upper-thresholds selecting the pulse height region of the amplifier output so that low- (1-3 V) or high-energy ions (7-10 V) or etc. were selected individually. Line 5 representing the minimized energy-loss events shows the narrowest peak, which remained when aluminum thin layers were inserted in front of the detector.

The profile measurements as well as a beam guiding experiment will be presented.

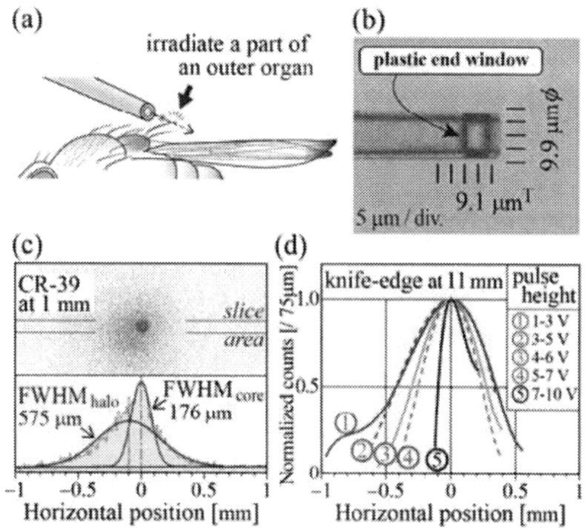

Figure 1. (a) Microbeam irradiation to a fruit fly, (b) The capillary outlet, (c) A 2.8 MeV H^+ microbeam spot on a CR-39 and the histogram along with slice area, (d) The profiles obtained with a knife-edge method according to the pulse heights from the PIN photodiode detector.

References

[1] Hasegawa J *et al* 2011 *J. Appl. Phys.* **110** 044913

* E-mail: tokihiro@riken.jp

Real-time trajectory control on a deterministic ion source

C Lopez, A Trimeche, D Comparat and Y J Picard*

Laboratoire Aimé Cotton, CNRS, Univ. Paris-Sud, ENS Paris Saclay, Université Paris-Saclay, Bât. 505, 91405 Orsay, France

Synopsis The major challenge to improve deterministic single ion sources is to control the position and momentum of each ion. Based on the extra information given by the electron created in a photoionization process, the trajectory of the correlated ion can be controlled and corrected using a fast real time feedback system. In this work, we report on a proof-of-principle experiment that demonstrates the performance of this feedback control with individual cesium ions.

Deterministic and high precision placement of individual atoms and ions offers new and exciting perspective for the realization of quantum based devices at the nanoscale. By using the correlation between ion and electron, produced by photoionization of an atomic beam, we developed a fast feedback system to control the ion trajectory [1]. As shown in figure 1, the produced electron is detected with a time and position sensitive detector, whose information are used to quickly infer the position of the corresponding ion. Then the feedback system drives the ion trajectory through steering plates, the individual ion can thus be sent to any dedicated location.

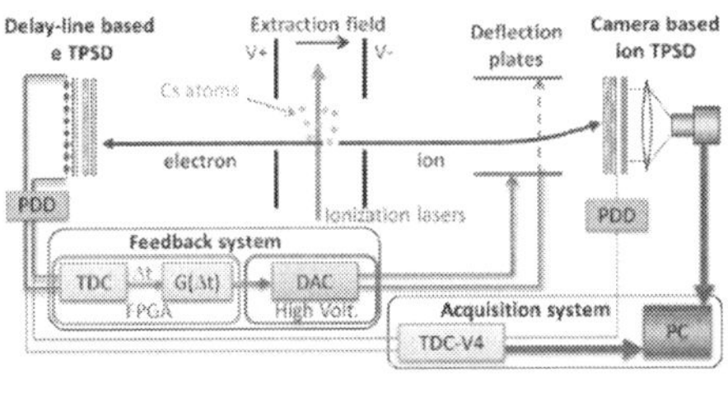

Figure 1. Scheme of the experimental setup. Electrons and ions, produced by photoionization of a cesium beam, are extracted in opposite directions and detected with two time and position sensitive detectors (TPSD). The acquisition system handles the information from the detectors and processes the data. The feedback system controls the ion trajectory by acting on the deflection plates with the appropriate voltages for each ion.

This feedback system enables us to add, to the deterministic character of the ion delivery, the control of its trajectory and complex patterning such as multi-spots and virtual diaphragm/mask (see figure 2) and to reach a factor 1000 improvement in spot area.

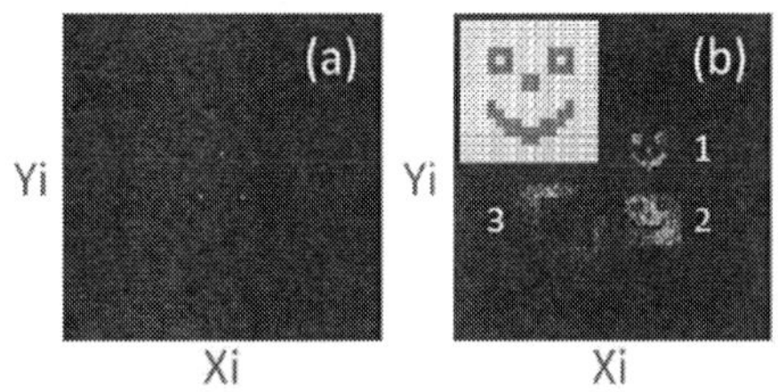

Figure 2. Raw images from the ion detector after: (a) application of a 2D correction in which the ions are carried on to three different spots at the target; (b): application of a virtual mask implemented into the feedback system.

The single ion feedback control is versatile and can be applied to different kind of ion sources. It provides a powerful tool to optimize the ion beam and offers new area for quantum systems and applications of materials science. This feedback system can correct more complex effects such as finite source size, electrostatic or magnetic lenses aberrations, stray fields, *etc...* Coupled to a FIB column, such system can probably produce sub-nm spots even for quite low beam energy —since we correct each ion independently —with deterministic and complex patterning property, making them very useful for focused ion beam studies such as for nano-structuring, microscopy and surface spectroscopy.

References

[1] Lopez C, Trimeche A, Comparat D and Picard Y J 2018 *e-Print: arXiv:1812.09039*

*E-mail: yan.picard@u-psud.fr

AUTHOR INDEX